Essential Formulas

1. $\int x^\alpha \, dx = \dfrac{x^{\alpha+1}}{\alpha+1} + C, \quad \alpha \neq -1$

2. $\int \dfrac{1}{x} \, dx = \ln|x| + C$

3. $\int e^x \, dx = e^x + C$

4. $\int \sin x \, dx = -\cos x + C$

5. $\int \cos x \, dx = \sin x + C$

6. $\int \tan x \, dx = \ln|\sec x| + C$

7. $\int \sec x \, dx = \ln|\sec x + \tan x| + C$

8. $\int \sec^2 x \, dx = \tan x + C$

9. $\int \sec x \tan x \, dx = \sec x + C$

10. $\int \dfrac{dx}{\sqrt{1-x^2}} = \sin^{-1}x + C$

11. $\int \dfrac{dx}{1+x^2} = \tan^{-1}x + C$

12. $\int \dfrac{dx}{x\sqrt{x^2-1}} = \sec^{-1}x + C$

Useful Formulas

13. $\int \cot x \, dx = \ln|\sin x| + C$

14. $\int \csc x \, dx = \ln|\csc x - \cot x| + C$

15. $\int \csc^2 x \, dx = -\cot x + C$

16. $\int \csc x \cot x \, dx = -\csc x + C$

17. $\int \dfrac{dx}{\sqrt{a^2-x^2}} = \sin^{-1}\dfrac{x}{a} + C, \quad a > 0$

18. $\int \dfrac{dx}{a^2+x^2} = \dfrac{1}{a}\tan^{-1}\dfrac{x}{a} + C, \quad a > 0$

19. $\int \dfrac{dx}{x\sqrt{x^2-a^2}} = \dfrac{1}{a}\sec^{-1}\dfrac{x}{a} + C, \quad a > 0$

20. $\int a^x \, dx = \dfrac{a^x}{\ln a} + C, \quad a > 0, \; a \neq 1$

21. $\int \sinh x \, dx = \cosh x + C$

22. $\int \cosh x \, dx = \sinh x + C$

23. $\int \operatorname{sech}^2 x \, dx = \tanh x + C$

24. $\int \operatorname{csch}^2 x \, dx = -\coth x + C$

25. $\int \operatorname{sech} x \tanh x \, dx = -\operatorname{sech} x + C$

26. $\int \operatorname{csch} x \coth x \, dx = -\operatorname{csch} x + C$

27. $\int \dfrac{dx}{a+bx} = \dfrac{1}{b}\ln|a+bx| + C$

28. $\int \dfrac{x \, dx}{a+bx} = \dfrac{1}{b^2}(a+bx - a\ln|a+bx|) + C$

29. $\int \dfrac{x \, dx}{(a+bx)^2} = \dfrac{a}{b^2(a+bx)} + \dfrac{1}{b^2}\ln|a+bx| + C$

30. $\int \dfrac{x^2 \, dx}{(a+bx)^2} = \dfrac{1}{b^3}\left(a+bx - \dfrac{a^2}{a+bx} - 2a\ln|a+bx|\right) + C$

31. $\int \dfrac{dx}{x(a+bx)^2} = \dfrac{1}{a(a+bx)} - \dfrac{1}{a^2}\ln\left|\dfrac{a+bx}{x}\right| + C$

32. $\int \dfrac{dx}{x^2(a+bx)} = -\dfrac{1}{ax} + \dfrac{b}{a^2}\ln\left|\dfrac{a+bx}{x}\right| + C$

33. $\int x(a+bx)^n \, dx = \dfrac{(a+bx)^{n+1}}{b^2}\left(\dfrac{a+bx}{n+2} - \dfrac{a}{n+1}\right) + C, \quad n \neq -1, -2$

34. $\int \dfrac{x \, dx}{(a+bx)(c+dx)} = \dfrac{1}{bc-ad}\left(-\dfrac{a}{b}\ln|a+bx| + \dfrac{c}{d}\ln|c+dx|\right) + C, \quad bc - ad \neq 0$

35. $\int \dfrac{x \, dx}{(a+bx)^2(c+dx)} = \dfrac{1}{bc-ad}\left[\dfrac{a}{b(a+bx)} + \dfrac{c}{bc-ad}\ln\left|\dfrac{a+bx}{c+dx}\right|\right] + C, \quad bc - ad \neq 0$

36. $\int \dfrac{dx}{a^2-x^2} = \dfrac{1}{2a}\ln\left|\dfrac{x+a}{x-a}\right| + C$

37. $\int \dfrac{dx}{x^2-a^2} = \dfrac{1}{2a}\ln\left|\dfrac{x-a}{x+a}\right| + C$

38. $\int \dfrac{dx}{(a^2 \pm x^2)^n} = \dfrac{1}{2(n-1)a^2}\left[\dfrac{x}{(a^2 \pm x^2)^{n-1}} + (2n-3)\int \dfrac{dx}{(a^2 \pm x^2)^{n-1}}\right], \quad n \neq 1$

39. $\int \dfrac{dx}{(x^2-a^2)^n} = \dfrac{1}{2(n-1)a^2}\left[-\dfrac{x}{(x^2-a^2)^{n-1}} - (2n-3)\int \dfrac{dx}{(x^2-a^2)^{n-1}}\right], \quad n \neq 1$

Forms Containing $\sqrt{a + bx}$

40. $\int x\sqrt{a + bx}\ dx = \dfrac{2}{15b^2}\ (3bx - 2a)(a + bx)^{3/2} + C$

41. $\int x^n\sqrt{a + bx}\ dx = \dfrac{2}{b(2n + 3)}\ [x^n(a + bx)^{3/2} - na\int x^{n-1}\sqrt{a + bx}\ dx]$

42. $\int \dfrac{x\ dx}{\sqrt{a + bx}} = \dfrac{2}{3b^2}(bx - 2a)\sqrt{a + bx} + C$

43. $\int \dfrac{x^2\ dx}{\sqrt{a + bx}} = \dfrac{2}{15b^3}\ (8a^2 + 3b^2x^2 - 4abx)\sqrt{a + bx} + C$

44. $\int \dfrac{x^n\ dx}{\sqrt{a + bx}} = \dfrac{2x^n\sqrt{a + bx}}{b(2n + 1)} - \dfrac{2na}{b(2n + 1)}\int \dfrac{x^{n-1}\ dx}{\sqrt{a + bx}}$

45. $\int \dfrac{dx}{x\sqrt{a + bx}} = \begin{cases} \dfrac{1}{\sqrt{a}}\ \ln\left|\dfrac{\sqrt{a + bx} - \sqrt{a}}{\sqrt{a + bx} + \sqrt{a}}\right| + C, & a > 0 \\[2em] \dfrac{2}{\sqrt{-a}}\ \tan^{-1}\sqrt{\dfrac{a + bx}{-a}} + C, & a < 0 \end{cases}$

46. $\int \dfrac{dx}{x^n\sqrt{a + bx}} = -\dfrac{\sqrt{a + bx}}{a(n - 1)x^{n-1}} - \dfrac{b(2n - 3)}{2a(n - 1)}\int \dfrac{dx}{x^{n-1}\sqrt{a + bx}}$

47. $\int \dfrac{\sqrt{a + bx}}{x}\ dx = 2\sqrt{a + bx} + a\int \dfrac{dx}{x\sqrt{a + bx}}$

48. $\int \dfrac{\sqrt{a + bx}}{x^2}\ dx = -\dfrac{\sqrt{a + bx}}{x} + \dfrac{b}{2}\int \dfrac{dx}{x\sqrt{a + bx}}$

Forms Containing $\sqrt{x^2 \pm a^2}$

49. $\int \sqrt{x^2 \pm a^2}\ dx = \dfrac{x}{2}\sqrt{x^2 \pm a^2} \pm \dfrac{a^2}{2}\ \ln|x + \sqrt{x^2 \pm a^2}| + C$

50. $\int x\sqrt{x^2 \pm a^2}\ dx = \frac{1}{3}(x^2 \pm a^2)^{3/2} + C$

51. $\int x^2\sqrt{x^2 \pm a^2}\ dx = \dfrac{x}{8}(2x^2 \pm a^2)\sqrt{x^2 \pm a^2} - \dfrac{a^4}{8}\ \ln|x + \sqrt{x^2 \pm a^2}| + C$

52. $\int \dfrac{\sqrt{x^2 + a^2}}{x}\ dx = \sqrt{x^2 + a^2} - a\ \ln\left|\dfrac{a + \sqrt{x^2 + a^2}}{x}\right| + C$

53. $\int \dfrac{\sqrt{x^2 - a^2}}{x}\ dx = \sqrt{x^2 - a^2} - a\ \sec^{-1}\left|\dfrac{x}{a}\right| + C$

54. $\int \dfrac{\sqrt{x^2 \pm a^2}}{x^2}\ dx = -\dfrac{\sqrt{x^2 \pm a^2}}{x} + \ln|x + \sqrt{x^2 \pm a^2}| + C$

55. $\int \dfrac{dx}{\sqrt{x^2 \pm a^2}} = \ln|x + \sqrt{x^2 \pm a^2}| + C$

56. $\int \dfrac{x^2\ dx}{\sqrt{x^2 \pm a^2}} = \dfrac{x}{2}\sqrt{x^2 \pm a^2} \mp \dfrac{a^2}{2}\ \ln|x + \sqrt{x^2 \pm a^2}| + C$

57. $\int \dfrac{dx}{x\sqrt{x^2 + a^2}} = -\dfrac{1}{a}\ \ln\left|\dfrac{a + \sqrt{x^2 + a^2}}{x}\right| + C$

58. $\int \dfrac{dx}{x\sqrt{x^2 - a^2}} = \dfrac{1}{a}\ \sec^{-1}\left|\dfrac{x}{a}\right| + C$

(continued inside back cover)

Second Edition

Calculus and
Analytic Geometry

Second Edition

Calculus and Analytic Geometry

Abe Mizrahi
Indiana University Northwest

Michael Sullivan
Chicago State University

Wadsworth Publishing Company
Belmont, California
A division of Wadsworth, Inc.

Cover: *Around a Point* by Frantisek Kupka. Photograph courtesy Musée National d'Art Moderne. © Copyright by ADAGP, Paris, 1986.

Credits: Page 469, Gateway Arch, St. Louis, Missouri, photo © UPI/Bettmann Newsphotos; page 652, Golden Gate Bridge, San Francisco, California, photo © Elizabeth Hamlin, provided by Stock, Boston, Inc.; page 654, photo courtesy of TS; page 662, Colosseum, Rome, photo by Bruce Kokernot; page 866, contour map of Menan Buttes, Idaho, courtesy of U.S. Geological Survey; page 867, annual average summer rainfall, redrawn from Chicago Tribune Graphic by Terry Volpp, copyrighted 1980, used with permission.

Signing Representative: Tom Braden
Mathematics Editor: Jim Harrison
Production Editor: Phyllis Niklas
Designers: Albert Burkhardt and Rick Chafian
Technical Illustrator: Scientific Illustrators
Print Buyer: Ruth Cole

Printed in the United States of America

2   3   4   5   6   7   8   9   10—90   89   88   87   86

Library of Congress Cataloging-in-Publication Data

Mizrahi, Abe.
  Calculus and analytic geometry.

  Includes index.
  1. Calculus. 2. Geometry, Analytic. I. Sullivan, Michael. II. Title.
  QA303.M6878   1986   515′.15   85-20199

ISBN 0-534-05454-4

To our wives,
Caryl and Mary

Contents

Preface to the Instructor

In our first edition we set out to write a straightforward, mathematically sound book that would nevertheless capture the interest of students. We felt—and still feel—that there is a place for a mature, mainstream text with the readability and practicality usually associated with lower-level books. In this edition we have sought to improve our text both mathematically and pedagogically, but our basic intentions remain unaltered.

Continuing Features

As before, we present concepts in both theorem–proof style and in the vernacular. We have often found that when students are presented with an alternative explanation, a mental block is unlocked, and better appreciation of the material results. Explanations in plain English also serve to provide the qualitative perspective from which analysis must at times be viewed.

We have retained and augmented our store of interesting applications. These applications are located at the most strategic places—where students are most likely to be fatigued by the presentation of difficult material or where an application can bring to life a particular aspect of mathematics.

We have also kept the historical sections in order to provide perspective on how calculus actually evolved. We have avoided the capsule biography approach to historical notes by including exercises in these optional sections. We want teachers to be able to give their students a taste of what it was like to invent the calculus. This Socratic style of instruction is, in our experience, more instructive than the simple reading of footnotes.

Each chapter still ends with a set of Miscellaneous Exercises. These are more difficult than the exercises at the ends of sections. They serve to extend the material of the chapter as well as to test the students' true grasp of the mechanics of the material.

We continue to provide several exercises taken directly from the best-selling texts in physics as well as many exercises that call for the use of a programmable calculator or computer.

New to This Edition

We have enlarged many problem sets with additional drill problems, challenging problems, and applied problems that allow the instructor to present certain topics in greater depth. In fact, over 700 exercises have been added to this edition, resulting in a total number of exercises in excess of 6000.

On the recommendation of reviewers and users, topics have been rewritten or reorganized and some new material has been added throughout the text. The main changes are as follows:

1. In Chapter 3 the sine and cosine functions now precede the chain rule so that trigonometric functions can be used in the examples and exercises.

2. Newton's method has been included in Chapter 3 so that instructors can, at their option, introduce numerical methods early in the course. A special section titled "Estimating Roots on a Computer" appears at the end of Chapter 3. Included here are sample programs in both BASIC and Pascal.

3. In Chapter 7 we have added considerably more coverage of differential equations.

4. In Chapter 9 a new section on integrals involving $ax^2 + bx + c$ has been added.

5. In Chapter 11 we have added a special review problem set that asks students to pick a test of convergence and then to apply it.

6. The section on quadric surfaces has been moved from Chapter 16 to Chapter 14.

7. The sections on differentials and on the chain rule have been moved to the end of Chapter 16.

8. In Chapter 17 many sections have been rewritten for greater clarity and new material added.

9. In Chapter 18 the material on triple integrals in cylindrical coordinates and on triple integrals in spherical coordinates has been expanded and now is presented in two sections.

10. In Chapter 19 some sections have been rewritten to tighten their logic and improve their readability. The theorem on conservative fields of force has been moved to earlier in the chapter where it is used extensively.

For those schools that prefer to have a single chapter devoted to differential equations, we have prepared a separate chapter on differential equations, available in bulk to adopters. This chapter was designed to provide a self-contained introduction to differential equations. It can work as a prospectus/advertisement for the innumerable uses of these powerful mathematical tools.

In addition to these substantive changes, we have replaced the original computer graphics in the text with new pieces commissioned for this edition. There are two types of three-dimensional computer-generated graphs: surfaces defined by equations of the form $z = f(x, y)$ and quadric surfaces. The graphs of the first type show lines of constant x and lines of constant y on the surface. This is done by drawing the portion of the graph nearest the viewer first, while maintaining a record of the outline of what is presently visible; further portions that are visible are then drawn and added to the outline, whereas new "back" portions that fall within the outline would be hidden and are not drawn. This technique is computationally more efficient than hidden-surface removal algorithms for arbitrary polygonal surfaces. Some of the quadric surfaces are drawn this way also; the rest are generated by "clipping" against the plane that separates the visible and invisible parts of a quadric when viewed from infinity. These graphics have been provided by Douglas Dunham and Robert Hapy of the University of Minnesota, Duluth.

Ancillaries

1. A *Student Solutions Manual*, prepared by Richard Fritz, accompanies this text. We have found the manual an invaluable aid for students who have weak backgrounds in algebra. The manual includes worked-out solutions to every other odd-numbered problem in the text. (The answers to odd-numbered problems appear in the back of the book.)

2. A complete *Solutions Manual* is also available to instructors.

3. A *Student Study Guide*, prepared by Richard St. André, accompanies the text. This study guide tests students on both key concepts and possible algebraic difficulties. It is a strategic aid and is not as comprehensive in solving remedial difficulties as the *Student Solutions Manual*. Our experience has been that the study guide is of great help at urban campuses where students may miss a few classes. This edition provides an abstract of the entire text.

4. We are publishing the *Calculus Blackboard*, by Lee Hill, Ed Lodi, and Wade Ellis, along with this edition. This software package for the APPLE II family and IBM computers is an extensive set of graphing and utility programs that covers the basic concepts of the first year of the course: composition of functions, Newton's method, stepwise functions, the derivative, the antiderivative, area between two curves, numerical integration, parametric equations, and limits. The program features a function plotter with the ability to zoom in on any portion of a given graph up to six times. The package is accompanied by a substantial student workbook with many exercises to help undergraduates fully utilize the power of the programs.

Acknowledgments

Books that are effective teaching tools (for us) and learning tools (for our students) are not so much written as they are developed out of classroom experience. The more experiences of colleagues an author is exposed to, the more the text is thoroughly developed into a reliable teaching tool. This work received an invaluable contribution through the suggestions, criticisms, and encouragement of the following colleagues:

Daniel D. Anderson
University of Iowa, Iowa City

Robert Anderson
University of Quebec at Montreal

William C. Belvel
Sierra College

William B. Bickford
Arizona State University

Robert Blatz
Marywood College

David Boyd
University of South Carolina, Sumter

Blaine Butler
Purdue University

Glenn Calkins
Western State College

Joseph Cannon
Elizabethtown College

Peter G. Casazza
University of Alabama

Benjamin R. Cato, Jr.
College of William and Mary

Charles Chapman
Cedar Crest College

Charles K. Chui
Texas A & M University

Philip S. Clarke
Los Angeles Valley College

Douglas B. Crawford
College of San Mateo

John Dinkins
Wallace Community College

Russell Euler
Northwest Missouri State University,
Maryville

Henri Feiner
Northrop Institute of Technology

Russell Floyd
Texas Wesleyan College

Richard A. Fritz
Moraine Valley Community College

Anton Glaser
Pennsylvania State University, Abington

Stuart Goldenberg
California Polytechnic State University,
San Luis Obispo

Michael B. Gregory
University of North Dakota, Grand Forks

Daniel R. Gustafson
Wayne State University

Alfred W. Hales
University of California, Los Angeles

Donald Hall
California State University, Sacramento

Ray Hamlett
East Central Oklahoma University

Vern Heeren
American River College

William Higgins
Wittenberg University

Kim Hughes
California State University,
San Bernardino

Glenn E. Johnston
Morehead State University

Ken Kalmanson
Montclair State College

Arthur Kaufman
College of Staten Island

William G. Koellner
Montclair State College

Steven Krantz
Pennsylvania State University

V. V. Krishnan
San Francisco State University

Thomas J. Kyrouz
Salem State College

Jeuel LaTorre
Clemson University

Stanley Lukawecki
Clemson University

George Luna
California Polytechnic State University,
San Luis Obispo

Thomas R. Lupton
University of North Carolina,
Wilmington

Thomas R. McCabe
Harper College

Mary McCammon
Pennsylvania State University,
University Park

Otis McCowan
Belmont College

Kendall McDonald
Northwest Missouri State University,
Marysville

Fred Martens
University of Alabama, Birmingham

John Mathews
California State College, Fullerton

Glenda Merhoff
Vanderbilt University

Eldon Miller
University of Mississippi

John Muth
Honolulu Community College

Thomas O'Neil
California Polytechnic State University

Thomas J. O'Reilly
St. Joseph's University

LeRoy Peterson
Indiana University Northwest

Perry G. Phillips
Pinebrook Junior College

Margaret S. Piedem
Somerset County College

Charles S. Rees
University of New Orleans

Dan Rinne
California State University,
San Bernardino

Richard St. André
Central Michigan University

Richard Savage, Jr.
Tennessee Technological University

John T. Scheick
Ohio State University

L. Schiefelbusch
Indiana University Northwest

Zeev Schuss
Tel Aviv University

Thomas S. Shores
University of Nebraska

Chanchal Singh
St. Lawrence University

Arthur G. Sparks
Georgia Southern College

David R. Stone
Georgia Southern College

Keith D. Stroyan
University of Iowa, Iowa City

Donna M. Szott
Community College of Allegheny

Willie E. Taylor, Jr.
Texas Southern University, Houston

Mel Tuscher
West Valley College

Paul Vicknair
California State University,
San Bernardino

Bob Allen Wake
University of California, Santa Cruz

J. Norman Wells
Georgia Southern College

W. Thurmon Whitley
University of New Haven

In addition, this text owes a considerable debt to Robert Brown, Richard Johnson-baugh, Marta Kongsle, Gloria Langer, Phyllis Marmont, Eldon Miller, John Muth, Jane Scott, David Wend, and Thurmon Whitley for its first edition. Special thanks for help in this revision go to Karin Breuer, Catherine Read, Anne Scanlan-Rohrer, Stuart Thomas, and Judith Weber. We would also like to acknowledge the special efforts of Thomas O'Neil, Richard Fritz, and Stanley Lukawecki, whose advice was invaluable.

The Historical Exercises in this text were contributed by Ken Abernethy, from an unpublished paper on the history of the calculus. This outstanding work is unique in that it consistently "reinvents" the calculus, rather than merely talking about the calculus. We hope that some of its contribution is evident in this textbook.

Many professors contributed their time and expertise to making this edition as error-free as the first edition. In particular, we would like to thank the following: Douglas B. Crawford, Richard Fritz, Stuart Goldenberg, Kim Hughes, Arthur Kaufman, Stanley Lukawecki, George Luna, Fred Martens, Thomas O'Neil, H. LeRoy Peterson, Perry G. Phillips, Richard Savage, Jr., Zeev Schuss, Mel Tuscher, and Henry Wyzinsky.

The authors wish to especially acknowledge the efforts of Tom Braden, who was instrumental in bringing us to Wadsworth; of Rich Jones, who orchestrated the development of the first edition; of Phyllis Niklas, whose expertise was invaluable in the production and editing of both the first and second editions; and of Jim Harrison, whose guidance and leadership made this edition possible.

Finally, we would like to thank the many teachers and students who wrote us un-solicited letters about our book. We have used many of their criticisms, but we would be more than human if we didn't admit we were most gratified by some of the comments of students who told us that their good experiences with the calculus made them eager to take additional mathematics courses. We don't pretend to take much credit for that, but it is a pleasure to imagine that we've had something to do with introducing a new generation to the power and beauty of the calculus.

Preface to the Student

I hear . . . . . and I forget
I see . . . . . and I remember
I do . . . . . and I understand

CHINESE SAYING

Second Edition

Calculus and Analytic Geometry

1

Introduction

This chapter contains material needed for the study of calculus. Most of this material is review, but some of it may be new to you.

1. Real Numbers

We begin with the idea of a *set*. A *set* is a collection of objects considered as a whole. The objects of a set S are called *elements* of S, or *members* of S. The set that has no elements, called the *empty set* or *null set*, is denoted by the symbol $\varnothing$.

If a is an element of the set S, we write $a \in S$, which is read "a is an element of S" or "a is in S." To indicate that a is not an element of S, we write $a \notin S$, which is read "a is not an element of S" or "a is not in S."

Ordinarily, a set S can be written in either of two ways. These two methods are illustrated by the following example: Consider the set D that has the elements

$$0, 1, 2, 3, 4, 5, 6, 7, 8, 9$$

In this case, we write

$$D = \{0, 1, 2, 3, 4, 5, 6, 7, 8, 9\}$$

This expression is read "D is the set consisting of the elements 0, 1, 2, 3, 4, 5, 6, 7, 8, 9." Here, we list or display the elements of the set D.

Another way of writing this same set D is to write

$$D = \{x \mid x \text{ is a nonnegative integer less than } 10\}$$

This is read "D is the set of all x such that x is a nonnegative integer less than 10." Here, we have described the set D by giving a property that every element of D has and that no element not in D can have.

Now consider the two sets

$$A = \{1, 2\} \qquad B = \{1, 2, 3, 4\}$$

where each element of A is also found in B. When the elements of A are also elements of B, we say A *is a subset of* B and use the notation $A \subseteq B$ to denote this fact. In addition, it is customary to consider $\varnothing \subseteq A$ for any set A. For example,

$$\{1, 3, 5\} \subseteq \{1, 3, 5, 7, 9\} \qquad \{1, 2, 3\} \subseteq \{1, 2, 3\} \qquad \varnothing \subseteq \{1, 2, 3\}$$

We now describe some important sets of numbers.

The set of *integers* is $\{0, 1, -1, 2, -2, \ldots\}$. The set of *nonnegative integers*, $\{0, 1, 2, \ldots\}$, forms a subset of the set of integers, as does the set of *negative integers*, $\{-1, -2, -3, \ldots\}$.

Rational numbers are ratios of integers. For a rational number a/b, the integer a is called the *numerator*, and the integer b, which cannot be zero, is called the *denominator*.

Figure 1

Although rational numbers occur frequently in nature, numbers that are not rational also occur. For example, the number $\sqrt{2}$ is not rational, but $\sqrt{2}$ is equal to the length of the hypotenuse of an isosceles right triangle with legs that are each of length 1 (see Fig. 1). As another example, the number π, which may be described as the ratio of the circumference of a circle to its diameter, is not a rational number (see Fig. 2). These numbers, $\sqrt{2}$ and π, are examples of *irrational numbers*—numbers that are not rational.

The irrational numbers and rational numbers together form the *set* $\mathbb{R}$ *of real numbers*. To represent each real number, we use what is commonly referred to as a *decimal representation*, or, simply, a *decimal*. For example, the decimal representations of the rational numbers $\frac{3}{4}$, $\frac{5}{2}$, $\frac{2}{3}$, and $\frac{7}{66}$ are $\frac{3}{4} = 0.75$, $\frac{5}{2} = 2.5$, $\frac{2}{3} = 0.666\ldots$, and $\frac{7}{66} = 0.10606\ldots$. We observe (and can prove) that the decimal representation of a rational number is always one of two types: (*1*) *terminating*, or ending ($\frac{3}{4}$, $\frac{5}{2}$, etc.); or (*2*) eventually *repeating* (in $\frac{2}{3}$, the 6's repeat; in $\frac{7}{66}$, the block 06 repeats).

At first, it may appear that these two types represent all possible decimals. However, it is relatively easy to construct a decimal that neither terminates nor eventually repeats. For example, the decimal $0.123456789101112\ldots$, where we write down the positive integers one after the other, will neither terminate nor eventually repeat. In fact, there are an infinite number of such decimals, and they represent the irrational

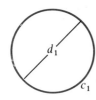

$$\frac{c_1}{d_1} = \frac{c_2}{d_2} = \pi$$

Figure 2

numbers. For example, the real numbers

$$\sqrt{2} = 1.414213\ldots \qquad \pi = 3.14159\ldots$$

are irrational, having decimals that neither terminate nor eventually repeat.

Thus, the set of real numbers may be thought of as the set of all possible decimals. This feature of real numbers gives them their practicality.* In the physical world, many changing magnitudes, like the length of a heated rod, the velocity of a particle, and so on, are assumed to pass through every possible magnitude from the initial one to the final one. Since the precise measurement of a magnitude is naturally given by a decimal, the equivalent of all possible magnitudes is all possible decimals (real numbers).

In practice, it is usually necessary to represent real numbers by approximations. For example, using the symbol $\approx$, read "approximately equal to," we can write

$$\sqrt{2} \approx 1.4142 \qquad \pi \approx 3.1416$$

PROPERTIES OF REAL NUMBERS

As an aid to your review of real numbers and their properties, we list several important rules and notations. The letters a, b, c, d represent real numbers; any exceptions will be noted as they occur.

1. *Commutative Laws*

    (a) $a + b = b + a$ (b) $a \cdot b = b \cdot a$

2. *Associative Laws*

    (a) $a + b + c = (a + b) + c = a + (b + c)$
    (b) $a \cdot b \cdot c = (a \cdot b) \cdot c = a \cdot (b \cdot c)$

3. *Distributive Law*

    $a \cdot (b + c) = (a \cdot b) + (a \cdot c)$

4. *Arithmetic of Ratios*

    (a) $\dfrac{a}{b} = a \cdot \dfrac{1}{b} \qquad b \neq 0$

    (b) $\dfrac{a}{b} + \dfrac{c}{d} = \dfrac{(a \cdot d) + (b \cdot c)}{b \cdot d} \qquad b \neq 0, \quad d \neq 0$

    (c) $\dfrac{a}{b} \cdot \dfrac{c}{d} = \dfrac{a \cdot c}{b \cdot d} \qquad b \neq 0, \quad d \neq 0$

    (d) $\dfrac{a}{b} \div \dfrac{c}{d} = \dfrac{a}{b} \cdot \dfrac{d}{c} = \dfrac{a \cdot d}{b \cdot c} \qquad b \neq 0, \quad c \neq 0, \quad d \neq 0$

---

* Other number systems exist that have many applications (such as the complex numbers). In this book, however, we limit our discussion to problems in which only real numbers are used.

5. *Rules for Division*

$$0 \div a = 0 \quad \text{or} \quad \frac{0}{a} = 0 \quad \text{for any real number } a \text{ different from } 0$$

$$a \div 0 \quad \text{or} \quad \frac{a}{0} \quad \text{is undefined for any real number } a; \text{ never divide by zero!}$$

$$a \div a = 1 \quad \text{for any real number } a \text{ different from } 0$$

6. *Rules of Signs*

(a) $a \cdot (-b) = -(a \cdot b)$ (b) $(-a) \cdot b = -(a \cdot b)$
(c) $(-a) \cdot (-b) = a \cdot b$ (d) $-(-a) = a$

7. *Agreements; Notations*

(a) Given $a \cdot b + c$ or $c + a \cdot b$, we agree to multiply $a \cdot b$ first, and then add c.

(b) A mixed number $3\frac{5}{8}$ means $3 + \frac{5}{8}$; 3 *times* $\frac{5}{8}$ is written as $3(\frac{5}{8})$ or $(3)(\frac{5}{8})$ or $3 \cdot \frac{5}{8}$.

8. *Exponents.* For any positive integer n and any real number x, we define

$$x^1 = x, \quad x^2 = x \cdot x, \quad \ldots, \quad x^n = \underbrace{x \cdot x \cdot \cdots \cdot x}_{n \text{ factors}}$$

For any real number $x \neq 0$, we define

$$x^0 = 1, \quad x^{-1} = \frac{1}{x}, \quad x^{-2} = \frac{1}{x^2}, \quad \ldots, \quad x^{-n} = \frac{1}{x^n}$$

9. *Laws of Exponents* (a, x are real numbers; n, m are integers)

(a) $x^n \cdot x^m = x^{n+m}$ (b) $(x^n)^m = x^{nm}$

(c) $(ax)^n = a^n \cdot x^n$ (d) $\left(\dfrac{x}{a}\right)^n = \dfrac{x^n}{a^n} \qquad a \neq 0$

(e) $\dfrac{x^n}{x^m} = x^{n-m} \qquad x \neq 0$

10. *Roots.* For a positive integer q, the qth root of x, $\sqrt[q]{x}$, is a symbol for the real number that, when raised to the power q, equals x. If q is even and x is positive, then $\sqrt[q]{x}$ is declared to be positive. For example,

$$\sqrt[3]{8} = 2 \quad \text{since} \quad 2^3 = 8 \qquad \sqrt[2]{64} = 8 \quad \text{since} \quad 8^2 = 64$$

Here, $\sqrt[3]{x}$ is called the *cube root* of x and $\sqrt[2]{x}$ is called the *square root* of x. Usually, we abbreviate square roots by $\sqrt{}$, dropping the ^{2}. *Be careful! No meaning is assigned to even roots of negative numbers*, since any real number raised to an even power is nonnegative. For example, $\sqrt{4} = 2$, whereas $\sqrt{-4}$ has no meaning in the set of real numbers. On the other hand, $\sqrt[3]{27} = 3$, while

$\sqrt[3]{-64} = -4$ since $(-4)^3 = -64$.   Thus, meaning is given to odd roots of negative numbers. Finally, following the usual convention, even roots of positive numbers are always positive. Thus, even though $(2)^2 = 4$ and $(-2)^2 = 4$, only the positive root 2 equals $\sqrt{4}$. That is, $\sqrt{4} = 2$.

11. *Rational Exponents.*   For any integer p and any positive integer q, we define $x^{1/q}$ and $x^{p/q}$ as

$$x^{1/q} = \sqrt[q]{x} \qquad x^{p/q} = (x^{1/q})^p = (\sqrt[q]{x})^p \quad \text{or} \quad x^{p/q} = (x^p)^{1/q} = \sqrt[q]{x^p}$$

For example,

$$27^{1/3} = \sqrt[3]{27} = 3 \qquad (-32)^{1/5} = \sqrt[5]{-32} = -2 \qquad x^{2/3} = \sqrt[3]{x^2} = (\sqrt[3]{x})^2$$

With this definition, the laws of exponents listed in 9 hold for rational exponents:

(a) $\sqrt{x} \cdot \sqrt[3]{x} = x^{1/2} \cdot x^{1/3} = x^{(1/2)+(1/3)} = x^{5/6} = \sqrt[6]{x^5}$

(b) $(\sqrt{x+2})^2 = [(x+2)^{1/2}]^2 = (x+2)^{(1/2)(2)} = (x+2)^1 = x+2$

(c) $\sqrt{20} = \sqrt{(4)(5)} = [(4)(5)]^{1/2} = (4^{1/2})(5^{1/2}) = (\sqrt{4})(\sqrt{5}) = 2\sqrt{5}$

(d) $\sqrt[3]{\dfrac{2}{27}} = \left(\dfrac{2}{27}\right)^{1/3} = \dfrac{2^{1/3}}{27^{1/3}} = \dfrac{\sqrt[3]{2}}{\sqrt[3]{27}} = \dfrac{\sqrt[3]{2}}{3}$

(e) $\dfrac{\sqrt{x}}{\sqrt[3]{x}} = \dfrac{x^{1/2}}{x^{1/3}} = x^{(1/2)-(1/3)} = x^{1/6} = \sqrt[6]{x}$

12. *Cancellation Laws*

(a)   If $a \cdot c = b \cdot c$ and c is not zero, then $a = b$.

(b)   If b and c are not zero, then $\dfrac{a \cdot c}{b \cdot c} = \dfrac{a}{b}$.

13. *Product Law.*   If $a \cdot b = 0$,   then either $a = 0$ or $b = 0$.

POSITIVE AND NEGATIVE NUMBERS

The real numbers can be divided into three nonempty subsets that have no elements in common: (*1*) the set of positive real numbers; (*2*) the set with just zero as a member; and (*3*) the set of negative real numbers. We list some familiar properties:

1. Any real number is either positive or negative or equal to zero.

2. The sum and the product of two positive numbers are positive.

3. The product of two negative numbers is positive.

4. The product of a positive number and a negative number is negative.

For real numbers a and b, a is less than b $(a < b)$ or b is greater than a $(b > a)$ if and only if the difference $b - a$ is a positive real number. For example,

$2 < 7$ and $-3 > -6$.   It is easy to conclude that

$$a \text{ is positive if and only if } \quad a > 0$$
$$a \text{ is negative if and only if } \quad a < 0$$

If a is less than or equal to b, we write $a \leq b$.   If a is greater than or equal to b, we write $a \geq b$.   If $a < c$ and $c < b$,   we write $a < c < b$.   This says that c is between a and b. Similarly, if $a \leq c$ and $c \leq b$,   we write $a \leq c \leq b$.   The notations $a \leq c < b$ and $a < c \leq b$ are given similar interpretations.

Inequalities obey the following laws:

1. *Addition Law.* If $a \leq b$,   then $a + c \leq b + c$ for any choice of c. That is, the addition of a number to each side of an inequality will not affect the sense or direction of the inequality.

2. *Multiplication Laws*

   (a) If $a \leq b$ and $c > 0$,   then $a \cdot c \leq b \cdot c$.
   (b) If $a \leq b$ and $c < 0$,   then $a \cdot c \geq b \cdot c$.

   When multiplying each side of an inequality by a number, the sense or direction of the inequality remains the same if we multiply by a positive number; it is reversed if we multiply by a negative number.

3. *Division Laws*

   (a) If $a > 0$,   then $\dfrac{1}{a} > 0$.   That is, the reciprocal of a positive number is positive.

   (b) If a, b are positive and if $a < b$,   then $\dfrac{1}{a} > \dfrac{1}{b}$.

4. *Trichotomy Law.*   For any two real numbers a, b,   one and only one of the following is true: $a < b$, $a = b$, $b < a$.

5. *Transitive Law.* If $a < b$ and $b < c$,   then $a < c$.

COORDINATES

Real numbers can be represented geometrically on a horizontal line. We begin by selecting an arbitrary point O, called the *origin*, and associate it with the real number 0. We then establish a scale by marking off line segments of equal length (units) on each side of 0. By agreeing that the positive direction is to the right of 0 and the negative direction is to the left of 0, we can successively associate the integers 1, 2, 3, ... with each mark to the right of 0 and the integers $-1, -2, -3, \ldots$ with each mark to the left of 0 (see Fig. 3). By subdividing these segments, we can locate rational numbers such as $\frac{1}{2}$ and $-\frac{3}{2}$. The irrational numbers are located by geometric construction (as in the case of $\sqrt{2}$) or by other means. In this way, every point P on the line is associated with a unique real number x, called the *coordinate of P* (see

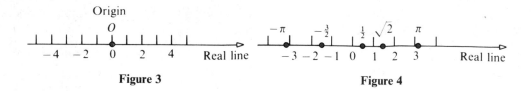

Figure 3   **Figure 4**

Fig. 4). Coordinates establish an ordering for the real numbers; that is, if a and b are coordinates of two points P and Q, respectively, then $a < b$ means that P lies to the left of Q on the line.

VARIABLES

A *variable* is a symbol (usually a letter, x, y, etc.) used to represent any real number. An *inequality* is a statement involving one or more variables and one of the inequality symbols ($<$, $\leq$, $>$, $\geq$). To *solve* an inequality means to find all possible numbers that the variables can assume to make the statement true. The set of all such numbers is called the *solution* (or *solution set*). Two inequalities with the same solution are called *equivalent*.

To find the solution of an inequality, we apply the laws of inequalities. For example, the statement $x + 2 \leq 3x - 5$ is an inequality with one variable; it can be solved by obtaining a series of equivalent inequalities until we get to an inequality with an obvious solution. Thus, for the inequality $x + 2 \leq 3x - 5$,   we have

$$x + 2 \leq 3x - 5$$

$$-2x \leq -7 \qquad \text{Subtract 2 and then } 3x \text{ from both sides}$$

$$x \geq \frac{7}{2} \qquad \text{Multiply by } -\tfrac{1}{2} \text{ and remember to reverse the inequality because we multiplied by a negative number}$$

The solution is the set of all real numbers to the right of $\frac{7}{2}$, including $\frac{7}{2}$ (Fig. 5).

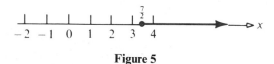

Figure 5

Solving other types of inequalities may be more difficult. Some examples are given below.

EXAMPLE 1   Solve the inequality: $x^2 + x - 12 > 0$

Solution   We factor the left side, obtaining

$$(x - 3)(x + 4) > 0$$

The product of two real numbers is positive either when both factors are positive or when both factors are negative.

Both Positive	or	*Both Negative*

$$x - 3 > 0 \quad \text{and} \quad x + 4 > 0 \qquad\qquad x - 3 < 0 \quad \text{and} \quad x + 4 < 0$$

$$x > 3 \quad \text{and} \quad x > -4 \qquad\qquad\qquad x < 3 \quad \text{and} \quad x < -4$$

The numbers x that are greater than 3 and -4 are simply

The numbers x that are less than 3 and -4 are simply

$$x > 3 \qquad\qquad or \qquad\qquad x < -4$$

The solution is $\{x \mid x > 3 \ \text{ or } \ x < -4\}$ (see Fig. 6).

Figure 6   ∎

We also can obtain the solution to the inequality of Example 1 by another method. The left-hand side of the inequality is factored so that it becomes $(x - 3)(x + 4) > 0$, as before. We then construct a graph (Fig. 7) that uses the numbers $x = 3$ and $x = -4$ as cutoff points.

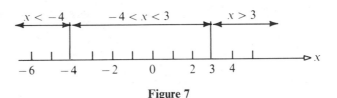

Figure 7

These numbers are the solutions to the equation

$$x^2 + x - 12 = (x - 3)(x + 4) = 0$$

and they separate the line into three parts: $x < -4$, $-4 < x < 3$, and $x > 3$. In the part of the line where $x < -4$, we deduce that both the quantities $(x - 3)$ and $(x + 4)$ are always negative, so their product must always be positive. Therefore, $x < -4$ is a solution of the inequality. In the part of the line where $-4 < x < 3$, we deduce that $(x - 3)$ is always negative and $(x + 4)$ is always positive, so their product is always negative. We conclude that the numbers between -4 and 3 are not solutions of the inequality. In the part of the line where $x > 3$, we deduce that both the quantities $(x - 3)$ and $(x + 4)$ are always positive, so their product is always positive. Hence, numbers greater than 3 are solutions of the inequality. Table 1 summarizes these results.

Table 1

	Sign of x − 3	Sign of x + 4	Sign of (x − 3)(x + 4)	Conclusion
$x < -4$	−	−	+	$x < -4$ is solution
$-4 < x < 3$	−	+	−	$-4 < x < 3$ is not solution
$x > 3$	+	+	+	$x > 3$ is solution

This method of solving inequalities is particularly appealing when the inequality contains more than two factors. It can also be used to solve inequalities that involve ratios.

EXAMPLE 2   Solve: $\dfrac{x}{x-4} < 3$

Solution   We begin by transforming the inequality to get 0 on the right side:

$$\frac{x}{x-4} - 3 < 0$$

$$\frac{x - 3(x-4)}{x-4} < 0$$

$$\frac{-2(x-6)}{x-4} < 0$$

$$\frac{x-6}{x-4} > 0$$

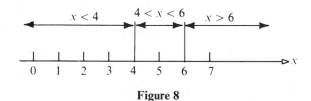

Figure 8

The quantities $(x - 4)$ and $(x - 6)$ supply two cutoff points, 4 and 6, as shown in Figure 8. Now we construct a table to find out where $\dfrac{x-6}{x-4}$ is positive (see Table 2).

Table 2

	Sign of x − 4	Sign of x − 6	Sign of $\dfrac{x-6}{x-4}$	Conclusion
$x < 4$	−	−	+	$x < 4$ is solution
$4 < x < 6$	+	−	−	$4 < x < 6$ is not solution
$x > 6$	+	+	+	$x > 6$ is solution

The solution is $\{x \mid x < 4 \quad \text{or} \quad x > 6\}$ (see Fig. 9).

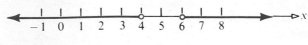

Figure 9   ■

ABSOLUTE VALUE

The absolute value of a number is the magnitude, or distance from 0, of that number. Thus, the absolute value of 5 is 5; and the absolute value of -6 is 6.

(1.1)   **DEFINITION**   *Absolute Value.*   **The *absolute value* of a real number x, denoted by $|x|$, is defined as**

$$|x| = \begin{cases} x & \text{if} \quad x \geq 0 \\ -x & \text{if} \quad x < 0 \end{cases}$$

For example,

$$|8| = 8 \qquad |0| = 0 \qquad |-4| = -(-4) = 4 \qquad |1 - \sqrt{2}| = -(1 - \sqrt{2}) = \sqrt{2} - 1$$

The definition of the absolute value of a number x may also be expressed as $|x| = \sqrt{x^2}$.

Since the absolute value of a negative number is positive, it follows that the absolute value of any number is either positive or zero. Some other properties of absolute value are listed below.

(1.2)   **If a and x are real numbers, then:**

(a) $|x - a| = |a - x|$ (b) $|ax| = |a| \, |x|$

(c) $-|x| \leq x \leq |x|$ (d) $\left| \dfrac{x}{a} \right| = \dfrac{|x|}{|a|} \qquad a \neq 0$

(e) $|x|^n = |x^n| \qquad n$ **an integer**

These properties are consequences of the definition.

A geometric interpretation of $|x - a|$, the absolute value of the difference between x and a, is that it equals the distance from a to x. For example, if $|x - 3| = 6$, then x lies either 6 units to the right of 3 or 6 units to the left of 3 (see Fig. 10). That is, the solution of $|x - 3| = 6$ is $\{-3, 9\}$.

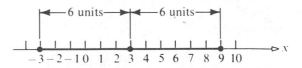

Figure 10

The general rule follows:

If the absolute value of a mathematical expression equals some positive number p, then the expression itself equals either p or $-p$.

For example,

$$|x| = 3 \quad \text{means} \quad x = 3 \quad \text{or} \quad x = -3$$

Similarly,

$$|x - 8| = 6 \quad \text{means} \quad x - 8 = 6 \qquad \text{or} \quad x - 8 = -6$$
$$x = 6 + 8 \qquad\qquad x = -6 + 8$$
$$= 14 \qquad\qquad\qquad = 2$$

Let's turn now to some examples of inequalities that involve absolute value.

EXAMPLE 3   Find all numbers x for which $\quad |x| < 7$.

Solution   Here we are asked to find all numbers x for which the distance from 0 is less than 7. Any x between -7 and 7 satisfies this condition. Consequently, the solution is $\{x \mid -7 < x < 7\}$.   ∎

In general, for any number $a > 0$:

(1.3) $-a < x < a \quad$ is equivalent to $\quad |x| < a$

(1.4) $-a \le x \le a \quad$ is equivalent to $\quad |x| \le a$

EXAMPLE 4   Find all x for which $\quad |x - 3| \le 6$.

Solution   Here, think of $(x - 3)$ as a single unknown quantity so that by applying expression (1.4), we find

$$-6 \le x - 3 \le 6$$

If we add 3 to each term in this inequality, we get

$$-6 + (3) \le (x - 3) + (3) \le 6 + (3)$$

which simplifies to $\quad -3 \le x \le 9$.   The solution is $\quad \{x \mid -3 \le x \le 9\}$.   ∎

We are now ready for an inequality involving the absolute value of a sum. This inequality will be useful in later chapters.

(1.5)   *Triangle Inequality.** **If x, y are real numbers, then**

$$|x + y| \le |x| + |y|$$

---

* This inequality is an algebraic expression of the fact that the length of any side of a triangle does not exceed the sum of the lengths of the other two sides.

Proof By property (c) in (1.2), $-|x| \leq x \leq |x|$ and $-|y| \leq y \leq |y|$. Adding these, we find

$$-(|x| + |y|) \leq x + y \leq |x| + |y|$$

Hence, by expression (1.4), $|x + y| \leq |x| + |y|$. ∎

Another useful property involving the absolute value of a difference is stated below. Its proof is left as an exercise.

(1.6)   **If x, y are real numbers, then**

$$|x - y| \geq |x| - |y|$$

INTERVALS

Let a and b be two real numbers with $a < b$. A *closed interval* $[a, b]$ is the set of all real numbers x from a to b, inclusive; that is,

$$[a, b] = \{x \mid a \leq x \leq b\}$$

An *open interval* (a, b) consists of all real numbers x between a and b, exclusive of both a and b; that is,

$$(a, b) = \{x \mid a < x < b\}$$

Finally, the *half-open* (*semi-open*) or *half-closed* (*semi-closed*) intervals are defined by

$$[a, b) = \{x \mid a \leq x < b\} \qquad (a, b] = \{x \mid a < x \leq b\}$$

In these definitions, a is the *left endpoint* and b is the *right endpoint* of each interval. In Figure 11 an open circle ○ is used to denote the fact that an endpoint is not included in the interval, while a filled circle ● is used when an endpoint is included in the interval.

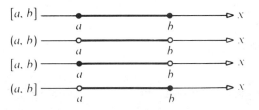

Figure 11

For a real number a, the notation $[a, +\infty)$ denotes the set of all real numbers greater than or equal to a; that is,

$$[a, +\infty) = \{x \mid x \geq a\}$$

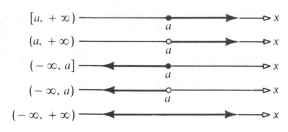

Figure 12

The symbol $+\infty$, read "plus infinity," is not a real number, but is merely a notational device. Similarly, we define

$$(a, +\infty) = \{x \,|\, x > a\}$$
$$(-\infty, a] = \{x \,|\, x \le a\}$$
$$(-\infty, a) = \{x \,|\, x < a\}$$
$$(-\infty, +\infty) = \mathbb{R} \text{ (set of real numbers)}$$

Figure 12 shows graphs of these intervals.

EXAMPLE 5   Find all x for which $\;|x - 2| \ge 3$.

Solution   Here we ask for all real numbers for which the distance from 2 is greater than or equal to 3. Using Figure 13 as an aid, we conclude that

$$x \le -1 \qquad \text{or} \qquad x \ge 5$$

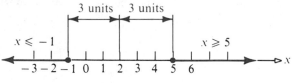

Thus, the solution is $(-\infty, -1]$ together with $[5, +\infty)$.   ∎

Figure 13

EXERCISE 1

In Problems 1–4 replace the * by $<$, $>$, or $=$, whichever gives a true statement.

1. $\frac{1}{3} * 0.33$ **2.** $\frac{1}{4} * 0.25$ **3.** $3 * \sqrt{9}$ **4.** $\pi * \frac{22}{7}$

In Problems 5–44 find the solution.

5. $3x + 5 \le 2$ **6.** $-3x + 5 \le 2$ **7.** $3x + 5 \ge 2$ **8.** $4 - 5x \ge 3$

9. $6x - 3 \ge 8x + 5$ **10.** $8 - 2x \le 5x - 6$ **11.** $14x - 21x + 16 \le 3x - 2$

12. $10x - 3x \le 2x + 5 - 15$ **13.** $x^2 - 5x + 6 \ge 0$ **14.** $x^2 + 2x \ge 0$

15. $x^2 + 7x < -12$ **16.** $x^2 - x < 12$ **17.** $|x| = 5$

18. $|x| = 6$ **19.** $|x| \le 3$ **20.** $|x| \le 4$

21. $|x - 3| < 4$ **22.** $|x + 2| \le 6$ **23.** $|2x - 4| + 5 \le 9$

24. $6 + |3x - 7| \le 10$ **25.** $|x - 3| > 4$ **26.** $|x + 2| > 3$

27. $|2x - 4| \ge -8$ **28.** $|3x + 4| \ge -5$ **29.** $|x - 3| < 0.01$

30. $|2x - 1| < 0.02$ **31.** $|\frac{1}{2} - 2x| \le 4$ **32.** $|4 - \frac{1}{2}x| < 5$

33. $\dfrac{1}{x} < 3$ **34.** $\dfrac{2}{x} \ge -5$ **35.** $\dfrac{2}{x - 2} \le -5$ **36.** $\dfrac{2}{x + 3} > 6$

37. $\dfrac{2x + 1}{x - 3} < 1$ **38.** $\dfrac{\frac{3}{2}x - 2}{x + 5} > 1$ **39.** $\dfrac{2}{3 - x} < 1$ **40.** $\dfrac{2}{3 - x} \le 4$

41. $\dfrac{3}{x} < \dfrac{2}{x-1}$ **42.** $\dfrac{3}{x+1} < \dfrac{2}{x-1}$ **43.** $\left|\dfrac{1}{x}\right| < 2$ **44.** $\left|\dfrac{2}{x}\right| < 3$

45. When does $|x| = x$ hold?   **46.** When does $|x| = -x$ hold?

In Problems 47 and 48 find all numbers x for which the given expression has meaning.

47. $\sqrt[4]{x^2 - 3x + 2}$ **48.** $\sqrt{x^2 - 4}$

49. Show that if $|x - a| < \frac{1}{3}$ and $|a - y| < \frac{1}{3}$, then $|x - y| < \frac{2}{3}$. [*Hint:* $|x - y| = |(x - a) + (a - y)|$]

50. Verify the triangle inequality (1.5) for:

(a) $x = 2,\quad y = 3$ (b) $x = -2,\quad y = 3$ (c) $x = 2,\quad y = -3$ (d) $x = -2,\quad y = -3$

51. If $a < b$, prove that $a < \dfrac{a+b}{2} < b$. The number $\dfrac{a+b}{2}$ is called the *arithmetic mean of a and b*. Show that the arithmetic mean is equidistant from a and b.

52. If $0 < a < b$, prove that $a < \sqrt{ab} < b$. The number $\sqrt{ab}$ is called the *geometric mean of a and b*. Show that the geometric mean is less than the arithmetic mean. [*Hint:* Use the fact that $(\sqrt{b/2} - \sqrt{a/2})^2 > 0$.]

53. If $0 < a < b$, prove that $a < h < b$, where h is defined by

$$\frac{1}{h} = \frac{1}{2}\left(\frac{1}{a} + \frac{1}{b}\right)$$

The number h is called the *harmonic mean of a and b*. Show that h equals the ratio of the geometric mean squared to the arithmetic mean.

54. Show that, for a fixed perimeter, a square encloses more area than any rectangle.

55. Show that, for a fixed perimeter, a circle encloses more area than a square.

56. Prove that $|x - y| \ge |x| - |y|$. [*Hint:* Show that $|x| = |x - y + y| \le |x - y| + |y|$.]

57. If $a \le b$ and $c < 0$, prove that $ac \ge bc$.

58. If a, b are positive and if $a < b$, prove that $1/a > 1/b$.

59. If $0 < a \le b$, prove that $a^2 \le b^2$.

60. If a, b are positive and if $a^2 \le b^2$, prove that $a \le b$.

2. Graphing

RECTANGULAR COORDINATES

Consider two lines, one horizontal and the other vertical. Call the horizontal line the *x-axis* and the vertical line the *y-axis*. Assign coordinates to points on these lines, as described previously, by using their point of intersection as the origin O and using a convenient scale on each. We follow the usual convention that points on the x-axis to the right of O are associated with positive real numbers, those to the left of O with negative numbers, those on the y-axis above O are associated with positive real numbers, and those below O with negative real numbers. This gives the origin a value of zero on both the x-axis and the y-axis.

Any point P in the plane formed by the x-axis and y-axis can then be located by using an *ordered pair* of real numbers. Let x denote the signed distance of P from the y-axis (signed in the sense that if P is to the right of the y-axis, then $x > 0$ and if P is to the left of the x-axis, then $x < 0$); and let y denote the signed distance of P from the x-axis. The ordered pair (x, y), the *coordinates of* P, then gives us enough information to locate the point P. We can assign ordered pairs of real numbers to every point P, as shown in Figure 14.

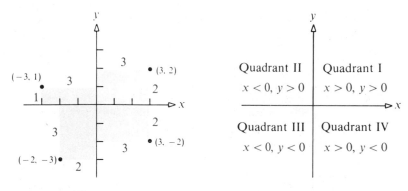

Quadrant II	Quadrant I
$x < 0, y > 0$	$x > 0, y > 0$
Quadrant III	Quadrant IV
$x < 0, y < 0$	$x > 0, y < 0$

Figure 14   **Figure 15**

If (x, y) are the coordinates of a point P, then x is called the *abscissa of* P and y is the *ordinate of* P. For example, the coordinates of the origin O are $(0, 0)$. The abscissa of any point on the y-axis is 0; the ordinate of any point on the x-axis is 0.

The coordinate system described here is a *rectangular* or *cartesian coordinate system* and divides the plane into four sections called *quadrants* (see Fig. 15). In quadrant I, for example, both the abscissa x and the ordinate y of all points are positive.

EQUATIONS

If an equation involves the variables x and y, then the *graph* of the equation consists of the set of points (x, y) in the plane with coordinates that satisfy the equation.

EXAMPLE 1   Graph the equation: $y = 2x + 5$

Solution   We want to find all points (x, y) for which the ordinate y equals twice the abscissa x plus 5. To locate some of these points (and thus get an idea of the pattern of the graph), let us assign some numbers to x and find corresponding values for y:

$$\text{If} \quad x = 0, \quad \text{then} \quad y = 2(0) + 5 = 5$$
$$\text{If} \quad x = 1, \quad \text{then} \quad y = 2(1) + 5 = 7$$
$$\text{If} \quad x = -5, \quad \text{then} \quad y = 2(-5) + 5 = -5$$
$$\text{If} \quad x = 10, \quad \text{then} \quad y = 2(10) + 5 = 25$$

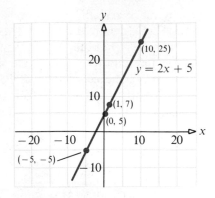

By connecting these points,* we obtain the graph of the equation (a straight line), as shown in Figure 16. ∎

Figure 16

EXAMPLE 2   Graph the equation: $y = x^2$

Solution   A table provides several points on the graph:

x	0	1	2	3	4	-1	-2	-3
y	0	1	4	9	16	1	4	9

In Figure 17 we list these points and, by connecting them with a smooth curve, we obtain the graph (a *parabola*). ∎

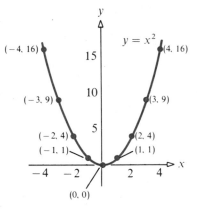

Figure 17

Two useful tools for obtaining the graph of an equation are *intercepts* and *symmetry*.

INTERCEPTS

The points at which a graph *intersects* the coordinate axes are called the *intercepts*. The abscissa of a point at which the graph crosses the x-axis is an *x-intercept*, and the ordinate of a point at which the graph crosses the y-axis is a *y-intercept* (see Fig. 18). To find the x-intercept(s) of an equation, we set $y = 0$ in the equation and solve the equation for x. To find the y-intercept(s), we set $x = 0$ and solve the equation for y. For example, to find the x-intercept(s) of $y = x^2 - 4$, we set $y = 0$. The resulting equation, $x^2 - 4 = 0$, has two solutions: $x = 2$, $x = -2$. Thus, the x-intercepts are 2 and -2. The y-intercept, found by setting $x = 0$ in the equation, is $y = -4$. The graph of $y = x^2 - 4$ thus has three intercepts: $(2, 0)$, $(-2, 0)$, and $(0, -4)$. See Figure 19.

---

* Merely connecting points will not give the complete picture for most problems. An important application of calculus is as an aid in graphing by identifying various characteristics of the graph (Chap. 4).

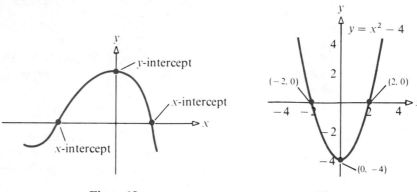

Figure 18   Figure 19

SYMMETRY

Another useful tool for graphing equations is *symmetry*, particularly symmetry with respect to the *x*-axis, *y*-axis, and origin.

1. **Symmetry with Respect to the x-Axis.**   For every point (x, y) on a graph, the point $(x, -y)$ is also on the graph.

2. **Symmetry with Respect to the y-Axis.**   For every point (x, y) on a graph, the point $(-x, y)$ is also on the graph.

3. **Symmetry with Respect to the Origin.**   For every point (x, y) on a graph, the point $(-x, -y)$ is also on the graph.

Figure 20 illustrates some of the possibilities that can occur.

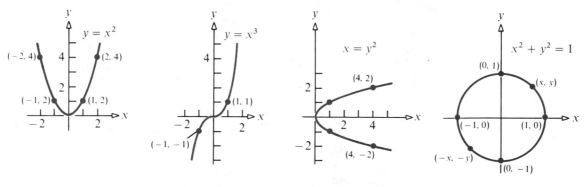

(a) Symmetry with respect to the *y*-axis

(b) Symmetry with respect to the origin

(c) Symmetry with respect to the *x*-axis

(d) Symmetry with respect to the *x*-axis, *y*-axis, and origin

Figure 20

EXAMPLE 3   Examine $xy^2 - x^4 + 5 = 0$ for symmetry with respect to the x-axis, y-axis, and origin.

Solution   For the graph of the equation $xy^2 - x^4 + 5 = 0$ to be symmetric with respect to the x-axis requires that whenever (x, y) is on the graph, then so is $(x, -y)$. This requires that

$$\text{if} \qquad xy^2 - x^4 + 5 = 0 \qquad \text{then} \qquad x(-y)^2 - x^4 + 5 = 0$$

This is the case, since $(-y)^2 = y^2$.   Hence, the graph of $xy^2 - x^4 + 5 = 0$ is symmetric with respect to the x-axis.

For symmetry with respect to the y-axis, we require that

$$\text{if} \qquad xy^2 - x^4 + 5 = 0 \qquad \text{then} \qquad (-x)y^2 - (-x)^4 + 5 = 0$$

This implication is not true, so the graph of $xy^2 - x^4 + 5 = 0$ is not symmetric with respect to the y-axis.

Similarly, we can show that the graph of $xy^2 - x^4 + 5 = 0$ is not symmetric with respect to the origin.   ■

Knowing that the graph of $xy^2 - x^4 + 5 = 0$ is symmetric with respect to the x-axis means that once we know its graph above the x-axis, we automatically know its graph below the x-axis.

Sometimes it is possible to obtain the graph of an equation by a simple *translation*. For example, the graph of the equation $y = x^2 + 1$ may be easily obtained by "lifting" the graph of $y = x^2$ one unit. Some examples are given in Figure 21, where we show translations of some of the graphs given in Figure 20.

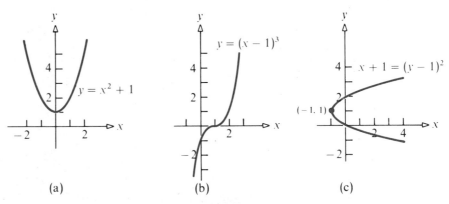

Figure 21

Another graphing technique is that of adding ordinates. To graph the equation $y = x^2 + x$,   we can graph the two equations $y = x^2$ and $y = x$ and then add the heights. Figure 22 illustrates this procedure.

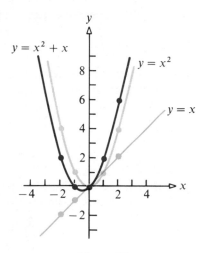

Figure 22

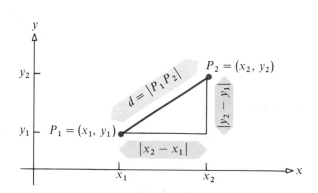

Figure 23

DISTANCE BETWEEN POINTS

Let (x_1, y_1) denote the coordinates of point P_1 and let (x_2, y_2) be the coordinates of point P_2. In moving from P_1 to P_2, the abscissa changes from x_1 to x_2. In calculus we denote this *change in x* by the symbol Δx (read "delta x"). That is, $\Delta x = x_2 - x_1$. Similarly, in moving from P_1 to P_2, the ordinate changes from y_1 to y_2, and this *change in y* is denoted by $\Delta y = y_2 - y_1$.* For example, if $P_1 = (5, -2)$ and $P_2 = (4, 7)$,   then the change in x from P_1 to P_2 is $\Delta x = 4 - 5 = -1$.   The change in y from P_1 to P_2 is $\Delta y = 7 - (-2) = 9$.

If the same scale is used on both the x-axis and the y-axis, then all distances in the plane can be measured by using this same scale. In fact, by using the theorem of Pythagoras, we find:

The distance between two points $P_1 = (x_1, y_1)$ and $P_2 = (x_2, y_2)$,   which we shall denote by $|P_1 P_2|$,   is

(1.7) $$|P_1 P_2| = \sqrt{(x_2 - x_1)^2 + (y_2 - y_1)^2} = \sqrt{(\Delta x)^2 + (\Delta y)^2}$$

See Figure 23.

Thus, to compute the distance between two points, find the change in their abscissas (Δx), square it, and add this to the square of the change in their ordinates (Δy). The square root of this sum is the distance. For example, to find the distance d between the points $(-2, 5)$ and $(3, 2)$, we compute

$$\Delta x = 3 - (-2) = 5 \qquad \text{and} \qquad \Delta y = 2 - 5 = -3$$

---

* The symbols Δx and Δy measure the change in x and the change in y, respectively, and do not mean the product of Δ by x or the product of Δ by y.

Then

$$d = \sqrt{(\Delta x)^2 + (\Delta y)^2} = \sqrt{(5)^2 + (-3)^2} = \sqrt{34}$$

The distance between two points is never a negative number. Furthermore, the only time the distance between two points is zero is when the two points are identical. Finally, it makes no difference whether the distance is computed from P_1 to P_2 or from P_2 to P_1; that is, $|P_1P_2| = |P_2P_1|$.

CIRCLES

Figure 24 is the graph of all points (x, y) that are a fixed distance R from a fixed point (h, k). We recognize this as the graph of a *circle* with its center at (h, k) and with radius R.

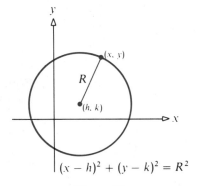

Figure 24

We can find the equation of this circle by using the distance formula (1.7). If (x, y) is any point on the circle, then

$$\sqrt{(x - h)^2 + (y - k)^2} = R$$

or, equivalently,

(1.8) $$(x - h)^2 + (y - k)^2 = R^2$$

This equation is referred to as the *standard equation* of a circle with radius R and center (h, k).

For example, the set of points defined by

$$(x - 2)^2 + (y + 3)^2 = 9$$

is a circle with center $(2, -3)$ and radius 3.

If the center is at the origin (that is, $h = 0$, $k = 0$) and the radius R is of unit length $(R = 1)$, then we obtain

$$x^2 + y^2 = 1$$

which is the equation of the *unit circle*.

EXAMPLE 4 Discuss the graph of $x^2 + y^2 + 4x - 6y + 12 = 0$.

Solution First, we group the terms as follows:

$$(x^2 + 4x) + (y^2 - 6y) = -12$$

We proceed to complete the square of each parenthetical term. Then

$$(x^2 + 4x + 4) + (y^2 - 6y + 9) = -12 + 4 + 9$$
$$(x + 2)^2 + (y - 3)^2 = 1$$

This is a circle with center $(-2, 3)$ and radius 1 (see Fig. 25).

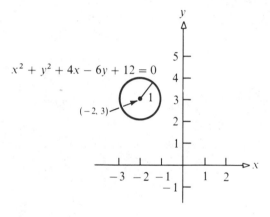

Figure 25   ■

It can be shown that any equation of the form

$$x^2 + y^2 + ax + by + c = 0$$

represents a circle, a point, or no graph at all. For example, the equation $x^2 + y^2 = 0$ represents the single point $(0, 0)$. As another example, the equation $x^2 + y^2 + 5 = 0$, or $x^2 + y^2 = -5$, represents no graph, since sums of squares are never negative. When it represents a circle, the equation

$$x^2 + y^2 + ax + by + c = 0$$

is referred to as the *general equation of a circle*.

MIDPOINT FORMULA

We close this section with the formula for the coordinates of the *midpoint of a line segment*. If $P_1 = (x_1, y_1)$ and $P_2 = (x_2, y_2)$ are the endpoints of a line segment, the point M that is equidistant from P_1 and P_2 has coordinates

$$\left(\frac{x_1 + x_2}{2}, \frac{y_1 + y_2}{2} \right)$$

To see how we obtained this result, look at Figure 26. There, triangles $P_1 A M$ and $M B P_2$ are congruent, since M is the midpoint and the three angles are congruent.

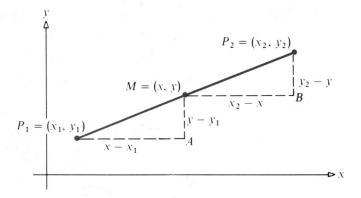

Figure 26

Hence, corresponding sides are equal in length. That is,

$$x - x_1 = x_2 - x \qquad \text{and} \qquad y - y_1 = y_2 - y$$

By solving for x and y, we obtain

(1.9) $$x = \frac{x_1 + x_2}{2} \qquad \text{and} \qquad y = \frac{y_1 + y_2}{2}$$

Thus, to find the midpoint of a line segment joining two points, we average the abscissas and ordinates of the points.

EXERCISE 2

In Problems 1–4 find the distance $|P_1 P_2|$ between the points P_1 and P_2.

1. $P_1 = (3, -4)$; $P_2 = (3, 1)$ 2. $P_1 = (-1, 0)$; $P_2 = (2, 1)$
3. $P_1 = (-0.6, 2)$; $P_2 = (-0.4, -0.2)$ 4. $P_1 = (-5, 1.2)$; $P_2 = (0.6, -0.5)$

In Problems 5–12 graph each equation.

5. $y = 3x$ 　　　　　　　**6.** $y = -2x$ 　　　　　　**7.** $y = 2x - 3$ 　　　　　**8.** $y = 2x - 4$

9. $3y + 2x + 1 = 0$ 　　　**10.** $4y - 2x - 5 = 0$ 　　**11.** $y = x^2 + 4$ 　　　　**12.** $y = 2x^2 - 4$

In Problems 13–16 find the general equation of the circle determined by the indicated center (h, k) and radius R.

13. $(h, k) = (2, -3);$ 　$R = 4$ 　　　　　　　　　　**14.** $(h, k) = (-1, 2);$ 　$R = 2$

15. $(h, k) = (1, -2);$ 　$R = 1$ 　　　　　　　　　　**16.** $(h, k) = (0, 3);$ 　$R = 3$

In Problems 17–20 find the center (h, k) and radius R of each circle.

17. $(x - 1)^2 + y^2 = 4$ 　　　　　　　　　　　　　**18.** $(x - 3)^2 + (y + 4)^2 = 9$

19. $x^2 + y^2 + 4x - 6y - 3 = 0$ 　　　　　　　　**20.** $x^2 + y^2 + 8y + 15 = 0$

In Problems 21–24 find the midpoint of the line segment joining the points P_1 and P_2.

21. $P_1 = (2, 3);$ 　$P_2 = (6, 5)$ 　　　　　　　　**22.** $P_1 = (1, -3);$ 　$P_2 = (3, 5)$

23. $P_1 = (0, 1);$ 　$P_2 = (1, 0)$ 　　　　　　　　**24.** $P_1 = (3, 0);$ 　$P_2 = (6, 2)$

In Problems 25–28 use the graph below.

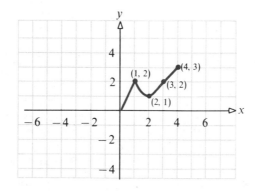

25. Add to the graph to make it symmetric with respect to the x-axis.

26. Add to the graph to make it symmetric with respect to the y-axis.

27. Add to the graph to make it symmetric with respect to the origin.

28. Add to the graph to make it symmetric with respect to the origin, x-axis, and y-axis.

In Problems 29–36 follow the steps of the solution for Example 3 to examine each equation for symmetry with respect to the x-axis, y-axis, and origin.

29. $x^2y + y^2x = 5$ 　　　　**30.** $x^2y^2 = 4$ 　　　　　**31.** $x^2 - 2xy = y^4$ 　　　**32.** $x^3 + y^3 = 1$

33. $3x^2 + 6y = 2$ 　　　　　**34.** $4y^2 - 6x^2 = x$ 　　　**35.** $y = -x^5 + 3x$ 　　　**36.** $y = (x^2 + 1)^2 - 1$

In Problems 37 and 38 find the general equation of each circle.

37. Center at $(1, -2)$ and passing through the point $(2, -1)$

38. Center at $(4, 1)$ and passing through the point $(5, 2)$

39. Find the lengths of the medians* of the triangle with vertices at $(0, 0)$, $(0, 6)$, and $(8, 0)$.

40. If two vertices of an equilateral triangle* are $(4, -3)$ and $(0, 0)$, find the third vertex. How many of these triangles are possible?

In Problems 41–44 find the length of each side of the triangle determined by the three points P_1, P_2, and P_3; and state whether the triangle is an isosceles triangle,* a right angle triangle,* neither of these, or both.

41. $P_1 = (2, 1)$; $P_2 = (-4, 1)$; $P_3 = (-4, -3)$ **42.** $P_1 = (-1, 4)$; $P_2 = (6, 2)$; $P_3 = (4, -5)$

43. $P_1 = (-2, -1)$; $P_2 = (0, 7)$; $P_3 = (3, 2)$ **44.** $P_1 = (7, 2)$; $P_2 = (-4, 0)$; $P_3 = (4, 6)$

45. In a study of conduction of heat through a wall, the ordered pair (x, y) represents the temperature x at each side of the wall and the corresponding thermal conductivity y. The formula for thermal resistance depends on the midpoint of the line segment determined by the temperatures and thermal conductivities on the two sides of the wall. If on one side of the wall the temperature is $300°F$ and the thermal conductivity is 0.22, while on the other side the temperature is $180°F$ and the thermal conductivity is 0.10, find the coordinates of the midpoint.

46. The earth is represented on a map of a portion of the solar system so that its surface is a circle with equation $x^2 + y^2 - 2x + 4y - 4091 = 0$.   A satellite circles 0.6 unit above the earth in a circular orbit with its center the center of the earth. Find the equation for the orbit of the satellite on this map.

47. If r is a real number, prove that the coordinates of the point $P = (x, y)$ that divides the line segment from $P_1 = (x_1, y_1)$ to $P_2 = (x_2, y_2)$ in the ratio r (that is, $|P_1P|/|P_1P_2| = r$)   are

$$x = (1 - r)x_1 + rx_2 \qquad y = (1 - r)y_1 + ry_2$$

[*Hint:*   Use similar triangles.]

In Problems 48–52 use the result of Problem 47.

48. Verify that the midpoint divides the line segment from $P_1 = (x_1, y_1)$ to $P_2 = (x_2, y_2)$ in the ratio $r = \frac{1}{2}$.

49. What point P divides the line segment from P_1 to P_2 in the ratio $r = 1$?

50. What point P divides the line segment from P_1 to P_2 in the ratio $r = 0$?

51. Find the point P on the line joining $P_1 = (1, 4)$ and $P_2 = (5, 6)$ that is twice as far from P_1 as P_2 is from P_1 and lies on the same side of P_1 as P_2 does.

52. Find the point(s) P on the line joining $P_1 = (0, 4)$ and $P_2 = (-1, 1)$ that is three times as far from P_1 as P_2 is from P_1.

3. The Straight Line

We begin with the result from plane geometry that there is one and only one line L containing two distinct points P_1 and P_2. If P_1 and P_2 are each represented by ordered pairs of real numbers, the following definition can be given:

---

* The *medians* of a triangle are the line segments from each vertex to the midpoint of the opposite side. An *equilateral triangle* is one in which all three sides are of equal length. An *isosceles triangle* is one in which two of the sides are of equal length. A *right angle triangle* is one in which one of the angles is $90°$. The Pythagorean theorem holds for all right triangles.

(1.10)   **DEFINITION**   *Slope.*   Let P_1 and P_2 be two distinct points with coordinates (x_1, y_1) and (x_2, y_2), respectively. The *slope m* of the line L containing P_1 and P_2 is defined by the formula*

$$m = \frac{y_2 - y_1}{x_2 - x_1} \qquad \text{if} \qquad x_1 \neq x_2$$

If $x_1 = x_2$,   the slope m of L is *undefined* (since this results in division by zero) and L is a *vertical line.*

Using the facts that $\Delta x = x_2 - x_1$ and $\Delta y = y_2 - y_1$, we can write the slope m of a nonvertical line as

$$m = \frac{\Delta y}{\Delta x}$$

That is, the slope m of a nonvertical line L is the ratio of the change in the ordinates from P_1 to P_2 to the change in the abscissas from P_1 to P_2 (see Fig. 27). Since

$$\frac{y_2 - y_1}{x_2 - x_1} = \frac{y_1 - y_2}{x_1 - x_2}$$

the result is the same whether the changes are computed from P_1 to P_2 or from P_2 to P_1. For example, the slope m of the line joining the points $(1, 2)$ and $(5, -3)$ may be computed as

$$m = \frac{-3 - 2}{5 - 1} = \frac{-5}{4} \qquad \text{or as} \qquad m = \frac{2 - (-3)}{1 - 5} = \frac{5}{-4} = \frac{-5}{4}$$

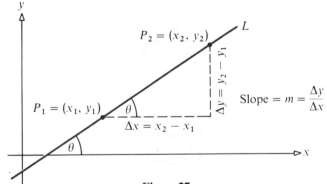

Figure 27

Observe in Figure 27 that the slope m of the nonvertical line L is a measure of the inclination of L to the x-axis. That is,

$$m = \tan \theta \qquad 0° \leq \theta \leq 180°, \qquad \theta \neq 90°$$

where θ is the angle between the positive direction of the x-axis and the line L. The angle θ is sometimes called the *angle of inclination* of the line L.

---

* The following argument, involving similar triangles, shows that the slope of a line L is the same no matter what two distinct points are used: Let L be a nonvertical line joining P_1 and P_2 and let X and Y be any other two distinct points on L. Construct the triangles depicted in the figure. Since triangle P_1P_2A is similar to triangle XYB (why?), it follows that the lengths of the corresponding sides are in proportion. That is, $|AP_2|/|BY| = |AP_1|/|BX|$ or $|AP_2|/|AP_1| = |BY|/|BX|$.   But the slope m of L is $|AP_2|/|AP_1|$, and by the foregoing equality, we see that $m = |BY|/|BX|$.   In other words, since X and Y are *any* two points, the slope m of a line L is the same no matter what points on L are used to compute m.

EXAMPLE 1   Compute the slopes and the angles of inclination of the lines L_1, L_2, L_3, and L_4 containing the given pairs of points. Graph each line.

$$L_1: \quad P = (2, 3) \qquad Q_1 = (-1, -2)$$
$$L_2: \quad P = (2, 3) \qquad Q_2 = (3, -1)$$
$$L_3: \quad P = (2, 3) \qquad Q_3 = (5, 3)$$
$$L_4: \quad P = (2, 3) \qquad Q_4 = (2, -2)$$

Solution   Let m_1 be the slope and let θ be the inclination of L_1, m_2 the slope and θ_2 the inclination of L_2, and so on. Then, with the help of a calculator, we find

$$m_1 = \frac{-2 - 3}{-1 - 2} = \frac{-5}{-3} \approx 1.66 \qquad \tan \theta_1 \approx 1.66 \qquad \text{so} \qquad \theta_1 \approx 59°$$

$$m_2 = \frac{-1 - 3}{3 - 2} = \frac{-4}{1} = -4 \qquad \tan \theta_2 = -4 \qquad \text{so} \qquad \theta_2 \approx 104°$$

$$m_3 = \frac{3 - 3}{5 - 2} = \frac{0}{3} = 0 \qquad \tan \theta_3 = 0 \qquad \text{so} \qquad \theta_3 = 0°$$

m_4 is undefined since $x_1 = x_2 = 2$; $\tan \theta_4$ is undefined, $\theta_4 = 90°$

These lines are graphed in Figure 28. Note that when the slope m is positive, the line slants upward from left to right (L_1); when m is negative, the line slants downward from left to right (L_2); when $m = 0$, the line is horizontal (L_3); and when m is undefined, the line is vertical (L_4).

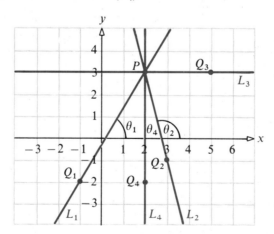

Figure 28   ∎

EQUATIONS OF LINES

A vertical line is given by the equation

(1.11)
$$x = a$$

where a is a given real number.

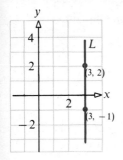

Figure 29

For example, the graph of the equation $x = 3$ is a vertical line (see Fig. 29). Now let L be a nonvertical line with slope m and containing (x_1, y_1). For (x, y) any other point on L, we have

$$m = \frac{y - y_1}{x - x_1} \qquad \text{or} \qquad y - y_1 = m(x - x_1)$$

(1.12)   *Point–Slope Form.*   **An equation of a nonvertical line of slope m that passes through the point (x_1, y_1) is**

$$y - y_1 = m(x - x_1)$$

For example, an equation of the line with slope 4 and passing through the point $(1, 2)$ is

$$y - 2 = 4(x - 1)$$
$$y = 4x - 2$$

EXAMPLE 2   Find an equation of the line passing through $(2, 3)$ and $(-4, 5)$.

Solution   Here, since two points are given, we first compute the slope of the line:

$$m = \frac{5 - 3}{-4 - 2} = \frac{2}{-6} = \frac{-1}{3}$$

Using the point $(2, 3)$ [we could just as well use $(-4, 5)$], we find the equation of the line to be

$$y - 3 = \left(\frac{-1}{3}\right)(x - 2)$$

$$y = \left(\frac{-1}{3}\right)x + \frac{11}{3} \qquad \blacksquare$$

Another useful equation of a line is obtained when the slope m and y-intercept b are known. Since in this event we know both the slope m of the line and a point $(0, b)$ on the line, we may use the point–slope form (1.12) to obtain the following equation:

$$y - b = m(x - 0) \qquad \text{or} \qquad y = mx + b$$

(1.13)   *Slope–Intercept Form.*   **An equation of a nonvertical line *L* with slope *m* and *y*-intercept *b* is**

$$y = mx + b$$

Sometimes, we write the equation of a line *L* in *general form*, namely,

$$Ax + By + C = 0$$

where *A*, *B*, and *C* are three real numbers with either $A \neq 0$ or $B \neq 0$.   This is referred to as the *general form* because every line has an equation that can be written this way.

EXAMPLE 3   Find the slope *m* and *y*-intercept *b* of the line *L* given by $2x + 4y - 8 = 0$.   Graph the line.

Solution   To obtain the slope and *y*-intercept, we transform the equation to its slope–intercept form. Thus, we need to solve for *y*:

$$2x + 4y - 8 = 0$$
$$4y = -2x + 8$$
$$y = \left(\frac{-1}{2}\right)x + 2$$

The coefficient of *x*, $-\frac{1}{2}$, is the slope, and the *y*-intercept is 2. To graph this line, we need two points. Normally, the easiest points to locate are the intercepts. Since the *y*-intercept is 2, we know one intercept is $(0, 2)$. To obtain the *x*-intercept, we set $y = 0$ and solve for *x*. When $y = 0$,   we have

$$2x - 8 = 0$$
$$x = 4$$

Thus, the intercepts are $(4, 0)$ and $(0, 2)$, as shown in Figure 30.

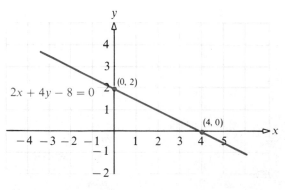

Figure 30   ■

INTERSECTING LINES; PARALLEL AND PERPENDICULAR LINES

Let L_1 and L_2 be two (distinct) lines. If L_1 and L_2 have exactly one point P in common, then L_1 and L_2 are said to *intersect* and the common point P is called the *point of intersection.*

EXAMPLE 4   Find the point of intersection of the lines

$$L_1: \quad x + y - 5 = 0 \quad \text{and} \quad L_2: \quad 2x + y - 6 = 0$$

Solution   Let the coordinates of the point P of intersection of L_1 and L_2 be (x_0, y_0). Since (x_0, y_0) is on both L_1 and L_2, then

$$x_0 + y_0 - 5 = 0 \qquad\qquad 2x_0 + y_0 - 6 = 0$$

so that

$$y_0 = 5 - x_0 \qquad\qquad y_0 = 6 - 2x_0$$

Setting these equal, we obtain

$$5 - x_0 = 6 - 2x_0$$
$$x_0 = 1$$

If $x_0 = 1$, then $y_0 = 5 - 1 = 4$. Thus, the point P of intersection of L_1 and L_2 is $(1, 4)$ (see Fig. 31).

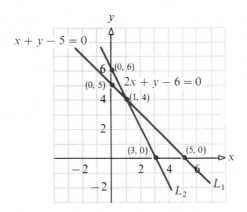

Figure 31   ∎

The technique given in Example 4 for finding the point P of intersection of two lines is sometimes called the *substitution technique* since in reality the value for y in one equation is substituted for the value of y in the other equation.

If two distinct lines (in a plane) do not intersect, they are said to be *parallel* and their angles of inclination are equal. Since the slope of a line is the tangent of the angle of inclination, we conclude:

(1.14) **Two nonvertical lines are parallel if and only if their slopes are equal.**

For example, the two lines

$$2x + 3y - 6 = 0 \quad \text{and} \quad 4x + 6y = 0$$

are parallel since each has slope $-\frac{2}{3}$ (see Fig. 32).

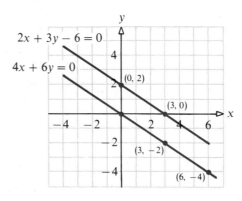

Figure 32

Slopes can also be used to determine whether two lines are *perpendicular*, that is, meet at a right angle.

(1.15) **Two nonvertical lines are perpendicular if and only if the product of their slopes is -1. Thus, two lines with slopes m_1 and m_2 are perpendicular if and only if**

$$m_1 = -\frac{1}{m_2}$$

Proof* Let θ_1 and θ_2 denote the inclinations of two nonvertical lines L_1 and L_2, respectively. We shall take θ_1 to be the smaller inclination. Then, as Figure 33 illustrates, if L_1 and L_2 are perpendicular, we have

$$\theta_2 = \theta_1 + \frac{\pi}{2}$$

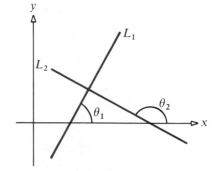

Figure 33

* Formulas that appear in this proof can be studied in the trigonometry review in Appendix I.

Thus,

$$m_2 = \tan\theta_2 = \tan\left(\theta_1 + \frac{\pi}{2}\right) = \frac{\sin[\theta_1 + (\pi/2)]}{\cos[\theta_1 + (\pi/2)]}$$

$$= \frac{\sin\theta_1 \cos(\pi/2) + \cos\theta_1 \sin(\pi/2)}{\cos\theta_1 \cos(\pi/2) - \sin\theta_1 \sin(\pi/2)} = -\frac{\cos\theta_1}{\sin\theta_1}$$

$$= -\cot\theta_1 = \frac{-1}{\tan\theta_1} = \frac{-1}{m_1}$$

The proof that if $m_1 = -1/m_2$ holds, then L_1 and L_2 are perpendicular is left as an exercise (see Problem 39). ∎

EXAMPLE 5   Show that the line L_1 through $(0, 0)$ and $(2, 4)$ is perpendicular to the line L_2 joining $(3, 1)$ and $(9, -2)$.

Solution   The slopes of these two lines are

$$L_1:\quad m_1 = \frac{4 - 0}{2 - 0} = 2 \qquad L_2:\quad m_2 = \frac{-2 - 1}{9 - 3} = \frac{-3}{6} = \frac{-1}{2}$$

Since $m_1 m_2 = 2(-\frac{1}{2}) = -1$, the lines are perpendicular. ∎

EXERCISE 3

In Problems 1–12 find the general equation of the line having the given properties.

1. Slope $= 2$; passing through $(-2, 3)$ **2.** Slope $= 3$; passing through $(4, -3)$

3. Slope $= -\frac{2}{3}$; passing through $(1, -1)$ **4.** Slope $= \frac{1}{2}$; passing through $(3, 1)$

5. Passing through $(1, 3)$ and $(-1, 2)$ **6.** Passing through $(-3, 4)$ and $(2, 5)$

7. Slope $= -3$; y-intercept $= 3$ **8.** Slope $= -2$; y-intercept $= -2$

9. x-intercept $= 2$; y-intercept $= -1$ **10.** x-intercept $= -4$; y-intercept $= 4$

11. Slope undefined; passing through $(1, 4)$ **12.** Slope undefined; passing through $(2, 1)$

In Problems 13–18 find the slope and y-intercept of the given line.

13. $3x - 2y = 6$ **14.** $4x + y = 2$ **15.** $x + 2y = 4$

16. $-x - y = 4$ **17.** $x = 4$ **18.** $y = 3$

19. Find the general equation of the line passing through $(1, 2)$ and parallel to $2x - y = 6$.

20. Find the general equation of the line passing through $(-1, 3)$ and parallel to $x + y = 4$.

In Problems 21–26 find the slope of a line perpendicular to the given line.

21. $3x - 2y = 6$ **22.** $3x + y = 4$ **23.** $x + 2y = -4$

24. $x - y = 1$ **25.** $x = 4$ **26.** $y = 2$

In Problems 27–32 determine whether the lines are parallel or intersecting. If they intersect, find the point of intersection.

27. L_1: $2x - 3y + 6 = 0$
     L_2: $4x - 6y + 7 = 0$

28. L_1: $4x - y + 2 = 0$
     L_2: $3x + 2y = 0$

29. L_1: $-x + 3y + 6 = 0$
     L_2: $x - 6y - 12 = 0$

30. L_1: $2x + 3y - 5 = 0$
     L_2: $x - 6y + 1 = 0$

31. L_1: $3x - 3y + 10 = 0$
     L_2: $x + y - 2 = 0$

32. L_1: $2x - 5y - 1 = 0$
     L_2: $x - 2y - 1 = 0$

33. Find an equation for the tangent line to the circle with center at $(2, -1)$ at the point $(3, 2)$.

34. Find an equation for the tangent line to the circle with center at $(4, 5)$ at the point $(6, -3)$.

35. Mr. Nicholson has just retired and needs $12,000 per year in income to live on. He has $100,000 to invest and can invest in AAA bonds at 14% interest annually or in savings and loan certificates at 10% a year. How much money should be invested in each so that he realizes exactly $12,000 in income per year?

36. One solution is 15% acid and another is 5% acid. How many cubic centimeters of each should be mixed to obtain 100 cubic centimeters of an 8% solution?

37. The relationship between the Celsius (°C) and Fahrenheit (°F) temperature scales is a straight line. Find the equation relating °C and °F if 0°C corresponds to 32°F and 100°C corresponds to 212°F. Use the equation to find the Celsius measure of 70°F.

38. The annual sales of Motors, Inc., for the past 5 years are given in the table.

Years	Units Sold (Thousands)
1981	2200
1982	2800
1983	3100
1984	3200
1985	3400

(a)   Graph this information using the x-axis for years and the y-axis for units sold. (For convenience, use a different scale on each axis.)

(b)   Draw a line L that passes through two of the points and comes close to passing through the remaining points.

(c)   Find the equation of the line L.

(d)   Using the equation of the line, what is your estimate for units sold in 1986?

39. Prove that if the product of the slopes of two nonvertical lines is -1, then the lines are perpendicular. [*Hint:*   One slope, say m_1, is positive and the other, m_2, is negative. Thus, if $m_1 = \tan \theta_1$ and $m_2 = \tan \theta_2$, then $0 < \theta_1 < \pi/2$ and $\pi/2 < \theta_2 < \pi$. Now use the identity $\tan[\theta_2 - (\pi/2)] = -1/\tan \theta_2$.]

40. Let L_1 and L_2 be two nonvertical, intersecting lines. If θ is the acute angle measured in the counterclockwise direction from L_1 to L_2, show that

$$\tan \theta = \frac{m_2 - m_1}{1 + m_1 m_2}$$

where m_1 and m_2 are the slopes of L_1 and L_2, respectively.

In Problems 41–44 use the result of Problem 40 to find the tangent of the acute angle from the line L_1 to the line L_2.

41. L_1: $2x + y = 2$
  L_2: $x - y = 4$

42. L_1: $x + y = 2$
  L_2: $2x - y = 0$

43. L_1: $2x + 3y = 4$
  L_2: $x - y = 6$

44. L_1: $x + 3y = 4$
  L_2: $2x - y = 5$

4. Functions and Their Graphs

In many applications, a correspondence often exists between two sets of numbers. For example, the volume V of a sphere of radius R is given by the formula (correspondence) $V = \frac{4}{3}\pi R^3$.

As another example, suppose a man standing on the moon throws a rock 20 meters (almost 22 yards) up and starts a stopwatch just as the rock begins to fall back down. Let x represent the number of seconds shown on the stopwatch and let y represent the height (in meters) of the rock above the surface of the moon. Then there is a correspondence between the time and the height, that is, between the numbers x and the numbers y. When the time is 0, the rock is at its highest point of 20 meters; therefore, $x = 0$ corresponds to $y = 20$. But to what heights do the numbers $x = 1$, $x = 2.5$, and $x = 5$ correspond? To find approximate answers to these questions without actually sending someone to the moon, we may use the following formula:

$$y = 20 - 0.8x^2$$

The height corresponding to $x = 1$ is found when we replace the letter x in the formula by the number 1, as follows:

$$y = 20 - 0.8(1)^2 = 19.2$$

Thus, when the stopwatch shows 1 second, the rock is still 19.2 meters above the surface of the moon. Similarly, when $x = 2.5$, the height is

$$y = 20 - 0.8(2.5)^2 = 15$$

When $x = 5$, the height is

$$y = 20 - 0.8(5)^2 = 0$$

and the rock has again reached the surface of the moon. (If you think that the rock falls to the moon more slowly than it would fall to the earth, you are right. See Problem 45.)

An important point made by this example is that if X is the set of times from 0 to 5 seconds and Y is the set of heights from 0 to 20 meters, then each element of X corresponds to one and only one element of Y. The correspondence $y = 20 - 0.8x^2$ is called a *function from X into Y*.

(1.16)   **DEFINITION**   *Function; Domain; Range.*   **Let X and Y be two sets of numbers. A *function from X into Y* is a correspondence that associates with each element of X a unique element of Y. The set X is called the *domain* of the function. For each element x in X, the corresponding element y in Y is called the *value* of the function at x, or the *image* of x. The set of all images of the elements of the domain is called the *range* of the function.**

Since there may be elements in Y that are the image of no x in X, it follows that the range of a function is a subset of Y.

Functions are often denoted by letters such as f, F, g, G, and so on. If f is a function from X into Y, then for each number x in X the corresponding image in the set Y is designated by the symbol $f(x)$ and is read "f of x." We refer to $f(x)$ as the *value of f at the number x*. For example, in the case of the falling rock, we may designate the function by the letter H (to remind us of the word height). Then for each x in X, $H(x)$ designates the value of H at x, that is, $H(x)$ designates the height of the rock at time x. In symbols, we write

$$H(x) = 20 - 0.8x^2$$

How do we designate the value of H at the times $x = 1$, $x = \frac{5}{4}$, $x = \sqrt{2}$? These are the heights $H(1)$, $H(\frac{5}{4})$, $H(\sqrt{2})$, and they may be computed by using the formula as follows:

$$H(1) = 20 - 0.8(1)^2 = 19.2$$
$$H(\tfrac{5}{4}) = 20 - 0.8(\tfrac{5}{4})^2 = 18.75$$
$$H(\sqrt{2}) = 20 - 0.8(\sqrt{2})^2 = 18.4$$

The expression $H(1) = 19.2$ is read as "the value of H at 1 is 19.2," or "19.2 meters corresponds to the time 1 second." Other ways to write this fact are by using *arrow notation*, thus,

$$1 \rightarrow 19.2$$

or *ordered-pair notation*,

$$(1, 19.2)$$

Each of these can be read "19.2 corresponds to 1."

It is convenient to use ordered-pair notation to show the difference between correspondences that *are* functions and a correspondence that is *not* a function. For example, let $X = \{1, 2\}$ and $Y = \{4, 5\}$.   The correspondence given by the pairs

$$(1, 5) \quad \text{and} \quad (2, 4)$$

is a function from X into Y because each element of X has a unique value in Y. Similarly, the correspondence given by

$$(1, 5) \quad \text{and} \quad (2, 5)$$

is a function. (Notice that 4 is *not* in the range of this function because the value of the function at 1 and 2 is the same, namely, 5. Nevertheless, the condition that each element of X has one and only one corresponding value is still satisfied.) The correspondence given by the pairs

$$(1, 4), \quad (1, 5), \quad \text{and} \quad (2, 5)$$

is *not* a function because the element 1 in X corresponds to *more than one value* in Y.

By using ordered-pair notation, we can also consider a function as a set of *ordered pairs* (x, y) in which no different pairs have the same first element. The set of all first elements is the *domain* and the set of all second elements is the *range* of the function. Thus, there is associated with each element x in the domain a unique element y in the range. An example is the set of all ordered pairs (x, y) such that $y = x^2$. Some of the pairs in this set are

$$(2, 2^2) = (2, 4) \qquad (0, 0^2) = (0, 0) \qquad (-2, (-2)^2) = (-2, 4) \qquad (\tfrac{1}{2}, (\tfrac{1}{2})^2) = (\tfrac{1}{2}, \tfrac{1}{4})$$

In this set no two different pairs have the same *first* element (even though there are different pairs that have the same *second* element). This set is the squaring function, which associates with each real number x the value x^2.

The ordered pairs (x, y) for which $y^2 = x$ do not represent a function because there are ordered pairs with the same first number but different second numbers. For example, $(1, 1)$ and $(1, -1)$ are ordered pairs obeying the relationship $y^2 = x$ with the same first number, but different second numbers.

The element x that appears in the first position of the ordered pair (x, y) is often called the *independent variable* since it can be assigned any of the permissible numbers from the domain; the second member of the pair is called the *dependent* variable since the value of y depends on the number x.

Another advantage of expressing a function (or any correspondence) as a set of ordered pairs is that we can then graph the set of pairs to make a "picture" of the function. For example, the graph of $y = x^2$ is depicted in Figure 20(a) on page 17.

To summarize, we have determined that a function f associates with real numbers x other real numbers y, and we now agree to use the notation

$$y = f(x)$$

to denote the rule that associates x and y. The set of all ordered pairs (x, y), where $y = f(x)$ is the ordinate and x is the abscissa, is called the *graph* of the function f.

Regardless of whether a function is described by a formula, by some rule, or by other means, it will always have a graph. However, not every collection of points is the graph of a function. In fact, a graph provides a visual technique for determining whether a collection of ordered pairs is a function. **If any vertical line intersects the graph in more than one point, the graph is not that of a function.** Compare Figures 20(a) and 20(c).

Up to this point we have discussed functions that could be described by a *single* formula or rule. For example, during the time $0 \leq x \leq 5$, the height of the falling rock is given by $H(x) = 20 - 0.8x^2$. But what is the height after the time $x = 5$ when the rock strikes the ground? Assuming that the rock does not bounce, we have a second rule that the height is 0 if $x > 5$. Therefore, to specify the height for *all* $x \geq 0$, we need a new function, say K, which incorporates *both* rules. The function K may be defined as follows:

$$K(x) = \begin{cases} 20 - 0.8x^2 & \text{if } 0 \leq x \leq 5 \\ 0 & \text{if } x > 5 \end{cases}$$

What is the difference between the functions H and K? The answer lies in the fact that they have different domains. The domain of H contains only those times from 0 to 5, while the domain of K contains all nonnegative times.

Here is another example of a function that is given by more than one rule.

EXAMPLE 1   Graph the function f given by the three rules:

$$f(x) = \begin{cases} x/2. & \text{if } -1 \leq x < 1 \\ 2 & \text{if } x = 1 \\ x + \frac{1}{2} & \text{if } x > 1 \end{cases}$$

Solution   Here, the domain of f is all real numbers $x \geq -1$. (Its graph is given in Figure 34.) We use a filled circle ● to indicate that at $x = 1$, the value of f is $f(1) = 2$; we use an open circle ○ to illustrate that the function does not assume either of the values $\frac{1}{2}$ or $\frac{3}{2}$ at $x = 1$.

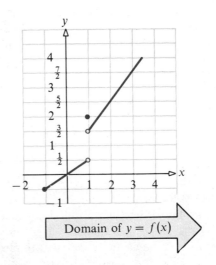

Figure 34   ■

In this book we will sometimes give directions such as "Graph the function $f(x) = x^2$." We actually mean "Graph the equation $y = x^2$, in which the numbers x are restricted to the numbers in the domain of f." But how can we graph a function if its domain is not specified? The answer is simply this: When the domain of a function is *not* specified but a rule of association is known, then we automatically assume that the domain is the largest set of real numbers for which the rule *makes sense* (or more precisely, for which we can compute $f(x)$ as a real number). For example, the operation of squaring can be performed on *any* real number x. Therefore, to associate x with x^2 makes sense for *every* real number x, and thus the domain of $y = f(x) = x^2$ is the set $\mathbb{R}$ of *all* real numbers.

What is the domain of $f(x) = 1/x$? We can divide any nonzero real number into 1. Hence, it makes sense to associate x with $1/x$ as long as $x \neq 0$. The domain of $y = f(x) = 1/x$ is therefore $\{x \mid x \in \mathbb{R} \text{ and } x \neq 0\}$, that is, all real numbers x except $x = 0$.

EXAMPLE 2   Find the domain and range, and graph the *square root function:* $y = f(x) = \sqrt{x}$

Solution   To find the domain D of f, we ask the question: "What are the numbers x for which we can compute $\sqrt{x}$?" Now, we know it is impossible (in the universe of real numbers) to find the square root of a negative number. Thus, we can only compute $\sqrt{x}$ if $x \geq 0$, and the domain is $\{x \mid x \geq 0\}$. Now we look for the range. To each number x in the domain, there is associated exactly one nonnegative number y. (This number y is nonnegative because of the definition of square root.) In fact, as x runs through the domain, $\sqrt{x}$ runs through *all* nonnegative numbers. Hence, the range is the set of nonnegative real numbers. The graph is given in Figure 35.

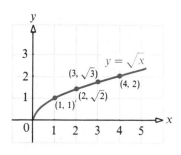

Figure 35   ∎

EXAMPLE 3   Find the domain and graph the function: $f(x) = \sqrt{x^2 + 2x - 3}$

Solution   For $\sqrt{x^2 + 2x - 3}$ to be a real number, the domain of f has to consist of all numbers x for which $x^2 + 2x - 3 = (x + 3)(x - 1)$ is zero or positive. The expression is zero for $x = -3$ and $x = 1$. What remains is to solve the inequality

$(x + 3)(x - 1) > 0$,   which we do by setting up a table (Table 3), using -3 and 1 as cutoff points, since $(x + 3)$ and $(x - 1)$ are the factors of $x^2 + 2x - 3$ (see Fig. 36). The preceding discussion shows that the domain of f consists of all real numbers in the intervals $(-\infty, -3]$ and $[1, +\infty)$. Figure 37 depicts the graph.

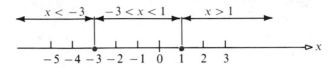

Figure 36

Table 3

	Sign of $x + 3$	Sign of $x - 1$	Sign of $x^2 + 2x - 3$	Conclusion
$x < -3$	$-$	$-$	$+$	$x < -3$ is solution
$-3 < x < 1$	$+$	$-$	$-$	$-3 < x < 1$ is not solution
$x > 1$	$+$	$+$	$+$	$x > 1$ is solution

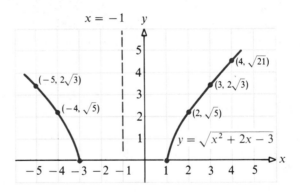

Figure 37   ∎

Observe the similarity between the graph of $y = \sqrt{x^2 + 2x - 3}$ in Figure 37 and that of the square root function in Figure 35. Note also the symmetry about the line $x = -1$ in Figure 37.

EXAMPLE 4   Graph the function:

(1.17)
$$f(x) = \begin{cases} x & \text{if } x \geq 0 \\ -x & \text{if } x < 0 \end{cases}$$

Solution If $x \geq 0$, then f is represented by the line $y = x$ (slope 1); when $x < 0$, then f is represented by the line $y = -x$ (slope -1). The graph of f is given in Figure 38. This function is called the *absolute value function* and is written as

$$f(x) = |x|$$

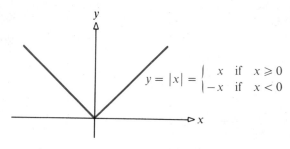

$$y = |x| = \begin{cases} x & \text{if } x \geq 0 \\ -x & \text{if } x < 0 \end{cases}$$

Figure 38   ■

The next function, which has the graph shown in Figure 39, occurs frequently enough in mathematics and in applications that it merits a special name, the *greatest integer function*.

(1.18)   DEFINITION *Greatest Integer Function.*   **The *greatest integer function*, denoted by $[\![x]\!]$ and read "bracket x," is defined as the greatest integer less than or equal to x.**

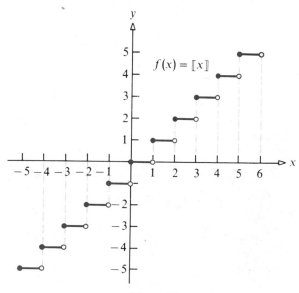

Figure 39

For example,

$$[\![5]\!] = 5 \qquad [\![4.9]\!] = 4 \qquad [\![-2.1]\!] = -3$$

The domain of this function is the set of real numbers, while its range is the set of integers (see Fig. 39). From the graph of the greatest integer function, we can see why it is sometimes referred to as a *step function*. Such functions exhibit what we shall refer to as a *discontinuity* at $x = \pm 1, \pm 2$, and so on, that is, points where the function suddenly jumps from one value to another without taking on any of the intermediate values. This occurs, for example, at $x = 3$, where to the left of 3, the y-values are at 2 and to the right of 3, the y-values are at 3.

EXAMPLE 5

Holders of credit cards issued by banks, department stores, oil companies, and so on, receive bills each month that state minimum amounts that must be paid by their due dates. The minimum depends on the total amount owed. For instance, for a bill of up to $10, the entire amount is due. For a bill of at least $10 but less than $500, the minimum is $10. There is a minimum of $15 on a bill of at least $500 but less than $1000; a minimum of $20 for $1000 up to $1500; and a minimum of $25 on bills of $1500 or more. The function f that describes the minimum payment on a bill of $\$x$ is

$$f(x) = \begin{cases} x & \text{if} \quad 0 \le x < 10 \\ 10 & \text{if} \quad 10 \le x < 500 \\ 15 & \text{if} \quad 500 \le x < 1000 \\ 20 & \text{if} \quad 1000 \le x < 1500 \\ 25 & \text{if} \quad 1500 \le x \end{cases}$$

The graph is given in Figure 40.

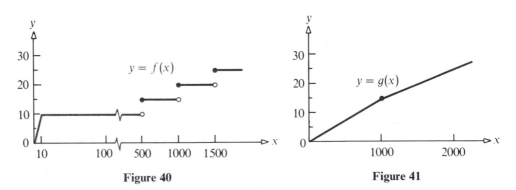

Figure 40   Figure 41

The card holder may pay any amount between the minimum and the total owed. The organization issuing the card charges the card holder interest of $1\frac{1}{2}\%$ a month for the first $1000 owed and 1% a month on any unpaid balance above $1000. Thus, if $g(x)$ is the amount of interest charged for a month on a balance of x, then $g(x) = 0.015x$ for $0 \le x \le 1000$. The amount of the unpaid balance above $1000 is $x - 1000$. If the balance due is $x > 1000$, then the interest is $0.015(1000) + 0.01(x - 1000)$, so

$$g(x) = \begin{cases} 0.015x & \text{if } 0 \le x \le 1000 \\ 5 + 0.01x & \text{if } x > 1000 \end{cases}$$

See Figure 41.   ∎

One important use of function notation is illustrated in the next two examples.

EXAMPLE 6   For the function $f(x) = 3x + 1$ find:

(a) $f(x + \Delta x)$ (b) $f(x + \Delta x) - f(x)$ (c) $\dfrac{f(x + \Delta x) - f(x)}{\Delta x}$ $\Delta x \ne 0$

Solution   (a)   The function $f(x) = 3x + 1$ tells us to multiply x by 3 and then add 1. To find $f(x + \Delta x)$, we multiply $(x + \Delta x)$ by 3 and then add 1. Thus,

$$f(x + \Delta x) = 3(x + \Delta x) + 1 = 3x + 3\Delta x + 1$$

Notice that x has been replaced by the quantity $(x + \Delta x)$.

(b) $f(x + \Delta x) - f(x) = (3x + 3\Delta x + 1) - (3x + 1) = 3\Delta x$

(c) $\dfrac{f(x + \Delta x) - f(x)}{\Delta x} = \dfrac{3\Delta x}{\Delta x} = 3 \qquad \Delta x \ne 0$ ∎

EXAMPLE 7   For the function $f(x) = 1/x$ find:

(a) $f(x + \Delta x)$ (b) $f(x + \Delta x) - f(x)$ (c) $\dfrac{f(x + \Delta x) - f(x)}{\Delta x}$ $\Delta x \ne 0$

Solution   (a)   The function $f(x) = 1/x$ tells us to find the reciprocal of x. Thus, for $f(x + \Delta x)$ we should find the reciprocal of $(x + \Delta x)$. That is,

$$f(x + \Delta x) = \frac{1}{x + \Delta x}$$

(b) $f(x + \Delta x) - f(x) = \dfrac{1}{x + \Delta x} - \dfrac{1}{x} = \dfrac{x - (x + \Delta x)}{x(x + \Delta x)} = \dfrac{-\Delta x}{x(x + \Delta x)}$

(c) $\dfrac{f(x + \Delta x) - f(x)}{\Delta x} = \dfrac{-\Delta x}{x(x + \Delta x)} \left(\dfrac{1}{\Delta x} \right) = \dfrac{-1}{x(x + \Delta x)}$ ∎

Many formulas that occur in mathematics and the sciences determine functions. For example, $A = \pi R^2$ is a formula that gives the area A of a circle in terms of its radius R. Similarly, if we know both the height h and the radius R of a right circular cylinder (such as a soup can), then we can find its volume V by the formula $V = \pi R^2 h$. In the formula $A = \pi R^2$, π is a constant (approximately 3.14159) and A is the dependent variable. Since its value depends only on the single independent variable R, A is called a *function of one variable*. In the formula $V = \pi R^2 h$, V is the dependent variable. Its value depends on the *two* independent variables R and h; therefore, V is a *function of two variables*.

We shall discuss functions of one variable in Chapters 2–13. In later chapters, we deal with functions of two or more variables.

EXERCISE 4

1. For the function $f(x) = 3x - 2$ find:

(a) $f(3)$ (b) $f(-2)$ (c) $f(0)$ (d) $f(x + 2)$ (e) $f(x + \Delta x)$ (f) $f\left(\dfrac{1}{x}\right)$

2. For the function $f(x) = 3x^2 + 1$ find:

(a) $f(1)$ (b) $f(-2)$ (c) $f(0)$ (d) $f(x + 4)$ (e) $f(x + \Delta x)$ (f) $f\left(\dfrac{1}{x}\right)$

In Problems 3–16 determine whether the given correspondence determines a function $y = f(x)$.

3. $y = x^2 + 2x + 1$ **4.** $y = x^3 - 3x$ **5.** $y = \dfrac{2}{x}$ **6.** $y = \dfrac{3}{x} - 4$

7. $y^2 = 1 - x^2$ **8.** $y = \pm\sqrt{1 - 2x}$ **9.** $x^2 + y = 1$ **10.** $x + y^2 = 1$

11. $x^2 y^2 = 5$ **12.** $x^2 y = 4$ **13.** $y = |x - 2|$ **14.** $y = \sqrt{x^2}$

15. $\{(1, 5), (2, 5), (5, 1)\}$ **16.** $\{(2, 2), (2, 3), (3, 4), (4, 5)\}$

In Problems 17–26 find the domain of the function f.

17. $f(x) = 3x + 5$ **18.** $f(x) = x^2 + 1$ **19.** $f(x) = \sqrt{x - 1}$ **20.** $f(x) = \sqrt{2x + 5}$

21. $f(x) = \sqrt{x^2 + 4}$ **22.** $f(x) = \sqrt{x^2 - 4}$ **23.** $f(x) = \dfrac{2x}{x - 2}$ **24.** $f(x) = \dfrac{x^2}{x^2 - 4}$

25. $f(x) = \sqrt{\dfrac{3}{x}}$ **26.** $f(x) = \dfrac{3x^2}{x^4 + 1}$

For the functions in Problems 27–36 find the domain, and graph each function.

27. $f(x) = \begin{cases} 2x - 3 & \text{if } x < 0 \\ x - 3 & \text{if } 0 \le x < 5 \end{cases}$ **28.** $f(x) = \begin{cases} 1 & \text{if } x \ge 0 \\ -1 & \text{if } x < 0 \end{cases}$

29. $f(x) = \begin{cases} 4x + 5 & \text{if } -2 \le x < 0 \\ 4 & \text{if } x = 0 \\ 2x & \text{if } x > 0 \end{cases}$ **30.** $f(x) = \begin{cases} 4 - x & \text{if } x \le 2 \\ x - 2 & \text{if } 2 < x \end{cases}$

31. $f(x) = \begin{cases} x^2 & \text{if } x \le 0 \\ \sqrt{x + 1} & \text{if } x > 0 \end{cases}$ **32.** $f(x) = \begin{cases} x^2 + 2 & \text{if } x \le 0 \\ \sqrt{x + 4} & \text{if } x > 0 \end{cases}$

33. $f(x) = x - [\![x]\!]$ **34.** $f(x) = x + [\![x]\!]$ **35.** $f(x) = |x + 4|$ **36.** $f(x) = |x - 2|$

For the functions in Problems 37–40 find:

(a) $f(x + \Delta x)$ (b) $f(x + \Delta x) - f(x)$ (c) $\dfrac{f(x + \Delta x) - f(x)}{\Delta x}$ $\Delta x \neq 0$

37. $f(x) = 2x + 5$ **38.** $f(x) = x^2 + 3$ **39.** $f(x) = x^2 + 3x + 4$ **40.** $f(x) = x + \dfrac{1}{x}$

41. For the function $f(x) = \sqrt{x}$ show that

$$\frac{f(x + \Delta x) - f(x)}{\Delta x} = \frac{1}{\sqrt{x + \Delta x} + \sqrt{x}} \qquad \Delta x \neq 0$$

42. For the function $f(x) = \sqrt{x + 3}$ show that

$$\frac{f(x + \Delta x) - f(x)}{\Delta x} = \frac{1}{\sqrt{x + 3 + \Delta x} + \sqrt{x + 3}} \qquad \Delta x \neq 0$$

43. If $f(x) = 2x^3 + Ax^2 + Bx - 5$ and if $f(2) = 3$ and $f(-2) = -37$,   what is the value of $A + B$?

44. If a rock falls from a height of 20 meters on the planet Jupiter, its height after x seconds is approximately

$$H(x) = 20 - 13x^2$$

(a) What is the height of the rock when $x = 1$ second? $x = 1.1$ seconds? $x = 1.2$ seconds? $x = 1.3$ seconds?

(b) When does the rock strike the ground?

(c) Compare these results with the results obtained at the beginning of this section for the rock falling on the moon.

(d) Write a function that gives the height of the rock for all times $x \geq 0$.

45. If a rock falls from a height of 20 meters here on the earth, the height H after x seconds is approximately

$$H(x) = 20 - 4.9x^2$$

Use this function to answer (a)–(d) of Problem 44.

46. Express the perimeter P and area A of a semicircle as a function of the diameter x.

47. A rectangular field requires 3000 feet of fence to enclose it. If the length of the field is x feet, express the area A as a function of x. What is the domain of A?

48. A trucking company transports goods between Chicago and New York, a distance of 960 miles. The company's policy is to charge, for each pound, $0.50 per mile for the first 100 miles, $0.40 per mile for the next 300 miles, $0.25 per mile for the next 400 miles, and no charge for the remaining 160 miles. Graph the relationship between the cost of transportation and mileage over the entire 960 mile route. Find the cost as a function of mileage for hauls between 100 and 400 miles from Chicago. Find the cost as a function of mileage for hauls between 400 and 800 miles from Chicago.

49. A page with dimensions 11 inches by 7 inches has a border of uniform width x surrounding the printed matter of the page. Write a formula for the area A of the printed part as a function of the width x of the border. Give the domain and range of A.

50. For the function

$$f(x) = \begin{cases} \dfrac{|x|}{x} & \text{if } x \neq 0 \\ 1 & \text{if } x = 0 \end{cases}$$

find:

(a) $f(1)$ (b) $f(-1)$ (c) $f(3)$ (d) $f(-3)$ (e) $f(5)$ (f) $f(-5)$

(g) $f(0 + \Delta x),\ \ \Delta x > 0$ (h) $f(0 + \Delta x),\ \ \Delta x < 0$

5. Operations on Functions;
Types of Functions

In this section we introduce some operations on functions. We shall see that functions, like numbers, can be added, subtracted, multiplied, and divided.

(1.19)   **DEFINITION**   **If f and g are functions, their *sum*, $f + g$; their *difference*, $f - g$; their *product*, $f \cdot g$; and their *quotient*, f/g, are defined by**

$$(f + g)(x) = f(x) + g(x) \qquad (f - g)(x) = f(x) - g(x)$$
$$(f \cdot g)(x) = f(x) \cdot g(x) \qquad (f/g)(x) = f(x)/g(x)$$

In each case, the *domain* of the resulting function consists of the numbers x that are common to the domains of f and g, but the numbers x for which $g(x) = 0$ must be excluded from the domain of the quotient f/g.

EXAMPLE 1   Let f and g be two functions defined as

$$f(x) = \sqrt{x + 2} \qquad \text{and} \qquad g(x) = \sqrt{x - 3}$$

Find the following, and, in each case, determine the domain:

(a) $(f + g)(x)$ (b) $(f - g)(x)$ (c) $(f \cdot g)(x)$ (d) $(f/g)(x)$

Solution   (a) $(f + g)(x) = \sqrt{x + 2} + \sqrt{x - 3}$

(b) $(f - g)(x) = \sqrt{x + 2} - \sqrt{x - 3}$

(c) $(f \cdot g)(x) = (\sqrt{x + 2})(\sqrt{x - 3}) = \sqrt{(x + 2)(x - 3)}$

(d) $(f/g)(x) = \dfrac{\sqrt{x + 2}}{\sqrt{x - 3}} = \sqrt{\dfrac{x + 2}{x - 3}}$

The domain of f is the interval $[-2, +\infty)$ and that of g is $[3, +\infty)$. The x common to both these domains is the interval $[3, +\infty)$; and, as a result, this is the domain of the sum $f + g$, the difference $f - g$, and the product $f \cdot g$. For part (d), the domain is the interval $(3, +\infty)$ since for $x = 3$ the denominator function g has the value zero.   ∎

TYPES OF FUNCTIONS

Many situations lead to functions that can be easily classified. We start with the *constant function*. By a *constant function* we mean a function $f(x) = A$ with domain the set of real numbers and range consisting of only one number, A. The graph of a constant function is a straight line parallel to the x-axis. For example, the function $f(x) = 3$ is a constant function; its graph is given in Figure 42.

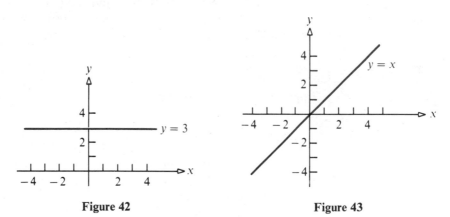

Figure 42   **Figure 43**

A function P is a *polynomial function* if

(1.20)
$$P(x) = a_n x^n + a_{n-1} x^{n-1} + \cdots + a_1 x + a_0$$

for all x, where the coefficients $a_0, a_1, \ldots, a_n$ are real numbers and the exponents are nonnegative integers. The domain of a polynomial function is the set of real numbers. If $a_n \neq 0$, then a_n is called the *leading coefficient of f*, and we say that the polynomial has *degree n*. For example, the function P defined by

(1.21)
$$P(x) = 2x^7 - 3x^2 + \tfrac{1}{2}x - 2$$

is a polynomial of degree 7 with leading coefficient 2. The coefficients of $P(x)$ are $a_7 = 2$, $a_6 = a_5 = a_4 = a_3 = 0$, $a_2 = -3$, $a_1 = \tfrac{1}{2}$, and $a_0 = -2$.

The constant function $f(x) = A$, $A \neq 0$, is a polynomial function of degree 0. The constant function $f(x) = 0$ is the *zero polynomial function* and has no degree.

If the degree of a polynomial function is 1, then the function is called a *linear function* and is of the form $P(x) = ax + b$, where $a \neq 0$. From the discussion in previous sections, we know that the graph of this function is a straight line with slope $m = a$ and y-intercept $= b$. If $a = 1$, $b = 0$, we get the linear function $P(x) = x$, which is known as the *identity function*. Its graph is given in Figure 43.

Any polynomial function P of degree 2 may be written as

$$P(x) = ax^2 + bx + c$$

where a, b, and c are constants and $a \neq 0$. Such polynomials are also called *quadratic functions*. The graph of a quadratic function is known as a *parabola*. Figure 17 (p. 16) illustrates a typical parabola.

Obtaining the graphs of most polynomials of degree 3 or higher is generally not easy with the tools we now have available. We would have to locate several well-chosen points on the graphs (this is easier said than done) and then hope that, by connecting them with a smooth curve, we would obtain an accurate picture. This tedious method is imprecise, and we will soon show how the power of calculus can be used to get accurate graphs without requiring a random selection of many points.

RATIONAL FUNCTIONS

A *rational function* is a function of the form

(1.22)
$$R(x) = \frac{P(x)}{Q(x)} = \frac{a_n x^n + \cdots + a_1 x + a_0}{b_m x^m + \cdots + b_1 x + b_0}$$

where P is a polynomial function of degree n and Q is a nonzero polynomial function of degree m.

To find the domain of a rational function, remember that the only time we will not be able to compute a value for $R(x)$ is when the x chosen gives a 0 in the denominator. Thus, the domain of R is $\{x \mid x \in \mathbb{R} \text{ and } Q(x) \neq 0\}$.

The graphs of most rational functions, like those of most polynomial functions, require the use of calculus. We postpone a discussion of their graphs to Chapter 4. As we continue in this book, other types of functions will be encountered, classified, and discussed. For most of them, calculus will not only be useful, but necessary, to obtain a complete description.

EXERCISE 5

For the functions f and g in Problems 1–6 find: (a) $f + g$ (b) $f - g$ (c) $f \cdot g$ (d) f/g
Determine the domains of f, g, $f + g$, $f \cdot g$, and f/g.

1. $f(x) = x - 1$; $g(x) = 2x^2$

2. $f(x) = \sqrt{x + 1}$; $g(x) = \sqrt{x^2 - 1}$

3. $f(x) = \sqrt{x + 1}$; $g(x) = x + 1$

4. $f(x) = |x|$; $g(x) = |x - 1|$

5. $f(x) = \dfrac{1}{x}$; $g(x) = \dfrac{1}{x} + 1$

6. $f(x) = (x^2 - 3x + 1)^5$; $g(x) = \sqrt{x^4 + 1}$

In Problems 7–12 indicate which are polynomial functions.

7. $f(x) = 2x^5 - 3x + 4$

8. $f(x) = \dfrac{1}{x^2}$

9. $f(x) = 2x^2 - \sqrt{x}$

10. $f(x) = 2x + \dfrac{3}{x} - 2$ 　　　　　**11.** $f(x) = \sqrt{x} - 2$ 　　　　　**12.** $f(x) = 3x^2 + 5x$

Problems 13–16 find the domain of each function.

13. $f(x) = \dfrac{3x}{x + 2}$ 　　　**14.** $f(x) = \dfrac{2x + 1}{3x^2 - 5x - 2}$ 　　　**15.** $f(x) = \dfrac{2}{x^2 - 4}$ 　　　**16.** $f(x) = \dfrac{x^4}{x^3 - 8}$

In Figure 17 (p. 16) the graph of $f(x) = x^2$ is given. In Problems 17–22 graph each function using Figure 17 as a guide.

17. $f(x) = x^2 + 4$ 　　　　　**18.** $f(x) = x^2 - 4$ 　　　　　**19.** $f(x) = x^2 + x$

20. $f(x) = x^2 - x$ 　　　　　**21.** $f(x) = |x^2 - 4|$ 　　　　　**22.** $f(x) = [\![x^2 - 4]\!]$

A function f is said to be *even* if $f(-x) = f(x)$ for every number x in the domain of f. A function f is said to be *odd* if $f(-x) = -f(x)$ for every x in the domain. (In both cases, it is understood that for each x in the domain, $-x$ must also be in the domain.)

In Problems 23–28 determine whether the given function f is even, odd, or neither.

23. $f(x) = 4x^3$ 　　　　　**24.** $f(x) = 3x^2 - 2x + 1$ 　　　　　**25.** $f(x) = |x|$

26. $f(x) = \dfrac{\sqrt{x^2 - 1}}{\sqrt{x^2 + 1}}$ 　　　**27.** $f(x) = \dfrac{x - 1}{x + 1}$ 　　　**28.** $f(x) = (x + 1)^2$

In Problems 29–34 give examples to illustrate each fact.

29. The sum of two odd functions is odd. 　　　**30.** The difference of two odd functions is odd.

31. The product of two even functions is even. 　　**32.** The product of two odd functions is even.

33. The function f/g is even if f is odd and g is odd. 　　**34.** The function f/g is even if f is even and g is even.

35. Given $f(x) = ax^2 + bx + c$, find numbers a, b, c such that

$$f(x + y) = f(x) + f(y)$$

36. Given $f(x) = 3x^2 + 2x - 1$ and $g(x) = (A + B)x^2 + Cx + D$, with A, B, C, D real numbers, under what conditions does $f(x) = g(x)$?

37. Given $f(x) = 3x + 1$ and $(f + g)(x) = 6 - \frac{1}{2}x$, find $g(x)$.

38. For what numbers a, b, c is the function $f(x) = ax^2 + bx + c$ even? Odd?

6. Composite Functions

Consider the function $y = (2x + 3)^2$. If we write $y = f(u) = u^2$ and $u = g(x) = 2x + 3$, then by a substitution process we can obtain the original function; namely, $y = f(u) = f(g(x)) = (2x + 3)^2$. This process is called *composition*. In general, suppose that f and g are two functions, and suppose that x is a number in the domain of g. By applying g to x, we get $g(x)$. If $g(x)$ is in the domain of f, then we may apply f to $g(x)$ and thereby obtain the value $f(g(x))$. If we do this for all such x's in the domain of g, the resulting correspondence is called a *composite function*.

(1.23)   **DEFINITION**   *Composite Function.*   **Given the two functions f and g, the *composite function*, denoted by $f \circ g$ (read "f circle g") is defined by**

$$(f \circ g)(x) = f(g(x))$$

where the domain of $f \circ g$ is the set of all numbers x in the domain of g such that $g(x)$ is in the domain of f.

Figures 44 and 45 illustrate the definition. Some examples will give you the idea.

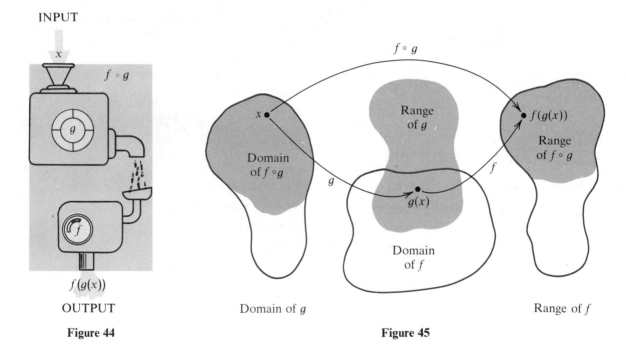

INPUT

OUTPUT

Figure 44   **Figure 45**

EXAMPLE 1   Suppose $f(x) = \sqrt{x}$ and $g(x) = x^3 - 1$.   Find the following composite functions, and then find the domain of each composite function:
(a) $f \circ g$ (b) $g \circ f$ (c) $f \circ f$ (d) $g \circ g$

Solution   (a)   The function f is the square root function, so the composite function $f \circ g = f(g(x))$ means to take the square root of $g(x)$; thus,

$$(f \circ g)(x) = f(g(x)) = \sqrt{g(x)} = \sqrt{x^3 - 1}$$

The domain of $f \circ g$ is the interval $[1, +\infty)$ and is found by determining those x in the domain of g for which $x^3 - 1 \geq 0$.

(b) The function g tells us to cube x and then subtract 1. The composite function $(g \circ f)(x) = g(f(x))$ tells us to cube $f(x)$ and then subtract 1. Thus,

$$(g \circ f)(x) = g(f(x)) = (f(x))^3 - 1 = (\sqrt{x})^3 - 1 = x^{3/2} - 1$$

The domain of $g \circ f$ is $[0, +\infty)$.

(c) $$(f \circ f)(x) = f(f(x)) = \sqrt{f(x)} = \sqrt{\sqrt{x}} = \sqrt[4]{x}$$

The domain of $f \circ f$ is $[0, +\infty)$.

(d) $$(g \circ g)(x) = g(g(x)) = (g(x))^3 - 1 = (x^3 - 1)^3 - 1$$

The domain of $g \circ g$,   evidently, is the set of all real numbers.   ■

Some techniques in calculus require that we be able to determine the components of a composite function. For example, the function $H(x) = \sqrt{x + 1}$ is the composition of the functions f and g, where $f(x) = \sqrt{x}$ and $g(x) = x + 1$,   because $H(x) = (f \circ g)(x) = f(g(x)) = \sqrt{g(x)} = \sqrt{x + 1}$.

EXAMPLE 2 Find functions f and g such that $f \circ g = H$ if $H(x) = (x^2 + 1)^{50}$.

Solution   Set $f(x) = x^{50}$ and $g(x) = x^2 + 1$.   Then

$$(f \circ g)(x) = f(g(x)) = (g(x))^{50} = (x^2 + 1)^{50} = H(x) \qquad ■$$

EXAMPLE 3 Find functions f and g so that $f \circ g = H$ and $H(x) = 1/(x + 1)$.

Solution   If we set $f(x) = 1/x$ and $g(x) = x + 1$,   we find that

$$(f \circ g)(x) = \frac{1}{g(x)} = \frac{1}{x + 1} = H(x) \qquad ■$$

Other functions f and g also have the above properties. For example, if $f(x) = 1/(x - 2)$ and $g(x) = x + 3$,   then

$$(f \circ g)(x) = \frac{1}{g(x) - 2} = \frac{1}{(x + 3) - 2} = \frac{1}{x + 1}$$

Although the answer to Example 3 and other problems involving composite functions is not unique, there is usually a "natural" selection for f and g—one that comes to mind first. More will be said later about this aspect of composition. In the meantime, it is sufficient to be able to write some functions f and g, whose composite is a given function H. You will most likely find that your selection is the "natural" one.

We end this section by describing two broad classes of functions—*algebraic* and *transcendental*. A function f is called *algebraic* if it can be expressed in terms of sums, differences, products, quotients, powers, or roots of polynomial functions. For example, the function f defined by

$$f(x) = \frac{3x^3 - x^2(x + 1)^{4/3}}{\sqrt{x^4 + 2}}$$

is an algebraic function. Functions that are not algebraic are termed *transcendental* functions. Examples of transcendental functions are trigonometric functions, logarithmic functions, and exponential functions. We study these functions in later chapters.

EXERCISE 6

In Problems 1–8 functions f and g are given. In each problem find:

(a) $f \circ g$ (b) $g \circ f$ (c) $g \circ g$ (d) $f \circ f$

1. $f(x) = 3x + 1$; $g(x) = x^2$

2. $f(x) = \sqrt{x + 1}$; $g(x) = \dfrac{1}{x^2}$

3. $f(x) = \sqrt{x}$; $g(x) = x^2 - 1$

4. $f(x) = \dfrac{1}{\sqrt{x - 1}}$; $g(x) = (x^2 + 1)^3$

5. $f(x) = \dfrac{x - 1}{x + 1}$; $g(x) = \dfrac{1}{x}$

6. $f(x) = \sqrt{x}$; $g(x) = \dfrac{1}{x}$

7. $f(x) = 3x^4 - 2x^2$; $g(x) = \dfrac{2}{\sqrt{x}}$

8. $f(x) = \dfrac{1}{3x + 2}$; $g(x) = \dfrac{3}{2x - 5}$

In Problems 9–14 find f and g such that $f \circ g = H$.

9. $H(x) = \sqrt{x^2 + x - 1}$

10. $H(x) = (1 + x^2)^{-3}$

11. $H(x) = (x^2 - 1)^7$

12. $H(x) = \left(1 - \dfrac{1}{x^2}\right)^2$

13. $H(x) = \dfrac{1}{(3x - 5)^2}$

14. $H(x) = \sqrt[3]{2 - 3x}$

15. If $f(x) = x^3$,   find a function g such that $f(g(x)) = x$ for every x in the domain of g.

16. If $f(x) = \sqrt{x}$,   find a function g such that $f(g(x)) = x$ for every x in the domain of g.

17. Let $f(x) = 3 - 2x$.   Find: (a) $f \circ f$ (b) $f^2 = f \cdot f$

18. Give an example of two functions f and g for which $f \circ g = g \circ f$.   Does $f \circ g = g \circ f$ for any choice of f and g?

19. If $f(x) = 2x^3 + 3x^2 + 4x + 5$ and $g(x) = 2$,   find $g(f(x))$ and $f(g(x))$.

7. Inverse Functions

Recall that a function f from X into Y is a correspondence that associates with each element of X a unique element of Y. The set X is called the *domain of f*. For each element x in X, the corresponding element y in Y is called the *value of f at x*. The set of all values is called the *range of f*. If a function f has the additional property that corresponding to each element in the range there is exactly one element in the domain, then f is called a *one-to-one function*. If the graph of a function f is known, there is a simple test to determine whether f is one-to-one. If any horizontal line of height h strikes the graph of f more than once, then the value of $y = h$ corresponds

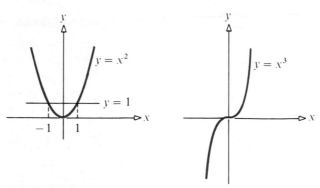

(a) Horizontal line strikes
graph twice

(b) Every horizontal line
strikes graph once

Figure 46

to more than one x and f is *not* one-to-one. Figure 46 illustrates this test for the functions $y = x^2$ and $y = x^3$. It is easy to see from these illustrations that $y = x^2$ is not one-to-one, while $y = x^3$ is one-to-one.

Analytically, this geometric interpretation means that when the equation of a one-to-one function $y = f(x)$ is solved for x in terms of y, then x is also a *function of y*. For example, the function $f(x) = 3x - 4$ is one-to-one since when we solve the equation $y = 3x - 4$ for x, then x is expressed as a function of y. In this case,

(1.24)
$$x = \tfrac{1}{3}(y + 4)$$

The function $f(x) = x^2 - 4$ is not one-to-one because when we solve the equation $y = x^2 - 4$ for x, we get

$$x = \pm\sqrt{y + 4}$$

This equation does not define x as a function of y—there are two numbers x for values of $y > -4$.

For the one-to-one function $f(x) = 3x - 4$, the function g defined by (1.24), namely, $g(y) = \tfrac{1}{3}(y + 4)$, is called the *inverse of f*. As the preceding examples illustrate, a function must be one-to-one to have an inverse. In fact, the inverse of a one-to-one function is unique. It is easy to prove that the inverse function $x = g(y)$ of a one-to-one function $y = f(x)$ is unique. If y is in the range of f, then f must take on the value y at some number x. Since f is one-to-one, this number x is unique. We have called this number $g(y)$ and hence g is unique.

(1.25)   **DEFINITION   *Inverse Function*.   Let f be a one-to-one function. The *inverse of f*, denoted by f^{-1}, is the unique function defined on the range of f for which**

$$x = f^{-1}(y) \quad \text{if and only if} \quad y = f(x)$$

In $x = f^{-1}(y)$ substitute $y = f(x)$.   Then

(1.26) $x = f^{-1}(f(x))$ for all x in the domain of f

Similarly,

$$y = f(f^{-1}(y)) \qquad \text{for all } y \text{ in the range of } f$$

As a result of (1.25), it follows that

Domain of f = Range of f^{-1} Range of f = Domain of f^{-1}

See Figure 47 for an illustration.

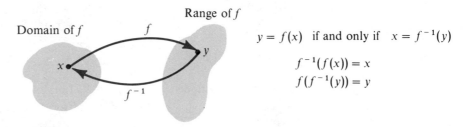

Figure 47

Example 1 demonstrates how to find the inverse of a given function f.

EXAMPLE 1   Find the inverse of $y = f(x) = 2x + 3$.

Solution   Since the graph of the line $y = 2x + 3$ is not horizontal, any horizontal line intersects it exactly once. Thus, the function is one-to-one. To find the inverse, we solve for x. Since $y = 2x + 3$,   we find

$$2x = y - 3$$
$$x = \tfrac{1}{2}(y - 3)$$

The inverse of $f(x) = 2x + 3$ is therefore $x = f^{-1}(y) = \tfrac{1}{2}(y - 3)$.   Since the symbol traditionally used to represent the independent variable of a function is x, it is convenient to replace y by x in f^{-1}. That is, $f^{-1}(x) = \tfrac{1}{2}(x - 3)$.   ∎

We can verify that the function f^{-1} is the inverse of the function f by checking to see that $f^{-1}(f(x)) = x$.   For the function in Example 1,

$$f^{-1}(f(x)) = \tfrac{1}{2}[f(x) - 3] = \tfrac{1}{2}[(2x + 3) - 3] = \tfrac{1}{2}(2x) = x$$

GEOMETRIC INTERPRETATION

If we sketch the graphs of f and f^{-1} of Example 1 (illustrated in Fig. 48), we notice an interesting fact:

The graphs of f and f^{-1} are symmetric with respect to the line $y = x$.

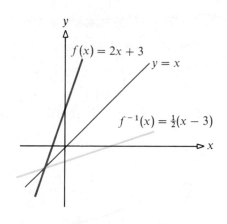

Figure 48   **Figure 49**

That is, the graph of f^{-1} is the reflection of the graph of f about the line $y = x$. (Each graph is a mirror image of the other, the mirror being the line $y = x$.)   See Figure 49 for a general demonstration of this situation.

EXAMPLE 2   Find the inverse of $y = f(x) = x^2$ if $x \geq 0$.

Solution   The function $f(x) = x^2$ is not one-to-one. However, if we restrict f to only that part of its domain for which $x \geq 0$,   we have a one-to-one function. If we solve for x, obtaining $x = \sqrt{y}$ (the minus sign is excluded since $x \geq 0$),   and replace y by x, we find the inverse of the new function to be

$$f^{-1}(x) = \sqrt{x}$$

Figure 50 illustrates the graphs of $f(x) = x^2$ and $f^{-1}(x) = \sqrt{x}$.

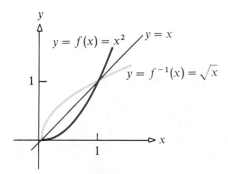

Figure 50   ■

The idea expressed in Example 2 of restricting the domain of a function that is not one-to-one is a common practice. We use this concept in Chapter 8 when inverse trigonometric functions are discussed.

EXERCISE 7

In Problems 1–8 use a graph of each function to determine whether it is one-to-one.

1. $y = 3x - 1$ **2.** $y = 5x + 3$ **3.** $y = x^2 + 3$ **4.** $y = x^2 - 4$

5. $y = x^3$ **6.** $y = x^3 + 1$ **7.** $y = x^n$, n even   **8.** $y = x^n$, n odd

In Problems 9–16 the function f is one-to-one. Find f^{-1}. For the functions in Problems 9–14, graph f and use it to get the graph of f^{-1}.

9. $f(x) = 3x - 1$ **10.** $f(x) = 5x + 3$ **11.** $f(x) = x^3$ **12.** $f(x) = \sqrt[3]{x}$

13. $f(x) = \dfrac{1}{x}$, $x > 0$ **14.** $f(x) = \dfrac{1}{1-x}$, $x > 1$ **15.** $f(x) = \dfrac{x}{x+1}$, $x > -1$ **16.** $f(x) = \dfrac{x}{x-1}$, $x > 1$

17. A function f has an inverse. If the graph of f lies in the first quadrant, in what quadrant does the graph of f^{-1} lie?

18. A function f has an inverse. If the graph of f lies in the second quadrant, in what quadrant does the graph of f^{-1} lie?

19. To convert from x degrees Celsius to y degrees Fahrenheit, we use the formula $y = f(x) = \frac{9}{5}x + 32$. To convert from u degrees Fahrenheit to v degrees Celsius, we use the formula $v = g(u) = \frac{5}{9}(u - 32)$. Show that f and g are inverse functions.

20. Prove that if f is a *periodic function*—that is, if there is a positive number a so that $f(x + a) = f(x)$, for all x in the domain of f—then f does not have an inverse.

21. If the function f is defined by $f(x) = x^5 - 1$, find f^{-1}.

Miscellaneous Exercises

In Problems 1–4 find the solution of each inequality.

1. $\left| \dfrac{2}{x+4} \right| > 3$ **2.** $\left| \dfrac{3}{x-2} \right| \leq 5$ **3.** $\dfrac{1}{x^2 - 4} < 0$ **4.** $\dfrac{1}{x(x^2 - 2x - 3)} > 0$

5. Plot the points $A = (1, 1)$, $B = (5, 3)$, $C = (3, 7)$, $D = (-1, 5)$.
   (a) Find the slopes of the lines containing A and B, containing B and C, containing C and D, and containing A and D.
   (b) Which of these lines are parallel? Which are perpendicular?
   (c) Find the tangent of the angle between the line containing AB and the line containing AD.
   (d) Find the point of intersection of the line containing AC and the line containing BD.
   (e) Find the distance between the line containing AB and the line containing CD.
   (f) Find the midpoint of the line segment joining AB.

6. Determine whether the points $(1, 8)$, $(2, 16)$, and $(-1, 2)$ are *collinear* (that is, lie on the same line) by:
   (a) Calculating slopes   (b) Using the distance formula

7. Show in two ways that the triangle with vertices at $A = (1, -6)$, $B = (8, 8)$, and $C = (-7, -2)$ is a right triangle.

8. Find the angle between the lines given by the equations $2x + 3y - 7 = 0$ and $3x - 5y - 2 = 0$.

9. Determine the coordinates of the point on the line $3x + 2y = 0$ that is equidistant from $(0, 0)$ and $(-2, 3)$.

10. Develop a formula for the distance between the parallel lines $y = mx + b_1$ and $y = mx + b_2$, where $b_1 \neq b_2$.

11. A point (a, b) lies at a distance 4 units from the line $15x + 8y - 34 = 0$. Find an equation involving a and b. (Two solutions.)

12. Find the coordinates of the point on the y-axis that is equidistant from the points $(-3, 5)$ and $(2, 4)$.

13. Express by an equation the fact that the point (x, y) is always at a distance 4 units from the point $(1, -2)$.

14. The point $(2, -5)$ is at a distance $\sqrt{65}$ from the midpoint of the segment joining $(4, 2)$ and $(x, 4)$. Find x.

15. If the line through $(x, 4)$ and $(-2, 1)$ is perpendicular to the line through $(2, 3)$ and $(-1, y)$, find an equation relating x and y.

16. The point $(2, 3)$ is a distance 4 units from the midpoint of the segment with endpoints $(-1, 4)$ and $(1, y)$. Find y.

17. Find the general equation of a circle tangent to the y-axis with center at $(2, 3)$.

18. Find the general equation of the circle with center at $(3, 1)$ passing through $(0, 0)$.

19. Find the general equation of the circle with center at $(-3, -2)$ and tangent to the line $y = 5$.

20. Find the general equation of the circle passing through the points $(2, 4)$, $(4, 8)$, and $(2, 8)$.

21. (a) Find the center and radius of the circle with equation

$$x^2 + ax + y^2 + by = 0$$

(b) Find a condition on a, b, and c so that

$$x^2 + ax + y^2 + by + c = 0$$

is the equation of a circle of positive radius.

22. The line with equation $y = \frac{1}{2}x + 2$ is tangent to a circle at $(0, 2)$. The line $y = 2x - 7$ is tangent to the same circle at $(3, -1)$. Find the center of this circle.

23. Develop a formula for the general equation of the tangent line to the circle $x^2 + y^2 = R^2$ at a point (x_0, y_0) on the circle.

24. Find the general equation of a line containing the centers of the circles

$$x^2 + y^2 + 4x - 8y + 4 = 0 \quad \text{and} \quad x^2 + y^2 - 2x - 2y = 0$$

25. Find the center of the circle circumscribing the triangle with sides $x + y = 4$, $y = x$, and $x + 2y = 9$.

26. Graph the functions: (a) $f(x) = |x^2 - 1|$ (b) $f(x) = x + |x - 1|$

27. If f denotes an even function and g denotes an odd function, determine whether the following functions are even, odd, or neither:
   (a) $f + g$ (b) $f \cdot g$ (c) $f \circ f$ (d) $g \circ g$ (e) $f \circ g$ (f) $g \circ f$

28. Sketch the graph of $f(x) = \text{Minimum}(x - [\![x]\!], \ 1 - x + [\![x]\!])$.

29. Let h be the function defined by $h(x) = x^n$, where n is an integer. What condition, if any, must be placed on n for h to have an inverse?

30. If $f(x) = x/(x + 1)$, find $f(1/x)$, $f(f(x))$, and $f(1/f(x))$.

31. Determine p so that $(f \circ g)(x) = (g \circ f)(x)$, where $f(x) = 3x + 2$ and $g(x) = 2x - p$.

32. If $f(x) = mx + b$ and $g(x) = sx + c$, where $m, b, s,$ and c are real numbers, show that $f \circ g$ and $g \circ f$ are linear functions.

33. A *fixed point* of f is a number x such that $f(x) = x$.
(a) If x_0 is a fixed point of f, show that $(f \circ f)(x_0) = x_0$.
(b) Find the fixed points of $f(x) = (2x + 3)/(x + 4)$.
(c) Show that, in general, $f(x) = (ax + b)/(cx + d)$, where $a, b, c,$ and d are constants, has at most two fixed points. Discuss.

34. If $f(x) = 3x/(x - 2)$, find f^{-1} and the domains of f and f^{-1}.

35. Suppose

$$f(x) = \begin{cases} -x & \text{if } x < 0 \\ -2 & \text{if } x = 0 \\ \dfrac{-1}{x + 1} & \text{if } x > 0 \end{cases}$$

(a) Show graphically that f has an inverse.
(b) Find the domains of f and f^{-1}.
(c) Graph f^{-1} and find equations for f^{-1} over various parts of its domain.

36. Find f^{-1} if $f(x) = (ax + b)/(cx + d)$. What happens if $ad - bc = 0$?

37. Show that $(f \circ g)^{-1} = g^{-1} \circ f^{-1}$ if the three inverses exist.

38. Solve: $|x| + |x - 3| < 4$.

39. If $|x - 2| < \frac{1}{5}$ and $|y - 3| < \frac{1}{10}$, show that $|xy - 6| < \frac{41}{50}$.

40. If $|x - 2| < \frac{1}{100}$, show that $|x^2 - 4| < \frac{1}{10}$ and $|x^3 - 8| < \frac{13}{100}$.

41. A rectangle is inscribed in a circle of radius R. Express the area A of the rectangle as a function of R and the length x of one of its sides.

42. A trapezoid is inscribed in a circle of radius 4 centimeters with one base coinciding with a diameter of the circle. Express the area of the trapezoid as a function of its altitude.

43. A strip of nickel 200 centimeters long and 16 centimeters wide is to be made into a rain gutter by turning up all four edges to form a trough with a rectangular cross section. If the height of the bent-up edge is x centimeters, express the volume of the trough as a function of x.

44. A gardener wishes to fence a rectangular garden along a straight river. No fence is required along the river. The gardener has enough wire to build a fence 200 meters long. If the length of the side bordering on the river is represented by the variable x, express the area of the garden as a function of x.

45. If a Norman window (a rectangle surmounted by a semicircle) has a perimeter of 100 centimeters and if the length of the side that is not the diameter of the semicircle is represented by x, express the area of the window as a function of x.

46. A wire 10 meters long is cut in two parts: one part is bent into the shape of a square and the other into the shape of the circumference of a circle. If the perimeter of the square is represented by p, express the sum of the areas of both the square and the circle as a function of p.

47. The modulus of elasticity of structural steel is 29×10^6 pounds per square inch at a temperature of $70°F$. At $900°F$ the modulus is 25×10^6 pounds per square inch. Between these temperatures, the change in modulus with temperature is known to be practically linear. Based on the above, give a quantitative formula for the modulus of elasticity of structural steel between the temperatures of 70°F and 900°F.

48. A right circular cylinder of radius x and height y just fits inside a right circular cone of radius b and height h. Draw a plane section through the common axis of the cone and cylinder, and from the geometry of the figure read off the relation that expresses y as a function of x (b and h are regarded as constants).

49. Graph any function f having the following properties: the domain of f is $(-1, 1)$; the range of f consists of exactly four numbers; $f(x) > 0$ when $x > 0$ and $f(x) < 0$ when $x < 0$.

50. Given $f(x) = 4x + 1$ and $g(x) = 2x - 5$, write each of the following in interval notation:
    (a) $f(x) > g(x)$ (b) $f(x) \le g(x)$ (c) $f(x) \ne g(x)$ (d) $|g(x) - f(x)| < 1$

51. If a^2 is an even integer, prove that a must also be even.   [*Hint:*   Assume the contrary.]

52. Prove that $\sqrt{2}$ is irrational.   [*Hint:*   Assume the contrary and use the result of Problem 51.]

53. A function f is defined on the closed interval from -3 to 3 and has the graph shown.
    (a) On the axes provided, sketch the entire graph of $y = |f(x)|$.
    (b) On the axes provided, sketch the entire graph of $y = f(|x|)$.
    (c) On the axes provided, sketch the entire graph of $y = f(-x)$.
    (d) On the axes provided, sketch the entire graph of $y = f(x - 1)$.

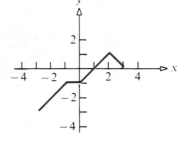

2

The Limit of a Function

The concept of the limit of a function is what bridges the gap between the mathematics of algebra and geometry and the mathematics of calculus. Although the idea of a limit is difficult to understand at first, it can be mastered. In fact, the *evaluation* of limits is fairly easy (after practice). For this reason, we will develop limits intuitively at first, then develop a technique for finding limits, and finally look at a precise formulation of limits and limit theorems.

1. Limits from an Intuitive Point of View

The development of calculus came as a result of the inability to answer certain questions in geometry. Two of these questions are:

1. Given a function f and a point P on its graph, what is the slope of the line tangent to the graph of the function at the point P? See Figure 1.

2. Given a nonnegative function f whose domain is the closed interval $[a, b]$, what is the area enclosed by the graph of f, the x-axis, and the vertical lines $x = a$ and $x = b$? See Figure 2.

These problems, traditionally called the *tangent problem* and the *area problem*, were solved by Gottfried Wilhelm von Leibniz (1646–1716) and Sir Isaac Newton (1642–1727).* Their solutions were shown to be intimately related not only to each

---

* The Historical Perspectives section at the end of this chapter traces the various solutions to the tangent problem from the Greeks to Fermat, who preceded Leibniz and Newton.

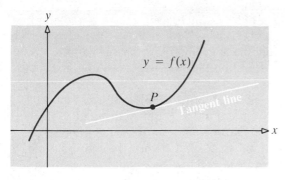

Figure 1   Tangent problem

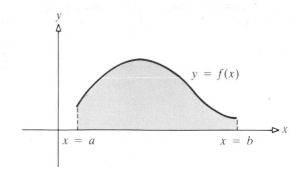

Figure 2   Area problem

other but to many other problems concerning the behavior of functions. Here, we will discuss the tangent problem; the area problem is postponed until Chapter 5.

THE TANGENT PROBLEM

Tangent line
to circle at P

Figure 3

Consider the function whose graph appears in Figure 1. Through the point P on the graph we have drawn a line that just touches the graph of f. We call this unique line the *tangent** line to the graph of f at P. Our problem is to give a satisfactory definition of a tangent line.

Let's start with some basics. In plane geometry, a tangent line to a circle is defined as a line having exactly one point in common with the circle (see Figure 3). However, this definition is not satisfactory for graphs in general. For example, consider the graph shown in Figure 4. There are many lines—for instance, L_1, L_2, L_3—that pass through the point P and have exactly one point in common with the graph, but they do not meet the requirement of just touching the graph at P. Furthermore,

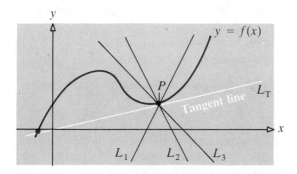

Figure 4

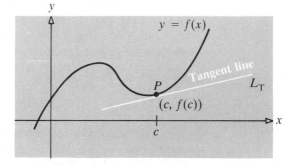

Figure 5

* *Tangere* is a Latin word meaning "to touch."

the line L_T, which does just touch the graph of f at P, also intersects the graph at another point.

It is evident that the definition from plane geometry for circles will not do for graphs in general. We need a different definition that will work not only for circles, but also for the graph of any function.

We begin with a function f and a point P on its graph (refer to Fig. 5). The tangent line L_T to f at P will necessarily pass through P. But to distinguish the tangent line L_T from all the other lines that pass through P, we will need to know its slope $m_{\tan}$. If we use the number c to represent the abscissa of P, then the coordinates of P are $(c, f(c))$. To get the slope $m_{\tan}$ of L_T, we need to know two points on L_T. However, we cannot find two points on L_T, so we look for another way to find $m_{\tan}$.

Suppose $Q = (x, f(x))$ is any point on the graph of f different from P. Then the line through P and Q is called a *secant line* of f (see Fig. 6).

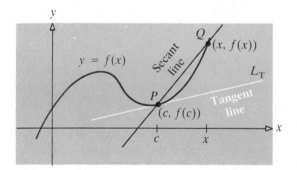

Figure 6

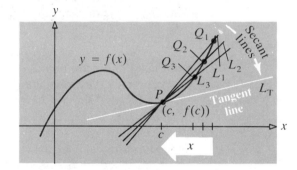

Figure 7

Now look at Figure 7, where we show three different points and the secant lines L_1, L_2, and L_3. As x gets closer to c, the secant lines L_1, L_2, L_3, ... tend to a limiting position. The line L_T that is the limiting position of these secant lines is the *tangent line to the graph of f at c*.

But what is the slope of the tangent line L_T? We know L_T passes through the point $(c, f(c))$. Furthermore, we know that the slope $m_{\sec}$ of each secant line joining the points $(c, f(c))$ and $(x, f(x))$ is given by

(2.1)
$$m_{\sec} = \frac{f(x) - f(c)}{x - c}$$

As x gets closer to c, the secant lines approach a limiting position—the tangent line L_T. So we may expect the limiting value of the slopes of the secant lines to equal the slope of the tangent line. That is, (2.1) suggests that

$$m_{\tan} = \begin{bmatrix} \text{Slope of tangent line} \\ \text{to } f \text{ at } c \end{bmatrix} = \begin{bmatrix} \text{Limiting value of } \dfrac{f(x) - f(c)}{x - c} \\ \text{as } x \text{ gets closer to } c \end{bmatrix}$$

In symbols, we write

(2.2)
$$m_{\tan} = \lim_{x \to c} \frac{f(x) - (c)}{x - c}$$

The notation "lim" with   "$x \to c$"   beneath it (or set to the side as   "$\lim_{x \to c}$")   is read "the limit as x approaches c of . . . ."

In order to study the tangent problem in more detail, the notion of limiting value or limit will have to be made mathematically precise. We will do this later. For the time being, all we need is an intuitive understanding of this concept. Let's look at an example.

EXAMPLE 1   Find an equation of the tangent line to the graph of $f(x) = x^2$ at the point $(2, 4)$.

Solution   The tangent line passes through the point $(2, 4)$. The slopes of the secant lines joining $(2, 4)$ to points $(x, f(x))$ on the graph of $y = x^2$ are

(2.3)
$$m_{\sec} = \frac{f(x) - f(2)}{x - 2} = \frac{f(x) - 4}{x - 2} = \frac{x^2 - 4}{x - 2}$$

(see Fig. 8). Since the point $(x, f(x))$ is taken to be different from the point $(2, 4)$, we conclude that $x \neq 2$.   Therefore, we may simplify (2.3) to get

$$m_{\sec} = \frac{(x - 2)(x + 2)}{x - 2} = x + 2$$

Now comes the important step in this procedure! As x gets closer to 2, the values of $m_{\sec} = x + 2$ get closer to 4. That is, the slope $m_{\tan}$ of the tangent line at $(2, 4)$

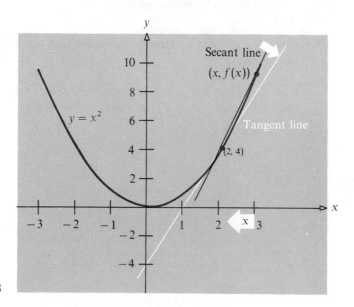

Figure 8

is

$$m_{\text{tan}} = \lim_{x \to 2} \frac{x^2 - 4}{x - 2} = \lim_{x \to 2} (x + 2) = 4$$

Therefore, an equation of this tangent line is

$$y - 4 = 4(x - 2)$$
$$y = 4x - 4 \quad \blacksquare$$

Now we can turn our attention to a more detailed description of the concept of limit. Our approach will proceed in stages that eventually lead to a precise definition of limit.

THE IDEA OF A LIMIT

We begin by asking a question: "What does it mean for a function f to have a limit L as x approaches some fixed number c?" To find an answer, we need to be more precise about f, L, and c. The function f must be defined in an open interval near the number c, but it does not have to be defined at c itself. The limit L is some number. With these restrictions in mind, we introduce the symbolism

(2.4) $$\lim_{x \to c} f(x) = L$$

which is read as "the limit of $f(x)$ as x approaches c equals the number L." This indicates that f has a limit L as x approaches c. We may describe statement (2.4) in two ways:

(2.5)   **For all x approximately equal to c, but $x \neq c$,   the value $f(x)$ is approximately equal to L.**

(2.6)   **For all x sufficiently close to c, but unequal to c, the value $f(x)$ can be made as close as we please to L.**

In Figure 9(a) we show the graph of the function f and observe that as x gets closer to c, the value of f, as measured by its height, gets closer to the number L. This

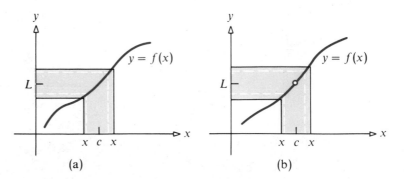

Figure 9   (a)   (b)

is the key idea behind the notion of a limit. Note that the value of f at c does not matter. As Figure 9(b) illustrates, even though f is not defined at c, it is still true that as x gets closer to c, the value of f gets closer to L.

NUMERICAL APPROACH

Based on descriptions (2.5) and (2.6), it is natural to try a numerical approach as an aid in calculating limits.

EXAMPLE 2   Let $f(x) = x^3$ and $c = 2$. The numerical approach uses a table and a calculator to guess $\lim_{x \to 2} x^3$:

x	1	1.5	1.6	1.75	1.8	1.9	1.99	1.999	1.9999
$f(x) = x^3$	1	3.375	4.096	5.359	5.832	6.859	7.8806	7.988	7.9988

x	3	2.5	2.4	2.25	2.2	2.1	2.01	2.001	2.0001
$f(x) = x^3$	27	15.625	13.824	11.3906	10.648	9.261	8.1206	8.012	8.0012

We infer that for x "sufficiently close" to 2, the value of $f(x) = x^3$ can be made "as close as we please" to 8. That is, $\lim_{x \to 2} x^3 = 8$. ∎

EXAMPLE 3   Use a numerical approach to find $\lim_{x \to 1} \dfrac{x^2 - 1}{x - 1}$.

Solution   As in Example 2, we construct a table. The choices for x, although arbitrary, are selected so that they are close to 1; some are chosen less than 1, and some greater than 1:

x	0	0.5	0.75	0.9	0.99	1.01	1.1	1.25	1.5	2
$\dfrac{x^2 - 1}{x - 1}$	1	1.5	1.75	1.9	1.99	2.01	2.1	2.25	2.5	3

We infer from the table that $\lim_{x \to 1} \dfrac{x^2 - 1}{x - 1} = 2$. ∎

You are right if you make the observation that as long as x is close to 1, but not equal to 1, f can be simplified as $(x^2 - 1)/(x - 1) = x + 1$. Now it is easy to see that for x close to 1, $f(x)$ will be close to $1 + 1 = 2$.

There are limits, though, for which the above type of simplification does not work.

EXAMPLE 4   Use a numerical approach to find $\displaystyle\lim_{x\to 0}\frac{\sin x}{x}$.

Solution   The functions $\sin x$ and x are familiar, and their quotient $(\sin x)/x$ can be calculated easily at every point except 0. At 0, the value of each function is 0, and this completely meaningless ratio (0/0) presents us with a problem. However, a careful calculation (using a calculator) for numbers x close to 0 yields the following table:

x (radian)	1	0.5	0.1	0.01	-0.01	-0.1	-0.5	-1
$\sin x$	0.8415	0.4794	0.0998	0.0099	-0.0099	-0.0998	-0.4794	-0.8415
$\dfrac{\sin x}{x}$	0.8415	0.9589	0.9983	**0.9999**	**0.9999**	0.9983	0.9589	0.8415

We infer that the ratio $(\sin x)/x$ approaches the value 1 as x tends to 0, and we write $\displaystyle\lim_{x\to 0}\frac{\sin x}{x}=1.$ ∎

The numerical approach we have been using involves guesses that are made based on a table of numerical values. It is possible that if we had used numbers x even closer to 0 in the table of Example 4, then the value of $(\sin x)/x$ might have been shown to equal some number that is not close to 1. Furthermore, since we cannot actually get "as close as we please to 0," this difficulty will remain with us. What we really need is a proof that $\lim_{x\to 0}(\sin x)/x = 1$. This proof is given in Chapter 3. For now, though, let's continue our intuitive development of limits. This time we use a graphical approach.

GRAPHICAL APPROACH

Let's look at some more examples. In these examples, we will graph each function and use the graph to find the limit.

EXAMPLE 5   Use the graph of the function below to find $\lim_{x\to 2} f(x)$.

$$f(x) = \begin{cases} 3x + 1 & \text{if } x \neq 2 \\ 3 & \text{if } x = 2 \end{cases}$$

Solution   First, we graph f in Figure 10. We observe that for x near 2, the value of the function f is near 7. In fact, by choosing x close enough to 2, we can force the value of f to get as close as we please to 7. We conclude that $\lim_{x\to 2} f(x) = 7$. Note that the value of f *at* 2 is $f(2) = 3$.

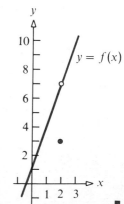

Figure 10

This example illustrates the following important principle about limits:

The limit L of a function $y = f(x)$ as x approaches the number c *does not* depend on the value of f at c.

If there is *no single* number that the value of f approaches, we say that f *has no limit as x approaches c*, or, more simply, that the *limit does not exist at c*. The next example illustrates this for $c = 0$.

EXAMPLE 6   Use the graph of the function below to find $\lim_{x \to 0} f(x)$,   if it exists.

$$f(x) = \begin{cases} -1 & \text{if} \quad x < 0 \\ 1 & \text{if} \quad x > 0 \end{cases}$$

Solution   Figure 11 illustrates the graph of f.

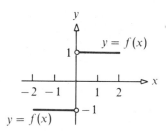

Figure 11

We observe that if x is close to 0 and negative, the value of f equals -1. On the other hand, if x is close to 0 and positive, the value of f equals 1. Since there is no single number that the values of f are close to when x is close to 0, we conclude that $\lim_{x \to 0} f(x)$ does not exist at 0.   ∎

In the next example we use the graphical approach to obtain two limits that we will rely on quite heavily in the next section.

EXAMPLE 7   Use a graphical approach to show that:

(2.7)   **(a)** $\lim_{x \to c} A = A,$ **A and c real numbers**   **(b)** $\lim_{x \to c} x = c,$ **c a real number**

Solution

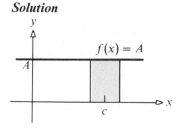

Figure 12

For x close to c, the value of f remains at A
$$\lim_{x \to c} A = A$$

(a) The function f is the constant function $f(x) = A,$ whose graph is a horizontal line (see Fig. 12). For any choice of c, it is clear that if x is close to c, the value of f remains at A. Therefore, we conclude that $\lim_{x \to c} A = A$.

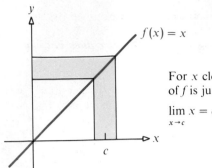

For x close to c, the value
of f is just as close to c

$$\lim_{x \to c} x = c$$

(b)   The function f is the identity function
      $f(x) = x$, whose graph is the straight
      line illustrated in Figure 13. For any
      choice of c, we see that if x is close to
      c, the value of f is just as close to c.
      We conclude that $\lim_{x \to c} x = c$.   ∎

Figure 13

Based on the important results (2.7), we can say that

$$\lim_{x \to -3} 4 = 4 \qquad \lim_{x \to 6} x = 6 \qquad \lim_{x \to -2} x = -2$$

In each of these examples, we draw conclusions about $\lim_{x \to c} f(x)$ by asking
whether the values of f can be forced to be "as close as we please" to a single number
L when x is "sufficiently close to" c. But the expressions "sufficiently close to" and
"as close as we please" are too vague for the purposes of mathematics. To be able to
treat the concept precisely, we need a *formal* definition of *limit*. For the sake of completeness and accuracy, we now give the definition, but we will postpone a detailed
discussion of it until Section 5. As is customary, the definition makes use of the Greek
letters ε (epsilon) and δ (delta) and is sometimes referred to as the ε, δ *definition of
a limit*.

(2.8)   **DEFINITION**   *Limit of a Function.*   **Let f be a function defined on an open interval
containing c, except possibly for c itself. Then the *limit of the function f* as x approaches c is the number L, written**

$$\lim_{x \to c} f(x) = L$$

if, for any given number $\varepsilon > 0$,   a number $\delta > 0$ exists such that

$$|f(x) - L| < \varepsilon \qquad \text{whenever} \qquad 0 < |x - c| < \delta \quad \text{and } x \text{ is in the domain of } f$$

EXERCISE 1

In Problems 1–8 complete each table and use it to evaluate the indicated limit. A calculator will be helpful.

1. x	0.9	0.99	0.999	1.001	1.01	1.1
$f(x) = 2x$						

$$\lim_{x \to 1} 2x = ?$$

2.

x	1.9	1.99	1.999	2.001	2.01	2.1
$f(x) = x + 3$						

$\lim\limits_{x \to 2}(x + 3) = ?$

3.

x	0.1	0.01	0.001	-0.001	-0.01	-0.1
$f(x) = x^2 + 2$						

$\lim\limits_{x \to 0}(x^2 + 2) = ?$

4.

x	-1.1	-1.01	-1.001	-0.999	-0.99	-0.9
$f(x) = x^2 - 2$						

$\lim\limits_{x \to -1}(x^2 - 2) = ?$

5.

x	-2.5	-2.9	-2.99	-3.01	-3.1	-3.5
$f(x) = \dfrac{x^2 - 9}{x + 3}$						

$\lim\limits_{x \to -3} \dfrac{x^2 - 9}{x + 3} = ?$

6.

x	-0.2	-0.1	-0.01	0.01	0.1	0.2
$f(x) = \dfrac{x^2 + x}{x}$						

$\lim\limits_{x \to 0} \dfrac{x^2 + x}{x} = ?$

7.

x (radian)	-0.2	-0.1	-0.01	0.01	0.1	0.2
$f(x) = \dfrac{\tan x}{x}$						

$\lim\limits_{x \to 0} \dfrac{\tan x}{x} = ?$

8.

x (radian)	−0.2	−0.1	−0.01	0.01	0.1	0.2
$f(x) = \dfrac{1 - \cos x}{x}$						

$$\lim_{x \to 0} \frac{1 - \cos x}{x} = ?$$

In Problems 9–16 use each graph to determine whether $\lim_{x \to c} f(x)$ exists at the number c.

9.
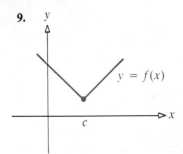

10.
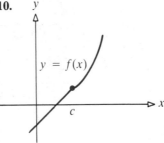

11.
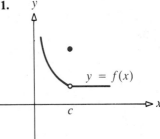

12.
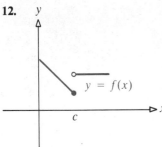

13.
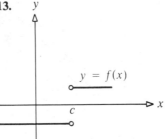

14.
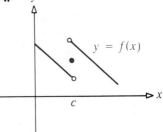

15.
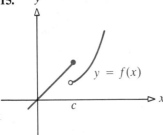

16.
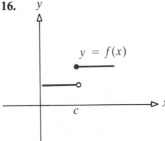

In Problems 17–24 use a graph to determine whether $\lim_{x \to c} f(x)$ exists at the given number c.

17. $f(x) = \begin{cases} 2x + 5 & \text{if } x \le 2 \\ 4x + 1 & \text{if } x > 2 \end{cases}$ $c = 2$

18. $f(x) = \begin{cases} 2x + 1 & \text{if } x \le 0 \\ 2x & \text{if } x > 0 \end{cases}$ $c = 0$

19. $f(x) = \begin{cases} 3x - 1 & \text{if } x < 1 \\ 4 & \text{if } x = 1 \\ 2x & \text{if } x > 1 \end{cases}$ $c = 1$

20. $f(x) = \begin{cases} 3x - 1 & \text{if } x < 1 \\ 2 & \text{if } x = 1 \\ 2x & \text{if } x > 1 \end{cases}$ $c = 1$

21. $f(x) = \begin{cases} 3x - 1 & \text{if } x < 1 \\ \text{Not defined} & \text{if } x = 1 \\ 2x & \text{if } x > 1 \end{cases}$ $c = 1$

22. $f(x) = \begin{cases} 3x - 1 & \text{if } x < 1 \\ 2 & \text{if } x = 1 \\ 3x & \text{if } x > 1 \end{cases}$ $c = 1$

23. $f(x) = \begin{cases} x^2 & \text{if } x \leq 0 \\ 2x + 1 & \text{if } x > 0 \end{cases}$ $c = 0$

24. $f(x) = \begin{cases} x^2 & \text{if } x < -1 \\ 2 & \text{if } x = -1 \\ -3x + 2 & \text{if } x > -1 \end{cases}$ $c = -1$

In Problems 25–28 use the results of (2.7) to find each limit.

25. $\lim\limits_{x \to 0}(-3)$ 26. $\lim\limits_{x \to -2}(-6)$ 27. $\lim\limits_{x \to -1} x$ 28. $\lim\limits_{x \to 0} x$

29. For $f(x) = 3x^2$:
   (a)   Find the slope of the secant line joining the points (2, 12) and (3, 27).
   (b)   Use the method illustrated in Example 1 to find an equation of the tangent line to the graph of f at 2.
   (c)   Graph f, show the tangent line at 2, and show the secant line from part (a).

30. For $f(x) = x^3$:
   (a)   Find the slope of the secant line joining the points (2, 8) and (3, 27).
   (b)   Use the method illustrated in Example 1 to find an equation of the tangent line to the graph of f at 2.
   (c)   Graph f, show the tangent line at 2, and show the secant line from part (a).

31. For $f(x) = \frac{1}{2}x^2 - 1$:
   (a)   Determine the slope m_{sec} of the secant line joining the points $P = (2, f(2))$ and $Q = (2 + h, f(2 + h))$.
   (b)   Use the result found in part (a) to determine the slopes of the secant lines for $h = -0.5$ and $h = 0.5$.
   (c)   Use the results found in parts (a) and (b) to complete the following table:

h	-0.5	-0.1	-0.001	0.001	0.1	0.5
m_{sec}						

   (d)   Determine the limiting value of the slope of the secant line found in part (a) as $h \to 0$.
   (e)   Find an equation for the tangent line to f at the point $P = (2, f(2))$.
   (f)   Draw the graph of the function f and the tangent line to f at $P = (2, f(2))$.

2. Algebraic Techniques for Finding Limits

The results (2.7) given in Section 1 and the algebraic properties of limits stated in the next theorem enable us to compute the limits of many functions we will encounter in calculus. Some of these properties are proved in Appendix II.

THEOREM   **Let f and g be two functions for which $\lim\limits_{x \to c} f(x) = L$ and $\lim\limits_{x \to c} g(x) = M$, L and M being two real numbers.**

(2.9)   *Limit of a Sum.*

$$\lim_{x \to c}[f(x) + g(x)] = \lim_{x \to c} f(x) + \lim_{x \to c} g(x) = L + M$$

That is, the limit of the sum of two functions equals the sum of their limits.

EXAMPLE 1
$$\lim_{x \to 1}(x + 5) = \lim_{x \to 1} x + \lim_{x \to 1} 5 = 1 + 5 = 6 \qquad \blacksquare$$

$\uparrow$ (2.9) $\qquad\qquad$ $\uparrow$ (2.7)

(2.10)   *Limit of a Difference.*

$$\lim_{x \to c}[f(x) - g(x)] = \lim_{x \to c} f(x) - \lim_{x \to c} g(x) = L - M$$

The limit of the difference of two functions equals the difference of their limits.

EXAMPLE 2
$$\lim_{x \to 3}(4 - x) = \lim_{x \to 3} 4 - \lim_{x \to 3} x = 4 - 3 = 1 \qquad \blacksquare$$

$\uparrow$ (2.10) $\qquad\qquad$ $\uparrow$ (2.7)

(2.11)   *If k is any real number,*

$$\lim_{x \to c}[kf(x)] = k \lim_{x \to c} f(x) = kL$$

The limit of a constant times a function equals the constant times the limit of the function.

EXAMPLE 3
$$\lim_{x \to -2}(3x + 5) = \lim_{x \to -2}(3x) + \lim_{x \to -2} 5 = 3 \lim_{x \to -2} x + \lim_{x \to -2} 5 = 3(-2) + 5 = -1 \qquad \blacksquare$$

$\uparrow$ (2.9) $\qquad\qquad$ $\uparrow$ (2.11) $\qquad\qquad$ $\uparrow$ (2.7)

(2.12)   *Limit of a Product.*

$$\lim_{x \to c}[f(x)g(x)] = \left[\lim_{x \to c} f(x)\right]\left[\lim_{x \to c} g(x)\right] = LM$$

The limit of the product of two functions equals the product of their limits.

EXAMPLE 4
$$\lim_{x \to 1}[(2x)(x + 4)] = \left[\lim_{x \to 1}(2x)\right]\left[\lim_{x \to 1}(x + 4)\right] = (2)(5) = 10 \qquad \blacksquare$$

$\uparrow$ (2.12) $\qquad\qquad$ $\uparrow$ (2.11), (2.9), (2.7)

(2.13)   *Limit of a Quotient.* If $\lim_{x \to c} g(x) = M \neq 0$,

$$\lim_{x \to c}\frac{f(x)}{g(x)} = \frac{\lim_{x \to c} f(x)}{\lim_{x \to c} g(x)} = \frac{L}{M}$$

If the limit of the denominator is not 0, the limit of the quotient of two functions equals the quotient of their limits.

EXAMPLE 5
$$\lim_{x \to 2}\frac{3x + 4}{x + 3} = \frac{\lim_{x \to 2}(3x + 4)}{\lim_{x \to 2}(x + 3)} = \frac{10}{5} = 2 \qquad \blacksquare$$

$\uparrow$ (2.13)

(2.14)   *Limit of* $[f(x)]^n$. **If n is a positive integer,**

$$\lim_{x \to c}[f(x)]^n = \left[\lim_{x \to c} f(x)\right]^n = L^n$$

EXAMPLE 6
$$\lim_{x \to 1}(2x - 3)^3 = \left[\lim_{x \to 1}(2x - 3)\right]^3 = (-1)^3 = -1 \qquad \blacksquare$$
$$\uparrow$$
$$(2.14)$$

(2.15)   *Limit of* $\sqrt[n]{f(x)}$. **If $n \geq 2$ is an integer,**

$$\lim_{x \to c} \sqrt[n]{f(x)} = \sqrt[n]{\lim_{x \to c} f(x)} = \sqrt[n]{L}$$

(Here we require $f(x) > 0$ and $L \geq 0$ if n is even.)

EXAMPLE 7
$$\lim_{x \to 4} \sqrt[3]{x^2 + 11} = \sqrt[3]{\lim_{x \to 4}(x^2 + 11)} = \sqrt[3]{16 + 11} = 3 \qquad \blacksquare$$
$$\uparrow$$
$$(2.15)$$

It is useful to observe in (2.14) the special case for $f(x) = x$,   namely,

(2.16)
$$\lim_{x \to c} x^n = c^n \qquad \textbf{\textit{n} a positive integer}$$

By using this result together with (2.11), we conclude

(2.17)
$$\lim_{x \to c} kx^n = kc^n \qquad \textbf{\textit{k} a real number, \textit{n} a positive integer}$$

These results demonstrate that *some* limits may be evaluated by merely substituting c for x. This is true for polynomial functions:

(2.18)   **For every polynomial $P(x) = a_n x^n + a_{n-1}x^{n-1} + \cdots + a_1 x + a_0$ of degree n, we have**

$$\lim_{x \to c} P(x) = P(c)$$

We can see that this is true because

$$\lim_{x \to c} P(x) = \lim_{x \to c}(a_n x^n + a_{n-1}x^{n-1} + \cdots + a_1 x + a_0)$$

$$= \lim_{x \to c}(a_n x^n) + \lim_{x \to c}(a_{n-1}x^{n-1}) + \cdots + \lim_{x \to c}(a_1 x) + \lim_{x \to c} a_0$$
$$\uparrow$$
By repeated use of (2.9)

$$= a_n c^n + a_{n-1}c^{n-1} + \cdots + a_1 c + a_0 = P(c)$$
$$\uparrow$$
By repeated use of (2.17)

EXAMPLE 8   (a) $\lim\limits_{x \to 3}(4x^2 - x + 2) = 4(3)^2 - 3 + 2 = 35$

(b) $\lim\limits_{x \to -1} (7x^5 + 4x^3 - 2x^2) = 7(-1)^5 + 4(-1)^3 - 2(-1)^2 = -13$

(c) $\lim\limits_{x \to 0}(10x^6 - 4x^5 + 20x^3 - 8x + 5) \doteq 10(0)^6 - 4(0)^5 + 20(0)^3 - 8(0) + 5 = 5$ ▪

(2.19)   **If c is in the domain of a rational function P/Q, so that $Q(c) \neq 0$, then**

$$\lim\limits_{x \to c} \frac{P(x)}{Q(x)} = \frac{P(c)}{Q(c)}$$

EXAMPLE 9   (a) $\lim\limits_{x \to 1} \dfrac{3x^3 - 2x + 1}{4x^2 + 5} = \dfrac{3 - 2 + 1}{4 + 5} = \dfrac{2}{9}$ (b) $\lim\limits_{x \to -2} \dfrac{2x + 4}{3x^2 - 1} = \dfrac{-4 + 4}{12 - 1} = 0$ ▪

These examples might lead one to conclude that the evaluation of limits is simply a question of substituting the number that x approaches into the function. Although this is often the case, the next few examples are a reminder that substitution cannot always be used.

EXAMPLE 10   Find: $\lim\limits_{x \to -2} \dfrac{x^2 + 5x + 6}{x^2 - 4}$

Solution $\lim\limits_{x \to -2} (x^2 - 4) = \lim\limits_{x \to -2} x^2 - \lim\limits_{x \to -2} 4 = 4 - 4 = 0$

Since the limit of the denominator function is 0, we cannot use (2.13). However, this does not mean that the limit does not exist! We can use the algebraic technique of factoring, and find

$$\frac{x^2 + 5x + 6}{x^2 - 4} = \frac{(x + 3)(x + 2)}{(x - 2)(x + 2)}$$

Recall that we are interested only in the limit as x approaches -2, and *not* in the value when x equals -2. Thus, the factor $(x + 2)$ is not 0 and we can cancel. Then

$$\lim\limits_{x \to -2} \frac{x^2 + 5x + 6}{x^2 - 4} = \lim\limits_{x \to -2} \frac{x + 3}{x - 2} \underset{\uparrow}{=} \frac{-2 + 3}{-2 - 2} = -\frac{1}{4}$$ ▪

By (2.13) or (2.19)

EXAMPLE 11   Find: $\lim\limits_{x \to 1} \dfrac{x^2 - 1}{x^2 - x}$

Solution $\lim\limits_{x \to 1} \dfrac{x^2 - 1}{x^2 - x} \underset{\uparrow}{=} \lim\limits_{x \to 1} \dfrac{(x - 1)(x + 1)}{x(x - 1)} \underset{\uparrow}{=} \lim\limits_{x \to 1} \dfrac{x + 1}{x} \underset{\uparrow}{=} \dfrac{1 + 1}{1} = 2$ ▪

Factor   Cancel   By (2.19)

EXAMPLE 12   For $f(x) = x^2 + 2x,$ find: $\lim\limits_{h \to 0} \dfrac{f(x+h) - f(x)}{h}$

Solution

$$f(x+h) = (x+h)^2 + 2(x+h) = x^2 + 2xh + h^2 + 2x + 2h$$

$$f(x+h) - f(x) = (x^2 + 2xh + h^2 + 2x + 2h) - (x^2 + 2x) = 2xh + h^2 + 2h$$

$$\frac{f(x+h) - f(x)}{h} = \frac{2xh + h^2 + 2h}{h} = \frac{h(2x + h + 2)}{h} = 2x + h + 2$$

$$\lim_{h \to 0} \frac{f(x+h) - f(x)}{h} = \lim_{h \to 0}(2x + h + 2) = 2x + 2 \qquad \blacksquare$$

These examples illustrate the usefulness of algebra for evaluating limits. However, as we shall see in later chapters, many limit problems arise that cannot be directly evaluated by algebraic techniques. These sometimes require geometric arguments [such as to prove* that $\lim_{x \to 0}(\sin x)/x = 1$]   and sometimes entirely new theories (such as for $\lim_{x \to 0} x^x,$ $x > 0$).

In many instances, the following result, called the *squeezing theorem*, may be helpful for evaluating limits. Its proof is given in Appendix II.

(2.20)   *Squeezing Theorem:* **Let f, g, and h be functions such that $f(x) \le g(x) \le h(x)$ for all numbers x in some open interval containing c, except possibly at c. If $\lim_{x \to c} f(x) = L$ and $\lim_{x \to c} h(x) = L,$ then $\lim_{x \to c} g(x) = L.$**

The idea behind this theorem is that it provides a technique for evaluating $\lim_{x \to c} g(x)$ if we know, or can find, simpler approximating functions f and h with g "sandwiched" between f and h for all x close to c. In this event, if f and h have the same limit L as x approaches c, then g is "squeezed" to this same limit L as x approaches c (see Fig. 14).

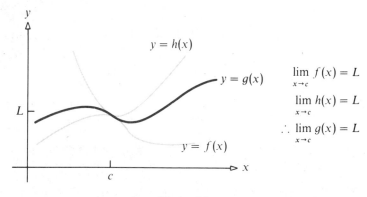

Figure 14

---

* Remember that the numerical argument given in Section 1 was not a proof.

For example, suppose we wish to find $\lim_{x \to 0} f(x)$ and it can be shown that $-x^2 \le f(x) \le x^2$ for all $x \ne 0$. Since $\lim_{x \to 0}(-x^2) = 0$ and $\lim_{x \to 0} x^2 = 0$, it follows from the squeezing theorem that $\lim_{x \to 0} f(x) = 0$ (see Fig. 15).

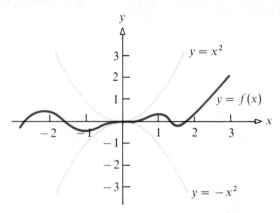

Figure 15

EXAMPLE 13   Show that: $\lim\limits_{x \to 0}\left(x \sin \dfrac{1}{x} \right) = 0$

Solution   If $x \ne 0$, we know from trigonometry that $\left| \sin(1/x) \right| \le 1$. Thus, for all $x \ne 0$, we have

$$\left| x \sin \frac{1}{x} \right| = |x| \left| \sin \frac{1}{x} \right| \le |x| \qquad \text{so that} \qquad \left| x \sin \frac{1}{x} \right| \le |x|$$

Consequently, by (1.4),

$$-|x| \le x \sin \frac{1}{x} \le |x|$$

Since $\lim_{x \to 0}(-|x|) = 0$ and $\lim_{x \to 0}|x| = 0$, it follows from (2.20) that

$$\lim_{x \to 0}\left(x \sin \frac{1}{x} \right) = 0 \qquad \blacksquare$$

We close this section with two results we need later in Chapter 4. Their proof, which uses the ε, δ definition of a limit, is given in Section 5 of this chapter.

(2.21)   If $\lim_{x \to c} f(x) > 0$, **then there is an open interval about c, possibly excluding c itself, on which $f(x) > 0$.**

(2.22)   If $\lim_{x \to c} f(x) < 0$, **then there is an open interval about c, possibly excluding c itself, on which $f(x) < 0$.**

These results may be summarized by saying that if the limit of a function is non-zero at c, then, for x sufficiently close to c, the value of the function will be nonzero and have the same sign as the limit of $f(x)$ at c.

EXERCISE 2

In Problems 1–40 evaluate each limit by using algebraic techniques.

1. $\lim\limits_{x\to 3} 2(x + 4)$

2. $\lim\limits_{x\to 4} 5x$

3. $\lim\limits_{x\to 1}(3x^2 - 2x + 4)$

4. $\lim\limits_{x\to 0}(-3x^4 + 2x + 1)$

5. $\lim\limits_{x\to -2}(x^4 + 2x)$

6. $\lim\limits_{x\to -1}(5x^6 - 2x^2 + x)$

7. $\lim\limits_{x\to 1/2}(2x^4 - 8x^3 + 4x - 5)$

8. $\lim\limits_{x\to -1/3}(27x^3 + 9x + 1)$

9. $\lim\limits_{x\to 4} 3\sqrt{x}$

10. $\lim\limits_{x\to 8} \frac{1}{4}\sqrt[3]{x}$

11. $\lim\limits_{x\to 3} \sqrt{x^2 + x + 4}$

12. $\lim\limits_{t\to -2} \sqrt{3t^2 + 4}$

13. $\lim\limits_{t\to 2} t\sqrt{t^3 - 4}$

14. $\lim\limits_{t\to -1} t^2\sqrt[3]{t}$

15. $\lim\limits_{x\to 2}(\sqrt{x^3 + 1} - \sqrt{x^2 + 5})$

16. $\lim\limits_{x\to 1}(\sqrt[3]{2 - x} - \sqrt[3]{1 - x})$

17. $\lim\limits_{x\to 2} \dfrac{x^2 + 4}{x}$

18. $\lim\limits_{x\to 5} \dfrac{x^2 + 5}{3x}$

19. $\lim\limits_{x\to -2} \dfrac{2x^3 + 5x}{3x - 2}$

20. $\lim\limits_{x\to 1} \dfrac{2x^4 - 1}{3x^3 + 2}$

21. $\lim\limits_{x\to 2} \dfrac{x^2 - 4}{x - 2}$

22. $\lim\limits_{x\to -1} \dfrac{x^3 - x}{x + 1}$

23. $\lim\limits_{x\to 2} \dfrac{x^3 - 8}{x^2 + x - 6}$

24. $\lim\limits_{x\to 1} \dfrac{x^3 - 3x^2 + 3x - 1}{x^3 - x}$

25. $\lim\limits_{x\to 0} \dfrac{3x^3 - 4x}{x^2 + x}$

26. $\lim\limits_{x\to -1} \dfrac{x^3 + x^2}{x^2 - 1}$

27. $\lim\limits_{x\to -3} \dfrac{x + 3}{x^2 - x - 12}$

28. $\lim\limits_{x\to 4} \dfrac{2x^2 - 32}{x^3 - 4x^2}$

29. $\lim\limits_{x\to -1} \dfrac{x^2 + 4x + 3}{x^2 + 5x + 4}$

30. $\lim\limits_{x\to 2} \dfrac{x^3 - 2x^2 + x - 2}{x - 2}$

31. $\lim\limits_{x\to -8}\left(\dfrac{2x}{x + 8} + \dfrac{16}{x + 8}\right)$

32. $\lim\limits_{x\to 2}\left(\dfrac{3x}{x - 2} - \dfrac{6}{x - 2}\right)$

33. $\lim\limits_{x\to -2} \dfrac{x^3 + 8}{x + 2}$

34. $\lim\limits_{x\to 1} \dfrac{x^4 - 1}{x - 1}$

35. $\lim\limits_{x\to 2} \dfrac{\sqrt{x} - \sqrt{2}}{x - 2}$

36. $\lim\limits_{x\to -2} \dfrac{x + 2}{x^2 - 4}$

37. $\lim\limits_{h\to 0} \dfrac{\frac{1}{x + h} - \frac{1}{x}}{h}$

38. $\lim\limits_{h\to 0} \dfrac{(x + h)^2 - x^2}{h}$

39. $\lim\limits_{h\to 0} \dfrac{\sqrt{x + h} - \sqrt{x}}{h}$

40. $\lim\limits_{h\to 0} \dfrac{\frac{1}{(x + h)^3} - \frac{1}{x^3}}{h}$

In Problems 41–44 use the facts that $\lim_{x\to c} f(x) = 5$ and $\lim_{x\to c} g(x) = 2$ to find each limit.

41. $\lim\limits_{x\to c}[2f(x)]$

42. $\lim\limits_{x\to c}[f(x) - g(x)]$

43. $\lim\limits_{x\to c} g(x)^3$

44. $\lim\limits_{x\to c} \dfrac{f(x)}{g(x) - f(x)}$

In Problems 45–50 find $\lim\limits_{h\to 0} \dfrac{f(x + h) - f(x)}{h}$ for the indicated function f.

45. $f(x) = 4x - 3$

46. $f(x) = 3x + 5$

47. $f(x) = 3x^2 + 4x + 1$

48. $f(x) = 2x^2 + x$

49. $f(x) = \dfrac{2}{x}$

50. $f(x) = \dfrac{3}{x^2}$

51. Find: $\lim\limits_{x\to 2} \dfrac{x^2 - 4}{3 - \sqrt{x^2 + 5}}$

52. Find: $\lim\limits_{x\to -1} \dfrac{2 - \sqrt{x^2 + 3}}{1 - x^2}$

53. Find: $\lim\limits_{x\to a} \dfrac{x^n - a^n}{x - a}$

54. Find: $\lim\limits_{x \to -a} \dfrac{x^n + a^n}{x + a}$ **55.** Find: $\lim\limits_{x \to 1} \dfrac{x^m - 1}{x^n - 1}$ **56.** Find: $\lim\limits_{x \to 0} \dfrac{\sqrt[3]{1 + x} - 1}{x}$

57. Find: $\lim\limits_{x \to 0} \dfrac{\sqrt{(1 + ax)(1 + bx)} - 1}{x}$ **58.** Find: $\lim\limits_{x \to 0} \dfrac{\sqrt{(1 + a_1 x)(1 + a_2 x) \cdots (1 + a_n x)} - 1}{x}$

In Problems 59–63 use the squeezing theorem (2.20).

59. If $1 - x^2 \le f(x) \le \cos x$ for all x in the interval $-\pi/2 < x < \pi/2$, show that $\lim\limits_{x \to 0} f(x) = 1$.

60. If $f(x) = \begin{cases} 1 & \text{if } x \text{ is rational} \\ 0 & \text{if } x \text{ is irrational} \end{cases}$ show that $\lim\limits_{x \to 0} [xf(x)] = 0$.

61. If $0 \le f(x) \le 1$ for every x, show that $\lim\limits_{x \to 0} [x^2 f(x)] = 0$.

62. If $0 \le f(x) \le M$ for every x, show that $\lim\limits_{x \to 0} [x^2 f(x)] = 0$.

63. Show that $\lim\limits_{x \to 0} [x^n \sin(1/x)] = 0$, n a positive integer.

3. One-Sided Limits

In defining $\lim\limits_{x \to c} f(x)$ in (2.8) we were careful to restrict x to an open interval containing c; that is, we studied the behavior of f on both sides of c. However, in some cases it is necessary to investigate *one-sided limits:* the *left-hand* limit, $\lim\limits_{x \to c^-} f(x)$, and the *right-hand* limit, $\lim\limits_{x \to c^+} f(x)$.

The idea behind the left-hand limit $\lim\limits_{x \to c^-} f(x) = L$, read "the limit of $f(x)$, as x approaches c from the left, equals L," is that for all x sufficiently close to c, but *less than c*, the value $f(x)$ can be made as close as we please to L. See Figure 16 for an illustration.

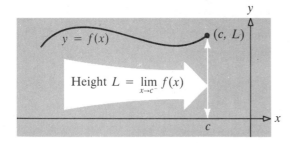

Figure 16

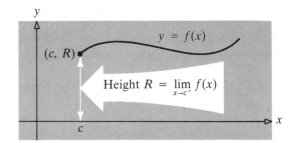

Figure 17

The idea behind the right-hand limit $\lim\limits_{x \to c^+} f(x) = R$, read "the limit of $f(x)$, as x approaches c from the right, equals R," is that for all x sufficiently close to c, but *greater than* c, the value of $f(x)$ can be made as close as we please to R. See Figure 17 for an illustration.

The rules for calculating left-hand and right-hand limits are the same as those used to calculate limits in the preceding section.

EXAMPLE 1   For the function below, calculate $\lim_{x\to 0^-} f(x)$ and $\lim_{x\to 0^+} f(x)$.

$$f(x) = \begin{cases} -1 & \text{if } x < 0 \\ 1 & \text{if } x > 0 \end{cases}$$

Solution   For $\lim_{x\to 0^-} f(x)$, x must remain negative. But, if $x < 0$, then $f(x)$ is constantly equal to -1 so that $\lim_{x\to 0^-} f(x) = \lim_{x\to 0^-}(-1) = -1$. Similar reasoning gives $\lim_{x\to 0^+} f(x) = \lim_{x\to 0^+} 1 = 1$. ∎

For the function in Example 1, the fact that $\lim_{x\to 0^-} f(x) = -1$ and $\lim_{x\to 0^+} f(x) = 1$ means that there can be no single number L that $f(x)$ is close to when x is close to 0. Consequently, $\lim_{x\to 0} f(x)$ does not exist (see Fig. 18).

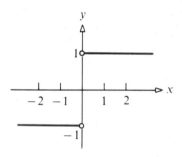

Figure 18

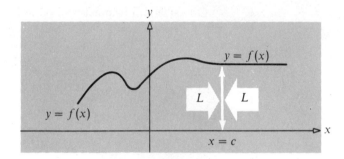

Figure 19

As Figure 19 illustrates, one-sided limits provide a way to determine whether the limit of a function exists. We state this criterion below:

(2.23) $\lim_{x\to c} f(x) = L$ if and only if $\lim_{x\to c^-} f(x) = L$ and $\lim_{x\to c^+} f(x) = L$

EXAMPLE 2   Determine whether $\lim_{x\to 2} f(x)$ and $\lim_{x\to 4} f(x)$ exist if

$$f(x) = \begin{cases} 2x + 1 & \text{if } 0 \le x \le 2 \\ 7 - x & \text{if } 2 < x < 4 \\ x & \text{if } 4 \le x \le 6 \end{cases}$$

Solution $\lim_{x\to 2} f(x)$:   Since the rule for $f(x)$ depends on whether $x < 2$ or $x > 2$, we need to look at the one-sided limits $\lim_{x\to 2^-} f(x)$ and $\lim_{x\to 2^+} f(x)$ to obtain information about $\lim_{x\to 2} f(x)$.

$$\lim_{x\to 2^-} f(x) = \lim_{x\to 2^-} (2x + 1) = 5 \qquad \lim_{x\to 2^+} f(x) = \lim_{x\to 2^+} (7 - x) = 5$$

Thus, by criterion (2.23), we conclude that $\lim_{x\to 2} f(x)$ exists and equals 5.

$\lim_{x \to 4} f(x)$:   The rule for $f(x)$ changes at $x = 4$.   Hence, to calculate $\lim_{x \to 4} f(x)$
we look at the one-sided limits

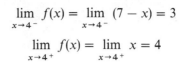

$$\lim_{x \to 4^-} f(x) = \lim_{x \to 4^-} (7 - x) = 3$$

$$\lim_{x \to 4^+} f(x) = \lim_{x \to 4^+} x = 4$$

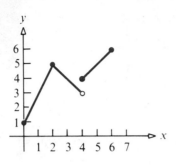

Figure 20

Since the limits are not equal, we conclude that $\lim_{x \to 4} f(x)$ does not exist. Figure 20 illustrates these conclusions.   ■

One-sided limits must be used to describe the behavior of functions like $f(x) = \sqrt{x - 1}$ near $x = 1$ since f is only defined for $x \geq 1$.   For this function, $\lim_{x \to 1^-} \sqrt{x - 1}$ makes no sense. However, the right-hand limit does make sense, and $\lim_{x \to 1^+} \sqrt{x - 1} = 0$ describes how $\sqrt{x - 1}$ behaves near and to the right of 1 (see Fig. 21).

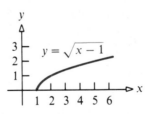

Figure 21

The ε, δ definitions of left-hand and right-hand limits are given in Appendix II.

EXERCISE 3

In Problems 1–20 evaluate each one-sided limit, if it exists.

1. $\lim_{x \to 3^-} (x^2 - 4)$

2. $\lim_{x \to 2^+} (3x^2 + x)$

3. $\lim_{x \to -2^+} (2x^3 + x - 1)$

4. $\lim_{x \to -4^-} (x^2 - 1)$

5. $\lim_{x \to -4^-} \dfrac{3x}{x - 4}$

6. $\lim_{x \to -4^+} \dfrac{3x}{x - 4}$

7. $\lim_{x \to 3^-} \dfrac{x^2 - 9}{x - 3}$

8. $\lim_{x \to 3^+} \dfrac{x^2 - 9}{x - 3}$

9. $\lim_{x \to 3^-} (\sqrt{9 - x^2} + x)$

10. $\lim_{x \to 2^+} (2\sqrt{x^2 - 4} + 3x)$

11. $\lim_{x \to 5^+} \dfrac{|x - 5|}{x - 5}$

12. $\lim_{x \to 5^-} \dfrac{|x - 5|}{x - 5}$

13. $\lim_{x \to 1/2^-} [\![2x]\!]$

14. $\lim_{x \to 1/2^+} [\![2x]\!]$

15. $\lim_{x \to 2/3^-} [\![2x]\!]$

16. $\lim_{x \to 2/3^+} [\![2x]\!]$

17. $\lim\limits_{x \to 2^+} \sqrt{|x| - x}$ **18.** $\lim\limits_{x \to 2^-} \sqrt{|x| - x}$ **19.** $\lim\limits_{x \to 2^+} \sqrt[3]{[\![x]\!] - x}$ **20.** $\lim\limits_{x \to 2^-} \sqrt[3]{[\![x]\!] - x}$

In Problems 21–32 evaluate $\lim_{x \to c^-} f(x)$ and $\lim_{x \to c^+} f(x)$ for the given number c. Based on your answer, determine whether $\lim_{x \to c} f(x)$ exists.

21. $f(x) = \begin{cases} 2x & \text{if } x \neq 0 \\ 1 & \text{if } x = 0 \end{cases}$ $c = 0$ **22.** $f(x) = \begin{cases} 2x & \text{if } x \neq 0 \\ 0 & \text{if } x = 0 \end{cases}$ $c = 0$

23. $f(x) = \begin{cases} \dfrac{x^2 - 9}{x - 3} & \text{if } x \neq 3 \\ 6 & \text{if } x = 3 \end{cases}$ $c = 3$ **24.** $f(x) = \begin{cases} \dfrac{x - 2}{x^2 - 4} & \text{if } x \neq 2 \\ 1 & \text{if } x = 2 \end{cases}$ $c = 2$

25. $f(x) = \begin{cases} 2x - 3 & \text{if } x \leq 1 \\ 3 - x & \text{if } x > 1 \end{cases}$ $c = 1$ **26.** $f(x) = \begin{cases} 5x + 2 & \text{if } x < -2 \\ 1 + 3x & \text{if } x \geq -2 \end{cases}$ $c = -2$

27. $f(x) = \begin{cases} 3x - 1 & \text{if } x < 1 \\ 4 & \text{if } x = 1 \\ 2x & \text{if } x > 1 \end{cases}$ $c = 1$ **28.** $f(x) = \begin{cases} 3x - 1 & \text{if } x < 1 \\ 2 & \text{if } x = 1 \\ 2x & \text{if } x > 1 \end{cases}$ $c = 1$

29. $f(x) = \begin{cases} 3x - 1 & \text{if } x < 1 \\ \text{Not defined} & \text{if } x = 1 \\ 2x & \text{if } x > 1 \end{cases}$ $c = 1$ **30.** $f(x) = \begin{cases} 3x - 1 & \text{if } x < 1 \\ 2 & \text{if } x = 1 \\ 3x & \text{if } x > 1 \end{cases}$ $c = 1$

31. $f(x) = \begin{cases} \sqrt{x^2 - 9} & \text{if } x \geq 3 \\ \sqrt{9 - x^2} & \text{if } x < 3 \end{cases}$ $c = 3$ **32.** $f(x) = \begin{cases} \dfrac{|x - 1|}{x - 1} & \text{if } x \neq 1 \\ 0 & \text{if } x = 1 \end{cases}$ $c = 1$

In Problems 33–38 use the function below to evaluate each limit, if it exists.

$$f(x) = \begin{cases} \sqrt{15 - 5x} & \text{if } x < 2 \\ \sqrt{5} & \text{if } x = 2 \\ \sqrt{9 - x^2} & \text{if } 2 < x < 3 \\ x - 2 & \text{if } 3 \leq x \end{cases}$$

33. $\lim\limits_{x \to 2^-} f(x)$ **34.** $\lim\limits_{x \to 2^+} f(x)$ **35.** $\lim\limits_{x \to 3^-} f(x)$

36. $\lim\limits_{x \to 3^+} f(x)$ **37.** $\lim\limits_{x \to 2} f(x)$ **38.** $\lim\limits_{x \to 3} f(x)$

For Problems 39 and 40 use the function below to evaluate each limit, if it exists.

$$f(x) = \begin{cases} 3x + 5 & \text{if } x \leq 2 \\ 13 - x & \text{if } x > 2 \end{cases}$$

39. $\lim\limits_{h \to 0^-} \dfrac{f(2 + h) - f(2)}{h}$ **40.** $\lim\limits_{h \to 0^+} \dfrac{f(2 + h) - f(2)}{h}$

For Problems 41–44 use the function f defined by

$$f(x) = \begin{cases} 1 & \text{if } x \text{ is an integer} \\ 0 & \text{if } x \text{ is not an integer} \end{cases}$$

41. Does $\lim_{x \to 2} f(x)$ exist?   **42.** Does $\lim_{x \to 1/2} f(x)$ exist?

43. Does $\lim_{x \to 3} f(x)$ exist?   **44.** Does $\lim_{x \to 0} f(x)$ exist?

45. Show by example that $\lim_{x \to c}[f(x) + g(x)]$ may exist even though $\lim_{x \to c} f(x)$ and $\lim_{x \to c} g(x)$ do not exist.

46. Show by example that $\lim_{x \to c}[f(x)g(x)]$ may exist even though $\lim_{x \to c} f(x)$ and $\lim_{x \to c} g(x)$ do not exist.

47. Show by example that $\lim_{x \to c}|f(x)|$ may exist even though $\lim_{x \to c} f(x)$ does not exist.

4. Continuous Functions

We have seen that sometimes $\lim_{x \to c} f(x)$ equals $f(c)$ and sometimes it does not. In fact, sometimes $f(c)$ is not even defined and $\lim_{x \to c} f(x)$ exists. Then, what is the relationship between $\lim_{x \to c} f(x)$ and $f(c)$? For the answer, we look at the possibilities (see Fig. 22):

(a) $\lim_{x \to c} f(x)$ exists and equals $f(c)$
(b) $\lim_{x \to c} f(x)$ exists and does not equal $f(c)$
(c) $\lim_{x \to c} f(x)$ exists and $f(c)$ is not defined
(d) $\lim_{x \to c} f(x)$ does not exist and $f(c)$ is defined
(e) $\lim_{x \to c} f(x)$ does not exist and $f(c)$ is not defined

Of these five situations, the "nicest" one is that given in (a). There, $\lim_{x \to c} f(x)$ both exists and is equal to $f(c)$. Functions that have these two qualities are said to be *continuous at c*. This appears to agree with the intuitive notion that "a function is continuous if its graph can be drawn without lifting the pencil." The functions in (b), (c), (d), and (e) are not continuous at c since each has a break in the graph at c. This leads us to definition (2.24).

(2.24)   DEFINITION *Continuous Function:* **Let** $y = f(x)$ **be a function defined on an open interval containing c. If**

1. $f(c)$ **is defined   and   2.** $\lim\limits_{x \to c} f(x)$ **exists   and   3.** $\lim\limits_{x \to c} f(x) = f(c)$

then the function is said to be *continuous at c*.

If *any one* of these three conditions is not satisfied, then the function is said to be *discontinuous at c*.

For example, the function $f(x) = 3x^3 - 5x + 4$ is continuous at 1 since

$$\lim_{x \to 1} f(x) = \lim_{x \to 1}(3x^3 - 5x + 4) = 2 \qquad \text{and} \qquad f(1) = 2$$

The function $f(x) = \dfrac{x^2 + 2}{x - 3}$ is discontinuous at 3 since $f(3)$ is not defined.

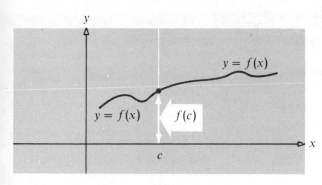

(a) $\lim\limits_{x\to c^-} f(x) = \lim\limits_{x\to c^+} f(x) = f(c)$

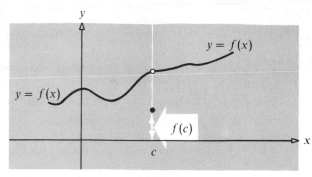

(b) $\lim\limits_{x\to c^-} f(x) = \lim\limits_{x\to c^+} f(x) \neq f(c)$

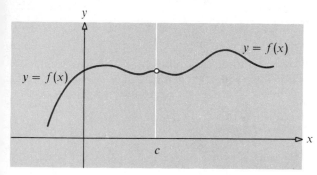

(c) $\lim\limits_{x\to c^-} f(x) = \lim\limits_{x\to c^+} f(x)$, $f(c)$ is not defined

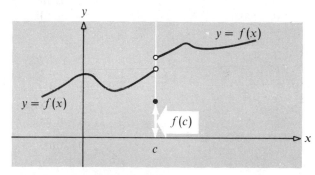

(d) $\lim\limits_{x\to c^-} f(x) \neq \lim\limits_{x\to c^+} f(x)$, $f(c)$ is defined

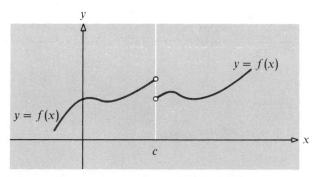

(e) $\lim\limits_{x\to c^-} f(x) \neq \lim\limits_{x\to c^+} f(x)$, $f(c)$ is not defined

Figure 22

In fact, because of (2.18), polynomial functions are everywhere continuous. And because of (2.19), it follows that rational functions are continuous where they are defined.

Let's look at some more examples.

EXAMPLE 1   Discuss the continuity of the function below at 3.

$$f(x) = \begin{cases} \dfrac{x^2 - 9}{x - 3} & \text{if } x \neq 3 \\ 6 & \text{if } x = 3 \end{cases}$$

Solution   The function f is defined at 3 since $f(3) = 6$.   Also,

$$\lim_{x \to 3^+} \frac{x^2 - 9}{x - 3} = \lim_{x \to 3^+} \frac{(x + 3)(x - 3)}{x - 3} = \lim_{x \to 3^+} (x + 3) = 6$$

$$\lim_{x \to 3^-} \frac{x^2 - 9}{x - 3} = \lim_{x \to 3^-} \frac{(x + 3)(x - 3)}{x - 3} = \lim_{x \to 3^-} (x + 3) = 6$$

Therefore $\lim_{x \to 3} f(x) = f(3) = 6$.   Thus, the three conditions in (2.24) are satisfied, and hence the function is continuous at 3 [see Fig. 23(a)].

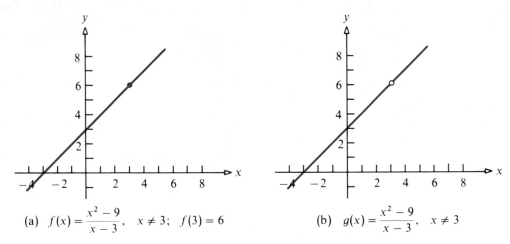

(a) $f(x) = \dfrac{x^2 - 9}{x - 3}$, $x \neq 3$; $f(3) = 6$
(b) $g(x) = \dfrac{x^2 - 9}{x - 3}$, $x \neq 3$

Figure 23   ■

Note that the function

$$g(x) = \frac{x^2 - 9}{x - 3} \qquad x \neq 3$$

is not continuous at 3 since condition 1 of (2.24) is not satisfied. We say, therefore, that g is discontinuous at 3 [see Fig. 23(b)].

EXAMPLE 2   Discuss the continuity of the function below at 0.

$$f(x) = \begin{cases} x^2 + 1 & \text{if} \quad x \neq 0 \\ 2 & \text{if} \quad x = 0 \end{cases}$$

Solution   The value of the function at 0 is $f(0) = 2$.   Also,

$$\lim_{x \to 0^-} f(x) = \lim_{x \to 0^-} (x^2 + 1) = 1$$

$$\lim_{x \to 0^+} f(x) = \lim_{x \to 0^+} (x^2 + 1) = 1$$

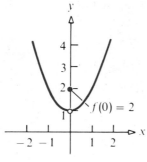

Thus, $\lim_{x \to 0} f(x) = 1$ and $f(0) = 2$.   Since condition 3 of (2.24) does not hold, the function is discontinuous at 0 (see Figure 24).   ■

Figure 24

EXAMPLE 3   Discuss the continuity of $f(x) = [\![x]\!]$ at 1.

Solution   The function f is defined at 1 and $f(1) = 1$.   But

$$\lim_{x \to 1^-} f(x) = 0$$

and

$$\lim_{x \to 1^+} f(x) = 1$$

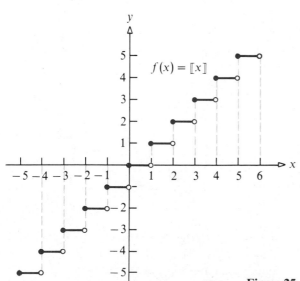

Therefore $\lim_{x \to 1} f(x)$ does not exist. Since the second condition of definition (2.24) is not met, we say that f is discontinuous at 1. In fact, as Figure 25 indicates, the function is discontinuous at each integer.   ■

Figure 25

The discontinuities encountered in Examples 2 and 3 illustrate a more general result. If a function f is defined on an open interval that contains c, then f will not be continuous at c for only one of two reasons: (1) $\lim_{x \to c} f(x)$ does not exist (Example 3) or (2) $\lim_{x \to c} f(x)$ exists but does not equal $f(c)$ (Example 2). The latter type of discontinuity is usually referred to as a *removable discontinuity* since we may redefine f at c to equal $\lim_{x \to c} f(x)$ and thus make f continuous at c. For instance, in Example 2, if we redefine f at 0 to be 1, then f will be continuous at 0.

There are more subtle types of discontinuity. For instance, the function below is nowhere continuous:

$$f(x) = \begin{cases} 1 & \text{if } x \text{ is rational} \\ 0 & \text{if } x \text{ is irrational} \end{cases}$$

Let's concentrate, however, on characteristics of functions that *are* continuous.

PROPERTIES OF CONTINUOUS FUNCTIONS

Because the definition of a continuous function is based on a limit, theorems that apply to limits also apply to continuous functions. We state these results below.

(2.25) **If f and g are continuous at c, then so are their sum $f + g$, difference $f - g$, and product $f \cdot g$.**

(2.26) **If f and g are continuous at c, and $g(c) \neq 0$, then the quotient f/g is also continuous at c.**

(2.27) **If f is continuous at $g(c)$ and g is continuous at c, then the composite function $f(g(x))$ is continuous at c.**

For example, the function $h(x) = \sqrt{3x^2 + 5}$ is continuous for all x. To see this, observe that $h = f \circ g$, where $f(x) = \sqrt{x}$ and $g(x) = 3x^2 + 5$. Now, g is continuous for all x and f is continuous for all $x \geq 0$. Since $g(x) > 0$ for all x, $f \circ g$ is defined for all x. It follows from (2.27) that $f \circ g = \sqrt{3x^2 + 5}$ is continuous for all x.

As another example, the function $h(x) = \sqrt{x^2/(x - 1)}$ is continuous for all $x > 1$. To see this, observe that $h = f \circ g$, where $f(x) = \sqrt{x}$ and $g(x) = x^2/(x - 1)$. Now g is continuous for all $x \neq 1$ and f is continuous for $x \geq 0$. However, since $g(x) \geq 0$ only for $x > 1$, $f \circ g$ is only defined for $x > 1$. Consequently, $f \circ g$ is continuous for $x > 1$.

CONTINUITY ON INTERVALS

In (2.24) we defined what is meant by a function f being continuous *at a number c*. Here, we define what is meant by a function f being continuous *on an interval*. As you will see, the definition requires that special attention be given to the endpoints, if they are contained in the interval. Figure 26 illustrates definition (2.28).

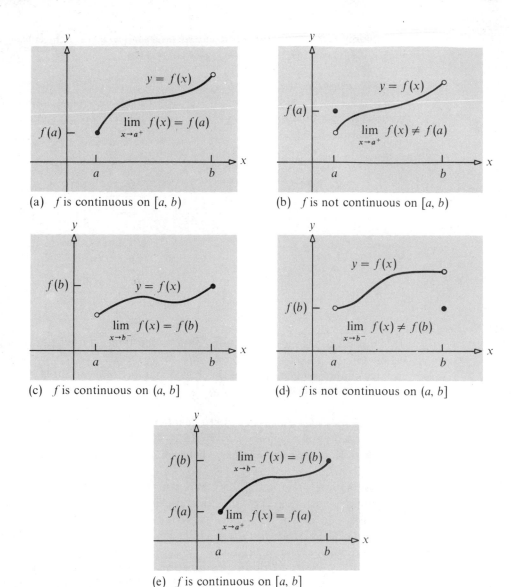

(a) f is continuous on $[a, b)$

(b) f is not continuous on $[a, b)$

(c) f is continuous on $(a, b]$

(d) f is not continuous on $(a, b]$

(e) f is continuous on $[a, b]$

Figure 26

(2.28)   DEFINITION   A function f is *continuous on an open interval* (a, b) if f is continuous at every number c between a and b.

A function f is *continuous on an interval* $[a, b)$ if f is continuous on (a, b) and $\lim_{x \to a^+} f(x) = f(a)$.

A function f is *continuous on an interval* $(a, b]$ if f is continuous on (a, b) and $\lim_{x \to b^-} f(b) = f(b)$.

A function f is *continuous on a closed interval* $[a, b]$ if f is continuous on (a, b), $\lim_{x \to a^+} f(x) = f(a)$, **and** $\lim_{x \to b^-} f(x) = f(b)$.

For example, the function $f(x) = \sqrt{x - 1}$ is continuous on $[1, +\infty)$, while $g(x) = \sqrt{2 - x}$ is continuous on $(-\infty, 2]$ (see Fig. 27). As Example 3 illustrates, the greatest integer function $f(x) = [\![x]\!]$ is continuous on every open interval $(n, n + 1)$, where n is an integer.

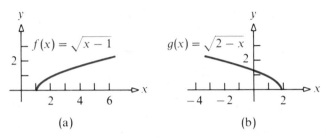

Figure 27

Functions that are continuous on a closed interval have many important properties. One of them is the *intermediate value theorem*, which we shall have use for in later chapters. (The proof of this result may be found in most books on advanced calculus.)

(2.29)   *Intermediate Value Theorem.*   **Let f denote a function that is continuous on the closed interval $[a, b]$ and suppose $f(a) \ne f(b)$. If N is any number between $f(a)$ and $f(b)$, then there is at least one number c between a and b so that $f(c) = N$.**

For an interpretation of this result, consider the following scenario: "If you climb a mountain, starting at 2000 meters and ending at 5000 meters, then no matter how many ups and downs you took in getting from the bottom to the top, there was some time when your altitude was 3765.6 meters—or any other number between 2000 and 5000 you may choose."

In other words, a continuous function whose domain is a closed interval $[a, b]$ must take on all values between $f(a)$ and $f(b)$. Figure 28 illustrates this situation and why the continuity of the function is crucial for the validity of the result.

One immediate application of the intermediate value theorem involves the location of the zeros* of a function. If a function f is continuous on the closed interval $[a, b]$ and if $f(a)$ and $f(b)$ are of opposite sign, there is at least one c between a and b so that $f(c) = 0$; that is, f has a zero between a and b. By "squeezing" the interval, better and better approximations of this zero may be obtained. For example, the function $f(x) = x^3 + x^2 - x - 2$ is continuous on the closed interval $[0, 2]$.

---

* r is a *zero* of f if $f(r) = 0$.

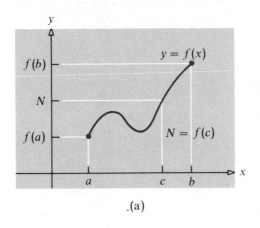

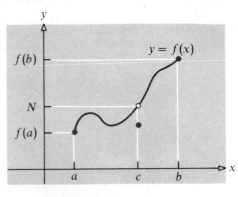

_(a)

(b) Discontinuity at c results in
     no number x in $[a, b]$ so that
     $f(x) = N$

Figure 28

Since $f(0) = -2$ and $f(2) = 8$ are of opposite sign, the intermediate value theorem implies $f(c) = 0$ for at least one number c in the interval $[0, 2]$. Therefore our first approximation to the zero is that it lies in the interval $[0, 2]$. We continue by testing the midpoint 1 of the interval $[0, 2]$. Since $f(1) = -1$, the function f must have a zero between 1 and 2. Our second approximation to the zero is that it lies in the interval $[1, 2]$. We could continue in this way (testing $\frac{3}{2}$ next) to obtain increasingly smaller intervals that contain the zero (see Fig. 29).

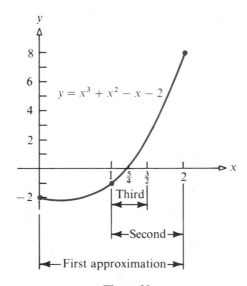

Figure 29

EXERCISE 4

In Problems 1–18 determine whether the function f is continuous at c.

1. $f(x) = 2x^2 + x - 5$ $c = 1$
2. $f(x) = 3 - 2x^3$ $c = -2$
3. $f(x) = \dfrac{x^2 - 9}{x - 3}$ $c = -3$

4. $f(x) = 1 + \dfrac{1}{x}$ $c = 1$
5. $f(x) = |x - 5|$ $c = 5$
6. $f(x) = \sqrt{x^2 + 9}$ $c = 0$

7. $f(x) = \begin{cases} 2x + 5 & \text{if } x \le 2 \\ 4x + 1 & \text{if } x > 2 \end{cases}$ $c = 2$
8. $f(x) = \begin{cases} 2x + 1 & \text{if } x \le 0 \\ 2x & \text{if } x > 0 \end{cases}$ $c = 0$

9. $f(x) = \begin{cases} 3x - 1 & \text{if } x < 1 \\ 4 & \text{if } x = 1 \\ 2x & \text{if } x > 1 \end{cases}$ $c = 1$
10. $f(x) = \begin{cases} 3x - 1 & \text{if } x < 1 \\ 2 & \text{if } x = 1 \\ 2x & \text{if } x > 1 \end{cases}$ $c = 1$

11. $f(x) = \begin{cases} 3x - 1 & \text{if } x < 1 \\ \text{Not defined} & \text{if } x = 1 \\ 2x & \text{if } x > 1 \end{cases}$ $c = 1$
12. $f(x) = \begin{cases} 3x - 1 & \text{if } x < 1 \\ 2 & \text{if } x = 1 \\ 3x & \text{if } x > 1 \end{cases}$ $c = 1$

13. $f(x) = \begin{cases} x^2 & \text{if } x \le 0 \\ 2x & \text{if } x > 0 \end{cases}$ $c = 0$
14. $f(x) = \begin{cases} x^2 & \text{if } x < -1 \\ 2 & \text{if } x = -1 \\ -3x + 2 & \text{if } x > -1 \end{cases}$ $c = -1$

15. $f(x) = \begin{cases} 4 - 3x^2 & \text{if } x < 0 \\ 4 & \text{if } x = 0 \\ \sqrt{16 - x^2} & \text{if } 0 < x < 4 \end{cases}$ $c = 0$
16. $f(x) = \begin{cases} \sqrt{4 + x} & \text{if } x \le 4 \\ \sqrt{\dfrac{x^2 - 16}{x - 4}} & \text{if } x > 4 \end{cases}$ $c = 4$

17. $f(x) = [\![x]\!]$ $c = 1$
18. $f(x) = [\![2x]\!]$ $c = \frac{1}{2}$

In Problems 19–22 the function f is not defined at c. For each function, decide how to define $f(c)$ so that f is continuous at c.

19. $f(x) = \dfrac{x^2 - 4}{x - 2}$ $c = 2$
20. $f(x) = \dfrac{x^2 + x - 12}{x - 3}$ $c = 3$

21. $f(x) = \begin{cases} 2x & \text{if } x > 1 \\ 1 + x & \text{if } x < 1 \end{cases}$ $c = 1$
22. $f(x) = \begin{cases} x^2 + 5x & \text{if } x < -1 \\ x - 3 & \text{if } x > -1 \end{cases}$ $c = -1$

In Problems 23–36 find all numbers x for which f is continuous.

23. $f(x) = 3x^5 - 2x^3 + x - 2$
24. $f(x) = -6x^3 + 4x$

25. $f(x) = \dfrac{x}{x^2 + 4}$
26. $f(x) = \dfrac{2 - x}{x^2 + x + 1}$
27. $f(x) = \dfrac{x}{x - 2}$
28. $f(x) = \dfrac{x^2 - 4}{x^2 - 1}$

29. $f(x) = \dfrac{x^2}{x^2 - 4}$
30. $f(x) = \dfrac{x}{x^3 - 8}$
31. $f(x) = |x|$
32. $f(x) = \dfrac{1}{x}$

33. $f(x) = \sqrt{x^2 + 1}$
34. $f(x) = \sqrt[3]{1 + x}$
35. $f(x) = \sqrt{\dfrac{x^2 + 1}{2 - x}}$
36. $f(x) = \sqrt{\dfrac{4}{x^2 - 1}}$

In Problems 37–40 use the function

$$f(x) = \begin{cases} \sqrt{15 - 3x} & \text{if } x < 2 \\ \sqrt{5} & \text{if } x = 2 \\ \sqrt{9 - x^2} & \text{if } 2 < x < 3 \\ [\![x - 2]\!] & \text{if } 3 \le x \end{cases}$$

37. Is f continuous at 0? Why or why not?   **38.** Is f continuous at 4? Why or why not?

39. Is f continuous at 2? Why or why not?   **40.** Is f continuous at 3? Why or why not?

In Problems 41–46 use the intermediate value theorem to determine which of the functions f must have zeros in the given intervals. Indicate those for which the theorem gives no information. Do not attempt to locate the zeros.

41. $f(x) = x^3 - 3x$ on $[-2, 2]$ **42.** $f(x) = x^4 - 1$ on $[-2, 2]$

43. $f(x) = \dfrac{x}{(x + 1)^2} - 1$ on $[10, 20]$ **44.** $f(x) = x^3 - 2x^2 - x + 2$ on $[3, 4]$

45. $f(x) = \sqrt{x^3 + 3} - \sqrt{x^3 - 1} - 1$ on $[1, 10]$ **46.** $f(x) = \sqrt{x^2 - 3x} - 2$ on $[3, 5]$

47. For the function below, find k so that f is continuous at 2.

$$f(x) = \begin{cases} \dfrac{\sqrt{2x + 5} - \sqrt{x + 7}}{x - 2} & \text{if } x \ne 2 \\ k & \text{if } x = 2 \end{cases}$$

48. Given the two functions f and h such that

$$f(x) = x^3 - 3x^2 - 4x + 12 \qquad h(x) = \begin{cases} \dfrac{f(x)}{x - 3} & \text{if } x \ne 3 \\ p & \text{if } x = 3 \end{cases}$$

(a)   Find all zeros of the function f.
(b)   Find the number p so that the function h is continuous at $x = 3$.   Justify your answer.
(c)   Use the number p found in part (b) to determine whether h is an even function. Justify your answer.

49. The function $f(x) = |x|/x$ is not defined at 0. Tell why it is impossible to define $f(0)$ so that f is continuous at 0.

50. If f and g are each continuous at c, prove that $f + g$ is continuous at c.

51. Discover two functions f and g that are each continuous at c, and yet f/g is not continuous at c.

5. Limit Theorems (If Time Permits)

In this section we take a closer look at the definition of a limit. Let's begin with an example.
       Consider the function f defined by

(2.30) $$f(x) = \begin{cases} 3x + 1 & \text{if } x \ne 2 \\ 3 & \text{if } x = 2 \end{cases}$$

Then, $\lim_{x \to 2} f(x) = 7$, as can be inferred from the graph (see Fig. 10, p. 64) or from the table below:

x	1.75	1.9	1.95	1.99	1.995	1.999	1.9999	2.0001	2.001	2.005	2.01	2.1	2.25	2.5
$f(x)$	6.25	6.7	6.85	6.97	6.985	6.997	6.9997	7.0003	7.003	7.015	7.03	7.3	7.75	8.5

Notice from the table that as x gets closer and closer to 2, $f(x)$ gets closer and closer to 7. In fact, we can make $f(x)$ as close to 7 as we please by taking x close enough to 2. For example, suppose we want $f(x)$ to differ from 7 by less than 0.3, that is,

$$7 - 0.3 < f(x) < 7 + 0.3$$
$$6.7 < f(x) < 7.3$$

First, we must require $x \neq 2$, because when $x = 2$, then $f(x) = f(2) = 3$ (by definition) and we obtain $6.7 < 3 < 7.3$, which is impossible.

From the table, we can see that 6.7 corresponds to 1.9, and 7.3 corresponds to 2.1. Thus,

$$6.7 < f(x) < 7.3 \qquad \text{whenever} \qquad x \neq 2 \quad \text{and} \quad 1.9 < x < 2.1$$

In other words, $f(x)$ differs from 7 by less than 0.3 whenever $x \neq 2$ and x differs from 2 by less than 0.1. Similarly, the table indicates that $f(x)$ differs from 7 by less than 0.003 whenever $x \neq 2$ and x differs from 2 by less than 0.001.

Now, let us ask a question that cannot be answered from the table. For $x \neq 2$, how close to 2 must x be in order to guarantee that $f(x)$ differs from 7 by less than some *arbitrary* positive number ε (ε might be extremely small)? Well, the words *$f(x)$ differs from 7 by less than ε* may be written

$$7 - \varepsilon < f(x) < 7 + \varepsilon$$

If $x \neq 2$, $f(x) = 3x + 1$. Hence,

$$7 - \varepsilon < 3x + 1 < 7 + \varepsilon$$

By subtracting 1 from all parts, we obtain $6 - \varepsilon < 3x < 6 + \varepsilon$. And, finally, we multiply by $\frac{1}{3}$ to obtain

$$2 - \tfrac{1}{3}\varepsilon < x < 2 + \tfrac{1}{3}\varepsilon$$

Thus, the answer to our question is $\frac{1}{3}\varepsilon$. That is, $f(x)$ differs from 7 by less than ε whenever $x \neq 2$ and x differs from 2 by less than $\frac{1}{3}\varepsilon$.

For example, $f(x)$ differs from 7 by less than $\varepsilon = \frac{1}{10}$ whenever $x \neq 2$ and x differs from 2 by less than $\frac{1}{3}\varepsilon = (\frac{1}{3})(\frac{1}{10}) = \frac{1}{30}$; $f(x)$ differs from 7 by less than $\varepsilon = 0.3$ whenever $x \neq 2$ and x differs from 2 by less than $\frac{1}{3}\varepsilon = \frac{1}{3}(0.3) = 0.1$; $f(x)$ differs from 7 by less than $\varepsilon = 0.003$ whenever $x \neq 2$ and x differs from 2 by less than $\frac{1}{3}\varepsilon = 0.001$. (As already noted, these last two statements are verified by the table.)

We now conjecture the following: If ε is any given positive number, then there is a positive number, say δ, such that

$$\left[\begin{array}{c} f(x) \text{ differs from } 7 \\ \text{by less than } \varepsilon \end{array}\right] \quad \text{whenever} \quad \left[\begin{array}{c} x \neq 2 \quad \text{and } x \text{ differs from } 2 \\ \text{by less than } \delta \end{array}\right]$$

(For the example just discussed, $\delta = \varepsilon/3$.)

This statement can be shortened by using the appropriate mathematical notation. We shorten the phrase "ε is any given positive number" by writing simply $\varepsilon > 0$. The difference between $f(x)$ and 7 is $|f(x) - 7|$. Thus, the statement "$f(x)$ differs from 7 by less than ε" may be written $|f(x) - 7| < \varepsilon$. Similarly, the difference between x and 2 is $|x - 2|$; and the statement "x differs from 2 by less than δ" may be written $|x - 2| < \delta$. To say that $x \neq 2$ is simply to say that the difference between them is positive, that is, $0 < |x - 2|$. Therefore, for the function (2.30), we have the following: If $\varepsilon > 0$ is given, then there is a $\delta > 0$ such that

$$|f(x) - 7| < \varepsilon \quad \text{whenever} \quad 0 < |x - 2| < \delta$$

In fact, $\delta = \frac{1}{3}\varepsilon$ works. Since this is possible for any $\varepsilon > 0$, we conclude that the limit of $f(x)$ as x approaches 2 is equal to 7, and we write $\lim_{x \to 2} f(x) = 7$.

The above discussion leads to the ε, δ definition of the limit of a function, which we now restate.

(2.31)   **DEFINITION**   *Limit of a Function.*   **Let f be a function defined on an open interval containing c, except possibly for c itself. Then $\lim_{x \to c} f(x) = L$ if, for any given number $\varepsilon > 0$, a number $\delta > 0$ exists such that**

$$|f(x) - L| < \varepsilon \quad \text{whenever} \quad 0 < |x - c| < \delta \quad \text{and } x \text{ is in the domain of } f$$

Figure 30 illustrates the definition for two choices of ε. Notice that in part (b), the smaller ε requires a smaller δ than in part (a).

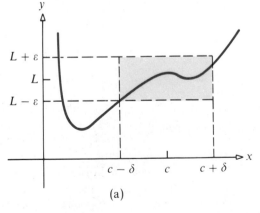
(a)

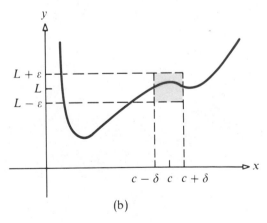
(b)

Figure 30

Figure 31 illustrates what happens if the δ is too large for the choice of ε; there are values of $f(x)$—for example, at x_1 and x_2—for which $|f(x) - L| \not< \varepsilon$.

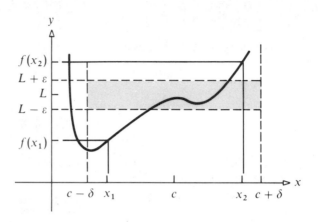

Figure 31

Let's look at our original example again to see how this definition of limit is applied.

EXAMPLE 1   Using the definition of limit, prove that $\lim_{x \to 2} f(x) = 7$ for the function

$$f(x) = \begin{cases} 3x + 1 & \text{if } x \neq 2 \\ 3 & \text{if } x = 2 \end{cases}$$

Solution   We seek a number $\delta > 0$ such that

$$|(3x + 1) - 7| < \varepsilon \qquad \text{whenever} \qquad 0 < |x - 2| < \delta$$

To find a connection between

$$|x - 2| \qquad \text{and} \qquad |(3x + 1) - 7|$$

we simplify the last expression, to get

$$|(3x + 1) - 7| = |3x - 6| = |3(x - 2)| = 3|x - 2|$$

We can make this expression less than ε by making $|x - 2|$ less than $\frac{1}{3}\varepsilon$. This suggests that we set $\delta = \frac{1}{3}\varepsilon$. Thus, we see that for any ε, we can find a δ so that whenever $0 < |x - 2| < \delta = \frac{1}{3}\varepsilon$, then $|(3x + 1) - 7| < \varepsilon$. Hence, we have established that $\lim_{x \to 2} f(x) = 7$. A geometrical interpretation is shown in Figure 32. We see that $f(x)$ on the vertical axis will lie between the horizontal lines $y = 7 + \varepsilon$ and $y = 7 - \varepsilon$, whenever x on the horizontal axis lies between $2 - \delta$ and $2 + \delta$. Thus, $\lim_{x \to 2} f(x) = 7$ describes the behavior of f near 2; namely, that the value of f is close to 7 when x is close to 2. Notice that $\lim_{x \to 2} f(x) \neq f(2)$.

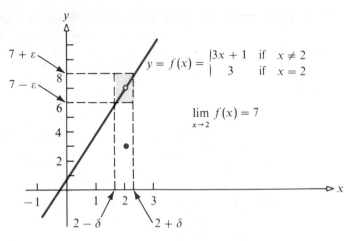

$$y = f(x) = \begin{vmatrix} 3x + 1 & \text{if} & x \neq 2 \\ 3 & \text{if} & x = 2 \end{vmatrix}$$

$$\lim_{x \to 2} f(x) = 7$$

Figure 32   ■

The next example proves the important results (2.7).

EXAMPLE 2   Use the definition of limit to prove that:
(a) $\lim_{x \to c} A = A$, A and c real numbers   (b) $\lim_{x \to c} x = c$, c a real number

Solution   (a) The function f is the constant function $f(x) = A$ whose graph is a horizontal line. Let $\varepsilon > 0$ be given. We must find $\delta > 0$ so that whenever
$0 < |x - c| < \delta$,   then $|f(x) - A| = |A - A| < \varepsilon$.   Since $|A - A| = 0$,   no
matter what δ is used, it always happens that $|f(x) - A| < \varepsilon$.   That is, any
choice of δ guarantees that whenever $0 < |x - c| < \delta$,   then $|f(x) - A| < \varepsilon$.
(b) The function f is the identity function $f(x) = x$.   Let $\varepsilon > 0$ be given. We
must find δ so that whenever $0 < |x - c| < \delta$,   then $|f(x) - c| = |x - c| < \varepsilon$.
The obvious choice for δ is ε itself. That is, whenever $0 < |x - c| < \delta = \varepsilon$,   it
follows that $|f(x) - c| = |x - c| < \varepsilon$.   ■

The following observations about the definition of a limit are important:

(2.32)   The limit of the function in no way depends on the value of the function at c.

Recall from Example 1 that $\lim_{x \to 2} f(x) = 7$,   and yet $f(2) \neq 7$.

(2.33)   In general, the size of δ depends on the size of ε.

For example, if $\delta = \varepsilon/3$ (as in Example 1), then δ grows larger or smaller as
ε does.

**(2.34)   For a given ε, if a suitable δ has been found, any *smaller positive number* will also work.
That is, δ is not uniquely determined when ε is given.**

Refer to our earlier discussion of the function (2.30) in Example 1, where $L = 7$ and $c = 2$. For $\varepsilon = \frac{1}{10}$, we found $\delta = \frac{1}{30}$, which is actually the *maximum* permissible δ for this ε. For $\varepsilon = 0.003$, we found $\delta = 0.001$, which again is the maximum permissible δ for this ε.

Let's look at some more examples.

EXAMPLE 3   Prove that $\lim_{x \to -1}(1 - 2x) = 3$.

Solution   To begin, we assume that a number $\varepsilon > 0$ is given. We must show that there exists a number $\delta > 0$ such that if $0 < |x - (-1)| < \delta$, then

$$|(1 - 2x) - 3| < \varepsilon$$

The idea is to find a connection between

$$|x - (-1)| \qquad \text{and} \qquad |(1 - 2x) - 3|$$

We begin by noting that

$$|x - (-1)| = |x + 1|$$

and

$$|(1 - 2x) - 3| = |-2(x + 1)| = 2|x + 1|$$

so that

$$|(1 - 2x) - 3| = 2|x - (-1)|$$

In other words,

$$|(1 - 2x) - 3| < \varepsilon \qquad \text{whenever} \qquad |x - (-1)| < \frac{\varepsilon}{2}$$

Thus, we may choose $\delta = \varepsilon/2$ (see Fig. 33). This completes the argument that $\lim_{x \to -1}(1 - 2x) = 3$.

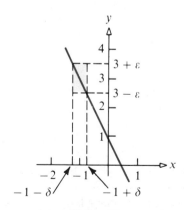

Figure 33   ∎

EXAMPLE 4   Prove that $\lim_{x \to 2} x^2 = 4$.

Solution   To begin, we assume that a number $\varepsilon > 0$ is given. We must show that there exists a number $\delta > 0$ such that

$$|x^2 - 4| < \varepsilon \qquad \text{whenever} \qquad 0 < |x - 2| < \delta$$

We need to establish a connection between $|x^2 - 4|$ and $|x - 2|$.   We note that

$$|x^2 - 4| = |(x + 2)(x - 2)| = |x + 2||x - 2|$$

If we can find a number K such that $|x + 2| < K$,   the choice of δ is clear, namely, $\delta < \varepsilon/K$.   If x is confined to some interval centered about 2, then K can be found. For example, suppose $|x - 2| < 1$—that is, $1 < x < 3$.   Then we add 2 to each part to get $1 + 2 < x + 2 < 3 + 2$;   and, in particular, $|x + 2| = x + 2 < 5$.   It then follows that whenever $|x - 2| < 1$,

$$|x^2 - 4| = |(x + 2)(x - 2)| < 5|x - 2|$$

If $|x - 2| < \varepsilon/5$ also, then

$$|x^2 - 4| < 5|x - 2| < 5\left(\frac{\varepsilon}{5}\right) = \varepsilon$$

as desired. But the quantity $|x - 2|$ now has two restrictions; namely,

$$|x - 2| < 1 \qquad \text{and} \qquad |x - 2| < \frac{\varepsilon}{5}$$

To ensure that both inequalities are obeyed, we select δ to be the smaller of 1 and $\varepsilon/5$, abbreviated $\delta \le \min(1, \varepsilon/5)$.   With this choice of δ, whenever

$$|x - 2| < \delta \le \min(1, \varepsilon/5)$$

we have $|x^2 - 4| < \varepsilon$.   ∎

In this example, the choice of restricting x so that $|x - 2| < 1$ is completely arbitrary. The reader should verify that if we had restricted x so that $|x - 2| < \frac{1}{3}$, then the choice for δ would be less than or equal to the smaller of $\frac{1}{3}$ and $3\varepsilon/13$; that is, $\delta \le \min(\frac{1}{3}, 3\varepsilon/13)$.

EXAMPLE 5   Prove that $\lim\limits_{x \to c} \dfrac{1}{x} = \dfrac{1}{c}$, $c > 0$.

Solution   For a given $\varepsilon > 0$,   we wish to find a δ such that $|(1/x) - (1/c)| < \varepsilon$ whenever $0 < |x - c| < \delta$.   Now, for $x \ne 0$,   we have

$$|f(x) - L| = \left|\frac{1}{x} - \frac{1}{c}\right| = \left|\frac{c - x}{xc}\right| = \frac{|x - c|}{|x||c|} = \frac{|x - c|}{c|x|}$$

The idea here is to find a connection between

$$|x - c| \quad \text{and} \quad \frac{|x - c|}{c|x|}$$

We proceed as in Example 4. Since we are only interested in x near c, we restrict x. so that $|x - c| < c/2$. It follows that $-c/2 < x - c < c/2$, or

$$\frac{c}{2} < x < \frac{3c}{2}$$

Therefore, if $|x - c| < c/2$, then $|x| > c/2$, so that $1/|x| < 2/c$, and

$$\left|\frac{1}{x} - \frac{1}{c}\right| = \frac{|x - c|}{c|x|} < \frac{2}{c^2}|x - c|$$

We can make $|(1/x) - (1/c)| < \varepsilon$ by choosing

$$|x - c| < \frac{\varepsilon}{2/c^2} = \frac{c^2\varepsilon}{2}$$

since then

$$\left|\frac{1}{x} - \frac{1}{c}\right| < \frac{2}{c^2}\left(\frac{c^2\varepsilon}{2}\right) = \varepsilon$$

But $|x - c|$ now has two restrictions: $|x - c| < c/2$ and $|x - c| < c^2\varepsilon/2$. Hence, given $\varepsilon > 0$, we choose $\delta \le \min(c/2, c^2\varepsilon/2)$. Then, whenever $|x - c| < \delta$, we have $|(1/x) - (1/c)| < \varepsilon$ (see Fig. 34).

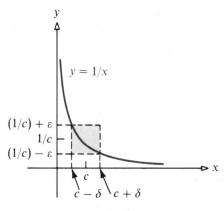

Figure 34   ■

The next example illustrates how a mistake in computing the limit can be discovered.

EXAMPLE 6   Prove that the statement $\lim_{x \to 3}(4x - 5) = 10$ is false.

Solution   This is a proof by contradiction. Thus, if we can show that for one ε, no number δ can be found that satisfies the definition of the limit, the proof will be complete. Suppose $\varepsilon = 1$.   If $\lim_{x \to 3}(4x - 5) = 10$,   then there is a $\delta > 0$ such that

$$|(4x - 5) - 10| < 1 \qquad \text{whenever} \qquad 0 < |x - 3| < \delta$$

The first inequality leads to $3.5 < x < 4$,   whenever $x \ne 3$ and $|x - 3| < \delta$. But the second inequality must be obeyed and, regardless of the value of δ, the interval $0 < |x - 3| < \delta$ contains a number x that is less than 3, in contradiction of the fact that $3.5 < x < 4$.   This impossibility shows us that $\lim_{x \to 3}(4x - 5) \ne 10$.   ∎

The next example uses an ε, δ proof to show that a limit does not exist.

EXAMPLE 7   Prove that for the function given below, $\lim_{x \to c} f(x)$ does not exist.

$$f(x) = \begin{cases} 1 & \text{if} \quad x \text{ is rational} \\ 0 & \text{if} \quad x \text{ is irrational} \end{cases}$$

Solution   Here, we use the technique of proof by contradiction. Suppose a limit does exist— that is, $\lim_{x \to c} f(x) = L$ for some number c. Let $\varepsilon = \frac{1}{2}$;   then there must exist a $\delta > 0$ such that

(2.35) $$|f(x) - L| < \tfrac{1}{2} \qquad \text{whenever} \qquad 0 < |x - c| < \delta$$

Now, let x_1 be a rational number satisfying $0 < |x_1 - c| < \delta$ and let x_2 be an irrational number satisfying $0 < |x_2 - c| < \delta$.   From the definition of the function, we know that

$$f(x_1) = 1 \qquad \text{and} \qquad f(x_2) = 0$$

Thus, by (2.35) we must have

$$|f(x_1) - L| = |1 - L| < \tfrac{1}{2} \qquad \text{and} \qquad |f(x_2) - L| = |0 - L| < \tfrac{1}{2}$$

From the first inequality, we have $L > \frac{1}{2}$.   From the second, we have $L < \frac{1}{2}$. But it is clearly impossible for both of these inequalities to be satisfied.   ∎

The next example proves (2.21), which was stated earlier on page 74. In Problem 24, Exercise 5, you are asked to prove (2.22).

EXAMPLE 8   Prove that if $\lim_{x \to c} f(x) > 0$,   then there is an open interval about c, possibly excluding c itself, on which $f(x) > 0$.

Solution   Suppose $\lim_{x \to c} f(x) = L > 0$.   Then, given any $\varepsilon > 0$,   there is a $\delta > 0$ such that

$$|f(x) - L| < \varepsilon \qquad \text{whenever} \qquad 0 < |x - c| < \delta$$

Choose $\varepsilon = L/2$.   Then, there is a $\delta > 0$ such that

$$|f(x) - L| < \frac{L}{2} \qquad \text{whenever} \qquad 0 < |x - c| < \delta$$

$$-\frac{L}{2} < f(x) - L < \frac{L}{2} \qquad \text{whenever} \qquad 0 < |x - c| < \delta$$

$$\frac{L}{2} < f(x) < \frac{3L}{2} \qquad \text{whenever} \qquad 0 < |x - c| < \delta$$

Since $L/2 > 0$,   the last statement proves our assertion that $f(x) > 0$ whenever $0 < |x - c| < \delta$.   ■

EXERCISE 5

1. For the function $f(x) = 4x - 1$, $\lim_{x \to 3} f(x) = 11$.   For each $\varepsilon > 0$,   find a $\delta > 0$ such that
$$|(4x - 1) - 11| < \varepsilon \qquad \text{whenever} \qquad 0 < |x - 3| < \delta$$
(a) $\varepsilon = 0.1$ (b) $\varepsilon = 0.01$ (c) $\varepsilon = 0.001$ (d) $\varepsilon > 0$ is arbitrary

2. For the function $f(x) = 2 - 5x$, $\lim_{x \to -2} f(x) = 12$.   For each $\varepsilon > 0$,   find a $\delta > 0$ such that
$$|(2 - 5x) - 12| < \varepsilon \qquad \text{whenever} \qquad 0 < |x + 2| < \delta$$
(a) $\varepsilon = 0.2$ (b) $\varepsilon = 0.02$ (c) $\varepsilon = 0.002$ (d) $\varepsilon > 0$ is arbitrary

3. For the function $f(x) = \dfrac{x^2 - 9}{x + 3}$, $\lim_{x \to -3} f(x) = -6$.   For each $\varepsilon > 0$,   find a $\delta > 0$ such that
$$\left| \frac{x^2 - 9}{x + 3} - (-6) \right| < \varepsilon \qquad \text{whenever} \qquad 0 < |x + 3| < \delta$$
(a) $\varepsilon = 0.1$ (b) $\varepsilon = 0.01$ (c) $\varepsilon > 0$ is arbitrary

4. For the function $f(x) = \dfrac{x^2 - 4}{x - 2}$, $\lim_{x \to 2} f(x) = 4$.   For each $\varepsilon > 0$,   find a $\delta > 0$ such that
$$\left| \frac{x^2 - 4}{x - 2} - 4 \right| < \varepsilon \qquad \text{whenever} \qquad 0 < |x - 2| < \delta$$
(a) $\varepsilon = 0.1$ (b) $\varepsilon = 0.01$ (c) $\varepsilon > 0$ is arbitrary

For the limits in Problems 5–8 find the largest δ that "works" for the given ε.

5. $\lim_{x \to 1}(2x) = 2$, $\varepsilon = 0.01$

6. $\lim_{x \to 2}(-3x) = -6$, $\varepsilon = 0.01$

7. $\lim_{x \to 2}(6x - 1) = 11$, $\varepsilon = \frac{1}{2}$

8. $\lim_{x \to -3}(2 - 3x) = 11$, $\varepsilon = \frac{1}{3}$

In Problems 9–20 give an ε, δ proof for each limit.

9. $\lim\limits_{x \to 2}(3x) = 6$

10. $\lim\limits_{x \to 3}(4x) = 12$

11. $\lim\limits_{x \to 0}(2x + 5) = 5$

12. $\lim\limits_{x \to -1}(2 - 3x) = 5$

13. $\lim\limits_{x \to -3}(-5x + 2) = 17$

14. $\lim\limits_{x \to 2}(2x - 3) = 1$

15. $\lim\limits_{x \to 2}(x^2 - 2x) = 0$

16. $\lim\limits_{x \to 0}(x^2 + 3x) = 0$

17. $\lim\limits_{x \to 1} \dfrac{1 + 2x}{3 - x} = \dfrac{3}{2}$

18. $\lim\limits_{x \to 2} \dfrac{2x}{4 + x} = \dfrac{2}{3}$

19. $\lim\limits_{x \to 0} \sqrt[3]{x} = 0$

20. $\lim\limits_{x \to 1} \sqrt{2 - x} = 1$

21. Show that the statement $\lim_{x \to 3}(3x - 1) = 12$ is false.

22. Show that the statement $\lim_{x \to -2}(4x) = -7$ is false.

23. Show that $\left| \dfrac{1}{x^2 + 9} - \dfrac{1}{18} \right| < \dfrac{7}{234}|x - 3|$ if $2 < x < 4$. Use this to show that $\lim\limits_{x \to 3} \dfrac{1}{x^2 + 9} = \dfrac{1}{18}$.

24. Prove that if $\lim_{x \to c} f(x) < 0$, then there is an open interval about c, possibly excluding c itself, on which $f(x) < 0$.

Historical Perspectives

Historically, one of the important problems for the development of the calculus is the *tangent problem*—the problem of finding a line tangent to the graph of a function at a given point on it. The tangent problem was solved for some special cases by the Greeks more than 2000 years ago. However, their methods were almost entirely geometric in nature, as opposed to the calculus method discussed in this chapter. After the decline of the Greek civilization, very little progress was made on the tangent problem until the invention of analytic geometry by two Frenchmen, René Descartes (1596–1650) and Pierre de Fermat (1601–1665).

One can hardly overestimate the importance of the new mathematical tool, analytic geometry, for the development of calculus. In the early seventeenth century, one great hindrance to solving some of the basic mathematical problems of the day was the need to translate these problems into geometric terms. The Greek methods, with their total dependence upon geometry (though modified considerably), still formed the basis for the mathematical analysis done prior to Descartes and Fermat.

Both Descartes and Fermat provided solutions to the tangent problem, working independently and employing quite dissimilar methods. Descartes' method was purely algebraic, and while it was a definite improvement over the Greek methods, its application was limited to a relatively small number of graphs. Fermat, on the other hand, developed a method that could be applied to a wide variety of graphs. In fact, his method was close in spirit to the method employed in this chapter. Fermat's method was to be improved later by one of the two principal founders of calculus, Sir Isaac

Newton (1642–1727) of England. And another Frenchman, Blaise Pascal (1623–1662), a contemporary of Descartes and Fermat, did work on the tangent problem that later inspired Gottfried Wilhelm von Leibniz (1646–1716) of Germany, who shares with Newton the honor of founding the calculus.

You are now invited to explore the history of the tangent problem in the following set of exercises.

GREEK METHODS FOR TANGENTS

1. The Greek method for finding the tangent line to a circle depended upon the fact that at any point on a circle the radius and the tangent lines are perpendicular. Use this method to find an equation of the tangent line to the circle $x^2 + y^2 = 4$ at the point $(1, \sqrt{3})$. See Figure 35. [*Hint:* First find the slope of L_2; then write the slope of L_1 and use the point–slope formula for the equation of L_1.]

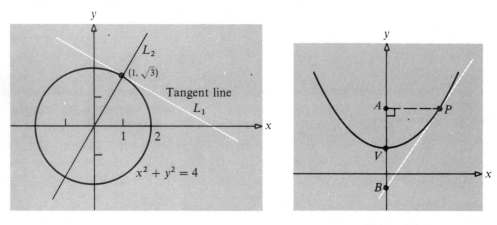

Figure 35   Figure 36

2. Use the method of Problem 1 to find an equation of the tangent line to the circle $(x - 1)^2 + (y - 2)^2 = 4$ at the point $(0, 2 + \sqrt{3})$.

3. The Greek method for finding a tangent line to a parabola is a bit more involved than that for a circle. If we seek the tangent line at P (see Fig. 36), we can show that it also passes through the point B on the axis of the parabola, where Distance BV = Distance AV. Since knowing P and V makes it easy to find AV, we can then get B and write the equation for L by using the two-point formula. Use this method to find an equation of the line tangent to $y = x^2$ at $(2, 4)$. Make a sketch.

4. Use the method of Problem 3 to find an equation of the tangent line to the graph of $x = y^2 + 1$ at the point $(2, 1)$.

5. Use the method of Problem 3 to find an equation of the tangent line to the graph of $y = x^2$ at *any* point (x_0, y_0).

Note: Similar methods for finding tangents to ellipses and hyperbolas are outlined in Figures 37 and 38. The Greeks also had some success with curves other than conics. For example, Archimedes (287–212 BC) developed a method for finding tangents to the Archimedean spiral.

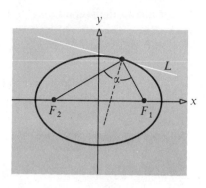

Figure 37 $L \perp$ to the line bisecting angle α

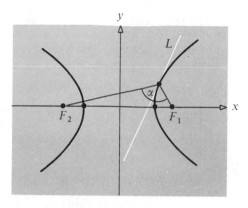

Figure 38 L bisects angle α

DESCARTES'* METHOD OF EQUAL ROOTS

6. Descartes' method for finding tangents depends upon the idea that for many graphs, the tangent line at a given point is the *unique* line that intersects the graph at that point only. We will apply his method to find an equation of the tangent line to the parabola $y = x^2$ at the point $(2, 4)$. First, we know the equation of the tangent line must be in the form $y = mx + b$. Using the fact that the point $(2, 4)$ is on the line, we can solve for b in terms of m and get the equation $y = mx + (4 - 2m)$. Now we want $(2, 4)$ to be the *unique* solution to the system

$$\begin{cases} y = x^2 \\ y = mx + 4 - 2m \end{cases}$$

From this system, we get $x^2 = mx + 4 - 2m$ or $x^2 - mx + (2m - 4) = 0$. By using the quadratic formula, we get

$$x = \frac{m \pm \sqrt{m^2 - 4(2m - 4)}}{2}$$

In order to obtain a *unique* solution for x, the *two roots must be equal*; in other words, the expression $m^2 - 4(2m - 4)$ must be 0. Complete the work to get m and write an equation of the tangent line. Compare this answer with the result of Problem 3.

7. Repeat Problem 4 using Descartes' method of equal roots.

8. Descartes' method of equal roots will not work for $y = x^3$. Prove this by finding the tangent line at the point $(2, 8)$ using the methods of this chapter and showing that it intersects the curve at *two* points, rather than at a single point.

---

* René Descartes, who came from a moderately wealthy family, was trained from an early age in a Jesuit school. Although Descartes showed academic promise early on, he did not settle down to scholarly work until he was past 30. Descartes moved to Holland at the age of 32, and for the next 20 years, he lived and worked there in relative obscurity, devoting most of his energy to the development of a deductive philosophical system. In 1649, Descartes was persuaded to go to Sweden as a tutor to Queen Christine. His duties there required him to begin his day at 5:00 AM in an unheated library. This schedule, so contrary to his lifelong habit of sleeping most of the morning, combined with a most severe winter in Sweden, was more than the frail Descartes could take, and he died of a lung inflammation a few months after his arrival in Sweden.

9. Use the method of equal roots to find an equation of the tangent line to the graph of $x^2 + y^2 = 25$ at the point $(3, 4)$.

FERMAT'S* METHOD FOR TANGENTS

10. We will illustrate Fermat's method by using it to find an equation of the tangent line to the graph of $y = x^2$ at $(2, 4)$. See Figure 39. Choose Q on the tangent line below P. Now, Q is outside the parabola, $y_0 < x_0^2$. (Why?) So we get $4/y_0 > 4/x_0^2$. But $y_0 = 4 - e$, so

$$\frac{4}{4 - e} > \frac{4}{x_0^2} = \left(\frac{2}{x_0}\right)^2$$

By using similar triangles, $2/x_0 = (4 + a)/(a + 4 - e)$, and combining, we get

$$\frac{4}{4 - e} > \left(\frac{4 + a}{a + 4 - e}\right)^2$$

$$\frac{4}{4 - e} > \frac{16 + 8a + a^2}{a^2 + 8a + 16 - 2ae - 8e + e^2}$$

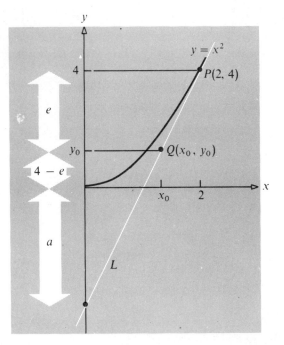

Figure 39

---

* Pierre Fermat ranks as the greatest "amateur" mathematician ever. We say "amateur" because Fermat was a lawyer by training and trade, and his considerable contributions to mathematics were made in his spare time! This fact becomes even more impressive when we consider those contributions. In addition to his work in analytic geometry, he was cofounder (with Pascal) of the theory of probability, an area of mathematics that has become increasingly important during the present century; he established many fundamental results in the field of number theory; and finally, since his work on calculus was the best done before Newton and Leibniz, he must be considered one of the principal founders of the subject. Besides his interest in mathematics he was a classical scholar, linguist, and poet.

Now, when Q is near P on the tangent line, it is not very far outside the parabola; thus, the inequality very nearly becomes an equality. In this case, we write

(2.36)
$$\frac{4}{4 - e} \approx \frac{16 + 8a + a^2}{a^2 + 8a + 16 - 2ae - 8e + e^2}$$

To complete the problem, we must calculate a (the y-intercept of the tangent line) and then write the equation of the tangent line. To do this, we need three steps: (*1*) cross-multiply in (2.36) to clear the fractions and then cancel like terms; (*2*) divide through by e; (*3*) set $e = 0$ and solve for a. Complete this procedure for this problem. Compare this answer with the results obtained in Problems 3 and 6.

11. Use Fermat's method to find an equation of the tangent line to the graph of $y = x^2$ at $(3, 9)$.

12. Fermat's method was criticized by many (especially by Descartes) because in one step he divides by e, and in the next step, he sets $e = 0$. You should be able to defend the method by pointing out that one does not really set $e = 0$. Elaborate. [*Hint:* What major concept of today's calculus is missing in Fermat's argument?]

13. Fermat's method is quite general. Apply it to find an equation of the tangent line to the graph of $y = x^3$ at the point $(2, 8)$. [*Hint:* Begin with $8/y_0 > 8/x_0^3 = (2/x_0)^3$. (Why is this true?)] Compare your answer with the result of Problem 8.

Note: Though Fermat's method is quite close in spirit to the limiting secant line concept developed in this chapter, it falls short in one fundamental way—it is basically *geometric* in nature. What was needed in order for calculus to develop fully was the analytic approach provided by limits. This approach was not perfected until 200 years after Fermat, with the work of Cauchy, Bolzano, and others.

Miscellaneous Exercises

In Problems 1–4 find the indicated limit.

1. $\lim\limits_{x \to 0} \dfrac{1}{x}\left(\dfrac{1}{4 + x} - \dfrac{1}{4}\right)$
  2. $\lim\limits_{x \to 0} \dfrac{1}{x}\left[\dfrac{1}{(2 + x)^2} - \dfrac{1}{4}\right]$
  3. $\lim\limits_{x \to 1} \dfrac{x^8 - 1}{x - 1}$
  4. $\lim\limits_{x \to 0} \dfrac{(x + 3)^2 - 9}{x}$

5. Graph the function $f(x) = x - [\![x]\!]$ on $[-3, 3]$ and discuss its discontinuities.

6. Suppose: $f(x) = \dfrac{x^2 - 6x - 16}{(x^2 - 7x - 8)\sqrt{x^2 - 4}}$

   (a) For what numbers x is f defined?
   (b) For what numbers x is f discontinuous?
   (c) Which of those numbers x found in part (b) are removable?

7. Let $f(x) = \dfrac{\sqrt{x} - 2}{x - 4}$. Use a calculator to complete the following table:

x	3.9	3.99	3.999	4.001	4.01	4.1
$f(x)$						

What is a reasonable guess for the value of $\lim\limits_{x \to 4} f(x)$? Defend it by evaluating the limit.

8. The graph of $y = \dfrac{x-3}{3-x}$ is a straight line with a point punched out. What straight line and what point?

9. Find $\lim\limits_{h \to 0} \dfrac{f(h) - f(0)}{h}$ if $f(x) = x|x|$.

10. Find $\lim\limits_{h \to 0} \dfrac{f(1+h) - f(1)}{h}$ if $f(x) = \sqrt{x}$.

11. Find $\lim\limits_{x \to 0^+} \dfrac{|x|}{x}(1-x)$ and $\lim\limits_{x \to 0^-} \dfrac{|x|}{x}(1-x)$. What can you say about $\lim\limits_{x \to 0} \dfrac{|x|}{x}(1-x)$?

12. Find $\lim\limits_{x \to 2}\left(\dfrac{x^2}{x-2} - \dfrac{2x}{x-2}\right)$. Then comment on the statement that this limit is given by

$$\lim_{x \to 2} \frac{x^2}{x-2} - \lim_{x \to 2} \frac{2x}{x-2}$$

13. Sketch the graph of the function below and determine where f is continuous.

$$f(x) = \begin{cases} 1 - x^2 & \text{if } |x| \leq 1 \\ x^2 - 1 & \text{if } |x| > 1 \end{cases}$$

14. Sketch the graph of the function below and determine where f is continuous.

$$f(x) = \begin{cases} \sqrt{4 - x^2} & \text{if } |x| \leq 2 \\ |x| - 2 & \text{if } |x| > 2 \end{cases}$$

15. Prove that the graphs of $y = x^3$ and $y = 1 - x^2$ intersect somewhere between $x = 0$ and $x = 1$.

16. Suppose that f and g are continuous in $[a, b]$, $f(a) < g(a)$, and $f(b) > g(b)$. Prove that the graphs of $y = f(x)$ and $y = g(x)$ intersect somewhere between $x = a$ and $x = b$.

17. Find constants A and B so that the function below is continuous for all x. Sketch the graph of the resulting function.

$$f(x) = \begin{cases} (x-1)^2 & \text{if } -\infty < x < 0 \\ A - x^2 & \text{if } 0 \leq x < 1 \\ x + B & \text{if } 1 \leq x < +\infty \end{cases}$$

18. Find constants A, B, C, and D so that the function below is continuous for all x. Sketch the graph of the resulting function.

$$f(x) = \begin{cases} \dfrac{x^2 + x - 2}{x - 1} & \text{if } -\infty < x < 1 \\ A & \text{if } x = 1 \\ B(x - C)^2 & \text{if } 1 < x < 4 \\ D & \text{if } x = 4 \\ 2x - 8 & \text{if } 4 < x < +\infty \end{cases}$$

19. Suppose f is defined on an interval (a, b) and there is a number K so that $|f(x) - f(c)| \leq K|x - c|$ for all c in (a, b) and x in (a, b). Such a constant K is called a *Lipschitz constant*. Find a Lipschitz constant for $f(x) = x^3$ on $(0, 2)$.

20. Let f be a function for which $0 \leq f(x) \leq 1$ for all x in $[0, 1]$. If f is continuous on $[0, 1]$, show that there exists at least one number c in $[0, 1]$ such that $f(c) = c$. [*Hint:* Let $g(x) = x - f(x)$ and apply the intermediate value theorem (2.29) to g.]

21. Show that $|(2 + x)^2 - 4| \le 5|x|$ if $-1 < x < 1$. Use this to prove that $\lim_{x \to 0}(2 + x)^2 = 4$. Use an ε, δ argument.

22. Prove that $\lim_{x \to 2}(x^2 + 3x) = 10$ by finding δ in terms of a given positive ε so that

$$|x^2 + 3x - 10| < \varepsilon \quad \text{if} \quad |x - 2| < \delta$$

23. Show that

$$\left| \frac{1}{x^2 + 9} - \frac{1}{13} \right| \le \frac{1}{26}|x - 2| \quad \text{if} \quad 1 < x < 3$$

Use this to prove that

$$\lim_{x \to 2} \frac{1}{x^2 + 9} = \frac{1}{13}$$

Use an ε, δ argument.

24. Suppose a function f is defined and continuous on the closed interval $[a, b]$. Is the domain of $h(x) = 1/f(x)$ also the closed interval $[a, b]$? What do you conclude about the continuity of h?

25. Use the definition of limit to show that $\lim_{x \to 1} x^2 \ne 1.31$. [*Hint:* Use $\varepsilon = 0.01$.]

26. Use the definition of limit to prove that no number L exists such that $\lim_{x \to 0}(1/x) = L$.

27. Use the definition of limit to prove that $\lim_{x \to 0}(4 - x^2) = 4$.

28. Use the definition of limit to prove that the linear function $f(x) = ax + b$ is continuous everywhere.

29. The definitions of $\sin x$ and $\cos x$ in trigonometry show that $|\sin x| \le |x|$ and $|\cos x - 1| \le |x|$ for all x. Why does it follow that $\lim_{x \to 0} \sin x = 0$ and $\lim_{x \to 0} \cos x = 1$?

30. Use Problem 29 to prove that $\sin x$ is continuous at every point c. [*Hint:* $\sin x = \sin[(x - c) + c] = \sin(x - c) \cos c + \cos(x - c) \sin c$.]

31. Use the same idea as in Problem 30 to prove that $\cos x$ is continuous at every point c.

3

The Derivative

In this chapter we use the idea of a limit to introduce the derivative of a function. The derivative of a function is intimately tied in with the notion of change, so we begin with a discussion of change and, in particular, rates of change.

1. Average Rate of Change

Recall that the symbol Δx denotes a change in x, while Δy denotes a change in y.

(3.1)   ***Average Rate of Change.***   **For a function** $y = f(x),$ **the *average rate of change* of y with respect to x is the ratio of the change in y to the change in x.**

In particular, if x changes from c to d, then

(3.2)

$$\left[\begin{array}{c} \textbf{Average rate of change} \\ \textbf{of } y \textbf{ with respect to } x \end{array}\right] = \frac{\Delta y}{\Delta x} = \frac{f(d) - f(c)}{d - c}$$

EXAMPLE 1   For the function $y = f(x) = x^2$, find the average rate of change of y with respect to x as x changes from $c = 0$ to $d = 3$.

Solution   First, compute the change in x: $\Delta x = 3 - 0 = 3$. The corresponding change in y is the difference between the values of $f(x) = x^2$ at $d = 3$ and at $c = 0$:

$$\Delta y = f(3) - f(0) = 3^2 - 0^2 = 9 - 0 = 9$$

The average rate of change of y with respect to x is

$$\frac{\Delta y}{\Delta x} = \frac{f(3) - f(0)}{3 - 0} = \frac{9}{3} = 3 \qquad \blacksquare$$

GEOMETRIC INTERPRETATION

A geometric interpretation of the average rate of change is provided in Figure 1. The line connecting the two points $(c, f(c))$ and $(d, f(d))$ on the graph of the function f is a secant line and its slope is

$$m_{\text{sec}} = \frac{f(d) - f(c)}{d - c}$$

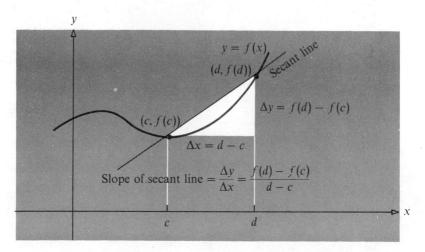

Figure 1

This is equal to the average rate of change of $y = f(x)$ with respect to x from c to d.

(3.3)   **Geometrically, the average rate of change of a function equals the slope of the secant line joining two points on the graph of the function.**

DIFFERENCE QUOTIENTS

The average rate of change $\Delta y/\Delta x$ of a function $y = f(x)$ from c to d is usually referred to as a *difference quotient*. We shall find it convenient to express the difference quotient in terms of c and Δx. We do this as follows: Let $\Delta x = d - c$.   Then $d = c + \Delta x$,   and if we substitute $c + \Delta x$ for d in (3.2), we find

(3.4)
$$\frac{\Delta y}{\Delta x} = \frac{f(d) - f(c)}{d - c} = \frac{f(c + \Delta x) - f(c)}{\Delta x}$$

EXAMPLE 2   Compute the difference quotient $\Delta y/\Delta x$ for the function $y = f(x) = x^2 + 5x$ where:

(a) $c = 2$, $\Delta x = 1$ (b) $c = 0$, $\Delta x = 0.5$
(c) c and $\Delta x \neq 0$ are any given real numbers

Solution   (a)   For $c = 2$ and $\Delta x = 1$,   we have

$$f(c) = f(2) = 2^2 + 5(2) = 14$$
$$f(c + \Delta x) = f(2 + 1) = f(3) = 3^2 + 5(3) = 24$$
$$\frac{\Delta y}{\Delta x} = \frac{f(c + \Delta x) - f(c)}{\Delta x} = \frac{24 - 14}{1} = 10$$

(b)   For $c = 0$ and $\Delta x = 0.5$,   we have

$$f(c) = f(0) = 0$$
$$f(c + \Delta x) = f(0 + 0.5) = f(0.5) = (0.5)^2 + 5(0.5) = 0.25 + 2.5 = 2.75$$
$$\frac{\Delta y}{\Delta x} = \frac{f(c + \Delta x) - f(c)}{\Delta x} = \frac{2.75}{0.5} = 5.5$$

(c)   For c and $\Delta x \neq 0$ any given real numbers, we have

$$f(c) = c^2 + 5c$$
$$f(c + \Delta x) = (c + \Delta x)^2 + 5(c + \Delta x) = c^2 + 2c\Delta x + (\Delta x)^2 + 5c + 5\Delta x$$
$$\Delta y = f(c + \Delta x) - f(c) = [c^2 + 2c\Delta x + (\Delta x)^2 + 5c + 5\Delta x] - (c^2 + 5c)$$
$$= 2c\Delta x + (\Delta x)^2 + 5\Delta x$$
$$\frac{\Delta y}{\Delta x} = \frac{f(c + \Delta x) - f(c)}{\Delta x} = \frac{2c\Delta x + (\Delta x)^2 + 5\Delta x}{\Delta x} = 2c + \Delta x + 5$$

Observe that we can use the answer to part (c) to check the answers found in parts (a) and (b).   ■

RECTILINEAR MOTION

Average rates of change are used extensively. For example, in physics we consider *rectilinear motion*, which is the motion of a particle along a straight line. It is often convenient to think of rectilinear motion along a line, usually horizontal, with the positive direction to the right. If we select the origin as the initial position of the particle and if we assume the distance of the particle from O is known for all times t (in seconds), then we can represent its motion by a function $s = f(t)$, where s is the signed or directed distance of the particle from O at t seconds of time (see Fig. 2).

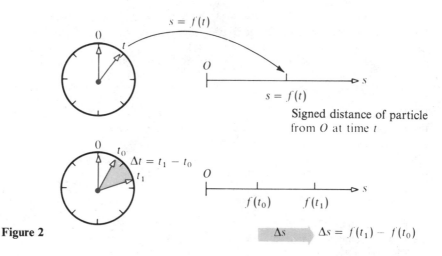

$s = f(t)$

Signed distance of particle from O at time t

$\Delta s = f(t_1) - f(t_0)$

Figure 2

If at time t_0 the particle is at $s_0 = f(t_0)$ and at time t_1 the particle is at the position $s_1 = f(t_1)$, then the change in time is $\Delta t = t_1 - t_0$ and the change in position is $\Delta s = s_1 - s_0 = f(t_1) - f(t_0)$. **The rate of change of position with respect to time, $\Delta s/\Delta t$, is called the** *average velocity* of the particle.*

For example, suppose the function $s = f(t) = 16t^2$ describes the distance s (in feet) an object dropped from a tall building has traveled after a time t (in seconds) for $0 \le t \le 6$. The average velocity of the object over the entire time interval (6 seconds) is

$$\frac{\Delta s}{\Delta t} = \frac{f(6) - f(0)}{6 - 0} = \frac{16(36) - 0}{6 - 0} = 96 \text{ feet per second}$$

More will be said about rectilinear motion in the next section. Now it is time for some exercises.

---

* *Speed* and *velocity* are not the same thing. Speed is defined as the *absolute value of velocity* and therefore is always nonnegative. Velocity, on the other hand, may be a positive or negative number, indicating direction to the right or left. Thus, speed is a measure of how fast a particle is moving, while velocity is a measure of both its speed and its direction.

EXERCISE 1

1. For the function $y = f(x) = 5x - 2$:
   (a) Find the change in y from $x = 0$ to $x = 4$.
   (b) Find the change in y from $x = -3$ to $x = 2$.
   (c) Find the average rate of change for part (a).
   (d) Find the average rate of change for part (b).
   (e) Compute the average rate of change from $x = c$ to $x = d$. What do you conclude? Why? [*Hint:* Graph the function.]

2. For the function $y = f(x) = x^2 - 1$:
   (a) Find the change in y from $x = 1$ to $x = 3$.
   (b) Find the change in y from $x = -1$ to $x = 1$.
   (c) Find the average rate of change in part (a).
   (d) Find the average rate of change in part (b).
   (e) Find the slope of the secant line from $(0, f(0))$ to $(4, f(4))$.

3. For the function $y = f(x) = \dfrac{x^2}{x + 3}$:
   (a) Find the change in y from $x = -1$ to $x = 1$.
   (b) Find the change in y from $x = 0$ to $x = 4$.
   (c) Find the average rate of change in part (a).
   (d) Find the average rate of change in part (b).
   (e) Find the slope of the secant line from $(-2, f(-2))$ to $(2, f(2))$.

4. For the function $y = f(x) = \dfrac{x}{x^2 - 1}$:
   (a) Find the change in y from $x = 2$ to $x = 3$.
   (b) Find the change in y from $x = -3$ to $x = -2$.
   (c) Find the average rate of change in part (a).
   (d) Find the average rate of change in part (b).
   (e) Find the slope of the secant line from $(-\frac{1}{2}, \frac{2}{3})$ to $(\frac{1}{2}, -\frac{2}{3})$.

In Problems 5–10 find the slope of the secant line from the point $(0, f(0))$ to the point $(2, f(2))$. Graph each function and show this secant line.

5. $f(x) = 3x - 2$

6. $f(x) = x^3 + 8$

7. $f(x) = x^2 + 4$

8. $f(x) = x$

9. $f(x) = 2 - x^2$

10. $f(x) = \dfrac{1}{x + 1}$

11. Suppose the function $s = f(t) = 16t^2$ relates the distance s (in feet) an object travels in time t (in seconds). Compute the average velocity,

$$\frac{\Delta s}{\Delta t} = \frac{f(t_0 + \Delta t) - f(t_0)}{\Delta t}$$

   when $t_0 = 3$, for:
   (a) $\Delta t = 0.5$ (b) $\Delta t = 0.1$

12. Follow the directions in Problem 11 for:
   (a) $\Delta t = 0.01$ (b) $\Delta t = 0.0001$

For the functions in Problems 13–22 compute the difference quotient below for the given number c:

$$\frac{\Delta y}{\Delta x} = \frac{f(c + \Delta x) - f(c)}{\Delta x} \qquad \Delta x \neq 0$$

13. $f(x) = 3x - 5, \quad c = 2$ **14.** $f(x) = 2x + 3, \quad c = 3$ **15.** $f(x) = 3x^2 - 2, \quad c = -1$

16. $f(x) = 2x^3 + 5, \quad c = 1$ **17.** $f(x) = x^3 + x^2, \quad c = 0$ **18.** $f(x) = x^3 - x^2, \quad c = -1$

19. $f(x) = x + \dfrac{1}{x}, \quad c = 1$ **20.** $f(x) = \dfrac{x}{x - 1}, \quad c = 2$ **21.** $f(x) = \sqrt{x}, \quad c = 4$ **22.** $f(x) = \dfrac{1}{x}, \quad c = 2$

23. What is the average velocity of an automobile traveling from New York City to Miami if half the distance is traversed at 45 miles per hour and the other half at 55 miles per hour? [*Hint:* The answer is not 50 miles per hour.]

24.[†] A body moves along a straight line, its distance from the origin at any instant being given by the equation $s = 8t - 3t^2$, where s is in centimeters and t is in seconds. Find the average velocity of the body in the interval from $t = 0$ to $t = 1$ second and in the interval from $t = 0$ to $t = 4$ seconds.

25. Suppose the distance s a person can walk in time t can be represented by $s = 4\sqrt{t}$, where s is measured in kilometers and t is measured in hours. What is the person's average velocity from $t = 0$ to $t = 16$ hours? From $t = 1$ to $t = 4$ hours? From $t = 1$ to $t = 2$ hours?

26. Consider the corn production data given in the table below. Graph the data. What is the average rate of increase in yield from $700 to $2,200? From $1,000 to $1,400? From $1,000 to $1,200?

Investment, x (dollars)	Corn Output, y (bushels)
700	12,700
1,000	13,100
1,200	14,700
1,400	17,100
1,600	19,000
2,000	22,100
2,200	21,700

27. Adding fertilizer to the soil can increase corn yields. The table below lists the corn yield obtained per acre by adding certain amounts of fertilizer. Graph the data and use the amount of fertilizer added as the x-axis. What is the average rate of increase in yield from 0 to 50 pounds? From 10 to 40 pounds? From 10 to 30 pounds?

Fertilizer (pounds)	Corn Yield per Acre (bushels)
0	45.0
10	46.2
20	48.0
30	48.2
40	47.6
50	47.3

[†] Adapted from F. W. Sears, M. W. Zemansky, and H. D. Young, *University Physics* (Reading, Mass.: Addison-Wesley Publishing Co., 1976), p. 63. Reprinted by permission.

28. A protein disintegrates into amino acids according to the formula $M = 28/(t + 2)$, where M is the mass of the protein (in grams) and t is time (in hours). Find the average reaction rate, $\Delta M/\Delta t$, from $t = 0$ to $t = 2$ hours.   Interpret your answer.

29. In a metabolic experiment, the mass M of glucose decreases according to the formula $M = 4.5 - 0.03t^2$, where M is measured in grams and t is time (in hours). Find the average reaction rate, $\Delta M/\Delta t$, from $t = 0$ to $t = 2$ hours.   Interpret your answer.

2. Instantaneous Rate of Change; the Derivative

At the conclusion of the last section we used the function $s = 16t^2$ to describe the distance s (in feet) an object dropped from a tall building has traveled after a time t (in seconds). When $0 \le t \le 6$, we found the average velocity of the object over the entire time interval of 6 seconds to be

$$\frac{\Delta s}{\Delta t} = \frac{f(6) - f(0)}{6 - 0} = 96 \text{ feet per second}$$

This average accurately describes the velocity of the object over the 6 second interval, but it gives no information at all about the actual velocity at any particular instant of time. We now wish to define the velocity of the object at a particular instant.

To see how this might be done, we will seek the exact velocity of the object at the instant when $t = 3$.   This velocity is called the *instantaneous velocity at 3*.

So far we have no *mathematical* method for finding instantaneous velocities. However, we can *estimate* the instantaneous velocity at $t = 3$ seconds   by computing the average velocities for some intervals of time beginning at $t = 3$.   For example, let's compute the average velocity for the 1 second interval beginning at $t = 3$ and ending at $t = 4$.   Here, $\Delta t = 1$.   At $t = 3$,   the distance of the object from the starting position is

$$s = f(3) = 16(9) = 144 \text{ feet}$$

At $t = 4$,

$$s = f(4) = 16(16) = 256 \text{ feet}$$

Thus, over the 1 second interval from $t = 3$ to $t = 4$,

$$\text{Average velocity} = \frac{\Delta s}{\Delta t} = \frac{f(4) - f(3)}{1} = \frac{256 - 144}{1} = 112 \text{ feet per second}$$

The average velocities for the smaller intervals of time $\Delta t = 0.5, 0.1, 0.01, 0.0001$ may be found similarly (see Problems 11 and 12 in Exercise 1); and Table 1 gives the five estimates obtained for the instantaneous velocity. We can see from the table that for this example, the larger the time interval Δt, the larger the average velocity $\Delta s/\Delta t$. The most accurate estimates for instantaneous velocity will correspond to very small time intervals Δt. For example, over the interval $\Delta t = 0.0001$ second,   we would not expect the velocity of the object to change very much. Thus, the average

Table 1

Start, t_0	End, t	Δt	Average Velocity, $\dfrac{\Delta s}{\Delta t} = \dfrac{f(t_0 + \Delta t) - f(t_0)}{\Delta t}$
3	4	1	$\dfrac{\Delta s}{\Delta t} = \dfrac{f(4) - f(3)}{1} = \dfrac{16(16)^2 - 16(9)}{1} = 112$
3	3.5	0.5	$\dfrac{\Delta s}{\Delta t} = \dfrac{f(3.5) - f(3)}{0.5} = \dfrac{16(3.5)^2 - 16(9)}{0.5} = 104$
3	3.1	0.1	$\dfrac{\Delta s}{\Delta t} = \dfrac{f(3.1) - f(3)}{0.1} = \dfrac{16(3.1)^2 - 16(9)}{0.1} = 97.6$
3	3.01	0.01	$\dfrac{\Delta s}{\Delta t} = \dfrac{f(3.01) - f(3)}{0.01} = \dfrac{16(3.01)^2 - 16(9)}{0.01} = 96.16$
3	3.0001	0.0001	$\dfrac{\Delta s}{\Delta t} = \dfrac{f(3.0001) - f(3)}{0.0001} = \dfrac{16(3.0001)^2 - 16(9)}{0.0001} = 96.0016$

velocity of 96.0016 feet per second during the very short time interval $\Delta t = 0.0001$ should be very close to the instantaneous velocity at $t = 3$.

But what is the exact or instantaneous velocity at $t = 3$? It must be close to 96.0016, but is it 96.0016? Or is it 96.0001? Or what?

To obtain the precise answer, we first use some algebra. Specifically, we find the average velocity for the object in the time interval that begins at $t_0 = 3$ and ends at $t = 3 + \Delta t$, where $\Delta t \neq 0$ represents any small interval of time:

$$f(3) = 16(9) = 144$$

$$f(3 + \Delta t) = 16(3 + \Delta t)^2 = 16[9 + 6\Delta t + (\Delta t)^2] = 144 + 96\Delta t + 16(\Delta t)^2$$

Thus,

$$\text{Average velocity} = \frac{\Delta s}{\Delta t} = \frac{f(3 + \Delta t) - f(3)}{\Delta t} = \frac{[144 + 96\Delta t + 16(\Delta t)^2] - 144}{\Delta t}$$

$$= \frac{16(\Delta t)^2 + 96\Delta t}{\Delta t} = \frac{16\Delta t(\Delta t + 6)}{\Delta t}$$

We can cancel the Δt's because the increment Δt does not equal 0. As a result,

$$\text{Average velocity} = \frac{\Delta s}{\Delta t} = 16(\Delta t + 6) = 96 + 16\Delta t$$

Now comes the important step in this procedure! As Δt gets closer and closer to 0, but not equal to 0, the values of $\Delta s / \Delta t = 96 + 16\Delta t$ get closer and closer to 96. The average velocity will never equal 96, because we must have a *nonzero* time interval Δt in order to compute an average velocity. Nevertheless, we can make $\Delta s / \Delta t = 96 + 16\Delta t$ as close as we please to 96 by taking Δt sufficiently close to 0.

Consistent with modern mathematical terminology (as introduced in Chapter 2), 96 is called the *limit* of the average velocity $\Delta s/\Delta t$ as Δt approaches 0. In symbols, we write

$$\lim_{\Delta t \to 0} \frac{\Delta s}{\Delta t} = \lim_{\Delta t \to 0} \frac{f(3 + \Delta t) - f(3)}{\Delta t} = \lim_{\Delta t \to 0} (96 + 16\Delta t) = 96$$

Intuition tells us that the limit 96 feet per second is what we mean by the (instantaneous) velocity of the object at time $t = 3$ seconds.   Thus, we are led to the following definition.

(3.5)   **In general, if $s = f(t)$ is a function that describes the distance s a particle travels in time t, the (*instantaneous*) *velocity* v of the particle at time t_0 is defined as the limit of the average rate of change $\Delta s/\Delta t$ as Δt approaches 0. Specifically, the velocity v at time t_0 is**

(3.6)
$$v = \lim_{\Delta t \to 0} \frac{\Delta s}{\Delta t} = \lim_{\Delta t \to 0} \frac{f(t_0 + \Delta t) - f(t_0)}{\Delta t}$$

provided this limit exists.

The limit in (3.6) has an important generalization:

Derivative of a Function.   **Let $y = f(x)$ be a function and let c be in the domain of f. The *derivative of f at c*, denoted by $f'(c)$, and read "f prime of c" is the number**

(3.7)
$$f'(c) = \lim_{\Delta x \to 0} \frac{f(c + \Delta x) - f(c)}{\Delta x}$$

provided this limit exists.

We calculate the derivative of a function f in four steps:

(3.8)   *Four-Step Method*

Step 1.   **Find $f(c + \Delta x)$.**

Step 2.   **Subtract $f(c)$ from $f(c + \Delta x)$ to get $\Delta y = f(c + \Delta x) - f(c)$.**

Step 3.   **Divide Δy by Δx to get the difference quotient $\dfrac{\Delta y}{\Delta x} = \dfrac{f(c + \Delta x) - f(c)}{\Delta x}$.**

Step 4.   **Find the limit (if it exists) of the difference quotient $\Delta y/\Delta x$ as Δx approaches 0.**

EXAMPLE 1   Find the derivative of $f(x) = x^3$ at 2 by using the four-step method.

Solution   *Step 1.* $f(2 + \Delta x) = (2 + \Delta x)^3 = 8 + 12\Delta x + 6(\Delta x)^2 + (\Delta x)^3$

Step 2.   Since $f(2) = 2^3 = 8$,

$$\Delta y = f(2 + \Delta x) - f(2) = [8 + 12\Delta x + 6(\Delta x)^2 + (\Delta x)^3] - 8$$
$$= 12\Delta x + 6(\Delta x)^2 + (\Delta x)^3$$

Step 3. $$\frac{\Delta y}{\Delta x} = \frac{12\Delta x + 6(\Delta x)^2 + (\Delta x)^3}{\Delta x} = \frac{\Delta x[12 + 6\Delta x + (\Delta x)^2]}{\Delta x}$$

$$= 12 + 6\Delta x + (\Delta x)^2$$

Step 4. $$f'(2) = \lim_{\Delta x \to 0} \frac{\Delta y}{\Delta x} = \lim_{\Delta x \to 0} [12 + 6\Delta x + (\Delta x)^2] = 12 \quad \blacksquare$$

In this example we calculated the derivative of $f(x) = x^3$ at 2. It is often just as easy to find the derivative at an arbitrary number c, as the next example illustrates.

EXAMPLE 2   Find the derivative of $f(x) = x^3$ at c.

Solution   We begin at Step 2 of the four-step method:

$$\Delta y = f(c + \Delta x) - f(c) = (c + \Delta x)^3 - c^3$$
$$= c^3 + 3c^2\Delta x + 3c(\Delta x)^2 + (\Delta x)^3 - c^3$$
$$= 3c^2\Delta x + 3c(\Delta x)^2 + (\Delta x)^3$$
$$= \Delta x[3c^2 + 3c\Delta x + (\Delta x)^2]$$

As a result,

$$\frac{\Delta y}{\Delta x} = \frac{\Delta x[3c^2 + 3c\Delta x + (\Delta x)^2]}{\Delta x} = 3c^2 + 3c\Delta x + (\Delta x)^2$$

The derivative is

$$f'(c) = \lim_{\Delta x \to 0} \frac{\Delta y}{\Delta x} = \lim_{\Delta x \to 0} [3c^2 + 3c\Delta x + (\Delta x)^2] = 3c^2 \quad \blacksquare$$

Notice that for $c = 2$, $f'(c) = 3c^2 = (3)(2^2) = 12$,   which agrees with the answer found in Example 1.

Since the limit

$$f'(c) = \lim_{\Delta x \to 0} \frac{\Delta y}{\Delta x} = 3c^2$$

exists for any choice of c, it is convenient to replace c by x, so that we can write $f'(x) = 3x^2$.   In general, the derivative $f'(x)$ of a function f at x is itself a function since it gives a rule for associating a number x with a number $f'(x)$. This function f', called the *derived function of f* or the *derivative of f*, is defined by

(3.9) $$f'(x) = \lim_{\Delta x \to 0} \frac{f(x + \Delta x) - f(x)}{\Delta x}$$

The domain of the derived function, which is always contained in the domain of the original function, consists of all numbers x for which the limit (3.9) exists. At such numbers, the function f is said to be *differentiable*. Thus, "differentiate f" means the same as "find the derivative of f."

EXAMPLE 3   Differentiate: $f(x) = x^2 + 2x$

Solution   To differentiate f, we compute

$$f'(x) = \lim_{\Delta x \to 0} \frac{\Delta y}{\Delta x} = \lim_{\Delta x \to 0} \frac{f(x + \Delta x) - f(x)}{\Delta x}$$

$$= \lim_{\Delta x \to 0} \frac{[(x + \Delta x)^2 + 2(x + \Delta x)] - (x^2 + 2x)}{\Delta x}$$

$$\underset{\substack{\uparrow \\ \text{Simplify}}}{= \lim_{\Delta x \to 0}} \frac{[x^2 + 2x\Delta x + (\Delta x)^2 + 2x + 2\Delta x] - x^2 - 2x}{\Delta x}$$

$$\underset{\substack{\uparrow \\ \text{Simplify}}}{= \lim_{\Delta x \to 0}} \frac{2x\Delta x + (\Delta x)^2 + 2\Delta x}{\Delta x}$$

By factoring Δx from the numerator and canceling it with the Δx in the denominator, we obtain

$$f'(x) = \lim_{\Delta x \to 0} \frac{\Delta x(2x + \Delta x + 2)}{\Delta x} = \lim_{\Delta x \to 0} (2x + \Delta x + 2) = 2x + 2$$

Thus, $f'(x) = 2x + 2$.   ■

EXAMPLE 4   Differentiate: $f(x) = \dfrac{1}{x^2}$

Solution $$f'(x) = \lim_{\Delta x \to 0} \frac{\Delta y}{\Delta x} = \lim_{\Delta x \to 0} \frac{f(x + \Delta x) - f(x)}{\Delta x} = \lim_{\Delta x \to 0} \frac{\dfrac{1}{(x + \Delta x)^2} - \dfrac{1}{x^2}}{\Delta x}$$

$$\underset{\substack{\uparrow \\ \text{Simplify}}}{= \lim_{\Delta x \to 0}} \frac{\dfrac{x^2 - (x + \Delta x)^2}{x^2(x + \Delta x)^2}}{\Delta x} \underset{\substack{\uparrow \\ \text{Simplify}}}{=\lim_{\Delta x \to 0}} \frac{-2x\Delta x - (\Delta x)^2}{x^2(x + \Delta x)^2\Delta x}$$

$$\underset{\substack{\uparrow \\ \text{Factor } \Delta x \text{ in numerator}}}{= \lim_{\Delta x \to 0}} \frac{(-2x - \Delta x)\Delta x}{x^2(x + \Delta x)^2\Delta x} \underset{\substack{\uparrow \\ \text{Cancel } \Delta x \neq 0}}{= \lim_{\Delta x \to 0}} \frac{-2x - \Delta x}{x^2(x + \Delta x)^2} = \frac{-2x}{x^4} = \frac{-2}{x^3}$$

Thus, $f'(x) = -2/x^3$.   ■

EXAMPLE 5   Find the derivative of $f(x) = \sqrt{x}$ and determine the domain of f'.

Solution   We first compute $\Delta y / \Delta x$:

$$\frac{\Delta y}{\Delta x} = \frac{f(x + \Delta x) - f(x)}{\Delta x} = \frac{\sqrt{x + \Delta x} - \sqrt{x}}{\Delta x}$$

In its present form, we cannot cancel the Δx's. Let's see what happens if we rationalize the numerator by multiplying the numerator and denominator by $\sqrt{x + \Delta x} + \sqrt{x}$:

$$\frac{\Delta y}{\Delta x} = \left(\frac{\sqrt{x + \Delta x} - \sqrt{x}}{\Delta x}\right)\left(\frac{\sqrt{x + \Delta x} + \sqrt{x}}{\sqrt{x + \Delta x} + \sqrt{x}}\right)$$

$$= \frac{(x + \Delta x) - x}{\Delta x(\sqrt{x + \Delta x} + \sqrt{x})} = \frac{\Delta x}{\Delta x(\sqrt{x + \Delta x} + \sqrt{x})}$$

Since $\Delta x \neq 0$, we can cancel the Δx's:

$$\frac{\Delta y}{\Delta x} = \frac{1}{\sqrt{x + \Delta x} + \sqrt{x}}$$

Now, when we let Δx approach 0, the quantity $\sqrt{x + \Delta x}$ approaches $\sqrt{x}$, so that

$$f'(x) = \lim_{\Delta x \to 0} \frac{\Delta y}{\Delta x} = \lim_{\Delta x \to 0} \frac{1}{\sqrt{x + \Delta x} + \sqrt{x}} = \frac{1}{2\sqrt{x}}$$

This limit does not exist when $x = 0$.   But for all other x in the domain of f $(x \geq 0)$,   the limit does exist. Hence, the domain of the function f' is all $x > 0$ and the function f has a derivative for all positive x.   ∎

ANOTHER FORMULATION OF THE DERIVATIVE

When it is required to calculate the derivative of a function f at a given number c, it is often advantageous to use an alternate formula for the derivative instead of (3.7).

Let $y = f(x)$ be a function, and let c be in the domain of f. The derivative of f at c is the number

(3.10) $$f'(c) = \lim_{x \to c} \frac{f(x) - f(c)}{x - c}$$

provided this limit exists.

We justify this alternate formula by setting $\Delta x = x - c$ in (3.7). Then $f(c + \Delta x) = f(x)$ and as $\Delta x \to 0$, $x \to c$. Consequently,

$$f'(c) = \lim_{\Delta x \to 0} \frac{f(c + \Delta x) - f(c)}{\Delta x} = \lim_{x \to c} \frac{f(x) - f(c)}{x - c}$$

EXAMPLE 6   Use (3.10) to find $f'(2)$ if $f(x) = x^2 + 2x$.

Solution $f(2) = 4 + 4 = 8$

$$f'(2) = \lim_{x \to 2} \frac{f(x) - f(2)}{x - 2} = \lim_{x \to 2} \frac{x^2 + 2x - 8}{x - 2} = \lim_{x \to 2} \frac{(x - 2)(x + 4)}{x - 2} = \lim_{x \to 2}(x + 4) = 6$$

∎

EXERCISE 2

In Problems 1–8 find the derivative of each function at the given number by using (3.10).

1. $f(x) = 2x + 3$,   at 1   **2.** $f(x) = 3x - 5$,   at 2   **3.** $f(x) = x^2 - 2$,   at 0

4. $f(x) = 2x^2 + 4$,   at 1   **5.** $f(x) = 3x^2 + x + 5$,   at -1 **6.** $f(x) = 2x^2 - x - 7$,   at -1

7. $f(x) = \sqrt{x}$,   at 4   **8.** $f(x) = \dfrac{1}{x^2}$,   at 2

In Problems 9–20 find the derivative of each function by using (3.9).

9. $f(x) = 2x + 3$ **10.** $f(x) = 3x - 5$ **11.** $f(x) = x^2 - 2$

12. $f(x) = 2x^2 + 4$ **13.** $f(x) = 3x^2 + x + 5$ **14.** $f(x) = 2x^2 - x - 7$

15. $f(x) = 5$ **16.** $f(x) = -2$ **17.** $f(x) = 5\sqrt{x}$

18. $f(x) = \dfrac{4}{x^3}$ **19.** $f(x) = mx + b$ **20.** $f(x) = ax^2 + bx + c$

21. The distance s (in meters) that a particle moves in time t (in seconds) is given by $s = f(t) = 3t^2 + 4t$.   Find the velocity at $t = 0$;   at $t = 2$;   at any time t.

22. The distance s (in meters) that a particle moves in time t (in seconds) is $s = f(t) = t^2 - 4t$.   Find the velocity at $t = 0$;   at $t = 3$;   at any time t.

23.*The motion of a certain body along the x-axis is described by the equation $s = 10t^2$ (s in centimeters). Compute the instantaneous velocity of the body at time $t = 3$ seconds   by letting Δt first equal 0.1 second, then 0.01 second, and finally 0.001 second. What limiting value do the results seem to be approaching?

24. At a certain instant the speedometer of an automobile reads V miles per hour. During the next $\frac{1}{4}$ second the automobile travels 20 feet. Estimate V from this information.

25. A ball is thrown upward. Let the height in feet of the ball be given by $s(t) = 100t - 16t^2$,   where t is the time elapsed in seconds. What is the velocity when $t = 0$, $t = 1$,   and $t = 4$?   At what time does the ball strike the ground? At what time does the ball reach its highest point? (At this time the ball should be "stationary" so that its velocity is 0.)

26. A rock is dropped from a height of 88.2 meters and falls toward earth in a straight line. In t seconds the rock falls $9.8t^2$ meters.
(a) How long does it take for the rock to hit the ground?
(b) What is the average velocity of the rock during the time it is falling?

---

* Adapted from F. W. Sears, M. W. Zemansky, and H. D. Young, *University Physics*, (Reading, Mass.: Addison-Wesley Publishing Co., 1976), p. 63. Reprinted by permission.

(c) What is the average velocity of the rock for the first 3 seconds?

(d) What is the velocity of the rock when it hits the ground?

27. If f is an even function that is differentiable at c, show that $f'(-c) = -f'(c)$.

28. If f is an odd function that is differentiable at c, show that $f'(-c) = f'(c)$.

29. Let f be a function defined for all x. Suppose f has the following properties: (1) $f(u + v) = f(u) \cdot f(v)$, (2) $f(0) = 1$,   (3) $f'(0)$ exists. Show that $f'(x)$ exists for all x. Also show that $f'(x) = f'(0)f(x)$.

30. Explain why $f(x) = |x|$ has no derivative at 0.   [*Hint:*   Use (3.10).]

3. Two Interpretations of the Derivative

GEOMETRIC INTERPRETATION

Let us now examine what the derivative of a function $y = f(x)$ means geometrically. Since the derivative is the limit of an average rate of change, we begin by recalling that the average rate of change $\Delta y/\Delta x$ equals the slope of the secant line joining two points on the graph of f. If we fix the initial point at $(c, f(c))$ and let the terminal point be at $(x, f(x))$ with $\Delta x \neq 0$,   the slope of this secant line is

$$m_{\text{sec}} = \frac{\Delta y}{\Delta x} = \frac{f(x) - f(c)}{x - c}$$

As x approaches c, the secant lines tend to a limiting position—the *tangent line to the graph of f at c*. Thus, in view of the discussion presented in Section 1 of Chapter 2, the slope m_{tan} of the tangent line to f at c is

$$m_{\text{tan}} = \lim_{x \to c} \frac{f(x) - f(c)}{x - c}$$

This limit, if it exists, is the derivative of f at c.

(3.11)   **In view of this, we define the *tangent line* to the graph of f at c to be the line passing through the point $(c, f(c))$ and having the slope**

$$f'(c) = \lim_{x \to c} \frac{f(x) - f(c)}{x - c}$$

provided this limit exists.

By using the point–slope form of the equation of a line, we find that an equation of the tangent line to the graph of f at $(c, f(c))$ is

(3.12) $$y - f(c) = f'(c)(x - c)$$

EXAMPLE 1   Find the slope of the tangent line to the graph of $f(x) = 2x^2$ at $(1, 2)$. What is its equation? Graph the function and show its tangent line.

Solution
$$f'(1) = \lim_{x \to 1} \frac{f(x) - f(1)}{x - 1} = \lim_{x \to 1} \frac{2x^2 - 2}{x - 1} = \lim_{x \to 1} \frac{2(x - 1)(x + 1)}{x - 1} = \lim_{x \to 1} 2(x + 1) = 4$$

By using (3.12), we find an equation of this tangent line to be

$$y - 2 = 4(x - 1)$$
$$y = 4x - 2$$

The graphs of $y = 2x^2$ and its tangent line at $(1, 2)$ are shown in Figure 3.

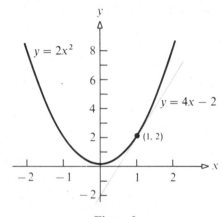

Figure 3   ■

RATES OF CHANGE

Our second interpretation of the derivative involves rates of change. Let x and y denote two physical quantities that are related by the function $y = f(x)$. We have defined the ratio $\Delta y/\Delta x$ of the corresponding changes in these quantities as the *average rate of change of y with respect to x over the interval* Δx. The limit of the ratio $\Delta y/\Delta x$ as Δx approaches 0, if it exists, is called the (*instantaneous*) *rate of change of y with respect to x.* Thus,

(3.13) $$f'(x) = \lim_{\Delta x \to 0} \frac{\Delta y}{\Delta x} = \textbf{(Instantaneous) rate of change of } y \textbf{ with respect to } x$$

In other words, the expression "rate of change of a function" means the derivative of the function.

EXAMPLE 2   Show that the rate of change of the area of a circle with respect to its radius is equal to its circumference.

Solution   The area A of a circle of radius R is $A = \pi R^2$.   Thus, the rate of change of area with respect to radius is

$$A'(R) = \lim_{\Delta R \to 0} \frac{\Delta A}{\Delta R} = \lim_{\Delta R \to 0} \frac{\pi(R + \Delta R)^2 - \pi R^2}{\Delta R}$$

$$= \lim_{\Delta R \to 0} \frac{\pi[R^2 + 2R(\Delta R) + (\Delta R)^2] - \pi R^2}{\Delta R}$$

$$= \lim_{\Delta R \to 0} \frac{2\pi R(\Delta R) + \pi(\Delta R)^2}{\Delta R}$$

$$= \lim_{\Delta R \to 0} \frac{\pi \Delta R(2R + \Delta R)}{\Delta R}$$

$$= \lim_{\Delta R \to 0} \pi(2R + \Delta R) = 2\pi R = \text{Circumference} \quad \blacksquare$$

EXAMPLE 3   In a metabolic experiment, the mass M of glucose decreases according to the formula $M(t) = 4.5 - 0.03t^2$,   where M is measured in grams and t is time (in hours). Find the reaction rate at 1 hour.

Solution   The reaction rate at $t = 1$ is $M'(1)$. Thus,

$$M'(1) = \lim_{t \to 1} \frac{M(t) - M(1)}{t - 1} = \lim_{t \to 1} \frac{(4.5 - 0.03t^2) - (4.5 - 0.03)}{t - 1}$$

$$= \lim_{t \to 1} \frac{(-0.03)(t^2 - 1)}{t - 1} = \lim_{t \to 1} \frac{(-0.03)(t + 1)(t - 1)}{t - 1}$$

$$= (-0.03)(2) = -0.06$$

The reaction rate at $t = 1$ is -0.06; that is, the mass M at $t = 1$ is decreasing at the rate of 0.06 gram per hour. $\blacksquare$

EXERCISE 3

In Problems 1–6 find an equation for the tangent line to the graph of each function at the indicated point. Graph each function and show this tangent line.

1. $f(x) = x^2$,   at $(3, 9)$ **2.** $f(x) = x^2$,   at $(-1, 1)$ **3.** $f(x) = x^2 + 2x + 1$,   at $(1, 4)$

4. $f(x) = x^3 + 1$,   at $(1, 2)$ **5.** $f(x) = \dfrac{1}{x}$,   at $(1, 1)$ **6.** $f(x) = \sqrt{x}$,   at $(4, 2)$

7. Does the tangent line to the graph of $y = x^2$ at $(1, 1)$ pass through the point $(2, 5)$?

8. Does the tangent line to the graph of $y = x^3$ at $(1, 1)$ pass through the point $(2, 5)$?

9. A dive bomber is flying from right to left along the graph of $y = x^2$. When a rocket bomb is released, it follows a path that is approximately along the tangent line. Where should the pilot release the bomb if the target is at $(1, 0)$?

10. Answer the question in Problem 9 if the plane is flying from right to left along the graph of $y = x^3$.

Normal Lines: The *normal line* to the graph of a function f at a point $(c, f(c))$ is defined as the line through $(c, f(c))$ and perpendicular to the tangent line to the graph of f at $(c, f(c))$. See Figure 4. Thus, if f is a function whose derivative at c is $f'(c) \neq 0$, the slope of the normal line to the graph of f at $(c, f(c))$ is the negative reciprocal of $f'(c)$, namely, $-1/f'(c)$, and an equation of the normal line to the graph of f at $(c, f(c))$ is

$$y - f(c) = \frac{-1}{f'(c)}(x - c)$$

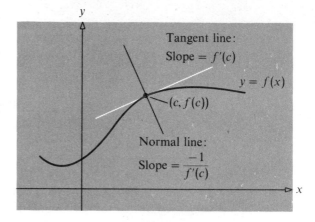

Figure 4

In Problems 11–16 find the slope of the normal line to the graph of each function at the indicated point. Graph each function and show this normal line.

11. $f(x) = x^2 + 1$, at $(1, 2)$ **12.** $f(x) = x^2 - 1$, at $(-1, 0)$ **13.** $f(x) = x^2 - 2x$, at $(-1, 3)$

14. $f(x) = 2x^2 + x$, at $(1, 3)$ **15.** $f(x) = \dfrac{1}{x}$, at $(1, 1)$ **16.** $f(x) = \sqrt{x}$, at $(4, 2)$

17. Atmospheric pressure decreases as the distance from the surface of the earth increases, and the rate of change of pressure with respect to height is proportional to the pressure. Express this law as an equation involving a derivative.

18. Under certain conditions, an electric current will die out at a rate (with respect to time) that is proportional to the current remaining. Express this law as an equation involving a derivative.

19. A circle of radius R has area $A = \pi R^2$ and circumference $C = 2\pi R$. If the radius changes from R to $R + \Delta R$, find the:
  (a) Change in area
  (b) Change in circumference
  (c) Average rate of change of area with respect to radius
  (d) Average rate of change of circumference with respect to radius
  (e) Rate of change of circumference with respect to radius

20. The volume V of a sphere of radius R is $V = 4\pi R^3/3$. If the radius changes from R to $R + \Delta R$, find the:
  (a) Change in volume
  (b) Average rate of change of volume with respect to radius
  (c) Rate of change of volume with respect to radius

21. A human being's respiration rate R (in breaths per minute) is given by $R = -10.35 + 0.59p$, where p is the partial pressure of carbon dioxide in the lungs. Find the rate of change in respiration when $p = 50$.

22. A metal cube with an edge length x is expanding uniformly as a consequence of being heated. Find the:
  (a) Average rate of change of volume of the cube with respect to an edge length as x increases from 2.00 to 2.01 centimeters
  (b) Instantaneous rate of change of volume of the cube with respect to an edge length at the instant when $x = 2$ centimeters

23. The equation of the tangent line to the graph of a function f at $(2, 6)$ is $y = -3x + 12$. What is $f'(2)$?

24. A simple model for population growth states that the rate of change of the population with respect to time is proportional to the population. Express this statement as an equation involving a derivative.

4. Formulas for Finding Derivatives

The derivative of a function f is the limit of a difference quotient, namely,

(3.14)
$$f'(x) = \lim_{\Delta x \to 0} \frac{\Delta y}{\Delta x} = \lim_{\Delta x \to 0} \frac{f(x + \Delta x) - f(x)}{\Delta x}$$

provided this limit exists. But calculating the derivative of a function by this definition can become a tedious chore, particularly if the function f is complicated. In this section and the next, we develop some formulas that make this calculation a relatively straightforward procedure for any algebraic function.

In the development of many of the formulas for derivatives, the required computations are easier to perform if Δx is replaced by h. Hence, we rewrite formula (3.14) as

(3.15)
$$f'(x) = \lim_{h \to 0} \frac{f(x + h) - f(x)}{h}$$

In terms of this notation the four-step method (3.8) developed earlier for finding the derivative of a function f takes the following form:

(3.16) *Step 1.* **Find** $f(x + h)$.

Step 2. **Subtract** $f(x)$ **from** $f(x + h)$ **to get** $f(x + h) - f(x)$.

Step 3. **Divide this result by** $h \neq 0$ **to get** $\dfrac{f(x + h) - f(x)}{h}$ **and simplify, if possible.**

Step 4. **Find the limit (if it exists) of** $\dfrac{f(x + h) - f(x)}{h}$ **as h approaches 0.**

CONSTANT FUNCTION

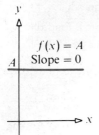

Figure 5

We begin with the constant function $f(x) = A$. A little geometry will tell us what to expect. Since the graph of the constant function f is a horizontal line (see Fig. 5), the tangent line to f at any point is also a horizontal line (slope equals 0). Since the derivative equals the slope of the tangent line to the graph of a function f at a point, then the derivative of f should be 0.

(3.17) **For the constant function $f(x) = A$, the derivative is $f'(x) = 0$.**

This result is sometimes restated to read "the derivative of a constant is 0." To prove this, we follow the four-step method in (3.16), using $f(x) = A$.

Step 1. $\qquad\qquad f(x + h) = A$

Step 2. $\qquad f(x + h) - f(x) = A - A = 0$

Step 3. $\qquad \dfrac{f(x + h) - f(x)}{h} = \dfrac{0}{h} = 0$

Step 4. $\qquad f'(x) = \lim_{h \to 0} \dfrac{f(x + h) - f(x)}{h} = \lim_{h \to 0} 0 = 0$

OTHER FORMS OF NOTATION

Besides the *prime notation f'*, there are several other ways to denote the derivative of a function $y = f(x)$. The most common ones are

$$y' \qquad \frac{dy}{dx} \qquad Df(x)$$

The notation dy/dx, often referred to as the *Leibniz notation*, may be interpreted as

$$\frac{dy}{dx} = \frac{d}{dx}(y) = \frac{d}{dx} f(x)$$

where d/dx is an instruction to compute the derivative (with respect to the independent variable x) of the function $y = f(x)$.

In the notation $Df(x)$, called the *operator notation*, D is said to *operate* on the function, and the result is the derivative of f. Sometimes, to emphasize that the operation is being performed with respect to the independent variable x, we write $Df(x) = D_x f(x)$.

In terms of these new forms of notation, the result (3.17) may also be stated as

(3.18) $$\frac{d}{dx} A = 0 \qquad \text{or} \qquad DA = 0$$

where A is a constant. For example,

$$\frac{d}{dx} 6 = 0$$

In subsequent work with derivatives we shall use the prime notation or the Leibniz notation, or sometimes a mixture of the two, depending on which is most convenient. We shall not use the operator notation in this book.

A BASIC FORMULA

The formula for the derivative of x^n, $n > 0$ an integer:

(3.19)   **For the function $f(x) = x^n$, n a positive integer, the derivative is $f'(x) = nx^{n-1}$. That is,**

$$\frac{d}{dx} x^n = nx^{n-1}$$

To prove this, we follow the four-step method in (3.16), using $f(x) = x^n$ for $n = 1,$ $n = 2,$ $n = 3,$ and $n > 3$.

Step 1. $n = 1$ $f(x) = x$ $f(x + h) = x + h$

  $n = 2$ $f(x) = x^2$ $f(x + h) = (x + h)^2 = x^2 + 2xh + h^2$

  $n = 3$ $f(x) = x^3$ $f(x + h) = (x + h)^3 = x^3 + 3x^2h + 3xh^2 + h^3$

  $n > 3$ $f(x) = x^n$ $f(x + h) = (x + h)^n$

By applying the binomial theorem (see Appendix I), we find the expansion of $(x + h)^n$ to be

$$(x + h)^n = x^n + nx^{n-1}h + \frac{n(n-1)}{2} x^{n-2}h^2 + \cdots + nxh^{n-1} + h^n$$

Step 2. $n = 1$ $f(x + h) - f(x) = (x + h) - x = h$

  $n = 2$ $f(x + h) - f(x) = (x^2 + 2xh + h^2) - x^2 = 2xh + h^2$

  $n = 3$ $f(x + h) - f(x) = (x^3 + 3x^2h + 3xh^2 + h^3) - x^3$

  $= 3x^2h + 3xh^2 + h^3$

  $n > 3$ $f(x + h) - f(x) = (x + h)^n - x^n$

$$= \left[x^n + nx^{n-1}h + \frac{n(n-1)}{2} x^{n-2}h^2 + \cdots \right.$$

$$\left. + nxh^{n-1} + h^n \right] - x^n$$

$$= nx^{n-1}h + (\text{Assorted terms})(h^2)$$

$\uparrow$

After subtracting the x^n terms,
we factor h^2 out of all terms, except the first

Step 3. $n = 1$ $\dfrac{f(x + h) - f(x)}{h} = \dfrac{h}{h} = 1$

$n = 2$ $\dfrac{f(x + h) - f(x)}{h} = \dfrac{2xh + h^2}{h} = \dfrac{h(2x + h)}{h} = 2x + h$

$n = 3$ $\dfrac{f(x + h) - f(x)}{h} = \dfrac{3x^2h + 3xh^2 + h^3}{h} = \dfrac{h(3x^2 + 3xh + h^2)}{h}$

$$= 3x^2 + 3xh + h^2$$

$n > 3$ $\dfrac{f(x + h) - f(x)}{h} = \dfrac{nx^{n-1}h + (\text{Assorted terms})h^2}{h}$

$$= nx^{n-1} + (\text{Assorted terms})(h)$$

Step 4. $n = 1$ $f'(x) = \lim\limits_{h \to 0} \dfrac{f(x + h) - f(x)}{h} = \lim\limits_{h \to 0} 1 = 1$

$n = 2$ $f'(x) = \lim\limits_{h \to 0} \dfrac{f(x + h) - f(x)}{h} = \lim\limits_{h \to 0}(2x + h) = 2x$

$n = 3$ $f'(x) = \lim\limits_{h \to 0} \dfrac{f(x + h) - f(x)}{h} = \lim\limits_{h \to 0}(3x^2 + 3xh + h^2) = 3x^2$

$n > 3$ $f'(x) = \lim\limits_{h \to 0} \dfrac{f(x + h) - f(x)}{h} = \lim\limits_{h \to 0}[nx^{n-1} + (\text{Assorted terms})(h)]$

$$= \lim\limits_{h \to 0} nx^{n-1} + \lim\limits_{h \to 0}[(\text{Assorted terms})(h)] = nx^{n-1}$$

This last formula for $n > 3$ agrees with the preceding three formulas, thus establishing the general formula (3.19) for n a positive integer; that is,

$$\frac{d}{dx} x^n = nx^{n-1}$$

This result tells us that the derivative of x raised to the power n is n times x raised to the power $n - 1$.

EXAMPLE 1   (a) $\dfrac{d}{dx} x^5 = 5x^4$ (b) $\dfrac{d}{dx} x^{13} = 13x^{12}$ (c) $\dfrac{d}{dx} x^{65} = 65x^{64}$ (d) $\dfrac{d}{dx} x = 1$

■

SOME GENERAL DIFFERENTIATION FORMULAS

The following theorem, which has a simple proof, is used often:

(3.20)   **THEOREM**   *Derivative of a Constant Times a Function.*   **If f is a differentiable function and if $F(x) = kf(x)$,   where k is a constant, then F is differentiable and**

$$F'(x) = kf'(x)$$

Proof We follow the four steps in (3.16) using the function $y = F(x) = kf(x)$.

Step 1. $\qquad\qquad F(x + h) = kf(x + h)$

Step 2. $\qquad F(x + h) - F(x) = kf(x + h) - kf(x) = k[f(x + h) - f(x)]$

Step 3. $\qquad \dfrac{F(x + h) - F(x)}{h} = k\dfrac{[f(x + h) - f(x)]}{h}$

Step 4. Since f is differentiable, the limit of $\dfrac{f(x + h) - f(x)}{h}$ as $h \to 0$ equals the derivative $f'(x)$. Taking the limit as $h \to 0$ and using the fact (2.11) that the limit of a constant times a function equals the constant times the limit of a function, we have

$$\lim_{h \to 0} \frac{F(x + h) - F(x)}{h} = \lim_{h \to 0}\left[k\frac{f(x + h) - f(x)}{h}\right]$$

$$= k\left[\lim_{h \to 0}\frac{f(x + h) - f(x)}{h}\right] = kf'(x) \qquad \blacksquare$$

Theorem (3.20) may be restated as follows:

(3.21) **The derivative of a constant times a differentiable function f equals the constant times the derivative of f. That is, if k is a constant, then**

(3.22)
$$\frac{d}{dx}[kf(x)] = k\left[\frac{d}{dx}f(x)\right]$$

EXAMPLE 2 (a) $\dfrac{d}{dx}(3x^2) = 3\dfrac{d}{dx}x^2 = (3)(2x) = 6x$ (b) $\dfrac{d}{dx}(-4x^3) = (-4)(3x^2) = -12x^2$

(c) $\dfrac{d}{dx}\left(\dfrac{1}{3}\right)x^3 = \left(\dfrac{1}{3}\right)(3x^2) = x^2$ (d) $\dfrac{d}{dx}\left(\dfrac{5}{2}x^8\right) = \left(\dfrac{5}{2}\right)(8x^7) = 20x^7$ $\qquad \blacksquare$

The next theorem provides a formula for finding the derivative of a function expressed as the sum of two functions whose derivatives are known.

(3.23) **THEOREM** *Derivative of a Sum.* **If f and g are differentiable functions and if $F(x) = f(x) + g(x)$,   then F is differentiable and**

$$F'(x) = f'(x) + g'(x)$$

Proof We follow the four steps in (3.16) using the function $F(x) = f(x) + g(x)$.

Step 1. $\qquad\qquad F(x + h) = f(x + h) + g(x + h)$

Step 2. $\qquad F(x + h) - F(x) = f(x + h) + g(x + h) - [f(x) + g(x)]$
$$= [f(x + h) - f(x)] + [g(x + h) - g(x)]$$

Step 3. $\dfrac{F(x + h) - F(h)}{h} = \dfrac{[f(x + h) - f(x)] + [g(x + h) - g(x)]}{h}$

$$= \dfrac{f(x + h) - f(x)}{h} + \dfrac{g(x + h) - g(x)}{h}$$

Step 4.   Taking the limit as $h \to 0$ and using the fact (2.9) that the limit of a sum equals the sum of the limits, we have

$$F'(x) = \lim_{h \to 0} \dfrac{F(x + h) - F(x)}{h} = \lim_{h \to 0} \dfrac{f(x + h) - f(x)}{h} + \lim_{h \to 0} \dfrac{g(x + h) - g(x)}{h}$$

The limits on the right are $f'(x)$ and $g'(x)$, respectively. Hence,

$$F'(x) = f'(x) + g'(x) \qquad \blacksquare$$

Theorem (3.23) may be restated as follows:

(3.24)   **The derivative of the sum of two differentiable functions equals the sum of their derivatives. In the Leibniz notation,**

(3.25)
$$\dfrac{d}{dx}[f(x) + g(x)] = \dfrac{d}{dx} f(x) + \dfrac{d}{dx} g(x)$$

EXAMPLE 3   Find the derivative of $f(x) = 3x^2 + 8$.

Solution   Here, f is the sum of $3x^2$ and 8. Hence,

$$f'(x) = \dfrac{d}{dx}(3x^2 + 8) = \dfrac{d}{dx}(3x^2) + \dfrac{d}{dx} 8 = 3\dfrac{d}{dx} x^2 + 0 = (3)(2x) = 6x \qquad \blacksquare$$

$\qquad\qquad\qquad\qquad\uparrow\qquad\qquad\qquad\qquad\uparrow$

$\qquad\qquad\qquad$ By (3.25)$\qquad\quad$ By (3.22) and (3.18)

(3.26)   **If f and g are two differentiable functions and if $F(x) = f(x) - g(x)$, then $F'(x) = f'(x) - g'(x)$. That is,**

(3.27)
$$\dfrac{d}{dx}[f(x) - g(x)] = \dfrac{d}{dx} f(x) - \dfrac{d}{dx} g(x)$$

The proof is left as an exercise (Problem 37 in Exercise 4).

The results given above extend to sums (or differences) of more than two differentiable functions. Thus, if $f_1, f_2, \ldots, f_n$ are n differentiable functions, we have the formula

(3.28)
$$\dfrac{d}{dx}[f_1(x) + f_2(x) + \cdots + f_n(x)] = \dfrac{d}{dx} f_1(x) + \dfrac{d}{dx} f_2(x) + \cdots + \dfrac{d}{dx} f_n(x)$$

Our final example for this section illustrates how these results may be combined.

EXAMPLE 4   Find the derivative of $f(x) = 6x^4 - 3x^2 + 2x - 5$.

Solution

$$f'(x) = \frac{d}{dx}(6x^4 - 3x^2 + 2x - 5)$$

$$= \frac{d}{dx}(6x^4) - \frac{d}{dx}(3x^2) + \frac{d}{dx}(2x) - \frac{d}{dx}5 = 24x^3 - 6x + 2 \quad \blacksquare$$

EXERCISE 4

In Problems 1–16 find the derivative of the function f by using the formulas of this section.

1. $f(x) = 3x^{15}$

2. $f(x) = \frac{1}{3}x^{12}$

3. $f(x) = 3x + 2$

4. $f(x) = 5x - \frac{1}{2}$

5. $f(x) = x^2 + 3x - 4$

6. $f(x) = 4x^4 + 2x^2 - 2$

7. $f(x) = 8x^5 - 5x + 1$

8. $f(x) = 9x^3 - 2x^2 + 4x + 4$

9. $f(x) = \frac{1}{3}x^4 - 3x + \frac{3}{2}$

10. $f(x) = -3x^4 - 2x^3$

11. $f(x) = \pi x^3 + \frac{3}{2}x^2$

12. $f(x) = 4 - \pi x^2$

13. $f(x) = \frac{1}{3}(x^5 - 8)$

14. $f(x) = \dfrac{x^3 + 2}{5}$

15. $f(x) = ax^2 + bx + c$

16. $f(x) = ax^3 + bx^2 + cx + d$

In Problems 17–22 find the indicated derivative.

17. $\dfrac{d}{dx}(\sqrt{3}x + \frac{1}{2})$

18. $\dfrac{d}{dx}\left(\dfrac{2x^4 - 5}{8}\right)$

19. $\dfrac{dA}{dR}$ if $A = \pi R^2$

20. $\dfrac{dC}{dR}$ if $C = 2\pi R$

21. $\dfrac{dV}{dR}$ if $V = \frac{4}{3}\pi R^3$

22. $\dfrac{dP}{dT}$ if $P = 0.2T$

In Problems 23 and 24 find the slope of the tangent line to the graph of the function f at the indicated point. What is an equation of the tangent line?

23. $f(x) = x^3 + 3x - 1$, at $(0, -1)$

24. $f(x) = x^4 + 2x - 1$, at $(1, 2)$

In Problems 25–28 find those x, if any, at which the graph of the function f has a horizontal tangent line—that is, at which $f'(x) = 0$.

25. $f(x) = 3x^2 - 12x + 4$

26. $f(x) = x^2 + 4x - 3$

27. $f(x) = x^3 - 3x + 2$

28. $f(x) = x^4 - 4x^3$

In Problems 29–32 find an equation for the tangent line(s) to the graph of the function f that is (are) parallel to the line L.

29. $f(x) = 3x^2 - x$, L: $y = 5x$

30. $f(x) = 2x^3 + 1$, L: $y = 6x - 1$

31. $f(x) = \frac{1}{3}x^3 - x^2$, L: $y - 3x + 2 = 0$

32. $f(x) = x^3 - x$, L: $x + y = 0$

33. In t seconds, the position of an object is a distance of s meters from the origin, where $s = t^3 - t + 1$. Find the velocity at $t = 0$; at $t = 5$.

34. In t seconds, the position of an object is a distance of s meters from the origin, where $s = t^4 - t^3 + t$. Find the velocity at $t = 0$; at $t = 1$.

35. What is $\lim\limits_{h\to 0} \dfrac{5(\frac{1}{2}+h)^8 - 5(\frac{1}{2})^8}{h}$?

36. What is $\lim\limits_{h\to 0} \dfrac{6(2+h)^5 - 6(2)^5}{h}$?

37. Prove that if f and g are differentiable functions and if $F(x) = f(x) - g(x)$, then $F'(x) = f'(x) - g'(x)$.

38. The velocity v of a liquid flowing through a cylindrical tube is given by the formula $v = k(R^2 - r^2)$, where R is the radius of the tube, k is a constant that depends on the length of the tube and the velocity of the liquid at its ends, and r is the variable distance of the liquid from the center of the tube. Find the rate of change of v with respect to r at the center of the tube. What is the rate halfway from the center to the wall of the tube? What is it at the wall of the tube?

39. Let $f(x) = x^n$, where n is a positive integer. Use (3.10) and a factoring principle to show that $f'(c) = nc^{n-1}$.

5. Formulas for Finding Derivatives (Continued)

In this section we will develop general formulas for differentiating functions that are products and quotients.

(3.29) **THEOREM** *Derivative of a Product.* **If f and g are differentiable functions and if $F(x) = f(x)g(x)$, then F is differentiable and**

$$F'(x) = f(x)g'(x) + f'(x)g(x)$$

Proof We proceed directly to Step 3 of (3.16):

$$\frac{F(x+h) - F(x)}{h} = \frac{f(x+h)g(x+h) - f(x)g(x)}{h}$$

Now, we subtract and add the term $f(x+h)g(x)$ in the numerator and factor (the reason for this will become apparent shortly):

$$\frac{F(x+h) - F(x)}{h} = \frac{f(x+h)g(x+h) - f(x+h)g(x) + f(x+h)g(x) - f(x)g(x)}{h}$$

$$= f(x+h)\frac{[g(x+h) - g(x)]}{h} + \frac{[f(x+h) - f(x)]}{h}g(x)$$

Then we take the limit as h approaches 0 and apply properties of limits:

$$F'(x) = \lim_{h\to 0} \frac{F(x+h) - F(x)}{h}$$

$$= \lim_{h\to 0} \left\{ f(x+h)\frac{[g(x+h) - g(x)]}{h} \right\} + \lim_{h\to 0} \left\{ \frac{[f(x+h) - f(x)]}{h}g(x) \right\}$$

$$= \left\{ \overset{①}{\lim_{h\to 0}} f(x+h) \right\} \left\{ \overset{②}{\lim_{h\to 0}} \frac{g(x+h) - g(x)}{h} \right\} + \left\{ \overset{③}{\lim_{h\to 0}} \frac{f(x+h) - f(x)}{h} \right\} \left\{ \overset{④}{\lim_{h\to 0}} g(x) \right\}$$

We have numbered the above limits for convenience. The limit ② is $g'(x)$, the derivative of g; the limit ③ is $f'(x)$, the derivative of f; limit ④ is the limit of $g(x)$, which equals $g(x)$ since x is fixed while $h \to 0$. The first limit requires a more careful look since, as we saw in Chapter 2, it may not be possible to merely replace h by zero.

$$\overset{①}{\underset{h \to 0}{\lim}} f(x + h) = \underset{h \to 0}{\lim}[f(x + h) - f(x) + f(x)] = \underset{h \to 0}{\lim}[f(x + h) - f(x)] + \underset{h \to 0}{\lim} f(x)$$

$$= \underset{h \to 0}{\lim}\left[h \frac{f(x + h) - f(x)}{h} \right] + f(x)$$

$$= \left[\underset{h \to 0}{\lim} h \right]\left[\underset{h \to 0}{\lim} \frac{f(x + h) - f(x)}{h} \right] + f(x) = (0)[f'(x)] + f(x) = f(x)$$

Hence, $\lim_{h \to 0} f(x + h) = f(x)$, and the four limits reduce to

$$F'(x) = f(x)g'(x) + f'(x)g(x) \qquad \blacksquare$$

In other words, the derivative of the product of two differentiable functions equals the first function times the derivative of the second plus the derivative of the first function times the second function. In the Leibniz notation

(3.30)
$$\frac{d}{dx}[f(x)g(x)] = f(x)\left[\frac{d}{dx} g(x) \right] + \left[\frac{d}{dx} f(x) \right]g(x)$$

Observe that, unlike the situation with limits, the derivative of a product does not equal the product of the derivatives.

EXAMPLE 1   Find the derivative of $F(x) = (x^2 + 2x - 5)(x^3 - 1)$.

Solution   The function F is the product of the two polynomial functions $f(x) = x^2 + 2x - 5$ and $g(x) = x^3 - 1$ so that, by (3.30), we have

$$F'(x) = (x^2 + 2x - 5)\left[\frac{d}{dx}(x^3 - 1) \right] + \left[\frac{d}{dx}(x^2 + 2x - 5) \right](x^3 - 1)$$

$$= (x^2 + 2x - 5)(3x^2) + (2x + 2)(x^3 - 1)$$

$$= 5x^4 + 8x^3 - 15x^2 - 2x - 2 \qquad \blacksquare$$

Now that you know the rule for the derivative of a product, be careful not to use it unnecessarily. When one of the factors is a constant, you should use (3.22). For example, it is easier to work

$$\frac{d}{dx}[5(x^2 + 1)] = 5\frac{d}{dx}(x^2 + 1) = (5)(2x) = 10x$$

than it is to work

$$\frac{d}{dx}[5(x^2+1)] = 5\left[\frac{d}{dx}(x^2+1)\right] + \left[\frac{d}{dx}5\right](x^2+1) = (5)(2x) + (0)(x^2+1) = 10x$$

The next theorem provides a formula for finding the derivative of the quotient of two functions.

(3.31)   **THEOREM** *Derivative of a Quotient.* **If f and $g \neq 0$ are differentiable functions and if $F(x) = f(x)/g(x)$, then F is differentiable and**

$$F'(x) = \frac{g(x)f'(x) - f(x)g'(x)}{[g(x)]^2}$$

Proof   We proceed directly to Step 3 of (3.16):

$$\frac{F(x+h) - F(x)}{h} = \frac{\dfrac{f(x+h)}{g(x+h)} - \dfrac{f(x)}{g(x)}}{h}$$

$$= \frac{g(x)f(x+h) - f(x)g(x+h)}{g(x)g(x+h)h}$$

We subtract and add $g(x)f(x)$ in the numerator and factor (the reason for doing this will become apparent shortly):

$$\frac{g(x)f(x+h) - g(x)f(x) + g(x)f(x) - f(x)g(x+h)}{g(x)g(x+h)h}$$

$$= \frac{g(x)[f(x+h) - f(x)] - f(x)[g(x+h) - g(x)]}{g(x)g(x+h)h}$$

$$= \left[\frac{g(x)}{g(x)g(x+h)}\right]\left[\frac{f(x+h) - f(x)}{h}\right] - \left[\frac{f(x)}{g(x)g(x+h)}\right]\left[\frac{g(x+h) - g(x)}{h}\right]$$

Taking the limit as h approaches 0 and applying properties of limits, gives

$$F'(x) = \frac{g(x)f'(x)}{[g(x)]^2} - \frac{f(x)g'(x)}{[g(x)]^2} = \frac{g(x)f'(x) - f(x)g'(x)}{[g(x)]^2}$$

Justification of this statement is left to you. [Refer to the proof of (3.29) for a hint.]   ∎

In the Leibniz notation, theorem (3.31) is stated as

(3.32)

$$\frac{d}{dx}\left[\frac{f(x)}{g(x)}\right] = \frac{g(x)\left[\dfrac{d}{dx}f(x)\right] - f(x)\left[\dfrac{d}{dx}g(x)\right]}{[g(x)]^2}$$

EXAMPLE 2   Find the derivative of $F(x) = \dfrac{x^2 + 1}{x - 3}$.

Solution   Here, F is the quotient of $f(x) = x^2 + 1$ over $g(x) = x - 3$. Thus,

$$F'(x) = \frac{d}{dx}\left(\frac{x^2 + 1}{x - 3}\right) = \frac{(x - 3)\left[\dfrac{d}{dx}(x^2 + 1)\right] - (x^2 + 1)\left[\dfrac{d}{dx}(x - 3)\right]}{(x - 3)^2}$$

$$= \frac{(x - 3)(2x) - (x^2 + 1)(1)}{(x - 3)^2}$$

$$= \frac{2x^2 - 6x - x^2 - 1}{(x - 3)^2} = \frac{x^2 - 6x - 1}{(x - 3)^2} \quad \blacksquare$$

A corollary of the theorem on the derivative of a quotient is that if $g \neq 0$ is a differentiable function, then

(3.33)
$$\frac{d}{dx}\left[\frac{1}{g(x)}\right] = \frac{-g'(x)}{[g(x)]^2}$$

For a proof, merely set $f(x) = 1$ in (3.32).

EXAMPLE 3   (a) $\dfrac{d}{dx}\left(\dfrac{1}{3x + 5}\right) = \dfrac{-3}{(3x + 5)^2}$ (b) $\dfrac{d}{dx}\left(\dfrac{1}{x^2 + 1}\right) = \dfrac{-2x}{(x^2 + 1)^2}$ $\quad \blacksquare$

We may use (3.33) to show that the formula $(d/dx)(x^n) = nx^{n-1}$ holds when n is a negative integer. Since in this event $-n$ is a positive integer, we see that

$$\frac{d}{dx}(x^n) \underset{\text{Definition}}{=} \frac{d}{dx}\left(\frac{1}{x^{-n}}\right) \underset{\text{By (3.33)}}{=} \frac{-\dfrac{d}{dx}(x^{-n})}{(x^{-n})^2} \underset{\text{By (3.19)}}{=} \frac{-(-n)x^{-n-1}}{x^{-2n}} \underset{\text{Simplify}}{=} nx^{n-1}$$

Now, we have the more general formula

$$\frac{d}{dx}(x^n) = nx^{n-1} \qquad \text{for } n \text{ any integer}$$

EXAMPLE 4   (a) $\dfrac{d}{dx}(x^{-1}) = -x^{-2} = -\dfrac{1}{x^2}$ (b) $\dfrac{d}{dx}\left(\dfrac{1}{x^2}\right) = \dfrac{d}{dx}(x^{-2}) = -2x^{-3} = -\dfrac{2}{x^3}$

(c) $\dfrac{d}{dx}\left(\dfrac{1}{x^3}\right) = \dfrac{d}{dx}(x^{-3}) = -3x^{-4} = -\dfrac{3}{x^4}$

(d) $\dfrac{d}{dx}\left(\dfrac{4}{x^5}\right) = \dfrac{d}{dx}(4x^{-5}) = 4(-5)x^{-6} = -\dfrac{20}{x^6}$ $\quad \blacksquare$

One last remark. A change in the symbol used for the independent variable does not affect the formula. For example,

$$\frac{d}{dt}(t^{-2}) = -2t^{-3} = -\frac{2}{t^3} \qquad \frac{d}{ds}\left(\frac{3}{s^4}\right) = 3(-4)s^{-5} = -\frac{12}{s^5}$$

$$\frac{d}{du}(6u^3 - 5u^{-2}) = 18u^2 + 10u^{-3} = 18u^2 + \frac{10}{u^3}$$

As a matter of fact, each of the derivative formulas given so far can be written without reference to the independent variable of the function by using the prime notation. That is, for differentiable functions f and g, we have

$$(f + g)' = f' + g' \qquad (fg)' = fg' + f'g \qquad \left(\frac{f}{g}\right)' = \frac{gf' - fg'}{g^2}$$

EXERCISE 5

In Problems 1–24, find the derivative of the function f by using the formulas of this section.

1. $f(x) = (x^2 + 1)(x^3 - 1)$

2. $f(x) = (x^4 - 2)(x + 5)$

3. $f(x) = (3x^2 - 5)(2x + 1)$

4. $f(x) = (3x - 2)(4x + 5)$

5. $f(t) = (2t^5 - t)(t^3 + 1)$

6. $f(u) = (u^4 - 3u^2 + 1)(u^2 - u + 2)$

7. $f(t) = t^{-3}$

8. $f(u) = u^{-4}$

9. $f(x) = \frac{10}{x^4} + \frac{3}{x^2}$

10. $f(x) = \frac{2}{x^5} - \frac{3}{x^3}$

11. $f(s) = \frac{2s}{s + 1}$

12. $f(z) = \frac{z + 1}{2z}$

13. $f(x) = \frac{4x^2 - 2}{3x + 4}$

14. $f(x) = \frac{-3x^3 - 1}{2x^2 + 1}$

15. $f(t) = 3t + \frac{1}{3t}$

16. $f(u) = 4u - \frac{1}{4u}$

17. $f(u) = \frac{1 - 2u}{1 + 2u}$

18. $f(w) = \frac{1 - w^2}{1 + w^2}$

19. $f(x) = 3x^3 - \frac{1}{3x^2}$

20. $f(x) = x^5 - \frac{5}{x^5}$

21. $f(t) = \frac{1}{t} - \frac{1}{t^2} + \frac{1}{t^3}$

22. $f(v) = \left(\frac{1 - v}{v}\right)(1 - v^2)$

23. $f(w) = \frac{1}{w^3 - 1}$

24. $f(v) = \frac{1}{v^2 + 5}$

In Problems 25 and 26 find the slope of the tangent line to the function f at the point indicated. What is an equation of the tangent line?

25. $f(x) = \frac{x^3}{x + 1}$,   at $(1, \frac{1}{2})$

26. $f(x) = \frac{x^2}{x - 1}$,   at $(-1, -\frac{1}{2})$

In Problems 27 and 28 find those x, if any, at which the function f has a horizontal tangent line—that is, at which $f'(x) = 0$.

27. $f(x) = \frac{x^2}{x + 1}$

28. $f(x) = \frac{x^2 + 1}{x}$

29. If $y = x^2(3x - 2)$,   find y' by:
  (a) Using the derivative of a product formula.
  (b) Multiplying the two factors first and then differentiating.
  (c) Compare the answers from parts (a) and (b).

30. If $y = (x^2 + 2)(x - 1)$,   find y' by:
  (a) Using the derivative of a product formula.
  (b) Multiplying the two factors first and then differentiating.
  (c) Compare the answers from parts (a) and (b).

31. The intensity of illumination I on a surface is inversely proportional to the square of the distance r from the surface to the source of light. If the intensity is 1000 units when the distance is 1 meter, find the rate of change of the intensity with respect to the distance when the distance is 10 meters.

32. Prove that if $g \neq 0$ is a differentiable function, then

$$\frac{d}{dx}\left[\frac{1}{g(x)}\right] = \frac{-g'(x)}{[g(x)]^2}$$

33. Prove that if f, g, and h are differentiable functions, then

$$\frac{d}{dx}\left[f(x)g(x)h(x)\right] = f(x)g(x)h'(x) + f(x)g'(x)h(x) + f'(x)g(x)h(x)$$

From this, deduce that

$$\frac{d}{dx}\left[f(x)\right]^3 = 3[f(x)]^2 f'(x)$$

In Problems 34–39 use the result of Problem 33 to find dy/dx.

34. $y = (x^2 + 1)(x - 1)(x + 5)$

35. $y = (x - 1)(x^2 + 5)(x^3 - 1)$

36. $y = (x^4 + 1)^3$

37. $y = (x^3 + 1)^3$

38. $y = (3x + 1)\left(1 + \frac{1}{x}\right)(x^{-5} + 1)$

39. $y = \left(1 - \frac{1}{x}\right)\left(1 - \frac{1}{x^2}\right)\left(1 - \frac{1}{x^3}\right)$

6. Higher-Order Derivatives

Earlier, we concluded that the derivative of a differentiable function $y = f(x)$ is also a function, called the *derivative function f'*.

The derivative (if there is one) of the function f' is called the *second derivative* of f and is denoted by f''. For example, if $f(x) = 6x^3 - 3x^2 + 2x - 5$,   then

$$f'(x) = 18x^2 - 6x + 2$$

$$f''(x) = \frac{d}{dx}f'(x) = \frac{d}{dx}(18x^2 - 6x + 2) = 36x - 6$$

By continuing in this fashion, we can find the *third derivative f'''*, the *fourth derivative $f^{(4)}$*, and so on, provided that these derivatives exist. These are collectively called *higher-order derivatives*. For example, the first, second, third, and fourth derivatives of

$$f(x) = x^4 + 3x^3 - 2x^2 + 5x - 6$$

are

$$f'(x) = 4x^3 + 9x^2 - 4x + 5$$

$$f''(x) = \frac{d}{dx} f'(x) = 12x^2 + 18x - 4$$

$$f'''(x) = \frac{d}{dx} f''(x) = 24x + 18$$

$$f^{(4)}(x) = 24$$

All derivatives of this function of order 5 or more equal 0.

The result obtained in this example can be generalized: For a polynomial function f of degree n, we have

$$f(x) = a_n x^n + a_{n-1} x^{n-1} + \cdots + a_1 x + a_0$$
$$f'(x) = na_n x^{n-1} + (n-1)a_{n-1} x^{n-2} + \cdots + a_1$$

Thus, the first derivative of a polynomial function of degree n is a polynomial function of degree $(n-1)$. By continuing the differentiation process, it follows that the nth-order derivative of f is

$$f^{(n)}(x) = n(n-1)(n-2) \cdots (3)(2)(1)a_n$$

a polynomial of degree 0—a constant function. Therefore, all derivatives of order greater than n will equal 0.

In some applications it is important to find both the first and second derivatives of a function and to solve for those numbers x that make these derivatives equal 0.

EXAMPLE 1   For $f(x) = 4x^3 - 12x^2 + 2$, find those x, if any, at which $f'(x) = 0$. For what numbers x will $f''(x) = 0$?

Solution $$f'(x) = 12x^2 - 24x = 12x(x-2) = 0 \quad \text{when} \quad x = 0 \quad \text{or} \quad x = 2$$
$$f''(x) = 24x - 24 = 24(x-1) = 0 \quad \text{when} \quad x = 1 \quad \blacksquare$$

OTHER FORMS OF NOTATION

The symbols f', f'', and so on, for higher-order derivatives of $y = f(x)$ have several parallel notations:

$$y' = f'(x) = \frac{dy}{dx} = \frac{d}{dx} f(x) = Dy$$

$$y'' = f''(x) = \frac{d^2y}{dx^2} = \frac{d^2}{dx^2} f(x) = D^2y$$

$$y''' = f'''(x) = \frac{d^3y}{dx^3} = \frac{d^3}{dx^3} f(x) = D^3y$$

$$\vdots$$

$$y^{(n)} = f^{(n)}(x) = \frac{d^ny}{dx^n} = \frac{d^n}{dx^n} f(x) = D^ny$$

ACCELERATION IN RECTILINEAR MOTION

Suppose the position of an object at time t is a distance s from the origin, where s is given as a function of t, say, as $s = f(t)$. Then from (3.6), the first derivative ds/dt is the velocity v of the particle.

(3.34) **The *acceleration a* of this particle is defined as the rate of change of velocity with respect to time. That is,**

$$a = \frac{dv}{dt} = \frac{d}{dt}v = \frac{d}{dt}\left(\frac{ds}{dt}\right) = \frac{d^2s}{dt^2}$$

In other words, acceleration is the second derivative of the function $s = f(t)$ with respect to time.

EXAMPLE 2   A ball is thrown vertically upward with an initial velocity of 19.6 meters per second. The distance s (in meters) of the ball above the ground is $s = -4.9t^2 + 19.6t$, where t is the number of seconds elapsed from the moment that the ball is thrown.
(a)   What is the velocity of the ball at the end of 1 second?
(b)   When will the ball reach its highest point?
(c)   What is the maximum height the ball reaches?
(d)   What is the acceleration of the ball at any time t?
(e)   How long is the ball in the air?
(f)   What is the velocity of the ball upon impact?
(g)   What is the total distance traveled by the ball?

Solution   (a) $v = \dfrac{ds}{dt} = -9.8t + 19.6$

At $t = 1$, $v = 9.8$ meters per second.
(b)   The ball will reach its highest point when $v = 0$.

$$v = -9.8t + 19.6 = 0 \qquad \text{when} \qquad t = 2 \text{ seconds}$$

(c)   At $t = 2$, $s = -4.9(4) + 19.6(2) = 19.6$ meters.

(d) $a = \dfrac{d^2s}{dt^2} = -9.8$ meters per second per second

(e)   We can answer this question in two ways. First, since it takes 2 seconds for the ball to reach its maximum height, it follows that it will take another 2 seconds to reach the ground, for a total time of 4 seconds in the air. The second way is to set $s = 0$ and solve for t:

$$-4.9t^2 + 19.6t = 0$$

$$t = 0 \qquad \text{or} \qquad t = \frac{19.6}{4.9} = 4$$

The ball is at ground level when $t = 0$ and when $t = 4$.

(f)   Upon impact, $t = 4$. Hence, when $t = 4$,

$$v = (-9.8)(4) + 19.6 = -19.6 \text{ meters per second}$$

The minus sign here indicates that the direction of the velocity is downward.

(g)   The total distance traveled is

$$\text{Distance up} + \text{Distance down} = 19.6 + 19.6 = 39.2 \text{ meters}$$

See Figure 6 for an illustration.

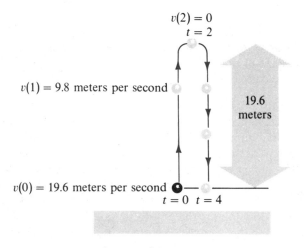

Figure 6   ∎

For Example 2, the acceleration of the ball is constant. This is approximately true for all falling bodies provided air resistance is ignored. In fact, the constant is even the same for all falling bodies, as Galileo (1564–1642) discovered in the sixteenth century. We can use calculus to see this. Galileo found by experimentation that all falling bodies obey the law that the distance they fall when dropped is proportional to the square of the time t it takes to fall that distance. Of great importance is the fact that the constant of proportionality c is the same for all bodies. Thus, Galileo's law states that the distance s a body falls in time t is given by

$$s = -ct^2$$

The reason for the minus sign is that the body is falling and we have chosen our coordinate system so that the positive s direction is up, along the y-axis.

The velocity v of this freely falling body is

$$v = \frac{ds}{dt} = -2ct$$

and its acceleration a is

$$a = \frac{dv}{dt} = \frac{d^2s}{dt^2} = -2c$$

Thus, the acceleration of a freely falling body is a constant. Usually, we denote this constant by $-g$ so that

$$a = -g$$

The number g is called the *acceleration of gravity*. For our planet, g may be approximated by 32 feet per second per second or 980 centimeters per second per second.* On the planet Jupiter, $g \approx 2600$ centimeters per second per second, and on our moon, $g \approx 160$ centimeters per second per second.

EXERCISE 6

In Problems 1–12 find f' and f'' for each function.

1. $f(x) = 2x + 5$ **2.** $f(x) = 3x + 2$ **3.** $f(x) = 3x^2 + x - 2$ **4.** $f(x) = -5x^2 - 3x$

5. $f(x) = x + \dfrac{1}{x}$ **6.** $f(x) = x - \dfrac{1}{x}$ **7.** $f(t) = \dfrac{t}{t+1}$ **8.** $f(u) = \dfrac{u+1}{u}$

9. $f(x) = \dfrac{x^2}{x+1}$ **10.** $f(x) = \dfrac{x^3}{x-1}$ **11.** $f(x) = \dfrac{4}{3x+5}$ **12.** $f(x) = \dfrac{10}{-2x+1}$

13. Find y' and y'' for the following:

(a) $y = 5x^4 - 2x^3 + x$ (b) $y = \dfrac{1}{x}$ (c) $y = (2x+1)(x^3+5)$ (d) $y = \dfrac{2x-5}{x}$

14. Find dy/dx and d^2y/dx^2 for the following:

(a) $y = -5x^3 + x^2 - 6$ (b) $y = \dfrac{5}{x^2}$ (c) $y = (3x-5)(x^2-2)$ (d) $y = \dfrac{2-3x}{x}$

15. Find y''' for the following:
(a) $y = 4x^3 - 3x^2 + x$ (b) $y = ax^3 + bx^2 + cx + d$

16. Find d^3y/dx^3 for the following:
(a) $y = -10x^4 + x^3 - 4$ (b) $y = ax^4 + bx^3 + cx^2 + dx + e$

In Problems 17–22 find the indicated derivative.

17. $f^{(4)}(x)$ if $f(x) = x^3 - 3x^2 + 2x - 5$ **18.** $f^{(5)}(x)$ if $f(x) = 4x^3 + x^2 - 1$

19. $\dfrac{d^{20}}{dx^{20}}(8x^{19} - 2x^{14} + 2x^5)$ **20.** $\dfrac{d^{14}}{dx^{14}}(x^{13} - 2x^{10} + 5x^3 - 1)$

21. $\dfrac{d^8}{dx^8}(\tfrac{1}{8}x^8 - \tfrac{1}{7}x^7 + x^5 - x^3)$ **22.** $\dfrac{d^6}{dx^6}(x^6 + 5x^5 - 2x + 4)$

In Problems 23–26 find the velocity v and acceleration a of an object whose position s at time t is given.

23. $s = 16t^2 + 20t$ **24.** $s = 16t^2 + 10t + 1$ **25.** $s = 4.9t^2 + 4t + 4$ **26.** $s = 4.9t^2 + 5t$

---

* The earth, as you know, is not perfectly round; it bulges slightly at the equator. But neither is it perfectly oval, and its mass is not distributed uniformly. As a result, the acceleration of any freely falling body varies slightly from these constants.

27. Find the second derivative of $f(x) = x^2g(x)$, where g' and g'' exist.

28. Find the second derivative of $f(x) = g(x)/x$, where g' and g'' exist.

29. An object is propelled vertically upward with an initial velocity of 39.2 meters per second. The distance s (in meters) of the object from the ground after t seconds is $s = -4.9t^2 + 39.2t$.
    (a) What is the velocity of the object at any time t?
    (b) When will the object reach its highest point?
    (c) What is the maximum height?
    (d) What is the acceleration of the object at any time t?
    (e) How long is the object in the air?
    (f) What is the velocity of the object upon impact?
    (g) What is the total distance traveled by the object?

30. A ball is thrown vertically upward with an initial velocity of 80 feet per second. The distance s (in feet) of the ball from the ground after t seconds is $s = 6 + 80t - 16t^2$.
    (a) What is the velocity of the ball after 2 seconds?
    (b) When will the ball reach its highest point?
    (c) What is the maximum height the ball reaches?
    (d) What is the acceleration of the ball at any time t?
    (e) How long is the ball in the air?
    (f) What is the velocity of the ball upon impact?
    (g) What is the total distance traveled by the ball?

31. Develop a formula for F'' if $F(x) = f(x)g(x)$. Assume that f', g', f'', and g'' exist.

32. Develop a formula for F'' if $F(x) = f(x)/g(x)$. Assume that f', g', f'', and g'' exist.

33. Show that if $f(x) = 1/x$, then

$$f^{(n)}(x) = \frac{(-1)^n(n)(n-1)(n-2)\cdots(3)(2)(1)}{x^{n+1}}$$

where $(-1)^n = 1$ if n is even, and $(-1)^n = -1$ if n is odd.

7. Derivatives of Trigonometric Functions

Appendix I gives a review of trigonometry that may be useful for understanding the material in this section.

In presenting the calculus of the trigonometric functions, the use of radian measure greatly simplifies the discussion. For this reason, we shall use radian measure and only mention degrees when it is convenient to do so.

We begin by establishing the following preliminary result, which we conjectured earlier in Chapter 2, since it is essential for finding the derivative of $f(x) = \sin x$.

THEOREM If θ is measured in radians, then

(3.35)
$$\lim_{\theta \to 0} \frac{\sin \theta}{\theta} = 1$$

Proof   To evaluate this limit, we employ a new technique. We start by taking θ to be a positive acute central angle of a circle with radius $R = 1$. As shown in Figure 7, $\widehat{OAB}$ represents a sector of the circle.

(3.36)
$$\text{Area of } \triangle OAB < \text{Area of } \widehat{OAB} < \text{Area } \triangle OAD$$

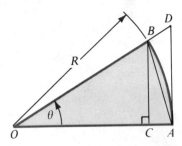

Figure 7

If we use vertical bars $|\ |$ to denote the length of a line segment, then, since the radius $R = 1$, we find that $|OB| = |OA| = 1$. Furthermore,

$$\sin \theta = \frac{|CB|}{|OB|} = \frac{|CB|}{1} = |CB| \qquad \tan \theta = \frac{|AD|}{|OA|} = \frac{|AD|}{1} = |AD|$$

Thus, in terms of θ, the areas in (3.36) may be expressed as

$$\text{Area of } \triangle OAB = \frac{1}{2}|OA||CB| = \frac{1}{2}(1)\sin\theta = \frac{1}{2}\sin\theta$$

(3.37)
$$\text{Area of } \widehat{OAB} = \frac{1}{2}R^2\theta = \frac{1}{2}(1)\theta = \frac{1}{2}\theta$$

$$\text{Area of } \triangle OAD = \frac{1}{2}|OA||AD| = \frac{1}{2}(1)\tan\theta = \left(\frac{1}{2}\right)\frac{\sin\theta}{\cos\theta}$$

By substituting (3.37) in (3.36), we obtain

$$\frac{1}{2}\sin\theta < \frac{\theta}{2} < \left(\frac{1}{2}\right)\frac{\sin\theta}{\cos\theta}$$

By dividing by $\frac{1}{2}\sin\theta$, we obtain

(3.38)
$$1 < \frac{\theta}{\sin\theta} < \frac{1}{\cos\theta} \qquad 0 < \theta < \frac{\pi}{2}$$

Since $0 < \theta < \pi/2$, the quantities $\theta/(\sin\theta)$ and $1/(\cos\theta)$ are positive. As a result, if we take the reciprocal of each expression in the inequality (3.38), the inequalities will be reversed. That is,

$$1 > \frac{\sin\theta}{\theta} > \cos\theta \qquad \text{or} \qquad \cos\theta < \frac{\sin\theta}{\theta} < 1$$

As θ tends to 0^+, $\cos \theta$ tends to 1.* Since the quantity $(\sin \theta)/\theta$ is between a quantity approaching 1 and 1 itself, by the squeezing theorem (2.20), it also must approach 1. Thus,

(3.39)
$$\lim_{\theta \to 0^+} \frac{\sin \theta}{\theta} = 1$$

Since $\sin(-\theta) = -\sin \theta$, we have

$$\frac{\sin(-\theta)}{-\theta} = \frac{-\sin \theta}{-\theta} = \frac{\sin \theta}{\theta}$$

Hence, as $\theta \to 0^-$, we must have

(3.40)
$$\lim_{\theta \to 0^-} \frac{\sin \theta}{\theta} = \lim_{\theta \to 0^+} \frac{\sin(-\theta)}{-\theta} = \lim_{\theta \to 0^+} \frac{\sin \theta}{\theta} = 1$$

Together, (3.39) and (3.40) yield

$$\lim_{\theta \to 0} \frac{\sin \theta}{\theta} = 1 \quad \blacksquare$$

The result just obtained states that for values of θ close to 0, $(\sin \theta)/\theta$ is close to 1.[†] This means that for values of θ close to 0, $\sin \theta \approx \theta$. Table 2 shows this.

Table 2

θ	$\sin \theta$	$(\sin \theta)/\theta$
0.1	0.0998334	0.998334
0.05	0.0499792	0.999583
0.04	0.0399893	0.999733
0.03	0.0299955	0.999850
0.02	0.0199987	0.999933
0.01	0.0099998	0.999983

---

* This is not obvious since we have not proved that $\cos \theta$ is continuous at 0. You are asked to prove this in Problem 61 in Exercise 7.

[†] In establishing the formula (3.35), the angle is measured in radians. If we measure the angle in degrees, we find that

$$\lim_{\alpha \to 0^\circ} \frac{\sin \alpha^\circ}{\alpha} = \lim_{\theta \to 0} \frac{\sin \theta}{180\theta/\pi} = \frac{\pi}{180} \lim_{\theta \to 0} \frac{\sin \theta}{\theta} = \frac{\pi}{180}$$

$$\text{Set } \alpha = \frac{180}{\pi} \theta$$

$$\sin \alpha^\circ = \sin \theta$$

The simplicity of the result obtained in formula (3.35) by using radian measure, compared to $\pi/180$ when degree measure is used, is another reason for the use of radian measure in calculus.

EXAMPLE 1   Evaluate: $\lim\limits_{\theta \to 0} \dfrac{\sin 3\theta}{\theta}$

Solution   Since $(\sin 3\theta)/\theta$ is not exactly in the form required by (3.35), we cannot use it immediately. However, by using the substitution $t = 3\theta$,   we find that

$$\lim_{\theta \to 0} \frac{\sin 3\theta}{\theta} = \lim_{t \to 0} \frac{\sin t}{t/3} = \lim_{t \to 0}\left(3\,\frac{\sin t}{t}\right) = 3 \lim_{t \to 0} \frac{\sin t}{t} = (3)(1) = 3 \qquad \blacksquare$$

Set $t = 3\theta$
As $\theta \to 0$,   then $t \to 0$

Apply (3.35)

EXAMPLE 2   Establish the formula

(3.41) $$\lim_{\theta \to 0} \frac{\cos \theta - 1}{\theta} = 0$$

Solution   To obtain this result, we proceed as follows:

$$\frac{\cos \theta - 1}{\theta} = \left(\frac{\cos \theta - 1}{\theta}\right)\left(\frac{\cos \theta + 1}{\cos \theta + 1}\right) = \frac{\cos^2\theta - 1}{\theta(\cos \theta + 1)} = \frac{-\sin^2\theta}{\theta(\cos \theta + 1)}$$

$$= (-\sin \theta)\left(\frac{\sin \theta}{\theta}\right)\left(\frac{1}{\cos \theta + 1}\right)$$

Thus,

$$\lim_{\theta \to 0} \frac{\cos \theta - 1}{\theta} = \lim_{\theta \to 0}\left[(-\sin \theta)\left(\frac{\sin \theta}{\theta}\right)\left(\frac{1}{\cos \theta + 1}\right)\right]$$

$$= \left[\lim_{\theta \to 0}(-\sin \theta)\right]\left[\lim_{\theta \to 0}\frac{\sin \theta}{\theta}\right]\left[\lim_{\theta \to 0}\frac{1}{\cos \theta + 1}\right]$$

$$= (0)(1)(\tfrac{1}{2}) = 0 \qquad \blacksquare$$

DERIVATIVES OF $y = \sin x$ AND $y = \cos x$

(3.42)   **THEOREM** *Derivative of* $y = \sin x$. **The derivative of** $y = \sin x$ **is** $y' = \cos x$. **That is,**

$$\frac{d}{dx}\sin x = \cos x$$

Proof   We follow the procedure established in (3.16) and calculate the difference quotient for $f(x) = \sin x$:

$$\frac{f(x + h) - f(x)}{h} = \frac{\sin(x + h) - \sin x}{h}$$

By using the fact that $\sin(x + h) = \sin x \cos h + \cos x \sin h,$ we have

$$\frac{f(x + h) - f(x)}{h} = \frac{\sin x \cos h + \cos x \sin h - \sin x}{h}$$

$$= \frac{\sin x \cos h - \sin x}{h} + \frac{\cos x \sin h}{h}$$

$$= (\sin x)\left(\frac{\cos h - 1}{h}\right) + (\cos x)\left(\frac{\sin h}{h}\right)$$

Now, using limit theorems for sums and products and formulas (3.35) and (3.41), we obtain the result

$$\frac{d}{dx}\sin x = \lim_{h \to 0}\frac{f(x + h) - f(x)}{h} = (\sin x)(0) + (\cos x)(1) = \cos x \qquad \blacksquare$$

EXAMPLE 3   Find y' if:

(a) $y = x + 4 \sin x$ (b) $y = x^2 \sin x$ (c) $y = \dfrac{\sin x}{x}$

Solution   (a)   We use the rule for differentiating a sum to get

$$y' = 1 + 4 \cos x$$

(b)   We use the rule for differentiating a product to get

$$y' = 2x \sin x + x^2 \cos x$$

(c)   We use the rule for differentiating a quotient to get

$$y' = \frac{x \cos x - \sin x}{x^2} \qquad \blacksquare$$

We obtain the derivative of $y = \cos x$ the same way we obtained the derivative of $y = \sin x$. Thus, if $f(x) = \cos x,$ then

$$\frac{f(x + h) - f(x)}{h} = \frac{\cos(x + h) - \cos x}{h}$$

$$= \frac{\cos x \cos h - \sin h \sin x - \cos x}{h}$$

$$= \cos x\left(\frac{\cos h - 1}{h}\right) - \sin x\left(\frac{\sin h}{h}\right)$$

Now,

$$\frac{d}{dx}\cos x = \lim_{h \to 0}\frac{f(x + h) - f(x)}{h} = (\cos x)(0) - (\sin x)(1)$$

$$= -\sin x$$

We have established the following result:

(3.43)   **THEOREM** *Derivative of* $y = \cos x$. **The derivative of** $y = \cos x$ **is** $y' = -\sin x$. **That is,**

$$\frac{d}{dx}\cos x = -\sin x$$

EXAMPLE 4   Find y' if:

(a) $y = x^2 \cos x$ (b) $y = \dfrac{\cos x}{1 - \sin x}$

Solution   (a) $y' = 2x \cos x + x^2(-\sin x)$

$\qquad\qquad = 2x \cos x - x^2 \sin x$

(b) $y' = \dfrac{(1 - \sin x)(-\sin x) - (\cos x)(-\cos x)}{(1 - \sin x)^2}$

$\qquad\quad = \dfrac{-\sin x + \sin^2 x + \cos^2 x}{(1 - \sin x)^2}$

$\qquad\quad = \dfrac{1 - \sin x}{(1 - \sin x)^2} = \dfrac{1}{1 - \sin x}$ ∎

DERIVATIVES OF OTHER TRIGONOMETRIC FUNCTIONS

EXAMPLE 5   Establish the formulas:

(3.44)   (a) $\dfrac{d}{dx}\tan x = \sec^2 x$ (b) $\dfrac{d}{dx}\sec x = \sec x \tan x$

Solution   (a) $\dfrac{d}{dx}\tan x = \dfrac{d}{dx}\dfrac{\sin x}{\cos x} = \dfrac{\cos x \cos x - \sin x(-\sin x)}{\cos^2 x}$

$\qquad\qquad\qquad = \dfrac{\cos^2 x + \sin^2 x}{\cos^2 x} = \dfrac{1}{\cos^2 x} = \sec^2 x$

(b) $\dfrac{d}{dx}\sec x = \dfrac{d}{dx}\dfrac{1}{\cos x} \underset{\underset{(3.33)}{\uparrow}}{=} \dfrac{-\dfrac{d}{dx}\cos x}{(\cos x)^2}$

$\qquad\qquad\qquad = \dfrac{\sin x}{\cos^2 x} = \left(\dfrac{1}{\cos x}\right)\left(\dfrac{\sin x}{\cos x}\right) = \sec x \tan x$ ∎

In Problem 64 in Exercise 7 you are asked to establish the remaining formulas:

(3.45) $\dfrac{d}{dx}\cot x = -\csc^2 x \qquad \dfrac{d}{dx}\csc x = -\csc x \cot x$

EXERCISE 7

In Problems 1–24 find y'.

1. $y = 3\sin x - 2\cos x$ **2.** $y = 4\tan x + \sin x$ **3.** $y = \sin x \cos x$

4. $y = x\cos x$ **5.** $y = \sec x \tan x$ **6.** $y = x\tan x$

7. $y = \dfrac{\cot x}{x}$ **8.** $y = \dfrac{\csc x}{x}$ **9.** $y = x^2 \tan x$

10. $y = x^2 \sin x$ **11.** $y = x\tan x - 3\sec x$ **12.** $y = x\sec x + 2\cot x$

13. $y = \dfrac{\sin x}{1 - \cos x}$ **14.** $y = \dfrac{x}{\cos x}$ **15.** $y = \dfrac{\sin x}{1 + x}$

16. $y = \dfrac{\tan x}{1 + x}$ **17.** $y = \dfrac{\sin x + \cos x}{\sin x - \cos x}$ **18.** $y = \dfrac{\sin x - \cos x}{\sin x + \cos x}$

19. $y = \dfrac{\sec x}{1 + x\sin x}$ **20.** $y = \dfrac{\csc x}{1 + x\cos x}$ **21.** $y = \csc x \cot x$

22. $y = \tan x \cos x$ **23.** $y = \dfrac{1 + \tan x}{1 - \tan x}$ **24.** $y = \dfrac{\csc x - \cot x}{\csc x + \cot x}$

In Problems 25–36 find y''.

25. $y = \sin x$ **26.** $y = \cos x$ **27.** $y = \sec x$

28. $y = \csc x$ **29.** $y = x\sin x$ **30.** $y = x\cos x$

31. $y = 2\sin x - 3\cos x$ **32.** $y = 3\sin x + 4\cos x$ **33.** $y = x^2 \sin x$

34. $y = x^2 \cos x$ **35.** $y = a\sin x + b\cos x$ **36.** $y = x\sec x$

In Problems 37–40 find the derivative of f at c.

37. $f(x) = 2\sin x + \cos x, \quad c = \dfrac{\pi}{2}$ **38.** $f(x) = x - \sin x, \quad c = \pi$

39. $f(x) = \dfrac{\cos x}{1 + \sin x}, \quad c = \dfrac{\pi}{3}$ **40.** $f(x) = \dfrac{\sin x}{1 + \cos x}, \quad c = \dfrac{5\pi}{6}$

In Problems 41–48 evaluate each limit.

41. $\displaystyle\lim_{x\to 0}\dfrac{\sin 7x}{x}$ **42.** $\displaystyle\lim_{x\to 0}\dfrac{\sin(x/3)}{x}$ **43.** $\displaystyle\lim_{x\to 0}\dfrac{\cos x}{1 + \sin x}$ **44.** $\displaystyle\lim_{x\to 0}\dfrac{\sin x}{1 + \cos x}$

45. $\displaystyle\lim_{x\to 0}\dfrac{\sin x^2}{x}$ **46.** $\displaystyle\lim_{x\to 0}\dfrac{\sin^2 2x}{x^2}$ **47.** $\displaystyle\lim_{x\to \pi}\dfrac{\sin x}{\pi - x}$ **48.** $\displaystyle\lim_{x\to 0}\dfrac{2x - 5\sin 3x}{x}$

In Problems 49–56 find the equation of the tangent line to the graph of f at the indicated point.

49. $f(x) = \sin x$,   at $(0, 0)$ **50.** $f(x) = \sin x$,   at $\left(\dfrac{\pi}{6}, \dfrac{1}{2}\right)$ **51.** $f(x) = \cos x$,   at $\left(\dfrac{\pi}{3}, \dfrac{1}{2}\right)$

52. $f(x) = \cos x$,   at $(0, 1)$ **53.** $f(x) = \tan x$,   at $(0, 0)$ **54.** $f(x) = \tan x$,   at $\left(\dfrac{\pi}{4}, 1\right)$

55. $f(x) = \sin x + \cos x$,   at $\left(\dfrac{\pi}{4}, \sqrt{2}\right)$ **56.** $f(x) = \sin x - \cos x$,   at $\left(\dfrac{\pi}{4}, 0\right)$

In Problems 57 and 58 find the nth derivative of each function.

57. $f(x) = \sin x$ **58.** $f(x) = \cos x$

59. What is $\displaystyle\lim_{h \to 0} \frac{\cos\left(\dfrac{\pi}{2} + h\right) - \cos\dfrac{\pi}{2}}{h}$?   **60.** What is $\displaystyle\lim_{h \to 0} \frac{\sin(\pi + h) - \sin \pi}{h}$?

61. Use Figure 7 to show that $|\sin \theta| \le |\theta|$.   Use this fact and the squeezing theorem (2.20) to show that $\lim_{\theta \to 0} \sin \theta = 0$.   Then, by means of the identity $\sin^2(\theta/2) = \frac{1}{2}(1 - \cos \theta)$,   show that $\lim_{\theta \to 0} \cos \theta = 1$.

62. If $y = \sin x$ and $y^{(n)}$ is the nth derivative of y with respect to x, find the smallest positive integer n for which $y^{(n)} = y$.

63. If $y = A \sin t + B \cos t$,   where A and B are constants, show that $y'' + y = 0$.

64. Establish the formulas (3.45):

$$\frac{d}{dx} \cot x = -\csc^2 x \qquad \frac{d}{dx} \csc x = -\csc x \cot x$$

65. Use the identity

$$\sin A - \sin B = 2 \cos \frac{A + B}{2} \sin \frac{A - B}{2}$$

with $A = x + h$, $B = x$ to prove that

$$\frac{d}{dx} \sin x = \lim_{h \to 0} \frac{\sin(x + h) - \sin x}{h} = \cos x$$

66. The function $f(x) = (\sin \pi x)/x$ is not defined at 0. Decide how to define $f(0)$ so that f is continuous at 0.

67. What is $\displaystyle\lim_{x \to 0} \frac{\sin ax}{\sin bx}$?

68. Let $f(x) = \sin x$.   Show that finding $f'(0)$ is the same as finding $\displaystyle\lim_{x \to 0} \frac{\sin x}{x}$.

69. Let $f(x) = \cos x$.   Show that finding $f'(0)$ is the same as finding $\displaystyle\lim_{x \to 0} \frac{\cos x - 1}{x}$.

70. How must we define $f(0)$ and $f(1)$ so that the function below is continuous on $0 \le x \le 1$?

$$f(x) = \frac{\sin \pi x}{x(1 - x)}$$

8. The Chain Rule

When the function to be differentiated is the composite of two differentiable functions, its derivative is found by using a formula called the *chain rule*. A complete proof that the composite of two differentiable functions is, in fact, differentiable is given in Appendix II.

THE POWER RULE

We begin this section by presenting a special case of the chain rule, called the *power rule*. Our purpose is first to learn the computational aspect of the power rule. Then, we state the chain rule and give examples to illustrate how it is employed.

Consider the function $y = (2x + 3)^2$. Here, y is a function of x that can be written as the composite of two simpler functions—the square function and a linear function. Specifically, if

$$f(x) = x^2 \quad \text{and} \quad g(x) = 2x + 3$$

then

$$y = (f \circ g)(x) = f(g(x)) = [g(x)]^2 = (2x + 3)^2$$

Other examples of composite functions are

$$y = (x^2 + 1)^4 \qquad \text{if} \qquad y = [g(x)]^4 \qquad \text{and} \qquad g(x) = x^2 + 1$$
$$y = (x^2 + 1)^{-3} \qquad \text{if} \qquad y = [g(x)]^{-3} \qquad \text{and} \qquad g(x) = x^2 + 1$$
$$y = (x^3 - x^2 + 1)^5 \qquad \text{if} \qquad y = [g(x)]^5 \qquad \text{and} \qquad g(x) = x^3 - x^2 + 1$$

To get some idea of how to find the derivative of a composite function of the form $[g(x)]^n$, with n an integer, let us see what happens when $n = 2$, $n = 3$, and $n = 4$. By the product formula (3.29),

$$n = 2 \qquad \frac{d}{dx}[g(x)]^2 = \frac{d}{dx}[g(x)g(x)] = g'(x)g(x) + g(x)g'(x) = 2g(x)g'(x)$$

$$n = 3 \qquad \frac{d}{dx}[g(x)]^3 = \frac{d}{dx}\{[g(x)]^2 g(x)\} = [g(x)]^2 g'(x) + g(x)\left\{\frac{d}{dx}[g(x)]^2\right\}$$

$$= [g(x)]^2 g'(x) + g(x)[2g(x)g'(x)] = 3[g(x)]^2 g'(x)$$

$$n = 4 \qquad \frac{d}{dx}[g(x)]^4 = \frac{d}{dx}\{[g(x)]^3 g(x)\} = [g(x)]^3 g'(x) + g(x)\left\{\frac{d}{dx}[g(x)]^3\right\}$$

$$= [g(x)]^3 g'(x) + g(x)\{3[g(x)]^2 g'(x)\} = 4[g(x)]^3 g'(x)$$

Let's summarize what we've found:

$$\frac{d}{dx}[g(x)]^2 = 2g(x)g'(x) \qquad \frac{d}{dx}[g(x)]^3 = 3[g(x)]^2 g'(x) \qquad \frac{d}{dx}[g(x)]^4 = 4[g(x)]^3 g'(x)$$

Based on these results, we conjecture the following formula, which is called the *power rule*. A proof of this rule is given a little later in the section.

(3.46)   **THEOREM**   *Power Rule.*   **If g is a differentiable function and n is any integer, then**

$$\frac{d}{dx}[g(x)]^n = n[g(x)]^{n-1}g'(x)$$

Notice the similarity between the power rule (3.46) and the formula (3.19) $(d/dx)(x^n) = nx^{n-1}$;   the main difference is the third factor, $g'(x)$.

EXAMPLE 1   Find the derivative of $f(x) = (x^2 + x + 1)^3$.

Solution   We could, of course, expand the right-hand side and proceed according to earlier techniques. However, the power rule lets us find derivatives of functions like this without resorting to tedious (and sometimes impossible) computation. The function $f(x) = (x^2 + x + 1)^3$ is the function $g(x) = x^2 + x + 1$ raised to the power $n = 3$.   By the power rule,

$$\frac{d}{dx}f(x) = \frac{d}{dx}(x^2 + x + 1)^3 = (3)(x^2 + x + 1)^2\left[\frac{d}{dx}(x^2 + x + 1)\right]$$

$$= 3(x^2 + x + 1)^2(2x + 1)   \blacksquare$$

Additional examples of the derivatives of composite functions are:
(a)   If $f(x) = (3 - x^3)^{-5}$,   then

$$f'(x) = (-5)(3 - x^3)^{-6}(-3x^2) = 15x^2(3 - x^3)^{-6}$$

(b)   If $f(x) = (2x + 3)^2$,   then

$$f'(x) = (2)(2x + 3)^1(2) = 4(2x + 3)$$

(c)   If $f(x) = (x^3 - 3x^2 + 1)^5$,   then

$$f'(x) = (5)(x^3 - 3x^2 + 1)^4(3x^2 - 6x) = 15x(x - 2)(x^3 - 3x^2 + 1)^4$$

Often, we must use at least one other rule along with the power rule to differentiate a function. Here is an example.

EXAMPLE 2   Find the derivative of $f(x) = x(x^2 + 1)^3$.

Solution   Here, the function is the product of x and $(x^2 + 1)^3$.   Thus, we begin by using the rule for differentiating a product. That is,

$$f'(x) = (x)\left[\frac{d}{dx}(x^2 + 1)^3\right] + (x^2 + 1)^3$$

By applying the power rule, we have

$$f'(x) = (x)[3(x^2 + 1)^2(2x)] + (x^2 + 1)^3$$
$$= 6x^2(x^2 + 1)^2 + (x^2 + 1)^3 = (x^2 + 1)^2(6x^2 + x^2 + 1)$$
$$= (x^2 + 1)^2(7x^2 + 1) \quad \blacksquare$$

EXAMPLE 3 Find the derivative of $f(x) = \left(\dfrac{3x + 2}{4x^2 - 5}\right)^5.$

Solution Here, f is the quotient $(3x + 2)/(4x^2 - 5)$ raised to the power 5. Thus, we begin by using the power rule and then use the rule for differentiating a quotient.

$$f'(x) \underset{\uparrow}{=} (5)\left(\frac{3x + 2}{4x^2 - 5}\right)^4\left[\frac{d}{dx}\left(\frac{3x + 2}{4x^2 - 5}\right)\right]$$

Apply power rule

$$\underset{\uparrow}{=} (5)\left(\frac{3x + 2}{4x^2 - 5}\right)^4\left[\frac{(4x^2 - 5)(3) - (3x + 2)(8x)}{(4x^2 - 5)^2}\right]$$

Apply quotient rule

$$= \frac{5(3x + 2)^4(-12x^2 - 16x - 15)}{(4x^2 - 5)^6} \quad \blacksquare$$

THE CHAIN RULE

We now turn our attention to composite functions in general. As we saw in Chapter 1, the composite of two functions f and g is defined by $(f \circ g)(x) = f(g(x))$. We may obtain the composite function $(f \circ g)(x) = f(g(x))$ by setting

$$y = f(u) \qquad \text{and} \qquad u = g(x) \qquad \text{to obtain} \qquad y = f(g(x))$$

That is, we regard y as a function of u and u as a function of x so that, after substitution, y becomes a function of x. To find the derivative of the composite function $y = f(g(x))$, we use the chain rule.

(3.47) **THEOREM** *Chain Rule.* **If f and g are differentiable functions, the composite function $f \circ g$ is also differentiable. Moreover, if $y = f(u)$ and $u = g(x)$, then the derivative of $y = (f \circ g)(x)$ is**

$$\frac{dy}{dx} = \left(\frac{dy}{du}\right)\left(\frac{du}{dx}\right)$$

Partial Proof of the Chain Rule To calculate the derivative of $y = f(g(x))$, we proceed directly to Step 3 of (3.16) and look at the difference quotient $\Delta y/\Delta x$. If $\Delta u \neq 0$, we may write the algebraic identity

$$\frac{\Delta y}{\Delta x} = \left(\frac{\Delta y}{\Delta u}\right)\left(\frac{\Delta u}{\Delta x}\right)$$

Since $u = g(x)$ is differentiable, it follows that $\Delta u \to 0$ when $\Delta x \to 0.$* Thus,

$$\lim_{\Delta x \to 0} \frac{\Delta y}{\Delta x} = \left(\lim_{\Delta u \to 0} \frac{\Delta y}{\Delta u} \right) \left(\lim_{\Delta x \to 0} \frac{\Delta u}{\Delta x} \right)$$

$$\frac{dy}{dx} = \left(\frac{dy}{du} \right) \left(\frac{du}{dx} \right)$$

This partial proof required that $\Delta u \neq 0$. In Appendix II we show that even if $\Delta u = 0$, the result (3.47) still holds. ∎

EXAMPLE 4   Find the derivative of the composite function $f \circ g$ for $y = f(u) = u^2 - 4$ and $u = g(x) = x^3$

Solution

$$\frac{dy}{du} = \frac{d}{du} f(u) = \frac{d}{du} (u^2 - 4) = 2u \underset{\uparrow}{=} 2x^3 \qquad \frac{du}{dx} = \frac{d}{dx} g(x) = \frac{d}{dx} x^3 = 3x^2$$

Remember, $u = x^3$

Thus, by the chain rule,

$$\frac{d}{dx} (f \circ g)(x) = \frac{dy}{dx} = \left(\frac{dy}{du} \right) \left(\frac{du}{dx} \right) = (2x^3)(3x^2) = 6x^5 \qquad ∎$$

Alternatively, we could have first formed the composite function $f \circ g$ and then differentiated. Using this approach, we find

$$(f \circ g)(x) = (x^3)^2 - 4 = x^6 - 4$$

Then

$$\frac{d}{dx} (f \circ g)(x) = 6x^5$$

Although the second approach appears shorter and easier here, the benefit of the chain rule will become apparent shortly. The following example illustrates a situation in which either the power rule or the chain rule can be used.

EXAMPLE 5   Find the derivative of $y = (3x^3 + 4x - 5)^{-5}$ by using:
(a) The power rule   (b) The chain rule

Solution   (a) *Power rule:* $y' = -5(3x^3 + 4x - 5)^{-6} \left[\frac{d}{dx} (3x^3 + 4x - 5) \right]$

$$= -5(3x^3 + 4x - 5)^{-6}(9x^2 + 4)$$

---

* Similar to part of the proof of (3.29),

$$\lim_{\Delta x \to 0} \Delta u = \lim_{\Delta x \to 0} \Delta x \left(\frac{\Delta u}{\Delta x} \right) = \left(\lim_{\Delta x \to 0} \Delta x \right) \left(\lim_{\Delta x \to 0} \frac{\Delta u}{\Delta x} \right) = (0) \left(\frac{du}{dx} \right) = 0$$

(b) *Chain rule:* Here y is the composite of $y = u^{-5}$ and $u = 3x^3 + 4x - 5$. Hence,

$$\frac{dy}{dx} = \left(\frac{dy}{du}\right)\left(\frac{du}{dx}\right) = (-5u^{-6})(9x^2 + 4) = -5(3x^3 + 4x - 5)^{-6}(9x^2 + 4) \quad \blacksquare$$

As this example illustrates, whenever a problem involves finding the derivative of a function raised to a power, either the power rule or the chain rule may be used. However, the power rule, as a special case of the chain rule, will not always work for composite functions involving nonalgebraic functions.

EXAMPLE 6   Find y' if: (a) $y = \sin x^2$ (b) $y = \sin^2 x$

Solution   (a)   We use the chain rule with $y = \sin u$ and $u = x^2$.   Then

$$y' = \frac{dy}{dx} = \left(\frac{dy}{du}\right)\left(\frac{du}{dx}\right) = (\cos u)(2x) = 2x \cos x^2$$

(b)   Since $y = \sin^2 x = (\sin x)^2$,   we may use either the power rule or the chain rule. First, we try the power rule:

$$y' = 2 \sin x \left(\frac{d}{dx} \sin x\right) = 2 \sin x \cos x$$

Using the chain rule, we set $y = u^2$ and $u = \sin x$:

$$y' = \left(\frac{dy}{du}\right)\left(\frac{du}{dx}\right) = 2u(\cos x) = 2 \sin x \cos x \quad \blacksquare$$

The above example is an important one. Although the solution in part (b) may be obtained by either the power rule or the chain rule, the solution in part (a) can only be obtained by using the chain rule. To summarize, whenever the power rule can be used, so can the chain rule; but there are many times when only the chain rule will do the job!

The formula

$$\frac{dy}{dx} = \left(\frac{dy}{du}\right)\left(\frac{du}{dx}\right) \qquad u = g(x)$$

can be extended. For example, if

$$y = f(u) \qquad u = g(v) \qquad v = h(x)$$

then the composite $y = (f \circ g \circ h)(x)$ is a function of x and

(3.48)
$$\frac{dy}{dx} = \left(\frac{dy}{du}\right)\left(\frac{du}{dv}\right)\left(\frac{dv}{dx}\right)$$

This "chain" of factors is the basis for the name *chain rule*.

EXAMPLE 7 Find dy/dx if $y = u^4$, $u = 4v^3 - 2$, and $v = 2/x^2$.

Solution
$$\frac{dy}{dx} = \left(\frac{dy}{du}\right)\left(\frac{du}{dv}\right)\left(\frac{dv}{dx}\right) = (4u^3)(12v^2)\left(-\frac{4}{x^3}\right) \underset{\underset{u = 4v^3 - 2}{\uparrow}}{=} 4(4v^3 - 2)^3(12v^2)\left(-\frac{4}{x^3}\right)$$

$$\underset{\underset{v = 2/x^2}{\uparrow}}{=} 4\left(\frac{32}{x^6} - 2\right)^3(12)\left(\frac{4}{x^4}\right)\left(-\frac{4}{x^3}\right) = \frac{-768}{x^7}\left(\frac{32}{x^6} - 2\right)^3 \quad \blacksquare$$

EXAMPLE 8 Find y' if: (a) $y = 5\cos^2(3x + 2)$ (b) $y = \sin^3 5x$

Solution (a) We use the chain rule, where $y = 5u^2$, $u = \cos v$, and $v = (3x + 2)$:

$$y' = \frac{dy}{dx} = \left(\frac{dy}{du}\right)\left(\frac{du}{dv}\right)\left(\frac{dv}{dx}\right) = 10u(-\sin v)(3)$$

$$= -30\cos v \sin v = -30\cos(3x + 2)\sin(3x + 2)$$

(b) We use the chain rule where $y = u^3$, $u = \sin v$, and $v = 5x$:

$$y' = \frac{dy}{dx} = \left(\frac{dy}{du}\right)\left(\frac{du}{dv}\right)\left(\frac{dv}{dx}\right) = (3u^2)(\cos v)(5) = 15\sin^2 v \cos v$$

$$= 15\sin^2 5x \cos 5x \quad \blacksquare$$

An important formula involving acceleration is derived by using the chain rule. We have seen that the acceleration of a body at time t is given by $a = dv/dt$, where v is the velocity of the object. Sometimes it is useful to express the velocity v as a function of the distance s. In this case, $v = v(s)$ and the acceleration a may be expressed as

$$a = \frac{dv}{dt} \underset{\underset{\substack{\text{Apply chain rule} \\ \text{to} \ v = v(s)}}{\equiv}}{} \left(\frac{dv}{ds}\right)\left(\frac{ds}{dt}\right) \underset{\underset{v = ds/dt}{\equiv}}{} v\frac{dv}{ds}$$

Thus, we have the following alternate formula for acceleration:

(3.49)
$$a = v\frac{dv}{ds}$$

We shall have occasion to refer to this formula later.

One final note before we prove the power rule. In terms of prime notation, the chain rule formula is

(3.50)
$$(f \circ g)'(x) = [f'g(x)]$$

The next example illustrates a use for (3.50).

EXAMPLE 9   Suppose $h = f \circ g$.   If

$$f(1) = 2 \qquad f'(1) = 3 \qquad f'(2) = -4$$
$$g(1) = 2 \qquad g'(1) = -3 \qquad g'(2) = 5$$

calculate $h'(1)$.

Solution   Based on (3.50),

$$h'(x) = [f'(g(x))][g'(x)]$$

When $x = 1$,

$$h'(1) = [f'(g(1))][g'(1)] = f'(2)g'(1) = (-4)(-3) = 12$$

Note that only a portion of the given information is required.   ∎

The chain rule is a powerful tool for establishing other differentiation formulas. We use it now to prove the power rule.

Proof of the Power Rule   The power rule states that if g is a differentiable function and n is an integer, then

$$\frac{d}{dx} [g(x)]^n = n[g(x)]^{n-1}g'(x)$$

To prove this, we use the chain rule. If

$$y = [g(x)]^n$$

we set

$$y = u^n \qquad \text{and} \qquad u = g(x)$$

Then

$$\frac{dy}{du} = nu^{n-1} = n[g(x)]^{n-1} \qquad \text{and} \qquad \frac{du}{dx} = g'(x)$$

By the chain rule,

$$\frac{d}{dx} [g(x)]^n = \frac{dy}{dx} = \left(\frac{dy}{du}\right)\left(\frac{du}{dx}\right) = n[g(x)]^{n-1}g'(x) \qquad ∎$$

EXERCISE 8

In Problems 1–22 find the derivative of the function f by using the power rule.

1. $f(x) = (3x + 5)^2$ **2.** $f(x) = (2x - 5)^3$ **3.** $f(x) = (6x - 5)^{-3}$

4. $f(x) = (4x + 1)^{-2}$ **5.** $f(x) = (x^2 + 5)^4$ **6.** $f(x) = (x^3 - 2)^5$

7. $f(t) = (t^5 - t^2 + t)^7$ **8.** $f(u) = (u^4 - u^2 + u - 1)^6$ **9.** $f(x) = \left(x - \dfrac{1}{x}\right)^3$

10. $f(x) = \left(x + \dfrac{1}{x}\right)^3$

11. $f(z) = \left(\dfrac{z}{z+1}\right)^3$

12. $f(w) = \dfrac{(3w+1)^4}{w}$

13. $f(x) = \tan^2 x$

14. $f(x) = \sec^3 x$

15. $f(t) = \sin^2 t - \cos^2 t$

16. $f(z) = (\sin z + \cos z)^2$

17. $f(x) = \left(\dfrac{3x^2+1}{x^3-1}\right)^2$

18. $f(x) = \left(\dfrac{x^3+4}{x^2-1}\right)^3$

19. $f(x) = (x^2+4)^2(2x^3-1)^3$

20. $f(x) = (x^2-2)^3(3x^4+1)^2$

21. $f(x) = [5x + (3x + 6x^2)^3]^4$

22. $f(x) = [2x - (3x^2 + 4x^3)^4]^2$

In Problems 23–30, for the functions f and g, find the derivative of $y = (f \circ g)(x)$ by using the chain rule.

23. $y = f(u) = u^5$, $u = g(x) = x^3 + 1$

24. $y = f(u) = u^3$, $u = g(x) = 2x + 5$

25. $y = f(u) = \dfrac{u}{u+1}$, $u = g(x) = x^2 + 1$

26. $y = f(u) = \dfrac{u-1}{u}$, $u = g(x) = x^2 - 1$

27. $y = f(u) = (u+1)^2$, $u = g(x) = \dfrac{1}{x}$

28. $y = f(u) = (u^2-1)^3$, $u = g(x) = \dfrac{1}{x+2}$

29. $y = f(u) = (u^3-1)^5$, $u = g(x) = x^{-2}$

30. $y = f(u) = (u^2+4)^4$, $u = g(x) = x^{-2}$

In Problems 31–48 find y' by using the chain rule.

31. $y = \sin 4x$

32. $y = \cos 5x$

33. $y = \tan 5x$

34. $y = \csc(x^3 + 1)$

35. $y = \sin(3x^2 + 4)$

36. $y = 5\cos(x^2 - 4)$

37. $y = 4\sin^2 3x$

38. $y = 2\cos^2(x^2)$

39. $y = 2\sin(x^2 + 2x - 1)$

40. $y = \frac{1}{2}\cos(x^3 - 2x + 5)$

41. $y = \csc^2(1 + 3x)$

42. $y = \cot^2(1 + 3x^2)$

43. $y = x^2 \sin 4x$

44. $y = x^2 \cos 4x$

45. $y = \sin\dfrac{1}{x}$

46. $y = \sin\dfrac{3}{x}$

47. $y = x^2 \sin^2 x$

48. $y = x\sin\dfrac{1}{x}$

49. Find the derivative y' of $y = (x^3 + 1)^2$ by:
 (a) Using the chain rule
 (b) Using the power rule
 (c) Expanding and then differentiating
 (d) Compare the answers from parts (a)–(c).

50. Follow the directions in Problem 49 for the function $y = (x^2 - 2)^3$.

In Problems 51–56 find dy/dx by using (3.48).

51. $y = u^3$, $u = 3v^2 + 1$, $v = \dfrac{4}{x^2}$

52. $y = 3u$, $u = 3v^2 - 4$, $v = \dfrac{1}{x}$

53. $y = u^2 + 1$, $u = \dfrac{4}{v}$, $v = x^2$

54. $y = u^3 - 1$, $u = -\dfrac{2}{v}$, $v = x^3$

55. $y = \tan^3 2x$

56. $y = \sec^2(x^3)$

In Problems 57 and 58 compute the indicated derivatives.

57. $\dfrac{d^2}{dx^2}\cos(x^5)$

58. $\dfrac{d^3}{dx^3}\sin^3 x$

59. Find the nth-order derivative of $f(x) = (2x + 3)^n$.

60. Find the nth-order derivative of $f(x) = 1/(3x - 4)$.

61. Prove that if a differentiable function f is odd, then f' is even.   [*Hint:* $f(x) = -f(-x)$]

62. Prove that if a differentiable function f is even, then f' is odd.

In Problems 63–70 find the indicated derivative.

63. $\dfrac{d}{dx} f(x^2 + 1)$ [*Hint:* Let $u = x^2 + 1$.]

64. $\dfrac{d}{dx} f(1 - x^2)$

65. $\dfrac{d}{dx} f\left(\dfrac{x+1}{x-1}\right)$

66. $\dfrac{d}{dx} f\left(\dfrac{1-x}{1+x}\right)$

67. $\dfrac{d}{dx} f(\sin x)$

68. $\dfrac{d}{dx} f(\tan x)$

69. $\dfrac{d^2}{dx^2} f(\cos x)$

70. $\dfrac{d^2}{dx^2} f(\sec x)$

71. Suppose $h = f \circ g$. If $f'(2) = 6$, $f(1) = 4$, $g(1) = 2$, and $g'(1) = -2$, calculate $h'(1)$.

72. Suppose $h = f \circ g$. If $f'(3) = 4$, $f(1) = 1$, $g(1) = 3$, and $g'(1) = 3$, calculate $h'(1)$.

73. If $y = u^5 + u$ and $u = 4x^3 + x - 4$, find dy/dx at $x = 1$.

74. The resistance R (measured in ohms) of an 80 meter long electric wire of radius x (in centimeters) is given by the formula $R = 0.0048/x^2$. In turn, the radius x varies with the absolute temperature T according to the rule $x = 0.1991 + 0.000003T$. How fast does R change with respect to T when $T = 320°$? (Do not eliminate x in obtaining your answer.)

75. A bullet is fired horizontally into a bale of paper. The distance s (in meters) the bullet travels in the bale of paper in t seconds is given by $s = 8 - (2 - t)^3$ for $0 \le t \le 2$. Find the velocity of the bullet after 1 second. Find the acceleration of the bullet at any time t.

76. If $y = A \sin \omega t + B \cos \omega t$, where A, B, and ω are constants, show that $y'' + \omega^2 y = 0$.

77. Find the acceleration of a car if its position s on a highway at time t is given by

$$ s = \frac{80}{3}\left[t + \left(\frac{3}{\pi}\right) \sin \frac{\pi}{6} t \right] $$

78. Find the nth derivative of: (a) $f(x) = \sin ax$ (b) $f(x) = \cos ax$

79. Use the chain rule and the fact that $\cos x = \sin\left(\dfrac{\pi}{2} - x\right)$ to show that $\dfrac{d}{dx} \cos x = -\sin x$.

9. Implicit Differentiation

So far we have considered only functions whose law of correspondence is expressed in the form $y = f(x)$. This expression of the relationship between x and y is said to be in *explicit form* because we have solved for the dependent variable y. If the functional relationship between the independent variable x and the dependent variable y is not of this form, we say that x and y are related *implicitly*. For example, x

and y are related implicitly in the expressions

$$3x + 4y - 5 = 0 \qquad 3y^2 + x^2 - 1 = 0, \quad y \geq 0 \qquad xy - 4 = 0$$

If x and y are related implicitly, and if it is possible to solve for y, say, as $y = f(x)$, then the replacement of y by $f(x)$ in the implicit equation results in an identity. For example, in the implicit equation $3x + 4y - 5 = 0$, the explicit form is given by $y = \frac{1}{4}(5 - 3x)$. This may be verified by showing that the replacement of y by $f(x)$ in $3x + 4y - 5 = 0$ results in an identity:

$$3x + 4[\tfrac{1}{4}(5 - 3x)] - 5 = 3x + 5 - 3x - 5 = 0$$

We will now develop a procedure for finding the derivative of y with respect to x when the functional relationship between x and y is given implicitly. This procedure, called *implicit differentiation*, is illustrated next.

EXAMPLE 1   Find dy/dx if $3x + 4y - 5 = 0$.

Solution   We begin by assuming that there is a differentiable function $y = f(x)$ implied by the above relationship. That is, the expression

$$3x + 4f(x) - 5 = 0$$

is an identity. We differentiate both sides of this identity with respect to x:

$$\frac{d}{dx}[3x + 4f(x) - 5] = \frac{d}{dx}(0)$$

$$\frac{d}{dx}(3x) + \frac{d}{dx}[4f(x)] - \frac{d}{dx}(5) = \frac{d}{dx}(0)$$

$$3 + (4)\left[\frac{d}{dx}f(x)\right] = 0$$

$$\frac{d}{dx}f(x) = -\frac{3}{4}$$

So, we have $dy/dx = -\frac{3}{4}$.   ■

In the above example the function $y = f(x)$, whose existence was assumed, can be found by algebraically solving for y. In this case,

$$y = \frac{1}{4}(5 - 3x) = \frac{5}{4} - \frac{3x}{4}$$

so that $dy/dx = -\frac{3}{4}$, which agrees with the result obtained using implicit differentiation.

Often, though, it is very difficult or even impossible to actually solve for y in terms of x. The next two examples illustrate this difficulty (try to solve for y and you will see). However, we still make the assumption that there is a differentiable function $y = f(x)$ that obeys the implicit equation.

In these examples we will use the power rule,

$$\frac{d}{dx}[f(x)]^n = n[f(x)]^{n-1}f'(x) \qquad n \text{ an integer}$$

which, if $y = f(x)$, looks like this:

(3.51)
$$\frac{d}{dx}y^n = ny^{n-1}\frac{dy}{dx}$$

For example, if $y = f(x)$, then

$$\frac{d}{dx}y^2 = 2y\frac{dy}{dx} \qquad \text{and} \qquad \frac{d}{dx}y^3 = 3y^2\frac{dy}{dx}$$

EXAMPLE 2   Find dy/dx if $3x^2 + 4y^2 = 2x$.

Solution   We again assume that y is a differentiable function $y = f(x)$ and that when y is replaced by $f(x)$ in the expression $3x^2 + 4y^2 = 2x$, we obtain an identity in x. We proceed to differentiate both sides of this identity with respect to x:

$$\frac{d}{dx}(3x^2 + 4y^2) = \frac{d}{dx}(2x)$$

$$\frac{d}{dx}(3x^2) + \frac{d}{dx}(4y^2) = 2$$

$$6x + 4\left[\frac{d}{dx}(y^2)\right] = 2$$

By using (3.51), we obtain

$$6x + 4\left(2y\frac{dy}{dx}\right) = 2$$

This is a linear equation in dy/dx. Solving for dy/dx, we have

$$8y\frac{dy}{dx} = 2 - 6x$$

$$\frac{dy}{dx} = \frac{1-3x}{4y} \qquad \text{provided} \quad y \neq 0 \qquad \blacksquare$$

We summarize the procedure of implicit differentiation as follows:

1. **To find dy/dx when x and y are related implicitly, assume that y is a differentiable function of x.**
2. **Differentiate both sides of the relationship with respect to x by employing the power rule or the chain rule and other properties of differentiation.**
3. **Solve the resulting equation, which is linear in dy/dx, for dy/dx.**

EXAMPLE 3   Find dy/dx if:

(a) $xy^2 - x + y^3x + 5y = 10$ (b) $(x^3 + y)^4 = x$ (c) $\cos(xy) = x$

Solution   (a)

$$\left[\frac{d}{dx}(xy^2)\right] - \frac{d}{dx}x + \left[\frac{d}{dx}(y^3x)\right] + \frac{d}{dx}(5y) = \frac{d}{dx}(10)$$

$$\left[x\frac{d}{dx}y^2 + y^2\frac{d}{dx}x\right] - 1 + \left[y^3\frac{d}{dx}x + x\frac{d}{dx}y^3\right] + 5\frac{dy}{dx} = 0$$

$$\left[x\left(2y\frac{dy}{dx}\right) + y^2\right] - 1 + \left[y^3 + x\left(3y^2\frac{dy}{dx}\right)\right] + 5\frac{dy}{dx} = 0$$

$$(2xy + 3xy^2 + 5)\frac{dy}{dx} = 1 - y^2 - y^3$$

Hence, if $2xy + 3xy^2 + 5 \neq 0$,   we find

$$\frac{dy}{dx} = \frac{1 - y^2 - y^3}{2xy + 3xy^2 + 5}$$

(b) $$\frac{d}{dx}(x^3 + y)^4 = 4(x^3 + y)^3\left[\frac{d}{dx}(x^3 + y)\right] = \frac{d}{dx}x$$

$$4(x^3 + y)^3\left(3x^2 + \frac{dy}{dx}\right) = 1$$

Then, if $(x^3 + y)^3 \neq 0$,

$$3x^2 + \frac{dy}{dx} = \frac{1}{4(x^3 + y)^3}$$

$$\frac{dy}{dx} = \frac{1}{4(x^3 + y)^3} - 3x^2$$

(c)   We use implicit differentiation and the chain rule to get

$$[-\sin(xy)]\left(y + x\frac{dy}{dx}\right) = 1$$

$$[-x\sin(xy)]\frac{dy}{dx} - y\sin(xy) = 1$$

$$\frac{dy}{dx} = \frac{1 + y\sin(xy)}{-x\sin(xy)}   \text{if}   x\sin(xy) \neq 0   ∎$$

The next example shows how implicit differentiation can be used to find the slope of the tangent line to the graph of a function that is defined implicitly.

EXAMPLE 4   Find the slope of the tangent line to the graph of $x^3 + xy + y^3 = 5$ at the point $(-1, 2)$.

Solution   We differentiate with respect to x, obtaining

$$3x^2 + x\frac{dy}{dx} + y + 3y^2\frac{dy}{dx} = 0$$

By solving for dy/dx,

$$\frac{dy}{dx} = \frac{-(3x^2 + y)}{x + 3y^2} \qquad \text{provided} \quad x + 3y^2 \neq 0$$

The derivative dy/dx equals the slope of the tangent line to the graph at any point (x, y) for which $x + 3y^2 \neq 0$. In particular, for $x = -1$ and $y = 2$, we find the slope of the tangent line to the graph at $(-1, 2)$ to be

$$\frac{dy}{dx} = \frac{-(3 + 2)}{-1 + 12} = -\frac{5}{11} \qquad \blacksquare$$

The prime notation y', y'', and so on, is usually used in finding higher-order derivatives for implicitly defined functions.

EXAMPLE 5   Using implicit differentiation, find y' and y'' in terms of x and y if $xy + y^2 - x^2 = 5$.

Solution

(3.52) $$y + xy' + 2yy' - 2x = 0$$
$$y'(x + 2y) = 2x - y$$

(3.53) $$y' = \frac{2x - y}{x + 2y} \qquad \text{provided} \quad x + 2y \neq 0$$

It is easier to find y'' by differentiating (3.52) than by using (3.53):

$$y' + y' + xy'' + 2y'(y') + 2yy'' - 2 = 0$$
$$y''(x + 2y) = 2 - 2y' - 2(y')^2$$
$$y'' = \frac{2 - 2y' - 2(y')^2}{x + 2y} \qquad \text{provided} \quad x + 2y \neq 0$$

To express y'' in terms of x and y, use (3.53). Then

$$y'' = \frac{2 - 2\left(\dfrac{2x - y}{x + 2y}\right) - 2\left(\dfrac{2x - y}{x + 2y}\right)^2}{x + 2y} = \frac{-10(x^2 - xy - y^2)}{(x + 2y)^3} \underset{\underset{x^2 - xy - y^2 = -5}{\uparrow}}{=} \frac{50}{(x + 2y)^3} \qquad \blacksquare$$

EXERCISE 9

In Problems 1–30 find dy/dx by using implicit differentiation.

1. $x^2 + y^2 = 4$ 2. $3x^2 + 2y^2 = 6$ 3. $x^2y = 5$ 4. $x^3y = 8$
5. $x^2 - y^2 - xy = 2$ 6. $x^2y + xy^2 = x + 1$ 7. $x^2 - 4xy + y^2 = y$ 8. $x^2 + 2xy + y^2 = x$

9. $3x^2 + y^3 = 4$

10. $y^4 - 4x^2 = 4$

11. $4x^3 + 2y^3 = x$

12. $5x^2 + xy - y^2 = 0$

13. $\dfrac{1}{x^2} - \dfrac{1}{y^2} = 1$

14. $\dfrac{1}{x^2} + \dfrac{1}{y^2} = 1$

15. $\dfrac{1}{x} + \dfrac{1}{y} = 1$

16. $\dfrac{1}{x} - \dfrac{1}{y} = 4$

17. $(x^2 + y)^3 = y$

18. $(x + y^2)^3 = 3x$

19. $\dfrac{x}{y} + \dfrac{y}{x} = 4$

20. $x^2 + y^2 = \dfrac{2y}{x}$

21. $x^2 = \dfrac{y^2}{y^2 - 1}$

22. $x^2 + y^2 = \dfrac{2y^2}{x^2}$

23. $y = x \sin y$

24. $y = x \cos y$

25. $\sin(xy) + xy = 0$

26. $x + y = \cos(x - y)$

27. $y = \tan(x - y)$

28. $y = \cos(x + y)$

29. $y = \sin(x + y) + \cos(x - y)$

30. $\cos(x + y) = y \sin x$

In Problems 31–34 find y' and y'' in terms of x and y.

31. $x^2 + y^2 = 4$

32. $x^2 - y^2 = 1$

33. $xy + yx^2 = 2$

34. $4xy = x^2 + y^2$

In Problems 35 and 36 find the slope of the tangent line at the indicated point. Write an equation for this tangent line.

35. $x^2 + y^2 = 5$,   at $(1, 2)$

36. $x^2 - y^2 = 8$,   at $(3, 1)$

37. Use implicit differentiation to show that the tangent line to a circle $x^2 + y^2 = R^2$ at any point P on the circle is perpendicular to OP, where O is the center of the circle.

38. For ideal gases, *Boyle's law* states that pressure is inversely proportional to volume. A more realistic relationship between pressure P and volume V is given by *van der Waals equation*

$$P + \frac{a}{V^2} = \frac{C}{V - b}$$

where C is the constant of proportionality, a is a constant that depends on molecular attraction, and b is a constant that depends on the size of the molecules. Find the compressibility of the gas, which is measured by dV/dP.

39. Given the equation $x + xy + 2y^2 = 6$:
   (a)   Find an expression for the slope of the tangent line at any point (x, y) on the graph.
   (b)   Write an equation for the line tangent to the graph at the point $(2, 1)$.
   (c)   Find the coordinates of all other points on this graph with slope equal to the slope at $(2, 1)$.

40. The graph of the function $(x^2 + y^2)^2 = x^2 - y^2$ contains exactly four points at which the tangent line is horizontal. Find them.

10. Derivative of an Inverse Function; Rational Exponents

There is a simple relationship between the derivative of a function and the derivative of the inverse function.* Let f be the function and let f^{-1} be its inverse function. Set $g = f^{-1}$ to simplify notation. Then, as a result of (1.26), we have

$$f^{-1}(f(x)) = x \qquad \text{or} \qquad g(f(x)) = x$$

---

* You may wish to review Section 7 of Chapter 1 before proceeding further.

If both g and f are differentiable, we may apply the chain rule (3.47) and conclude that

$$[g'(f(x))][f'(x)] = 1$$

Since the product is not 0, this shows that each function has a nonzero derivative.

It turns out that, conversely, if one of the functions has a nonzero derivative, the other also has a nonzero derivative. (See Appendix II for a proof.) More precisely, we have the following theorem.

THEOREM Let $y = f(x)$ and $x = g(y)$ be inverse functions. Assume that f is differentiable on an open interval containing x_0 and that $y_0 = f(x_0)$. If $f'(x_0) \neq 0$, then g is differentiable at y_0 and

(3.54)
$$g'(y_0) = \frac{1}{f'(x_0)}$$

DERIVATIVE OF $\sqrt[n]{x}$

We now use (3.54) to find the derivative of $\sqrt[n]{x}$. The function $y = f(x) = x^n$, n a positive integer, has the derivative nx^{n-1}. Hence, if $x \neq 0$, then $f'(x) \neq 0$. The inverse function of f, namely, $x = g(y) = \sqrt[n]{y}$, is defined for all y if n is odd, and for all $y \geq 0$ if n is even. By (3.54), this inverse function is differentiable for all $y \neq 0$ and

$$g'(y) = \frac{d}{dy} \sqrt[n]{y} = \frac{1}{f'(x)} = \frac{1}{nx^{n-1}}$$

But $nx^{n-1} = n(\sqrt[n]{y})^{n-1} = ny^{(n-1)/n} = ny^{1-(1/n)}$. Therefore,

$$\frac{d}{dy} \sqrt[n]{y} = \frac{d}{dy} y^{1/n} = \frac{1}{ny^{1-(1/n)}} = \frac{1}{n} y^{(1/n)-1}$$

Remembering that the symbol used to represent the variable is immaterial, we may replace y by x. Thus, we have the formula

(3.55)
$$\frac{d}{dx} x^{1/n} = \frac{1}{n} x^{(1/n)-1}$$

EXAMPLE 1   (a) $\dfrac{d}{dx} \sqrt{x} = \dfrac{d}{dx} x^{1/2} = \dfrac{1}{2} x^{(1/2)-1} = \dfrac{1}{2} x^{-1/2} = \dfrac{1}{2x^{1/2}} = \dfrac{1}{2\sqrt{x}}$

(b) $\dfrac{d}{dx} x^{1/5} = \dfrac{1}{5} x^{-4/5}$

(c) $\dfrac{d}{dx} \sqrt[3]{x} = \dfrac{d}{dx} x^{1/3} = \dfrac{1}{3} x^{-2/3} = \dfrac{1}{3\sqrt[3]{x^2}}$ ∎

DERIVATIVE OF $x^{p/q}$

We can apply (3.55) to get a formula for the derivative of $x^{p/q}$, where p/q is a *rational* exponent. We find the derivative of $x^{p/q}$, where $q > 0$ and p are integers, as follows:

$$\frac{d}{dx} \, x^{p/q} = \underset{\uparrow}{\frac{d}{dx}} \, (x^{1/q})^p = \underset{\uparrow}{p(x^{1/q})^{p-1}} \left[\frac{d}{dx} \, x^{1/q} \right]$$

$$\qquad\qquad \text{Definition} \qquad \text{Power rule (3.46)}$$

$$= \underset{\uparrow}{px^{(p-1)/q}} \left[\frac{1}{q} \, x^{(1/q)-1} \right] = \underset{\uparrow}{\frac{p}{q} \, x^{p/q - 1/q + 1/q - 1}} = \frac{p}{q} \, x^{(p/q)-1}$$

$$\qquad\quad \text{Formula (3.55)} \qquad\qquad \text{Simplify}$$

Thus, we have the formula*

(3.56)
$$\frac{d}{dx} \, x^{p/q} = \frac{p}{q} \, x^{(p/q)-1}$$

EXAMPLE 2   (a) $\dfrac{d}{dx} \, x^{5/2} = \dfrac{5}{2} x^{3/2}$ (b) $\dfrac{d}{dx} \, x^{2/3} = \dfrac{2}{3} x^{-1/3}$

(c) $\dfrac{d}{dx} \, x^{-3/2} = -\dfrac{3}{2} x^{-5/2}$ (d) $\dfrac{d}{dx} \, x^{-4/3} = -\dfrac{4}{3} x^{-7/3}$ ∎

Because of (3.56), we may restate the power rule as follows:

(3.57)   **If g is a differentiable function and r is any rational number, then**

$$\frac{d}{dx} \, [g(x)]^r = r[g(x)]^{r-1} g'(x)$$

EXAMPLE 3   For the function $f(x) = \sqrt{x^2 + 4}$ find those x, if any, where $f'(x) = 0$.   At which x will $f''(x) = 0$?

Solution   Use the power rule (3.46) with $y = \sqrt{g(x)} = [g(x)]^{1/2}$ and $g(x) = x^2 + 4$:

$$f'(x) = \tfrac{1}{2}(x^2 + 4)^{-1/2}(2x) = \frac{1}{2\sqrt{x^2 + 4}} \, (2x) = \frac{x}{\sqrt{x^2 + 4}}$$

---

* The formula (3.56) is valid for *any real exponent* α (the proof is given in Chap. 7). That is

$$\frac{d}{dx} \, x^\alpha = \alpha x^{\alpha - 1} \qquad \alpha \text{ a real number}$$

Using this fact,

$$\frac{d}{dx} \, x^{\sqrt{2}} = \sqrt{2} x^{\sqrt{2} - 1} \qquad \text{and} \qquad \frac{d}{dx} \, x^\pi = \pi x^{\pi - 1}$$

Now, $f'(x) = 0$ if $x/\sqrt{x^2 + 4} = 0$, or $x = 0$. Next, by using the rule for differentiating a quotient,

$$f''(x) = \frac{(1)\sqrt{x^2 + 4} - x(x/\sqrt{x^2 + 4})}{x^2 + 4} = \frac{x^2 + 4 - x^2}{(x^2 + 4)^{3/2}} = \frac{4}{(x^2 + 4)^{3/2}}$$

Notice that $f''(x)$ is never 0; in fact, $f''(x) > 0$ for all x.   ■

Because of (3.57), we can now find the derivative of any algebraic function—that is, any function composed of a finite number of sums, products, quotients, powers, and roots of x.

Let's look at another use of formula (3.54). In the Leibniz notation, this formula assumes the simple form

$$\frac{dx}{dy} = \frac{1}{dy/dx}$$

An advantage of (3.54) is that it enables us to find the derivative of the inverse function at y_0, where $y_0 = f(x_0)$, without explicitly knowing a formula for $g = f^{-1}$, provided we can determine x_0 and thus calculate $f'(x_0)$.

Let's look at an example.

EXAMPLE 4   The function $f(x) = x^5 + x$ has an inverse function g. Find $g'(2)$.

Solution   To find $g'(2)$, we utilize (3.54). Then

$$g'(2) = \frac{1}{f'(x_0)} \qquad \text{where} \quad 2 = f(x_0)$$

By inspection, we find that a solution of the equation

$$f(x_0) = x_0^5 + x_0 = 2$$

is $x_0 = 1$.   Since $f'(x) = 5x^4 + 1$,   it follows that

$$g'(2) = \frac{1}{f'(1)} = \frac{1}{(5)(1^4) + 1} = \frac{1}{6}$$ ■

Observe in Example 4 that we calculated the derivative of the inverse g without actually knowing a formula for g.

APPLICATION: CALCULATING THE RATE OF SPREADING OF AN OIL SPILL (IF TIME PERMITS)

The situation we wish to analyze is that of an oil spill from a tanker at sea. Our question is "At what rate will the oil spread?" We make several assumptions to simplify and clarify the situation. First, we assume that a fixed volume V of oil is spilled and that the spill occurs in a short period of time. Second, we assume that the sea is calm and that the oil spreads in the shape of a circle with uniform thickness that may vary over time. Figure 8 illustrates the situation.

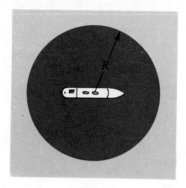

Figure 8

Let

$$V = \text{Fixed volume of oil spilled}$$
$$R = R(t), \quad \text{the radius of the circle at time } t$$
$$h = h(t), \quad \text{the thickness of the oil at time } t$$

Since the thickness of the oil is uniform, the spill takes on the shape of a cylinder. Thus,

(3.58)
$$V = \pi R^2(t)h(t)$$

In the early stages of an oil spill, the thickness h of the oil may be calculated from the formula

$$h(t) = \frac{k}{t^{1/2}} \quad k \text{ a constant}$$

Replacing $h(t)$ by $k/t^{1/2}$ in (3.58) gives

(3.59)
$$V = \pi R^2(t)\frac{k}{t^{1/2}}$$

$$Vt^{1/2} = \pi k R^2(t)$$

We wish to find the rate of spreading of the oil. Since this is measured by the time rate of change of the radius R, we differentiate (3.59) with respect to t to get dR/dt:

$$\frac{V}{2}t^{-1/2} = 2\pi k R(t)\frac{dR}{dt}$$

By solving for dR/dt and using (3.59), we find the rate of oil spreading to be given by the formula

$$\frac{dR}{dt} = \frac{1}{4}\left(\frac{V}{\pi k R}\right)t^{-1/2} \overset{\scriptstyle=}{\underset{\uparrow}{}} \frac{1}{4}\frac{R}{t}$$

$$\frac{V}{\pi k R} = Rt^{-1/2}$$

EXERCISE 10

In Problems 1–40 find dy/dx.

1. $y = x^{2/3} + 4$

2. $y = x^{1/3} - 1$

3. $y = \sqrt[3]{x^2}$

4. $y = \sqrt[4]{x^5}$

5. $y = \dfrac{2}{x^{1/2}} + \dfrac{3}{x^{1/3}} - \dfrac{4}{x^{3/2}} + \dfrac{8}{x^{3/4}}$

6. $y = 2x^{1/2} + 3x^{1/3} - 4x^{3/2}$

7. $y = \sqrt[3]{x} - \dfrac{1}{\sqrt[3]{x}}$

8. $y = 3\sqrt{x^3} + 4\sqrt[3]{x}$

9. $y = \sqrt{\dfrac{1}{3x}}$

10. $y = \sqrt{x} + \dfrac{1}{\sqrt{x}}$

11. $y = (x^3 - 1)^{1/2}$

12. $y = (x^2 - 1)^{1/3}$

13. $y = \sqrt{\sin x}$

14. $y = \sqrt[3]{\tan x}$

15. $y = \sec\sqrt{x}$

16. $y = \cos\sqrt{1 + x}$

17. $y = x\sqrt{x^2 - 1}$

18. $y = (\sqrt{x^2 + 4})^{2/3}$

19. $y = x\sqrt{x^3 + 1}$

20. $y = x^2\sqrt{x + 1}$

21. $y = \sqrt{3 - x^2} + \sqrt{4 - x^2}$

22. $y = \sqrt{x^2 + 2x - 1}$

23. $y = \sqrt{x^2 + 1}$

24. $y = \dfrac{\sqrt{x^2 + 1}}{x}$

25. $y = \sqrt{\dfrac{x + 1}{x - 1}}$

26. $y = \sqrt[3]{\dfrac{x}{x + 1}}$

27. $y = \sqrt{x^3(8x + 1)}$

28. $y = \dfrac{\sqrt[3]{3x + 1}}{\sqrt{3x + 2}}$

29. $y = \sqrt{4 + \sqrt[3]{x}}$

30. $y = \sqrt{4 + \sqrt{x}}$

31. $y = (x^2 \cos x)^{3/2}$

32. $y = (x^2 \sin x)^{3/2}$

33. $y = \sin(\cos\sqrt{x^2 + 1})$

34. $y = \cos(\sin\sqrt{x^2 - 1})$

35. $y = (x^2 - 3)^{1/2}(6x + 1)^{1/3}$

36. $y = (3x + 4)^{3/2}(x^3 - 4)^{2/3}$

37. $y = \dfrac{(2x^3 - 1)^{2/3}}{(3x + 4)^{1/2}}$

38. $y = \dfrac{(4x^2 - 1)^{1/4}}{(3x + 5)^{3/2}}$

39. $y = \sqrt[3]{2x^3 + x} \, \sqrt[4]{x^2 + 1}$

40. $y = \sqrt[3]{5x - 4} \, \sqrt[4]{2x^2 + x}$

In Problems 41 and 42 find y' and y'' for the given function.

41. $y = \sqrt{x^2 + 1}$

42. $y = \sqrt{4 - x^2}$

In Problems 43 and 44 find an equation of the tangent line at the given point.

43. $x^{1/3} + y^{1/3} = 1$,   at $(8, -1)$

44. $x^{2/3} + y^{2/3} = \frac{1}{2}$,   at $(\frac{1}{8}, \frac{1}{8})$

45. Show that the slope of the tangent line to $x^{2/3} + y^{2/3} = a^{2/3}$, $a > 0$,   at any point is $-y^{1/3}/x^{1/3}$.

46. Use implicit differentiation to find dy/dx if $x^2 + y^2 = 1$. Show that dy/dx equals the derivative of each of the functions $y = \sqrt{1 - x^2}$ and $y = -\sqrt{1 - x^2}$.

In Problems 47–50 the functions f and g are inverse functions. Calculate the indicated derivative.

47. If $f(0) = 4$, $f'(0) = -2$, find $g'(4)$.   **48.** If $f(1) = -2$, $f'(1) = 4$, find $g'(-2)$.

49. If $g(3) = -2$, $g'(3) = \frac{1}{2}$, find $f'(-2)$.   **50.** If $g(-1) = 0$, $g'(-1) = -\frac{1}{3}$, find $f'(0)$.

51. The function $f(x) = x^3 + 2x$ has an inverse function g. Find $g'(0)$ and $g'(3)$.

52. The function $f(x) = 2x^3 + x - 3$ has an inverse function g. Find $g'(-3)$ and $g'(0)$.

53. For a freely falling object that is falling from rest (in a vacuum), the distance s the object falls in time t is given by $s = \frac{1}{2}gt^2$. Its velocity v after time t is $v = gt$. Find dv/ds without eliminating t from the given equation.

11. Newton's Method of Solving Equations

Newton's method will enable us to find, to any desired degree of accuracy, the real roots of many equations.

Suppose we let $y = f(x)$ denote a function whose derivative f' is continuous, and we wish to find the real roots of the equation $f(x) = 0$. Graphically, this means that we are to find the x-intercepts of the graph. Now suppose that, from a graph or from tables or by trial calculations, we have found that the equation $f(x) = 0$ has a real root in some open interval containing the number $x = c_1$. We draw the tangent line to the graph of f at the point $P_1 = (c_1, f(c_1))$ and let P_2 be the point where this tangent line intersects the x-axis (see Fig. 9). Then, in general, if $x = c_1$

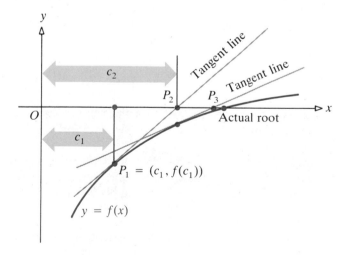

Figure 9

is a fair *first approximation* to the required root of the given equation $f(x) = 0$, then the x-intercept $c_2 = \overline{OP_2}$ of the tangent line will give a better, or *second*, *approximation* to the root. Graphically, this is the idea behind Newton's method of solving equations.

We may derive a formula for calculating the approximation c_2 as follows: The coordinates of P_1 are $(c_1, f(c_1))$, and the slope of the tangent line to the graph of f at P_1 is $f'(c_1)$. Therefore, the equation of this tangent line at P_1 is

$$y - f(c_1) = f'(c_1)(x - c_1)$$

By putting $y = 0$ in this equation and solving for x in order to find the point of intersection of the tangent line with the x-axis, we obtain

$$x = c_2 = c_1 - \frac{f(c_1)}{f'(c_1)}$$

Hence:

If $x = c_1$ is a sufficiently close first approximation to a real root of the equation $f(x) = 0$, then the formula

(3.60)
$$c_2 = c_1 - \frac{f(c_1)}{f'(c_1)}$$

gives a second approximation to the root.

We may now use c_2 in formula (3.60) in place of c_1 and get a third, and possibly closer, approximation to the required root of the equation; namely,

$$c_3 = c_2 - \frac{f(c_2)}{f'(c_2)}$$

The process may be repeated as often as required to give the desired degree of accuracy.*

EXAMPLE 1   Use Newton's method to find a third approximation to the real positive root r of the equation $x^3 - 2x - 5 = 0$, where $2 \leq r \leq 3$.

Solution   Let $f(x) = x^3 - 2x - 5$. We find $f(2) = -1$ and $f(3) = 16$, which indicates that the given equation has a root between 2 and 3, probably nearer 2 than 3. Therefore, we take $c_1 = 2$ and apply Newton's formula. Then, $f(c_1) = f(2) = -1$,

---

* Successive approximations may not get successively closer to the actual solution, even though they do approach the solution (as in a limit process). In addition, there are instances in which successive approximations do not approach the solution. Consult a numerical analysis book for details.

and since $f'(x) = 3x^2 - 2$, we have $f'(c_1) = f'(2) = 10$. By formula (3.60),

(3.61)
$$c_2 = 2 - \frac{(-1)}{10} = 2.1$$

Now we apply formula (3.60) again, with $c_1 = 2.1$ We find $f(2.1) = 0.061$ and $f'(2.1) = 11.23$.   Hence, a third approximation to the root is

$$c_3 = 2.1 - \frac{0.061}{11.23} = 2.1 - 0.0054 = 2.0946$$

To see how good this approximation is, we use a calculator and find that $f(2.0946) = 0.0005416$.   ∎

EXAMPLE 2   Use Newton's method to find a third approximation to the real root of the equation $\sin x + x - 1 = 0$.

Solution   To get the first approximation c_1, we note that the real root of $\sin x + x - 1 = 0$ occurs at the intersection of $y = \sin x$ and $y = 1 - x$, which is some number between 0 and 1 (see Fig. 10). We begin with $c_1 = 0$. To use (3.60), we set $f(x) = \sin x + x - 1$.   Then $f'(x) = \cos x + 1$ and

$$c_2 = c_1 - \frac{f(c_1)}{f'(c_1)} = 0 - \frac{f(0)}{f'(0)} = -\frac{-1}{2} = 0.5$$

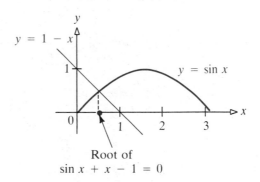

Root of
$\sin x + x - 1 = 0$

Figure 10

Since

$$f(0.5) = \sin \frac{1}{2} + \frac{1}{2} - 1 \approx -0.0206$$

Use a calculator

and

$$f'(0.5) = \cos \frac{1}{2} + 1 \approx 1.8776$$

we find the third approximation to be

$$c_3 = 0.5 - \frac{-0.0206}{1.8776} \approx 0.5 + 0.0110 = 0.5110$$

Using a calculator, we find that $f(0.5110) = -0.0095745.$ ∎

EXERCISE 11

Solve Problems 1–8 by using Newton's method. In each equation find a third approximation to the root indicated.

1. $x^3 + 3x - 5 = 0$;   root between 1 and 2
2. $x^3 - 4x + 2 = 0$;   root between 1 and 2
3. $2x^3 + 3x^2 + 4x - 1 = 0$;   root between 0 and 1
4. $x^3 - x^2 - 2x + 1 = 0$;   root between 0 and 1
5. $x^3 - 6x - 12 = 0$;   root between 3 and 4
6. $3x^3 + 5x - 40 = 0$;   root between 2 and 3
7. $x^4 - 2x^3 + 21x - 23 = 0$;   root between 1 and 2
8. $x^4 - x^3 + x - 2 = 0$;   root between 1 and 2

9. Use Newton's method to find a second approximation to the positive root of the equation $2x - 3 \sin x = 0$. Use $c_1 = 1.5$ as your first approximation.

10. Follow the directions of Problem 9 for the equation $x^2 = 2 \cos x$,   using $c_1 = \pi/4$ as your first approximation.

11. The volume of a spherical segment is given by $V = \frac{1}{3}\pi h^2(3R - h)$,   where R is the radius of the sphere and h is the height of the segment. If $R = 4$ feet   and $V = 12$ cubic feet,   find a second approximation to h.

12. A solid wooden sphere of diameter d and specific gravity S sinks in water to a depth h, which is determined by the equation $2x^3 - 3x^2 + S = 0$,   where $x = h/d$.   Find a second approximation to h for a maple ball of diameter 6 inches for which $S = 0.786$.

13. The equation $x - p \sin x = M$,   called *Kepler's equation*, occurs in astronomy. Find a second approximation to x when $p = 0.2$ and $M = 0.85$.   Use $c_1 = 1$ as your first approximation.

12. Functions that Are Not Differentiable at c

Thus far we have been totally concerned with functions f that are differentiable at a number c. In this section we look at some functions that are not differentiable at c—that is, functions f for which

(3.62)
$$\lim_{x \to c} \frac{f(x) - f(c)}{x - c}$$

does not exist. Two* of the most common ways that (3.62) may fail to exist are:

1. When f has a corner at c (see Fig. 11)
2. When f is not continuous at c (see Fig. 12)

---

* There are other ways, but they will be discussed later.

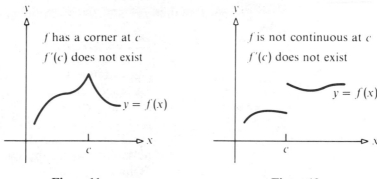

Figure 11   **Figure 12**

The first of these situations can be described most easily in terms of one-sided derivatives.

ONE-SIDED DERIVATIVES

The derivative of a function f at c is given by

$$f'(c) = \lim_{x \to c} \frac{f(x) - f(c)}{x - c}$$

From (2.23), a criterion for this limit to exist is that the one-sided limits

$$\lim_{x \to c^-} \frac{f(x) - f(c)}{x - c} \quad \text{and} \quad \lim_{x \to c^+} \frac{f(x) - f(c)}{x - c}$$

exist and are equal. These limits, when they exist, are referred to as the *left derivative of f at c* and the *right derivative of f at c*, respectively. Collectively, these are called *one-sided derivatives of f at c.*

EXAMPLE 1   For the function below, determine whether $f'(1)$ exists.

$$f(x) = \begin{cases} -2x^2 + 4 & \text{if} \quad x < 1 \\ x^2 + 1 & \text{if} \quad x \geq 1 \end{cases}$$

Solution   We need to find the limit

$$\lim_{x \to 1} \frac{f(x) - f(1)}{x - 1} = \lim_{x \to 1} \frac{f(x) - 2}{x - 1}$$

But we face a difficulty. If $x < 1$, then $f(x) = -2x^2 + 4$; if $x \geq 1$, then $f(x) = x^2 + 1$.   Consequently, we calculate the one-sided derivatives of f at 1:

$$\lim_{x \to 1^-} \frac{f(x) - f(1)}{x - 1} = \lim_{x \to 1^-} \frac{(-2x^2 + 4) - 2}{x - 1} = \lim_{x \to 1^-} \frac{-2(x^2 - 1)}{x - 1} = -4$$

$$\lim_{x \to 1^+} \frac{f(x) - f(1)}{x - 1} = \lim_{x \to 1^+} \frac{(x^2 + 1) - 2}{x - 1} = \lim_{x \to 1^+} \frac{(x - 1)(x + 1)}{x - 1} = 2$$

The left derivative of f at 1 is -4 and the right derivative of f at 1 is 2. Thus, $f'(1)$ does not exist.   ∎

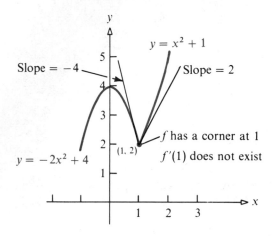

Figure 13

Figure 13 illustrates the graph of the function f in Example 1. At 1, where the derivative does not exist (and hence there is no tangent line), the graph of f has a *corner*. We usually say that the graph of f is not *smooth* at a corner.

Functions defined by more than one rule may be smooth where the split occurs. Let's look at an example.

EXAMPLE 2   For the function below, determine whether $f'(1)$ exists.

$$f(x) = \begin{cases} 2x^2 + 1 & \text{if } x \le 1 \\ 4x - 1 & \text{if } x > 1 \end{cases}$$

Solution   Since the rule for f changes at 1, we calculate the one-sided derivatives of f at 1:

$$\lim_{x \to 1^-} \frac{f(x) - f(1)}{x - 1} = \lim_{x \to 1^-} \frac{(2x^2 + 1) - (3)}{x - 1} = \lim_{x \to 1^-} \frac{2(x^2 - 1)}{x - 1} = \lim_{x \to 1^-} 2(x + 1) = 4$$

$$\lim_{x \to 1^+} \frac{f(x) - f(1)}{x - 1} = \lim_{x \to 1^+} \frac{(4x - 1) - (3)}{x - 1} = \lim_{x \to 1^+} \frac{4(x - 1)}{x - 1} = 4$$

Thus, $f'(1) = 4$. The graph of f is smooth at 1; that is, the graph of f has a tangent line at 1 and its slope is $f'(1) = 4$ (see Fig. 14).   ∎

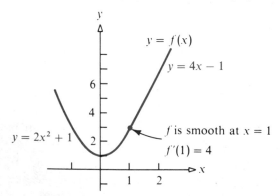

Figure 14

Finally, we mention that if f is defined only for $x \geq a$ or only for $x \leq b$, the slope of the tangent line to the graph of f at a would be given by the right derivative of f at a, and the slope of the tangent line to the graph of f at b would be given by the left derivative of f at b, provided these one-sided derivatives exist.

DERIVATIVES AND CONTINUITY

The next result states a relationship between a continuous function and a differentiable function.

(3.63)   **THEOREM**   **If a function f has a derivative at c, then it is continuous at c.**

Proof   To show that f is continuous at c, we need to verify that $\lim_{x \to c} f(x) = f(c)$. We begin by observing that $x \neq c$, so

$$f(x) - f(c) = \left[\frac{f(x) - f(c)}{x - c} \right](x - c)$$

Now we take the limit of both sides as $x \to c$ and use the fact that the limit of a product equals the product of the limits:

$$\lim_{x \to c}[f(x) - f(c)] = \lim_{x \to c}\left\{ \left[\frac{f(x) - f(c)}{x - c} \right](x - c) \right\}$$

$$= \left[\lim_{x \to c} \frac{f(x) - f(c)}{x - c} \right]\left[\lim_{x \to c}(x - c) \right]$$

Since f has a derivative at c, we know that

$$\lim_{x \to c} \frac{f(x) - f(c)}{x - c} = f'(c)$$

is a number. Furthermore, since $\lim_{x \to c}(x - c) = 0$, we find

$$\lim_{x \to c}[f(x) - f(c)] = [f'(c)](0) = 0$$

That is, $\lim_{x \to c} f(x) = f(c)$, so f is continuous at c.   ∎

Thus, every differentiable function is continuous. Or, to put it another way:

(3.64)   **If a function f is not continuous at a number c, then it has no derivative at c.**

For example, the greatest integer function $f(x) = [\![x]\!]$ has no derivative for x an integer since it is not continuous for x an integer.

The converse of (3.63) is not true. There are functions that are continuous at c and yet have no derivative at c. The function from Example 1 is continuous at 1, but it is not differentiable at 1.

Result (3.64) may save some time if you are asked to find the derivative of a function you suspect is not continuous. For, if the function is indeed not continuous at a number c, then by (3.64) the function f will have no derivative at c.

EXERCISE 12

In Problems 1–8 determine whether the given function f has a derivative at c. Graph each function.

1. $f(x) = \begin{cases} 2x + 3 & \text{if } x < 1 \\ x^2 + 4 & \text{if } x \geq 1 \end{cases} \quad c = 1$

2. $f(x) = \begin{cases} 3 - 4x & \text{if } x < -1 \\ 2x + 9 & \text{if } x \geq -1 \end{cases} \quad c = -1$

3. $f(x) = \begin{cases} -4 + 2x & \text{if } x \leq \frac{1}{2} \\ 4x^2 - 4 & \text{if } x > \frac{1}{2} \end{cases} \quad c = \frac{1}{2}$

4. $f(x) = \begin{cases} 2x^2 + 1 & \text{if } x < -1 \\ -1 - 4x & \text{if } x \geq -1 \end{cases} \quad c = -1$

5. $f(x) = |x^2 - 4| \qquad c = 2$

6. $f(x) = |x^2 - 4| \qquad c = -2$

7. $f(x) = \begin{cases} 2x^2 + 1 & \text{if } x < -1 \\ 2 + 2x & \text{if } x \geq -1 \end{cases} \quad c = -1$

8. $f(x) = \begin{cases} 5 - 2x & \text{if } x < 2 \\ x^2 & \text{if } x \geq 2 \end{cases} \quad c = 2$

9. For the function

$$f(x) = \begin{cases} x^3 & \text{if } x \leq 0 \\ x^2 & \text{if } x > 0 \end{cases}$$

determine whether: (a) f is continuous at 0   (b) $f'(0)$ exists
(c) Graph the function and give a geometric interpretation to the answers found in parts (a) and (b).

10. Repeat Problem 9 for the function

$$f(x) = \begin{cases} 2x & \text{if } x \leq 0 \\ x^2 & \text{if } x > 0 \end{cases}$$

11. Calculate analytically the velocity (in feet per second) of an automobile whose position is given by

$$s = \begin{cases} t^3 & \text{if } 0 \leq t < 5 \\ 125 & \text{if } 5 \leq t \end{cases}$$

This could represent a crash test in which a vehicle is accelerated into a brick wall. Find the velocity just before and just after impact. Are the formulas quoted accurate during impact?

Estimating Roots on a Computer

In Section 11 of this chapter, we saw how to use Newton's method to approximate a real root for the equation $f(x) = 0$. To approximate a root, Newton's method requires an initial estimate (c_1) for the root and repeated (frequently time-consuming) applications of the formula

$$c_{n+1} = c_n - \frac{f(c_n)}{f'(c_n)}$$

The successive applications of this formula produce the sequence of values c_1, c_2, c_3, . . . , which usually will approach the desired root for $f(x) = 0$. That is, $|f(c_n)|$ will approach 0 as n increases.

A number of problems are associated with Newton's method:

1. The initial estimate c_1 may not be "good enough." A poor first value can invalidate the whole process (and produce a really strange sequence of values).

2. A poor first value may also cause the sequence of values to approach the root so slowly that we have to repeat the formula calculation hundreds of times before we get close enough to the root to be satisfied.

3. One of the values in the sequence may fall at a place where the derivative is 0 (i.e., a horizontal tangent line). We must stop the process at that point since the formula is not defined for $f'(c_n) = 0$.

To see a few examples of these problems, use a hand calculator and Newton's formula to calculate a few steps in the sequence of estimates for each of the following functions:

(a) $f(x) = -x^3 + 6x^2 - 9x + 6$ with an initial estimate of $c_1 = 2.9$ and then with $c_1 = 3.0$ (The root is near $x = 4.2$.)

(b) $f(x) = x^8 - 1$ with an initial estimate of $c_1 = 0.1$

(c) $f(x) = (x - 1)^{1/3}$ with an initial estimate of $c_1 = 2.0$

Although the problems listed above are not the only problems that can occur, they are sufficient to discourage the use of Newton's method even when the calculations are done on a hand calculator. However, now that microcomputers are readily available, we can have a computer do all the tedious calculations, while we just sit back and observe how well Newton's method works.

Programming a computer to repeat a sequence of calculations is easily done in languages such as BASIC and Pascal. Each step (calculating the next estimate) is just another application of Newton's formula. However, we need a set of conditions that tell the computer when to stop doing the calculations. There are many different stopping conditions that we could choose. The set of conditions we will use is one that seems at first to be the most intuitively acceptable (and one that we might use if we had to do the calculations by hand).

We will instruct the computer to stop if:

(a) $|f(c_n)| <$ eps, where eps(ilon) is a measure of how close $f(c_n)$ is to 0^*

(b) The number of estimates (steps in the sequence) calculated so far reaches the maximum allowed (max)

(c) The derivative is 0, that is, $f'(c_n) = 0$

The algorithm (procedure, or list of instructions) we will use when we write the program that tells the computer what to do consists of a very simple loop:

> if (while) no stopping condition is satisfied
>   calculate the next estimate
>   print the values of c_n and $f(c_n)$
>   check the stopping conditions

Sample computer programs in BASIC and Pascal are shown here, which implement the strategy outlined above. Both were run on a standard microcomputer using the MS

---

*The primary stopping condition $|f(c_n)| <$ eps does not give us much information about the size of the actual error ($|c_n -$ Root$|$). Thus, in most implementations of Newton's method, the primary condition for terminating the loop is to stop when $|c_{n+1} - c_n| < \delta$. Here, δ is a user-supplied value that specifies how close we want our estimates to be to the root, instead of how close we want the $f(c_n)$ value to 0.

DOS operating system. Find a microcomputer and try to get one of these programs running. Then go back and redo some of the problems in Section 11. See how fast (or how slowly) Newton's method approaches each root.

Another alternative is to see if your school already owns a software package (i.e., computer program) that will do Newton's method for functions defined by the user. Often, these programs will offer a variety of different methods using other techniques (bisection, secants, etc.) for finding roots. Also, many of these programs are designed to show you graphically what each method does as it approximates a root.

PROGRAMS

The programs given here were not written for speed or efficiency. We simply want to display the basic algorithm as clearly as possible. We anticipate that the version of BASIC you use will be an interpreted form of the language and that Pascal will probably have to be a compiled language. Therefore, the BASIC version is a stripped-down, but perfectly functional, version of the Pascal program.

Both programs are currently set up to apply Newton's method to the function $f(x) = x^3 - 2x - 5$. In order to apply these programs to other functions, the function definitions in the programs for $f(x)$ and $f'(x)$ must be changed.

The identifiers used in the programs are:

c1   = Initial estimate for the desired root
cn   = Current estimate for the root
eps   = Tolerance limit for how close we want $|f(c_n)|$ to 0
max   = Maximum number of estimation steps that will be done
n   = Current step in the sequence of estimates
abs( ) = Absolute value function

In the BASIC program, the values for c_1, eps, and max are assigned in lines 120–140. In the Pascal program, the user must enter these values at the computer keyboard after the program has begun.

```
100  DEF FNF(X)=X^3-2*X-5
110  DEF FNP(X)=3*X^2-2
120  CN=.1
130  EPS=.0001
140  MAX=20
150  PRINT "STEP","X","F(X)"
160  N=1
170  PRINT N,CN,FNF(CN)
180  IF (ABS(FNF(CN)) < EPS) OR (FNP(CN) = 0) OR (N >= MAX) THEN GOTO 500
190      N=N+1
200      CN=CN-FNF(CN)/FNP(CN)
210      PRINT N,CN,FNF(CN)
220  GOTO 180
500  IF ABS(FNF(CN)) < EPS THEN PRINT "F(X) WITHIN TOLERANCE" : END
510  IF FNP(CN) = 0  THEN PRINT "DERIVATIVE = 0 - LOOP TERMINATED" : END
520  IF N >= MAX THEN PRINT "MAXIMUM NUMBER OF STEPS"
```

BASIC Program

```
program newton_method;

function f(x: real) : real;
begin (* f *)
     f := x * x * x  - 2.0 * x - 5.0
end; (* f *)

function fp(x: real) : real;
begin (* fprime *)
     fp := 3.0 * x * x  - 2.0
end; (* fp = f prime *)

var
     cn, eps : real;
     n, max : integer;

begin (* main *)
     write('Enter the initial estimate for the root --> ');
     readln(cn);
     write('Enter the maximum number of approximation steps --> ');
     readln(max);
     write('Enter the tolerance limit for  ! f(cn) !  --> ');
     readln(eps);
     writeln('step','x':10,'f(x)':20);
     n := 1;
     writeln(n:3, cn:18:7, f(cn):18:7);
     while (abs(f(cn)) >= eps) and (fp(cn) <> 0.0) and (n < max) do
     begin
          cn := cn - f(cn) / fp(cn);
          n := n + 1;
          writeln(n:3, cn:18:7, f(cn):18:7);
     end; (* while *)
     if abs(f(cn)) < eps then
          writeln('f(x) value within tolerance of 0')
     else if fp(cn) = 0.0 then
          writeln('derivative = 0 -- cannot continue')
     else
          writeln('maximum number of steps allowed')
end. (* main *)
```

Pascal Program

Miscellaneous Exercises

In Problems 1–42 find the derivative. When a or b appears, it is a constant.

1. $y = (ax + b)^n$

2. $y = \sqrt{2ax}$

3. $y = x\sqrt{1 - x}$

4. $y = \dfrac{1}{\sqrt{x^2 + 1}}$

5. $f(x) = \dfrac{x^2}{x + 1}$

6. $F(z) = \dfrac{2}{\sqrt{z}}$

7. $u = \dfrac{r}{a^2\sqrt{a^2 - r^2}}$

8. $s = \dfrac{t^3}{t - 2}$

9. $y = (x^2 + 4)^{3/2}$

10. $y = x(a^2 + x^2)\sqrt{a^2 - x^2}$

11. $y = 3x^{-2} + 2x^{-1} + 1$

12. $w = (z^2 + 1)^{5/2}$

13. $u = \dfrac{1}{z^2 + 1}$

14. $s = (t - 1)^{5/2}$

15. $f(x) = \dfrac{x^2}{\sqrt{x^2 - 1}}$

16. $\phi(x) = \dfrac{\sqrt{x + 1}}{x}$

17. $q = \dfrac{\sqrt{p^2 - 1}}{a^2 p}$

18. $r = (x + \sqrt{x^2 - 1})^n$

19. $y = x^2(a^2 + x^2)^{3/2}$

20. $z = \dfrac{\sqrt{2ax - x^2}}{x}$

21. $y = \sqrt{x + 2}$

22. $y = \sqrt{x} + \sqrt[3]{x}$

23. $u = \dfrac{y}{\sqrt{y^2 + 9}}$

24. $w = \dfrac{v - 1}{v^2 + 1}$

25. $f(x) = \sqrt{1 - x^2}$

26. $F(x) = \dfrac{2}{(a^2 - x^2)^{3/2}}$

27. $w = \dfrac{1}{1 - z + z^2}$

28. $r = (t - 1)\sqrt{t^2 + 1}$

29. $y = (1 - x^3)^{1/2}$

30. $\phi(x) = \dfrac{(x^2 - a^2)^{3/2}}{x^3}$

31. $v = (1 + u)^{3/2}$

32. $y = x\sqrt{3 - x}$

33. $f(x) = \dfrac{x^2}{(x - 1)^2}$

34. $g(x) = x^2\sqrt{2ax - x^2}$

35. $u = (a^{1/2} - x^{1/2})^2$

36. $v = (2r - 1)^{5/2}$

37. $y = x \sin 2x$

38. $u = \cos^3 x$

39. $v = \tan u + \sec u$

40. $y = \sqrt{a^2 \sin(x/a)}$

41. $\phi(z) = \sqrt{1 + \sin z}$

42. $u = \sin v - \frac{1}{3} \sin^3 v$

43. If $f(x) = (x - 1)/(x + 1)$ for all $x \neq -1$, find $f'(1)$.

44. If $f(x) = x^{1/2}(x - 2)^{3/2}$ for all $x \geq 2$, find the domain of f'.

45. If $f(x) = 2 + |x - 3|$ for all x, determine whether the derivative $f'(x)$ exists at $x = 3$.

46. If $\tan(xy) = x$, find dy/dx.

47. If $y = x + \sin(xy)$, find dy/dx.

48. Let $f(x) = 4x^3 - 3x - 1$.
   (a) Find the x-intercepts of the graph of f.
   (b) Write an equation for the tangent line to the graph of f at $x = 2$.
   (c) Write an equation of the graph that is the reflection across the y-axis of the graph of f.

49. Let f be the function defined by $f(x) = \sqrt{1 + 6x}$.
   (a) Give the domain and range of f.
   (b) Determine the slope of the line tangent to the graph of f at $x = 4$.
   (c) Determine the y-intercept of the line tangent to the graph of f at $x = 4$.
   (d) Give the coordinates of the point on the graph of f where the tangent line is parallel to $y = x + 12$.

50. If the line $3x - 4y = 0$ is tangent to the graph of $y = x^3 + k$ in the first quadrant, find k.

51. For what nonnegative number b is the line given by $y = -\frac{1}{3}x + b$ normal to the graph of $y = x^3$?

52. If

$$\frac{d}{dx} f(x) = g(x) \quad \text{and} \quad h(x) = x^2$$

find $(d/dx)f(h(x))$.

53. What is the nth derivative of $f(x) = 1/(2 + x)$ at 0?

54. Let $f(x) = x^3 + x$. If h is the inverse function of f, find $h'(2)$.

55. Write a formula for the derivative of the product of four differentiable functions. That is, find a formula for $(d/dx)[f_1(x)f_2(x)f_3(x)f_4(x)]$.

56. If f and g are differentiable functions, prove that if $k(x) = 1/[f(x)g(x)]$, then

$$k'(x) = -k(x)\left[\frac{f'(x)}{f(x)} + \frac{g'(x)}{g(x)}\right]$$

57. If f and g are twice differentiable functions, find formulas for the second derivative of:

(a) $f(x)g(x)$ (b) $\dfrac{1}{g(x)}$

58. Prove that if y is a differentiable function of x such that x and y are related by the equation $x^n y^m + x^m y^n = k$ (k a constant), then

$$\frac{dy}{dx} = -\frac{y(nx^r + my^r)}{x(mx^r + ny^r)} \qquad \text{where} \quad r = n - m$$

59. A particle moves in a straight line according to the law $s = \frac{1}{8}\cos 4\pi t$. Find its velocity and acceleration.

60. As a particle of mass m moves along the x-axis, its position s and velocity $v = ds/dt$ obey the equation

$$m(v^2 - v_0^2) = k(s_0^2 - s^2)$$

where k is a positive constant and v_0, s_0 are the initial velocity and position, respectively. Show that if $v > 0$, then

$$ma = -ks$$

where $a = d^2x/dt^2$ is the acceleration of the particle. [*Hint:* Differentiate the expression $m(v^2 - v_0^2) = k(s_0^2 - s^2)$ with respect to t.]

61. Find the constants a, b, c so that the graph of $y = ax^2 + bx + c$ passes through the point $(-1, 1)$ and is tangent to the line $y = 2x$ at $(0, 0)$.

62. At what point does the graph of $y = 1/\sqrt{x}$ have a tangent line parallel to the line $x + 16y = 5$?

63. If f and g are differentiable functions, find the derivative of $fg/(f + g)$.

64. If a function f has the properties that   (1) $f(u + v) = f(u)f(v)$ for all choices of u and v, and   (2) $f(x) = 1 + xg(x)$,   where $\lim_{x \to 0} g(x) = 1$,   show that $f' = f$.

65. Find y' and y'' at the point $(-1, 1)$ if $3x^2 y + 2y^3 = 5x^2$.

66. If n is an odd positive integer, show that the tangent lines to the graph of $y = x^n$ at $(1, 1)$ and at $(-1, -1)$ are parallel.

67. If n is an even positive integer, show that the tangent line to the graph of $y = \sqrt[n]{x}$ at $(1, 1)$ is perpendicular to the tangent line to the graph of $y = x^n$ at $(-1, 1)$.

68. Show that the tangent line to the graph of $y = x^n$ at $(1, 1)$ has y-intercept $1 - n$.

69. If $f(x) = \sin x$ and $F(t) = f(t^2 - 1)$, find $F'(1)$.

70. Find y'' in terms of y alone if $b^2 x^2 + a^2 y^2 = a^2 b^2$.

71. Find a, b, c, d so that the tangent line to the graph of the cubic $y = ax^3 + bx^2 + cx + d$ at the point $(1, 0)$ is $y = 3x - 3$ and at the point $(2, 9)$ is $y = 18x - 27$.

72. The graphs of two functions are said to be *orthogonal* (perpendicular to each other) if their tangent lines are perpendicular at each point of intersection.
  (a) Show that the graphs of $xy = c_1$ and $-x^2 + y^2 = c_2$ are orthogonal, where c_1 and c_2 are constants.
  (b) Plot the graphs on one coordinate system for $c_1 = 1, 2, 3$ and $c_2 = 1, 9, 25$.

73. Show that the parabolas $y^2 = 2ax + a^2$ and $y^2 = a^2 - 2ax$ intersect at right angles.

74. If $f(x) = Ax^2 + B$, then:
  (a) Find c in terms of A such that the tangent lines to the graph of f at $(c, f(c))$ and $(-c, f(-c))$ are perpendicular.
  (b) Find the slopes of the tangent lines in part (a).
  (c) Find the coordinates, in terms of A and B, of the point of intersection of the tangent lines in part (a).

75. A function f is defined for all real numbers and has the following three properties: (*1*) $f(1) = 5$,   (*2*) $f(3) = 21$,   and   (*3*) for all real a and b, $f(a + b) - f(a) = kab + 2b^2$, where k is a fixed real number independent of a and b.
  (a) Use $a = 1$ and $b = 2$ to find k.   (b) Find $f'(3)$.   (c) Find $f'(x)$ and $f(x)$ for all real x.

76. For a differentiable function f, let $f*$ be the function defined by

$$f*(x) = \lim_{h \to 0} \frac{f(x + h) - f(x - h)}{h}$$

  (a) Determine $f*(x)$ for $f(x) = x^2 + x$.
  (b) Determine $f*(x)$ for $f(x) = \cos x$.
  (c) Write an equation that expresses the relationship between the functions $f*$ and f', where f' denotes the derivative of f. Justify your answer.

77. The line $x = c$, where $c > 0$, intersects the cubic $y = 2x^3 + 3x^2 - 9$ at point P and the parabola $y = 4x^2 + 4x + 5$ at point Q.
  (a) If a line tangent to the cubic at point P is parallel to the line tangent to the parabola at point Q, find the number c.
  (b) Write the equations of the two tangent lines in part (a).

78. A function f is periodic if there is a positive number p so that $f(x + p) = f(x)$ for all x. Show that if f is a differentiable function of period p, then f' is also of period p.

79. If the position s of an object at time t is $s = t/(t^2 - 1)$, find its velocity and its acceleration.

80. If $f(x) = 1/(1 - x)$, find a formula for the nth derivative—that is, $f^{(n)}(x)$.

81. Let f and g be two differentiable functions at c. State the relationship between their tangent lines at c if:

  (a) $f'(c) = g'(c)$ (b) $f'(c) = -\dfrac{1}{g'(c)}$

82. A particle moves in a straight line according to the equation $s = 2t^3 - 15t^2 + 24t + 3$, where t is measured in minutes and s in meters. Determine:
  (a) When the particle is at rest   (b) Acceleration when $t = 3$

83. Find the value of the limit below and specify the function f for which this is the derivative.

$$\lim_{\Delta x \to 0} \frac{[4 - 2(x + \Delta x)]^2 - (4 - 2x)^2}{\Delta x}$$

84. Find equations of the tangent and normal lines to the graph of $y = x\sqrt{x} + (x - 1)^2$ at the point $(2, 2\sqrt{3})$.

85. Let T be the line tangent to the graph of $y = x^3$ at the point $(\frac{1}{2}, \frac{1}{8})$. At what point Q does the line T intersect the graph? What is the angle between the tangent line at Q and the line T?

86. Let N be the line normal to the graph of $y = x^2$ at the point $(-2, 4)$. At what point Q does N meet the graph?

87. At what point(s), if any, is the line $y = x - 1$ tangent to the graph of $y = \sqrt{25 - x^2}$?

88. If $f(x) = \sqrt{x - 1}$, let A and B be points on the graph of f with $x = 2$ and $x = 5$, respectively. A line is moved upward on the graph such that it remains parallel to the secant line AB. Find the coordinates of the last point on the graph of f before the secant line loses contact with the graph.

4

Applications of the Derivative

In this chapter we present important applications of the techniques developed in Chapters 2 and 3. The first part gives applications of the derivative to related rates and differentials. Then we consider maxima and minima of functions and the related subject of graphing functions. Finally, we introduce the concept of antiderivatives, which is the basis for our later study of integral calculus.

1. Related Rates

In all of the natural sciences and many of the social and behavioral sciences, quantities that are related, but that vary with time, are encountered. For example, the pressure of an ideal gas of fixed volume is proportional to temperature, yet each of these quantities may change over a period of time. Problems involving rates of related variables are referred to as *related rate problems*. In such problems we normally want to find the rate at which one of the variables is changing at a certain time, while the rates at which the other variables are changing are known.

The usual procedure in such problems is to write an equation that relates all the time-dependent variables involved. Such a relationship is often obtained by investigating the geometric and/or physical conditions imposed by the problem. When this relationship is differentiated with respect to the time t, a new equation that involves the variables and their rates of change with respect to time is obtained.

For example, suppose x and y are two differentiable functions of time t; that is, $x = x(t)$, $y = y(t)$.   And suppose they obey the equation

$$x^3 - y^3 + 2y - x - 99 = 0$$

If we differentiate with respect to time t (remember x and y are functions of time t), we obtain

$$3x^2 \frac{dx}{dt} - 3y^2 \frac{dy}{dt} + 2 \frac{dy}{dt} - \frac{dx}{dt} = 0$$

This equation is valid for all times t under consideration and involves the derivatives of x and y with respect to t, as well as the variables themselves. Because the derivatives are related by this equation, we call them *related rates*. We can solve for one of these rates once the value of the other rate and the values of the variables are known. For example, if in the above equation at a specific time t, we know that $x = 5$, $y = 3$, and $dx/dt = 2$,   then by direct substitution we find that $dy/dt = \frac{148}{25}$.

EXAMPLE 1   A child throws a stone into a still millpond causing a circular ripple to spread. If the radius of the circle increases at the constant rate of 0.5 meter per second, how fast is the area of the ripple increasing when the radius of the ripple is 20 meters?

Figure 1

Solution   The variables involved are:

t = Time (in seconds) elapsed from the time the throw hits the water

R = Radius of the ripple (in meters) after t seconds

A = Area of the ripple (in square meters) after t seconds

The rates involved are:

$$\frac{dR}{dt} = \text{Rate at which the radius is increasing at each instant}$$

$$\frac{dA}{dt} = \text{Rate at which the area is increasing with time}$$

We wish to find dA/dt when $R = 20$; that is, the rate at which the area of the ripple is increasing at the instant when $R = 20$. The relationship between A and R is given by the formula for the area of a circle:

(4.1)
$$A = \pi R^2$$

Since A and R are functions of t, we differentiate both sides of (4.1) with respect to t to obtain

(4.2)
$$\frac{dA}{dt} = 2\pi R \frac{dR}{dt}$$

Since the radius increases at the rate of 0.5 meter per second, we know that

(4.3)
$$\frac{dR}{dt} = 0.5$$

By substituting (4.3) into (4.2), we get

$$\frac{dA}{dt} = 2\pi R(0.5) = \pi R$$

Thus, when $R = 20$, the area of the ripple is increasing at the rate

$$\frac{dA}{dt} = \pi(20) = 20\pi \approx 62.8 \text{ square meters per second} \qquad \blacksquare$$

Example 1 illustrates some general guidelines that will prove helpful for solving related rate problems:

1. **If possible, draw a picture illustrating the problem.**
2. **Identify the variables and assign symbols to them.**
3. **Identify and interpret rates of change as derivatives.**
4. **Express all relationships among the variables by equations.**
5. **Obtain additional relationships among the variables and their derivatives by differentiating.**
6. **Substitute numerical values for the variables and the derivatives. Solve for the unknown rate.**

Note: It is important to remember that the substitution of numerical values must occur after the differentiation process (Step 5).

EXAMPLE 2   A balloon in the form of a sphere is being inflated at the rate of 10 cubic meters per minute. Find the rate at which the surface area of the sphere is increasing at the instant when the radius of the sphere is 3 meters.

Solution   The variables of the problem are:

t = Time (in minutes) measured from the moment inflation of the balloon begins

R = Length (in meters) of the radius of the balloon at time t

V = Volume (in cubic meters) of the balloon at time t

S = Surface area (in square meters) of the balloon at time t

The rates of change are:

$$\frac{dR}{dt} = \text{Rate of change of radius with respect to time}$$

$$\frac{dV}{dt} = \text{Rate of change of volume with respect to time}$$

$$\frac{dS}{dt} = \text{Rate of change of surface area with respect to time}$$

We are given that $dV/dt = 10$ cubic meters per minute, and we seek dS/dt when $R = 3$ meters. At any time t the volume V of the balloon (a sphere) is $V = \frac{4}{3}\pi R^3$ and the surface area S of the balloon is $S = 4\pi R^2$. By differentiating each of these equations with respect to the time t, we have

$$\frac{dV}{dt} = 4\pi R^2 \frac{dR}{dt} \qquad \text{and} \qquad \frac{dS}{dt} = 8\pi R \frac{dR}{dt}$$

In the equation for dV/dt, we solve for dR/dt and substitute this quantity into the equation for dS/dt. Then

$$\frac{dS}{dt} = 8\pi R \frac{dV/dt}{4\pi R^2} = \frac{2}{R}\frac{dV}{dt}$$

At $R = 3$ and $dV/dt = 10$, we have

$$\frac{dS}{dt} = \left(\frac{2}{3}\right)(10) \approx 6.67 \text{ square meters per minute}$$

Thus, the surface area is increasing at the rate of 6.67 square meters per minute when the radius is 3 meters. ∎

EXAMPLE 3   A rectangular swimming pool 10 meters long and 5 meters wide is 3 meters deep at one end and 1 meter deep at the other. (A cross-sectional view of the pool is illustrated in Fig. 2.) If water is pumped into the pool at the rate of 300 liters per minute, at what rate is the water level rising when it is 1.5 meters deep at the deep end? (*Note:*   1 liter of water $= 10^{-3}$ cubic meter.)

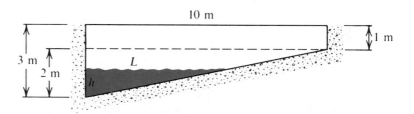

Figure 2

Solution   First, the variables involved are:

$t =$ Time (in minutes) measured from the moment water begins to flow into the pool

$h =$ Water level (in meters) measured at the deep end

$L =$ Distance (in meters) from the deep end toward the short end measured at water level

$V =$ Volume (in cubic meters) of water in the pool

The rates of change are:

$$\frac{dV}{dt} = \text{Rate of increase in volume at a given instant}$$

$$\frac{dh}{dt} = \text{Rate of increase in height at a given instant}$$

The volume V is related to L and h by the formula

(4.4) $V = (\text{Cross-sectional triangular area})(\text{Width}) = (\tfrac{1}{2}Lh)(5)$ cubic meters

Using similar triangles, we see from Figure 2 that L and h are related by the equation

$$\frac{L}{h} = \frac{10}{2}$$

$$L = 5h$$

By replacing L by $5h$ in (4.4), we have

(4.5) $V = \tfrac{1}{2}(5h)(h)(5) = \tfrac{25}{2}h^2$ cubic meters

Here, V and h are each functions of time t. By differentiating (4.5) with respect to t, we obtain

$$\frac{dV}{dt} = 25h\frac{dh}{dt} \text{ cubic meters per minute}$$

We seek the rate at which the water level is rising, dh/dt, when $h = 1.5$ meters and the rate of water pumped into the pool is $dV/dt = 300$ liters per minute $= 300(10^{-3})$ cubic meter per minute. Thus, the water level is rising at a rate of

$$\frac{dh}{dt} = \frac{(300)(10^{-3})}{25(1.5)} = 0.008 \text{ meter per minute} \quad \blacksquare$$

EXAMPLE 4   A person is standing on a pier and pulling a boat inward by pulling a rope at the rate of 2 meters per second. The end of the rope is 3 meters above water level. (See Fig. 3.) How fast is the boat approaching the base of the pier when 5 meters of rope are left to pull in? Disregard sagging of the rope and assume the rope is attached to the boat at water level.

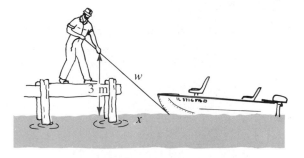

Figure 3

Solution   The variables of the problem are:

$t = $ Time (in seconds)

$x = $ Distance (in meters) from the boat to the base of the pier

$w = $ Distance (in meters) from the boat to the person
(that is, the length of rope)

The rates of change are:

$$\frac{dx}{dt} = \text{Rate at which the boat is approaching the pier}$$

$$\frac{dw}{dt} = \text{Rate at which the rope is being pulled}$$

From Figure 3 we see that

(4.6) $$w^2 = 9 + x^2$$

The variables w and x are functions of time t. By differentiating (4.6) with respect to time t, we obtain

$$2w \frac{dw}{dt} = 2x \frac{dx}{dt}$$

(4.7)

$$w \frac{dw}{dt} = x \frac{dx}{dt}$$

We seek to find dx/dt, the rate at which the boat is approaching the pier, when $w = 5$ meters and $dw/dt = -2$ meters per second. (The negative sign is used to indicate that the length of the rope is *decreasing* at the rate of 2 meters per second.) The value of x when $w = 5$ is found from equation (4.6) to be $x = \sqrt{25 - 9} = 4$ meters. By substituting these values into (4.7), we find

$$4 \frac{dx}{dt} = 5(-2)$$

$$\frac{dx}{dt} = -2.5 \text{ meters per second}$$

Thus, the boat is approaching the pier at the rate of 2.5 meters per second.   ∎

EXAMPLE 5   A revolving light located 5 kilometers from a straight shoreline has a constant angular velocity of 3 radians per minute.* With what velocity does the spot of light move along the shore when the beam makes an angle of 60° with the shoreline?

Solution   Using Figure 4, we find that the variables are:

Figure 4

$$t = \text{Time (in minutes)}$$
$$x = \text{Distance (in kilometers) of the beam of light from } B$$
$$\theta = \text{Angle (in radians) the beam of light makes with } AB$$

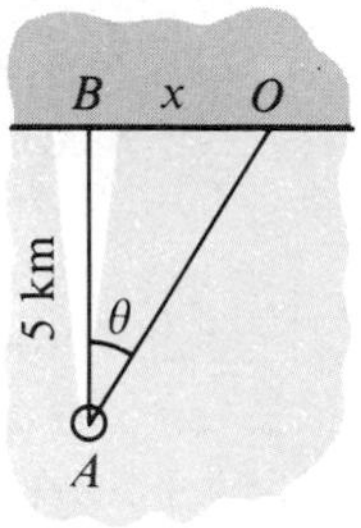

The rates of change are:

$$\frac{dx}{dt} = \text{Velocity of the spot of light along the shore}$$

$$\frac{d\theta}{dt} = \text{Angular velocity of the beam of light}$$

From Figure 4, we see that

$$\tan \theta = \frac{x}{5}$$

$$x = 5 \tan \theta$$

---

* *Angular velocity* ω is defined as the rate of change of angle with respect to time, that is, $\omega = \lim_{\Delta t \to 0}(\Delta\theta/\Delta t) = d\theta/dt$, with θ in radians.

Hence,

(4.8)
$$\frac{dx}{dt} = 5 \sec^2\theta \, \frac{d\theta}{dt}$$

We are given that $d\theta/dt = 3$ radians per minute. We seek dx/dt when angle $AOB = 60°$ ($\theta = 30° = \pi/6$ radian). From (4.8),

$$\frac{dx}{dt} = 5 \sec^2\theta \, \frac{d\theta}{dt} = \frac{5}{\cos^2\theta} \frac{d\theta}{dt} = \frac{5(3)}{(\cos \pi/6)^2} = \frac{15}{\frac{3}{4}} = 20 \text{ kilometers per minute}$$

Thus, when $\theta = 30°$, the velocity of the light along the shore is 20 kilometers per minute. ∎

EXERCISE 1

In Problems 1–4 assume x and y are differentiable functions of t. Find dx/dt when $x = 3$, $y = 4$, and $dy/dt = 2$.

1. $x^2 + y^2 = 25$ **2.** $x^2 - y^2 = -7$ **3.** $x^3y^2 = 432$ **4.** $x^2y^3 = 576$

5. Suppose h is a differentiable function of t and suppose that when $h = 3$, $dh/dt = \frac{1}{12}$. Find dV/dt if $V = 80h^2$.

6. Suppose x is a differentiable function of t and suppose that when $x = 15$, $dx/dt = 3$. Find dy/dt if $y^2 = 625 - x^2$.

7. Suppose h is a differentiable function of t and suppose that $dh/dt = \frac{5}{16}\pi$ when $h = 8$. Find dV/dt if $V = \frac{1}{12}\pi h^3$.

8. Suppose x and y are differentiable functions of t and suppose that when $t = 20$, $dx/dt = 5$, $dy/dt = 4$, $x = 150$, and $y = 80$. Find ds/dt if $s^2 = x^2 + y^2$.

9. If each edge of a cube is increasing at the constant rate of 3 centimeters per second, how fast is the volume increasing when x, the length of an edge, is 10 centimeters long?

10. If the radius of a sphere is increasing at 1 centimeter per second, find the rate of change of its volume when the radius is 6 centimeters.

11. Consider a right triangle with hypotenuse of (fixed) length 45 centimeters and variable legs of lengths x and y, respectively. If the leg of length x increases at the rate of 2 centimeters per minute, how fast is y changing when x is 4 centimeters long?

12. Air is pumped into a balloon with a spherical shape at the rate of 80 cubic centimeters per second. How fast is the surface area of the balloon increasing when the radius is 10 centimeters?

13. A spherical balloon filled with gas has a leak that permits the gas to escape at a rate of 1.5 cubic meters per minute. How fast is the surface area of the balloon shrinking when the radius is 4 meters?

14. When a metal plate is heated, it expands. If the shape of the metal is circular and if its radius, as a result of expansion, increases at the rate of 0.02 centimeter per second, at what rate is the area of the top surface increasing when the radius is 3 centimeters?

15. A public swimming pool has a rectangular shape with the following dimensions: length 30 meters, width 15 meters, depth 3 meters at the adult side and 1 meter at the children's side. If water is pumped into the pool at the rate of 15 cubic meters per minute, how fast is the water level rising when it is 2 meters deep at the adult side?

16. A radar antenna, making one revolution every 5 seconds, is located on a ship that is 6 kilometers from a straight shore. How fast is the radar beam moving along the shoreline when the beam makes an angle of 45° with the shore?

17. A light in a lighthouse 2000 meters from a straight shoreline is rotating at 2 revolutions per minute. How fast is the beam moving along the shore when it passes a point 500 meters from the point on shore opposite the lighthouse?   (1 revolution = 2π radians.)

18. Water is flowing into a vertical cylindrical tank of diameter 6 meters at the rate of 5 cubic meters per minute. Find the rate at which the depth of the water is rising.

19. Consider a container in the form of a right circular cone (vertex down) with radius $R = 4$ meters   and height $h = 16$ meters.   If water is poured into the container at the constant rate of 16 cubic meters per minute, how fast is the water level rising when the water is 8 meters deep? [*Hint:*   The volume V of a cone of radius R and height h is $V = \frac{1}{3}\pi R^2 h$.]

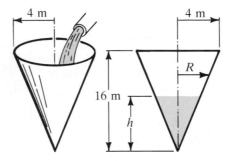

20. Sand is being poured on the ground, forming a conical pile with height equal to one-fourth of the diameter of the base. If the sand is falling at the rate of 20 cubic centimeters per second, how fast is the height increasing when it is 3 centimeters?

21. A cistern in the shape of a cone 4 meters deep and 2 meters in diameter at the top, is being filled with water at the rate of 3 cubic meters per minute. If at the time the water is 3 meters deep, it is observed to be rising 0.5 meter per minute, at what rate is the water leaking away?

22. A conical funnel 15 centimeters in diameter and 15 centimeters deep is filled with a liquid that runs out at the rate of 5 cubic centimeters per minute. How fast is the surface falling when the depth of the liquid is 8 centimeters?

23. An 8 meter ladder is leaning against a vertical wall. If a person pulls the base of the ladder away from the wall at the rate of 0.5 meter per second, how fast is the top going down the wall when the base of the ladder is:
    (a)   3 meters from the wall?
    (b)   4 meters from the wall?
    (c)   6 meters from the wall?

24. A boy is walking toward the base of a pole 20 meters high at the rate of 4 kilometers per hour. At what rate (in meters per second) is the distance from his feet to the top of the pole changing when he is 5 meters from the pole?

25. A girl flies a kite at a height of 30 meters above her hand. If the kite flies horizontally away from the girl at the rate of 2 meters per second, at what rate is the string being let out when the length of the string released is 70 meters? Assume that the string remains taut.

26. A street light hangs 6 meters high. A child (1 meter tall) is walking directly under it. If the child walks away from the light at the rate of 40 meters per minute, how fast is the child's shadow lengthening?

27. An isosceles triangle has equal sides 4 centimeters long and the included angle is θ. If θ increases at the rate of 2° per minute, how fast is the area of the triangle changing when θ is 30°?

28. A particle P is moving along the parabola $y^2 = 4(3 - x)$. When P passes the point $(-1, 4)$, its y-coordinate is increasing at the rate of 3 units per second. How fast is the distance from P to the origin changing at that instant?

29. A gas is said to be compressed adiabatically if there is no gain or loss of heat. When such a gas is diatomic (has two atoms per molecule), it satisfies the equation $PV^{1.4} = k$, where k is a constant, P is the pressure, and V is the volume. At a given instant, the pressure is 20 kilograms per square centimeter, the volume is 32 cubic centimeters, and the volume is decreasing at the rate of 2 cubic centimeters per minute. At what rate is the pressure changing?

30. Two cars approach an intersection, one heading east at the rate of 30 kilometers per hour and the other heading south at the rate of 40 kilometers per hour. At what rate are the two cars approaching each other at the instant when the first car is 100 meters from the intersection and the second car is 75 meters from the intersection? Assume the cars maintain their respective speeds.

31. In order to lift a container to the third floor, which is 10 meters above the top of the container, a rope is attached to the container and, with the help of a pulley, hoists the container up. If a person holds the end of the rope and walks away from beneath the pulley at the rate of 2 meters per second, how fast is the container rising when the person is 5 meters away? We assume that the end of the rope in the person's hand was originally at the same height as the top of the container.

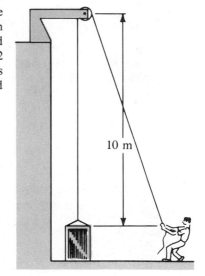

10 m

32. In a rectangle with a diagonal 15 centimeters long, one side is increasing at the rate of $2\sqrt{5}$ centimeters per second. Find the rate of change of the area when that side is 10 centimeters long.

33. Consider the following situation: An elevator in a building is located on the fifth floor, which is 25 meters above the ground. A delivery truck is positioned directly beneath the elevator at street level. If, simultaneously, the elevator goes up at a speed of 5 meters per second and the truck pulls away at a speed of 8 meters per second, how fast will the elevator and the truck be separating 1 second later? Assume the speeds remain constant at all times.

34. A soldier at an antiaircraft battery observes an airplane flying toward him at an altitude of 4500 feet. When the angle of elevation of the battery is 30°, the soldier must increase the angle of elevation by 1° per second to keep the plane in sight. What is the ground speed of the plane?

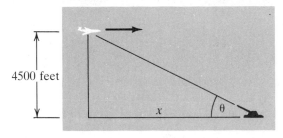

35. An airplane 2000 meters high is flying horizontally with a speed of 300 meters per second. How fast is the angle of elevation of the plane changing when this angle is 45°?

36. In a certain piston engine the distance x in meters between the center of the driving shaft and the head of the piston is given by $x = \cos\theta + \sqrt{16 - \sin^2\theta}$,   where θ is the angle between the crank and the path of the piston head. If θ increases at the constant rate of 45 radians per second, what is the speed of the piston head when θ is $\pi/6$?

37. An object that weighs K pounds on the surface of the earth weighs approximately

$$W(R) = K\left(\frac{4000}{4000 + R}\right)^2$$

pounds when it is a distance of R miles from the earth's surface. Find the rate at which the weight of an object weighing 1000 pounds on the earth's surface is changing when it is 50 miles above the earth's surface and is being lifted at the rate of 10 miles per second.

2. Differentials

In studying the derivative of a function $y = f(x)$ we use the notation dy/dx to represent the derivative. The symbols dy and dx, called *differentials*, which appear in this notation may also be given their own meanings.

   To pursue this, recall that for a differentiable function f, the derivative is defined as

$$\frac{dy}{dx} = f'(x) = \lim_{\Delta x \to 0} \frac{\Delta y}{\Delta x} = \lim_{\Delta x \to 0} \frac{f(x + \Delta x) - f(x)}{\Delta x}$$

That is, the derivative f' is the limit of the ratio of the change in y to the change in x as Δx tends to 0, but $\Delta x \neq 0$.   In other words, for Δx sufficiently close to 0, we can make $\Delta y/\Delta x$ as close as we please to $f'(x)$. We express this fact by writing

(4.9) $$\frac{\Delta y}{\Delta x} \approx f'(x) \quad \text{when} \quad \Delta x \approx 0 \quad (\Delta x \neq 0)$$

Another way of writing (4.9) is

(4.10) $$\Delta y \approx f'(x)\Delta x \quad \text{when} \quad \Delta x \approx 0 \quad (\Delta x \neq 0)$$

The quantity $f'(x)\Delta x$ is given a special name, the *differential of y.*

(4.11)   DEFINITION   *Differential.*   Let f denote a differentiable function and let Δx denote a change in x.
(a)   The *differential of y*, denoted by dy, is defined as $dy = f'(x)\Delta x$.
(b)   The *differential of x*, denoted by dx, is defined as $dx = \Delta x \neq 0$.

Thus, using the notation of differentials, we have

(4.12) $$dy = f'(x)\,dx$$

Since $dx \neq 0$,   (4.12) can be written as

(4.13) $$\frac{dy}{dx} = f'(x)$$

The expression in (4.13) should look very familiar. Interestingly enough, we have given an independent meaning to the symbols dy and dx in such a way that, when dy is divided by dx, their *quotient* will be equal to the derivative. That is, the differential of y divided by the differential of x is equal to the derivative $f'(x)$. For this reason, **we may formally regard the derivative as a quotient of differentials.**

Note that the differential dy is a function of both x and dx. For example, the differential dy of the function $y = x^3$ is

$$dy = 3x^2\,dx$$

so that

$$\begin{array}{lllllll}
\text{if} & x = 1 & \text{and} & dx = 2, & \text{then} & dy = 3(1)^2(2) = 6 \\
\text{if} & x = 0.5 & \text{and} & dx = 0.1, & \text{then} & dy = 3(0.5)^2(0.1) = 0.075 \\
\text{if} & x = 2 & \text{and} & dx = -5, & \text{then} & dy = 3(2)^2(-5) = -60
\end{array}$$

EXAMPLE 1   (a)   If $y = x^2 + 3x - 5$,   then $dy = (2x + 3)\,dx$.
(b)   If $y = x \sin x$,   then $dy = (\sin x + x \cos x)\,dx$.

(c)   If $y = \sqrt{x^2 + 4}$,   then $dy = \dfrac{x}{\sqrt{x^2 + 4}}\,dx$.   ∎

GEOMETRIC INTERPRETATION

We use Figure 5 to arrive at a geometric interpretation of the differentials dx and dy and their relationship to Δx and Δy. From the definition, the differential dx and the change Δx are equal. Therefore, we concentrate on the relationship between dy and Δy.

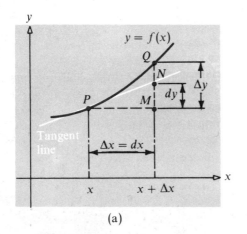

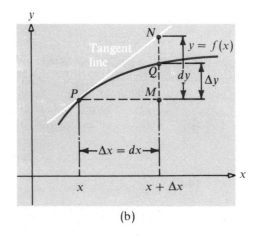

Figure 5   (a)   (b)

In Figure 5(a), $P = (x, y)$ is a point on the graph of $y = f(x)$ and $Q = (x + \Delta x, y + \Delta y)$ is a nearby point that is also on the graph of f. The slope of the tangent line to the graph of f at P is $f'(x)$. From the figure, it follows that

$$f'(x) = \frac{dy}{\Delta x} = \frac{dy}{dx} \qquad \text{or} \qquad dy = f'(x)\, dx$$

Figure 5(a) illustrates the case for which $dy < \Delta y$ (concave up) and $\Delta x > 0$. The case for which $dy > \Delta y$ and $\Delta x > 0$ is illustrated in Figure 5(b). The remaining cases, in which $\Delta x = dx < 0$, have similar graphical representations.

LINEAR APPROXIMATIONS

Let us now establish a relationship between Δy and dy. In Figure 5(a), $\Delta x = dx$ and the increment Δy is represented by the length of the line segment $|MQ|$. Thus, $\Delta y - dy$ is just the length of the line segment $|NQ|$. The size of $|NQ|$ equals the amount by which the graph departs from its tangent line. In fact, for $dx = \Delta x$ sufficiently small, the graph does not depart very much from its tangent line. As a result, the function whose graph is this tangent line is referred to as the *linear approximation to f near P*.

(4.14)   **Thus, for $dx = \Delta x$ sufficiently small, the differential dy is a good approximation to Δy, in the sense that dy differs from Δy by a small percentage of dx. That is,**

$$\Delta y \approx dy \qquad \text{if} \quad \Delta x \approx 0$$

We can use (4.14) to obtain the linear approximation to a function f near a point $P = (x_0, y_0)$ on f. Since $dy = f'(x_0)\,dx = f'(x_0)\,\Delta x = f'(x_0)(x - x_0)$, we find from (4.14) that

$$\Delta y \approx dy$$
$$f(x) - f(x_0) \approx f'(x_0)(x - x_0)$$
$$f(x) \approx f(x_0) + f'(x_0)(x - x_0)$$

(4.15)   **Thus, for $dx = \Delta x$ sufficiently small, that is, for x close to x_0,**

$$f(x) \approx f(x_0) + f'(x_0)(x - x_0)$$

The function $y = f(x_0) + f'(x_0)(x - x_0)$ is the linear approximation to f near x_0.

EXAMPLE 2   Find the linear approximation to $f(x) = x^2 + 2x$ near $x = 1$.   Graph f and the linear approximation.

Solution   First, $f(1) = 3$ and $f'(1) = 2(1) + 2 = 4$.   By (4.15), the linear approximation to f near $x = 1$ is $f(1) + f'(1)(x - 1)$.   Therefore,

$$f(x) \approx 3 + 4(x - 1) = 4x - 1$$

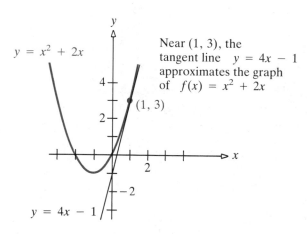

Near $(1, 3)$, the tangent line $y = 4x - 1$ approximates the graph of $f(x) = x^2 + 2x$

Figure 6   ■

Although the practicality of the next example is lessened due to the availability of hand calculators, whenever $f(x)$ is difficult to compute and $f(x_0)$ and $f'(x_0)$ are easy to compute, (4.15) may be used to obtain a numerical approximation.

EXAMPLE 3   Use differentials to approximate $\sqrt{123}$.

Solution   The closest perfect square to 123 is 121, and $\sqrt{121} = 11$.   What we wish to know is the value of $y = f(x) = \sqrt{x}$ when $x = 123$.   The change in x from a square root we know $(x_0 = 121)$ to the square root we seek $(x = 123)$ is $x - x_0 = 2$.

At $x_0 = 121$,

$$f(x_0) = \sqrt{121} = 11 \quad \text{and} \quad f'(x_0) = \frac{1}{2\sqrt{121}} = \frac{1}{22}$$

From (4.15),

$$f(x) = \sqrt{123} \approx f(x_0) + f'(x_0)(x - x_0) = 11 + \frac{1}{22}(2) = 11 + 0.0909 = 11.0909$$

(On a calculator, $\sqrt{123} = 11.090537$.)   ∎

The next two examples use (4.14).

EXAMPLE 4   A bearing with a spherical shape has a radius of 3 centimeters when it is new. Find the approximate volume of the metal lost after it wears down to a radius of 2.971 centimeters.

Solution   The exact volume of metal lost equals the change, ΔV, in volume V of the sphere, where $V = \frac{4}{3}\pi R^3$. The change in radius R is $\Delta R = -0.029$ centimeter. Since the change ΔR is small, we can use the differential dV of volume to approximate the change ΔV in volume. Therefore,

$$\Delta V \approx dV = 4\pi R^2 \, dR = (4\pi)(9)(-0.029) \approx -3.28$$

$$\underbrace{}$$
$$R = 3$$
$$dR = \Delta R = -0.029$$

The approximate loss in volume is 3.28 cubic centimeters.   ∎

The use of dy to approximate Δy when dx is small may also be helpful in approximating *errors*. If Q is the quantity to be measured and if ΔQ is the change in Q, we define

Relative error in $Q = \dfrac{|\Delta Q|}{Q}$ **Percentage error in** $Q = \dfrac{|\Delta Q|}{Q}$ **(100%)**

For example, if $Q = 50$ units and the change ΔQ in Q is measured to be 5 units, then

Relative error in $Q = \frac{5}{50} = 0.10$ Percentage error in $Q = 10\%$

EXAMPLE 5   Suppose a company manufactures spherical ball bearings with radius 3 centimeters, and the percentage error in the radius must be no more than 1%. What is the approximate percentage error for the surface area of the ball bearing?

Solution   If S is the surface area of a sphere of radius R, then $S = 4\pi R^2$. The actual error $\Delta S/S$ we seek may be approximated by the use of differentials. That is,

$$\frac{\Delta S}{S} \approx \frac{dS}{S} = \frac{8\pi R \, dR}{4\pi R^2} = \frac{2 \, dR}{R} = \frac{2\Delta R}{R} = 2(0.01) = 0.02$$

The percentage error in the surface area is 2%.   ∎

In Example 5 the percentage error of 1% in the radius of the sphere means the radius will lie somewhere between 2.97 and 3.03 centimeters. But the percentage error of 2% in the surface area means the surface area lies within a factor of $\pm(0.02)$ of $S = 4\pi R^2 = 36\pi;$ that is, it lies between $35.28\pi = 110.84$ and $36.72\pi = 115.36$ square centimeters.   A rather small error in the radius results in a more significant range of possibilities for the surface area!

DIFFERENTIAL FORMULAS

All the formulas derived earlier for finding derivatives carry over to differentials. We use the symbol df to indicate the differential of the function f. Thus, $d(x^2 + 2) = 2x\,dx$.   The list below gives formulas for differentials next to the corresponding derivative formulas.

	Derivative		*Differential*
(1)	$\dfrac{d}{dx}c = 0$	(1′)	$dc = 0,$ if c is constant
(2)	$\dfrac{d}{dx}(kx) = k$	(2′)	$d(kx) = k\,dx,$ if k is constant
(3)	$\dfrac{d}{dx}(u + v) = \dfrac{du}{dx} + \dfrac{dv}{dx}$	(3′)	$d(u + v) = du + dv$
(4)	$\dfrac{d}{dx}(uv) = u\dfrac{dv}{dx} + v\dfrac{du}{dx}$	(4′)	$d(uv) = u\,dv + v\,du$
(5)	$\dfrac{d}{dx}\left(\dfrac{u}{v}\right) = \dfrac{v\dfrac{du}{dx} - u\dfrac{dv}{dx}}{v^2}$	(5′)	$d\left(\dfrac{u}{v}\right) = \dfrac{v\,du - u\,dv}{v^2}$
(6)	$\dfrac{d}{dx}x^r = rx^{r-1}$	(6′)	$d(x^r) = rx^{r-1}\,dx,$ r a rational number
(7)	$\dfrac{d}{dx}\sin x = \cos x$	(7′)	$d(\sin x) = (\cos x)\,dx$
(8)	$\dfrac{d}{dx}\cos x = -\sin x$	(8′)	$d(\cos x) = (-\sin x)\,dx$
(9)	$\dfrac{d}{dx}\tan x = \sec^2 x$	(9′)	$d(\tan x) = (\sec^2 x)\,dx$

From now on, to find the differential of a function $y = f(x),$ either find the derivative dy/dx and then multiply by dx or employ formulas (1′)–(9′). For example, if $y = x^3 + 2x + 1$, then

$$\frac{dy}{dx} = 3x^2 + 2 \qquad \text{so that} \qquad dy = (3x^2 + 2)\, dx$$

By using the differential formulas,

$$dy = d(x^3 + 2x + 1) = d(x^3) + d(2x) + d(1) = 3x^2\, dx + 2\, dx + 0 = (3x^2 + 2)\, dx$$

But be careful! The use of dy on the left side of an equation requires dx on the right side. Thus, $dy = 3x^2 + 2$ is incorrect.

The symbol d is an instruction to take the differential.

EXAMPLE 6   (a) $d(x^2 - 3x) = (2x - 3)\, dx$ (b) $d(3y^4 - 2y + 4) = (12y^3 - 2)\, dy$

(c) $d(\sqrt{z^2 + 1}) = \dfrac{z}{\sqrt{z^2 + 1}}\, dz$ ■

The differential can be used as an alternative to implicit differentiation to find the derivative of a function that is defined implicitly.

EXAMPLE 7   Find dy/dx and dx/dy if $x^2 + y^2 = 2xy^2$

Solution   We take the differential of each side:

$$d(x^2 + y^2) = d(2xy^2)$$
$$2x\, dx + 2y\, dy = 2(y^2\, dx + 2xy\, dy)$$
$$(y - 2xy)\, dy = (y^2 - x)\, dx$$

$$\frac{dy}{dx} = \frac{y^2 - x}{y - 2xy} \qquad \text{provided} \quad y - 2xy \neq 0$$

$$\frac{dx}{dy} = \frac{y - 2xy}{y^2 - x} \qquad \text{provided} \quad y^2 - x \neq 0 \qquad ■$$

EXERCISE 2

In Problems 1–6 find the differential dy.

1. $y = x^3 - 2x + 1$ 2. $y = 4(x^2 + 1)^{3/2}$ 3. $y = \dfrac{x - 1}{x^2 + 2x - 8}$ 4. $y = \sqrt{x^2 - 1}$

5. $y = 3 \sin 2x + x$ 6. $y = \cos^2 3x - x$

In Problems 7–14 find dy/dx and dx/dy by means of differentials.

7. $xy = 7$ 8. $3x^2y + 2x - 9 = 0$ 9. $x^2 + y^2 = 4$

10. $4xy^2 + yx^2 + 2 = 0$ 11. $x^3 + y^3 = 3x^2y$ 12. $2x^2 + y^3 = xy^2$

13. $\sin 3y = 2x$ 14. $y \sin 2x + x \cos 2y = 1$

In Problems 15–18 find the indicated differential.

15. $d(\sqrt{x} + 2)$ **16.** $d\left(\dfrac{1 + x}{1 - x}\right)$ **17.** $d(x^3 + x - 2)$ **18.** $d(x^2 + 2)^{2/3}$

In Problems 19–24 find the linear approximation to f near x_0. Graph f and the linear approximation.

19. $f(x) = x^2 - 2x + 1, \quad x_0 = 2$ **20.** $f(x) = x^3 - 1, \quad x_0 = 0$ **21.** $f(x) = \sqrt{x}, \quad x_0 = 4$

22. $f(x) = x^{2/3}, \quad x_0 = 1$ **23.** $f(x) = \sin x, \quad x_0 = \dfrac{\pi}{6}$ **24.** $f(x) = \cos x, \quad x_0 = \dfrac{\pi}{3}$

25. Use (4.15) to approximate:

     (a) $\sqrt{35}$ (b) $\sqrt{26.2}$ (c) $\dfrac{1}{\sqrt{1.2}}$ (d) $\sin 29°$ (Use radians.)

26. Use (4.15) to approximate:

     (a) $\sqrt[3]{126}$ (b) $\sqrt[3]{123}$ (c) $\sqrt[4]{15}$ (d) $\cos 31°$

27. Use (4.15) to find the approximate change in:
     (a) $y = f(x) = x^2$ as x changes from 3 to 3.001
     (b) $y = f(x) = 1/(x + 2)$ as x changes from 2 to 1.98

28. Use (4.15) to find the approximate change in:
     (a) $y = x^3$ as x changes from 3 to 3.01   (b) $y = 1/(x - 1)$ as x changes from 2 to 1.98

29. A circular plate is heated and expands. If the radius of the plate increases from $R = 10$ centimeters to $R = 10.1$ centimeters, find the approximate increase in area of the top surface.

30. In a wooden block 3 centimeters thick, an existing circular hole with a radius of 2 centimeters is enlarged to a hole with a radius of 2.2 centimeters. Approximately what volume of wood is removed?

31. Find the approximate change in volume of a spherical balloon of radius 3 meters as the balloon swells to a radius of 3.1 meters.

32. A bee flies around the circumference of a circle traced on a ball with a radius of 7 centimeters at a constant distance of 2 centimeters from the ball. An ant travels along the circumference of the same circle on the ball. Approximately how many more centimeters does the bee travel in one trip around than does the ant?

33. If the percentage error in measuring the edge of a cube is 2%, what is the percentage error in computing its volume?

34. The radius of a spherical ball is computed by measuring the volume of the sphere (by finding how much water it displaces). The volume is found to be 40 cubic centimeters, with a percentage error of 1%. Compute the corresponding percentage error in the radius (due to the error in measuring the volume).

35. A manufacturer produces paper cups in the shape of a right circular cone with radius equal to one-fourth its height. Specifications call for the cups to have a diameter of 4 centimeters. After production, it is discovered that the diameters measure only 3 centimeters. Assuming that the radius is still one-fourth of the height, what is the approximate loss in the capacity of the cup?

36. The oil pan of a car is shaped in the form of a hemisphere with a radius of 8 centimeters. The depth h of the oil is found to be 3 centimeters, with a percentage error of 10%. Approximate the percentage error in the volume. [*Hint:* The volume V for a spherical segment is $V = \frac{1}{3}\pi h^2(3R - h)$, where R is the radius.]

37. To find the height of a building, the length of the shadow of a 3 meter pole placed 9 meters from the building is measured. This measurement is found to be 1 meter, with a percentage error of 1%. What is the estimated height of the building? What is the percentage error in the estimate?

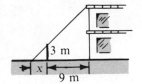

38. The period of the pendulum of a grandfather clock is $T = 2\pi\sqrt{l/g}$, where l is the length (in meters) of the pendulum, T is the period (in seconds), and g is the acceleration due to gravity (9.8 meters per second per second). Suppose the length of the pendulum, a thin wire, increases by 1% due to an increase in temperature. What is the corresponding percentage error in the period? How much time will the clock lose each day?

39. Refer to Problem 38. If the pendulum of a grandfather clock is normally 1 meter long and the length is increased by 10 centimeters, how many minutes will the clock lose each day?

40. What is the approximate volume enclosed by a hollow sphere if its inner radius is 2 meters and its outer radius is 2.1 meters?

41. (a) If $y = f(x)$, $y' = f'(x)$, and so on, why does $dy' = y'' \, dx$?
  (b) What are dy'' and $d(y'^2)$ when expressed with dx as a factor?

3. Maxima and Minima*

Consider the function f defined on the closed interval $[a, b]$, whose graph appears in Figure 7. Note the behavior of the graph of the function f at the numbers x_1, x_2, x_3, and x_4. In a sufficiently small open interval surrounding x_1, the value of the function is greatest at x_1; the same remark holds true for x_3 and x_4. We say that at x_1, x_3, and x_4, f has *local maxima*—local in the sense that the value of f is

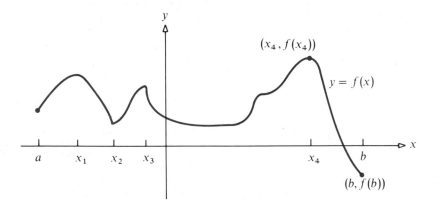

Figure 7

---

* *Maxima* is the plural of *maximum; minima* is the plural of *minimum.*

greatest (maximum) in some open interval about x_1, x_3, x_4. Similarly, f has a *local minimum* at x_2.

On the closed interval $[a, b]$, the largest value of f is $f(x_4)$, while the smallest value of f is $f(b)$. These are called, respectively, the *absolute maximum* and *absolute minimum* of f on $[a, b]$.

(4.16)   **DEFINITION**   *Absolute Maximum and Absolute Minimum.*   **Let f denote a function defined on some interval I.**

If there is a number u in I for which $f(u) \geq f(x)$ for all x in I, then $f(u)$ is the *absolute maximum* of f on I and we say that the absolute maximum of f occurs at u.

If there is a number v in I for which $f(v) \leq f(x)$ for all x in I, then $f(v)$ is the *absolute minimum* of f on I and we say that the absolute minimum of f occurs at v.

The values $f(u)$ and $f(v)$ are sometimes referred to as the *absolute extrema** or the *extreme values* of f on I (see Fig. 8).

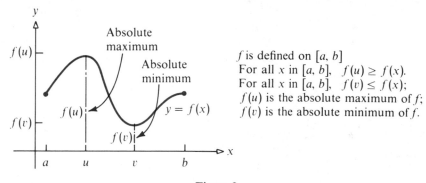

f is defined on $[a, b]$
For all x in $[a, b]$, $f(u) \geq f(x)$.
For all x in $[a, b]$, $f(v) \leq f(x)$;
$f(u)$ is the absolute maximum of f;
$f(v)$ is the absolute minimum of f.

Figure 8

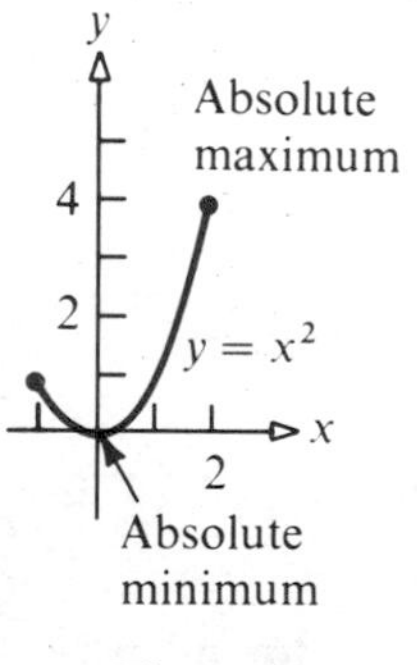

For example, the function $f(x) = x^2$ on the closed interval $[-1, 2]$ has 0 as its absolute minimum and 4 as its absolute maximum (see Fig. 9).

It is possible for a function to have an absolute maximum but no absolute minimum. Some functions may have neither. Figure 10 (p. 202) illustrates these possibilities. Of course, it may also happen that a function has an absolute minimum but no absolute maximum. For example, consider the graph of the function $f(x) = 1/x$ on the interval $(0, 4]$, as illustrated in Figure 11 (p. 202). This function has an absolute minimum $\frac{1}{4}$, but no absolute maximum.

The *absolute* maximum and *absolute* minimum, if they exist, are the largest and smallest values, respectively, that a function f assumes *on the entire interval* over which f is defined. Contrast this idea with that of a *local* maximum and a *local* minimum. These are the largest and smallest values a function f assumes in *some*

Figure 9

* *Extrema* is the plural of *extremum*.

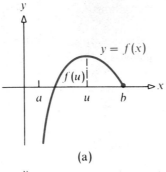

f is defined on $(a, b]$
For all x in $(a, b]$, $f(u) \geq f(x)$;
  $f(u)$ is the absolute maximum of f.
There is no v in $(a, b]$ so that $f(v) \leq f(x)$ for all x;
  thus, f has no absolute minimum on $(a, b]$.

(a)

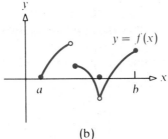

f is defined on $[a, b]$
There is no u nor any v in $[a, b]$ so that
  $f(u) \geq f(x)$ or $f(v) \leq f(x)$ for all x;
  hence, f has neither an absolute maximum
  nor an absolute minimum on $[a, b]$.

(b)

Figure 10

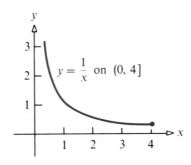

Figure 11

open interval contained in the interval over which f is defined. The next definition makes this precise.

(4.17)   **DEFINITION   *Local Maximum; Local Minimum.*   Let f be a function defined on some interval I and let u and v be numbers in I.**

If there is some open interval in I containing u so that $f(u) \geq f(x)$ for all x in this open interval, then f has a *local maximum* at u.

If there is some open interval in I containing v so that $f(v) \leq f(x)$ for all x in this open interval, then f has a *local minimum* at v.

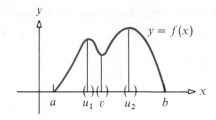

Figure 12

Figure 12 illustrates (4.17) for a function f defined on $[a, b]$. There is an open interval containing u_1 so that $f(u_1) \geq f(x)$ for all x in this open interval. There is also an open interval containing u_2 so that $f(u_2) \geq f(x)$ for all x in this open interval. Then, f has a local maximum at u_1 and a local maximum at u_2. There is an open interval containing v so that $f(v) \leq f(x)$ for all x in this open interval; thus, f has a local minimum at v. Note that the endpoints a and b are not even considered since there is no open interval containing them on which f is defined.

We call attention to the fact that the interval that contains u in the above definition is required to be *open* and that the value $f(u)$ must be larger than or equal to *all* the other values in this open interval. The word *local* is used to emphasize that $f(u)$ is larger than other values of $f(x)$ "around u" or in some (possibly small) open interval containing u. Of course, similar remarks hold for a local minimum.

We shall use the term *local extremum* to describe either a local maximum of f or a local minimum of f.

Of course, a function may not have any local extrema; or a function may have no local maximum, but have local minima (Fig. 13 illustrates these possibilities); or a function may have no local minimum, but have local maxima.

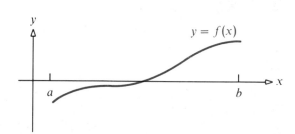

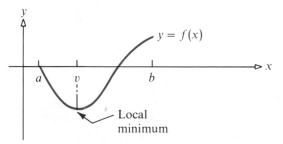

(a)   There is no u in (a, b) so that $f(u) \geq f(x)$ for some open interval containing u. There is no v in (a, b) so that $f(v) \leq f(x)$ for some open interval containing v. Thus, f has no local extrema.

(b)   There is no u in (a, b) so that $f(u) \geq f(x)$ for some open interval containing u. Therefore, f has no local maximum. For the number v, we see that $f(v) \leq f(x)$ for some open interval containing v. Hence, f has a local minimum at v.

Figure 13

Let's now see how we can use the material of Chapter 3 to study functions that do have local extrema.

(4.18)   *A Necessary Condition for Local Extrema.*   **If a function f has a local maximum or a local minimum at the number c, then either $f'(c) = 0$ or $f'(c)$ does not exist.**

Proof   Suppose f has a local maximum at c. Then, by definition, $f(x) \le f(c)$ for all x in some open interval containing c. As a result, $f(x) - f(c) \le 0$.   The derivative of f at c may be written as

$$\lim_{x \to c} \frac{f(x) - f(c)}{x - c}$$

If this limit does not exist, then $f'(c)$ does not exist and there is nothing further to prove. If this limit does exist, then

$$\lim_{x \to c^-} \frac{f(x) - f(c)}{x - c} = \lim_{x \to c^+} \frac{f(x) - f(c)}{x - c}$$

In the limit on the left side, $x < c$ and $f(x) - f(c) \le 0$.   Hence, the quantity

$$\frac{f(x) - f(c)}{x - c} \ge 0$$

and so

(4.19)
$$\lim_{x \to c^-} \frac{f(x) - f(c)}{x - c} \ge 0^*$$

In the limit on the right side, $x > c$ and $f(x) - f(c) \le 0$.   Hence, the quantity

$$\frac{f(x) - f(c)}{x - c} \le 0$$

and so

(4.20)
$$\lim_{x \to c^+} \frac{f(x) - f(c)}{x - c} \le 0$$

Since the limits (4.19) and (4.20) are required to be equal, we must have

$$\lim_{x \to c} \frac{f(x) - f(c)}{x - c} = f'(c) = 0$$

The proof when f has a local minimum at c is similar and is left as an exercise.   ∎

---

*If $\displaystyle\lim_{x \to c^-} \frac{f(x) - f(c)}{x - c} < 0$ there would be some left-sided interval about c on which $\displaystyle\frac{f(x) - f(c)}{x - c} < 0$, which is not possible.

For differentiable functions, (4.18) has the form:

(4.21)   **If a differentiable function f has a local maximum or a local minimum at c, then $f'(c) = 0$.**

In other words, for differentiable functions, a local maximum or local minimum occurs at a point where the tangent line to the graph of f is horizontal. We shall make use of this fact a little later.

As (4.18) shows, the numbers at which a function f has a 0 derivative, or at which f' does not exist, provide a clue for locating where f has local extrema. Unfortunately, the fact that the derivative is 0 at a number will not guarantee a local extremum at this number. Nor will the nonexistence of the derivative guarantee a local extremum. For example, in Figure 14 $f'(x_3) = 0$,   but f has neither a local maximum nor a local minimum at x_3. Similarly, $f'(x_4)$ does not exist, but f has neither a local maximum nor a local minimum at x_4.

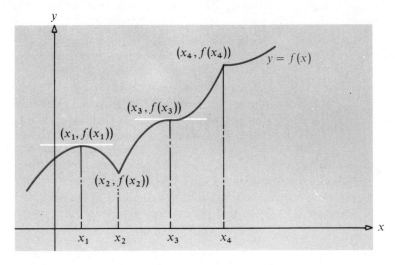

$f'(x_1) = 0$; f has local maximum at x_1

$f'(x_2)$ does not exist; f has local minimum at x_2

$f'(x_3) = 0$; f has neither a local maximum nor a local minimum

$f'(x_4)$ does not exist; f has neither a local maximum nor a local minimum

Figure 14

In spite of the fact that local extrema are not found at all numbers x for which $f'(x) = 0$ or $f'(x)$ does not exist, these numbers do provide *all* the candidates at which f *might* have local extrema. For this reason, we give a special name to the numbers x at which $f'(x) = 0$ or $f'(x)$ does not exist—they are called *critical numbers.*

(4.22)   **DEFINITION** *Critical Number.* **A number in the domain of a function f for which either** $f'(c) = 0$ **or** $f'(c)$ **does not exist is called a** *critical number* c **of the function** f.

EXAMPLE 1   Find the critical numbers of the following functions:

(a) $f(x) = 6 - x^2$ (b) $f(x) = x^3 - 6x^2 + 9x + 2$

(c) $f(x) = \dfrac{1}{x - 2}$ (d) $f(x) = \dfrac{(x - 2)^{2/3}}{x}$

Solution   (a) For $f(x) = 6 - x^2$, the derivative $f'(x) = -2x$ exists everywhere. Since $f'(x) = 0$ only at $x = 0$, then 0 is the only critical number.

(b) For $f(x) = x^3 - 6x^2 + 9x + 2$, the derivative $f'(x) = 3x^2 - 12x + 9 = 3(x - 1)(x - 3)$ exists everywhere. Since $f'(x) = 0$ at $x = 1$ and $x = 3$, then 1 and 3 are the only critical numbers.

(c) For $f(x) = 1/(x - 2)$, the derivative $f'(x) = -1/(x - 2)^2$ exists throughout the domain of f and is never 0. (Note that 2 is not in the domain of f and hence is excluded.) Therefore, there are no critical numbers.

(d) For $f(x) = (x - 2)^{2/3}/x$, the derivative is

$$f'(x) = \frac{(x)(\frac{2}{3})(x - 2)^{-1/3} - (x - 2)^{2/3}}{x^2} = \frac{2x - 3(x - 2)}{3x^2(x - 2)^{1/3}} = \frac{6 - x}{3x^2(x - 2)^{1/3}}$$

Clearly, 6 is a critical number since $f'(6) = 0$. The derivative f' does not exist for those numbers x that give a 0 denominator. There are two such numbers, 0 and 2. However, of these, we must exclude 0 since 0 is not in the domain of f. Thus, f has only two critical numbers, 6 and 2.   ∎

We are not yet ready to give a procedure for determining whether a function f has a local maximum or a local minimum or neither at a critical number. This requires the mean value theorem, which is the subject of the next section. However, the critical numbers do help us find the extreme values of a function f.

FINDING EXTREME VALUES

A function f may have neither an absolute maximum nor an absolute minimum, or it may have one without the other. The next theorem provides a condition under which a function f will always possess extreme values. (A proof of this theorem may be found in most advanced calculus books.)

(4.23)   *Extreme Value Theorem.* **If f is a continuous function defined on a closed interval** $[a, b]$**, then f has an absolute maximum and an absolute minimum on** $[a, b]$**.**

The absolute maximum (and absolute minimum) of a continuous function f defined on a closed interval $[a, b]$ will occur either at an endpoint of the interval $[a, b]$ or else at a critical number in the open interval (a, b). Hence, we have the following test:

(4.24)   *Test for Absolute Maximum and Absolute Minimum.* **If a continuous function f is defined on a closed interval $[a, b]$, the absolute maximum and the absolute minimum of f are, respectively, the largest and the smallest values found among the following:**

1. The values of f at the critical numbers in the open interval (a, b)

2. $f(a)$ and $f(b)$, the values of f at the endpoints a and b

Let's use (4.24) to find the extreme values of some specific functions.

EXAMPLE 2   Find the absolute maximum and absolute minimum of the functions below.

(a) $f(x) = x^3 - 6x^2 + 9x + 2$,   on $[0, 2]$ (b) $f(x) = \dfrac{(x - 2)^{2/3}}{x}$,   on $[1, 10]$

Solution   (a)   The function f is continuous on $[0, 2]$. Thus, by (4.23), f has an absolute maximum and an absolute minimum. From Example 1, part (b), the critical numbers of f are 1 and 3. However, we exclude 3 since we are interested only in the extrema of f on $[0, 2]$. We calculate

$$f(1) = 6$$

At the endpoints 0 and 2, we have

$$f(0) = 2 \quad \text{and} \quad f(2) = 4$$

Thus, the absolute maximum of f, which occurs at 1, is 6; and the absolute minimum of f, which occurs at 0, is 2.

(b)   The function f is continuous on $[1, 10]$. Thus, by (4.23), f has an absolute maximum and an absolute minimum. From Example 1, part (d), the critical numbers of f are 6 and 2. We calculate

$$f(6) = \frac{4^{2/3}}{6} \approx 0.42 \qquad f(2) = 0$$

At the endpoints 1 and 10, we have

$$f(1) = 1 \qquad f(10) = \tfrac{4}{10} = 0.4$$

Thus, the absolute maximum of f, which occurs at 1, is 1; and the absolute minimum of f, which occurs at 2, is 0.   ∎

For functions f given by more than one rule, we need to be careful at the number where the split occurs. Let's look at an example.

EXAMPLE 3   Find the absolute maximum and absolute minimum of the function

$$f(x) = \begin{cases} 2x - 1 & \text{if } 0 \le x \le 2 \\ x^2 - 5x + 9 & \text{if } 2 < x \le 3 \end{cases}$$

Solution   The function f is continuous on $[0, 3]$. (You should check this.) To find the critical numbers of f on $[0, 3]$, we observe that for $0 < x < 2$, $f'(x) = 2$, and for $2 < x < 3$, $f'(x) = 2x - 5$. Thus, $f'(x)$ is never 0 on $(0, 2)$, while $f'(x) = 0$ at $\frac{5}{2}$ on $(2, 3)$. At 2, the derivative does not exist (check this), so 2 is a critical number. Hence, the critical numbers of f are $\frac{5}{2}$ and 2. We list the values of $f(x)$ at critical numbers and the endpoints of the interval:

$$f(2) = (2)(2) - 1 = 3$$
$$f(\tfrac{5}{2}) = (\tfrac{5}{2})^2 - 5(\tfrac{5}{2}) + 9 = \tfrac{11}{4}$$
$$f(0) = (2)(0) - 1 = -1$$
$$f(3) = 9 - 15 + 9 = 3$$

The absolute maximum is 3 and occurs at 2 and at 3; the absolute minimum is -1 and occurs at 0. The graph of f is given in Figure 15.

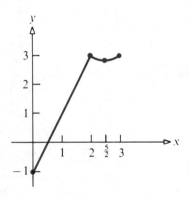

Figure 15

■

EXERCISE 3

In Problems 1–22 find all the critical numbers for the given functions.

1. $f(x) = x^2 + 2x$

2. $f(x) = x^2 - 8x$

3. $f(x) = 1 - 6x + x^2$

4. $f(x) = 4 - 2x - x^2$

5. $f(x) = x^3 - 3x^2$

6. $f(x) = x^3 - 6x$

7. $f(x) = x^4 - 2x^2 + 1$

8. $f(x) = 3x^4 - 4x^3$

9. $f(x) = x^{2/3}$

10. $f(x) = x^{1/3}$

11. $f(x) = 2\sqrt{x}$

12. $f(x) = 4 - \sqrt{x}$

13. $f(x) = x\sqrt{1 - x^2}$

14. $f(x) = x^2\sqrt{2 - x}$

15. $f(x) = \dfrac{x^2}{x - 1}$

16. $f(x) = \dfrac{x}{x^2 - 1}$

17. $f(x) = (x + 3)^2(x - 1)^{2/3}$

18. $f(x) = (x - 1)^2(x + 1)^{1/3}$

19. $f(x) = \dfrac{(x-3)^{1/3}}{x-1}$
 20. $f(x) = \dfrac{(x+3)^{2/3}}{x+1}$
 21. $f(x) = \dfrac{\sqrt[3]{x^2-9}}{x}$

22. $f(x) = \dfrac{\sqrt[3]{4-x^2}}{x}$

In Problems 23–48 find the absolute maximum and absolute minimum of each function f on the indicated interval. Note that the functions in Problems 23–44 are the same as those in Problems 1–22.

23. $f(x) = x^2 + 2x$,   on $[-3, 3]$
 24. $f(x) = x^2 - 8x$,   on $[-1, 10]$

25. $f(x) = 1 - 6x + x^2$,   on $[0, 4]$
 26. $f(x) = 4 - 2x - x^2$,   on $[-2, 2]$

27. $f(x) = x^3 - 3x^2$,   on $[1, 4]$
 28. $f(x) = x^3 - 6x$,   on $[-1, 1]$

29. $f(x) = x^4 - 2x^2 + 1$,   on $[0, 2]$
 30. $f(x) = 3x^4 - 4x^3$,   on $[-2, 0]$

31. $f(x) = x^{2/3}$,   on $[-1, 1]$
 32. $f(x) = x^{1/3}$,   on $[-1, 1]$

33. $f(x) = 2\sqrt{x}$,   on $[1, 4]$
 34. $f(x) = 4 - \sqrt{x}$,   on $[0, 4]$

35. $f(x) = x\sqrt{1 - x^2}$,   on $[-1, 1]$
 36. $f(x) = x^2\sqrt{2 - x}$,   on $[0, 2]$

37. $f(x) = \dfrac{x^2}{x-1}$,   on $[-1, \frac{1}{2}]$
 38. $f(x) = \dfrac{x}{x^2-1}$,   on $[-\frac{1}{2}, \frac{1}{2}]$

39. $f(x) = (x+3)^2(x-1)^{2/3}$,   on $[-4, 5]$
 40. $f(x) = (x-1)^2(x+1)^{1/3}$,   on $[-2, 7]$

41. $f(x) = \dfrac{(x-3)^{1/3}}{x-1}$,   on $[2, 11]$
 42. $f(x) = \dfrac{(x+3)^{2/3}}{x+1}$,   on $[-4, -2]$

43. $f(x) = \dfrac{\sqrt[3]{x^2-9}}{x}$,   on $[3, 6]$
 44. $f(x) = \dfrac{\sqrt[3]{4-x^2}}{x}$,   on $[-4, -1]$

45. $f(x) = \begin{cases} 2x + 1 & \text{if } 0 \le x < 1 \\ 3x & \text{if } 1 \le x \le 3 \end{cases}$
 46. $f(x) = \begin{cases} x + 3 & \text{if } -1 \le x \le 2 \\ 2x + 1 & \text{if } 2 < x \le 4 \end{cases}$

47. $f(x) = \begin{cases} x^2 & \text{if } -2 \le x < 1 \\ x^3 & \text{if } 1 \le x \le 2 \end{cases}$
 48. $f(x) = \begin{cases} x + 2 & \text{if } -1 \le x < 0 \\ 2 - x & \text{if } 0 \le x \le 1 \end{cases}$

49. A truck has a top speed of 75 miles per hour and, when traveling at the rate of x miles per hour, it consumes gasoline at the rate of $\frac{1}{200}(1600/x + x)$ gallon per mile. If the length of the trip is 200 miles and the price of gasoline is \$1.60 per gallon, the cost $C(x)$ (in dollars) is

$$C(x) = (1.60)\left(\frac{1600}{x} + x\right)$$

What is the most economical speed for the truck? Use the interval $[10, 75]$.

50. If the driver of the truck in Problem 49 is paid \$8.00 per hour so that his salary is added to the cost, what is the most economical speed for the truck?

51. The function $f(x) = Ax^2 + Bx + C$ has a local minimum at 0, and its graph passes through the points $(0, 2)$ and $(1, 8)$. Find A, B, and C.

52. Let $f(x) = \sqrt{1 + x^2} + |x - 2|$. Find the absolute maximum and absolute minimum of f on $[0, 3]$, and determine where each occurs.

53. Prove that if f has local minimum at c, then either $f'(c) = 0$ or $f'(c)$ does not exist.

54. Discuss the domain of the function $f(x) = [(16 - x^2)(x^2 - 9)]^{1/2}$ and find the absolute maximum of f in its domain.

4. Rolle's Theorem; Mean Value Theorem

The next theorem, which is due to the French mathematician Michel Rolle (1652–1719), is important because of its theoretical value. We shall use it to prove the mean value theorem, a result needed to derive tests for locating local extrema.

(4.25)   *Rolle's Theorem.*   **Let f be a function defined on a closed interval $[a, b]$. If   (1) f is continuous on $[a, b]$,   (2) f is differentiable on (a, b), and   (3) $f(a) = f(b)$,   then there is at least one number c in the open interval (a, b) for which $f'(c) = 0$.**

We defer an analytic proof of Rolle's theorem to the end of this section. A geometric interpretation is provided by Figure 16. The graphs presented in parts (a) and (b) meet conditions (1), (2), and (3) of the theorem. The existence of at least one number c at which $f'(c) = 0$ is apparent. The graphs in parts (c), (d), and (e) demonstrate that the conclusion of Rolle's theorem may not hold when one or more of the conditions (1), (2), and (3) are not met.

Let's look at a specific function to see whether it satisfies conditions (1), (2), and (3) of Rolle's theorem.

EXAMPLE 1   For the function $f(x) = x^2 - 5x + 6$,   find the two x-intercepts and show that $f'(x) = 0$ for some number between these two intercepts.

Solution   To find the x-intercepts, we set $f(x) = 0$ to get

$$f(x) = x^2 - 5x + 6 = (x - 2)(x - 3) = 0$$

Thus, $f(2) = 0$ and $f(3) = 0$. Since f satisfies the three conditions of Rolle's theorem (f is a polynomial), it guarantees the existence of a number c in the interval $(2, 3)$ such that $f'(c) = 0$. We obtain c as follows:

$$f'(x) = 2x - 5 = 0$$
$$x = \tfrac{5}{2}$$

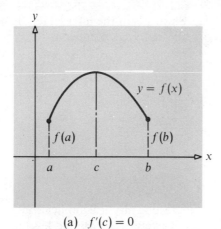

(a) $f'(c) = 0$

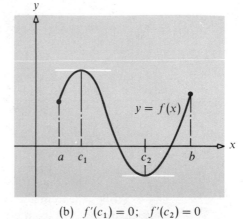

(b) $f'(c_1) = 0$; $f'(c_2) = 0$

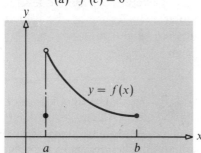

(c) f is not continuous at a
  f is differentiable on (a, b)
  $f(a) = f(b)$
  No c in (a, b) at which $f'(c) = 0$

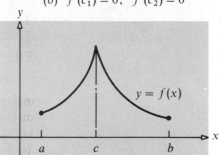

(d) f is continuous on $[a, b]$
  f is not differentiable on (a, b),
    no derivative at c
  $f(a) = f(b)$
  No c in (a, b) at which $f'(c) = 0$

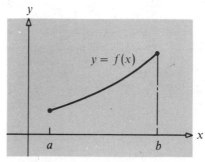

(e) f is continuous on $[a, b]$
  f is differentiable on (a, b)
  $f(a) \neq f(b)$
  No c in (a, b) at which $f'(c) = 0$

Figure 16

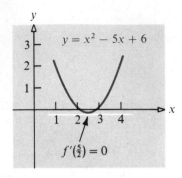

Figure 17

Since $\frac{5}{2}$ is in the interval $(2, 3)$ and since $f'(\frac{5}{2}) = 0$, the required number is $c = \frac{5}{2}$ (see Fig. 17). ∎

Rolle's theorem also has an interesting physical application. If an object is thrown vertically upward from ground level, it returns to ground level at a later time. Rolle's theorem predicts that at some intermediate time t_0, the velocity of the object must be 0. Of course, this time is when the height of the object is greatest.

MEAN VALUE THEOREM

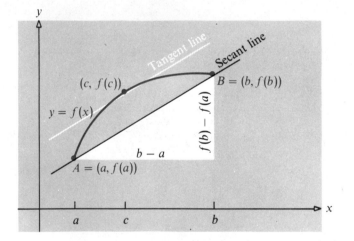

Figure 18

The significance of Rolle's theorem is its theoretical value in obtaining other results, many of which have wide-ranging application. Perhaps the most important of these is the *mean value theorem*, which is sometimes referred to as the *theorem of the mean for derivatives*. This theorem asserts the geometric condition that if f is continuous on $[a, b]$ and differentiable on (a, b), then there is at least one point $(c, f(c))$ between $(a, f(a))$ and $(b, f(b))$ on the graph of f at which the slope $f'(c)$ of the tangent line equals the slope of the secant line joining $A = (a, f(a))$ and $B = (b, f(b))$. That is, as Figure 18 illustrates,

$$f'(c) = \text{Slope of } AB = \frac{f(b) - f(a)}{b - a}$$

(4.26)   *Mean Value Theorem.* **Let f be a function defined on a closed interval $[a, b]$. If (1) f is continuous on $[a, b]$ and (2) f is differentiable on (a, b), then there is at least one number c in the open interval (a, b) at which**

$$f'(c) = \frac{f(b) - f(a)}{b - a}$$

An analytic proof of this theorem is deferred to the end of this section.

Let's look at a specific function to see whether it satisfies conditions (*1*) and (*2*) of the mean value theorem.

EXAMPLE 2   Verify that the function $f(x) = x^3 - 3x + 5$ on the interval $[-1, 1]$ satisfies the conditions of the mean value theorem, and find the number(s) in the interval $(-1, 1)$ at which

$$f'(c) = \frac{f(1) - f(-1)}{1 - (-1)}$$

Solution   Since f is a polynomial function, we know that f is continuous on $[-1, 1]$ and differentiable on $(-1, 1)$. Thus,

$$f'(x) = 3x^2 - 3 \qquad f'(c) = 3c^2 - 3$$

and

$$f(1) = 3 \qquad f(-1) = 7$$

By the mean value theorem, there is a number c in $(-1, 1)$ so that

$$f'(c) = \frac{f(1) - f(-1)}{1 - (-1)} \qquad \text{or} \qquad 3c^2 - 3 = \frac{3 - 7}{2}$$

$$3c^2 - 3 = -2$$

$$3c^2 = 1$$

There are two numbers in the interval $(-1, 1)$ that satisfy the requirement, namely, $-1/\sqrt{3} = -\sqrt{3}/3$ and $1/\sqrt{3} = \sqrt{3}/3$ (see Fig. 19).

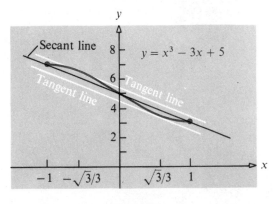

Figure 19   ■

PHYSICAL INTERPRETATION OF MEAN VALUE THEOREM

Suppose $s = f(t)$ denotes the distance a particle has traveled at time t. If f is continuous on $[a, b]$ and differentiable on (a, b), the mean value theorem tells us that at some time t_0, $a < t_0 < b$, the velocity $ds/dt = f'(t_0)$ equals the average velocity

$[f(b) - f(a)]/(b - a)$ from a to b. Thus, for example, if a car travels 300 kilometers in 5 hours, then at some time its speedometer has to read 60 kilometers per hour.

Proof of Rolle's Theorem   Because f is continuous on a closed interval $[a, b]$, we know that f has an absolute maximum and an absolute minimum [the extreme value theorem, (4.23)]. We look at the possibilities:

If f is a constant function on $[a, b]$, then $f(x) = f(a)$ for all x in $[a, b]$, and, hence, $f'(x) = 0$ for all x in (a, b).

If f is not constant on $[a, b]$, then, because $f(a) = f(b)$,   either the absolute maximum or the absolute minimum must be attained in the open interval (a, b), say, at c. But at c, f has a local maximum (or a local minimum). From (4.18), we know that a necessary condition for a local extremum is that $f'(c) = 0$ or $f'(c)$ does not exist. Since f is differentiable on (a, b) and c is in (a, b), then we must have $f'(c) = 0$.   ■

Proof of Mean Value Theorem   We begin by constructing a function g that satisfies the conditions of Rolle's theorem. We define g as

$$g(x) = f(x) - \left[f(a) + \frac{f(b) - f(a)}{b - a} (x - a) \right]$$

for all x on $[a, b]$. This function g has geometric significance—its value equals the directed vertical distance from the graph of $y = f(x)$ to the secant line joining $(a, f(a))$ to $(b, f(b))$. (Refer back to Figure 19.)

Since f is continuous on $[a, b]$ and is differentiable on (a, b), it follows that g is also continuous on $[a, b]$ and is differentiable on (a, b). Furthermore,

$$g(a) = f(a) - \left[f(a) + \frac{f(b) - f(a)}{b - a} (a - a) \right] = 0$$

$$g(b) = f(b) - \left[f(a) + \frac{f(b) - f(a)}{b - a} (b - a) \right] = 0$$

Since g satisfies conditions (1), (2), and (3) of Rolle's theorem, there is a number c in (a, b) at which $g'(c) = 0$.   Also, since $f(a)$ and $[f(b) - f(a)]/(b - a)$ are numbers, we calculate $g'(x)$ to be

$$g'(x) = f'(x) - \frac{f(b) - f(a)}{b - a}$$

Hence,

$$g'(c) = f'(c) - \left[\frac{f(b) - f(a)}{b - a} \right] = 0$$

so that

$$f'(c) = \frac{f(b) - f(a)}{b - a}$$ ■

EXERCISE 4

In Problems 1–10 the function f defined on the indicated interval $[a, b]$ satisfies the conditions of Rolle's theorem. Find all numbers c in (a, b) at which $f'(c) = 0$.

1. $f(x) = x^2 - 3x$, on $[0, 3]$ **2.** $f(x) = x^2 + 2x$, on $[-2, 0]$

3. $f(x) = x^2 - 2x - 2$, on $[0, 2]$ **4.** $f(x) = x^2 + 1$, on $[-1, 1]$

5. $f(x) = x^3 - x$, on $[-1, 0]$ **6.** $f(x) = x^3 - 4x$, on $[-2, 2]$

7. $f(x) = x^3 - x + 2$, on $[-1, 1]$ **8.** $f(x) = x^4 - 3$, on $[-2, 2]$

9. $f(x) = x^4 - 2x^2 + 1$, on $[-2, 2]$ **10.** $f(x) = x^4 + x^2$, on $[-2, 2]$

State why Rolle's theorem cannot be applied to the following functions:

11. $f(x) = x^2 - 2x + 1$, on $[-2, 1]$ **12.** $f(x) = x^3 - 3x$, on $[2, 4]$

13. $f(x) = x^{1/3} - x$, on $[-1, 1]$ **14.** $f(x) = x^{2/5}$, on $[-1, 1]$

In Problems 15–24 the function f defined on the indicated interval $[a, b]$ satisfies the conditions of the mean value theorem. Find all numbers c in (a, b) for which $f'(c) = \dfrac{f(b) - f(a)}{b - a}$.

15. $f(x) = x^2$, on $[-1, 2]$ **16.** $f(x) = x^2 + 1$, on $[0, 2]$

17. $f(x) = \sqrt{x}$, on $[0, 4]$ **18.** $f(x) = x^4 + 5$, on $[0, 1]$

19. $f(x) = x^3 - 5x^2 + 4x - 2$, on $[1, 3]$ **20.** $f(x) = x^3 - 7x^2 + 5x$, on $[-2, 2]$

21. $f(x) = \dfrac{x + 1}{x}$, on $[1, 3]$ **22.** $f(x) = \dfrac{x^2}{x + 1}$, on $[0, 1]$

23. $f(x) = \sqrt[3]{x^2}$, on $[1, 8]$ **24.** $f(x) = \sqrt{x - 2}$, on $[2, 4]$

25. Consider $f(x) = |x|$ on the interval $[-1, 1]$. Here, $f(1) = f(-1) = 1$, but there is no c in $(-1, 1)$ at which $f'(c) = 0$. Explain why this does not contradict Rolle's theorem.

26. Consider $f(x) = x^{2/3}$ on the interval $[-1, 1]$. Verify that there is no c in $(-1, 1)$ for which

$$f'(c) = \frac{f(1) - f(-1)}{1 - (-1)}$$

Explain why this does not contradict the mean value theorem.

27. To demonstrate that the conclusion of the mean value theorem may not hold, draw the graph of a function f that is continuous on $[a, b]$ and not differentiable on (a, b).

28. Repeat Problem 27 for a function f that is differentiable on (a, b) but not continuous on $[a, b]$.

29. Use the mean value theorem to verify that $\frac{1}{9} < \sqrt{66} - 8 < \frac{1}{8}$. [*Hint:* Consider $f(x) = \sqrt{x}$ on the interval $[64, 66]$.]

30. Prove that there is no k for which the function $f(x) = x^3 - 3x + k$ has two distinct zeros in the interval $[0, 1]$.

31. Apply Rolle's theorem to the function $f(x) = (x - 1) \sin x$ on $[0, 1]$. Thereby conclude that the equation $\tan x + x = 1$ has a solution in $(0, 1)$.

32. Explain why the equation $ax^4 + bx^3 + cx^2 + dx + e = 0$ must have a root between 0 and 1 if

$$\frac{a}{5} + \frac{b}{4} + \frac{c}{3} + \frac{d}{2} + e = 0$$

33. Explain why the equation $x^n + ax + b = 0$, n a positive even integer, has at most two distinct real roots.

34. Explain why the equation $x^n + ax + b = 0$, n a positive odd integer, has at most three distinct real roots.

35. Explain why the equation $x^n + ax^2 + b = 0$, n a positive odd integer, has at most three distinct real roots.

36. Explain why the equation $x^n + ax^2 + b = 0$, n a positive even integer, has at most four distinct real roots.

37. Let f be a function that is continuous on $[a, b]$ and differentiable on (a, b). If $f(x) = 0$ for three different numbers x in (a, b), show that there must be at least two numbers in (a, b) at which f has a zero derivative.

5. Increasing and Decreasing Functions;
First Derivative Test

In Section 3 we learned that all local extrema of a function f occur at critical numbers. Each of these critical numbers is a candidate for locating a local extremum for f. But how do we sift out those that locate local extrema from those that do not? And then how do we determine whether each local extremum thus located is a local maximum or is a local minimum? A study of Figure 20 will provide a clue. If you look from left to right along the graph of the continuous function f, you will notice that parts of the graph are increasing and parts are decreasing. The function f is increasing to the left of x_1, where a local maximum occurs, and is decreasing to its right; the function f is decreasing to the left of x_2, where a local minimum occurs, and is increasing to its right. Apparently, knowing when a function f is increasing and when it is decreasing will enable us to identify the local maxima and the local minima.

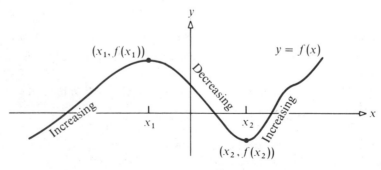

Figure 20

(4.27)   **DEFINITION**   *Increasing Function; Decreasing Function.*
   (a)   The graph of f is increasing on an interval I if for any choice of x_1, x_2 in I with $x_1 < x_2$,   we have $f(x_1) < f(x_2)$.
   (b)   The graph of f is decreasing on an interval I if for any choice of x_1, x_2 in I with $x_1 < x_2$,   we have $f(x_1) > f(x_2)$.

The terms *increasing* and *decreasing* describe the possible behavior of the graph of a function as we examine it from left to right [see Fig. 21(a), (b)]. If there is an interval on which f remains constant, we say that the graph of f is *stationary on the interval* [see Fig. 21(c)].

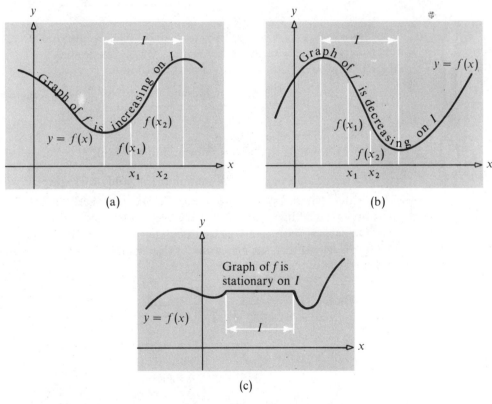

Figure 21

The calculations to determine whether a given function f is increasing or decreasing may be tedious and cumbersome since it is required that a certain inequality must hold for *every* choice of numbers x_1, x_2 in an interval. However, for a function that is differentiable, a relatively straightforward test is available.

(4.28)   *Test for Increasing and Decreasing Function.*   **Let f be a function defined on a closed interval $[a, b]$. Suppose f is continuous on $[a, b]$ and is differentiable on (a, b).**
(a)   If $f'(x) > 0$ throughout (a, b), then f is increasing on $[a, b]$.
(b)   If $f'(x) < 0$ throughout (a, b), then f is decreasing on $[a, b]$.

Proof
(a)   Here $f'(x) > 0$ throughout (a, b). Let x_1, x_2, with $x_1 < x_2$,   denote any two distinct numbers in $[a, b]$. We want to show that $f(x_1) < f(x_2)$.   The function f satisfies the conditions of the mean value theorem on the interval $[x_1, x_2]$. Hence, there is a number c in the interval (x_1, x_2) for which

$$f'(c) = \frac{f(x_2) - f(x_1)}{x_2 - x_1}$$

Now, $x_2 - x_1$ is positive, and since $f'(c)$ is also positive, it follows that $f(x_2) - f(x_1) > 0$;   that is, $f(x_2) > f(x_1)$,   as desired.
(b)   This proof is left as an exercise.   ∎

Let's look at a specific function and determine where it is increasing and where it is decreasing.

EXAMPLE 1   Determine   where   the   function $f(x) = 2x^3 - 9x^2 + 12x - 5$ is   increasing   and where it is decreasing. Use this information to graph f.

Solution $f'(x) = 6x^2 - 18x + 12 = 6(x - 2)(x - 1)$

The critical numbers of f are 1 and 2. We use these as cutoffs, which in turn partition the real line into three intervals as illustrated in Figure 22 and Table 1.

On each of these intervals we calculate the sign of the factors of $f'(x)$, which by (4.28) tells us whether f is increasing or decreasing. Thus, f is increasing on $(-\infty, 1]$ and on $[2, +\infty)$; and f is decreasing on $[1, 2]$.

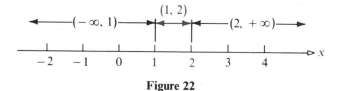

Figure 22

Table 1

Interval	Sign of $x - 1$	Sign of $x - 2$	Sign of $f'(x) = 6(x - 2)(x - 1)$	Conclusion
$(-\infty, 1)$	Negative $(-)$	Negative $(-)$	Positive $(+)$	f is increasing
$(1, 2)$	Positive $(+)$	Negative $(-)$	Negative $(-)$	f is decreasing
$(2, +\infty)$	Positive $(+)$	Positive $(+)$	Positive $(+)$	f is increasing

Since f is increasing to the left of 1 and decreasing to the right of 1, we conclude that at 1, f has a local maximum. Similarly, at 2, f has a local minimum. See Figure 23 for the graph of f.

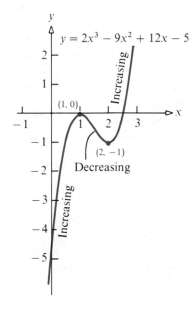

Figure 23   ■

EXAMPLE 2   Show that $f(x) = x^3 + 3x^2 + 3x$ is increasing for all x.

Solution $$f'(x) = 3x^2 + 6x + 3 = 3(x^2 + 2x + 1) = 3(x + 1)^2$$

This function f has only one critical number, and it is at -1. However, since

$$f'(x) > 0 \qquad \text{for} \quad x < -1$$

and

$$f'(x) > 0 \qquad \text{for} \quad x > -1$$

we conclude that f is increasing for all x. Figure 24 illustrates the graph of

$$y = x^3 + 3x^2 + 3x$$

Note the horizontal tangent line at the point $(-1, -1)$.   ■

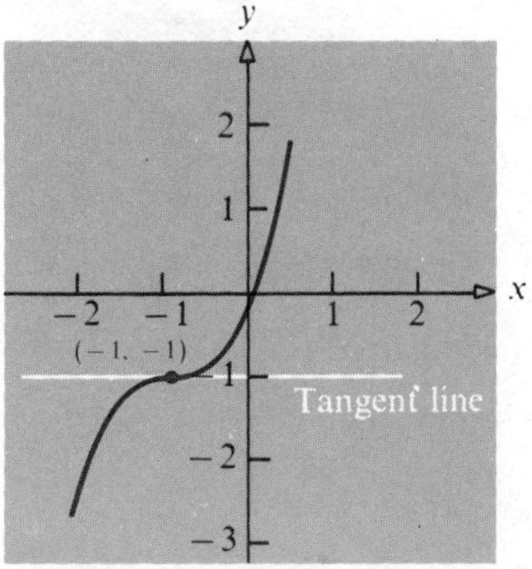

Figure 24

The following result is usually referred to as the *first derivative test* since it relies on information obtained from the first derivative of a function.

(4.29)   *First Derivative Test.*   **Let c be a critical number of a function f and let (a, b) denote an open interval containing c. The function f is assumed to be continuous on the closed interval $[a, b]$ and to be differentiable on the open interval (a, b), except possibly at c.**
(a)   **If $f'(x)$ is positive for $a < x < c$ and is negative for $c < x < b$, then f has a local maximum at c.**
(b)   **If $f'(x)$ is negative for $a < x < c$ and is positive for $c < x < b$, then f has a local minimum at c.**
(c)   **If $f'(x)$ is positive on both sides of c, or is negative on both sides of c, then f has neither a local maximum nor a local minimum at c.**

Proof
(a)   The function f is increasing on $a \le x \le c$ since $f'(x) > 0$ for $a < x < c$; in addition, f is decreasing on $c \le x \le b$ since $f'(x) < 0$ on $c < x < b$. Hence, for all x in $[a, b]$, we have $f(x) \le f(c)$. That is, f has a local maximum at c. See Figure 25(a).
(b)   This is left as an exercise. See Figure 25(b).
(c)   This is also left as an exercise.

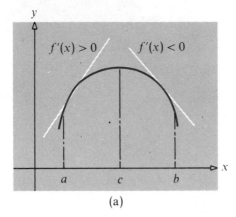

(a)

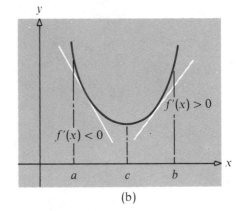

(b)

Figure 25   ∎

EXAMPLE 3   Find the local extrema of $f(x) = x^4 - 4x^3$ and sketch a graph of f.

Solution
$$f'(x) = 4x^3 - 12x^2 = 4x^2(x - 3)$$

The critical numbers are 0 and 3. As in Example 1, we use Figure 26 and Table 2 to help determine where f is increasing and where it is decreasing. Thus, by the first

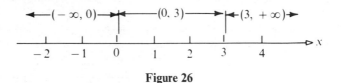

Figure 26

Table 2

Interval	Sign of x^2	Sign of $x - 3$	Sign of $f'(x) = 4x^2(x - 3)$	Conclusion
$(-\infty, 0)$	$+$	$-$	$-$	f is decreasing
$(0, 3)$	$+$	$-$	$-$	f is decreasing
$(3, +\infty)$	$+$	$+$	$+$	f is increasing

derivative test, f has neither a local maximum nor a local minimum at 0, and f has a local minimum at 3.

It is advisable to locate the intercepts and test for possible symmetries before sketching the graph of f. By inspection, the y-intercept is $y = 0$. The x-intercepts, which obey the equation $x^4 - 4x^3 = 0$, are $x = 0$ and $x = 4$. The graph is not symmetric with respect to the x-axis, the y-axis, or the origin. See Figure 27 for a sketch of the graph. Note the horizontal tangent line at $(0, 0)$.

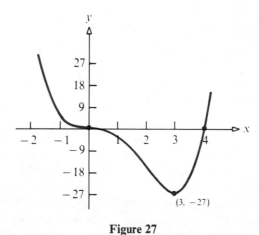

Figure 27   ■

EXAMPLE 4   Find the local extrema of $f(x) = x^{2/3}(x - 5)$ and sketch the graph of f.

Solution $f'(x) = x^{2/3} + \left(\dfrac{2}{3}\right)x^{-1/3}(x - 5) = \dfrac{3x + 2(x - 5)}{3x^{1/3}} = \dfrac{5}{3}\left(\dfrac{x - 2}{x^{1/3}}\right)$

Table 3

Interval	Sign of $x - 2$	Sign of $x^{1/3}$	Sign of $f'(x) = \dfrac{5}{3}\left(\dfrac{x - 2}{x^{1/3}}\right)$	Conclusion
$(-\infty, 0)$	$-$	$-$	$+$	f is increasing
$(0, 2)$	$-$	$+$	$-$	f is decreasing
$(2, +\infty)$	$+$	$+$	$+$	f is increasing

The critical numbers are 0 and 2. Table 3 summarizes the behavior of f. By the first derivative test, f has a local maximum at 0 and has a local minimum at 2.

At the point $(2, -3\sqrt[3]{4})$, the tangent line is horizontal; at 0, the derivative does not exist. The x-intercepts are

$$x = 0 \qquad \text{and} \qquad x = 5$$

The y-intercept is 0. There is no symmetry. See Figure 28 for a sketch of the graph.   ∎

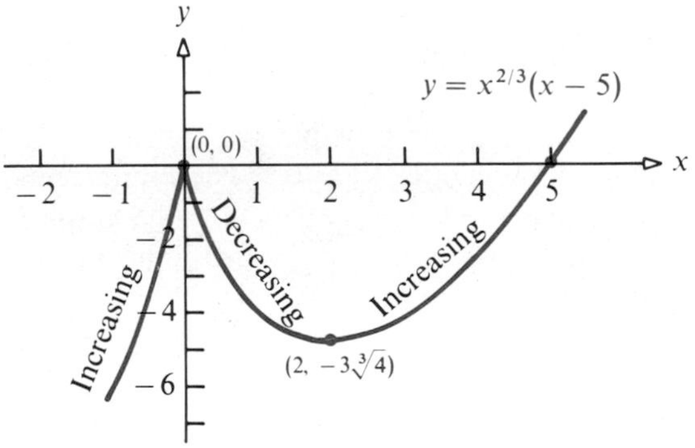

Figure 28

RECTILINEAR MOTION

It is advantageous to use the notion of increasing and decreasing functions in discussing the motion of a particle. Suppose the directed distance s of a particle from the origin at time t is given by $s = f(t)$. As is customary, we think of the motion as along a horizontal line with the positive direction to the right. The velocity v of this particle is $v = ds/dt$. If $v = ds/dt > 0$, the directed distance s is increasing as time t increases, and hence the particle will move to the right. If $v = ds/dt < 0$, the directed distance s is decreasing as time t increases, and the particle will move to the left. This information may be used with the first derivative test to find the local extrema of f and determine at what times the direction of motion of the particle changes.

The acceleration of the particle is $a = dv/dt$. If $a = dv/dt > 0$, the velocity is increasing; if $a = dv/dt < 0$, the velocity is decreasing. This information, along with the use of the first derivative test, can be used to find the local extrema of the velocity.

If both $a < 0$ and $v < 0$, the velocity is decreasing (becoming more negative), and the particle is moving to the left. However, the speed $|v|$ actually may be increasing. The next example illustrates this phenomenon.

EXAMPLE 5

Suppose the directed distance s of a particle from the origin at time t is given by

$$s = t^3 + 3t^2 - 9t + 3$$

Determine the time interval during which the particle is moving to the right. When is it moving to the left? When does it reverse direction? When is its velocity increasing? When is it decreasing? Draw a figure to illustrate the motion of the particle.

Solution

The velocity v of the particle is

$$v = \frac{ds}{dt} = 3t^2 + 6t - 9 = 3(t^2 + 2t - 3) = 3(t + 3)(t - 1)$$

Table 4 summarizes the motion of the particle. The particle reverses direction when $t = -3$ and when $t = 1$. The acceleration a of the particle is

$$a = \frac{dv}{dt} = 6t + 6 = 6(t + 1)$$

When $-\infty < t < -1$, then $a < 0$, and the velocity is decreasing. When $-1 < t < +\infty$, then $a > 0$, and the velocity is increasing.

Table 4

Time Interval	Velocity, v	Motion of Particle
$(-\infty, -3)$	$+$	To the right
$(-3, 1)$	$-$	To the left
$(1, +\infty)$	$+$	To the right

Note that for $-3 < t < -1$, we have $a < 0$ and $v < 0$. Yet the speed $|v| = 3|(t + 3)(t - 1)|$ over this time interval is increasing. Figure 29 illustrates the motion of the particle.

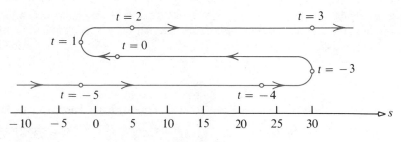

Figure 29

EXERCISE 5

1. Show that the function $f(x) = 2x^3 - 6x^2 + 6x - 5$ is increasing for all x.
2. Show that the function $f(x) = x^3 - 3x^2 + 3x$ is increasing for all x.
3. Show that the function $f(x) = x/(x + 1)$ is increasing on any interval not containing $x \neq -1$.
4. Show that the function $f(x) = (x + 1)/x$ is decreasing on any interval not containing $x \neq 0$.

In Problems 5–32 find the critical numbers of f. Determine the intervals on which f is increasing or decreasing, and find the local extrema of f.

5. $f(x) = x^2 - x - 2$
6. $f(x) = -x^2 + 4x - 3$
7. $f(x) = -2x^2 + 4x - 5$
8. $f(x) = -3x^2 - 12x + 2$
9. $f(x) = 2x^3 + 3x^2 + 4$
10. $f(x) = x^3 - 6x^2 + 2$
11. $f(x) = x^3 + 6x^2 + 12x + 1$
12. $f(x) = -x^3 - 3x^2 + 4$
13. $f(x) = 3x^4 - 12x^3 + 5$
14. $f(x) = x^4 - 4x + 2$
15. $f(x) = 3x^4 - 4x^3$
16. $f(x) = x^4 + 2x^3 - 3$
17. $f(x) = x^3 + 3x^2 - 9x + 1$
18. $f(x) = 2x^3 + 3x^2 - 36x + 4$
19. $f(x) = x^2(x + 1)$
20. $f(x) = x(x^2 - 1)$
21. $f(x) = x^4 + 2x^2$
22. $f(x) = x^4 - 2x^2$
23. $f(x) = x^{2/3} + x^{1/3}$
24. $f(x) = \frac{1}{2}x^{2/3} - x^{1/3}$
25. $f(x) = x^{2/3}(x - 10)$
26. $f(x) = x^{1/3}(x - 8)$
27. $f(x) = x^{2/3}(x^2 - 4)$
28. $f(x) = x^{1/3}(x^2 - 7)$
29. $f(x) = |x^2 - 1|$
30. $f(x) = |x^2 - 4|$
31. $f(x) = 3 \sin x$
32. $f(x) = \cos 2x$

In Problems 33–40 the distance s of a particle from the origin at time t is given. Determine the time interval during which the particle is moving to the right. When is it moving to the left? When does it reverse direction? When is its velocity increasing? When is it decreasing? Draw a figure to illustrate the motion of the particle.

33. $s = t^2 - 2t + 3$
34. $s = 2t^2 + 8t - 7$
35. $s = 2t^3 + 6t^2 - 18t + 1$
36. $s = 3t^4 - 16t^3 + 24t^2$
37. $s = 2t - \dfrac{6}{t}, \quad t > 0$
38. $s = 3\sqrt{t} - \dfrac{1}{\sqrt{t}}, \quad t > 0$
39. $s = 2 \sin 3t, \quad 0 \leq t \leq \dfrac{2\pi}{3}$
40. $s = 3 \cos \pi t, \quad 0 \leq t \leq 2$

41. If $f(x) = ax^2 + bx + c$, $a \neq 0$, prove that f has a local maximum at $-b/(2a)$ if $a < 0$ and has a local minimum at $-b/(2a)$ if $a > 0$.
42. If $f(x) = ax^3 + bx^2 + cx + d$, $a \neq 0$, how does the quantity $b^2 - 3ac$ determine the number of potential local extrema?
43. If $f(x) = ax^3 + bx^2 + cx + d$, $a \neq 0$, determine a, b, c, and d so that f has a local minimum at 0, has a local maximum at 4, and has a graph that passes through the points $(0, 5)$ and $(4, 33)$.
44. Prove part (b) of (4.28).
45. Prove part (b) of (4.29).
46. Prove part (c) of (4.29).
47. Show that 0 is the only critical number of $f(x) = \sqrt[3]{x}$ and that f has no local extrema.
48. Show that 0 is the only critical number of $f(x) = \sqrt[3]{x^2}$ and that f has a local minimum at 0.

49. Find a number x, $0 < x < 1$,   so that the function below is as small as possible.

$$f(x) = \frac{2}{x} + \frac{8}{1 - x}$$

50. Show that $\sin x \le x$, $0 \le x \le 2\pi$.   [*Hint:*   Let $f(x) = x - \sin x$;   find x for which f is minimum.]

51. Show that $1 - (x^2/2) \le \cos x$, $0 \le x \le 2\pi$.   [*Hint:*   Use the result of Problem 50.]

6. Concavity; Second Derivative Test

We begin by looking at the graphs of two familiar functions: $y = x^2$, $x \ge 0$ and $y = \sqrt{x}$ (see Fig. 30). Each graph starts at the origin, passes through the point $(1, 1)$, and is increasing. But there is a noticeable difference. The graph of $y = x^2$ opens up—is *concave up*—while the graph of $y = \sqrt{x}$ opens down—is *concave down*.

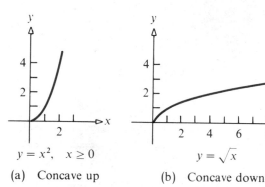

(a)   Concave up   (b)   Concave down

Figure 30

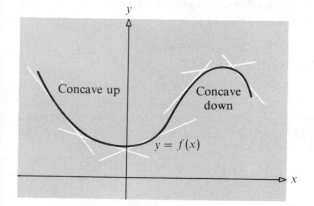

Figure 31

In this section we discuss a test to determine where a function is concave up and where it is concave down. First, though, we need a definition. The definition is based on the illustration given in Figure 31. We observe in this figure that when tangent lines are drawn to the graph where it opens up (is concave up), the graph lies above the tangent lines (except at the point of tangency). Similarly, where the graph opens down (is concave down), the graph lies below the tangent lines (except at the point of tangency). This leads us to formulate definition (4.30).

(4.30)   **DEFINITION**   *Concave Up; Concave Down.*   **Let f be a function that is continuous on a closed interval $[a, b]$ and is differentiable on the open interval (a, b).**

(a)   **f is *concave up* on $[a, b]$ if, throughout (a, b), the graph of f lies above the tangent lines to f.**

(b)   **f is *concave down* on $[a, b]$ if, throughout (a, b), the graph of f lies below the tangent lines to f.**

We may formulate a useful test for determining where a function f is concave up or concave down, providing f'' exists. Since f'' equals the rate of change of f', it follows that if in some open interval, we have f'' positive, then f' will be increasing on that interval. But f' equals the slope of the tangent line to the graph of f. Hence, if the slope is increasing, the tangent line will turn in a counterclockwise sense, as indicated in Figure 32. The graph of f will therefore lie above the tangent line, and f will be concave up. These observations lead to the test for concavity.

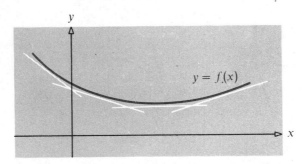

Figure 32

(4.31)   *Test for Concavity.*   **Let f denote a function that is continuous on a closed interval $[a, b]$ and is twice differentiable on the open interval (a, b).**

(a)   **If $f''(x) > 0$ throughout (a, b), then f is concave up on $[a, b]$.**

(b)   **If $f''(x) < 0$ throughout (a, b), then f is concave down on $[a, b]$.**

Proof

(a)   We need to show that the graph of f throughout (a, b) lies above the tangent lines to f. Let c be any fixed number in (a, b). The equation of the tangent line to f at the point $(c, f(c))$ is

$$y = f(c) + f'(c)(x - c)$$

We must establish that

$$f(x) \geq f(c) + f'(c)(x - c) \qquad \text{for all } x \text{ in } (a, b)$$

If $x = c$, this is obviously true. If x is different from c, then, by applying the mean value theorem to the function f, there is a number x_1 between c and x so that

(4.32) $$f'(x_1) = \frac{f(x) - f(c)}{x - c}$$ where x_1 is between c and x

(4.33) $$f(x) = f(c) + f'(x_1)(x - c)$$

There are two possibilities: either $c < x_1 < x$ or $x < x_1 < c$. Suppose $c < x_1 < x$. Since f'' is positive throughout (a, b), it follows from (4.28) that f' is increasing on (a, b). For $x_1 > c$, this means that $f'(x_1) > f'(c)$. Therefore, we may write (4.33) as

$$f(x) > f(c) + f'(c)(x - c)$$

That is, the graph of f lies above its tangent line to the right of c in (a, b). The case where $x < x_1 < c$ is left as an exercise.

(b)   We also leave this part of the proof as an exercise.   ■

Let's look at a specific function and determine where it is concave up and where it is concave down.

EXAMPLE 1   Determine where $f(x) = x^3 - 6x^2 + 9x + 30$ is concave up and where it is concave down.

Solution $$f'(x) = 3x^2 - 12x + 9 \qquad f''(x) = 6x - 12$$

$f''(x) < 0$ if $x < 2$, so f is concave down on $(-\infty, 2]$.
$f''(x) > 0$ if $x > 2$, so f is concave up on $[2, +\infty)$.   ■

Inflection Points.   **An *inflection point* of a function f is a point on the graph of f at which the concavity of f changes.**

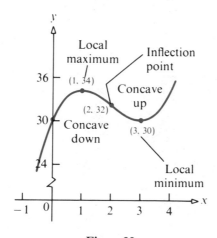

The function f in Example 1 has an inflection point at $(2, 32)$ since f is concave down to the left of 2 and is concave up to the right of 2. You may verify that the sketch of the graph of f is as given in Figure 33.

Figure 33

If f has an inflection point at c* and if $f'(c)$ exists, then the function f' must have a local maximum or minimum at c. In either case, it follows that $f''(c) = 0$ or $f''(c)$ does not exist. This test is stated more formally next.

(4.34) ***Test for Inflection Point.*** **Let f denote a function that is continuous on a closed interval $[a, b]$ and is differentiable on the open interval (a, b). If f has an inflection point at the number c in (a, b), then either $f''(c) = 0$ or f'' does not exist at c.**

Note the wording. If you *know* there is an inflection point at c, the second derivative at c is 0 or does not exist. The converse is not necessarily true. In other words, numbers at which $f''(x)$ is 0 or does not exist will not always determine points of inflection. Thus, to find the points of inflection of a function f:

1. Find all numbers in the domain of f at which $f''(x) = 0$ or $f''(x)$ does not exist.

2. Use the test for concavity to determine the concavity on either side of each of these numbers.

3. If the concavity changes, there is an inflection point; otherwise, there is not.

For the function given in Example 1, $f''(x) = 6x - 12 = 0$ at 2. Since the concavity changes at 2, f has an inflection point at 2 (refer to Fig. 33).

EXAMPLE 2 Find all inflection points of $f(x) = x^{5/3}$. Sketch the graph of f.

Solution
$$f'(x) = \tfrac{5}{3}x^{2/3} \qquad f''(x) = \tfrac{10}{9}x^{-1/3} = \frac{10}{9x^{1/3}}$$

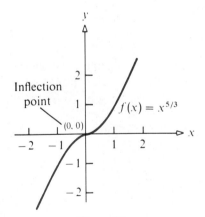

Inflection point

$f(x) = x^{5/3}$

$(0, 0)$

Figure 34

The second derivative of f is never 0 and does not exist when $x = 0$. Thus, the only candidate for an inflection point is at 0. Now,

if $x < 0$, then $f''(x) < 0$ so f is concave down on $(-\infty, 0]$

if $x > 0$, then $f''(x) > 0$ so f is concave up on $[0, +\infty)$

Hence, f has an inflection point at 0. The sketch of the graph of f is given in Figure 34. ∎

---

* By this we mean "at the point $(c, f(c))$ on the graph of f."

The fact that a change in concavity occurs is not, of itself, a guarantee that there is an inflection point. For example, the function $f(x) = 1/x$ is concave down on $(-\infty, 0)$ and concave up on $(0, +\infty)$ since $f''(x) = 2/x^3 < 0$ if $x < 0$ and $f''(x) = 2/x^3 > 0$ if $x > 0$. Yet f has no inflection points. This is due to the fact that f is not defined at 0.

SECOND DERIVATIVE TEST

The second derivative can be used to find local extrema of a function f.

(4.35) *Second Derivative Test.* **Let f be a function that is twice differentiable on an open interval (a, b) containing a critical number c. Suppose further that $f''(c)$ exists.**
(a)   If $f''(c) < 0$,   then f has a local maximum at c.
(b)   If $f''(c) > 0$,   then f has a local minimum at c.

Proof
(a)   Since $f''(c)$ exists and is negative, we have

$$f''(c) = \lim_{x \to c} \frac{f'(x) - f'(c)}{x - c} < 0$$

Referring back to (2.22) (p. 74), on some interval about c, we must have

$$\frac{f'(x) - f'(c)}{x - c} < 0$$

But c is a critical number, so $f'(c) = 0$. Hence,

$$\frac{f'(x)}{x - c} < 0$$

Now for $x < c$ on this interval, $f'(x) > 0$,   and for $x > c$ on this interval, $f'(x) < 0$.   By the first derivative test, f has a local maximum at c.
(b)   This proof is left as an exercise.   ∎

The second derivative test should be used when the second derivative of f is easily calculated. If the second derivative equals 0 or does not exist, then the test gives no information. In such cases, the first derivative test must be used.

EXAMPLE 3   Find all local extrema of $f(x) = x^3 - 6x^2 + 9x + 1$ by using the second derivative test. Sketch the graph of f.

Solution
$$f'(x) = 3x^2 - 12x + 9 = 3(x^2 - 4x + 3) = 3(x - 1)(x - 3)$$

The critical numbers are 1 and 3.

$$f''(x) = 6x - 12$$

At the critical numbers, the second derivatives are $f''(1) < 0$ and $f''(3) > 0$. Thus, f has a local maximum at 1 and a local minimum at 3. Figure 35 illustrates the graph of f.

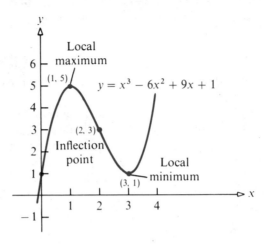

Figure 35   ■

EXERCISE 6

In Problems 1–12 determine the intervals on which the function f is concave up and on which f is concave down. Find all points of inflection.

1. $f(x) = x^2 - 2x + 5$

2. $f(x) = x^2 + 4x - 2$

3. $f(x) = x^3 - 9x^2 + 2$

4. $f(x) = x^3 - 6x^2 + 9x + 1$

5. $f(x) = x^4 - 4x^3 + 10$

6. $f(x) = 3x^4 - 8x^3 + 6x + 1$

7. $f(x) = x + \dfrac{1}{x}$

8. $f(x) = 2x^2 - \dfrac{1}{x}$

9. $f(x) = 3x^{1/3} + 2x$

10. $f(x) = (x - 1)^{3/2}$

11. $f(x) = 3 - \dfrac{4}{x} + \dfrac{4}{x^2}$

12. $f(x) = x^{4/3} - 8x^{1/3}$

In Problems 13–36 use the second derivative test, if possible, to find the local extrema of f. Determine the intervals on which f is concave up and on which f is concave down. Find all points of inflection. Graph each function.

13. $f(x) = 2x^3 - 6x^2 + 6x - 3$

14. $f(x) = 2x^3 + 9x^2 + 12x - 4$

15. $f(x) = -2x^3 + 15x^2 - 36x + 7$

16. $f(x) = x^3 + 10x^2 + 25x - 25$

17. $f(x) = x^4 - 4x$

18. $f(x) = x^4 + 4x$

19. $f(x) = 5x^4 - x^5$

20. $f(x) = 4x^6 + 6x^4$

21. $f(x) = 3x^5 - 20x^3$

22. $f(x) = 3x^5 + 5x^3$

23. $f(x) = 6x^{4/3} - 3x^{1/3}$

24. $f(x) = x^{2/3} - x^{1/3}$

25. $f(x) = x^{2/3}(x - 10)$

26. $f(x) = x^{1/3}(x - 3)$

27. $f(x) = x^{2/3}(x^2 - 8)$

28. $f(x) = x^{1/3}(x^2 - 2)$

29. $f(x) = \dfrac{x^2}{1 + x^2}$ **30.** $f(x) = \dfrac{x}{(1 + x^2)^{5/2}}$ **31.** $f(x) = \dfrac{\sqrt{x}}{1 + x}$

32. $f(x) = x^2\sqrt{1 - x^2}$ **33.** $f(x) = \sqrt{x - x^2}$ **34.** $f(x) = x\sqrt{1 - x}$

35. $f(x) = \dfrac{x^2}{3x + 1}$ **36.** $f(x) = \dfrac{2x}{x^2 + 2}$

In Problems 37–48 sketch the graph of a continuous function f that has the given properties. More than one correct sketch is possible.

37. The graph of f is concave up and increasing on $(-\infty, 0)$ and is concave up and decreasing on $(0, +\infty)$; $f(0) = 1$.

38. The graph of f is concave up and decreasing on $(-\infty, 0)$ and is concave down and increasing on $(0, +\infty)$; $f(0) = 1$.

39. The graph of f is concave down and decreasing on $(-\infty, 0)$, concave down and increasing on $(0, 1)$, and concave up and increasing on $(1, +\infty)$; $f(0) = 1$; $f(1) = 2$.

40. The graph of f is concave down and increasing on $(-\infty, 0)$, concave up and increasing on $(0, 1)$, and concave up and increasing on $(1, +\infty)$; $f(0) = 1$; $f(1) = 2$.

41. $f'(x) > 0$ if $x < 0$; $f'(x) < 0$ if $x > 0$; $f''(x) > 0$ if $x < 0$; $f''(x) > 0$ if $x > 0$; $f(0) = 1$

42. $f'(x) > 0$ if $x < 0$; $f'(x) > 0$ if $x > 0$; $f''(x) > 0$ if $x < 0$; $f''(x) < 0$ if $x > 0$; $f(0) = 1$

43. $f''(0) = 0$; $f'(0) = 0$; $f''(x) > 0$ if $x < 0$; $f''(x) > 0$ if $x > 0$; $f(0) = 1$

44. $f''(0) = 0$; $f'(x) > 0$ if $x \neq 0$; $f''(x) < 0$ if $x < 0$; $f''(x) > 0$ if $x > 0$; $f(0) = 1$

45. $f'(0) = 0$; $f'(x) < 0$ if $x \neq 0$; $f''(x) > 0$ if $x < 0$; $f''(x) < 0$ if $x > 0$; $f(0) = 1$

46. $f''(0) = 0$; $f'(0) = \frac{1}{2}$; $f''(x) > 0$ if $x < 0$; $f''(x) < 0$ if $x > 0$; $f(0) = 1$

47. $f'(0)$ does not exist; $f''(x) > 0$ if $x < 0$; $f''(x) > 0$ if $x > 0$; $f(0) = 1$

48. $f'(0)$ does not exist; $f''(0)$ does not exist; $f''(x) < 0$ if $x < 0$; $f''(x) > 0$ if $x > 0$; $f(0) = 1$

49. For the function $f(x) = ax^3 + bx^2$, determine a and b so that the point $(1, 6)$ is a point of inflection of f.

50. For the cubic polynomial function $f(x) = ax^3 + bx^2 + cx + d$, determine a, b, c, and d so that the point $(0, 4)$ is a critical point and the point $(1, -2)$ is a point of inflection.

51. Use calculus to show that $x^2 - 8x + 21 > 0$ for all x.

52. Use calculus to show that $3x^4 - 4x^3 - 12x^2 + 40 > 0$ for all x.

53. Show that the function $f(x) = ax^2 + bx + c$, $a \neq 0$, has no inflection points. For what values of a is f concave up? For what values of a is f concave down?

54. Show that the function $f(x) = (ax + b)/(cx + d)$ has no critical points and no inflection points.

55. Find the local extrema and the points of inflection of $y = \sqrt{3} \sin x + \cos x$ on the interval $(0, 2\pi)$.

56. Show that every polynomial of degree 3, $f(x) = ax^3 + bx^2 + cx + d$, $a \neq 0$, has exactly one inflection point.

57. Prove that a polynomial of degree $n \geq 3$ has at most $(n - 1)$ critical numbers and at most $(n - 2)$ inflection points.

58. Show that the function $f(x) = (x - a)^n$, a a constant, has exactly one point of inflection if n is odd.

59. Show that the function $f(x) = (x - a)^n$, a a constant, has no points of inflection if n is even.

60. Complete the proof of part (a) of the test for concavity (4.31).

61. Prove part (b) of the test for concavity.

62. Prove part (b) of the second derivative test (4.35).

7. Limits at Infinity; Infinite Limits; Asymptotes

In Chapter 2 we describe $\lim_{x \to c} f(x) = L$ by saying that the value of $f(x)$ can be made as close as we please to L by choosing numbers x sufficiently close to c. It is understood that L and c are numbers. In this section we extend the language of limits to allow c to be $+\infty$ or $-\infty$ (*limits at infinity*) and to allow L to be $+\infty$ or $-\infty$ (*infinite limits*).* These limits, it turns out, are useful for locating *asymptotes* and hence aid in obtaining the graph of a function.

We begin with limits at infinity.

LIMITS AT INFINITY

Let's look at a familiar function, $f(x) = 1/x$, whose domain is $x \neq 0$ (see Fig. 36). This function has the property that the value $f(x)$ can be made as close as we

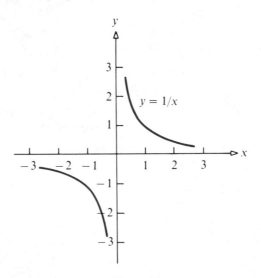

Figure 36

---

* You are reminded that the symbols $+\infty$ (plus infinity) and $-\infty$ (minus infinity) are *not numbers*. Plus infinity expresses the idea of unboundedness in the positive direction; minus infinity expresses the idea of unboundedness in the negative direction.

please to 0 when the number x is sufficiently positive. The table illustrates this fact for selected numbers x:

x	1	10	100	1,000	10,000	100,000
$f(x) = 1/x$	1	0.1	0.01	0.001	0.0001	0.00001

This phenomenon is expressed by saying that $f(x) = 1/x$ has the limit 0 as x approaches $+\infty$ and is symbolized by writing

(4.36)
$$\lim_{x \to +\infty} \frac{1}{x} = 0$$

Informally, this statement means that $1/x$ can be made as close to 0 as we please by taking x sufficiently positive. For example, to make $1/x < 0.0001$, we take $x > 10,000$; to make $1/x < 0.00001$, we take $x > 100,000$. In an analogous way, we write

(4.37)
$$\lim_{x \to -\infty} \frac{1}{x} = 0$$

to indicate that $1/x$ can be made as close as we please to 0 by selecting numbers x sufficiently negative. We summarize statements (4.36) and (4.37) by saying that $f(x) = 1/x$ has *limits at infinity*.*

The statement $\lim_{x \to +\infty}(1/x) = 0$ may be extended to $\lim_{x \to +\infty}(1/x^p) = 0$ for $p > 0$ a real number. In fact,

(4.38)
$$\lim_{x \to +\infty} \frac{A}{x^p} = 0 \qquad p > 0, \quad A \text{ any real number}$$

As examples,

$$\lim_{x \to +\infty} \frac{4}{x^2} = 0 \qquad \lim_{x \to +\infty} \frac{-10}{\sqrt{x}} = 0$$

Limits as $x \to -\infty$ are handled in the same way as limits as $x \to +\infty$. For example, if p is a positive number for which x^p, $x < 0$, is defined and A is any real number, then

(4.39)
$$\lim_{x \to -\infty} \frac{A}{x} = 0 \qquad \lim_{x \to -\infty} \frac{A}{x^p} = 0$$

We shall use (4.38) and (4.39) in the examples that follow.

---

* A definition of limits at infinity is given in Appendix II.

EXAMPLE 1   Find: $\displaystyle\lim_{x \to +\infty} \frac{3x - 2}{4x - 1}$

Solution   We evaluate this limit by first dividing each term of both the numerator and the denominator by the highest power of x that appears in the denominator (in this case x). Then

$$\lim_{x \to +\infty} \frac{3x - 2}{4x - 1} = \lim_{x \to +\infty} \frac{3 - (2/x)}{4 - (1/x)} = \frac{\lim_{x \to +\infty}[3 - (2/x)]}{\lim_{x \to +\infty}[4 - (1/x)]}$$

$$= \frac{\lim_{x \to +\infty} 3 - \lim_{x \to +\infty}(2/x)}{\lim_{x \to +\infty} 4 - \lim_{x \to +\infty}(1/x)} = \frac{3 - 0}{4 - 0} = \frac{3}{4} \qquad \blacksquare$$

EXAMPLE 2   Find: $\displaystyle\lim_{x \to -\infty} \frac{4x^2 - 5x}{3x^3 - 2x + 9}$

Solution   We divide the numerator and the denominator by x^3:

$$\lim_{x \to -\infty} \frac{4x^2 - 5x}{3x^3 - 2x + 9} = \lim_{x \to -\infty} \frac{(4x^2/x^3) - (5x/x^3)}{(3x^3/x^3) - (2x/x^3) + (9/x^3)}$$

$$= \frac{\lim_{x \to -\infty}[(4/x) - (5/x^2)]}{\lim_{x \to -\infty}[3 - (2/x^2) + (9/x^3)]} = \frac{0 - 0}{3 - 0 + 0} = 0 \qquad \blacksquare$$

From the above two examples, we see that the idea of dividing the numerator and the denominator by the highest power of x appearing in the denominator reduces the problem to looking at just the term in the numerator with the highest exponent and the term in the denominator with the highest exponent. These terms are said to *dominate* the other terms *near* $-\infty$ and $+\infty$.* For example, the limit in Example 1 can be found as

$$\lim_{x \to +\infty} \frac{3x - 2}{4x - 1} = \lim_{x \to +\infty} \frac{3x}{4x} = \frac{3}{4}$$

since

$$\frac{3x - 2}{4x - 1} \approx \frac{3x}{4x} \qquad \text{for } x \text{ sufficiently positive}$$

The limit in Example 2 can be found as

$$\lim_{x \to -\infty} \frac{4x^2 - 5x}{3x^3 - 2x + 9} = \lim_{x \to -\infty} \frac{4x^2}{3x^3} = \lim_{x \to -\infty} \frac{4}{3x} = 0$$

---

* The expression "near $+\infty$" means "for all numbers x greater than some positive number," and "near $-\infty$" means "for all numbers x less than some negative number."

since

$$\frac{4x^2 - 5x}{3x^3 - 2x + 9} \approx \frac{4x^2}{3x^3} \quad \text{for } x \text{ sufficiently negative}$$

Let's look at some other examples:

$$\lim_{x \to +\infty} \frac{2x^3 - 5x + 4}{3x^3 + 2x^2 - 10} = \lim_{x \to +\infty} \frac{2x^3}{3x^3} = \frac{2}{3}$$

$$\lim_{x \to -\infty} \frac{5x^4 - 10x^3 + 5}{2x^3 + 10x - 3} = \lim_{x \to -\infty} \frac{5x^4}{2x^3} = \lim_{x \to -\infty} \frac{5x}{2} = -\infty$$

$$\lim_{x \to +\infty} \frac{-10x^2 + 5x + 2}{5x^3 + 2x - 1} = \lim_{x \to +\infty} \frac{-10x^2}{5x^3} = \lim_{x \to +\infty} \frac{-2}{x} = 0$$

$$\lim_{x \to -\infty} \frac{5x^4 - 10x^2 + 1}{-3x^3 + 10x^2 + 50} = \lim_{x \to -\infty} \frac{5x^4}{-3x^3} = \lim_{x \to -\infty} \frac{5x}{-3} = +\infty$$

$$\lim_{x \to +\infty} \frac{10x^{3/2} + 50x - 2}{2x^2 + 1} = \lim_{x \to +\infty} \frac{10x^{3/2}}{2x^2} = \lim_{x \to +\infty} \frac{5}{x^{1/2}} = 0$$

$$\lim_{x \to -\infty} \frac{2 - 3x}{\sqrt{3 + 4x^2}} = \lim_{x \to -\infty} \frac{-3x}{\sqrt{4x^2}} \underset{\underset{\sqrt{x^2}\,=\,|x|}{\uparrow}}{=} \lim_{x \to -\infty} \frac{-3x}{2|x|} \underset{\underset{\substack{|x|\,=\,-x \\ \text{if } x < 0}}{\uparrow}}{=} \lim_{x \to -\infty} \frac{-3x}{2(-x)} = \frac{3}{2}$$

INFINITE LIMITS

Let's again use the function $f(x) = 1/x$, whose graph was given earlier in Figure 36, to introduce the idea of *infinite limits*. We construct a table illustrating values of $f(x)$ for selected numbers x that are close to 0:

x	1	0.1	0.01	0.001	0.0001	0.00001
$f(x) = 1/x$	1	10	100	1,000	10,000	100,000

Here we see that as x gets closer to 0 from the right, the value of $f(x) = 1/x$ can be made as positive as we please—that is, $1/x$ becomes unbounded in the positive direction. We express this fact by writing

(4.40)
$$\lim_{x \to 0^+} \frac{1}{x} = +\infty$$

Similarly, we use the notation

(4.41)
$$\lim_{x \to 0^-} \frac{1}{x} = -\infty$$

to indicate that $1/x$ can be made as negative as we please by selecting numbers x sufficiently close to 0, but less than 0. We summarize (4.40) and (4.41) by saying that $f(x) = 1/x$ has *one-sided infinite limits* at 0.*

As another example, consider $f(x) = 1/x^2$ as x approaches 0. When x approaches 0 from the left or when x approaches 0 from the right, the value of $1/x^2$ becomes positively infinite so that

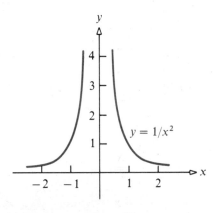

$$\lim_{x \to 0^-} \frac{1}{x^2} = +\infty$$

$$\lim_{x \to 0^+} \frac{1}{x^2} = +\infty$$

Even though we shall write $\lim_{x \to 0}(1/x^2) = +\infty$, we still say that the *limit as $x \to 0$ of $1/x^2$ does not exist†* because $+\infty$ is not a real number. Figure 37 illustrates the graph of $f(x) = 1/x^2$.

Figure 37

We postpone a discussion of other examples of infinite limits for now, and apply the ideas of limits at infinity and infinite limits to the problem of finding *horizontal and vertical asymptotes.*

HORIZONTAL ASYMPTOTES

Limits at infinity have an interesting geometric interpretation. When $\lim_{x \to +\infty} f(x) = L$, it means that as x becomes sufficiently positive, the value of $f(x)$ can be made as close as we please to L; that is, the graph of $y = f(x)$ for x sufficiently positive is as close as we please to the horizontal line $y = L$. Similarly, $\lim_{x \to -\infty} f(x) = M$ means that the values of $f(x)$ can be made as close as we please to M for x sufficiently negative. Thus, the graph of $y = f(x)$ for x sufficiently negative is as close as we please to the horizontal line $y = M$. These lines are called

---

* A definition of infinite limits is given in Appendix II.

† So far, we have seen two conditions under which $\lim_{x \to c} f(x)$ will not exist: (*1*) $\lim_{x \to c^-} f(x) \neq \lim_{x \to c^+} f(x)$ or (*2*) $\lim_{x \to c} f(x) = +\infty$ or $\lim_{x \to c} f(x) = -\infty$. This list of conditions is by no means complete. Other examples include functions that exhibit a highly oscillatory behavior, such as

$$f(x) = \begin{cases} \sin(1/x) & \text{if } x \neq 0 \\ 0 & \text{if } x = 0 \end{cases} \quad \text{at } x = 0$$

and $\quad f(x) = \begin{cases} 1 & \text{if } x \text{ is rational} \\ 0 & \text{if } x \text{ is irrational} \end{cases} \quad$ whose limit does not exist for any choice of x

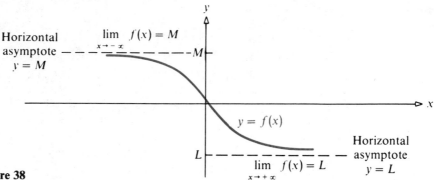

Figure 38

horizontal asymptotes of the graph of f (see Fig. 38). We find horizontal asymptotes by calculating the limits at infinity.

EXAMPLE 3   Find the horizontal asymptotes of

$$y = f(x) = \frac{x}{\sqrt{x^2 + 4}}$$

Solution   For x sufficiently positive, $\sqrt{x^2 + 4} \approx \sqrt{x^2} = x$. Thus,

$$\lim_{x \to +\infty} \frac{x}{\sqrt{x^2 + 4}} = \lim_{x \to +\infty} \frac{x}{\sqrt{x^2}} = \lim_{x \to +\infty} \frac{x}{x} = 1$$

Thus, $y = 1$ is a horizontal asymptote for x sufficiently positive. Similarly, for x sufficiently negative, $\sqrt{x^2 + 4} \approx \sqrt{x^2} = -x$. Thus,

$$\lim_{x \to -\infty} \frac{x}{\sqrt{x^2 + 4}} = \lim_{x \to -\infty} \frac{x}{\sqrt{x^2}} = \lim_{x \to -\infty} \frac{x}{-x} = -1$$

Thus, $y = -1$ is a horizontal asymptote for x sufficiently negative (see Fig. 39).

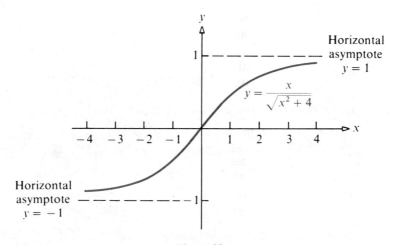

Figure 39

VERTICAL ASYMPTOTES

Infinite limits are used to find vertical asymptotes. Figure 40 illustrates the possibilities that can occur when a function has an infinite limit. Whenever

$$\lim_{x \to c^-} f(x) = \pm \infty \qquad \text{or} \qquad \lim_{x \to c^+} f(x) = \pm \infty$$

we call the line $x = c$ a *vertical asymptote* of the graph of f.

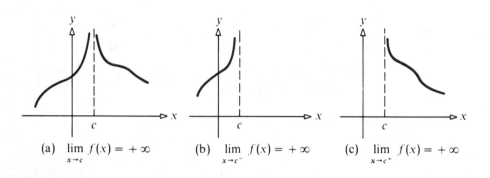

(a) $\lim_{x \to c} f(x) = +\infty$ (b) $\lim_{x \to c^-} f(x) = +\infty$ (c) $\lim_{x \to c^+} f(x) = +\infty$

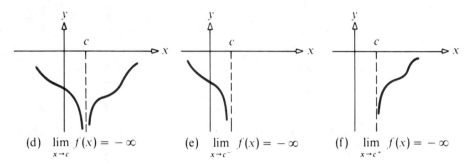

(d) $\lim_{x \to c} f(x) = -\infty$ (e) $\lim_{x \to c^-} f(x) = -\infty$ (f) $\lim_{x \to c^+} f(x) = -\infty$

Figure 40

We state three useful results for locating vertical asymptotes:

1. If n is an even positive integer, then $\lim_{x \to c} \dfrac{1}{(x - c)^n} = +\infty.$

2. If n is an odd positive integer, then $\lim_{x \to c^-} \dfrac{1}{(x - c)^n} = -\infty.$

3. If n is an odd positive integer, then $\lim_{x \to c^+} \dfrac{1}{(x - c)^n} = +\infty.$

EXAMPLE 4   Locate all vertical and horizontal asymptotes of $f(x) = x/(x - 3)$ and graph f.

Solution   Since $x = 3$ is the only number for which the denominator of f equals 0, we evaluate the limit of f as $x \to 3$ to determine whether $x = 3$ is a vertical asymptote:

$$\lim_{x \to 3^-} \frac{x}{x - 3} = -\infty \quad \text{and} \quad \lim_{x \to 3^+} \frac{x}{x - 3} = +\infty$$

Hence, the line $x = 3$ is a vertical asymptote for the graph. To locate the horizontal asymptotes, if any, we look at the limits at infinity of f:

$$\lim_{x \to +\infty} \frac{x}{x - 3} = \lim_{x \to +\infty} \frac{x}{x} = 1 \quad \text{and} \quad \lim_{x \to -\infty} \frac{x}{x - 3} = \lim_{x \to -\infty} \frac{x}{x} = 1$$

Thus, the line $y = 1$ is a horizontal asymptote for x sufficiently positive and for x sufficiently negative. At $x = 0$, we have $f(0) = 0$, and this is the only x-intercept. Putting all this information together, we obtain the graph of f depicted in Figure 41.

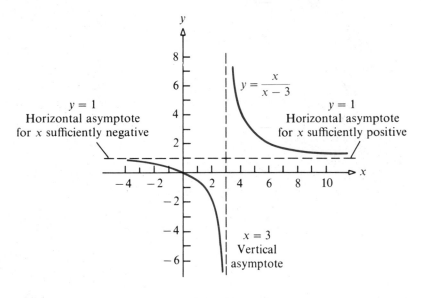

Figure 41   ■

VERTICAL TANGENT LINES

Infinite limits are also used to locate vertical tangent lines to the graph of a function.

(4.42)   **DEFINITION**   **The graph of a function f is said to have a *vertical tangent line* at c if f is continuous at c and**

$$\lim_{x \to c} |f'(x)| = +\infty$$

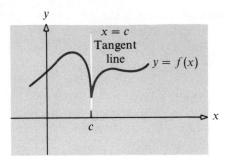

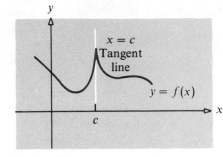

(a) $\lim_{x \to c^-} f'(x) = -\infty$, $\lim_{x \to c^+} f'(x) = +\infty$ (b) $\lim_{x \to c^-} f'(x) = +\infty$, $\lim_{x \to c^+} f'(x) = -\infty$

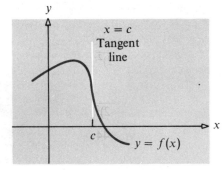

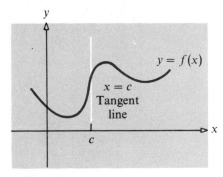

(c) $\lim_{x \to c^-} f'(x) = -\infty$, $\lim_{x \to c^+} f'(x) = -\infty$ (d) $\lim_{x \to c^-} f'(x) = +\infty$, $\lim_{x \to c^+} f(x) = +\infty$

Figure 42

Vertical tangent lines can occur in a variety of ways, as Figure 42 illustrates. Let's look at a specific example.

EXAMPLE 5   Show that the graph of $f(x) = (x - 2)^{2/3}$ has a vertical tangent line at 2.

Solution   The function f is continuous at 2, and

$$f'(x) = \left(\frac{2}{3}\right)(x - 2)^{-1/3} = \frac{2}{3(x - 2)^{1/3}}$$

At 2,

$$\lim_{x \to 2}|f'(x)| = \lim_{x \to 2}\left|\frac{2}{3(x - 2)^{1/3}}\right| = +\infty$$

Hence, f has a vertical tangent line at 2. See Figure 43 for a sketch of the graph of $f(x) = (x - 2)^{2/3}$.

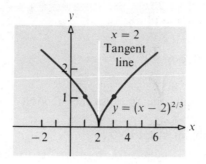

Figure 43   ■

EXAMPLE 6   Show that the graph of $f(x) = (x - 2)^{1/3}$ has a vertical tangent line at 2.

Solution   The function f is continuous at 2, and

$$f'(x) = \frac{1}{3(x - 2)^{2/3}}$$

At 2,

$$\lim_{x \to 2} |f'(x)| = \lim_{x \to 2} \frac{1}{3(x - 2)^{2/3}} = +\infty$$

Thus, f has a vertical tangent at 2. See Figure 44 for a sketch of $f(x) = (x - 2)^{1/3}$.

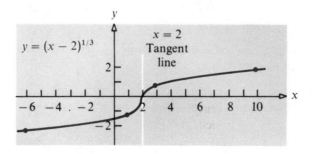

Figure 44   ■

SKETCHING GRAPHS

The next example summarizes how some of the concepts we have discussed in this and earlier chapters can be used to obtain a quite accurate sketch of the graph of a function, while computing only a few points. For the purpose of this example and for Problems 53–68 in Exercise 7 we define "investigate fully the graph of the function" to mean: Find the following:

1. Intercepts of f

2. Symmetries of f

3. Asymptotes of f

4. Critical numbers of f

5. Local extrema of f
6. Concavity of f
7. Inflection points of f

EXAMPLE 7   Investigate fully the graph of the function $f(x) = \dfrac{x^2}{x^2 - 1}$

Solution   1. The only intercept is $(0, 0)$.
2. The graph is symmetric with respect to the y-axis.
3. Since

$$\lim_{x \to +\infty} \frac{x^2}{x^2 - 1} = 1 \quad \text{and} \quad \lim_{x \to -\infty} \frac{x^2}{x^2 - 1} = 1$$

the line $y = 1$ is a horizontal asymptote for x sufficiently positive and for x sufficiently negative. The lines $x = -1$ and $x = 1$ are vertical asymptotes since

$$\lim_{x \to -1^-} \frac{x^2}{x^2 - 1} = +\infty \qquad \lim_{x \to -1^+} \frac{x^2}{x^2 - 1} = -\infty$$

$$\text{and}$$

$$\lim_{x \to 1^-} \frac{x^2}{x^2 - 1} = -\infty \qquad \lim_{x \to 1^+} \frac{x^2}{x^2 - 1} = +\infty$$

4. $$f'(x) = \frac{(x^2 - 1)(2x) - x^2(2x)}{(x^2 - 1)^2} = \frac{-2x}{(x^2 - 1)^2}$$

There is a critical number at $x = 0$ ($x = -1$ and $x = 1$ are not in the domain of f).

5. Since we shall require f'' below, we attempt to use the second derivative test on this critical number:

$$f''(x) = (-2) \left[\frac{(x^2 - 1)^2(1) - (x)(2)(x^2 - 1)(2x)}{(x^2 - 1)^4} \right] = \frac{2(3x^2 + 1)}{(x^2 - 1)^3}$$

Since $f''(0) < 0$,   it follows that f has a local maximum at 0.

6. The table provides an analysis of the concavity of f:

Interval	Sign of f''	Concavity
$(-\infty, -1)$	$+$	Up
$(-1, 1)$	$-$	Down
$(1, +\infty)$	$+$	Up

7. Since $f''(x)$ exists for all $x \neq \pm 1$ and is never 0, we conclude that f has no inflection points.

Figure 45 illustrates the graph of f.

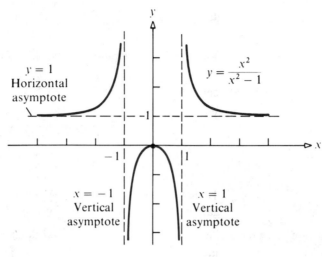

Figure 45   ■

EXERCISE 7

In Problems 1–30 find the indicated limit.

1. $\displaystyle\lim_{x\to+\infty} \frac{x^3 + x^2 + 2x - 1}{x^3 + x + 1}$

2. $\displaystyle\lim_{x\to+\infty} \frac{2x^2 - 5x + 2}{5x^2 + 7x - 1}$

3. $\displaystyle\lim_{x\to+\infty} \frac{2x + 4}{x - 1}$

4. $\displaystyle\lim_{x\to+\infty} \frac{x + 1}{x}$

5. $\displaystyle\lim_{x\to+\infty} \frac{3x^2 - 1}{x^2 + 4}$

6. $\displaystyle\lim_{x\to-\infty} \frac{x^2 - 2x + 1}{x^3 + 5x + 4}$

7. $\displaystyle\lim_{x\to-\infty} \frac{5x^3 - 1}{x^2 + 1}$

8. $\displaystyle\lim_{x\to-\infty} \frac{x^2 + 1}{x^3 - 1}$

9. $\displaystyle\lim_{x\to+\infty} \frac{x^{3/2} + 1}{x - 2}$

10. $\displaystyle\lim_{x\to+\infty} \frac{x^2 - 1}{x^{1/2} + 4}$

11. $\displaystyle\lim_{x\to+\infty} \frac{\sqrt{x^2 + 4}}{3x - 1}$

12. $\displaystyle\lim_{x\to+\infty} \frac{\sqrt[3]{x^3 - 4}}{2x + 5}$

13. $\displaystyle\lim_{x\to2^-} \frac{3x}{x - 2}$

14. $\displaystyle\lim_{x\to2^+} \frac{x}{x^2 - 4}$

15. $\displaystyle\lim_{x\to-1^+} \frac{5x + 3}{x(x + 1)}$

16. $\displaystyle\lim_{x\to-3^-} \frac{1}{x^2 - 9}$

17. $\displaystyle\lim_{x\to2^+} \frac{x + 4}{\sqrt{x - 2}}$

18. $\displaystyle\lim_{x\to3^+} \frac{1 - x}{(3 - x)^2}$

19. $\displaystyle\lim_{t\to-\infty} \frac{3 - t}{\sqrt{4 + 5t^2}}$

20. $\displaystyle\lim_{t\to-\infty} \frac{3 + t}{\sqrt{4 + 5t^2}}$

21. $\displaystyle\lim_{x\to-\infty} \frac{1 + \sqrt[3]{x}}{1 - \sqrt[3]{x}}$

22. $\displaystyle\lim_{x\to+\infty} \frac{\sqrt[5]{x}}{x^4 + 4}$

23. $\displaystyle\lim_{x\to+\infty} (x - \sqrt{x^2 + 4})$
[*Hint:*   Rationalize.]

24. $\displaystyle\lim_{x\to+\infty} (x - \sqrt{x^2 - x})$

25. $\lim\limits_{x \to 0^+} \left(\dfrac{1}{x+1} - \dfrac{1}{x} \right)$

26. $\lim\limits_{x \to 0^-} \left(\dfrac{1}{x+1} - \dfrac{1}{x} \right)$

27. $\lim\limits_{x \to +\infty} \dfrac{\sin x}{x}$

28. $\lim\limits_{x \to +\infty} \dfrac{\cos x}{x}$

29. $\lim\limits_{x \to 0^+} \cot x$

30. $\lim\limits_{x \to \pi/2^-} \tan x$

In Problems 31–46 locate all horizontal and vertical asymptotes, if any, of the function f.

31. $f(x) = 3 + \dfrac{1}{x}$

32. $f(x) = 2 - \dfrac{1}{x^2}$

33. $f(x) = \dfrac{2}{(x-1)^2}$

34. $f(x) = \dfrac{5}{(x+2)^2}$

35. $f(x) = \dfrac{3x-1}{x+1}$

36. $f(x) = \dfrac{x^2}{x^2-4}$

37. $f(x) = \dfrac{x}{x^2-1}$

38. $f(x) = \dfrac{x^2}{x^2+1}$

39. $f(x) = \dfrac{x^2+4}{x^2+1}$

40. $f(x) = \dfrac{x^4}{x^3-1}$

41. $f(x) = \dfrac{3x^4+1}{x^3}$

42. $f(x) = \dfrac{x^3-1}{x^4+1}$

43. $f(x) = \dfrac{x^5}{x^2+1}$

44. $f(x) = \dfrac{2x^2-1}{x^2-1}$

45. $f(x) = \dfrac{ax+b}{cx+d}$

46. $f(x) = \dfrac{x^n}{(x-a)^n}$

In Problems 47–52 verify that the graph of each function has a vertical tangent line at the indicated number.

47. $f(x) = x^{1/3}$, at 0

48. $f(x) = x^{1/4} + 2$, at 0

49. $f(x) = \sqrt{x+4}$, at -4

50. $f(x) = x + x^{1/3}$, at 0

51. $f(x) = (x-3)^{2/3} + 2$, at 3

52. $f(x) = \sqrt[5]{x} - 7$, at 0

In Problems 53–68 investigate fully the graph of the function f (see pp. 241–242 for directions).

53. $f(x) = \dfrac{2}{x^2-4}$

54. $f(x) = \dfrac{1}{x^2-1}$

55. $f(x) = \dfrac{2x-1}{x+1}$

56. $f(x) = \dfrac{x-2}{x}$

57. $f(x) = \dfrac{x}{x^2+1}$

58. $f(x) = \dfrac{2x}{x^2-4}$

59. $f(x) = \dfrac{8}{x^2-16}$

60. $f(x) = \dfrac{x^2}{4-x^2}$

61. $f(x) = \dfrac{x^2+1}{x}$

62. $f(x) = \dfrac{x^2+4}{x^2-1}$

63. $f(x) = \dfrac{x^2}{x+3}$

64. $f(x) = \dfrac{3x^2-1}{x-1}$

65. $f(x) = \dfrac{x^{2/3}}{x-1}$

66. $f(x) = \dfrac{x^{1/3}}{x-1}$

67. $f(x) = 1 + \dfrac{1}{x} + \dfrac{1}{x^2}$

68. $f(x) = \dfrac{2}{x} + \dfrac{1}{x^2}$

69. Explain why a rational function will have vertical asymptotes at each point of discontinuity.

70. Explain why a nonconstant polynomial function cannot have any asymptotes.

71. Show that the tangent line to the circle $x^2 + y^2 = 1$ is vertical at $x = 1$ and at $x = -1$.

72. Sketch the graph of a function f defined and continuous for $-1 \le x \le 2$, which satisfies the following conditions:

$$f(-1) = 1, \quad f(1) = 2, \quad f(2) = 3, \quad f(0) = 0, \quad f(\tfrac{1}{2}) = 3,$$

$$\lim\limits_{x \to -1^+} f'(x) = -\infty, \quad \lim\limits_{x \to 1^-} f'(x) = -1, \quad \lim\limits_{x \to 1^+} f'(x) = +\infty,$$

$$f \text{ has a local minimum at } 0, \quad f \text{ has a local maximum at } \tfrac{1}{2}$$

73. If P and Q are polynomials of degrees m and n, respectively, discuss $\lim\limits_{x \to +\infty}[P(x)/Q(x)]$ when:
(a) $m > n$ (b) $m = n$ (c) $m < n$

8. Applied Extrema Problems

We begin with an example.

EXAMPLE 1   A farmer with 4000 meters of available fencing wishes to enclose a rectangular plot that borders on a straight river (see Figure 46). If the farmer does not fence the side along the river, what is the largest area that can be enclosed?

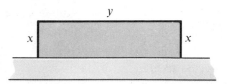

Figure 46

Solution   The quantity to be maximized is the area. We denote it by A, and we denote the dimensions of the rectangle by x and y, with y the length of the side parallel to the river. The area A is therefore

$$A = xy$$

But we need to express A in terms of a single variable. Since the length of available fence is 4000 meters, the variables x and y are related by the equation

$$x + y + x = 4000$$
$$y = 4000 - 2x$$

Thus, the area A is

$$A = x(4000 - 2x) = 4000x - 2x^2$$

The restrictions on x are $x \geq 0$ and $x \leq 2000$ (if $x > 2000$,   then $x + y + x$ is greater than 4000 meters). The problem therefore is to maximize $A = 4000x - 2x^2$ on the closed interval $[0, 2000]$. The critical numbers obey

$$A'(x) = 4000 - 4x = 0$$
$$x = 1000$$

Since this is the only critical number, we calculate

$$A(1000) = 2,000,000 \qquad A(0) = 0 \qquad A(2000) = 0$$

The maximum area that can be enclosed is 2,000,000 square meters.   ∎

In general, each problem we discuss in this section requires that some quantity be minimized or maximized. We assume that this quantity can be represented by a function. Once this function is determined, the problem can be reduced to the question of determining at what number the function assumes its absolute maximum or absolute minimum. Even though each applied problem has its unique features, it is possible to outline in a rough way a procedure for obtaining a solution. This five-step procedure is given below:

1. Identify the quantity for which a maximum or a minimum value is to be found and assign a symbol to represent it.

2. Assign symbols to represent other variables in the problem. If possible, use a picture to assist you.

3. Determine the relationships among these variables.

4. Express the quantity to be maximized or minimized as a function of one of the variables, and determine the domain of meaningful numbers for this variable.

5. Apply the techniques of the previous sections to this function to determine the absolute maximum or absolute minimum relative to the domain found in Step 4.

The following examples illustrate this procedure.

EXAMPLE 2   From each corner of a square piece of sheet metal 18 centimeters on a side, remove a small square and turn up the edges to form an open box. What should be the dimensions of the box so as to maximize its volume?

Solution   The quantity to be maximized is the volume of the box. Denote the volume by V, and denote the dimension of each side of the small square by x, as shown in Figure 47. Although the total area of the piece of sheet metal is fixed, the length of the sides of the squares can be changed, and this is treated as a variable. If y denotes the length left after cutting out the squares, we have

$$y = 18 - 2x$$

The height of the box is x, while the area of the base of the box is y^2. The volume

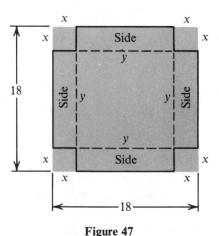

Figure 47

V (height times area of base) is therefore

$$V = x(y^2)$$

This function of two variables can be expressed in terms of one variable by substituting for y in the formula for the volume. This gives

$$V = x(18 - 2x)^2$$

which is the function to be maximized. The only real numbers x that make sense in this case are those between 0 and 9. Thus, we want to find the absolute maximum of

$$V = x(18 - 2x)^2 \qquad 0 \le x \le 9$$

To find the x that maximizes V, we differentiate and find the critical numbers:

$$V'(x) = (18 - 2x)^2 + 2x(18 - 2x)(-2) = (18 - 2x)(18 - 6x) = 12(9 - x)(3 - x)$$

Now, set $V'(x) = 0$ and solve for x. The solutions are

$$x = 9 \qquad \text{or} \qquad x = 3$$

The only critical number in the open interval $(0, 9)$ is $x = 3$.   Thus, we calculate

$$V(0) = 0 \qquad V(3) = 3(18 - 6)^2 = 432 \qquad V(9) = 0$$

The maximum volume is 432 cubic centimeters, and the dimensions of the box that yield the maximum volume are

$$\text{Height} = x = 3 \text{ centimeters}$$
$$\text{Base} = y^2 = 12^2 = 144 \text{ square centimeters} \qquad \blacksquare$$

EXAMPLE 3*   A certain manufacturer makes a flexible square playpen that can be opened at a corner and attached at right angles to a wall (the side of a house, for example). Each side is 1 unit long, so that when the playpen is used in its square shape, its area is 1 square unit. When the playpen is placed as in Figure 48, the area enclosed is 2 square units, which doubles the child's play area. Is there a configuration that will do better than double the child's play area?

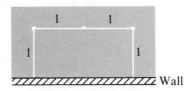

Figure 48

---

* Adapted from *Proceedings, Summer Conference for College Teachers on Applied Mathematics* (University of Missouri–Rolla), 1971.

Solution

Since the playpen must be attached at right angles to the wall, the possible configurations depend on the amount of wall used as a fifth side for the playpen (see Fig. 49). Let x represent half the length of wall used as a fifth side. The area A is a function of x and is the sum of two rectangles (with sides 1 and x) and two right triangles (with hypotenuse 1 and base x). Thus, the quantity to be maximized is

$$A(x) = 2x + x\sqrt{1 - x^2} \qquad 0 \leq x \leq 1$$

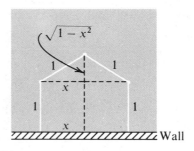

Figure 49

To compute the maximum area, we note that

$$A'(x) = 2 + \sqrt{1 - x^2} - \frac{x^2}{\sqrt{1 - x^2}} = \frac{2\sqrt{1 - x^2} + 1 - 2x^2}{\sqrt{1 - x^2}}$$

The critical numbers obey

$$2\sqrt{1 - x^2} + 1 - 2x^2 = 0$$
$$2\sqrt{1 - x^2} = 2x^2 - 1$$
$$4(1 - x^2) = 4x^4 - 4x^2 + 1$$
$$4x^4 = 3$$
$$x = \sqrt[4]{\tfrac{3}{4}} \approx 0.931$$

Thus, the only critical number in the open interval $(0, 1)$ is $\sqrt[4]{\tfrac{3}{4}} \approx 0.931$. Now, compute $A(x)$ at the endpoints $x = 0$ and $x = 1$ and at the critical number $x \approx 0.931$. The results are

$$A(0) = 0 \qquad A(1) = 2 \qquad A(0.931) \approx 2.20$$

Thus, a wall of length approximately $2x = 1.862$ will maximize the area, and a configuration like the one in Figure 49 increases the play area by about 10% (from 2 to 2.20 square units).   ∎

EXAMPLE 4

A can company wishes to produce a cylindrical container with a capacity of 1000 cubic centimeters. The top and bottom of the container must be made of material that costs $0.05 per square centimeter, while the sides of the container can be made

of material costing $0.03 per square centimeter. Find the dimensions that will mini-
mize the total cost of the container.

Solution   Figure 50 shows a cylindrical container and the area of its top, bottom, and lateral
surface. As indicated in the figure, if we let h stand for the height of the can and R
for the radius, then the total area of the bottom and top is $2\pi R^2$, and the area of the
lateral surface of the can is $2\pi Rh$. The total cost C (in cents) of manufacturing the can
is therefore

$$C = (5)(2\pi R^2) + (3)(2\pi Rh) = 10\pi R^2 + 6\pi Rh$$

This is the function we wish to minimize. The cost function is a function of two
variables, h and R; but there is a relationship between h and R since the volume of
the cylinder is fixed at 1000 cubic centimeters. That is,

$$V = 1000 = \pi R^2 h$$

$$h = \frac{1000}{\pi R^2}$$

By substituting for h in the cost function C, we obtain

$$C = 10\pi R^2 + (6\pi R)\left(\frac{1000}{\pi R^2}\right) = 10\pi R^2 + \frac{6000}{R}.$$

The only restriction on R is that $R > 0$. To find the R that gives minimum cost,
we differentiate C with respect to R. Thus,

$$\frac{dC}{dR} = C'(R) = 20\pi R - \frac{6000}{R^2} = \frac{20\pi R^3 - 6000}{R^2}$$

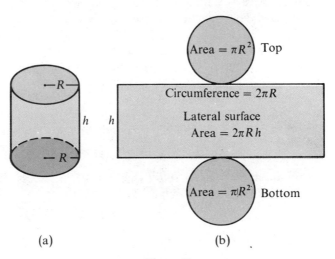

(a)   (b)

Figure 50

The critical numbers obey $C'(R) = 0,$ or

$$20\pi R^3 - 6000 = 0$$

$$R^3 = \frac{300}{\pi}$$

$$R = \sqrt[3]{\frac{300}{\pi}} \approx 4.57 \text{ centimeters}$$

By using the second derivative test (4.35), we have

$$C''(R) = 20\pi + \frac{12{,}000}{R^3}$$

$$C''\left(\sqrt[3]{\frac{300}{\pi}}\right) = 20\pi + \frac{12{,}000\pi}{300} > 0$$

Thus, for $R = \sqrt[3]{300/\pi} \approx 4.57$ centimeters,   the cost has a local minimum. Since $C''(R)$ is clearly positive for all positive R, it follows that the graph of C is everywhere concave up, and hence the local minimum is the absolute minimum. The corresponding height of this can is

$$h = \frac{1000}{\pi R^2} \approx \frac{1000}{20.89\pi} \approx 15.24 \text{ centimeters}$$

These are the dimensions that will minimize the cost of the material.   ■

If the costs of the materials for the top, bottom, and lateral surfaces of a cylindrical container are all the same, then the minimum total cost occurs when the surface area is minimum. It can be shown (see Problem 14 in Exercise 8) that for any fixed volume, the minimum surface area is obtained when the height equals twice the radius.

EXAMPLE 5   A company charges $200 for each set of tools on orders of 150 or fewer sets. The cost to the buyer on every set is reduced by $1 for each set in excess of 150. For what size order is revenue maximum?

Solution   For an order of exactly 150 sets, the company's revenue is

$$\$200(150) = \$30{,}000$$

For an order of 160 sets (which is 10 in excess of 150), the per set charge is $200 - 10(\$1) = \$190$ and the revenue is

$$\$190(160) = \$30{,}400$$

To solve the problem, let x denote the number of sets sold. The revenue R is

$$R = (\text{Number of sets})(\text{Cost per set}) = x(\text{Cost per set})$$

The cost per set is

$200 − $1(Number of sets in excess of 150) = 200 − 1(x − 150) = 350 − x

Hence, the revenue R is

$$R = x(350 − x) = 350x − x^2$$

The only meaningful values for x are $150 \leq x \leq 350$.   To find the **number of** sets leading to maximum revenue, we find the critical numbers of R:

$$R'(x) = 350 − 2x = 0 \qquad \text{when} \quad x = 175$$

We evaluate the revenue R at the critical number and at the endpoints:

$$R(175) = 175(175) = \$30,625 \qquad R(150) = 150(200) = \$30,000 \qquad R(350) = 350(0) = 0$$

The company's revenue is a maximum when 175 sets are sold. (Of course, the company would set this figure as the most it would allow anyone to purchase on this plan, since revenue to the company starts to decrease for orders in excess of 175.)

■

EXAMPLE 6   If a rectangle is inscribed in a semicircle of radius 2, find the dimensions of the rectangle that will have the maximum area.

Solution   Locate the semicircle so that its center is at the origin and its diameter is along the x-axis. Then the length of the inscribed rectangle is $2x$ and its height is y (see Fig. 51). We shall give two methods of solution; the first one is algebraic, while the second uses trigonometry.

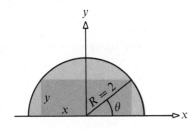

Figure 51

Method I.   From Figure 51, it follows that

$$A = 2xy \qquad \text{and} \qquad x^2 + y^2 = 4 \qquad y \geq 0$$

We solve for y in $x^2 + y^2 = 4$ to get $y = \sqrt{4 − x^2}$,   so we can use this in $A = 2xy$.   As a result, the area A to be maximized can be expressed in terms of x alone as

$$A = 2x\sqrt{4 − x^2} \qquad 0 \leq x \leq 2$$

The critical numbers of A obey

$$A'(x) = 2\left[\sqrt{4 - x^2} + \frac{x(\frac{1}{2})(-2x)}{\sqrt{4 - x^2}}\right] = 2\frac{(4 - x^2) - x^2}{\sqrt{4 - x^2}}$$

$$= \frac{-4(x^2 - 2)}{\sqrt{4 - x^2}} = 0$$

The only critical number in the open interval $(0, 2)$ is $\sqrt{2}$, so we list the values of $A(x)$ at the endpoints and at the critical number:

$$A(2) = 0 \qquad A(0) = 0 \qquad A(\sqrt{2}) = 4$$

The maximum area is 4, and the corresponding dimensions of the rectangle are

$$\text{Length} = 2x = 2\sqrt{2} \qquad \text{Height} = y = \sqrt{2}$$

Method II.   From Figure 51 we see that $x = 2 \cos \theta$ and $y = 2 \sin \theta$ with $0 \le \theta \le \pi/2$. The area A of the rectangle is

$$A = 2xy = 2(2 \cos \theta)(2 \sin \theta) = 8 \cos \theta \sin \theta = 4 \sin 2\theta$$

$$\uparrow$$
$$\sin 2\theta = 2 \sin \theta \cos \theta$$

To obtain the critical numbers, we differentiate $A = A(\theta)$ with respect to θ and set the result equal to 0:

$$A'(\theta) = 8 \cos 2\theta = 0$$

$$\cos 2\theta = 0$$

$$\theta = \frac{\pi}{4}$$

By using the second derivative test (4.35), we get

$$A''(\theta) = -16 \sin 2\theta$$

$$A''\left(\frac{\pi}{4}\right) = -16 \sin \frac{\pi}{2} = -16$$

Thus, at $\theta = \pi/4$ the area is maximum, and $A = 4 \sin 2(\pi/4) = 4$. The dimensions of the rectangle are: Length $= 2\sqrt{2}$, Height $= \sqrt{2}$.

∎

EXERCISE 8

1. A farmer with 3000 meters of available fencing wishes to enclose a rectangular plot that borders on a straight highway. If the farmer does not fence the side along the highway, what is the largest area that can be enclosed?

2. If the farmer in Problem 1 decides also to fence the side along the highway, what is the largest area that can be enclosed?

3. Find the dimensions of the rectangle with the largest area that can be enclosed on all sides by L meters of fencing.

4. A builder wishes to fence in 60,000 square meters of land in a rectangular shape. Because of security reasons, the fence along the front part of the land will cost $2.00 per meter, while the fence for the other three sides will cost $1.00 per meter. How much of each type of fence will the builder have to buy in order to minimize the cost of the fence? What is the minimum cost?

5. A gardener with 20,000 meters of available fencing wishes to enclose a rectangular field and then divide it into two plots with a fence parallel to one of the sides, as shown in the figure. What is the largest area that can be enclosed?

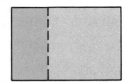

6. A realtor wishes to enclose 600 square meters of land in a rectangular plot and then divide it into two plots with a fence parallel to one of the sides. What are the dimensions of the rectangular plot that require the least amount of fence?

7. An isosceles triangle has a perimeter of fixed length L. What should the dimensions of the triangle be if its area is to be a maximum?

8. An open box with a square base is to be made from a square piece of cardboard that measures 12 centimeters on each side. A square will be cut out from each corner of the cardboard and the sides will be turned up to form the box. Find the dimensions that yield the maximum volume of the box.

9. An open box with a square base is to be made from a square piece of cardboard 24 centimeters on a side by cutting out a square from each corner and turning up the sides. Find the dimensions that yield the maximum volume.

10. An open box with a square base is to have a volume of 2000 cubic centimeters. What should the dimensions of the box be if the amount of material used is to be a minimum?

11. If the box in Problem 10 is to be closed on top, what should the dimensions of the box be if the amount of material used is to be a minimum?

12. A cylindrical container that has a capacity of 10 cubic meters is to be produced. The top and bottom of the container are to be made of a material that costs $2 per square meter, while the side of the container is to be made of material costing $1.50 per square meter. Find the dimensions that will minimize the total cost of the container.

13. A cylindrical container that has a capacity of 4000 cubic centimeters is to be produced. The top and bottom of the container are to be made of material that costs $0.50 per square centimeter, while the side of the container is to be made of material costing $0.40 per square centimeter. Find the dimensions that will minimize the total cost of the container.

14. Prove that a cylindrical container of fixed volume V requires the least material (minimum surface area) when its height is twice its radius.

15. A car rental agency has 24 cars (identical model). The owner of the agency finds that at a price of $10 per day, all the cars can be rented; however, for each $1 increase in rental, one of the cars is not rented. What should the agency charge to maximize income?

16. A charter flight club charges its members $200 per year. But for each new member in excess of 60, the charge for every member is reduced by $2. What number of members leads to a maximum revenue?

17. A heavy object of mass m is to be dragged along a horizontal surface by a rope making an angle θ with the horizontal. The force F required to move the object is given by the formula

$$F = \frac{mc}{c \sin \theta + \cos \theta}$$

where c is the *coefficient of friction* of the surface. Show that the force is least when $\tan \theta = c$.

18. A self-catalytic chemical reaction results in the formation of a product that causes its formation rate to increase. The reaction rate V of many self-catalytic chemicals obeys the relationship

$$V = kx(a - x) \qquad 0 \leq x \leq a$$

where k is a positive constant, a is the initial amount of the chemical, and x is the variable amount of the chemical. For what value of x is the reaction rate a maximum?

19. A truck has a top speed of 75 miles per hour and, when traveling at the rate of x miles per hour, consumes gasoline at the rate of $\frac{1}{200}[(1600/x) + x]$ gallon per mile. This truck is to be taken on a 200 mile trip by a driver who is to be paid at the rate of $\$b$ per hour plus a commission of $\$c$. Since the time required for this trip at x miles per hour is $200/x$, the total cost, if gasoline costs $\$a$ per gallon, is

$$C(x) = \left(\frac{1600}{x} + x\right)a + \frac{200}{x}b + c$$

Find the most economical possible speed under each of the following sets of conditions:
(a) $a = \$1.50$, $b = 0$, $c = 0$ (b) $a = \$1.50$, $b = \$8.00$, $c = \$500$
(c) $a = \$1.60$, $b = \$10.00$, $c = 0$

20. Find the largest area of a rectangle with one vertex on the parabola $y = 9 - x^2$,   another at the origin, and the remaining two on the positive x-axis and positive y-axis, respectively.

21. A telephone company is asked to provide telephone service to a customer whose house is located 2 kilometers away from the road along which the telephone lines run. The nearest telephone box is located 5 kilometers down the road. As shown in the figure, let $5 - x$ denote the distance from the box to the connection so that x is the distance from this point to the point on the road closest to the house. If the cost to connect the telephone line is $50 per kilometer along the road and $60 per kilometer away from the road, where along the road from the box should the company connect the telephone line so as to minimize construction cost?

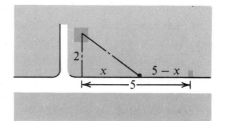

22. A small island is 3 kilometers from the nearest point P on the straight shoreline of a large lake. A town is 12 kilometers down the shore from P. If a person on the island can row a boat 2.5 kilometers per hour and can walk 4 kilometers per hour, where should the boat be landed so that the person arrives in town in the shortest time?

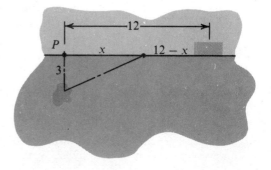

23. Two houses A and B on the same side of a road are a distance p apart, with distances q and r, respectively, from the road. Find the length of the shortest path that goes from A to the road and then on to the other house B.
   (a)   Use the derivative techniques introduced in this section.
   (b)   Use only elementary geometry.
   [*Hint:*   Introduce an imaginary house C on the other side of the road such that the midpoint between B and C is on the road and the segment BC is perpendicular to the road; that is, "reflect" B across the road to become C.]

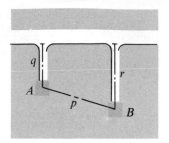

24. The strength of a rectangular beam is proportional to the product of the width and the cube of its depth. Find the dimensions of the strongest beam that can be cut from a log whose cross section has the form of a circle of fixed radius R.

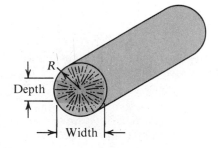

25. If the strength of a rectangular beam is proportional to the product of its width and the square of its depth, find the dimensions of the strongest beam that can be cut from a log whose cross section has the form of the ellipse $10x^2 + 9y^2 = 90$.   [*Hint:*   Choose   Width $= 2x$ and   Depth $= 2y$.]

26. The strength of a beam made from a certain wood is proportional to the product of its width and the cube of its depth. Find the dimensions of the rectangular cross section of the beam with maximum strength that can be cut from a log whose original cross section is in the form of the ellipse $b^2x^2 + a^2y^2 = a^2b^2$, $a \geq b$.

27. The diagram shows two corridors meeting at a right angle. One has width 1 meter, and the other, width 8 meters. Find the length of the longest pipe that can be carried horizontally from one hall, around the corner, and into the other hall.

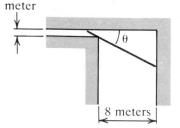

28. Find the length of the shortest beam that can be used to brace a wall if the beam is to pass over a second wall 2 meters high and 5 meters from the first wall.

29. The sides of a V-shaped trough are 28 centimeters wide. Find the angle between the sides of the trough that results in maximum capacity.

30. A metal rain gutter is to have 10 centimeter sides and a 10 centimeter horizontal bottom, with the sides making equal angles with the bottom. How wide should the opening across the top be for maximum carrying capacity?

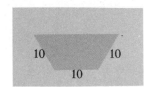

31. An open irrigation ditch of given (fixed) cross-sectional area is to be lined with concrete to prevent seepage. If the two equal sides are perpendicular to the flat bottom, find the relative dimensions that require the least amount of concrete.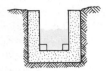

32. The intensity of illumination at a point varies inversely as the square of the distance between the point and the light source. Two lights, one having an intensity 8 times that of the other, are 6 meters apart. At what point between the two lights is the total illumination least?

33. A proposed tunnel of given (fixed) cross-sectional area is to have a horizontal floor, vertical walls of equal height, and a ceiling that is a semicircular cylinder. If the ceiling costs 3 times as much per square meter to build as the vertical walls and the floor, find the most economical ratio of the diameter of the semicircular cylinder to the height of the vertical walls.

34. An observatory is to be in the form of a right circular cylinder surmounted by a hemispherical dome. If the hemispherical dome costs 3 times as much per square meter as the cylindrical wall, what are the most economical dimensions for a given volume? Neglect the floor.

35. A wire is to be cut into two pieces. One piece will be bent into a square, and the other piece will be bent into a circle. If the total area enclosed by the two pieces is to be 64 square centimeters, what is the minimum length of wire that can be used? What is the maximum length of wire that can be used?

36. A wire is to be cut into two pieces. One piece will be bent into an equilateral triangle, and the other piece will be bent into a circle. If the total area enclosed by the two pieces is to be 64 square centimeters, what is the minimum length of wire that can be used? What is the maximum length of wire that can be used?

37. A wire 35 centimeters long is cut into two pieces. One piece is bent into the shape of a square, and the other piece is bent into the shape of a circle. How should the wire be cut so that the area enclosed is minimum? How should it be cut to maximize the area?

38. A wire 35 centimeters long is cut into two pieces. One piece is bent into the shape of an equilateral triangle, and the other piece is bent into the shape of a circle. How should the wire be cut so that the area enclosed is maximum? How should it be cut to minimize the area?

39. A Norman window has the shape of a rectangle surmounted by a semicircle of diameter equal to the width of the rectangle (see the figure). If the perimeter of the window is 10 meters, what dimensions will admit the most light?

9. Antiderivatives

We have already learned that for each differentiable function f there is a corresponding derivative function f'. We now ask the following question: "If a function f is given, can we find a function F whose derivative is f?" That is, is it possible to find a function F so that $F' = dF/dx = f$? If such a function F can be found, it is called an *antiderivative of f*.

(4.43)   **DEFINITION**   *Antiderivative.*   **A function F is called an *antiderivative* of the function f if $F' = f$.**

For example, an antiderivative of $2x$ is x^2 since

$$\frac{d}{dx} x^2 = 2x$$

Another function whose derivative is $2x$ is $x^2 + 3$ since

$$\frac{d}{dx} (x^2 + 3) = 2x$$

This leads us to suspect that the function $f(x) = 2x$ has an unlimited number of antiderivatives. Indeed, any of the functions x^2, $x^2 + \frac{1}{2}$, $x^2 + 2$, $x^2 + \sqrt{5}$, $x^2 - \pi$, $x^2 - 1$, $\ldots$, $x^2 + C$,   where C is any constant, has the property that its derivative is $2x$. We conjecture that all the antiderivatives of $2x$ are given by $x^2 + C$, where C is any constant. The following theorem assures us that this conjecture is correct.

(4.44)   **THEOREM**   **If f and g are differentiable functions, and**

$$f'(x) = g'(x) \qquad \text{for all } x \text{ in } [a, b]$$

then there is a constant C so that

$$f(x) = g(x) + C$$

for all x in $[a, b]$.

Before we prove the theorem, let's interpret it geometrically. The hypotheses of the theorem state that at corresponding points in $[a, b]$ the slopes of the tangent lines to the graphs of f and g must be the same (see Fig. 52). If we plot the graphs

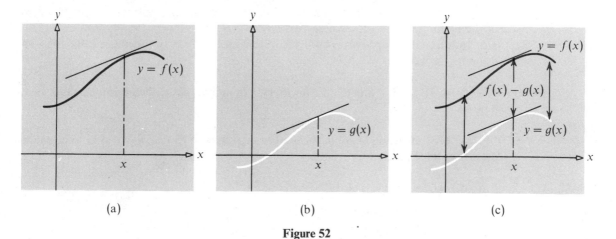

(a)   (b)   (c)

Figure 52

of f and g from parts (a) and (b) of the figure, along with their tangent lines, on the same set of coordinate axes, we obtain Figure 52(c). Then we can see that the values $f(x)$ and $g(x)$ differ by a constant amount as x varies. This constant amount is simply the vertical difference between the two graphs.

Proof   Define the function h by

$$h(x) = f(x) - g(x)$$

By differentiating both sides with respect to x and using the fact that $f'(x) = g'(x)$ for all x, we find

$$h'(x) = f'(x) - g'(x) = 0 \qquad \text{for all } x \text{ in } [a, b]$$

and hence

$$h'(x) = 0 \qquad \text{for all } x \text{ in } [a, b]$$

Now, we let x be any number for which $a < x \le b$. By the mean value theorem (4.26) there exists a number c between a and x such that

$$h(x) - h(a) = (x - a)h'(c) = (x - a)0 = 0$$

Thus, $h(x) = h(a)$, a constant, for all x in $[a, b]$. By setting $C = h(a)$, we get

$$C = f(x) - g(x)$$
$$f(x) = g(x) + C \qquad \text{for all } x \text{ in } [a, b] \qquad \blacksquare$$

As a result of (4.44), we have the following:

(4.45)   **If F is an antiderivative of f, then any other antiderivative has the form $F(x) + C$, where C is an (arbitrary) constant.**

All the antiderivatives of f may be obtained from the expression $F(x) + C$ by letting C range over all real numbers. For example, all the antiderivatives of x^5 are of the form $(x^6/6) + C$, where C is a constant.

In the course of proving (4.44), the following corollary of the mean value theorem was also proved:

(4.46)   **COROLLARY**   **If $h'(x) = 0$ for every x in an interval I, then h is a constant function on I.**

We shall be using (4.46) in this and later chapters. For now, we give some examples of antidifferentiation.

EXAMPLE 1   Find all the antiderivatives of $f(x) = x^{1/2}$.

Solution   Recall that the derivative of $\frac{2}{3}x^{3/2}$ is

$$(\tfrac{2}{3})(\tfrac{3}{2}x^{1/2}) = x^{1/2}$$

Hence, all the antiderivatives of $f(x) = x^{1/2}$ are of the form

$$\tfrac{2}{3}x^{3/2} + C \qquad \text{where } C \text{ is a constant} \qquad \blacksquare$$

In the above example, you may ask how we knew to choose $\tfrac{2}{3}x^{3/2}$. We arrived at the answer in two stages. First we know that

$$\frac{d}{dx}\, x^r = rx^{r-1}$$

That is, differentiation reduces the exponent by 1. Antidifferentiation is the inverse process, so it should increase the exponent by 1. This is how we obtain the $x^{3/2}$ part of $\tfrac{2}{3}x^{3/2}$. Second, the $\tfrac{2}{3}$ factor is needed so that when we differentiate $x^{3/2}$, we get $x^{1/2}$ and not $\tfrac{3}{2}x^{1/2}$.

Because of the differentiation formulas

1. $\dfrac{d}{dx}\, x^{r+1} = (r+1)x^r,\quad r$ a rational number

2. $\dfrac{d}{dx}\, \sin x = \cos x$

3. $\dfrac{d}{dx}\, \cos x = -\sin x$

we have the corresponding antidifferentiation formulas:

(4.47)   **All the antiderivatives of x^r are** $\dfrac{x^{r+1}}{r+1} + C,\quad r \neq -1,^*\quad C$ **any constant.**

(4.48)   **All the antiderivatives of $\cos x$ are** $\sin x + C,\quad C$ **any constant.**

(4.49)   **All the antiderivatives of $\sin x$ are** $-\cos x + C,\quad C$ **any constant.**

Because of the close relationship between the processes of differentiation and finding antiderivatives, we are able to state the following results:

(4.50)   **If functions F_1 and F_2 are antiderivatives of f_1 and f_2, respectively, then $F_1 + F_2$ is an antiderivative of $f_1 + f_2$.**

(4.51)   **If k is a constant and if F is an antiderivative of f, then kF is an antiderivative of kf.**

These statements are a consequence of the properties of derivatives. They may be more easily remembered in the following forms:

An antiderivative of a sum of functions equals the sum of the antiderivatives of the functions.

---

* The case for which $r = -1$ requires special attention and is taken up in Chapter 7.

An antiderivative of a constant times a function equals the constant times an antiderivative of the function.

The result (4.50) can be extended to any finite sum of functions. A similar result is also true for differences. Let's see how these are used by looking at an example.

EXAMPLE 2   If $f(x) = x^5 - x^{1/2} + (6/x^2) - \sin x$,   find all its antiderivatives.

Solution   We have already found that $x^6/6$ and $\frac{2}{3}x^{3/2}$ are antiderivatives of x^5 and $x^{1/2}$, respectively. For the term $6/x^2$, we write $6/x^2 = 6x^{-2}$.   By using (4.51) and (4.47), an antiderivative of $6x^{-2}$ is

$$\frac{6x^{-1}}{-1} = \frac{-6}{x}$$

Finally, an antiderivative of $\sin x$ is $-\cos x$. Hence, all the antiderivatives of the function f are given by

(4.52)
$$\frac{x^6}{6} - \frac{2}{3}x^{3/2} - \frac{6}{x} + \cos x + C \qquad C \text{ a constant} \qquad \blacksquare$$

Note that the constant C in the above example is the sum of the constants resulting from each term since the antiderivatives of x^5 are $(x^6/6) + C_1$,   the antiderivatives of $x^{1/2}$ are $\frac{2}{3}x^{3/2} + C_2$,   the antiderivatives of $6/x^2$ are $(-6/x) + C_3$, and the antiderivatives of $\sin x$ are $-\cos x + C_4$.   Combining these expressions, we get result (4.52) by letting

$$C_1 - C_2 + C_3 - C_4 = C$$

DIFFERENTIAL EQUATIONS

In scientific studies of physical, chemical, biological, and other phenomena, attempts are made, on the basis of long observation, to deduce mathematical laws that describe and predict natural behavior. Such laws often involve the derivatives of some unknown function F, which is to be found. For example, it may be required to find all functions $y = F(x)$ so that $dy/dx = f(x)$.   This equation is an example of what is called a *differential equation*, and a function $y = F(x)$ for which $dy/dx = f(x)$ is a *solution* of the differential equation. The *general solution* of $dy/dx = f(x)$ consists of all the antiderivatives of f.

For example, the general solution of the differential equation

(4.53)
$$\frac{dy}{dx} = 5x^2 + 2$$

is

$$y = F(x) = \tfrac{5}{3}x^3 + 2x + C$$

A *particular solution* of $dy/dx = f(x)$ occurs when C is assigned a particular value. When a particular solution is required, we use a *boundary condition on F*. For example, in the differential equation (4.53) we might require the general solution to obey the condition that $y = 5$ when $x = 3$. Then

$$5 = (\tfrac{5}{3})(27) + (2)(3) + C$$
$$C = -46$$

The particular solution of (4.53) with the boundary condition that $y = 5$ when $x = 3$ is therefore

$$y = \tfrac{5}{3}x^3 + 2x - 46$$

EXAMPLE 3 Solve the differential equation below with the boundary condition that $y = -1$ when $x = 3$.

$$\frac{dy}{dx} = x^2 + 2x + 1$$

Solution The general solution of the differential equation is

$$y = \frac{x^3}{3} + x^2 + x + C$$

To determine the number C, we use the boundary condition. Then

$$-1 = \frac{3^3}{3} + 3^2 + 3 + C$$

$$C = -22$$

The particular solution of the differential equation with the boundary condition that $y = -1$ when $x = 3$ is

$$y = \frac{x^3}{3} + x^2 + x - 22 \qquad \blacksquare$$

RECTILINEAR MOTION

Suppose the functions $s = s(t)$, $v = v(t)$, and $a = a(t)$ represent the position, velocity, and acceleration, respectively, of a particle at time t. The three quantities s, v, and a are related by the differential equations

(4.54)
$$\frac{ds}{dt} = v(t) \qquad \text{and} \qquad \frac{dv}{dt} = a(t)$$

Thus, if the acceleration $a = a(t)$ of a particle is a known function of the time t, its velocity may be found by solving the differential equation $dv/dt = a(t)$. Similarly, if the velocity $v = v(t)$ of a particle is a known function of t, its position $s = s(t)$ at time t is the solution of the differential equation $ds/dt = v(t)$.

In the next example we illustrate how to determine the position $s = s(t)$ of a particle once the velocity or the acceleration function and some boundary conditions are given. In a physical problem, boundary conditions are often the values of the velocity v and position s at time $t = 0$. In such cases, $v(0)$ and $s(0)$ are referred to as *initial conditions*. They can also be written as v_0 and s_0.

EXAMPLE 4   Find the position of a particle at any time t if its acceleration a is known to be

$$a(t) = 8t - 3 \text{ meters per second per second}$$

and the initial conditions are given as

$$v_0 = v(0) = 4 \text{ meters per second} \qquad \text{and} \qquad s_0 = s(0) = 1 \text{ meter}$$

Solution   Since $dv/dt = a(t) = 8t - 3$, we have $v(t) = 4t^2 - 3t + C_1$ for some constant C_1. The initial conditions state that when $t = 0$, then $v_0 = v(0) = 4$. Thus,

$$v_0 = v(0) = 4(0)^2 - 3(0) + C_1 = 4$$
$$C_1 = 4$$

The velocity at any time t is therefore

$$v(t) = 4t^2 - 3t + 4$$

The position of the particle at time t obeys the differential equation

$$\frac{ds}{dt} = v(t) = 4t^2 - 3t + 4$$

Hence,

$$s(t) = \frac{4}{3}t^3 - \frac{3}{2}t^2 + 4t + C_2 \qquad \text{for some constant } C_2$$

Applying the initial condition that $s_0 = s(0) = 1$, we find

$$s_0 = s(0) = 0 - 0 + 0 + C_2 = 1$$
$$C_2 = 1$$

Hence,

$$s(t) = \frac{4}{3}t^3 - \frac{3}{2}t^2 + 4t + 1$$

is the position of the particle at any time t.   ∎

EXAMPLE 5   When the brakes of a car are applied, they produce a deceleration at the constant rate of 10 meters per second per second. If the car is to stop within 20 meters after the brakes are applied, what is the maximum allowable speed?

Solution   Let $s(t)$ represent the distance in meters the car has traveled t seconds after the brakes are applied. Let v_0 be the speed of the car at the time the brakes are applied $(t = 0)$.

Since the car decelerates at the rate of 10 meters per second per second, its acceleration a is

$$a = -10 \quad \text{or} \quad \frac{dv}{dt} = -10$$

By solving the differential equation for v, we find

$$v(t) = -10t + C_1$$

At $t = 0$, $v(0) = v_0$, the speed of the car when the brakes are applied. Thus,

$$v(t) = -10t + v_0 \quad \text{or} \quad \frac{ds}{dt} = -10t + v_0$$

By solving the differential equation for s, we get

$$s(t) = -5t^2 + v_0 t + C_2 \qquad C_2 \text{ a constant}$$

At $t = 0$, $s(0) = 0$, since we start measuring distance at the point at which the brakes are applied. Hence,

(4.55) $$s(t) = -5t^2 + v_0 t$$

The car stops completely when the speed is equal to 0; that is, when

$$\frac{ds}{dt} = -10t + v_0 = 0$$

$$t = \frac{v_0}{10}$$

This is the time that must elapse for the car to come to rest. If we substitute $v_0/10$ for t in (4.55), we get the total distance the car has traveled:

$$s\left(\frac{v_0}{10}\right) = -5\left(\frac{v_0}{10}\right)^2 + v_0\left(\frac{v_0}{10}\right) = \frac{v_0^2}{20}$$

But, according to our original conditions, this distance cannot exceed 20 meters; that is, $v_0^2/20 \leq 20$. Thus, the maximum allowable speed v_0 for the car is

$$v_0^2 = 400$$

$$v_0 = 20 \text{ meters per second}$$

$$= \left(\frac{20 \text{ meters}}{\text{second}}\right)\left(\frac{1 \text{ kilometer}}{1000 \text{ meters}}\right)\left(\frac{3600 \text{ seconds}}{1 \text{ hour}}\right)$$

$$= 72 \text{ kilometers per hour}$$

$$\approx \left(\frac{72 \text{ kilometers}}{\text{hour}}\right)\left(\frac{1 \text{ mile}}{1.6 \text{ kilometer}}\right) = 45 \text{ miles per hour} \quad \blacksquare$$

FREELY FALLING BODIES

A common example of motion with (nearly) constant acceleration is that of a body falling toward the earth. In the absence of air resistance we find that all bodies, regardless of their size, weight, or composition, fall with the same acceleration at the same point of the earth's surface, and if the distance covered is not too great, the acceleration remains constant throughout the fall. This ideal motion, in which air resistance and the small change in acceleration with altitude are neglected, is called *free fall*. The acceleration of a freely falling body is called the *acceleration due to gravity* and is denoted by the symbol g. Near the earth's surface its magnitude is approximately 32 feet per second per second or 9.8 meters per second per second, and it is directed down toward the center of the earth.

If F is the weight of a body of mass m, then according to Galileo, assuming air resistance is negligible, a freely falling body obeys the relationship

(4.56) $$F = -mg$$

The minus sign is chosen since the body is falling, and we choose upward movement to be positive. Also, according to Newton, $F = ma$ (this is called *Newton's law of motion*), so (4.56) can be written as

(4.57) $$ma = -mg$$
$$a = -g$$

where a is acceleration.

We wish to obtain formulas for the velocity v and position s of a falling body at time t. To fix our ideas, suppose an object at $t = 0$ has speed v_0 and is at a height s_0 above the ground. By solving the differential equation $dv/dt = a = -g$ for v, we get

$$v = -gt + v_0$$

Since $v = ds/dt$,   we find the distance s above ground level of the object to be

$$s = (-\tfrac{1}{2})gt^2 + v_0 t + s_0$$

The next example illustrates how problems involving falling bodies are solved in practice.

EXAMPLE 6   A rock is thrown straight up with an initial velocity of 9.8 meters per second from the roof of a building 14.7 meters above ground level. How long does it take the rock to reach its maximum height? What is the maximum height of the rock? If the rock misses the edge of the building on the way down and eventually strikes the ground, what is the total time the rock is in the air?

Solution   In Figure 53 we start measuring time at the moment the rock is released. If s is the distance in meters of the rock from the ground, then, since the rock is released at a height of 14.7 meters, we have

(4.58) $$s_0 = s(0) = 14.7$$

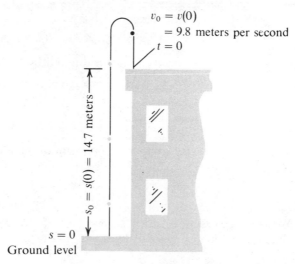

Figure 53

The initial velocity of the rock is

(4.59)
$$v_0 = v(0) = 9.8$$

If we ignore air resistance, the only force acting on the rock is gravity. Since gravity $(g = 9.8$ meters per second per second$)$ is acting in a direction opposite to that of the motion of the rock, the acceleration a of the rock is

$$a = -9.8$$

But $a = dv/dt$, so that

$$\frac{dv}{dt} = -9.8 \qquad v(t) = -9.8t + v_0$$

By using the initial condition (4.59), we find the velocity of the rock at any time t:

(4.60)
$$v(t) = -9.8t + 9.8$$

But $v = ds/dt$, so

$$\frac{ds}{dt} = -9.8t + 9.8 \qquad s(t) = -4.9t^2 + 9.8t + s_0$$

By using the initial condition (4.58), the distance of the rock from the ground at any time t is therefore

(4.61)
$$s(t) = -4.9t^2 + 9.8t + 14.7$$

The rock reaches its maximum height when its velocity is 0. From (4.60), this happens when $t = 1$ second. To obtain the maximum height, we use (4.61) with $t = 1$.

The maximum height of the rock is

$$s(1) = -4.9 + 9.8 + 14.7 = 19.6 \text{ meters above the ground}$$

The total time the rock is in the air is found by setting $s = 0$. From (4.61), we find

$$-4.9t^2 + 9.8t + 14.7 = 0$$
$$t^2 - 2t - 3 = 0$$
$$(t - 3)(t + 1) = 0$$

The only meaningful solution (since t cannot be negative) is $t = 3$. Thus, the rock is in the air for 3 seconds.   ■

LAW OF INERTIA

We close this section with a special case of the *law of inertia*, which was originally stated by Galileo.

(4.62)   **A body acted upon by no force remains at rest or in a state of uniform motion along a straight line.**

The force F acting upon a body of mass m is given by Newton's law of motion $F = ma$, where a is the acceleration of the body. If there is no force acting upon the body, then $F = 0$. In this case, the acceleration a must be 0. But $a = dv/dt$, where v is the speed of the body. Hence,

$$\frac{dv}{dt} = 0 \qquad \text{or} \qquad v = \text{Constant}$$

That is, the body is at rest $(v = 0)$ or else in a state of uniform motion (v is a non-zero constant).

EXERCISE 9

In Problems 1–18 find all the antiderivatives of each function.

1. $f(x) = 4x^5$

2. $f(x) = x^{4/3}$

3. $f(x) = 5x^{3/2}$

4. $f(x) = x^{5/2}$

5. $f(x) = 2x^{-2}$

6. $f(x) = 3x^{-3}$

7. $f(x) = \sqrt{x}$

8. $f(x) = \dfrac{1}{\sqrt{x}}$

9. $f(x) = 4x^3 - 3x^2 + 1$

10. $f(x) = x^2 - x$

11. $f(x) = (2 - 3x)^2$

12. $f(x) = (3x - 1)^2$

13. $f(x) = \dfrac{3x - 2}{\sqrt{x}}$

14. $f(x) = \dfrac{4x^{3/2} - 1}{x^2}$

15. $f(x) = \dfrac{x^2 + 10x + 21}{3x + 9}$

16. $f(x) = \dfrac{x^3 - 5x + 8}{x^5}$

17. $f(x) = 2x - 3\cos x$

18. $f(x) = \sin x - \cos 2x$

In Problems 19–26 find the solution of each differential equation having the given boundary conditions.

19. $\dfrac{dy}{dx} = 3x^2 - 2x + 1$, if $y = 1$ when $x = 0$

20. $\dfrac{dy}{dx} = x^{1/3} + x\sqrt{x} - 2$, if $y = 2$ when $x = 0$

21. $\dfrac{dv}{dt} = 3t^2 - 2t + 1$, if $v = 5$ when $t = 1$

22. $\dfrac{ds}{dt} = t^4 + 4t^3 - 5$, if $s = 5$ when $t = 2$

23. $\dfrac{ds}{dt} = t^3 + \dfrac{1}{t^2}$, if $s = 2$ when $t = 1$

24. $\dfrac{dy}{dx} = \sqrt{x} - x\sqrt{x} + 1$, if $y = 0$ when $x = 1$

25. $\dfrac{dy}{dx} = x - 2\sin x$, if $y = 0$ when $x = 0$

26. $\dfrac{dy}{dx} = x^2 - \sin 2x$, if $y = 0$ when $x = 0$

In Problems 27–30 find the distance $s = s(t)$ under the stated conditions.

27. $a = -32$ feet per second per second, $s(0) = 0$ feet, $v(0) = 128$ feet per second

28. $a = -980$ centimeters per second per second, $s(0) = 5$ centimeters, $v(0) = 1980$ centimeters per second

29. $a = 3t$ meters per second per second, $s(0) = 2$ meters, $v(0) = 18$ meters per second

30. $a = 5t - 2$ feet per second per second, $s(0) = 0$ feet, $v(0) = 8$ feet per second

31. Using the fact that

$$\frac{d}{dx}(x \cos x + \sin x) = -x \sin x + 2 \cos x$$

find F if

$$\frac{dF}{dx} = -x \sin x + 2 \cos x \qquad \text{and} \qquad F(0) = 1$$

32. Using the fact that

$$\frac{d}{dx}\sin x^2 = 2x \cos x^2$$

find h if

$$\frac{dh}{dx} = x \cos x^2 \qquad \text{and} \qquad h(0) = 2$$

33. A car decelerates at a constant rate of 10 meters per second per second when its brakes are applied. If the car must stop within 15 meters after applying the brakes, what is the maximum allowable speed for the car?

34. A car can accelerate from 0 to 60 kilometers per hour in 60 seconds. If the acceleration is constant, how far does the car travel during this time?

35. The 2 meter high jump is rather commonplace today. If this event were held on the moon, where the acceleration due to gravity is 1.6 meters per second per second, what height would be attained? Assume the athlete can propel herself with the same force on the moon as on the earth.

36. A ball is thrown straight up from ground level with an initial velocity of 19.6 meters per second. How high is the ball thrown? How long will it take the ball to return to ground level?

37. A child throws a ball straight up. If the ball is to reach a height of 9.8 meters, what is the minimum velocity that must be imparted to the ball? (Assume the initial height of the ball is 1 meter.)

38. A ball thrown directly down from a roof 49 meters high reaches the ground in 3 seconds. What is the initial velocity?

39. A constant force is applied to a particle that is initially at rest. If the mass of the particle is 4 grams and if its velocity after 6 seconds is 12 centimeters per second, determine the force applied to it.

40. Starting from rest, with what constant acceleration must a car proceed to go 2 kilometers in 2 minutes? (Give your answer in centimeters per second per second.)

41. A child on top of a building 24 meters high drops a rock and then 1 second later throws another rock straight down. What initial velocity must the second rock be given so that the dropped rock and the thrown rock hit the ground at the same time?

10. Application to Economics (If Time Permits)

In general, a company that produces items for sale can lose money by producing either too few items or too many items. *Marginal analysis* provides a means to answer the question, "What level of production will result in maximum profits?"

If P is profit, R is revenue received, and C is cost of production, then

$$\text{Profit} = \text{Revenue} - \text{Cost}$$

(4.63)
$$P = R - C$$

Revenue is the amount of money derived from the sale of a product and equals the sales price of the product times the quantity of the product that is actually sold. But the sales price and the quantity sold are not always independent. As the price (p) falls, the demand for the product usually increases, and the quantity sold increases (see Fig. 54). When the price rises, the opposite occurs.

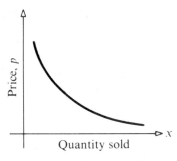

Figure 54

The relationship between the price p of a product and the quantity x demanded in the marketplace is called the *demand equation*. If in this equation we solve for p, we have

$$p = d(x)$$

The function d is called the *price function*, and $d(x)$ is the price per unit, when x units are demanded.

Even though in normal economic transactions x can assume only positive integer values, in order to apply the calculus we assume that x is a real number so that the price function may be treated as a continuous function.

In practice, a price function is found through surveys, analysis of data, history, and other sources available to the economist. The next example illustrates how a linear price function can be constructed. It is important to observe that the fundamental assumption made here is the linear nature of the price function.

EXAMPLE 1   Dan's Toy Store has observed that each week 1000 electric trucks are sold at a price of $5 per truck. When the store has a sale, the trucks sell for $4 each and 1200 per week are sold. Construct the price function, assuming it is linear.

Solution   Let p be the price of each truck and let x be the number sold. If the price function $p = d(x)$ is linear, then we know that (1000, 5) and (1200, 4) are two points on the line $p = d(x)$. Thus, the slope is $-\frac{1}{200}$ and the price function is

$$p - 4 = \frac{-1}{200}(x - 1200)$$

$$p = \frac{-1}{200}x + 10 \quad \blacksquare$$

The price function obtained in the above example is not meant to reflect extreme situations. For example, we do not expect to sell $x = 0$ trucks nor do we expect to sell too many trucks in excess of 1500 since even during a special only 1200 are sold. The price function does represent the relationship between price and quantity in a certain range—in this case, perhaps $500 < x < 1500$. Our next example illustrates a constant price function.

EXAMPLE 2   A farmer can expect to sell wheat at $2 per bushel, no matter how much of it is grown. Find the price function.

Solution   Since the price per bushel is fixed at $2 per bushel, the price function is

$$p = 2 \quad \blacksquare$$

One way economists describe the behavior of a demand function is by a term called the *price elasticity of demand*. It describes the relative responsiveness of consumers to a change in the price of an item. If $p = d(x)$ is a differentiable demand function, then the price elasticity of demand is given by

$$\eta = \frac{p/x}{dp/dx}$$

For a given price, if $|\eta| < 1$, the demand is said to be *inelastic*; if $|\eta| > 1$, the demand is said to be *elastic*.

EXAMPLE 3   Show that the demand function $p = 10x^{-1/2}$ is elastic.

Solution
$$\eta = \frac{p/x}{dp/dx} = \frac{(10x^{-1/2})/x}{-5x^{-3/2}} = \frac{10x^{-3/2}}{-5x^{-3/2}} = -2$$

Since $|\eta| = 2$, we conclude that the demand is elastic for any price.   ∎

REVENUE FUNCTION

If x is the number of units sold and $d(x)$ is the price for each unit, the *revenue function* R is defined as
$$R(x) = xd(x)$$

A measure of the rate of change of revenue as output changes is called the *marginal revenue*, MR. If we know the revenue function R, then the marginal revenue function is defined as the derivative of the revenue function, that is,
$$MR = \lim_{\Delta x \to 0} \frac{\Delta R}{\Delta x} = R'(x)$$

MR is frequently described by economists as the additional revenue from selling an additional unit of output.

COST FUNCTION

We consider now the cost C of producing goods. The cost to a firm is composed of two parts: *fixed costs*, FC, and *variable costs*, VC. The cost C is the total of these two parts:
$$C = FC + VC$$

Fixed costs remain the same no matter how much product is manufactured. Examples of fixed costs are mortgage payments, interest, insurance, real estate taxes, and salaries of people employed even during a time of no production. Variable costs change as production changes. Examples are costs of raw material, salaries of people directly related to production, and so on. Fixed costs are constant, while VC is a function of the quantity produced. The variable cost is 0 when 0 quantity is produced.

To increase profits, a company may decide to increase its production. Associated with this increase are added costs, such as new capital investments, additional labor costs, increased storage facilities, added sales and advertising expenditures, and the like. The question that concerns management is "How will the cost be affected by an increase in production?" What we need to measure is the rate of change of cost as output changes. This measure is called the *marginal cost*, MC. If we know the cost function C, the marginal cost function is the derivative of the cost function. That is,
$$MC = \lim_{\Delta x \to 0} \frac{\Delta C}{\Delta x} = C'(x)$$

MC is frequently described by economists as the additional cost of producing an additional unit of output.

MAXIMIZING PROFITS

We now seek the level of production that will maximize profits. Suppose that x units are produced and that all of these are sold. Then the profit is

$$P(x) = R(x) - C(x)$$

where R and C are the revenue and cost functions, respectively. To maximize the profit P, we find the critical numbers:

$$\frac{d}{dx}P = \frac{d}{dx}(R - C) = 0$$

$$\frac{d}{dx}R - \frac{d}{dx}C = 0$$

$$MR - MC = 0$$

Thus, critical numbers of P occur when marginal revenue equals marginal cost.

The equality $MR = MC$ is the basis for the classical economic criterion for maximum profit.

When we apply the second derivative test (4.35) to the function P,

$$P''(x) = R''(x) - C''(x)$$

we find that P has a local maximum at a critical number x if $P''(x) < 0$. This will occur at any x for which the marginal revenue function equals the marginal cost and $R''(x) < C''(x)$. These x are restricted to a closed interval in which the endpoints should be tested separately.

Suppose that the wheat farmer in Example 2, who can sell wheat at a fixed price of \$2 per bushel, can produce anywhere from 0 to 15,000 bushels. Suppose the cost function in dollars is

$$C = \frac{x^2}{10,000} + 500$$

where x represents the number of bushels produced. We interpret this cost function as consisting of total fixed costs of \$500 and total variable costs of $x^2/10,000$ dollars. Here, the total fixed cost of \$500 is due to costs of land, equipment, and the like. The total variable cost represents the cost of planting, fertilizing, and harvesting the crop. For example, the cost of producing 15,000 bushels is

$$C = \frac{(15,000)(15,000)}{10,000} + 500 = 22,500 + 500 = \$23,000$$

The marginal cost is of particular interest; it is

$$MC = \frac{dC}{dx} = \frac{x}{5000}$$

For example, for $x = 5000$ bushels, the marginal cost is

$$MC = 1 \text{ dollar per bushel}$$

and for $x = 6000$ bushels, the marginal cost is

$$MC = 1.2 \text{ dollars per bushel}$$

This difference in marginal cost means that the cost of producing each bushel is increasing from \$1 per bushel to \$1.20 per bushel. This increase is tolerable if it is not detrimental to the total profit picture. That is, the combination of cost and revenue is what is critical—not cost alone. We need to ask how the revenue function changes relative to the quantity produced. Comparing this to the marginal cost will provide valuable information to the farmer.

Assuming the farmer can sell all the wheat that is produced, how much wheat should be produced to maximize profit? This is a classic example of free competition. The maximum profit occurs when marginal cost equals marginal revenue. Since

$$C = \frac{x^2}{10,000} + 500 \quad \text{and} \quad R = 2x$$

we find

$$MC = \frac{x}{5000} \quad \text{and} \quad MR = 2$$

These are equal when

$$2 = \frac{x}{5000}$$

$$x = 10,000$$

Since

$$P(x) = R(x) - C(x) = 2x - \frac{x^2}{10,000} - 500$$

it follows that

$$P'(x) = 2 - \frac{2x}{10,000} \quad \text{and} \quad P''(x) = \frac{-1}{5000} < 0$$

Thus, at $x = 10,000$, there is a local maximum. Since

$$P(0) = -500$$

$$P(10,000) = 2(10,000) - \frac{(10,000)^2}{10,000} - 500 = 20,000 - 10,500 = \$9,500$$

$$P(15,000) = 2(15,000) - 23,000 = \$7,000$$

maximum profit occurs for a production that is 5000 bushels under the maximum output of 15,000 bushels.

EXERCISE 10

1. If the cost function is $C(x) = 2x + 5$ and the revenue function is $R(x) = 8x - x^2$, where x is the number of units produced (in thousands) and R and C are measured in millions of dollars, find the following:
   (a) Marginal revenue
   (b) Marginal revenue at $x = 3$, $x = 4$
   (c) Marginal cost
   (d) Fixed cost
   (e) Variable cost at $x = 4$
   (f) *Break-even points;* that is, $R(x) = C(x)$; interpret your answer
   (g) Profit function
   (h) Most profitable output
   (i) Maximum profit
   (j) Marginal revenue at the most profitable output
   (k) Revenue at the most profitable output
   (l) Variable cost at the most profitable output
   (m) Graph MR and MC.
   (n) Graph R and C.

2. Suppose the cost function is given by $C(x) = x^2 + 5$ and the price function is $p = 12 - 2x$, where p is the price in dollars and x is the number of units produced (in thousands). Answer the questions asked in Problem 1.

3. During a fixed period, a retail store can sell x units of a product at a price of p cents per unit, where $p = 20 - 0.03x$. The cost of making x units is C cents, where $C = 3 + 0.02x$. What number of units will lead to maximum profit?

4. A certain item can be produced at a unit cost of $10 for all possible outputs. The price function for the product is $p = 90 - 0.02x$, where p is the price in dollars and x is the number of units.
   (a) How many units should be produced to maximize profit?
   (b) What is the price that gives maximum profit?
   (c) What is the maximum profit?

5. A coal company can produce x tons of coal at a daily cost of $\$C$, where $C = 200 + 35x + 0.02x^2$. The coal can be sold at a price of $39 per ton. How many tons should be produced each day so as to maximize profit?

6. A tractor company can manufacture at most 1000 heavy-duty tractors per year. Furthermore, from past demand data, the company knows that the number of heavy-duty tractors it can sell depends only on the price p of each unit. The company also knows that the cost to produce the units is a function of the number x of units sold. Assume that the price function is $p = 19,000 - 2x$ and the cost function is $C = 1,000,000 + 10,000x + 3x^2$. How many units should be produced to maximize profit?

7. A manufacturer estimates that 500 articles can be sold each week if the unit price is $20 and that weekly sales will rise by 50 with each $0.50 reduction in price. The cost of producing and selling x articles a week is $C(x) = 4200 + 5.10x + 0.0001x^2$. Find:
   (a) Price function
   (b) Level of weekly production for maximum profit
   (c) Price per article at the maximum profit level

8. Let x be the total number of items produced per year and let p be the fixed price at which each is sold. If the cost for producing x items is $C = ax^2 + bx + c$ and $p = r - sx$, what value of x produces maximum profit? Here, $a, b, c, r,$ and s are nonnegative constants, with $a \neq 0$.

9. The quantity $C(x)/x$ is called the *average cost* of producing x units. If $C''(x) > 0$, show that the average cost is a minimum when it equals the marginal cost.

10. The quantity $R(x)/x$ is called the *average revenue* from selling x units. If $R''(x) < 0$, show that the average revenue is a maximum when it equals the marginal revenue.

Miscellaneous Exercises

1. Without finding them, explain why the function $f(x) = \sqrt{x(2 - x)}$ must have an absolute maximum and an absolute minimum. Then find them in two ways (with and without the help of calculus).

2. Without finding the derivative, prove that if $f(x) = (x^2 - 4x + 3)(x^2 + x + 1)$, then $f'(x) = 0$ for at least one number x between 1 and 3. Check by finding the derivative and applying the intermediate value theorem.

3. Does the mean value theorem (4.26) apply to the function $f(x) = \sqrt{x}$ in the interval $[0, 9]$? If not, why not? If so, find the number c referred to in the theorem.

4. Show that when the mean value theorem is applied to the function $f(x) = Ax^2 + Bx + C$ in the interval $[a, b]$, the number c referred to in the theorem is the midpoint of the interval.

5. Suppose that the domain of f is an open interval and $f'(x) > 0$ for all x in the interval. Prove that f cannot have an extreme value.

6. Sketch the graph of $y = x\sqrt{6 - x}$ after answering all the questions below. Also, show the graph of $y^2 = x^2(6 - x)$ on the same drawing.
   (a) What is the domain?
   (b) Where does the graph intersect the x-axis?
   (c) Find all extreme values of y; identify all horizontal and vertical tangents.
   (d) Show that $y'' = 3(x - 8)/4(6 - x)^{3/2}$. Comment on the statement that an inflection point occurs at $x = 8$ and discuss the concavity of the graph.

7. Discuss and sketch the graph of $y = x - 2\sqrt{x}$.

8. Discuss and sketch the graph of $y^2 = x^2(4 - x^2)$.

9. Let g be a continuous function on the closed interval $[0, 1]$. Let $g(0) = 1$ and $g(1) = 0$. Which of the following is *not* necessarily true?
   (a) There exists a number u in $[0, 1]$ such that $g(u) \geq g(x)$ for all x in $[0, 1]$.
   (b) For all a and b in $[0, 1]$, if $a = b$, then $g(a) = g(b)$.
   (c) There exists a number c in $[0, 1]$ such that $g(c) = \frac{1}{2}$.
   (d) There exists a number c in $[0, 1]$ such that $g(c) = \frac{3}{2}$.
   (e) For all c in the open interval $(0, 1)$, $\lim_{x \to c} g(x) = g(c)$.

10. If f is a continuous function on the closed interval $[a, b]$, which of the following is necessarily true?
   (a) f' exists on (a, b).
   (b) If $f(u)$ is a maximum of f, then $f'(u) = 0$.
   (c) $\lim_{x \to c} f(x) = f(\lim_{x \to c} x)$, for $a < c < b$.
   (d) $f'(x) = 0$, for some x, $a \leq x \leq b$.
   (e) The graph of f' is a straight line.

11. If y is a function of x such that $y' > 0$ for all x and $y'' < 0$ for all x, which of the following could be part of the graph of $y = f(x)$?

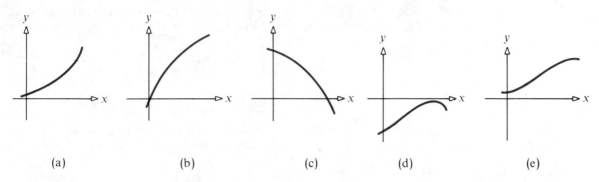

(a)   (b)   (c)   (d)   (e)

12. If a function f is continuous for all x and if f has a local maximum at $(-1, 4)$ and a local minimum at $(3, -2)$, which of the following statements must be true?
(a) The graph of f has a point of inflection somewhere between $x = -1$ and $x = 3$.
(b) $f'(-1) = 0$.
(c) The graph of f has a horizontal asymptote.
(d) The graph of f has a horizontal tangent line at $x = 3$.
(e) The graph of f intersects both axes.

13. Sketch the graph of a function f that has the following properties:

$$f(-3) = 2 \qquad f(-1) = 5 \qquad f(2) = -4 \qquad f(6) = -1$$
$$f'(-1) = f'(2) = 0 \qquad f''(-3) = f''(6) = 0$$

$$\lim_{x \to -\infty} f(x) = 1 \qquad \lim_{x \to 0^-} f(x) = -\infty \qquad \lim_{x \to 0^+} f(x) = +\infty \qquad \lim_{x \to +\infty} f(x) = 0$$

$f''(x) > 0$ if $x < -3$ or $0 < x < 6$ $\qquad f''(x) < 0$ if $-3 < x < 0$ or $6 < x$

14. Sketch the graph of a function f that has the following properties:

$$f(-2) = 2 \qquad f(5) = 1 \qquad f(0) = 0$$

$f'(x) > 0$ if $x < -2$ or $5 < x$ $\qquad f'(x) < 0$ if $-2 < x < 2$ or $2 < x < 5$ $\qquad f'(5) = 0$
$f''(x) > 0$ if $x < 0$ or $2 < x$ $\qquad f''(x) < 0$ if $0 < x < 2$

$$\lim_{x \to 2^-} f(x) = -\infty \qquad \lim_{x \to 2^+} f(x) = +\infty \qquad \lim_{x \to -\infty} f(x) = 0$$

15. Sketch the graph of $f(x) = x(x^2 - 4)^{1/3}$.

16. Sketch the graph of $f(x) = (x^2 + 4)/(x^2 - 9)$.

17. Suppose that the volume of a spherical ball of ice decreases (by melting) at a rate proportional to its surface area. Show that its radius decreases at a constant rate. (Volume and surface area of a sphere of radius R are $V = \frac{4}{3}\pi R^3$ and $S = 4\pi R^2$, respectively.)

18. If p is the period of a pendulum of length L, the acceleration due to gravity may be computed by the formula $g = 4\pi^2 L/p^2$. If L is measured with negligible error, but a 2% error may occur in the measurement of p, what is the (approximate) percentage error in the computation of g?

19. A motorcycle accelerates (at a constant rate) from 0 to 72 kilometers per hour in 10 seconds. How far has it traveled in that time?

20. Suppose that in a town with population P the number of people with an infectious disease increases at a rate proportional to both the number who have the disease and the number who do not.
    (a) Write the above statement in the form of a differential equation.
    (b) If N is the number of people who have the disease at time t, show that the graph of N as a function of t has an inflection point when half the town is infected. Sketch such a graph, showing its asymptote.

21. Find the point on the graph of $2y = x^2$ nearest to (4, 1).

22. At what number x does the *derivative* of $f(x) = (x^4/3) - (x^5/5)$ attain its maximum value?

23. *Calculator problem.*   Compute each function f for several large numbers x. Based on this experimental evidence, what would you guess is $\lim_{x \to +\infty} f(x)$ for each function $f(x)$?

    (a) $f(x) = \left(1 + \dfrac{1}{x}\right)^x$
    (b) $f(x) = \left(1 + \dfrac{1}{x}\right)^{x^2}$
    (c) $f(x) = \left(1 - \dfrac{1}{x}\right)^x$

    (d) $f(x) = \dfrac{\sin x}{x}$
    (e) $f(x) = |\sin x|^x$

24. For the function $f(x) = x\sqrt{x + 1}$, $0 \le x \le b$,   the number c, $0 < c < b$,   satisfying the mean value theorem is $c = 3$.   Find b.

25. Use differentials to approximate the square root of your birth month (not the day or year). If that month is a perfect square, add 1 and approximate the new number. Check your answer with a calculator.

26. A box moves down an inclined plane at an acceleration of $t^2(t - 3)$ centimeters per second per second.   It covers a distance of 10 centimeters in 2 seconds. What was the original velocity?

27. A lamp is on a post 10 meters high. A ball is thrown straight up at a distance of 5 meters from the lamp and 20 meters from a wall with initial velocity 19.6 meters per second. The acceleration due to gravity is $a = -9.8$ meters per second per second.   If the ball is thrown up from an initial height of 1 meter above ground, how fast is the shadow of the ball moving on the wall 3 seconds after the ball is released? Is the ball moving up or down? How far is the ball above ground at $t = 3$ seconds?

28. How fast is the maximum volume of a cone inscribed in a sphere of (variable) radius R decreasing, when the volume of the sphere is decreasing at a rate of 9 cubic meters per minute?

29. How fast is the maximum volume of a cylinder inscribed in a sphere of (variable) radius R decreasing, when the volume of the sphere decreases at a rate of 27 cubic meters per minute?

30. Which point of the semicircle $y = \sqrt{25 - x^2}$ is farthest from the chord AB if $A = (0, 5)$ and $B = (3, 4)$?

31. If $f'(x)$ and $g'(x)$ exist and $f'(x) > g'(x)$ for all real x, then which of the following statements must be true about the graph of $y = f(x)$ and the graph of $y = g(x)$?
    (a)   They intersect exactly once.
    (b)   They intersect no more than once.
    (c)   They do not intersect.
    (d)   They could intersect more than once.
    (e)   They have a common tangent at each point of intersection.

32. Find where the general cubic $f(x) = ax^3 + bx^2 + cx + d$ is increasing and where it is decreasing by considering cases depending on the value of $b^2 - 3ac$.

33. (a)   Prove that a rational function of the form $f(x) = (ax^{2n} + b)/(cx^n + d)$ has at most five critical numbers.
    (b)   Give an example of such a rational function with exactly five critical numbers.

34. Given

$$f(x) = \frac{ax^n + b}{cx^n + d} \qquad \text{with} \qquad ad - bc \neq 0$$

determine the critical numbers and where the function is increasing and where it is decreasing.

35.*Two bodies begin a free fall from rest from the same height 1.0 second apart. How long after the first body begins to fall will the two bodies be 10 meters apart?

36.[†]The engineer of a train moving at a speed v_1 sights a freight train a distance d ahead of him on the same track moving in the same direction with a slower speed v_2. He puts on the brakes and gives his train a constant deceleration a. Show that:

$$\text{If} \qquad d > \frac{(v_1 - v_2)^2}{2a} \qquad \text{there will be no collision}$$

$$\text{If} \qquad d < \frac{(v_1 - v_2)^2}{2a} \qquad \text{there will be a collision}$$

37.[‡]Two cars, A and B, travel in a straight line. The distance of A from the starting point is given as a function of time by $s_A = 4t + t^2$, and the distance of B from the starting point is $s_B = 2t^2 + 2t^3$.
(a) Which car is ahead just after they leave the starting point?
(b) At what times are the cars at the same point?
(c) At what times is the velocity of B relative to A zero?
(d) At what times is the distance from A to B neither increasing nor decreasing?

38. Let A_n be the area bounded by a regular n-sided polygon inscribed in a circle of radius R. Show that:

(a) $A_n = \left(\dfrac{n}{2}\right)R^2 \sin\left(\dfrac{2\pi}{n}\right)$ (b) $\displaystyle\lim_{n \to +\infty} A_n = \lim_{n \to +\infty} \left(\dfrac{n}{2}\right)R^2 \sin\left(\dfrac{2\pi}{n}\right) = \pi R^2$ (the area of a circle of radius R)

39. Let P_n be the perimeter of a regular n-sided polygon inscribed in a circle of radius R. Show that:

(a) $P_n = 2nR \sin\left(\dfrac{\pi}{n}\right)$ (b) $\displaystyle\lim_{n \to +\infty} P_n = 2\pi R$ (the perimeter of a circle of radius R)

---

* Adapted from D. Halliday and R. Resnick, *Physics*, part I (New York: John Wiley & Sons, 1977), p. 52. Reprinted by permission.

[†] Ibid., p. 51. Reprinted by permission.

[‡] Adapted from F. W. Sears, M. W. Zemansky, and H. D. Young, *University Physics* (Reading, Mass.: Addison-Wesley Publishing Co., 1976), p. 67. Reprinted by permission.

5

The Definite Integral

The development of the integral, like that of the derivative, was motivated to a large extent by attempts to solve a basic problem in geometry—namely, the *area problem*. As we stated in Chapter 2, the question is: "Given a nonnegative function f whose domain is the closed interval $[a, b]$, what is the area enclosed by the graph of f, the x-axis, and the vertical lines $x = a$ and $x = b$?" Figure 1 illustrates the area to be found.

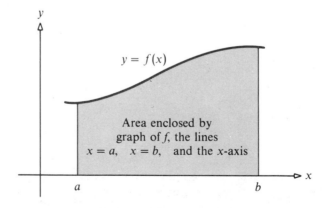

Figure 1

In Sections 1–3, we show how the concept of the integral develops from the area problem. At first glance, the area problem and the tangent problem, which helped motivate the invention of the derivative (see Chap. 2), may look quite dissimilar. However, much of calculus is built upon a surprising relationship between the two problems and their associated concepts. This relationship is the basis for the *fundamental theorem of calculus.*

1. Area

In this section we discuss the problem of finding area. To be more precise, we wish to compute the area of the region enclosed by the graph of a nonnegative function $y = f(x)$, the lines $x = a$ and $x = b$, and the x-axis (see Fig. 1).

If the graph of $y = f(x)$ is a horizontal line, say, $f(x) = h$, with h positive, the region is a rectangle and its area (A) is the product of the height (h) by the width $(b - a)$, as shown in Figure 2. If the graph of $y = f(x)$ consists of three horizontal lines, each of positive height (see Fig. 3), the area A of the region enclosed by the graph of $y = f(x)$, the lines $x = a$ and $x = b$, and the x-axis may be computed by adding up the rectangular areas A_1, A_2, and A_3. In general, the procedure for computing the area of any region enclosed by the graph of a function $y = f(x)$, the lines $x = a$ and $x = b$, and the x-axis is based on this idea of adding up rectangular areas. For convenience, we shall refer to this region as the *area under the graph of f from a to b.*

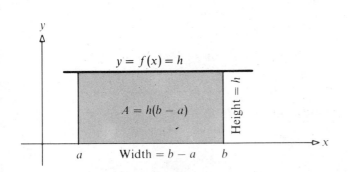

Figure 2

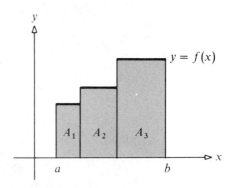

Figure 3

To fix our ideas, we make two assumptions:

1. The function f is continuous on the closed interval $[a, b]$.
2. The function f is nonnegative on the closed interval $[a, b]$.

Since we know how to compute the area of a rectangle, we approximate the area under the graph of f from a to b by using rectangles. We do this in the following

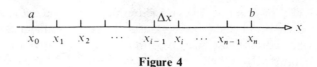

Figure 4

way: We pick $(n-1)$ numbers between a and b and label them $x_1, x_2, \ldots, x_{n-1}$, with $a < x_1 < x_2 < \cdots < x_{n-1} < b$ (see Fig. 4). For convenience, we set $a = x_0$ and $b = x_n$.   The numbers thus selected divide, or *partition*, the interval $[a, b]$ into n subintervals:

$$[x_0, x_1], \quad [x_1, x_2], \quad \ldots, \quad [x_{n-1}, x_n]$$

The selection of numbers between a and b is arbitrary, but for simplicity, we select them so that the length of each subinterval is the same. If we denote this common length by Δx, so that $\Delta x = x_1 - x_0 = x_2 - x_1 = \cdots = x_n - x_{n-1}$,   it follows that

$$\Delta x = \frac{b - a}{n}$$

Since f is continuous on the closed interval $[a, b]$, it is continuous on every subinterval $[x_{i-1}, x_i]$ of $[a, b]$. Because of the extreme value theorem (4.23), there is a number in each of these subintervals at which f attains its absolute minimum. We denote these numbers by $c_1, c_2, c_3, \ldots, c_n$,   so that $f(c_i)$ is the absolute minimum of f in the subinterval $[x_{i-1}, x_i]$. We now construct n rectangles, each having a subinterval as a base and $f(c_i)$ as an altitude (see Fig. 5). In doing this, we obtain n thin strips of uniform width $\Delta x = (b - a)/n$ and altitudes $f(c_1), f(c_2), \ldots, f(c_n)$, respectively, so that their areas may be calculated as follows:

$$\text{Area of first rectangle} = f(c_1)(x_1 - x_0) = f(c_1)\Delta x$$
$$\text{Area of second rectangle} = f(c_2)(x_2 - x_1) = f(c_2)\Delta x$$
$$\vdots$$
$$\text{Area of } n\text{th (and last) rectangle} = f(c_n)(x_n - x_{n-1}) = f(c_n)\Delta x$$

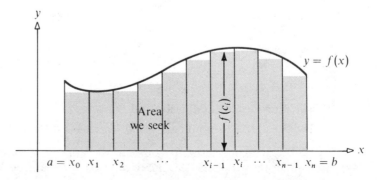

Figure 5

The sum s_n of the areas of these n inscribed rectangles gives an approximation of the area A we seek. That is, the area A is approximately equal to

$$s_n = f(c_1)\Delta x + f(c_2)\Delta x + \cdots + f(c_i)\Delta x + \cdots + f(c_n)\Delta x$$

Since the rectangles we used to approximate the area A are *inscribed rectangles*, the sum s_n *underestimates* the area A of the region. Thus, we conclude that

$$s_n \leq A$$

Let's look at an example before continuing the discussion.

EXAMPLE 1   Approximate the area under the graph of $f(x) = 3x$ from 0 to 10, by computing s_n for $n = 2$, $n = 5$, $n = 10$.

Solution   For $n = 2$,   we partition the closed interval $[0, 10]$ into two subintervals of equal length, $[0, 5]$ and $[5, 10]$, as shown in Figure 6(a). The length of each of these sub-intervals is $\Delta x = (10 - 0)/2 = 5$.   To compute s_2, we need to know where f attains its minimum in each subinterval. Since f is increasing, the minimum is attained at the left endpoint of each subinterval. Thus, for $n = 2$,   the minimum of f on $[0, 5]$ occurs at 0 and the minimum of f on $[5, 10]$ occurs at 5. The value of s_2 is therefore

$$s_2 = f(c_1)\Delta x + f(c_2)\Delta x = f(0)(5) + f(5)(5) = (0)(5) + (15)(5) = 75$$

For $n = 5$,   we partition $[0, 10]$ into five subintervals of equal length; namely, $[0, 2], [2, 4], [4, 6], [6, 8]$, and $[8, 10]$, as shown in Figure 6(b). The length of each of these is $\Delta x = (10 - 0)/5 = 2$.   The sum s_5 is

$$\begin{aligned} s_5 &= f(c_1)\Delta x + f(c_2)\Delta x + f(c_3)\Delta x + f(c_4)\Delta x + f(c_5)\Delta x \\ &= f(0)(2) + f(2)(2) + f(4)(2) + f(6)(2) + f(8)(2) \\ &= (0)(2) + (6)(2) + (12)(2) + (18)(2) + (24)(2) = 120 \end{aligned}$$

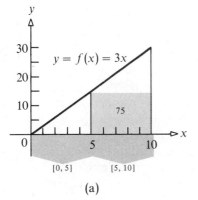

[0, 5]   [5, 10]

(a)

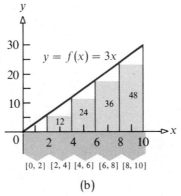

[0, 2] [2, 4] [4, 6] [6, 8] [8, 10]

(b)

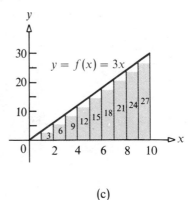

(c)

Figure 6

For $n = 10$, we partition $[0, 10]$ into 10 subintervals of equal length; namely, $[0, 1]$, $[1, 2]$, $[2, 3]$, . . . , $[9, 10]$, as shown in Figure 6(c). The length of each subinterval is $\Delta x = (10 - 0)/10 = 1$. The sum s_{10} is

$$s_{10} = f(0)(1) + f(1)(1) + f(2)(1) + \cdots + f(9)(1)$$
$$= 0 + 3 + 6 + 9 + 12 + 15 + 18 + 21 + 24 + 27 = 135 \qquad \blacksquare$$

The region under the graph of $f(x) = 3x$ from 0 to 10 is, of course, a triangle with base 10 and height 30. Thus, the *actual area* is

$$A = (\tfrac{1}{2})(10)(30) = 150$$

The following table summarizes the results of Example 1:

n	2	5	10
s_n	75	120	135

Observe that as the number n of subintervals increases, the estimates s_n of the area get closer to the actual area. For $n = 2$, the error in estimating A is 75 square units, while for $n = 10$, the error has been reduced to 15 square units.

In general, the error due to using inscribed rectangles occurs when a portion of the region lies outside the inscribed rectangles (see Fig. 7). It is this error that makes the sum s_n less than the actual area A. To get a better approximation of the area A, we must decrease such errors, and we can usually do this by increasing the number of subintervals. For example, suppose the number n of subintervals is doubled; that is, each interval of the first subdivision is itself subdivided to give a finer subdivision. In doing so we double the number of inscribed rectangles; for example, compare Figures 6(b) and 6(c). The result is that a greater portion of the region is covered, and the error is smaller than it was in the first subdivision. By further subdivision, we can reduce the error even more. Thus, by taking a finer and finer subdivision of the interval $[a, b]$ (we do this by increasing n without bound), we can make the sum of the areas of the inscribed rectangles as close as we please to the area A. (You can find the proof to this statement in books on advanced calculus.) With this analysis in mind, we now give the definition of the area under the graph of a function f from a to b.

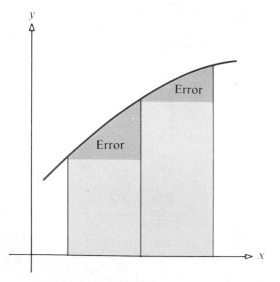

Figure 7

(5.1)   **DEFINITION**   *Area under a Graph.*   **Let f denote a function that is both continuous and nonnegative on a closed interval $[a, b]$. Divide the interval $[a, b]$ into n subintervals $[x_0, x_1], [x_1, x_2], [x_2, x_3], \ldots, [x_{n-1}, x_n]$,   each of length**

$$\Delta x = \frac{b - a}{n}$$

In each subinterval $[x_{i-1}, x_i]$, let $f(c_i)$ denote the absolute minimum of f on this subinterval. Form the sum

$$s_n = f(c_1)\Delta x + \cdots + f(c_n)\Delta x$$

The area A under the graph of $y = f(x)$ from a to b is the number

(5.2) $$A = \lim_{n \to +\infty} s_n$$

provided this limit exists.

   In the above discussion we calculated the area A of a region by the use of inscribed rectangles. By a parallel argument, we can use circumscribed rectangles to compute the area A, as shown in Figure 8. In this case, the values $C_1, C_2, \ldots, C_n$ are chosen so that the altitude $f(C_i)$ of the ith rectangle is the absolute maximum of f on the ith subinterval. The corresponding sum S_n of the areas of the *circumscribed rectangles* is an *overestimate* of the area A. In this case, $S_n \geq A$.   It can be shown that as n increases without bound, the limit of the sum S_n approaches the same limit as the one obtained from the inscribed rectangles; that is,

$$\lim_{n \to +\infty} s_n = \lim_{n \to +\infty} S_n = A$$

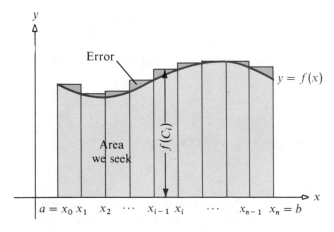

Figure 8

Expressions such as $\lim_{n \to +\infty} s_n$ (and $\lim_{n \to +\infty} S_n$) require some further comment. In limits like this, where s_n is defined only for positive integers n, $\lim_{n \to +\infty} s_n = A$ means that for any given $\varepsilon > 0$, there is a positive number N so that $|s_n - A| < \varepsilon$ whenever $n > N$, n being a positive integer. Based on the similarity between this definition and the preceding discussion (in Chap. 4) on limits at infinity, the procedures adopted for evaluating limits at infinity will be used to evaluate limits such as (5.2).

EXERCISE 1

In Problems 1–4 partition the given interval into n subintervals of equal length.

1. $[1, 4]$; $n = 3$ **2.** $[0, 9]$; $n = 9$ **3.** $[-1, 4]$; $n = 10$ **4.** $[-4, 4]$; $n = 16$

In Problems 5–10 approximate the area under the graph of each function from a to b by computing s_n and S_n for $n = 4$ and $n = 8$.

5. $f(x) = 2x + 5$,   on $[2, 6]$ **6.** $f(x) = -x + 10$,   on $[0, 8]$ **7.** $f(x) = x^2$,   on $[0, 2]$

8. $f(x) = x^3$,   on $[0, 8]$ **9.** $f(x) = \dfrac{1}{x}$,   on $[1, 5]$ **10.** $f(x) = \dfrac{1}{x^2}$,   on $[2, 6]$

2. Evaluation of Area

To evaluate the area A under the graph of a continuous nonnegative function $y = f(x)$ defined on a closed interval $[a, b]$, we need to find the limit of a certain sum, namely,

(5.3) $$A = \lim_{n \to +\infty} s_n = \lim_{n \to +\infty} [f(c_1)\Delta x + f(c_2)\Delta x + \cdots + f(c_n)\Delta x]$$

where Δx is the length of each subinterval $[x_{i-1}, x_i]$ of $[a, b]$ and $f(c_i)$ is the minimum of f on $[x_{i-1}, x_i]$, $i = 1, 2, \ldots, n$.

In order to calculate the limit in (5.3), and hence find the area A, we need to express the sum

(5.4) $$f(c_1)\Delta x + \cdots + f(c_n)\Delta x$$

as a function of n. As a start, it is convenient to introduce *summation notation* in order to write (5.4) concisely:

(5.5) $$f(c_1)\Delta x + \cdots + f(c_n)\Delta x = \sum_{i=1}^{n} f(c_i)\Delta x$$

The symbol $\sum$ (the Greek letter sigma) is simply an instruction to add; the integer i is called the *index* of the sum. The expression on the right in equation (5.5) should be read "the sum of $f(c_i)\Delta x$ from $i = 1$ through $i = n$." For example, we can use summation notation to represent the sum $1^2 + 2^2 + \cdots + 20^2$ concisely as

$$\sum_{i=1}^{20} i^2 = 1^2 + 2^2 + \cdots + 20^2$$

Letters other than i can be used as the index. For example,

$$\sum_{j=1}^{20} j^2 \quad \text{and} \quad \sum_{k=1}^{20} k^2$$

each represent the same sum. Also, for convenience, the symbols $\sum_{j=1}^{20} j^2$ and $\sum_{k=1}^{20} k^2$ may be used to represent these sums. .

To evaluate the limit in (5.3), we require that the sum $\sum_{i=1}^{n} f(c_i)\Delta x$ be expressed as a function of n. This, in turn, requires a knowledge of some of the properties of sums, which we now take up.

We begin with a reminder. If $f(i)$ denotes some formula involving the integer i, then

$$\sum_{i=1}^{n} f(i) = f(1) + f(2) + \cdots + f(n)$$

Now for the rules that sums obey.

If f and g are functions, k is a real number, and n is an integer, we have the following rules for sums:

(5.6)
$$\sum_{i=1}^{n} kf(i) = k \sum_{i=1}^{n} f(i)$$

(5.7)
$$\sum_{i=1}^{n} [f(i) + g(i)] = \sum_{i=1}^{n} f(i) + \sum_{i=1}^{n} g(i)$$

(5.8)
$$\sum_{i=1}^{n} [f(i) - g(i)] = \sum_{i=1}^{n} f(i) - \sum_{i=1}^{n} g(i)$$

(5.9)
$$\sum_{i=1}^{n} f(i) = \sum_{i=1}^{j} f(i) + \sum_{i=j+1}^{n} f(i) \quad \text{when} \quad 1 < j < n$$

In addition to these four rules, we need to know some summation formulas. Listed below are the ones we shall be using. They may be verified for any positive integer n by mathematical induction.

(5.10)
$$\sum_{i=1}^{n} 1 = \underbrace{1 + 1 + \cdots + 1}_{n \text{ times}} = n$$

(5.11)
$$\sum_{i=1}^{n} i = 1 + 2 + 3 + \cdots + n = \frac{n(n+1)}{2}$$

$$(5.12) \qquad \sum_{i=1}^{n} i^2 = 1^2 + 2^2 + 3^2 + \cdots + n^2 = \frac{n(n+1)(2n+1)}{6}$$

$$(5.13) \qquad \sum_{i=1}^{n} i^3 = 1^3 + 2^3 + \cdots + n^3 = \frac{n^2(n+1)^2}{4}$$

Let's look at two examples of how these rules and formulas are used in practice.

EXAMPLE 1   Express the sum $\sum_{i=1}^{n}(i-1)$ as a function of n.

Solution
$$\sum_{i=1}^{n} (i-1) \underset{\underset{\text{By (5.8)}}{\uparrow}}{=} \sum_{i=1}^{n} i - \sum_{i=1}^{n} 1 \underset{\underset{\text{By (5.11), (5.10)}}{\uparrow}}{=} \frac{n(n+1)}{2} - n = \frac{1}{2}(n^2 - n) \qquad \blacksquare$$

EXAMPLE 2   Express the sum $\sum_{i=1}^{n}(i-1)^2 \left(\dfrac{4}{n}\right)^3$ as a function of n.

Solution
$$\sum_{i=1}^{n} (i-1)^2 \left(\frac{4}{n}\right)^3 \underset{\underset{\text{By (5.6)}}{\uparrow}}{=} \frac{64}{n^3} \sum_{i=1}^{n} (i-1)^2 = \frac{64}{n^3} \sum_{i=1}^{n} (i^2 - 2i + 1)$$

$$\underset{\underset{\text{By (5.7), (5.8)}}{\uparrow}}{=} \frac{64}{n^3} \left(\sum_{i=1}^{n} i^2 - 2 \sum_{i=1}^{n} i + \sum_{i=1}^{n} 1 \right)$$

Now, by using (5.12), (5.11), and (5.10), we get

$$\sum_{i=1}^{n} (i-1)^2 \left(\frac{4}{n}\right)^3 = \frac{64}{n^3} \left[\frac{n(n+1)(2n+1)}{6} - \frac{2n(n+1)}{2} + n \right]$$

$$= \frac{64}{n^3} (n) \left[\frac{(n+1)(2n+1) - 6(n+1) + 6}{6} \right]$$

$$= \frac{64}{n^2} \left(\frac{2n^2 - 3n + 1}{6} \right) = \frac{32}{3} \left(\frac{2n^2 - 3n + 1}{n^2} \right) \qquad \blacksquare$$

Now let's do some area problems.

EXAMPLE 3   Find the area under the graph of $f(x) = 3x$ from 0 to 10 by using the definition of area, $A = \lim_{n \to +\infty} S_n$ (inscribed rectangles).

Solution   The region whose area is to be computed is illustrated in Figure 9. We divide the closed interval $[0, 10]$ into n equal subintervals

$$[x_0, x_1], [x_1, x_2], \ldots, [x_{n-1}, x_n]$$

where

$$0 = x_0 < x_1 < x_2 < \cdots < x_{n-1} < x_n = 10$$

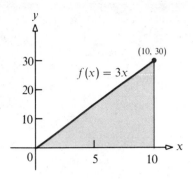

Figure 9

and

$$x_1 - x_0 = x_2 - x_1 = \cdots = x_n - x_{n-1} = \Delta x$$

Since the length of each subinterval is

$$\Delta x = \frac{10 - 0}{n} = \frac{10}{n}$$

the coordinates of the endpoints (as Fig. 10 illustrates) can be expressed in terms of n as follows:

$$x_0 = 0, \quad x_1 = \frac{10}{n}, \quad x_2 = 2\left(\frac{10}{n}\right), \quad x_3 = 3\left(\frac{10}{n}\right), \quad \ldots, \quad x_{i-1} = (i-1)\left(\frac{10}{n}\right),$$

$$x_i = i\left(\frac{10}{n}\right), \quad \ldots, \quad x_{n-1} = (n-1)\left(\frac{10}{n}\right), \quad x_n = n\left(\frac{10}{n}\right) = 10$$

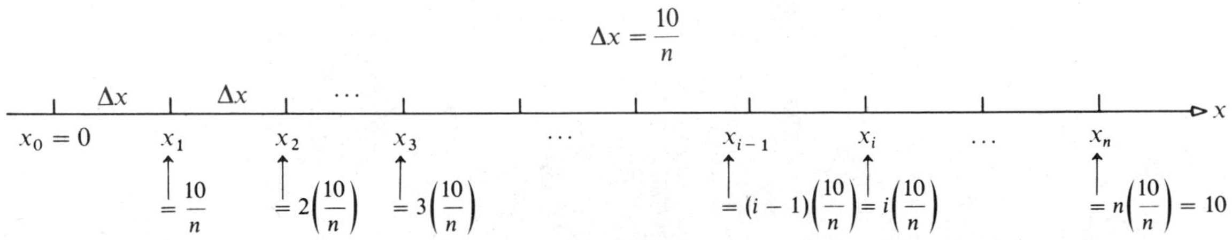

$$\Delta x = \frac{10}{n}$$

Figure 10

To find area A by using the limit of s_n, we need to compute the absolute minimum of f on each subinterval $[x_{i-1}, x_i]$, $i = 1, 2, \ldots, n$.   In this example, since the function $f(x) = 3x$ is an increasing function, it will assume its absolute minimum at the left endpoint x_{i-1} of each subinterval. Since $x_{i-1} = (i-1)(10/n)$, the absolute minimum of f on $[x_{i-1}, x_i]$ is

$$f(x_{i-1}) = f\left[(i-1)\left(\frac{10}{n}\right)\right] = 3(i-1)\left(\frac{10}{n}\right)$$

Thus,

$$S_n = \sum_{i=1}^{n} f(x_{i-1})\Delta x = \sum_{i=1}^{n} 3(i-1)\left(\frac{10}{n}\right)\left(\frac{10}{n}\right) = \frac{300}{n^2}\sum_{i=1}^{n}(i-1)$$

$$\Delta x = \frac{10}{n}$$

Using the result obtained in Example 1, we find

$$S_n = \frac{300}{n^2}\left(\frac{n^2-n}{2}\right) = 150\left(\frac{n^2-n}{n^2}\right) = 150\left(1 - \frac{1}{n}\right)$$

By letting $n \to +\infty$, the area A of the region under the graph of $f(x) = 3x$ from 0 to 10 is

$$A = \lim_{n\to+\infty} S_n = \lim_{n\to+\infty} 150\left(1 - \frac{1}{n}\right) = 150 \text{ square units}$$

This answer is in agreement with the answer obtained earlier (in Section 1) by using the formula for the area of a triangle, namely, $A = (\frac{1}{2})bh = (\frac{1}{2})(10)(30) = 150$. ∎

EXAMPLE 4   Find the area A of the region under the graph of $y = f(x) = 16 - x^2$ from 0 to 4 by using $A = \lim_{n\to+\infty} S_n$ (circumscribed rectangles).

Solution   Figure 11 illustrates the region and a typical circumscribed rectangle with base of length $\Delta x = x_i - x_{i-1}$. We divide the closed interval $[0, 4]$ into n subintervals of

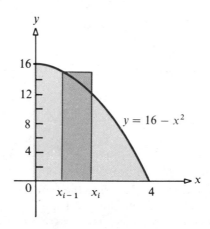

Figure 11

equal length,

$$[x_0, x_1], \quad [x_1, x_2], \quad \ldots, \quad [x_{n-1}, x_n]$$

where

$$0 = x_0 < x_1 < \cdots < x_{n-1} < x_n = 4$$

and

$$x_1 - x_0 = x_2 - x_1 = \cdots = x_n - x_{n-1} = \Delta x$$

The length of each subinterval is $\Delta x = (4 - 0)/n = 4/n$. The coordinates of the endpoints (as Fig. 12 illustrates) can be expressed in terms of n as follows:

$$x_0 = 0, \quad x_1 = \frac{4}{n}, \quad x_2 = 2\left(\frac{4}{n}\right), \quad \ldots, \quad x_{i-1} = (i-1)\left(\frac{4}{n}\right),$$

$$x_i = i\left(\frac{4}{n}\right), \quad \ldots, \quad x_{n-1} = (n-1)\left(\frac{4}{n}\right), \quad x_n = n\left(\frac{4}{n}\right) = 4$$

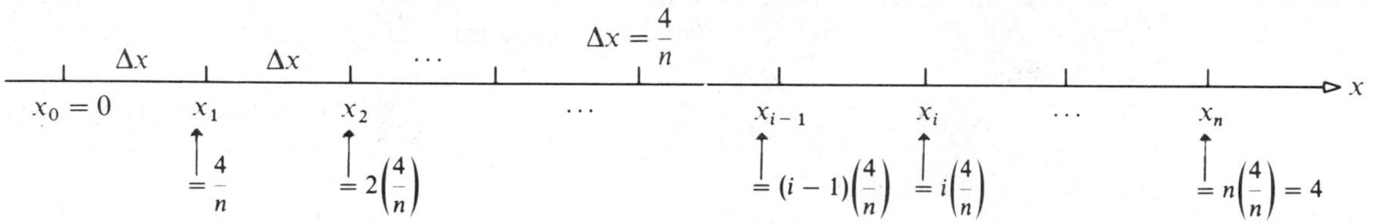

Figure 12

To use circumscribed rectangles, we need to compute the absolute maximum of f on each subinterval $[x_{i-1}, x_i]$, $i = 1, 2, \ldots, n$. In this example, since the function f is a decreasing function, it will assume its absolute maximum at the left endpoint x_{i-1} of each subinterval. Thus,

$$S_n = \sum_{i=1}^{n} f(x_{i-1})\Delta x$$

Since $x_{i-1} = (i-1)(4/n)$, $f(x) = 16 - x^2$, and $\Delta x = 4/n$, we calculate S_n to be

$$S_n = \sum_{i=1}^{n} f(x_{i-1})\Delta x = \sum_{i=1}^{n} \left\{ 16 - \left[(i-1)\left(\frac{4}{n}\right)\right]^2 \right\}\left(\frac{4}{n}\right)$$

$$= \sum_{i=1}^{n} \frac{64}{n} - \sum_{i=1}^{n} (i-1)^2\left(\frac{64}{n^3}\right) = \frac{64}{n}\sum_{i=1}^{n} 1 - \frac{64}{n^3}\sum_{i=1}^{n} (i-1)^2$$

By using the result obtained in Example 2 and formula (5.10), we find

$$S_n = \frac{64}{n}(n) - \frac{64}{n^3}(n)\left(\frac{2n^2 - 3n + 1}{6}\right) = 64 - \frac{32}{3}\left(2 - \frac{3}{n} + \frac{1}{n^2}\right) = \frac{128}{3} + \frac{32}{n} - \frac{32}{3n^2}$$

Taking the limit as $n \to +\infty$, we get

$$A = \lim_{n \to +\infty} S_n = \lim_{n \to +\infty} \left(\frac{128}{3} + \frac{32}{n} - \frac{32}{3n^2} \right) = \frac{128}{3}$$

The area of the region is therefore $\frac{128}{3}$ square units. ■

If we had used $A = \lim_{n \to +\infty} s_n$ (inscribed rectangles) in the above example, we would have obtained the same answer. Of course, in using $A = \lim_{n \to +\infty} s_n$, we must find the number c_i in $[x_{i-1}, x_i]$ at which the function f attains its absolute minimum. In this example, since $f(x) = 16 - x^2$ is decreasing on $[0, 4]$, we find $c_i = x_i = i(4/n)$.

The next example illustrates how to obtain the coordinates of the partition when the initial left endpoint is other than 0.

EXAMPLE 5   Find the area A of the region under the graph of $y = f(x) = 4x$ from 1 to 3 by using $A = \lim_{n \to +\infty} S_n$ (circumscribed rectangles).

Solution   Figure 13 illustrates the region and a typical circumscribed rectangle with base of length $\Delta x = x_i - x_{i-1}$. We divide the closed interval $[1, 3]$ into n subintervals of equal length

$$[x_0, x_1], \quad [x_1, x_2], \quad \dots, \quad [x_{n-1}, x_n]$$

where

$$1 = x_0 < x_1 < \cdots < x_{n-1} < x_n = 3$$

and

$$x_1 - x_0 = x_2 - x_1 = \cdots = x_n - x_{n-1} = \Delta x$$

The length of each subinterval is $\Delta x = (3 - 1)/n = 2/n$. The coordinates of the

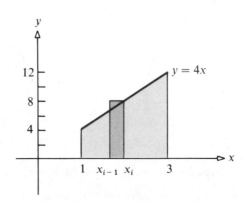

Figure 13

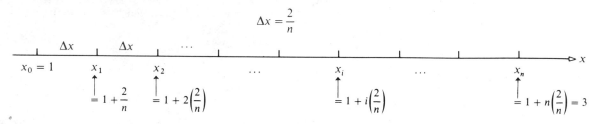

Figure 14

endpoints, as Figure 14 illustrates, can be expressed in terms of n as follows:

$$x_0 = 1, \quad x_1 = 1 + \frac{2}{n}, \quad x_2 = 1 + 2\left(\frac{2}{n}\right), \quad \ldots,$$

$$x_i = 1 + i\left(\frac{2}{n}\right), \quad \ldots, \quad x_n = 1 + n\left(\frac{2}{n}\right) = 1 + 2 = 3$$

To use circumscribed rectangles, we need to compute the absolute maximum of f on each subinterval $[x_{i-1}, x_i]$, $i = 1, 2, \ldots, n$. In this example, since f is increasing, it assumes its absolute maximum at the right endpoint of each subinterval. Thus,

$$S_n = \sum_{i=1}^{n} f(x_i)\Delta x$$

Since $x_i = 1 + (i)(2/n)$, $f(x) = 4x$, and $\Delta x = 2/n$, we find S_n to be

$$S_n = \sum_{i=1}^{n} f(x_i)\Delta x = \sum_{i=1}^{n} 4\left[1 + (i)\left(\frac{2}{n}\right)\right]\left(\frac{2}{n}\right)$$

$$= \left(\frac{8}{n}\right)\left[\sum_{i=1}^{n} 1 + \left(\frac{2}{n}\right)\sum_{i=1}^{n} i\right]$$

$$= \left(\frac{8}{n}\right)\left[n + \left(\frac{2}{n}\right)\frac{n(n+1)}{2}\right] = \left(\frac{8}{n}\right)[n + (n+1)] = 16 + \frac{8}{n}$$

The area A is therefore

$$A = \lim_{n \to +\infty} S_n = 16 \text{ square units} \quad \blacksquare$$

EXERCISE 2

In Problems 1–8 express each sum as a function of n.

1. $\displaystyle\sum_{i=1}^{n} 3i(i + 1)$ 　　　　 **2.** $\displaystyle\sum_{i=1}^{n} 3i(i - 1)$ 　　　　 **3.** $\displaystyle\sum_{i=1}^{n} (i^3 - 2i + 1)$ 　　　　 **4.** $\displaystyle\sum_{i=1}^{n} (i^3 + i + 1)$

5. $\displaystyle\sum_{i=1}^{n} \left(\frac{i}{n}\right)^3\left(\frac{1}{n}\right)$ 　　 **6.** $\displaystyle\sum_{i=1}^{n} \left(\frac{i}{n}\right)^2\left(\frac{1}{n}\right)$ 　　 **7.** $\displaystyle\sum_{i=1}^{n} \left[\left(\frac{i}{n}\right)^2 - \frac{i}{n}\right]\left(\frac{1}{n}\right)$ 　　 **8.** $\displaystyle\sum_{i=1}^{n} \left(\frac{i+1}{2}\right)^2\left(\frac{1}{n}\right)$

In Problems 9–16 express each sum by using summation notation in two ways in which the index begins at
(a) $i = 0$ and (b) $i = 1$.

9. $1^2 + 2^2 + 3^2 + \cdots + n^2$

10. $1^3 + 2^3 + 3^3 + 4^3 + \cdots + n^3$

11. $1 + 2 + 4 + 8 + 16 + \cdots + 2^n$

12. $1 + 3 + 9 + 27 + 81 + \cdots + 3^n$

13. $1 + \dfrac{1}{2} + \dfrac{1}{4} + \dfrac{1}{8} + \dfrac{1}{16} + \cdots + \dfrac{1}{2^n}$

14. $1 + 3 + 5 + 7 + 9 + \cdots + (2n - 1)$

15. $1 \cdot 2 + 2 \cdot 3 + 3 \cdot 4 + \cdots + n(n + 1)$

16. $\dfrac{1}{2} + \dfrac{2}{3} + \dfrac{3}{4} + \cdots + \dfrac{n}{n + 1}$

17. Use circumscribed rectangles to find the area under the graph of $f(x) = 3x$ from 0 to 10.

18. Use inscribed rectangles to find the area under the graph of $f(x) = 16 - x^2$ from 0 to 4.

In Problems 19–28 use inscribed rectangles to find the area under the graph of $y = f(x)$ from a to b. A
graph will be helpful.

19. $f(x) = 2x + 1, \quad a = 0, \quad b = 4$

20. $f(x) = 3x + 1, \quad a = 0, \quad b = 4$

21. $f(x) = x^2, \quad a = 0, \quad b = 2$

22. $f(x) = x^2 + 2, \quad a = 0, \quad b = 3$

23. $f(x) = 4 - x^2, \quad a = 0, \quad b = 2$

24. $f(x) = 12 - x^2, \quad a = 0, \quad b = 3$

25. $f(x) = x^3, \quad a = 0, \quad b = 2$

26. $f(x) = x^3 + 3, \quad a = 0, \quad b = 1$

27. $f(x) = 2x + 1, \quad a = 1, \quad b = 3$

28. $f(x) = 4 + 3x, \quad a = 1, \quad b = 2$

In Problems 29–38 use circumscribed rectangles to find the area under the graph of $y = f(x)$ from a to b.

29. $f(x) = 2x + 1, \quad a = 0, \quad b = 4$

30. $f(x) = 3x + 1, \quad a = 0, \quad b = 4$

31. $f(x) = x^2, \quad a = 0, \quad b = 2$

32. $f(x) = x^2 + 2, \quad a = 0, \quad b = 3$

33. $f(x) = 4 - x^2, \quad a = 0, \quad b = 2$

34. $f(x) = 12 - x^2, \quad a = 0, \quad b = 3$

35. $f(x) = x^3, \quad a = 0, \quad b = 2$

36. $f(x) = x^3 + 3, \quad a = 0, \quad b = 1$

37. $f(x) = 2x + 1, \quad a = 1, \quad b = 3$

38. $f(x) = 4 + 3x, \quad a = 1, \quad b = 2$

39. Show that $\displaystyle\sum_{i=1}^{n} [(i + 1)^2 - i^2] = (n + 1)^2 - 1.$

40. Show that $\displaystyle\sum_{i=1}^{n} [(i + 1)^4 - i^4] = n^4 + 4n^3 + 6n^2 + 4n.$

41. Show that $\displaystyle\sum_{i=1}^{n} \left(\dfrac{1}{i} - \dfrac{1}{i + 1}\right) = 1 - \dfrac{1}{n + 1}.$ [*Hint:* Write out the terms of the sum.]

42. Show that $\displaystyle\sum_{i=1}^{n} (2^{i+1} - 2^i) = 2^{n+1} - 2.$ [*Hint:* Write out the terms of the sum.]

43. Use a method of this section to find the area of a right triangle of height H and base B.

44. Use a method of this section to find the area of a trapezoid of height H and bases B_1 and B_2.

45. Find the area under the graph of $y = f(x) = x^2$ from $a = 0$ to $b = 3$ by evaluating

$$\lim_{n \to +\infty} \sum_{i=1}^{n} f(m_i)\Delta x$$

where $\Delta x = 3/n$ and m_i is the midpoint of the ith subinterval.

46. Follow the instructions of Problem 45 to find the area under the graph of $f(x) = 9 - x^2$ from $a = 0$ to $b = 2$.

47. Approximate the area under the graph of $f(x) = x$ from a to b by computing s_n and S_n for a partition of $[a, b]$ into n subintervals of equal length. Show that

$$s_n < \frac{b^2 - a^2}{2} < S_n$$

48. Approximate the area under the graph of $f(x) = x^2$ from $a \geq 0$ to b by computing s_n and S_n for a partition of $[a, b]$ into n subintervals of equal length. Show that

$$s_n < \frac{b^3 - a^3}{3} < S_n$$

49. Derive formula (5.11) by using the fact that the sum can be written in two ways as:

$$\sum_{i=1}^{n} i = 1 + 2 + 3 + \cdots + (n - 2) + (n - 1) + n$$

$$\sum_{i=1}^{n} i = n + (n - 1) + (n - 2) + \cdots + 3 + 2 + 1$$

[*Hint:* Add the two equations.]

50. Derive a formula for $\sum_{i=1}^{n} r^i$ by using the fact that

$$\sum_{i=1}^{n} r^i = r + r^2 + r^3 + \cdots + r^n$$

$$r \sum_{i=1}^{n} r^i = r^2 + r^3 + r^4 + \cdots + r^{n+1}$$

3. The Definite Integral

We have defined the area A under the graph of $y = f(x)$ from a to b as the limit

(5.14) $$A = \lim_{n \to +\infty} \sum_{i=1}^{n} f(c_i) \Delta x$$

where the following assumptions were made:

I. The function f is continuous on $[a, b]$.
II. The function f is nonnegative on $[a, b]$.
III. The subintervals of $[a, b]$ are all of equal length, $\Delta x = (b - a)/n$.
IV. In each subinterval $[x_{i-1}, x_i]$ the number c_i in $[x_{i-1}, x_i]$ is the one at which f has its absolute minimum.

We also noted that

$$A = \lim_{n \to +\infty} \sum_{i=1}^{n} f(C_i) \Delta x$$

where in each subinterval $[x_{i-1}, x_i]$ the number C_i in $[x_{i-1}, x_i]$ is the one at which f has its absolute maximum.

This procedure for calculating area involves partitioning an interval into n subintervals, finding the area of each inscribed (or circumscribed) rectangle, adding them up, and taking the limit as $n \to +\infty$. Interestingly enough, similar processes enable us to find the volume of a solid of revolution, the length of a graph, the work done by a variable force, and other quantities. For this reason, we now study sums of the form

(5.15)
$$\sum_{i=1}^{n} f(u_i)\Delta x_i$$

using the following more general assumptions:

  I. The function f is not necessarily continuous on $[a, b]$.

  II. The function f does not have to be nonnegative on $[a, b]$.

  III. The lengths Δx_i of the subintervals $[x_{i-1}, x_i]$ of $[a, b]$ do not have to be equal.

  IV. The number u_i may be any number in $[x_{i-1}, x_i]$; it is not necessarily the number at which f has its absolute minimum (or maximum).

Observe that the sum in (5.14) is a special case of the sums in (5.15), which we will study in this section. The new sums are called *Riemann sums* for f on $[a, b]$; they are named after the German mathematician Georg Friedrich Bernhard Riemann (1826–1866).

We begin with a function f defined on a closed interval $[a, b]$, and we divide the interval $[a, b]$ into n subintervals

$$[x_0, x_1], \quad [x_1, x_2], \quad [x_2, x_3], \quad \ldots, \quad [x_{n-1}, x_n]$$

where

$$a = x_0 < x_1 < x_2 < \cdots < x_{n-1} < x_n = b$$

These subintervals are not necessarily of the same length. As a result, we denote the length of the first interval by $\Delta x_1 = x_1 - x_0$, the length of the second interval by $\Delta x_2 = x_2 - x_1$, and so on. In general, the length of the ith subinterval is

$$\Delta x_i = x_i - x_{i-1}, \qquad i = 1, 2, \ldots, n$$

The set of all such subintervals of the interval $[a, b]$ is called a *partition* P of $[a, b]$. Three examples of possible partitions of the interval $[0, 2]$ are:

(a) $[0, 1], \quad [1, 2]$

(b) $[0, \frac{1}{2}], \quad [\frac{1}{2}, 1], \quad [1, \frac{3}{2}], \quad [\frac{3}{2}, 2]$

(c) $[0, \frac{1}{4}], \quad [\frac{1}{4}, \frac{1}{3}], \quad [\frac{1}{3}, \frac{1}{2}], \quad [\frac{1}{2}, \frac{7}{8}], \quad [\frac{7}{8}, 1], \quad [1, \frac{5}{4}], \quad [\frac{5}{4}, \frac{3}{2}], \quad [\frac{3}{2}, \frac{7}{4}], \quad [\frac{7}{4}, 2]$

The length of the largest subinterval in a partition P is called the *norm* of the partition and is denoted by $\|P\|$. For example, $\|P\| = 1$ in (a); $\|P\| = \frac{1}{2}$ in (b); $\|P\| = \frac{3}{8}$ in (c).

Suppose in each subinterval $[x_{i-1}, x_i]$ of a partition P we choose a number u_i, $i = 1, 2, \ldots, n,$ and form the Riemann sums

(5.16)
$$\sum_{i=1}^{n} f(u_i)\Delta x_i = f(u_1)\Delta x_1 + f(u_2)\Delta x_2 + \cdots + f(u_n)\Delta x_n$$

Notice that once the interval $[a, b]$ and the function f have been chosen, then the choices for the partition P are unlimited. Furthermore, once the partition has been chosen, the choices for the u_i's are unlimited. The value of the Riemann sums (5.16) depends on *all* of these choices.

Now, suppose P is a partition for which the norm $\|P\|$ is close to 0. Then, since $\|P\|$ is the length of the largest subinterval, the lengths of *all* subintervals in P will be close to 0. In particular, when the norm $\|P\|$ approaches 0, the effect is similar to that produced by repeatedly refining the original partition.

For many functions, when $\|P\|$ approaches 0, the Riemann sums approach a limit, say, I. In this event, we write

(5.17)
$$\lim_{\|P\| \to 0} \sum_{i=1}^{n} f(u_i)\Delta x_i = I$$

In words, this means that values of the Riemann sums can be made as close to I as we please by choosing a partition P whose norm $\|P\|$ is sufficiently close to 0. This choice of P is made independently of the choice of u_i.

We now state the meaning of (5.17) more precisely.

(5.18)   **DEFINITION**   **Let f be a function defined on a closed interval $[a, b]$ and I be a number. The statement**

(5.19)
$$\lim_{\|P\| \to 0} \sum_{i=1}^{n} f(u_i)\Delta x_i = I$$

means that for any given $\varepsilon > 0$,   there is a positive number δ so that if P is a partition of $[a, b]$ for which $\|P\| < \delta$,   then

$$\left| \sum_{i=1}^{n} f(u_i)\Delta x_i - I \right| < \varepsilon$$

for any choice of numbers u_i in the subintervals $[x_{i-1}, x_i]$ of P.

The number I in (5.19) plays such a major role in mathematics that a special name and symbol is given to it.

(5.20)   **DEFINITION**   *Definite Integral.*   **Let f be a function defined on the closed interval $[a, b]$. If the limit in (5.19) exists, then the number I is called the *definite integral* of f from a to b and is denoted by $\int_a^b f(x)\,dx$.   That is,**

(5.21)
$$\int_a^b f(x)\,dx = \lim_{\|P\| \to 0} \sum_{i=1}^{n} f(u_i)\Delta x_i$$

For the definite integral $\int_a^b f(x)\,dx$, the number a is called the *lower limit of integration*, the number b is called the *upper limit of integration*,* the symbol $\int$ (an elongated S to remind you of summation) is called the *integral sign*, and $f(x)$ is called the *integrand*. The variable used in the definite integral is an *artificial*, or *dummy*, variable because it may be replaced by any other letter. Thus, for example,

$$\int_a^b f(x)\,dx \qquad \int_a^b f(t)\,dt \qquad \int_a^b f(s)\,ds$$

all denote the definite integral of f from a to b, and if any of them exist, they are all equal to the same number.

In defining the definite integral $\int_a^b f(x)\,dx$, we have assumed that $a < b$. To remove this restriction, we give the following definition.

(5.22)   **DEFINITION**   **If $f(a)$ is defined, then**

$$\int_a^a f(x)\,dx = 0$$

If $a > b$ and if $\int_b^a f(x)\,dx$ exists, then

$$\int_a^b f(x)\,dx = -\int_b^a f(x)\,dx$$

Thus, interchanging the limits of integration will reverse the sign of the integral. Here are some specific examples to illustrate the definition:

$$\int_1^1 x^2\,dx = 0 \qquad \int_3^2 x^2\,dx = -\int_2^3 x^2\,dx$$

We can specify a condition on the function f that will guarantee that the limit in (5.21) exists. (The proof of this result may be found in advanced calculus texts.)

(5.23)   **THEOREM**   **If a function f is continuous on a closed interval $[a, b]$, then the definite integral**

(5.24) $$\int_a^b f(x)\,dx = \lim_{\|P\| \to 0} \sum_{i=1}^n f(u_i)\Delta x_i$$

exists.

Two items deserve special mention here. First, f is defined on a *closed* interval, and, second, f is *continuous* on that interval. There are some functions that are continuous on an open interval (or even a semi-open interval) for which the limit in (5.24) does not exist. For example, the definite integral $\int_0^1 (1/x^2)\,dx$ does not exist, yet $f(x) = 1/x^2$ is continuous on $(0, 1)$ (and on $(0, 1]$). Also, there are examples of functions that are discontinuous at some numbers in the closed interval $[a, b]$ and yet the limit in (5.24) will exist (see Problems 17 and 18 in Exercise 3).† To summarize,

---

* The terms *upper limit* and *lower limit* of integration simply refer to the endpoints of the closed interval $[a, b]$ and have nothing to do with the highly specialized concept of limit as introduced in Chapter 2.

† Additional discussion of the definition and existence of $\int_a^b f(x)\,dx$, where f is not continuous on $[a, b]$ is given in Chapter 10.

theorem (5.23) states that if f is continuous on $[a, b]$, then we are guaranteed that $\int_a^b f(x)\,dx$ exists.

In the event that $\int_a^b f(x)\,dx$ exists, the limit in (5.21) will exist for any choice of u_i in the ith subinterval. This means that we are free to choose the u_i in any manner we please—such as the left endpoint of each subinterval, or the right endpoint, or the midpoint, or in any other way. Furthermore, if the limit in (5.21) exists, it is independent of the partitions P of the closed interval $[a, b]$, provided $\|P\|$ is close to 0. It is this feature that enables the definite integral to play such an important role in applications to engineering, physics, chemistry, geometry, and economics.

In evaluating the limit in (5.21), we will usually use a partition that divides the interval $[a, b]$ into n subintervals of the same length. We refer to such a partition as a *regular partition*. For a regular partition, the norm is

$$\|P\| = \frac{b - a}{n}$$

From this relationship, it follows that for a regular partition, the two statements

$$\|P\| \to 0 \qquad \text{and} \qquad n \to +\infty$$

are interchangeable. As a result, we may write

$$\int_a^b f(x)\,dx = \lim_{\|P\|\to 0} \sum_{i=1}^n f(u_i)\Delta x_i = \lim_{n\to +\infty} \sum_{i=1}^n f(u_i)\Delta x$$

where $\Delta x = \Delta x_i = (b - a)/n$.

EXAMPLE 1   Evaluate: $\int_0^2 (3x - 8)\,dx$

Solution   Since the integrand $f(x) = 3x - 8$ is continuous on the closed interval $[0, 2]$, we know from (5.23) that the definite integral exists. To evaluate it, we may use any partition of $[0, 2]$ whose norm can be made as close to 0 as we please, and we may choose any u_i in each subinterval. We elect to use a regular partition of $[0, 2]$, and we will choose the u_i as the right endpoint of each subinterval. As a result, we partition $[0, 2]$ into n subintervals,

$$[0, x_1], \quad [x_1, x_2], \quad \ldots, \quad [x_{i-1}, x_i], \quad \ldots, \quad [x_{n-1}, 2]$$

each of length $\Delta x = 2/n$. The coordinates of the partition, in terms of n, are

$$x_0 = 0, \quad x_1 = \frac{2}{n}, \quad x_2 = 2\left(\frac{2}{n}\right), \quad \ldots, \quad x_i = i\left(\frac{2}{n}\right), \quad \ldots, \quad x_n = n\left(\frac{2}{n}\right) = 2$$

The Riemann sum of $f(x) = 3x - 8$ from 0 to 2, using $u_i = x_i = i(2/n)$, is

$$\sum_{i=1}^n f(u_i)\Delta x_i = \sum_{i=1}^n f(x_i)\Delta x_i = \sum_{i=1}^n (3x_i - 8)\Delta x = \sum_{i=1}^n \left[3i\left(\frac{2}{n}\right) - 8\right]\left(\frac{2}{n}\right)$$

$$= \left(\frac{12}{n^2}\right) \sum_{i=1}^n i - \left(\frac{16}{n}\right) \sum_{i=1}^n 1 = \left(\frac{12}{n^2}\right)\frac{n(n+1)}{2} - \left(\frac{16}{n}\right)(n) = -10 + \frac{6}{n}$$

Therefore,

$$\int_0^2 (3x - 8)\, dx = \lim_{||P|| \to 0} \sum_{i=1}^{n} f(u_i)\Delta x_i = \lim_{n \to +\infty} \left(-10 + \frac{6}{n} \right) = -10 \quad \blacksquare$$

Observe that the integrand $f(x) = 3x - 8$ in Example 1 is negative on the interval $[0, 2]$. As a result, we may *not* interpret $\int_0^2 (3x - 8)\, dx$ as representing an area. The fact that our answer (-10) is not positive is further evidence that we do not have an area problem here. Thus, when looking at a definite integral, do not presume it represents area. As you will see later in Chapter 6, the definite integral may have many interpretations.

EXERCISE 3

In Problems 1–6 evaluate each definite integral. Use a regular partition and choose u_i any way you wish.

1. $\int_0^1 (x - 4)\, dx$
2. $\int_0^3 (3x - 1)\, dx$
3. $\int_0^{-4} (2x^2)\, dx$

4. $\int_0^{-1} (x^2 + 1)\, dx$
5. $\int_{-2}^1 (3x^2 - x)\, dx$
6. $\int_{-1}^1 (2x^2 + x)\, dx$

In Problems 7–10 use (5.16) to calculate each Riemann sum of f for the partition P and the numbers u_i listed.

7. $f(x) = x$, $[0, 2]$; for P: $x_0 = 0$, $x_1 = \frac{1}{4}$, $x_2 = \frac{1}{2}$, $x_3 = \frac{3}{4}$, $x_4 = 1$, $x_5 = \frac{5}{4}$, $x_6 = \frac{3}{2}$, $x_7 = \frac{7}{4}$, $x_8 = 2$; $u_1 = \frac{1}{8}$, $u_2 = \frac{3}{8}$, $u_3 = \frac{5}{8}$, $u_4 = \frac{7}{8}$, $u_5 = \frac{9}{8}$, $u_6 = \frac{11}{8}$, $u_7 = \frac{13}{8}$, $u_8 = \frac{15}{8}$

8. $f(x) = x$, $[0, 2]$; for P: $x_0 = 0$, $x_1 = \frac{1}{2}$, $x_2 = 1$, $x_3 = \frac{3}{2}$, $x_4 = 2$; $u_1 = \frac{1}{2}$, $u_2 = 1$, $u_3 = \frac{3}{2}$, $u_4 = 2$

9. $f(x) = x^2$, $[-2, 1]$; for P: $x_0 = -2$, $x_1 = -1$, $x_2 = 0$, $x_3 = 1$; $u_1 = -\frac{3}{2}$, $u_2 = -\frac{1}{2}$, $u_3 = \frac{1}{2}$

10. $f(x) = x^2$, $[1, 2]$; for P: $x_0 = 1$, $x_1 = \frac{5}{4}$, $x_2 = \frac{3}{2}$, $x_3 = \frac{7}{4}$, $x_4 = 2$; $u_1 = \frac{5}{4}$, $u_2 = \frac{3}{2}$, $u_3 = \frac{7}{4}$, $u_4 = 2$

In Problems 11–16 find the Riemann sum for the given function and interval. Use a regular partition to divide the interval $[a, b]$ into n subintervals, and always choose u_i as the right endpoint of the ith subinterval $[x_{i-1}, x_i]$. Leave your answer in summation notation.

11. $f(x) = \sqrt{x}$, $[0, 1]$
12. $f(x) = \sqrt{x - 1}$, $[1, 2]$
13. $f(x) = \frac{1}{x}$, $[2, 3]$

14. $f(x) = \frac{1}{x + 3}$, $[-2, 1]$
15. $f(x) = \frac{2}{x^2}$, $[1, 4]$
16. $f(x) = x^{1/3}$, $[0, 8]$

17. The function $f(x) = [\![x]\!]$ is not continuous on $[0, 4]$. Show that $\int_0^4 f(x)\, dx$ exists.

18. Consider the function f, where

$$f(x) = \begin{cases} 0 & \text{if } x \text{ is rational} \\ 1 & \text{if } x \text{ is irrational} \end{cases}$$

Show that $\int_0^1 f(x)\, dx$ does not exist. [*Hint:* Evaluate the Riemann sum in two different ways: first by using rational numbers for u_i and then by using irrational numbers for u_i.]

19. Use the definition of a definite integral in terms of Riemann sums to prove that $\int_a^b k\, dx = k(b - a)$, where k is a constant.

4. The Fundamental Theorem of Calculus

The following properties of the definite integral are needed for the discussion of the fundamental theorem of calculus:

(5.25) **THEOREM   If a function f is continuous on an interval containing the numbers a, b, and c, then**

$$\int_a^b f(x)\,dx = \int_a^c f(x)\,dx + \int_c^b f(x)\,dx$$

A proof of (5.25) is given in Appendix II.

In particular, if f is continuous and nonnegative on $[a, b]$ and if c is between a and b, then (5.25) has a simple geometric interpretation, as seen in Figure 15.

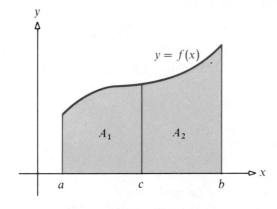

$A = $ Area under f from a to $b = A_1 + A_2$

$$\int_a^b f(x)\,dx = \int_a^c f(x)\,dx + \int_c^b f(x)\,dx$$

Figure 15

(5.26) **THEOREM   If a function f is continuous on a closed interval $[a, b]$ and if m and M denote the absolute minimum and absolute maximum of f on $[a, b]$, respectively, then**

(5.27) $$m(b - a) \le \int_a^b f(x)\,dx \le M(b - a)$$

A proof of (5.26) is given in Appendix II. If f is nonnegative on $[a, b]$, then (5.26) may be illustrated geometrically. In Figure 16 (p. 300) the area of the shaded region is $\int_a^b f(x)\,dx$.   The area of the smaller rectangle, which has   width $= (b - a)$ and height $= m$,   is $m(b - a)$;   the area of the larger rectangle, which has   width $= (b - a)$ and   height $= M$,   is $M(b - a)$.   These three areas are numerically related by the inequalities in (5.27).

Interestingly enough, the area under the graph of f from a to b, namely, $\int_a^b f(x)\,dx$,   *is equal to* the area of a certain rectangle of width $(b - a)$ and height

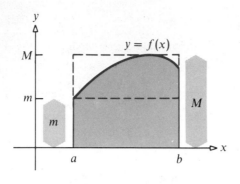

$$f(u)(b - a) = \int_a^b f(x)\, dx$$

Figure 16   **Figure 17**

$f(u)$ for a special choice (or choices) of u in $[a, b]$. (See Figure 17.) In terms of definite integrals, this result is known as the *mean value theorem for integrals*.

(5.28)   ***Mean Value Theorem for Integrals.***   **If f is continuous on a closed interval $[a, b]$, there exists a number u, $a \le u \le b$,   so that**

$$\int_a^b f(x)\, dx = f(u)(b - a)$$

Note the similarity between (5.28) and (4.26), the mean value theorem for derivatives. The proof of (5.28), which follows, is independent of the geometric interpretation.

Proof   If f is a constant function on $[a, b]$, the result is true for any choice of u in $[a, b]$. Suppose f is not identically constant on $[a, b]$. Since f is continuous on the closed interval $[a, b]$, by the extreme value theorem (4.23), f attains its extreme values on $[a, b]$. Suppose f assumes its absolute minimum m at the number c so that $f(c) = m$;   and suppose f assumes its absolute maximum M at the number C so that $f(C) = M$.   Then for all x in $[a, b]$, by (5.26), we have

(5.29) $$m(b - a) \le \int_a^b f(x)\, dx \le M(b - a)$$

If we divide (5.29) by $(b - a)$ and replace m by $f(c)$ and M by $f(C)$, then

$$f(c) \le \frac{1}{b - a} \int_a^b f(x)\, dx \le f(C)$$

Since $[1/(b - a)] \int_a^b f(x)\, dx$ is a number between $f(c)$ and $f(C)$, it follows from the intermediate value theorem (2.28) that there is a number u between c and C, and hence in $[a, b]$, so that

$$f(u) = \frac{1}{b - a} \int_a^b f(x)\, dx$$

That is, there is a u, $a \le u \le b$, so that

$$\int_a^b f(x)\, dx = f(u)(b - a) \qquad \blacksquare$$

The number u is not necessarily unique. However, the value $f(u)$ in (5.28), which is unique, is called the *average value* or *mean value* of f on $[a, b]$. The reasons for this name shall be made clear in Chapter 6.

A RELATIONSHIP BETWEEN THE DERIVATIVE AND THE DEFINITE INTEGRAL

We begin with a function f that is continuous on a closed interval $[a, b]$. As a result of theorem (5.23), the definite integral $\int_a^b f(t)\, dt$ exists and is a real number. If x denotes any number in $[a, b]$, the definite integral $\int_a^x f(t)\, dt$ will also exist, and it will depend on x. That is, $\int_a^x f(t)\, dt$ is a function, say, I, of x, namely,

$$I(x) = \int_a^x f(t)\, dt$$

whose domain is the closed interval $[a, b]$. Note that the integral I above has a *variable upper limit* x; the t that appears is a dummy variable.

The relationship between the functions I and f on $[a, b]$ is stated in the next theorem.

(5.30)   **THEOREM**   **Let f be a continuous function defined on the closed interval $[a, b]$. The function I defined by**

$$I(x) = \int_a^x f(t)\, dt$$

has the property that

$$I'(x) = \frac{d}{dx}\left[\int_a^x f(t)\, dt\right] = f(x) \qquad \text{for all } x \text{ in } [a, b]$$

Proof If x and $(x + h)$, $h \ne 0$, are in $[a, b]$, form the difference

$$I(x + h) - I(x) = \int_a^{x+h} f(t)\, dt - \int_a^x f(t)\, dt = \underset{\underset{\text{By (5.22)}}{\uparrow}}{\int_a^{x+h} f(t)\, dt} + \int_x^a f(t)\, dt = \underset{\underset{\text{By (5.25)}}{\uparrow}}{\int_x^{x+h} f(t)\, dt}$$

To form the difference quotient, we divide the last equality by $h \ne 0$.   Then

(5.31)
$$\frac{I(x + h) - I(x)}{h} = \frac{1}{h}\int_x^{x+h} f(t)\, dt$$

We apply (5.28), the mean value theorem for integrals, to the integral on the right. There are two possibilities: $h > 0$ or $h < 0$. If $h > 0$, there exists a u,

$x \le u \le x + h$, so that

$$\int_x^{x+h} f(t)\, dt = f(u)h \qquad \text{or} \qquad \frac{1}{h}\int_x^{x+h} f(t)\, dt = f(u)$$

Therefore, from (5.31),

(5.32)
$$\frac{I(x+h) - I(x)}{h} = f(u)$$

Suppose we let $h \to 0^+$ in (5.32). Since $x \le u \le x + h$, as $h \to 0^+$, then u must tend to x^+. Thus, $\lim_{h \to 0^+} f(u) = \lim_{u \to x^+} f(u)$. But f is continuous. Therefore $\lim_{u \to x^+} f(u) = f(x)$. Hence

(5.33)
$$\lim_{h \to 0^+} \frac{I(x+h) - I(x)}{h} = f(x)$$

Similarly, if $h < 0$, then

(5.34)
$$\lim_{h \to 0^-} \frac{I(x+h) - I(x)}{h} = f(x)$$

Since the two one-sided limits (5.33) and (5.34) are equal, we conclude that

$$\lim_{h \to 0} \frac{I(x+h) - I(x)}{h} = f(x)$$

We recognize this limit as the derivative of the function I. Thus, $I'(x) = f(x)$ for all x in $[a, b]$. ∎

If f is nonnegative, a geometric justification of theorem (5.30) may be given, as shown in Figure 18.

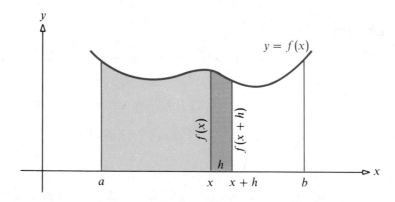

$$I(x + h) = \text{Area from } a \text{ to } x + h$$
$$I(x) = \text{Area from } a \text{ to } x$$
$$I(x + h) - I(x) = \text{Area from } x \text{ to } x + h$$
$$\frac{I(x + h) - I(x)}{h} = \frac{\text{Area from } x \text{ to } x + h}{h} \approx f(x) \quad \text{if } h \text{ is small}$$

Figure 18

We can summarize theorem (5.30) as follows: Let f be continuous on $[a, b]$. If $I(x) = \int_a^x f(t)\, dt$,   then $I'(x) = f(x)$, $a \le x \le b$.   Stated another way,

(5.35)
$$\frac{d}{dx}[I(x)] = \frac{d}{dx}\left[\int_a^x f(t)\, dt\right] = f(x)$$

EXAMPLE 1   (a) $\dfrac{d}{dx} \displaystyle\int_0^x \sqrt{t^2 + 1}\, dt = \sqrt{x^2 + 1}$ (b) $\dfrac{d}{dx} \displaystyle\int_2^x \dfrac{s^3 - 1}{2s^2 + s + 1}\, ds = \dfrac{x^3 - 1}{2x^2 + x + 1}$ ∎

Note that in each of the above examples the differentiation is with respect to the variable upper limit. Furthermore, observe that the dummy variable plays no role—the answer is a function of the variable upper limit; in fact, the answer is the same function as the integrand.

Since the function I defined in (5.30) is differentiable, it follows by (3.63) on page 173, that I is also continuous. But theorem (5.30) does more than relate the derivative and the definite integral—it also provides a method for calculating definite integrals without using definition (5.21). This method is so valuable it is referred to as the *fundamental theorem of calculus.*

(5.36)   **Fundamental Theorem of Calculus.   Let f be a continuous function defined on the closed interval $[a, b]$. If F is any antiderivative of f on $[a, b]$, then**

$$\int_a^b f(x)\, dx = F(b) - F(a)$$

Proof   By theorem (5.30), we know that $\int_a^x f(t)\, dt$ is an antiderivative of f. Since F is any antiderivative of f, we may write

(5.37)
$$F(x) = \int_a^x f(t)\, dt + C$$

where C is some constant. (Here we use the fact that if two functions have the same derivative, they differ by a constant.) Now we replace x by a and then replace x by b in (5.37). The results are

$$F(a) = \int_a^a f(t)\, dt + C \qquad F(b) = \int_a^b f(t)\, dt + C$$

But by (5.22), $\int_a^a f(t)\, dt = 0$.   Hence, by subtracting $F(a)$ from $F(b)$ we obtain

$$F(b) - F(a) = \int_a^b f(t)\, dt$$

Since t is an artificial variable (see p. 296), we may replace t by x, and the result follows.   ∎

Thus, if F is an antiderivative of f, we have the formula

$$\int_a^b f(x)\, dx = F(b) - F(a)$$

As an aid in computation, we introduce the notation

(5.38)
$$\int_a^b f(x)\, dx = F(x)\Big|_a^b = F(b) - F(a)$$

This notation suggests that we first find an antiderivative $F(x)$ of $f(x)$. Then we write $F(x)\big|_a^b$ as an aid in computing $F(b) - F(a)$. Specifically, we first replace x by the upper limit b, obtaining $F(b)$, and from this subtract the value $F(a)$ obtained by setting $x = a$.

EXAMPLE 2

(a) $x^2\big|_{-2}^1 = 1^2 - (-2)^2 = 1 - 4 = -3$

(b) $(1 - 3x^2)\big|_{-1}^5 = [1 - 3(5)^2] - [1 - 3(-1)^2] = -74 - (-2) = -72$

(c) $\sin^2 x\big|_0^{\pi/6} = \left(\sin\frac{\pi}{6}\right)^2 - (\sin 0)^2 = \left(\frac{1}{2}\right)^2 - 0 = \frac{1}{4}$ ∎

We observe that for any constant,

$$[F(x) + C]\big|_a^b = [F(b) + C] - [F(a) + C] = F(b) - F(a)$$

In other words, it does not matter which of the antiderivatives of f is chosen in applying the fundamental theorem of calculus since the same answer is obtained for every antiderivative. Let's look at some more examples.

EXAMPLE 3   Use the fundamental theorem of calculus to evaluate:

(a) $\displaystyle\int_{-2}^1 x^2\, dx$ (b) $\displaystyle\int_0^{\pi/6} \cos x\, dx$

(c) $\displaystyle\int_{-2}^1 (x^2 - 2x + 1)\, dx$ (d) $\displaystyle\int_1^2 \frac{x^3 - 4}{x^2}\, dx$

Solution

(a)   An antiderivative of x^2 is $x^3/3$. By applying (5.38) with $F(x) = x^3/3$,   we get

$$\int_{-2}^1 x^2\, dx = \frac{x^3}{3}\bigg|_{-2}^1 = \frac{1}{3} - \frac{(-2)^3}{3} = \frac{1}{3} + \frac{8}{3} = \frac{9}{3} = 3$$

(b)   An antiderivative of $\cos x$ is $\sin x$. By applying (5.38), we get

$$\int_0^{\pi/6} \cos x\, dx = \sin x\bigg|_0^{\pi/6} = \sin\frac{\pi}{6} - \sin 0 = \frac{1}{2}$$

(c)   An antiderivative of $x^2 - 2x + 1$ is $(x^3/3) - x^2 + x$.   By applying (5.38), we get

$$\int_{-2}^1 (x^2 - 2x + 1)\, dx = \left[\frac{x^3}{3} - x^2 + x\right]\bigg|_{-2}^1$$

$$= \left[\frac{(1)^3}{3} - (1)^2 + 1\right] - \left[\frac{(-2)^3}{3} - (-2)^2 - 2\right]$$

$$= \frac{1}{3} + \frac{26}{3} = \frac{27}{3} = 9$$

(d) $\int_1^2 \dfrac{x^3 - 4}{x^2}\, dx = \int_1^2 \left(x - \dfrac{4}{x^2}\right) dx = \int_1^2 (x - 4x^{-2})\, dx = \left(\dfrac{x^2}{2} - \dfrac{4x^{-1}}{-1}\right)\Big|_1^2$

$$= \left(\dfrac{x^2}{2} + \dfrac{4}{x}\right)\Big|_1^2 = \left[\dfrac{(2)^2}{2} + \dfrac{4}{2}\right] - \left(\dfrac{1}{2} + \dfrac{4}{1}\right) = 4 - \dfrac{9}{2} = -\dfrac{1}{2} \qquad \blacksquare$$

We have just seen that the fundamental theorem of calculus provides a short way of evaluating the limit of some Riemann sums, namely,

$$\lim_{||P|| \to 0} \sum_{i=1}^{n} f(u_i)\Delta x_i = \int_a^b f(x)\, dx = F(b) - F(a)$$

provided f has an antiderivative F. This is what is so astonishing about the fundamental theorem and the reason for its name: Evaluating the limit of a sum and the inverse of the process of differentiation (antidifferentiation) are intimately related.

EXERCISE 4

In Problems 1–22 apply the fundamental theorem of calculus to evaluate each definite integral (a and b are constants).

1. $\int_1^2 (3x - 1)\, dx$

2. $\int_1^2 (2x + 1)\, dx$

3. $\int_0^1 3x^2\, dx$

4. $\int_{-2}^0 (x + x^2)\, dx$

5. $\int_0^1 \sqrt{u}\, du$

6. $\int_1^8 \sqrt[3]{y}\, dy$

7. $\int_0^{\pi/3} \sin x\, dx$

8. $\int_0^{\pi/2} \cos x\, dx$

9. $\int_0^1 (t^2 - t^{3/2})\, dt$

10. $\int_1^4 (\sqrt{x} - a^2 x)\, dx$

11. $\int_{-2}^3 (x - 1)(x + 3)\, dx$

12. $\int_0^1 (z^2 + 1)^2\, dz$

13. $\int_1^2 \dfrac{x^2 - 12}{x^4}\, dx$

14. $\int_1^3 \dfrac{2 - x^2}{x^4}\, dx$

15. $\int_0^1 (\sqrt[5]{t^2} + 1)\, dt$

16. $\int_1^4 (\sqrt{u} + a)\, du$

17. $\int_1^4 \dfrac{x + 1}{\sqrt{x}}\, dx$

18. $\int_1^9 \dfrac{\sqrt{x} + 1}{x^2}\, dx$

19. $\int_0^1 (ax^4 + b)\, dx$

20. $\int_{-1}^1 (x + 1)^3\, dx$

21. $\int_a^b (x + 2)^2\, dx$

22. $\int_{-1}^0 (x^2 + 4a)\, dx$

In Problems 23–24 find the area under the graph of $y = f(x)$ from a to b. Draw a graph first.

23. $f(x) = x^2 + 9$ from $a = -2$ to $b = 2$ 24. $f(x) = x^3 + 4$ from $a = -1$ to $b = 3$

In Problems 25–32 find the indicated derivative.

25. $\dfrac{d}{dx} \int_1^x \sqrt{t^2 + 1}\, dt$

26. $\dfrac{d}{dx} \int_3^x \dfrac{t + 1}{t}\, dt$

27. $\dfrac{d}{dt} \left[\int_0^t (3 + x^2)^{3/2}\, dx\right]$

28. $\dfrac{d}{dx} \left[\int_{-4}^x (t^3 + 8)^{1/3}\, dt\right]$

29. $\dfrac{d}{dx}\left[\displaystyle\int_1^x f(u)\,du\right]$

30. $\dfrac{d}{dt}\left[\displaystyle\int_4^t g(x)\,dx\right]$

31. $\dfrac{d}{dx}\left[\displaystyle\int_1^{2x^3}\sqrt{t^2+1}\,dt\right]$

32. $\dfrac{d}{dx}\left[\displaystyle\int_1^{\sqrt{x}}\sqrt{t^4+5}\,dt\right]$

33. Given that $f(x)=(2x^3-3)^2$ and $f'(x)=12x^2(2x^3-3)$, use the fundamental theorem of calculus to find $\int_0^2 12x^2(2x^3-3)\,dx$.

34. Given that $f(x)=(x^2+5)^3$ and $f'(x)=6x(x^2+5)^2$, use the fundamental theorem of calculus to find $\int_{-1}^2 6x(x^2+5)^2\,dx$.

In Problems 35–38 find the number(s) u referred to in the mean value theorem for integrals (5.28).

35. $\displaystyle\int_0^3 6x^2\,dx$

36. $\displaystyle\int_0^2 4x^3\,dx$

37. $\displaystyle\int_0^2 (x^2-4)\,dx$

38. $\displaystyle\int_0^4 (4-x)\,dx$

In Problems 39–44 use theorem (5.26) to obtain a lower estimate and an upper estimate for each of the given integrals.

39. $\displaystyle\int_1^3 (5x+1)\,dx$

40. $\displaystyle\int_0^1 (1-x)\,dx$

41. $\displaystyle\int_{\pi/4}^{\pi/2}\sin x\,dx$

42. $\displaystyle\int_{\pi/6}^{\pi/3}\cos x\,dx$

43. $\displaystyle\int_0^1 \sqrt{1+x^2}\,dx$

44. $\displaystyle\int_{-1}^1 \sqrt{1+x^4}\,dx$

45. If f is continuous on $[a,b]$, show that the functions defined by

$$F(x)=\int_c^x f(t)\,dt \qquad G(x)=\int_d^x f(t)\,dt$$

for any choice of c and d in (a,b), always differ by a constant, and also show that

$$F(x)-G(x)=\int_c^d f(t)\,dt$$

46. If f'' is continuous on $[a,b]$, show that

$$\int_a^b xf''(x)\,dx = bf'(b)-af'(a)-f(b)+f(a)$$

[*Hint:*   Look at the derivative of the function $F(x)=xf'(x)-f(x)$.]

47. If f' is continuous in $[a,b]$, show that

$$\int_a^b f(x)f'(x)\,dx = \frac{1}{2}\{[f(b)]^2-[f(a)]^2\}$$

48. If u and v are differentiable functions and f is a continuous function, develop a formula for

$$\frac{d}{dx}\left[\int_{u(x)}^{v(x)} f(t)\,dt\right]$$

5. Properties of the Definite Integral

In this section we list several properties of the definite integral that are used when working with integrals. Although you already know some of them, we list them again for completeness.

(5.39)   **If a function f is continuous on an interval containing a, b, and c, then**

$$\int_a^b f(x)\, dx = \int_a^c f(x)\, dx + \int_c^b f(x)\, dx$$

This, of course, is a duplication of (5.25).

EXAMPLE 1   If f is continuous on the closed interval $[2, 7]$, then

$$\int_2^7 f(x)\, dx = \int_2^4 f(x)\, dx + \int_4^7 f(x)\, dx \qquad \blacksquare$$

The choice of c in (5.39) does not have to lie between a and b.

EXAMPLE 2   If g is continuous on the closed interval $[3, 25]$, then

$$\int_3^{10} g(x)\, dx = \int_3^{25} g(x)\, dx + \int_{25}^{10} g(x)\, dx \qquad \blacksquare$$

(5.40)   **If the functions f and g are continuous on the closed interval $[a, b]$, then**

$$\int_a^b [f(x) + g(x)]\, dx = \int_a^b f(x)\, dx + \int_a^b g(x)\, dx$$

EXAMPLE 3 $\displaystyle\int_0^1 (x^2 + x)\, dx = \int_0^1 x^2\, dx + \int_0^1 x\, dx \qquad \blacksquare$

(5.41)   **If a function f is continuous on a closed interval $[a, b]$ and if k is a constant, then**

$$\int_a^b kf(x)\, dx = k \int_a^b f(x)\, dx$$

EXAMPLE 4 $\displaystyle\int_0^1 3x^2\, dx = 3 \int_0^1 x^2\, dx \qquad \blacksquare$

By means of repeated use of (5.40) and (5.41), we have the extended result:

(5.42)   **If the functions $f_1, f_2, \ldots, f_n$ are continuous on a closed interval $[a, b]$ and if $k_1, k_2, \ldots, k_n$ are constants, then**

$$\int_a^b [k_1 f_1(x) + k_2 f_2(x) + \cdots + k_n f_n(x)]\, dx$$

$$= k_1 \int_a^b f_1(x)\, dx + k_2 \int_a^b f_2(x)\, dx + \cdots + k_n \int_a^b f_n(x)\, dx$$

(5.43)   **If a function f is continuous on a closed interval $[a, b]$ and if $f(x) \geq 0$ on $[a, b]$, then**

$$\int_a^b f(x)\, dx \geq 0$$

This property asserts the geometric fact that the area under the graph of a non-negative function is never negative.

(5.44)   If the functions f and g are continuous on a closed interval $[a, b]$ and if $f(x) \le g(x)$ on $[a, b]$, then

$$\int_a^b f(x)\, dx \le \int_a^b g(x)\, dx$$

A proof of (5.39)–(5.44) will require the use of definition (5.20). [See Appendix II for such a proof of (5.39).] However, if we assume that each of the integrands have antiderivatives, then we can use theorem (5.23) and the fundamental theorem of calculus (5.36). We shall prove (5.40) and (5.43) here with this added assumption.

Proof of (5.40)   Since f and g each have antiderivatives F and G on $[a, b]$, then $F' = f$ and $G' = g$ on $[a, b]$. Further, $F + G$ is an antiderivative of $f + g$ on $[a, b]$ since $(F + G)' = F' + G' = f + g$. Consequently,

$$\int_a^b [f(x) + g(x)]\, dx = [F(x) + G(x)]\Big|_a^b = [F(b) + G(b)] - [F(a) + G(a)]$$

$$= [F(b) - F(a)] + [G(b) - G(a)]$$

$$= \int_a^b f(x)\, dx + \int_a^b g(x)\, dx \quad \blacksquare$$

Proof of (5.43)   Since $f(x) \ge 0$ on $[a, b]$ and f has an antiderivative F, then

$$F'(x) = f(x) \ge 0 \qquad \text{on } [a, b]$$

Thus, F is an increasing function on $[a, b]$, which implies that $F(b) \ge F(a)$. Consequently,

$$\int_a^b f(x)\, dx = F(b) - F(a) \ge 0 \quad \blacksquare$$

EXERCISE 5

In Problems 1–4 use properties of the definite integral to verify each statement. Assume that all integrals involved exist.

1. $\int_3^{11} f(x)\, dx - \int_7^{11} f(x)\, dx = \int_3^7 f(x)\, dx$ **2.** $\int_{-2}^6 f(x)\, dx - \int_3^6 f(x)\, dx = \int_{-2}^3 f(x)\, dx$

3. $\int_0^4 f(x)\, dx - \int_6^4 f(x)\, dx = \int_0^6 f(x)\, dx$ **4.** $\int_{-1}^3 f(x)\, dx - \int_5^3 f(x)\, dx = \int_{-1}^5 f(x)\, dx$

In Problems 5–10 evaluate each definite integral if it is known that $\int_1^3 f(x)\, dx = 5$, $\int_1^3 g(x)\, dx = -2$, $\int_3^5 f(x)\, dx = 2$, $\int_3^5 g(x)\, dx = 1$.

5. $\int_1^3 [f(x) + g(x)]\, dx$ **6.** $\int_1^3 [f(x) - g(x)]\, dx$ **7.** $\int_1^3 [5f(x) - 3g(x)]\, dx$

8. $\int_1^3 [3f(x) + 4g(x)]\, dx$ **9.** $\int_1^5 [2f(x) - 3g(x)]\, dx$ **10.** $\int_1^5 [f(x) - g(x)]\, dx$

11. Without evaluating the definite integral, show that $\int_0^1 x\, dx \ge \int_0^1 x^3\, dx$.

12. Without evaluating the definite integral, show that $\int_1^2 x^3\, dx \ge \int_1^2 x\, dx$.

13. Prove (5.41). Assume f has an antiderivative on $[a, b]$.

14. Prove (5.42). Assume $f_1, f_2, \ldots, f_n$ have antiderivatives on $[a, b]$.

15. Prove (5.44). Assume f and g have an antiderivative on $[a, b]$.

16. Prove that if f is continuous on $[a, b]$, then

$$\left| \int_a^b f(x)\, dx \right| \le \int_a^b |f(x)|\, dx$$

What is the geometric interpretation?

17. If $f(x) = \begin{cases} x + 1 & x < 0 \\ x^2 + 1 & x \ge 0 \end{cases}$, find $\int_{-1}^1 f(x)\, dx$.   **18.** If $f(x) = \begin{cases} 1 & x < 0 \\ x^2 + 1 & x \ge 0 \end{cases}$, find $\int_{-1}^1 f(x)\, dx$.

6. The Indefinite Integral; Method of Substitution

Through the fundamental theorem of calculus, we discovered an intimate relationship between definite integrals and antiderivatives. Namely, the definite integral $\int_a^b f(x)\, dx$ can be evaluated in many cases by finding an antiderivative of f. Because of this, it has become customary to use the integral symbol $\int$ as a direction to find all antiderivatives of a function. That is:

The notation $\int f(x)\, dx$ **is used to denote all antiderivatives of** f**, and we refer to** $\int f(x)\, dx$ **as the *indefinite integral of* f.**

For example,

$$\int x^2\, dx = \frac{x^3}{3} + C \qquad \int (x^2 + 1)\, dx = \frac{x^3}{3} + x + C$$

where C is a constant called the *constant of integration*.

The process of evaluating either the indefinite integral $\int f(x)\, dx$ or the definite integral $\int_a^b f(x)\, dx$ is called *integration* and the function f is called the *integrand*. It is important to distinguish between the definite integral $\int_a^b f(x)\, dx$ and the indefinite integral $\int f(x)\, dx$. The definite integral is a number that depends on the limits of integration a and b. On the other hand, the indefinite integral of f is defined as a collection of functions $F(x) + C$, C a constant, such that $F'(x) = f(x)$. For example,

$$\int_0^2 x^2\, dx = \frac{x^3}{3}\Big|_0^2 = \frac{8}{3} \qquad \int x^2\, dx = \frac{x^3}{3} + C$$

Again, we remind you that an antiderivative F of f and the definite integral $\int_a^b f(x)\, dx$ of f are related by the fundamental theorem of calculus, namely,

$$\int_a^b f(x)\, dx = F(x)\Big|_a^b = F(b) - F(a)$$

PROPERTIES OF THE INDEFINITE INTEGRAL

We now state several rules concerning indefinite integrals. The first is a consequence of the definition of $\int f(x)\,dx$ as all antiderivatives of f:

(5.45)
$$\frac{d}{dx}\int f(x)\,dx = f(x)$$

For example,

$$\frac{d}{dx}\int \sqrt{x^2+1}\,dx = \sqrt{x^2+1}$$

Also, we have the following property:

(5.46)
$$\int kf(x)\,dx = k\int f(x)\,dx \qquad \textbf{where } k \textbf{ is a real number}$$

This formula states that to find the indefinite integral of a constant k times a function f, we first find the indefinite integral of f and then multiply by k. To prove this statement, we differentiate the right side of (5.46):

$$\frac{d}{dx}\left[k\int f(x)\,dx\right] = k\left[\frac{d}{dx}\int f(x)\,dx\right] = kf(x)$$

Another property is

(5.47)
$$\int [f(x)+g(x)]\,dx = \int f(x)\,dx + \int g(x)\,dx$$

This formula states that the indefinite integral of a sum equals the sum of the indefinite integrals.

The next formula is a consequence of (3.56):

(5.48)
$$\int x^r\,dx = \frac{x^{r+1}}{r+1} + C \qquad r \neq -1 \text{ a rational number}$$

This is easily verified since

$$\frac{d}{dx}\left(\frac{x^{r+1}}{r+1} + C\right) = x^r$$

EXAMPLE 1
$$\int (2x^{1/3}+5x^{2/3})\,dx = \int 2x^{1/3}\,dx + \int 5x^{2/3}\,dx = 2\int x^{1/3}\,dx + 5\int x^{2/3}\,dx$$

$$= \frac{2x^{4/3}}{\frac{4}{3}} + \frac{5x^{5/3}}{\frac{5}{3}} + C = \frac{3x^{4/3}}{2} + 3x^{5/3} + C \qquad \blacksquare$$

Sometimes an appropriate algebraic manipulation is required before formula (5.48) can be used.

EXAMPLE 2
$$\int \left(\frac{12}{x^5} + \frac{1}{\sqrt{x}}\right) dx = 12 \int \frac{1}{x^5} \, dx + \int \frac{1}{\sqrt{x}} \, dx = 12 \int x^{-5} \, dx + \int x^{-1/2} \, dx$$

$$= 12\left(\frac{x^{-4}}{-4}\right) + \frac{x^{1/2}}{\frac{1}{2}} + C = \frac{-3}{x^4} + 2x^{1/2} + C \qquad \blacksquare$$

Now, since the differentiation formulas for $\sin x$ and $\cos x$ are

$$\frac{d}{dx} \sin x = \cos x \qquad \text{and} \qquad \frac{d}{dx} \cos x = -\sin x$$

we have the following integral formulas:

(5.49)
$$\int \cos x \, dx = \sin x + C \qquad \int \sin x \, dx = -\cos x + C$$

METHOD OF SUBSTITUTION; CHANGE OF VARIABLE

Indefinite integrals that cannot be evaluated directly by using formulas (5.48) and (5.49) may sometimes be evaluated by the *substitution technique*. This technique involves the introduction of a function that changes the integrand into one to which the formulas apply. For example, to evaluate $\int (x^2 + 5)^3 2x \, dx$, we use the substitution $u = x^2 + 5$ so that $du = 2x \, dx$:

$$\int (x^2 + 5)^3 2x \, dx = \int u^3 \, du = \frac{u^4}{4} + C = \frac{(x^2 + 5)^4}{4} + C$$

Note that to make the substitution method work, the differentials must also be substituted for. Of course, once a substitution is chosen, its differential is determined. We may verify the correctness of the answer as follows:

$$\frac{d}{dx} \left[\frac{(x^2 + 5)^4}{4} + C\right] = \frac{4(x^2 + 5)^3(2x)}{4} = (x^2 + 5)^3 2x$$

EXAMPLE 3   Evaluate:

(a) $\int (2x + 1)^2 \, dx$ (b) $\int \frac{dx}{\sqrt{5 - 2x}}$ (c) $\int \sin(3x + 2) \, dx$

Solution   (a) Let $u = 2x + 1$. Then $du = 2 \, dx$, so that $dx = du/2$. Hence,

$$\int (2x + 1)^2 \, dx = \int u^2 \left(\frac{du}{2}\right) = \frac{1}{2} \int u^2 \, du = \frac{1}{2}\left(\frac{u^3}{3}\right) + C = \frac{(2x + 1)^3}{6} + C$$

(b) Let $u = 5 - 2x$. Then $du = -2 \, dx$, so that $dx = du/(-2)$. Hence,

$$\int \frac{dx}{\sqrt{5 - 2x}} = \int \frac{du/(-2)}{\sqrt{u}} = -\frac{1}{2} \int \frac{du}{\sqrt{u}}$$

$$= -\frac{1}{2} \int u^{-1/2} \, du = -\frac{1}{2}\left(\frac{u^{1/2}}{\frac{1}{2}}\right) + C = -\sqrt{5 - 2x} + C$$

(c) Let $u = 3x + 2$. Then $du = 3\, dx$, so that $dx = du/3$. Hence,

$$\int \sin(3x + 2)dx = \int (\sin u)\frac{du}{3} = \frac{1}{3}\int \sin u\, du$$

$$= \left(\frac{1}{3}\right)(-\cos u) + C = \left(-\frac{1}{3}\right)\cos(3x + 2) + C \qquad \blacksquare$$

EXAMPLE 4   Evaluate: $\int x\sqrt{x^2 + 1}\, dx$

Solution   We use the substitution $u = x^2 + 1$. Then $du = 2x\, dx$, so that $x\, dx = du/2$. Hence,

$$\int x\sqrt{x^2 + 1}\, dx = \int \sqrt{x^2 + 1}x\, dx = \int \sqrt{u}\,\frac{du}{2} = \frac{1}{2}\int u^{1/2}\, du$$

$$= \frac{1}{2}\left(\frac{u^{3/2}}{\frac{3}{2}}\right) + C = \frac{(x^2 + 1)^{3/2}}{3} + C \qquad \blacksquare$$

Note that in using the substitution $u = x^2 + 1$ in the above example we must substitute not only for the integrand $x\sqrt{x^2 + 1}$, but also for the differential dx. In addition, the existence of x as part of the integrand makes the substitution $u = x^2 + 1$ work. For example, if we used this same substitution to evaluate $\int \sqrt{x^2 + 1}\, dx$, we would obtain $u = x^2 + 1$, $du = 2x\, dx$. Since $x = \sqrt{u - 1}$, we have

$$\int \sqrt{x^2 + 1}\, dx = \int \sqrt{u}\,\frac{du}{2\sqrt{u - 1}} = \int \frac{\sqrt{u}}{2\sqrt{u - 1}}\, du$$

In this case, the substitution we used results in an integrand that does not conform to formula (5.48). In fact, it results in an integrand that is *more* complicated than the original one.

Thus, the idea behind the substitution method is to obtain an integral $\int h(u)\, du$ that is simpler than the original integral $\int f(x)\, dx$. When a substitution does not simplify the integral, other substitutions should be tried. If these do not work, other integration methods should be applied (we will discuss some of these other techniques in Chap. 9). Thus, since integration—unlike differentiation—has no prescribed method, some ingenuity and a lot of practice is required.

To illustrate that the integral has no prescribed method for evaluation, we use two different substitutions in the next example.

EXAMPLE 5   Evaluate: $\int x\sqrt{4 + x}\, dx$

Solution   *Substitution I:*   Let $u = 4 + x$. Then $du = dx$ and $x = u - 4$,   so that

$$\int x\sqrt{4 + x}\, dx = \int (u - 4)\sqrt{u}\, du = \int (u^{3/2} - 4u^{1/2})\, du$$

$$= \frac{u^{5/2}}{\frac{5}{2}} - \frac{4u^{3/2}}{\frac{3}{2}} + C = \frac{2(4 + x)^{5/2}}{5} - \frac{8(4 + x)^{3/2}}{3} + C$$

Substitution II: Let $u = \sqrt{4+x}$, so that $u^2 = 4 + x$. Then we have $2u\,du = dx$, $x = u^2 - 4$, and

$$\int x\sqrt{4 + x}\,dx = \int (u^2 - 4)(u)(2u\,du) = 2\int (u^4 - 4u^2)\,du$$

$$= \frac{2u^5}{5} - \frac{8u^3}{3} + C = \frac{2(4 + x)^{5/2}}{5} - \frac{8(4 + x)^{3/2}}{3} + C \quad \blacksquare$$

EXAMPLE 6   Evaluate: $\int_0^2 x\sqrt{4 - x^2}\,dx$

Solution   First we evaluate the indefinite integral $\int x\sqrt{4 - x^2}\,dx$, using the substitution $u = 4 - x^2$. Then $du = -2x\,dx$, and

$$\int x\sqrt{4 - x^2}\,dx = \int \sqrt{4 - x^2}\,x\,dx = \int \sqrt{u}\left(-\frac{du}{2}\right)$$

$$= -\frac{1}{2}\left(\frac{u^{3/2}}{\frac{3}{2}}\right) + C = \frac{-(4 - x^2)^{3/2}}{3} + C$$

By the fundamental theorem, we find

$$\int_0^2 x\sqrt{4 - x^2}\,dx = \frac{-(4 - x^2)^{3/2}}{3}\bigg|_0^2 = 0 + \frac{4^{3/2}}{3} = \frac{8}{3} \quad \blacksquare$$

CHANGING LIMITS OF INTEGRATION

When the substitution method is used to evaluate definite integrals, it is sometimes easier to change the limits of integration. For example, in order to evaluate $\int_0^2 x\sqrt{4 - x^2}\,dx$ in Example 6, we let $u = 4 - x^2$ and $du = -2x\,dx$. But now we use the equation $u = 4 - x^2$ to change the limits of integration. Thus, when $x = 0$ (the *old* lower limit of integration), we have $u = 4 - 0^2 = 4$ (the *new* lower limit of integration). Similarly, when $x = 2$ (the old upper limit of integration), we have $u = 0$ (the new upper limit of integration). By using this technique,

$$u = 4 - x^2 = 4 - 2^2 = 0$$

$$\int_0^2 x\sqrt{4 - x^2}\,dx = \int_4^0 \sqrt{u}\left(-\frac{du}{2}\right) = -\frac{1}{2}\left(\frac{u^{3/2}}{\frac{3}{2}}\right)\bigg|_4^0 = -\frac{1}{3}(0 - 4^{3/2}) = \frac{8}{3}$$

$$u = 4 - x^2 = 4 - 0 = 4$$

A word to the wise! This second way of evaluating a definite integral may be faster, but it also requires more care. Be certain not to forget to change the limits of integration when using this method.

EXAMPLE 7   Evaluate: $\int_0^{\pi/2} \frac{1 - \cos 2\theta}{2}\,d\theta$

Solution
$$\int_0^{\pi/2} \frac{1 - \cos 2\theta}{2}\, d\theta = \frac{1}{2}\int_0^{\pi/2}(1 - \cos 2\theta)\, d\theta = \frac{1}{2}\int_0^{\pi/2} d\theta - \frac{1}{2}\int_0^{\pi/2}\cos 2\theta\, d\theta$$

In the second integral on the right, make the substitution $u = 2\theta$. Then $du = 2\, d\theta$, and

$$\int_0^{\pi/2} \frac{1 - \cos 2\theta}{2}\, d\theta = \frac{1}{2}\theta\Big|_0^{\pi/2} - \frac{1}{2}\int_0^{\pi}(\cos u)\frac{du}{2}$$

$$= \left(\frac{1}{2}\right)\left(\frac{\pi}{2}\right) - \left(\frac{1}{4}\right)\sin u\Big|_0^{\pi} = \frac{\pi}{4} \qquad \blacksquare$$

A JUSTIFICATION OF THE SUBSTITUTION METHOD

A justification for the substitution method is based on the chain rule, which states that

$$f'(g(x))g'(x) = \frac{d}{dx} f(g(x))$$

This formula provides us with a rule for integration, namely,

$$\int f'(g(x))g'(x)\, dx = f(g(x)) + C \qquad C \text{ a constant}$$

If we let $u = g(x)$ in the above, and note that $du = g'(x)\, dx$, we have

$$\int f'(g(x))g'(x)\, dx = \int f'(u)\, du = f(u) + C$$

In other words, if we replace $g(x)$ by u and $g'(x)\, dx$ by du in $\int f'(g(x))g'(x)\, dx$, we obtain an easily evaluated integral, $\int f'(u)\, du$. This replacement of $g(x)$ by u is called the *substitution method* since we substitute u for $g(x)$. Sometimes this technique is referred to as the *change of variable method* since the variable u has replaced the variable x.

EXERCISE 6

In Problems 1–46 evaluate each indefinite integral.

1. $\int 6\, dx$,

2. $\int 3\, dx$

3. $\int 3x\, dx$

4. $\int 6x^2\, dx$

5. $\int t^{-4}\, dt$

6. $\int u^{-1/2}\, du$

7. $\int (x^2 + 2)\, dx$

8. $\int (3x^3 - 1)\, dx$

9. $\int (x^7 + 1)\, dx$

10. $\int (4\sqrt{x} + 1)\, dx$

11. $\int (4x^3 - 3x^2 + 5x - 2)\, dx$

12. $\int (3x^5 - 2x^4 - x^2 - 1)\, dx$

13. $\int (3\sqrt{z} + z) \, dz$

14. $\int (4t^{3/2} + t^{1/2}) \, dt$

15. $\int \left(x - \dfrac{1}{x^2} \right) dx$

16. $\int \left(3x^2 - \dfrac{1}{\sqrt{x}} \right) dx$

17. $\int u(u - 1) \, du$

18. $\int t^2(t + 1) \, dt$

19. $\int \dfrac{3x^5 + 1}{x^2} \, dx$

20. $\int \dfrac{x^2 + 2x + 1}{x^4} \, dx$

21. $\int \dfrac{t^2 - 4}{t - 2} \, dt$

22. $\int \dfrac{z^3 - 8}{z - 2} \, dz$

23. $\int (2x + 1)^5 \, dx$

24. $\int (1 - 3x)^3 \, dx$

25. $\int \sqrt{x^2 - 9} \, x \, dx$

26. $\int (1 - t^2)^6 t \, dt$

27. $\int \dfrac{x}{\sqrt{1 + x^2}} \, dx$

28. $\int \dfrac{x}{\sqrt[5]{1 - 3x^2}} \, dx$

29. $\int x\sqrt{x + 3} \, dx$

30. $\int x\sqrt{4 - x} \, dx$

31. $\int \sin 3x \, dx$

32. $\int (1 - \cos 4x) \, dx$

33. $\int x \sin x^2 \, dx$

34. $\int \sin x \cos^2 x \, dx$

35. $\int x^2\sqrt{x + 1} \, dx$

36. $\int x^2\sqrt{3x - 1} \, dx$

37. $\int \dfrac{x}{\sqrt{1 + x}} \, dx$

38. $\int \dfrac{x}{\sqrt{x + 2}} \, dx$

39. $\int (s - 5)^{1/2} \, ds$

40. $\int \dfrac{2x + 1}{(x^2 + x - 5)^2} \, dx$

41. $\int \dfrac{1}{\sqrt{x}(1 + \sqrt{x})^4} \, dx$

42. $\int \dfrac{x + 4x^3}{\sqrt{x}} \, dx$

43. $\int (x - 5)\sqrt{x + 1} \, dx$

44. $\int \dfrac{x^3}{\sqrt{2x + 1}} \, dx$

45. $\int \sqrt{t}\sqrt{4 + t\sqrt{t}} \, dt$

46. $\int \dfrac{z \, dz}{z + \sqrt{z^2 + 4}}$

In Problems 47–58 evaluate each definite integral.

47. $\int_{-2}^{0} \dfrac{x}{(x^2 + 3)^2} \, dx$

48. $\int_{-1}^{1} (s^2 - 1)^5 s \, ds$

49. $\int_{-1/3}^{0} x\sqrt[3]{3x + 1} \, dx$

50. $\int_{0}^{26} x\sqrt[3]{x + 1} \, dx$

51. $\int_{6}^{1} x\sqrt{x + 3} \, dx$

52. $\int_{6}^{2} x^2\sqrt{x - 2} \, dx$

53. $\int_{2}^{6} \dfrac{4}{\sqrt{3x - 2}} \, dx$

54. $\int_{0}^{4} \dfrac{x}{\sqrt{x^2 + 9}} \, dx$

55. $\int_{-1}^{7} x(1 + x)^{1/3} \, dx$

56. $\int_{1}^{3} \dfrac{1}{x^2}\sqrt{1 - \dfrac{1}{x}} \, dx$

57. $\int_{0}^{\pi/3} \sin \dfrac{x}{2} \, dx$

58. $\int_{0}^{\pi/6} \cos 2x \, dx$

59. Find the area under the graph of $f(x) = \sin x$ from 0 to $\pi/2$.

60. Find the area under the graph of $f(x) = \cos x$ from $-\pi/2$ to $\pi/2$.

61. Find the area under the graph of $f(x) = \sqrt{x + 4}$ from 0 to 5.

62. Find the area under the graph of $f(x) = x^2/\sqrt{2x + 1}$ from 0 to 4.

63. Find the area under the graph of $f(x) = \sqrt{x-1}$ from 1 to 2.

64. Find the area under the graph of $f(x) = x/(x^2 + 1)^2$ from 0 to 2.

In Problems 65 and 66 evaluate each indefinite integral both by (a) using the substitution method and (b) expanding the integrand. Compare the two results. How do the constants of integration differ?

65. $\int (x + 1)^2 \, dx$

66. $\int (x^2 + 1)^2 x \, dx$

67. Verify that $\int x\sqrt{x} \, dx \neq (\int x \, dx)(\int \sqrt{x} \, dx)$.

68. Verify that $\int x(x^2 + 1) \, dx \neq x \int (x^2 + 1) \, dx$.

69. Verify that $\int \dfrac{x^2 - 1}{x - 1} \, dx \neq \dfrac{\int (x^2 - 1) \, dx}{\int (x - 1) \, dx}$.

70. The electric field strength a distance z out along the axis of a ring of radius R carrying a charge Q is given by the formula

$$E(z) = \frac{Qz}{(R^2 + z^2)^{3/2}}$$

If the electric potential V is related to E by $E = -dV/dz$, what is $V(z)$?

71. For the function given below find $\int_{-1}^{1} f(x) \, dx$.

$$f(x) = \begin{cases} x + 1 & \text{if } x < 0 \\ \cos \pi x & \text{if } x \geq 0 \end{cases}$$

72. If f is continuous on $[a, b]$, show that $\int_a^b f(x) \, dx = \int_a^b f(a + b - x) \, dx$.

73. Use an appropriate substitution to show that $\int_0^{\pi/2} \sin^n \theta \, d\theta = \int_0^{\pi/2} \cos^n \phi \, d\phi$.

74. Use an appropriate substitution to show that $\int_0^1 x^m (1 - x)^n \, dx = \int_0^1 x^n (1 - x)^m \, dx$.

75. If $\int_0^2 f(x - 3) \, dx = 8$, find $\int_{-3}^{-1} f(x) \, dx$.

76. If f is a continuous function defined on the interval $[0, 1]$, show that

$$\int_0^\pi x f(\sin x) \, dx = \frac{\pi}{2} \int_0^\pi f(\sin x) \, dx$$

77. If f is an even function, show that $\int_{-a}^{a} f(x) \, dx = 2 \int_0^a f(x) \, dx$.

78. If f is an odd function, show that $\int_{-a}^{a} f(x) \, dx = 0$.

Historical Perspectives

The fundamental idea of integral calculus can be traced back over 2000 years to the Greek method of "exhaustion," which is described in Euclid's *Tenth Book*. The method is illustrated in Problems 6–8 at the end of this section. The great Greek mathematician Archimedes (287–212 BC) made extensive use of this method to establish many theorems on areas and volumes, but the method could not be fully developed until the discovery of algebra and its relationship to geometry, which was not to come until several hundred years later in the late sixteenth and early seventeenth centuries.

One of the first men to rediscover and exploit the method of exhaustion was the Italian mathematician Bonaventura Cavalieri (1598–1647), who developed an *infinitesimal* method (characterized by the use of infinitesimal, or infinitely small, length, area, and volume units) for finding areas and volumes (see Problems 4 and 5). At about the same time, many others, including the German astronomer Johannes Kepler (1571–1630), the French mathematicians Fermat and Roberval, and the Italian mathematician and astronomer Evangelista Torricelli (1608–1647), were developing their own infinitesimal methods for finding areas and volumes. All of these methods built upon the Greek geometric method of exhaustion by adding analytic or algebraic features to the analysis.

However, old prejudices die hard and not every mathematician of the seventeenth century was willing to abandon the classical Greek methods and embrace the new algebraic methods. An important example is provided by Newton's teacher, Sir Isaac Barrow (1630–1677). Barrow vigorously opposed the new approach and pushed the old geometric methods further than anyone before him. In fact, he discovered the fundamental theorem of calculus in geometric form, and if he had been more receptive to the new methods, he might well have gained recognition as the founder of calculus. But, in 1669, he gave up his notes and his position as professor at Cambridge to Newton in order to devote his time to the study of theology, leaving that distinction to his successor.

HISTORICAL EXERCISES

The problem of finding the area of a plane region was suggested to Cavalieri by his teacher, Galileo. In his work on falling objects, Galileo had recognized that the distance an object fell could be interpreted as the area under the graph of its velocity function. In Problems 1–3 we explore in some simple cases the relationship between velocity and area. (Keep in mind the intimate relationship between velocity and the derivative.) Note that Problems 2 and 3 provide a kind of "proof" of the fundamental theorem for the special case $\int_0^{t_0} kt\, dt$.

1. Suppose an object moves in a straight line with constant velocity (speed) of 4 feet per second.
   (a) How far does the object travel in 5 seconds? Graph the velocity function v over the interval $0 \le t \le 5$ and interpret the distance traveled as an area.
   (b) Shade the area that represents the distance traveled during the third second.

2. Suppose an object that was originally at rest is moving in a straight line and is accelerating uniformly (a falling object, for example), so that its velocity is being increased by a constant amount each second; that is, the velocity function is of the form $v(t) = kt$.
   (a) Graph the particular velocity function $v(t) = 3t$ over the interval $0 \le t \le 4$. What is the velocity at the start of the interval? At the end? At the midpoint? What is the acceleration, that is, how much does the velocity increase each second?
   (b) It can be shown that under the assumption of a uniform rate of acceleration, the average velocity over any interval equals the velocity at the midpoint of the interval. That is, the object will end up at the same point it would have if it had had a constant velocity over the entire interval equal to its velocity at the midpoint. Verify this for the velocity function in part (a).
   (c) Graph the velocity function $v_1 = 6t$ over $0 \le t \le 4$. On the same axes graph the constant velocity function v_2, which represents the average velocity over this interval. Calculate the areas under the two velocity graphs. How far did the object travel during the interval?

3. Consider the accompanying figure.
   (a) Verify that (no matter what k and t_0 are) the area of tri-
       angle ABE equals the area of $ABCD$, by directly calcu-
       lating each area.
   (b) Verify the equality in part (a) by integrating to get the
       area of the triangle.

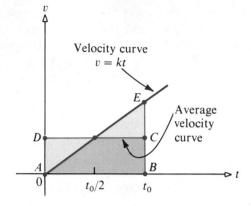

Cavalieri studied under Galileo and became a professor at Bologna in 1629, a position he held until his death.
Credit for the first general theorem of calculus goes to Cavalieri. He derived (in our notation) the theorem

$$\int_0^a x^n \, dx = \frac{a^{n+1}}{n+1}$$

for positive integers n, a result that had been demonstrated for some special cases as far back as Archimedes.
Although his theories were not complete or rigorously defensible, in 1635 Cavalieri published the first textbook
in what we would now call *integral calculus*.

4. In this exercise we give a simple application of Cavalieri's method
   of *indivisibles* to "prove" a theorem from elementary geometry. Consider the parallelogram shown here. The theorem from geometry is:
   Area of $\triangle ABC$ = Area of $\triangle ADC$. Cavalieri would have considered
   $\triangle ABC$ to be generated by the infinite set of lines (like OP) parallel

   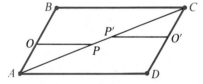

   to AD that we get by letting point O move from A to B. He would have called these lines the *indivisibles* that
   make up the triangle. State an argument to "prove" the theorem after considering the diagram and thinking
   about what can be said about OP and $O'P'$ when $AO = CO'$. (Prove your conclusion concerning OP and
   $O'P'$.)

5. A well-known theorem of elementary geometry is due
   to Cavalieri and is often called *Cavalieri's principle*.
   Referring to the figure, the theorem asserts that if
   $f(x) - g(x)$ and $h(x) - k(x)$ are in a constant proportion, say, $f(x) - g(x) = M[h(x) - k(x)]$, for all x
   between a and b, then Area A = M(Area B). Cavalieri's proof of this theorem occupies several pages in
   his textbook on his method of indivisibles. Give a one-
   line proof using integration.

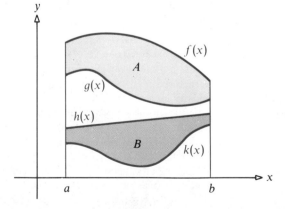

To illustrate the Greek method of exhaustion, we sketch the proof of a theorem in geometry by using that method. Problems 6–8 complete the proof and provide a proof of a corollary to the theorem. The method of exhaustion was usually used in giving indirect proofs, which managed to avoid any need for an argument involving infinite processes. This kind of indirect proof, in which we argue to a contradiction from an assumption and hence conclude that the assumption is false, is called *reductio ad absurdum.*

THEOREM   The areas of two circles, A_1 and A_2, are in the same proportion as the squares of their diameters, d_1^2 and d_2^2. That is, $A_1/A_2 = d_1^2/d_2^2$.

Proof   Consider the following figures:

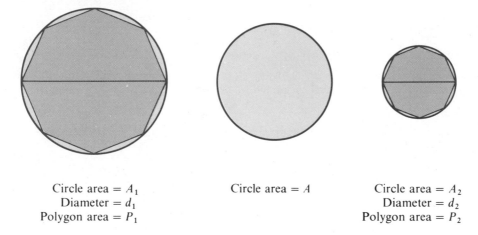

Circle area $= A_1$
Diameter $= d_1$
Polygon area $= P_1$

Circle area $= A$

Circle area $= A_2$
Diameter $= d_2$
Polygon area $= P_2$

Let us assume that $A_1/A_2 \neq d_1^2/d_2^2$.   We will show that this assumption leads to a contradiction. Now, consider a circle, different from the first circle, with area A so that $A/A_2 = d_1^2/d_2^2$,   and suppose $A < A_1$ (a similar argument will work if we suppose $A_1 < A$).   Then we inscribe in the circle with area A_1, a polygon with area P_1, so that $A < P_1 < A_1$.   We can do this because we can make P_1 as close to A_1 as we wish by doubling the number of sides of the polygon as many times as needed (this is the core idea of the method of exhaustion—notice the lack of the need for an infinite process here). Now, if we inscribe a similar polygon with area P_2 inside the circle with area A_2, we can show that $P_1/P_2 = d_1^2/d_2^2$ (see Problem 7). But $d_1^2/d_2^2 = A/A_2$ and so $P_1/P_2 = A/A_2$. However, $P_1 > A$,   so what can we conclude about the relative sizes of P_2 and A_2 (see Problem 6)?   ■

6. Draw the appropriate conclusion about the relative sizes of P_2 and A_2. Then look again at the figure and state the resulting contradiction that completes the proof.

7. (a)   Consider two *similar rectangles*, as shown here.
      Show that

$$\frac{\text{Area } R_1}{\text{Area } R_2} = \frac{a^2}{A^2} = \frac{b^2}{B^2}$$

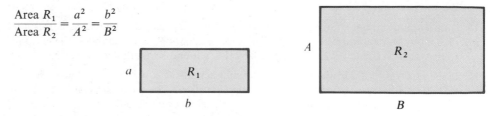

(b) Consider two *similar triangles*, as shown here. Show that

$$\frac{\text{Area } T_1}{\text{Area } T_2} = \frac{b^2}{B^2} = \frac{h^2}{H^2}$$

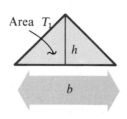

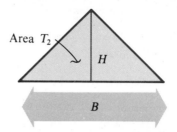

(c) Use part (b) to argue that $P_1/P_2 = d_1^2/d_2^2$ in the above proof.

8. Use the theorem proved above to prove that for any circle, the area equals a constant C times the radius squared, and that C is the same constant for all circles. (Of course, we usually call C by the name π.)

Miscellaneous Exercises

In Problems 1–6 evaluate each integral.

1. $\int \left[\sqrt{(z^2 + 1)^4 - 3}\,\right][(z^2 + 1)^3]z\,dz$

2. $\int_0^1 \dfrac{(z^2 + 5)(z^3 + 15z - 3)\,dz}{\sqrt{194 - (z^3 + 15z - 3)^2}}$

3. $\int \left(2\sqrt{x^2 + 3} - \dfrac{4}{x} + 9\right)^6 \left(\dfrac{x}{\sqrt{x^2 + 3}} + \dfrac{2}{x^2}\right) dx$

4. $\int \dfrac{x^3\,dx}{\sqrt{x^2 - 9}}$

5. $\int_2^{17} \dfrac{dx}{\sqrt{\sqrt{x - 1} + (x - 1)^{5/4}}}$

6. $\int \dfrac{y\,dy}{(y - 2)^3}$

7. If $\int_1^b t^2(5t^3 - 1)^{1/2}\,dt = \frac{38}{45}$,   find b.

8. If $\int_a^3 t\sqrt{9 - t^2}\,dt = 6$,   find a.

9. Find $\int_{\pi^2/4}^{4\pi^2} \dfrac{1}{\sqrt{x}} \sin \sqrt{x}\,dx$.

10. Find $\int_0^1 \dfrac{1}{\sqrt{x}} \cos \dfrac{\pi\sqrt{x}}{2}\,dx$.

11. Find $\int_1^{\sqrt[3]{1/2}} \dfrac{[1 - (1/t^3)]^3}{t^4}\,dt$.

12. The evaluation of $\int_0^2 \sqrt{4 - x^2}\,dx$ by Riemann sums or by the fundamental theorem is difficult. However, this integral may be interpreted as a certain area, and evaluated by elementary geometry. What is the result?

13. Find an approximate value of $\int_1^2 dx/x$ by computing the Riemann sums corresponding to a partition of $[1, 2]$ into four subintervals of equal length and evaluating the integrand at the midpoint of each subinterval. Compare with the true value, which is $0.6931 \ldots$ (found by methods not yet discussed).

14. If the interval $[1, 5]$ is divided into eight subintervals of equal length, what is the largest Riemann sum of $f(x) = x^2$ that can be computed using this partition? The smallest? Compute the average of these sums. What integral has been approximated and what is its exact value?

15. Suppose that P is a partition of $[0, \pi/2]$ into n subintervals and u_i is an arbitrary point of the typical subinterval $[x_{i-1}, x_i]$, $i = 1, 2, \ldots, n$. Explain why

$$\lim_{\|P\| \to 0} \sum_{i=1}^{n} (\cos u_i)\Delta x_i = 1$$

16. The interval $[0, 4]$ is divided into n subintervals of equal length Δx and a point u_i is chosen in the typical subinterval $[x_{i-1}, x_i]$, $i = 1, 2, \ldots, n$. What is the limit of $\sum_{i=1}^{n} \sqrt{u_i}\, \Delta x$ as $n \to +\infty$?

17. Find k, $0 \le k \le 3$, so that $\displaystyle\int_0^3 \frac{x}{\sqrt{x^2 + 16}}\, dx = \frac{3k}{\sqrt{k^2 + 16}}$.

18. For all real b find $\int_0^b |2x|\, dx$.

19. If n is a nonnegative integer, for what n does $\int_0^1 x^n\, dx = \int_0^1 (1 - x)^n\, dx$?

20. Given the function f defined for all real numbers x by $f(x) = 2|x - 1|x^2$:
    (a) What is the range of the function? (b) For what x is the function continuous?
    (c) For what x is the derivative of f continuous? (d) Determine the value of $\int_0^1 f(x)\, dx$.

21. Use integration to find the area enclosed by the graph of $y = 3 - |x|$ and the x-axis. Check by elementary geometry.

22. Evaluate: $\displaystyle\int_{-2}^{3} (x + |x|)\, dx$ 23. Evaluate: $\displaystyle\int_0^3 |x - 1|\, dx$

24. If F is a function whose derivative is continuous for all real x, find

$$\lim_{h \to 0} \frac{1}{h} \int_c^{c+h} F'(x)\, dx$$

25. Let $f(x) = k \sin(kx)$, where k is a positive constant.
    (a) Find the area of the region enclosed by one arch of the graph of f and the x-axis.
    (b) Find the area of the triangle formed by the x-axis and the tangents to one arch of f at the points where the graph of f crosses the x-axis.

26. If $f(x) = \int_0^x 1/\sqrt{t^3 + 2}\, dt$, which of the following is *false*?
    (a) $f(0) = 0$ (b) f is continuous at x for all $x \ge 0$
    (c) $f(1) > 0$ (d) $f'(1) = 1/\sqrt{3}$ (e) $f(-1) > 0$

27. Find $(d/dx) \int_x^1 (t - 1)^2\, dt$ without integrating. Then check by integrating before differentiating.

28. Find $f''(x)$ if $f(x) = \int_0^x \sqrt{1 - t^2}\, dt$.

29. Suppose that $F(x) = \int_0^x \sqrt{t}\, dt$ and $G(x) = \int_1^x \sqrt{t}\, dt$. Explain why $F(x) - G(x)$ is constant and find the constant.

30. For each $x > 0$ let $F(x) = \int_1^x dt/t$ and $G(x) = \int_1^{ax} dt/t$, where a is a positive constant.
    (a) Show that $F'(x) = G'(x)$ and use the result to obtain $F(ax) = F(a) + F(x)$.
    (b) Recall from your precalculus study that the logarithm of a product is the sum of the logarithms of the factors. Do you see any resemblance to that idea in part (a)? (This suggests that F is a logarithmic function of some kind; see Chap. 7.)

31. Given $y = \sqrt{x^2 - 1}(4 - x)$, $1 \le x \le a$, for what number a will the area under the graph have a maximum value?

32. If $\int_1^2 f(x - c)\, dx = 5$, where c is a constant, find $\int_{1-c}^{2-c} f(x)\, dx$.

33. If the substitution $\sqrt{x} = \sin y$ is made in the integrand of $\int_0^{1/2} (\sqrt{x}/\sqrt{1 - x})\, dx$, what is the resulting integral? Do not integrate.

34. Suppose that the graph of $y = f(x)$ contains the points $(0, 1)$ and $(2, 5)$. Find $\int_0^2 f'(x)\,dx$. (Assume that f' is continuous.)

35. If f is continuous for all x, which of the following integrals necessarily have the same value?

I. $\int_a^b f(x)\,dx$ **II.** $\int_0^{b-a} f(x + a)\,dx$ **III.** $\int_{a+c}^{b+c} f(x + c)\,dx$

36. Let $a < c < b$ and let f be differentiable on $[a, b]$. Which of the following is *not* necessarily true?

(a) $\int_a^b f(x)\,dx = \int_a^c f(x)\,dx + \int_c^b f(x)\,dx$

(b) There exists d in $[a, b]$ such that $f'(d) = \dfrac{f(b) - f(a)}{b - a}$.

(c) $\int_a^b f(x)\,dx \geq 0$

(d) $\lim_{x \to c} f(x) = f(c)$

(e) If k is a real number, then $\int_a^b kf(x)\,dx = k\int_a^b f(x)\,dx$.

37. Find the area under the graph of $y = 1/\sqrt{x}$ from $x = 1$ to $x = r$ (where $r > 1$). Then examine the behavior of this area as $r \to +\infty$.

38. Find the area under the graph of $y = 1/x^2$ from $x = 1$ to $x = r$ (where $r > 1$). Then examine the behavior of this area as $r \to +\infty$.

In Problems 39 and 40 use (5.27) to find lower and upper estimates for each integral.

39. $\int_{-4}^{-1} (2x^3 + 9x^2 + 12x + 32)\,dx$ **40.** $\int_0^3 x^2(x^2 - 1)^{1/3}\,dx$

41. If n is a known positive integer, for what number c is $\int_1^c x^{n-1}\,dx = 1/n$?

42. Approximate $\int_{-1}^3 (x^3 - 3x^2 + 3)\,dx$ using a regular partition with four subintervals. Choose u_i as the number at which $(x^3 - 3x^2 + 3)$ assumes its minimum value in the ith subinterval.

43. Use the definition of average value of a function from page 301 to find the *average slope* of the graph of $y = f(x)$, $a \leq x \leq b$. (Assume that f' is continuous.) What is the geometric interpretation?

44. What theorem guarantees that the average slope found in Problem 43 is equal to $f'(u)$ for some u in $[a, b]$? What *different* theorem guarantees the same thing? (The connection between these theorems should now be apparent!)

45. The formula $(d/dx)\int f(x)\,dx = f(x)$ says that if a function is integrated and the result is differentiated, the original function is returned. What about the other way around? Is the formula $\int f'(x)\,dx = f(x)$ correct?

46. Prove that if f is continuous in $[a, b]$ and $\int_a^b f(x)\,dx = 0$, there is at least one number u in $[a, b]$ such that $f(u) = 0$.

47. Give a counterexample to the statement in Problem 46 if f is not required to be continuous.

48. Prove that if f is continuous and even in $[-a, a]$, then $\int_{-a}^a f(x)\,dx = 2\int_0^a f(x)\,dx$.

49. Prove that if f is continuous and odd in $[-a, a]$, then $\int_{-a}^a f(x)\,dx = 0$.

50. Suppose that F is an antiderivative of f in $[a, b]$ and partition $[a, b]$ into n subintervals.
   (a) Apply the mean value theorem for derivatives to F in each subinterval $[x_{i-1}, x_i]$ to show that there is a point u_i in the subinterval such that $F(x_i) - F(x_{i-1}) = f(u_i)\Delta x_i$.
   (b) Show that $\sum_{i=1}^n [F(x_i) - F(x_{i-1})] = F(b) - F(a)$.
   (c) Use parts (a) and (b) to explain why $\int_a^b f(x)\,dx = F(b) - F(a)$.

51. Give reasons for the steps in the following argument, which yields an estimate for the error in approximating an integral by a Riemann sum: Let f' be continuous on the closed interval $[a, b]$ and let M be the maximum of $|f'(x)|$ on $[a, b]$. Also, let P be a regular partition of $[a, b]$ and let $R = \sum_{i=1}^{n} f(u_i)\Delta x_i$ be a Riemann sum. Then,

$$R - \int_a^b f(x)\,dx = R - \sum_{i=1}^{n} \int_{x_{i-1}}^{x_i} f(x)\,dx \qquad \text{(a) Why?}$$

$$= R - \sum_{i=1}^{n} f(t_i)\Delta x_i \qquad \text{(b) Why?}$$

$$= \sum_{i=1}^{n} f'(c_i)(u_i - t_i)\Delta x_i \qquad \text{(c) Why?}$$

Therefore,

$$\left| R - \int_a^b f(x)\,dx \right| \le \sum_{i=1}^{n} |f'(c_i)|\,|u_i - t_i|\Delta x_i \qquad \text{(d) Why?}$$

$$\le M \sum_{i=1}^{n} (\Delta x_i)^2 \qquad \text{(e) Why?}$$

$$= Mn\left(\frac{b-a}{n}\right)^2 \qquad \text{(f) Why?}$$

$$= \frac{M(b-a)^2}{n}$$

52. Let $f(x) = 2x$ on the closed interval $[1, 3]$. Let P be the regular partition consisting of 10 subintervals. Let u_i be the left endpoint of the ith interval. Compare the actual error, $\left| \sum_{i=1}^{n} f(u_i)\Delta x_i - \int_a^b f(x)\,dx \right|$, with the estimate given in Problem 51.

53. Repeat Problem 52 with $f(x) = x^2$ on the closed interval $[0, 1]$.

6

Applications of the Integral

In the applications that follow, we rely on two basic facts from Chapter 5. The first is that for a continuous function f on a closed interval $[a, b]$, the limit of the Riemann sum—the definite integral—exists. That is,

(6.1)
$$\lim_{\|P\| \to 0} \sum_{i=1}^{n} f(u_i)\Delta x_i = \int_a^b f(x)\, dx = \text{Some number}$$

The second basic fact is that if F is an antiderivative of a continuous function f defined on $[a, b]$, then, by the fundamental theorem of calculus,

$$\int_a^b f(x)\, dx = F(b) - F(a)$$

When $f(x) \geq 0$ on $[a, b]$, we can use this result to calculate the area under the graph of $y = f(x)$ from a to b. In this chapter we extend this result to calculate the area enclosed by the graphs of two or more functions. Furthermore, we show how formula (6.1) can be applied to find the volume of a solid, to compute work and fluid pressure, to find the length of a graph, and to calculate average values. For each of these applications, we can approximate the required quantity by a Riemann sum and then use (6.1).

1. Area

Suppose we desire to calculate the area enclosed by the graphs of $y = f(x)$ and $y = g(x)$ and the lines $x = a$ and $x = b$. As indicated in Figure 1, we assume that f and g are continuous on $[a, b]$ and that $f(x) \geq g(x)$ for all x in $[a, b]$.

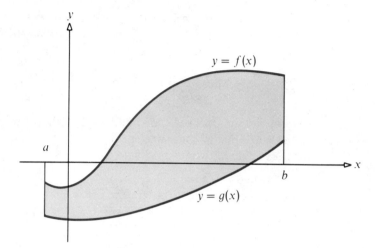

Figure 1

We begin by partitioning the interval $[a, b]$ into n subintervals,

$$[a, x_1], \quad [x_1, x_2], \quad \ldots, \quad [x_{i-1}, x_i], \quad \ldots, \quad [x_{n-1}, b]$$

We denote the length of the ith subinterval by $\Delta x_i = x_i - x_{i-1}$. For each i, $1 \leq i \leq n$, we select a number u_i in the ith subinterval $[x_{i-1}, x_i]$ and construct n rectangles, each of width Δx_i and height $f(u_i) - g(u_i)$. The area of each rectangle is then $[f(u_i) - g(u_i)]\Delta x_i$ (see Fig. 2). The sum of the areas of these n rectangles,

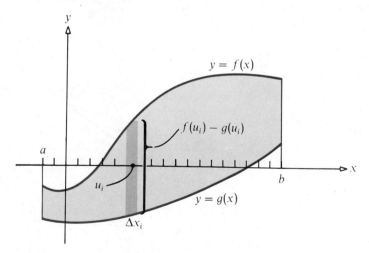

Figure 2

$\sum_{i=1}^{n} [f(u_i) - g(u_i)]\Delta x_i$, gives an approximation to the area we seek. As the length of each subinterval gets smaller and smaller, that is, as the norm $\|P\|$ of the partition approaches 0, the sum $\sum_{i=1}^{n} [f(u_i) - g(u_i)]\Delta x_i$ is a better and better approximation to the area we seek. However, this sum is a Riemann sum, so, as $\|P\| \to 0$, its limit is a definite integral. This suggests the following definition:

(6.2)   *Area.* **The area A enclosed by the graphs of $y = f(x)$ and $y = g(x)$ and the lines $x = a$ and $x = b$, where f and g are continuous on $[a, b]$ and $f(x) \geq g(x)$ on $[a, b]$, is**

$$A = \lim_{\|P\| \to 0} \sum_{i=1}^{n} [f(u_i) - g(u_i)]\Delta x_i = \int_a^b [f(x) - g(x)]\, dx$$

An interesting aspect of the above result is that it works whether the graphs lie above the x-axis, below the x-axis, or partially above and partially below the x-axis. The next example illustrates this fact.

EXAMPLE 1   Find the area enclosed by the graphs of $y = f(x) = 10x - x^2$ and $y = g(x) = 3x - 8$.

Solution   First we graph the two functions (see Fig. 3). Before we can compute the area, we need to locate the points of intersection of the two graphs. Thus, we need to solve

$$f(x) = g(x)$$
$$10x - x^2 = 3x - 8$$
$$x^2 - 7x - 8 = 0 \qquad \text{so that} \qquad x = -1, \quad x = 8$$

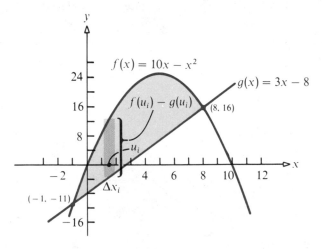

Figure 3

From Figure 3, we note $f(x) \geq g(x)$ on $[-1, 8]$ so that the area we seek is given by

$$\lim_{||P|| \to 0} \sum_{i=1}^{n} [f(u_i) - g(u_i)] \Delta x_i = \int_{-1}^{8} [f(x) - g(x)] \, dx = \int_{-1}^{8} [(10x - x^2) - (3x - 8)] \, dx$$

$$= \int_{-1}^{8} (-x^2 + 7x + 8) \, dx = \left(\frac{-x^3}{3} + \frac{7x^2}{2} + 8x \right) \Big|_{-1}^{8}$$

$$= \left(\frac{-512}{3} + 224 + 64 \right) - \left(\frac{1}{3} + \frac{7}{2} - 8 \right)$$

$$= -171 + 296 - \frac{7}{2} = \frac{243}{2} \qquad \blacksquare$$

The application of (6.2) requires that one graph lies above the other graph on $[a, b]$. The next example illustrates how to proceed when this is not the case.

EXAMPLE 2   Find the area enclosed by the graphs of $f(x) = x^3$ and $g(x) = x$.

Solution   First we graph the two functions (see Fig. 4). The points of intersection obey

$$x^3 = x$$
$$x^3 - x = 0$$
$$x(x^2 - 1) = 0$$

so they occur at

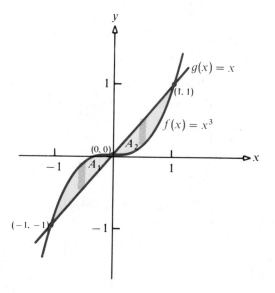

Figure 4

$$x = -1 \qquad x = 0 \qquad \text{and} \qquad x = 1$$

On $[-1, 0]$, $f(x) \geq g(x)$, while on $[0, 1]$, we have $g(x) \geq f(x)$. Thus, the area A_1 enclosed by the graphs of f and g from $x = -1$ to $x = 0$ is

$$A_1 = \int_{-1}^{0} [f(x) - g(x)] \, dx$$

$$= \int_{-1}^{0} (x^3 - x) \, dx$$

$$= \left(\frac{x^4}{4} - \frac{x^2}{2} \right) \Big|_{-1}^{0}$$

$$= 0 - \left(\frac{1}{4} - \frac{1}{2} \right) = \frac{1}{4}$$

The area A_2 enclosed by the graphs of f and g from $x = 0$ to $x = 1$ is

$$A_2 = \int_0^1 [g(x) - f(x)]\, dx = \int_0^1 (x - x^3)\, dx$$

$$= \left(\frac{x^2}{2} - \frac{x^4}{4} \right)\Big|_0^1 = \left(\frac{1}{2} - \frac{1}{4} \right) - 0 = \frac{1}{4}$$

The area we seek is therefore $A_1 + A_2 = \frac{1}{2}$. ∎

The result of Example 2 can be more easily obtained by using symmetry. By examining Figure 4, we can see that $A_1 = A_2$. Thus,

$$A = A_1 + A_2 = 2A_2 = 2 \int_0^1 (x - x^3)\, dx = 2\left(\frac{1}{4} \right) = \frac{1}{2}$$

The importance of graphing when doing an area problem cannot be over-emphasized. For the functions in Example 2, a thoughtless application of (6.2) would give

$$\int_{-1}^1 (x^3 - x)\, dx = \left(\frac{x^4}{4} - \frac{x^2}{2} \right)\Big|_{-1}^1 = \left(\frac{1}{4} - \frac{1}{2} \right) - \left(\frac{1}{4} - \frac{1}{2} \right) = 0$$

which is a ridiculous answer. A graph will prevent this kind of mistake.

Finding the area A enclosed by the graph of $y = g(x)$, the x-axis, and the lines $x = a$ and $x = b$, when $g(x) \leq 0$ on $[a, b]$, is the same as finding the area enclosed by the graphs of $y = f(x) = 0$ (the x-axis) and $y = -g(x)$ and the lines $x = a$ and $x = b$ (see Fig. 5). Then

$$A = \int_a^b [0 - g(x)]\, dx = \int_a^b -g(x)\, dx = -\int_a^b g(x)\, dx$$

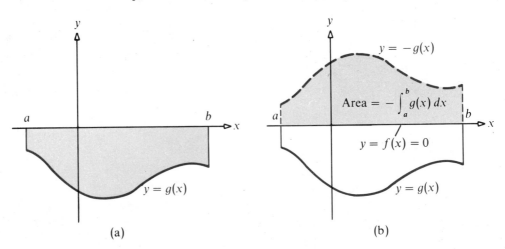

(a)   (b)

Figure 5

We use this fact in the next example.

EXAMPLE 3   Find the area enclosed by the graph of $f(x) = x^2 - 4$, the x-axis, and the lines $x = 0$ and $x = 4$.

Solution   On the interval $[0, 4]$, the graph crosses the x-axis at $x = 2$. As we can see from Figure 6, $f(x) \leq 0$ from $x = 0$ to $x = 2$ and $f(x) \geq 0$ from $x = 2$ to $x = 4$. Thus, the area A_1 enclosed by the graph of $f(x) = x^2 - 4$ and the x-axis from $x = 0$ to $x = 2$ is

$$A_1 = -\int_0^2 (x^2 - 4)\, dx = -\left(\frac{x^3}{3} - 4x\right)\Big|_0^2$$

$$= -\left(\frac{8}{3} - 8\right) = \frac{16}{3}$$

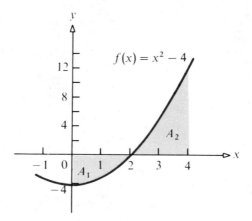

Figure 6

The area A_2 enclosed by the graph of $f(x) = x^2 - 4$ and the x-axis from $x = 2$ to $x = 4$ is

$$A_2 = \int_2^4 (x^2 - 4)\, dx = \left(\frac{x^3}{3} - 4x\right)\Big|_2^4$$

$$= \left(\frac{64}{3} - 16\right) - \left(\frac{8}{3} - 8\right) = \frac{56}{3} - 8 = \frac{32}{3}$$

The total area is therefore

$$A = A_1 + A_2 = \frac{16}{3} + \frac{32}{3} = \frac{48}{3} = 16 \quad \blacksquare$$

Sometimes it is better to partition the y-axis instead of the x-axis. The next example illustrates this.

EXAMPLE 4   Find the area enclosed by the graphs of $y = f(x) = \sqrt{4 - 4x}$, $y = g(x) = \sqrt{4 - x}$, and the x-axis.

Solution   First, we shall solve the problem by partitioning the x-axis, and then we will solve it by partitioning the y-axis.

Partition of x-axis: Refer to Figure 7. There, we see that the area we seek is the sum of two areas:

$$A_1 = \text{Area enclosed by} \quad y = \sqrt{4 - x}, \quad y = \sqrt{4 - 4x}, \quad \text{and} \quad x = 1$$
$$A_2 = \text{Area enclosed by} \quad y = \sqrt{4 - x}, \quad \text{the } x\text{-axis, and} \quad x = 1$$

The area A we seek is therefore,

$$A = A_1 + A_2 = \int_0^1 (\sqrt{4 - x} - \sqrt{4 - 4x})\, dx + \int_1^4 \sqrt{4 - x}\, dx$$

$$= \int_0^1 \sqrt{4 - x}\, dx - \int_0^1 \sqrt{4 - 4x}\, dx + \int_1^4 \sqrt{4 - x}\, dx$$

$$= \int_0^4 \sqrt{4 - x}\, dx - \int_0^1 \sqrt{4 - 4x}\, dx$$

Now,

$$\int_0^4 \sqrt{4 - x}\, dx = -\int_4^0 u^{1/2}\, du = \frac{-2}{3} u^{3/2}\Big|_4^0 = \frac{-2}{3}(0 - 8) = \frac{16}{3}$$

Set $u = 4 - x$

$$\int_0^1 \sqrt{4 - 4x}\, dx = -\frac{1}{4}\int_4^0 u^{1/2}\, du = -\frac{2u^{3/2}}{4(3)}\Big|_4^0 = \frac{-1}{6}(0 - 8) = \frac{4}{3}$$

Set $u = 4 - 4x$

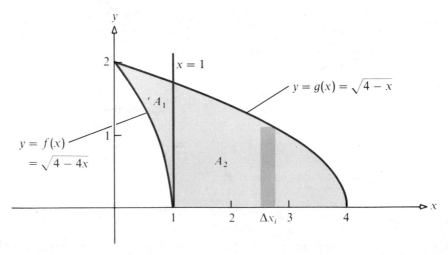

Figure 7

The required area A is

$$A = \frac{16}{3} - \frac{4}{3} = 4$$

Partition of y-axis: When we partition the y-axis, we obtain horizontal rectangles of width Δy and length $x_2 - x_1$, where x_2 is the distance from the y-axis to $y = g(x)$ and x_1 is the distance from the y-axis to $y = f(x)$ (refer to Fig. 8). The area A we seek is

$$A = \lim_{||P|| \to 0} \sum_{i=1}^{n} (x_2 - x_1)\Delta y_i = \int_0^2 \left[(4 - y^2) - \frac{1}{4}(4 - y^2) \right] dy$$

$$= \int_0^2 \frac{3}{4}(4 - y^2) \, dy = \frac{3}{4}\left(4y - \frac{y^3}{3} \right)\Big|_0^2 = \frac{3}{4}\left(8 - \frac{8}{3} \right) = 4$$

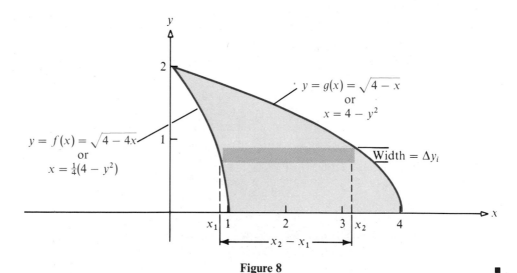

Figure 8

As the example illustrates, a partition of the y-axis is simpler. But how would you have known this in advance? Usually the configuration of the graph will make it clear whether to partition the x-axis or the y-axis. However, as is so often the case, there is no substitute for experience.

EXERCISE 1

In Problems 1–24 find the area enclosed by the graphs of the given functions and lines. Draw a sketch first.

1. $f(x) = x$, $g(x) = 2x$, $x = 0$, $x = 1$

2. $f(x) = x$, $g(x) = 3x$, $x = 0$, $x = 3$

3. $f(x) = x^2$, $g(x) = x$

4. $f(x) = x^2$, $g(x) = 4x$

5. $f(x) = x^2 + 1$, $g(x) = x + 1$

6. $f(x) = x^2 + 1$, $g(x) = 4x + 1$

7. $f(x) = \sqrt{x}, \quad g(x) = x^3$

8. $f(x) = x^2, \quad g(x) = x^3$

9. $f(x) = x^2, \quad g(x) = x^4$

10. $f(x) = \sqrt{x}, \quad g(x) = x^2$

11. $f(x) = x^2 - 4x, \quad g(x) = -x^2$

12. $f(x) = x^2 - 8x, \quad g(x) = -x^2$

13. $f(x) = 4 - x^2, \quad g(x) = x + 2$

14. $f(x) = 2 + x - x^2, \quad g(x) = -x - 1$

15. $f(x) = x^3, \quad g(x) = 4x$

16. $f(x) = x^3, \quad g(x) = 16x$

17. $y = x^2, \quad y = x, \quad y = -x$

18. $y = x^2 - 1, \quad y = x - 1, \quad y = -x - 1$

19. $y = \sqrt{9 - x}, \quad y = \sqrt{9 - 3x}, \quad x\text{-axis}$

20. $y = \sqrt{16 - 2x}, \quad y = \sqrt{16 - 4x}, \quad x\text{-axis}$

21. $y^2 = x, \quad y = x - 6$

22. $y^2 = x + 16, \quad y = -x - 4$

23. $y = \cos x, \quad y = 1, \quad x = \pi/6$

24. $y = \sin x, \quad y = 1, \quad x = 0$

In Problems 25–30 find the area by using a partition of the y-axis.

25. Area enclosed by $\quad y = \sqrt{9 - x}, \quad y = \sqrt{9 - 3x}, \quad x\text{-axis}$

26. Area enclosed by $\quad y = \sqrt{16 - 2x}, \quad y = \sqrt{16 - 4x}, \quad x\text{-axis}$

27. Area enclosed by $\quad y^2 = x, \quad x + y = 2$

28. Area enclosed by $\quad y^2 = x, \quad x + y = 6$

29. Area enclosed by $\quad y^2 = 4x, \quad 4x - 3y - 4 = 0$

30. Area enclosed by $\quad y^2 = 4x + 1, \quad x = y + 1$

31. Find the area enclosed by $\quad x^{1/2} + y^{1/2} = 1,\quad$ the x-axis, and the y-axis.

32. Find the area of the "triangle" in the first quadrant enclosed by $\quad y = \sin x, \quad y = \cos x, \quad$ and the y-axis.

33. Find c, $\quad 0 < c < 1,\quad$ so that the area under the graph of $\quad y = x^2\quad$ from 0 to c equals the area under the same graph from c to 1.

34. Find the area enclosed by $\quad x = y^2,\quad$ the y-axis, $\quad y = 1,\quad$ and $\quad y = 2\quad$ by using:
    (a)   A partition of the x-axis   (b)   A partition of the y-axis

35. Show that the shaded area in the figure is $\frac{2}{3}$ of the area of the parallelogram $ABCD$. (This illustrates a result due to Archimedes concerning sectors of parabolas.)

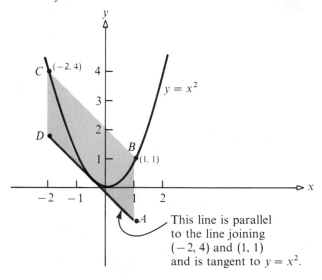

This line is parallel to the line joining $(-2, 4)$ and $(1, 1)$ and is tangent to $y = x^2$.

2. Volume of a Solid of Revolution: Disk Method

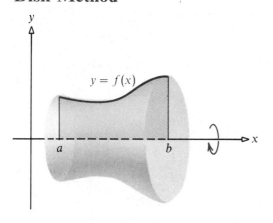

Figure 9

A solid of revolution may be generated by revolving the area under the graph of a continuous, nonnegative function $y = f(x)$ from a to b about the x-axis (see Fig. 9). A common example of a solid of revolution is a *cone*, which may be generated by revolving the area under a line that passes through the origin from 0 to h about the x-axis (see Fig. 10). Another common example of a solid of revolution is a *cylinder*, which may be generated by revolving the area under a horizontal line of positive height from a to b about the x-axis (see Fig. 11).

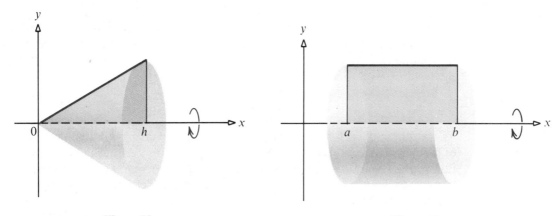

Figure 10   **Figure 11**

We seek a formula for finding the volume V of a solid of revolution. To fix our ideas, suppose the area under the graph of $y = f(x)$, $f(x) \geq 0$, from a to b is revolved about the x-axis, thus generating a solid of revolution. If we cut this solid with planes perpendicular to it and the x-axis, we find that a typical cross section is a circle and a typical slice is a *disk*. For this reason, the method developed in this section is referred to as the *disk method*.

We begin by partitioning the interval $[a, b]$ into n subintervals,

$$[a, x_1], \quad [x_1, x_2], \quad \ldots, \quad [x_{n-1}, b]$$

As before, we denote the length of the ith subinterval by $\Delta x_i = x_i - x_{i-1}$, and for each i, $1 \leq i \leq n$, we select a number u_i in the ith subinterval $[x_{i-1}, x_i]$ (see Fig.

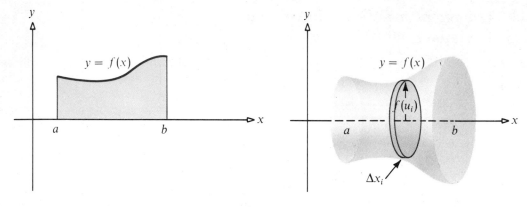

Area of circle $= \pi[f(u_i)]^2$
Width of disk $= \Delta x_i$
Volume of disk $= \pi[f(u_i)]^2\Delta x_i$

(a)   (b)

Figure 12

12). The circle obtained by cutting the solid at $x = u_i$ has radius $f(u_i)$, and its area is $\pi[f(u_i)]^2$. The solid of revolution obtained from the ith subinterval is a disk; its volume is $\pi[f(u_i)]^2\Delta x_i$. Hence, an approximation to the desired volume V of the solid of revolution is

$$V \approx \sum_{i=1}^{n} \pi[f(u_i)]^2\Delta x_i$$

As the length of each subinterval gets smaller and smaller—that is, as the norm $\|P\|$ of the partition approaches 0—the sum $\sum_{i=1}^{n} \pi[f(u_i)]^2\Delta x_i$ is a better approximation to the volume we seek. However, this sum is a Riemann sum. Hence, if f is continuous on $[a, b]$, then as $\|P\| \to 0$, its limit is a definite integral. This suggests the following definition:

(6.3)   *Volume.* **The volume V of the solid of revolution obtained by revolving the area under the graph of a continuous nonnegative function $y = f(x)$ from a to b about the x-axis is**

(6.4) $$V = \lim_{\|P\| \to 0} \sum_{i=1}^{n} \pi[f(u_i)]^2\Delta x_i = \pi \int_{a}^{b} [f(x)]^2 \, dx$$

EXAMPLE 1   Find the volume of the solid of revolution generated by revolving the area under the graph of $f(x) = \sqrt{x}$ from $x = 0$ to $x = 5$ about the x-axis.

Solution   First, we graph the area to be revolved and the resultant solid of revolution, as shown in Figure 13. The desired volume V is

$$V = \int_{0}^{5} \pi[f(x)]^2 \, dx = \int_{0}^{5} \pi(\sqrt{x})^2 \, dx = \int_{0}^{5} \pi x \, dx = \frac{\pi x^2}{2}\Big|_{0}^{5} = \frac{25\pi}{2}$$

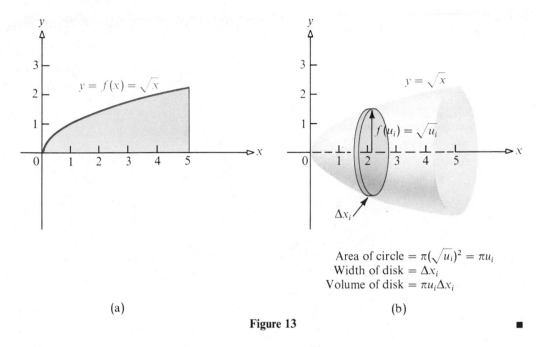

Area of circle $= \pi(\sqrt{u_i})^2 = \pi u_i$
Width of disk $= \Delta x_i$
Volume of disk $= \pi u_i \Delta x_i$

(a)   (b)

Figure 13   ■

The function f in (6.3) does not have to be nonnegative. As Figure 14 illustrates, the volume V of the solid of revolution obtained by revolving about the x-axis the area enclosed by the graph of a continuous function f, the x-axis, $x = a$,   and $x = b$ [Fig. 14(a)] will equal the volume of the solid of revolution obtained by revolving about the x-axis the area under the graph of $y = |f(x)|$ from a to b

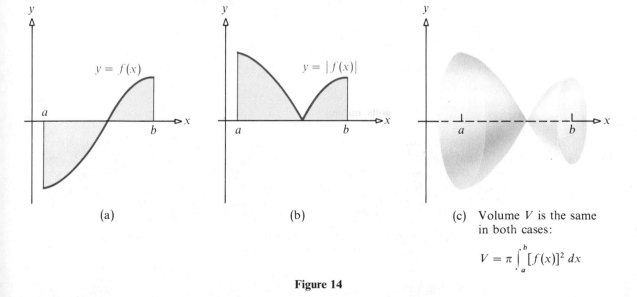

(a)   (b)   (c)   Volume V is the same
                                                                            in both cases:

$$V = \pi \int_a^b [f(x)]^2 \, dx$$

Figure 14

[Fig. 14(b)]. Since $|f(x)|^2 = [f(x)]^2$,   the formula for V is the same as the one given by (6.4).

EXAMPLE 2   Find the volume of the solid of revolution generated by revolving the area enclosed by $y = x^3$,   the x-axis, $x = -1$,   and $x = 2$ about the x-axis.

Solution   Figure 15 illustrates the area to be revolved and the resultant solid of revolution. The desired volume V is

$$V = \pi \int_{-1}^{2} (x^3)^2 \, dx = \pi \left.\frac{x^7}{7}\right|_{-1}^{2} = \frac{\pi}{7}(128 + 1) = \frac{129}{7}\pi$$

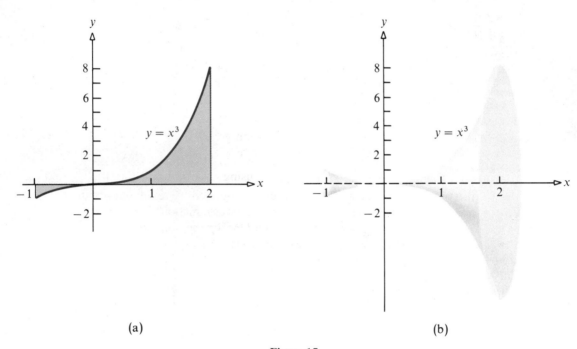

(a)   (b)

Figure 15   ∎

REVOLVING ABOUT THE y-AXIS

A solid of revolution may be generated by revolving an area about any line. In particular, the volume V of the solid of revolution generated by revolving the area enclosed by the graph of $x = g(y)$, $g(y) \geq 0$,   the y-axis, $y = c$ and $y = d$ about the y-axis may be obtained by partitioning the y-axis and taking slices of width Δy_i perpendicular to the y-axis (see Fig. 16). In this case, the area of a typical slice is $\pi x_i^2 = \pi[g(v_i)]^2$,   and the volume of a typical disk is $\pi[g(v_i)]^2\Delta y_i$.   By summing all the volumes and taking the limit, we find the required volume V to be

(6.5) $$V = \pi \int_{c}^{d} [g(y)]^2 \, dy$$

We use (6.5) in the next example.

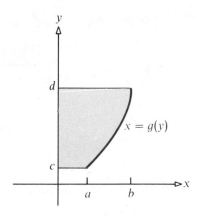

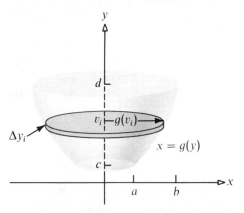

Area of circle $= \pi[g(v_i)]^2$
Width of disk $= \Delta y_i$
Volume of disk $= \pi[g(v_i)]^2 \Delta y_i$

Figure 16   (a)   (b)

EXAMPLE 3   Find the volume of the solid of revolution generated by revolving the area enclosed
by the graphs of $y = x^3$,   the y-axis, $y = 1$,   and $y = 8$ about the y-axis.

Solution   Figure 17 illustrates the situation. By using $x = g(y) = y^{1/3}$ in (6.5), we find the
volume V to be

$$V = \int_1^8 \pi[g(y)]^2 \, dy = \pi \int_1^8 y^{2/3} \, dy = \frac{\pi y^{5/3}}{\frac{5}{3}} \Big|_1^8 = \frac{3\pi}{5}(32 - 1) = \frac{93}{5}\pi$$

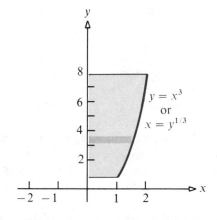

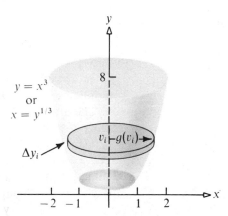

Area of circle $= \pi(y^{1/3})^2$
Width of disk $= \Delta y_i$
Volume of disk $= \pi(y^{1/3})^2 \, \Delta y_i$

(a)   (b)

Figure 17   ∎

WASHERS

We seek a formula for the volume of a solid of revolution generated by revolving about the x-axis the area enclosed by the graphs of two continuous functions $y = f(x)$ and $y = g(x)$, $f(x) \geq g(x) \geq 0$, and the lines $x = a$ and $x = b$. Figure 18(a) illustrates the situation.

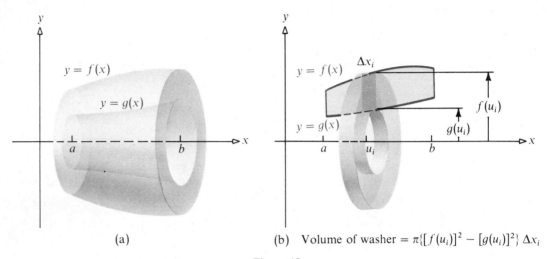

(a)   (b)   Volume of washer $= \pi\{[f(u_i)]^2 - [g(u_i)]^2\}\,\Delta x_i$

Figure 18

As has been our practice, we begin by partitioning the interval $[a, b]$ into n subintervals,

$$[a, x_1], \quad [x_1, x_2], \quad \ldots, \quad [x_{n-1}, b]$$

The length of the ith subinterval is $\Delta x_i = x_i - x_{i-1}$. For each i, $1 \leq i \leq n$, select a number u_i in the ith subinterval $[x_{i-1}, x_i]$ [see Fig. 18(b)].

The area obtained by cutting the solid at $x = u_i$ is the difference of the area of the outer circle, $\pi[f(u_i)]^2$, and the area of the inner circle, $\pi[g(u_i)]^2$, namely, $\pi[f(u_i)]^2 - \pi[g(u_i)]^2$. The solid of revolution obtained from the ith subinterval is called a *washer*, and its volume is $\{\pi[f(u_i)]^2 - \pi[g(u_i)]^2\}\Delta x_i$. By adding up the volumes of all the washers and taking the limit as the norm of the partition approaches 0, we obtain the following definition:

(6.6)   **The volume V of the solid of revolution obtained by revolving about the x-axis the area enclosed by the graphs of two continuous functions $y = f(x)$ and $y = g(x)$, $f(x) \geq g(x) \geq 0$, and the lines $x = a$ and $x = b$ is**

(6.7) $$V = \lim_{\|P\| \to 0} \sum_{i=1}^{n} \{\pi[f(u_i)]^2 - \pi[g(u_i)]^2\}\Delta x_i = \int_a^b \pi\{[f(x)]^2 - [g(x)]^2\}\,dx$$

Because of the way we obtained definition (6.6), its use is sometimes referred to as the *washer method*. In using formula (6.7), it may be helpful to remember the general formula

(6.8)   **Volume of a washer $= \pi[(\text{Outer radius})^2 - (\text{Inner radius})^2](\text{Thickness})$**

EXAMPLE 4   Find the volume of the solid of revolution generated by revolving the area enclosed by the graphs of $y = 2/x$ and $y = 3 - x$ about the x-axis.

Solution   Figure 19 illustrates the situation. The points of intersection of the graphs obey

$$\frac{2}{x} = 3 - x$$

$$x^2 - 3x + 2 = 0$$
$$(x - 2)(x - 1) = 0$$

So, the area to be revolved lies between $x = 1$ and $x = 2$. Partitioning the x-axis gives the volume of a typical washer:

$$\pi\{[f(u_i)]^2 - [g(u_i)]^2\}\Delta x_i = \pi\left[(3 - u_i)^2 - \left(\frac{2}{u_i}\right)^2\right]\Delta x_i$$

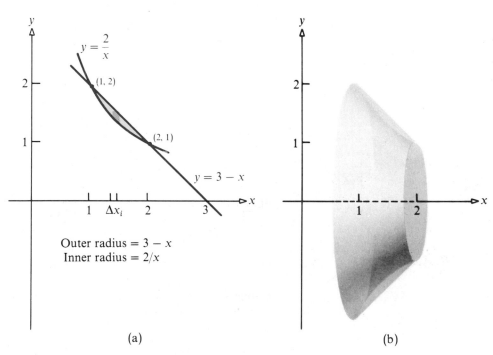

Outer radius $= 3 - x$
Inner radius $= 2/x$

(a)   (b)

Figure 19

The desired volume V is

$$V = \int_1^2 \pi\left[(3-x)^2 - \left(\frac{2}{x}\right)^2\right]dx = \pi\int_1^2\left(9 - 6x + x^2 - \frac{4}{x^2}\right)dx$$

$$= \pi\left(9x - 3x^2 + \frac{x^3}{3} + \frac{4}{x}\right)\Big|_1^2 = \frac{\pi}{3} \quad \blacksquare$$

If the solid of revolution is obtained by revolving about the y-axis, the partition will be along the y-axis and the thickness of the washer in (6.8) will be Δy_i. Let's look at an example.

EXAMPLE 5   Find the volume of the solid of revolution generated by revolving about the y-axis the area enclosed by the graphs of $y = 2x$ and $y = x^2$.

Solution   Figure 20 illustrates the situation. The points of intersection of the graphs obey

$$x^2 = 2x$$
$$x^2 - 2x = 0$$
$$x(x-2) = 0$$

So, the area to be revolved lies between $x = 0$ and $x = 2$, or between $y = 0$ and $y = 4$. Partitioning the y-axis gives the volume of a typical washer as

$$\pi[(\text{Outer radius})^2 - (\text{Inner radius})^2]\Delta y_i = \pi\left[(\sqrt{v_i})^2 - \left(\frac{v_i}{2}\right)^2\right]\Delta y_i$$

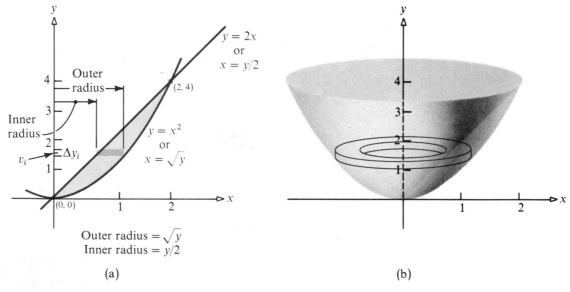

Outer radius $= \sqrt{y}$
Inner radius $= y/2$

(a)   (b)

Figure 20

The desired volume V is

$$V = \int_0^4 \pi \left[(\sqrt{y})^2 - \left(\frac{y}{2}\right)^2 \right] dy = \pi \int_0^4 \left(y - \frac{y^2}{4} \right) dy = \pi \left(\frac{y^2}{2} - \frac{y^3}{12} \right)\Big|_0^4 = \frac{8\pi}{3} \quad \blacksquare$$

APPLICATION: COOLING TOWERS*

A proposed design for the cooling towers at a nuclear power plant is a branch of a hyperbola† rotated about an axis (see Fig. 21). The equation of the branch of the hyperbola—assuming it is rotated about the y-axis, its vertex is on the x-axis, and its center is at the origin—is $x = \sqrt{a^2 + by^2}$, where $a = 147$ and $b = 0.16$.

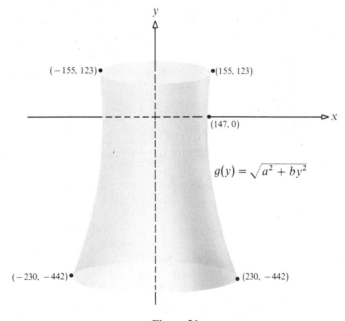

Figure 21

The base of the tower is 442 feet below the vertex, and the top of the tower is 123 feet above the vertex. Then, using (6.5), $V = \pi \int_c^d [g(y)]^2 \, dy$, we find the volume of the cooling tower to be

$$V = \pi \int_{-442}^{123} (\sqrt{a^2 + by^2})^2 \, dy = \pi \int_{-442}^{123} (a^2 + by^2) \, dy$$

$$= \pi \left(a^2 y + \frac{by^3}{3} \right)\Big|_{-442}^{123} = \pi(16{,}913{,}712) \approx 53{,}135{,}993 \text{ cubic feet}$$

---

* Data obtained from U.S. Nuclear Regulatory Commission.

† Refer to Section 4 of Chapter 12, where a detailed discussion of the hyperbola is given.

Now, let's assume that the walls of this cooling tower are a constant 5 inches (≈ 0.42 foot) thick. Then the interior hyperbola (Fig. 22) has as its equation $x = \sqrt{q^2 + ry^2}$, where $q = 146.58$ and $r = 0.16$. Then, again using

$$V = \pi \int_c^d [g(y)]^2 \, dy$$

we find that the volume of the space between the interior walls—that is, the space inside the cooling tower—is

$$V = \pi \int_{-442}^{123} (\sqrt{q^2 + ry^2})^2 \, dy$$

$$= \pi \int_{-442}^{123} (q^2 + ry^2) \, dy$$

$$= \pi \left(q^2 y + \frac{ry^3}{3} \right) \Big|_{-442}^{123}$$

$$= \pi(16,844,045) \approx 52,917,129 \text{ cubic feet}$$

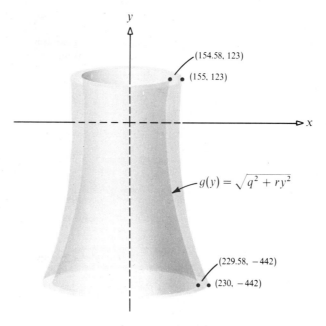

Figure 22

By combining this result with the previous result, we find that the volume of the walls themselves is approximately 218,864 cubic feet, so 218,864 cubic feet of reinforced concrete are needed to construct the walls of the tower.

EXERCISE 2

In Problems 1–12 find the volume of the solid of revolution generated by revolving each enclosed area about the indicated axis. Draw a sketch first.

1. Enclosed by the graphs of $y = x^2$, the x-axis, $x = 0$, and $x = 2$; about the x-axis

2. Enclosed by the graphs of $y = 2x^2$, the x-axis, $x = 0$, and $x = 1$; about the x-axis

3. Enclosed by the graphs of $y = x^3$, the x-axis, $x = 0$, and $x = 2$; about the x-axis

4. Enclosed by the graphs of $y = 2x^4$, the x-axis, $x = 0$, and $x = 1$; about the x-axis

5. Enclosed by the graphs of $y = 2\sqrt{x}$, the x-axis, $x = 1$, and $x = 4$; about the x-axis

6. Enclosed by the graphs of $y = \sqrt{x}$, the x-axis, $x = 4$, and $x = 9$; about the x-axis

7. Enclosed by the graphs of $y = 1/x$, the x-axis, $x = 1$, and $x = 2$; about the x-axis

8. Enclosed by the graphs of $y = x^{2/3}$, the x-axis, $x = 0$, and $x = 8$; about the x-axis

9. Enclosed by the graphs of $y = x^2$, the y-axis, $y = 1$, and $y = 4$; about the y-axis

10. Enclosed by the graphs of $y = 2\sqrt{x}$, the y-axis, $y = 1$, and $y = 4$; about the y-axis

11. Enclosed by the graphs of $y = 1/x$, the y-axis, $y = 1$, and $y = 4$; about the y-axis

12. Enclosed by the graphs of $y = x^{2/3}$, the y-axis, $y = 1$, and $y = 4$; about the y-axis

In Problems 13-16 find the volume of the solid of revolution generated by revolving the area enclosed by the graphs $y = f(x)$ and $y = g(x)$ and the two vertical lines about the x-axis.

13. $f(x) = 3x$, $g(x) = x^3$, $x = 0$, $x = 1$

14. $f(x) = 2x + 1$, $g(x) = x$, $x = 0$, $x = 3$

15. $f(x) = -x$, $g(x) = x^2$, $x = -1$, $x = 0$

16. $f(x) = \cos x$, $g(x) = \sin x$, $x = 0$, $x = \pi/4$

17. Find the volume of the solid of revolution generated by revolving the area enclosed by the graphs of $y = x$ and $y = x^2$ about the x-axis.

18. Find the volume of the solid of revolution generated by revolving the area enclosed in the first quadrant by the graphs of $y = x$ and $y = x^3$ about the x-axis.

In Problems 19–24 find the volume of the solid of revolution generated by revolving each area about the indicated axis.

19. Enclosed by the graphs of $y = x^2$, the y-axis, and $y = 4$; about the x-axis

20. Enclosed by the graphs of $y = 2x^2$, the y-axis, and $y = 2$; about the x-axis

21. Enclosed by the graphs of $y = x^3$, the x-axis, and $x = 2$; about the y-axis

22. Enclosed by the graphs of $y = 2x^4$, the x-axis, and $x = 1$; about the y-axis

23. Enclosed by the graphs of $y = 2\sqrt{x}$, the y-axis, and $y = 4$; about the x-axis

24. Enclosed by the graphs of $y = \sqrt{x}$, the y-axis, and $y = 9$; about the x-axis

In Problems 25–28 find the volume of the solid of revolution generated by revolving each area about the line indicated.

25. Enclosed by the graphs of $y = x^2$, the x-axis, $x = 0$, and $x = 1$; about $x = 1$

26. Enclosed by the graphs of $y = x^3$, the x-axis, $x = 0$, and $x = 1$; about $x = 1$

27. Enclosed by the graphs of $y = \sqrt{x}$, the x-axis, $x = 0$, and $x = 4$; about $x = 4$

28. Enclosed by the graphs of $y = 1/\sqrt{x}$, the x-axis, $x = 1$, and $x = 4$; about $x = 4$

3. Volume of a Solid of Revolution: Shell Method

In this section we approach the problem of finding volumes of solids of revolution by using *cylindrical shells*. A cylindrical shell is the solid region between two concentric cylinders. For example, the water pipes in your house are cylindrical shells (see Fig. 23). If the inner radius is r_1 and the outer radius is r_2, the volume V of a cylindrical shell of height h is

$$V = \pi r_2^2 h - \pi r_1^2 h$$

We shall find it convenient to write this formula as

$$V = \pi h(r_2^2 - r_1^2) = \pi h(r_2 + r_1)(r_2 - r_1)$$

$$= 2\pi h\left(\frac{r_2 + r_1}{2}\right)(r_2 - r_1)$$

As an aid to remembering this formula, it helps to state it as

(6.9) $V = 2\pi(\textbf{Height})(\textbf{Average radius})(\textbf{Thickness})$

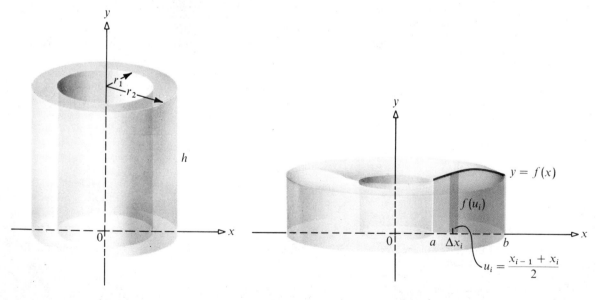

Figure 23   Figure 24

We use formula (6.9) to find the volume of the solid generated by revolving the area under the graph of a function $y = f(x)$ from $x = a$ to $x = b$ about the y-axis. We assume f is nonnegative and continuous on $[a, b]$ and $a \geq 0$ (see Fig. 24). To find the volume of such a solid, we partition the interval $[a, b]$ into n subintervals

$$[a, x_1], \quad [x_1, x_2], \quad \ldots, \quad [x_{i-1}, x_i], \quad \ldots, \quad [x_{n-1}, b]$$

The length of the ith subinterval is $\Delta x_i = x_i - x_{i-1}$, $i = 1, 2, \ldots, n$. Now, we concentrate on the rectangle whose base is the subinterval $[x_{i-1}, x_i]$, and whose height is $f(u_i)$, where $u_i = (x_{i-1} + x_i)/2$ is the midpoint of the subinterval $[x_{i-1}, x_i]$. When this rectangle is revolved about the y-axis, it generates a cylindrical shell of inner radius x_{i-1}, outer radius x_i, and height $f(u_i)$. From (6.9), the volume of this cylindrical shell is

$$V = 2\pi(\text{Height})(\text{Average radius})(\text{Thickness}) = 2\pi f(u_i) \frac{(x_{i-1} + x_i)}{2} (x_i - x_{i-1})$$

$$= 2\pi f(u_i) u_i \Delta x_i = 2\pi u_i f(u_i) \Delta x_i$$

The sum of all the volumes due to each subinterval is

$$\sum_{i=1}^{n} 2\pi u_i f(u_i) \Delta x_i$$

This sum represents an approximation to the volume V of the solid generated by revolving the area under the graph of $y = f(x)$ from $x = a$ to $x = b$ about the y-axis. As the length of each subinterval gets smaller and smaller—that is, as the norm $\|P\|$ approaches 0—the sum $\sum_{i=1}^{n} 2\pi u_i f(u_i) \Delta x_i$ is a better and better approximation to the volume V of the solid. But this sum is a Riemann sum, so that, as $\|P\| \to 0$, its limit is a definite integral. This suggests the following definition:

(6.10)   *Volume.*   **The volume V of the solid generated by revolving the area under the graph of a continuous nonnegative function $y = f(x)$ from $x = a$ to $x = b$ about the y-axis is**

(6.11)
$$V = \lim_{\|P\| \to 0} \sum_{i=1}^{n} 2\pi u_i f(u_i) \Delta x_i = \int_a^b 2\pi x f(x) \, dx$$

Because of the way we obtained definition (6.10), finding the volume of a solid of revolution by using formula (6.11) is referred to as the *shell method*.

It can be shown that this definition and the washer method of Section 2 each lead to the same answer.* The advantage of having two equivalent, yet different, formulas is that on occasion one might be easier to use. The next example illustrates just such a situation.

---

* This topic is discussed in detail in an article by Charles A. Cable, "The Disk and Shell Method," *American Mathematical Monthly* 91, no. 2 (Feb. 1984): 139.

EXAMPLE 1   Find the volume of the solid generated by revolving the area under the graph
of $y = x^2 + 2x$ from $x = 0$ to $x = 1$ about the y-axis.

Solution   *By shell method:*   Figure 25 illustrates the situation. Note that for the shell method,
a revolution about the y-axis requires a partition of the x-axis. The height of the
rectangle is $f(u_i)$, where $f(x) = x^2 + 2x$. The required volume V is

$$V = 2\pi \int_0^1 x(x^2 + 2x)\, dx = 2\pi \int_0^1 (x^3 + 2x^2)\, dx = 2\pi \left(\frac{x^4}{4} + \frac{2x^3}{3} \right)\Big|_0^1 = \frac{11\pi}{6}$$

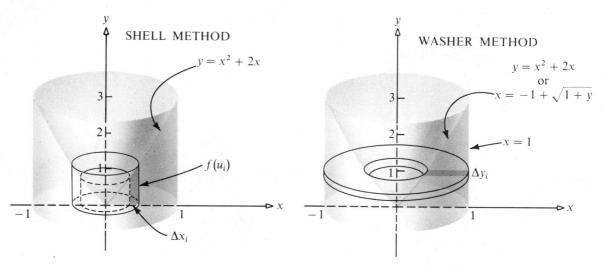

Figure 25   **Figure 26**

By washer method: Figure 26 illustrates the situation. Note that for the washer
method, a revolution about the y-axis requires a partition of the y-axis. To use the
washer method, we need to solve for x in the expression $y = x^2 + 2x$. We treat
this as a quadratic equation in the variable x and use the quadratic formula* with
$a = 1$, $b = 2$, $c = -y$ to obtain $x = -1 \pm \sqrt{1 + y}$. Since we require $x > 0$,
we use only the $+$ sign. Then

$$V = \pi \int_0^3 (1)^2\, dy - \pi \int_0^3 (-1 + \sqrt{1 + y})^2\, dy$$

$$= \pi \int_0^3 [1 - (1 - 2\sqrt{1 + y} + 1 + y)]\, dy$$

$$= \pi \int_0^3 [2\sqrt{1 + y} - 1 - y]\, dy = 2\pi \int_0^3 \sqrt{1 + y}\, dy - \pi \int_0^3 (1 + y)\, dy$$

---

* If $ax^2 + bx + c = 0$, then $x = \dfrac{-b \pm \sqrt{b^2 - 4ac}}{2a}$.

Now

$$2\pi \int_0^3 \sqrt{1+y}\, dy = 2\pi \int_1^2 u(2u\, du) = 4\pi \left.\frac{u^3}{3}\right|_1^2 = \frac{28\pi}{3}$$

Set $u^2 = 1 + y$

$$\pi \int_0^3 (1+y)\, dy = \pi\left.\left(y + \frac{y^2}{2}\right)\right|_0^3 = \frac{15\pi}{2}$$

so the required volume V is

$$V = \frac{28\pi}{3} - \frac{15\pi}{2} = \frac{11\pi}{6} \qquad \blacksquare$$

This example gives a clue as to when the shell method is preferable to the washer method, namely, when it may be difficult to solve $y = f(x)$ for x in terms of y. For example, if the function given in Example 1 had been $y = x^5 + x^2 + 1$, the only practical choice would have been the shell method.

The next example illustrates the importance of sketching a graph before blindly using a formula. Note especially the limits of integration and how they come about when the washer method is used.

EXAMPLE 2   Find the volume of the solid generated by revolving the area enclosed by $y = x^2$ and $y = 12 - x$ to the right of $x = 1$ about the y-axis.

Solution   *By shell method:*   Figure 27 illustrates the situation. The height of a typical rectangle is $(12 - x) - (x^2) = 12 - x - x^2$, and the integration is with respect to x from 1

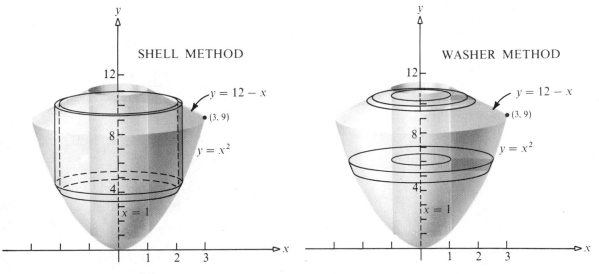

Figure 27   Figure 28

to 3. The desired volume V is

$$V = 2\pi \int_1^3 x(12 - x - x^2)\,dx = 2\pi \int_1^3 (12x - x^2 - x^3)\,dx$$

$$= 2\pi \left(6x^2 - \frac{x^3}{3} - \frac{x^4}{4} \right)\Big|_1^3 = 2\pi \left[\left(54 - 9 - \frac{81}{4} \right) - \left(6 - \frac{1}{3} - \frac{1}{4} \right) \right] = \frac{116\pi}{3}$$

By washer method: Figure 28 illustrates the situation. The integration is with respect to y from 1 to 11, with a change occurring at $y = 9$.   The required volume V is obtained by subtracting the inner volume from the outer volume. Thus,

$$V = \pi \int_1^9 (\sqrt{y})^2\,dy + \pi \int_9^{11} (12 - y)^2\,dy - \pi \int_1^{11} (1)^2\,dy$$

$$= \pi \frac{y^2}{2}\Big|_1^9 + \pi \left[144y - 24\left(\frac{y^2}{2}\right) + \frac{y^3}{3} \right]\Big|_9^{11} - \pi y\Big|_1^{11}$$

$$= 40\pi + \pi \left[288 - (12)(40) + \frac{602}{3} \right] - \pi(10) = -162\pi + \frac{602\pi}{3} = \frac{116\pi}{3} \qquad \blacksquare$$

REVOLVING ABOUT THE x-AXIS

The shell method can also be used when the solid is generated by a revolution about the x-axis, as shown in Figure 29.

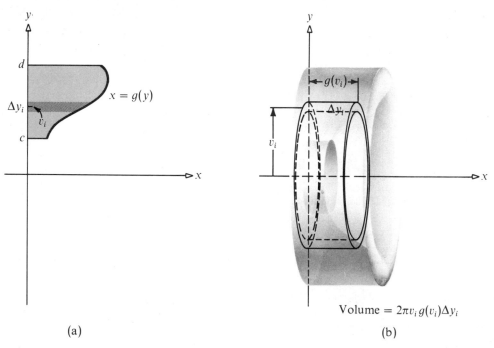

(a)

(b)

Figure 29

(6.12)   *Volume.*   **The volume V of the solid generated by revolving the area enclosed by the graph of $x = g(y)$, $g(y) \geq 0$,   the y-axis, $y = c$,   and $y = d$ about the x-axis is**

(6.13) $$V = 2\pi \int_c^d yg(y)\, dy$$

Figure 29 illustrates how (6.13) is obtained.

EXAMPLE 3   Find the volume generated by revolving the area enclosed in the first quadrant by $x^2/a^2 + y^2/b^2 = 1$,   about the x-axis. Here, a and b are positive constants.

Solution   *By shell method:*   Figure 30 illustrates the situation. The integration is with respect to y from 0 to b and the function g is $g(y) = (a/b)\sqrt{b^2 - y^2}$.   The required volume V is

$$V = 2\pi \int_0^b yg(y)\, dy = 2\pi \int_0^b y\left(\frac{a}{b}\sqrt{b^2 - y^2}\right) dy$$

$$= \frac{2\pi a}{b} \int_b^0 u(-u\, du) = \frac{2\pi a}{b}\left(\frac{-u^3}{3}\right)\Big|_b^0 = \frac{2\pi a}{b}\left(\frac{b^3}{3}\right) = \frac{2\pi ab^2}{3}$$

Set $u^2 = b^2 - y^2$

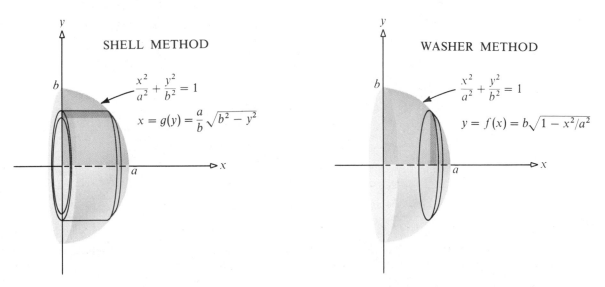

Figure 30   **Figure 31**

By washer method: Figure 31 illustrates the situation. The integration is with respect to x from 0 to a. The required volume V is

$$V = \pi \int_0^a b^2\left(1 - \frac{x^2}{a^2}\right) dx = \pi b^2\left(x - \frac{x^3}{3a^2}\right)\Big|_0^a = \pi b^2\left(a - \frac{a}{3}\right) = \frac{2\pi ab^2}{3} \qquad \blacksquare$$

If $a = b$ in Example 3, the solid generated is a hemisphere and its volume is $2\pi a^3/3$. The volume of a sphere of radius R is therefore $4\pi R^3/3$.

Table 1 summarizes the washer and shell methods.

Table 1

	Washer Method	Shell Method
Revolution about x-axis	Partition the x-axis; use vertical strips	Partition the y-axis; use horizontal strips
Revolution about y-axis	Partition the y-axis; use horizontal strips	Partition the x-axis; use vertical strips

EXERCISE 3

In Problems 1–12 use the shell method to find the volume of the solid of revolution generated by revolving each area about the indicated axis.

1. Enclosed by the graphs of $y = x^2 - 1$, the x-axis, $x = 1$, and $x = 3$; about the y-axis
2. Enclosed by the graphs of $y = x^3 + x$, the x-axis, $x = 0$, and $x = 1$; about the y-axis
3. Enclosed by the graphs of $y = \sqrt{x} + x$, the x-axis, $x = 1$, and $x = 4$; about the y-axis
4. Enclosed by the graphs of $y = x^{2/3} + x^{1/3}$, the x-axis, $x = 1$, and $x = 8$; about the y-axis
5. Enclosed by the graphs of $y = x^3$ and $y = x^2$; about the y-axis
6. Enclosed by the graphs of $y = \sqrt{x}$ and $y = x^2$; about the y-axis
7. Enclosed by the graphs of $y = x^3$, the y-axis, and $y = 8$; about the x-axis
8. Enclosed by the graphs of $y = \sqrt{x}$, the y-axis, and $y = 2$; about the x-axis
9. Enclosed by the graphs of $x = \sqrt{y}$, the y-axis, and $y = 1$; about the x-axis
10. Enclosed by the graphs of $x = 4\sqrt{y}$, the y-axis, and $y = 4$; about the x-axis
11. Enclosed by the graphs of $y = x$ and $y = x^2$; about the x-axis
12. Enclosed in the first quadrant by the graphs of $y = x$ and $y = x^3$; about the x-axis

In Problems 13–22 use either the shell method or the washer method to find the volume of the solid of revolution generated by revolving each area about the indicated axis.

13. Enclosed by the graphs of $y = \sqrt{x}$, the y-axis, and $y = 4$; about the x-axis
14. Enclosed by the graphs of $y = 1/x$, the x-axis, $x = 1$, and $x = 4$; about the y-axis
15. Enclosed in the first quadrant by the graphs of $y = x^3$ and $y = x$; about the y-axis
16. Enclosed in the first quadrant by the graphs of $y = x^3$ and $y = x^2$; about the x-axis
17. Enclosed by the graphs of $y = 3x^2$ and $y = 30 - x$ to the right of $x = 1$; about the y-axis
18. Enclosed by the graphs of $y = 3x^2$ and $y = 30 - x$ to the right of $x = 1$; about the x-axis

19. Enclosed by the graphs of $y = x^2$ and $y = 8 - x^2$ to the right of $x = 1$; about the y-axis

20. Enclosed by the graphs of $y = x^2$ and $y = 8 - x^2$ to the right of $x = 1$; about the x-axis

21. Enclosed by the graphs of $y = \sqrt{x}$ and $y = 18 - x^2$ to the right of $x = 1$; about the y-axis

22. Enclosed by the graphs of $y = \sqrt{x}$ and $y = 18 - x^2$ to the right of $x = 1$; about the x-axis

23. Use the shell method to find the outside volume of the cooling tower described in the application at the end of Section 2. [*Hint:* The equation $x = \sqrt{a^2 + by^2}$ can be rewritten as $y = \pm\sqrt{(x^2 - a^2)/b}$.]

24. Let A be the area of the first quadrant enclosed by the x-axis and the graph of $y = 2x - x^2$.
    (a)   Find the volume produced when A is revolved about the x-axis.
    (b)   Find the volume produced when A is revolved about the y-axis.

4. Volume by Slicing

In this section we consider the problem of computing the volume of a solid by the *method of slicing*. The idea is to cut the solid into thin slices using planes perpendicular to the x-axis and then add up these slices to obtain the volume (see Fig. 32).

We begin by partitioning the interval $[a, b]$ into n subintervals

$$[a, x_1], \quad [x_1, x_2], \quad \ldots, \quad [x_{i-1}, x_i], \quad \ldots, \quad [x_{n-1}, b]$$

The length of each subinterval is

$$\Delta x_i = x_i - x_{i-1} \qquad i = 1, \ldots, n$$

In the ith subinterval, we pick a number u_i and consider the slice (cross section) cut from the solid at $x = u_i$ by a plane perpendicular to the x-axis. We denote the area of this cross section by $A(u_i)$, as indicated in Figure 33. The volume of the thin slice from $x = x_{i-1}$ to $x = x_i$ may then be approximated by $A(u_i)\Delta x_i$, and the sum of all the volumes due to each subinterval is

(6.14)

$$\sum_{i=1}^{n} A(u_i)\Delta x_i$$

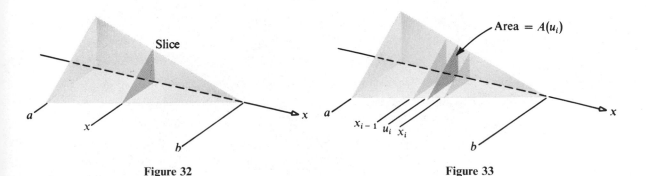

Figure 32   Figure 33

This sum represents an approximation to the volume V of the solid from $x = a$ to $x = b$.   As the length of each subinterval gets smaller and smaller, that is, as the norm $\|P\|$ approaches 0, the sum $\sum_{i=1}^{n} A(u_i)\Delta x_i$ is a better and better approximation to the volume V of the solid. But the sum (6.14) is a Riemann sum. Hence, if the cross sections $A(x)$ vary continuously with x, then, as $\|P\| \to 0$,   the limit of this sum is a definite integral. This suggests the following definition:

(6.15)   *Volume.*   **If for each x in $[a, b]$ the area $A(x)$ of the cross section of a solid is known and is continuous on $[a, b]$, then the volume V of the solid is**

(6.16)
$$V = \lim_{\|P\| \to 0} \sum_{i=1}^{n} A(u_i)\Delta x_i = \int_{a}^{b} A(x)\,dx$$

Formula (6.16) may be used if the solid whose volume we seek is not a solid of revolution. Although the volume of such a solid usually requires the use of a double integral or a triple integral (the subjects of Chap. 18), in certain situations a single definite integral will suffice. For example, this is the case whenever parallel cross sections of the solid all have the same simple geometric configuration (all are semi-circles, or triangles, or squares, etc.). The result is that the area $A(x)$ of the cross section is easy to calculate, and formula (6.16) can then be used to obtain the volume of the solid.

We use (6.16) when the area of the cross section varies (that is, when the areas of the slices are not all equal). For solids of constant cross-sectional areas, such as the one in Figure 34, the volume is simply the area of the cross section times its thickness, that is,

$$V = \textbf{(Area of cross section)}(b - a)$$

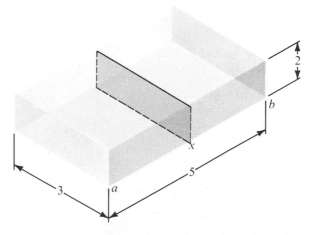

Figure 34

For example, the volume of the solid in Figure 34 is $(6)(5) = 30$ since the cross section $A(x)$ is always equal to $(2)(3) = 6$ and the length $b - a$ is 5.

The use of (6.15) and (6.16) to calculate the volume of a solid is usually referred to as the *slicing method*. We illustrate its use in the next example.

EXAMPLE 1   Find the volume of a right circular cone having radius R and height h.

Solution   We position the cone in such a way that its vertex is at the origin and its axis coincides with the x-axis (see Fig. 35). Therefore, the cone extends from $x = 0$ to $x = h$. The cross section at any number x is a circle. To obtain its area $A(x)$, we must find its radius $r(x)$, which depends on x. Because we have similar triangles, as shown in Figure 35(b),

$$\frac{r(x)}{x} = \frac{R}{h}$$

$$r(x) = \frac{xR}{h}$$

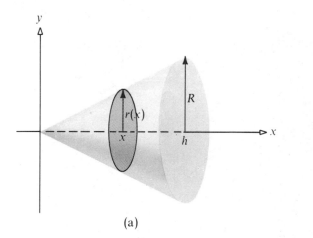

(a)

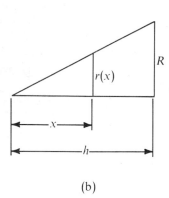

(b)

Figure 35

so that

$$A(x) = \pi[r(x)]^2 = \pi\left(\frac{xR}{h}\right)^2$$

Using (6.16), the volume of the cone is

$$V = \int_a^b A(x)\,dx = \int_0^h \frac{\pi x^2 R^2}{h^2}\,dx = \frac{\pi R^2}{h^2}\left(\frac{x^3}{3}\right)\Big|_0^h = \frac{\pi R^2 h}{3} \qquad \blacksquare$$

Of course, a right circular cone is a solid of revolution, and its volume can also be found by the disk method (see Problem 11 in Exercise 4).

The above example illustrates that the way in which the solid is positioned relative to the x-axis is important if the slice $A(x)$ is to be easily found and integrated. The next two examples further illustrate the importance of good positioning.

EXAMPLE 2   A solid has a circular base of radius 3 units. Find the volume of the solid if every plane cross section that is perpendicular to a fixed diameter is an equilateral triangle.

Solution   Position the circular base as shown in Figure 36(a), with the x-axis as the fixed diameter. The equation of the circle is then $x^2 + y^2 = 9$.   The cross section of the solid is an equilateral triangle of side $2y$ and area $A(x) = \sqrt{3}y^2$,   as indicated in Figure 36(b). Since $y^2 = 9 - x^2$,   we have $A(x) = \sqrt{3}(9 - x^2)$,   so that the volume is

$$V = \int_a^b A(x)\, dx = 2\int_0^3 \sqrt{3}(9 - x^2)\, dx = 2\sqrt{3}\left(9x - \frac{x^3}{3}\right)\Bigg|_0^3 = 36\sqrt{3}$$

Use symmetry

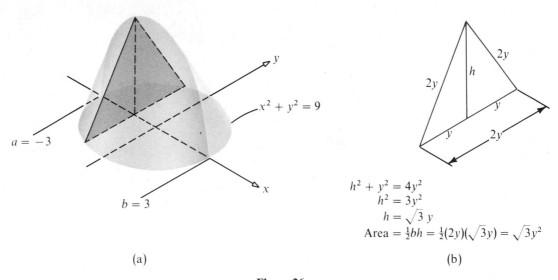

$$h^2 + y^2 = 4y^2$$
$$h^2 = 3y^2$$
$$h = \sqrt{3}\, y$$
$$\text{Area} = \tfrac{1}{2}bh = \tfrac{1}{2}(2y)(\sqrt{3}y) = \sqrt{3}y^2$$

(a)   (b)

Figure 36   ∎

EXAMPLE 3   Find the volume of a pyramid having a height of length h and a square base, with each side of length b.

Solution   Position the pyramid so that its vertex is at the origin and its height is along the positive x-axis. Then a typical cross section $A(x)$ is a square with side s, where s is

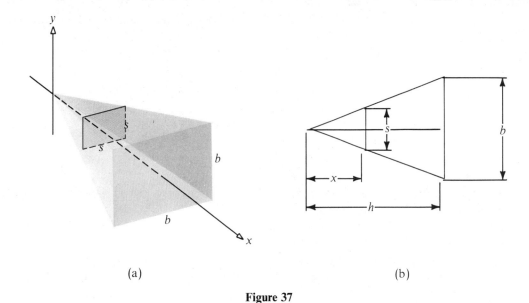

(a)   (b)

Figure 37

a function of x (see Fig. 37). The volume of the pyramid is

$$V = \int_0^h s^2 \, dx$$

In order to evaluate the integral, we must express s in terms of x. By using similar triangles, we find that

$$\frac{s}{x} = \frac{b}{h}$$

$$s = \frac{b}{h} x$$

As a result,

$$V = \int_0^h \frac{b^2}{h^2} x^2 \, dx = \frac{b^2}{h^2} \left(\frac{x^3}{3} \right) \Big|_0^h = \frac{1}{3} b^2 h \qquad \blacksquare$$

EXERCISE 4

In Problems 1–10 compute the volume of each solid by the method of slicing.

1. The base is a square with each side of length 2; cross sections taken perpendicular to the base are rectangles of height 1

2. The base is a square with each side of length 1; cross sections taken perpendicular to the base are rectangles of height 2

3. The volume of a solid whose base is the area enclosed by the circle $x^2 + y^2 = 1$; cross sections taken perpendicular to the base and parallel to the y-axis are squares

4. Rework Problem 3 if the cross sections are isosceles triangles of constant altitude h. [*Hint:* $\int_{-1}^{1} \sqrt{1 - x^2}\, dx$ is the area of the upper half of the circle $x^2 + y^2 = 1$.]

5. The volume of a solid whose base is the area enclosed by $y = x^2$, $x = 1$, and $y = 0$; cross sections taken perpendicular to the base and parallel to the y-axis are semicircles

6. The base is a circle of radius 2; cross sections taken perpendicular to the base are isosceles right triangles with one leg on the base

7. The volume of a pyramid 40 meters high whose horizontal cross section h meters from the top is a square with sides of length $2h$ meters

8. The volume of a pyramid 20 meters high whose horizontal cross section h meters from the top is a rectangle with sides of length $2h$ and h meters

9. The volume of a horn-shaped region; cross sections taken perpendicular to the x-axis are circles whose diameters extend from the graph of $y = x^{1/2}$ to the graph of $y = \frac{4}{3}x^{1/3}$ from $x = 0$ to $x = 1$

10. The volume of a horn-shaped region; cross sections taken perpendicular to the x-axis are circles whose diameters extend from the graph of $y = x^{1/3}$ to $y = \frac{3}{2}x^{1/3}$ from $x = 0$ to $x = 1$

11. Use the disk method to verify that the volume V of a right circular cone having radius R and height h is $V = \pi R^2 h/3$ (see Example 1).

12. Find the volume of a parallelepiped with edge lengths a, b, c such that the edges having lengths a and b make an acute angle θ with each other, and the edge of length c makes an acute angle of ϕ with the diagonal of the parallelogram formed by a and b.

13. A hemispherical bowl of radius R contains water to the depth h. Find the volume of the water in the bowl.

14. Suppose a cylindrical glass full of water is tipped until the water level bisects the base and touches the rim. What is the volume of the water remaining? Set up the integral; do not evaluate.

15. Suppose a wedge is cut from a solid right circular cylinder (like a wedge cut in a tree by an axe) such that one side of the wedge is horizontal and the other is inclined at 30°. Assuming the wedged sides meet in a diameter of a circular cross section and the diameter of the cylinder is 10 meters, find the volume of the wedge removed. [*Hint:* The vertical cross sections of the wedge are right triangles.]

16. A hole of radius 2 centimeters is bored completely through a solid metal sphere of radius 5 centimeters, such that the axis of the hole passes through the center of the sphere. Find the volume of the metal removed by the drilling.

5. Arc Length

In this section we find a formula for measuring the length of the graph of a function (*arc length*). As the Greeks discovered, the formula for the circumference of a circle is $C = \pi d$, where d is the diameter. The Greeks arrived at this by the use of inscribed polygons; that is, they picked appropriate points on a circle and connected

(a)   6 points

(b)   12 points

Figure 38

them with straight lines of equal length (see Fig. 38). By adding up the lengths of these line segments, they obtained an approximation to the length (circumference) of the circle. And by choosing more and more points on the circle, they obtained a better and better approximation to the actual length, so they eventually obtained the formula $C = \pi d$.

To find a formula for the length of the graph of $y = f(x)$ from $x = a$ to $x = b$,   where the function f has a continuous derivative on $[a, b]$,* we use the same approach. First, we partition the closed interval $[a, b]$ into n subintervals,

$$[a, x_1], \quad [x_1, x_2], \quad \ldots, \quad [x_{i-1}, x_i], \quad \ldots, \quad [x_{n-1}, b]$$

We denote the length of the ith subinterval by $\Delta x_i = x_i - x_{i-1}$.   Corresponding to each number $a, x_1, x_2, \ldots, x_{n-1}, b$ in the partition, there is a succession of points $P_0, P_1, P_2, \ldots, P_{n-1}, P_n$ on the graph (see Fig. 39). When we join each

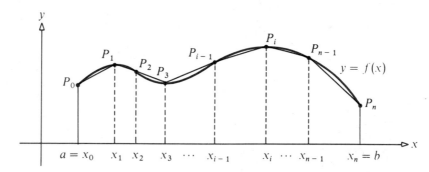

Figure 39

point to its successor by a line segment, the sum L of the lengths of these line segments provides an approximation to the length of the graph of $y = f(x)$ from $x = a$ to $x = b$.   This sum may be written as

(6.17) $$L = d(P_0, P_1) + d(P_1, P_2) + \cdots + d(P_{n-1}, P_n) = \sum_{i=1}^{n} d(P_{i-1}, P_i)$$

where $d(P_{i-1}, P_i)$ is the length of the line segment joining P_{i-1} to P_i.

By the formula for the distance between two points, it follows that the length of the ith line segment is

(6.18) $$d(P_{i-1}, P_i) = \sqrt{(x_i - x_{i-1})^2 + (y_i - y_{i-1})^2} = \sqrt{(\Delta x_i)^2 + (\Delta y_i)^2} = \sqrt{1 + \left(\frac{\Delta y_i}{\Delta x_i}\right)^2}\, \Delta x_i$$

---

* We restrict our discussion of arc length to functions for which f' is continuous on $[a, b]$ in order to eliminate some complications that would otherwise occur.

where $\Delta x_i = x_i - x_{i-1}$ and $\Delta y_i = y_i - y_{i-1} = f(x_i) - f(x_{i-1})$. By combining (6.17) and (6.18), we find

(6.19)
$$L = \sum_{i=1}^{n} \sqrt{1 + \left(\frac{\Delta y_i}{\Delta x_i}\right)^2} \, \Delta x_i$$

Ordinarily, we would now let the norm $\|P\|$ of the partition approach 0. However, this will not be helpful at this stage because the sum (6.19) is not in the form (5.15)—it is not a Riemann sum; that is, taking the limit will not result in a definite integral. However, the sum (6.19) can be put in a proper form as follows: Since we assume the function f has a derivative on $[a, b]$, it follows that f has a derivative on each subinterval $[x_{i-1}, x_i]$. As a result, we can apply the mean value theorem (4.26). For each subinterval $[x_{i-1}, x_i]$, there is a number u_i between x_{i-1} and x_i such that

$$f(x_i) - f(x_{i-1}) = f'(u_i)(x_i - x_{i-1})$$
$$\Delta y_i = f'(u_i)\Delta x_i$$
$$\frac{\Delta y_i}{\Delta x_i} = f'(u_i)$$

Thus, the sum L of the lengths of the line segments in (6.19) becomes

(6.20)
$$L = \sum_{i=1}^{n} \sqrt{1 + [f'(u_i)]^2} \, \Delta x_i$$

where u_i is some number in the subinterval $[x_{i-1}, x_i]$.

The sum (6.20) gives an approximation to the length of the graph of $y = f(x)$ from $x = a$ to $x = b$. As the length of each subinterval gets smaller and smaller, that is, as the norm $\|P\|$ of the partition approaches 0, the sum (6.20) is a better and better approximation to the length of the graph of $y = f(x)$ from $x = a$ to $x = b$. However, the sum (6.20) is now a Riemann sum, so that, as $\|P\| \to 0$, its limit is a definite integral. This suggests the following definition:

(6.21) *Length of a Graph.* **Consider a function f, which has a continuous derivative on $[a, b]$. The length s of the graph of $y = f(x)$ from $x = a$ to $x = b$ is**

(6.22)
$$s = \int_a^b \sqrt{1 + [f'(x)]^2} \, dx$$

We refer to the length s of the graph of $y = f(x)$ from $x = a$ to $x = b$ as *the arc length s of $y = f(x)$ from $x = a$ to $x = b$.*

EXAMPLE 1   Find the arc length of $y = 4x + 5$ from $x = 0$ to $x = 1$.

Solution   Here $f'(x) = 4$ and the arc length s is

$$s = \int_0^1 \sqrt{1 + 16} \, dx = \sqrt{17}x \Big|_0^1 = \sqrt{17} \qquad \blacksquare$$

For this function, whose graph is a straight line, we can verify our answer by using the distance formula. Using this formula, we find that the distance from $(0, 5)$ to $(1, 9)$ is $d = \sqrt{4^2 + 1^2} = \sqrt{17}$.

EXAMPLE 2   Find the arc length of $y = x^{2/3}$ from $x = 1$ to $x = 8$.

Solution   The derivative of $f(x) = x^{2/3}$ is $f'(x) = \frac{2}{3}x^{-1/3} = 2/(3x^{1/3})$.   The arc length from $x = 1$ to $x = 8$ is

$$\int_1^8 \sqrt{1 + \frac{4}{9x^{2/3}}} \, dx$$

Since

$$\sqrt{1 + \frac{4}{9x^{2/3}}} = \sqrt{\frac{9x^{2/3} + 4}{9x^{2/3}}} = \frac{1}{3}\sqrt{9x^{2/3} + 4} \, x^{-1/3}$$

we use the substitution $u = 9x^{2/3} + 4$, $du = 6x^{-1/3} \, dx$.   Also, $u = 13$ when $x = 1$,   and $u = 40$ when $x = 8$.   Thus,

$$\int_1^8 \sqrt{1 + \frac{4}{9x^{2/3}}} \, dx = \int_1^8 \frac{1}{3}\sqrt{9x^{2/3} + 4} \, x^{-1/3} \, dx = \int_{13}^{40} \frac{1}{3}\sqrt{u} \, \frac{du}{6}$$

$$= \frac{1}{18}\left(\frac{u^{3/2}}{\frac{3}{2}}\right)\Big|_{13}^{40} = \frac{u^{3/2}}{27}\Big|_{13}^{40} = \frac{1}{27}(80\sqrt{10} - 13\sqrt{13}) \qquad \blacksquare$$

Figure 40 illustrates the graph of $y = x^{2/3}$.   Observe that if we had been asked to find the arc length of $y = x^{2/3}$ from, say, -1 to 8, we would not have been able to use (6.21) since the derivative does not exist when $x = 0$.   However, in such cases we may be able to do the problem by using a partition of the y-axis. We give a precise statement of the formula below, and then we use it in Example 3 to solve this problem.

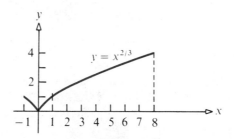

Figure 40

(6.23)   If x is a function of y, $x = g(y)$, $c \leq y \leq d$,   and g' is continuous on $[c, d]$, the arc length s of the graph of $x = g(y)$ from c to d is given by the formula

(6.24)
$$s = \int_c^d \sqrt{1 + [g'(y)]^2} \, dy$$

EXAMPLE 3   Compute the arc length of $y = x^{2/3}$ from $x = -1$ to $x = 8$.

Solution   The derivative of $y = x^{2/3}$ is $y' = 2/(3x^{1/3})$. Since the derivative does not exist at $x = 0$, we cannot use (6.22) to calculate the arc length (see Fig. 40). In this situation we apply (6.24) twice: first, to calculate the arc length of

$$x_1 = g_1(y) = -y^{3/2}$$

from $y = 0$ to $y = 1$; and second, to calculate the arc length of

$$x_2 = g_2(y) = y^{3/2}$$

from $y = 0$ to $y = 4$. The sum of these lengths is the arc length of $y = x^{2/3}$ from $x = -1$ to $x = 8$. By (6.24), the arc length s_1 of $x_1 = g_1(y) = -y^{3/2}$ from $y = 0$ to $y = 1$ is

$$s_1 = \int_0^1 \sqrt{1 + [g_1'(y)]^2}\, dy = \int_0^1 \sqrt{1 + \frac{9y}{4}}\, dy = \frac{1}{2} \int_0^1 \sqrt{4 + 9y}\, dy$$

Use the substitution $u = 4 + 9y$ so that $du = 9\, dy$. Then, since $u = 4$ when $y = 0$, and $u = 13$ when $y = 1$, we have

$$s_1 = \frac{1}{2} \int_4^{13} \sqrt{u}\, \frac{du}{9} = \frac{1}{18} \left(\frac{u^{3/2}}{\frac{3}{2}} \right)\Bigg|_4^{13} = \frac{1}{27}(13\sqrt{13} - 8)$$

Similarly, the arc length s_2 of $x_2 = g_2(y) = y^{3/2}$ from $y = 0$ to $y = 4$ is

$$s_2 = \int_0^4 \sqrt{1 + [g'(y)]^2}\, dy = \int_0^4 \sqrt{1 + \frac{9y}{4}}\, dy = \frac{1}{2} \int_0^4 \sqrt{4 + 9y}\, dy$$

$$\underset{u = 4 + 9y}{=} \frac{1}{2} \int_4^{40} \sqrt{u}\, \frac{du}{9} = \frac{1}{18} \left(\frac{u^{3/2}}{\frac{3}{2}} \right)\Bigg|_4^{40} = \frac{1}{27}(80\sqrt{10} - 8)$$

Thus, the total length s of $y = x^{2/3}$ from $x = -1$ to $x = 8$ is the sum

$$s = s_1 + s_2 = \tfrac{1}{27}(80\sqrt{10} + 13\sqrt{13} - 16) \quad \blacksquare$$

As Example 3 illustrates, when one or more vertical tangents occur between the ends of a portion of a graph whose length is to be found, we use (6.24) to calculate the lengths of the portions of the graph between the vertical tangents and add them.

EXERCISE 5

In Problems 1–4 find the arc length of each line between the points indicated. Verify your answer by using the distance formula.

1. $y = 3x - 1$, from (1, 2) to (3, 8)

2. $y = -4x + 1$, from $(-1, 5)$ to $(1, -3)$

3. $2x - 3y + 4 = 0$, from (1, 2) to (4, 4)

4. $3x + 4y - 12 = 0$, from (0, 3) to (4, 0)

In Problems 5–20 find the indicated arc length.

5. $y = x^{2/3} + 1$, from $x = 1$ to $x = 8$

6. $y = x^{2/3} + 6$, from $x = 1$ to $x = 8$

7. $y = x^{3/2}$, from $x = 0$ to $x = 4$

8. $y = x^{3/2} + 4$, from $x = 1$ to $x = 4$

9. $9y^2 = 4x^3$, from $x = 0$ to $x = 1$

10. $y = \dfrac{x^3}{6} + \dfrac{1}{2x}$, from $x = 1$ to $x = 3$

11. $y = \frac{2}{3}(x^2 + 1)^{3/2}$, from $x = 1$ to $x = 4$

12. $y = \frac{1}{3}(x^2 + 2)^{3/2}$, from $x = 2$ to $x = 4$

13. $y = \frac{2}{9}\sqrt{3}(3x^2 + 1)^{3/2}$, from $x = -1$ to $x = 2$

14. $y = (1 - x^{2/3})^{3/2}$, from $x = \frac{1}{8}$ to $x = 1$

15. $8y = x^4 + \dfrac{2}{x^2}$, from $x = 1$ to $x = 2$

16. $9y^2 = 4(1 + x^2)^3$, from $x = 0$ to $x = 2\sqrt{2}$

17. $y = x^{2/3}$, from $x = 0$ to $x = 1$

18. $y = x^{2/3}$, from $x = -1$ to $x = 0$

19. $(x + 1)^2 = 4y^3$, from $y = 0$ to $y = 1$

20. $x = \frac{2}{3}(y - 5)^{3/2}$, from $y = 5$ to $y = 6$

21. Find the total length of the *hypocycloid* $x^{2/3} + y^{2/3} = a^{2/3}$, $a > 0$.

22. Find the distance between $(0, 1)$ and $(3, 7)$ along $2x - y + 1 = 0$:
   (a) By the distance formula   (b) By formula (6.22)

23. Find the distance between $(1, 1)$ and $(3, 3\sqrt{3})$ along $y^2 = x^3$.

24. Find the perimeter of the area enclosed by $y^3 = x^2$ and $x + 3y - 4 = 0$.

25. Find the perimeter of the area enclosed by $y = 3(x - 1)^{3/2}$ and $y = 6(x - 1)$.

26. Find the length of one loop of $9y^2 = x(x - 3)^2$, $0 \leq x \leq 3$.

27. Find the length of $6xy = y^4 + 3$ from $y = 1$ to $y = 2$.

28. Find the length of one loop of $9y^2 = x^2(2x + 3)$, $-\frac{3}{2} \leq x \leq 0$.

In Problems 29–32 use (6.22) to set up the integral for the arc length. Do not attempt to integrate. (Techniques for evaluating these integrals are given in Chap. 9.)

29. $y = x^2$, from $x = 0$ to $x = 2$

30. $x = y^2$, from $y = 1$ to $y = 3$

31. $y = \sqrt{25 - x^2}$, from $x = 0$ to $x = 4$

32. $x = \sqrt{4 - y^2}$, from $y = 0$ to $y = 1$

33. The graph whose equation is given by $(x^2/a^2) + (y^2/b^2) = 1$ is called an *ellipse*. Use (6.22) to set up the arc length of this ellipse from $x = 0$ to $x = a/2$. This integral, which is approximated by numerical techniques (see Chap. 9), is called an *elliptic integral of the second kind*.

34. The arc length formula does not apply to the function f given below. Why not?

$$f(x) = \begin{cases} \sin 1/x & \text{if } 0 < x \leq 1 \\ 0 & \text{if } x = 0 \end{cases}$$

35. In each case below, $P_1 = (x_1, y_1)$ and $P_2 = (x_2, y_2)$ are points on the circle $x^2 + y^2 = 1$, with neither coordinate zero. Express the length of the counterclockwise arc $P_1 P_2$ in terms of integrals of the form

$$\int_u^v \frac{1}{\sqrt{1 - t^2}} \, dt \qquad -1 < u < v < 1$$

   (a) When P_1 is in quadrant I and P_2 is in quadrant II
   (b) When P_1 and P_2 are both in quadrant III and $y_1 < y_2$
   (c) When P_1 is in quadrant II and P_2 is in quadrant IV

6. Work

The work W done by a *constant force* F in moving an object a distance x along a straight line in the direction of F is defined to be

(6.25)
$$W = Fx$$

One unit of work is the work done by a unit force in moving an object a unit distance in the direction of the force. In the International System of metric units (abbreviated SI, for Système International d'Unités), the unit of work is 1 newton-meter, which is generally called 1 joule. In terms of customary U.S. units, the unit of work is the foot-pound (ft-lb).* These terms are summarized in Table 2.

Table 2

	Work	=	Force	×	Distance
SI	joule (J)		newton (N)		meter (m)
U.S.	foot-pound (ft-lb)		pound (lb)		foot (ft)

For example, the work required to lift an object weighing 80 pounds a distance of 5 feet would be $(80)(5) = 400$ foot-pounds.

When the force F acts in the same direction as the motion, the work done is positive; if the force F acts in a direction opposite to the motion, the work is negative. In some cases, a force F acts along the line of motion of an object, but the magnitude of the force varies depending on the position of the object. For example, the force required to stretch a spring depends on how far the spring has already been stretched from its normal length. This is an example of a *variable* force. Similarly, suppose a cylindrical tank full of water is to be emptied by pumping the water over the top. The work required to do this depends on the weight of the column of water, a variable, and the distance it is to be lifted.

In general, suppose the force F is given by a continuous function $F = F(x)$, $a \leq x \leq b$,   of the position x of the object it acts upon. We seek a procedure for calculating the work done by a variable force F in moving an object from one position to another. We begin by partitioning $[a, b]$ into n subintervals

$$[a, x_1], \quad [x_1, x_2], \quad \dots, \quad [x_{i-1}, x_i], \quad \dots, \quad [x_{n-1}, b]$$

The length of the ith subinterval is $\Delta x_i = x_i - x_{i-1}, \quad i = 1, 2, \dots, n.$

---

* 1 newton is the force required to accelerate a 1 kilogram mass at 1 meter per second per second; 1 joule = 0.7376 foot-pound;   1 foot-pound = 1.356 joules.

We now consider the ith subinterval $[x_{i-1}, x_i]$ and choose a number u_i in $[x_{i-1}, x_i]$. If the length of this subinterval is small enough, the force $F = F(x)$ will not change too much, so that the work done by F as the object moves from x_{i-1} to x_i is, by (6.25), approximately $F(u_i)(x_i - x_{i-1}) = F(u_i)\Delta x_i$ (see Fig. 41). Thus, a good approximation to the total work W done by F as the object moves from a to b may be found by adding up the work done by F on each subinterval. That is,

(6.26)
$$W \approx F(u_1)\Delta x_1 + F(u_2)\Delta x_2 + \cdots + F(u_n)\Delta x_n = \sum_{i=1}^{n} F(u_i)\Delta x_i$$

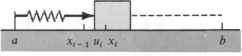

(a)   Constant force
$W = F(b - a)$

(b)   Variable force
$$W = \int_a^b F(x)\, dx$$

Figure 41

As the length of each subinterval gets smaller and smaller, that is, as the norm $\|P\|$ approaches 0, the sum $\sum_{i=1}^{n} F(u_i)\Delta x_i$ is a better and better approximation to the work done by the force $F = F(x)$ in moving an object along a line from $x = a$ to $x = b$ in the direction of F. However, this sum is a Riemann sum, so that, as $\|P\| \to 0$, its limit is a definite integral. This suggests the following definition:

(6.27)   **Work.   The work W done by a continuously varying force $F = F(x)$ acting upon an object, as that object moves along a straight line in the direction of F from $x = a$ to $x = b$, is**

(6.28)
$$W = \int_a^b F(x)\, dx$$

HOOKE'S LAW

A common example of the work done by a variable force is found in the extension or compression of an elastic *spring*. In the case of an elastic spring of *stiffness k** and negligible mass, the force F supported by the spring at any deformation x–either compression or extension—is given by *Hooke's law:*

(6.29)
$$F(x) = kx$$

where the constant k, measured in newtons per meter (N/m, in SI units) or pounds

---

* Sometimes called the *spring constant k*.

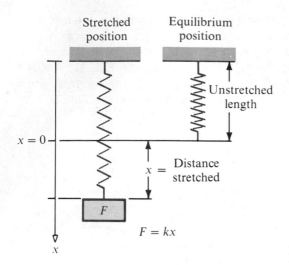

Stretched position

Equilibrium position

Unstretched length

$x = 0$

$x = $ Distance stretched

F

$F = kx$

x

Figure 42

per foot (lb/ft, in U.S. units), depends on the type of spring. As Figure 42 illustrates, the distance x in (6.29) is measured from the equilibrium, or unstretched, position of the spring.

As an example, if a certain spring without any force applied to it is 0.8 meter long, and a force of 2 newtons stretches the spring to a length of 1.2 meters, the constant k for this spring is

$$2 = k(1.2 - 0.8)$$

$$\frac{2}{0.4} = k$$

$$k = 5 \text{ newtons per meter}$$

Thus, the force F required to hold the spring stretched to a length of 3 meters is

$$F = kx = (5)(3 - 0.8) = (5)(2.2) = 11 \text{ newtons}$$

EXAMPLE 1   Consider a spring of stiffness $k = 5$ newtons per meter,   whose unstretched length is 0.8 meter.

(a)   How much work is required to stretch the spring to a length of 1.4 meters?

(b)   How much work is required to stretch it from 2 to 3 meters?

Solution   (a)   By using (6.28) and (6.29) and remembering that our coordinate system sets $x = 0$ at the unstretched length, the work W required to stretch the spring from $x = 0$ (equilibrium) to $x = 1.4 - 0.8 = 0.6$ is

$$W = \int_0^{0.6} 5x \, dx = \tfrac{5}{2}x^2 \Big|_0^{0.6} = \tfrac{5}{2}(0.36) = 0.9 \text{ joule}$$

(b)   The work required to stretch the spring from 2 to 3 meters equals the work required to stretch it to 3 meters from equilibrium less the work required to stretch it to 2 meters from equilibrium. The work W required is therefore,

$$W = \int_0^{3 - 0.8} 5x \, dx - \int_0^{2 - 0.8} 5x \, dx = \int_{1.2}^{2.2} 5x \, dx = \tfrac{5}{2}x^2 \Big|_{1.2}^{2.2}$$

$$= \tfrac{5}{2}(4.84 - 1.44) = 8.5 \text{ joules}   \blacksquare$$

PUMPING LIQUIDS

As a second example of a situation in which force is variable, we take up the problem of the work needed to pump a liquid out of a tank. We rely on the basic idea that the work needed to lift an object a given distance is the product of the weight (force) of the object times the distance it is lifted. That is,

(6.30)   **Work = (Weight of object)(Distance lifted)**

Let's see how this formula works with a liquid in a cylindrical container.

EXAMPLE 2   An oil tank in the shape of a right circular cylinder, with height 40 meters and radius 5 meters, is half full of oil. How much work is required to pump all the oil over the top of the tank?

Solution   Position the cylinder as illustrated in Figure 43. Note that we position the top of the tank at $x = 0$ so that the bottom of the tank is at $x = 40$. The work required to pump a certain volume of the oil over the top depends on the weight of this amount of oil and its distance from the top. The oil fills the tank from $x = 20$ to $x = 40$. Therefore, we partition the interval $[20, 40]$ into n subintervals and let Δx_i denote the length of the ith subinterval. In the ith subinterval, we view the oil as consisting of a

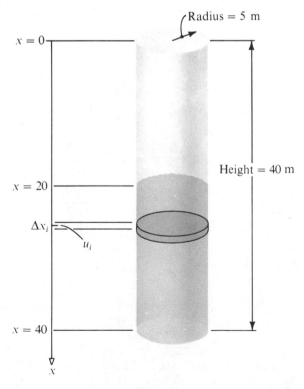

Figure 43

thin layer of height Δx_i. Now we choose a number u_i in the ith subinterval. Then,

$$\text{Area of } i\text{th layer} = \pi(\text{Radius})^2 = 25\pi$$
$$\text{Volume of } i\text{th layer} = (\text{Area of layer})(\text{Height}) = 25\pi\Delta x_i$$
$$\text{Weight* of } i\text{th layer} = \rho g(\text{Volume}) = \rho g(25\pi\Delta x_i)$$
$$\text{Distance } i\text{th layer is lifted} = u_i$$
$$\text{Work done in lifting } i\text{th layer} = 25\pi\rho g u_i \Delta x_i$$

The layers of oil occur from 20 to 40. Therefore, the work W required to pump all the oil over the top is

$$W = \int_{20}^{40} 25\pi\rho gx \, dx = 25\pi\rho g \left.\frac{x^2}{2}\right|_{20}^{40} = 25\pi\rho g(800 - 200) = 15{,}000\pi\rho g \text{ joules} \quad \blacksquare$$

Let's look at another example.

EXAMPLE 3   A water tank in the shape of a hemisphere of radius 2 meters is full of water. How much work is required to pump all the water to a level 3 meters above the tank?

Solution   We choose our coordinate system so that the top of the tank is positioned at $x = 0$ and the bottom is at $x = 2$ (see Fig. 44). The work required to pump a certain volume of the water to a level 3 meters above the top of the tank will depend on the

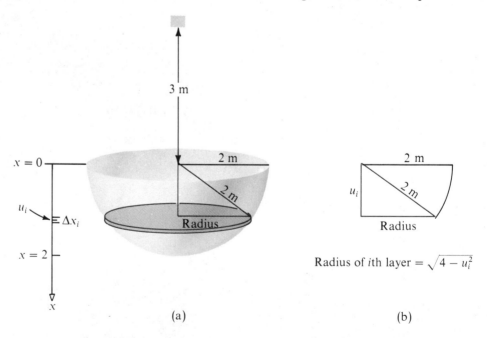

3 m

$x = 0$

2 m

2 m

u_i

$\equiv \Delta x_i$

Radius

$x = 2$

Radius

x

Radius of ith layer $= \sqrt{4 - u_i^2}$

Figure 44   (a)   (b)

---

* Weight $= \rho g(\text{Volume})$, where ρ is the mass density (mass per unit volume, a constant that depends on the type of liquid involved) and $g \approx 9.8 \text{ m/sec}^2$ (the acceleration of gravity).

weight of this amount of water and its distance from the level 3 meters above the tank. The water fills the container from $x = 0$ to $x = 2$. We partition the interval $[0, 2]$ into n subintervals, and let Δx_i be the length of the ith interval. We may view the water in the tank as composed of circular layers, each of height Δx_i. We choose a number u_i in the ith subinterval. As Figure 44 illustrates, the area of the circular layer u_i meters from the top of the tank is

$$\text{Area of } i\text{th layer} = \pi(\text{Radius})^2 = \pi(4 - u_i^2)$$

Then,

$$\text{Volume of } i\text{th layer} = (\text{Area of layer})\Delta x_i = \pi(4 - u_i^2)\Delta x_i$$

The mass density of water is $\rho = 1000$ kilograms per cubic meter; hence,

$$\text{Weight of } i\text{th layer} = \rho g(\text{Volume}) = 1000 g\pi(4 - u_i^2)\Delta x_i$$
$$\text{Distance } i\text{th layer is lifted} = u_i + 3$$
$$\text{Work done in lifting } i\text{th layer} = 1000 g\pi(4 - u_i^2)(u_i + 3)\Delta x_i$$

The work W required to lift all the water is thus

$$W = \int_0^2 1000 g\pi(4 - x^2)(x + 3)\, dx$$

$$= 1000 g\pi \int_0^2 (-x^3 - 3x^2 + 4x + 12)\, dx = 1000 g\pi(20) \approx 615{,}752 \text{ joules} \quad \blacksquare$$

$$g \approx 9.8$$

In general, to compute the work required to pump liquid from a container, think of the liquid as composed of thin layers of height Δx_i and area $A(x)$. If the liquid is to be lifted a height h above the tank, the work required is

(6.31) $$W = \int_a^b \rho g A(x)(x + h)\, dx$$

where ρ is the mass density of the liquid and the liquid fills the container from $x = a$ to $x = b$ (see Fig. 45).

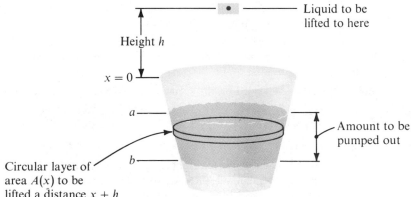

Figure 45

Liquid to be lifted to here

Height h

$x = 0$

a

Amount to be pumped out

b

Circular layer of area $A(x)$ to be lifted a distance $x + h$

GRAVITATIONAL FORCES

One of the conclusions of Newton's law of universal gravitation is that the force required to move an object, say, a rocket, of mass m (in kilograms) at a point x meters above the center of the earth is $F(x) = GMm/x^2$, where G is the gravitational constant of the earth and M is the mass (in kilograms) of the earth. Hence, using (6.28), we see that the work required to move an object of mass m from the surface of the earth to a distance r meters from the center of the earth (where R is the radius of the earth in meters) is

(6.32)
$$W = \int_R^r \frac{GMm}{x^2}\,dx = -\frac{GMm}{x}\bigg|_R^r = \frac{GMm}{R} - \frac{GMm}{r} = GMm\left(\frac{1}{R} - \frac{1}{r}\right)$$

Physicists know that $GM = gR^2$, where $g \approx 9.8$ meters per second per second is the acceleration of gravity of the earth and $R \approx 6.37 \times 10^6$ meters. Further, if d is the distance the object is to be moved above the surface of the earth, then $r = R + d$. Consequently, from (6.32), we find that the work required to move a body of mass m to a distance d meters above the surface of the earth is

$$W = gRm\left(1 - \frac{R}{R + d}\right) \text{joules}$$

Observe that even though d may be extremely large, W will never be bigger than $gRm \approx 62.4 \times 10^6 m$ joules.

EXERCISE 6

1. A spring, whose unstretched length is 1 meter, stretches to a length of 3 meters when a force of 3 newtons is applied. Find the work needed to stretch the spring to a length of 2 meters from its natural length.

2. How much work is required to stretch the spring in Problem 1 to a length of 4 meters?

3. A spring, whose unstretched length is 2 meters, is compressed to a length of $\frac{1}{2}$ meter when a force of 10 newtons is applied. Find the work required to compress the spring to a length of 1 meter.

4. How much work is required to compress the spring in Problem 3 to a length of 1.5 meters?

5. A spring, whose unstretched length is 4 feet, stretches to a length of 8 feet when a force of 2 pounds is applied. If 9 foot-pounds of work is expended to stretch this spring from an unstretched position, what is its total length?

6. If 8 foot-pounds of work is expended on the spring in Problem 5, how far is it stretched?

7. How much work is required to pump all the water over the top of a full cylindrical tank that is 4 meters in diameter and 6 meters high?

8. How much work is required to pump half the water over the top of the tank in Problem 7?

9. A full water tank in the shape of an inverted right circular cone is 8 meters across the top and 4 meters high. How much work is required to pump all the water over the top of the tank?

10. If the surface of the water in the tank of Problem 9 is 2 meters below the top of the tank, how much work is required to pump all the water over the top of the tank?

11. A water tank in the shape of a hemispherical bowl of radius 4 meters is filled with water to a depth of 2 meters. How much work is required to pump all the water over the top of the tank?

12. If the water tank in Problem 11 is completely filled with water, how much work is required to pump all the water to a height 2 meters above the tank?

13. A swimming pool is in the shape of a rectangular parallelepiped 6 feet deep, 30 feet long, and 20 feet wide. It is filled with water to a depth of 5 feet. How much work is required to pump all the water over the top? [*Hint:* $\rho g = 62.5$ pounds per cubic foot.]

14. A 1 horsepower motor can do 550 foot-pounds of work per second. Using this motor, how long does it take to pump all the water out of a swimming pool 5 feet deep, 25 feet long, and 15 feet wide if the pool is filled to a depth of 4 feet? The pool is in the shape of a rectangular parallelepiped.

Use (6.32) to do Problems 15 and 16.

15. The minimum energy required to move an object of mass 30 kilograms a distance 500 kilometers above the surface of the earth is equal to the work required to accomplish this. Find the work required. (The radius of the earth is approximately 6370 kilometers.)

16. The minimum energy required to move a rocket of mass 1000 kilograms a distance of 800 kilometers above the surface of the earth is equal to the work required to do this. Find the work required.

17. By *Coulomb's law*, a positive charge m of electricity repels a unit of positive charge at a distance x with the force m/x^2. What is the work done when the unit charge is carried from $x = 2a$ to $x = a$?

18. In raising a leaky bucket from the bottom of a well 25 feet deep, one-fourth of the water is lost. If the bucket weighs 1.5 pounds, the water in the bucket at the start weighs 20 pounds, and the amount that has leaked out is assumed to be proportional to the displacement, find the work done in raising the bucket.

19. The stiffness of a spring on a bumping post in a freight yard is 300,000 newtons per meter. Find the work done in compressing the spring 0.1 meter.

20. A cable of uniform linear mass $\rho = 9$ kilograms per meter is being unwound from a cylindrical drum. If 15 meters are already unwound, what is the work done by gravity in unwinding 60 meters more?

21. A uniform chain 30 feet long and weighing 30 pounds is hanging from the top of a building 30 feet high. If a bucket filled with cement weighing 150 pounds is attached to the end of the chain, how much work is required to pull the bucket and chain to the top of the building?

22. In Problem 21, if a uniform cable 30 feet long and weighing 20 pounds is used instead, how much work is required to pull the bucket and chain to the top of the building?

The pressure p (in pounds per square inch, lb/in.²) and the volume v (in cubic inches) of an adiabatic expansion of a gas are related by $pv^k = c$, where k and c are constants that depend on the gas. If the gas expands from $v = a$ to $v = b$, the work done (in inch-pounds, in.-lb) is $W = \int_a^b p \, dv$. Use this fact in Problems 23 and 24.

23. The pressure of 1 pound of a gas is 100 pounds per square inch and the volume is 2 cubic feet. Find the work done by the gas in expanding to double its volume according to the law $pv^{1.4} = c$ (in inch-pounds).

24. The pressure and volume of a certain gas obey the law $pv^{1.2} = 120$ (in inch-pounds). Find the work done when the gas expands from $v = 2.4$ to $v = 4.6$ cubic inches.

7. Liquid Pressure and Force

When containers are built to hold liquids in place, it is important to know the force F caused by liquid pressure on the sides of the containers. The pressure P exerted by a liquid of mass density ρ at a depth h below the surface of the liquid is defined as

(6.33)
$$P = \rho g h$$

where $g \approx 9.8 \text{ m/sec}^2 \approx 32.2 \text{ ft/sec}^2$ is the acceleration of gravity. Table 3 summarizes the units of measure needed for these calculations.

Table 3

	Pressure	=	Mass Density	×	g	×	Depth
SI	newton/meter² (N/m²)		kilogram/meter³ (kg/m³)		9.8 m/sec²		meter (m)
U.S.	pound/foot² (lb/ft²)		slug/foot³ (slug/ft³)		32.2 ft/sec²		foot (ft)

If a flat plate of area A is suspended horizontally in a liquid at a depth h, the force F (weight) caused by the liquid on one face of the plate is

(6.34)
$$F = PA = \rho g h A$$

For example, the mass density of water is about $\rho = 1.94$ slugs per cubic foot, so the pressure due to water on a plate suspended horizontally at a depth of 4 feet is $(4)(1.94)g = 4(62.5) = 250$ pounds per square foot. If the plate has an area of 2 square feet, the force (weight) exerted on one side of the plate is $(2)(250) = 500$ pounds.

On the other hand, if a plate is suspended vertically in a liquid, the force caused by the liquid on one face of the plate is more difficult to find because the force at the bottom of the plate is greater than that at the top. To solve this problem, we proceed as follows: Suppose a plate is suspended vertically in a liquid of mass density ρ. Suppose further that the plate is bounded by $y = c$, $y = d$, $x = g(y)$, and $x = f(y)$, where f and g are continuous functions on $c \leq y \leq d$ and $f(y) \geq g(y)$ on $c \leq y \leq d$ (see Fig. 46). The surface of the liquid is the line $y = H$, where $H \geq d$, so that the top of the plate is at a depth of $(H - d)$ and the bottom of the plate is at a depth of $(H - c)$.

We partition the interval $[c, d]$ into n subintervals

$$[c, y_1], \quad [y_1, y_2], \quad \ldots, \quad [y_{i-1}, y_i], \quad \ldots, \quad [y_{n-1}, d]$$

The length of the ith subinterval is $\Delta y_i = y_i - y_{i-1}$. And, as usual, we select a number u_i in the ith subinterval $[y_{i-1}, y_i]$. If the length Δy_i of the ith subinterval is small, then all points in the ith portion of the plate are roughly the same distance $(H - u_i)$ from the surface, so that the pressure P of the liquid on this portion of the

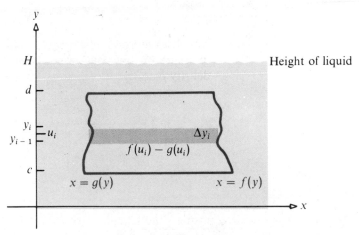

Figure 46

plate is approximately $P = \rho g(H - u_i)$. By (6.34), the force due to liquid pressure on this portion of the plate is approximately

$$\rho g(H - u_i)[f(u_i) - g(u_i)]\Delta y_i$$

A good approximation to the force F on the plate can be found by adding up the forces on each subinterval. That is,

(6.35)
$$F \approx \sum_{i=1}^{n} \rho g(H - u_i)[f(u_i) - g(u_i)]\Delta y_i$$

As the length of each subinterval gets smaller and smaller, that is, as the norm $\|P\|$ approaches 0, the sum (6.35) becomes a better and better approximation of the force due to liquid pressure on the plate. However, the sum (6.35) is a Riemann sum, so that, as $\|P\| \to 0$, its limit is a definite integral. This suggests the following definition:

(6.36) *Liquid Pressure.* **The force F due to a liquid of mass density ρ on a plate of the type illustrated in Figure 46, where f and g are continuous on $[c, d]$, is**

(6.37)
$$F = \int_{c}^{d} \rho g(H - y)[f(y) - g(y)]\,dy$$

As an aid in using (6.37), it is useful to remember that

Force on ith rectangle = Mass density $\times g \times$ Depth $\times$ Area of ith rectangle

EXAMPLE 1   A trough, whose cross section is a trapezoid, is 2 meters across at the bottom, 4 meters across at the top, and 2 meters deep. If the trough is filled with a liquid of mass density ρ, what is the force due to liquid pressure on one end of the trough?

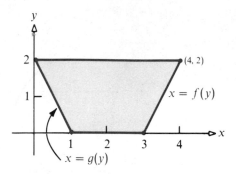

Figure 47

Solution   It is convenient to position the trough as shown in Figure 47. The sides of the end of the trough are lines that pass through the points $(0, 2)$, $(1, 0)$ and $(3, 0)$, $(4, 2)$, respectively. Their equations are

$$y - 0 = \frac{-2}{1}(x - 1) \qquad \text{and} \qquad y - 0 = \frac{2}{1}(x - 3)$$

$$x = g(y) = \frac{-1}{2}y + 1 \qquad\qquad x = f(y) = \frac{1}{2}y + 3$$

The height of the liquid is $H = 2$, so by (6.37), the force F due to the liquid on an end of the trough is

$$F = \int_0^2 \rho g(2 - y)[f(y) - g(y)]\, dy = \int_0^2 \rho g(2 - y)(y + 2)\, dy$$

$$= \rho g \int_0^2 (-y^2 + 4)\, dy = \rho g \left(\frac{-y^3}{3} + 4y\right)\Bigg|_0^2$$

$$= \rho g \left(\frac{-8}{3} + 8\right) = \frac{16}{3}\rho g \text{ newtons} \qquad \blacksquare$$

In the development of the definition of liquid pressure, we positioned the coordinate system so that the submerged plate was located in the first quadrant. As the next example illustrates, the coordinates may be placed in any convenient position. Keep in mind that the essential idea behind the formula for force due to liquid pressure is that   Force $= \rho g \times$ Depth $\times$ Area.

EXAMPLE 2   A cylindrical sewer pipe, of radius 2 meters, is half full of water. Find the force exerted on one side of a gate that is used to seal off the sewer.

Solution   Figure 48 illustrates the situation. The equation of the circle with center at $(0, 0)$ and radius 2 is $x^2 + y^2 = 4$.   Thus, we have

$$x = g(y) = -\sqrt{4 - y^2} \qquad \text{and} \qquad x = f(y) = \sqrt{4 - y^2}$$

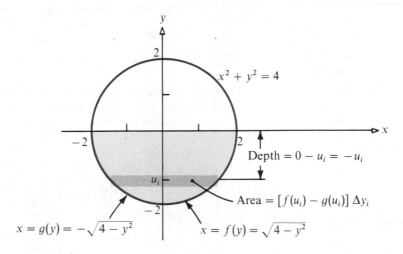

Figure 48

The force F exerted on one side of the gate by the water is

$$F = \int_{-2}^{0} \rho g(0 - y)[2\sqrt{4 - y^2}]\, dy$$

Now we use the substitution $u = 4 - y^2$ so that $du = -2y\, dy$. Since $u = 0$ when $y = -2$, and $u = 4$ when $y = 0$, we find that

$$F = \rho g \int_{0}^{4} 2\sqrt{u}\, \frac{du}{2} = \rho g \left. \frac{u^{3/2}}{\frac{3}{2}} \right|_{0}^{4} = \rho g \frac{16}{3} \approx 52{,}266.7 \text{ newtons} \qquad \blacksquare$$

$$\rho = 1000 \text{ kg/m}^3$$
$$g = 9.8 \text{ m/sec}^2$$

EXERCISE 7

1. A rectangular plate of width 2 meters and depth 6 meters is suspended vertically in a pool of water so that the top of the plate is even with the surface of the water. What is the force due to water pressure on one side of the plate?

2. If the plate in Problem 1 is suspended in the water so that the top of the plate is 1 meter below the water surface, what is the force due to water pressure on one side of the plate?

3. A swimming pool is in the shape of a rectangular parallelepiped 6 feet deep, 30 feet long, and 20 feet wide. If the pool is full of water, what is the force due to water pressure on one short side of the pool?   ($\rho \approx 1.94$ slugs per cubic foot.)

4. For the pool in Problem 3, what is the force due to water pressure on one long side of the pool?

5. A trough, whose cross section is a trapezoid, is 1 meter across at the bottom, 5 meters across at the top, and 2 meters deep. If the trough is filled with water, what is the force due to water pressure on one end of the trough?

6. A trough, whose cross section is an equilateral triangle with side 2 meters long, is filled with water. What is the force due to water pressure on one end of the trough?

7. A trough, whose cross section is a semicircle of radius 2 meters, is filled with water. What is the force due to water pressure on one end of the trough?

8. A viewing plate in a submarine is a circle of radius 1 foot. If the depth of the center of the viewing plate is 5 feet below the surface of the water, what is the force due to water pressure on one end of the plate? Assume the viewing plate is vertical. (See the comment to Problem 4 in Exercise 4, p 356.)

9. Find the force on the face of a vertical floodgate in the shape of an isosceles triangle whose base is 1.5 meters and whose altitude is 1 meter, if its base is on the surface of the water.

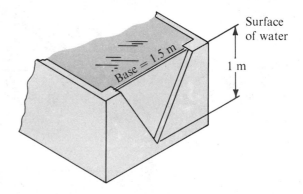

10. A vertical masonry dam in the form of an isosceles trapezoid is 200 meters long at the surface of the water, 150 meters long at the bottom, and 60 meters high. What force must it withstand? Express the result in newtons.   [*Hint:*   Use similar triangles.]

11. A tanker truck carrying a half load of oil is in the shape of a right circular cylinder. If the radius of the cylinder is 4 feet, what is the force due to the pressure of the oil on one end of the cylinder? Assume that the mass density of the oil is $\rho = 1.86$ slugs per cubic foot.

12. The gas tank in a sports car is a cylinder lying on its side. If the diameter of the tank is 1 meter and if the tank is filled with gasoline to within 0.5 meter of the top, find the force on one end of the tank. (The mass density of gasoline is 600 kilograms per cubic meter.)

8. Average Value of a Function

In this section we examine how the definite integral can be applied to calculate averages. For example, at the U.S. Weather Bureau, a continuous reading of the temperature over a 24 hour period is taken daily. To obtain the average daily temperature, 12 readings may be taken at 2 hour intervals beginning at midnight: $f(0), f(2), f(4), \ldots, f(20), f(22)$.   The average temperature is then calculated as

$$\frac{f(0) + f(2) + f(4) +\; \cdots \;+ f(20) + f(22)}{12}$$

This number represents a good approximation to the true average as long as there are no drastic temperature changes over short periods of time. To improve the approximation, readings may be taken every hour. The average in this case would be

$$\frac{f(0) + f(1) + \cdots + f(22) + f(23)}{24}$$

An even better approximation would be obtained if readings were recorded every half hour.

In general, if $y = f(x)$ is a continuous function defined on the interval $[a, b]$, we can obtain the *average of f on* $[a, b]$ as follows: We partition the closed interval $[a, b]$ into n subintervals

$$[a, x_1], \quad [x_1, x_2], \quad \ldots, \quad [x_{i-1}, x_i], \quad \ldots, \quad [x_{n-1}, b]$$

each of length $\Delta x = (b - a)/n$. We pick a number u_i in the ith subinterval $[x_{i-1}, x_i]$. An approximation of the average value of f over the interval $[a, b]$ is then the sum

(6.38)
$$\frac{f(u_1) + f(u_2) + \cdots + f(u_n)}{n}$$

If we multiply and divide the expression in (6.38) by $(b - a)$, we get

$$\frac{f(u_1) + f(u_2) + \cdots + f(u_n)}{n} = \frac{1}{b - a}\left[f(u_1)\frac{b - a}{n} + f(u_2)\frac{b - a}{n} + \cdots + f(u_n)\frac{b - a}{n} \right]$$

(6.39)
$$= \frac{1}{b - a}\left[f(u_1)\Delta x + f(u_2)\Delta x + \cdots + f(u_n)\Delta x \right]$$

$$= \frac{1}{b - a}\sum_{i=1}^{n} f(u_i)\Delta x$$

The sum obtained gives an approximation to the average value. As the length of each subinterval gets smaller and smaller, this sum becomes a better and better approximation to the average value of f on $[a, b]$. However, this sum is a Riemann sum, so that its limit is a definite integral. This suggests the following definition:

(6.40)   *Average Value of a Function over an Interval.*   **The average value $\bar{y}$ of a continuous function f over $[a, b]$ is**

(6.41)
$$\bar{y} = \frac{1}{b - a}\int_a^b f(x)\, dx$$

EXAMPLE 1   The average value of $f(x) = x^3$ over the interval $[0, 2]$ is

$$\bar{y} = \frac{\displaystyle\int_0^2 x^3\, dx}{2 - 0} = \frac{4}{2} = 2 \quad \blacksquare$$

The average value $\bar{y}$ of a function f, as defined in (6.41), equals the value $f(u)$ referred to in (5.28), the mean value theorem for integrals. Let's review the geometric interpretation.

We begin by rearranging the formula for $\bar{y}$ as

(6.42)
$$\bar{y}(b - a) = \int_a^b f(x)\, dx$$

If $f(x) \geq 0$ on $[a, b]$, the right side of (6.42) represents the area enclosed by the graph of $y = f(x)$, the x-axis, the line $x = a$, and the line $x = b$. The left side of the equation can be interpreted as the area of a rectangle of height $\bar{y}$ and base $(b - a)$. Hence, (6.42) asserts that $\bar{y}$, the average value of the function, is the height of a rectangle whose base is $(b - a)$ and whose area is equal to the area under the graph of f (see Fig. 49).

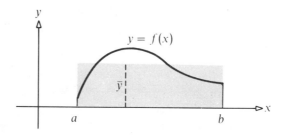

Figure 49

The next example illustrates that in computing averages it is important to indicate the variable with respect to which the average is to be computed.

EXAMPLE 2   For a freely falling body starting from rest, calculate:
(a)   The average velocity $\bar{v}_t$ with respect to time over $[0, t_1]$
(b)   The average velocity $\bar{v}_s$ with respect to distance over $[0, s_1]$, where s_1 is the distance the body falls in time t_1

Solution   (a)   Since we wish to compute the average velocity $\bar{v}_t$ relative to time t, we need to express the velocity v as a function of t. The distance is $s = \frac{1}{2}gt^2$, so it follows that

$$v(t) = \frac{ds}{dt} = gt$$

Then

$$\bar{v}_t = \frac{1}{t_1 - 0}\int_0^{t_1} v(t)\, dt = \frac{1}{t_1}\int_0^{t_1} gt\, dt = \left(\frac{1}{t_1}\right)\frac{1}{2}gt^2\bigg|_0^{t_1} = \frac{1}{2}gt_1 = \frac{1}{2}v(t_1)$$

(b) To find the average velocity $\bar{v}_s$ relative to distance s, we note that $\;\; s = \frac{1}{2}gt^2,\;\;$ so that $\;\; t = \sqrt{2s/g}\;\;$ and $\;\; v(s) = g\sqrt{2s/g} = \sqrt{2gs}.\;\;$ Then,

$$\bar{v}_s = \frac{1}{s_1 - 0} \int_0^{s_1} v(s)\, ds = \frac{1}{s_1} \int_0^{s_1} \sqrt{2g}\sqrt{s}\, ds$$

$$= \frac{\sqrt{2g}}{s_1}\left.\left(\frac{s^{3/2}}{\frac{3}{2}}\right)\right|_0^{s_1} = \frac{2\sqrt{2g}}{3}s_1^{1/2} = \frac{2}{3}\sqrt{2gs_1} = \frac{2}{3}v(s_1) \qquad \blacksquare$$

EXERCISE 8

In Problems 1–10 find the average value of each function f over the given interval.

1. $f(x) = x^2,$ over $[0, 1]$ 2. $f(x) = 2x^2,$ over $[-4, 2]$ 3. $f(x) = 1 - x^2,$ over $[-1, 1]$

4. $f(x) = 16 - x^2,$ over $[-4, 4]$ 5. $f(x) = 3x,$ over $[1, 5]$ 6. $f(x) = 4x,$ over $[-5, 5]$

7. $f(x) = -5x^4 + 4x - 10,$ over $[-2, 2]$ 8. $f(x) = 10x^4 - 2x + 7,$ over $[-1, 2]$

9. $f(x) = \sin x,$ over $[0, \pi/2]$ 10. $f(x) = \cos x,$ over $[0, \pi/2]$

11. Find the average velocity with respect to time of a freely falling object that falls from rest for 5 seconds. What is its average velocity with respect to distance? Use SI units.

12. If the object in Problem 11 falls from rest for 3 seconds, what is its average velocity with respect to time? What is its average velocity with respect to distance? Use SI units.

13. A rod 3 meters long is heated to $25x$ degrees Celsius, where x is the distance (in meters) from one end of the rod. Calculate the average temperature of the rod.

14. The rainfall per day, x days after the beginning of the year, is $\;\;0.00002(6511 + 366x - x^2),\;\;$ measured in centimeters. By integration, estimate the average daily rainfall for the first 180 days of the year.

15. A car starting from rest accelerates at the rate of 3 meters per second per second. Find its average speed over the first 8 seconds.

16. What is the average area of all circles whose radii are between 1 and 3 meters?

17. The mass density of a metal bar of length 3 meters is given by $\;\;\rho(x) = 1 + x - \sqrt{x}\;\;$ kilograms per cubic meter, where x is the distance in meters from one end of the bar. What is the average mass density over the length of the entire bar?

18. The acceleration at time t of a particle moving on the x-axis is $\;\;4\pi \cos t.\;\;$ If the velocity is 0 at $\;\;t = 0,\;\;$ what is the average velocity of the particle over the interval $\;\;0 \le t \le \pi$?

Miscellaneous Exercises

In Problems 1–3 find the total area enclosed by the graphs. Use any convenient method.

1. $y = x\sqrt{4 - x^2}$ and $y = -x,\;\; 0 \le x \le 2$ 2. $y = x^2$ and $y = 18 - x^2$

3. $x = 2(y - 1)^2$ and $(y - 1)^2 = x - 1$

4. Find b such that the area under the graph of $\;\;y = (x + 1)\sqrt{x^2 + 2x + 4}\;\;$ is $\frac{56}{3}$ for $\;\;0 \le x \le b$.

5. Find h such that the area under the graph of $y^2 = x^3$, $0 \leq x \leq 64$, $y \geq 0$, is equal to that of a rectangle of base 4 and height h.

6. Find h if the area enclosed by the graphs of $y = x$, $y = 8x$, and $y = 1/x^2$ is equal to that of an isosceles triangle of base 1 and height h.

7. In the figure, the shaded area A is enclosed by the graphs of $xy = 1$, $x = 1$, $x = 2$, and $y = 0$. Find the volume of the solid generated by revolving the area A about the x-axis.

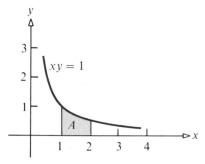

8. Let A be the area in the first quadrant enclosed by the x-axis and the graph of $y = kx - x^2$, where $k > 0$.
   (a) In terms of k, find the volume produced when A is revolved around the x-axis.
   (b) In terms of k, find the volume produced when A is revolved around the y-axis.
   (c) Find the number k for which the volumes found in parts (a) and (b) are equal.

9. Use integration to find the area of the triangle enclosed by the lines $x - y + 1 = 0$, $3x + y - 13 = 0$, and $x + 3y - 7 = 0$.

10. If P is a polynomial that is positive for $x > 0$ and for each $k > 0$ the area under the graph of P from $x = 0$ to $x = k$ is equal to $k^3 + 3k^2 - 6k$, find P.

11. Find the area enclosed by the coordinate axes and the graph of $\sqrt{x} + \sqrt{y} = \sqrt{a}$.

12. Find $b > 0$ so that the area in the first quadrant enclosed by the graph of $y = 1 + b - bx^2$ and the coordinate axes is a minimum.

13. (a) Find all numbers b for which the graphs of $y = 2x + b$ and $y^2 = 4x$ intersect in two distinct points.
    (b) If $b = -4$, find the area enclosed by the graphs of $y = 2x - 4$ and $y^2 = 4x$.
    (c) If $b = 0$, find the volume of the solid generated by revolving about the x-axis the area enclosed by the graphs of $y = 2x$ and $y^2 = 4x$.

14. Find the volume of the solid generated by revolving the area enclosed by the graph of $y = x^2 - 4$ and the x-axis about the line $y = -4$.

15. Find the volume of the solid generated by revolving the area enclosed by the graph of $y = -x^2 + 6x - 8$ and the x-axis about the y-axis. Use both the washer method and the shell method.

16. Suppose $f(x) \geq 0$ for $x \geq 0$ and the area below f from $x = 0$ to $x = k$ is revolved about the x-axis. If for each $k > 0$ the volume of the resulting solid is $\frac{1}{5}k^5 + k^4 + \frac{4}{3}k^3$, find f.

17. Suppose a plane region of area A to the right of the y-axis is revolved about the y-axis, generating a solid of volume V. If this same area is revolved about the line $x = -k$, $k > 0$, show that the solid thus generated has volume $V + 2\pi kA$.

18. A container is formed by revolving the area under the graph of $y = x^2$, $0 \leq x \leq 2$, about the y-axis. How much work is required to fill the container with water from a source 2 units below the x-axis by pumping through a hole in the bottom of the container?

19. A spring of unstretched length 0.6 meter would be stretched to a length of 0.8 meter by a force of 4 newtons. If the upper end of the spring is attached to the point $(0, 0.6)$, how much work is done in moving the lower end from the origin to $(0.8, 0)$?

20. Find the unstretched length of a spring if the work required to stretch the spring from 1.0 meter to 1.4 meters is half the work required to stretch it from 1.2 meters to 1.8 meters.

21. Find the point P on the graph of $y = \frac{2}{3}x^{3/2}$ to the right of the y-axis such that the length of the graph $y = \frac{2}{3}x^{3/2}$ from 0 to P is equal to $\frac{52}{3}$.

22. Consult the accompanying figure. The area enclosed by the graphs of $y_1 = g(x)$, $y_2 = f(x)$, and the y-axis is to be revolved about the y-axis.

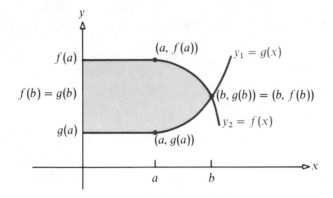

Show that the resulting volume V is given by:

(a) $V = \pi a^2 [f(a) - g(a)] + 2\pi \int_a^b x[f(x) - g(x)]\, dx$ if the shell method is used

(b) $V = \pi \int_{g(a)}^{g(b)} [g^{-1}(y)]^2\, dy + \pi \int_{g(b)}^{f(a)} [f^{-1}(y)]^2\, dy$ if the disk method is used

23. *The force in newtons required to stretch a certain spring a distance of x meters beyond its unstretched length is given by $F = 100x$.
(a)   What force will stretch the spring 0.1 meter? 0.2 meter? 0.4 meter?
(b)   How much work is required to stretch the spring 0.1 meter? 0.2 meter? 0.4 meter?

24. *Calculator problem.*   Inscribe a regular polygon of 2^n sides in a circle of radius 1. The figure illustrates the situation for $n = 3$.   Show that the indicated angle $\alpha = 45/2^{n-2}$ degrees.   Show that the formula $2^{n+1} \sin(45/2^{n-2})$ gives a good approximation to the arc length (circumference) of the circle. Evaluate this approximation for several values of n. Compare your answers with the exact value.

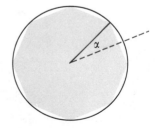

* Adapted from F. W. Sears, M. W. Zemansky, and H. D. Young, *University Physics* (Reading, Mass.: Addison-Wesley Publishing Co., 1976), p. 132. Reprinted by permission.

25. *Calculator problem.*   By considering Problem 24 and the half-angle formula, show that $2^{n+2}a_n$ gives an approximation to π, where

$$a_0 = \frac{1}{\sqrt{2}} \quad \text{and} \quad a_{n+1} = \sqrt{\frac{1 - \sqrt{1 - a_n^2}}{2}} \qquad n = 0, 1, 2, \ldots$$

Evaluate $2^{n+2}a_n$ for several values of n and compare your answer with the exact value of π.

26. A continuous function f on $[a, b]$ satisfying   (i) $f(x) \geq 0$ for x in $[a, b]$ and   (ii) $\int_a^b f(x)\,dx = 1$ is called a *probability density function* on $[a, b]$. If $a \leq c < d \leq b$,   the probability of obtaining a value between c and d is defined as $\int_c^d f(x)\,dx$.
   (a)   Find a constant c so that $f(x) = cx$ is a probability density function on $[0, 2]$.
   (b)   Find the probability of obtaining a value between 1 and 1.5.

27. Refer to Problem 26. If f is a probability density function on $[a, b]$, the *distribution function* F for f is defined as

$$F(x) = \int_a^x f(t)\,dt \qquad a \leq x \leq b$$

Find the distribution function F for the probability density function f of Problem 26, part (a).

28. For the distribution function $F(x) = x - 1$ on $[1, 2]$, find:
   (a)   The probability density function f corresponding to F
   (b)   The probability of obtaining a value between 0.5 and 0.7

7

Exponential and Logarithmic Functions

1. Introduction

In this section we review the laws for exponents and logarithms, as presented in pre-calculus mathematics, in preparation for the definitions of the logarithmic and exponential functions given in Sections 2 and 3.

Let a be a positive number. The following definitions should be familiar:

$$a^n = \underbrace{a \cdot a \cdots \cdot a}_{n \text{ factors}} \qquad \text{for } n \text{ a positive integer}$$

$$a^0 = 1$$

$$a^{n/m} = (\sqrt[m]{a})^n = \sqrt[m]{a^n} \qquad \text{for } m \text{ and } n \text{ positive integers}$$

$$a^{-r} = \frac{1}{a^r} \qquad \text{for } r \text{ a rational number}$$

From these definitions we can obtain the following laws, which hold for any positive real numbers a and b and any rational numbers r and s:

(7.1)
$$a^r a^s = a^{r+s} \qquad (a^r)^s = a^{rs} \qquad (ab)^r = a^r b^r$$

At this point in most precalculus courses, the student is asked to *assume* that if $a > 0$, then a^x can be defined for *any* real number x. However, making this assumption causes difficulty for two reasons: First, the development is not rigorous; and second, we have stepped out beyond our intuition. For example, what is meant by $2^{\sqrt{5}}$ or 3^π?

But if we do not object to this "definition" of a^x, $a > 0$, then it can be used to "define" the logarithm as follows: Consider the exponential relationship $x = a^y$, $a > 0$, $a \neq 1$. The name we give the solution for y when x is given is the *logarithm to the base a of x*, and we symbolize it by $y = \log_a x$. That is,

$$y = \log_a x \qquad \text{if and only if} \qquad x = a^y$$

For example,

$$
\begin{array}{lllll}
y = \log_2 1 & \text{means} & 2^y = 1, & \text{that is,} & y = 0 \\
y = \log_2 4 & \text{means} & 2^y = 4, & \text{that is,} & y = 2 \\
y = \log_3 \tfrac{1}{3} & \text{means} & 3^y = \tfrac{1}{3}, & \text{that is,} & y = -1 \\
y = \log_a 1 & \text{means} & a^y = 1, & \text{that is,} & y = 0 \\
y = \log_a a & \text{means} & a^y = a, & \text{that is,} & y = 1
\end{array}
$$

The last two results may be rewritten as

(7.2)
$$\log_a 1 = 0 \qquad \log_a a = 1 \qquad \text{for} \qquad a > 0, \quad a \neq 1$$

For positive real numbers a, M, and N, with $a \neq 1$, the following laws for logarithms may be derived from the laws for exponents:

$$\log_a(MN) = \log_a M + \log_a N$$

(7.3)
$$\log_a \frac{M}{N} = \log_a M - \log_a N$$

$$\log_a M^r = r \log_a M \qquad \text{for } r \text{ any real number}$$

Here are some examples of (7.3):

$$\log_a[x(x + 1)] = \log_a x + \log_a(x + 1)$$

$$\log_a \frac{x^2 + 4}{x + 1} = \log_a(x^2 + 4) - \log_a(x + 1)$$

$$\log_a x^{3/2} = \tfrac{3}{2} \log_a x$$

Until recently, *common* logarithms—that is, logarithms to the base 10—were used to facilitate arithmetic computations. However, the development of hand calculators has made this particular use of logarithms less important than it once was.

On the other hand, *natural* logarithms—that is, logarithms to the base $e = 2.718\ldots$ —remain important for two major reasons: First, they arise naturally in the study of many real-world phenomena; and, second, their use simplifies many mathematical calculations.

Since we now have calculus as a tool, the approach we shall take to logarithms will be based on the definite integral. Although this approach is different from the approach taken in precalculus courses, it has the advantage that it avoids the lack of rigor of the precalculus approach.

2. The Natural Logarithm as an Integral

We shall begin with a definition of the natural logarithm, which we choose to symbolize as $\ln x$ (instead of $\log_e x$, $e = 2.718\ldots$).* The basis for the definition is the integration problem

$$\int \frac{1}{t}\, dt$$

Recall that the integration formula $\int t^n\, dt = t^{n+1}/(n+1) + C$, does not hold when $n = -1$, in which event $t^n = t^{-1} = 1/t$. However, since the function $f(t) = 1/t$ is continuous when $t > 0$, it follows from (5.23) that the definite integral $\int_1^x (1/t)\, dt$ exists for all positive x. This integral therefore defines a function whose domain is $x > 0$.

(7.4)   **DEFINITION**   *Natural Logarithm.*   **For** $x > 0$, **the** *natural logarithm of x* **is**

$$\ln x = \int_1^x \frac{1}{t}\, dt$$

As it turns out, this function $\int_1^x (1/t)\, dt$ has all the properties of a logarithm function, as stated in (7.2) and (7.3), along with those familiar from precalculus mathematics.

(7.5)   **For positive real numbers a and b and any rational number r:**

(a) $\ln 1 = 0$ (b) $\ln(ab) = \ln a + \ln b$

(c) $\ln a^r = r \ln a$ (d) $\ln \dfrac{a}{b} = \ln a - \ln b$

Proof
(a)   Replace x by 1 in (7.4) to get $\ln 1 = \int_1^1 (1/t)\, dt = 0$.
(b)   Replace x by ab in (7.4). The result is

(7.6)
$$\ln(ab) = \int_1^{ab} \frac{1}{t}\, dt = \int_1^a \frac{1}{t}\, dt + \int_a^{ab} \frac{1}{t}\, dt = \ln a + \int_a^{ab} \frac{1}{t}\, dt$$

---

* The letters ln, read "ell-en," in $\ln x$ are an abbreviation for the Latin *logarithmus naturalis*.

In $\int_a^{ab} (1/t) \, dt$,   we use the substitution $t = au$ so that $dt = a \, du$.   When $t = a$, $u = 1$;   and when $t = ab$, $u = b$.   Thus,

$$\int_a^{ab} \frac{1}{t} \, dt = \int_1^b \frac{1}{au} \, a \, du = \int_1^b \frac{1}{u} \, du = \ln b$$

By (7.4)

By using this result in (7.6), we have

$$\ln(ab) = \ln a + \ln b$$

(c)   We replace x by a^r in (7.4). The result is

$$\ln a^r = \int_1^{a^r} \frac{1}{t} \, dt$$

If $r = 0$,   the result is true. If $r \neq 0$,   we use the substitution $t = u^r$, so   that $dt = ru^{r-1} \, du$.   When $t = 1$, $u = 1$,   and when $t = a^r$, $u = a$. Hence,

$$\ln a^r = \int_1^{a^r} \frac{1}{t} \, dt = \int_1^a \frac{1}{u^r} \, ru^{r-1} \, du = r \int_1^a \frac{du}{u} = r \ln a$$

(d)   The proof of part (d) is left as an exercise.   ∎

THE DERIVATIVE OF ln x

A straightforward application of (5.35) yields

$$\frac{d}{dx} \ln x = \frac{d}{dx} \int_1^x \frac{1}{t} \, dt = \frac{1}{x}$$

(7.7)   **THEOREM**   *Derivative of* $y = \ln x$.   **The derivative of the logarithm function** $y = \ln x$ **is** $dy/dx = 1/x$.   **That is,**

$$\frac{d}{dx} \ln x = \frac{1}{x}$$

EXAMPLE 1   Find y' if: (a) $y = x \ln x$ (b) $\ln x + \ln y = 2x$

Solution   (a)   We use the rule for differentiating a product to get

$$y' = (x)\left(\frac{1}{x}\right) + (1)(\ln x) = 1 + \ln x$$

(b)   We use implicit differentiation to get

$$\frac{1}{x} + \left(\frac{1}{y}\right)y' = 2$$

$$y' = y\left(2 - \frac{1}{x}\right) = \frac{y(2x - 1)}{x}$$ ∎

The chain rule is frequently used when differentiating logarithms. For example, if we apply the chain rule to the composite function $y = \ln u$, where $u = g(x) > 0$ is a differentiable function, we find

$$\frac{dy}{dx} = \left(\frac{dy}{du}\right)\left(\frac{du}{dx}\right) = \frac{1}{u}\left(\frac{du}{dx}\right) = \frac{1}{g(x)}\,g'(x)$$

Thus, if $g > 0$ is a differentiable function,

(7.8)
$$\frac{d}{dx}\ln g(x) = \frac{g'(x)}{g(x)}$$

EXAMPLE 2   Find y' if:
(a) $y = \ln(x^2 + 1)$ (b) $y = \ln\sqrt{x^2 + 1}$ (c) $y = (\ln x)^2$ (d) $y = \sin(\ln x)$

Solution   (a) We may use either (7.8) or the chain rule. By (7.8), $g(x) = x^2 + 1$ and $g'(x) = 2x$, so that

$$y' = \frac{2x}{x^2 + 1}$$

By the chain rule, we set $y = \ln u$ and $u = x^2 + 1$. Then

$$y' = \left(\frac{dy}{du}\right)\left(\frac{du}{dx}\right) = \left(\frac{1}{u}\right)(2x) = \frac{2x}{x^2 + 1}$$

(b) Always simplify first if possible! Here we may use part (c) of (7.5) to write $y = \ln\sqrt{x^2 + 1}$ as $y = \frac{1}{2}\ln(x^2 + 1)$. Then we may use the solution to part (a) above to get $y' = x/(x^2 + 1)$.

(c) We may use either the power rule or the chain rule.

Power rule: $y = [g(x)]^2$ and $g(x) = \ln x$

$$y' = 2\ln x\left(\frac{d}{dx}\ln x\right) = \frac{2\ln x}{x}$$

Chain rule: Set $y = u^2$ and $u = \ln x$; then

$$y' = \left(\frac{dy}{du}\right)\left(\frac{du}{dx}\right) = (2u)\left(\frac{1}{x}\right) = \frac{2\ln x}{x}$$

(d) We use the chain rule with $y = \sin u$ and $u = \ln x$. Then

$$y' = \left(\frac{dy}{du}\right)\left(\frac{du}{dx}\right) = (\cos u)\left(\frac{1}{x}\right) = \frac{\cos(\ln x)}{x} \qquad\blacksquare$$

Sometimes the chain rule must be used with three factors to get a derivative.

EXAMPLE 3   Find y' if $y = \cos[\ln(4x - 3)]$, $x > \frac{3}{4}$.*

---

* In subsequent examples, we do not explicitly state the domain of a function containing a natural logarithm. Instead, we assume that the variable is restricted so that the given expression makes sense.

Solution   We use the chain rule with $y = \cos u$, $u = \ln v$,   and $v = 4x - 3$.   Then

$$y' = \left(\frac{dy}{du}\right)\left(\frac{du}{dv}\right)\left(\frac{dv}{dx}\right) = (-\sin u)\left(\frac{1}{v}\right)(4)$$

$$= -4[\sin(\ln v)]\frac{1}{v} = \frac{-4\sin[\ln(4x - 3)]}{4x - 3} \qquad \blacksquare$$

The properties of logarithms (7.5) can sometimes be used to simplify the work needed to find the derivative of certain algebraic functions.

EXAMPLE 4   Find the derivative of $y = \ln[(2x - 1)^3(2x + 1)^5]$.

Solution   Rather than attempt to use (7.8) or the chain rule, we first use the fact that a logarithm transforms products into sums. That is, we may write

$$y = \ln(2x - 1)^3 + \ln(2x + 1)^5 = 3\ln(2x - 1) + 5\ln(2x + 1)$$

$$\uparrow \qquad\qquad\qquad\qquad \uparrow$$

By (7.5(b))   By (7.5(c))

Now we differentiate by using (7.8):

$$y' = (3)\left(\frac{2}{2x - 1}\right) + (5)\left(\frac{2}{2x + 1}\right) = \frac{6}{2x - 1} + \frac{10}{2x + 1} = \frac{4(8x - 1)}{4x^2 - 1} \qquad \blacksquare$$

As Example 4 illustrates, some thought should be given to the possibility of simplification before differentiating. The next example illustrates a somewhat more subtle procedure.

EXAMPLE 5   Find the derivative of $y = x^2\sqrt{5x + 1}/(3x - 2)^3$.

Solution   As you will see, it is easier to take the natural logarithm of both sides before differentiating. That is, look instead at

$$\ln y = \ln\frac{x^2\sqrt{5x + 1}}{(3x - 2)^3}$$

By using some of the properties (7.5), we may write the above expression as

$$\ln y = \ln x^2 + \ln\sqrt{5x + 1} - \ln(3x - 2)^3$$

$$= 2\ln x + \frac{1}{2}\ln(5x + 1) - 3\ln(3x - 2)$$

We now use implicit differentiation and (7.8) to find y':

$$\frac{y'}{y} = \frac{2}{x} + \frac{5}{2(5x + 1)} - \frac{(3)(3)}{3x - 2}$$

$$y' = \frac{x^2\sqrt{5x + 1}}{(3x - 2)^3}\left[\frac{2}{x} + \frac{5}{2(5x + 1)} - \frac{9}{3x - 2}\right] \qquad \blacksquare$$

We refer to the procedure used in Example 5 as *logarithmic differentiation*. This procedure was first used in 1697 by Johann Bernoulli (1667–1748).*

Here are the basic steps used in logarithmic differentiation:

Comment

Step 1. $y = f(x)$ The function f consists of products and quotients

Step 2. $\ln y = \ln f(x)$ Take the natural logarithm of both sides and simplify by using (7.5)

Step 3. $\dfrac{y'}{y} = \dfrac{d}{dx} \ln f(x)$ Differentiate by using (7.8)

Step 4. $y' = f(x) \dfrac{d}{dx} \ln f(x)$ Multiply both sides by y

Let's do another example to illustrate the procedure.

EXAMPLE 6   Find the derivative of $y = \sqrt{4x + 3}/(2x - 5)^3$.

Solution
$$\ln y = \ln\left[\frac{(4x + 3)^{1/2}}{(2x - 5)^3}\right] = \ln(4x + 3)^{1/2} - \ln(2x - 5)^3$$

$$= \frac{1}{2}\ln(4x + 3) - 3\ln(2x - 5)$$

$$\frac{y'}{y} = \frac{1}{2}\left(\frac{4}{4x + 3}\right) - 3\left(\frac{2}{2x - 5}\right)$$

$$y' = \frac{\sqrt{4x + 3}}{(2x - 5)^3}\left(\frac{2}{4x + 3} - \frac{6}{2x - 5}\right) \quad \blacksquare$$

THE GRAPH OF $f(x) = \ln x$

We now proceed to obtain the graph of the natural logarithm function. From (7.4), the domain of $f(x) = \ln x$ is the set of *positive real numbers*, and because of (5.30) (p. 301), $f(x) = \ln x$ is both continuous and differentiable. Since $(d/dx) \ln x = 1/x > 0$, we conclude that $f(x) = \ln x$ is an increasing function. Moreover, since $f''(x) = -1/x^2 < 0$, it follows that $y = \ln x$ is concave down on its domain.

Since $y = \ln x$ is increasing on its domain and $\ln 1 = 0$, we conclude that $\ln x > 0$ for $x > 1$, and $\ln x < 0$ for $0 < x < 1$. This conclusion can also be obtained by using a geometric (area) argument, which we now give:

If $x > 1$, then

$$\ln x = \int_1^x \frac{1}{t}\, dt = \text{Area under graph of } y = \frac{1}{x} \text{ from 1 to } x$$

---

* Johann was only one of the famous Bernoulli brothers. See the Historical Perspectives section in Chapter 11 for a discussion of their work.

See Figure 1(a). Hence, for $x > 1$, $\ln x > 0$ and the graph of $y = \ln x$ lies above the x-axis.

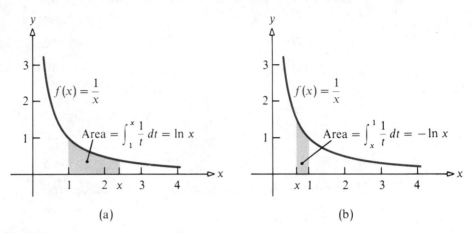

(a)   (b)

Figure 1

If $0 < x < 1$,   then

$$\ln x = \int_1^x \frac{1}{t}\, dt = -\int_x^1 \frac{1}{t}\, dt = -\text{Area under graph of }\; y = \frac{1}{x}\; \text{ from } x \text{ to } 1$$

See Figure 1(b). Hence, for $0 < x < 1$, $\ln x < 0$ and the graph of $y = \ln x$ lies below the x-axis.

Let's summarize what we know about the graph of $y = \ln x$:

1. Domain: positive real numbers
2. $y = \ln x$ is increasing and continuous
3. $y = \ln x$ is concave down

4. $y = \ln x < 0$ if $0 < x < 1$
5. $\ln x = 0$ if $x = 1$
6. $y = \ln x > 0$ if $x > 1$

We now proceed to get additional information about $y = \ln x$.   First, let's find out how to locate other points on the graph. One way is by approximating $\ln 2$. As Figure 2 illustrates, the areas A_1, A_2, and A_3 obey the inequality

$$A_1 < A_2 < A_3$$

These areas are computed to be

$$(1)\left(\frac{1}{2}\right) < \int_1^2 \frac{1}{t}\, dt < (1)(1)$$

(7.9)
$$\frac{1}{2} < \ln 2 < 1$$

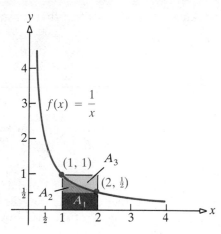

Figure 2

Inequality (7.9) establishes that $\ln 2$ lies between $\frac{1}{2}$ and 1. More precisely, we show in Chapter 11 that

$$\ln 2 = 0.6931 \ldots$$

Using this approximation and part (c) of (7.5), we can get estimates of $\ln x$ when x is any rational power of 2. As examples,

$$\ln \tfrac{1}{2} = \ln 2^{-1} = -\ln 2 \approx -0.6931$$
$$\ln \tfrac{1}{4} = -\ln 4 = -\ln 2^2 = -2 \ln 2 \approx -1.3862$$
$$\ln \tfrac{1}{8} = -\ln 8 = -\ln 2^3 = -3 \ln 2 \approx -2.0793$$
$$\ln 4 = \ln 2^2 = 2 \ln 2 \approx 1.3862$$
$$\ln 8 = \ln 2^3 = 3 \ln 2 \approx 2.0793$$

We now use inequality (7.9) to show that:

(7.10)   (a) $\quad \lim_{x \to +\infty} \ln x = +\infty \quad$ (b) $\quad \lim_{x \to 0^+} \ln x = -\infty$

Proof

(a) Since $\quad f(x) = \ln x \quad$ is an increasing function, it follows that

$$\ln x > \ln 4^n \qquad \text{whenever} \qquad x > 4^n$$

Since $\quad \ln 4^n = n \ln 4, \quad$ this becomes

$$\ln x > n \ln 4 \qquad \text{whenever} \qquad x > 4^n$$

But $\quad \ln 4 > 1 \quad$ since $\quad \ln 4 = 2 \ln 2 \quad$ and $\quad \ln 2 > \tfrac{1}{2}.$ Thus,

$$\ln x > n \qquad \text{whenever} \qquad x > 4^n$$

By taking $N = 4^n$, we find that for any $n > 0$,

$$\ln x > n \qquad\qquad \text{whenever} \qquad x > N = 4^n$$

In other words, $\lim_{x \to +\infty} \ln x = +\infty$.

(b) To determine $\lim_{x \to 0^+} \ln x$, we set $x = 1/n$. Then, as $x \to 0^+$, n approaches $+\infty$ and

$$\lim_{x \to 0^+} \ln x = \lim_{n \to +\infty} \ln \frac{1}{n} = \lim_{n \to +\infty} (\ln 1 - \ln n) = - \lim_{n \to +\infty} \ln n = -\infty \qquad \blacksquare$$

$$\underset{\text{By (7.5(d))}}{\uparrow} \qquad\qquad \underset{\ln 1 = 0}{\uparrow} \qquad \underset{\text{By (7.10(a))}}{\uparrow}$$

We conclude from part (b) of (7.10) that the line $x = 0$ is a vertical asymptote to $y = \ln x$. We further conclude that the range of the natural logarithm function is $(-\infty, +\infty)$. Putting all of the foregoing facts together, we obtain the graph of $f(x) = \ln x$, as shown in Figure 3.

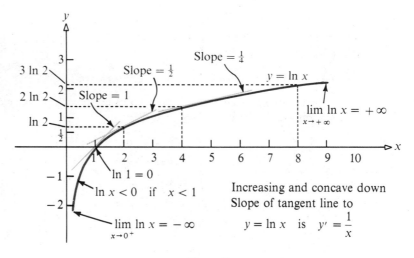

Figure 3

THE NUMBER e

We have already mentioned that the natural logarithm function $y = \ln x$ is continuous and increasing on its domain, $x > 0$, and that its range is $-\infty < y < +\infty$. By the intermediate value theorem (2.28), it follows that the graph of $y = \ln x$ must cross each horizontal line $y = c$ at a point to the right of the y-axis (see Fig. 4). Moreover, because $f(x) = \ln x$ is an increasing function, it cannot cross any horizontal line $y = c$ more than once. Thus, we have the following theorem:

(7.11)   **THEOREM** **For every real number c there is a unique positive real number x such that**

$$\ln x = c$$

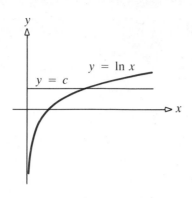

Figure 4   **Figure 5**

In particular, there is a unique number x for which $\ln x = 1$.   This number is designated by e (see Fig. 5). Thus,

$$e \text{ is that unique number for which } \ln e = 1$$

We have already discovered that $\ln 2 < 1 < \ln 4$,   so that $\ln 2 < \ln e < \ln 4$. Hence, we conclude that

(7.12) $$2 < e < 4$$

In Section 6 we show that e may be expressed as the limit

(7.13) $$e = \lim_{n \to +\infty} \left(1 + \frac{1}{n} \right)^n$$

and we will use this formula to approximate e as

$$e = 2.71828 \ldots$$

EXERCISE 2

In Problems 1–8 let $a = \ln 2$ and $b = \ln 5$.   Express each logarithm in terms of a and b only.

1. $\ln 10$ **2.** $\ln 25$ **3.** $\ln \sqrt[2]{5}$ **4.** $\ln \sqrt[3]{10}$

5. $\ln 0.05$ **6.** $\ln(0.05)^3$ **7.** $\ln \frac{5}{2}$ **8.** $\ln \frac{1}{25}$

In Problems 9—12 sketch the graphs.

9. $y = -\ln x$ **10.** $y = \ln x - x$

11. $y = \ln 3x$ **12.** $y = \ln(x + 2)$ $x > -2$

In Problems 13–44 find the derivative of each function.

13. $y = \ln 3x$ **14.** $y = 5 \ln x$ **15.** $y = x \ln(x^2 + 4)$

16. $y = x \ln(x^2 + 5x + 1)$ **17.** $y = x \ln \sqrt{x^2 + 1}$ **18.** $y = x \ln(\sqrt[3]{3x + 1})$

19. $y = \dfrac{1}{2} \ln\left(\dfrac{1 + x}{1 - x}\right)$　　　**20.** $y = \dfrac{1}{2} \ln\left(\dfrac{1 + x^2}{1 - x^2}\right)$　　　**21.** $y = \ln(\ln x)$

22. $y = \ln\left(\ln \dfrac{1}{x}\right)$　　　**23.** $y = \ln\left(\dfrac{x}{\sqrt{x^2 + 1}}\right)$　　　**24.** $y = \ln\left(\dfrac{4x^3}{\sqrt{x^2 + 4}}\right)$

25. $y = \ln\left[\dfrac{(x^2 + 1)^2}{x\sqrt{x^2 - 1}}\right]$　　　**26.** $y = \ln\left[\dfrac{x\sqrt{3x - 1}}{(x^2 + 1)^3}\right]$　　　**27.** $y = \ln(\sin x)$

28. $y = \ln(\cos x)$　　　**29.** $y = (\ln x)^{1/2}$　　　**30.** $y = (\cos x)(\ln x)$

31. $y = x \ln \sqrt{\cos 2x}$　　　**32.** $y = x^2 \ln \sqrt{\sin 2x}$　　　**33.** $y = \ln(x + \sqrt{x^2 + 4})$

34. $y = \ln(\sqrt{x + 1} + \sqrt{x})$　　　**35.** $y = \ln(x + \sqrt{x^2 + a^2})$　　　**36.** $y = \ln(\sqrt{x + a} + \sqrt{x})$

[*Hint:* In Problems 37–44 use logarithmic differentiation.]

37. $y = (x^2 + 1)^2(2x^3 - 1)^4$　　　　　　　**38.** $y = (3x^2 + 4)^3(x^2 + 1)^4$

39. $y = (x^3 + 1)(x - 1)(x^4 + 5)$　　　　　　**40.** $y = \sqrt{x^2 + 1}(x^3 - 5)(3x + 4)$

41. $y = \dfrac{x^2(x^3 + 1)}{\sqrt{x^2 + 1}}$　　　　　　　**42.** $y = \dfrac{\sqrt{x}(x^3 + 2)^2}{\sqrt[3]{3x + 4}}$

43. $y = \dfrac{x \cos x}{(x^2 + 1)^3 \sin x}$　　　　　**44.** $y = (x \sin x)(\cos x)(\ln x)$

In Problems 45–48 use implicit differentiation to find dy/dx.

45. $x \ln y + y \ln x = 2$　　　　　　　**46.** $\ln(x^2 + y^2) = x + y$

47. $\ln\left(\dfrac{y}{x}\right) - \ln\left(\dfrac{x}{y}\right) = 1$　　　　**48.** $\ln(x^2 - y^2) = x - y$

49. Find an equation of the tangent line to　$y = \ln 5x$　at $(\tfrac{1}{5}, 0)$.

50. Find an equation of the tangent line to　$y = x \ln x$　at $(1, 0)$.

51. Show that for　$x > 1$,　$\ln x < 2(\sqrt{x} - 1)$.　[*Hint:*　Compare $\int_1^x (1/t)\, dt$　with　$\int_1^x (1/\sqrt{t})\, dt$　for　$t > 1$.]

52. Show that　$0 < \ln x < x$　for　$x > 1$.

53. Use Problem 51 to evaluate　$\lim_{x \to +\infty}(\ln x/x)$.

54. Use Problem 52 to evaluate　$\lim_{x \to 0^+} x(\ln x)$.　[*Hint:*　Let　$x = 1/u$.]

55. Prove that　$x \le \ln(1 + x)^{1+x}$　for all　$x > -1$.　[*Hint:*　Consider　$f(x) = -x + \ln(1 + x)^{1+x}$.]

56. Show that　$1 + \ln x < x$　whenever　$x > 1$.

In Problems 57 and 58 find all local maxima and local minima. Graph each function.

57. $y = x \ln x$　　　　　　　　　　**58.** $y = \left(\dfrac{1}{x}\right) \ln x$

59. If　$y = \ln(x^2 + y^2)$,　find the value of dy/dx at the point $(1, 0)$.

60. If　$y = \tan u$,　$u = v - (1/v)$,　and　$v = \ln x$,　what is the value of dy/dx at　$x = e$?

61. A point moves on the x-axis in such a way that its velocity at time t　$(t > 0)$　is given by　$v = (\ln t)/t$.　At what time t does v attain its maximum?

62. Find a formula for the nth derivative of $y = \ln(ax)$, where a is a positive constant.

63. A telephone cable is made up of a core of copper wires covered by an insulating material. If x is the ratio of the radius of the core to the thickness of the insulating material, the speed v of signaling is

$$v = kx^2 \ln \frac{1}{x}$$

where k is a constant. Determine the ratio x that results in maximum speed.

64. Use (7.5(b)) and (7.5(c)) to show that

$$\ln \frac{a}{b} = \ln a - \ln b \qquad \text{where } a, b \text{ are positive real numbers}$$

3. The Exponential Function

Let r be any rational number. Using the facts that $\ln e^r = r \ln e$ and $\ln e = 1$, we have

$$\ln e^r = r$$

Consistent with this equation and the fact that the range of $\ln x$ is all real numbers, we state the following definition:

(7.14) **If α is a real number, we define e^α to be that (unique) number for which $\ln e^\alpha = \alpha$.**

With this definition, we have given meaning to expressions such as $e^{\sqrt{2}}$, which is the unique number for which $\ln e^{\sqrt{2}} = \sqrt{2}$ (see Fig. 6).

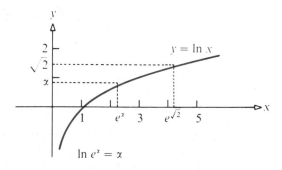

Figure 6

The function $y = e^x$ is called the *exponential function.** Since, by definition,

(7.15) $$\ln e^x = x \qquad \text{for all real } x$$

---

* Some books use the notation $y = \exp x$ to represent the exponential function.

it follows that:

(7.16)   **The exponential function $y = e^x$ and the natural logarithm function $y = \ln x$ are inverses.**

We may also write this fact as

(7.17)
$$e^{\ln x} = x \qquad \text{for all} \quad x > 0$$

We use the fact that $y = \ln x$ and $y = e^x$ are inverses to get information about the graph of $y = e^x$. Indeed, based on this, we can obtain the graph $y = e^x$ by simply reflecting the graph of $y = \ln x$ about the line $y = x$ (see Fig. 7). Because $y = \ln x$ and $y = e^x$ are inverses, we conclude that:

Domain of $y = e^x$ equals range of $y = \ln x$ equals all real numbers.

Range of $y = e^x$ equals domain of $y = \ln x$ equals the positive real numbers.

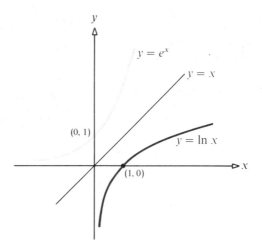

Figure 7

Furthermore, since $y = \ln x$ crosses the x-axis at $(1, 0)$, then $y = e^x$ crosses the y-axis at $(0, 1)$. Also, $y = \ln x$ is increasing, concave down, and continuous on its domain. Hence, $y = e^x$ is increasing, concave up, and continuous on its domain (see Appendix II).

Finally, based on (7.10) and Figure 7, we have the important limits

(7.18)
$$\lim_{x \to +\infty} e^x = +\infty \qquad \lim_{x \to -\infty} e^x = 0$$

EXAMPLE 1   Sketch the graph of $y = e^{-x}$.

Solution   The replacement of x by $-x$ in the equation $y = e^x$ results in the reflection of the graph of this equation about the y-axis. Thus, the graph of this equation is as shown in Figure 8.

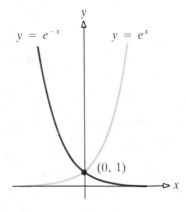

Figure 8   ■

The following limits, which are consequences of (7.18), are evident from the graph of e^{-x}:

$$\lim_{x \to +\infty} e^{-x} = \lim_{x \to +\infty} \frac{1}{e^x} = 0 \qquad \lim_{x \to -\infty} e^{-x} = \lim_{x \to -\infty} \frac{1}{e^x} = +\infty$$

The familiar laws of exponents hold for e^x.

(7.19)   **For any real numbers α and β:**

$$\textbf{(a)} \quad e^\alpha e^\beta = e^{\alpha + \beta} \qquad \textbf{(b)} \quad \frac{e^\alpha}{e^\beta} = e^{\alpha - \beta} \qquad \textbf{(c)} \quad (e^\alpha)^\beta = e^{\alpha\beta}$$

Proof   To prove part (a), we use the properties of the natural logarithm function. Since

$$\ln(e^\alpha e^\beta) = \ln e^\alpha + \ln e^\beta = \alpha + \beta = \ln e^{\alpha + \beta}$$

and since $\ln x$ is a one-to-one function, it follows that

$$e^\alpha e^\beta = e^{\alpha + \beta}$$

The proofs of parts (b) and (c) are left as exercises (see Problems 89 and 90 in Exercise 3).   ■

DERIVATIVE OF $y = e^x$

We are now ready to find the derivative of $y = e^x$.   Although the formula is quite simple, it is one of the most important results in calculus.

(7.20) **THEOREM** *Derivative of* $y = e^x$. **The derivative of the exponential function** $y = e^x$ **is** $dy/dx = e^x$. **That is,**

$$\frac{d}{dx} e^x = e^x$$

Proof Since the natural logarithm function is differentiable and its derivative is never 0, it follows by (3.54) (p. 162) that its inverse, the exponential function, is also differentiable. By (7.15), we know that

$$\ln e^x = x \qquad \text{for all real } x$$

We differentiate both sides with respect to x, applying (7.8) to the left side, to get

$$\frac{\frac{d}{dx} e^x}{e^x} = 1$$

Therefore,

$$\frac{d}{dx} e^x = e^x \qquad \blacksquare$$

Let's now look at some derivative problems.

EXAMPLE 2   Find y' if:

(a) $y = x^3 e^x$ (b) $y = e^{x^2}$ (c) $e^y + e^x = 4x$ (d) $y = \cos e^x$

(e) $y = \sin e^{\sqrt{x^2 + 4}}$

Solution   (a) We use the rule for differentiating a product to get

$$y' = x^3 \left(\frac{d}{dx} e^x \right) + \left(\frac{d}{dx} x^3 \right) e^x = (x^3)(e^x) + (3x^2)(e^x) = x^2 e^x (x + 3)$$

(b) We use the chain rule with $y = e^u$ and $u = x^2$ to get

$$y' = \left(\frac{dy}{du} \right)\left(\frac{du}{dx} \right) = (e^u)(2x) = 2x e^{x^2}$$

(c) We use implicit differentiation to get

$$e^y y' + e^x = 4$$

$$y' = \frac{4 - e^x}{e^y}$$

(d) We use the chain rule with $y = \cos u$ and $u = e^x$ to get

$$y' = \left(\frac{dy}{du} \right)\left(\frac{du}{dx} \right) = (-\sin u)(e^x) = -e^x \sin e^x$$

(e)   We use the chain rule with $y = \sin u$, $u = e^v$,   and $v = \sqrt{x^2 + 4}$ to get

$$y' = \left(\frac{dy}{du}\right)\left(\frac{du}{dv}\right)\left(\frac{dv}{dx}\right) = (\cos u)(e^v)\left(\frac{x}{\sqrt{x^2 + 4}}\right)$$

$$= \frac{xe^{\sqrt{x^2+4}} \cos e^{\sqrt{x^2+4}}}{\sqrt{x^2 + 4}} \quad ■$$

EXAMPLE 3   Graph the function: $f(x) = xe^{-x}$

Solution   The domain of this function is all real numbers, and it is continuous on its domain. Since e^{-x} is always positive, then $f(x) < 0$ if $x < 0$ and $f(x) > 0$ if $x > 0$. Also, $f(0) = 0$.   The first derivative is

$$f'(x) = x(-e^{-x}) + (1)e^{-x} = (1 - x)e^{-x}$$

Again, since e^{-x} is always positive, $f'(x)$ has the same sign as $(1 - x)$.   Hence, we conclude that the function is decreasing for $x > 1$ and increasing for $x < 1$. Also, $x = 1$ is a critical number. The second derivative is

$$f''(x) = (1 - x)(-e^{-x}) + (-1)(e^{-x}) = (x - 2)e^{-x}$$

To find the local extrema, we use the second derivative test. At the critical number $x = 1$,   the sign of the second derivative is negative, so that f has a local maximum at $x = 1$.   The value of the function at $x = 1$ is $f(1) = e^{-1} \approx 0.368$.   The lone inflection point occurs at $x = 2$.   If we put all this information together, we obtain the graph in Figure 9.

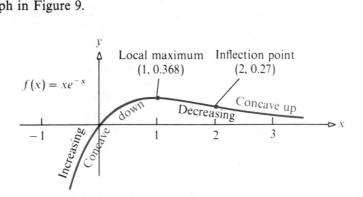

Figure 9   ■

EXAMPLE 4   The damped vibration curve $y = e^{-ax} \sin bx$, a, b positive constants, is of importance in many applications, such as the motion of musical strings, the vibration of a pendulum, and the current in a radio circuit. Graph this function.

Solution   We seek the intercepts first. If $x = 0$,   then $y = 0$.   If $y = 0$,   then $\sin bx = 0$ since e^{-ax} can never be 0. From $\sin bx = 0$,   we find $bx = n\pi$,   or $x = n\pi/b$, where n is any integer. Thus, the x-intercepts are a distance π/b apart. As x increases, $\sin bx$ oscillates regularly between -1 and 1, but e^{-ax} decreases and approaches 0 as $x \to +\infty$. That is, $\lim_{x \to +\infty} e^{-ax} \sin bx = 0$. Hence, the curve oscillates regularly, but with a constantly decreasing amplitude.

When $\sin bx = 1$ or $x = \pi/2b, 5\pi/2b, 9\pi/2b$, and so on, we have $y = e^{-ax}$, so that the damped vibration curve is in contact with the exponential curve $y = e^{-ax}$ at these x's. Similarly, at $x = 3\pi/2b, 7\pi/2b, 11\pi/2b$, and so on, the damped vibration curve is in contact with the exponential curve $y = -e^{-ax}$. The damped vibration curve therefore oscillates between the two exponential guiding curves.

We find $y' = e^{-ax}(b \cos bx - a \sin bx)$. If $y' = 0$, then, since e^{-ax} cannot be 0, we must have $b \cos bx - a \sin bx = 0$, or $\sin bx/\cos bx = b/a$, or $\tan bx = b/a$; these numbers x, at which alternate maximum and minimum points of the damped vibration curve occur, are to the left of the points of contact with the guiding curves, $y = \pm e^{-ax}$.

Making use of all these facts, we obtain the graph shown in Figure 10.

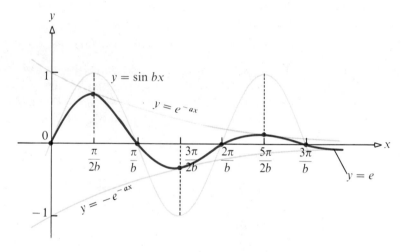

Figure 10　　　　　　　■

EXERCISE 3

In Problems 1–10 use (7.17) and (7.19) to simplify each expression as far as possible.

1. $\ln e^{x - x^2}$

2. $e^{-\ln(1/x)}$

3. $e^{\ln(1/x)}$

4. $e^{\ln 5 - \ln 2}$

5. $\ln \dfrac{1}{e^{x^2}}$

6. $e^{\ln x^2}$

7. $e^{x + \ln x}$

8. $e^{\ln x - 10 \ln y}$

9. $\ln e^{\ln 2}$

10. $e^{-\ln(xy^{-2})}$

In Problems 11–14 solve for x.

11. $\ln e^{-\ln x^2} = 5$

12. $\ln x^{5/2} + \ln \sqrt{x} = 3$

13. $e^{2x} - 3e^x + 2 = 0$

14. $\ln(1 + x) = 5$

In Problems 15–22 use (7.18) to find the indicated limit.

15. $\lim\limits_{x \to +\infty} 3e^{3x}$

16. $\lim\limits_{x \to -\infty} 3e^{3x}$

17. $\lim\limits_{x \to +\infty} 3e^{-3x}$

18. $\lim\limits_{x \to -\infty} 3e^{-3x}$

19. $\lim\limits_{x \to +\infty} (1 - e^{-x^3})$

20. $\lim\limits_{x \to -\infty} (1 - e^{-x^3})$

21. $\lim\limits_{x \to +\infty} e^{1/x}$

22. $\lim\limits_{x \to -\infty} e^{1/x}$

In Problems 23–52 find y'.

23. $y = 5e^{3x}$

24. $y = xe^{2x}$

25. $y = \dfrac{e^x + e^{-x}}{2}$

26. $y = \dfrac{e^x - e^{-x}}{2}$

27. $y = e^{-3x} \ln(2x)$

28. $y = \frac{1}{2} \ln(4x) - 5e^{2x}$

29. $y = e^x \sin x$

30. $y = e^{2x} \cos(3x)$

31. $y = e^{ax} \sin bx$

32. $y = \dfrac{e^x + e^{-x}}{e^x - e^{-x}}$

33. $y = e^{\sqrt{x^2 - 9}}$

34. $y = e^{\cos(4x)}$

35. $y = e^{1/x}$

36. $y = \sqrt{e^x}$

37. $y = \dfrac{e^{ax} - 1}{e^{ax} + 1}$

38. $y = \ln(e^{ax} + e^{-ax})$

39. $e^{x+y} = y$

40. $x^2 y = e^{xy}$

41. $ye^x = y - x$

42. $e^{x+y} = x^2$

43. $y = \dfrac{100}{1 + 99e^{-x}}$

44. $y = \dfrac{1}{1 + 2e^{-x}}$

45. $y = \cos e^{x^2}$

46. $y = \dfrac{\cos(e^{\sqrt{x}})}{\sqrt{x}}$

47. $y = \ln(\sin e^x)$

48. $y = \sin e^{\ln x}$

49. $\ln(\ln y) = e^x$

50. $e^y = \sin x$

51. $e^x \sin y + e^y \cos x = 4$

52. $e^y \cos x + e^{-x} \sin y = 10$

In Problems 53 and 54 find a general formula for the nth derivative of y.

53. $y = e^{ax}$

54. $y = e^{-ax}$

55. If $y = e^{2x}$, show that $y'' - 4y = 0$.

56. If $y = e^{-2x}$, show that $y'' - 4y = 0$.

57. If $y = Ae^{2x} + Be^{-2x}$, where A and B are constants, show that $y'' - 4y = 0$.

58. If $y = Ae^{ax} + Be^{-ax}$, where A, B, and a are constants, show that $y'' - a^2 y = 0$.

59. If $y = Ae^{2x} + Be^{3x}$, where A and B are constants, show that $y'' - 5y' + 6y = 0$.

60. If $y = Ae^{-2x} + Be^{-x}$, where A and B are constants, show that $y'' + 3y' + 2y = 0$.

61. If $y = e^{-at}(A \sin \omega t + B \cos \omega t)$, where A, B, a, and ω are constants, find y'.

62. If $\ln T = kt$, where k is a constant, show that $dT/dt = kT$.

In Problems 63–68 graph the given function, and label local extrema, concavity, and inflection points.

63. $y = 3e^{3x}$

64. $y = 3e^{-3x}$

65. $y = e^{-x^2}$

66. $y = e^{-x} \cos x, \quad 0 \le x \le 2\pi$

67. $y = e^{1/x}$

68. $y = e^{|x|}$

69. Find the absolute maximum and absolute minimum of $f(x) = e^x - 3x$ on $[0, 1]$.

70. Find the local maximum, local minimum, absolute maximum, and absolute minimum of $y = e^{\cos x}$ on $[-\pi, 2\pi]$.

71. Let $y = 2e^{\cos x}$.
   (a) Calculate dy/dx and $d^2 y/dx^2$.

(b) If x and y both vary with time in such a way that y increases at a steady rate of 5 units per second, at what rate is x changing when $x = \pi/2$?

72. A particle moves along the x-axis in such a way that at time $t > 0$ its position is $x = \sin e^t$.
(a) Find the velocity and acceleration of the particle at time t.
(b) At what time does the particle first have 0 velocity?
(c) What is the acceleration of the particle at the time determined in part (b)?

73. The formula $i = I_0 e^{(-R/L)t}$ measures the current i in amperes after t seconds in an electric circuit containing no capacitors, a resistance of R ohms, and an inductance of L henrys. The current in amperes at time $t = 0$ (when the electromotive force is cut off) is represented by I_0. Show that the rate of change of the current is proportional to the current.

74. Use Newton's method to solve $e^{-x} = \ln x$ correct to one decimal place.

75. Show that $e^x > 1 + x$ for all $x > 0$. [*Hint:* Show that $f(x) = e^x - 1 - x$ is an increasing function for $x > 0$.]

76. Find the triangle of largest area that has two sides along the positive coordinate axes and its hypotenuse tangent to the function $f(x) = 3e^{-x}$.

77. Show that $2x - \ln(3 + 6e^x + 3e^{2x}) = c - 2\ln(1 + e^{-x})$ for some constant c. [*Hint:* Show that both sides have the same derivative.]

78. Use Newton's method to estimate the value of e by finding the root of the equation $\ln x - 1 = 0$. Start with $c_1 = 3$, and calculate c_4. Compare your results with the approximation 2.71828 for e.

79. Show that the line perpendicular to the x-axis passing through the point (x, y) on the graph of $y = e^x$ and the tangent to $y = e^x$ at the point (x, y) intersect the x-axis 1 unit apart. See the figure.

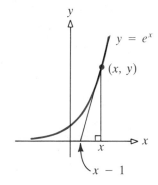

80. A pipe in the shape of an hourglass is formed by revolving the function given by $y = e^{3x}$, $-10 \le x \le 10$, about the line whose equation is $y = x$. What is the radius of the largest ball bearing that can pass through this hourglass pipe? [*Hint:* The radius will be equal to the minimum distance from the function to the line $y = x$.]

81. Define f by

$$f(x) = \begin{cases} e^{-1/x^2} & \text{if } x \neq 0 \\ 0 & \text{if } x = 0 \end{cases}$$

Show that f is differentiable on $(-\infty, \infty)$ and find $f'(x)$ for each value of x. [*Hint:* For $f'(0)$ you must use the definition of the derivative. Show that $1 < x^2 e^{1/x^2}$ for $x \neq 0$.]

82. The function $f(x) = (1/\sqrt{2\pi})e^{-x^2/2}$, encountered in probability theory, is called the *standard normal density function*. Determine where this function is increasing and decreasing, find all local maxima and local minima, find all inflection points, and determine intervals of concavity. Then graph the function.

83. The vertices of a rectangle are at $(0, 0)$, $(0, e^x)$, $(x, 0)$, and (x, e^x). If x increases at the rate of 1 unit per second, at what rate is the area increasing when $x = 10$?

84. The atmospheric pressure at a height of x meters above sea level is $P(x) = 10^4 e^{-0.00012x}$ kilograms per square meter. What is the rate of change of the pressure with respect to the height at $x = 500$ meters? What is it at $x = 750$ meters?

85. Show that the largest rectangle that can be inscribed under the graph of $y = e^{-x^2}$ has two of its vertices at the points of inflection of the graph.

86. The sales of a new car model over a period of time are expected to follow the so-called logistic curve,

$$f(x) = \frac{20{,}000}{1 + 50e^{-x}} \qquad x \geq 0$$

where x is measured in months. Determine the month in which the sales rate is a maximum. Graph the function.

87. The sales of a new stereo system over a period of time are expected to follow the so-called logistic curve,

$$f(x) = \frac{5000}{1 + 5e^{-x}} \qquad x \geq 0$$

where x is measured in years. Determine the year in which the sales rate is a maximum. Graph the function.

88. In a town of 50,000 people, the number of people at time t who have influenza is

$$N(t) = \frac{10{,}000}{1 + 9999e^{-t}}$$

where t is measured in days. Note that the flu is spread by the one person who has it at $t = 0$. At what time t is the rate of spreading of flu the greatest? Graph the function.

89. Prove part (b) of (7.19).

90. Prove part (c) of (7.19). [*Hint:* $\ln(e^x)^\beta = \beta \ln e^x$]

4. Related Integrals

Corresponding to every differentiation formula there is a related integration formula. Thus, as a consequence of (7.20), $(d/dx)e^x = e^x$, we have the formula

(7.21)
$$\int e^x\, dx = e^x + C \qquad \text{where } C \text{ is a constant}$$

Let's look at the differentiation formula (7.7), namely,

(7.22)
$$\frac{d}{dx} \ln x = \frac{1}{x}$$

which holds when $x > 0$. On the other hand, if $x < 0$, we may use (7.8) to obtain the formula

(7.23)
$$\frac{d}{dx} \ln(-x) = \left(\frac{1}{-x}\right)(-1) = \frac{1}{x}$$

By combining (7.22) and (7.23) into one formula, we get

(7.24)
$$\frac{d}{dx}\ln|x| = \frac{1}{x} \qquad x \neq 0$$

The differentiation formula (7.24) yields the related integration formula

(7.25)
$$\int \frac{1}{x}\,dx = \ln|x| + C \qquad \text{where } C \text{ is a constant}$$

Let's look at some examples that require the use of (7.21) or (7.25).

EXAMPLE 1   Evaluate each integral:

(a) $\displaystyle\int e^{3x+4}\,dx$ (b) $\displaystyle\int \frac{5x^2\,dx}{4x^3-1}$ (c) $\displaystyle\int_0^1 \frac{e^x\,dx}{e^x+4}$

(d) $\displaystyle\int_1^4 \frac{e^{\sqrt{x}}\,dx}{\sqrt{x}}$ (e) $\displaystyle\int e^{3x}\cos e^{3x}\,dx$

Solution   (a)   We use the substitution $u = 3x + 4$.   Then $du = 3\,dx$,   and

$$\int e^{3x+4}\,dx = \int e^u\left(\frac{du}{3}\right) = \frac{1}{3}e^u + C = \frac{1}{3}e^{3x+4} + C$$

(b)   We note that the numerator, except for a constant factor, is the derivative of the denominator. Hence, we use the substitution $u = 4x^3 - 1$,   so that $du = 12x^2\,dx$ or $5x^2\,dx = \frac{5}{12}\,du$.   Then,

$$\int \frac{5x^2\,dx}{4x^3-1} = \int \frac{5\,du}{u\,12} = \frac{5}{12}\ln|u| + C = \frac{5}{12}\ln|4x^3-1| + C$$

(c) $\displaystyle\int_0^1 \frac{e^x\,dx}{e^x+4} = \int_5^{e+4} \frac{du}{u} = \ln|u|\Big|_5^{e+4} = \ln(e+4) - \ln 5 \approx 1.9048 - 1.6094 = 0.2954$

Set $u = e^x + 4$ Use a calculator

(d) $\displaystyle\int_1^4 \frac{e^{\sqrt{x}}\,dx}{\sqrt{x}} = \int_1^2 e^u(2\,du) = 2e^u\Big|_1^2 = 2(e^2 - e) \approx 2(7.389 - 2.718) = 9.342$

Set $u = \sqrt{x}$ Use a calculator

(e) $\displaystyle\int e^{3x}\cos e^{3x}\,dx = \int (\cos u)\frac{du}{3} = \frac{1}{3}\sin u + C = \frac{1}{3}\sin e^{3x} + C$ ■

Set $u = e^{3x}$

In parts (b) and (c) of Example 1, you may have noticed a general integration rule that evolves from (7.8), namely,

(7.26)
$$\int \frac{g'(x)}{g(x)}\,dx = \ln|g(x)| + C$$

In other words, whenever the numerator is the derivative of the denominator, the integral results in a logarithm. Note in both (7.25) and in (7.26) that we must use absolute value bars since the domain of the logarithm function is the positive real numbers. Of course, if in (7.26) the function $g(x)$ is known to be positive, we need not use the absolute value bars.

In Example 2 we use (7.26) to get two basic integration formulas.

EXAMPLE 2   Use (7.26) to show that:

(7.27)   (a) $\displaystyle\int \tan x \, dx = \ln|\sec x| + C$ (b) $\displaystyle\int \cot x \, dx = \ln|\sin x| + C$

Solution   (a) $\displaystyle\int \tan x \, dx = \int \frac{\sin x}{\cos x} \, dx = \int -\frac{du}{u}$

                              Set $u = \cos x$

$$= -\ln|u| + C = \ln|u|^{-1} + C = \ln|\cos x|^{-1} + C = \ln|\sec x| + C$$

                 (b) $\displaystyle\int \cot x \, dx = \int \frac{\cos x}{\sin x} \, dx = \int \frac{du}{u} = \ln|u| + C = \ln|\sin x| + C$ ∎

                              Set $u = \sin x$

EXAMPLE 3   Find the volume of the solid of revolution generated by revolving the area under the graph of $y = e^x$ from $x = 0$ to $x = 1$ about the x-axis.

Solution   Figure 11 illustrates the situation. We choose to use the disk method because the shell method leads to an integral we cannot yet evaluate. (Check this out yourself.*)

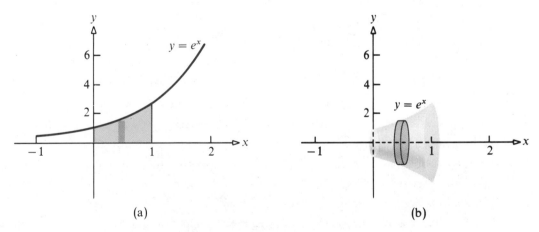

(a)   (b)

Figure 11

* The integral $\int \ln x \, dx$ is discussed in Chapter 9.

Thus, by the disk method,

$$V = \pi \int_0^1 (e^x)^2 \, dx = \pi \int_0^1 e^{2x} \, dx = \pi \left. \frac{e^{2x}}{2} \right|_0^1 = \frac{\pi}{2} (e^2 - 1) \approx \frac{\pi}{2} (6.389) = 10.036 \qquad \blacksquare$$

Use a calculator

EXERCISE 4

In Problems 1–34 evaluate each integral.

1. $\int e^{3x} \, dx$

2. $\int e^{-2x} \, dx$

3. $\int e^{3x+1} \, dx$

4. $\int e^{bx+1} \, dx$

5. $\int (x + e^{7x}) \, dx$

6. $\int (x^2 + e^{-7x}) \, dx$

7. $\int \frac{dx}{3x - 1}$

8. $\int \frac{dx}{1 - 5x}$

9. $\int \frac{x \, dx}{x^2 - 1}$

10. $\int \frac{5x \, dx}{1 - x^2}$

11. $\int_0^2 \frac{e^{2x}}{e^{2x} + 1} \, dx$

12. $\int_1^3 \frac{e^{3x}}{e^{3x} - 1} \, dx$

13. $\int_0^1 x^2 e^{x^3+1} \, dx$

14. $\int_0^1 x e^{x^2-2} \, dx$

15. $\int \frac{e^{1/x}}{x^2} \, dx$

16. $\int \frac{e^{\sqrt[3]{x}}}{\sqrt[3]{x^2}} \, dx$

17. $\int \frac{e^x + e^{-x}}{e^x - e^{-x}} \, dx$

18. $\int \frac{e^{-x}}{6 + e^{-x}} \, dx$

19. $\int \frac{3e^x}{\sqrt[4]{e^x - 1}} \, dx$

20. $\int \frac{e^x}{\sqrt{1 + e^x}} \, dx$

21. $\int_2^3 \frac{dx}{x \ln x}$

22. $\int_2^3 \frac{dx}{x(\ln x)^2}$

23. $\int \frac{dx}{x(\ln x)^n}$, $n > 0$, n an integer

24. $\int \frac{(\ln 5x)^3}{x} \, dx$

25. $\int_0^\pi e^x \cos e^x \, dx$

26. $\int_0^\pi e^{-x} \cos e^{-x} \, dx$

27. $\int \frac{4}{1 + e^x} \, dx$

28. $\int \frac{1}{2 + 3e^{-x}} \, dx$

29. $\int_1^2 \frac{x + 2}{3x^2} \, dx$

30. $\int_1^2 \frac{x - 4}{x^2} \, dx$

31. $\int \frac{dx}{\sqrt{x}(1 + \sqrt{x})}$

32. $\int \frac{dx}{(x - 2)^{1/2} + (x - 2)}$

33. $\int \frac{\cos x \, dx}{2 \sin x - 1}$

34. $\int \frac{1 - \cos 3x}{3x - \sin 3x} \, dx$

35. Find the area enclosed by the graphs of $y = e^x$, $y = e^{-x}$, and $x = \ln 2$.

36. What is the area enclosed by the graph of $y = e^{2x}$ and the lines $x = 1$ and $y = 1$?

37. Find the area under the graph of $y = (\sin x)/(2 - \cos x)$ from $x = 0$ to $x = \pi/2$.

38. Find the area under the graph of $y = x/(2 - x^2)$ from $x = 0$ to $x = 1$.

39. Find the area enclosed by the graphs of $y = e^x$, $y = e^{3x}$, and $x = 2$.

40. Find the area under the graph of $y = 1/(4 - x)$ from $x = 0$ to $x = 3$.

41. Find the volume of a solid generated by revolving the area under the graph of $y = e^{-x}$ from $x = 0$ to $x = 2$ about the x-axis.

42. Find the volume of a solid generated by revolving the area under the graph of $y = e^{-x^2}$ from $x = 0$ to $x = 2$ about the y-axis.

43. Find the average value of $f(x) = e^x$ on $[0, 1]$.

44. Find the average value of $f(x) = 1/x$ on $[1, e]$.

5. Exponentials and Logarithms to Other Bases

THE FUNCTION $f(x) = x^\alpha$, α **REAL**

In our development thus far, we have given meaning to expressions such as x^2, x^{-4}, $x^{1/2}$, and $x^{8/3}$, but we have not given meaning to expressions such as $x^{\sqrt 2}$, x^π, or, in general, x^α, where α is irrational. Just as we used the logarithm function to give meaning to e^x, we use it to give meaning to x^α, where α is irrational. This definition of x^α, α irrational, avoids the difficulty mentioned in the opening of this chapter and is motivated by the property

(7.28)
$$x^r = (e^{\ln x})^r = e^{r \ln x} \qquad r \text{ rational}$$

Consistent with this equation and the fact that the domain of $\ln x$ is $x > 0$, we define

(7.29)
$$x^\alpha = e^{\alpha \ln x} \qquad \alpha \text{ real}, \quad x > 0$$

For example,

$$x^{\sqrt 2} = e^{\sqrt 2 \ln x} \qquad x^\pi = e^{\pi \ln x} \qquad 3^\pi = e^{\pi \ln 3}$$

The familiar properties of exponents (7.1) hold for x^α, α real. For example, if α and β are real, then

$$x^\alpha x^\beta = x^{\alpha + \beta}$$

because

$$x^\alpha x^\beta = e^{\alpha \ln x} e^{\beta \ln x} \underset{\underset{\text{By (7.19(a))}}{\uparrow}}{=} e^{\alpha \ln x + \beta \ln x} = e^{(\alpha + \beta) \ln x} \underset{\underset{\text{By (7.29)}}{\uparrow}}{=} x^{\alpha + \beta}$$

The proofs of the other properties of exponents are left as exercises.

DERIVATIVE OF x^α

To obtain the derivative of x^α, α a real number,* we use (7.29). Then,

$$\frac{d}{dx} x^\alpha \underset{\underset{\text{By (7.29)}}{\uparrow}}{=} \frac{d}{dx} e^{\alpha \ln x} \underset{\underset{\text{Chain rule}}{\uparrow}}{=} e^{\alpha \ln x} \frac{d}{dx} (\alpha \ln x) = x^\alpha \frac{\alpha}{x} = \alpha x^{\alpha - 1}$$

---

* Recall from (3.56) that thus far, we have found the derivative of x^r only for r a *rational number*.

Thus,

(7.30)
$$\frac{d}{dx} x^{\alpha} = \alpha x^{\alpha-1} \qquad \alpha \text{ a real number}$$

Another way of deriving this result is to use logarithmic differentiation on $y = x^{\alpha} = e^{\alpha \ln x}$. Then

$$\ln y = \ln e^{\alpha \ln x} = \alpha \ln x$$

$$\frac{y'}{y} = \frac{\alpha}{x}$$

$$y' = \frac{\alpha y}{x} = \frac{\alpha x^{\alpha}}{x} = \alpha x^{\alpha-1}$$

Let's look at an example of how (7.30) can be used to find the derivative of $y = x^x$, $x > 0$.

EXAMPLE 1   Find the derivative of $y = x^x$, $x > 0$.

Solution   Since $y = x^x = e^{x \ln x}$, we have

$$y' = e^{x \ln x} \frac{d}{dx} (x \ln x) = e^{x \ln x} \left[x\left(\frac{1}{x}\right) + \ln x \right] = x^x\left(1 + \ln x\right) \qquad \blacksquare$$

Another way of solving Example 1 is to use logarithmic differentiation. Then
$$\ln y = \ln x^x = \ln e^{x \ln x} = x \ln x$$

We differentiate with respect to x; then

$$\frac{1}{y}\left(\frac{dy}{dx}\right) = x\left(\frac{1}{x}\right) + \ln x$$

$$\frac{dy}{dx} = y(1 + \ln x) = x^x(1 + \ln x)$$

THE FUNCTION $f(x) = a^x$, $a > 0$, $a \neq 1$

Since $e^{x \ln a} = (e^{\ln a})^x = a^x$, we define

(7.31)
$$a^x = e^{x \ln a} \qquad a > 0, \quad a \neq 1$$

We use the chain rule on (7.31) to get

(7.32)
$$\frac{d}{dx} a^x = a^x(\ln a) \qquad a > 0, \quad a \neq 1$$

EXAMPLE 2   Find y' if: (a) $y = 2^x$ (b) $y = 3^{x^2}$

Solution   (a)   By (7.32), $y' = 2^x \ln 2$.
             (b)   We use the chain rule with $y = 3^u$ and $u = x^2$ to get

$$y' = \left(\frac{dy}{du}\right)\left(\frac{du}{dx}\right) = (3^u \ln 3)(2x) = (2 \ln 3)x \ 3^{x^2} \quad \blacksquare$$

In view of (7.32), we have the related integration formula

(7.33) $$\int a^x \, dx = \frac{a^x}{\ln a} + C \quad a > 0, \quad a \neq 1$$

EXAMPLE 3 $$\int 3^{2x+1} \, dx = \int 3^u \frac{du}{2} = \frac{1}{2}\left(\frac{3^u}{\ln 3}\right) + C = \frac{3^{2x+1}}{2 \ln 3} + C \quad \blacksquare$$

Set $u = 2x + 1$

THE GRAPH OF $y = a^x$

From (7.32), it follows that when $a > 1$,

$$\frac{d}{dx} a^x > 0$$

and when $0 < a < 1$,

$$\frac{d}{dx} a^x < 0$$

Hence, the function $y = a^x$, $a > 1$,   is increasing, while $y = a^x$, $0 < a < 1$,   is decreasing. Furthermore, $a^x > 0$ for all x, so that its graph lies above the x-axis. The y-intercept is $(0, 1)$. Figure 12 illustrates the graph of the function $y = a^x$ for selected values of a.

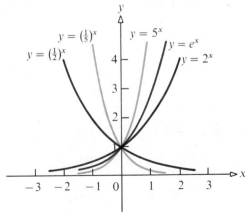

Figure 12   Slopes to $y = a^x$ for $0 < a < 1$ are negative
                     Slopes to $y = a^x$ for $a > 1$ are positive

THE FUNCTION $f(x) = \log_a x, \quad a > 0, \quad a \neq 1$

We define $\ f(x) = \log_a x\ $ as the inverse of the function $\ y = a^x$. See Figure 13 for the graph for selected values of a.

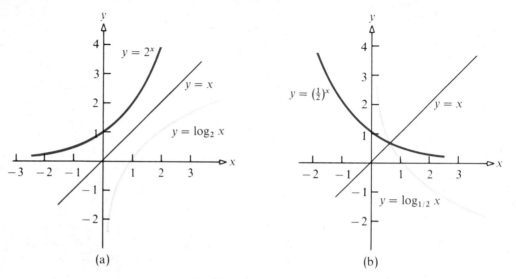

(a)   (b)

Figure 13

The function $\ f(x) = \log_a x\ $ can always be calculated from the natural logarithms of a and x by the following formula:

(7.34)
$$\log_a x = \frac{\ln x}{\ln a} \qquad a > 0, \quad a \neq 1$$

This formula can be derived in the following way. If $\ y = \log_a x,\ $ then $\ a^y = x$. Therefore,

$$\ln a^y = \ln x$$
$$y \ln a = \ln x$$
$$y = \frac{\ln x}{\ln a}$$

which is (7.34). It follows that

(7.35)
$$\frac{d}{dx} \log_a x = \frac{1}{x \ln a} \qquad a > 0, \quad a \neq 1$$

Notice that when $\ a = e,\ \log_e x = \ln x,\ $ and we have the simpler formula

$$\frac{d}{dx} \ln x = \frac{1}{x}$$

From (7.34), it also follows that

$$\log_a a^x = \frac{\ln a^x}{\ln a} = \frac{x \ln a}{\ln a} = x$$

EXERCISE 5

In Problems 1–28 find y'.

1. $y = x^{\sqrt{2}}$ **2.** $y = x^{\pi}$ **3.** $y = (1 + x^2)^{\sqrt{2}}$

4. $y = (x^2 - 1)^{\sqrt{3}}$ **5.** $y = (\tfrac{1}{2})^x$ **6.** $y = 2^{-x}$

7. $y = \log_2 x$ **8.** $y = \log_3 x$ **9.** $y = \log_2(1 + x^2)$

10. $y = \log_2(x^2 - 1)$ **11.** $y = 2^{-x} \sin x$ **12.** $y = (\tfrac{1}{2})^x \cos x$

13. $y = (3x)^x, \quad x > 0$ **14.** $y = (x^2 + 1)^{2x}, \quad x > 0$ **15.** $y = x^{\ln x}$

16. $y = x^{x^2}$ **17.** $y = (3x)^{\sqrt{x}}, \quad x > 0$ **18.** $y = (x^2 + 1)^{x/2}, \quad x > 0$

19. $y = x^{e^x}$ **20.** $y = (x^2 + 1)^{e^x}$ **21.** $y = 2^{\sin x}$

22. $y = 2^{\cos x}$ **23.** $y = x^{\sin x}, \quad x > 0$ **24.** $y = (\sin x)^x, \quad \sin x > 0$

25. $y = (\sin x)^{\cos x}, \quad \sin x > 0$ **26.** $y = (\sin x)^{\tan x}, \quad \sin x > 0$ **27.** $2^{xy} = x$

28. $x^y = 4$

In Problems 29–38 evaluate each integral.

29. $\displaystyle\int 2^x \, dx$ **30.** $\displaystyle\int 3^x \, dx$ **31.** $\displaystyle\int 2^{3x+5} \, dx$ **32.** $\displaystyle\int 3^{2x+7} \, dx$

33. $\displaystyle\int_0^1 (1 + x)^{3\pi} \, dx$ **34.** $\displaystyle\int_0^1 x \, 10^{-x^2} \, dx$ **35.** $\displaystyle\int_3^9 \frac{dx}{x \log_3 x}$ **36.** $\displaystyle\int_{10}^{100} \frac{dx}{x \log_{10} x}$

37. $\displaystyle\int_{10}^{100} \frac{dx}{x \ln x}$ **38.** $\displaystyle\int_{10}^{100} \frac{dx}{x \log_5 x}$

39. Prove that $\log_b a = 1/\log_a b$, $\quad a > 0, \quad a \neq 1, \quad b > 0, \quad b \neq 1$.

40. Prove that $\log_a x^{\alpha} = \alpha \log_a x$, $\quad \alpha$ a real number.

41. Estimate $\ln a$ in terms of x_1 and x_2 if $\quad e^{x_1} < a < e^{x_2}$.

42. Estimate e^a in terms of x_1 and x_2 if $\quad \ln x_1 < a < \ln x_2$.

43. If f and g are differentiable functions, and if $\quad f(x) > 0, \quad$ show that

$$\frac{d}{dx} f(x)^{g(x)} = g(x) f(x)^{g(x) - 1} f'(x) + f(x)^{g(x)} \ln f(x) g'(x)$$

44. Use (7.29) to show that $\quad (x^{\alpha})^{\beta} = x^{\alpha\beta}, \quad \alpha$ and β real.

45. Use (7.29) to show that $\quad (xy)^{\alpha} = x^{\alpha} y^{\alpha}, \quad \alpha$ real.

46. If α and β are real, prove that $\quad x^{\alpha}/x^{\beta} = x^{\alpha - \beta}$.

6. The Number e as a Limit; Continuously Compounded Interest

In this section, we show that the number e defined in (7.11) can also be expressed as a limit. In fact, there are two equivalent ways of writing e as a limit, each of which we shall have occasion to refer to later.

(7.36)   **THEOREM**   (a) $\lim_{h \to 0}(1 + h)^{1/h} = e$ (b) $\lim_{n \to +\infty}\left(1 + \dfrac{1}{n}\right)^n = e$

Proof

(a) We use the definition of a derivative to calculate the derivative of $f(x) = \ln x$ at $x = 1$.   Of course, we already know that the answer is $f'(1) = \frac{1}{1} = 1$.

$$1 = f'(1) = \lim_{h \to 0}\frac{\ln(1 + h) - \ln 1}{h} = \lim_{h \to 0}\left[\left(\frac{1}{h}\right)\ln(1 + h)\right] = \lim_{h \to 0}\left[\ln(1 + h)^{1/h}\right]$$

$$\underset{\ln 1 = 0}{\uparrow} \qquad\qquad \underset{(7.5(c))}{\uparrow}$$

Now we use the fact that $y = \ln x$ is continuous, so that

$$\lim_{x \to c}[\ln f(x)] = \ln[\lim_{x \to c} f(x)]$$

Also, since $\ln e = 1$,

$$1 = \ln\left[\lim_{h \to 0}(1 + h)^{1/h}\right] = \ln e$$

Since $y = \ln x$ is one-to-one, we conclude that

$$e = \lim_{h \to 0}(1 + h)^{1/h}$$

(b) The limit derived in part (a) is valid when $h \to 0^+$.   Hence, if we set $n = 1/h$, then $h \to 0^+$ implies $n \to +\infty$, and

$$e = \lim_{h \to 0^+}(1 + h)^{1/h} = \lim_{n \to +\infty}\left(1 + \frac{1}{n}\right)^n \qquad \blacksquare$$

Table 1 lists some approximate values for e that were calculated by using (7.36(b)).

Table 1

$n = 100$	$\left(1 + \dfrac{1}{n}\right)^n = \left(1 + \dfrac{1}{100}\right)^{100}$	≈ 2.704814
$n = 10{,}000$	$\left(1 + \dfrac{1}{n}\right)^n = \left(1 + \dfrac{1}{10{,}000}\right)^{10{,}000}$	≈ 2.718146
$n = 100{,}000$	$\left(1 + \dfrac{1}{n}\right)^n = \left(1 + \dfrac{1}{100{,}000}\right)^{100{,}000}$	≈ 2.718268237

Correct to nine decimal places, it turns out that

$$e = 2.718281828 \ldots$$

We hasten to point out that this pattern for e does not repeat. In fact, it can be shown that e is an irrational number.

Theorem (7.36) can be used to evaluate certain limits in terms of the number e.

EXAMPLE 1   Express each limit in terms of the number e.

(a) $\lim\limits_{h \to 0}(1 + 2h)^{1/h}$ (b) $\lim\limits_{n \to +\infty} \left(1 + \dfrac{3}{n}\right)^{2n}$

Solution   (a) Observe the resemblance of the limit in question to that of (7.36(a)). In fact, note that

$$(1 + 2h)^{1/h} = [(1 + 2h)^{1/2h}]^2$$

Since $h \to 0$ is equivalent to $2h = k \to 0$,   we discover that

$$\lim_{h\to 0}(1 + 2h)^{1/h} = \lim_{h\to 0}[(1 + 2h)^{1/2h}]^2 \underset{\underset{\text{Set}\ \ 2h = k}{\uparrow}}{=} \lim_{k\to 0}[(1 + k)^{1/k}]^2$$

$$= [\lim_{k\to 0}(1 + k)^{1/k}]^2 = e^2 \approx 7.389$$

(b) Observe the resemblance of the limit in question to that of (7.36(b)). Furthermore,

$$\left(1 + \frac{3}{n}\right)^{2n} = \left[\left(1 + \frac{3}{n}\right)^{n/3}\right]^6 = \left[\left(1 + \frac{1}{n/3}\right)^{n/3}\right]^6$$

Since $n \to +\infty$ is equivalent to $n/3 = m \to +\infty$,   we find that

$$\lim_{n\to +\infty} \left(1 + \frac{3}{n}\right)^{2n} = \lim_{n\to +\infty}\left[\left(1 + \frac{1}{n/3}\right)^{n/3}\right]^6$$

$$\underset{\underset{\text{Set}\ \ m = n/3}{\uparrow}}{=} \lim_{m\to +\infty}\left[\left(1 + \frac{1}{m}\right)^{m}\right]^6 = \left[\lim_{m\to +\infty}\left(1 + \frac{1}{m}\right)^{m}\right]^6 = e^6 \approx 403.429 \qquad \blacksquare$$

CONTINUOUSLY COMPOUNDED INTEREST

One use of the fact that $e = \lim_{h\to 0^+}(1 + h)^{1/h}$ is found in finance. Suppose a principal P is to be invested at an annual rate of interest r, which is compounded n times per year for t years. The interest earned on the principal P at each compounding period is then $P(r/n)$.   The amount A after one compounding period is

$$A = P + P\left(\frac{r}{n}\right) = P\left(1 + \frac{r}{n}\right)$$

After 2 compoundings,* $\quad A = P\left(1 + \dfrac{r}{n}\right) + P\left(1 + \dfrac{r}{n}\right)\left(\dfrac{r}{n}\right) = P\left(1 + \dfrac{r}{n}\right)^2$

After 3 compoundings, $\quad A = P\left(1 + \dfrac{r}{n}\right)^2 + P\left(1 + \dfrac{r}{n}\right)^2\left(\dfrac{r}{n}\right) = P\left(1 + \dfrac{r}{n}\right)^3$

...   ...

After k compoundings, $\quad A = P\left(1 + \dfrac{r}{n}\right)^k$

In t years there are nt compounding periods.

The amount A after t years accrued on a principal P when it is invested at an annual rate of interest r and is compounded n times per year is

(7.37)
$$A = P\left(1 + \dfrac{r}{n}\right)^{nt}$$

Formula (7.37) is usually referred to as the *compound interest formula*.

Table 2 lists the results of investing \$1,000 at an annual rate of 10% for 1 year for various compounding periods.

Table 2

$P = Principal = \$1{,}000;$ $r = Annual\ Rate\ of\ Interest = 10\% = 0.10;$ $t = 1\ Year = 1$

$n = Number\ of\ Times$ Compounded per Year	$A = Amount\ after\ 1\ Year$
1 Annual compounding	$A = P(1 + r) = 1{,}000(1 + 0.1) = \$1{,}100.00$
2 Semiannual compounding	$A = P\left(1 + \dfrac{r}{2}\right)^2 = 1{,}000(1 + 0.05)^2 = \$1{,}102.50$
4 Quarterly compounding	$A = P\left(1 + \dfrac{r}{4}\right)^4 = 1{,}000(1 + 0.025)^4 = \$1{,}103.81$
12 Monthly compounding	$A = P\left(1 + \dfrac{r}{12}\right)^{12} = 1{,}000(1 + 0.00833)^{12} = \$1{,}104.71$
365 Daily compounding†	$A = P\left(1 + \dfrac{r}{365}\right)^{365} = 1{,}000(1 + 0.000274)^{365} = \$1{,}105.16$

Now we ask what happens to the amount after t years as the number of times n that the interest is compounded per year gets larger and larger. In other words, we seek to calculate $\lim_{n \to +\infty}[P(1 + r/n)^{nt}]$.

* Remember, the new principal is now $\quad P(1 + r/n)$.

† Often, banks use 360 days for daily compounding.

$$\lim_{n \to +\infty} \left[P \left(1 + \frac{r}{n} \right)^{nt} \right] = P \lim_{n \to +\infty} \left[\left(1 + \frac{r}{n} \right)^n \right]^t = P \lim_{n \to +\infty} \left[\left(1 + \frac{r}{n} \right)^{n/r} \right]^{rt}$$

$$= P \left[\lim_{n \to +\infty} \left(1 + \frac{r}{n} \right)^{n/r} \right]^{rt} = P \left[\lim_{h \to 0} (1 + h)^{1/h} \right]^{rt} = Pe^{rt}$$

$$\underset{h = r/n}{\uparrow}$$

Thus, no matter how often the interest is compounded during the year, the amount after t years has a definite ceiling, Pe^{rt}. When interest is compounded so that the amount after t years is Pe^{rt}, we say that the interest is *compounded continuously*.

For example, the amount A due to investing $1,000 for 1 year at an annual rate of 10% compounded continuously is

(7.38)
$$A = 1,000e^{0.1} = \$1,105.17$$

The phrase *effective rate of interest* is often used. This is the equivalent annual rate of interest with compounding. When interest is compounded annually, there is no difference between the annual rate and the effective rate; however, when interest is compounded more than once a year, the effective rate always exceeds the annual rate. For example, by using the results in Table 2 and equation (7.38), we find that when the annual rate is 10%, the effective rates are those listed in Table 3. It is worth noting that although the difference between a bank's paying interest yearly (almost none do now) versus compounding quarterly or monthly is fairly substantial, the difference between daily and continuous compounding is practically negligible.

Table 3

	Annual Rate (%)	Effective Rate (%)
Annual compounding	10	10
Semiannual compounding	10	10.25
Quarterly compounding	10	10.381
Monthly compounding	10	10.471
Daily compounding	10	10.516
Continuous compounding	10	10.517

EXERCISE 6

In Problems 1–4 express each limit in terms of e.

1. $\lim_{n \to +\infty} \left(1 + \frac{1}{n} \right)^{2n}$ **2.** $\lim_{n \to +\infty} \left(1 + \frac{1}{n} \right)^{n/2}$ **3.** $\lim_{n \to +\infty} \left(1 + \frac{1}{3n} \right)^n$ **4.** $\lim_{n \to +\infty} \left(1 + \frac{4}{n} \right)^n$

5. Find the amount after 1 year if $500 is invested at 6% compounded continuously for 1 year. Use a calculator to obtain the answer if the rate is $6\frac{1}{4}$% compounded quarterly. Which is better?

6. Find the amount after 2 years if $1,000 is invested at 8% compounded continuously. Use a calculator to obtain the amount if the rate is $8\frac{1}{2}$% compounded quarterly. Which is better?

7. What principal P should be invested at 6% compounded continuously in order to have $1,000 after 1 year? What principal is required if the interest is compounded quarterly?

8. What principal P should be invested at 10% compounded continuously in order to have $2,000 after 3 years? What principal is required if the interest is compounded quarterly?

9. How long (in months) does it take for a principal P to double if it is invested at 10% compounded continuously? How long does it take if the compounding is quarterly?

10. Rework Problem 9 if the rate of interest is 8%.

7. Separable Differential Equations

We have noted (in Section 9 of Chapter 4) that a differential equation is an equation involving a function and its derivatives, such as

(1) $\quad \dfrac{dy}{dx} = 3(y - 1)$ (2) $\quad \dfrac{dy}{dx} + y = 2e^x$

(3) $\quad \dfrac{d^2y}{dx^2} - 8\dfrac{dy}{dx} + 4y = 0$ (4) $\quad \dfrac{d^3y}{dx^3} + x\dfrac{d^2y}{dx^2} + x = \left(\dfrac{d^2y}{dx^2}\right)^2$

The *order* of a differential equation is the highest order of the derivatives that appear in it. In the list above, (1) and (2) are first-order differential equations, (3) is a second-order differential equation, and (4) is a third-order differential equation. In this section we consider first-order differential equations of the following special type:

(7.39) **A *separable* differential equation is one that can be written in the form**

$$\frac{dy}{dx} = \frac{f(x)}{g(y)}$$

where f and g are continuous functions.

Such an equation can be solved by *separating the variables;* that is, by writing it in the form

$$g(y)\, dy = f(x)\, dx$$

so that the variables x and y appear on opposite sides of the equation. A differential equation written in this way is solved by integrating both sides:

$$\int g(y)\, dy = \int f(x)\, dx + C$$

where C is an arbitrary constant. (Any convenient antiderivative can be used on the left side; the constant of integration is retained on the right side.)

The following examples illustrate the technique.

EXAMPLE 1   (a)   Find a general solution of the differential equation $dy/dx = 2x/5y$.
                (b)   Solve the differential equation in part (a) with the boundary condition* that $y = 2$ when $x = 1$.

Solution   (a)   Separate the variables to obtain

$$5y\, dy = 2x\, dx$$

$$\int 5y\, dy = \int 2x\, dx + C$$

$$\frac{5y^2}{2} = x^2 + C$$

Here, y is defined implicitly as described in Section 9, Chapter 3. Implicit differentiation can be used to verify that y is a solution of the differential equation.

(b)   The general solution is $5y^2/2 = x^2 + C$.   We use the boundary condition (substituting 2 for y and 1 for x) to determine the particular solution:

$$\frac{5(4)}{2} = 1 + C$$

$$9 = C$$

Thus, the solution of the given differential equation with the boundary condition that $y = 2$ when $x = 1$ is

$$\frac{5y^2}{2} = x^2 + 9 \qquad \blacksquare$$

EXAMPLE 2   Solve the differential equation $dy/dx = 3y/x$.

Solution   If we rewrite the equation as

$$\frac{dy}{dx} = \frac{3/x}{1/y} \qquad y \neq 0$$

we see that it is separable:

$$\frac{1}{y}\, dy = \frac{3}{x}\, dx$$

$$\int \frac{1}{y}\, dy = \int \frac{3}{x}\, dx + C$$

$$\ln|y| = 3 \ln|x| + C$$

We solve for y explicitly by taking exponents on both sides,

$$e^{\ln|y|} = e^{3 \ln|x| + C}$$

---

* You may wish to review Section 9 in Chapter 4, where these concepts are introduced. Also, "boundary conditions" are frequently called "initial conditions" in other textbooks.

or, by (7.17),

$$|y| = |x|^3 e^C$$

Since C is an arbitrary constant, we may write

$$y = kx^3$$

where $k = \pm e^C$ is an arbitrary constant.   ∎

DIFFERENTIAL EQUATIONS FOR GROWTH AND DECAY

In studies of physical, chemical, biological, and other phenomena, scientists attempt, on the basis of long observation or by other means, to deduce mathematical laws that describe or predict nature's behavior. In many situations the amount A of a substance varies with time t in such a way that the time rate of change of A is proportional to A itself. We may state this in the form of the differential equation

(7.40)
$$\frac{dA}{dt} = kA$$

where $k \neq 0$ is a real number. If $k > 0$, then (7.40) asserts that the time rate of change of A is positive, so that the amount A of the substance is increasing; if $k < 0$, then (7.40) asserts that the time rate of change of A is negative, so that the amount A of the substance is decreasing.

We assume that the initial amount A_0 is known, giving us the boundary condition $A = A_0$ when $t = 0$. By separating the variables and integrating,

$$\frac{dA}{A} = k \, dt$$

$$\int \frac{dA}{A} = \int k \, dt + C$$

$$\ln A = kt + C$$

The boundary condition requires that $A = A_0$ when $t = 0$. Thus, $C = \ln A_0$, so that

$$\ln A = kt + \ln A_0 = \ln A_0 + kt$$

or

(7.41)
$$A = A_0 e^{kt}$$

Therefore, this is the solution to the differential equation (7.40), where A_0 is the original amount.

When a function $A = A(t)$ varies according to the law (7.40), or its equivalent (7.41), it is said to follow the *exponential law*, or the *law of uninhibited growth or decay*, or the *law of continuously compounded interest*. Figure 14 illustrates the graphs of (7.41) for both $k > 0$ and $k < 0$.

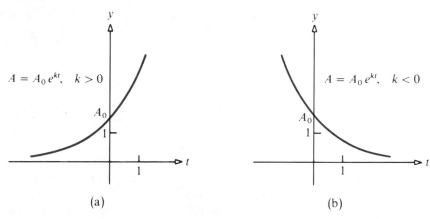

Figure 14

BACTERIAL GROWTH

EXAMPLE 3   Assume that a colony of bacteria increases at a rate proportional to the number present.* If the number of bacteria doubles in 5 hours, how long will it take for the bacteria to triple?

Solution   Let $N(t)$ be the number of bacteria present at time t. Then the assumption that this colony of bacteria increases at a rate proportional to the number present can be mathematically written as the differential equation

(7.42)
$$\frac{dN}{dt} = kN$$

where k is a positive constant of proportionality. The solution of (7.42) is

$$N(t) = N_0 e^{kt}$$

where N_0 is the initial number of bacteria in this colony. Since the number of bacteria doubles in 5 hours, we have

$$N(5) = 2N_0$$
$$N_0 e^{5k} = 2N_0$$
$$e^{5k} = 2$$
$$k = (\tfrac{1}{5}) \ln 2$$

---

* This is a model of uninhibited growth. However, after enough time has passed, growth will not continue at a rate proportional to the number present. Other factors, such as lack of living space, dwindling food supply, and so on, will start to afffect the rate of growth. The model presented accurately reflects the way growth occurs in the early stages.

The time t required for this colony to triple must satisfy the equation

$$N(t) = 3N_0$$
$$N_0 e^{kt} = 3N_0$$
$$e^{kt} = 3$$

$$t = \left(\frac{1}{k}\right) \ln 3 = 5\frac{\ln 3}{\ln 2} = 5\frac{1.0986}{0.6931} = 7.925 \text{ hours} \quad \blacksquare$$

RADIOACTIVE DECAY

Our next application is related to the problem of dating archaeological specimens, organic fossils, and so on by measuring radioactive decay.

For a radioactive substance, the *rate of decay is proportional to the amount present at a given time t*. That is, if A represents the amount of a radioactive substance at time t, we have

(7.43)
$$\frac{dA}{dt} = kA$$

where the constant k is negative and depends on the atomic structure of the substance. The *half-life* of a radioactive substance is the time required for half of the substance to decay.

In carbon dating, we use the fact that all living organisms contain two kinds of carbon, carbon-12 (a stable carbon) and carbon-14 (a radioactive carbon). As a result, when an organism dies, the amount of carbon-12 present remains unchanged while the amount of carbon-14 begins to decrease. This change in the amount of carbon-14 present relative to the amount of carbon-12 present makes it possible to calculate the time at which the organism lived.

EXAMPLE 4   In the skull of an animal found in an archaeological dig, it was determined that about 20% of the original amount of carbon-14 was still present. If the half-life of carbon-14 is 5730 years, find the approximate age of the animal.

Solution   Let A be the amount of carbon-14 present in the skull at time t. Then A satisfies the differential equation $dA/dt = kA$,   whose solution is

$$A = A_0 e^{kt}$$

where A_0 is the amount of carbon-14 present at time $t = 0$.   To determine the constant k, we use the fact that when $t = 5730$,   half of the original amount A_0 will remain. Thus,

$$\tfrac{1}{2}A_0 = A_0 e^{5730k}$$
$$\tfrac{1}{2} = e^{5730k}$$
$$5730k = \ln \tfrac{1}{2}$$
$$k = -0.000121$$

The relationship between the amount A of carbon-14 and time t is therefore

$$A = A_0 e^{-(0.000121)t}$$

If the amount A of carbon-14 is 20% of the original amount A_0, we have

$$0.2A_0 = A_0 e^{-(0.000121)t}$$
$$0.2 = e^{-(0.000121)t}$$

By taking the natural logarithm of both sides, we have

$$-(0.000121)t = \ln 0.2$$
$$-(0.000121)t = -1.6094$$

$$t = \frac{1.6094}{0.000121} \approx 13,300 \text{ years}$$

Thus, the animal lived approximately 13,000 years ago.   ■

NEWTON'S LAW OF COOLING

Newton's *law of cooling* states that the rate of decrease of temperature of an object is continuous and proportional to the difference between the temperature of the object and that of the surrounding medium. That is, if u is the temperature of the body at time t and if T $(u > T)$ is the (constant) temperature of the surrounding medium, then

(7.44)
$$\frac{du}{dt} = k(u - T)$$

where k is a negative constant* that depends on the object.

We seek u as a function of t, that is, we seek a rule for finding the temperature of the object at any time t. Using an argument similar to the one used to solve (7.40), we separate the variables in order to rewrite (7.44) as

$$\frac{du}{u - T} = k \, dt$$

Then

$$\int \frac{du}{u - T} = \int k \, dt$$

from which

(7.45)
$$\ln(u - T) = kt + C$$

---

* If u is the temperature of a body at time t and if T $(u < T)$ is the constant temperature of the surrounding medium, the *law of heating* states that $du/dt = k(T - u)$,   where k is a positive constant.

We use the boundary condition that when $t = 0$, the initial temperature of the body is u_0. As a result, $C = \ln(u_0 - T)$. By substituting this into (7.45) and simplifying, we obtain

$$\ln(u - T) = kt + \ln(u_0 - T)$$

$$\ln(u - T) - \ln(u_0 - T) = kt$$

$$\ln \frac{u - T}{u_0 - T} = kt$$

(7.46) $$u - T = (u_0 - T)e^{kt}$$

Let's look at a specific example.

EXAMPLE 5   Suppose an object is heated to 90°C and allowed to cool in a room with air temperature at 20°C. If after 10 minutes the temperature of the object is 60°C, what will be its temperature after 20 minutes?

Solution   When $t = 0$, $u_0 = 90°C$, and when $t = 10$, $u = 60°C$. The temperature T of the medium is 20°C. By using this information in equation (7.46), we get

$$(60 - 20) = (90 - 20)e^{10k}$$

$$\tfrac{40}{70} = e^{10k}$$

$$10k = \ln \tfrac{4}{7} = -0.5596$$

$$k = -0.05596$$

The relationship between the temperature u and time t is therefore

$$u - 20 = 70e^{-0.05596t}$$

When $t = 20$, the temperature u of the object is

$$u = 70e^{-0.05596(20)} + 20 = 70(0.3265) + 20 = 22.86 + 20 = 42.86°C \quad \blacksquare$$

EXERCISE 7

In Problems 1–20 determine whether or not the equation is separable. If it is, find a general solution; if it is not, leave it alone. Check each solution by substituting back into the differential equation.

1. $\dfrac{y^2}{x}\left(\dfrac{dy}{dx}\right) = 1 + x^2$

2. $\sec x \left(\dfrac{dy}{dx}\right) = \tan y$

3. $\cos y \left(\dfrac{dy}{dx}\right) = \sin(x + y)$

4. $\dfrac{dy}{dx} = \dfrac{(x + 1)^2 - 2y}{y}$

5. $(x + y)\left(\dfrac{dy}{dx}\right) = 2$

6. $\dfrac{dy}{dx} = \dfrac{\cos y}{e^x \sin y}$

7. $\ln y^x \left(\dfrac{dy}{dx}\right) = 3x^2 y$

8. $(x + y)^2 \left(\dfrac{dy}{dx}\right) = 2xy$

9. $\dfrac{dy}{dx} = \dfrac{8x + y^2 - 1}{2yx}$

10. $y(2 + x)\dfrac{dy}{dx} = \dfrac{4}{x} + x + 4$

11. $\dfrac{dy}{dx} = y\left(\dfrac{x^2 - 2x + 1}{y + 3}\right)$

12. $[(x + y)^2 - (x^2 + y^2)]\dfrac{dy}{dx} = 1$

Problems 13–22 contain a separable differential equation, together with a boundary condition. Find a general solution and the solution with the given boundary condition. Check your solutions.

13. $\dfrac{dy}{dx} = 3x^2(y + 2);\quad y = 8$ when $x = 0$

14. $x\left(\dfrac{dy}{dx}\right) = y^2;\quad y = 5$ when $x = 3$

15. $\dfrac{dy}{dx} = \dfrac{x^2 + 2}{y};\quad y = 1$ when $x = 1$

16. $y\left(\dfrac{dy}{dx}\right) = \dfrac{x^2}{y + 4};\quad y = 2$ when $x = 8$

17. $\dfrac{dy}{dx} = \dfrac{x - 1}{y + 2};\quad y = 6$ when $x = -1$

18. $x^2\left(\dfrac{dy}{dx}\right) = \dfrac{1}{y};\quad y = 2$ when $x = 4$

19. $\dfrac{dy}{dx} = \dfrac{e^y}{x};\quad y = 0$ when $x = 1$

20. $\dfrac{dy}{dx} = \dfrac{-2\sin x}{y^2};\quad y = 4$ when $x = \pi$

21. $\dfrac{dy}{dx} = \dfrac{-3x}{y + 4};\quad y = 3$ when $x = 2$

22. $y\left(\dfrac{dy}{dx}\right) = \dfrac{8x + 1}{y^2};\quad y = -1$ when $x = 1$

23. The half-life of radium is 1690 years. If 8 grams of radium is present now, how much will be present in 100 years?

24. If 25% of a radioactive substance disappears in 10 years, what is the half-life of the substance?

25. A piece of charcoal is found to contain 30% of the carbon-14 it originally had. When did the tree from which the charcoal came die? Use 5730 years as the half-life of carbon-14.

26. A fossilized leaf contains 70% of a normal amount of carbon-14. How old is the fossil?

27. The population growth of a colony of mosquitoes obeys the uninhibited growth equation (7.42). If there are 1500 mosquitoes initially, and there are 2500 mosquitoes after 24 hours, what is the size of the mosquito population after 3 days?

28. The population of a suburb doubled in size in an 18 month period. If this growth continues and the current population is 8000, what will the population be in 4 years?

29. The number of bacteria in a culture is growing at a rate of $3000e^{2t/5}$ per unit of time t. At $t = 0$, the number of bacteria present is 7500. Find the number present at $t = 5$.

30. At any time t, the rate of increase in the area of a culture of bacteria is twice the area of the culture. If the initial area of the culture is 10, then what is the area at time t?

31. The rate of change in the number of bacteria in a culture is proportional to the number present. In a certain laboratory experiment, a culture has 10,000 bacteria initially, 20,000 bacteria at time t_1 minutes, and 100,000 bacteria at $(t_1 + 10)$ minutes.
   (a) In terms of t only, find the number of bacteria in the culture at any time t minutes $(t \ge 0)$.
   (b) How many bacteria are there after 20 minutes?
   (c) At what time are 20,000 bacteria observed? That is, find the value of t_1.

32. Salt (NaCl) dissociates in water into sodium (Na^+) and chloride (Cl^-) ions at a rate proportional to its mass. If the initial amount of salt is 25 kilograms, and after 10 hours, 15 kilograms are left:
   (a) How much salt would be left after 1 day?
   (b) After how many hours would there be less than $\frac{1}{2}$ kilogram of salt left?

33. Radioactive beryllium is sometimes used to date fossils found in deep-sea sediment. The decay of beryllium satisfies the equation $dA/dt = -\alpha A$, where $\alpha = 1.5 \times 10^{-7}$, and t is measured in years. What is the half-life of beryllium?

34. The voltage of a certain condenser decreases at a rate proportional to the voltage. If the initial voltage is 20, and 2 seconds later it is 10, what is the voltage at time t? When will the voltage be 5?

35. A thermometer reading 4°C is brought into a room where the temperature reading is 30°C. If the thermometer reads 10°C after 2 minutes, determine the temperature reading 5 minutes after the thermometer is first brought into the room.

36. A thermometer reading 70°F is taken outside where the temperature is 22°F. Four minutes later the reading is 32°F. Find:
(a)   The thermometer reading 7 minutes after the thermometer was brought outside
(b)   The time taken for the reading to change from 70°F to within $\frac{1}{2}$°F of the air temperature

37. Atmospheric pressure is a function of altitude above sea level and is given by the equation $dP/da = \beta P$, where β is a constant. The pressure is measured in millibars (mb). At sea level $(a = 0)$, $P(0)$ is 1013.25 mb, which means that the atmosphere at sea level will support a column of mercury 1013.25 millimeters high at a standard temperature of 15°C. At an altitude of $a = 1500$ meters,   the pressure is 845.6 millibars.
(a)   What is the pressure at $a = 4000$ meters?
(b)   What is the pressure at 10 kilometers?
(c)   In California, the highest and lowest points are Mount Whitney (4418 meters) and Death Valley (86 meters below sea level). What is the difference in their atmospheric pressures?
(d)   What is the atmospheric pressure at Mount Everest (elevation 8848 meters)?
(e)   At what elevation is the atmospheric pressure equal to 1 millibar?

8. Linear Differential Equations

A first-order differential equation is said to be *linear* if we can write it in the form

(7.47)
$$\frac{dy}{dx} + P(x)y = Q(x)$$

where P and Q are continuous functions. [Note that (7.47) is not separable unless $P(x) = 0$ or $Q(x) = 0$ or $Q(x)$ is a multiple of $P(x)$ for all x.] Some examples of linear first-order differential equations are given below.

$$\frac{dy}{dx} - 2y = e^x \qquad \text{Here,}\quad P(x) = -2, \quad Q(x) = e^x$$

$$\frac{dy}{dx} - \frac{3}{x}y = x^4 \qquad \text{Here,}\quad P(x) = -\frac{3}{x}, \quad Q(x) = x^4$$

$$\frac{dy}{dx} + y\tan x = \sin x \qquad \text{Here,}\quad P(x) = \tan x, \quad Q(x) = \sin x$$

Solutions to linear first-order differential equations of this type may be found by finding a function $u = u(x)$ such that

(7.48)
$$\frac{du}{dx} = u(x)P(x)$$

Suppose we have found such a function. Then, by multiplying both sides of (7.47) by u, we would obtain the differential equation

(7.49)
$$u(x)\frac{dy}{dx} + u(x)P(x)y = u(x)Q(x)$$

From (7.48), $du/dx = u(x)P(x)$.   Thus,

(7.50)
$$u(x)\frac{dy}{dx} + \frac{du}{dx}y = u(x)Q(x)$$

$$\frac{d}{dx}\left[u(x)y\right] = u(x)Q(x)$$

(7.51)
$$u(x)y = \int u(x)Q(x)\,dx$$

Thus, if a function $u = u(x)$ can be found that obeys the differential equation (7.48), then the linear first-order differential equation (7.47) has the solution (7.51).

We find the function $u = u(x)$,   called an *integrating factor*, by solving the differential equation (7.48), $du/dx = u(x)P(x)$.   This is a separable differential equation so that

$$\frac{du}{u} = P(x)\,dx$$

Integrating both sides, we get

$$\ln u = \int P(x)\,dx$$

$$u = e^{\int P(x)\,dx}$$

Note that in computing $u = u(x)$,   we do not include a constant of integration since any function u that obeys (7.48) will work as an integrating factor.

The steps to use to solve a linear first-order differential equation in the form (7.47) are listed below:

1. Find an integrating factor $u = u(x) = e^{\int P(x)\,dx}$.

2. Multiply the linear first-order differential equation by the integrating factor $u = u(x)$.

3. Integrate both sides, remembering that the left side will integrate to $u(x)y$ [see (7.50)].

EXAMPLE 1   Solve: $\dfrac{dy}{dx} - 2y = e^x$

Solution   This is a linear first-order differential equation $(P = -2, Q = e^x)$.   An integrating factor is

$$u = e^{\int -2\,dx} = e^{-2x}$$

By multiplying by u and integrating, we get

$$e^{-2x}\frac{dy}{dx} - 2e^{-2x}y = e^{-2x}e^x = e^{-x}$$

$$e^{-2x}y = \int e^{-x}\,dx$$

$$e^{-2x}y = -e^{-x} + C$$

$$y = e^{2x}(-e^{-x} + C) = -e^x + Ce^{2x}$$

This is the general solution.   ■

EXAMPLE 2   Solve

$$x\frac{dy}{dx} - y = x^2 \qquad x > 0$$

with the boundary condition that $y = -2$ when $x = 1$.

Solution   As written, the differential equation to be solved is not of the form (7.47). However, if we divide by x, we get

$$\frac{dy}{dx} - \frac{1}{x}y = x$$

which is a linear first-order differential equation. An integrating factor is

$$u = e^{\int(-1/x)\,dx} = e^{-\ln|x|} = \frac{1}{|x|} \underset{\substack{\uparrow \\ x > 0}}{=} \frac{1}{x}$$

Thus,

$$\frac{1}{x}\left(\frac{dy}{dx}\right) - \frac{1}{x^2}y = \frac{1}{x}x = 1$$

$$\frac{1}{x}y = \int dx = x + C$$

$$y = x(x + C) = x^2 + Cx$$

When $x = 1$ and $y = -2$,   we get $-2 = 1 + C$ or $C = -3$.   Thus,   the particular solution is

$$y = x^2 - 3x \qquad ■$$

EXAMPLE 3   *Economics*   Assume that the rate of change of the price P of a commodity is proportional to the difference between the demand D and the supply S at any time t. If $D = a - bP$ and $S = -c + dP$,   where a, b, c, and d are constants, find an expression for P.

Solution

(7.52)
$$\frac{dP}{dt} = K(D - S)$$

where K is the constant of proportionality.

$$D - S = (a - bP) - (-c + dP) = (a + c) - (b + d)P$$

Hence, (7.52) can be written as

$$\frac{dP}{dt} + K(b + d)P = K(a + c)$$

By using the integrating factor $e^{K(b + d)t}$,   we have

$$e^{K(b + d)t} \frac{dP}{dt} + K(b + d)e^{K(b + d)t}P = K(a + c)e^{K(b + d)t}$$

Now we integrate both sides to get

$$e^{K(b + d)t}P = \frac{a + c}{b + d} e^{K(b + d)t} + C$$

The price P is therefore

(7.53)
$$P = \frac{a + c}{b + d} + Ce^{-K(b + d)t}$$

Notice that after a sufficient amount of time, that is, as $t \to +\infty$,   the price P gets closer to $(a + c)/(b + d)$.   For $P = (a + c)/(b + d)$,   we find that

$$D = a - bP = a - b\left(\frac{a + c}{b + d}\right) = \frac{ad - bc}{b + d}$$

and

$$S = -c + dP = -c + d\left(\frac{a + c}{b + d}\right) = \frac{ad - bc}{b + d}$$

That is, supply and demand are eventually equal. In this case, P is called the *equilibrium price*.   ■

EXAMPLE 4   *Mixtures*   A large tank contains 81 gallons of brine in which 20 pounds of salt are dissolved. Brine containing 3 pounds of dissolved salt per gallon runs into the tank at the rate of 5 gallons per minute. The mixture, kept uniform by stirring, runs out of the tank at the rate of 2 gallons per minute. How much salt is in the tank at the end of 37 minutes?

Solution   Let $y(t)$ be the amount of salt in the tank at any time t. Then $y'(t)$ is the rate of change of the salt in the tank at time t. Clearly,

$$y'(t) = \text{Rate in} - \text{Rate out}$$

where "rate in" is the rate at which salt runs into the tank at time t, and "rate out" is the rate at which salt runs out of the tank at time t. Here

$$\text{Rate in} = (3 \text{ lb/gal})(5 \text{ gal/min}) = 15 \text{ lb/min}$$

that is, 15 pounds of salt per minute flows into the tank. To compute the rate out, we first find the *concentration* of salt at time t, that is, the amount of salt per gallon of brine at time t. Since

$$\text{Concentration} = \frac{\text{Pounds of salt in tank at time } t}{\text{Gallons of brine in tank at time } t}$$

$$= \frac{y(t)}{81 + (5 - 2)t}$$

we have

$$\text{Rate out} = \left[\frac{y(t)}{81 + 3t} \text{ lb/gal} \right](2 \text{ gal/min}) = \frac{2y(t)}{81 + 3t} \text{ lb/min}$$

Thus, the differential equation describing this mixture problem is

(7.54) $$y' = 15 - \frac{2y}{81 + 3t}$$

We also have the boundary condition that $y = 20$ when $t = 0$. The differential equation in (7.54) is linear and can be written in the form

$$y' + \frac{2}{81 + 3t} y = 15$$

By multiplying both sides by the integrating factor

$$e^{\int [2/(81 + 3t)] \, dt} = e^{(2/3) \int [dt/(27 + t)]} = e^{(2/3)[\ln(27 + t)]} = (27 + t)^{2/3}$$

we obtain

$$\frac{d}{dt} \left[y(t)(27 + t)^{2/3} \right] = 15(27 + t)^{2/3}$$

By integrating both sides with respect to t, we obtain

$$y(t)(27 + t)^{2/3} = 15(27 + t)^{5/3} \frac{3}{5} + C$$

$$y(t) = 9(27 + t) + C(27 + t)^{-2/3}$$

By using the boundary condition, we find that

$$20 = 9(27) + C(27)^{-2/3}$$

and so $C = -2007$.   Thus,

$$y(t) = 9(27 + t) - 2007(27 + t)^{-2/3}$$

and the amount of salt in the tank at the end of 37 minutes is

$$y(37) = 9(27 + 37) - 2007(27 + 37)^{-2/3} = 450.6 \text{ pounds} \quad\blacksquare$$

SECOND-ORDER LINEAR DIFFERENTIAL EQUATIONS

Another type of differential equation that is useful in many applications is

(7.55) $$y'' + by' + cy = 0 \qquad \text{where } b \text{ and } c \text{ are constants}$$

It is called the *second-order homogeneous linear differential equation with constant coefficients*. In order to solve such an equation, we assume that there is an exponential solution

$$y = e^{mx}$$

for some number m. By substituting this hypothetical solution and its derivatives into (7.55), we have

$$m^2 e^{mx} + bm e^{mx} + c e^{mx} = 0$$
$$(m^2 + bm + c) e^{mx} = 0$$

Since $e^{mx} > 0$ for all x, we conclude that

$$m^2 + bm + c = 0$$

This quadratic equation is called the *characteristic equation* for the differential equation (7.55). The general solution for (7.55) depends on whether the characteristic equation has two real roots $(b^2 - 4c > 0)$,   a single root $(b^2 - 4c = 0)$,   or two complex roots $(b^2 - 4c < 0)$.

In texts on differential equations, the following result is proved:

The general solution of a second-order homogeneous linear differential equation with constant coefficients,

(7.56) $$y'' + by' + cy = 0$$

is obtained by solving the characteristic equation

$$m^2 + bm + c = 0$$

1. If the characteristic equation has two distinct real roots, m_1 and m_2, the general solution of (7.56) is

$$y = C_1 e^{m_1 x} + C_2 e^{m_2 x}$$

where C_1 and C_2 are arbitrary constants.

2. **If the characteristic equation has one real root, say m, the general solution of (7.56) is**

$$y = (C_1 + C_2 x)e^{mx}$$

where C_1 and C_2 are arbitrary constants.

3. **If the characteristic equation has no real roots, the general solution of (7.56) is**

$$y = C_1 e^{\lambda x} \sin \omega x + C_2 e^{\lambda x} \cos \omega x$$

where C_1 and C_2 are arbitrary constants and $\lambda \pm i\omega$ are the two complex roots of the characteristic equation.

Let's look at some examples that illustrate the three possibilities.

EXAMPLE 5   Solve: $y'' + 2y' - 3y = 0$

Solution   This differential equation is of the form (7.56). Its characteristic equation is

$$m^2 + 2m - 3 = 0$$
$$(m + 3)(m - 1) = 0$$

The two real roots are $m_1 = -3$ and $m_2 = 1$.   Then the general solution of the differential equation is

$$y = C_1 e^{-3x} + C_2 e^x$$

where C_1 and C_2 are arbitrary constants.   ∎

EXAMPLE 6   Solve: $4y'' + 12y' + 9y = 0$

Solution   To put this differential equation in the form (7.56), we need to divide by 4:

$$4y'' + 12y' + 9y = 0$$
$$y'' + 3y' + \tfrac{9}{4}y = 0$$

The characteristic equation is

$$m^2 + 3m + \tfrac{9}{4} = 0$$
$$(m + \tfrac{3}{2})^2 = 0$$

The single real root is $-\tfrac{3}{2}$. Thus, the general solution of the differential equation is

$$y = (C_1 + C_2 x)e^{(-3/2)x}$$

where C_1 and C_2 are arbitrary constants.   ∎

EXAMPLE 7   Solve: $\dfrac{d^2 y}{dx^2} + 4y = 0$

Solution   This differential equation is of the form (7.56). Its characteristic equation is

$$m^2 + 4 = 0$$

This equation has two complex roots $m = \pm 2i$. The general solution of the differential equation is

$$y = C_1 \sin 2x + C_2 \cos 2x$$

where C_1 and C_2 are arbitrary constants.   ■

EXERCISE 8

In Problems 1–6 solve each differential equation.

1. $\dfrac{dy}{dx} + \dfrac{y}{x} = 3; \quad y = 0$ when $x = 1$

2. $x\dfrac{dy}{dx} + y = 3x^2; \quad y = 1$ when $x = 1$

3. $\dfrac{dy}{dx} + 2xy = e^{-x^2}$

4. $x\dfrac{dy}{dx} - y = 2x^2, \quad x > 0$

5. $\dfrac{dy}{dx} + (\tan x)y = \sin x, \quad -\dfrac{\pi}{2} < x < \dfrac{\pi}{2}$

6. $x\dfrac{dy}{dx} + 6y = 3x + 1$

In Problems 7–12 verify that the given functions are solutions of the differential equation.

7. $y'' + 4y' - 12y = 0; \quad y_1 = e^{2x}, \quad y_2 = e^{-6x}$

8. $y'' + 11y' + 24y = 0; \quad y_1 = e^{-8x}, \quad y_2 = e^{-3x}$

9. $y'' + 8y' + 16y = 0; \quad y_1 = e^{-4x}, \quad y_2 = xe^{-4x}$

10. $y'' - 10y' + 25y = 0; \quad y_1 = e^{5x}, \quad y_2 = xe^{5x}$

11. $y'' + 2y = 0; \quad y_1 = \cos(\sqrt{2}x), \quad y_2 = \sin(\sqrt{2}x)$

12. $y'' - 4y' + 5y = 0; \quad y_1 = e^{2x} \cos x, \quad y_2 = e^{2x} \sin x$

In Problems 13–20 find the general solution of the differential equation.

13. $y'' + 22y' + 121y = 0$

14. $y'' + 6y' + 9y = 0$

15. $y'' + 3y' - 4y = 0$

16. $y'' + 5y' + 6y = 0$

17. $y'' + 25y = 0$

18. $y'' + 2y' + 2y = 0$

19. $y'' + 6y' = 0$

20. $y'' - 2y' - 8y = 0$

In Problems 21–24 solve each differential equation with the given boundary condition.

21. $y'' + 4y' + 4y = 0; \quad y = 1$ when $x = 0; \quad y' = 2$ when $x = 0$

22. $y'' - 7y' + 12y = 0; \quad y = 7$ when $x = 0; \quad y' = 24$ when $x = 0$

23. $y'' + 9y = 0; \quad y = 2$ when $x = 3\pi/2; \quad y' = 6$ when $x = 3\pi/2$

24. $y'' + y = 0; \quad y = 1$ when $x = 0; \quad y' = 2$ when $x = 0$

25. Show that if $y_1(x)$ and $y_2(x)$ are solutions of the differential equation

$$y'' + by' + cy = 0$$

and C_1 and C_2 are constants, then $y(x) = C_1 y_1(x) + C_2 y_2(x)$ is also a solution.

26. The basic equation governing the amount of current I (in amperes) in a simple RL circuit consisting of a resistance R (in ohms), an inductance L (in henrys), and an electromotive force E (in volts) is

$$\frac{dI}{dt} + \frac{R}{L}I = \frac{E}{L}$$

where t is the time in seconds (see the figure). Solve the differential equation assuming that E, R, and L are constants and that $I = 0$ when $t = 0$.

27. The equation governing the amount of electrical charge q (in coulombs) of an RC circuit consisting of a resistance R, a capacitance C (in farads), an electromotive force E, and no inductance is

$$\frac{dq}{dt} + \frac{1}{RC}q = \frac{E}{R}$$

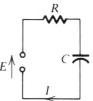

where t is the time in seconds (see the figure). Solve the differential equation assuming E, R, and C are constants and that $q = 0$ when $t = 0$.

28. A large tank contains 40 gallons of brine in which 10 pounds of salt are dissolved. Brine containing 2 pounds of dissolved salt per gallon runs into the tank at the rate of 4 gallons per minute. The mixture, kept uniform by stirring, runs out of the tank at the rate of 3 gallons per minute.
(a)   How much salt is in the tank at any time t?
(b)   Find the amount of salt in the tank at the end of 1 hour.

29. A tank initially contains 10 gallons of pure water. Starting at time $t = 0$, brine containing 3 pounds of salt per gallon flows into the tank at the rate of 2 gallons per minute. The mixture is kept uniform by stirring, and the well-stirred mixture flows out of the tank at the same rate as the inflow. How much salt is in the tank after 5 minutes? How much salt is in the tank after a very long time?

Miscellaneous Exercises

In Problems 1–12 differentiate and simplify each expression (a is a positive constant when it appears).

1. $v = \ln(y^2 + 1)$

2. $z = \ln(u - \sqrt{u^2 + a^2})$

3. $y = x^{1/a} + a^{1/x}$

4. $y = x^a + a^x$

5. $y = \ln(\sin 2x)$

6. $f(y) = e^{-y}\sin y$

7. $g(x) = \ln(x^2 - 2x)$

8. $y = x \ln e^x$

9. $y = \ln \dfrac{x^2 + 1}{x^2 - 1}$

10. $y = e^{-x} \ln x$

11. $w = \ln(\sqrt{x + a} - \sqrt{x})$

12. $y = \dfrac{1}{a} \ln \dfrac{x}{x + \sqrt{a^2 - x^2}}$

13. $f(x) = (x^2 + 1)^{(2 - 3x)}$,   find $f'(1)$.

14. Find: $\displaystyle\lim_{h \to 0}\left[\frac{1}{h}\ln\left(\frac{2 + h}{2}\right)\right]$

In Problems 15–18 evaluate each integral.

15. $\int \dfrac{e^x + 1}{e^x - 1}\, dx$ **16.** $\int \dfrac{dx}{\sqrt{x}(1 - 2\sqrt{x})}$ **17.** $\int_{1/5}^{3} \dfrac{\log_3(5x)}{x}\, dx$ **18.** $\int_{-1}^{1} \dfrac{5^{-x}}{2^x}\, dx$

19. Give a different proof of the formula $\ln(ab) = \ln a + \ln b$ (where a and b are positive) as follows:
   (a) Confirm that $\ln ax$ and $\ln x$ have the same derivative for all $x > 0$.
   (b) Why does it follow that $\ln ax - \ln x = \ln a$ for all 'x > 0$? Let $x = b$ to finish the proof.

20. Give a different proof of the formula $\ln a^r = r \ln a$ (where $a > 0$ and r is rational) as follows:
   (a) Confirm that $\ln x^r$ and $r \ln x$ have the same derivative for all $x > 0$.
   (b) Why does it follow that $\ln x^r - r \ln x = 0$ for all $x > 0$? Let $x = a$ to finish the proof.

21. Sketch the graph of $y = x^3 - 3 \ln x$ after discussing domain, asymptotes, extreme values, and concavity.

22. Suppose that a, b, and c are positive constants. Show that the minimum value of $ae^{cx} + be^{-cx}$ is $2\sqrt{ab}$.

23. If $f'(x) = -f(x)$ and $f(1) = 1$, find $f(x)$.

24. Prove that $\pi^e < e^\pi$. [*Hint:* What is the maximum value of $\ln x/x$?]

25. Find $y = f(x)$ if the arc length s of $y = f(x)$ from 0 to x satisfies $s = e^x - y$ and $f(0) = 1$.

26. A function $P(x) = y = kx^2$ is symmetric with respect to the y-axis and passes through $(0, 0)$ and (b, e^{-b^2}), where $b > 0$.
   (a) Write an equation for P.
   (b) The area enclosed by P and the line $y = e^{-b^2}$ is revolved about the y-axis to form a solid. Compute its volume.
   (c) For what number b is the volume of the solid in part (b) a maximum? Justify your answer.

27. A function f has the following properties: (i) $f(x + h) = e^h f(x) + e^x f(h)$ for all real numbers x and h; (ii) $f(x)$ has a derivative for all real numbers x; (iii) $f'(0) = 2$.
   (a) Show that $f(0) = 0$.
   (b) Using the definition of $f'(0)$, find $\lim_{x\to 0}[f(x)/x]$.
   (c) Prove that there is a real number p such that $f'(x) = f(x) + pe^x$ for all real numbers x.
   (d) What is the number p that is described in part (c)?

28. At $x = 0$, which of the following is true of the function f defined by $f(x) = x^2 + e^{-2x}$?
   (a) f is increasing   (b) f is decreasing   (c) f is discontinuous
   (d) f has a local minimum   (e) f has a local maximum

29. Let A be the area in the first quadrant that lies below both of the graphs of $y = 3x^2$ and $y = 3/x$ and to the left of the line $x = k$, where $k > 1$.
   (a) Find the area A as a function of k.
   (b) When the area is 7, what is k?
   (c) If the area A is increasing at the constant rate of 5 square units per second, at what rate is k increasing when $k = 15$?

30. (a) For what number m is the line $y = mx$ tangent to the graph of $y = \ln x$?
   (b) Prove that the graph of $y = \ln x$ lies entirely below the graph of the line found in part (a).
   (c) Use the results of part (b) to show that $e^x \geq x^e$ for $x > 0$.

31. Find the area of the largest rectangle in the fourth quadrant that has three vertices on the coordinate axes and the fourth vertex on the graph of $y = \ln x$.

32. Find a so that the area under the graph of $y = x + (1/x)$ from a to $(a + 1)$ is minimum.

33. Find the volume of the solid generated by revolving about the y-axis the area under the graph of $y = e^{-x^2}$ from $x = 0$ to the abscissa of the point of inflection of $y = e^{-x^2}$.

34. Let A be the area enclosed by the graph of $y = 1/x$, the x-axis, the line $x = m$, and the line $x = 2m$, $m > 0$. Which of the following is true about the area A?
    (a) Independent of m (b) Increases as m increases   (c) Decreases as m increases
    (d) Decreases as m increases when $m < \frac{1}{2}$; increases as m increases when $m > \frac{1}{2}$
    (e) Increases as m increases when $m < \frac{1}{2}$; decreases as m increases when $m > \frac{1}{2}$

35. Which of the following is true about the graph of $y = \ln|x^2 - 1|$ in the interval $(-1, 1)$?
    (a) Increasing   (b) Attains a local minimum at $(0, 0)$ (c) Has a range of all real numbers
    (d) Concave down   (e) Has an asymptote of $x = 0$

36. (a) Sketch the graph of $y = \frac{1}{2}(e^x + e^{-x})$.
    (b) Let R be a point on the graph and let the x-coordinate of R be r $(r \neq 0)$. The tangent line to the graph at R crosses the x-axis at a point Q. Find the coordinates of Q.
    (c) If P is the point $(r, 0)$, find the length of PQ as a function of r and the limiting value of this length as r increases without bound.

37. Apply the mean value theorem (4.26) to $y = \ln x$ to show that $a \ln(b/a) < b - a < b \ln(b/a)$ if $0 < a < b$.

38. Prove *Bernoulli's inequality* (Jakob Bernoulli): $(1 + x)^p \geq 1 + px$ for $x \geq 0$ and $p > 1$. [*Hint:* Consider $y = (1 + x)^p - 1 - px$ and y'.]

39. Consider $f(x) = (1/x) + \ln x$, defined only on the closed interval $1/e \leq x \leq e$.
    (a) Show your reasoning to determine the number(s) x at which f has its absolute maximum and absolute minimum.
    (b) For what numbers x is the curve concave up?
    (c) Sketch the graph of f over the interval $1/e \leq x \leq e$.

40. The number y of bacteria in a culture at time t is given approximately by

$$y = 1000(25 + te^{-t/20}) \qquad \text{for} \quad 0 \leq t \leq 100$$

    (a) Find the largest number and the smallest number of bacteria in the culture during the interval.
    (b) At what time during the interval is the rate of change in the number of bacteria a minimum?

41. Find: $\lim\limits_{x \to 2} \dfrac{\ln x - \ln 2}{x - 2}$ 42. Find: $\lim\limits_{h \to 0} \dfrac{e^h - 1}{h}$ 43. Find: $\lim\limits_{x \to c} \dfrac{e^x - e^c}{x - c}$

44. Show that $\lim\limits_{x \to +\infty} xe^{-x} = 0$. [*Hint:* Use the result that $0 < \ln x < x$ for $x > 1$.]

In Problems 45 and 46 use the result in Problem 44 to find the limits.

45. $\lim\limits_{x \to +\infty} xe^{-2x}$ and $\lim\limits_{x \to -\infty} xe^{-2x}$ 46. $\lim\limits_{x \to +\infty} x^2e^{2x}$ and $\lim\limits_{x \to -\infty} x^2e^{2x}$

47. Sketch the graph of $y = xe^x$, taking into account concavity, local extrema, and inflection points.

48. Sketch the graph of $y = x^2e^{2x}$, taking into account concavity, local extrema, and inflection points.

49. Find a function that is equal to its own derivative and such that $f(3) = 10$.

50. Use Newton's method to solve $e^{-x} = x - 4$.

51. Use Newton's method to solve $10x = e^x$.

52. *Calculator problem.* Use a calculator to evaluate $(3^{0.003})^{1001}$ and $3^{(0.003 \times 1001)}$ by computing the expressions within parentheses first. Did you get the same results? If not, can you explain the difference? What

does this tell you about the laws of exponents (7.1) as applied to calculator arithmetic? Experiment with other values and some of the other laws of exponents.

53. *Calculator problem.* Use a calculator to evaluate $\log_{10}(2^{0.0001})$ and $0.0001 \times (\log_{10} 2)$ by computing the expressions within parentheses first. Did you get the same results? If not, can you explain the difference? What does this tell you about the laws of logarithms (7.3) as applied to calculator arithmetic? Experiment with other values and some of the other laws of logarithms.

54. *Calculator problem.* Verify that $2^{10} \approx 10^3$. By taking $\log_{10}$ of both sides of this approximate equation, show that $\log_{10} 2 \approx 0.3$. (In fact, $\log_{10} 2 = 0.30103\ldots$.) Can you find other powers like 2^{10} that are approximately 10^n for some n? If so, you can approximate other logarithms.

55. *The force exerted by a gas in a cylinder on a piston whose area is A is given by $F = pA$, where p is the force per unit area, or *pressure*. The work W in a displacement of the piston from x_1 to x_2 is

$$W = \int_{x_1}^{x_2} F\, dx = \int_{x_1}^{x_2} pA\, dx = \int_{V_1}^{V_2} p\, dV$$

where dV is the accompanying infinitesimal change of volume of the gas.

(a) During expansion of a gas at constant temperature (isothermal), the pressure depends on the volume according to the relation

$$p = \frac{nRT}{V}$$

where n and R are constants and T is the constant temperature. Calculate the work in expanding the gas isothermally from volume V_1 to volume V_2.

(b) During expansion of a gas at constant entropy (adiabatic), the pressure depends on the volume according to the relation

$$p = \frac{K}{V^\gamma}$$

where K and $\gamma \neq 1$ are constants. Calculate the work in expanding the gas adiabatically from V_1 to V_2.

In Problems 56–63 solve the differential equations.

56. $\dfrac{dy}{dx} = e^{-2x}$; $y = 0$ when $x = 4$

57. $\dfrac{1}{y+2}\dfrac{dy}{dx} = \dfrac{1}{2x}$; $x > 0$; $y = 2$ when $x = 1$

58. $2y'' + 3y' + y = 0$

59. $2y'' + 3y' - 2y = 0$

60. $y'' - 9y = 0$

61. $y'' - 4y = 0$

62. $y'' - 5y' + 6y = 0$

63. $2y'' - 4y' - y = 0$

* Adapted from F. W. Sears, M. W. Zemansky, and H. D. Young, *University Physics*, (Reading, Mass.: Addison-Wesley Publishing Co., 1976), p. 131. Reprinted by permission.

8

Trigonometric and Hyperbolic Functions

1. Derivatives of Trigonometric Functions

We begin by reproducing two basic formulas from Chapter 3, namely,

$$(8.1) \qquad \frac{d}{dx} \sin x = \cos x$$

$$(8.2) \qquad \frac{d}{dx} \cos x = -\sin x$$

From these, it is fairly easy to obtain formulas for the derivatives of the remaining trigonometric functions. The idea is to use the identities

$$\tan x = \frac{\sin x}{\cos x} \qquad \cot x = \frac{\cos x}{\sin x}$$

$$\sec x = \frac{1}{\cos x} \qquad \csc x = \frac{1}{\sin x}$$

in conjunction with (8.1), (8.2), and various differentiation formulas. A list of these formulas follows:

(8.3)
$$\frac{d}{dx}\tan x = \sec^2 x$$

(8.4)
$$\frac{d}{dx}\cot x = -\csc^2 x$$

(8.5)
$$\frac{d}{dx}\sec x = \sec x \tan x$$

(8.6)
$$\frac{d}{dx}\csc x = -\csc x \cot x$$

Formulas (8.3) and (8.5) were obtained in Example 5 of Section 7, Chapter 3 (p. 145). We shall show the derivation of (8.4) and leave (8.6) as an exercise.

$$\frac{d}{dx}\cot x = \frac{d}{dx}\frac{\cos x}{\sin x} = \frac{(\sin x)[(d/dx)\cos x] - (\cos x)[(d/dx)\sin x]}{(\sin x)^2}$$

$$= \frac{(\sin x)(-\sin x) - (\cos x)(\cos x)}{(\sin x)^2}$$

$$= \frac{-(\sin^2 x + \cos^2 x)}{(\sin x)^2} = \frac{-1}{\sin^2 x} = -\csc^2 x$$

Notice that formulas (8.2), (8.4), and (8.6) for the derivatives of cos x, cot x, and csc x (sometimes referred to as the *cofunctions*) can be obtained from formulas (8.1), (8.3), and (8.5) for the derivatives of sin x, tan x, and sec x by: (*1*) introducing a minus sign and (*2*) replacing each function by its cofunction. For example, from (8.5) we know that

$$\frac{d}{dx}\sec x = \sec x \tan x$$

If we replace sec x by its cofunction csc x and replace tan x by its cofunction cot x and put in a minus sign, we obtain

$$\frac{d}{dx}\csc x = -\csc x \cot x$$

which is formula (8.6).

We suggest that formulas (8.1), (8.3), and (8.5) be memorized and that the rule above be applied to get (8.2), (8.4), and (8.6).

Now let's see how the chain rule is used with formulas (8.1)–(8.6).

EXAMPLE 1   Find y' if:
(a) $y = \sin e^x$ (b) $y = e^{\tan x}$
(c) $y = \ln(\sec x + \tan x)$ (d) $y = \sec^2(3x + 2)$

Solution   (a)   Set $y = \sin u$ and $u = e^x$.   Then

$$y' = \left(\frac{dy}{du}\right)\left(\frac{du}{dx}\right) = (\cos u)(e^x) = e^x \cos e^x$$

(b)   Set $y = e^u$ and $u = \tan x$.   Then

$$y' = \left(\frac{dy}{du}\right)\left(\frac{du}{dx}\right) = (e^u)(\sec^2 x) = e^{\tan x} \sec^2 x$$

(c)   Set $y = \ln u$ and $u = \sec x + \tan x$.   Then

$$y' = \left(\frac{dy}{du}\right)\left(\frac{du}{dx}\right) = \left(\frac{1}{u}\right)(\sec x \tan x + \sec^2 x) = \frac{(\sec x + \tan x)(\sec x)}{\sec x + \tan x} = \sec x$$

(d)   Set $y = u^2$, $u = \sec v$,   and $v = 3x + 2$.   Then

$$y' = \left(\frac{dy}{du}\right)\left(\frac{du}{dv}\right)\left(\frac{dv}{dx}\right) = (2u)(\sec v \tan v)(3) = 6 \sec^2(3x + 2) \tan(3x + 2) \qquad \blacksquare$$

In the next example, we review graphing techniques.

EXAMPLE 2   Sketch the graph of $f(x) = \cos^2 x - \sin x$.

Solution   Because $\sin x$ and $\cos x$ are periodic functions with period 2π, so is f; that is, $f(x + 2\pi) = f(x)$.   As a result, we need only sketch the graph of f on an interval of length 2π, say, the interval $[0, 2\pi]$, and the rest of the graph will be a repetition of this part of the graph.

The derivative of f is

$$f'(x) = -2 \cos x \sin x - \cos x = -\cos x(2 \sin x + 1)$$

The critical numbers obey

$$\cos x = 0 \qquad \text{or} \qquad \sin x = -\tfrac{1}{2}$$

Now,

$$\cos x = 0 \qquad \text{on } [0, 2\pi] \text{ only if } \quad x = \frac{\pi}{2} \quad \text{or} \quad x = \frac{3\pi}{2}$$

$$\sin x = -\tfrac{1}{2} \qquad \text{on } [0, 2\pi] \text{ only if } \quad x = \frac{7\pi}{6} \quad \text{or} \quad x = \frac{11\pi}{6}$$

Hence, the critical numbers of f on $[0, 2\pi]$ are

$$\frac{\pi}{2}, \quad \frac{3\pi}{2}, \quad \frac{7\pi}{6}, \quad \frac{11\pi}{6}$$

We choose to use the second derivative test on these critical numbers. The second derivative of f is given by

$$f''(x) = 2(\sin^2 x - \cos^2 x) + \sin x$$
$$= 2(\sin^2 x - 1 + \sin^2 x) + \sin x$$
$$= 4 \sin^2 x + \sin x - 2$$

By the second derivative test, we verify that

$$f''\left(\frac{\pi}{2}\right) = 4 + 1 - 2 = 3 > 0 \qquad \text{so that } f \text{ has a local minimum at } \frac{\pi}{2}$$

$$f''\left(\frac{3\pi}{2}\right) = 4 - 1 - 2 = 1 > 0 \qquad \text{so that } f \text{ has a local minimum at } \frac{3\pi}{2}$$

$$f''\left(\frac{7\pi}{6}\right) = 1 - \frac{1}{2} - 2 = -\frac{3}{2} < 0 \qquad \text{so that } f \text{ has a local maximum at } \frac{7\pi}{6}$$

$$f''\left(\frac{11\pi}{6}\right) = 1 - \frac{1}{2} - 2 = -\frac{3}{2} < 0 \qquad \text{so that } f \text{ has a local maximum at } \frac{11\pi}{6}$$

By using the above information and the computations from Table 1, we are able to sketch the graph, as shown in Figure 1.

Table 1

x	0	$\pi/2$	π	$7\pi/6$	$3\pi/2$	$11\pi/6$	2π
$\sin x$	0	1	0	$-\frac{1}{2}$	-1	$-\frac{1}{2}$	0
$\cos^2 x$	1	0	1	$\frac{3}{4}$	0	$\frac{3}{4}$	1
$\cos^2 x - \sin x$	1	-1	1	$\frac{5}{4}$	1	$\frac{5}{4}$	1

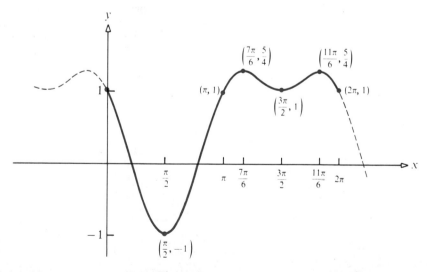

Figure 1

In the final example of this section, we prove *Snell's law of refraction*.

EXAMPLE 3   Light travels at different speeds in different media (such as air, water, glass, and so on). Suppose that light travels from a point A in one medium, where its speed is c_1, to a point B in another medium, where its speed is c_2 (see Fig. 2). Use Fermat's principle that light always travels along the path requiring least time to prove Snell's law of refraction, namely, that

$$\frac{\sin \theta_1}{c_1} = \frac{\sin \theta_2}{c_2}$$

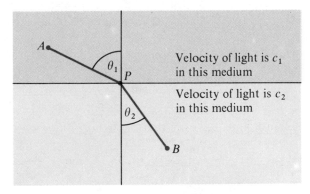

Figure 2

Solution   Let the light pass from one medium to the other at the point P. Then the path taken by the light is made up of two line segments—from A to P and from P to B—since the shortest distance between two points is a line. We position our coordinate system as illustrated in Figure 3. By using the formula

$$\text{Time} = \frac{\text{Distance}}{\text{Speed}}$$

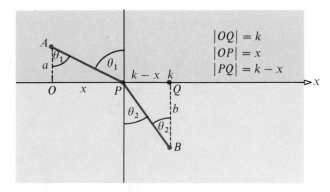

Figure 3

we see from Figure 3 that the travel time from A to P is

$$t_1 = \frac{\sqrt{x^2 + a^2}}{c_1}$$

and the travel time from P to B is

$$t_2 = \frac{\sqrt{(k - x)^2 + b^2}}{c_2}$$

The total time is therefore given by the formula

$$t = t_1 + t_2 = \frac{\sqrt{x^2 + a^2}}{c_1} + \frac{\sqrt{(k - x)^2 + b^2}}{c_2}$$

To find the least time, we compute dt/dx and set it equal to 0:

(8.7) $$\frac{dt}{dx} = \frac{x}{c_1 \sqrt{x^2 + a^2}} - \frac{k - x}{c_2 \sqrt{(k - x)^2 + b^2}} = 0$$

From Figure 3, we see that

$$\frac{x}{\sqrt{x^2 + a^2}} = \sin \theta_1 \qquad \text{and} \qquad \frac{k - x}{\sqrt{(k - x)^2 + b^2}} = \sin \theta_2$$

Hence,

$$\frac{dt}{dx} = \frac{\sin \theta_1}{c_1} - \frac{\sin \theta_2}{c_2} = 0$$

$$\frac{\sin \theta_1}{c_1} = \frac{\sin \theta_2}{c_2}$$

To ensure that a minimum occurs when $dt/dx = 0$, we need to show that $d^2t/dx^2 > 0$. By using (8.7), we find that

$$\frac{d^2t}{dx^2} = \frac{d}{dx}\left[\frac{x}{c_1\sqrt{x^2 + a^2}}\right] - \frac{d}{dx}\left[\frac{k - x}{c_2\sqrt{(k - x)^2 + b^2}}\right]$$

$$= \frac{a^2}{c_1(x^2 + a^2)^{3/2}} + \frac{b^2}{c_2[(k - x)^2 + b^2]^{3/2}} > 0 \qquad \blacksquare$$

EXERCISE 1

In Problems 1–38 find $y' = dy/dx$.

1. $y = \sec 4x$ **2.** $y = \cot 5x$ **3.** $y = \tan 5x$

4. $y = \csc(x^3 + 1)$ **5.** $y = e^x \sec x$ **6.** $y = e^x \cot x$

$2x \sin(x^2) x$

7. $y = e^{-4x} \sin 3x$

8. $y = e^{-2x} \cos 3x$

9. $y = e^{\pi x} \tan \pi x$

10. $y = x^\pi \cot \pi x$

11. $y = \sin(x^2)$

12. $y = \cos(x^2)$

13. $y = \sec^4 x - \tan^4 x$

14. $y = \cot^2 x - \csc^2 x$

15. $y = \sec(x^2 + 2x - 1)$

16. $y = \cot(x^3 - 2x + 5)$

17. $y = \csc^2(1 + 2x)$

18. $y = \cot^2(1 + 3x^2)$

19. $y = x^2 \sin 4x$

20. $y = x^2 \cos 4x$

21. $y = \sqrt{\sin(1/x)}$

22. $y = \sqrt{\cos(3/x)}$

23. $y = x \tan x$

24. $y = x \cot x$

25. $y = x^2 \sin^2 x$

26. $y = x \sin \dfrac{1}{x}$

27. $y = \sec x \tan x$

28. $y = \sec^2 2x \tan^3 2x$

29. $y = \dfrac{\sec x}{1 + x}$

30. $y = \dfrac{\tan x}{1 + x}$

31. $x \sec y + y \tan x = x$

32. $\tan(xy) = x$

33. $\sec^2(xy) = 3x$

34. $\csc(x + y) = 2y$

35. $y = \sec^2(x^3 + x)$

36. $y = \cot^2(3x + 1)$

37. $y = e^{\csc^2 x}$

38. $y = \ln(\cot^2 x)$

In Problems 39–44 graph each function.

39. $f(x) = \sin x - \cos x$

40. $f(x) = \sin x + \tan x$

41. $f(x) = \sin^2 x - \cos x$

42. $f(x) = \cos^2 x + \sin x$

43. $f(x) = \sin x - \tan x$

44. $f(x) = \sec x - \tan x$

45. Find the dimensions of the right circular cone of maximum volume having slant height 4 units.

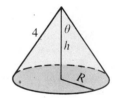

46. A 6 foot fence is erected 4 feet from a house. What is the shortest ladder that can be stood outside the fence and leaned against the wall? What is its angle of inclination θ?

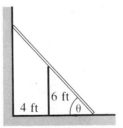

47. The sides of an isosceles triangle (see the figure) are sliding outward with velocity 1 centimeter per minute. At what rate is the area enclosed by the triangle changing when $\theta = 30°$?

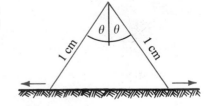

48. A weight hangs on a spring that is 2 meters long when it is stretched out (see the figure). The weight is pulled down and then released. The weight then oscillates up and down, and the length x of the spring when t seconds have elapsed is given by the formula $x = 2 + \cos 2\pi t$. Find:

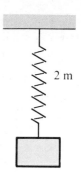

2 m

   (a)   Length of the spring at times $t = 0, \frac{1}{2}, 1, \frac{3}{2}, \frac{5}{8}$
   (b)   Velocity of the weight at time $t = \frac{1}{4}$
   (c)   Acceleration of the weight at time $t = \frac{1}{4}$
   (d)   Time intervals during which the weight is moving down

49. The hands of a clock are 2 inches and 3 inches long (see the figure). As the hands move around the clock, they sweep out the triangle OAB. At what rate is the area of the triangle changing at time 12:10?

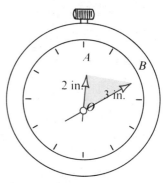

50. A searchlight is following a plane flying at an altitude of 3000 feet in a straight line over the light; the plane's velocity is 500 miles per hour. At what rate is the searchlight turning when the distance between the light and the plane is 5000 feet?

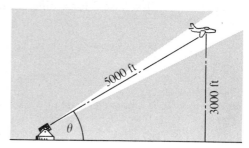

51. Verify (8.6).

2. Related Integrals

Each of the six differentiation formulas developed thus far leads to a corresponding integration formula:

$$\int \sin x \, dx = -\cos x + C \qquad \int \cos x \, dx = \sin x + C$$

(8.8)

$$\int \sec^2 x \, dx = \tan x + C \qquad \int \csc^2 x \, dx = -\cot x + C$$

$$\int \sec x \tan x \, dx = \sec x + C \qquad \int \csc x \cot x \, dx = -\csc x + C$$

To this list we add four more formulas:

(8.9) $$\int \tan x \, dx = \ln|\sec x| + C$$

(8.10) $$\int \sec x \, dx = \ln|\sec x + \tan x| + C$$

(8.11) $$\int \cot x \, dx = \ln|\sin x| + C$$

(8.12) $$\int \csc x \, dx = \ln|\csc x - \cot x| + C$$

Formulas (8.9) and (8.11) are a reproduction of (7.27), derived in Chapter 7 (p. 403). We derive (8.10) by multiplying by $(\sec x + \tan x)/(\sec x + \tan x)$:

$$\int \sec x \, dx = \int \frac{(\sec x)(\sec x + \tan x)}{\sec x + \tan x} \, dx = \int \frac{\sec^2 x + \sec x \tan x}{\sec x + \tan x} \, dx$$

The substitution $u = \sec x + \tan x$ results in $du = (\sec x \tan x + \sec^2 x) \, dx$, so that

$$\int \sec x \, dx = \int \frac{du}{u} = \ln|u| + C = \ln|\sec x + \tan x| + C$$

If you are wondering how we decided to multiply by $(\sec x + \tan x)/(\sec x + \tan x)$ in the first place, look at Example 1(c) in Section 1 (p. 436). Also, see Problem 28 in Exercise 2 for a different formula for $\int \sec x \, dx$.

The derivation of (8.12) is similar and is left as an exercise (Problem 27 in Exercise 2).

EXAMPLE 1   Evaluate each integral:

(a) $\int \tan(3x + 1) \, dx$ (b) $\int_0^{\sqrt{\pi/2}} x \sec x^2 \tan x^2 \, dx$ (c) $\int \cos^2 x \, dx$

Solution   (a)   We use the substitution $u = 3x + 1$. Then $du = 3 \, dx$, so that

$$\int \tan(3x + 1) \, dx = \frac{1}{3} \int \tan u \, du = \frac{1}{3} \ln|\sec u| + C = \frac{1}{3} \ln|\sec(3x + 1)| + C$$

By (8.9)

(b)   Let $u = x^2$. Then $du = 2x \, dx$. Also, $u = 0$ when $x = 0$, and $u = \pi/4$ when $x = \sqrt{\pi}/2$. Consequently,

$$\int_0^{\sqrt{\pi/2}} x \sec x^2 \tan x^2 \, dx = \int_0^{\pi/4} \sec u \tan u \, \frac{du}{2} = \frac{1}{2} \sec u \Big|_0^{\pi/4} = \frac{1}{\sqrt{2}} - \frac{1}{2} \approx 0.207$$

By (8.8)

(c)   The substitution approach will not work on this integral. However, the use of a certain trigonometric identity does work. By using the identity

$$\cos^2 x = \tfrac{1}{2}(1 + \cos 2x)$$

we find that

$$\int \cos^2 x \, dx = \frac{1}{2} \int (1 + \cos 2x) \, dx = \frac{1}{2} \int dx + \frac{1}{2} \int \cos 2x \, dx$$

$$= \frac{x}{2} + \frac{1}{4} \sin 2x + C \quad \blacksquare$$

EXAMPLE 2   Find the volume of the solid of revolution generated by revolving the area under the graph of $y = \cos x$ from $x = 0$ to $x = \pi/2$ about the x-axis.

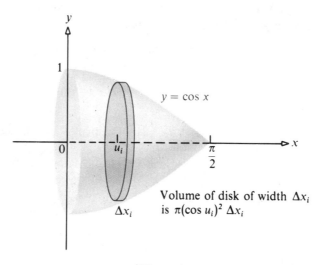

$y = \cos x$

Volume of disk of width Δx_i is $\pi (\cos u_i)^2 \, \Delta x_i$

Figure 4

Solution   Figure 4 illustrates the situation. We choose to use the disk method here because the shell method requires that we solve for x in the equation $y = \cos x$, the solution of which we will not study until the next section. Thus, we partition along the x-axis, obtaining disks of radius $\cos u_i$ and width Δx_i. The required volume V is

$$V = \pi \int_0^{\pi/2} \cos^2 x \, dx = \pi \left(\frac{x}{2} + \frac{\sin 2x}{4} \right) \Big|_0^{\pi/2} = \frac{\pi^2}{4} \approx 2.467 \quad \blacksquare$$

See Example 1(c)

EXERCISE 2

In Problems 1–22 evaluate each integral.

1. $\int_0^{\pi/4} \sec^2 x \, dx$ **2.** $\int_{-1}^{1} \sec x \tan x \, dx$ **3.** $\int \sec 5x \, dx$ **4.** $\int \csc 3x \, dx$

5. $\int \tan 2x \, dx$ **6.** $\int \cot 4x \, dx$ **7.** $\int \sec 4x \tan 4x \, dx$ **8.** $\int \csc 3x \cot 3x \, dx$

9. $\int \sec^2 2x \, dx$

10. $\int \csc^2 4x \, dx$

11. $\int \sqrt{\tan x} \, \sec^2 x \, dx$

12. $\int (2 + 3 \cot x)^{3/2} \csc^2 x \, dx$

13. $\int \dfrac{1}{\sec 5x} \, dx$

14. $\int \dfrac{1}{\csc 3x} \, dx$

15. $\int \sec(3x - 1) \tan(3x - 1) \, dx$

16. $\int \csc^2 4x \, dx$

17. $\int \dfrac{\sin x}{\cos^2 x} \, dx$

18. $\int \dfrac{\cos x}{\sin^2 x} \, dx$

19. $\int_0^{\pi/4} (1 + \sec^2 x) \, dx$

20. $\int_{\pi/4}^{\pi/2} \dfrac{dx}{\sin^2 x}$

21. $\int \sin x \, e^{\cos x} \, dx$

22. $\int \sec^2 x \, e^{\tan x} \, dx$

23. Find the volume of the solid of revolution generated by revolving the area under the graph of $y = \csc x$ from $\pi/2$ to $3\pi/4$ about the x-axis.

24. Find the volume of the solid of revolution generated by revolving the area under the graph of $y = \sec x$ from 0 to $\pi/4$ about the x-axis.

25. Find the average value of $y = \tan x$ on the interval $[0, \pi/4]$.

26. Find the average value of $y = \sec x$ on the interval $[0, \pi/4]$.

27. Prove that $\int \csc x \, dx = \ln|\csc x - \cot x| + C$. [*Hint:* Multiply and divide by $(\csc x - \cot x)$.]

28. Find y' if $y = \ln[\tan(x/2 + \pi/4)]$. Use the result to show that

(8.13) $$\int \sec x \, dx = \ln\left[\tan\left(\frac{x}{2} + \frac{\pi}{4}\right)\right] + C$$

Also, show that (8.13) and (8.10) are identities.

29. Show that $\int \sin x \cos x \, dx$ can be given in three different ways as

$$\int \sin x \cos x \, dx = \frac{1}{2} \sin^2 x + C_1 = -\frac{1}{2} \cos^2 x + C_2 = -\frac{1}{4} \cos 2x + C_3$$

Find the relationship between the constants C_1 and C_2 and the constants C_1 and C_3.

30. Use the identity below to obtain a formula for $\int \csc x \, dx$.

$$\csc x = \frac{1}{\sin x} = \frac{1}{2 \sin(x/2) \cos(x/2)}$$

31. Use the identity below and the result of Problem 30 to obtain formula (8.13).

$$\sec x = \frac{1}{\cos x} = \frac{1}{\sin(\pi/2 - x)}$$

32. Use the identities $\cos^2 x = \frac{1}{2}(1 + \cos 2x)$ and $\sin^2 x = \frac{1}{2}(1 - \cos 2x)$ to show that:

(a) $\int \sin^2 x \, dx = \frac{1}{2}x - \frac{1}{4} \sin 2x + C$

(b) $\int \cos^2 4x \, dx = \frac{1}{2}x + \frac{1}{16} \sin 8x + C$

(c) $\int \dfrac{\sin^2 \sqrt{x}}{\sqrt{x}} \, dx = \sqrt{x} - \frac{1}{2} \sin(2\sqrt{x}) + C$

3. Inverse Trigonometric Functions

The graph of $y = \sin x$ is given in Figure 5(a). There we observe that every horizontal line between -1 and $+1$ inclusive intersects the graph of $y = \sin x$ at infinitely many points. Thus, the function $y = \sin x$ is not one-to-one and consequently has no inverse. However, if we agree to restrict $y = \sin x$ to only that part of its domain for which $-\pi/2 \le x \le \pi/2$, we have a one-to-one function. With this restricted domain,* the function $y = \sin x$ has an inverse. To obtain the inverse function, we interchange x and y and solve for y. This value of y is called the *inverse sine of* x and we write

$$y = \sin^{-1}x$$

which we read as "*y is the angle whose sine is x*," or "*y is the inverse sine of x.*"†

(8.14) $y = \sin^{-1}x$ means $x = \sin y$ where $-\dfrac{\pi}{2} \le y \le \dfrac{\pi}{2},\quad -1 \le x \le 1$

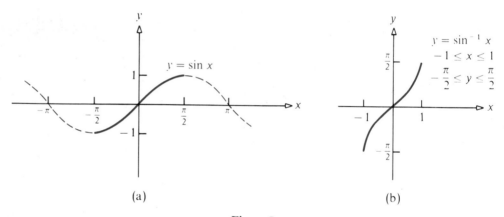

(a)   (b)

Figure 5

The domain of $y = \sin^{-1}x$ is the closed interval $[-1, 1]$, and the range is the closed interval $[-\pi/2, \pi/2]$. The graph of $y = \sin^{-1}x$ is the reflection about the line $y = x$ of the restricted portion of the graph of $y = \sin x$, as shown in Figure 5(b).

Let's look at an example that reviews the basic idea behind the definition of $y = \sin^{-1}x$.

---

* Although there are other restrictions that work just as well, mathematicians have agreed to be consistent by using the interval $-\pi/2 \le x \le \pi/2$ for $y = \sin x$.

† In some books the inverse sine function is referred to as the *arcsine function*, written arcsin x.

EXAMPLE 1   Find:
(a) $\sin^{-1}(\sqrt{3}/2)$ (b) $\sin^{-1}(-\sqrt{2}/2)$

Solution   (a) Let $y = \sin^{-1}(\sqrt{3}/2)$. From (8.14), $\sqrt{3}/2 = \sin y$, where $-\pi/2 \le y \le \pi/2$.
Thus, $y = \pi/3$. [*Note:* $\sin y = \sqrt{3}/2$ has infinitely many solutions, but
$y = \pi/3$ is the only solution in $-\pi/2 \le y \le \pi/2$.]

(b) Let $y = \sin^{-1}(-\sqrt{2}/2)$. From (8.14), $-\sqrt{2}/2 = \sin y$, where $-\pi/2 \le y \le \pi/2$.
Thus, $y = -\pi/4$.   ∎

Do not misinterpret the use of the minus 1 in $y = \sin^{-1}x$. This symbolism is
used to remind you that it is the inverse of the sine function that is being discussed.
If it is desired to discuss the reciprocal of the sine function, then write $y = (\sin x)^{-1}$
or $y = \csc x$.

In a similar manner, we can define inverse functions for the remaining trigonometric functions. In each case we suitably restrict the domain so that the function
is one-to-one.

For example, the inverse tangent (arctangent) function, denoted by $\tan^{-1}$, is the
angle between $-\pi/2$ and $\pi/2$ whose tangent is x. Thus,

(8.15) $y = \tan^{-1}x$ means $x = \tan y$ where $-\dfrac{\pi}{2} < y < \dfrac{\pi}{2}, \quad -\infty < x < +\infty$

See Figures 6(a) and 6(b) for the graphs of the tangent function and its inverse. Note
that the values $y = \pm\pi/2$ are excluded from (8.15) since the tangent is not defined
at $\pm\pi/2$.

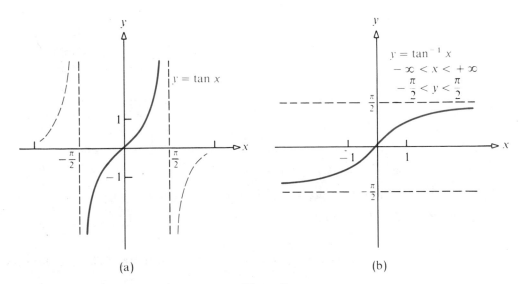

(a)   (b)

Figure 6

The *inverse secant* (arcsecant) function is denoted by $\sec^{-1}$. Similar to previous definitions,

(8.16) $y = \sec^{-1} x$ means $\sec y = x$ where $|x| \geq 1, \quad 0 \leq y < \dfrac{\pi}{2}$ or $\pi \leq y < \dfrac{3\pi}{2}$

See Figures 7(a) and 7(b). We have restricted y to $0 \leq y < \pi/2$ or $\pi \leq y < 3\pi/2$, instead of using the restrictions $0 \leq y < \pi/2$ or $\pi/2 < y \leq \pi$,   in order to make $\tan y \geq 0$.   This, in turn, will make the differentiation formula for the inverse secant simpler.

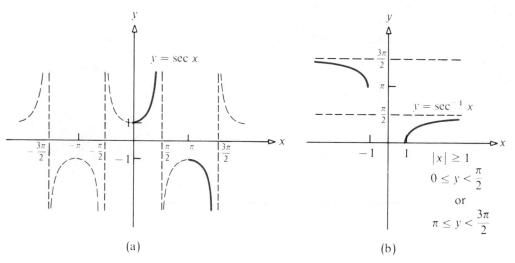

(a)   (b)

Figure 7

The remaining three inverse trigonometric functions are defined as follows:

Inverse cosine function:

$y = \cos^{-1} x$ means $x = \cos y$ where $0 \leq y \leq \pi$ and $-1 \leq x \leq 1$

Inverse cotangent function:

$y = \cot^{-1} x$ means $x = \cot y$ where $0 < y < \pi$ and $-\infty < x < +\infty$

Inverse cosecant function:

$y = \csc^{-1} x$ means $x = \csc y$ where $-\pi < y \leq \dfrac{\pi}{2}$ or $0 < y \leq \dfrac{\pi}{2}$

and $|x| \geq 1$

See Figures 8, 9, and 10 for these graphs.

The inverse trigonometric functions satisfy the following three identities:

(8.17)

$$\cos^{-1} x = \frac{\pi}{2} - \sin^{-1} x$$

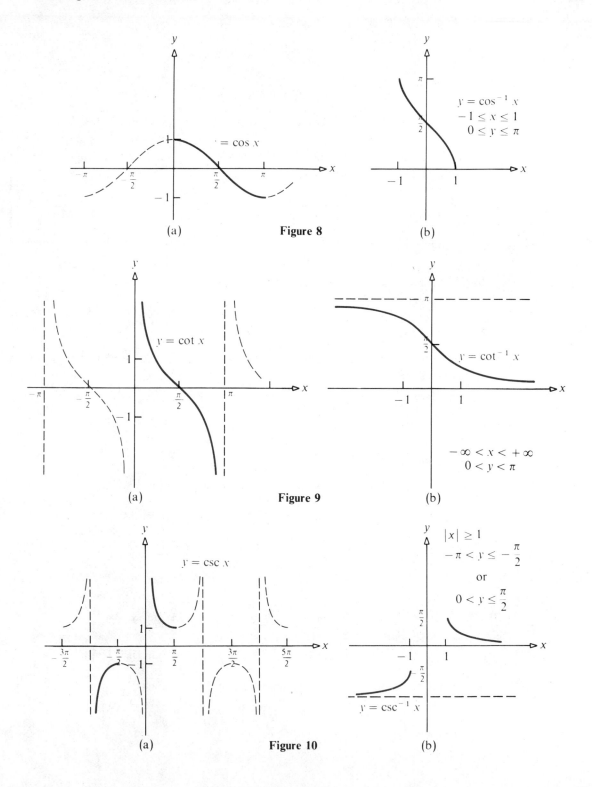

(a)   **Figure 8**   (b)

$y = \cos^{-1} x$
$-1 \le x \le 1$
$0 \le y \le \pi$

$y = \cos x$

(a)   **Figure 9**   (b)

$y = \cot x$

$y = \cot^{-1} x$

$-\infty < x < +\infty$
$0 < y < \pi$

(a)   **Figure 10**   (b)

$y = \csc x$

$|x| \ge 1$
$-\pi < y \le -\dfrac{\pi}{2}$
or
$0 < y \le \dfrac{\pi}{2}$

$y = \csc^{-1} x$

(8.18)
$$\cot^{-1}x = \frac{\pi}{2} - \tan^{-1}x$$

(8.19)
$$\csc^{-1}x = \frac{\pi}{2} - \sec^{-1}x$$

We shall only give the following restricted proof for (8.17):

Proof   Consider the right triangle given in Figure 11, in which the angles α and β are acute and complementary. Then

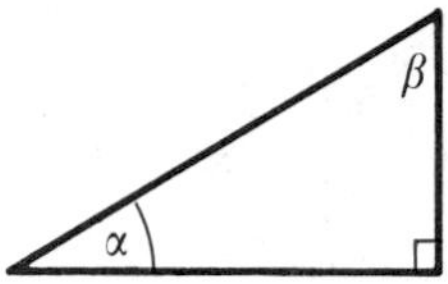

$$\alpha + \beta = \frac{\pi}{2}$$

(8.20)
$$\beta = \frac{\pi}{2} - \alpha$$

Figure 11

Since $\sin \alpha = \cos \beta$,   we may set $\sin \alpha = \cos \beta = x$ to get

(8.21)
$$\alpha = \sin^{-1}x \quad \text{and} \quad \beta = \cos^{-1}x$$

By combining (8.20) and (8.21), we get (8.17).   ∎

In Problems 29 and 30 (Exercise 3) you are asked to give similar derivations of (8.18) and (8.19).

EXERCISE 3

In Problems 1–20 find the exact value of each expression.

1. $\sin^{-1}\left(\dfrac{\sqrt{2}}{2}\right)$

2. $\sin^{-1}\left(\dfrac{-1}{2}\right)$

3. $\sin^{-1}(-1)$

4. $\cot^{-1}(-1)$

5. $\tan^{-1}(0)$

6. $\cos^{-1}(-1)$

7. $\sin^{-1}\left(\dfrac{\sqrt{2}}{2}\right) + \sin^{-1}\left(-\dfrac{\sqrt{2}}{2}\right)$

8. $\tan^{-1}\left(\tan\dfrac{\pi}{4}\right)$

9. $\sin\left[\sin^{-1}\left(\dfrac{1}{2}\right)\right]$

10. $\cos^{-1}\left[\cos\left(-\dfrac{\pi}{4}\right)\right]$

11. $\tan^{-1}(\sin 0)$

12. $\tan^{-1}(\cos 0)$

13. $\sin\left[\cos^{-1}\left(\dfrac{3}{5}\right)\right]$

14. $\sin^{-1}\left(\sin\dfrac{1}{2}\right)$

15. $\tan(\sec^{-1}5)$

16. $\tan[\sec^{-1}(-3)]$

17. $\sin^{-1}(\sin 2)$

18. $\sin\left[\sin^{-1}\left(\dfrac{1}{3}\right)\right]$

[*Hint:*   For Problems 19 and 20 use the formula for the sine of the sum of two angles.]

19. $\sin\left[\cos^{-1}\left(\dfrac{1}{3}\right) + \sin^{-1}\left(\dfrac{2}{3}\right)\right]$

20. $\sin\left[\cos^{-1}\left(\dfrac{1}{2}\right) + \sin^{-1}\left(\dfrac{-3}{4}\right)\right]$

21. Show that $\cos(\sin^{-1}x) = \sqrt{1 - x^2}$.

22. Show that $\tan(\sin^{-1}x) = x/\sqrt{1 - x^2}$.

23. Show that $\sin(2\cos^{-1}x) = 2x\sqrt{1 - x^2}$.

24. Show that $\cos(2\sin^{-1}x) = 1 - 2x^2$.

25. Show that $\sin^{-1}(-x) = -\sin^{-1}x$.

26. Show that $\tan^{-1}(-x) = -\tan^{-1}x$.

27. Show that $\cos^{-1}(-x) = \pi - \cos^{-1}x$.

28. Given that $x = \sin^{-1}(\frac{1}{2})$,   find $\cos x$, $\tan x$, $\cot x$, $\sec x$, and $\csc x$.

29. Give a restricted proof of (8.18).

30. Give a restricted proof of (8.19).

4. Derivatives of Inverse Trigonometric Functions; Related Integrals

To find the derivative of the function

$$y = \sin^{-1}x \qquad -\frac{\pi}{2} \le y \le \frac{\pi}{2}, \quad -1 \le x \le 1$$

we write $\sin y = x$ and differentiate implicitly with respect to x:

$$\cos y \frac{dy}{dx} = 1$$

$$\frac{dy}{dx} = \frac{1}{\cos y} \qquad \text{provided} \quad y \ne \pm\frac{\pi}{2}$$

Since y is the angle whose sine is x and $-\pi/2 < y < \pi/2$,   we have

$$\cos y = \sqrt{1 - x^2}$$

In Figure 12 we can "see" this formula for $0 < y < \pi/2$.

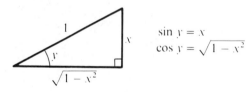

$\sin y = x$

$\cos y = \sqrt{1 - x^2}$

Figure 12

(8.22)   **THEOREM**   *Derivative of* $y = \sin^{-1}x$.   **The derivative of the inverse sine function** $y = \sin^{-1}x$ **is** $dy/dx = 1/\sqrt{1 - x^2}$, $-1 < x < 1$.   **That is,**

$$\frac{d}{dx}\sin^{-1}x = \frac{1}{\sqrt{1 - x^2}} \qquad -1 < x < 1$$

EXAMPLE 1   Find y' if:

(a) $y = \sin^{-1}(4x^2)$ (b) $y = e^{\sin^{-1}x}$ (c) $y = [\sin^{-1}(3x + 5)]^{1/2}$

Solution   (a)   We use the chain rule with $y = \sin^{-1}u$ and $u = 4x^2$.   Then

$$y' = \left(\frac{dy}{du}\right)\left(\frac{du}{dx}\right) = \left(\frac{1}{\sqrt{1 - u^2}}\right)(8x) = \frac{8x}{\sqrt{1 - 16x^4}}$$

(b)   We use the chain rule with $y = e^u$ and $u = \sin^{-1}x$.   Then

$$y' = \left(\frac{dy}{du}\right)\left(\frac{du}{dx}\right) = (e^u)\left(\frac{1}{\sqrt{1-x^2}}\right) = \frac{e^{\sin^{-1}x}}{\sqrt{1-x^2}}$$

(c)   We use the chain rule with $y = u^{1/2}$, $u = \sin^{-1}v$,   and $v = 3x + 5$.   Then

$$y' = \left(\frac{dy}{du}\right)\left(\frac{du}{dv}\right)\left(\frac{dv}{dx}\right) = \left(\frac{1}{2u^{1/2}}\right)\left(\frac{1}{\sqrt{1-v^2}}\right) \quad (3)$$

$$= \frac{3}{2[\sin^{-1}(3x+5)]^{1/2}\sqrt{1-(3x+5)^2}} \qquad \blacksquare$$

To get a formula for the derivative of the function $y = \tan^{-1}x$,   we write $\tan y = x$ and differentiate implicitly with respect to x. Then

$$\sec^2 y \,\frac{dy}{dx} = 1$$

$$\frac{dy}{dx} = \frac{1}{\sec^2 y}$$

Since y is an angle whose tangent is x and $-\pi/2 < y < \pi/2$,   we have $\sec y > 0$. It follows that $\sec y = \sqrt{1 + x^2}$.   Figure 13 illustrates this result. Thus,

$$\frac{1}{\sec^2 y} = \frac{1}{1 + x^2}$$

and we have the following theorem:

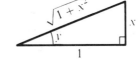

$\tan y = x$
$\sec y = \sqrt{1 + x^2}$

Figure 13

(8.23)   **THEOREM** *Derivative of* $y = \tan^{-1}x$. **The derivative of the inverse tangent function** $y = \tan^{-1}x$ **is** $dy/dx = 1/(1 + x^2)$. **That is,**

$$\frac{d}{dx} \tan^{-1}x = \frac{1}{1 + x^2}$$

EXAMPLE 2   Find y' if:
(a) $y = \tan^{-1}4x$ (b) $y = \sin(\tan^{-1}x)$ (c) $y = \tan^{-1}e^{2x}$

Solution   (a)   Set $y = \tan^{-1}u$ and $u = 4x$.   Then

$$y' = \left(\frac{dy}{du}\right)\left(\frac{du}{dx}\right) = \left(\frac{1}{1+u^2}\right)(4) = \frac{4}{1 + 16x^2}$$

(b)   Set $y = \sin u$ and $u = \tan^{-1}x$.   Then

$$y' = \left(\frac{dy}{du}\right)\left(\frac{du}{dx}\right) = (\cos u)\left(\frac{1}{1+x^2}\right) = \frac{\cos(\tan^{-1}x)}{1 + x^2}$$

(c) Set $y = \tan^{-1} u$, $u = e^v$, and $v = 2x$. Then

$$y' = \left(\frac{dy}{du}\right)\left(\frac{du}{dv}\right)\left(\frac{dv}{dx}\right) = \left(\frac{1}{1+u^2}\right)(e^v)(2) = \frac{2e^{2x}}{1+e^{4x}} \qquad \blacksquare$$

Finally, we derive the derivative of $y = \sec^{-1} x$. If we write $x = \sec y$ and differentiate implicitly, we find

$$1 = \sec y \tan y \frac{dy}{dx}$$

By the definition of $y = \sec^{-1} x$, we know that $0 \le y < \pi/2$ or $\pi \le y < 3\pi/2$. We can solve for dy/dx provided $\sec y \ne 0$ and $\tan y \ne 0$. Thus,

$$\frac{dy}{dx} = \frac{1}{\sec y \tan y} \qquad \text{provided } 0 < y < \frac{\pi}{2} \text{ or } \pi < y < \frac{3\pi}{2}$$

With these restrictions on y, we have $\tan y > 0$.* Thus, as Figure 14 illustrates,

$$\sec y = x \qquad \tan y = \sqrt{x^2 - 1}$$

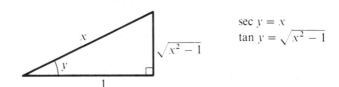

$$\sec y = x$$
$$\tan y = \sqrt{x^2 - 1}$$

Figure 14

Hence, we have the formula

$$\frac{d}{dx} \sec^{-1} x = \frac{1}{x\sqrt{x^2 - 1}}$$

(8.24)   THEOREM *Derivative of $y = \sec^{-1} x$.* **The derivative of the inverse secant function $y = \sec^{-1} x$ is $dy/dx = 1/(x\sqrt{x^2 - 1})$, $|x| > 1$; that is,**

$$\frac{d}{dx} \sec^{-1} x = \frac{1}{x\sqrt{x^2 - 1}} \qquad |x| > 1$$

Note that $y = \sec^{-1} x$ is not differentiable when $x = \pm 1$. In fact, as Figure 7 (p. 448) illustrates, at $x = \pm 1$, the graph of $y = \sec^{-1} x$ has a vertical tangent.

---

* Now the reason for the previous restrictions on $y = \sec^{-1} x$ [see (8.16)] should be apparent. Any other choice of restrictions would complicate the formula for $(d/dx) \sec^{-1} x$.

The three remaining formulas for the derivatives can be obtained by using the identities (8.17), (8.18), and (8.19):

Since $\cos^{-1}x = \dfrac{\pi}{2} - \sin^{-1}x,$ $\dfrac{d}{dx}\cos^{-1}x = \dfrac{-1}{\sqrt{1-x^2}}$ $|x| < 1$

Since $\cot^{-1}x = \dfrac{\pi}{2} - \tan^{-1}x,$ $\dfrac{d}{dx}\cot^{-1}x = \dfrac{-1}{1+x^2}$

Since $\csc^{-1}x = \dfrac{\pi}{2} - \sec^{-1}x,$ $\dfrac{d}{dx}\csc^{-1}x = \dfrac{-1}{x\sqrt{x^2-1}}$ $|x| > 1$

RELATED INTEGRALS

Now we turn our attention to the integrals that correspond to each of the derivative formulas just obtained:

(8.25)
$$\int \frac{dx}{\sqrt{1-x^2}} = \sin^{-1}x + C$$

(8.26)
$$\int \frac{dx}{1+x^2} = \tan^{-1}x + C$$

(8.27)
$$\int \frac{dx}{x\sqrt{x^2-1}} = \sec^{-1}x + C$$

EXAMPLE 3   Evaluate each integral:

(a) $\displaystyle\int \frac{dx}{\sqrt{4-x^2}}$ (b) $\displaystyle\int \frac{dx}{9+4x^2}$ (c) $\displaystyle\int \frac{dx}{x^2+x+1}$ (d) $\displaystyle\int \frac{x\,dx}{x^2+x+1}$

Solution   (a)   This resembles (8.25). We rewrite the original problem as follows:

$$\int \frac{dx}{\sqrt{4-x^2}} = \int \frac{dx}{2\sqrt{1-(x/2)^2}}$$

Now we use the substitution $u = x/2$.   Then $du = dx/2$,   and

$$\int \frac{dx}{\sqrt{4-x^2}} = \int \frac{2\,du}{2\sqrt{1-u^2}} = \sin^{-1}u + C = \sin^{-1}\left(\frac{x}{2}\right) + C$$

(b)   This resembles (8.26). We rewrite the original problem as follows:

$$\int \frac{dx}{9+4x^2} = \int \frac{dx}{9[1+(2x/3)^2]}$$

Let $u = 2x/3$.   Then $du = 2\,dx/3$,   and

$$\int \frac{dx}{9+4x^2} = \int \frac{3\,du/2}{9(1+u^2)} = \frac{1}{6}\tan^{-1}u + C = \frac{1}{6}\tan^{-1}\left(\frac{2x}{3}\right) + C$$

(c) We complete the square to get

$$\int \frac{dx}{x^2 + x + 1} = \int \frac{dx}{(x + \frac{1}{2})^2 + \frac{3}{4}}$$

Let $u = x + \frac{1}{2}$. Then $du = dx$, and

$$\int \frac{dx}{x^2 + x + 1} = \int \frac{du}{u^2 + \frac{3}{4}} = \int \frac{4\,du}{4u^2 + 3} = \int \frac{4\,du}{3[(2u/\sqrt{3})^2 + 1]}$$

Set $2u/\sqrt{3} = v$. Then $2\,du/\sqrt{3} = dv$, and

$$\int \frac{dx}{x^2 + x + 1} = \int \frac{4\sqrt{3}\,dv/2}{3(v^2 + 1)} = \frac{2\sqrt{3}}{3}\tan^{-1}v + C = \frac{2\sqrt{3}}{3}\tan^{-1}\left[\frac{2u}{\sqrt{3}}\right] + C$$

$$= \frac{2\sqrt{3}}{3}\tan^{-1}\left[\frac{2(x + \frac{1}{2})}{\sqrt{3}}\right] + C$$

$$= \frac{2\sqrt{3}}{3}\tan^{-1}\left[\frac{2x + 1}{\sqrt{3}}\right] + C$$

(d) Note how we force the derivative of the denominator to appear in the numerator:

$$\int \frac{x\,dx}{x^2 + x + 1} = \frac{1}{2}\int \frac{2x\,dx}{x^2 + x + 1} = \frac{1}{2}\int \frac{[(2x + 1) - 1]\,dx}{x^2 + x + 1}$$

$$= \frac{1}{2}\int \frac{(2x + 1)\,dx}{x^2 + x + 1} - \frac{1}{2}\int \frac{dx}{x^2 + x + 1}$$

$$= \frac{1}{2}\ln(x^2 + x + 1) - \frac{\sqrt{3}}{3}\tan^{-1}\left(\frac{2x + 1}{\sqrt{3}}\right) + C \qquad \blacksquare$$

↑

See part (c)

The integrals in (8.25), (8.26), and (8.27) have a more general form.

EXAMPLE 4 Verify that

(8.28)

$$\int \frac{dx}{a^2 + x^2} = \frac{1}{a}\tan^{-1}\left(\frac{x}{a}\right) + C \qquad a \neq 0$$

Solution

$$\int \frac{dx}{a^2 + x^2} = \frac{1}{a^2}\int \frac{dx}{1 + (x/a)^2}$$

We use the substitution $u = x/a$. Then $du = dx/a$, and

$$\int \frac{dx}{a^2 + x^2} = \frac{1}{a^2}\int \frac{a\,du}{1 + u^2} = \frac{1}{a}\tan^{-1}u + C = \frac{1}{a}\tan^{-1}\frac{x}{a} + C \qquad \blacksquare$$

In Problems 71 and 72 in Exercise 4 you are asked to show that

(8.29)

$$\int \frac{dx}{\sqrt{a^2 - x^2}} = \sin^{-1}\left(\frac{x}{a}\right) + C \qquad a \neq 0$$

(8.30)
$$\int \frac{dx}{x\sqrt{x^2 - a^2}} = \frac{1}{a} \sec^{-1}\left(\frac{x}{a}\right) + C \qquad a \neq 0$$

The next example is reproduced from *Dynamics*,* as an illustration of the occurrence in dynamics of integrals that lead to inverse trigonometric functions.

EXAMPLE 5   Consult Figure 15.

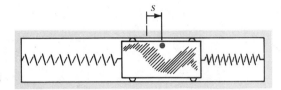

Figure 15

The spring-mounted slider moves in the horizontal guide with negligible friction and has a velocity v_0 in the s direction as it crosses the midposition where $s = 0$ and $t = 0$. The two springs together exert a retarding force to the motion of the slider, which gives it an acceleration proportional to the displacement but oppositely directed and equal to $a = -k^2 s$ where k is constant. (The constant is arbitrarily squared for later convenience in the form of the expressions.) Determine the expressions for the displacement s and velocity v as functions of the time t.

Solution   Since the acceleration is specified in terms of the displacement, the differential relation $v\,dv = a\,ds$ may be integrated. Thus,

$$\int v\,dv = \int -k^2 s\,ds + C_1 \text{ a constant} \qquad \text{or} \qquad \frac{v^2}{2} = -\frac{k^2 s^2}{2} + C_1$$

When $s = 0$, $v = v_0$, so that $C_1 = v_0^2/2$, and the velocity becomes

$$v = +\sqrt{v_0^2 - k^2 s^2}$$

The plus sign of the radical is taken when v is positive (in the plus s direction). This last expression may be integrated by substituting $v = ds/dt$. Thus,

$$\int \frac{ds}{\sqrt{v_0^2 - k^2 s^2}} = \int dt + C_2 \text{ a constant} \qquad \text{or} \qquad \frac{1}{k} \sin^{-1} \frac{ks}{v_0} = t + C_2$$

With the requirement of $t = 0$ when $s = 0$, the constant of integration becomes $C_2 = 0$, and we may solve the equation for s, so that

$$s = \frac{v_0}{k} \sin kt \qquad\qquad\qquad\qquad\qquad \textit{Answer}$$

The velocity is $v = \dot{s}[= ds/dt]$, which gives

$$v = v_0 \cos kt \qquad\qquad\qquad\qquad\qquad \textit{Answer}$$

         ■

---

* J. L. Meriam, *Dynamics*, 2d ed. (New York: John Wiley & Sons, 1975), p. 23. Reprinted by permission.

EXERCISE 4

In Problems 1–34 find $y' = dy/dx$.

1. $y = \sin^{-1}4x$

2. $y = \tan^{-1}5x$

3. $y = \cos^{-1}x^2$

4. $y = \sec^{-1}3x$

5. $y = \csc^{-1}(5x + 1)$

6. $y = \cot^{-1}x^2$

7. $y = \sin^{-1}x + \cos^{-1}x$

8. $y = \tan^{-1}x + \cot^{-1}x$

9. $y = \tan^{-1}\left(\dfrac{1}{x}\right)$

10. $y = \sin^{-1}\left(\dfrac{1}{x}\right)$

11. $y = \tan^{-1}\left(\dfrac{2x - 1}{2x}\right)$

12. $y = \sec^{-1}\sqrt{x}$

13. $y = \csc^{-1}\sqrt{x}$

14. $y = x(\sin^{-1}x)$

15. $y = \sin^{-1}(1 - x^2)$

16. $y = x\tan^{-1}(x + 1)$

17. $y = (1 + \tan^{-1}x)^2$

18. $y = \sin^{-1}(x^2 + 2x)$

19. $y = \sin^{-1}(1 - x^2)^{1/2}$

20. $y = \sqrt{1 - x^2}\,\sin^{-1}x$

21. $y = x\cot^{-1}(1 + x^2)$

22. $y = \cos^{-1}x + \dfrac{1}{1 - x^2}$

23. $y = \sin^{-1}\left(\dfrac{x - 1}{x + 1}\right)$

24. $y = x\cot^{-1}x + (1 + x)^{3/2}$

25. $y = \sin^{-1}(\cos x)$

26. $y = \tan^{-1}\sqrt{x}$

27. $y = \tan^{-1}(\ln x)$

28. $y = \sin^{-1}(\ln x)$

29. $y = \ln(\tan^{-1}x)$

30. $y = \ln\sqrt{\sin^{-1}x}$

31. $y = \tan^{-1}(\sin x)$

32. $y = \sin(\tan^{-1}x)$

33. $y = x\tan^{-1}\left(\dfrac{x}{a}\right) - \dfrac{1}{2}a\ln(x^2 + a^2)$

34. $y = x\sin^{-1}\left(\dfrac{x}{a}\right) + a\ln\sqrt{a^2 - x^2}$

In Problems 35 and 36 find dy/dx in terms of x and y.

35. $\sin^{-1}y + \cos^{-1}x = y$

36. $\tan^{-1}y = 3x + y$

In Problems 37–56 evaluate each integral.

37. $\displaystyle\int_0^1 \dfrac{5\,dx}{1 + x^2}$

38. $\displaystyle\int_0^1 \dfrac{x\,dx}{1 + x^4}$

39. $\displaystyle\int \dfrac{dx}{x^2 + 25}$

40. $\displaystyle\int \dfrac{dx}{x\sqrt{x^2 - 4}}$

41. $\displaystyle\int \dfrac{dx}{\sqrt{9 - x^2}}$

42. $\displaystyle\int_0^1 \dfrac{dx}{\sqrt{4 - x^2}}$

43. $\displaystyle\int \dfrac{dx}{\sqrt{16 - 9x^2}}$

44. $\displaystyle\int \dfrac{dx}{x\sqrt{9x^2 - 16}}$

45. $\displaystyle\int \dfrac{\cos x}{1 + \sin^2 x}\,dx$

46. $\displaystyle\int_0^1 \dfrac{e^x}{1 + e^{2x}}\,dx$

47. $\displaystyle\int \dfrac{\sin x}{\sqrt{4 - \cos^2 x}}\,dx$

48. $\displaystyle\int \dfrac{\sec^2 x\,dx}{\sqrt{1 - \tan^2 x}}$

49. $\displaystyle\int \dfrac{8x\,dx}{\sqrt{1 - (2x^2 - 1)^2}}$

50. $\displaystyle\int \dfrac{5\,dx}{\sqrt{e^{2x} - 16}}$

51. $\displaystyle\int \dfrac{2\,dx}{3 + 2x + 2x^2}$

52. $\displaystyle\int \dfrac{3\,dx}{x^2 + 6x + 10}$

53. $\displaystyle\int \dfrac{x\,dx}{2x^2 + 2x + 3}$

54. $\displaystyle\int \dfrac{3x\,dx}{x^2 + 6x + 10}$

55. $\displaystyle\int \dfrac{dx}{\sqrt{2x - x^2}}$

56. $\displaystyle\int_{\sqrt{2}}^2 \dfrac{dx}{x\sqrt{x^2 - 1}}$

In Problems 57–60 find the area under the graph of each function.

57. Under $y = \dfrac{1}{x^2 + 1}$ from $x = 0$ to $x = \sqrt{3}$

58. Under $y = \dfrac{1}{\sqrt{1 - x^2}}$ from $x = 0$ to $x = \frac{1}{2}$

59. Under $y = \dfrac{1}{5x^2 + 1}$ from $x = 0$ to $x = 1$

60. Under $y = \dfrac{1}{x\sqrt{x^2 - 4}}$ from $x = 3$ to $x = 4$

61. Show that $\dfrac{d}{dx} \cot^{-1}x = \dfrac{d}{dx} \tan^{-1}\left(\dfrac{1}{x}\right)$ for all $x \neq 0$.

62. We might try to infer from Problem 61 that $\cot^{-1}x = \tan^{-1}(1/x) + C$ for all $x \neq 0$, where C is a constant. Show, however, that

$$\cot^{-1}x = \begin{cases} \tan^{-1}(1/x) & \text{if } x > 0 \\ \tan^{-1}(1/x) + \pi & \text{if } x < 0 \end{cases}$$

What is the explanation of the incorrect inference?

63. Show that $\dfrac{d}{dx} \tan^{-1}(\cot x) = -1$.

64. A ladder 5 meters long is leaning against a wall. If the lower end of the ladder slides away from the wall at the rate of 0.5 meter per second, at what rate is the inclination of the ladder with respect to the ground changing when the lower end is 4 meters from the wall?

65. A man 2 meters tall walks horizontally at a constant rate of 1 meter per second toward the base of a tower 25 meters high. When the man is 10 meters from the tower, how fast is the angle of elevation changing if that angle is measured from the horizontal to the line joining the top of the man's head to the top of the tower?

66. A picture 4 meters in height is hung on a wall with the lower edge 3 meters above the level of an observer's eye. How far from the wall should the observer stand in order to obtain the most favorable view? (That is, the picture should subtend a maximum angle.)

67. Establish the identity $\sin^{-1}x + \cos^{-1}x = \pi/2$ by showing that the derivative of $y = \sin^{-1}x + \cos^{-1}x$ is 0. Then use the fact that when $x = 0$, $y = \pi/2$.

68. Establish the identity $\tan^{-1}x + \cot^{-1}x = \pi/2$ by showing that the derivative of $y = \tan^{-1}x + \cot^{-1}x$ is 0. Then use the fact that when $x = 1$, $y = \pi/2$.

69. The mean value theorem (4.26) guarantees that there is a real number N in $(0, 1)$ such that $f'(N) = f(1) - f(0)$ if f is continuous in $[0, 1]$ and differentiable in $(0, 1)$. Find N if $f(x) = \sin^{-1}x$.

70. A hot air balloon is rising at a speed of 100 meters per minute. What is the rate of change of the angle of elevation of an observer's line of sight if the observer is standing 200 meters from the lift-off point when the balloon is 600 meters high?

71. Verify (8.29).　　　　　　　　　　　**72.** Verify (8.30).

5. Harmonic Motion (If Time Permits)

The importance of the sine and cosine functions in mathematics and science derives primarily from the differential equation

(8.31)
$$\frac{d^2x}{dt^2} + \omega^2 x = 0 \qquad \omega > 0$$

This differential equation occurs in the study of oscillations, such as with a coiled spring or in wave theory (sound waves, radio waves, lightwaves, and so on).

The general solution of the differential equation (8.31) (see Section 8 of Chapter 7, pp. 428–429) is

(8.32) $$x = a \cos \omega t + b \sin \omega t$$

where a and b are constants.

We shall study the derivation of the differential equation (8.31) as it relates to a coiled spring. Suppose a spring of natural length L is placed in a vertical position, as shown in Figure 16(a). If a body of mass m is attached to the end of the spring, the spring is stretched beyond its natural length and comes to rest in a new equilibrium position of length $L + d$, as shown in Figure 16(b). The force of gravity that acts on the body of mass m is mg, and, by Hooke's law, the resisting force is kd. For equilibrium to occur, we must have

(8.33) $$mg = kd$$

Suppose we pull the weight down a distance A from this new equilibrium position and then release it [see Fig. 16(c)]. We want to find an expression for the displacement of the body from its equilibrium position (length $L + d$) as a function of the time t after the motion begins.

We take the positive direction of x to be down. If we assume there is no friction in the spring and no air resistance to its motion, the forces acting on the body are

$$\begin{bmatrix} \text{Force due to} \\ \text{gravity} \end{bmatrix} - \begin{bmatrix} \text{Force due to} \\ \text{resisting force} \\ \text{of spring (Hooke's law)} \end{bmatrix} = mg - k(d + x)$$

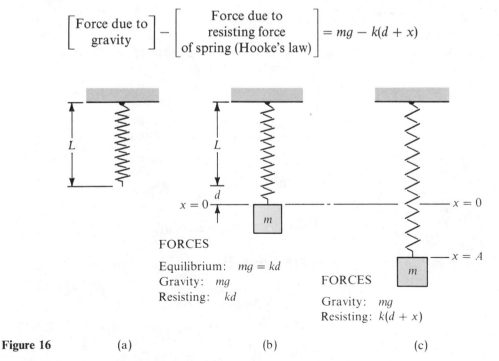

FORCES

Equilibrium: $mg = kd$
Gravity: mg
Resisting: kd

FORCES

Gravity: mg
Resisting: $k(d + x)$

Figure 16   (a)   (b)   (c)

By Newton's second law of motion, the sum of these forces, F, equals ma, where $a = d^2x/dt^2$ is the acceleration of the body. That is,

$$F = ma$$

or

$$mg - k(d + x) = ma$$

(8.34)
$$mg - kd - kx = m\frac{d^2x}{dt^2}$$

But $mg = kd$, so (8.34) becomes

$$m\frac{d^2x}{dt^2} = -kx$$

$$m\frac{d^2x}{dt^2} + kx = 0$$

(8.35)
$$\frac{d^2x}{dt^2} + \frac{k}{m}x = 0$$

The initial conditions require that

(8.36)
$$\text{at} \quad t = 0, \quad x = A \quad \text{and} \quad \frac{dx}{dt} = 0$$

The first condition arises since the body is pulled down a distance A when it is released. The second condition arises because when $t = 0$, the body has zero speed.

Since k and m are positive real numbers, we set $k/m = \omega^2$, and the differential equation (8.35) may be written as

(8.37)
$$\frac{d^2x}{dt^2} + \omega^2 x = 0 \qquad \omega = \sqrt{k/m}$$

As we pointed out in (8.32), the general solution of this differential equation is

(8.38)
$$x = a\cos\omega t + b\sin\omega t$$

The initial conditions (8.36) require that at $t = 0$, $x = A$ and $dx/dt = 0$. So we have

$$A = a\cos 0 + b\sin 0 \qquad \text{and} \qquad \frac{dx}{dt} = -a\omega\sin\omega t + b\omega\cos\omega t$$

$$A = a \qquad\qquad\qquad\qquad 0 = -a\omega\sin 0 + b\omega\cos 0$$

$$0 = b\omega$$

$$0 = b$$

The solution of the differential equation (8.37) with the initial conditions (8.36) is therefore

$$x = A\cos\omega t \qquad \omega = \sqrt{k/m}$$

This is an equation describing the motion of a body and is an example of *simple harmonic motion*. The maximum displacement A is called the *amplitude*. The motion is *periodic* with *period* $T = 2\pi/\omega$. This means that if time is measured in seconds, then every $2\pi/\omega$ seconds the motion repeats itself. The reciprocal $1/T$ of the period T is called the *frequency* $f = \omega/2\pi$, the number of complete cycles per second. In terms of the spring constant k and the mass m of the body, the period T and frequency f are

$$T = 2\pi \sqrt{\frac{m}{k}} \qquad \text{and} \qquad f = \frac{1}{2\pi} \sqrt{\frac{k}{m}}$$

ANOTHER FORM OF THE GENERAL SOLUTION

Another form of the general solution (8.32) to the differential equation (8.31) can be obtained if we write (8.32) as

(8.39)
$$x = \sqrt{a^2 + b^2} \left(\frac{a}{\sqrt{a^2 + b^2}} \cos \omega t + \frac{b}{\sqrt{a^2 + b^2}} \sin \omega t \right)$$

Now we set

$$\sin \phi = \frac{a}{\sqrt{a^2 + b^2}} \qquad \text{and} \qquad \cos \phi = \frac{b}{\sqrt{a^2 + b^2}}$$

(see Fig. 17). An application of the identity

$$\sin(\omega t + \phi) = \sin \omega t \cos \phi + \cos \omega t \sin \phi$$

allows us to write (8.32) as

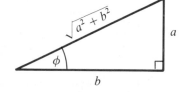

Figure 17

(8.40)
$$x = A \sin(\omega t + \phi) \qquad A = \sqrt{a^2 + b^2}$$

This is the equation of simple harmonic motion with amplitude $A = \sqrt{a^2 + b^2}$ and period $2\pi/\omega$. The angle ϕ is referred to as the *phase angle* (see Fig. 18).

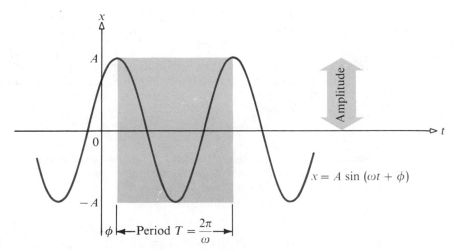

Figure 18

Any motion that obeys the differential equation (8.31) is referred to as *harmonic oscillation*. Harmonic oscillation occurs not only in elastic bodies but also in vibrating tuning forks, in vibrating strings, inside an atom, and elsewhere. This phenomenon explains why watches keep time and why atoms emit light of a particular frequency.

EXERCISE 5

In Problems 1–5 a particle of mass m is attached to one end of a vertical elastic spring whose spring constant is k. The other end of the spring is fixed. If the particle is pulled down a distance A (the amplitude) from the equilibrium position, the displacement of the particle from its equilibrium position at time t is denoted by $x = x(t)$, where at $t = 0$, $x = A$.

1. Suppose that $m = 15$ grams and $x = 12 \cos(t/3)$. Find the spring constant.

2. Suppose that $m = 12$ grams and $k = 18$ grams per second per second. Suppose also that, at time $t = 0$ second, the spring is stretched 2 centimeters from its equilibrium position and the particle is at rest. Find $x(t)$. Find the amplitude.

3. Suppose that $m = 1$ gram and $k = 4$ grams per second per second. Suppose also that, at time $t = 0$ second, the spring is compressed 1 centimeter from its equilibrium position. Find an expression for the velocity of the particle.

4. Suppose that $m = 2$ grams and $k = 18$ grams per second per second. Suppose also that, at time $t = 0$ second, the spring is compressed 2 centimeters from its equilibrium position. Find the velocity of the particle at time $t = \pi/6$ second.

5. Suppose that $m = 1$ gram and $k = 0.25$ gram per second per second. Suppose also that, at time $t = 0$ second, the spring is stretched 1 centimeter from its equilibrium position. Find $x(t)$.

6. Consider a simple pendulum whose length in centimeters is l. If the pendulum makes an angle θ with the vertical, its motion (assuming no friction or air resistance) obeys the differential equation

$$\frac{d^2\theta}{dt^2} = -\frac{g}{l}\sin\theta \qquad g \approx 980 \text{ centimeters per second per second}$$

If the angle θ is sufficiently small ($<5°$), then $\sin\theta \approx \theta$, and we may replace $\sin\theta$ by θ (θ in radians). Under this assumption, find the equation of the motion, its period, and its frequency. Use the initial conditions that when $t = 0$, $\theta = \theta_0$ and $d\theta/dt = 0$.

7. If the disk in the figure is rotated about the vertical through an angle θ, torsion in the wire will attempt to turn the disk in the opposite direction. The motion (assuming no friction or air resistance) obeys the differential equation

$$\frac{1}{2}mR^2\frac{d^2\theta}{dt^2} = -k\theta$$

where m is the mass of the disk, R is the radius of the disk, and k is the coefficient of torsion of the wire. Find the equation of the motion and its period. Use the initial conditions that when $t = 0$, $\theta = \theta_0$ and $d\theta/dt = 0$.

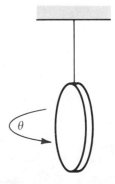

6. Hyperbolic Functions

Certain combinations of the functions e^x and e^{-x} occur so frequently in applied mathematics that they warrant special study. This collection of functions, called *hyperbolic functions,* has some properties that bear a close resemblance to those of the trigonometric functions. In fact, hyperbolic functions are so named because they are related to a hyperbola in much the same way that the trigonometric functions (sometimes called *circular functions*) are related to a circle.* Because of this, we give them names such as *hyperbolic sine* (*sinh*) and *hyperbolic cosine* (*cosh*). These functions are defined as

(8.41)
$$\sinh x = \frac{e^x - e^{-x}}{2} \qquad \cosh x = \frac{e^x + e^{-x}}{2}$$

The remaining hyperbolic functions are combinations of the first two; they should be easy to remember since they are analogous to the relationships among the trigonometric functions.

(8.42)

Hyperbolic tangent: $\qquad \tanh x = \dfrac{\sinh x}{\cosh x} = \dfrac{e^x - e^{-x}}{e^x + e^{-x}}$

Hyperbolic cotangent: $\qquad \coth x = \dfrac{\cosh x}{\sinh x} = \dfrac{e^x + e^{-x}}{e^x - e^{-x}}$

Hyperbolic secant: $\qquad \operatorname{sech} x = \dfrac{1}{\cosh x} = \dfrac{2}{e^x + e^{-x}}$

Hyperbolic cosecant: $\qquad \operatorname{csch} x = \dfrac{1}{\sinh x} = \dfrac{2}{e^x - e^{-x}}$

There are also identities involving the hyperbolic functions that are strikingly reminiscent of the familiar trigonometric identities. Some of the basic ones are

(8.43)
$$\cosh^2 x - \sinh^2 x = 1$$

(8.44)
$$\tanh^2 x + \operatorname{sech}^2 x = 1$$

(8.45)
$$\coth^2 x - \operatorname{csch}^2 x = 1$$

The first of these is derived directly from the definition, as follows:

$$\cosh^2 x - \sinh^2 x = \left(\frac{e^x + e^{-x}}{2}\right)^2 - \left(\frac{e^x - e^{-x}}{2}\right)^2$$

$$= \frac{e^{2x} + 2e^0 + e^{-2x}}{4} - \frac{e^{2x} - 2e^0 + e^{-2x}}{4} = \frac{2+2}{4} = 1$$

---

* More on this at the end of the section.

The second identity (8.44) is obtained from the first by dividing each term by $\cosh^2 x$. Identity (8.45) is derived similarly.

Numerous other identities may be derived. We list some of them below:

$$(8.46) \qquad\qquad \sinh(A + B) = \sinh A \cosh B + \cosh A \sinh B$$

$$(8.47) \qquad\qquad \cosh(A + B) = \cosh A \cosh B + \sinh A \sinh B$$

$$(8.48) \qquad\qquad \sinh(-A) = -\sinh A$$

$$(8.49) \qquad\qquad \cosh(-A) = \cosh A$$

The derivations of these identities are left as exercises.

The differentiation formulas for the hyperbolic functions can be obtained readily:

$$\frac{d}{dx} \sinh x = \frac{d}{dx} \left[\frac{1}{2} (e^x - e^{-x}) \right] = \frac{e^x + e^{-x}}{2} = \cosh x$$

$$\frac{d}{dx} \cosh x = \frac{d}{dx} \left[\frac{1}{2} (e^x + e^{-x}) \right] = \frac{e^x - e^{-x}}{2} = \sinh x$$

Thus,

$$(8.50) \qquad\qquad \frac{d}{dx} \sinh x = \cosh x \qquad \frac{d}{dx} \cosh x = \sinh x$$

The formulas for differentiating the rest of the hyperbolic functions are

$$(8.51) \qquad\qquad \frac{d}{dx} \tanh x = \operatorname{sech}^2 x \qquad\qquad \frac{d}{dx} \coth x = -\operatorname{csch}^2 x$$

$$\frac{d}{dx} \operatorname{sech} x = -\operatorname{sech} x \tanh x \qquad\qquad \frac{d}{dx} \operatorname{csch} x = -\operatorname{csch} x \coth x$$

THE GRAPHS OF THE HYPERBOLIC FUNCTIONS

Since $\cosh x = \frac{1}{2}(e^x + e^{-x})$, it follows that, at $x = 0$, $y = \frac{1}{2}(e^0 + e^0) = 1$. Thus, the graph of $y = \cosh x$ has its y-intercept at $(0, 1)$. Since $\cosh(-x) = \cosh x$, the cosh function is an *even function*, and hence its graph is symmetric with respect to the y-axis. Also, since $\frac{1}{2}(e^x + e^{-x})$ is positive for all x, the points on the graph of $y = \cosh x$ are always above the x-axis.

Since $\lim_{x \to +\infty} [\frac{1}{2}(e^x + e^{-x})] = +\infty$, the graph is not bounded above. If $y = \cosh x$, then $dy/dx = \sinh x$. Now, $\sinh x > 0$ for $x > 0$. Hence, cosh x is increasing for $x > 0$. Also, since $\sinh x < 0$ for $x < 0$, cosh x is decreasing for $x < 0$. Furthermore, at $x = 0$, $dy/dx = 0$. Hence, there is a horizontal tangent line at the point $(0, 1)$.

Also, $d^2y/dx^2 = \cosh x$, which is always positive. Hence, the point $(0, 1)$ is a local minimum, and the graph is concave upward.

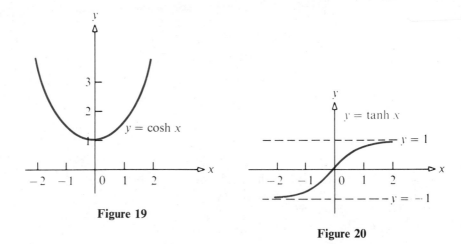

Figure 19

Figure 20

Using all this information, we are able to sketch the graph of cosh x, as shown in Figure 19.

A similar analysis will yield enough information to graph tanh x, sinh x, coth x, sech x, and csch x, as shown in Figures 20–24.

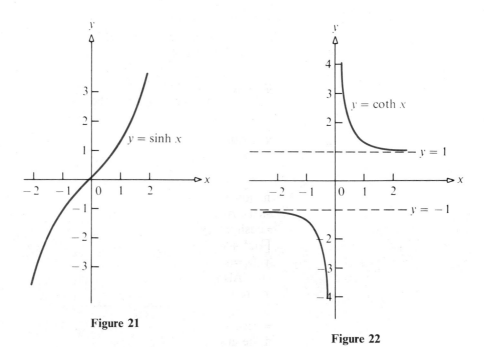

Figure 21

Figure 22

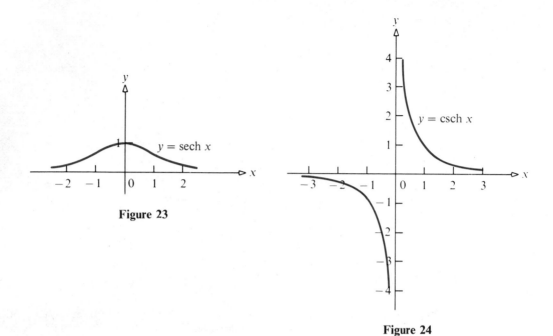

Figure 23

Figure 24

HANGING CHAINS

The hyperbolic cosine has an interesting physical property. If a cable or chain of uniform density is suspended at its ends, it will assume the shape of the graph of a hyperbolic cosine.

If we fix our coordinate system (as in Fig. 25) so that the cable lies in the xy-plane, with the y-axis vertical and the lowest point of the cable on the y-axis at the point

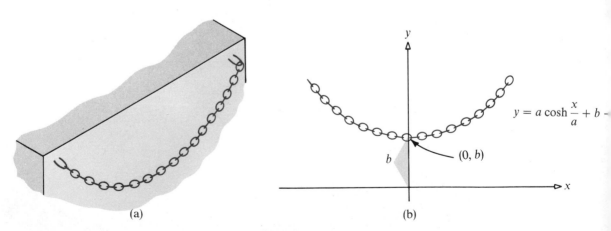

(a)   (b)

Figure 25

(0, b), the shape of the chain will be given by the equation

(8.52)
$$y = a \cosh \frac{x}{a} + b - a$$

where a is a constant that depends on the linear density or weight per unit length of the chain and the tension or horizontal force holding the ends of the chain apart. The graph of this equation is called a *catenary*.*

RELATED INTEGRALS

Formulas (8.50) and (8.51) give rise to the following integration formulas:

(8.53)
$$\int \sinh x \, dx = \cosh x + C \qquad \int \cosh x \, dx = \sinh x + C$$
$$\int \operatorname{sech}^2 x \, dx = \tanh x + C \qquad \int \operatorname{csch}^2 x \, dx = -\coth x + C$$
$$\int \operatorname{sech} x \tanh x \, dx = -\operatorname{sech} x + C \qquad \int \operatorname{csch} x \coth x \, dx = -\operatorname{csch} x + C$$

WHY THE NAME HYPERBOLIC?

The trigonometric functions are sometimes referred to as *circular functions* because any point P on the unit circle $x^2 + y^2 = 1$ has coordinates ($\cos t$, $\sin t$), as shown in Figure 26. The hyperbolic functions are so named because any point P on the hyperbola $x^2 - y^2 = 1$ has coordinates ($\cosh t$, $\sinh t$), as in Figure 27.

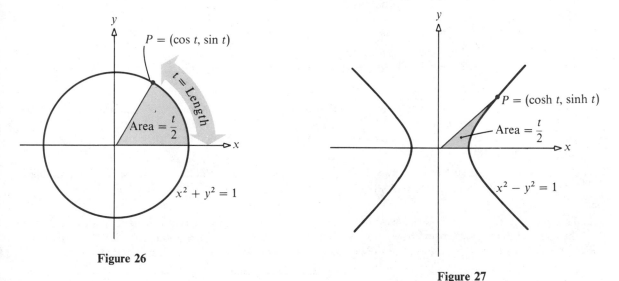

Figure 26

Figure 27

---

* From the Latin word *catena*, meaning "chain."

The relationship is even deeper, though. From elementary geometry it is known that the area of the sector shaded in Figure 26 equals $t/2$. And it can be shown that the area shaded in Figure 27 also equals $t/2$.

EXERCISE 6

In Problems 1–8 derive each identity.

1. $\tanh^2 x + \text{sech}^2 x = 1$

2. $\coth^2 x - \text{csch}^2 x = 1$

3. $\sinh(-A) = -\sinh A$

4. $\cosh(-A) = \cosh A$

5. $\sinh(A + B) = \sinh A \cosh B + \cosh A \sinh B$

6. $\cosh(A + B) = \cosh A \cosh B + \sinh A \sinh B$

7. $\cosh 3x = 4 \cosh^3 x - 3 \cosh x$

8. $\tanh 2x = \dfrac{2 \tanh x}{1 + \tanh^2 x}$

In Problems 9–20 find y'.

9. $y = \sinh 3x$

10. $y = \cosh \dfrac{x}{2}$

11. $y = \cosh(x^2 + 1)$

12. $y = \cosh(2x^3 - 1)$

13. $y = \sinh x \cosh 4x$

14. $y = \cosh^2 x$

15. $y = \tanh^2 x$

16. $y = e^x \cosh x$

17. $y = \coth \dfrac{1}{x}$

18. $y = e^x(\cosh x + \sinh x)$

19. $y = \tanh x^2$

20. $y = x^2 \text{sech } x$

In Problems 21–28 evaluate each integral.

21. $\displaystyle\int \frac{\sinh \sqrt{x}}{\sqrt{x}} \, dx$

22. $\displaystyle\int \frac{1}{\cosh^2 2x} \, dx$

23. $\displaystyle\int \sinh^2 x \, dx$

24. $\displaystyle\int \cosh^4 8x \, dx$

25. $\displaystyle\int \sinh x \cosh x \, dx$

26. $\displaystyle\int \text{sech}^2 x \tanh x \, dx$

27. $\displaystyle\int \sqrt{\cosh x + 1} \, dx$

28. $\displaystyle\int \cosh^2 x \, dx$

29. Find the area under the graph of $y = \cosh x$ from $x = -1$ to $x = 1$.

30. A cable hangs in the shape of a catenary between two supports at the same height 20 meters apart. The slope of the cable at the right-hand support is $\tfrac{3}{4}$.
   (a) What is the height of the supports above the lowest point in the middle of the cable?
   (b) Find the length of the cable.

31. A rope of length L is supported at two points, (c, d) and $(-c, d)$. The middle of the rope is located at $(0, b)$. Find the constant a in the formula (8.52) for the shape of the rope. [*Hint:* Show that $(d - b + a)^2 - a^2 = L^2/4$.]

32. The famous Gateway Arch to the West in St. Louis, Missouri, is constructed in the shape of a catenary (see the figure at the top of the next page). For this arch, the number a in (8.52) is about 127.7 feet. Find an equation of the arch. (Assume that the coordinate system has been positioned in such a way that $b = a$.)

33. Find the length of the catenary (8.52) from the point $(0, b)$ to a point (x, y), $x > 0$.

34. Find the area under the graph of the catenary (8.52) from $x = 0$ to $x = c$.

35. Find the volume of the solid generated by revolving the area under the graph of the catenary in (8.52) from $x = 0$ to $x = 1$ about the x-axis.

Gateway Arch to the West,
St. Louis, Missouri

36. Show that formula (8.52) obeys the differential equation $d^2y/dx^2 = (1/a^2)(y + a - b)$.

37. Establish the identity $(\cosh x + \sinh x)^n = \cosh nx + \sinh nx$ for any integer n.

38. Establish formulas (8.51).

39. Show that the catenary function (8.52) is a solution of the differential equation

$$\frac{d^2y}{dx^2} = \frac{1}{a}\sqrt{1 + \left(\frac{dy}{dx}\right)^2}$$

7. Inverse Hyperbolic Functions

The graph of $y = \sinh x$ is given in Figure 21 (p. 465). From the graph, we observe that every horizontal line intersects the graph of $y = \sinh x$ in exactly one point. The function $y = \sinh x$ is therefore one-to-one and, hence, has an inverse. We denote the inverse by $\sinh^{-1}$ and define it as

(8.54) $y = \sinh^{-1}x$ means $x = \sinh y$

The domain of $y = \sinh^{-1}x$ is the set of real numbers, and the range is also the set of real numbers.

The graph of $y = \cosh x$ is given in Figure 19 (p. 465). From the graph, we observe that every horizontal line above 1 intersects the graph of $y = \cosh x$ at two points. Thus, the function is not one-to-one. However, if we restrict $y = \cosh x$

to the nonnegative values of x, we have a one-to-one function that, therefore, has an inverse. We denote the inverse by $\cosh^{-1}$ and define it as

(8.55) $$y = \cosh^{-1}x \quad \text{means} \quad x = \cosh y \quad y \geq 0$$

The domain of $y = \cosh^{-1}x$ is $x \geq 1$, and the range is $y \geq 0$.

The other inverse hyperbolic functions are defined similarly.

The inverse hyperbolic functions may be expressed in terms of natural logarithms as follows:

(8.56) $$\sinh^{-1}x = \ln(x + \sqrt{x^2 + 1}) \quad \text{for all } x$$

(8.57) $$\cosh^{-1}x = \ln(x + \sqrt{x^2 - 1}) \quad x \geq 1$$

(8.58) $$\tanh^{-1}x = \frac{1}{2}\ln\left(\frac{1 + x}{1 - x}\right) \quad -1 < x < 1$$

We will prove (8.56) here and leave (8.57) and (8.58) as exercises.

Proof Let $y = \sinh^{-1}x$. Then $x = \sinh y = (e^y - e^{-y})/2$. Multiplying by $2e^y$, we get

$$2xe^y = (e^y)^2 - 1$$
$$(e^y)^2 - 2xe^y - 1 = 0$$

This is a quadratic equation in e^y. By solving it for e^y, we find

$$e^y = \tfrac{1}{2}(2x \pm \sqrt{4x^2 + 4}) = x \pm \sqrt{x^2 + 1}$$

Since $e^y > 0$ and $x < \sqrt{x^2 + 1}$ for all x, the minus sign on the right is not possible. As a result,

$$e^y = x + \sqrt{x^2 + 1} \quad \text{or} \quad y = \ln(x + \sqrt{x^2 + 1}) \quad \blacksquare$$

The derivatives of $y = \sinh^{-1}x$, $y = \cosh^{-1}x$, and $y = \tanh^{-1}x$ are obtained by a straightforward calculation using formulas (8.56), (8.57), and (8.58):

$$\frac{d}{dx}\sinh^{-1}x = \frac{1}{\sqrt{x^2 + 1}} \qquad \frac{d}{dx}\cosh^{-1}x = \frac{1}{\sqrt{x^2 - 1}} \qquad \frac{d}{dx}\tanh^{-1}x = \frac{1}{1 - x^2}$$

EXERCISE 7

In Problems 1–10 find y'.

1. $y = \sinh^{-1}3x$

2. $y = \cosh^{-1}4x$

3. $y = \tanh^{-1}(x^2 - 1)$

4. $y = x\sinh^{-1}x$

5. $y = \cosh^{-1}(2x + 1)$

6. $y = x^2\cosh^{-1}x$

7. $y = \tanh^{-1}(\cos 2x)$

8. $y = \cosh^{-1}(\sqrt{x^2 - 1}), \quad x > 0$

9. $y = \tanh^{-1}(\tan x)$

10. $y = \sinh^{-1}(\sin x)$

11. Show that $\cosh^{-1}x = \ln(x + \sqrt{x^2 - 1}), \quad x \geq 1.$

12. Show that $\tanh^{-1}x = \frac{1}{2}\ln\left(\frac{1 + x}{1 - x}\right), \quad -1 < x < 1.$

13. Show that $\dfrac{d}{dx}\sinh^{-1}x = \dfrac{1}{\sqrt{x^2 + 1}}.$

14. Show that $\dfrac{d}{dx}\cosh^{-1}x = \dfrac{1}{\sqrt{x^2 - 1}}.$

15. Show that $\dfrac{d}{dx}\tanh^{-1}x = \dfrac{1}{1 - x^2}.$

16. Obtain the graph of $y = \sinh^{-1}x.$

17. Obtain the graph of $y = \cosh^{-1}x.$

18. Obtain the graph of $y = \tanh^{-1}x.$

19. Evaluate: $\displaystyle\int_2^3 \frac{1}{\sqrt{x^2 - 1}}\,dx$

20. Derive the formula $\displaystyle\int\frac{dx}{\sqrt{x^2 + a^2}} = \sinh^{-1}\left(\frac{x}{a}\right) + C.$

Miscellaneous Exercises

In Problems 1–24 differentiate and simplify.

1. $w = z \sinh z$

2. $y = \tanh\left(\dfrac{x}{2}\right) + \dfrac{2x}{4 + x^2}$

3. $y = \tanh\dfrac{1}{x}$

4. $u = x^2 \tanh^{-1}x$

5. $w = \sqrt{1 - z^2}(\sin^{-1}z)$

6. $y = \sqrt{\sinh x}$

7. $y = \sin^{-1}(x - 1) + (x - 1)\sqrt{2x - x^2}$

8. $y = \cosh e^x$

9. $y = 2\sqrt{x} - 2\tan^{-1}\sqrt{x}$

10. $z = \ln(\sinh^2 x)$

11. $y = \dfrac{8x}{x^2 + 4} - 4\tan^{-1}\left(\dfrac{1}{2}x\right) + x$

12. $u = \sin^{-1}(2x - 1)$

13. $w = z^2 \sin^{-1}z$

14. $y = \tan^{-1}\sqrt{v}$

15. $y = \sin^{-1}\left(\dfrac{x}{a}\right) + \dfrac{\sqrt{a^2 - x^2}}{x}$

16. $y = \sinh^{-1}e^x$

17. $y = \cos^{-1}\left(1 - \dfrac{x}{a}\right)$

18. $y = x^2 \tan^{-1}\left(\dfrac{1}{x}\right)$

19. $y = x \tan^{-1}x - \ln\sqrt{1 + x^2}$

20. $y = x^2 \sin^{-1}\left(\dfrac{2}{x}\right)$

21. $y = \ln(\tan^2 v)$

22. $y = \dfrac{-\sqrt{a^2 - x^2}}{x} + \cos^{-1}\left(\dfrac{x}{a}\right)$

23. $y = \sin^{-1}e^x$

24. $y = \sin^{-1}(x^2)$

25. Find: $\displaystyle\int \sin^2 3x \, dx$

26. Find: $\displaystyle\int \frac{1}{\sqrt{x}}\cos^2\sqrt{x}\,dx$

27. Find: $\displaystyle\lim_{x \to 0}\left(\frac{\sinh x}{x}\right)$

28. Find: $\displaystyle\lim_{x \to 0}\left(\frac{\cosh x - 1}{x}\right)$

29. Find y' if $x = \ln(\csc y + \cot y).$

30. Find the value of y' at $x = \pi/2$, $y = \pi$ if $x \sin y + y \cos x = 0$.

31. Find the minimum value of $y = x - \cosh^{-1} x$.

32. Find y' if $y = \ln \sqrt{1 - x^2} + x \tanh^{-1} x$.

33. Use differentiation to verify the integration formula

$$\int \sin^{-1} x \, dx = x \sin^{-1} x + \sqrt{1 - x^2} + C$$

34. Use differentiation to verify the integration formula

$$\int \sqrt{a^2 - x^2} \, dx = \frac{1}{2} x \sqrt{a^2 - x^2} + \frac{1}{2} a^2 \sin^{-1}\left(\frac{x}{a}\right) + C$$

35. Use Problem 34 to find the area enclosed by the graph of $y = \sqrt{a^2 - x^2}$ and the x-axis. Check by elementary geometry.

36. Find: $\displaystyle\int_0^1 \frac{x \, dx}{\sqrt{2 - x^4}}$ **37.** Find: $\displaystyle\int_2^4 \frac{dx}{\sqrt{8x - x^2}}$ **38.** Find: $\displaystyle\int_0^1 \frac{x + 1}{x^2 + 3} \, dx$

39. Find: $\displaystyle\int_{2/\sqrt{3}}^2 \frac{dx}{x \sqrt{x^2 - 1}}$ **40.** Find: $\displaystyle\int_{3\sqrt{2}}^6 \frac{dx}{x \sqrt{x^2 - 9}}$

41. Evaluate: $\displaystyle\int \frac{e^{4x} \, dx}{\sqrt{16 - e^{8x}}}$ **42.** Evaluate: $\displaystyle\int \frac{x^2 \, dx}{\sqrt{1 - x^6}}$

43. Find the area under the graph of $y = 1/(1 + x^2)$ from $x = -r$ to $x = r$, where $r > 0$. Then examine the behavior of the area as $r \to +\infty$.

44. Find the length of the graph of $y = \sqrt{a^2 - x^2}$ in two ways (by calculus and by elementary geometry).

45. If $f(x) = \int_0^x dt/\sqrt{1 - t^2}$, what is $f^{-1}(x)$?

46. Express the length of the graph of $y = \ln(\sec x)$ from $x = 0$ to $x = b$, where $0 < b < \pi/2$, as an integral. Do not evaluate the integral.

47. Express the length of the graph of $y = \tan x$ from $x = a$ to $x = b$, where $0 < a < b < \pi/2$, as an integral. Do not evaluate the integral.

48. (a) Express the area A of the region in the first quadrant enclosed by the y-axis and the graphs of $y = \tan x$ and $y = k$ for $k > 0$ as a function of k.
    (b) What is the value of A when $k = 1$?
    (c) If the line $y = k$ is moving upward at the rate of $\frac{1}{10}$ unit per second, at what rate is A changing when $k = 1$?

49. What are the coordinates of the inflection point on the graph of $y = (x + 1) \tan^{-1} x$?

50. Find an equation for the tangent line to the graph of $y = \sin^{-1}(x/2)$ at the origin.

51. The region enclosed by the x-axis and the part of the graph of $y = \cos x$ from $x = -\pi/2$ to $x = \pi/2$ is separated into two regions by the line $x = k$. If the area of the region for $-\pi/2 \le x \le k$ is three times the area of the region for $k \le x \le \pi/2$, find k.

52. Let $f(x) = x \sin^{-1} x$.
    (a) Show that f is an even function.   (b) Find $f'(x)$ and $f''(x)$.   (c) Sketch the graph of f.

53. What happens when we try to find the derivative of $f(x) = \sin^{-1}(\cosh x)$? What is the explanation?

54. Since $(d/dx) \tanh^{-1} x = 1/(1 - x^2)$, we might expect that $\int_2^3 dx/(1 - x^2) = \tanh^{-1}(3) - \tanh^{-1}(2)$. Why is this incorrect? What is the correct result?

55. (a) Sketch the graph of $y = \text{gd}(x) = \tan^{-1}(\sinh x)$; this is called the *gudermannian of* x (named after Christoph Gudermann).

   (b) If $y = \text{gd}(x)$, show that $\cos y = \text{sech } x$ and $\sin y = \tanh x$.

56. In Section 6 we said that the shaded area in Figure 27 (p. 467) is $t/2$. Prove this as follows:

   (a) Let A be twice the area outlined in Figure 27, and let (x, y) be the coordinates of P. Explain why
   $A = x\sqrt{x^2 - 1} - 2\int_1^x \sqrt{u^2 - 1} \, du$.

   (b) Differentiate in part (a) to show that $dA/dx = 1/\sqrt{x^2 - 1}$.

   (c) Show that $A = \cosh^{-1} x + C$, and explain why $C = 0$.

   (d) Why does it follow that $A = t$?

57. *The general equation of simple harmonic motion,

$$x = A \sin(\omega t + \theta_0)$$

can be written in the equivalent form

$$x = B \sin \omega t + C \cos \omega t$$

Find the expressions for the amplitudes B and C in terms of the amplitude A and the initial phase angle θ_0.

58.† A body is vibrating with simple harmonic motion of amplitude 15 centimeters and frequency 4 hertz. Compute:

   (a) The maximum values of the acceleration and velocity

   (b) The acceleration and velocity when the coordinate is 9 centimeters

   (c) The time required to move from the equilibrium position to a point 12 centimeters distant from it

59.† A body of mass 10 grams moves with simple harmonic motion of amplitude 24 centimeters and period 4 seconds. The coordinate is $+24$ centimeters when $t = 0$. Compute:

   (a) The position of the body when $t = 0.5$ second

   (b) The magnitude and direction of the force acting on the body when $t = 0.5$ second

   (c) The minimum time required for the body to move from its initial position to the point where $x = -12$ centimeters

   (d) The velocity of the body when $x = -12$ centimeters

60.† The motion of the piston of an automobile engine is approximately simple harmonic.

   (a) If the stroke of an engine (twice the amplitude) is 4 inches and the angular velocity is 3600 revolutions per minute, compute the acceleration of the piston at the end of its stroke.

   (b) If the piston weighs 1 pound, what resultant force must be exerted on it at this point?

---

* Adapted from F. W. Sears, M. W. Zemansky, and H. D. Young, *University Physics* (Reading, Mass.: Addison-Wesley Publishing Co., 1976), p. 206. Reprinted by permission.

† Ibid., p. 208. Reprinted by permission.

9

Techniques of Integration

1. Basic Integration Formulas

In the preceding chapters we developed a collection of basic integration formulas. For convenience, a list of those that are *essential* is given here and inside the covers of this book. If you have not already committed these to memory, then you should do so now.

1. $\int x^\alpha \, dx = \dfrac{x^{\alpha+1}}{\alpha+1} + C, \quad \alpha \neq -1$

2. $\int \dfrac{1}{x} \, dx = \ln|x| + C$

3. $\int e^x \, dx = e^x + C$

4. $\int \sin x \, dx = -\cos x + C$

5. $\int \cos x \, dx = \sin x + C$

6. $\int \tan x \, dx = \ln|\sec x| + C$

7. $\int \sec x \, dx = \ln|\sec x + \tan x| + C$

8. $\int \sec^2 x \, dx = \tan x + C$

9. $\int \sec x \tan x \, dx = \sec x + C$

10. $\int \dfrac{dx}{\sqrt{1 - x^2}} = \sin^{-1}x + C$

11. $\int \dfrac{dx}{1 + x^2} = \tan^{-1}x + C$

12. $\int \dfrac{dx}{x\sqrt{x^2 - 1}} = \sec^{-1}x + C$

The next list contains formulas that are *useful*. For short-term use (a test, for example), you may wish to learn them. For long-term or occasional use, you can look them up (inside the covers of this book) or derive them.

13. $\int \cot x \, dx = \ln|\sin x| + C$

14. $\int \csc x \, dx = \ln|\csc x - \cot x| + C$

15. $\int \csc^2 x \, dx = -\cot x + C$

16. $\int \csc x \cot x \, dx = -\csc x + C$

17. $\int \dfrac{dx}{\sqrt{a^2 - x^2}} = \sin^{-1}\left(\dfrac{x}{a}\right) + C, \quad a > 0$

18. $\int \dfrac{dx}{a^2 + x^2} = \dfrac{1}{a}\tan^{-1}\left(\dfrac{x}{a}\right) + C, \quad a > 0$

19. $\int \dfrac{dx}{x\sqrt{x^2 - a^2}} = \dfrac{1}{a}\sec^{-1}\left(\dfrac{x}{a}\right) + C, \quad a > 0$

20. $\int a^x \, dx = \dfrac{a^x}{\ln a} + C, \quad a > 0, \quad a \neq 1$

21. $\int \sinh x \, dx = \cosh x + C$

22. $\int \cosh x \, dx = \sinh x + C$

23. $\int \text{sech}^2 x \, dx = \tanh x + C$

24. $\int \text{csch}^2 x \, dx = -\coth x + C$

25. $\int \text{sech } x \tanh x \, dx = -\text{sech } x + C$

26. $\int \text{csch } x \coth x \, dx = -\text{csch } x + C$

When evaluating an integral, it is best first to compare it with the above lists. If an expression is identical to one of these, its integral is known. However, quite often it is necessary to employ some special techniques in order to express the integrand in a form that is found in the above lists. For example, one *technique of integration*— the substitution method—has been studied in Chapter 5. We review this technique in the example below.

EXAMPLE 1   Evaluate each integral: (a) $\displaystyle\int \frac{e^x \, dx}{e^x + 4}$ (b) $\displaystyle\int \frac{5 \, dx}{\sqrt{1 - 4x^2}}$

Solution   (a)   We use the substitution $u = e^x + 4$.   Then $du = e^x \, dx$,   and

$$\int \frac{e^x}{e^x + 4} \, dx = \int \frac{1}{u} \, du = \ln|u| + C = \ln(e^x + 4) + C$$

$$\uparrow$$
$$\text{Formula 2}$$

(b)   We use the substitution $u = 2x$.   Then $du = 2 \, dx$,   and

$$\int \frac{5 \, dx}{\sqrt{1 - 4x^2}} = \left(\frac{5}{2}\right) \int \frac{2 \, dx}{\sqrt{1 - (2x)^2}} = \left(\frac{5}{2}\right) \int \frac{du}{\sqrt{1 - u^2}}$$

$$= \frac{5}{2} \sin^{-1} u + C = \frac{5}{2} \sin^{-1}(2x) + C$$

$$\uparrow$$
$$\text{Formula 10}$$

Alternatively, we could have used formula 17 from the list of useful integral formulas:

$$\int \frac{5 \, dx}{\sqrt{1 - 4x^2}} = 5 \int \frac{dx}{2\sqrt{\frac{1}{4} - x^2}} = \frac{5}{2} \sin^{-1}\left(\frac{x}{\frac{1}{2}}\right) + C = \frac{5}{2} \sin^{-1}(2x) + C$$

$$\uparrow$$
$$a = \tfrac{1}{2}$$

In the remaining sections of this chapter we detail procedures for identifying the most appropriate substitution to be made to simplify the integral. In addition, we study the techniques of integration referred to as *integration by parts* and *partial fractions*.

Many of the integration formulas that appear inside the covers of this book are derived in the coming sections. A more comprehensive list of integration formulas is found in William H. Beyer, ed., *Standard Mathematical Tables* (Cleveland, Ohio: CRC Press, Inc.).

EXERCISE 1

In Problems 1–36 evaluate each integral by using the formulas listed in this section.

1. $\int \sqrt{3x + 1}\, dx$

2. $\int \sqrt{5x - 7}\, dx$

3. $\int x(3x^2 - 5)^5\, dx$

4. $\int x^2 \sqrt{5x^3 - 2}\, dx$

5. $\int \dfrac{\sin x}{1 + \cos x}\, dx$

6. $\int \dfrac{\sin x}{1 - \cos x}\, dx$

7. $\int \dfrac{x\, dx}{\sqrt{1 - 9x^2}}$

8. $\int \dfrac{x\, dx}{\sqrt{1 - x^2}}$

9. $\int \dfrac{1}{\sqrt{4 - x^2}}\, dx$

10. $\int \dfrac{1}{x\sqrt{9x^2 - 1}}\, dx$

11. $\int \dfrac{e^x}{1 + 2e^x}\, dx$

12. $\int \dfrac{e^x + 1}{e^x}\, dx$

13. $\int \dfrac{e^{\sqrt{x}}}{\sqrt{x}}\, dx$

14. $\int \dfrac{e^{\sqrt{3x+1}}}{\sqrt{3x + 1}}\, dx$

15. $\int x e^{x^2}\, dx$

16. $\int x e^{-x^2}\, dx$

17. $\int \cos x\, e^{\sin x}\, dx$

18. $\int \sin x\, e^{\cos x}\, dx$

19. $\int \sin x \cos x\, dx$

20. $\int \dfrac{\sec^2 u}{1 + \tan u}\, du$

21. $\int \dfrac{dx}{x^2 + 16}$

22. $\int \dfrac{dx}{x^2 + 4}$

23. $\int \dfrac{1 + \cos(2x)}{\sin^2(2x)}\, dx$

24. $\int \dfrac{\sin^2(2x)}{1 + \cos(2x)}\, dx$

25. $\int x \csc x^2 \cot x^2\, dx$
   [*Hint:* Let $u = x^2$.]

26. $\int \dfrac{\csc v}{\tan v}\, dv$
   [*Hint:* $\cot v = 1/(\tan v)$]

27. $\int \dfrac{\tan t}{\cos t}\, dt$
   [*Hint:* $\sec t = 1/(\cos t)$]

28. $\int \dfrac{dx}{x\sqrt{x^2 - 4}}$

29. $\int \dfrac{dx}{x\sqrt{x^2 - 9}}$

30. $\int \dfrac{dx}{(x + 2)^2 + 1}$

31. $\int \dfrac{\sin^2(3x)}{1 + \cos(3x)}\, dx$

32. $\int \dfrac{\cos^2(5x)}{1 + \sin(5x)}\, dx$

33. $\int 5^x\, dx$

34. $\int 9^{5x}\, dx$

35. $\int \dfrac{\sinh x\, dx}{1 - \cosh x}$

36. $\int \dfrac{\operatorname{csch}^2 x\, dx}{1 - \coth x}$

2. Integration by Parts

Integration by parts is based on the product rule for differentiation and is an effective and versatile technique for integration.

Recall that if u and v are differentiable functions of x, then

$$\frac{d}{dx}(uv) = v\frac{du}{dx} + u\frac{dv}{dx}$$

Integrating both sides gives

$$uv = \int v \frac{du}{dx} \, dx + \int u \frac{dv}{dx} \, dx$$

$$\int u \frac{dv}{dx} \, dx = uv - \int v \frac{du}{dx} \, dx$$

We may write this in the abbreviated form

(9.1)
$$\int u \, dv = uv - \int v \, du$$

This formula is usually referred to as the *by-parts formula*.

The corresponding formula for definite integrals is

$$\int_a^b u(x) \, dv(x) = u(x)v(x) \Big|_a^b - \int_a^b v(x) \, du(x)$$

where $dv(x) = v'(x) \, dx$ and $du(x) = u'(x) \, dx$.

To apply (9.1), we separate the integrand into two parts. We call one u and the other dv. We differentiate u to obtain du and integrate dv to obtain v. If we can then integrate $\int v \, du$, the problem is solved. The goal of this procedure, then, is to choose u and dv so that the term $\int v \, du$ is easier to solve than the original problem. As the examples will illustrate, this usually happens when u is simplified by differentiation.

EXAMPLE 1   Evaluate: $\int xe^x \, dx$

Solution   To use the by-parts formula, we choose u and dv so that

$$\int u \, dv = \int xe^x \, dx$$

and $\int v \, du$ is easier to evaluate than $\int u \, dv$. In this example we decide to choose

$$u = x \quad \text{and} \quad dv = e^x \, dx$$

As a result of this choice,

$$du = dx \quad \text{and} \quad v = \int dv = \int e^x \, dx = e^x$$

Note that we only require a particular antiderivative of dv at this stage; we will add the constant of integration later. Substitution in (9.1) results in

$$\int \overset{u}{x} \, \overset{dv}{e^x \, dx} = \overset{u}{x} \, \overset{v}{e^x} - \int \overset{v}{e^x} \, \overset{du}{dx} = xe^x - e^x + C = e^x(x - 1) + C \quad \blacksquare$$

Let's look once more at Example 1. Suppose we had chosen u and dv differently as

$$u = e^x \quad \text{and} \quad dv = x \, dx$$

This choice would have resulted in

$$du = e^x \, dx \qquad \text{and} \qquad v = \frac{x^2}{2}$$

and equation (9.1) would have yielded

$$\int xe^x \, dx = \frac{x^2}{2} e^x - \int \frac{x^2 e^x}{2} \, dx$$

As you can see, instead of obtaining an integral that is easier to evaluate, we obtain one that is more complicated than the original. This means that an unwise choice of u and dv has been made.

Unfortunately, there are no general directions for choosing u and dv except that:

1. dx is always a part of dv.
2. It must be possible to integrate dv.
3. u and dv are chosen so that $\int v \, du$ is easier to evaluate than the original integral $\int u \, dv$; this often happens when u is simplified by differentiation.

In making an initial choice for u and dv, a certain amount of trial and error is used. If a selection appears to hold little promise, abandon it and try some other choice. If no choices work, it may be that some other technique of integration should be tried.

Let's look at some more examples.

EXAMPLE 2   Derive the formula* $\int \ln x \, dx = x \ln x - x + C$.

Solution   We choose

$$u = \ln x \qquad \text{and} \qquad dv = dx$$

Then

$$du = \frac{1}{x} \, dx \qquad \text{and} \qquad v = \int dx = x$$

By substituting in (9.1),

$$\int \ln x \, dx = x \ln x - \int (x) \left(\frac{1}{x} \, dx \right) = x \ln x - \int dx = x \ln x - x + C \qquad \blacksquare$$

EXAMPLE 3   Evaluate: $\int x \sin x \, dx$

Solution   We choose

$$u = x \qquad \text{and} \qquad dv = \sin x \, dx$$

---

* This is formula 118 in the Table of Integrals (inside the covers of this book).

Then

$$du = dx \qquad \text{and} \qquad v = \int \sin x \, dx = -\cos x$$

By substituting in (9.1),

$$\int x \sin x \, dx = -x \cos x + \int \cos x \, dx = -x \cos x + \sin x + C \qquad \blacksquare$$

EXAMPLE 4 Show that* $\int \tan^{-1}x \, dx = x \tan^{-1}x - \frac{1}{2}\ln(1 + x^2) + C.$

Solution We choose

$$u = \tan^{-1}x \qquad \text{and} \qquad dv = dx$$

Then

$$du = \frac{1}{1 + x^2}\, dx \qquad \text{and} \qquad v = \int dx = x$$

By substituting in (9.1),

$$\int \tan^{-1}x \, dx = x \tan^{-1}x - \int \frac{x}{1 + x^2}\, dx$$

To evaluate the integral $\int [x/(1 + x^2)] \, dx,$ we use the substitution $t = 1 + x^2.$ Then $dt = 2x \, dx,$ and

$$\int \frac{x}{1 + x^2} \, dx = \frac{1}{2} \int \frac{dt}{t} = \frac{1}{2}\ln|t| = \frac{1}{2}\ln(1 + x^2)$$

As a result,

$$\int \tan^{-1}x \, dx = x \tan^{-1}x - \frac{1}{2}\ln(1 + x^2) + C \qquad \blacksquare$$

EXAMPLE 5 Evaluate: $\int_1^2 x \ln x \, dx$

Solution First, we calculate the indefinite integral $\int x \ln x \, dx$ by using the by-parts formula (9.1). We choose u and dv as

$$u = \ln x \qquad \text{and} \qquad dv = x \, dx$$

Then

$$du = \frac{1}{x}\, dx \qquad \text{and} \qquad v = \int x \, dx = \frac{x^2}{2}$$

Thus,

$$\int x \ln x \, dx = \frac{x^2}{2}\ln x - \int \frac{x^2}{2}\left(\frac{1}{x}\, dx\right) = \frac{x^2}{2}\ln x - \frac{x^2}{4} + C = \frac{x^2}{2}\left(\ln x - \frac{1}{2}\right) + C$$

---

* This is formula 108 in the Table of Integrals (inside the covers of this book).

It follows that

$$\int_1^2 x \ln x \, dx = \frac{x^2}{2}\left(\ln x - \frac{1}{2}\right)\Big|_1^2 = 2\left(\ln 2 - \frac{1}{2}\right) - \frac{1}{2}\left(-\frac{1}{2}\right) = 2\ln 2 - \frac{3}{4} \approx 0.6363$$

■

Sometimes it may be necessary to integrate by parts more than once to solve a particular problem. The next example is a case in point.

EXAMPLE 6   Evaluate: $\int x^2 e^x \, dx$

Solution   We choose

$$u = x^2 \qquad \text{and} \qquad dv = e^x \, dx$$

Then

$$du = 2x \, dx \qquad \text{and} \qquad v = \int e^x \, dx = e^x$$

Thus,

$$\int x^2 e^x \, dx = x^2 e^x - 2\int xe^x \, dx$$

We must still evaluate $\int xe^x \, dx$.   In Example 1 we found that

$$\int xe^x \, dx = xe^x - e^x + C$$

As a result,

$$\int x^2 e^x \, dx = x^2 e^x - 2xe^x + 2e^x - 2C = e^x(x^2 - 2x + 2) + C_1 \qquad ■$$

The next example illustrates how integration by parts can be used to establish a general integration formula.

EXAMPLE 7   Show that*

(9.2)
$$\int e^{ax} \cos bx \, dx = \frac{e^{ax}(b \sin bx + a \cos bx)}{a^2 + b^2} + C \qquad b \neq 0$$

Solution   Let

$$u = e^{ax} \qquad \text{and} \qquad dv = \cos bx \, dx$$

Then

$$du = ae^{ax} \, dx \qquad \text{and} \qquad v = \frac{1}{b} \sin bx$$

---

* This is formula 126 in the Table of Integrals (inside the covers of this book).

Thus,

(9.3)
$$\int e^{ax} \cos bx \, dx = \frac{e^{ax} \sin bx}{b} - \frac{a}{b} \int e^{ax} \sin bx \, dx$$

Now, apply the by-parts formula (9.1) to $\int e^{ax} \sin bx \, dx$. We choose

$$u = e^{ax} \qquad \text{and} \qquad dv = \sin bx \, dx$$

Then

$$du = ae^{ax} \, dx \qquad \text{and} \qquad v = -\frac{1}{b} \cos bx$$

Thus,

(9.4)
$$\int e^{ax} \sin bx \, dx = -\left(\frac{1}{b}\right) e^{ax} \cos bx + \frac{a}{b} \int e^{ax} \cos bx \, dx$$

From (9.3) and (9.4), we get

$$\int e^{ax} \cos bx \, dx = \frac{1}{b} e^{ax} \sin bx - \frac{a}{b} \left[-\frac{1}{b} e^{ax} \cos bx + \frac{a}{b} \int e^{ax} \cos bx \, dx \right]$$

$$= \frac{1}{b} e^{ax} \sin bx + \frac{a}{b^2} e^{ax} \cos bx - \frac{a^2}{b^2} \int e^{ax} \cos bx \, dx$$

Finally, collecting terms, we get

$$\int e^{ax} \cos bx \, dx + \frac{a^2}{b^2} \int e^{ax} \cos bx \, dx = \frac{1}{b} e^{ax} \sin bx + \frac{a}{b^2} e^{ax} \cos bx$$

$$\left(1 + \frac{a^2}{b^2} \right) \int e^{ax} \cos bx \, dx = \frac{1}{b^2} e^{ax}(b \sin bx + a \cos bx)$$

$$\int e^{ax} \cos bx \, dx = \frac{e^{ax}(b \sin bx + a \cos bx)}{a^2 + b^2} + C \qquad ■$$

As an illustration of (9.2) for $a = 4$ and $b = 5$, we have

$$\int e^{4x} \cos 5x \, dx = \frac{e^{4x}(5 \sin 5x + 4 \cos 5x)}{41} + C$$

Let's derive another general formula.

EXAMPLE 8   Derive the formula*

(9.5)
$$\int \sec^n x \, dx = \frac{\sec^{n-2}x \tan x}{n - 1} + \frac{n - 2}{n - 1} \int \sec^{n-2}x \, dx \qquad n > 1$$

---

* This is formula 94 in the Table of Integrals (inside the covers of this book).

Solution   We write $\sec^n x = \sec^{n-2} x \sec^2 x$,   and choose

$$u = \sec^{n-2} x \qquad\qquad\qquad \text{and} \qquad dv = \sec^2 x \, dx$$

Then,

$$du = (n - 2) \sec^{n-3} x \sec x \tan x \, dx$$
$$= (n - 2) \sec^{n-2} x \tan x \, dx \qquad \text{and} \qquad v = \tan x$$

Hence,

$$\int \sec^n x \, dx = \sec^{n-2} x \tan x - (n - 2) \int \sec^{n-2} x \tan^2 x \, dx$$

We replace $\tan^2 x$ by $\sec^2 x - 1$ to get

$$\int \sec^n x \, dx = \sec^{n-2} x \tan x - (n - 2) \int \sec^{n-2} x (\sec^2 x - 1) \, dx$$

$$= \sec^{n-2} x \tan x - (n - 2) \int \sec^n x \, dx + (n - 2) \int \sec^{n-2} x \, dx$$

By transposing the second term on the right to the left, we get

$$(n - 1) \int \sec^n x \, dx = \sec^{n-2} x \tan x + (n - 2) \int \sec^{n-2} x \, dx$$

Dividing both sides by $n - 1$ produces the result:

$$\int \sec^n x \, dx = \frac{\sec^{n-2} x \tan x}{n - 1} + \frac{n - 2}{n - 1} \int \sec^{n-2} x \, dx \qquad \blacksquare$$

Formula (9.5) is referred to as a *reduction formula* because repeated applications eventually lead to an elementary integral. Consider, for example, the reduction formula for $\int \sec^n x \, dx$. When n is even, repeated applications lead to $\int \sec^2 x \, dx = \tan x + C$. When n is odd, repeated applications lead to $\int \sec x \, dx = \ln|\sec x + \tan x| + C$.

As an illustration of (9.5), for $n = 3$,   we have

(9.6) $$\int \sec^3 x \, dx = \frac{\sec x \tan x}{2} + \frac{1}{2} \int \sec x \, dx = \frac{\sec x \tan x}{2} + \frac{1}{2} \ln|\sec x + \tan x| + C$$

Below, we outline a few of the types of integrals that usually are evaluated by using integration by parts (n is always a positive integer):

I. $\int x^n e^{ax} \, dx \qquad \int x^n \cos ax \, dx, \qquad \int x^n \sin ax \, dx$

Here, let $u = x^n$,   and $dv = $ what remains.

II. $\int x^n \sin^{-1} x \, dx, \qquad \int x^n \cos^{-1} x \, dx \qquad \int x^n \tan^{-1} x \, dx$

Here, set $u = \sin^{-1} x$,   or $u = \cos^{-1} x$,   or $u = \tan^{-1} x$,   and $dv = x^n \, dx$.

III. $\int x^m (\ln x)^n \, dx, \qquad m \neq -1$

Here, set $u = (\ln x)^n$, and $dv = x^m \, dx$.

EXERCISE 2

In Problems 1–32 use integration by parts to evaluate each integral.

1. $\int x e^{2x} \, dx$ **2.** $\int x e^{-3x} \, dx$ **3.** $\int x \cos x \, dx$ **4.** $\int x \sin(3x) \, dx$

5. $\int \sqrt{x} \ln x \, dx$ **6.** $\int x^{-2} \ln x \, dx$ **7.** $\int \cot^{-1} x \, dx$ **8.** $\int \sin^{-1} x \, dx$

9. $\int (\ln x)^2 \, dx$ **10.** $\int x(\ln x)^2 \, dx$ **11.** $\int e^x \sin x \, dx$ **12.** $\int_0^\pi e^x \cos x \, dx$

13. $\int_0^1 x^2 e^{-x} \, dx$ **14.** $\int x^2 e^x \, dx$ **15.** $\int x^2 \sin x \, dx$ **16.** $\int x^2 \cos x \, dx$

17. $\int x \cos^2 x \, dx$ **18.** $\int x \sin^2 x \, dx$ **19.** $\int_0^2 x^2 e^{-3x} \, dx$ **20.** $\int_0^{\pi/4} x \tan^2 x \, dx$

21. $\int x \sinh x \, dx$ **22.** $\int \sinh^{-1} x \, dx$ **23.** $\int x^n \ln x \, dx,$ **24.** $\int x^3 e^{x^2} \, dx$
                      $n > 1$ is an integer   [*Hint:* Let $u = x^2$,
                                      $dv = xe^{x^2} dx.$]

25. $\int (\ln x)^3 \, dx$ **26.** $\int (\ln x)^4 \, dx$ **27.** $\int x e^x \cos x \, dx$ **28.** $\int x e^x \sin x \, dx$

29. $\int x^2 (\ln x)^2 \, dx$ **30.** $\int x^3 (\ln x)^2 \, dx$ **31.** $\int x^2 \tan^{-1} x \, dx$ **32.** $\int x \tan^{-1} x \, dx$

33. Derive the formula $\int \ln(x + \sqrt{x^2 + a^2}) \, dx = x \ln(x + \sqrt{x^2 + a^2}) - \sqrt{x^2 + a^2} + C.$

34. Derive the formula $\int e^{ax} \sin bx \, dx = \dfrac{e^{ax}(a \sin bx - b \cos bx)}{a^2 + b^2} + C.$

In Problems 35–40 establish each reduction formula where $n > 1$ is an integer.

35. $\int x^n \sin^{-1} x \, dx = \dfrac{x^{n+1}}{n+1} \sin^{-1} x - \dfrac{1}{n+1} \int \dfrac{x^{n+1}}{\sqrt{1 - x^2}} \, dx$

36. $\int x^n \tan^{-1} x \, dx = \dfrac{x^{n+1}}{n+1} \tan^{-1} x - \dfrac{1}{n+1} \int \dfrac{x^{n+1}}{1 + x^2} \, dx$

37. $\int x^n (ax + b)^{1/2} \, dx = \dfrac{2x^n (ax + b)^{3/2}}{(2n + 3)a} - \dfrac{2bn}{(2n + 3)a} \int x^{n-1} (ax + b)^{1/2} \, dx$

38. $\int \dfrac{dx}{(x^2 + 1)^{n+1}} = \left(1 - \dfrac{1}{2n}\right) \int \dfrac{dx}{(x^2 + 1)^n} + \dfrac{x}{2n(x^2 + 1)^n}$

39. $\int \sin^n x \, dx = -\dfrac{\sin^{n-1} x \cos x}{n} + \dfrac{n-1}{n} \int \sin^{n-2} x \, dx$

40. $\int \sin^n x \cos^m x \, dx = -\dfrac{\sin^{n-1} x \cos^{m+1} x}{n + m} + \dfrac{n-1}{n + m} \int \sin^{n-2} x \cos^m x \, dx, \quad m \neq -n$

41. Find the area enclosed by $y = \ln x$, the x-axis, and the line $x = e$.

42. Find the area enclosed by $y = x \sin x$, $y = x$, $x = 0$, and $x = \pi/2$.

43. Find the volume of the solid generated by revolving the area under the graph of $y = \cos x$ from $x = 0$ to $x = \pi/2$ about the y-axis.

44. Find the volume of the solid generated by revolving the area under the graph of $y = \sin x$ from $x = 0$ to $x = \pi/2$ about the y-axis.

45. Find the volume of the solid generated by revolving the area under the graph of $y = x\sqrt{\sin x}$ from $x = 0$ to $x = \pi/2$ about the x-axis.

46. Show that the following formula is true:

(9.7) $$\int_0^{\pi/2} \sin^n x \, dx = \int_0^{\pi/2} \cos^n x \, dx = \begin{cases} \dfrac{(n-1)(n-3)\cdots(4)(2)}{n(n-2)\cdots(5)(3)(1)} & \text{if } n > 1 \text{ is odd} \\[4mm] \dfrac{(n-1)(n-3)\cdots(5)(3)(1)}{n(n-2)\cdots(4)(2)}\left(\dfrac{\pi}{2}\right) & \text{if } n > 1 \text{ is even} \end{cases}$$

This is called *Wallis' formula*. [*Hint:* Use the result of Problem 39.]

47. Use Wallis' formula (9.7) to find:

(a) $\displaystyle\int_0^{\pi/2} \sin^6 x \, dx$ (b) $\displaystyle\int_0^{\pi/2} \sin^5 x \, dx$ (c) $\displaystyle\int_0^{\pi/2} \cos^8 x \, dx$ (d) $\displaystyle\int_0^{\pi/2} \cos^6 x \, dx$

3. Trigonometric Integrals

In this section we consider the problem of evaluating certain trigonometric integrals. In studying the techniques presented here, you should follow the procedures outlined in the examples rather than try to memorize the results.

We confine our discussion to four commonly encountered types of trigonometric integrals:

I. $\displaystyle\int \sin^n x \, dx, \qquad \int \cos^n x \, dx$

II. $\displaystyle\int \sin^m x \cos^n x \, dx$

III. $\displaystyle\int \tan^m x \sec^n x \, dx, \qquad \int \cot^m x \csc^n x \, dx$

IV. $\displaystyle\int \sin mx \sin nx \, dx, \qquad \int \sin mx \cos nx \, dx, \qquad \int \cos mx \cos nx \, dx$

TYPE I

$$\int \sin^n x \, dx \qquad \int \cos^n x \, dx \qquad n \text{ is a positive integer}$$

Although we could use integration by parts to obtain a reduction formula for each of these integrals (see Problem 39 in Exercise 2), they can also be evaluated in another—usually easier—way.

We consider two cases: (a) n is an odd positive integer, and   (b) n is an even positive integer.

(a)   When n is an odd positive integer, we begin by writing

$$\int \sin^n x \, dx = \int \sin^{n-1} x \, \sin x \, dx$$

Since n is odd, $(n-1)$ is even. Using the fact that $\sin^2 x = 1 - \cos^2 x$,   we obtain an integral that can be evaluated by using the substitution $u = \cos x$.   The example below illustrates this technique.

EXAMPLE 1   Evaluate: $\int \sin^5 x \, dx$

Solution
$$\int \sin^5 x \, dx = \int \sin^4 x \, \sin x \, dx = \int (1 - \cos^2 x)^2 \, \sin x \, dx$$

$$= \int (1 - 2\cos^2 x + \cos^4 x) \, \sin x \, dx$$

Use the substitution $u = \cos x$.   Then $du = -\sin x \, dx$,   so that

$$\int \sin^5 x \, dx = -\int (1 - 2u^2 + u^4) \, du = -u + \tfrac{2}{3}u^3 - \tfrac{1}{5}u^5 + C$$

$$= -\cos x + \tfrac{2}{3}\cos^3 x - \tfrac{1}{5}\cos^5 x + C \quad \blacksquare$$

A similar technique may be used to evaluate $\int \cos^n x \, dx$ when n is an odd positive integer. In this case, we write

$$\int \cos^n x \, dx = \int \cos^{n-1} x \, \cos x \, dx$$

and use the fact that $\cos^2 x = 1 - \sin^2 x$,   along with the substitution $u = \sin x$.

(b)   To evaluate $\int \sin^n x \, dx$ when n is an even positive integer, we use the identities (half-angle formulas)

$$\sin^2 x = \frac{1 - \cos 2x}{2} \qquad \text{and} \qquad \cos^2 x = \frac{1 + \cos 2x}{2}$$

to obtain an integrand that may be simpler.

EXAMPLE 2   Evaluate: $\int \sin^2 x \, dx$

Solution
$$\int \sin^2 x \, dx = \tfrac{1}{2}\int (1 - \cos 2x) \, dx = \tfrac{1}{2}x - \tfrac{1}{2}\int \cos 2x \, dx$$

$$= \tfrac{1}{2}x - \tfrac{1}{4}\sin 2x + C \quad \blacksquare$$

EXAMPLE 3   Evaluate: $\int \cos^4 x \, dx$

Solution

$$\int \cos^4 x \, dx = \int (\cos^2 x)^2 \, dx = \int \tfrac{1}{4}(1 + \cos 2x)^2 \, dx$$

$$= \tfrac{1}{4} \int (1 + 2 \cos 2x + \cos^2 2x) \, dx$$

You should be able to supply the necessary details to show that

$$\int \cos^4 x \, dx = \tfrac{3}{8}x + \tfrac{1}{4} \sin 2x + \tfrac{1}{32} \sin 4x + C \qquad \blacksquare$$

TYPE II

$$\int \sin^m x \, \cos^n x \, dx$$

Again, we consider two cases: (a) at least one of the exponents m or n is an odd positive integer; and (b) both are even positive integers.

     Integrals of Type II may be evaluated by using variations of previous techniques.

EXAMPLE 4   Evaluate: $\int \cos^{1/5} x \, \sin^5 x \, dx$

Solution   Here, $\sin x$ is raised to an odd power and $\cos x$ to the number $\tfrac{1}{5}$. We write

$$\int \cos^{1/5} x \, \sin^5 x \, dx = \int \cos^{1/5} x \, \sin^4 x \, \sin x \, dx$$

But

$$\sin^4 x = (\sin^2 x)^2 = (1 - \cos^2 x)^2$$

This leads us to make the substitution $u = \cos x$. Then $du = -\sin x \, dx$, and

$$\int \cos^{1/5} x \, \sin^5 x \, dx = \int \cos^{1/5} x (1 - \cos^2 x)^2 \sin x \, dx$$

$$= \int u^{1/5}(1 - u^2)^2(-du) = -\int (u^{1/5} - 2u^{11/5} + u^{21/5}) \, du$$

$$= -\tfrac{5}{6}u^{6/5} + \tfrac{5}{8}u^{16/5} - \tfrac{5}{26}u^{26/5} + C$$

$$= -\tfrac{5}{6} \cos^{6/5} x + \tfrac{5}{8} \cos^{16/5} x - \tfrac{5}{26} \cos^{26/5} x + C \qquad \blacksquare$$

     Note that if either m or n is an odd positive integer, the other exponent may be *any real number*.

     If m and n are both even positive integers, then we may use the identity $\sin^2 x + \cos^2 x = 1$ to obtain a sum of integrals, each one of which involves only powers of either $\sin x$ or $\cos x$. For example,

$$\int \sin^2 x \, \cos^4 x \, dx = \int (1 - \cos^2 x) \cos^4 x \, dx = \int \cos^4 x \, dx - \int \cos^6 x \, dx$$

The two integrals on the right are now of Type I.

Table 1 summarizes the methods for evaluating $\int \sin^m x \cos^n x \, dx$.

Table 1

Case	Method	Useful Identities
n odd	Substitute $u = \sin x$	$\cos^2 x = 1 - \sin^2 x$
m odd	Substitute $u = \cos x$	$\sin^2 x = 1 - \cos^2 x$
m and n even	Reduce to smaller powers of m or n	$\sin x \cos x = \frac{1}{2} \sin 2x$
		$\sin^2 x = \dfrac{1 - \cos 2x}{2}$
		$\cos^2 x = \dfrac{1 + \cos 2x}{2}$

TYPE III

$$\int \tan^m x \sec^n x \, dx \qquad \int \cot^m x \csc^n x \, dx$$

This time, we consider three cases: (a) m is an odd positive integer;   (b) n is an even positive integer; and   (c) m is an even positive integer and n is an odd positive integer.

(a)   If m is an odd positive integer, we write $\int \tan^m x \sec^n x \, dx$ as

$$\int \tan^{m-1} x \sec^{n-1} x \sec x \tan x \, dx$$

Since $(m - 1)$ is even, we can use the identity $\tan^2 x = \sec^2 x - 1$ to express $\tan^{m-1} x$ in terms of $\sec x$. The substitution $u = \sec x$ will then lead to a simpler integral since $du = \sec x \tan x \, dx$.

EXAMPLE 5   Evaluate: $\int \tan^3 x \sec^3 x \, dx$

Solution   Here $m = 3$ is odd and

$$\int \tan^3 x \sec^3 x \, dx = \int \tan^2 x \sec^2 x (\sec x \tan x \, dx) = \int (\sec^2 x - 1) \sec^2 x (\sec x \tan x \, dx)$$

Let $u = \sec x$, so that $du = \sec x \tan x \, dx$.   Then

$$\int \tan^3 x \sec^3 x \, dx = \int (u^2 - 1) u^2 \, du = \int (u^4 - u^2) \, du$$

$$= \frac{u^5}{5} - \frac{u^3}{3} + C = \frac{\sec^5 x}{5} - \frac{\sec^3 x}{3} + C \qquad \blacksquare$$

(b)   If n is an even positive integer, we write $\int \tan^m x \sec^n x \, dx$ as

$$\int \tan^m x \sec^{n-2} x \sec^2 x \, dx$$

We now express $\sec^{n-2} x$ in terms of tan x by using the identity $\sec^2 x = 1 + \tan^2 x$. The substitution $u = \tan x$ will lead to a simpler integral. (Do you see why?)

EXAMPLE 6   Evaluate: $\int \tan^2 x \sec^4 x \, dx$

Solution

$$\int \tan^2 x \sec^4 x \, dx = \int \tan^2 x \sec^2 x (\sec^2 x) \, dx$$

$$= \int \tan^2 x (1 + \tan^2 x) \sec^2 x \, dx$$

Let $u = \tan x$,   so that $du = \sec^2 x \, dx$.   Then

$$\int \tan^2 x \sec^4 x \, dx = \int u^2 (1 + u^2) \, du = \int (u^2 + u^4) \, du$$

$$= \frac{u^3}{3} + \frac{u^5}{5} + C = \frac{\tan^3 x}{3} + \frac{\tan^5 x}{5} + C \quad \blacksquare$$

(c)   When m is an even positive integer and n is an odd positive integer, we express the integrand of $\int \tan^m x \sec^n x \, dx$ in terms of sec x by using the identity $\tan^2 x = \sec^2 x - 1$ and then use the technique of integration by parts.

EXAMPLE 7   Evaluate: $\int \tan^2 x \sec x \, dx$

Solution   Here

$$\int \tan^2 x \sec x \, dx = \int (\sec^2 x - 1) \sec x \, dx = \int \sec^3 x \, dx - \int \sec x \, dx$$

By using result (9.6) for $\int \sec^3 x \, dx$ (obtained in Section 2), we have

$$\int \tan^2 x \sec x \, dx = \tfrac{1}{2} \sec x \tan x - \tfrac{1}{2} \ln|\sec x + \tan x| + C \quad \blacksquare$$

To evaluate integrals of the form $\int \cot^m x \csc^n x \, dx$,   we follow the same procedures, except that the identity $\csc^2 x = 1 + \cot^2 x$ is employed.

Table 2 summarizes the methods for evaluating $\int \tan^m x \sec^n x \, dx$ $(m \geq 0,$ $n \geq 0)$.

Table 2

Case	Method	Useful Identities
m odd	Substitute $u = \sec x$	$\tan^2 x = \sec^2 x - 1$
n even	Substitute $u = \tan x$	$\sec^2 x = 1 + \tan^2 x$
m even, n odd	Reduce to powers of sec x alone	$\tan^2 x = \sec^2 x - 1$

TYPE IV

$$\int \sin mx \sin nx \, dx \qquad \int \sin mx \cos nx \, dx \qquad \int \cos mx \cos nx \, dx$$

Trigonometric integrals of Type IV are handled with the aid of the following trigonometric identities (product-to-sum formulas):

$$2 \sin A \sin B = \cos(A - B) - \cos(A + B)$$
$$2 \sin A \cos B = \sin(A + B) + \sin(A - B)$$
$$2 \cos A \cos B = \cos(A - B) + \cos(A + B)$$

EXAMPLE 8   Evaluate: $\int \sin 3x \sin 2x \, dx$

Solution   We set $A = 3x$ and $B = 2x$.   Then

$$\int \sin 3x \sin 2x \, dx = \tfrac{1}{2} \int (\cos x - \cos 5x) \, dx$$
$$= \tfrac{1}{2} \sin x - \tfrac{1}{10} \sin 5x + C \qquad \blacksquare$$

EXERCISE 3

In Problems 1–38 evaluate each integral.

1. $\int \sin^2 x \cos x \, dx$

2. $\int \sin^3 x \cos x \, dx$

3. $\int \dfrac{\sin x \, dx}{\cos^2 x}$

4. $\int \dfrac{\cos x \, dx}{\sin^4 x}$

5. $\int \sin^3 x \, dx$

6. $\int \cos^3 x \, dx$

7. $\int \sin^2 3x \, dx$

8. $\int \sin^4 x \, dx$

9. $\int \cos^5 x \, dx$

10. $\int \cos^7 x \, dx$

11. $\int \sin^5 3x \, dx$

12. $\int \cos^3 3x \, dx$

13. $\int \sin^3 x \cos^2 x \, dx$

14. $\int \sin^4 x \cos^3 x \, dx$

15. $\int \sin^3 x \cos^5 x \, dx$

16. $\int \sin^3 x \cos^3 x \, dx$

17. $\int \sin^{1/3} x \cos x \, dx$

18. $\int \dfrac{\sin^2 x}{\cos^3 x} \, dx$

19. $\int \sin^{1/2} x \cos^3 x \, dx$

20. $\int \sin^2 x \cos^4 x \, dx$

21. $\int \sin^2 x \cos^2 x \, dx$

22. $\int \sin^2 3x \cos^2 3x \, dx$

23. $\int \tan^3 x \, dx$

24. $\int \tan^3 x \sec^2 x \, dx$

25. $\int \tan^2 x \sec^2 x \, dx$

26. $\int \tan^4 x \sec^2 x \, dx$

27. $\int \csc^3 x \cot^5 x \, dx$

28. $\int \cot 2x \csc^4 2x \, dx$

29. $\int \tan^2 x \sec^3 x \, dx$

30. $\int \tan^4 x \sec^4 x \, dx$

31. $\int \sin 3x \cos x \, dx$

32. $\int \sin x \cos 3x \, dx$

33. $\int \sin 2x \sin 4x \, dx$

34. $\int \cos x \cos 3x \, dx$

35. $\int \cos 2x \cos x \, dx$

36. $\int \sin 2x \sin x \, dx$

37. $\int \sin \dfrac{x}{2} \cos \dfrac{3x}{2} \, dx$

38. $\int \sin \dfrac{x}{2} \sin \dfrac{3x}{2} \, dx$

39. Evaluate: $\int \sec^n x \, dx$, n a positive even integer, by the use of the identity

$$\sec^n x = \sec^{n-2} x \sec^2 x = (\tan^2 x + 1)^{(n-2)/2} \sec^2 x$$

40. Evaluate: $\int \cot^5 x \csc^2 x \, dx$

41. Find the volume of the solid generated by revolving the area under the graph of $y = \sin x$ from $x = 0$ to $x = \pi$ about the x-axis.

42. Find the volume of the solid generated by revolving the area enclosed by $y = \cos x$, $y = \sin x$, $x = 0$, and $x = \pi/4$ about the x-axis.

43. The acceleration a of a particle at time t is given by $a = \cos^2 t \sin t$ meters per second per second. At $t = 0$ the particle is at the origin and its velocity is 5 meters per second. Find its position at any time t.

4. Integration by Trigonometric Substitution: Integrands Containing $\sqrt{a^2 - x^2}$, $\sqrt{a^2 + x^2}$, or $\sqrt{x^2 - a^2}$

When an integrand contains a square root of one of the forms $\sqrt{a^2 - x^2}$, $\sqrt{a^2 + x^2}$, or $\sqrt{x^2 - a^2}$, where $a > 0$, an appropriate trigonometric substitution will eliminate the radical and sometimes transform the integrand into a trigonometric integral like those studied in the previous section. The three cases and the suggested substitutions are listed below.

	Integrand	*Substitution*	*Derived from the Identity*
Case 1:	$\sqrt{a^2 - x^2}$	$x = a \sin \theta$	$1 - \sin^2\theta = \cos^2\theta$
Case 2:	$\sqrt{a^2 + x^2}$	$x = a \tan \theta$	$1 + \tan^2\theta = \sec^2\theta$
Case 3:	$\sqrt{x^2 - a^2}$	$x = a \sec \theta$	$\sec^2\theta - 1 = \tan^2\theta$

The substitutions can be memorized, but you will probably find it easier to draw a right triangle and derive them as needed. Each substitution derives from the Pythagorean theorem. If we place the sides a and x on the triangle appropriately, we can make the third side of the triangle represent any of the cases above. For example, the substitution $x = a \sin \theta$ is used in $\sqrt{a^2 - x^2}$ since it eliminates the radical (see Fig. 1). That is,

$$\sqrt{a^2 - x^2} = \sqrt{a^2 - a^2 \sin^2\theta}$$
$$= a\sqrt{1 - \sin^2\theta}$$
$$= a\sqrt{\cos^2\theta}$$
$$= a|\cos \theta|$$

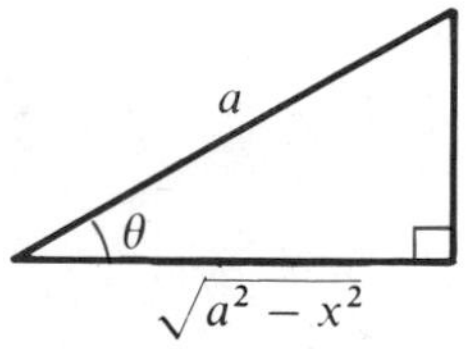

Figure 1

With the restriction $-\pi/2 \le \theta \le \pi/2$, $\cos \theta \ge 0$ and, thus, $\sqrt{a^2 - x^2} = a \cos \theta$.
If $\sqrt{a^2 - x^2}$ occurs in the denominator of the integrand, we let $x = a \sin \theta$, with the restriction $-\pi/2 < \theta < \pi/2$. Let's look at an example.

EXAMPLE 1   Evaluate: $\displaystyle\int \frac{dx}{x^2\sqrt{4 - x^2}}$

Solution   Let $x = 2 \sin \theta,$ $-\pi/2 < \theta < \pi/2.$ Then $dx = 2 \cos \theta \, d\theta,$ and

$$\sqrt{4 - x^2} = \sqrt{4 - 4 \sin^2 \theta} = 2\sqrt{\cos^2 \theta} = 2 \cos \theta$$

Therefore,

$$\int \frac{dx}{x^2 \sqrt{4 - x^2}} = \int \frac{2 \cos \theta \, d\theta}{(4 \sin^2 \theta)(2 \cos \theta)} = \frac{1}{4} \int \csc^2 \theta \, d\theta = -\frac{1}{4} \cot \theta + C$$

The integral has now been found in terms of θ. In order to express $\cot \theta$ in terms of x, we draw a right triangle, as shown in Figure 2, and use the fact that $\sin \theta = x/2$, $-\pi/2 < \theta < \pi/2.$ Using the Pythagorean theorem, we conclude that $\cot \theta = \sqrt{4 - x^2}/x.$ Thus,

$$\int \frac{dx}{x^2 \sqrt{4 - x^2}} = -\frac{1}{4} \cot \theta + C = -\frac{1}{4} \frac{\sqrt{4 - x^2}}{x} + C$$

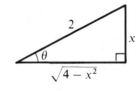

Figure 2   ■

For integrands involving $\sqrt{a^2 + x^2}$, $a > 0$, we use the substitution $x = a \tan \theta$, $-\pi/2 < \theta < \pi/2$ (see Fig. 3). Since $\sec \theta > 0$ when $-\pi/2 < \theta < \pi/2$, we have

$$\sqrt{a^2 + x^2} = \sqrt{a^2 + a^2 \tan^2 \theta}$$
$$= a\sqrt{1 + \tan^2 \theta}$$
$$= a\sqrt{\sec^2 \theta}$$
$$= a \sec \theta$$

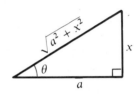

Figure 3

EXAMPLE 2   Evaluate: $\displaystyle \int \frac{dx}{(x^2 + 9)^{3/2}}$

Solution   The given integrand involves a power of $\sqrt{x^2 + 9}$. Hence, we use the substitution $x = 3 \tan \theta$, $-\pi/2 < \theta < \pi/2$. Then $dx = 3 \sec^2 \theta \, d\theta$, and

$$\int \frac{dx}{(x^2 + 9)^{3/2}} = \int \frac{3 \sec^2 \theta \, d\theta}{(9 \tan^2 \theta + 9)^{3/2}} = \int \frac{3 \sec^2 \theta \, d\theta}{27(1 + \tan^2 \theta)^{3/2}} = \frac{1}{9} \int \frac{\sec^2 \theta \, d\theta}{\sec^3 \theta}$$

$$= \frac{1}{9} \int \cos \theta \, d\theta = \frac{1}{9} \sin \theta + C$$

Since $\tan \theta = x/3$, $-\pi/2 < \theta < \pi/2$, from Figure 4 we see that $\sin \theta = x/\sqrt{x^2 + 9}$. Thus,

$$\int \frac{dx}{(x^2 + 9)^{3/2}} = \frac{x}{9\sqrt{x^2 + 9}} + C$$

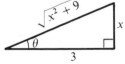

Figure 4   ■

In evaluating integrals involving $\sqrt{a^2 + x^2}$, $a > 0$, we may also use the substitution $x = a \sinh \theta$. Then $dx = a \cosh \theta \, d\theta$, and

$$\sqrt{a^2 + x^2} = \sqrt{a^2 + a^2 \sinh^2 \theta} = a\sqrt{1 + \sinh^2 \theta} = a\sqrt{\cosh^2 \theta} = a \cosh \theta$$

For example,

$$\int \frac{dx}{\sqrt{x^2 + a^2}} = \int \frac{a \cosh \theta}{a \cosh \theta} \, d\theta = \int d\theta = \theta + C = \sinh^{-1}\left(\frac{x}{a}\right) + C$$

$$\underset{\underset{\text{By (8.56)}}{\uparrow}}{=} \ln\left(\frac{x}{a} + \frac{\sqrt{x^2 + a^2}}{a}\right) + C$$

For integrands involving $\sqrt{x^2 - a^2}$, $a > 0$, we may use the substitution $x = a \sec \theta$, $0 < \theta < \pi/2$ or $\pi < \theta < 3\pi/2$.* Since $\tan \theta > 0$ for $0 < \theta < \pi/2$ or $\pi < \theta < 3\pi/2$, we have (see Fig. 5)

$$\sqrt{x^2 - a^2} = \sqrt{a^2(\sec^2\theta - 1)}$$
$$= a\sqrt{\tan^2\theta}$$
$$= a \tan \theta$$

Figure 5

EXAMPLE 3   Evaluate: $\displaystyle\int \frac{\sqrt{x^2 - 4}}{x} \, dx$

Solution   Let $x = 2 \sec \theta$. Then $dx = 2 \sec \theta \tan \theta \, d\theta$, and

$$\sqrt{x^2 - 4} = \sqrt{4 \sec^2\theta - 4} = 2\sqrt{\sec^2\theta - 1} = 2 \tan \theta$$

Therefore,

$$\int \frac{\sqrt{x^2 - 4}}{x} \, dx = \int \frac{(2 \tan \theta)(2 \sec \theta \tan \theta \, d\theta)}{2 \sec \theta} = 2 \int \tan^2\theta \, d\theta$$

$$= 2 \int (\sec^2\theta - 1) \, d\theta = 2(\tan \theta - \theta) + C$$

$$\underset{\underset{\text{Use Fig. 6}}{\uparrow}}{=} \sqrt{x^2 - 4} - 2 \sec^{-1}\left(\frac{x}{2}\right) + C$$

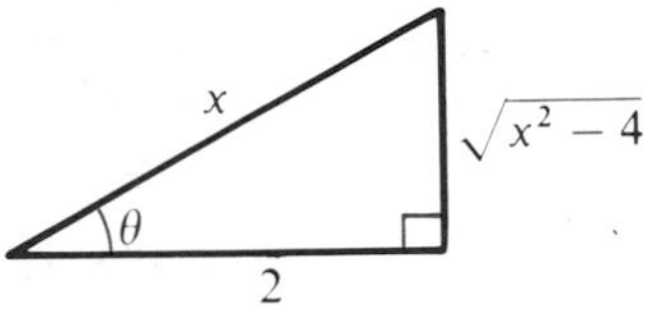

Figure 6   ■

It may be required to complete the square of the expression under the radical before the trigonometric substitution is made. For example, if the integrand contains a square root of the form $\sqrt{x^2 + 2x - 3}$, rewrite it as $\sqrt{(x + 1)^2 - 4}$ and use the substitution $x + 1 = 2 \sec \theta$.

EXAMPLE 4   Find the area enclosed by the ellipse: $\dfrac{x^2}{4} + \dfrac{y^2}{9} = 1$

---

* We may also use the substitution $x = a \cosh \theta$.

Solution   The total area (as Fig. 7 illustrates) is four times the area in the first quadrant, namely, $\int_0^2 y\, dx$,   where

$$y = 3\sqrt{1 - \frac{x^2}{4}} = \left(\frac{3}{2}\right)\sqrt{4 - x^2}$$

Hence, the total area is

$$A = (4)\left(\frac{3}{2}\right)\int_0^2 \sqrt{4 - x^2}\, dx$$

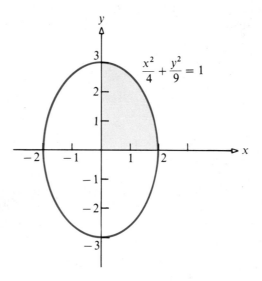

Figure 7

We make the substitution $x = 2\sin\theta$. Now, when $x = 0$, $2\sin\theta = 0$ and $\theta = 0$; when $x = 2$, $2\sin\theta = 2$ and so $\sin\theta = 1$ and $\theta = \pi/2$. Thus,

$$A = 6\int_0^2 \sqrt{4 - x^2}\, dx = 6\int_0^{\pi/2} \sqrt{4 - 4\sin^2\theta}\; 2\cos\theta\, d\theta$$

$$= 24\int_0^{\pi/2} \cos^2\theta\, d\theta = \frac{24}{2}\int_0^{\pi/2}(1 + \cos 2\theta)\, d\theta$$

$$= 12\left(\theta + \frac{\sin 2\theta}{2}\right)\Bigg|_0^{\pi/2}$$

$$= 12\left(\frac{\pi}{2} + 0\right) - 12(0) = 6\pi \qquad \blacksquare$$

In Example 4 the new limits of integration were easy to obtain. This is not always the case.

EXAMPLE 5   Evaluate: $\int_1^3 \sqrt{x^2 - 1}\, dx$

Solution   Here we use the substitution $x = \sec\theta$.   Then $dx = \sec\theta \tan\theta\, d\theta$.   Since the upper limit $x = 3$ does not result in a nice angle $(\theta = \sec^{-1}3)$,   we evaluate the indefinite integral and do not change the limits of integration. Then

$$\int \sqrt{x^2 - 1}\, dx = \int \tan\theta \sec\theta \tan\theta\, d\theta$$

$$= \int \tan^2\theta \sec\theta\, d\theta = \tfrac{1}{2} \sec\theta \tan\theta - \tfrac{1}{2} \ln|\sec\theta + \tan\theta| + C$$

See Example 7,
Section 3

As Figure 8 illustrates,

$$\sec\theta = x$$

and

$$\tan\theta = \sqrt{x^2 - 1}$$

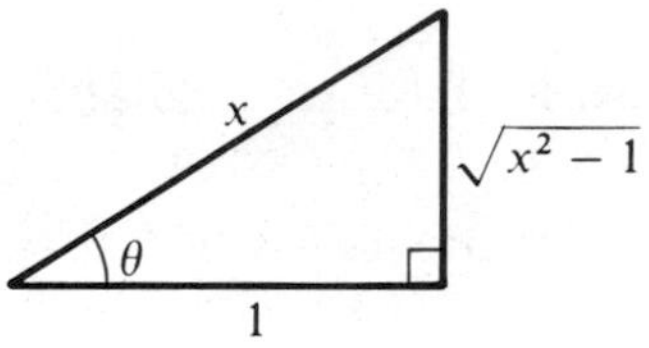

Figure 8

Hence

$$\int_1^3 \sqrt{x^2 - 1}\, dx = \left[\left(\frac{x}{2}\right) \sqrt{x^2 - 1} - \frac{1}{2} \ln\left|x + \sqrt{x^2 - 1}\right| \right]\Bigg|_1^3$$

$$= \tfrac{3}{2}\sqrt{8} - \tfrac{1}{2}\ln|3 + \sqrt{8}| \approx (1.5)(2.828) - (0.5)(1.763) = 3.361 \quad \blacksquare$$

EXERCISE 4

In Problems 1–32 evaluate each integral.

1. $\displaystyle\int \frac{dx}{\sqrt{4 - x^2}}$

2. $\displaystyle\int x\sqrt{x^2 - 4}\, dx$

3. $\displaystyle\int \frac{\sqrt{9 - x^2}}{x^2}\, dx$

4. $\displaystyle\int \frac{x^2}{\sqrt{9 - x^2}}\, dx$

5. $\displaystyle\int \frac{x^2}{\sqrt{x^2 - 1}}\, dx$

6. $\displaystyle\int \frac{\sqrt{x^2 - 1}}{x^2}\, dx$

7. $\displaystyle\int \frac{dx}{(x^2 + 4)^{3/2}}$

8. $\displaystyle\int \frac{x^3}{(x^2 + 4)^{1/2}}\, dx$

9. $\displaystyle\int \frac{dx}{x^2\sqrt{x^2 + 4}}$

10. $\displaystyle\int \sqrt{4 - x^2}\, dx$

11. $\displaystyle\int \sqrt{16 - x^2}\, dx$

12. $\displaystyle\int x^2\sqrt{4 - x^2}\, dx$

13. $\displaystyle\int \frac{dx}{(4 - x^2)^{3/2}}$

14. $\displaystyle\int \frac{x^2\, dx}{(x^2 + 9)^{3/2}}$

15. $\displaystyle\int \frac{x^2}{\sqrt{16 - x^2}}\, dx$

16. $\displaystyle\int \frac{dx}{x^2\sqrt{x^2 - 9}}$

17. $\displaystyle\int \frac{\sqrt{x^2 - 1}}{x}\, dx$

18. $\displaystyle\int \frac{\sqrt{x^2 + 1}}{x^2}\, dx$

19. $\displaystyle\int \frac{dx}{(x^2 - 9)^{3/2}}$

20. $\displaystyle\int \frac{x^2 + 9}{x^6}\, dx$

21. $\displaystyle\int \frac{dx}{\sqrt{1 - (x - 2)^2}}$

22. $\int \sqrt{4 - (x + 2)^2} \, dx$

23. $\int \dfrac{dx}{\sqrt{(x - 1)^2 - 4}}$

24. $\int \dfrac{dx}{(x - 2)\sqrt{(x - 2)^2 + 9}}$

25. $\int \dfrac{dx}{(x^2 - 2x + 10)^{3/2}}$

26. $\int \dfrac{dx}{\sqrt{x^2 - 2x + 10}}$

27. $\int \dfrac{dx}{\sqrt{x^2 + 2x - 3}}$

28. $\int \sqrt{5 + 4x - x^2} \, dx$

29. $\int_0^2 \dfrac{x^2 \, dx}{(16 - x^2)^{3/2}}$

30. $\int_1^2 \dfrac{dx}{x^2 \sqrt{1 + 9x^2}}$

31. $\int_1^2 \dfrac{x^3 \, dx}{\sqrt{1 + 4x^2}}$

32. $\int_0^3 \dfrac{x^2 \, dx}{9 + x^2}$

33. Evaluate $\int \sqrt{a^2 - x^2} \, dx$ and use it to find the area enclosed by the ellipse $\dfrac{x^2}{a^2} + \dfrac{y^2}{b^2} = 1.$

34. Find the area enclosed by the hyperbola $\dfrac{x^2}{9} - \dfrac{y^2}{16} = 1$ and the line $x = 6$.

35. Derive the formula: $\int \dfrac{dx}{\sqrt{x^2 - a^2}} = \ln \left| \dfrac{x}{a} + \dfrac{\sqrt{x^2 - a^2}}{a} \right| + C$

36. Evaluate: $\int \sqrt{x^2 + a^2} \, dx$
  (a) By using a trigonometric substitution   (b) By using a hyperbolic substitution

37. Derive the formula for the area of a circle.

38. Find the area under the graph of $y = x^3/\sqrt{9 - x^2}$ from $x = 0$ to $x = 2$.

39. Find the length of the arch of the parabola $y = 5x - x^2$ above the x-axis.

40. Find the volume of the solid generated by revolving the area under the graph of $y = 1/(x^2 + 4)$ from $x = 0$ to $x = 1$ about the x-axis.

In Problems 41 and 42 use integration by parts and then the method of this section to evaluate each integral.

41. $\int x \sin^{-1} x \, dx$

42. $\int x \cos^{-1} x \, dx$

5. Integrals Involving $ax^2 + bx + c$

Integrals involving $ax^2 + bx + c$,   where $a \ne 0$, b, c are real numbers, can often be handled by the technique of completing the square:

$$ax^2 + bx + c = a\left(x^2 + \frac{b}{a} x + \qquad \right) + c$$

$$= a\left(x^2 + \frac{b}{a} x + \frac{b^2}{4a^2} \right) + c - \frac{b^2}{4a}$$

$$= a\left(x + \frac{b}{2a} \right)^2 + \left(c - \frac{b^2}{4a} \right)$$

The substitution

$$u = x + \frac{b}{2a}$$

reduces the original expression $ax^2 + bx + c$ to the simpler form $au^2 + r$, where $r = c - (b^2/4a)$.

EXAMPLE 1   Evaluate: $\displaystyle\int \frac{dx}{x^2 + 6x + 10}$

Solution   We complete the square by writing

$$x^2 + 6x + 10 = (x^2 + 6x + 9) + 1 = (x + 3)^2 + 1$$

Now we can use the substitution $u = x + 3$. Then $du = dx$, and

$$\int \frac{dx}{x^2 + 6x + 10} = \int \frac{dx}{(x + 3)^2 + 1} = \int \frac{du}{u^2 + 1}$$

$$= \tan^{-1}u + C = \tan^{-1}(x + 3) + C \qquad \blacksquare$$
$$\uparrow$$
$$\text{Formula 11}$$

EXAMPLE 2   Evaluate: $\displaystyle\int \frac{(x - 1)\, dx}{x^2 + x + 1}$

Solution   Complete the square:

$$x^2 + x + 1 = (x^2 + x + \tfrac{1}{4}) + 1 - \tfrac{1}{4} = (x + \tfrac{1}{2})^2 + \tfrac{3}{4}$$

Now let $u = x + \tfrac{1}{2}$. Then $du = dx$, $x = u - \tfrac{1}{2}$, and

$$\int \frac{(x - 1)\, dx}{x^2 + x + 1} = \int \frac{(x - 1)\, dx}{(x + \tfrac{1}{2})^2 + \tfrac{3}{4}} = \int \frac{(u - \tfrac{3}{2})\, du}{u^2 + \tfrac{3}{4}}$$

$$= \int \frac{u\, du}{u^2 + \tfrac{3}{4}} - \frac{3}{2} \int \frac{du}{u^2 + \tfrac{3}{4}}$$

$$= \frac{1}{2} \ln\left(u^2 + \frac{3}{4}\right) - \sqrt{3} \tan^{-1}\left(\frac{2u}{\sqrt{3}}\right) + C$$
$$\uparrow$$
$$\text{Formulas 2 and 18}$$

$$= \frac{1}{2} \ln(x^2 + x + 1) - \sqrt{3} \tan^{-1}\left(\frac{2x + 1}{\sqrt{3}}\right) + C \qquad \blacksquare$$

EXAMPLE 3   Evaluate: $\displaystyle\int \frac{dx}{\sqrt{2x - x^2}}$

Solution   We complete the square by writing

$$-x^2 + 2x = -(x^2 - 2x + 1) + 1 = 1 - (x - 1)^2$$

We let $u = x - 1$.   Then $du = dx$ and

$$\int \frac{dx}{\sqrt{2x - x^2}} = \int \frac{dx}{\sqrt{1 - (x - 1)^2}} = \int \frac{du}{\sqrt{1 - u^2}}$$

$$= \sin^{-1}u + C = \sin^{-1}(x - 1) + C \qquad \blacksquare$$

Formula 10

EXERCISE 5

In Problems 1–14 evaluate each integral.

1. $\displaystyle\int \frac{dx}{x^2 + 4x + 5}$
  2. $\displaystyle\int \frac{dx}{x^2 + 2x + 5}$
  3. $\displaystyle\int \frac{dx}{\sqrt{8 + 2x - x^2}}$
  4. $\displaystyle\int \frac{dx}{\sqrt{5 - 4x - 2x^2}}$

5. $\displaystyle\int \frac{dx}{\sqrt{4x - x^2}}$
  6. $\displaystyle\int \frac{dx}{\sqrt{x^2 - 6x + 10}}$
  7. $\displaystyle\int \frac{dx}{\sqrt{24 - 2x - x^2}}$
  8. $\displaystyle\int \frac{dx}{\sqrt{9x^2 + 6x + 10}}$

9. $\displaystyle\int \frac{x\,dx}{\sqrt{x^2 - 2x + 5}}$
  10. $\displaystyle\int \frac{(x + 1)\,dx}{x^2 - 4x + 3}$
  11. $\displaystyle\int_1^3 \frac{dx}{\sqrt{x^2 - 2x + 5}}$
  12. $\displaystyle\int_{1/2}^1 \frac{x^2\,dx}{\sqrt{2x - x^2}}$

13. $\displaystyle\int \frac{e^x\,dx}{\sqrt{e^{2x} + e^x + 1}}$
  14. $\displaystyle\int \frac{\cos x\,dx}{\sqrt{\sin^2 x + 4\sin x + 3}}$

15. Show that if $k > 0$,   then

$$\int \frac{dx}{\sqrt{(x + h)^2 + k}} = \ln\big[\sqrt{(x + h)^2 + k} + x + h\big] + C \qquad \text{for all } x.$$

16. Show that if $a > 0$ and $b^2 - 4ac > 0$,   then

$$\int \frac{dx}{\sqrt{ax^2 + bx + c}} = \frac{1}{\sqrt{a}} \ln\left(\sqrt{ax^2 + bx + c} + \sqrt{a}\,x + \frac{b}{2\sqrt{a}}\right) + C \qquad \text{for all } x.$$

6. Integration of Rational Functions by Partial Fractions

In many cases, the integration of *rational functions* can be accomplished by using certain standard integral forms together with the algebraic method of *partial fractions*.

Recall that, by definition, a rational function is the quotient of two polynomials, P and $Q \neq 0$,   with no common factors. A rational function in which the polynomial in the numerator is of lower degree than the polynomial in the denominator is called a *proper rational function*; otherwise it is an *improper rational function*. Any improper rational function can be reduced by division to a mixed form, consisting of the sum of a polynomial and a proper rational function.

For example, the rational function $x^2/(x - 1)$ is improper. By long division, we obtain

(9.8)
$$\frac{x^2}{x - 1} = x + 1 + \frac{1}{x - 1}$$

To evaluate

$$\int \frac{x^2}{x - 1}\, dx$$

we integrate both sides of (9.8), obtaining

$$\int \frac{x^2}{x - 1}\, dx = \int \left(x + 1 + \frac{1}{x - 1} \right) dx = \int (x + 1)\, dx + \int \frac{1}{x - 1}\, dx$$

$$= \frac{x^2}{2} + x + \ln|x - 1| + C$$

Thus, because an improper rational function can always be written as the sum of a polynomial plus a proper rational function, we restrict our discussion to proper rational functions.

In general, a proper rational function P/Q may be written as the sum of simpler fractions, called *partial fractions*. The method of integration by partial fractions involves separating the given rational function into a sum of partial fractions. These partial fractions are then integrated by standard integration formulas. The result is that *the integral of every rational function can be expressed in terms of algebraic, logarithmic, and/or inverse trigonometric expressions*.

When a rational function is separated into partial fractions, the result is an *identity*; that is, it is true for all values of the variable for which the expressions involved have meaning. The evaluation of the coefficients of the partial fractions is based on the following theorem from algebra:

(9.9)   **THEOREM   If two polynomials are equal for all values of the variable, then the polynomials have the same degree and the coefficients of like powers of the variable in both polynomials must be equal.**

For example, if

$$ax^2 + bx + c = 3x^2 - 5x + 2 \qquad \text{for all } x$$

then

$$a = 3 \qquad b = -5 \qquad c = 2$$

As it turns out, the partial fraction decomposition of P/Q depends on the nature of the factors of the denominator Q. According to the fundamental theorem of algebra, any polynomial in x with real coefficients can be expressed as a product of factors of one or both of the following types:

1. *Linear factors* of the form $ax + b$,   where a and b are real numbers
2. *Irreducible quadratic factors* of the form $ax^2 + bx + c$,   where a, b, and c are real numbers and $ax^2 + bx + c$ cannot be factored into real linear factors

We begin with the case for which Q contains only nonrepeated linear factors.

Case 1. Q has only nonrepeated linear factors

The polynomial Q may be written as

$$Q(x) = (x - a_1)(x - a_2) \cdots (x - a_n)$$

where none of the real numbers $a_1, a_2, \ldots, a_n$ are the same. In this case, P/Q may be written as the identity

(9.10)
$$\frac{P(x)}{Q(x)} = \frac{A_1}{x - a_1} + \frac{A_2}{x - a_2} + \cdots + \frac{A_n}{x - a_n}$$

in which $A_1, A_2, \ldots, A_n$ are numbers to be found. By integrating both sides of (9.10), we find

$$\int \frac{P(x)}{Q(x)} \, dx = \int \left(\frac{A_1}{x - a_1} + \cdots + \frac{A_n}{x - a_n} \right) dx$$

$$= \int \frac{A_1}{x - a_1} \, dx + \cdots + \int \frac{A_n}{x - a_n} \, dx$$

$$= A_1 \ln|x - a_1| + \cdots + A_n \ln|x - a_n| + C$$

The procedure for finding the numbers $A_1, \ldots, A_n$ is illustrated in Example 1.

EXAMPLE 1   Evaluate: $\displaystyle\int \frac{x \, dx}{x^2 - 5x + 6}$

Solution   First we factor $x^2 - 5x + 6 = (x - 2)(x - 3)$.   Then we write the identity

(9.11)
$$\frac{x}{(x - 2)(x - 3)} = \frac{A}{x - 2} + \frac{B}{x - 3}$$

Now we multiply both sides by $(x - 2)(x - 3)$ to clear fractions, giving

$$x = A(x - 3) + B(x - 2)$$

(9.12)
$$x = (A + B)x - 3A - 2B$$

This is an identity in x. Now, using theorem (9.9), we may conclude:

$$\underset{\substack{\text{Coefficient of } x \\ \text{on left side of (9.12)}}}{1} = \underset{\substack{\text{Coefficient of } x \\ \text{on right side of (9.12)}}}{A + B}$$

$$\underset{\substack{\text{Coefficient of } x^0 \\ \text{on left side of (9.12)} \\ \text{(the constant terms)}}}{0} = \underset{\substack{\text{Coefficient of } x^0 \\ \text{on right side of (9.12)} \\ \text{(the constant terms)}}}{-3A - 2B}$$

The solution of this system of equations is $A = -2$ and $B = 3$.   By substituting into (9.11), we find

$$\frac{x}{(x-2)(x-3)} = \frac{-2}{x-2} + \frac{3}{x-3}$$

so that

$$\int \frac{x}{(x-2)(x-3)}\, dx = \int \frac{-2}{x-2}\, dx + \int \frac{3}{x-3}\, dx = -2\ln|x-2| + 3\ln|x-3| + C$$

$$= \ln\left|\frac{(x-3)^3}{(x-2)^2}\right| + C \quad \blacksquare$$

Case 2. Q has repeated linear factors

If the polynomial Q has a factor $(x-a)^n$, $n \ge 2$ an integer,   then in the partial fraction decomposition of P/Q we must allow for the terms

$$\frac{A_1}{x-a} + \frac{A_2}{(x-a)^2} + \cdots + \frac{A_n}{(x-a)^n}$$

where $A_1, \ldots, A_n$ are numbers to be found.

EXAMPLE 2   Evaluate: $\displaystyle\int \frac{dx}{x(x-1)^2}$

Solution   Since x is a distinct linear factor of the denominator Q, by Case 1, we allow for the term A/x in the decomposition of P/Q. Since $(x-1)^2$ is a repeated linear factor, we allow for the terms $B/(x-1)$ and $C/(x-1)^2$ in the decomposition of P/Q. Hence, we write the identity

(9.13)
$$\frac{1}{x(x-1)^2} = \frac{A}{x} + \frac{B}{x-1} + \frac{C}{(x-1)^2}$$

As in Example 1, we clear fractions:

$$1 = A(x-1)^2 + Bx(x-1) + Cx$$
$$1 = (A+B)x^2 + (C-B-2A)x + A$$

By equating coefficients, we obtain the system of equations

$$A + B = 0$$
$$-2A - B + C = 0$$
$$A = 1$$

By solving this system of equations, we find $A = 1$, $B = -1$, and $C = 1$. Substituting into (9.13), we get

$$\frac{1}{x(x-1)^2} = \frac{1}{x} + \frac{-1}{x-1} + \frac{1}{(x-1)^2}$$

so that

$$\int \frac{dx}{x(x-1)^2} = \int \frac{dx}{x} - \int \frac{dx}{x-1} + \int \frac{dx}{(x-1)^2}$$

$$= \ln|x| - \ln|x-1| - \frac{1}{x-1} + C = \ln\left|\frac{x}{x-1}\right| - \frac{1}{x-1} + C \quad \blacksquare$$

The numbers to be found in the partial fraction decomposition of P/Q can often be obtained more easily by substituting convenient values of x into the identity obtained after clearing fractions. In the example above, after clearing fractions, we have the identity

$$1 = A(x-1)^2 + Bx(x-1) + Cx$$

If we set $x = 0$, the terms involving B and C drop out, leaving $1 = A(1)$, so that $A = 1$. If we set $x = 1$, the terms involving A and B drop out, leaving $1 = C(1)$, so that $C = 1$. To get B, we can use any x other than $x = 0$ or $x = 1$ with $A = 1$ and $C = 1$. Using $x = 2$, we get

$$1 = 1(1)^2 + B2(1) + 1(2)$$
$$2B = -2$$
$$B = -1$$

The advantage of this method is evident from the next example.

EXAMPLE 3   Evaluate: $\int \frac{x^3 - 8}{x^2(x-1)^3} dx$

Solution   Due to the x^2 factor, we allow for the terms A/x and B/x^2; and due to the $(x-1)^3$ term, we allow for the terms $C/(x-1)$, $D/(x-1)^2$, and $E/(x-1)^3$. We write the identity

(9.14) $$\frac{x^3 - 8}{x^2(x-1)^3} = \frac{A}{x} + \frac{B}{x^2} + \frac{C}{x-1} + \frac{D}{(x-1)^2} + \frac{E}{(x-1)^3}$$

As usual, we clear fractions and simplify:

(9.15) $\qquad x^3 - 8 = Ax(x - 1)^3 + B(x - 1)^3 + Cx^2(x - 1)^2 + Dx^2(x - 1) + Ex^2$

We set $\;x = 0.\;$ Then

$$-8 = B(-1)$$
$$B = 8$$

We set $\;x = 1.\;$ Then

$$1 - 8 = E$$
$$E = -7$$

To find the other coefficients, we replace B by 8 and E by -7 in (9.15) and collect like terms:

$$x^3 - 8 - 8(x - 1)^3 + 7x^2 = Ax(x - 1)^3 + Cx^2(x - 1)^2 + Dx^2(x - 1)$$
$$-7x^3 + 31x^2 - 24x = Ax(x - 1)^3 + Cx^2(x - 1)^2 + Dx^2(x - 1)$$
$$-x(x - 1)(7x - 24) = x(x - 1)[A(x - 1)^2 + Cx(x - 1) + Dx]$$

By dividing by $\;x(x - 1)$, we obtain

(9.16) $\qquad -(7x - 24) = A(x - 1)^2 + Cx(x - 1) + Dx$

In (9.16), we set $\;x = 0.\;$ Then

$$24 = A$$

In (9.16), we set $\;x = 1.\;$ Then

$$17 = D$$

In (9.16), we replace A by 24 and D by 17. Then

$$-(7x - 24) = 24(x - 1)^2 + Cx(x - 1) + 17x$$

Now we set $\;x = 2\;$ (any choice other than 0 or 1 could have been used.) Then

$$10 = 24 + 2C + 34$$
$$C = -24$$

Using the values found for A, B, C, D, and E, in (9.14), we have

$$\frac{x^3 - 8}{x^2(x - 1)^3} = \frac{24}{x} + \frac{8}{x^2} - \frac{24}{x - 1} + \frac{17}{(x - 1)^2} - \frac{7}{(x - 1)^3}$$

Thus,

$$\int \frac{x^3 - 8}{x^2(x - 1)^3}\, dx = 24\ln|x| - \frac{8}{x} - 24\ln|x - 1| - \frac{17}{x - 1} + \frac{7}{2(x - 1)^2} + C$$

$$= 24\ln\left|\frac{x}{x - 1}\right| - \frac{8}{x} - \frac{17}{x - 1} + \frac{7}{2(x - 1)^2} + C \quad \blacksquare$$

Although this procedure may seem tedious, it is much faster than solving five equations in five unknowns by equating coefficients in (9.15).*

We close this section by deriving two useful formulas.

EXAMPLE 4   Derive the formulas:

(9.17)

(a) $\displaystyle \int \frac{dx}{x^2 - a^2} = \frac{1}{2a} \ln \left| \frac{x - a}{x + a} \right| + C \qquad a \neq 0$

(b) $\displaystyle \int \frac{dx}{a^2 - x^2} = \frac{1}{2a} \ln \left| \frac{x + a}{x - a} \right| + C \qquad a \neq 0$

Solution   (a)

$$\frac{1}{x^2 - a^2} = \frac{1}{(x - a)(x + a)} = \frac{A}{x - a} + \frac{B}{x + a}$$

By solving for A and B, we find that $A = 1/(2a)$ and $B = -1/(2a)$,   so that

$$\int \frac{dx}{x^2 - a^2} = \frac{1}{2a} \int \frac{dx}{x - a} - \frac{1}{2a} \int \frac{dx}{x + a} = \frac{1}{2a} \left(\int \frac{dx}{x - a} - \int \frac{dx}{x + a} \right)$$

$$= \frac{1}{2a} (\ln|x - a| - \ln|x + a|) + C$$

$$= \frac{1}{2a} \ln \left| \frac{x - a}{x + a} \right| + C$$

(b) $\displaystyle \int \frac{dx}{a^2 - x^2} = - \int \frac{dx}{x^2 - a^2} = -\frac{1}{2a} \ln \left| \frac{x - a}{x + a} \right| + C = \frac{1}{2a} \ln \left| \frac{x - a}{x + a} \right|^{-1} + C$

$$= \frac{1}{2a} \ln \left| \frac{x + a}{x - a} \right| + C \qquad \blacksquare$$

EXERCISE 6

In Problems 1–4 evaluate each integral by expressing the integrand as the sum of a polynomial plus a proper rational function.

1. $\displaystyle \int \frac{x^2 + 1}{x + 1} \, dx$

2. $\displaystyle \int \frac{x^2 + 4}{x - 2} \, dx$

3. $\displaystyle \int \frac{x^3 + 3x - 4}{x - 2} \, dx$

4. $\displaystyle \int \frac{x^3 - 3x^2 + 4}{x + 3} \, dx$

In Problems 5–22 evaluate each integral.

5. $\displaystyle \int \frac{dx}{(x - 2)(x + 1)}$

6. $\displaystyle \int \frac{dx}{(x + 4)(x - 1)}$

7. $\displaystyle \int \frac{x \, dx}{(x - 1)(x - 2)}$

8. $\displaystyle \int \frac{3x \, dx}{(x + 2)(x - 4)}$

9. $\displaystyle \int \frac{x^2 \, dx}{(x - 1)^2(x + 1)}$

10. $\displaystyle \int \frac{(x + 1) \, dx}{x^2(x - 2)}$

---

* The method used in this example is discussed in more detail in H. J. Straight and R. Dowds, *American Mathematical Monthly*, 91, no. 6 (June–July 1984): 365.

11. $\int \dfrac{(x-3)\,dx}{(x+2)(x+1)^2}$

12. $\int \dfrac{(x^2+x)\,dx}{(x+2)(x-1)^2}$

13. $\int \dfrac{x\,dx}{(3x-2)(2x+1)}$

14. $\int \dfrac{dx}{(2x+3)(4x-1)}$

15. $\int \dfrac{x\,dx}{x^2+2x-3}$

16. $\int \dfrac{x^2-x-8}{(x+1)(x^2+5x+6)}\,dx$

17. $\int \dfrac{7x+3}{x^3-2x^2-3x}\,dx$

18. $\int \dfrac{x^5+1}{x^6-x^4}\,dx$

19. $\int \dfrac{x^2}{x^3-4x^2+5x-2}\,dx$

20. $\int \dfrac{x^2+1}{x^3+x^2-5x+3}\,dx$

21. $\int \dfrac{\cos\theta\,d\theta}{\sin^2\theta+\sin\theta-6}$

22. $\int \dfrac{e^t\,dt}{e^{2t}+e^t-2}$

In Problems 23–26 evaulate each definite integral using part (a) or (b) of formula (9.17).

23. $\int_0^1 \dfrac{dx}{x^2-9}$

24. $\int_2^4 \dfrac{dx}{x^2-25}$

25. $\int_{-2}^3 \dfrac{dx}{16-x^2}$

26. $\int_1^2 \dfrac{dx}{9-x^2}$

27. Find the area under the graph of $y=4/(x^2-4)$ from $x=3$ to $x=5$.

28. Find the area under the graph of $y=(x-4)/(x+3)^2$ from $x=4$ to $x=6$.

29. Find the volume of the solid generated by revolving the area under the graph of $y=x/(x^2-4)$ from $x=3$ to $x=5$ about the x-axis.

7. Integration of Rational Functions
by Partial Fractions (Continued)

In this section we treat proper rational functions P/Q, where one of the factors of Q is an irreducible quadratic. *A quadratic factor is called irreducible if it cannot be factored into real linear factors.* Thus, ax^2+bx+c is an irreducible quadratic factor if $b^2-4ac<0$. For example, x^2+x+1 and x^2+4 are irreducible.

Case 3. Q contains a nonrepeated irreducible quadratic factor

If the polynomial Q contains a nonrepeated irreducible quadratic factor ax^2+bx+c, then in the partial fraction decomposition of P/Q we must allow for the terms

$$\frac{Ax+B}{ax^2+bx+c}$$

where A and B are numbers to be found.

EXAMPLE 1   Evaluate: $\int \dfrac{5x^2+3x-2}{x^3-1}\,dx$

Solution   Here, $Q(x)=x^3-1=(x-1)(x^2+x+1)$. Since $(x-1)$ is a nonrepeated linear factor, by Case 1 we allow for the term $A/(x-1)$ in the decomposition of P/Q. Since x^2+x+1 is a nonrepeated irreducible quadratic factor, we allow for the term $(Bx + C)/(x^2 + x + 1)$ in the decomposition of P/Q. Hence, we write the identity

(9.18)
$$\frac{5x^2 + 3x - 2}{x^3 - 1} = \frac{A}{x - 1} + \frac{Bx + C}{x^2 + x + 1}$$

We multiply both sides by $x^3 - 1 = (x - 1)(x^2 + x + 1)$ to get

(9.19)
$$5x^2 + 3x - 2 = A(x^2 + x + 1) + (Bx + C)(x - 1)$$

To determine the coefficients A, B, and C, we collect terms:

$$5x^2 + 3x - 2 = (A + B)x^2 + (A - B + C)x + A - C$$

By equating coefficients of like powers of x, we obtain the system of equations

$$A + B = 5$$
$$A - B + C = 3$$
$$A - C = -2$$

A little effort* yields $A = 2$, $B = 3$, and $C = 4$. By now substituting in (9.18), we get

$$\frac{5x^2 + 3x - 2}{x^3 - 1} = \frac{2}{x - 1} + \frac{3x + 4}{x^2 + x + 1}$$

so that

(9.20)
$$\int \frac{5x^2 + 3x - 2}{x^3 - 1}\, dx = \int \frac{2}{x - 1}\, dx + \int \frac{3x + 4}{x^2 + x + 1}\, dx = 2\ln|x - 1| + \int \frac{3x + 4}{x^2 + x + 1}\, dx$$

For the integral on the right, we complete the square in the denominator and make a substitution (see Section 5):

(9.21)
$$\int \frac{3x + 4}{x^2 + x + 1}\, dx = \int \frac{3x + 4}{(x + \tfrac{1}{2})^2 + \tfrac{3}{4}}\, dx = \int \frac{3u + \tfrac{5}{2}}{u^2 + \tfrac{3}{4}}\, du$$
$$\underset{u = x + \frac{1}{2}}{\uparrow}$$

$$= 3\int \frac{u}{u^2 + \tfrac{3}{4}}\, du + \frac{5}{2}\int \frac{du}{u^2 + \tfrac{3}{4}}$$

$$= \frac{3}{2}\ln\left(u^2 + \frac{3}{4}\right) + \frac{5}{2}\left(\frac{2}{\sqrt{3}}\right)\tan^{-1}\frac{2}{\sqrt{3}}u + C$$

$$= \frac{3}{2}\ln(x^2 + x + 1) + \frac{5\sqrt{3}}{3}\tan^{-1}\frac{2x + 1}{\sqrt{3}} + C$$
$$\underset{u = x + \frac{1}{2}}{\uparrow}$$

---

* An alternative here is to set $x = 1$ in (9.19), which results in finding $A = 2$. By resubstituting $A = 2$ into (9.19) and collecting terms, we get $(3x + 4)(x - 1) = (Bx + C)(x - 1)$, from which we conclude that $B = 3$ and $C = 4$. The effort is somewhat less if the alternative is used.

Hence, from (9.20) and (9.21), we find that

$$\int \frac{5x^2 + 3x - 2}{x^3 - 1} dx = 2 \ln|x - 1| + \frac{3}{2} \ln(x^2 + x + 1) + \frac{5\sqrt{3}}{3} \tan^{-1} \frac{2x + 1}{\sqrt{3}} + C \quad \blacksquare$$

Note that we could have saved some computation time if, in place of (9.18), we had written the partial fraction decomposition as

(9.22)
$$\frac{5x^2 + 3x - 2}{x^3 - 1} = \frac{D}{x - 1} + \frac{E(2x + 1) + F}{x^2 + x + 1}$$

This immediately places the derivative of $x^2 + x + 1$ in the decomposition. Following the usual procedure, we find the numbers D, E, and F to be $D = 2$, $E = \frac{3}{2}$, and $F = \frac{5}{2}$. Integrating (9.22) saves a few steps.

Case 4. Q has repeated irreducible quadratic factors

If the polynomial Q contains a repeated irreducible quadratic factor $(x^2 + bx + c)^n$, $n \geq 2$ an integer, then in the partial fraction decomposition of P/Q we must allow for the terms

$$\frac{A_1 x + B_1}{x^2 + bx + c} + \frac{A_2 x + B_2}{(x^2 + bx + c)^2} + \cdots + \frac{A_n x + B_n}{(x^2 + bx + c)^n}$$

where $A_1, B_1, A_2, B_2, \ldots, A_n, B_n$ are numbers to be found.

EXAMPLE 2 Evaluate: $\int \dfrac{x^3 + 1}{(x^2 + 4)^2} dx$

Solution The partial fraction decomposition is

$$\frac{x^3 + 1}{(x^2 + 4)^2} = \frac{Ax + B}{x^2 + 4} + \frac{Cx + D}{(x^2 + 4)^2}$$

Upon clearing fractions and combining terms, we arrive at

$$x^3 + 1 = Ax^3 + Bx^2 + (4A + C)x + 4B + D$$

By equating coefficients, we find

$$A = 1 \quad B = 0 \quad 4A + C = 0 \quad 4B + D = 1$$
$$C = -4 \quad\quad D = 1$$

Therefore,

$$\frac{x^3 + 1}{(x^2 + 4)^2} = \frac{x}{x^2 + 4} + \frac{-4x + 1}{(x^2 + 4)^2}$$

so that

$$\int \frac{x^3 + 1}{(x^2 + 4)^2} dx = \int \frac{x}{x^2 + 4} dx + \int \frac{-4x + 1}{(x^2 + 4)^2} dx$$

$$= \frac{1}{2} \ln(x^2 + 4) + \int \frac{-4x}{(x^2 + 4)^2} dx + \int \frac{dx}{(x^2 + 4)^2}$$

In the first integral on the right side, we use the substitution $u = x^2 + 4$.   Then $du = 2x\,dx$,   and

$$\int \frac{-4x}{(x^2 + 4)^2}\,dx = \int \frac{-2\,du}{u^2} = \frac{2}{u} = \frac{2}{x^2 + 4}$$

In the second integral on the right, we use the trigonometric substitution $x = 2\tan\theta$. Then $dx = 2\sec^2\theta\,d\theta$,   and

$$\int \frac{dx}{(x^2 + 4)^2} = \int \frac{2\sec^2\theta\,d\theta}{16\sec^4\theta} = \frac{1}{8}\int \cos^2\theta\,d\theta$$

$$= \frac{1}{8}\int \frac{1 + \cos 2\theta}{2}\,d\theta$$

$$\underset{\underset{\text{From Fig. 9}}{\uparrow}}{=} \frac{1}{16}\left(\theta + \frac{1}{2}\sin 2\theta\right) = \frac{1}{16}(\theta + \sin\theta\cos\theta)$$

Figure 9

$$= \frac{1}{16}\left(\tan^{-1}\frac{x}{2} + \frac{2x}{x^2 + 4}\right)$$

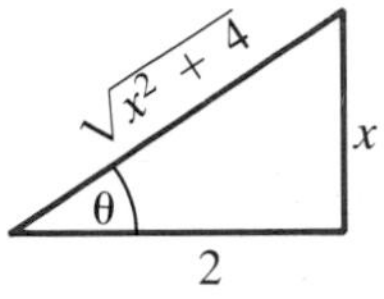

By combining these results, we find

$$\int \frac{x^3 + 1}{(x^2 + 4)^2}\,dx = \frac{1}{2}\ln(x^2 + 4) + \frac{2}{x^2 + 4} + \frac{1}{16}\tan^{-1}\frac{x}{2} + \frac{x}{8(x^2 + 4)} + C \quad\blacksquare$$

EXERCISE 7

In Problems 1–14 evaluate each integral.

1. $\displaystyle\int \frac{dx}{x(x^2 + 1)}$

2. $\displaystyle\int \frac{dx}{(x + 1)(x^2 + 4)}$

3. $\displaystyle\int \frac{(2x + 1)\,dx}{x^3 - 1}$

4. $\displaystyle\int \frac{dx}{x^3 - 8}$

5. $\displaystyle\int \frac{x + 4}{x^2(x^2 + 4)}\,dx$

6. $\displaystyle\int \frac{10x^2 + 2x}{(x - 1)^2(x^2 + 2)}\,dx$

7. $\displaystyle\int \frac{x^2 + 2x + 3}{(x + 1)(x^2 + 2x + 4)}\,dx$

8. $\displaystyle\int \frac{x^2 - 11x - 18}{x(x^2 + 3x + 3)}\,dx$

9. $\displaystyle\int \frac{x^2 + 2x + 3}{(x^2 + 4)^2}\,dx$

10. $\displaystyle\int \frac{2x + 1}{(x^2 + 16)^2}\,dx$

11. $\displaystyle\int \frac{x^3\,dx}{(x^2 + 16)^3}$

12. $\displaystyle\int \frac{x^2\,dx}{(x^2 + 4)^3}$

13. $\displaystyle\int \frac{\sin\theta\,d\theta}{\cos^3\theta + \cos\theta}$

14. $\displaystyle\int \frac{dt}{e^{2t} + 1}$

15. Find the area under the graph of $y = 8/(x^3 + 1)$ from $x = 0$ to $x = 2$.

16. Find the volume of the solid generated by revolving the area under the graph of $y = 1/(x^2 + 1)^2$ from $x = 0$ to $x = 1$ about the y-axis.

8. Integration of Rational Functions of Sine and Cosine

If an integrand is a rational expression of $\sin x$ or $\cos x$ or both, the substitution

$$z = \tan \frac{x}{2} \qquad -\frac{\pi}{2} < \frac{x}{2} < \frac{\pi}{2}$$

will transform the integrand into a rational function of z. To see why this substitution works, we rely on Figure 10 and the double-angle formulas from trigonometry. In the figure we see that

$$\sin \frac{x}{2} = \frac{z}{\sqrt{1 + z^2}} \qquad \cos \frac{x}{2} = \frac{1}{\sqrt{1 + z^2}} \qquad -\frac{\pi}{2} < \frac{x}{2} < \frac{\pi}{2}$$

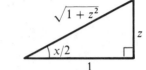

Figure 10

By using the double-angle formulas, we find that

$$\sin x = \sin 2\left(\frac{x}{2}\right) = 2 \sin \frac{x}{2} \cos \frac{x}{2} = 2\left(\frac{z}{\sqrt{1 + z^2}}\right)\left(\frac{1}{\sqrt{1 + z^2}}\right) = \frac{2z}{1 + z^2}$$

$$\cos x = \cos 2\left(\frac{x}{2}\right) = 2 \cos^2 \frac{x}{2} - 1 = \frac{2}{1 + z^2} - 1 = \frac{1 - z^2}{1 + z^2}$$

Since both $\sin x$ and $\cos x$ are rational expressions in z, any rational expression of $\sin x$ and $\cos x$ will also be a rational expression in z.

In making the substitution $z = \tan(x/2)$, we need to know dx. Since

$$x = 2 \tan^{-1} z$$

we calculate

(9.23)
$$dx = \frac{2\,dz}{1 + z^2}$$

To summarize, for rational expressions involving $\sin x$ or $\cos x$, we use the substitution

$$z = \tan \frac{x}{2} \qquad -\frac{\pi}{2} < \frac{x}{2} < \frac{\pi}{2}$$

which requires the use of

(9.24)
$$\sin x = \frac{2z}{1 + z^2} \qquad \cos x = \frac{1 - z^2}{1 + z^2} \qquad dx = \frac{2\,dz}{1 + z^2}$$

Let's look at an example.

EXAMPLE 1   Evaluate: $\displaystyle\int \frac{3}{2 \cos x + 1}\, dx$

Solution   This is a rational expression in cos x. Hence, we use (9.24) to get

$$\int \frac{3}{2\cos x + 1}\, dx = \int \frac{3}{2\left(\dfrac{1 - z^2}{1 + z^2}\right) + 1}\left(\frac{2\, dz}{1 + z^2}\right) = \int \frac{6\, dz}{2 - 2z^2 + 1 + z^2} = 6\int \frac{dz}{3 - z^2}$$

By part (b) of formula (9.17) in Section 6, we find

$$6\int \frac{dz}{3 - z^2} = (6)\left(\frac{1}{2\sqrt{3}}\right)\ln\left|\frac{z + \sqrt{3}}{z - \sqrt{3}}\right| + C$$

By replacing z by $\tan(x/2)$, we get

$$\int \frac{3}{2\cos x + 1}\, dx = \sqrt{3}\,\ln\left|\frac{\tan(x/2) + \sqrt{3}}{\tan(x/2) - \sqrt{3}}\right| + C \qquad ■$$

EXAMPLE 2   Evaluate: $\displaystyle\int \frac{dx}{\cos x - \sin x + 1}$

Solution   This is a rational expression in sin x and cos x. Hence, we use (9.24) to get

$$\int \frac{dx}{\cos x - \sin x + 1} = \int \frac{\dfrac{2\, dz}{1 + z^2}}{\dfrac{1 - z^2 - 2z + 1 + z^2}{1 + z^2}} = \int \frac{dz}{1 - z}$$

$$= -\ln|1 - z| + C = -\ln\left|1 - \tan\frac{x}{2}\right| + C \qquad ■$$

We can use the technique of this section to obtain an alternate formula for $\int \sec x\, dx$.   Since

$$\int \sec x\, dx = \int \frac{dx}{\cos x}$$

we use (9.24), and get

$$\int \sec x\, dx = 2\int \frac{dz}{1 - z^2} = \ln\left|\frac{1 + z}{1 - z}\right| + C = \ln\left|\frac{1 + \tan(x/2)}{1 - \tan(x/2)}\right| + C$$

In Problem 17 (Exercise 8) you are asked to verify that the above formula and formula 7 (in the Table of Integrals) are equivalent.

EXERCISE 8

In Problems 1–16 evaluate each integral.

1. $\displaystyle\int \frac{dx}{1 - \sin x}$ 2. $\displaystyle\int \frac{dx}{1 + \sin x}$ 3. $\displaystyle\int \frac{dx}{1 - \cos x}$ 4. $\displaystyle\int \frac{dx}{3 + 2\cos x}$

5. $\int \dfrac{2\,dx}{\sin x + \cos x}$

6. $\int \dfrac{dx}{1 - \sin x + \cos x}$

7. $\int \dfrac{\sin x\,dx}{3 + \cos x}$

8. $\int \dfrac{dx}{\tan x - 1}$

9. $\int \dfrac{dx}{\tan x - \sin x}$

10. $\int \dfrac{\sec x\,dx}{\tan x - 2}$

11. $\int \dfrac{\cot x\,dx}{1 + \sin x}$

12. $\int \dfrac{\sec x\,dx}{1 + \sin x}$

13. $\int_0^{\pi/2} \dfrac{dx}{\sin x + 1}$

14. $\int_{\pi/4}^{\pi/3} \dfrac{\csc x}{3 + 4\tan x}\,dx$

15. $\int_0^{\pi/2} \dfrac{\cos x}{2 - \cos x}\,dx$

16. $\int_0^{\pi/4} \dfrac{4\,dx}{\tan x + 1}$

17. Show that the two formulas below are equivalent.

$$\int \sec x\,dx = \ln|\sec x + \tan x| + C \qquad \int \sec x\,dx = \ln\left|\dfrac{1 + \tan(x/2)}{1 - \tan(x/2)}\right| + C$$

$$\left[\textit{Hint:}\quad \tan\left(\dfrac{x}{2}\right) = \dfrac{\sin(x/2)}{\cos(x/2)} = \dfrac{\sin^2(x/2)}{\sin(x/2)\cos(x/2)} = \dfrac{1 - \cos x}{\sin x} \right]$$

18. Using the technique of this section, show that

$$\int \csc x\,dx = \ln\sqrt{\dfrac{1 - \cos x}{1 + \cos x}} + C$$

19. Show that the result obtained in Problem 18 is equivalent to the formula

$$\int \csc x\,dx = \ln|\csc x - \cot x| + C$$

9. Miscellaneous Substitutions

If the integrand involves fractional powers $x^{p/q}$, $x^{r/s}$, and so forth, the substitution $x = u^n$, where n is the least common denominator of p/q, r/s, and so forth, will transform the integrand into a rational function of u.

EXAMPLE 1   Evaluate: $\displaystyle\int \dfrac{x^{1/2}\,dx}{x^{1/3} + 4}$

Solution   The least common denominator of $\frac{1}{2}$ and $\frac{1}{3}$ is 6, so we use the substitution $x = u^6$. Then $dx = 6u^5\,du$, $x^{1/2} = u^3$, and $x^{1/3} = u^2$, so that*

$$\int \dfrac{x^{1/2}\,dx}{x^{1/3} + 4} = 6\int \dfrac{u^3 u^5\,du}{u^2 + 4} = 6\int \dfrac{u^8\,du}{u^2 + 4} = 6\int\left(u^6 - 4u^4 + 16u^2 - 64 + \dfrac{256}{u^2 + 4}\right)du$$

$$= 6\left(\dfrac{u^7}{7} - \dfrac{4u^5}{5} + \dfrac{16u^3}{3} - 64u + 128\tan^{-1}\dfrac{u}{2}\right) + C$$

$$= 6\left(\dfrac{x^{7/6}}{7} - \dfrac{4}{5}x^{5/6} + \dfrac{16}{3}x^{1/2} - 64x^{1/6} + 128\tan^{-1}\dfrac{x^{1/6}}{2}\right) + C \qquad \blacksquare$$

---

* Now you should see why we use the substitution $x = u^6$. It enables us to express each fractional exponent of x as a positive integer exponent of u.

If the integrand is a rational function of x and $\sqrt[n]{a + bx}$, we use the substitution $u^n = a + bx$.

EXAMPLE 2   Evaluate: $\displaystyle\int \frac{2 + x}{\sqrt[3]{2 - x}}\, dx$

Solution   Let $u^3 = 2 - x$, so that $3u^2\, du = -dx$. Then

$$\int \frac{2 + x}{\sqrt[3]{2 - x}}\, dx = \int \frac{(4 - u^3)(-3u^2\, du)}{u} = \int(-12u + 3u^4)\, du = -6u^2 + \frac{3}{5} u^5 + C$$

$$= -6(2 - x)^{2/3} + \frac{3}{5}(2 - x)^{5/3} + C \quad \blacksquare$$

EXERCISE 9

In Problems 1–14 evaluate each integral.

1. $\displaystyle\int \frac{x\, dx}{3 + \sqrt{x}}$

2. $\displaystyle\int \frac{dx}{\sqrt{x} + 2}$

3. $\displaystyle\int \frac{dx}{x - \sqrt[3]{x}}$

4. $\displaystyle\int \frac{x\, dx}{\sqrt[3]{x} - 1}$

5. $\displaystyle\int \frac{dx}{\sqrt{x} + \sqrt[3]{x}}$

6. $\displaystyle\int \frac{dx}{\sqrt{x} - \sqrt[3]{x}}$

7. $\displaystyle\int \frac{dx}{\sqrt[3]{2 + 3x}}$

8. $\displaystyle\int \frac{dx}{\sqrt[4]{1 + 2x}}$

9. $\displaystyle\int \frac{x\, dx}{(1 + x)^{3/4}}$

10. $\displaystyle\int \frac{dx}{(1 + x)^{2/3}}$

11. $\displaystyle\int \frac{dx}{\sqrt{x} + 1}$

12. $\displaystyle\int \frac{dx}{\sqrt{x}(1 + \sqrt[3]{x})^2}$

13. $\displaystyle\int \frac{\sqrt[3]{x} + 1}{\sqrt[3]{x} - 1}\, dx$

14. $\displaystyle\int \frac{x\, dx}{\sqrt[5]{x + 4}}$

15. Find the volume of the solid generated by revolving the area under the graph of $y = \sqrt{x + 1} + x$ from $x = 0$ to $x = 3$ about the x-axis.

16. Rework Problem 15 if the revolution is about the y-axis.

10. Integrals that Are Not Expressible in Terms of Elementary Functions

We have seen that the derivatives of all elementary functions are also elementary functions,* but integrals of elementary functions are *not* all expressible (in finite form) in terms of elementary functions. Some examples of integrals of elementary functions

---

*The elementary functions are those that we have been discussing, such as polynomials, exponential, logarithmic, trigonometric, inverse trigonometric, hyperbolic, and so on.

that are not expressible in terms of elementary functions are

$$\int e^{x^2}\,dx \qquad \int e^{-x^2}\,dx \qquad \int \frac{\sin x}{x}\,dx \qquad \int \frac{\cos x}{x}\,dx \qquad \int \frac{e^x\,dx}{x} \qquad \int \frac{dx}{\sqrt{1-x^3}}$$

An important class of integral forms that are not expressible in terms of elementary functions are the *elliptic integrals*. An elliptic integral has the form

$$\int R(x, \sqrt{P(x)})\,dx$$

where P is a polynomial function of degree 3 or 4 with nonrepeated roots, and R denotes a rational function.

Numerous problems in geometry, mechanics, and other subjects lead to elliptic integrals. For example, the problems of finding the length of arc of an ellipse and of finding the time of oscillation of a pendulum (for arbitrary angle of oscillation) lead to elliptic integrals.

In the next section we give two techniques for approximating definite integrals. This is especially useful where the integral is not expressible in terms of elementary functions.

11. Numerical Techniques

So far in this chapter, we have dealt mainly with various techniques for evaluating indefinite integrals, that is, finding the antiderivative of a function. The reason for this study is that to evaluate the definite integral $\int_a^b f(x)\,dx$, the fundamental theorem of calculus requires that we know an antiderivative of the integrand f. When it is not possible to find an antiderivative of the integrand f or when the integrand f is defined by an empirical table of values or by an empirical graph, we turn to numerical techniques to approximate the value of the definite integral. To aid in the following discussion, we shall assume that the integrand f is nonnegative and continuous on the closed interval $[a, b]$.

Most of the methods of approximate integration are based on the fact that a definite integral equals the area under a graph, so that any method of approximating this area will also give an approximation to the integral. Two of the most widely used numerical techniques of approximate integration are the *trapezoidal rule* and *Simpson's rule*. Another useful method is based on the use of *series* (see Chap. 11).

(9.25)   *Trapezoidal Rule.*   **If a function f is continuous on the closed interval $[a, b]$, then**

(9.26) $$\int_a^b f(x)\,dx \approx \left(\frac{1}{2}\right)\left(\frac{b-a}{n}\right)[f(x_0) + 2f(x_1) + 2f(x_2) + \cdots + 2f(x_{n-1}) + f(x_n)]$$

where the closed interval $[a, b]$ has been partitioned into n subintervals $[x_0, x_1]$, $[x_1, x_2], \ldots, [x_{n-1}, x_n]$, each of length $(b-a)/n$.

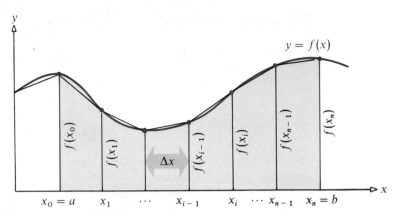

Figure 11   Area of trapezoid in ith subinterval $= \frac{1}{2}[f(x_{i-1}) + f(x_i)]\,\Delta x$

Proof   The trapezoidal rule is based on the idea of representing a definite integral by an area under a graph and on approximating this area by a collection of trapezoids obtained by replacing the graph by a set of chords. Suppose that we are to evaluate $\int_a^b f(x)\,dx$ approximately. The integral is then equal to the area enclosed by the graph of f, the x-axis, and the lines $x = a$ and $x = b$ (see Fig. 11). Now, partition the interval $[a, b]$ into n subintervals

$$[a, x_1], \quad [x_1, x_2], \quad \ldots, \quad [x_{i-1}, x_i], \quad \ldots, \quad [x_{n-1}, b]$$

each of length $\Delta x = (b - a)/n$. The ordinates corresponding to $x_0 = a$, $x_1, x_2, \ldots, x_{n-1}, x_n = b$ are $f(x_0), f(x_1), f(x_2), \ldots, f(x_{n-1}), f(x_n)$. When we join consecutive points on the graph by straight line segments (chords), trapezoids are formed, and the sum of the areas of the trapezoids is taken as an approximation to the area under the graph. Since the area of a trapezoid is equal to half the sum of the length of the parallel sides times the altitude, we have the following equation for the sum of the areas of the trapezoids:

$$\begin{aligned}
\text{Area} &= \tfrac{1}{2}[f(x_0) + f(x_1)]\Delta x + \tfrac{1}{2}[f(x_1) + f(x_2)]\Delta x + \cdots + \tfrac{1}{2}[f(x_{n-1}) + f(x_n)]\Delta x \\
&= \tfrac{1}{2}\Delta x[f(x_0) + 2f(x_1) + 2f(x_2) + \cdots + 2f(x_{n-1}) + f(x_n)]
\end{aligned}$$

By setting $\Delta x = (b - a)/n$, we have (9.26).   ∎

(9.27)   *Error.*   **The difference between the exact value of the integral $\int_a^b f(x)\,dx$ and the approximate value given by the trapezoidal rule is called the *error*. It may be estimated by the formula**

$$\text{Error} \le \frac{(b - a)^3 M}{12n^2}$$

where M is the largest value of $|f''(x)|$ on the closed interval $[a, b]$.*

---

* The verification of this formula, which uses an extension of the mean value theorem, is found in advanced calculus books.

EXAMPLE 1   Use the trapezoidal rule with $n = 4$ to approximate $\int_0^1 \dfrac{dx}{1 + x^2}$. Estimate the error in using this approximation.

Solution   We partition the interval $[0, 1]$ into four subintervals of equal length, namely, $[0, \frac{1}{4}]$, $[\frac{1}{4}, \frac{1}{2}]$, $[\frac{1}{2}, \frac{3}{4}]$, $[\frac{3}{4}, 1]$. The corresponding values of f are

$$f(0) = \frac{1}{1 + 0} = 1 \qquad f\left(\frac{1}{4}\right) = \frac{1}{1 + (\frac{1}{4})^2} = \frac{16}{17} \approx 0.94117 \qquad f\left(\frac{1}{2}\right) = \frac{1}{1 + (\frac{1}{2})^2} = \frac{4}{5} = 0.8$$

$$f\left(\frac{3}{4}\right) = \frac{1}{1 + (\frac{3}{4})^2} = \frac{16}{25} = 0.64 \qquad f(1) = \frac{1}{1 + 1} = \frac{1}{2} = 0.5$$

It is convenient to set up a table, as shown in Table 3. The sum of the entries in the bottom row of the table is 6.26234, so that by the trapezoidal rule (9.25), we get

$$\int_0^1 \frac{dx}{1 + x^2} \approx \left(\frac{1}{8}\right)(6.26234) \approx 0.78279$$

Table 3

	$x = 0$	$x = \frac{1}{4}$	$x = \frac{1}{2}$	$x = \frac{3}{4}$	$x = 1$
$f(x) = 1/(1 + x^2)$	1	0.94117	0.8	0.64	0.5
Factor	×1	×2	×2	×2	×1
Product	1	1.88234	1.6	1.28	0.5

We now use (9.27) to estimate the error. For this, we need to find the maximum value of $|f''(x)|$, which in turn requires that we find $f'''(x)$. Some calculations will lead to

$$f(x) = \frac{1}{1 + x^2} \qquad f'(x) = \frac{-2x}{(1 + x^2)^2} \qquad f''(x) = \frac{2(3x^2 - 1)}{(1 + x^2)^3}$$

$$f'''(x) = 24x(1 - x^2)(1 + x^2)^{-4}$$

Because $f'''(x) > 0$ for $0 < x < 1$, f'' has no critical numbers in $(0, 1)$. The largest value for $|f''(x)|$ occurs at the endpoint 0. For $M = |f''(0)| = 2$, $b = 1$, $a = 0$, and $n = 4$, an upper estimate to the error is

$$\text{Error} \leq \frac{(1 - 0)^3(2)}{(12)(4^2)} = \frac{1}{96} \approx 0.0104 \qquad \blacksquare$$

As a result of Example 1, we have

$$0.78279 - 0.0104 < \int_0^1 \frac{dx}{1 + x^2} < 0.78279 + 0.0104$$

Since the exact value of the integral $\int_0^1 dx/(1 + x^2)$ is $\tan^{-1}x\big|_0^1 = \pi/4$, we conclude that

$$0.77239 < \frac{\pi}{4} < 0.79319$$

$$3.08956 < \pi < 3.17276$$

In Chapter 11 we give another way of approximating π, which equals 3.14159 correct to five decimal places.

The next example illustrates how the trapezoidal rule is used when only discrete information is known.

EXAMPLE 2   A tree trunk is 140 feet long. At a distance x feet from one end, its sectional area A is given in square feet by the following table at intervals of 20 feet:

x	0	20	40	60	80	100	120	140
A	120	124	128	130	132	136	144	158

Find the approximate volume of the tree trunk.

Solution   From Chapter 6 we know that the volume is

$$V = \int_0^{140} A \, dx$$

Since $n = 7$, $a = 0$, and $b = 140$, by the trapezoidal rule, we find

$$V \approx \tfrac{140}{14}[120 + 2(124) + 2(128) + 2(130) + 2(132) + 2(136) + 2(144) + 158]$$
$$= 18{,}660 \text{ cubic feet} \quad \blacksquare$$

(9.28)   *Simpson's Rule.* **If a function f is continuous on the closed interval $[a, b]$, then**

(9.29) $$\int_a^b f(x) \, dx \approx \frac{b - a}{3n}[f(x_0) + 4f(x_1) + 2f(x_2) + 4f(x_3) + 2f(x_4) + \cdots$$

$$+ 2f(x_{n-2}) + 4f(x_{n-1}) + f(x_n)]$$

where the closed interval $[a, b]$ has been partitioned into an even number n of subintervals $[x_0, x_1], [x_1, x_2], \ldots, [x_{n-1}, x_n]$,   each of length $(b - a)/n$.

Simpson's rule (named after the English mathematician, Thomas Simpson, 1710–1761) is obtained by interpreting the definite integral as an area under a graph and by approximating the graph as a collection of parabolic arcs. By using parabolic arcs instead of chords, as in the derivation of the trapezoidal rule, we often get a closer approximation to the area. Thus, as a preliminary to the derivation of Simpson's rule, we need to find a formula for the area under a parabolic arc.

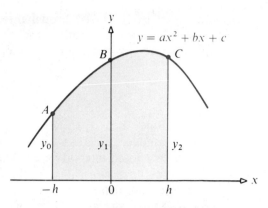

Figure 12

Let the graph in Figure 12 represent the parabola $y = ax^2 + bx + c$. Draw the lines $x = -h$ and $x = h$, and denote the ordinates at $x = -h$, $x = 0$, and $x = h$ by y_0, y_1, and y_2, respectively. The area enclosed by the parabola, the x-axis, and the lines $x = -h$ and $x = h$ is given by

(9.30)
$$\text{Area} = \int_{-h}^{h} y \, dx = \int_{-h}^{h} (ax^2 + bx + c) \, dx = \left(a\frac{x^3}{3} + b\frac{x^2}{2} + cx \right)\Bigg|_{-h}^{h}$$

$$= \frac{2}{3} ah^3 + 2ch = \frac{h}{3}(2ah^2 + 6c)$$

Since the three points $A = (-h, y_0)$, $B = (0, y_1)$, and $C = (h, y_2)$ lie on the parabola, their coordinates satisfy the equation $y = ax^2 + bx + c$, and we have the system of equations

(9.31)
$$y_0 = ah^2 - bh + c$$
$$y_1 = c$$
$$y_2 = ah^2 + bh + c$$

Then

$$y_0 + y_2 = 2ah^2 + 2c$$

and, therefore,

$$y_0 + y_2 + 4y_1 = 2ah^2 + 6c$$

By substituting this in the area formula (9.30), we get

(9.32)
$$\text{Area} = \frac{h}{3}(y_0 + 4y_1 + y_2)$$

This formula depends only upon the three ordinates and the distance h, and so it is independent of the position of the y-axis. We may state the result as follows: If a parabola with vertical axis is passed through three points (x_0, y_0), (x_1, y_1), and (x_2, y_2), with the distance h between consecutive abscissas, the area enclosed by the parabola, the x-axis, and the lines $x = x_0$ and $x = x_2$ is given by

$$\text{Area} = \frac{h}{3}(y_0 + 4y_1 + y_2)$$

Proof of Simpson's Rule   Let $\int_a^b f(x)\, dx$ be the integral to be approximated; it then equals the area under the graph of $y = f(x)$ from $x = a$ to $x = b$ (see Fig. 13). Partition the closed interval $[a, b]$ into an even number n of subintervals, each of length $\Delta x = (b - a)/n$. The ordinates corresponding to $x_0 = a$, x_1, $x_2, \ldots, x_{n-1}$, $x_n = b$ are denoted by $y_0, y_1, y_2, \ldots, y_{n-1}, y_n$. Arrange these ordinates in groups of three, the last ordinate of each group being the same as the first ordinate of the next group. Through each group of three ordinates, pass an arc of a parabola with vertical axis.

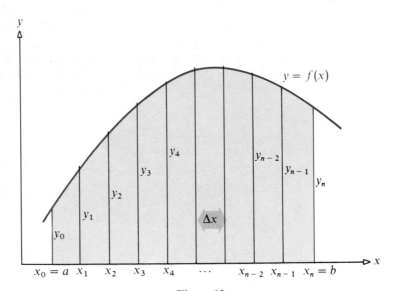

Figure 13

By (9.32), the area under the graph $y = f(x)$ from $x = x_0$ to $x = x_2$ may be approximated by the area under the parabola passing through (x_0, y_0), (x_1, y_1), and (x_2, y_2); namely,

$$A_1 = \frac{b - a}{3n}(y_0 + 4y_1 + y_2)$$

Similarly, the area under the graph of $y = f(x)$ from $x = x_2$ to $x = x_4$ may be approximated by the area under the parabola passing through (x_2, y_2), (x_3, y_3), and (x_4, y_4); namely,

$$A_2 = \frac{b-a}{3n}(y_2 + 4y_3 + y_4)$$

Continuing in this way (now you should realize why the number of subintervals must be even), we can approximate $\int_a^b f(x)\,dx$ by

$$\int_a^b f(x)\,dx \approx A_1 + A_2 + \cdots + A_n$$

$$= \frac{b-a}{3n}(y_0 + 4y_1 + y_2) + \frac{b-a}{3n}(y_2 + 4y_3 + y_4)$$

$$+ \frac{b-a}{3n}(y_4 + 4y_5 + y_6) + \cdots + \frac{b-a}{3n}(y_{n-2} + 4y_{n-1} + y_n)$$

$$= \frac{b-a}{3n}(y_0 + 4y_1 + 2y_2 + 4y_3 + 2y_4 + \cdots + 2y_{n-2} + 4y_{n-1} + y_n)$$

Thus, we have (9.29).   ■

(9.33) *Error.* **The difference between the exact value of the integral $\int_a^b f(x)\,dx$ and the approximate value given by Simpson's rule is called the *error*. It may be estimated by the formula**

$$\text{Error} \leq \frac{M(b-a)^5}{180n^4}$$

where M is the greatest value of $|f^{(4)}(x)|$ on the closed interval $[a, b]$.*

EXAMPLE 3   Use Simpson's rule with $n = 4$ to approximate $\int_0^1 \dfrac{dx}{1+x}$. Estimate the error in using this approximation.

Table 4

	$x = 0$	$x = \frac{1}{4}$	$x = \frac{1}{2}$	$x = \frac{3}{4}$	$x = 1$
$f(x) = 1/(1 + x)$	1	0.8	0.66666	0.57143	0.5
Factor	$\times 1$	$\times 4$	$\times 2$	$\times 4$	$\times 1$
Product	1	3.2	1.33333	2.28572	0.5

Solution   For $n = 4$, the partition of $[0, 1]$ is $[0, \frac{1}{4}]$, $[\frac{1}{4}, \frac{1}{2}]$, $[\frac{1}{2}, \frac{3}{4}]$, $[\frac{3}{4}, 1]$. Table 4 summarizes the information we need. Thus,

$$\int_0^1 \frac{dx}{1+x} \approx \left(\frac{1}{12}\right)(1 + 3.2 + 1.33333 + 2.28572 + 0.5) = 0.69325$$

---

* The verification of this formula, which uses the extension of the mean value theorem, is found in advanced calculus books.

To estimate the error, we need to find the maximum value of $|f^{(4)}(x)| = |24(1+x)^{-5}|$. On $[0, 1]$, the largest value of $|24(1+x)^{-5}|$ occurs at $x = 0$. For $M = |f^{(4)}(0)| = 24$, $b = 1$, $a = 0$, and $n = 4$, an upper estimate to the error is

$$\text{Error} \leq \frac{24(1-0)^5}{(180)(4^4)} = \frac{1}{1920} \approx 0.00052 \qquad \blacksquare$$

As a result of Example 3, we have

$$0.69325 - 0.00052 < \int_0^1 \frac{dx}{1+x} < 0.69325 + 0.00052$$

Since the exact value of $\int_0^1 dx/(1+x)$ is $\ln(x+1)|_0^1 = \ln 2$, we conclude that

$$0.69273 < \ln 2 < 0.69377$$

In Chapter 11 we give a different technique for approximating $\ln 2$, which equals 0.69314 correct to five decimal places.

EXERCISE 11

In Problems 1–10 use the trapezoidal rule to approximate each integral.

1. $\int_0^4 x^2 \, dx; \quad n = 8$

2. $\int_0^3 x^3 \, dx; \quad n = 6$

3. $\int_1^2 \frac{dx}{x}; \quad n = 4$

4. $\int_0^1 \frac{dx}{1+x}; \quad n = 6$

5. $\int_0^\pi \sin x \, dx; \quad n = 6$

6. $\int_{-\pi/2}^{\pi/2} \cos x \, dx; \quad n = 6$

7. $\int_{\pi/2}^\pi \frac{\sin x}{x} \, dx; \quad n = 3$

8. $\int_{3\pi/2}^{2\pi} \frac{\cos x}{x} \, dx; \quad n = 3$

9. $\int_0^1 e^{-x^2} \, dx; \quad n = 4$

10. $\int_0^1 e^{x^2} \, dx; \quad n = 4$

In Problems 11–20 use Simpson's rule to approximate each integral.

11. $\int_0^4 x^2 \, dx; \quad n = 8$

12. $\int_0^3 x \, dx; \quad n = 6$

13. $\int_1^2 \frac{dx}{x^2}; \quad n = 4$

14. $\int_0^2 \sqrt{x} \, dx; \quad n = 6$

15. $\int_0^2 \frac{1}{\sqrt{1+x}} \, dx; \quad n = 6$

16. $\int_0^\pi \sin x \, dx; \quad n = 6$

17. $\int_0^\pi \sqrt{\sin x} \, dx; \quad n = 6$

18. $\int_{-\pi/2}^{\pi/2} \sqrt{\cos x} \, dx; \quad n = 6$

19. $\int_0^1 e^{-x^2} \, dx; \quad n = 4$

20. $\int_0^1 e^{x^2} \, dx; \quad n = 4$

In Problems 21–24 approximate each integral by both the trapezoidal rule and Simpson's rule.

21. $\int_3^6 \dfrac{x \, dx}{4 + x^2}$;   use six subintervals

22. $\int_0^1 \dfrac{dx}{1 + x^2}$;   use ten subintervals

23. $\int_4^7 \sqrt{9 + x^2} \, dx$;   use six subintervals

24. $\int_0^{\pi/2} \sqrt{\sin x} \, dx$;   use four subintervals

25. Show that $\int_1^2 dx/x = \ln 2$. Then use the trapezoidal rule with $n = 5$ to approximate $\int_1^2 dx/x$ and hence obtain an approximation to $\ln 2$.

26. Use Simpson's rule with $n = 6$ to approximate $\int_1^2 dx/x$, and hence obtain an approximation to $\ln 2$.

27. In the table, S is the area in square meters of the cross section of a railroad cutting, and x meters is the corresponding distance along the line:

x	0	25	50	75	100	125	150
S	105	118	142	120	110	90	78

Use the trapezoidal rule to calculate the number of cubic meters of earth removed to make the cutting from $x = 0$ to $x = 150$. Do not attempt to compute an error since a function f for the area is not known.

28. A series of soundings taken across a river channel is given in the table, where x is the distance from one shore and y is the corresponding depth:

x	0	10	20	30	40	50	60	70	80
y	5	10	13.2	15	15.6	12	6	4	0

Draw the section and find its area by the trapezoidal rule.

29. In the table, F is the force in pounds acting on a body in its direction of motion and s is the displacement in feet.

s	0	5	10	15	20	25	30	35	40	45	50
F	100	80	66	56	50	45	40	36	33	30	28

Use the trapezoidal rule to calculate the total work done by the force from $s = 0$ to $s = 50$.

30. Use Simpson's rule to find the area enclosed by the pairs of rectangular coordinates in the table, the x-axis, and the lines $x = 2$ and $x = 4.4$.

x	2.0	2.4	2.8	3.2	3.6	4.0	4.4
y	3.03	4.61	5.80	6.59	7.76	8.46	9.19

31. The area of the horizontal section of a reservoir is A square meters at a height x meters from the bottom; corresponding values of x and A are given in the table:

x	0	2.5	5	7.5	10	12.5	15	17.5	20	22.5	25
A	0	2510	3860	4870	5160	5590	5810	6210	6890	7680	8270

Find the volume of water in the reservoir by use of the trapezoidal rule and also by Simpson's rule.

32. A gas expands from a volume of 1 cubic inch to 2.5 cubic inches; values of the volume (v) and pressure (p, in pounds per square inch) during the expansion are given in the table:

v	1	1.25	1.5	1.75	2	2.25	2.5
p	68.7	55.0	45.8	39.3	34.4	30.5	27.5

Calculate the total work W done in the expansion by using Simpson's rule $(W = \int_a^b p\, dv)$.

33. Use Simpson's rule to approximate the area of the pond pictured in the figure.

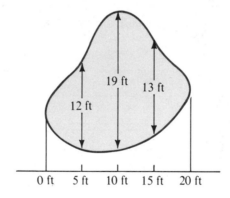

34. Use the trapezoidal rule to find the arc length of the ellipse $9x^2 + 100y^2 = 900$ in the first quadrant from $x = 0$ to $x = 8$.   Divide the interval into four equal parts, and obtain an answer to three decimal places.

35. Use (9.27) to estimate the error that can result in Problem 1.

36. Use (9.27) to estimate the error that can result in Problem 2.

37. Use (9.33) to estimate the error that can result in Problem 13.

38. Use (9.33) to estimate the error that can result in Problem 14.

39. Approximate the arc length of $y = \sin x$ from $x = 0$ to $x = \pi/2$ by using the trapezoidal rule with $n = 3$.

40. Rework Problem 39 by using Simpson's rule with $n = 4$.

41. Approximate the volume of the solid generated by revolving the area under the graph of $y = \sin x$ from $x = 0$ to $x = \pi/2$ about the y-axis by using the trapezoidal rule with $n = 3$.

42. Rework Problem 41 by using Simpson's rule with $n = 4$.

43. For the function below, use the trapezoidal rule with $n = 6$ to approximate $\int_0^\pi f(x)\,dx$.

$$f(x) = \begin{cases} (\sin x)/x & \text{if } x \neq 0 \\ 1 & \text{if } x = 0 \end{cases}$$

44. Use Simpson's rule with $n = 6$ to rework Problem 43.

45. The velocity (v, in meters per second) of a particle at time t is given by the table:

t	0	0.5	1	1.5	2	2.5	3
v	5.1	5.3	5.6	6.1	6.8	6.7	6.5

Use the trapezoidal rule to approximate the distance traveled from $t = 0$ to $t = 3$.

46. Use Simpson's rule to rework Problem 45.

Miscellaneous Exercises

In Problems 1–60 evaluate each integral.

1. $\int x^2(x^2 - 2)\,dx$

2. $\int \sqrt{ax + b}\,dx$

3. $\int \dfrac{dx}{x^2 + 4x + 20}$

4. $\int x\sqrt{x + 2}\,dx$

5. $\int \sec^2 2\theta\,d\theta$

6. $\int \dfrac{y^3 + y}{y + 1}\,dy$

7. $\int \dfrac{\sqrt{x}\,dx}{1 + x}$

8. $\int \sec^3\phi \tan\phi\,d\phi$

9. $\int \dfrac{t\,dt}{9 + t^2}$

10. $\int \dfrac{dt}{9 + t^2}$

11. $\int \cot^2\theta\,d\theta$

12. $\int \dfrac{x\,dx}{(1 + x)^4}$

13. $\int e^{\cos x} \sin x\,dx$

14. $\int \sin^3\phi\,d\phi$

15. $\int x \sin 2x\,dx$

16. $\int \dfrac{(y - 2)\,dy}{y^2 - 4y + 2}$

17. $\int (z^2 + 4)^2\,dz$

18. $\int v \csc^2 v\,dv$

19. $\int t^3\sqrt{2 - t}\,dt$

20. $\int x(1 - x^2)\,dx$

21. $\int \sin^2 x \cos^3 x\,dx$

22. $\int (4 - x^2)^{3/2}\,dx$

23. $\int \dfrac{dy}{\sqrt{2y + 1}}$

24. $\int \dfrac{3x^2 + 1}{x^3 + 2x^2 - 3x}\,dx$

25. $\int \dfrac{e^{2t}\,dt}{e^t - 2}$

26. $\int \dfrac{2z + 3}{\sqrt{1 + 2z}}\,dz$

27. $\int \tanh 2v\,dv$

28. $\int \dfrac{dy}{5 + 4y + 4y^2}$

29. $\int \dfrac{\sin x + \cos x}{\tan x}\,dx$

30. $\int \dfrac{e^{2x}\,dx}{e^{2x} + 1}$

31. $\int \dfrac{x\,dx}{x^4 - 4}$

32. $\int x^3 e^{x^2}\,dx$

33. $\int \dfrac{dx}{x \ln x}$

34. $\int \dfrac{y^2\,dy}{(y + 1)^3}$

35. $\int \tan\left(\dfrac{1}{4}\pi - \theta\right)d\theta$

36. $\int \dfrac{x + 1}{x\sqrt{x - 2}}\,dx$

37. $\int a^x b^x\,dx$

38. $\int \sin^2\theta \csc^2 2\theta\,d\theta$

39. $\int \sqrt{\dfrac{a + x}{a - x}}\,dx$

40. $\int \dfrac{dx}{\sqrt{16 + 4x - 2x^2}}$

41. $\int \dfrac{x^3 - 2x}{x - 1}\, dx$

42. $\int \ln(1 - y)\, dy$

43. $\int (3y^2 - 6y)^3 (y - 1)\, dy$

44. $\int \cos^n\theta \sin\theta\, d\theta$

45. $\int \dfrac{3x^2\, dx}{1 - x}$

46. $\int x^2 e^x\, dx$

47. $\int \cos^3 3x\, dx$

48. $\int \dfrac{dy}{\sqrt{2 + 3y^2}}$

49. $\int \dfrac{x^2\, dx}{4 - x^2}$

50. $\int \csc^4 x\, dx$

51. $\int x^2 \sin^{-1}x\, dx$

52. $\int \dfrac{\sec^2 z\, dz}{a + b \tan z}$

53. $\int \dfrac{(3t + 2)\, dt}{t\sqrt{t + 1}}$

54. $\int e^{y/2}\, dy$

55. $\int x \cos^2 x\, dx$

56. $\int \dfrac{dv}{\sqrt{3v - v^2}}$

57. $\int \dfrac{dx}{x^2 + ax}$

58. $\int \sin^4 y \cos^4 y\, dy$

59. $\int \dfrac{w^3\, dw}{1 - w^2}$

60. $\int \dfrac{\cos^2 mx\, dx}{\sin^3 mx}$

In Problems 61–65 use a trigonometric substitution to derive each formula.

61. $\int \dfrac{dx}{\sqrt{a^2 - x^2}} = \sin^{-1}\left(\dfrac{x}{a}\right) + C$

62. $\int \dfrac{dx}{a^2 + x^2} = \dfrac{1}{a} \tan^{-1}\left(\dfrac{x}{a}\right) + C$

63. $\int \dfrac{dx}{x\sqrt{x^2 - a^2}} = \dfrac{1}{a} \sec^{-1}\left(\dfrac{x}{a}\right) + C$

64. $\int \sqrt{x^2 - a^2}\, dx = \dfrac{1}{2} x\sqrt{x^2 - a^2} - \dfrac{1}{2} a^2 \ln\left|x + \sqrt{x^2 - a^2}\right| + C$

65. $\int \dfrac{dx}{\sqrt{x^2 + a^2}} = \ln(x + \sqrt{x^2 + a^2}) + C$

66. Find $\int dx/\sqrt{x^2 + a^2}$, $a > 0$, by using the substitution $u = \sinh^{-1}(x/a)$. Express your answer in logarithmic form.

67. Find $\int dx/\sqrt{x^2 - a^2}$, $a > 0$, using an appropriate hyperbolic function substitution. Express your answer in logarithmic form.

68. Suppose $F(x) = \int_0^x t g'(t)\, dt$ for all $x \geq 0$. Show that $F(x) = x g(x) - \int_0^x g(t)\, dt$.

69. If $\int x^2 \cos x\, dx = f(x) - \int 2x \sin x\, dx$, find f.

70. If the graph of $y = f(x)$ contains the point $(0, 2)$, $dy/dx = -x/ye^{x^2}$, and $f(x) > 0$ for all x, find f.

71. Let f be a function defined for all $x > -5$ with the following properties. (i) $f''(x) = 1/(3\sqrt{x + 5})$ for all x in the domain of f; (ii) the line tangent to the graph of f at $(4, 2)$ has an angle of inclination of $45°$. Find an expression for f.

72. (a) Determine $\int x^2 e^{5x}\, dx$.
   (b) Using integration by parts, derive a general formula for $\int x^n e^{kx}\, dx$, $k \neq 0$, in which the resulting integrand involves x^{n-1}.

73. The area in the first quadrant enclosed by the graph of $y = \sec x$, $x = \pi/4$, and the axes is rotated about the x-axis. What is the volume of the solid generated?

74. Let A be the area enclosed by the graph of $y = \ln x$, the line $x = e$, and the x-axis.
   (a) Find the volume generated by revolving A about the x-axis.
   (b) Find the volume generated by revolving A about the y-axis.

75. Let A be the area in the first quadrant enclosed by $y = \sec x$, $y = 2 \sin x$, and the y-axis.
   (a) Find A.   (b) Find the volume of the solid formed when A is revolved about the x-axis.

76. Find the length of the graph $y = \ln x$ from $x = \sqrt{3}/3$ to $x = \sqrt{3}$.

77. Let T_n be the approximation to $\int_a^b f(x)\, dx$ given by the trapezoidal rule with n subintervals. Without appealing to the error formula given in the text, prove that $\lim_{n \to +\infty} T_n = \int_a^b f(x)\, dx$.

78. Show that if $f(x) = Ax^3 + Bx^2 + Cx + D$, then Simpson's rule gives the exact value of $\int_a^b f(x)\, dx$.

79. When integration by parts is used to find $\int e^x \cosh x\, dx$, what happens? Explain. Find $\int e^x \cosh x\, dx$ without using integration by parts.

80. Assume that a population grows according to *Verhurst's logistic law of population growth*, that is, $dN/dt = AN - BN^2$, where $N = N(t)$ is the population at time t (in years), and the constants A and B are the *vital coefficients* of the population. Assume that the vital coefficients have the values $A = 10^5$ and $B = 0.01$. If initially the size of this population was 12×10^5, what will its size be after 10 years? What will the population be after a very long time?

81. The *White Lightning* (the first human-powered vehicle to exceed 55 miles per hour, built by Northrop University students) is hypothesized to accelerate according to the function $a(t) = t^2/(20\sqrt{t^3 + 4})$ meters per second per second. What is its velocity at time t if the initial velocity is 0?

82. *Computer problem.* Another approach to approximating an integral is called the *Monte Carlo method*. Suppose we have a dart board consisting of a circle inscribed in a square that measures 2 meters on each side (see the figure). Suppose we randomly throw 10,000 darts. (By "randomly" we mean that a dart is equally likely to land anywhere within the square.) If we hit the circle 8567 times, it is reasonable to approximate the area of the circle as

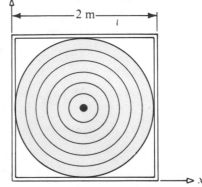

$$\frac{\text{Hits}}{\text{Total}} \times \text{Area of square} = \frac{8567}{10,000} \times 4 = 3.4268$$

(Note that the area of the circle is $\pi R^2 = \pi = 3.14159\ldots$, so that we have an approximation to π.)

It would be rather tedious actually to construct the dart board and throw 10,000 darts, and it would be impossible to throw the darts in a truly random manner. But, fortunately, we can simulate dart throwing on a computer. All that is needed is a random number generator and a few lines of computer coding. Suppose the command RND(1) returns a random number between 0 and 1. By multiplying by 2, we obtain a random number between 0 and 2. If we call RND(1) 20,000 times, we will obtain 10,000 pairs of numbers, or 10,000 points in our square. Then all we need to do is determine which points are within the circle, and keep track of the number of hits. A portion of the computer program in the BASIC language* might be

$$\text{HIT} = \emptyset$$
$$\text{FOR I} = 1 \text{ TO } 1\emptyset,\emptyset\emptyset\emptyset$$

---

* The variable names in this program may have to be adjusted to be compatible with your particular system.

$$X = 2 * \text{RND}(1)$$
$$Y = 2 * \text{RND}(1)$$
$$\text{IF } (X - 1) \wedge 2 + (Y - 1) \wedge 2 < 1 \text{ THEN}$$
$$\text{HIT} = \text{HIT} + 1$$
$$\text{NEXT I}$$

Write a program to estimate the area of the circle using the Monte Carlo method.

83. *Computer problem.* Use the Monte Carlo method (see Problem 82) to approximate the integrals:

(a) $\int_0^1 \dfrac{dx}{1 + x^2}$ (b) $\int_0^1 \dfrac{dx}{1 + x}$

Compare your results to the approximations of Examples 1 and 3 in Section 11.

84. *Buffon needle problem.* A floor pattern consists of lines 1 meter apart. Find the probability that a meter stick thrown at random hits a crack (see the figure).

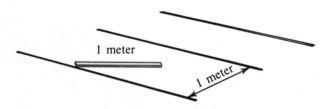

[*Hint:* Let d = Distance from bottom of stick to next higher line   and let θ = Angle from horizontal to stick,   as indicated:

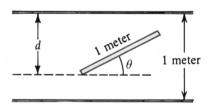

The inequalities $0 \le d \le 1,\ \ 0 \le \theta \le \pi$ represent all possibilities:

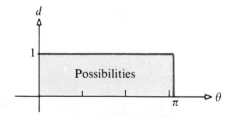

Show that the meter stick hits a crack if and only if $d \leq \sin \theta$, so that the shaded area in the figure represents the hits. Deduce that the desired probability is $\int_0^\pi \sin \theta \, d\theta / \pi$.]

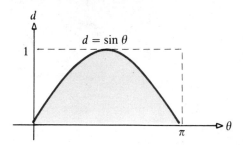

85. *Computer problem.* Simulate Problem 84 on a computer. Compare the approximation obtained with the exact result obtained in Problem 84.

10

Indeterminate Forms; Improper Integrals; Taylor Polynomials

1. The Indeterminate Forms 0/0 and ∞/∞
2. Other Indeterminate Forms
3. Improper Integrals
4. Taylor Polynomials
   Miscellaneous Exercises

1. The Indeterminate Forms 0/0 and ∞/∞

To evaluate

$$\lim_{x \to c} \frac{f(x)}{g(x)}$$

we usually try to use the rule

(10.1)
$$\lim_{x \to c} \frac{f(x)}{g(x)} = \frac{\lim_{x \to c} f(x)}{\lim_{x \to c} g(x)}$$

However, in many cases, this result cannot be applied.

For example, to find

$$\lim_{x \to 2} \frac{x^2 - 4}{x - 2}$$

we do not use (10.1) since the numerator and the denominator each approach 0. That is,

$$\frac{\lim_{x \to 2}(x^2 - 4)}{\lim_{x \to 2}(x - 2)} \qquad \text{leads to} \qquad \frac{0}{0}$$

Instead, to find this limit, we rely on algebra to get

$$\lim_{x \to 2} \frac{x^2 - 4}{x - 2} = \lim_{x \to 2} \frac{(x + 2)(x - 2)}{x - 2} = \lim_{x \to 2}(x + 2) = 4$$

To find

$$\lim_{x \to +\infty} \frac{3x - 2}{x + 5}$$

we do not use (10.1) since the numerator and the denominator each approach infinity. That is,

$$\frac{\lim_{x \to +\infty}(3x - 2)}{\lim_{x \to +\infty}(x + 5)} \qquad \text{leads to} \qquad \frac{\infty}{\infty}$$

Instead, to find this limit, we divide the numerator and the denominator by x and find

$$\lim_{x \to +\infty} \frac{3 - (2/x)}{1 + (5/x)} = 3$$

As another example, to find

$$\lim_{x \to 0} \frac{\sin x}{x}$$

we do not use (10.1) since it leads to 0/0. Instead, we use a geometric argument (see Section 7, Chap. 3) to show that

$$\lim_{x \to 0} \frac{\sin x}{x} = 1$$

Whenever the application of (10.1) leads to 0/0 or ∞/∞, we say that $f(x)/g(x)$ is *an indeterminate form*. More precisely:

The quotient $f(x)/g(x)$ is an *indeterminate form at c* if

(10.2) $$\lim_{x \to c} f(x) = 0 \qquad \text{and} \qquad \lim_{x \to c} g(x) = 0 \qquad \text{Type } \frac{0}{0}$$

or

(10.3) $$\lim_{x \to c} f(x) = \pm\infty \qquad \text{and} \qquad \lim_{x \to c} g(x) = \pm\infty \qquad \text{Type } \frac{\infty}{\infty}$$

The word *indeterminate* conveys the idea that the limit cannot be determined without additional work. However, a remarkable theorem, named after the French mathematician Guillaume Francois de L'Hospital* (1661–1704), provides a simple and general method for finding the limit of an indeterminate form.

---

* Pronounced "lōpital" and sometimes spelled L'Hôpital.

(10.4)   **THEOREM**   *L'Hospital's Rule.**   **Suppose f and g are two functions that are differentiable at every point in an open interval I containing c, except possibly at c itself, and suppose $g'(x) \neq 0$ for all $x \neq c$ in I. Let L denote either a real number or $+\infty$ or $-\infty$, and suppose $f(x)/g(x)$ is an indeterminate form at c.**

$$\text{If} \quad \lim_{x \to c} \frac{f'(x)}{g'(x)} = L, \quad \text{then} \quad \lim_{x \to c} \frac{f(x)}{g(x)} = L$$

A partial proof of L'Hospital's rule is given at the end of this section.

Under the conditions stated in the theorem, the limit of a quotient of two functions equals the limit of the quotient of their derivatives. Let's look at some examples to see how the rule is applied.

EXAMPLE 1   Use L'Hospital's rule to verify that $\lim\limits_{x \to 0} \dfrac{\sin x}{x} = 1$.

Solution   Let $f(x) = \sin x$ and $g(x) = x$.   Since $\lim_{x \to 0} f(x) = \lim_{x \to 0} \sin x = 0$ and $\lim_{x \to 0} g(x) = \lim_{x \to 0} x = 0$,   we find that $f(x)/g(x) = (\sin x)/x$ is an indeterminate form at 0 of the type 0/0. Since

$$\lim_{x \to 0} \frac{f'(x)}{g'(x)} = \lim_{x \to 0} \frac{\cos x}{1} = 1$$

it follows by L'Hospital's rule that

$$\lim_{x \to 0} \frac{\sin x}{x} = 1 \quad \blacksquare$$

In the solution of Example 1 we were careful to determine that the limit of the ratio of the derivatives, namely,

$$\lim_{x \to 0} \frac{\cos x}{1}$$

existed or became infinite *before* applying L'Hospital's rule. However, to avoid complicating things, the usual practice is to proceed directly as follows:

$$\lim_{x \to 0} \frac{\sin x}{x} = \lim_{x \to 0} \frac{\cos x}{1} = 1$$

Apply   Evaluate
L'Hospital's   limit
rule

A word of caution! L'Hospital's rule is applied by taking the derivative of the numerator and the derivative of the denominator *separately*. Be careful not to take the derivative of the quotient.

---

* The result was actually discovered by his teacher, Johann Bernoulli, who sent L'Hospital his proof in a letter in 1694.

The next example illustrates that sometimes we may have to apply L'Hospital's rule more than once.

EXAMPLE 2   Evaluate: $\lim\limits_{x\to 0}\dfrac{\sin x - x}{x^2}$

Solution   Since $\lim_{x\to 0}(\sin x - x) = 0$ and $\lim_{x\to 0} x^2 = 0$,   we have an indeterminate form at 0 of the type 0/0. Applying L'Hospital's rule, we find

$$\lim_{x\to 0}\frac{\sin x - x}{x^2} = \lim_{x\to 0}\frac{\cos x - 1}{2x}$$

Since $\lim_{x\to 0}(\cos x - 1) = 0$ and $\lim_{x\to 0} 2x = 0$,   we still have an indeterminate form at 0 of the type 0/0. Hence, we apply the rule once more. Then

$$\lim_{x\to 0}\frac{\cos x - 1}{2x} = \lim_{x\to 0}\frac{-\sin x}{2}$$

Since the latter limit equals 0, we conclude that

$$\lim_{x\to 0}\frac{\sin x - x}{x^2} = 0 \quad\blacksquare$$

The statement (10.4) is also true if $c = +\infty$ or $c = -\infty$.   The next examples illustrate the use of (10.4) for $c = +\infty$ as it is applied to indeterminate forms at $+\infty$ of the type ∞/∞.

EXAMPLE 3   Evaluate: (a) $\lim\limits_{x\to +\infty}\dfrac{\ln x}{x}$ (b) $\lim\limits_{x\to +\infty}\dfrac{x}{e^x}$

Solution   (a)   Since $\lim_{x\to +\infty} \ln x = +\infty$ and $\lim_{x\to +\infty} x = +\infty$,   we have an indeterminate form at $+\infty$ of the type ∞/∞. By L'Hospital's rule, we have

$$\lim_{x\to +\infty}\frac{\ln x}{x} = \lim_{x\to +\infty}\frac{1/x}{1} = \lim_{x\to +\infty}\frac{1}{x} = 0$$

(b)   Since $\lim_{x\to +\infty} x = +\infty$ and $\lim_{x\to +\infty} e^x = +\infty$,   we have an indeterminate form at $+\infty$ of the type ∞/∞. Hence, by L'Hospital's rule, we have

$$\lim_{x\to +\infty}\frac{x}{e^x} = \lim_{x\to +\infty}\frac{1}{e^x} = 0 \quad\blacksquare$$

The results obtained in this example tell us that x grows faster than $\ln x$, while e^x grows faster than x. In fact, this is true for any *power* of x. That is, for α a positive number,

(10.5) $$\lim_{x\to +\infty}\frac{\ln x}{x^\alpha} = 0 \qquad \lim_{x\to +\infty}\frac{x^\alpha}{e^x} = 0$$

You are asked to verify these results in Problems 33 and 35 of Exercise 1.

EXAMPLE 4   Evaluate: $\displaystyle\lim_{x \to -1} \frac{3x^2 + 2x - 1}{x^2 + x}$

Solution   Since $\lim_{x \to -1}(3x^2 + 2x - 1) = 0$ and $\lim_{x \to -1}(x^2 + x) = 0$, we have an inde-terminate form at -1 of the type 0/0. Hence, we apply L'Hospital's rule:

$$\lim_{x \to -1} \frac{3x^2 + 2x - 1}{x^2 + x} = \lim_{x \to -1} \frac{6x + 2}{2x + 1} = 4 \quad \blacksquare$$

Before applying L'Hospital's rule, it is important to check on the indeterminacy of the expression since a routine and thoughtless application may yield an incorrect result. For instance, in Example 4, note that in obtaining the last equality, we use the fact that the quotient $(6x + 2)/(2x + 1)$ is continuous at $x = -1$, so that a direct substitution yields 4. Be careful not to apply L'Hospital's rule to expressions such as $(6x + 2)/(2x + 1)$. In this case, you would obtain $\frac{6}{2} = 3$, which is a wrong answer.

The next example illustrates how an early simplification may reduce the effort needed to solve the problem.

EXAMPLE 5   Evaluate: $\displaystyle\lim_{x \to 0} \frac{\tan x - \sin x}{x^2 \tan x}$

Solution   This is an example of an indeterminate form of the type 0/0. However, due to the complicated nature of the denominator, we attempt to simplify using the fact that $\tan x = (\sin x)/(\cos x)$ *before* applying L'Hospital's rule. The result is

$$\frac{\tan x - \sin x}{x^2 \tan x} = \frac{\dfrac{\sin x}{\cos x} - \sin x}{x^2 \dfrac{\sin x}{\cos x}} = \frac{\dfrac{\sin x - \sin x \cos x}{\cos x}}{\dfrac{x^2 \sin x}{\cos x}} = \frac{\sin x(1 - \cos x)}{x^2 \sin x} = \frac{1 - \cos x}{x^2}$$

Now,

$$\lim_{x \to 0} \frac{\tan x - \sin x}{x^2 \tan x} = \lim_{x \to 0} \frac{1 - \cos x}{x^2} \underset{\substack{\uparrow \\ \text{Apply} \\ \text{L'Hospital's rule}}}{=} \lim_{x \to 0} \frac{\sin x}{2x} = \frac{1}{2}$$

Compare this solution to one that applies L'Hospital's rule to the original expression.
                                                                                                          $\blacksquare$

L'Hospital's rule also may be used for one-sided limits.

EXAMPLE 6   Evaluate: $\displaystyle\lim_{x \to 0^+} \frac{\cot x}{\ln x}$

Solution   Since $\lim_{x \to 0^+} \cot x = +\infty$ and $\lim_{x \to 0^+} \ln x = -\infty$,   we have an indeterminate form at 0 of the type ∞/∞. Hence,

$$\lim_{x \to 0^+} \frac{\cot x}{\ln x} \underset{\substack{\uparrow \\ \text{L'Hospital's} \\ \text{rule}}}{=} \lim_{x \to 0^+} \frac{-\csc^2 x}{1/x} = -\lim_{x \to 0^+} \frac{x}{\sin^2 x}$$

$$\underset{\substack{\uparrow \\ \text{L'Hospital's} \\ \text{rule}}}{=} -\lim_{x \to 0^+} \frac{1}{2 \sin x \cos x} = -\infty \quad \blacksquare$$

PARTIAL PROOF OF L'HOSPITAL'S RULE

To prove L'Hospital's rule, we use a formula that bears the name of the French mathematician Augustin Cauchy (1789–1857) and is an extension of the mean value theorem (4.26).

(10.6)   **Cauchy's Mean Value Theorem.**   **Let f and g be continuous on the closed interval $[a, b]$ and differentiable on the open interval (a, b). If g' is never 0 on (a, b), there exists a number c in (a, b) such that**

$$\frac{f'(c)}{g'(c)} = \frac{f(b) - f(a)}{g(b) - g(a)}$$

Remark: Under the conditions stated, $g(b)$ and $g(a)$ cannot be equal, because if $g(b) = g(a)$, then Rolle's theorem (4.25) would assert that g' takes on the value 0 somewhere in (a, b). Since g' is never 0 on (a, b), we conclude that $g(b) \neq g(a)$.

Proof of Cauchy's Mean Value Theorem   We prove this result by applying the mean value theorem to the function h defined by

$$h(x) = [g(b) - g(a)][f(x) - f(a)] - [g(x) - g(a)][f(b) - f(a)] \qquad a \le x \le b$$

We note that h is continuous on $[a, b]$, is differentiable on (a, b), and that $h(a) = h(b) = 0$.   Applying Rolle's theorem to h produces a number c in (a, b) such that $h'(c) = 0$.   That is,

$$h'(c) = [g(b) - g(a)]f'(c) - g'(c)[f(b) - f(a)] = 0$$

Thus,

(10.7) $$\frac{f'(c)}{g'(c)} = \frac{f(b) - f(a)}{g(b) - g(a)} \qquad \blacksquare$$

If we let $g(x) = x$ in (10.7), we obtain the familiar mean value theorem. A partial proof of L'Hospital's rule follows.

Proof of L'Hospital's Rule   Suppose that $f(x)/g(x)$ is an indeterminate form at c of the type 0/0, and suppose $\lim_{x \to c}[f'(x)/g'(x)] = L$,   where L is a real number.

We wish to prove that $\lim_{x \to c}[f(x)/g(x)] = L$.   First, we introduce the functions F and G as follows:

(10.8)
$$F(x) = \begin{cases} f(x) & \text{if } x \neq c \\ 0 & \text{if } x = c \end{cases} \qquad G(x) = \begin{cases} g(x) & \text{if } x \neq c \\ 0 & \text{if } x = c \end{cases}$$

Both F and G are continuous at c since

$$\lim_{x \to c} F(x) = \lim_{x \to c} f(x) = 0 = F(c) \qquad \lim_{x \to c} G(x) = \lim_{x \to c} g(x) = 0 = G(c)$$

Moreover,

$$F'(x) = f'(x) \qquad G'(x) = g'(x)$$

for all x on the given interval I, except possibly at c. Since the conditions for Cauchy's mean value theorem are met by the functions F and G in either $[x, c]$ or $[c, x]$, there is a number u between c and x so that

(10.9)
$$\frac{F(x) - F(c)}{G(x) - G(c)} = \frac{F'(u)}{G'(u)} = \frac{f'(u)}{g'(u)}$$

By using (10.8), this simplifies to

$$\frac{f(x)}{g(x)} = \frac{f'(u)}{g'(u)}$$

Since u is between c and x, it follows from the squeezing theorem (2.20) that

$$\lim_{x \to c} \frac{f(x)}{g(x)} = \lim_{u \to c} \frac{f'(u)}{g'(u)} = L$$

A similar argument may be given if L is infinite. The proof when $f(x)/g(x)$ is an indeterminate form at $+\infty$ of the type ∞/∞ is omitted here, but it may be found in books on advanced calculus. ■

We justify the use of L'Hospital's rule when $c = +\infty$ for an indeterminate form of the type 0/0 by the following argument: In $\lim_{x \to +\infty}[f(x)/g(x)]$, let $x = 1/u$.   Then, as $x \to +\infty$, $u \to 0^+$, and

$$\lim_{x \to +\infty} \frac{f(x)}{g(x)} = \lim_{u \to 0^+} \frac{f\left(\dfrac{1}{u}\right)}{g\left(\dfrac{1}{u}\right)} = \lim_{u \to 0^+} \frac{\dfrac{d}{du} f\left(\dfrac{1}{u}\right)}{\dfrac{d}{du} g\left(\dfrac{1}{u}\right)}$$

$$= \underset{\uparrow}{\lim_{u \to 0^+}} \frac{-\dfrac{1}{u^2} f'\left(\dfrac{1}{u}\right)}{-\dfrac{1}{u^2} g'\left(\dfrac{1}{u}\right)} \underset{\uparrow}{=} \lim_{x \to +\infty} \frac{f'(x)}{g'(x)} = L$$

Chain rule $x = 1/u$

A similar argument handles the case $c = -\infty$ (see Problem 67 in Exercise 1).

EXERCISE 1

In Problems 1–56 evaluate each limit.

1. $\lim\limits_{x \to 2} \dfrac{x^2 + x - 6}{x^2 - 3x + 2}$

2. $\lim\limits_{x \to -1} \dfrac{x^2 + 5x + 4}{x^2 - 4x - 5}$

3. $\lim\limits_{x \to 1} \dfrac{2x^3 + 5x^2 - 4x - 3}{x^3 + x^2 - 10x + 8}$

4. $\lim\limits_{x \to 0} \dfrac{x^3 - 3x^2 + 5x}{x^3 - x}$

5. $\lim\limits_{x \to +\infty} \dfrac{3x^2 + 7}{4x^2 - 5}$

6. $\lim\limits_{x \to +\infty} \dfrac{4x^3 + 7}{3x^5 + 6}$

7. $\lim\limits_{x \to 1} \dfrac{\ln x}{x^2 - 1}$

8. $\lim\limits_{x \to 1} \dfrac{\sin \pi x}{x - 1}$

9. $\lim\limits_{x \to 0} \dfrac{e^x - e^{-x}}{\sin x}$

10. $\lim\limits_{x \to 0} \dfrac{\tan 2x}{\ln(1 + x)}$

11. $\lim\limits_{x \to \pi} \dfrac{1 + \cos x}{\sin 2x}$

12. $\lim\limits_{x \to 0} \dfrac{\ln(1 - x)}{e^x - 1}$

13. $\lim\limits_{x \to +\infty} \dfrac{x^2}{e^x}$

14. $\lim\limits_{x \to +\infty} \dfrac{x^4 + x^3}{e^x + 1}$

15. $\lim\limits_{x \to 0^+} \dfrac{\cot x}{\cot 2x}$

16. $\lim\limits_{x \to +\infty} \dfrac{\ln x}{a^x}$

17. $\lim\limits_{x \to +\infty} \dfrac{x^2 + x - 1}{e^x + e^{-x}}$

18. $\lim\limits_{x \to +\infty} \dfrac{x + \ln x}{x \ln x}$

19. $\lim\limits_{x \to 0} \dfrac{\sin^2 x}{x}$

20. $\lim\limits_{x \to 0} \dfrac{\tan x - x}{\sin x}$

21. $\lim\limits_{x \to 0} \dfrac{e^x + e^{-x} - 2}{x^2}$

22. $\lim\limits_{x \to 0} \dfrac{e^x + e^{-x} - 2}{\sin^2 x}$

23. $\lim\limits_{x \to 0} \dfrac{\cos x - 1}{\cos 2x - 1}$

24. $\lim\limits_{x \to 0} \dfrac{e^x - 1 - \sin x}{1 - \cos x}$

25. $\lim\limits_{x \to 0} \dfrac{\sin x - x}{x^3}$

26. $\lim\limits_{x \to 0} \dfrac{\tan x - x}{\cos x - 1}$

27. $\lim\limits_{x \to 0} \dfrac{e^x - e^{-x} - 2 \sin x}{3x^3}$

28. $\lim\limits_{x \to 0} \dfrac{\tan x - \sin x}{x^3}$

29. $\lim\limits_{x \to +\infty} \dfrac{\ln(\ln x)}{\ln x}$

30. $\lim\limits_{x \to 1/2^-} \dfrac{\ln(1 - 2x)}{\tan \pi x}$

31. $\lim\limits_{x \to 1^-} \dfrac{\ln(1 - x)}{\cot \pi x}$

32. $\lim\limits_{x \to 1^+} \dfrac{\ln(\ln x)}{1/(1 - x)}$

33. $\lim\limits_{x \to +\infty} \dfrac{\ln x}{x^\alpha} \quad (\alpha > 0)$

34. $\lim\limits_{x \to +\infty} \dfrac{\ln x}{e^x}$

35. $\lim\limits_{x \to +\infty} \dfrac{x^\alpha}{e^x} \quad (\alpha > 0)$

36. $\lim\limits_{x \to +\infty} \dfrac{\cosh x}{x}$

37. $\lim\limits_{x \to 0} \dfrac{xe^{4x} - x}{1 - \cos 2x}$

38. $\lim\limits_{x \to 0} \dfrac{x \tan x}{1 - \cos x}$

39. $\lim\limits_{x \to \pi/4} \dfrac{1 + \cos 4x}{\sec^2 x - 2 \tan x}$

40. $\lim\limits_{x \to 0} \dfrac{\sec x - 1}{x^2 \sec x}$

41. $\lim\limits_{x \to 0} \dfrac{2^x - 1}{3^x - 1}$

42. $\lim\limits_{x \to 0} \dfrac{a^x - b^x}{x}$

43. $\lim\limits_{x \to \pi/4} \dfrac{\cos^2 2x}{1 - \tan x}$

44. $\lim\limits_{x \to 0} \dfrac{\tan^{-1} x}{x}$

45. $\lim\limits_{x \to 0} \dfrac{e^x - \ln(1 + x) - 1}{x^2}$

46. $\lim\limits_{x \to 0} \dfrac{2x - \sin^{-1} x}{2 \tan^{-1} x - x}$

47. $\lim\limits_{x \to 0} \dfrac{x - \tan^{-1} x}{x^3}$

48. $\lim\limits_{x \to 0} \dfrac{\tan 3x}{x \cos x}$

49. $\lim\limits_{x \to 0} \dfrac{e^{2x} - 1}{x^2 - \sin x}$

50. $\lim\limits_{x \to 0} \dfrac{\sin 2x - \sin x}{\tan 3x}$

51. $\lim\limits_{x \to 0} \dfrac{\tan^{-1} x}{\sin^{-1} x}$

52. $\lim\limits_{x \to 0} \dfrac{\ln(\sec 2x)}{\ln(\sec x)}$

53. $\lim\limits_{x \to 0} \dfrac{\sinh x}{x}$

54. $\lim\limits_{x \to 0} \dfrac{\cosh x - 1}{x^2}$

55. $\lim\limits_{x \to 0} \dfrac{\ln(\cosh x)}{x}$

56. $\lim\limits_{x \to 0} \dfrac{\tanh^{-1}x}{x}$

57. Discuss the behavior of $\dfrac{\sin x - \tan x}{e^x + e^{-x} - 2}$ as $x \to 0$.

58. Discuss the behavior of $\dfrac{1 - \cos x - x \sin x}{2 - 2 \cos x - \sin^2 x}$ as $x \to 0$.

59. If α and β are positive real numbers, show that $\lim\limits_{x \to +\infty} \dfrac{(\ln x)^\beta}{x^\alpha} = 0$.

60. If α and β are positive real numbers, show that $\lim\limits_{x \to +\infty} \dfrac{x^\alpha}{e^{\beta x}} = 0$.

61. Evaluate: $\lim\limits_{x \to 0^+} \dfrac{\int_0^x \cos t^2 \, dt}{\int_0^x e^{t^2} \, dt}$

62. Evaluate: $\lim\limits_{x \to 0^+} \dfrac{\int_0^x \sin t^3 \, dt}{x^4}$

63. If α, $\beta \neq 0$, and $c > 0$ are real numbers, show that $\lim\limits_{x \to c} \dfrac{x^\alpha - c^\alpha}{x^\beta - c^\beta} = \dfrac{\alpha}{\beta} c^{\alpha - \beta}$.

64. The equation governing the amount of current I (in amperes) in a simple RL circuit consisting of a resistance R (in ohms), an inductance L (in henrys), and an electromotive force E (in volts) is

$$I = \frac{E}{R}(1 - e^{-Rt/L})$$

Find $\lim\limits_{t \to +\infty} I(t)$ and $\lim\limits_{R \to 0^+} I(t)$.

65. The sum S_n of the first n terms of a geometric series is given by the formula

$$S_n = a + ar + ar^2 + \cdots + ar^{n-1} = \frac{a(r^n - 1)}{r - 1}$$

Evaluate $\lim\limits_{r \to 1} \dfrac{a(r^n - 1)}{r - 1}$, and compare the result with a geometric series in which $r = 1$.

66. Suppose f is a continuous function. Evaluate $\lim\limits_{x \to 0} \left[\dfrac{1}{x} \int_0^x f(t) \, dt \right]$.

67. Prove L'Hospital's rule when $f(x)/g(x)$ is an indeterminate form at $-\infty$ of the type 0/0.

2. Other Indeterminate Forms

L'Hospital's rule only applies to indeterminate forms of the types 0/0 and ∞/∞. By algebraic rearrangement, indeterminate forms of the types $0 \cdot \infty$, $\infty - \infty$,* 0^0, 1^∞, and ∞^0 often may be transformed into the type 0/0 or the type ∞/∞, and then L'Hospital's rule may be applied.

---

* The indeterminate form of the type $\infty - \infty$ is a convenient notation for any of the following: $(+\infty) - (+\infty)$; $(-\infty) - (-\infty)$; $(+\infty) + (-\infty)$.   Note that $(+\infty) + (+\infty) = +\infty$ and $(-\infty) + (-\infty) = -\infty$ are not indeterminate forms.

If a function $F(x) = f(x)g(x)$ results in the indeterminate form $0 \cdot \infty$, we may write the product $f(x)g(x)$ in quotient form as

$$f(x)g(x) = \frac{f(x)}{1/g(x)} \qquad \text{or} \qquad f(x)g(x) = \frac{g(x)}{1/f(x)}$$

which will produce one of the indeterminate forms 0/0 or ∞/∞. This may then be handled by L'Hospital's rule. The choice of the quotient form is usually dictated by the ease of differentiation. Let's look at two examples.

EXAMPLE 1   Evaluate: $\lim\limits_{x \to 0^+} (x \ln x)$

Solution   This is an indeterminate form of the type $\infty \cdot 0$ since $\lim_{x \to 0^+} x = 0$ and $\lim_{x \to 0^+} \ln x = -\infty$. To transform this to the type 0/0 or ∞/∞, we write

$$\lim_{x \to 0^+} (x \ln x) = \lim_{x \to 0^+} \frac{\ln x}{1/x}$$

Since the limit on the right is an indeterminate form of the type ∞/∞, we may apply L'Hospital's rule. Then

$$\lim_{x \to 0^+} (x \ln x) \underset{\substack{\uparrow \\ \text{Algebra}}}{=} \lim_{x \to 0^+} \frac{\ln x}{1/x} \underset{\substack{\uparrow \\ \text{L'Hospital's} \\ \text{rule}}}{=} \lim_{x \to 0^+} \frac{1/x}{-1/x^2} \underset{\substack{\uparrow \\ \text{Algebra}}}{=} \lim_{x \to 0^+} (-x) = 0 \qquad \blacksquare$$

The idea behind the solution given in Example 1 is to replace a product by a quotient. We chose to replace $x \ln x$ by $\dfrac{\ln x}{1/x}$. Why did we make this choice instead of $\dfrac{x}{1/\ln x}$? The answer lies in the subsequent step, where we apply L'Hospital's rule. The choice we made results in a limit that is easy to find; the other choice results in a limit that is more complicated than the original one. Again, experience is the best teacher.

You might be tempted to argue that $0 \cdot \infty$ is 0 since "anything" times 0 is 0. This is not true, because $0 \cdot \infty$ is not a product of numbers—it is a statement about limits. The next example illustrates an indeterminate form of the type $0 \cdot \infty$ that yields a nonzero limit.

EXAMPLE 2   Evaluate: $\lim\limits_{x \to +\infty} \left(x \sin \dfrac{1}{x} \right)$

Solution   This is an indeterminate form of the type $0 \cdot \infty$ since $\lim_{x \to +\infty} x = +\infty$ and $\lim_{x \to +\infty} \sin(1/x) = 0$. To transform this, we write

$$\lim_{x \to +\infty} \left(x \sin \frac{1}{x} \right) \underset{\substack{\uparrow \\ \text{Algebra}}}{=} \lim_{x \to +\infty} \frac{\sin(1/x)}{1/x} \underset{\substack{\uparrow \\ \text{Set} \\ t = 1/x}}{=} \lim_{t \to 0^+} \frac{\sin t}{t} = 1 \qquad \blacksquare$$

If the limit of a function results in the indeterminate form $\infty - \infty$, it is generally possible to rewrite the function (usually by using algebra or trigonometry) so that it changes to a quotient, producing one of the indeterminate forms $0/0$ or ∞/∞. Let's look at an example.

EXAMPLE 3   Evaluate: $\displaystyle\lim_{x \to 0^+} \left(\frac{1}{x} - \frac{1}{\sin x} \right)$

Solution   This is an indeterminate form of the type $\infty - \infty$, since $\lim_{x \to 0^+} (1/x) = +\infty$ and $\lim_{x \to 0^+} (1/\sin x) = +\infty$. To transform this, we write the difference as a single fraction:

$$\lim_{x \to 0^+} \left(\frac{1}{x} - \frac{1}{\sin x} \right) = \lim_{x \to 0^+} \frac{\sin x - x}{x \sin x}$$

The limit on the right is an indeterminate form of the type $0/0$. By applying L'Hospital's rule twice, we get

$$\lim_{x \to 0^+} \left(\frac{1}{x} - \frac{1}{\sin x} \right) \underset{\underset{\text{Algebra}}{\uparrow}}{=} \lim_{x \to 0^+} \frac{\sin x - x}{x \sin x} \underset{\underset{\substack{\text{L'Hospital's} \\ \text{rule}}}{\uparrow}}{=} \lim_{x \to 0^+} \frac{\cos x - 1}{\sin x + x \cos x}$$

$$\underset{\underset{\substack{\text{L'Hospital's} \\ \text{rule}}}{\uparrow}}{=} \lim_{x \to 0^+} \frac{-\sin x}{\cos x + \cos x - x \sin x}$$

$$= \lim_{x \to 0^+} \frac{-\sin x}{2 \cos x - x \sin x} = 0 \qquad \blacksquare$$

A function of the form $f(x)^{g(x)}$ may result in one of the indeterminate forms 1^∞, 0^0, or ∞^0. To handle these, we set $y = f(x)^{g(x)}$ and take logarithms to obtain

$$\ln y = \ln f(x)^{g(x)} = g(x) \ln f(x)$$

The expression on the right is then an indeterminate form of the type $0 \cdot \infty$, which can be evaluated by techniques already discussed.

The four steps listed below are followed when we want to evaluate $\lim_{x \to c} f(x)^{g(x)}$ and this limit results in one of the indeterminate forms 0^0, 1^∞, or ∞^0:

(10.10)   *Step 1.*   **Set $y = f(x)^{g(x)}$.**

Step 2.   **Take logarithms: $\ln y = g(x) \ln f(x)$.**

Step 3.   **Evaluate $\displaystyle\lim_{x \to c} \ln y$,   if it exists.**

Step 4.   **If $\displaystyle\lim_{x \to c} \ln y = L$,   then $\displaystyle\lim_{x \to c} y = e^L$.***

---

* If $\lim_{x \to c} \ln y = L$,   then $\lim_{x \to c} e^{\ln y} = \lim_{x \to c} y = e^L$.

These four steps may be used if $x \to +\infty$ or $x \to -\infty$, or if one-sided limits are involved.

EXAMPLE 4   Evaluate: $\lim\limits_{x \to 0^+} x^x$

Solution   This expression is of the form 0^0. We follow (10.10):

Step 1.   Let $y = x^x$.

Step 2. $\ln y = x \ln x$

Step 3. $\lim_{x \to 0^+} \ln y = \lim_{x \to 0^+} x \ln x = 0$ (by Example 1)

Step 4.   Since $\lim_{x \to 0^+} \ln y = 0$, $\lim_{x \to 0^+} y = e^0 = 1$,   so that $\lim_{x \to 0^+} x^x = 1$.
           ∎

Caution: Do not stop after showing that $\lim_{x \to c} \ln y = L$ and conclude that $\lim_{x \to c} f(x)^{g(x)} = L$. What we are looking for is $\lim_{x \to c} y$, which equals e^L.

EXAMPLE 5   Evaluate: $\lim\limits_{x \to 0^+} (1 + x)^{1/x}$

Solution   This expression is of the form 1^∞. We follow (10.10):

Step 1.   Let $y = (1 + x)^{1/x}$.

Step 2. $\ln y = \ln(1 + x)^{1/x} = (1/x) \ln(1 + x)$

Step 3. $\lim\limits_{x \to 0^+} \ln y = \lim\limits_{x \to 0^+} \dfrac{\ln(1 + x)}{x} = \lim\limits_{x \to 0^+} \dfrac{1/(1 + x)}{1} = 1$

                                   L'Hospital's
                                    rule

Step 4.   Since $\lim_{x \to 0^+} \ln y = 1$, $\lim_{x \to 0^+} y = e^1 = e$,   so that
$\lim_{x \to 0^+} (1 + x)^{1/x} = e$.   ∎

We close this section with some important limits that we shall have occasion to refer to in Chapter 11.

EXAMPLE 6   Show that:

     (a) $\lim\limits_{n \to +\infty} \sqrt[n]{\alpha} = 1$ α **any positive real number**

(10.11)   (b) $\lim\limits_{n \to +\infty} \sqrt[n]{n} = 1$

     (c) $\lim\limits_{n \to +\infty} \left(1 + \dfrac{\alpha}{n}\right)^n = e^\alpha$ α **any real number**

Solution   (a)   Set $y = \sqrt[n]{\alpha} = \alpha^{1/n}$.   Then $\ln y = \ln \alpha^{1/n} = (1/n) \ln \alpha$,   so that

$$\lim_{n \to +\infty} \ln y = \lim_{n \to +\infty} \frac{\ln \alpha}{n} = 0$$

Hence,

$$\lim_{n \to +\infty} y = \lim_{n \to +\infty} \sqrt[n]{\alpha} = e^0 = 1$$

(b)   Set $y = \sqrt[n]{n} = n^{1/n}$.   Then $\ln y = \ln n^{1/n} = (\ln n)/n$,   so that

$$\lim_{n \to +\infty} \ln y = \lim_{n \to +\infty} \frac{\ln n}{n} = \lim_{n \to +\infty} \frac{1/n}{1} = 0$$

Hence,

$$\lim_{n \to +\infty} y = \lim_{n \to +\infty} \sqrt[n]{n} = e^0 = 1$$

(c)   Set $y = (1 + \alpha/n)^n$.   Then $\ln y = n \ln(1 + \alpha/n)$,   so that

$$\lim_{n \to +\infty} \ln y = \lim_{n \to +\infty} n \ln\left(1 + \frac{\alpha}{n}\right) = \lim_{n \to +\infty} \frac{\ln(1 + \alpha/n)}{1/n}$$

$$= \lim_{n \to +\infty} \frac{\dfrac{-\alpha/n^2}{1 + \alpha/n}}{-1/n^2} = \lim_{n \to +\infty} \frac{\alpha}{1 + \alpha/n} = \alpha$$

Hence,

$$\lim_{n \to +\infty} y = \lim_{n \to +\infty} \left(1 + \frac{\alpha}{n}\right)^n = e^{\alpha} \quad \blacksquare$$

EXERCISE 2

In Problems 1–36 evaluate each limit.

1. $\lim_{x \to 0}(x \cot x)$

2. $\lim_{x \to \pi/2} (1 - \sin x) \tan x$

3. $\lim_{x \to +\infty} \left[x(e^{1/x} - 1)\right]$

4. $\lim_{x \to 0^+} (x^2 \ln x)$

5. $\lim_{x \to 0^+} \left[x \ln(\sin x)\right]$

6. $\lim_{x \to +\infty} (xe^{-x})$

7. $\lim_{x \to \pi/2} \left[\tan x \ln(\sin x)\right]$

8. $\lim_{x \to 0}[\csc x \ln(x + 1)]$

9. $\lim_{x \to 0}(x^2 \csc x \cot x)$

10. $\lim_{x \to \pi/4} (1 - \tan x) \sec 2x$

11. $\lim_{x \to -\infty} (x^2 e^x)$

12. $\lim_{x \to +\infty} (x - 1)e^{-x^2}$

13. $\lim_{x \to 0^+} (x^{\alpha} \ln x), \quad \alpha > 0$

14. $\lim_{x \to a}(a^2 - x^2) \tan \dfrac{\pi x}{2a}$

15. $\lim_{x \to 1^+} (1 - x) \tan \dfrac{1}{2} \pi x$

16. $\lim_{x \to 0^+} \left[\sin x \ln(\sin x)\right]$

17. $\lim_{x \to \pi/2} (\sec x - \tan x)$

18. $\lim_{x \to 0}\left(\csc \dfrac{1}{2} x - \cot \dfrac{1}{2} x\right)$

19. $\lim\limits_{x \to 0} \left(\cot x - \dfrac{1}{x} \right)$

20. $\lim\limits_{x \to 1} \left(\dfrac{1}{\ln x} - \dfrac{x}{\ln x} \right)$

21. $\lim\limits_{x \to 0} \left(\dfrac{1}{x} - \dfrac{1}{e^x - 1} \right)$

22. $\lim\limits_{x \to 1} \left(\dfrac{x}{x - 1} - \dfrac{1}{\ln x} \right)$

23. $\lim\limits_{x \to 1} \left(\dfrac{1}{\ln x} - \dfrac{1}{x - 1} \right)$

24. $\lim\limits_{x \to \pi/2} \left(x \tan x - \dfrac{\pi}{2} \sec x \right)$

25. $\lim\limits_{x \to 0^+} (2x)^{3x}$

26. $\lim\limits_{x \to 0} (1 + x^2)^{1/x^2}$

27. $\lim\limits_{x \to \pi/2} (\sin x)^{\tan x}$

28. $\lim\limits_{x \to 0^+} (\csc x)^{\sin x}$

29. $\lim\limits_{x \to 0^+} (1 + \sin x)^{\cot x}$

30. $\lim\limits_{x \to 0^+} \left(\dfrac{1}{x} \right)^{\sin x}$

31. $\lim\limits_{x \to +\infty} (1 + x^2)^{1/x}$

32. $\lim\limits_{x \to 0} (\cos x)^{1/x}$

33. $\lim\limits_{x \to 1^-} (1 - x)^{\tan \pi x}$

34. $\lim\limits_{x \to 0} (e^x + x)^{1/x}$

35. $\lim\limits_{x \to 0} \left(\dfrac{\sin x}{x} \right)^{1/x}$

36. $\lim\limits_{x \to \pi/2} (\csc x)^{\tan^2 x}$

37. Show that $\lim\limits_{x \to 0^+} (\cos x + 2 \sin x)^{\cot x} = e^2$.

38. Show that $\lim\limits_{x \to +\infty} \left(\dfrac{x + a}{x - a} \right)^x = e^{2a}$.

39. Find $\lim\limits_{x \to +\infty} \dfrac{P(x)}{e^x}$, where P is a polynomial function.

40. Find $\lim\limits_{x \to +\infty} [\ln(x + 1) - \ln(x - 1)]$.

41. Show that $\lim\limits_{x \to 0^+} \dfrac{e^{-1/x^2}}{x} = 0$. $\left[\textit{Hint:} \text{ Write } \dfrac{e^{-1/x^2}}{x} = \dfrac{1/x}{e^{1/x^2}}. \right]$

42. If n is an integer, show that $\lim\limits_{x \to 0^+} \dfrac{e^{-1/x^2}}{x^n} = 0$.

43. Show that the function below has a derivative at 0. What is $f'(0)$?　[*Hint:*　Use Problem 41.]

$$ f(x) = \begin{cases} e^{-1/x^2} & \text{if } x \neq 0 \\ 0 & \text{if } x = 0 \end{cases} $$

44. Sketch the graph of $y = \dfrac{1}{x} \tan x$, $-\pi/2 < x < \pi/2$.

3. Improper Integrals

　In previous discussions on the definite integral $\int_a^b f(x)\, dx$, we required that a and b both be numbers and that f be continuous throughout the closed interval $[a, b]$. However, in many situations, one or more of these assumptions are not met. For example, one of the limits of integration might be infinite; or the function f might not be defined at some number in $[a, b]$. In such cases, $\int_a^b f(x)\, dx$ is called an *improper integral*. The purpose of this section is to define improper integrals and to determine when they have a value.

We begin by discussing the situation in which one of the limits of integration is infinite.

If f is continuous for all $x \geq a$,   then

(10.12)
$$\int_a^{+\infty} f(x) \, dx = \lim_{b \to +\infty} \left[\int_a^b f(x) \, dx \right]$$

provided this limit exists. If the limit exists, the improper integral $\int_a^{+\infty} f(x) \, dx$ is said to *converge*. If the limit does not exist,* the improper integral $\int_a^{+\infty} f(x) \, dx$ is said to *diverge*.

If f is continuous for all $x \leq b$,   then

(10.13)
$$\int_{-\infty}^b f(x) \, dx = \lim_{a \to -\infty} \left[\int_a^b f(x) \, dx \right]$$

provided this limit exists. The terms *converge* and *diverge* are defined similarly.

EXAMPLE 1   Determine whether the following improper integrals converge or diverge:

(a) $\int_1^{+\infty} \frac{1}{x} \, dx$ (b) $\int_{-\infty}^0 e^x \, dx$ (c) $\int_1^{+\infty} \sin \frac{\pi}{2} x \, dx$

Solution   (a) $\int_1^{+\infty} \frac{1}{x} \, dx = \lim_{b \to +\infty} \left(\int_1^b \frac{1}{x} \, dx \right) = \lim_{b \to +\infty} \ln x \Big|_1^b = \lim_{b \to +\infty} \ln b = +\infty$

Thus, $\int_1^{+\infty} (1/x) \, dx$ diverges.

(b) $\int_{-\infty}^0 e^x \, dx = \lim_{a \to -\infty} \left(\int_a^0 e^x \, dx \right) = \lim_{a \to -\infty} e^x \Big|_a^0 = \lim_{a \to -\infty} (1 - e^a) = 1 - \lim_{a \to -\infty} e^a = 1$

Thus, $\int_{-\infty}^0 e^x \, dx$ converges and equals 1.

(c) $\int_1^{+\infty} \sin \frac{\pi}{2} x \, dx = \lim_{b \to +\infty} \left(\int_1^b \sin \frac{\pi}{2} x \, dx \right) = \lim_{b \to +\infty} \frac{-\cos(\pi/2)x}{\pi/2} \Big|_1^b$

$\qquad = \lim_{b \to +\infty} \left(-\frac{2}{\pi} \cos \frac{\pi}{2} b \right) = -\frac{2}{\pi} \lim_{b \to +\infty} \cos \frac{\pi}{2} b$

This limit does not exist since as $b \to +\infty$,   the value of $\cos(\pi/2)b$ oscillates between -1 and 1. Hence, $\int_1^{+\infty} \sin(\pi/2)x \, dx$ diverges.   ∎

GEOMETRIC INTERPRETATION

If f is continuous and nonnegative for $x \geq a$,   then $\int_a^{+\infty} f(x) \, dx$ may be interpreted geometrically. For each number $b > a$,   the definite integral $\int_a^b f(x) \, dx$

---

* Remember, a limit exists only when it equals a number.

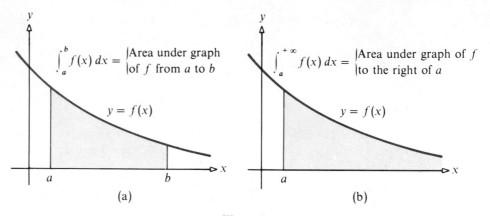

Figure 1

represents the area under the graph of $y = f(x)$ from a to b, as shown in Figure 1(a). As we let $b \to +\infty$, this area approaches the area under the graph of $y = f(x)$ over the infinite interval $[a, +\infty)$, as shown in Figure 1(b). We define the area under the graph of $y = f(x)$ to the right of a to be $\int_a^{+\infty} f(x)\, dx$, provided the improper integral converges. If it diverges, there is no defined area.

EXAMPLE 2   Find the area under the graph of $y = 1/x^2$ to the right of $x = 1$.

Solution   Look at the graph in Figure 2. Is there a finite number that measures the indicated area? At first, we may be tempted to answer "no," but, in fact, it turns out that this area is finite. To find the area, we proceed as follows:

$$\int_1^{+\infty} \frac{1}{x^2}\, dx = \lim_{b \to +\infty} \left(\int_1^b \frac{1}{x^2}\, dx \right) = \lim_{b \to +\infty} \left(-\frac{1}{x} \right)\Big|_1^b = \lim_{b \to +\infty} \left(-\frac{1}{b} + 1 \right) = 1$$

Thus, the area under the graph of $f(x) = 1/x^2$ to the right of 1, is 1 square unit.

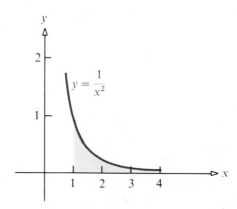

Figure 2   ∎

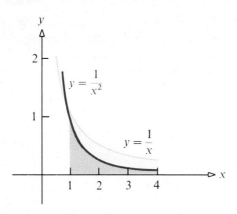

Figure 3

In Example 1(a) it was shown that $\int_1^{+\infty} (1/x)\, dx$ diverges, whereas in Example 2 it was shown that $\int_1^{+\infty} (1/x^2)\, dx$ converges. Yet, as Figure 3 illustrates, the graphs of $y = 1/x$ and $y = 1/x^2$ are similar on the interval $[1, +\infty)$. The explanation for the difference is that $y = 1/x^2$ approaches 0 more rapidly than $y = 1/x$ as $x \to +\infty$. The difference in the areas under the graphs is large enough to make $\int_1^{+\infty} (1/x)\, dx$ diverge and $\int_1^{+\infty} (1/x^2)\, dx$ converge.

EXAMPLE 3   Find the volume of the solid generated by revolving the area under the graph of $y = 1/x$ to the right of 1·about the x-axis. Use the disk method.

Solution   Figure 4 illustrates the situation. The volume V is

$$V = \int_1^{+\infty} \pi \left(\frac{1}{x}\right)^2 dx = \pi \int_1^{+\infty} \frac{dx}{x^2} = \pi \lim_{b \to +\infty} \left(\int_1^b \frac{dx}{x^2}\right) = \pi \lim_{b \to +\infty} \left(-\frac{1}{x}\right)\Big|_1^b$$

$$= \pi \lim_{b \to +\infty} \left(-\frac{1}{b} + 1\right) = \pi \text{ cubic units}$$

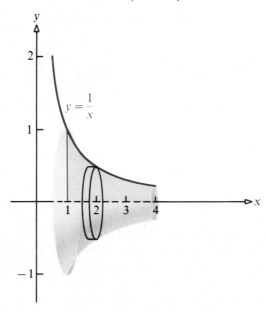

Figure 4   ∎

It is interesting to note that the area under the graph of $y = 1/x$ to the right of 1 is not defined, as shown in Example 1(a), while the volume obtained by revolving this same area about the x-axis equals π cubic units.

The result obtained in the next example will be put to good use in the following chapter.

EXAMPLE 4   Show that $\int_1^{+\infty} \dfrac{dx}{x^p}$ converges if $p > 1$ and diverges if $p \le 1$.

Solution $p \ne 1$: $\int_1^{+\infty} \dfrac{dx}{x^p} = \lim_{b \to +\infty} \left(\int_1^b \dfrac{dx}{x^p} \right) = \lim_{b \to +\infty} \dfrac{x^{-p+1}}{-p+1} \bigg|_1^b$

$$= \lim_{b \to +\infty} \dfrac{b^{-p+1} - 1}{-p+1} = \begin{cases} \dfrac{-1}{-p+1} & \text{if } p > 1 \\ +\infty & \text{if } p < 1 \end{cases}$$

$p = 1$: $\int_1^{+\infty} \dfrac{dx}{x^p} = \int_1^{+\infty} \dfrac{dx}{x} = \lim_{b \to +\infty} \left(\int_1^b \dfrac{dx}{x} \right)$

$$= \lim_{b \to +\infty} \ln x \bigg|_1^b = \lim_{b \to +\infty} \ln b = +\infty \quad \blacksquare$$

In case both limits of integration are infinite, we use the following definition:

(10.14)   **If f is continuous for all x and if the two improper integrals $\int_{-\infty}^0 f(x)\,dx$ and $\int_0^{+\infty} f(x)\,dx$ both converge, then we say that $\int_{-\infty}^{+\infty} f(x)\,dx$ converges and we define**

(10.15)
$$\int_{-\infty}^{+\infty} f(x)\,dx = \int_{-\infty}^0 f(x)\,dx + \int_0^{+\infty} f(x)\,dx$$

If *either* or *both* of the integrals *diverge*, then we say that $\int_{-\infty}^{+\infty} f(x)\,dx$ diverges.*

EXAMPLE 5   Determine whether $\int_{-\infty}^{+\infty} 4x^3\,dx$ converges or diverges.

Solution   We look first at the improper integral $\int_{-\infty}^0 4x^3\,dx$:

$$\int_{-\infty}^0 4x^3\,dx = \lim_{a \to -\infty} \left(\int_a^0 4x^3\,dx \right) = \lim_{a \to -\infty} x^4 \bigg|_a^0 = \lim_{a \to -\infty} (-a^4) = -\infty$$

There is no need to go on: $\int_{-\infty}^{+\infty} 4x^3\,dx$ is divergent. $\blacksquare$

A word of caution! The definition given in (10.14) requires that *two* improper integrals each converge in order for $\int_{-\infty}^{+\infty} f(x)\,dx$ to converge. We cannot set $\int_{-\infty}^{+\infty} f(x)\,dx = \lim_{a \to +\infty} [\int_{-a}^a f(x)\,dx]$. If we had done this in Example 5, the result would have been $\int_{-a}^a 4x^3\,dx = x^4 \big|_{-a}^a = a^4 - a^4 = 0$, from which we would have incorrectly concluded that $\int_{-\infty}^{+\infty} f(x)\,dx$ converges and equals 0.

---

* In (10.15), the number 0 is not crucial; any real number can be used.

ANOTHER TYPE OF IMPROPER INTEGRAL

An integral $\int_a^b f(x)\,dx$ is also improper when f is continuous on $(a, b]$ but is not defined at a. In this case, we define

(10.16)
$$\int_a^b f(x)\,dx = \lim_{t \to a^+}\left[\int_t^b f(x)\,dx\right]$$

provided this limit exists.

Similarly, if f is continuous in $[a, b)$ but is not defined at b, the integral $\int_a^b f(x)\,dx$ is improper and

(10.17)
$$\int_a^b f(x)\,dx = \lim_{t \to b^-}\left[\int_a^t f(x)\,dx\right]$$

provided this limit exists.

When the limit exists, the improper integral is said to *converge;* otherwise, it is said to *diverge.*

Figure 5 may help you remember the correct one-sided limit to use.

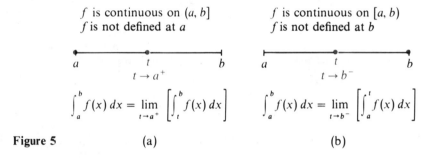

f is continuous on $(a, b]$
f is not defined at a

f is continuous on $[a, b)$
f is not defined at b

$$\int_a^b f(x)\,dx = \lim_{t \to a^+}\left[\int_t^b f(x)\,dx\right] \qquad \int_a^b f(x)\,dx = \lim_{t \to b^-}\left[\int_a^t f(x)\,dx\right]$$

Figure 5 (a) (b)

If f is continuous and nonnegative on the interval $(a, b]$, then the improper integral $\int_a^b f(x)\,dx$ may be interpreted geometrically. The integral $\int_t^b f(x)\,dx$ for each number t, $a < t \le b$, represents the area under the graph of $y = f(x)$ over the interval $[t, b]$, as shown in Figure 6(a). As we let $t \to a^+$, this area tends to

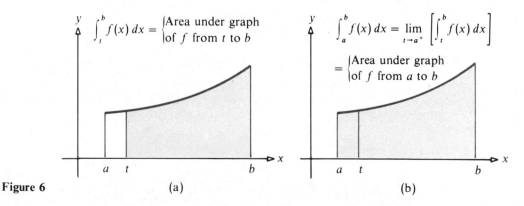

$$\int_t^b f(x)\,dx = \left\{ \begin{matrix}\text{Area under graph}\\ \text{of } f \text{ from } t \text{ to } b\end{matrix}\right.$$

$$\int_a^b f(x)\,dx = \lim_{t \to a^+}\left[\int_t^b f(x)\,dx\right]$$
$$= \left\{\begin{matrix}\text{Area under graph}\\ \text{of } f \text{ from } a \text{ to } b\end{matrix}\right.$$

Figure 6 (a) (b)

approach the entire area under the graph of $y = f(x)$ over $(a, b]$, as shown in Figure 6(b). Hence, if $\int_a^b f(x)\,dx$ converges, we shall define its value to be the area under the graph of f from a to b.

EXAMPLE 6   Determine whether $\int_0^4 \dfrac{1}{\sqrt{x}}\,dx$ converges or diverges.

Solution   This is an improper integral since $f(x) = 1/\sqrt{x}$ is not defined at 0. We use (10.16) to get

$$\int_0^4 \frac{1}{\sqrt{x}}\,dx = \lim_{t \to 0^+}\left(\int_t^4 \frac{1}{\sqrt{x}}\,dx\right) = \lim_{t \to 0^+} \frac{x^{1/2}}{\frac{1}{2}}\bigg|_t^4$$

$$= \lim_{t \to 0^+}(2\sqrt{4} - 2\sqrt{t}) = 4 - \lim_{t \to 0^+} 2\sqrt{t} = 4$$

We conclude that $\int_0^4 (1/\sqrt{x})\,dx$ converges.   ■

EXAMPLE 7   Determine whether $\int_0^{\pi/2} \tan x\,dx$ converges or diverges.

Solution   Since $\tan x$ is not defined at $\pi/2$, $\int_0^{\pi/2} \tan x\,dx$ is an improper integral. Thus, we use (10.17) to get

$$\int_0^{\pi/2} \tan x\,dx = \lim_{t \to \pi/2^-}\left(\int_0^t \tan x\,dx\right) = \lim_{t \to \pi/2^-} \ln(\sec x)\bigg|_0^t = \lim_{t \to \pi/2^-} \ln(\sec t) = +\infty$$

We conclude that $\int_0^{\pi/2} \tan x\,dx$ diverges.   ■

Another condition under which $\int_a^b f(x)\,dx$ is an improper integral is when the integrand is not defined at some number c in the closed interval $[a, b]$.

If f is continuous on $[a, b]$ except at c, $a < c < b$, where f is not defined, the integral $\int_a^b f(x)\,dx$ is improper and

(10.18)
$$\int_a^b f(x)\,dx = \int_a^c f(x)\,dx + \int_c^b f(x)\,dx$$

provided that each of the improper integrals on the right converges.

If either of the improper integrals on the right side of (10.18) diverge, then the improper integral on the left side of (10.18) also diverges.

EXAMPLE 8   Determine whether $\int_0^2 \dfrac{1}{(x - 1)^2}\,dx$ converges or diverges.

Solution   Since $f(x) = 1/(x-1)^2$ is not defined at 1, $0 < 1 < 2$, we look at the two improper integrals $\int_0^1 [1/(x-1)^2]\, dx$ and $\int_1^2 [1/(x-1)^2]\, dx$:

$$\int_0^1 \frac{1}{(x-1)^2}\, dx = \lim_{t \to 1^-} \left[\int_0^t \frac{1}{(x-1)^2}\, dx \right] = \lim_{t \to 1^-} \frac{-1}{x-1} \Big|_0^t = \lim_{t \to 1^-} \left(\frac{-1}{t-1} + 1 \right) = +\infty$$

Since $\int_0^1 [1/(x-1)^2]\, dx$ diverges, there is no need to investigate the second integral, and we conclude that $\int_0^2 [1/(x-1)^2]\, dx$ diverges.   ∎

A word of caution! It is a common mistake to look at an integral like $\int_0^2 dx/(x-1)^2$ and not notice that the integrand is undefined at 1. Then, if you merely plow ahead, using the fundamental theorem, you get

$$\int_0^2 \frac{dx}{(x-1)^2} = \frac{-1}{x-1} \Big|_0^2 = -1 - 1 = -2$$

which is an incorrect answer. Look before you leap!

EXERCISE 3

In Problems 1–8 determine whether each integral is improper. For those that are improper, state the reason.

1. $\int_0^{+\infty} x^2\, dx$ **2.** $\int_0^5 x^3\, dx$ **3.** $\int_2^3 \frac{dx}{x-1}$ **4.** $\int_1^2 \frac{dx}{x-1}$

5. $\int_0^1 \frac{1}{x}\, dx$ **6.** $\int_{-1}^1 \frac{x}{x^2+1}\, dx$ **7.** $\int_0^1 \frac{x}{x^2-1}\, dx$ **8.** $\int_0^{+\infty} e^{-2x}\, dx$

In Problems 9–60 determine whether the given improper integral converges or diverges and evaluate each convergent improper integral.

9. $\int_{10}^{+\infty} \frac{dx}{x^2}$ **10.** $\int_0^{+\infty} e^{-x}\, dx$ **11.** $\int_0^{+\infty} e^{2x}\, dx$ **12.** $\int_1^{+\infty} \frac{dx}{\sqrt{x}}$

13. $\int_{-\infty}^{-1} \frac{4}{x}\, dx$ **14.** $\int_3^{+\infty} \frac{dx}{(x-1)^4}$ **15.** $\int_0^1 \frac{dx}{\sqrt{x}}$ **16.** $\int_0^1 \frac{dx}{x^3}$

17. $\int_0^1 \frac{dx}{x}$ **18.** $\int_4^6 \frac{dx}{\sqrt{x-4}}$ **19.** $\int_0^a \frac{dx}{\sqrt{a-x}}$ **20.** $\int_1^5 \frac{x\, dx}{\sqrt{5-x}}$

21. $\int_0^{+\infty} xe^{-x^2}\, dx$ **22.** $\int_0^{+\infty} \cos x\, dx$ **23.** $\int_0^1 \frac{x\, dx}{(1-x^2)^2}$ **24.** $\int_0^{+\infty} \frac{x\, dx}{1+x^2}$

25. $\int_0^{+\infty} \frac{x\, dx}{\sqrt{x+1}}$ **26.** $\int_1^{+\infty} \frac{dx}{x(1+x^2)}$ **27.** $\int_0^{\pi/4} \tan 2x\, dx$ **28.** $\int_0^{\pi/2} \csc x\, dx$

29. $\int_{-1}^1 \frac{dx}{\sqrt[3]{x}}$ **30.** $\int_0^3 \frac{dx}{(x-2)^2}$ **31.** $\int_0^{2a} \frac{dx}{(x-a)^2}$ **32.** $\int_a^{3a} \frac{2x\, dx}{(x^2-a^2)^{3/2}}$

33. $\int_{-\infty}^0 \frac{dx}{x^2+4}$ **34.** $\int_{-\infty}^2 \frac{dx}{\sqrt{4-x}}$ **35.** $\int_{-\infty}^1 \frac{x\, dx}{\sqrt{2-x}}$ **36.** $\int_{-\infty}^0 e^x\, dx$

37. $\int_{-\infty}^{+\infty} \dfrac{dx}{x^2 + a^2}$ **38.** $\int_0^4 \dfrac{2x \, dx}{\sqrt[3]{x^2 - 4}}$ **39.** $\int_0^1 \dfrac{dx}{1 - x^2}$ **40.** $\int_1^2 \dfrac{dx}{\sqrt{x^2 - 1}}$

41. $\int_0^a \dfrac{x^2 \, dx}{\sqrt{a^2 - x^2}}$ **42.** $\int_0^3 \dfrac{x \, dx}{(9 - x^2)^{3/2}}$ **43.** $\int_0^{+\infty} \sin \pi x \, dx$ **44.** $\int_0^1 \dfrac{\ln x \, dx}{x}$

45. $\int_1^{+\infty} \dfrac{dx}{x^{2/3}}$ **46.** $\int_{-\infty}^{+\infty} \dfrac{dx}{x^2 + 4x + 5}$ **47.** $\int_{2a}^{+\infty} \dfrac{dx}{x^2 - a^2}$ **48.** $\int_2^{+\infty} \dfrac{dx}{x\sqrt{x^2 - 1}}$

49. $\int_1^{+\infty} \dfrac{dx}{x^4 + x^2}$ **50.** $\int_{-\infty}^{+\infty} \dfrac{dx}{e^x + e^{-x}}$ **51.** $\int_0^a \dfrac{dx}{\sqrt{a^2 - x^2}}$ **52.** $\int_a^{2a} \dfrac{x \, dx}{\sqrt{x^2 - a^2}}$

53. $\int_1^2 \dfrac{dx}{(2 - x)^{3/4}}$ **54.** $\int_0^4 \dfrac{dx}{\sqrt{8x - x^2}}$ **55.** $\int_0^\pi \dfrac{1}{1 - \cos x} \, dx$ **56.** $\int_0^{\pi/2} \dfrac{1}{1 - \sin x} \, dx$

57. $\int_{-1}^1 \dfrac{1}{x^3} \, dx$ **58.** $\int_0^2 \dfrac{dx}{x - 1}$ **59.** $\int_0^2 \dfrac{dx}{(x - 1)^{1/3}}$ **60.** $\int_{-1}^1 \dfrac{dx}{x^{5/3}}$

61. Find the area, if it exists, under the graph of $y = 1/(1 + x^2)$ to the right of 0.

62. Find the volume, if it exists, of the solid generated by revolving the area under the graph of $y = e^{-x}$ to the right of 0 about the x-axis.

63. Find the volume, if it exists, of the solid generated by revolving the area under the graph of $y = 1/\sqrt{x}$ to the right of 1 about the x-axis.

64. For what α does $\int_0^1 x^\alpha \, dx$ converge?

65. Show that $\int_0^{+\infty} \sin x \, dx$ and $\int_{-\infty}^0 \sin x \, dx$ each diverge, yet $\lim_{t \to +\infty}(\int_{-t}^t \sin x \, dx) = 0$.

66. Find a function f for which $\int_0^{+\infty} f(x) \, dx$ and $\int_{-\infty}^0 f(x) \, dx$ each diverge, yet $\lim_{t \to +\infty}[\int_{-t}^t f(x) \, dx] = 1$.

In Problems 67–70 use integration by parts and perhaps L'Hospital's rule to evaluate each integral.

67. $\int_0^{+\infty} xe^{-x} \, dx$ **68.** $\int_0^1 (x \ln x) \, dx$ **69.** $\int_0^{+\infty}(e^{-x} \cos x) \, dx$ **70.** $\int_0^{+\infty} (\tan^{-1} x) \, dx$

71. The rate of reaction to a given dose of a drug at time t hours after administration is given by $r(t) = te^{-t^2}$ (measured in appropriate units). Why is it reasonable to define the total *reaction* as the area under the graph of $y = r(t)$ from $t = 0$ to $t = +\infty$? Evaluate the total reaction to the given dose of the drug.

72. The present value (PV) of a capital asset that provides a perpetual stream of revenues that flows continuously at a rate of $R(t)$ dollars per year is given by

$$PV = \int_0^{+\infty} R(t)e^{-rt} \, dt$$

where r is the per annum rate of interest compounded continuously.
(a) Find the present value of an asset if it provides a constant return of \$100 per year and r is 8%.
(b) Find the present value of an asset if it provides a return of $R(t) = 1000 + 80t$ dollars per year and $r = 7\%$.

73. In a problem in electrical theory, the following integral occurs:

$$\int_0^{+\infty} Ri^2 \, dt$$

where the current $i = Ie^{-Rt/L}$, t is time, and R, I, and L are constants. Evaluate this integral.

74. The magnetic potential at a point on the axis of a circular coil is given by

$$u = \frac{2\pi N I r}{10} \int_x^{+\infty} \frac{dy}{(r^2 + y^2)^{3/2}}$$

where N, I, r, and x are constants. Evaluate this expression.

75. The field intensity around a long ("infinite") straight wire carrying electric current is given by the integral

$$F = \frac{r I m}{10} \int_{-\infty}^{+\infty} \frac{dy}{(r^2 + y^2)^{3/2}}$$

where r, I, and m are constants. Evaluate this expression.

4. Taylor Polynomials

Of the functions studied thus far, the polynomial functions

$$P_n(x) = a_0 + a_1 x + a_2 x^2 + \cdots + a_n x^n$$

stand out as the simplest. Such functions are easy to evaluate since the only operations needed are addition and multiplication. Such is not the case with, for example, the exponential function or the sine function if it is desired to calculate e^π or sin 0.1. Because of the simple nature of polynomials, we are interested in *approximating functions by a special class of polynomials known as the Taylor polynomials.*[*]

We begin by recalling that the linear approximation of f (studied in Section 2 of Chapter 4) is a polynomial of degree 1 that serves as an approximation to f. In fact, based on (4.15), the first-degree polynomial $f(c) + f'(c)(x - c)$ approximates f for x close to c. That is,

$$f(x) \approx f(c) + f'(c)(x - c)$$

We define the first-degree polynomial P_1 as

$$P_1(x) = f(c) + f'(c)(x - c)$$

The graph of this approximating polynomial for f is the tangent line to f at $x = c$ (see Fig. 7, p. 552).

An improvement on this approximation would be a second-degree polynomial whose graph not only has the same tangent line at c as f, but whose second derivative at c equals that of f. Such a polynomial is

$$P_2(x) = f(c) + f'(c)(x - c) + \frac{f''(c)}{2}(x - c)^2$$

---

[*] Named for the English mathematician Brook Taylor (1681–1731).

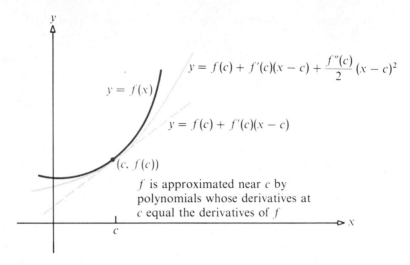

Figure 7

because

1. $P_2(c) = f(c)$ $P_2(x)$ has the same value at c as f

2. $P_2'(x) = f'(c) + f''(c)(x - c)$

    $P_2'(c) = f'(c) + f''(c)(c - c) = f'(c)$ The slope of the tangent line to P_2 at c is the same as the slope of the tangent line to f at c

3. $P_2''(x) = f''(c)$ The second derivative of P_2 at c is the same as

    $P_2''(c) = f''(c)$ that of f at c

See Figure 7.

    A third-degree polynomial whose first three derivatives at c are equal to those of f at c should provide an even better approximation to f for x near c. As we shall verify, such a polynomial is

$$P_3(x) = f(c) + f'(c)(x - c) + \frac{f''(c)}{2!}(x - c)^2 + \frac{f'''(c)}{3!}(x - c)^3$$

For

1. $P_3(c) = f(c)$

2. $P_3'(x) = f'(c) + f''(c)(x - c) + \frac{f'''(c)}{2}(x - c)^2$ $P_3'(c) = f'(c)$

3. $P_3''(x) = f''(c) + f'''(c)(x - c)$ $P_3''(c) = f''(c)$

4. $P_3'''(x) = f'''(c)$ $P_3'''(c) = f'''(c)$

If continued, this same reasoning would, after n steps, bring us to a polynomial of degree n:

(10.19) $$P_n(x) = f(c) + f'(c)(x - c) + \frac{f''(c)}{2!}(x - c)^2 + \cdots + \frac{f^{(n)}(c)}{n!}(x - c)^n$$

This polynomial P_n is called a *Taylor polynomial of degree n for f at c.*

EXAMPLE 1   Write the Taylor polynomial of degree 3 for $f(x) = \sqrt{x}$ at 1.

Solution   We use (10.19) with $n = 3$, which requires that we calculate the derivatives of f at 1:

$$f(x) = \sqrt{x} \qquad f(1) = 1$$

$$f'(x) = \frac{1}{2\sqrt{x}} \qquad f'(1) = \frac{1}{2}$$

$$f''(x) = -\frac{1}{4x^{3/2}} \qquad f''(1) = -\frac{1}{4}$$

$$f'''(x) = \frac{3}{8x^{5/2}} \qquad f'''(1) = \frac{3}{8}$$

The Taylor polynomial $P_3(x)$ of $f(x) = \sqrt{x}$ at 1 is

$$P_3(x) = f(1) + f'(1)(x - 1) + \frac{f''(1)}{2!}(x - 1)^2 + \frac{f'''(1)}{3!}(x - 1)^3$$

$$= 1 + \frac{x - 1}{2} - \frac{(x - 1)^2}{8} + \frac{(x - 1)^3}{16} \qquad \blacksquare$$

EXAMPLE 2   Write the Taylor polynomial of degree 7 for $f(x) = \sin x$ at 0.

Solution   The successive derivatives of $f(x) = \sin x$ at 0 are

$$f(x) = \sin x \qquad f(0) = 0$$
$$f'(x) = \cos x \qquad f'(0) = 1$$
$$f''(x) = -\sin x \qquad f''(0) = 0$$
$$f'''(x) = -\cos x \qquad f'''(0) = -1$$
$$f^{(4)}(x) = \sin x \qquad f^{(4)}(0) = 0$$

The pattern now begins to repeat: even derivatives of f at 0 equal 0; and odd derivatives of f at 0 alternate between 1 and -1. Therefore, the Taylor polynomial of degree 7 is

$$P_7(x) = f(0) + f'(0)x + \frac{f''(0)x^2}{2!} + \frac{f'''(0)x^3}{3!} + \cdots + \frac{f^{(7)}(0)x^7}{7!}$$

$$= x - \frac{x^3}{3!} + \frac{x^5}{5!} - \frac{x^7}{7!}$$

Figure 8 illustrates the graph of P_7 superimposed on the graph of sin x.

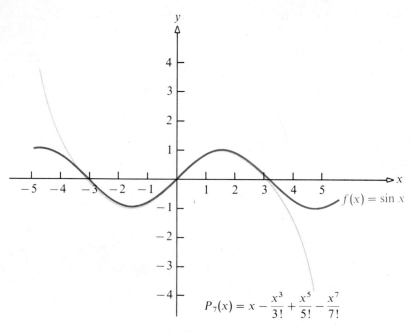

$$P_7(x) = x - \frac{x^3}{3!} + \frac{x^5}{5!} - \frac{x^7}{7!}$$

Figure 8   ∎

EXAMPLE 3   Write the Taylor polynomial of degree n for $f(x) = e^x$ at 0.

Solution   The successive derivatives of $f(x) = e^x$ at 0 are

$$
\begin{aligned}
f(x) &= e^x & f(0) &= 1 \\
f'(x) &= e^x & f'(0) &= 1 \\
f''(x) &= e^x & f''(0) &= 1
\end{aligned}
$$

As is evident, the derivative of any order of $f(x) = e^x$ equals 1 at 0. Thus, the Taylor polynomial of degree n of $f(x) = e^x$ at 0 is

$$P_n(x) = f(0) + f'(0)x + \frac{f''(0)x^2}{2!} + \cdots + \frac{f^{(n)}(0)x^n}{n!}$$

$$= 1 + x + \frac{x^2}{2!} + \cdots + \frac{x^n}{n!}$$

Figure 9 illustrates how good the Taylor polynomials P_1, P_2, P_3 of $f(x) = e^x$ are at approximating e^x near 0.

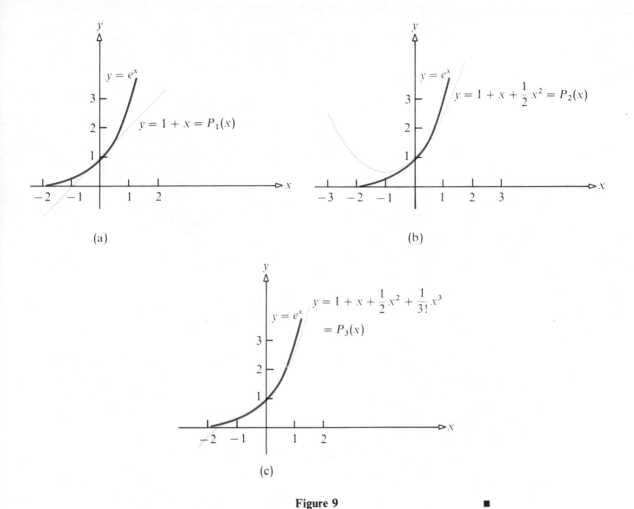

(a)

(b)

(c)

Figure 9   ■

The accuracy of the polynomial P_n as an approximation to the function f for x close to c is provided by the next result, which is called *Taylor's formula with remainder.*

(10.20) *Taylor's Formula with Remainder.* **Let f be a function whose first $(n + 1)$ derivatives exist and are continuous on some interval I containing c. Then, for every x in I, there exists a number u between x and c such that**

(10.21)
$$f(x) = f(c) + f'(c)(x - c) + \frac{f''(c)}{2!}(x - c)^2$$

$$+ \frac{f'''(c)}{3!}(x - c)^3 + \cdots + \frac{f^{(n)}(c)}{n!}(x - c)^n + R_{n+1}(x)$$

where

(10.22)
$$R_{n+1}(x) = \frac{f^{(n+1)}(u)}{(n+1)!}(x-c)^{n+1}$$

A proof of (10.20) is postponed to the end of the section.

In the expression (10.21), the sum of the first $n+1$ terms on the right is the Taylor polynomial of degree n of f at c. The expression R_{n+1} is called the *remainder after $n+1$ terms* and, as it is formulated in (10.22), it is referred to as *Lagrange's form of the remainder*.

Since we have

$$P_n(x) = f(c) + f'(c)(x-c) + \frac{f''(c)}{2!}(x-c)^2 + \cdots + \frac{f^{(n)}(c)}{n!}(x-c)^n$$

and

$$R_{n+1}(x) = \frac{f^{(n+1)}(u)}{(n+1)!}(x-c)^{n+1} \qquad u \text{ between } x \text{ and } c$$

then Taylor's formula with remainder asserts that

(10.23)
$$f(x) = P_n(x) + R_{n+1}(x)$$

where P_n is a polynomial of degree n in $(x-c)$, whose first n derivatives at c coincide with those of f at c.

The Taylor polynomial P_n may be considered an approximating polynomial for the function f for x close to c. The remainder R_{n+1} measures the accuracy of this approximation. Thus:

If the numerical value of R_{n+1} is small, then

$$f(x) \approx P_n(x)$$

and the error involved in using the approximation equals $|R_{n+1}(x)|$.

Of course, if we knew the value of $R_{n+1}(x)$, we would know $f(x)$ exactly. The reason we do not know $R_{n+1}(x)$ is that its value depends on the number u and we do not know u. However, we can usually get estimates of the remainder $R_{n+1}(x)$ and thereby determine just how good an approximation we have.

For instance, using the result obtained in Example 1, we can approximate $\sqrt{1.2}$ by $P_3(1.2)$. That is,

$$\sqrt{1.2} \approx P_3(1.2) = 1 + \frac{0.2}{2} - \frac{(0.2)^2}{8} + \frac{(0.2)^3}{16} = 1 + 0.1 - 0.005 + 0.0005 = 1.0955$$

The error of this approximation for $\sqrt{1.2}$ is obtained from (10.22):

$$R_4(x) = \frac{f^{(4)}(u)}{4!}(x-1)^4$$

where $f^{(4)}(u) = -15/(16u^{7/2})$, $x = 1.2$,   and $1 < u < 1.2$.   Since $u > 1$,   it follows that $1/u < 1$.   Therefore, an upper estimate to the error is

$$|R_4(1.2)| = \frac{15}{(4!)(16u^{7/2})}(0.2)^4 < \frac{15(0.2)^4}{384} = 0.0000625$$

By using a hand calculator, we compute the actual error to be

$$|\sqrt{1.2} - 1.0955| \approx |1.0954451 - 1.0955| = 0.0000549$$

EXAMPLE 4   Write the polynomial $f(x) = x^3 - 3x^2 + 6x - 9$ as a polynomial in $(x - 2)$.

Solution   We seek the Taylor polynomial of f at 2:

$$f(x) = x^3 - 3x^2 + 6x - 9 \qquad f(2) = -1$$
$$f'(x) = 3x^2 - 6x + 6 \qquad f'(2) = 6$$
$$f''(x) = 6x - 6 \qquad f''(2) = 6$$
$$f'''(x) = 6 \qquad f'''(2) = 6$$

Since $f^{(4)}(x) = 0$,   it follows that $R_4(x) = 0$ and $f(x) = P_3(x)$.   That is,

$$f(x) = P_3(x) = -1 + 6(x - 2) + \frac{6(x - 2)^2}{2!} + \frac{6(x - 2)^3}{3!}$$

Thus,

$$x^3 - 3x^2 + 6x - 9 = (x - 2)^3 + 3(x - 2)^2 + 6(x - 2) - 1 \qquad ■$$

Example 4 illustrates a general result. If f is a polynomial of degree n, it follows that the remainder R_{n+1} must equal 0 since $f^{(n+1)}(u) = 0$.   Hence, a polynomial of degree n is equal to a Taylor polynomial of degree n at c.

Let's look at another example of approximation.

EXAMPLE 5   Approximate sin 1 using a Taylor polynomial of degree 7 for $f(x) = \sin x$ at 0. Estimate the error obtained in using this Taylor polynomial.

Solution   Using the result of Example 2, we have

$$\sin 1 \approx 1 - \frac{(1)^3}{3!} + \frac{(1)^5}{5!} - \frac{(1)^7}{7!}$$

$$\approx 1 - 0.1666666 + 0.0083333 - 0.0001984 = 0.8414683$$

The error obtained in using this Taylor polynomial is

$$|R_8(1)| = \left| \frac{f^{(8)}(u)(1)^8}{8!} \right| = \frac{|\sin u|}{8!}$$

where u is between 0 and 1. Since $|\sin u| \leq 1$, we have

$$|R_8(1)| \leq \frac{1}{8!} \approx 0.0000248 \qquad \blacksquare$$

Let's look at another example and approximate the number e.

EXAMPLE 6   Use a Taylor polynomial to approximate the number e correct to within 0.0001.

Solution   From Example 3, the Taylor polynomial of degree n at 0 of $f(x) = e^x$ is

(10.24)
$$P_n(x) = 1 + \frac{x}{1!} + \frac{x^2}{2!} + \cdots + \frac{x^n}{n!}$$

and

$$e^x = P_n(x) + R_{n+1}(x) = 1 + \frac{x}{1!} + \frac{x^2}{2!} + \cdots + \frac{x^n}{n!} + \frac{e^u}{(n+1)!}x^{n+1}$$

where u is a number between 0 and x. If we set $x = 1$, we get

$$e^1 = P_n(1) + R_{n+1}(1)$$

$$e = 1 + \frac{1}{1!} + \frac{1}{2!} + \cdots + \frac{1}{n!} + \frac{e^u}{(n+1)!} \qquad 0 < u < 1$$

We need to find an integer n such that the remainder $e^u/(n+1)!$ is less than 0.0001. Our problem in finding such an n is that we do not know exactly what e^u is. However, since we set $x = 1$, we do know that $0 < u < 1$. In addition, since it is common knowledge that $e < 3$, we also know that

$$\frac{e^u}{(n+1)!} < \frac{e}{(n+1)!} < \frac{3}{(n+1)!}$$

The smallest n for which $3/(n+1)! < 0.0001$ is $n = 7$. Thus, for $n = 7$, $e^u/(n+1)! < 0.0001$. Hence, we use $n = 7$ in (10.24) to find that

$$P_7(1) = 1 + 1 + \frac{1}{2!} + \frac{1}{3!} + \frac{1}{4!} + \frac{1}{5!} + \frac{1}{6!} + \frac{1}{7!}$$

$$\approx 1 + 1 + 0.500000 + 0.166667 + 0.041667 + 0.008333 + 0.001389 + 0.000198$$

$$= 2.718254$$

approximates e to within 0.0001. In other words, $|e - 2.718254| < 0.0001$ or $2.718154 < e < 2.718354$. $\blacksquare$

In approximating e by using a Taylor polynomial of degree n, the error introduced equals $e^u/(n+1)!$. Evidently this error can be made as small as we please by choosing n sufficiently large. However, by taking n larger and larger, we need to add more and more terms to the Taylor polynomial. In fact, we show in the next chapter that e actually equals an infinite sum of numbers. This leads us to the subject of the next chapter—*infinite series*.

Proof of Taylor's Formula with Remainder   For a fixed x in I there is a number L (depending on x) for which

(10.25)
$$f(x) = f(c) + \frac{f'(c)}{1!}(x - c) + \frac{f''(c)}{2!}(x - c)^2 + \cdots + \frac{f^{(n)}(c)}{n!}(x - c)^n$$

$$+ \frac{L}{(n + 1)!}(x - c)^{n+1}$$

We define the function F so that

(10.26)
$$F(t) = f(x) - f(t) - \frac{f'(t)}{1!}(x - t) - \frac{f''(t)}{2!}(x - t)^2 - \cdots - \frac{f^{(n)}(t)}{n!}(x - t)^n$$

$$- \frac{L}{(n + 1)!}(x - t)^{n+1}$$

The domain of F is $c \leq t \leq x$ if $x > c$ and $x \leq t \leq c$ if $x < c$. Clearly, since $f(t), f'(t), \ldots, f^{(n)}(t)$ are each continuous, F is continuous on its domain. Furthermore, F is differentiable and

$$\frac{dF}{dt} = F'(t) = -f'(t) + \left[f'(t) - \frac{f''(t)}{1!}(x - t) \right] + \left[\frac{f''(t)}{1!}(x - t) - \frac{f'''(t)}{2!}(x - t)^2 \right] + \cdots$$

$$+ \left[\frac{f^{(n)}(t)}{(n - 1)!}(x - t)^{n-1} - \frac{f^{(n+1)}(t)}{n!}(x - t)^n \right] + \frac{L}{n!}(x - t)^n$$

$$= \frac{-f^{(n+1)}(t)}{n!}(x - t)^n + \frac{L}{n!}(x - t)^n \qquad \text{for all } t \text{ between } c \text{ and } x$$

In view of (10.25) and (10.26),

$$F(c) = f(x) - f(c) - \frac{f'(c)}{1!}(x - c) - \frac{f''(c)}{2!}(x - c)^2 - \cdots - \frac{f^{(n)}(c)}{n!}(x - c)^n$$

$$- \frac{L}{(n + 1)!}(x - c)^{n+1} = 0$$

and, by direct substitution in (10.26),

$$F(x) = f(x) - f(x) - \frac{f'(x)}{1!}(x - x) - \cdots - \frac{f^{(n)}(x)}{n!}(x - x)^n$$

$$- \frac{L}{(n + 1)!}(x - x)^{n+1} = 0$$

Thus, we may apply Rolle's theorem (4.25) to F. Then there is a number u between c and x so that

$$F'(u) = \frac{-f^{(n+1)}(u)}{n!}(x-u)^n + \frac{L(x-u)^n}{n!} = 0$$

Solving this equation for L, we find $L = f^{(n+1)}(u)$. Then, setting $t = c$ and $L = f^{(n+1)}(u)$ in (10.26) and solving for $f(x)$ gives the desired result. ■

EXERCISE 4

In Problems 1–4 write each polynomial as powers of $(x - 1)$.

1. $f(x) = 3x^2 - 6x + 4$ **2.** $f(x) = 4x^2 - x + 1$ **3.** $f(x) = x^3 + x^2 - 8$ **4.** $f(x) = x^4 + 1$

In Problems 5–22 write the Taylor polynomial of degree n about c for the given c and n. Also, write an expression for the remainder $R_{n+1}(x)$.

5. $f(x) = 3x^3 + 2x^2 - 6x + 5$; $c = 1$, $n = 3$ **6.** $f(x) = 4x^3 - 2x^2 - 4$; $c = 1$, $n = 3$

7. $f(x) = 2x^4 - 6x^3 + x$; $c = -1$, $n = 4$ **8.** $f(x) = -3x^4 + 2x^2 - 5$; $c = -1$, $n = 4$

9. $f(x) = x^5$; $c = 2$, $n = 4$ **10.** $f(x) = x^6$; $c = 3$, $n = 5$

11. $f(x) = \ln x$; $c = 1$, $n = 5$ **12.** $f(x) = \ln(1 + x)$; $c = 0$, $n = 5$

13. $f(x) = \dfrac{1}{x}$; $c = 1$, $n = 5$ **14.** $f(x) = \sqrt[3]{x}$; $c = 1$, $n = 5$

15. $f(x) = \cos x$; $c = 0$, $n = 6$ **16.** $f(x) = \sin x$; $c = \dfrac{\pi}{4}$, $n = 7$

17. $f(x) = \cosh x$; $c = 0$, $n = 6$ **18.** $f(x) = e^{-x}$; $c = 0$, $n = 5$

19. $f(x) = \dfrac{1}{1-x}$; $c = 0$, $n = 4$ **20.** $f(x) = \dfrac{1}{1+x}$; $c = 0$, $n = 4$

21. $f(x) = \dfrac{1}{(1+x)^2}$; $c = 0$, $n = 3$ **22.** $f(x) = \dfrac{1}{1+x^2}$; $c = 0$, $n = 2$

In Problems 23–32 use a Taylor polynomial to approximate the given number with an error less than that shown.

23. $\ln 1.1$;   Error < 0.00001 **24.** $\ln 1.1$;   Error < 0.01 **25.** $\dfrac{1}{1.1}$;   Error < 0.001

26. $\sqrt[3]{0.9}$;   Error < 0.001 **27.** $\cos 1°$;   Error < 0.0001 **28.** $\sin 46°$;   Error < 0.0001
                                        [*Hint:* $1° = 2\pi/360$ radian]

29. e^{-1};   Error < 0.01 **30.** $e^{2/3}$;   Error < 0.01 **31.** $e^{1/2}$;   Error < 0.01

32. e^{-2};   Error < 0.1

In Problems 33–40 write the Taylor polynomial of degree n for each function about c using the indicated values of c and n.

33. $f(x) = x \ln x$; $n = 4$, $c = 1$ **34.** $f(x) = \sqrt{3 + x^2}$; $n = 3$, $c = 1$

35. $f(x) = e^{-x^2}$; $n = 2$, $c = 0$

36. $f(x) = \sin x^2$; $n = 3$, $c = 0$

37. $f(x) = \tan x$; $n = 4$, $c = \dfrac{\pi}{4}$

38. $f(x) = \sec x$; $n = 3$, $c = 0$

39. $f(x) = \sqrt{1 + x}$; $n = 3$, $c = 0$

40. $f(x) = \tan^{-1} x$; $n = 3$, $c = 0$

41. The graphs of $y = \sin x$ and $y = \lambda x$ intersect near $x = \pi$ if λ is small. Set $f(x) = \sin x - \lambda x$. Write the Taylor polynomial of degree 2 for f about π and use it to show that an approximate solution of the equation $\sin x = \lambda x$ is $x = \pi/(1 + \lambda)$.

42. Proceed as in Problem 41 and obtain an approximate solution of the equation $\cot x = \lambda x$. [*Hint:* Assume that λ is small and use $n = 2$ and $c = \pi/2$.]

43. Approximate $\int_0^{0.01} e^{x^2}\, dx$ by using a Taylor polynomial of degree 4 for $f(x) = e^{x^2}$ at $c = 0$.

Miscellaneous Exercises

In Problems 1–4 evaluate each limit.

1. $\displaystyle \lim_{x \to +\infty} (1 + 4x)^{2/x}$

2. $\displaystyle \lim_{x \to 0} \frac{xe^{3x} - x}{1 - \cos 2x}$

3. $\displaystyle \lim_{x \to \pi/2} \frac{\sec^2 x}{\sec^2 3x}$

4. $\displaystyle \lim_{x \to 0} \left(\frac{1}{x^2} - \frac{1}{x^2 \sec x} \right)$

In Problems 5–14 determine whether the improper integral converges or diverges and evaluate each convergent improper integral.

5. $\displaystyle \int_1^{+\infty} \frac{e^{-\sqrt[3]{x}}}{\sqrt[3]{x^2}}\, dx$

6. $\displaystyle \int_1^{+\infty} \frac{e^{-\sqrt{x}}}{\sqrt{x}}\, dx$

7. $\displaystyle \int_0^1 \frac{\sin \sqrt{x}}{\sqrt{x}}\, dx$

8. $\displaystyle \int_0^{\pi/2} \frac{x\, dx}{\sin x^2}$

9. $\displaystyle \int_0^2 \frac{x\, dx}{(x^2 - 1)^{2/3}}$

10. $\displaystyle \int_{-a}^a \frac{dx}{\sqrt{a^2 - x^2}}$

11. $\displaystyle \int_0^1 \frac{x\, dx}{\sqrt{1 - x^2}}$

12. $\displaystyle \int_0^{+\infty} e^{-x} \sin x\, dx$

13. $\displaystyle \int_{-\infty}^0 xe^x\, dx$

14. $\displaystyle \int_0^{+\infty} x^2 e^{-x}\, dx$

15. If n is a positive integer, show that $\int_0^{+\infty} x^n e^{-x}\, dx = n \int_0^{+\infty} x^{n-1} e^{-x}\, dx$.

16. If n is a positive integer, show that $\int_0^{+\infty} x^n e^{-x}\, dx = n!$.

17. Show that $\int_e^{+\infty} dx/[x(\ln x)^p]$ converges if $p > 1$, and diverges if $p \le 1$.

18. Show that $\int_0^{\pi/2} \sin x\, dx/\sqrt{\cos x}$ converges.

19. Show that $\int_1^{+\infty} (\sqrt{1 + x^{1/8}}/x^{3/4})\, dx$ diverges.

20. Show that the area under the graph of $y = x^{-2/3}$ from $x = 0$ to $x = 1$ is defined.

21. Show that the volume of the solid generated by revolving the area in Problem 20 about the x-axis is not defined.

22. Sketch the graph of $f(x) = x^x$, $x > 0$.

23. Show that $\lim_{n \to +\infty} n(\sqrt[n]{x} - 1) = \ln x$.

24. Find the limiting area between $y = 8a^3/(x^2 + 4a^2)$ and its asymptote.

25. Find the area enclosed by $y = 1/(x + 1)$ and $y = 1/(x + 2)$ on the interval $[0, +\infty)$.

26. Suppose f and g denote continuous functions defined on the interval $[a, +\infty)$, for which $0 \le f(x) \le g(x)$ for $x \ge a$. If $\int_a^{+\infty} f(x)\, dx = +\infty$, show that $\int_a^{+\infty} g(x)\, dx = +\infty$.

27. Use the result obtained in Problem 26 to show that $\int_0^{+\infty} (1/\sqrt{2 + \sin x})\, dx$ diverges.

28. Use the result obtained in Problem 26 to show that $\int_2^{+\infty} [(\ln x)/\sqrt{x^2 - 1}]\, dx$ diverges.

29. Evaluate: $\int_{-\infty}^a e^{(x - e^x)}\, dx$

30. Evaluate: $\int_{-\infty}^{+\infty} e^{(x - e^x)}\, dx$

31. Explain why L'Hospital's rule does not apply to $\lim\limits_{x \to 0} \dfrac{x^2 \sin(1/x)}{\sin x}$.

32. Find: $\lim\limits_{x \to \pi/2^-} \cos x[\ln(\cos x)]$

33. (a) Prove that the area in the first quadrant enclosed by the graph of $y = e^{-x}$ and the x-axis is divided into two equal parts by the line $x = \ln 2$.

    (b) If the two equal areas described in part (a) are rotated about the x-axis, are the resulting volumes equal? If so, prove it. If not, determine which one is larger and by how much.

34. Find constants a, b, c, and d so that: $\lim\limits_{x \to 0} \dfrac{\sin ax + bx + cx^2 + dx^3}{x^5} = \dfrac{4}{15}$

35. (a) Find: $\lim\limits_{h \to 0} \dfrac{f(x + 2h) - 2f(x + h) + f(x)}{h^2}$

    (b) Find: $\lim\limits_{h \to 0} \dfrac{f(x + 3h) - 3f(x + 2h) + 3f(x + h) - f(x)}{h^3}$

    (c) Generalize parts (a) and (b).

36. *Calculator problem.* The formulas in Problem 35 can be used to approximate derivatives. Approximate $f'(2)$, $f''(2)$, and $f'''(2)$ from the following table:

x	2.0	2.1	2.2	2.3	2.4
$f(x)$	0.6931	0.7419	0.7885	0.8329	0.8755

The data are for $f(x) = \ln x$. Compare the exact values with your approximations.

37. Find the length of the curve $y = \sqrt{x - x^2} - \sin^{-1}\sqrt{x}$.

38. The force of gravitational attraction between two point masses m and M that are r units apart is $F = GmM/r^2$, where G is a constant. Find the work done in moving the mass m along a straight-line path from $r = 1$ unit from M to $r = +\infty$.

Laplace transforms are useful in solving a special class of differential equations. The Laplace transform $L\{f(x)\}$ of a function f is defined as

(10.27) $$L\{f(x)\} = \int_0^{+\infty} e^{-sx} f(x)\, dx$$

In Problems 39–44 compute the Laplace transform of the given function.

39. $f(x) = 1$ 40. $f(x) = x$ 41. $f(x) = \cos x$ 42. $f(x) = \sin x$

43. $f(x) = e^x$ 44. $f(x) = e^{ax}$

A *probability density function* is a function f whose domain is the set of all real numbers such that   (*1*) $f(x) \geq 0$ for all x and   (*2*) $\int_{-\infty}^{+\infty} f(x)\,dx = 1$.

45. Show that the function below is a probability density function. (This is called the *uniform density function.*)

$$f(x) = \begin{cases} 0 & \text{if } x < a \\[2mm] \dfrac{1}{b-a} & \text{if } a \leq x \leq b \\[2mm] 0 & \text{if } x > b \end{cases}$$

46. Show that the function below is a probability density function for $a > 0$.   (This is called the *exponential density function.*)

$$f(x) = \begin{cases} \dfrac{1}{a}\,e^{-x/a} & \text{if } x \geq 0 \\[2mm] 0 & \text{if } x < 0 \end{cases}$$

The *expected value* or *mean* μ associated with the density function f is defined by

$$\mu = \int_{-\infty}^{+\infty} xf(x)\,dx$$

We can think of the expected value as a weighted average of various probabilities.

47. Calculate the expected value associated with the probability density function given in Problem 45.

48. Calculate the expected value associated with the probability density function given in Problem 46.

The *variance* σ^2 associated with the probability density function f is defined by

$$\sigma^2 = \int_{-\infty}^{+\infty} (x - \mu)^2 f(x)\,dx$$

We can think of the variance as the average square of the deviation from the mean. We define the *standard deviation* σ associated with the function f to be the square root of the variance σ^2.

49. Calculate the variance and standard deviation associated with the probability density function given in Problem 45.

50. Calculate the variance and standard deviation associated with the probability density function given in Problem 46.

11

Infinite Series

Historical Perspectives

Both Newton and Leibniz realized that the foundation of the theory they had helped create was poorly and inadequately laid. Newton especially seemed to be aware of the seriousness of the difficulties and inconsistencies contained in his efforts to explain the logical basis of calculus. Although there were mathematicians in the eighteenth century who made attempts to supply the missing rigor for the subject, most of the mathematical talent of the day was busy at the task of expanding and refining the methodology of calculus rather than worrying about the solidity of its logical foundation. Much of this expanded methodology depended directly upon the use of infinite series. The missing rigor was finally supplied in the nineteenth century—largely in response to inconsistencies and paradoxes in the theory of infinite series that could no longer be ignored if the development of calculus was to proceed.

Among the most important contributors to the development of calculus methods were the three Swiss mathematicians Jakob Bernoulli (1654–1705), his brother Johann (1667–1748), and Leonhard Euler (1707–1783). Much of their work involved the representation of functions by infinite series for the purposes of integration and differentiation. In particular, this method was the standard way of treating the trigonometric, exponential, and logarithmic functions at the time.

Both of the Bernoulli brothers were in regular correspondence with Leibniz, and each made a major contribution in interpreting and completing the details in Leibniz's sketchy papers on calculus. A famous dispute developed between the ambitious brothers when Johann presented some results that Jakob had communicated to him as if he had discovered them himself. When Jakob learned of this, he reciprocated and took credit for some of Johann's work. Perhaps justice was served in the end though, because, as we mentioned in Chapter 10, one of Johann's nicest discoveries is now known as L'Hospital's rule—credit for it wrongly going to Johann's benefactor, Guillaume L'Hospital.

Euler was one of the most prolific mathematicians ever (his collected works fill more than 70 volumes!) and one of the most capable. He had the good fortune to be educated in mathematics by Johann Bernoulli. Euler was noted for his remarkable memory, and although he was totally blind for the last 17 years of his life, he employed a secretary to take down his discoveries and over 400 of his research papers were written during those years. The range of Euler's work covered all the mathematics known in his day. In 1755, he wrote the first reasonably complete textbook on differential calculus, which he followed in 1768–1770 with a three-volume text on integral calculus. Through these popular books, his numerous research papers, and correspondence, Euler exerted an inestimable influence on the development of calculus.

Throughout most of the eighteenth century, calculus was regarded as essentially an extension of algebra, expanded through the use and manipulation of infinite series. However, even Euler, who accomplished this manipulation with greater success than anyone, had no method for analyzing the convergence or divergence of a series, and confusion about the proper role of infinite series abounded (see Problems 1–10 in the Miscellaneous Exercises at the end of this chapter).

The present-day concepts of convergence and divergence were not clearly defined until the early nineteenth century in the works of the great French mathematician Augustin Cauchy and the Czechoslovakian Bernhard Bolzano. Although others made important contributions, these two men were the first to put forth the concept of the sum of a series being the limit of its sequence of partial sums. The fact that over 150 years elapsed between Newton's use of the binomial theorem to expand a function into an infinite series and integrate it term-by-term, and an acceptable definition of the sum of a series, indicates how hard-won the concepts of convergence and divergence were.

1. Sequences

The study of infinite sums of numbers has important applications in physics and engineering since it provides an alternate way of representing functions. In particular, *infinite series* may be used to approximate numbers such as e, π, $\ln 2$, and so on.

The theory of infinite series is developed through the use of a special kind of function called a *sequence*.

A *sequence* is a function whose domain is the set of positive integers.

Let us *see* what this means. The graph in Figure 1(a) is the graph of the function $f(x) = 1/x$ for $x > 0$. If all of the graph were erased *except* the points $(1, 1)$, $(2, \frac{1}{2})$, $(3, \frac{1}{3})$, and so on, as shown in Figure 1(b), then these points would form the graph of a sequence. In the graph of a sequence, there is *one point* for every positive integer.

A sequence is often represented by listing its values in order. For example, for the above sequence we write

$$f(1), \quad f(2), \quad f(3), \quad f(4), \quad f(5), \quad \ldots$$

or

$$1, \quad \tfrac{1}{2}, \quad \tfrac{1}{3}, \quad \tfrac{1}{4}, \quad \tfrac{1}{5}, \quad \ldots$$

The list never ends, as indicated by the three dots at the end. The numbers in the list are called the *terms* of the sequence.

It is common to use a letter with a subscript, such as s_1, to stand for the first term, s_2 for the second term, s_3 for the third, and so on. Thus, for the above sequence, we have

$$s_1 = f(1) = 1$$
$$s_2 = f(2) = \tfrac{1}{2}$$
$$s_3 = f(3) = \tfrac{1}{3}$$
$$\vdots$$
$$s_n = f(n) = \frac{1}{n}$$
$$\vdots$$

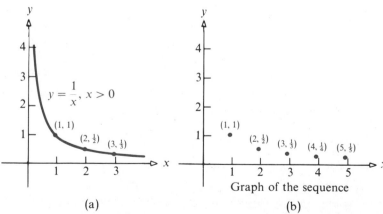

(a)

(b)

Graph of the sequence

Figure 1

It is easy to find any term of this particular sequence because it has the simple rule $s_n = f(n) = 1/n$. In general, if a rule for the nth term of a sequence is known, then any term of that sequence may be found.

For example, suppose the nth term of a sequence is

$$b_n = g(n) = \frac{2n - 1}{n^3}$$

Then

$$b_1 = g(1) = \frac{2(1) - 1}{1^3} = 1$$

$$b_2 = g(2) = \frac{2(2) - 1}{2^3} = \frac{3}{8}$$

$$b_3 = g(3) = \frac{2(3) - 1}{3^3} = \frac{5}{27}$$

$$\vdots$$

When a rule for the nth term of a sequence is known, there is yet another advantage; namely, we may shorten our notation. For example, we may write $\{1/n\}$ for the sequence $1, \frac{1}{2}, \frac{1}{3}, \frac{1}{4}, \frac{1}{5}, \frac{1}{6}, \ldots$. In words, $\{1/n\}$ is the sequence whose nth term is $1/n$. Similarly, $\{2n/(n + 1)\}$ is the sequence whose nth term is $2n/(n + 1)$, namely,

$$\frac{2(1)}{1 + 1}, \quad \frac{2(2)}{2 + 1}, \quad \frac{2(3)}{3 + 1}, \quad \cdots, \quad \frac{2n}{n + 1}, \quad \cdots$$

or

$$1, \quad \frac{4}{3}, \quad \frac{3}{2}, \quad \cdots, \quad \frac{2n}{n + 1}, \quad \cdots$$

Consider another example. If the nth term of a sequence is

$$c_n = (-1)^n (\tfrac{1}{2})^n$$

then

$$c_1 = (-1)^1 (\tfrac{1}{2})^1 = -\tfrac{1}{2}$$
$$c_2 = (-1)^2 (\tfrac{1}{2})^2 = \tfrac{1}{4}$$
$$c_3 = (-1)^3 (\tfrac{1}{2})^3 = -\tfrac{1}{8}$$

(11.1)

$$c_4 = \tfrac{1}{16}$$
$$c_5 = -\tfrac{1}{32}$$
$$\vdots$$

In this sequence the signs of the terms *alternate* because the factor $(-1)^n$ equals -1 when n is odd and equals 1 when n is even. Factors that have just the reverse behavior are $(-1)^{n+1}$ and $(-1)^{n-1}$. These equal 1 when n is odd and equal -1 when n is even. For example, the first five terms of the sequence $\{(-1)^{n+1}(\tfrac{1}{2})^n\}$ are $\tfrac{1}{2}, -\tfrac{1}{4}, \tfrac{1}{8}, -\tfrac{1}{16}, \tfrac{1}{32}$.

Compare these with the first five terms of the sequence $\{c_n\} = \{(-1)^n(\tfrac{1}{2})^n\}$, listed in (11.1).

The rule defining a sequence is often expressed by an explicit formula for its nth term in terms of the index n. Sometimes, a sequence is indicated by an observed pattern in the first few terms so that a natural choice for the nth term is suggested. A sufficient number of terms is given for each sequence listed below, so you should be able to figure out the nth term at the right.

(a) $\quad e, \dfrac{e^2}{2}, \dfrac{e^3}{3}, \ldots$ $\qquad\qquad\qquad\qquad\qquad a_n = \dfrac{e^n}{n}$

(b) $\quad 1, \tfrac{1}{3}, \tfrac{1}{9}, \tfrac{1}{27}, \ldots$ $\qquad\qquad\qquad\qquad b_n = (\tfrac{1}{3})^{n-1}$

(c) $\quad 1, 3, 5, 7, \ldots$ $\qquad\qquad\qquad\qquad\qquad c_n = 2n - 1$

(d) $\quad 1, 4, 9, 16, 25, \ldots$ $\qquad\qquad\qquad\qquad d_n = n^2$

(e) $\quad \tfrac{2}{2}, \tfrac{4}{3}, \tfrac{6}{4}, \tfrac{8}{5}, \ldots$ $\qquad\qquad\qquad\qquad e_n = \dfrac{2n}{n+1}$

(f) $\quad 1, -\tfrac{1}{2}, \tfrac{1}{3}, -\tfrac{1}{4}, \tfrac{1}{5}, \ldots$ $\qquad\qquad\qquad f_n = \dfrac{(-1)^{n+1}}{n}$

(g) $\quad 1, \tfrac{1}{2}, 1, \tfrac{1}{4}, 1, \tfrac{1}{6}, \ldots$ $\qquad\qquad\qquad g_n = \begin{cases} 1 & \text{if } n \text{ is odd} \\ \dfrac{1}{n} & \text{if } n \text{ is even} \end{cases}$

(h) $\quad 0, -1, -1, -2, -2, -3, -3, \ldots$ $\qquad h_n = \begin{cases} -\dfrac{n-1}{2} & \text{if } n \text{ is odd} \\ -\dfrac{n}{2} & \text{if } n \text{ is even} \end{cases}$

The graphs of sequences (e)–(h) are given in Figure 2.

CONVERGENT SEQUENCES

Consider the sequence $\{e_n\} = \{2n/(n+1)\}$, whose graph is given in Figure 2. As we look to the right in this graph, we see that the points get closer and closer to the line $y = 2$. In fact, if we could look far enough to the right, the points would eventually *appear* to lie on the line $y = 2$ (although in actuality they would always be some very small distance below). This is because the nth term $2n/(n+1)$ becomes a closer and closer approximation to 2 as n grows larger. In fact, as the following table suggests, $2n/(n+1)$ can be made as close as we please to 2 by making n sufficiently large:

n	1	9	99	999	9,999	999,999
$2n/(n+1)$	1	1.8	1.98	1.998	1.9998	1.999998

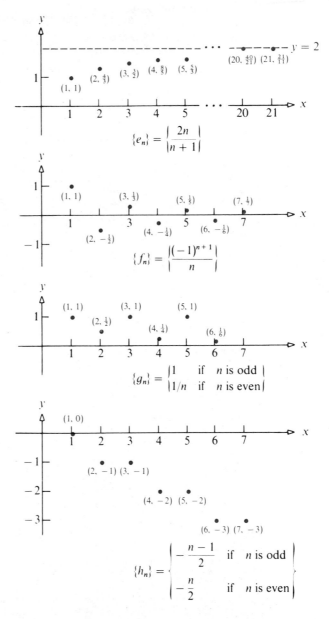

Figure 2

We describe this behavior by saying that the sequence $\{2n/(n + 1)\}$ *converges to 2.*
In this case, we write

$$\lim_{n \to +\infty} \frac{2n}{n + 1} = 2$$

We now state the definition of convergence.

(11.2)   **DEFINITION**   **Let L be a real number and let $\{s_n\}$ be a sequence. The statement $\{s_n\}$ *converges to* L means that for any given $\varepsilon > 0$, there exists a positive integer N such that**

$$|s_n - L| < \varepsilon \qquad \text{for all integers} \quad n > N$$

We use the notation $\lim_{n \to +\infty} s_n = L$ to mean that $\{s_n\}$ converges to L. The real number L is called the *limit* of the sequence $\{s_n\}$.

In general, if a sequence converges to some number, the sequence is said to be *convergent;* otherwise, it is said to be *divergent.*

Figure 3 provides a geometric interpretation of the statement $\lim_{n \to +\infty} s_n = L$. The figure also shows that the convergence (or divergence) of a sequence is in no way affected by the beginning terms of the sequence. In fact, we may ignore the beginning terms of a sequence in making a determination of convergence or divergence.

Notice the similarity between the statements

$$\lim_{n \to +\infty} \frac{2n}{n+1} = 2 \qquad \text{and} \qquad \lim_{x \to +\infty} \frac{2x}{x+1} = 2$$

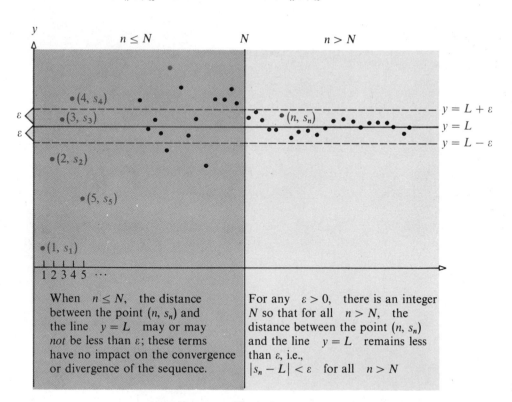

When $n \leq N$, the distance between the point (n, s_n) and the line $y = L$ may or may *not* be less than ε; these terms have no impact on the convergence or divergence of the sequence.

For any $\varepsilon > 0$, there is an integer N so that for all $n > N$, the distance between the point (n, s_n) and the line $y = L$ remains less than ε, i.e., $|s_n - L| < \varepsilon$ for all $n > N$

Figure 3

In general, there is a connection between the convergence of certain sequences $\{s_n\}$ and the behavior of certain functions f. The following result explains this connection:

(11.3)   **Suppose $\{s_n\}$ is a sequence, f is a function defined for all $x \geq 1$, and $s_n = f(n)$ for all positive integers n. If L is a number and**

$$\text{if} \qquad \lim_{x \to +\infty} f(x) = L, \qquad \text{then also} \qquad \lim_{n \to +\infty} s_n = L$$

Because of this, many of the results from Chapter 4 concerning limits at infinity are also valid for sequences. For example, *the limit of a sequence, if it converges, is unique.* We state below some limit theorems for convergent sequences.

(11.4)   **THEOREM**   **If $\{s_n\}$ and $\{t_n\}$ are convergent sequences and if c is a number, then:**

(a) $\displaystyle \lim_{n \to +\infty} (cs_n) = c \lim_{n \to +\infty} s_n$

(b) $\displaystyle \lim_{n \to +\infty} (s_n \pm t_n) = \lim_{n \to +\infty} s_n \pm \lim_{n \to +\infty} t_n$

(c) $\displaystyle \lim_{n \to +\infty} (s_n t_n) = \left(\lim_{n \to +\infty} s_n \right) \left(\lim_{n \to +\infty} t_n \right)$

(d) $\displaystyle \lim_{n \to +\infty} \frac{s_n}{t_n} = \frac{\lim_{n \to +\infty} s_n}{\lim_{n \to +\infty} t_n}$ provided $\displaystyle \lim_{n \to +\infty} t_n \neq 0$

EXAMPLE 1

$$\lim_{n \to +\infty} \frac{3n^2 + 5n - 2}{6n^2 - 6n + 5} = \lim_{n \to +\infty} \frac{3 + 5/n - 2/n^2}{6 - 6/n + 5/n^2}$$

$$= \underset{\uparrow}{\frac{\lim_{n \to +\infty} (3 + 5/n - 2/n^2)}{\lim_{n \to +\infty} (6 - 6/n + 5/n^2)}} = \frac{(3 + 0 - 0)}{(6 - 0 + 0)} = \frac{1}{2}$$

Use (11.4(d))

The sequence $\{(3n^2 + 5n - 2)/(6n^2 - 6n + 5)\}$ converges to $\frac{1}{2}$.   ∎

It may be possible to use L'Hospital's rule (10.4) to evaluate the limit of a sequence $\{s_n\}$, provided the corresponding function f meets the necessary requirements. We demonstrate such a use of L'Hospital's rule in part (a) of the next example. In part (b) we give an example of a sequence that converges, but whose corresponding function f does not converge.

EXAMPLE 2   Show that the following sequences converge: (a) $\left\{ \dfrac{n}{e^n} \right\}$ (b) $\left\{ \dfrac{1}{n} + \sin n\pi \right\}$

Solution   (a)   We employ L'Hospital's rule by treating n as a real variable:

$$\lim_{n \to +\infty} \frac{n}{e^n} \underset{\substack{\uparrow \\ \text{Type } \infty/\infty; \\ \text{use L'Hospital's} \\ \text{rule (10.4)}}}{=} \lim_{n \to +\infty} \frac{\dfrac{d}{dn} n}{\dfrac{d}{dn} e^n} = \lim_{n \to +\infty} \frac{1}{e^n} = 0$$

(b)   The terms of the given sequence are $1, \frac{1}{2} + 0, \frac{1}{3} + 0, \ldots, 1/n + 0, \ldots$. As $n \to +\infty$, the terms converge to 0. Hence,

$$\lim_{n \to +\infty} \left(\frac{1}{n} + \sin n\pi \right) = 0 \qquad \blacksquare$$

Let's look at the function f corresponding to the sequence $\{1/n + \sin n\pi\}$, namely,

$$f(x) = \frac{1}{x} + \sin \pi x \qquad x \text{ a real number}$$

As $x \to +\infty$, the values of f oscillate back and forth between -1 and 1. Therefore, $\lim_{x \to +\infty} f(x)$ does not exist. Thus, the function corresponding to a sequence may not have a limit at infinity, while the sequence itself may converge. However, if the function does have a limit at infinity, then so will the corresponding sequence.

To summarize, suppose f is any function corresponding to the sequence $\{s_n\}$:

If $\lim_{x \to +\infty} f(x) = L$ (a number), then $\lim_{n \to +\infty} s_n = L$ and $\{s_n\}$ converges.

If $\lim_{x \to +\infty} f(x) = +\infty$ (or $-\infty$), then $\{s_n\}$ diverges.

If $\lim_{x \to +\infty} f(x)$ does not exist, then there is no information gained about $\{s_n\}$.

DIVERGENT SEQUENCES

The sequence $\{s_n\}$ has the limit $+\infty$, written

$$\lim_{n \to +\infty} s_n = +\infty$$

if corresponding to any real number K (however large), there is a positive integer N such that $s_n > K$ for all integers $n > N$.

Similarly, $\{s_n\}$ has the limit $-\infty$ if corresponding to any real number K (however large in magnitude and negative), there is a positive integer N such that $s_n < K$ for all integers $n > N$.

Although the word *limit* is applied to both the finite and infinite cases, the word *converge* is used only for finite limits. Thus, a sequence that tends toward $+\infty$ or toward $-\infty$ diverges. For example, the sequence $2, 4, 8, \ldots, 2^n, \ldots$ diverges since $\lim_{n \to +\infty} 2^n = +\infty$.

In the next example, we illustrate how the definition can be used to show that a sequence diverges.

EXAMPLE 3   Show that the sequence $\{s_n\}$ whose terms are   1, 2, 1, 2, 1, 2, . . .   diverges.

Solution   Let us *assume* that $\{s_n\}$ *converges* to some number L. From this we will derive a contradiction, which will prove the falseness of the assumption and show that the sequence *diverges*.

Take $\varepsilon = \frac{1}{2}$.   (Actually, any $\varepsilon \le \frac{1}{2}$ can be used. Look at Fig. 4 to see why.)

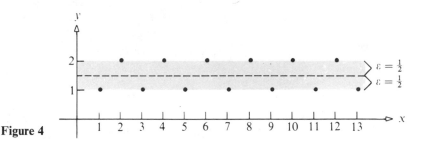

Figure 4

By the definition of convergence, there is a positive integer N such that

$$|s_n - L| < \tfrac{1}{2} \qquad \text{for all integers} \quad n > N$$

If $n > N$ and n is odd, then $s_n = 1$.   This means that

$$|1 - L| < \tfrac{1}{2} \quad \text{or} \quad -\tfrac{1}{2} < L - 1 < \tfrac{1}{2} \quad \text{or} \quad \tfrac{1}{2} < L < \tfrac{3}{2}$$

In particular, $L < \frac{3}{2}$.   If $n > N$ and n is even, then $s_n = 2$.   This means that

$$|2 - L| < \tfrac{1}{2} \quad \text{or} \quad -\tfrac{1}{2} < L - 2 < \tfrac{1}{2} \quad \text{or} \quad \tfrac{3}{2} < L < \tfrac{5}{2}$$

In particular, $L > \frac{3}{2}$.   We clearly have a contradiction because L cannot be both less than $\frac{3}{2}$ *and* greater than $\frac{3}{2}$. Therefore, the sequence diverges.   ∎

This example illustrates an important point: Sequences may diverge without tending to $+\infty$ or to $-\infty$. The sequence discussed in Example 3 has another important characteristic—its value is never more than 2 nor less than 1.

BOUNDED SEQUENCES

If every point of the graph of a sequence $\{s_n\}$ lies between the horizontal lines $y = -K$ and $y = K$,   we say that $\{s_n\}$ is *bounded by K*. Figure 5 (p. 574) illustrates the idea.

(11.5)   **DEFINITION**   *Bounded Sequence.*   **A sequence $\{s_n\}$ is *bounded* if and only if there exists a positive number K such that**

$$|s_n| \le K \qquad \text{for all integers } n$$

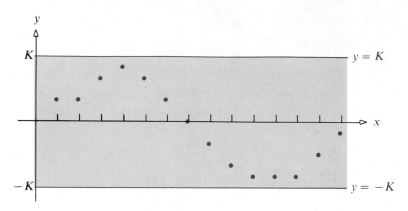

Figure 5

The sequence $\{1/n\}$ is bounded since $|1/n| \le 1$ for all n. The sequence $\{n^2\}$ is *unbounded* since for any choice of K, there is an integer n so that $n^2 > K$.

The following theorem gives a general tool for finding bounded sequences:

(11.6)   **THEOREM**. **Any convergent sequence is bounded.**

Proof   Let $\{s_n\}$ denote a convergent sequence. There is a number L so that $\lim_{n \to +\infty} s_n = L$. We use (11.2) and, for simplicity, we take $\varepsilon = 1$. Then we know that there exists a positive integer N such that

$$|s_n - L| < 1 \qquad \text{for all} \quad n > N$$

This means that

$$|s_n| < 1 + |L| \qquad \text{for all} \quad n > N$$

If we choose K to be the largest number in the finite collection

$$|s_1|, \quad |s_2|, \quad |s_3|, \quad \ldots, \quad |s_N|, \quad 1 + |L|$$

it will follow that $|s_n| \le K$ for *all* n. Consequently, the sequence $\{s_n\}$ is bounded.
∎

The contrapositive of theorem (11.6) provides a test for divergent sequences.

(11.7)   *Test for Divergence.*   **If a sequence is unbounded, then it diverges.**

For example, the sequences $\{n - 1/n\}$, $\{n/3\}$, and $\{e^n\}$ are unbounded and are therefore divergent.

Warning: The converse of (11.7) is not true. Bounded sequences may converge or they may diverge. Example 3 demonstrates a bounded sequence that diverges.

MONOTONIC SEQUENCES

(11.8)   DEFINITION   A sequence $\{s_n\}$ is said to be:

(a)   *Increasing* **if and only if** $s_n < s_{n+1}$ **for each** n

(b)   *Nondecreasing* **if and only if** $s_n \leq s_{n+1}$ **for each** n

(c)   *Decreasing* **if and only if** $s_n > s_{n+1}$ **for each** n

(d)   *Nonincreasing* **if and only if** $s_n \geq s_{n+1}$ **for each** n

A sequence that has one of these properties is called *monotonic*. Table 1 lists three ways to show whether a given sequence $\{s_n\}$ is increasing or decreasing.

Table 1

	To Show $\{s_n\}$ Is Decreasing	To Show $\{s_n\}$ Is Increasing
Algebraic Difference	Show $s_{n+1} - s_n < 0$ for all n	Show $s_{n+1} - s_n > 0$ for all n
Algebraic Ratio	If $s_n > 0$ for all n, show $\dfrac{s_{n+1}}{s_n} < 1$ for all n	If $s_n > 0$ for all n, show $\dfrac{s_{n+1}}{s_n} > 1$ for all n
Derivative	Show $\dfrac{d}{dn}(s_n) < 0$ for all n (Here, n is treated as though it were a real variable.)	Show $\dfrac{d}{dn}(s_n) > 0$ for all n (Here, n is treated as though it were a real variable.)

EXAMPLE 4   Show that each of the following sequences is monotonic by telling whether it is increasing, nondecreasing, decreasing, or nonincreasing.

(a) $\{s_n\} = \left\{\dfrac{n}{n+1}\right\}$ (b) $\{s_n\} = \left\{\dfrac{e^n}{n!}\right\}$ (c) $\{s_n\} = \{\ln n\}$

Solution   (a)   We choose the algebraic difference test. Since $s_n = n/(n+1)$ and $s_{n+1} = (n+1)/(n+2)$, we find

$$s_{n+1} - s_n = \frac{n+1}{n+2} - \frac{n}{n+1} = \frac{n^2 + 2n + 1 - n^2 - 2n}{(n+2)(n+1)}$$

$$= \frac{1}{(n+2)(n+1)} > 0 \qquad \text{for all } n$$

Hence, $\{n/(n+1)\}$ is an increasing sequence.

(b) The presence of a factorial usually means the algebraic ratio test applies. Since $s_n = e^n/n!$ and $s_{n+1} = e^{n+1}/(n+1)!$, we find

$$\frac{s_{n+1}}{s_n} = \frac{e^{n+1}/(n+1)!}{e^n/n!} = \left(\frac{e^{n+1}}{e^n}\right)\frac{n!}{(n+1)!} = \frac{e}{n+1} < 1 \qquad \text{for all}\quad n > 1$$

Hence, after the first term, $\{e^n/n!\}$ is a decreasing sequence.

(c) This is an easy function to differentiate. By using the derivative test, we find

$$\frac{d}{dn}\ln n = \frac{1}{n} > 0 \qquad \text{for all } n$$

Hence, the sequence $\{\ln n\}$ is increasing.   ∎

Not all sequences are monotonic. For example, the sequences $\{s_n\} = \{\sin(\pi/2)n\}$ and $\{t_n\} = \{1 + (-1)^n/n^2\}$ are not monotonic (see Figs. 6 and 7). Although both s_n and t_n are bounded sequences, s_n diverges and t_n converges.

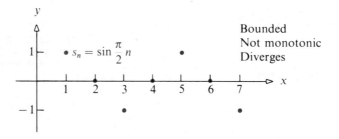

Figure 6

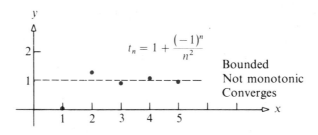

Figure 7

The sequence $\{n\}$ is monotonic, not bounded, and diverges.

Thus, we have examples of monotonic sequences that diverge and bounded sequences that diverge. However, when a sequence is both monotonic and bounded, it will always converge. We take this property as an axiom.

(11.9)   AXIOM   A bounded monotonic sequence converges.

Figure 8 provides an illustration for this statement. The next example shows how (11.9) is applied.

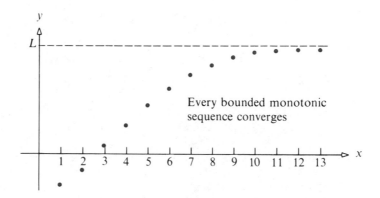

Figure 8

EXAMPLE 5   Show that the sequence $\{(\ln n)/n\}$ converges.

Solution   To use (11.9), we need to show that $\{(\ln n)/n\}$ is both bounded and monotonic. We note that $(\ln n)/n \geq 0$ for all $n \geq 1$,   from which we conclude that $\{(\ln n)/n\}$ is bounded (from below) by 0. Treating n as though it were a real number, we may apply the derivative test to find that

$$\frac{d}{dn}\frac{\ln n}{n} = \frac{1 - \ln n}{n^2} < 0 \qquad \text{for all} \quad n \geq 3$$

Consequently, except for the first two terms, $\{(\ln n)/n\}$ is decreasing. By (11.9), we conclude that the sequence $\{(\ln n)/n\}$ converges.   ∎

For the particular sequence in Example 5 we can apply L'Hospital's rule (10.4) and discover to what number the sequence converges. Here, too, we treat n as though it were a real number:

$$\lim_{n \to +\infty} \frac{\ln n}{n} = \lim_{n \to +\infty} \frac{\dfrac{d}{dn}\ln n}{\dfrac{d}{dn}n} = \lim_{n \to +\infty} \frac{1/n}{1} = 0$$

Thus, the sequence $\{(\ln n)/n\}$ converges to 0.

We close this section by summarizing the methods we have for showing that a sequence converges:

1. Look at the first few terms of the sequence to see if a trend is developing. For example, the first few terms of the sequence $\{1 + (-1)^n/n^2\}$ are $1 - 1$, $1 + \frac{1}{4}$, $1 - \frac{1}{9}$, $1 + \frac{1}{16}$, $1 - \frac{1}{25}$.   The pattern makes it clear that the sequence converges to 1.

2. Calculate the limit of the nth term, using any available limit technique. For example, for the sequence $\{[n^2/(n-1)] - [n^2/(n+1)]\}$ we find

$$\lim_{n \to +\infty} \left(\frac{n^2}{n-1} - \frac{n^2}{n+1} \right) = \lim_{n \to +\infty} \frac{2n^2}{(n-1)(n+1)} = 2$$

Thus, $\{[n^2/(n-1)] - [n^2/(n+1)]\}$ converges to 2.

3. Show that the sequence is both bounded and monotonic.

EXERCISE 1

In Problems 1–8 the nth term of a sequence is given. Write the first four terms of the sequence.

1. $s_n = \dfrac{1}{n}$

2. $s_n = \dfrac{2}{n^2}$

3. $s_n = \ln n$

4. $s_n = (-1)^{n+1}$

5. $s_n = \dfrac{(-1)^{n+1}}{2n+1}$

6. $s_n = \dfrac{n}{\ln(n+1)}$

7. $s_n = \dfrac{1-(-1)^n}{2}$

8. $s_n = \begin{cases} n^2 + n & \text{if } n \text{ is even} \\ 4n+1 & \text{if } n \text{ is odd} \end{cases}$

In Problems 9–40 tell whether the given sequence converges or diverges. If it converges, determine to what number.

9. $\left\{ \dfrac{5}{n} \right\}$

10. $\left\{ \dfrac{n}{5} \right\}$

11. $\left\{ 2 - \dfrac{4}{n} \right\}$

12. $\left\{ \left(4 + \dfrac{2}{n} \right)^2 \right\}$

13. $\left\{ 1 + \dfrac{(-1)^n}{n} \right\}$

14. $\left\{ \dfrac{1 + (-1)^n}{n} \right\}$

15. $\{(0.5)^n\}$

16. $\left\{ \left(\dfrac{1}{3} \right)^n \right\}$

17. $\{1 + (-1)^n\}$

18. $\left\{ \dfrac{(-1)^n}{2n} \right\}$

19. $\left\{ \dfrac{n + (-1)^n}{n} \right\}$

20. $\left\{ \left(-\dfrac{1}{3} \right)^n \right\}$

21. $\left\{ \left(1 - \dfrac{1}{n} \right) \left(1 - \dfrac{1}{n^2} \right) \right\}$

22. $\left\{ \left(1 - \dfrac{1}{n} \right) \left(1 - \dfrac{1}{n^2} \right) \left(1 - \dfrac{1}{n^3} \right) \right\}$

23. $\left\{ \dfrac{n^2}{2n+1} - \dfrac{n^2}{2n-1} \right\}$

24. $\left\{ \dfrac{6n^4 - 5}{7n^4 + 3} \right\}$

25. $\left\{ \dfrac{n}{\sqrt{n^2 + 1}} \right\}$

26. $\left\{ \dfrac{\sqrt{n} + 2}{\sqrt{n} + 5} \right\}$

27. $\left\{ 2 - \dfrac{1}{2^n} \right\}$

28. $\left\{ \dfrac{n^2}{3^n} \right\}$

29. $\left\{ \dfrac{1}{ne^{-n}} \right\}$

30. $\left\{ \dfrac{n}{e^n} \right\}$

31. $\left\{ \cos \left(n\pi + \dfrac{\pi}{2} \right) \right\}$

32. $\left\{ \sin \dfrac{n\pi}{2} \right\}$

33. $\left\{ n \sin \dfrac{1}{n} \right\}$

34. $\left\{ \dfrac{\sqrt{n}}{e^n} \right\}$

35. $\{\ln n - \ln(n+1)\}$

36. $\left\{\dfrac{\ln(n+1)}{n+1}\right\}$ 　　　　　 **37.** $\{\tan^{-1}n\}$ 　　　　　 **38.** $\left\{\dfrac{n\tan^{-1}n}{n^2+1}\right\}$

39. $\left\{\dfrac{n+\sin n}{n+\cos 4n}\right\}$ 　　　　 **40.** $\left\{\dfrac{n^2}{2n+1}\sin\dfrac{1}{n}\right\}$

In Problems 41–62 the nth term of a sequence is given. Tell whether the sequence converges or diverges. Give a reason for your decision.

41. $s_n = \dfrac{3}{n}$ 　　　　　 **42.** $s_n = \dfrac{(-1)^n}{n}$ 　　　　　 **43.** $s_n = \dfrac{3^n+1}{4^n}$

44. $s_n = \dfrac{(n-1)^2}{e^n}$ 　　　　 **45.** $s_n = \sqrt{n}$ 　　　　　 **46.** $s_n = 1 - \left(\dfrac{1}{2}\right)^n$

47. $s_n = (0.8)^n$ 　　　　 **48.** $s_n = \dfrac{n+1}{n^2}$ 　　　　 **49.** $s_n = \dfrac{\ln(n+1)}{\sqrt{n}}$

50. $s_n = \ln\left(\dfrac{n+1}{3n}\right)$ 　　　 **51.** $s_n = \cos n\pi$ 　　　　 **52.** $s_n = ne^{-n}$

53. $s_n = \dfrac{\sqrt{n+1}}{n}$ 　　　　 **54.** $s_n = \dfrac{\sin n}{n}$ 　　　　 **55.** $s_n = \dfrac{(-3)^n}{n^{100}}$

56. $s_n = \dfrac{2^n}{(2)(4)(6)\cdots(2n)}$ 　　 **57.** $s_n = \dfrac{3^{n+1}}{(3)(6)(9)\cdots(3n)}$ 　　 **58.** $s_n = (-1)^n\sqrt{n}$

59. $s_n = \sqrt{8-(1/n)}$ 　　　 **60.** $s_n = \dfrac{\sqrt{2n+7}}{n+7}$ 　　　 **61.** $s_n = \dfrac{5^n}{(n+1)^2}$

62. $s_n = 1 + \dfrac{1}{n^2+n\cos n+1}$

　　[*Hint:* Show that the derivative of $x^2 + x\cos x + 1$ is positive for $x > 1$.]

63. For what integers n is $1/n$ within $1/100$ of $1/50$?

64. For what integers n is $1/n$ within $1/100$ of 0?

65. For what integers n is $(n-1)/n$ within $1/100$ of 0.9?

66. For what integers n is $(n-1)/n$ within $1/100$ of 1.1?

67. Prove that if $0 < r < 1$, then $\lim_{n\to+\infty} r^n = 0$. [*Hint:* Let $r = 1/(1+p)$, where $p > 0$; then, by the binomial theorem, $r^n = 1/(1+p)^n = 1/[1 + np + n(n-1)p^2/2 + \cdots + p^n] < 1/np$.]

68. Use the result of Problem 67 to prove that if $-1 < r < 0$, then $\lim_{n\to+\infty} r^n = 0$.

69. Prove that if $r > 1$, then $\lim_{n\to+\infty} r^n = +\infty$. [*Hint:* Let $r = 1 + p$, where $p > 0$; then by the binomial theorem, $r^n = (1+p)^n = 1 + np + n(n-1)p^2/2 + \cdots + p^n > np$.]

70. Use the result of Problem 69 to show that if $r < -1$, then $\lim_{n\to+\infty} r^n$ does not exist. [*Hint:* r^n oscillates between positive and negative values.]

71. Prove that if the sequence $\{s_n\}$ is convergent and $\lim_{n\to+\infty} s_n = L$, then the sequence $\{s_n^2\}$ is also convergent and $\lim_{n\to+\infty} s_n^2 = L^2$.

72. A sequence $\{s_n\}$ is said to be a *Cauchy sequence* if and only if for each $\varepsilon > 0,$ there exists a positive integer N such that

$$|s_n - s_m| < \varepsilon \qquad \text{for all} \quad n, m > N$$

Show that every convergent sequence is a Cauchy sequence.

73. Prove that if $\lim_{n \to +\infty} s_n = L,$ then $\lim_{n \to +\infty} |s_n|$ exists and $\lim_{n \to +\infty} |s_n| = |L|.$ Is the converse true?

74. Let $a_1 > 0$ and $b_1 > 0$ be two real numbers, for which $a_1 > b_1.$ Define sequences $\{a_n\}$ and $\{b_n\}$ as

$$a_{n+1} = \frac{a_n + b_n}{2} \qquad b_{n+1} = \sqrt{a_n b_n}$$

(a) Show that $b_n < b_{n+1} < a_1$ for all $n.$
(b) Show that $b_1 < a_{n+1} < a_n$ for all $n.$

(c) Show that $0 < a_{n+1} - b_{n+1} < \dfrac{a_1 - b_1}{2^n}.$

(d) Show that $\lim_{n \to +\infty} a_n$ and $\lim_{n \to +\infty} b_n$ each exist and are equal.

75. The famous *Fibonacci sequence* $\{u_n\}$ is defined recursively by the formula

$$u_{n+2} = u_{n+1} + u_n, \qquad u_1 = 1, \quad u_2 = 1$$

(a) Write out the first eight terms of the Fibonacci sequence.
(b) Verify that a closed form representation for u_n is given by

$$u_n = \frac{(1 + \sqrt{5})^n - (1 - \sqrt{5})^n}{2^n \sqrt{5}}$$

[*Hint:* Show that $u_1 = 1,$ $u_2 = 1,$ and $u_{n+2} = u_{n+1} + u_n.$]

2. Infinite Series

Is it logical for the sum of an infinite collection of nonzero numbers to be finite? Consider Figure 9. The square in the figure has each side of length 1 unit, and hence its area is 1 square unit. When we divide the box into two boxes of equal area, each half has an area of $\frac{1}{2}$ square unit. If we continue this process indefinitely, we obtain a decomposition of the area 1 into boxes of area $\frac{1}{2}, \frac{1}{4}, \frac{1}{8}, \frac{1}{16},$ and so forth. Therefore,

$$1 = \tfrac{1}{2} + \tfrac{1}{4} + \tfrac{1}{8} + \tfrac{1}{16} + \cdots$$

Surprised?

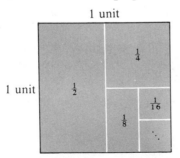

Figure 9

Let's look at this result from a different point of view by starting with the infinite sum

(11.10)
$$\tfrac{1}{2} + \tfrac{1}{4} + \tfrac{1}{8} + \tfrac{1}{16} + \cdots$$

One way we might add this up is by the use of *partial sums* in order to see if a trend develops. The first four partial sums are

$$\tfrac{1}{2} + \tfrac{1}{4} = \tfrac{3}{4} = 0.75$$
$$\tfrac{1}{2} + \tfrac{1}{4} + \tfrac{1}{8} = \tfrac{3}{4} + \tfrac{1}{8} = \tfrac{7}{8} = 0.875$$
$$\tfrac{1}{2} + \tfrac{1}{4} + \tfrac{1}{8} + \tfrac{1}{16} = \tfrac{7}{8} + \tfrac{1}{16} = \tfrac{15}{16} = 0.9375$$
$$\tfrac{1}{2} + \tfrac{1}{4} + \tfrac{1}{8} + \tfrac{1}{16} + \tfrac{1}{32} = \tfrac{15}{16} + \tfrac{1}{32} = \tfrac{31}{32} = 0.96875$$

Each of these sums uses more of the terms in (11.10) and each sum seems to be getting closer to 1.

In general, if $a_1, a_2, \ldots, a_n, \ldots$ is some infinite collection of numbers, the expression

(11.11)
$$\sum_{k=1}^{+\infty} a_k = a_1 + a_2 + \cdots + a_n + \cdots$$

is called an *infinite series*, or, simply, a *series*. The numbers $a_1, a_2, \ldots, a_n, \ldots$ are called *terms* and the number a_n is called the *nth term* or the *general term*. As before, the symbol $\sum$ stands for *summation; k* is the *index of summation*.

The index of summation is not fixed for a given series. For example, the series in (11.10), which may be written as

$$\frac{1}{2} + \frac{1}{2^2} + \frac{1}{2^3} + \cdots + \frac{1}{2^n} + \cdots = \sum_{k=1}^{+\infty} \frac{1}{2^k}$$

also may be written in any of the following equivalent ways:

$$\sum_{k=1}^{+\infty} \frac{1}{2}\left(\frac{1}{2^{k-1}}\right) \qquad \sum_{k=0}^{+\infty} \frac{1}{2^{k+1}} \qquad \sum_{k=4}^{+\infty} \frac{1}{2^{k-3}}$$

In most of our work, the index of summation will begin at 1.

To define a sum of an infinite series $\sum_{k=1}^{+\infty} a_k$, we make use of the sequence $\{S_n\}$ defined by

$$S_1 = a_1$$

$$S_2 = a_1 + a_2 = \sum_{k=1}^{2} a_k$$

$$\vdots$$

$$S_n = a_1 + a_2 + \cdots + a_n = \sum_{k=1}^{n} a_k$$

This sequence $\{S_n\}$ is called the *sequence of partial sums* of the series $\sum_{k=1}^{+\infty} a_k$.

For example, consider again the series

$$\sum_{k=1}^{+\infty} \frac{1}{2^k} = \frac{1}{2} + \frac{1}{2^2} + \frac{1}{2^3} + \frac{1}{2^4} + \cdots = \frac{1}{2} + \frac{1}{4} + \frac{1}{8} + \frac{1}{16} + \cdots$$

As it turns out, the partial sums can each be written as 1 minus a power of $\frac{1}{2}$, as follows:

$$S_1 = \frac{1}{2} = 1 - \frac{1}{2}$$

$$S_2 = \frac{1}{2} + \frac{1}{4} = \frac{3}{4} = 1 - \frac{1}{4} = 1 - \frac{1}{2^2}$$

$$S_3 = \frac{1}{2} + \frac{1}{4} + \frac{1}{8} = \frac{7}{8} = 1 - \frac{1}{8} = 1 - \frac{1}{2^3}$$

$$S_4 = \frac{15}{16} = 1 - \frac{1}{16} = 1 - \frac{1}{2^4}$$

$$S_5 = \frac{31}{32} = 1 - \frac{1}{2^5}$$

$$\vdots$$

$$S_n = 1 - \frac{1}{2^n}$$

The nth partial sum is $S_n = 1 - (1/2^n)$, and, as n gets larger and larger, the sequence $\{S_n\}$ of partial sums approaches a limit, namely,

$$\lim_{n \to +\infty} S_n = \lim_{n \to +\infty} \left(1 - \frac{1}{2^n}\right) = \lim_{n \to +\infty} 1 - \lim_{n \to +\infty} \frac{1}{2^n} = 1 - 0 = 1$$

We agree to call this limit the *sum of the series*, and we write

(11.12) $$\sum_{k=1}^{+\infty} \frac{1}{2^k} = \frac{1}{2} + \frac{1}{4} + \frac{1}{8} + \frac{1}{16} + \cdots = 1$$

(11.13) **If the sequence $\{S_n\}$ of partial sums of an infinite series $\sum_{k=1}^{+\infty} a_k$ has a limit S, then the series *converges* and is said to have the *sum S*. That is, if $\lim_{n \to +\infty} S_n = S$, then**

$$\sum_{k=1}^{+\infty} a_k = a_1 + a_2 + \cdots + a_n + \cdots = S$$

An infinite series *diverges* if the sequence of partial sums diverges.

Remember that if an infinite series does not converge, then the symbol $a_1 + a_2 + \cdots = \sum_{k=1}^{+\infty} a_k$ is *not* the name of a number.

Let's look at some examples.

EXAMPLE 1   Show that: $\displaystyle\sum_{k=1}^{+\infty}\frac{1}{k(k+1)}=\frac{1}{1(2)}+\frac{1}{2(3)}+\frac{1}{3(4)}+\cdots=\frac{1}{2}+\frac{1}{6}+\frac{1}{12}+\cdots=1$

Solution   We begin with the sequence $\{S_n\}$ of partial sums, namely,

$$S_1=\frac{1}{1(2)}$$

$$S_2=\frac{1}{1(2)}+\frac{1}{2(3)}$$

$$S_3=\frac{1}{1(2)}+\frac{1}{2(3)}+\frac{1}{3(4)}$$

$$\vdots$$

$$S_n=\frac{1}{1(2)}+\frac{1}{2(3)}+\frac{1}{3(4)}+\cdots+\frac{1}{n(n+1)}$$

To find the sum of the infinite series requires that we find the limit of the sequence $\{S_n\}$. This is usually difficult, because to find $\lim_{n\to+\infty} S_n$ requires that S_n be expressed as a function of n. In this example, though, since

$$\frac{1}{n(n+1)}=\frac{1}{n}-\frac{1}{n+1}$$

we can express S_n as

$$S_n=\left(\frac{1}{1}-\frac{1}{2}\right)+\left(\frac{1}{2}-\frac{1}{3}\right)+\cdots+\left(\frac{1}{n-1}-\frac{1}{n}\right)+\left(\frac{1}{n}-\frac{1}{n+1}\right)$$

Note that all the terms except the first and last cancel,* so that

$$S_n=1-\frac{1}{n+1}$$

It follows that

$$\lim_{n\to+\infty} S_n=\lim_{n\to+\infty}\left(1-\frac{1}{n+1}\right)=1$$

Thus, $\sum_{k=1}^{+\infty}[1/k(k+1)]$ converges, and its sum is 1.   ∎

EXAMPLE 2   Show that the series $\sum_{k=1}^{+\infty}(-1)^k=-1+1-1+\cdots$ diverges.

---

* Sequences in which the middle terms cancel in this manner are sometimes called *telescoping sequences*.

Solution   The sequence $\{S_n\}$ of partial sums for this series is

$$S_1 = -1$$
$$S_2 = -1 + 1 = 0$$
$$S_3 = -1 + 1 - 1 = -1$$
$$S_4 = -1 + 1 - 1 + 1 = 0$$
$$\vdots$$
$$S_n = \begin{cases} -1 & \text{if } n \text{ is odd} \\ 0 & \text{if } n \text{ is even} \end{cases}$$

Since the sequence $\{S_n\}$ diverges, the infinite series diverges.   ■

EXAMPLE 3   Show that the series $\sum_{k=1}^{+\infty} k = 1 + 2 + 3 + \cdots$ diverges.

Solution   The sequence $\{S_n\}$ of partial sums is

$$S_1 = 1$$
$$S_2 = 1 + 2$$
$$S_3 = 1 + 2 + 3$$
$$\vdots$$
$$S_n = 1 + 2 + 3 + \cdots + n$$

We seek to express S_n in a way that will make it easy to find $\lim_{n \to +\infty} S_n$. Using (5.11), we write

$$S_n = 1 + 2 + \cdots + n = \frac{n(n + 1)}{2}$$

Since $\lim_{n \to +\infty} S_n = \lim_{n \to +\infty} [n(n + 1)/2] = +\infty$, $\{S_n\}$ diverges. Hence, the series $\sum_{k=1}^{+\infty} k$ diverges.   ■

GEOMETRIC SERIES

A special infinite series that is useful in many applied problems is the *geometric series*:

(11.14)
$$\sum_{k=0}^{+\infty} ar^k = \sum_{k=1}^{+\infty} ar^{k-1} = a + ar + ar^2 + \cdots + ar^{n-1} + \cdots \qquad a \neq 0$$

In this series, the ratio of each term to its predecessor is r, where r is some fixed real number.

To determine the conditions for convergence of the geometric series, we examine the nth partial sum

(11.15)
$$S_n = a + ar + ar^2 + \cdots + ar^{n-1}$$

We look at various cases for r.

If $r = 1$, the series becomes $\sum_{k=1}^{+\infty} a = a + a + \cdots + a + \cdots$ and the nth partial sum is

$$S_n = a + a + \cdots + a = na$$

In this case, $\{S_n\}$ diverges (since $a \neq 0$), so that for $r = 1$, the geometric series diverges.

If $r = -1$, the series becomes $\sum_{k=1}^{+\infty} a(-1)^{k-1} = a - a + a - a + \cdots$ and the nth partial sum is

$$S_n = \begin{cases} 0 & \text{if } n \text{ is even} \\ a & \text{if } n \text{ is odd} \end{cases}$$

Since $\{S_n\}$ diverges, we conclude that for $r = -1$, the geometric series diverges.

To see what happens for other choices of r, we multiply both sides of (11.15) by r, obtaining

$$rS_n = ar + ar^2 + \cdots + ar^n$$

By subtracting this from the expression for S_n in (11.15), we have

$$S_n - rS_n = (a + ar + ar^2 + \cdots + ar^{n-1}) - (ar + ar^2 + \cdots + ar^{n-1} + ar^n)$$

$$= a - ar^n$$

$$S_n(1 - r) = a(1 - r^n)$$

If $r \neq 1$, we can express the nth partial sum of the geometric series as

$$S_n = \frac{a(1 - r^n)}{1 - r}$$

Now,

$$\lim_{n \to +\infty} S_n = \lim_{n \to +\infty} \frac{a - ar^n}{1 - r} = \frac{a}{1 - r} - \frac{a}{1 - r} \lim_{n \to +\infty} r^n$$

Combining the results obtained in Problems 67 and 68 of Exercise 1, we conclude that:

For $-1 < r < 1$, $\displaystyle\lim_{n \to +\infty} r^n = 0$ so that $\displaystyle\lim_{n \to +\infty} S_n = \frac{a}{1 - r}$

Hence, for $-1 < r < 1$, the geometric series converges and its sum is $a/(1 - r)$.

Combining the results obtained in Problems 69 and 70 of Exercise 1, we have:

For $r > 1$ or $r < -1$, $\displaystyle\lim_{n \to +\infty} r^n$ does not exist

Hence, for $r > 1$ or $r < -1$, the geometric series diverges.

We summarize these results in the following theorem:

(11.16)   **THEOREM**

(a)   If $-1 < r < 1$,   the geometric series $\sum_{k=1}^{+\infty} ar^{k-1}$ **converges, and its sum is**

$$\sum_{k=1}^{+\infty} ar^{k-1} = \frac{a}{1-r}$$

(b)   If $|r| \geq 1$,   the geometric series $\sum_{k=1}^{+\infty} ar^{k-1}$ **diverges.**

EXAMPLE 4   Determine whether each geometric series converges or diverges. If it converges, find its sum.

(a) $\sum_{k=1}^{+\infty} 8\left(\frac{2}{5}\right)^{k-1}$ (b) $\sum_{k=1}^{+\infty} \left(-\frac{5}{9}\right)^{k-1}$ (c) $\sum_{k=1}^{+\infty} 3\left(\frac{3}{2}\right)^{k-1}$ (d) $\sum_{k=1}^{+\infty} \frac{1}{2^k}$

Solution

(a)   We have $a = 8$, $r = \frac{2}{5}$.   Since $|r| < 1$,   the series converges and

$$\sum_{k=1}^{+\infty} 8\left(\frac{2}{5}\right)^{k-1} = \frac{8}{1 - \frac{2}{5}} = 8\left(\frac{5}{3}\right) = \frac{40}{3}$$

(b)   We have $a = 1$, $r = -\frac{5}{9}$.   Since $|r| < 1$,   the series converges and

$$\sum_{k=1}^{+\infty} \left(-\frac{5}{9}\right)^{k-1} = \frac{1}{1 + \frac{5}{9}} = \frac{9}{14}$$

(c)   We have $a = 3$, $r = \frac{3}{2}$.   Since $|r| > 1$,   the series diverges.

(d)   The given series is not exactly in the form $\sum_{k=1}^{+\infty} ar^{k-1}$.   However, since

$$\sum_{k=1}^{+\infty} \frac{1}{2^k} = \sum_{k=1}^{+\infty} \left(\frac{1}{2}\right)^k = \sum_{k=1}^{+\infty} \frac{1}{2}\left(\frac{1}{2}\right)^{k-1}$$

we conclude that $\sum_{k=1}^{+\infty} (1/2^k)$ is a geometric series with $a = \frac{1}{2}$, $r = \frac{1}{2}$. Hence, the series converges and its sum is

$$\sum_{k=1}^{+\infty} \frac{1}{2^k} = \frac{\frac{1}{2}}{1 - \frac{1}{2}} = 1$$

which, of course, verifies the solution given earlier in (11.12).   ∎

Geometric series can be used to express a nonterminating repeating decimal as a rational number. For example, to express the repeating decimal $0.33333\ldots$ as a quotient of two integers, we write the infinite series

$$0.333\ldots = 0.3 + 0.03 + 0.003 + 0.0003 + \cdots = \frac{3}{10} + \frac{3}{10^2} + \frac{3}{10^3} + \cdots$$

$$= \sum_{k=1}^{+\infty} \frac{3}{10^k} = \sum_{k=1}^{+\infty} \frac{3}{10}\left(\frac{1}{10}\right)^{k-1}$$

This is a geometric series with $a = \frac{3}{10}$ and $r = \frac{1}{10}$. Since $|r| < 1$, the series converges and its sum is

$$\sum_{k=1}^{+\infty} \frac{3}{10^k} = \frac{\frac{3}{10}}{1 - \frac{1}{10}} = \frac{3}{9} = \frac{1}{3}$$

Hence, $0.333\ldots = \frac{1}{3}$.

EXAMPLE 5   A ball is dropped from a height of 12 meters. Each time it strikes the ground, it bounces back to a height three-fourths the distance from which it fell. Find the total distance traveled by the ball before it comes to rest. (See Fig. 10.)

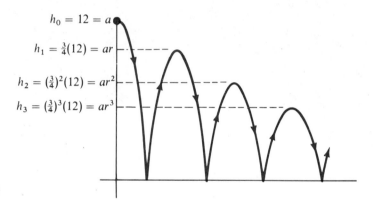

Figure 10

Solution   Let h_n denote the height the ball reaches after the nth bounce. Then

$$h_0 = 12$$
$$h_1 = \tfrac{3}{4}(12)$$
$$h_2 = \tfrac{3}{4}[\tfrac{3}{4}(12)] = (\tfrac{3}{4})^2(12)$$
$$\vdots$$
$$h_k = (\tfrac{3}{4})^k(12)$$

After the first bounce, the ball travels up a distance $h_1 = \tfrac{3}{4}(12)$ and then the same distance back down. Between the first and second bounce, the total distance traveled is therefore $h_1 + h_1 = 2h_1$. The *total* distance H traveled by the ball is

$$H = h_0 + 2h_1 + 2h_2 + 2h_3 + \cdots = h_0 + \sum_{k=1}^{+\infty} 2h_k = 12 + \sum_{k=1}^{+\infty} 2\left[12\left(\frac{3}{4}\right)^k\right]$$

$$= 12 + \sum_{k=1}^{+\infty} 24\left[\frac{3}{4}\left(\frac{3}{4}\right)^{k-1}\right]$$

$$= 12 + \sum_{k=1}^{+\infty} 18\left(\frac{3}{4}\right)^{k-1}$$

The last expression is a geometric series with $a = 18$, $r = \frac{3}{4}$. Hence, it converges and

$$H = 12 + \sum_{k=1}^{+\infty} 18\left(\frac{3}{4}\right)^{k-1} = 12 + \frac{18}{1 - \frac{3}{4}} = 84 \text{ meters}$$

The ball travels a total distance of 84 meters.   ■

We postpone a rather detailed application of geometric series in biology until the end of this section in order to continue our study of series, but if you wish to look at it now, see page 591.

HARMONIC SERIES

The infinite series

$$\sum_{k=1}^{+\infty} \frac{1}{k} = 1 + \frac{1}{2} + \frac{1}{3} + \cdots$$

is called the *harmonic series*. This series diverges.

To show this, we look at the partial sums whose index is a power of 2:

$$S_1 = 1 > 1(\tfrac{1}{2})$$
$$S_2 = 1 + \tfrac{1}{2} > \tfrac{1}{2} + \tfrac{1}{2} = 2(\tfrac{1}{2})$$
$$\uparrow$$

Replace 1 by $\tfrac{1}{2}$,
which is smaller

$$S_4 = 1 + \tfrac{1}{2} + (\tfrac{1}{3} + \tfrac{1}{4}) > 2(\tfrac{1}{2}) + (\tfrac{1}{4} + \tfrac{1}{4}) = 3(\tfrac{1}{2})$$
$$S_8 = 1 + \tfrac{1}{2} + (\tfrac{1}{3} + \tfrac{1}{4}) + (\tfrac{1}{5} + \tfrac{1}{6} + \tfrac{1}{7} + \tfrac{1}{8}) > 3(\tfrac{1}{2}) + (\tfrac{1}{8} + \tfrac{1}{8} + \tfrac{1}{8} + \tfrac{1}{8}) = 4(\tfrac{1}{2})$$
$$\vdots$$
$$S_{2^{n-1}} > n(\tfrac{1}{2})$$

We conclude that $\{S_{2^{n-1}}\}$ is unbounded, and, hence, so is $\{S_n\}$,* so that $\{S_n\}$ diverges. Thus, the harmonic series $\sum_{k=1}^{+\infty}(1/k)$ diverges.

_n_th TERM TEST

For most series, it is not possible to find a single compact expression for the sequence $\{S_n\}$ of partial sums as a function of n. Thus, obtaining the sum of a convergent infinite series is often very difficult. However, in many applications involving series, the significant point is to know whether or not the series converges. Knowing the sum of a convergent series, although desirable, is not always necessary. We are there-

---

* Do you see why? The sequence of partial sums is increasing, so that once we know that $\{S_{2^{n-1}}\}$ is unbounded, we also know that $\{S_n\}$ itself is unbounded.

fore interested in finding tests for convergence or divergence. To be useful, these tests should only depend on a knowledge of the form of the general term of the series.

(11.17)   **THEOREM**   **If the series** $\sum_{k=1}^{+\infty} a_k$ **converges, then** $\lim_{n \to +\infty} a_n = 0.$

Proof   The nth partial sum of $\sum_{k=1}^{+\infty} a_k$ is $S_n = \sum_{k=1}^{n} a_k$. Since $S_{n-1} = \sum_{k=1}^{n-1} a_k$, it follows that

$$a_n = S_n - S_{n-1}$$

But $\sum_{k=1}^{+\infty} a_k$ converges. Hence, the sequence $\{S_n\}$ has a limit, say, S. Since $\lim_{n \to +\infty} S_n = S$ and $\lim_{n \to +\infty} S_{n-1} = S$, we find that

$$\lim_{n \to +\infty} a_n = \lim_{n \to +\infty} (S_n - S_{n-1}) = \lim_{n \to +\infty} S_n - \lim_{n \to +\infty} S_{n-1} = S - S = 0 \qquad \blacksquare$$

Theorem (11.17) provides a *necessary condition* for convergence since the condition $\lim_{n \to +\infty} a_n = 0$ necessarily holds for *any convergent series*. However, the condition is *not a sufficient condition* for convergence since there are many divergent series $\sum_{k=1}^{+\infty} a_k$ for which $\lim_{n \to +\infty} a_n = 0$.

For example, consider the harmonic series $\sum_{k=1}^{+\infty}(1/k)$. The limit of the nth term is $\lim_{n \to +\infty}(1/n) = 0$, and yet the series diverges. For the geometric series $\sum_{k=1}^{+\infty}(1/2^{k-1})$, the limit of the nth term is $\lim_{n \to +\infty}(1/2^{n-1}) = 0$, and the series converges $(r = \frac{1}{2})$.

Theorem (11.17) is most useful as a test for divergence.

(11.18)   *Test for Divergence.*   **If the nth term of an infinite series** $\sum_{k=1}^{+\infty} a_k$ **does not approach 0 as** $n \to +\infty$, **the series is divergent.**

This test tells us immediately that the following series diverge:

$$\sum_{k=1}^{+\infty} 87 \qquad \sum_{k=1}^{+\infty} k \qquad \sum_{k=1}^{+\infty} (-1)^k \qquad \sum_{k=1}^{+\infty} 2^k$$

OTHER REMARKS

Some general remarks about series follow.

(11.19)   **If two infinite series are identical after a certain term, then either both converge or both diverge.***

To verify this statement, consider the two series

$$\sum_{k=1}^{+\infty} a_k = a_1 + a_2 + \cdots + a_p + a_{p+1} + \cdots + a_n + \cdots$$

$$\sum_{k=1}^{+\infty} b_k = b_1 + b_2 + \cdots + b_p + a_{p+1} + \cdots + a_n + \cdots$$

---

* If both series converge, they will not necessarily have the same sum.

in which after the first p terms of each series the remaining terms are identical. The sequence $\{S_n\}$ of partial sums of $\sum_{k=1}^{+\infty} a_k$ and the sequence $\{T_n\}$ of partial sums of $\sum_{k=1}^{+\infty} b_k$ are given by

$$S_n = a_1 + a_2 + \cdots + a_p + a_{p+1} + \cdots + a_n$$
$$T_n = b_1 + b_2 + \cdots + b_p + a_{p+1} + \cdots + a_n$$

Now,

$$S_n - T_n = (a_1 + \cdots + a_p) - (b_1 + \cdots + b_p)$$
$$S_n = T_n + (a_1 + \cdots + a_p) - (b_1 + \cdots + b_p)$$
$$\lim_{n \to +\infty} S_n = \lim_{n \to +\infty} T_n + (a_1 + \cdots + a_p) - (b_1 + \cdots + b_p)$$

Consequently, either both limits exist (both series converge) or both limits do not exist (both series diverge). No other possibility can occur.

(11.20)   **Convergent series can be added. If $\sum_{k=1}^{+\infty} a_k = S$ and $\sum_{k=1}^{+\infty} b_k = T$ are two convergent series, then the sum is a convergent series and**

$$\sum_{k=1}^{+\infty} (a_k + b_k) = S + T$$

The proof is left as an exercise (see Problem 47 in Exercise 2).

(11.21)   **Let c be any nonzero constant. If the series $\sum_{k=1}^{+\infty} a_k$ is convergent to the sum S, then the series $\sum_{k=1}^{+\infty} ca_k$ is convergent to the sum cS. If the series $\sum_{k=1}^{+\infty} a_k$ diverges, then so does the series $\sum_{k=1}^{+\infty} ca_k$.**

The proof is left as an exercise (see Problem 48 in Exercise 2).
Let's look at an example to illustrate these remarks.

EXAMPLE 6   Determine whether each series converges or diverges. If it converges, find its sum.

(a) $\displaystyle\sum_{k=1}^{+\infty} \frac{2}{k}$ (b) $\displaystyle\sum_{k=1}^{+\infty} \left(\frac{1}{2^{k-1}} + \frac{1}{3^{k-1}}\right)$

Solution   (a)   Except for a factor of 2, this is a harmonic series, which diverges. Hence, by (11.21) $\sum_{k=1}^{+\infty}(2/k)$ diverges.

(b)   The series $\sum_{k=1}^{+\infty}(1/2^{k-1})$ is a convergent geometric series, as is the series $\sum_{k=1}^{+\infty}(1/3^{k-1})$. Hence, by (11.20), $\sum_{k=1}^{+\infty}[(1/2^{k-1}) + (1/3^{k-1})]$ converges, and its sum is

$$\sum_{k=1}^{+\infty} \left(\frac{1}{2^{k-1}} + \frac{1}{3^{k-1}}\right) = \sum_{k=1}^{+\infty} \frac{1}{2^{k-1}} + \sum_{k=1}^{+\infty} \frac{1}{3^{k-1}} = \frac{1}{1 - \frac{1}{2}} + \frac{1}{1 - \frac{1}{3}} = 2 + \frac{3}{2} = 3.5 \quad \blacksquare$$

We conclude this section with a summary of several key points:

1. A series converges if and only if its sequence of partial sums converges.

2. The geometric series $\sum_{k=1}^{+\infty} ar^{k-1}$, $a \neq 0$, converges for $|r| < 1$ and diverges for $|r| \geq 1$. If it converges, its sum is $a/(1 - r)$.

3. The harmonic series $\sum_{k=1}^{+\infty} (1/k)$ diverges.

4. Divergence test: If $\lim_{n \to +\infty} a_n \neq 0$, then the series $\sum_{k=1}^{+\infty} a_k$ diverges.

APPLICATION IN BIOLOGY*

In this application we study the rate of occurrence of *retinoblastoma*, a rare type of eye cancer in children, which, at the turn of this century, was nearly always fatal.

To begin, we need a term from biology. An *allele* (*allelomorph*) is a gene that gives rise to one of a pair of contrasting characteristics, such as smooth or rough, tall or short, and so on. Each person normally has two such genes for each characteristic. An individual may have two "tall" genes, two "short" genes, or one of each. In reproduction, each parent gives one of the two types to the child.

The tendency to develop retinoblastoma apparently depends upon a single dominant allele, say, A. If the corresponding normal allele is represented by a, the mutation rate from a to A in each generation is approximately $m = 0.00002 = 2 \times 10^{-5}$. In this example, we ignore the very unlikely possibility of mutation from A to a. With the medical care available in the early 1950's, approximately 70% of those affected with the disease survived, although they usually became blind in one or both eyes. Let us assume that survivors reproduce at about half the normal rate. (This assumption is based on scientific guesswork.) Then the productive proportion of persons affected with the disease is $r = 0.35$. This rate is remarkable, considering that around 1900, r was approximately 0.

Starting with 0 inherited cases in an early generation, for the nth consecutive generation, we obtain a rate of

m due to mutation in the nth generation

mr due to mutation in the $(n - 1)$st generation

mr^2 due to mutation in the $(n - 2)$nd generation

$\vdots$

mr^n due to mutation in the zero (original) generation

Hence, the total rate of occurrence of the disease in the nth generation is

$$p_n = m + mr + \cdots + mr^n = \frac{m(1 - r^{n+1})}{1 - r}$$

---

* Adapted from J. V. Neel and W. J. Schull, *Human Heredity*, 3rd impression (Chicago: University of Chicago Press, 1958), pp. 333–334. Reprinted by permission.

from which

$$p = \lim_{n \to +\infty} p_n = \frac{m}{1-r} = 3.08 \times 10^{-5}$$

Thus, the total rate of persons affected with the disease will be slightly more than 50% higher than the mutation rate.

Observe that if $r = 0$, then $p = m$. Thus, retinoblastoma has become more frequent with better medical care. As medical care improves, the rate of occurrence of the disease can be expected to become even greater. Neel and Schull point out that, with improved medical care, the frequency of the gene A increases rapidly at first, then more slowly, until an equilibrium point is reached. This equilibrium point is closely approximated after about eight generations.

EXERCISE 2

In Problems 1–4 find the fourth partial sum for each series.

1. $\sum_{k=1}^{+\infty} \left(\frac{3}{4}\right)^{k-1}$ **2.** $\sum_{k=1}^{+\infty} \frac{(-1)^{k+1}}{3^{k-1}}$ **3.** $\sum_{k=1}^{+\infty} k$ **4.** $\sum_{k=1}^{+\infty} \ln k$

In Problems 5–14 determine whether each geometric series converges or diverges. If it converges, find its sum.

5. $1 + \frac{1}{3} + \frac{1}{9} + \cdots + \left(\frac{1}{3}\right)^n + \cdots$ **6.** $1 + \frac{1}{4} + \frac{1}{16} + \cdots + \left(\frac{1}{4}\right)^n + \cdots$

7. $1 + 2 + 4 + \cdots + 2^n + \cdots$ **8.** $1 - \frac{1}{2} + \frac{1}{4} - \frac{1}{8} + \cdots + \frac{(-1)^{n-1}}{2^{n-1}} + \cdots$

9. $\left(\frac{1}{7}\right)^2 + \left(\frac{1}{7}\right)^3 + \cdots + \left(\frac{1}{7}\right)^n + \cdots$ **10.** $\left(\frac{3}{4}\right)^5 + \left(\frac{3}{4}\right)^6 + \cdots + \left(\frac{3}{4}\right)^n + \cdots$

11. $\sum_{k=1}^{+\infty} (\sqrt{2})^{k-1}$ **12.** $\sum_{k=1}^{+\infty} (0.33)^{k-1}$ **13.** $\sum_{k=1}^{+\infty} 7\left(\frac{1}{3}\right)^k$ **14.** $\sum_{k=1}^{+\infty} \left(\frac{7}{4}\right)^k$

In Problems 15–34 use the examples and theorems of this section to determine whether each series converges or diverges. If it converges, find its sum.

15. $\sum_{k=1}^{+\infty} \frac{1}{100^k}$ **16.** $\sum_{k=0}^{+\infty} e^{-k}$ **17.** $\sum_{k=1}^{+\infty} \frac{10}{k}$ **18.** $\sum_{k=1}^{+\infty} \frac{2^k 3^k}{4^k}$

19. $\sum_{k=1}^{+\infty} \frac{k^2+1}{4k+1}$ **20.** $\sum_{k=1}^{+\infty} \frac{k^3}{k^3+3}$ **21.** $\sum_{k=1}^{+\infty} \left(\frac{1}{k+2} - \frac{1}{k+3}\right)$ **22.** $\sum_{k=1}^{+\infty} \left[\frac{1}{k^2} - \frac{1}{(k+1)^2}\right]$

23. $\sum_{k=1}^{+\infty} \left(k + \frac{1}{k}\right)$ **24.** $\sum_{k=1}^{+\infty} \left(\frac{1}{3^k} - \frac{1}{4^k}\right)$ **25.** $\sum_{k=1}^{+\infty} \left(\frac{1}{3k} - \frac{1}{4k}\right)$ **26.** $\sum_{k=1}^{+\infty} \left(k - \frac{10}{k}\right)$

27. $\sum_{k=1}^{+\infty} (2^{1/3})^k$ **28.** $\sum_{k=1}^{+\infty} (3^{1/2})^k$ **29.** $\sum_{k=1}^{+\infty} \sin \frac{\pi}{2} k$ **30.** $\sum_{k=1}^{+\infty} \sec \pi k$

31. $\sum_{k=3}^{+\infty} \frac{k+1}{k-2}$ **32.** $\sum_{k=5}^{+\infty} \frac{2k^5+3}{k^5-4k^4}$

33. $\displaystyle\sum_{k=1}^{+\infty} \frac{1}{4k^2 - 1}$ $\left[Hint: \quad \frac{1}{4k^2 - 1} = \frac{1}{2}\left(\frac{1}{2k-1} - \frac{1}{2k+1} \right) \right]$

34. $\displaystyle\sum_{k=1}^{+\infty} \frac{1}{k(k+1)(k+2)}$ $\left[Hint: \quad \frac{1}{k(k+1)(k+2)} = \frac{1}{2}\left[\frac{1}{k(k+1)} - \frac{1}{(k+1)(k+2)} \right] \right]$

In Problems 35–38 express each repeating decimal as a rational number by using a geometric series.

35. $0.555\ldots$

36. $0.999\ldots$

37. $4.28555\ldots$

38. $7.162162\ldots$

[*Hint:* Write the number as $4.28 + 0.00555\ldots$]

In Problems 39 and 40 use a geometric series to prove the given statement.

39. $\displaystyle\frac{x}{x-1} = \sum_{k=1}^{+\infty} \frac{1}{x^{k-1}}$; for $|x| > 1$

40. $\displaystyle\frac{1}{1+x} = \sum_{k=0}^{+\infty} (-1)^k x^k$; for $|x| < 1$

41. A ball is dropped from a height of 18 feet. Each time it strikes the ground, it bounces back to two-thirds of the previous height. Find the total distance traveled by the ball before it comes to rest.

42. A rich man promises to give you \$1000 on January 1, 1986. Each day thereafter he will give you $\frac{9}{10}$ of what he gave you the previous day. What is the total amount you will receive? What is the first date on which the amount you receive is less than 1¢?

43. A coin-flipping game involves two people who successively flip a coin. The first person to obtain a head is the winner. In probability, it turns out that the person who flips first has the probability of winning given by the series below. Find this number.

$$\frac{1}{2} + \frac{1}{8} + \frac{1}{32} + \cdots + \frac{1}{2^{2n-1}} + \cdots$$

44. Use a hand calculator to find the smallest number n for which $\sum_{k=1}^{n}(1/k) \geq 3$.

45. Answer Problem 44 for $\sum_{k=1}^{n}(1/k) \geq 4$.

46. Let $S = 1 + 2 + 4 + 8 + \cdots$. Then

$$2S = 2 + 4 + 8 + 16 + \cdots = -1 + (1 + 2 + 4 + \cdots) = -1 + S$$

Therefore,

$$S = 1 + 2 + 4 + 8 + \cdots = -1$$

What did we do wrong here?

47. Prove that the sum of two convergent series is a convergent series.

48. Prove that if $\sum_{k=1}^{+\infty} a_k = S$, then $\sum_{k=1}^{+\infty} ca_k = cS$ for any real number c.

49. Suppose $\sum_{k=N+1}^{+\infty} a_k = S$ and suppose $a_1 + a_2 + \cdots + a_N = K$. Prove that $\sum_{k=1}^{+\infty} a_k$ converges and its sum is $S + K$.

50. Zeno's paradox is about a race between Achilles and a tortoise, in which the tortoise is allowed a certain lead at the start of the race. Zeno claimed the tortoise must win such a race. He reasoned that in order for Achilles to overtake the tortoise, at some time he must cover $\frac{1}{2}$ of the distance that originally separated them. Then, when he covers $\frac{1}{2}$ of the new distance separating them, he will still have $\frac{1}{4}$ of the original distance remaining. And so on. Therefore, by Zeno's reasoning, Achilles never catches the tortoise. Use a series argument to explain this paradox. Assume that the difference in speed between Achilles and the tortoise is v meters per second.

3. Series of Positive Terms

The definition of convergence of an infinite series requires that we know the sequence $\{S_n\}$ of partial sums of the series. In many cases it is difficult to obtain a formula for $\{S_n\}$ and therefore it is desirable to have tests for convergence that bypass the sequence of partial sums. For series whose terms are all positive, it is possible to construct tests for convergence that require the use of the *nth term of the series* and do not depend on a knowledge of the form of the sequence $\{S_n\}$ of partial sums.

Consider an infinite series

$$\sum_{k=1}^{+\infty} a_k = a_1 + a_2 + \cdots + a_n + \cdots$$

in which each term is positive. Because each term is positive, the sequence $\{S_n\}$ of partial sums will be increasing.* If the sequence of partial sums is bounded (from above), it follows from axiom (11.9) that it will converge. Hence, we have the *general convergence test:*

(11.22)   **An infinite series of positive terms will converge if and only if its sequence of partial sums is bounded. The sum of such an infinite series will not exceed an upper bound.**

We use this fact in developing three standard types of tests for convergence of positive series: the *comparison tests*, the *integral test*, and the *ratio test*.

COMPARISON TESTS FOR CONVERGENCE AND DIVERGENCE

Series may be tested for convergence or divergence by comparing them with series whose behavior is known. Suppose $\sum_{k=1}^{+\infty} b_k$ is a series of positive terms that is known to converge, and suppose $\sum_{k=1}^{+\infty} a_k$ is the series of positive terms to be tested. If S_n is the *n*th partial sum of $\sum_{k=1}^{+\infty} b_k$ and if B is its sum, then we have $S_n < B$. If term-by-term $a_k \le b_k$, that is, if

$$a_1 \le b_1, \quad a_2 \le b_2, \quad \ldots, \quad a_n \le b_n, \quad \ldots$$

then it follows that

$$\left(n\text{th partial sum of } \sum_{k=1}^{+\infty} a_k \right) \le S_n < B$$

and hence, by (11.22), $\sum_{k=1}^{+\infty} a_k$ must also converge. Thus, we have the first comparison test.

(11.23)   *Comparison Test I.* **A series $\sum_{k=1}^{+\infty} a_k$ of positive terms is convergent if each of its terms is less than or equal to the corresponding term of a known convergent series $\sum_{k=1}^{+\infty} b_k$ of positive terms.**

---

* Do you see why? $S_{n+1} = a_{n+1} + S_n > S_n$

It is appropriate to recall at this point that the early terms in a series have no effect on the convergence or divergence of the series. In fact, comparison test I is true if $0 < a_n \leq b_n$ for all $n \geq N$, where N is some suitably selected integer. We use this in the next example by ignoring the first term.

EXAMPLE 1   Prove that the series given below converges.

$$\sum_{k=1}^{+\infty} \frac{1}{k^k} = 1 + \frac{1}{2^2} + \frac{1}{3^3} + \cdots + \frac{1}{n^n} + \cdots$$

Solution   We have already seen that the geometric series $\sum_{k=1}^{+\infty}(1/2^k)$ converges to 1. Therefore, since $1/n^n \leq 1/2^n$ for $n \geq 2$,* each term of the given series is less than or equal to each corresponding term of a convergent series (except for the first term). Therefore, by comparison test I, we conclude that $\sum_{k=1}^{+\infty}(1/k^k)$ converges.   ∎

In the above example, you may ask "How did you know that the given series should be compared to $\sum_{k=1}^{+\infty}(1/2^k)$?" The answer is that only practice and experience will guide you.

Now, let's look at a comparison test for divergence.

Let $\sum_{k=1}^{+\infty} c_k$ be a known divergent series of positive terms. If S_n is the nth partial sum $\sum_{k=1}^{+\infty} c_k$, then we know that $\{S_n\}$ is unbounded. Suppose $\sum_{k=1}^{+\infty} a_k$ is the series of positive terms to be tested, and suppose that term-by-term we have

$$a_1 \geq c_1, \quad a_2 \geq c_2, \quad \ldots, \quad a_n \geq c_n, \quad \ldots$$

Then the nth partial sum of $\sum_{k=1}^{+\infty} a_k$ is greater than or equal to the nth partial sum of $\sum_{k=1}^{+\infty} c_k$, which is unbounded. Hence, the nth partial sum of $\sum_{k=1}^{+\infty} a_k$ is unbounded, so that $\sum_{k=1}^{+\infty} a_k$ diverges.

(11.24)   **Comparison Test II.   A series $\sum_{k=1}^{+\infty} a_k$ of positive terms is divergent if each of its terms is greater than or equal to the corresponding term of a known divergent series $\sum_{k=1}^{+\infty} c_k$ of positive terms.**

EXAMPLE 2   Show that the following series diverges: $\displaystyle\sum_{k=1}^{+\infty} \frac{k+3}{k(k+2)}$

Solution   We choose to compare the given series to the harmonic series $\sum_{k=1}^{+\infty}(1/k)$, which diverges.

$$\frac{n+3}{n(n+2)} = \underbrace{\left(\frac{n+3}{n+2}\right)}_{\substack{\uparrow \\ \text{This number is always} \\ \text{greater than 1}}}\left(\frac{1}{n}\right) > \frac{1}{n}$$

It follows from (11.24) that $\sum_{k=1}^{+\infty}[(k+3)/k(k+2)]$ diverges.   ∎

---

* Do you see why? $n \geq 2 \Rightarrow n^n \geq 2^n \Rightarrow 1/n^n \leq 1/2^n$

p-Series. **An important class of infinite series are the *p-series*,**

(11.25)
$$\sum_{k=1}^{+\infty} \frac{1}{k^p} = 1 + \frac{1}{2^p} + \frac{1}{3^p} + \cdots + \frac{1}{n^p} + \cdots$$

where *p* is a real number. This series:

(a)   Converges if $p > 1$ (b)   Diverges if $p \le 1$

These are sometimes referred to as the *hyperharmonic series* since they include the harmonic series as a special case when $p = 1$.

EXAMPLE 3   Use comparison tests to show that the *p*-series:
(a)   Converges if $p > 1$ (b)   Diverges if $p \le 1$

Solution   (a)   We group the terms of the series as follows:

$$1 + \frac{1}{2^p} + \frac{1}{3^p} + \cdots = 1 + \left(\frac{1}{2^p} + \frac{1}{3^p} \right) + \left(\frac{1}{4^p} + \frac{1}{5^p} + \frac{1}{6^p} + \frac{1}{7^p} \right)$$

$$+ \left(\frac{1}{8^p} + \frac{1}{9^p} + \frac{1}{10^p} + \cdots + \frac{1}{15^p} \right)$$

$$+ \left(\frac{1}{16^p} + \cdots + \frac{1}{31^p} \right) + \cdots$$

in which each group contains twice as many terms as the preceding group. Since $p > 1$,   we have

$$\frac{1}{2^p} + \frac{1}{3^p} < \frac{1}{2^p} + \frac{1}{2^p} = \frac{2}{2^p} = \frac{1}{2^{p-1}}$$

$$\frac{1}{4^p} + \frac{1}{5^p} + \frac{1}{6^p} + \frac{1}{7^p} < \frac{1}{4^p} + \frac{1}{4^p} + \frac{1}{4^p} + \frac{1}{4^p} = \frac{4}{4^p} = \frac{1}{4^{p-1}} = \frac{1}{(2^{p-1})^2}$$

$$\frac{1}{8^p} + \cdots + \frac{1}{15^p} < \frac{1}{8^p} + \cdots + \frac{1}{8^p} = \frac{8}{8^p} = \frac{1}{8^{p-1}} = \frac{1}{(2^{p-1})^3}$$

and so on. Therefore,

$$1 + \frac{1}{2^p} + \frac{1}{3^p} + \cdots < 1 + \frac{1}{2^{p-1}} + \frac{1}{(2^{p-1})^2} + \frac{1}{(2^{p-1})^3} + \cdots$$

The series on the right is a geometric series whose ratio is $1/2^{p-1}$. This ratio is less than 1 since $p > 1$.   It therefore converges. Hence, by Test I, the *p*-series $\sum_{k=1}^{+\infty}(1/k^p)$ converges if $p > 1$.

(b)   If $p = 1$,   the *p*-series is the harmonic series, which has been shown to be divergent. Suppose $p < 1$.   Then each term $1/n^p$ of the *p*-series is greater than or equal to the corresponding term $1/n$ of the harmonic series. That is,

$$\frac{1}{n^p} \ge \frac{1}{n} \qquad \text{for all}   n \ge 1$$

Since the harmonic series diverges, it follows from Test II that the p-series $\sum_{k=1}^{+\infty}(1/k^p)$ diverges if $p < 1$.   ∎

Here are two examples:

$$\sum_{k=1}^{+\infty} \frac{1}{k^{3/2}} \quad \text{converges} \quad (p = \tfrac{3}{2}) \qquad \sum_{k=1}^{+\infty} \frac{1}{k^{1/2}} \quad \text{diverges} \quad (p = \tfrac{1}{2})$$

Tests I and II are algebraic tests that require certain inequalities to hold. The next comparison test is analytic; it requires that certain conditions on the limit of a ratio occur. You may find this test the easiest to use when a comparison test is called for.

(11.26)   *Comparison Test III: Limit Comparison Test.* **Suppose $\sum_{k=1}^{+\infty} a_k$ and $\sum_{k=1}^{+\infty} b_k$ are each positive series. If L is a positive real number and if $\lim_{n \to +\infty}(a_n/b_n) = L$, then both series converge or both diverge.**

Proof   If $\lim_{n \to +\infty}(a_n/b_n) = L > 0$, there is a number N so that

$$\frac{L}{2} < \frac{a_n}{b_n} < \frac{3L}{2} \qquad \text{for all} \quad n > N^*.$$

Since $b_n > 0$, this is the same as

$$\frac{L}{2} b_n < a_n < \frac{3L}{2} b_n \qquad \text{for all} \quad n > N$$

If $\sum_{k=1}^{+\infty} a_n$ converges, then so does $\sum_{k=1}^{+\infty}(L/2)b_n$ and hence, by (11.21), so does

$$\sum_{k=1}^{+\infty} b_n = \frac{2}{L} \sum_{k=1}^{+\infty} \frac{L}{2} b_n$$

If $\sum_{k=1}^{+\infty} b_n$ converges, then so does $\sum_{k=1}^{+\infty}(3L/2)b_n$ and, by Test I, so does $\sum_{k=1}^{+\infty} a_n$. Thus, $\sum_{k=1}^{+\infty} a_n$ and $\sum_{k=1}^{+\infty} b_n$ converge together. Consequently, they also diverge together.   ∎

The limit comparison test (11.26) is quite versatile for comparing algebraically complex series to a p-series. *The correct choice of the p-series to use in this comparison is obtained by disregarding all terms but the highest powers of n in both the numerator and the denominator of the series to be tested.* For example:

1. To test $\sum_{k=1}^{+\infty} \frac{1}{3k^2 + 5k + 2}$ use the p-series $\sum_{k=1}^{+\infty} \frac{1}{k^2}$ because for large n,

$$\frac{1}{3n^2 + 5n + 2} \approx \left(\frac{1}{3}\right)\frac{1}{n^2}$$

---

* To see this, use (11.2) with $\varepsilon = L/2$.

2. To test $\displaystyle\sum_{k=1}^{+\infty}\frac{2k^2+5}{3k^3-5k^2+2}$ use the *p*-series $\displaystyle\sum_{k=1}^{+\infty}\frac{1}{k}$ because for large *n*,

$$\frac{2n^2+5}{3n^3-5n^2+2}\approx\frac{2n^2}{3n^3}=\left(\frac{2}{3}\right)\frac{1}{n}$$

3. To test $\displaystyle\sum_{k=1}^{+\infty}\frac{\sqrt{3k+1}}{\sqrt{4k^2-2k+1}}$ use the *p*-series $\displaystyle\sum_{k=1}^{+\infty}\frac{1}{k^{1/2}}$ because for large *n*,

$$\frac{\sqrt{3n+1}}{\sqrt{4n^2-2n+1}}\approx\frac{\sqrt{3}\sqrt{n}}{2\sqrt{n^2}}=\left(\frac{\sqrt{3}}{2}\right)\frac{1}{n^{1/2}}$$

EXAMPLE 4   Test the series $\displaystyle\sum_{k=1}^{+\infty}[1/(2k^{3/2}+5)]$ for convergence or divergence, using the limit comparison test.

Solution   We compare this series to the convergent *p*-series $\displaystyle\sum_{k=1}^{+\infty}(1/k^{3/2})$.   Thus, we evaluate

$$\lim_{n\to+\infty}\frac{\dfrac{1}{2n^{3/2}+5}}{\dfrac{1}{n^{3/2}}}=\lim_{n\to+\infty}\frac{n^{3/2}}{2n^{3/2}+5}=\lim_{n\to+\infty}\frac{n^{3/2}}{2n^{3/2}}=\frac{1}{2}$$

By the limit comparison test, since $\displaystyle\sum_{k=1}^{+\infty}(1/k^{3/2})$ converges, $\displaystyle\sum_{k=1}^{+\infty}[1/(2k^{3/2}+5)]$ does also.   ∎

EXAMPLE 5   Test the series below for convergence or divergence using the limit comparison test.

$$\sum_{k=1}^{+\infty}\frac{3\sqrt{k}+2}{\sqrt{k^3-3k^2+1}}$$

Solution   We choose for (limit) comparison the *p*-series $\displaystyle\sum_{k=1}^{+\infty}(1/k)$ since

$$\frac{3\sqrt{n}+2}{\sqrt{n^3-3n^2+1}}\approx\frac{3\sqrt{n}}{n^{3/2}}\approx\frac{3}{n}$$

Then

$$\lim_{n\to+\infty}\frac{\dfrac{3\sqrt{n}+2}{\sqrt{n^3-3n^2+1}}}{\dfrac{1}{n}}=\lim_{n\to+\infty}\frac{(3\sqrt{n}+2)n}{\sqrt{n^3-3n^2+1}}=\lim_{n\to+\infty}\frac{3n^{3/2}+2n}{\sqrt{n^3-3n^2+1}}$$

$$=\lim_{n\to+\infty}\frac{3n^{3/2}}{n^{3/2}}=3$$

By the limit comparison test, since $\sum_{k=1}^{+\infty}(1/k)$ diverges, the series to be tested does also.   ■

The comparison tests for convergence and divergence require a knowledge of convergent and divergent series. We list below some series we have already encountered:

	Convergent	Divergent
1. The geometric series $\sum_{k=1}^{+\infty} ar^{k-1}$	$\lvert r\rvert < 1$	$r \geq 1$ or $r \leq -1$
2. The harmonic series $\sum_{k=1}^{+\infty} (1/k)$		Divergent
3. The p-series $\sum_{k=1}^{+\infty} (1/k^p)$	$p > 1$	$p \leq 1$
4. The series $\sum_{k=1}^{+\infty} (1/k^k)$	Convergent	

THE INTEGRAL TEST

When a comparison test is not readily applicable, it is sometimes possible to use the *integral test*. For this test, the series to be tested is compared with an improper integral.

(11.27)   *Integral Test.* **Let f be a continuous, positive, nonincreasing function defined for all real numbers $x \geq 1$. Let $a_k = f(k)$ for all positive integers k. Then the series $\sum_{k=1}^{+\infty} a_k$ converges if and only if the improper integral $\int_1^{+\infty} f(x)\, dx$ converges.**

The proof of this statement is motivated by Figures 11 and 12 (p. 600).

Proof   Suppose $\int_1^{+\infty} f(x)\, dx$ converges. Then $\lim_{n \to +\infty} \int_1^n f(x)\, dx$ exists and equals some number L. Since $f(x) > 0$, we must have

$$\int_1^n f(x)\, dx < L \quad \text{for any } n$$

But f is nonincreasing on the interval $[1, n]$. Thus, the $(n-1)$ rectangles indicated in Figure 11 have a total area less than or equal to the area under the graph of $y = f(x)$ from 1 to n. That is,

$$f(2) + f(3) + \cdots + f(n) \leq \int_1^n f(x)\, dx < L$$

Since $a_1 = f(1)$, $a_2 = f(2)$, $\ldots$, $a_n = f(n)$, upon adding $f(1) = a_1$ to both sides, we have

$$a_1 + a_2 + a_3 + \cdots + a_n = f(1) + f(2) + \cdots + f(n) \leq f(1) + \int_1^n f(x)\, dx < f(1) + L$$

This means that the partial sums of $\sum_{k=1}^{+\infty} a_k$ are bounded, and, hence, $\sum_{k=1}^{+\infty} a_k$ converges.

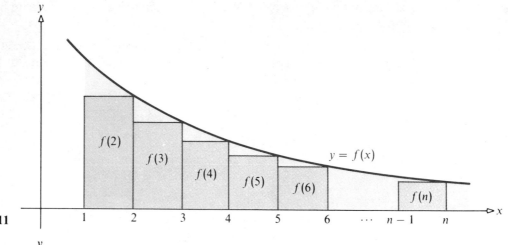

Figure 11

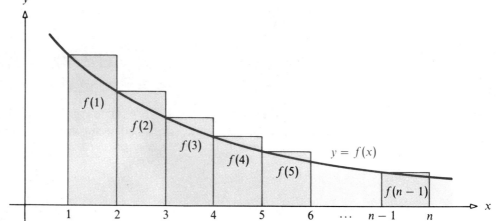

Figure 12

Now suppose $\int_1^{+\infty} f(x)\,dx$ diverges. Since f is nonincreasing on $[1, n]$, the $(n-1)$ rectangles indicated in Figure 12 have a total area greater than or equal to the area under the graph of $y = f(x)$ from 1 to n. That is,

$$f(1) + f(2) + \cdots + f(n-1) \geq \int_1^n f(x)\,dx$$

But $\int_1^n f(x)\,dx \to +\infty$ as $n \to +\infty$. Hence, the nth partial sum of $\sum_{k=1}^{+\infty} a_k$ diverges so that $\sum_{k=1}^{+\infty} a_k$ diverges. ∎

EXAMPLE 6   Use the integral test to show that the p-series

$$\sum_{k=1}^{+\infty} \frac{1}{k^p} = 1 + \frac{1}{2^p} + \frac{1}{3^p} + \cdots + \frac{1}{n^p} + \cdots$$

converges for $p > 1$ and diverges for $p \le 1$.   In addition, show that for $p > 1$,

(11.28)
$$\frac{1}{p-1} < \sum_{k=1}^{+\infty} \frac{1}{k^p} < 1 + \frac{1}{p-1}$$

Solution

The function f defined by $f(x) = 1/x^p$, where p is a nonnegative constant, is continuous, positive, and nonincreasing for $x \ge 1$.   Hence, by the integral test, the series $\sum_{k=1}^{+\infty}(1/k^p)$ converges if and only if the improper integral $\int_1^{+\infty} (1/x^p)\, dx$ converges. Now,

$$\int_1^{+\infty} \frac{1}{x^p}\, dx = \int_1^{+\infty} x^{-p}\, dx = \lim_{b \to +\infty} \int_1^b x^{-p}\, dx$$

$$= \begin{cases} \lim_{b \to +\infty} \dfrac{x^{-p+1}}{-p+1} \Big|_1^b & \text{if } p \neq 1 \\[2ex] \lim_{b \to +\infty} \ln x \Big|_1^b & \text{if } p = 1 \end{cases}$$

$$= \begin{cases} \lim_{b \to +\infty} \dfrac{b^{1-p}-1}{1-p} & \begin{cases} \text{converges if } p > 1 \;\; (1-p < 0) \\ \text{diverges if } p < 1 \;\; (1-p > 0) \end{cases} \\[2ex] \lim_{b \to +\infty} \ln b & \text{diverges } \;\; (p = 1) \end{cases}$$

Hence, $\sum_{k=1}^{+\infty}(1/k^p)$ converges if $p > 1$ and diverges if $p \le 1$.
   To establish (11.28), we refer to the shaded areas in Figures 13 and 14. By combining the inequalities given in these figures, we obtain

$$\int_1^{+\infty} \frac{dx}{x^p} < \sum_{k=1}^{+\infty} \frac{1}{k^p} < 1 + \int_1^{+\infty} \frac{dx}{x^p}$$

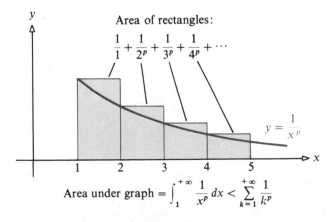

y

Area of rectangles:

$$\frac{1}{1} + \frac{1}{2^p} + \frac{1}{3^p} + \frac{1}{4^p} + \cdots$$

$$y = \frac{1}{x^p}$$

x

$$\text{Area under graph} = \int_1^{+\infty} \frac{1}{x^p}\, dx < \sum_{k=1}^{+\infty} \frac{1}{k^p}$$

Figure 13

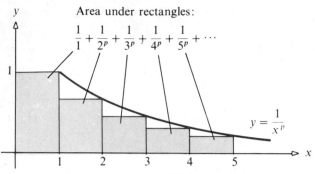

Figure 14 $1 +$ Area under graph $= 1 + \int_{1}^{+\infty} \frac{1}{x^p}\, dx > \sum_{k=1}^{+\infty} \frac{1}{k^p}$

For $p > 1$,   we have

$$\int_{1}^{+\infty} \frac{dx}{x^p} = \lim_{b \to +\infty} \int_{1}^{b} \frac{dx}{x^p} = \frac{1}{p-1}$$

and, therefore, it follows that

$$\frac{1}{p-1} < \sum_{k=1}^{+\infty} \frac{1}{k^p} < 1 + \frac{1}{p-1} \qquad \blacksquare$$

The inequality (11.28) establishes a pair of bounds for the sum of a convergent *p*-series. What the actual sum of most convergent *p*-series is remains an open question.*

EXAMPLE 7   Test the series $\sum_{k=2}^{+\infty}[1/k(\ln k)^2]$ for convergence or divergence by using the integral test.

Solution   The function $f(x) = 1/x(\ln x)^2$ is continuous, positive, and nonincreasing for all $x \geq 2$.   Using the integral test, we proceed to investigate the improper integral $\int_{2}^{+\infty} dx/x(\ln x)^2$.   In $\int dx/x(\ln x)^2$ make the substitution $u = \ln x$, $du = dx/x$. Then

$$\int \frac{dx}{x(\ln x)^2} = \int \frac{du}{u^2} = \frac{u^{-1}}{-1} + C = \frac{-1}{\ln x} + C$$

Hence,

$$\int_{2}^{+\infty} \frac{dx}{x(\ln x)^2} = \lim_{b \to +\infty} \frac{-1}{\ln x}\Big|_{2}^{b} = \left(\lim_{b \to +\infty} \frac{-1}{\ln b} \right) + \frac{1}{\ln 2} = \frac{1}{\ln 2}$$

Hence, $\sum_{k=2}^{+\infty}[1/k(\ln k)^2]$ converges. $\blacksquare$

---

* Since $\sum_{k=1}^{+\infty}(1/k^p)$ converges for $p > 1$,   it has a sum. In particular, when $p > 1$ is an integer, it has a sum. However, this exact sum has only been found for p any positive even integer, and the exact sum remains unknown for p a positive odd integer. In 1752, Euler was the first to show that $\sum_{k=1}^{+\infty}(1/k^2) = \pi^2/6$.

The integral test is usually applied when the nth term of the series has the form of a positive, nonincreasing function whose antiderivative can be readily found.

THE RATIO TEST

One of the most practical tests for convergence of a positive series makes use of the ratio of consecutive terms.

(11.29) *Ratio Test.* Let $\sum_{k=1}^{+\infty} a_k$ be a series of positive terms, and suppose that $\lim_{n \to +\infty}(a_{n+1}/a_n) = \lambda$, where $0 \le \lambda \le +\infty$.

(a) If $0 \le \lambda < 1$, then $\sum_{k=1}^{+\infty} a_k$ converges.

(b) If $1 < \lambda \le +\infty$, then $\sum_{k=1}^{+\infty} a_k$ diverges.

(c) If $\lambda = 1$, the test fails to indicate either convergence or divergence.

A proof is given at the end of this section. Let's look at some examples of how (11.29) is applied.

EXAMPLE 8   Use the ratio test to test each series for convergence or divergence.

(a) $\displaystyle\sum_{k=1}^{+\infty} \frac{k}{4^k}$ (b) $\displaystyle\sum_{k=1}^{+\infty} \frac{3k+1}{k^2}$ (c) $\displaystyle\sum_{k=1}^{+\infty} \frac{2^k}{k}$ (d) $\displaystyle\sum_{k=0}^{+\infty} \frac{1}{k!}$

Solution   (a) For $\sum_{k=1}^{+\infty}(k/4^k)$, we have $a_{n+1} = (n+1)/4^{n+1}$ and $a_n = n/4^n$. We form their ratio:

$$\frac{a_{n+1}}{a_n} = \frac{(n+1)/4^{n+1}}{n/4^n} = \frac{n+1}{4n} = \frac{1 + (1/n)}{4}$$

Since

$$\lambda = \lim_{n \to +\infty} \frac{a_{n+1}}{a_n} = \lim_{n \to +\infty} \frac{1 + (1/n)}{4} = \frac{1}{4} < 1$$

the series converges.

(b) For $\sum_{k=1}^{+\infty}[(3k+1)/k^2]$, we have $a_{n+1} = (3n+4)/(n+1)^2$ and $a_n = (3n+1)/n^2$. Their ratio is

$$\frac{a_{n+1}}{a_n} = \left[\frac{3n+4}{(n+1)^2}\right]\left(\frac{n^2}{3n+1}\right) = \left(\frac{3n+4}{3n+1}\right)\left(\frac{n^2}{n^2 + 2n + 1}\right)$$

$$= \left[\frac{3 + (4/n)}{3 + (1/n)}\right]\left[\frac{1}{1 + (2/n) + (1/n^2)}\right]$$

Then

$$\lambda = \lim_{n \to +\infty} \frac{a_{n+1}}{a_n} = \lim_{n \to +\infty} \left\{ \left[\frac{3 + (4/n)}{3 + (1/n)} \right] \left[\frac{1}{1 + (2/n) + (1/n^2)} \right] \right\} = 1$$

The ratio test gives no information about this series.*

(c) For $\sum_{k=1}^{+\infty} (2^k/k)$, we have $a_{n+1} = 2^{n+1}/(n+1)$ and $a_n = 2^n/n$, and their ratio is

$$\frac{a_{n+1}}{a_n} = \frac{2^{n+1}}{n+1} \left(\frac{n}{2^n} \right) = \frac{2n}{n+1} = \frac{2}{1 + (1/n)}$$

Then

$$\lambda = \lim_{n \to +\infty} \frac{a_{n+1}}{a_n} = \lim_{n \to +\infty} \frac{2}{1 + (1/n)} = 2$$

The series diverges.

(d) For $\sum_{k=0}^{+\infty} (1/k!)$, we have $a_{n+1} = 1/(n+1)!$ and $a_n = 1/n!$, so

$$\frac{a_{n+1}}{a_n} = \frac{n!}{(n+1)!} = \frac{1}{n+1}$$

Then

$$\lambda = \lim_{n \to +\infty} \frac{a_{n+1}}{a_n} = \lim_{n \to +\infty} \frac{1}{n+1} = 0$$

Hence, $\sum_{k=0}^{+\infty} (1/k!)$ converges.† ∎

As Example 8 illustrates, the ratio test is especially useful in handling series containing factorials and/or powers.

EXAMPLE 9   Test the series $\sum_{k=1}^{+\infty} (k!/k^k)$ for convergence or divergence.

Solution   Since $a_n = n!/n^n$ and $a_{n+1} = (n+1)!/(n+1)^{n+1}$, we find that

$$\frac{a_{n+1}}{a_n} = \frac{(n+1)!/(n+1)^{n+1}}{n!/n^n} = \frac{(n+1)!n^n}{n!(n+1)^{n+1}} = \frac{n^n}{(n+1)^n} = \frac{1}{[(n+1)/n]^n} = \frac{1}{[1 + (1/n)]^n}$$

Hence,

$$\lambda = \lim_{n \to +\infty} \frac{1}{[1 + (1/n)]^n} = \frac{1}{\lim_{n \to +\infty} [1 + (1/n)]^n} = \frac{1}{e}$$

By (7.36)

Since $1/e < 1$, the series converges. ∎

---

* You should verify that the series diverges by comparing it to the harmonic series $\sum_{k=1}^{+\infty} (1/k)$.

† We shall discover in Section 7 that $\sum_{k=0}^{+\infty} (1/k!) = e$.

Caution: Here we have found that the ratio a_{n+1}/a_n converges to $1/e$; this does not mean that $\sum_{k=1}^{+\infty}(k!/k^k)$ converges to $1/e$. In fact, it is not known what the sum of the series is; all we know is that it converges.

We conclude our discussion of the ratio test with some observations:

1. To test $\sum_{k=1}^{+\infty} a_k$ for convergence, it is important to check whether the *limit of the ratio* a_{n+1}/a_n—not the ratio itself—is less than 1 for all n. For example, the ratio a_{n+1}/a_n for the harmonic series $\sum_{k=1}^{+\infty}(1/k)$, which diverges, is $n/(n+1) < 1$, but its limit is not less than 1.

2. For divergence, it is sufficient that the ratio a_{n+1}/a_n itself is greater than 1 for all n.

3. The ratio test is not conclusive when $\lim_{n \to +\infty}(a_{n+1}/a_n) = 1$. It is also not conclusive when $\lim_{n \to +\infty}(a_{n+1}/a_n) \neq +\infty$ does not exist.

4. If the general term of an infinite series involves the index n, either exponentially or factorially, the ratio test may be expected to yield an answer to the question of convergence or divergence.

Proof of the Ratio Test

Part (a): $\mathbf{0 \leq \lambda < 1.}$ Let r be any number such that $\lambda < r < 1$. Since $\lim_{n \to +\infty}(a_{n+1}/a_n) = \lambda < r$, by the definition of limit, we can find a number N so that for $n > N$, the ratio a_{n+1}/a_n will differ from λ by as little as we please, and will therefore be less than r. Then

$$\frac{a_{N+1}}{a_N} < r \qquad \text{or} \qquad a_{N+1} < ra_N$$

$$\frac{a_{N+2}}{a_{N+1}} < r \qquad \text{or} \qquad a_{N+2} < ra_{N+1} < r^2 a_N$$

$$\frac{a_{N+3}}{a_{N+2}} < r \qquad \text{or} \qquad a_{N+3} < ra_{N+2} < r^3 a_N$$

Thus, each term of the series $a_{N+1} + a_{N+2} + \cdots$ is less than the corresponding term of the geometric series $a_N r + a_N r^2 + a_N r^3 + \cdots$, which is convergent since $r < 1$. By comparison test I (11.23), the series $a_{N+1} + a_{N+2} + \cdots$ is also convergent and, hence, the series $\sum_{k=1}^{+\infty} a_k$ is convergent.

Part (b): $\mathbf{1 < \lambda \leq +\infty.}$ Let r be any number such that $1 < r < \lambda$. Since $\lim_{n \to +\infty}(a_{n+1}/a_n) = \lambda$, we can find a number N so that for $n > N$, the ratio a_{n+1}/a_n will differ from λ by as little as we please, and will therefore be greater than r. That is, for all $n > N$, we have $a_{n+1}/a_n > r > 1$ so that $a_{n+1} > a_n$. That is, after the Nth term, the terms are positive and increasing. Hence, $\lim_{n \to +\infty} a_n \neq 0$ and therefore by the test for divergence (11.24), the series diverges.

Part (c): $\mathbf{\lambda = 1.}$ To show that the test fails for $\lambda = 1$, we merely check two series—one that diverges and another that converges—to show that no conclusion can be drawn. Consider $\sum_{k=1}^{+\infty}(1/k)$ and $\sum_{k=1}^{+\infty}(1/k^2)$. The first is the harmonic series and we know it diverges; the second is a p-series with $p > 1$, and it converges. It is left to you to show that $\lambda = 1$ in each case.   ∎

EXERCISE 3

In Problems 1–6 use (11.23) or (11.24) to test each series for convergence or divergence.

1. Test $\sum_{k=1}^{+\infty} \dfrac{1}{k(k+1)}$ by comparing it with $\sum_{k=1}^{+\infty} \dfrac{1}{k^2}$.

2. Test $\sum_{k=1}^{+\infty} \dfrac{1}{(k+2)^2}$ by comparing it with $\sum_{k=1}^{+\infty} \dfrac{1}{k^2}$.

3. Test $\sum_{k=2}^{+\infty} \dfrac{\sqrt{k}}{k-1}$ by comparing it with $\sum_{k=2}^{+\infty} \dfrac{1}{\sqrt{k}}$.

4. Test $\sum_{k=1}^{+\infty} \dfrac{1}{(2k-1)(2^k)}$ by comparing it with $\sum_{k=1}^{+\infty} \dfrac{1}{2^k}$.

5. Test $\sum_{k=2}^{+\infty} \dfrac{1}{\sqrt{k(k-1)}}$ by comparing it with $\sum_{k=2}^{+\infty} \dfrac{1}{k}$.

6. Test $\sum_{k=1}^{+\infty} \dfrac{1}{k(k+1)(k+2)}$ by comparing it with $\sum_{k=1}^{+\infty} \dfrac{1}{k^3}$.

In Problems 7–22 use (11.26) to test each series for convergence or divergence.

7. $\sum_{k=1}^{+\infty} \dfrac{1}{(k+1)(k+2)}$ **8.** $\sum_{k=1}^{+\infty} \dfrac{1}{k^2+1}$ **9.** $\sum_{k=1}^{+\infty} \dfrac{1}{\sqrt{k^2+1}}$ **10.** $\sum_{k=1}^{+\infty} \dfrac{\sqrt{k}}{k+4}$

11. $\sum_{k=1}^{+\infty} \dfrac{3\sqrt{k}+2}{2k^2+5}$ **12.** $\sum_{k=2}^{+\infty} \dfrac{3\sqrt{k}+2}{2k-3}$ **13.** $\sum_{k=2}^{+\infty} \dfrac{1}{k\sqrt{k^2-1}}$ **14.** $\sum_{k=1}^{+\infty} \dfrac{k}{(2k-1)^2}$

15. $\sum_{k=1}^{+\infty} \dfrac{3k+4}{k2^k}$ **16.** $\sum_{k=1}^{+\infty} \dfrac{1}{2^k+1}$ **17.** $\sum_{k=1}^{+\infty} \dfrac{5}{3^k+2}$ **18.** $\sum_{k=2}^{+\infty} \dfrac{k-1}{k2^k}$

19. $\sum_{k=1}^{+\infty} \dfrac{k+5}{k^{k+1}}$ **20.** $\sum_{k=1}^{+\infty} \dfrac{5}{k^k+1}$ **21.** $\sum_{k=1}^{+\infty} \sin\dfrac{1}{k}$ **22.** $\sum_{k=1}^{+\infty} \tan\dfrac{1}{k}$

In Problems 23–36 test each series for convergence or divergence by using the integral test (11.27).

23. $\sum_{k=1}^{+\infty} \dfrac{1}{k^{1.01}}$ **24.** $\sum_{k=1}^{+\infty} \dfrac{1}{k^{0.9}}$ **25.** $\sum_{k=1}^{+\infty} \dfrac{\ln k}{k}$ **26.** $\sum_{k=1}^{+\infty} \dfrac{1}{\sqrt{k^2+25}}$

27. $\sum_{k=1}^{+\infty} ke^{-k}$ **28.** $\sum_{k=1}^{+\infty} ke^{-k^2}$ **29.** $\sum_{k=1}^{+\infty} \dfrac{1}{k^2+1}$ **30.** $\sum_{k=1}^{+\infty} \dfrac{1}{k(k+1)}$

31. $\sum_{k=2}^{+\infty} \dfrac{1}{k(\ln k)}$ **32.** $\sum_{k=2}^{+\infty} \dfrac{1}{k(\ln k)^3}$ **33.** $\sum_{k=2}^{+\infty} \dfrac{1}{k\sqrt{k^2-1}}$ **34.** $\sum_{k=1}^{+\infty} \ln\dfrac{k+2}{k}$

35. $\sum_{k=1}^{+\infty} \dfrac{\ln k}{k^2}$ **36.** $\sum_{k=2}^{+\infty} \dfrac{1}{k\sqrt{\ln k}}$

In Problems 37–54 test for convergence or divergence by using the ratio test (11.29).

37. $\sum_{k=1}^{+\infty} \dfrac{(2k-1)(2k+1)}{2^k}$

38. $\sum_{k=1}^{+\infty} \dfrac{1}{(2k+1)2^k}$

39. $\sum_{k=1}^{+\infty} k\left(\dfrac{2}{3}\right)^k$

40. $\sum_{k=1}^{+\infty} \dfrac{(k+1)!}{3^k}$

41. $\sum_{k=1}^{+\infty} \dfrac{10^k}{(2k)!}$

42. $\sum_{k=1}^{+\infty} \dfrac{5^k}{k^2}$

43. $\sum_{k=1}^{+\infty} \dfrac{k}{(2k-2)!}$

44. $\sum_{k=1}^{+\infty} \dfrac{(2k)!}{5^k 3^{k-1}}$

45. $\sum_{k=1}^{+\infty} \dfrac{2^k}{k(k+1)}$

46. $\sum_{k=1}^{+\infty} \dfrac{k!}{k^2(k+1)^2}$

47. $\sum_{k=1}^{+\infty} \dfrac{k^3}{k!}$

48. $\sum_{k=1}^{+\infty} \dfrac{k!}{k^{k+1}}$

49. $\sum_{k=2}^{+\infty} \dfrac{3^{k-1}}{k2^k}$

50. $\sum_{k=1}^{+\infty} \dfrac{k(k+2)}{3^k}$

51. $\sum_{k=1}^{+\infty} \dfrac{k}{e^k}$

52. $\sum_{k=1}^{+\infty} \dfrac{e^k}{k^3}$

53. $\sum_{k=1}^{+\infty} k2^k$

54. $\sum_{k=1}^{+\infty} \dfrac{1}{k}2^k$

55. Use the integral test to prove that the series $\sum_{k=2}^{+\infty}[1/k(\ln k)^p]$ is convergent if and only if $p>1$.

56. Use the integral test to prove that the series $\sum_{k=3}^{+\infty}[1/k(\ln k)[\ln(\ln k)]^p]$ is convergent if and only if $p>1$.

57. Use the ratio test to find the positive numbers x for which the series $\sum_{k=1}^{+\infty}(x^k/k^2)$ converges or diverges.

58. Prove that any series of the form $\sum_{k=1}^{+\infty}(d_k/10^k)$, where the d_k are digits $(0, 1, 2, \ldots, 9)$, converges.

59. For the series $\sum_{k=1}^{+\infty}(1/k)$, show that $\lim_{n\to+\infty}(a_{n+1}/a_n)=1$, yet $\sum_{k=1}^{+\infty}(1/k)$ diverges.

60. For the series $\sum_{k=1}^{+\infty}(1/k^2)$, show that $\lim_{n\to+\infty}(a_{n+1}/a_n)=1$, yet $\sum_{k=1}^{+\infty}(1/k^2)$ converges.

61. Prove that $\lim_{n\to+\infty}(n!/n^n)=0$, where n denotes a positive integer.

62. Prove that if $\sum_{k=1}^{+\infty}a_k$ is a positive series to be tested and if $\sum_{k=1}^{+\infty}c_k$ is a known convergent, positive series and if $\lim_{n\to+\infty}(a_n/c_n)$ exists, then the series $\sum_{k=1}^{+\infty}a_k$ converges.

63. Prove that if $\sum_{k=1}^{+\infty}a_k$ is a positive series to be tested and if $\sum_{k=1}^{+\infty}d_k$ is a known divergent, positive series and if $\lim_{n\to+\infty}(a_n/d_n)$ exists and is nonzero, then the series $\sum_{k=1}^{+\infty}a_k$ diverges. Also, if $\lim_{n\to+\infty}(a_n/d_n)=+\infty$, then $\sum_{k=1}^{+\infty}a_k$ diverges.

In Problems 64 and 65 use the results of Problems 62 and 63 to test each series for convergence or divergence.

64. $\sum_{k=2}^{+\infty} \dfrac{\ln k}{k+3}$

65. $\sum_{k=2}^{+\infty} \dfrac{\ln(k+1)}{k^3-4}$

Root Test. **Suppose** $\sum_{k=1}^{+\infty}a_k$ **is a series of positive terms to be tested, and suppose** $\lim_{n\to+\infty}\sqrt[n]{a_n}=\lambda$:

(a) If $\lambda<1$, then $\sum_{k=1}^{+\infty}a_k$ converges.

(b) If $\lambda>1$, then $\sum_{k=1}^{+\infty}a_k$ diverges.

(c) If $\lambda=1$, the test is inconclusive.

66. Use the root test to show that $\sum_{k=2}^{+\infty}[1/(\ln k)^k]$ converges.

67. Use the root test to show that $\sum_{k=1}^{+\infty}[1/(2k)^k]$ converges.

68. Show that the root test is inconclusive for $\sum_{k=1}^{+\infty}(1/k)$.

4. Alternating Series and Absolute Convergence

In Section 3 we studied series with positive terms; we now discuss series that have both positive and negative terms. Even though the behavior of such series is markedly different from that of series with positive terms, we shall find that we can make good use of the latter to clarify the former. By far the most frequently encountered series with both positive and negative terms is the *alternating series*, in which the terms are alternately positive and negative.

An *alternating series* is a series either of the form

$$\sum_{k=1}^{+\infty} (-1)^{k+1} a_k = a_1 - a_2 + a_3 - a_4 + \cdots$$

or of the form

$$\sum_{k=1}^{+\infty} (-1)^{k} a_k = -a_1 + a_2 - a_3 + a_4 - \cdots$$

where each number a_k is positive for every k.

For example, the series

$$1 - \frac{1}{2} + \frac{1}{3} - \frac{1}{4} + \cdots = \sum_{k=1}^{+\infty} \frac{(-1)^{k+1}}{k}$$

and

$$-1 + \frac{1}{3!} - \frac{1}{5!} + \frac{1}{7!} - \cdots = \sum_{k=1}^{+\infty} \frac{(-1)^{k}}{(2k-1)!}$$

are alternating series.

A test for convergence of alternating series is given next.

(11.30) *Alternating Series Test.* **If the numbers a_k of the alternating series**

$$\sum_{k=1}^{+\infty} (-1)^{k+1} a_k = a_1 - a_2 + a_3 - a_4 + \cdots \qquad a_k > 0$$

satisfy the following two conditions:

(a) **the a_k tend to 0; that is, $\lim_{n \to +\infty} a_n = 0$**

(b) **the a_k are nonincreasing; that is, $a_1 \geq a_2 \geq a_3 \geq \cdots \geq a_n \geq a_{n+1} \geq \cdots$**

then the alternating series converges.

Proof We first consider the partial sums with an even number of terms:

$$S_{2n} = a_1 - a_2 + a_3 - a_4 + \cdots + a_{2n-1} - a_{2n}$$
$$= (a_1 - a_2) + (a_3 - a_4) + \cdots + (a_{2n-1} - a_{2n})$$

Since $a_1 \geq a_2$, $a_3 \geq a_4$, ..., $a_{2n-1} \geq a_{2n}$, the difference shown within each pair of parentheses is either 0 or positive. Hence, the sequence of partial sums S_{2n} is increasing, that is,

(11.31)
$$S_2 \leq S_4 \leq S_6 \leq \cdots \leq S_{2n} \leq S_{2n+2} \leq \cdots$$

For the partial sums with an odd number of terms, we group the terms a little differently and write

$$S_{2n+1} = a_1 - a_2 + a_3 - a_4 + a_5 - \cdots - a_{2n} + a_{2n+1}$$
$$= a_1 - (a_2 - a_3) - (a_4 - a_5) - \cdots - (a_{2n} - a_{2n+1})$$

Here again, the difference within each pair of parentheses is either positive or 0, and it is to be subtracted from the previous term. Hence, the sequence of partial sums S_{2n+1} is decreasing, that is,

(11.32)
$$S_1 \geq S_3 \geq S_5 \geq \cdots \geq S_{2n-1} \geq S_{2n+1} \geq \cdots$$

Figure 15 illustrates what happens.

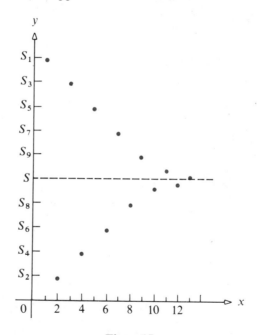

Figure 15

Finally, we note that

(11.33)
$$S_{2n+1} - S_{2n} = a_{2n+1} > 0 \qquad \text{so that} \qquad S_{2n} < S_{2n+1}$$

We combine the three statements (11.31), (11.32), and (11.33) as

$$S_2 \leq S_4 \leq S_6 \leq \cdots \leq S_{2n} < S_{2n+1} \leq \cdots \leq S_5 \leq S_3 \leq S_1$$

The sequence $S_2, S_4, S_6, \ldots$ is increasing and bounded above by S_1. By axiom (11.9), this sequence has a limit, which we call T. Similarly, the sequence $S_1, S_3, S_5, \ldots$ is decreasing and bounded below, so it has a limit, which we call S. Furthermore, by (11.33),

$$S - T = \lim_{n \to +\infty} S_{2n+1} - \lim_{n \to +\infty} S_{2n} = \lim_{n \to +\infty} (S_{2n+1} - S_{2n}) = \lim_{n \to +\infty} a_{2n+1}$$

Therefore, since $\lim_{n \to +\infty} a_{2n+1} = 0$, we have $S - T = 0$. This means $S = T$ and, hence, the sequences $\{S_{2n}\}$ and $\{S_{2n+1}\}$ both converge to S. It follows that

$$\lim_{n \to +\infty} S_n = S$$

and so, the alternating series converges.   ∎

Let's now look at an example of how (11.30) is used.

EXAMPLE 1   Test the alternating harmonic series given below for convergence or divergence.

$$\sum_{k=1}^{+\infty} \frac{(-1)^{k+1}}{k} = 1 - \frac{1}{2} + \frac{1}{3} - \frac{1}{4} + \cdots$$

Solution   Since

$$\lim_{n \to +\infty} a_n = \lim_{n \to +\infty} \frac{1}{n} = 0 \quad \text{and} \quad a_{n+1} = \frac{1}{n+1} < \frac{1}{n} = a_n$$

the conditions of the test are satisfied. Hence, the series converges.*   ∎

As the above example illustrates, to use the alternating series test on $\sum_{k=1}^{+\infty} (-1)^k a_k$, we need to determine whether the a_k are nonincreasing. Therefore, we reproduce here the three ways to make this verification:

1. Show that $a_{n+1} - a_n \le 0$ for all n.
2. Show that $a_{n+1}/a_n \le 1$ for all n.
3. Show that the derivative of the function representing a_n is always negative or 0.

For example, we can show that the a_k in the series $\sum_{k=1}^{+\infty} [(-1)^{k+1}/k]$ are decreasing by observing that $(d/dn)(1/n) = -1/n^2 < 0$ for $n \ge 1$.

In applying the alternating series test, it is always best to check on the limit of the nth term first. If it is not 0, the test for divergence (11.18) tells us immediately that the series diverges. If it equals 0, then we should continue on to determine whether the a_n are nonincreasing. For example, in testing the alternating series

$$\sum_{k=2}^{+\infty} (-1)^k \frac{k}{k-1} = 2 - \frac{3}{2} + \frac{4}{3} - \frac{5}{4} + \cdots$$

---

* We shall discover in Section 7 that the sum of this alternating harmonic series is ln 2.

we find out that $\lim_{n \to +\infty} a_n = \lim_{n \to +\infty}[n/(n-1)] = 1$. There is no need to look further. The series diverges by (11.18)

ESTIMATING THE SUM OF AN ALTERNATING SERIES

An important characteristic of a convergent alternating series is that it is possible to estimate quite easily the error made by taking a partial sum as an approximation to the sum of the series.

(11.34) *Error Estimate.* **For a convergent alternating series $\sum_{k=1}^{+\infty}(-1)^{k+1}a_k$, the error E_n made in taking the sum of the first n terms as an approximation to the sum S of the series is numerically less than the $(n+1)$st term of the series. That is,**

$$E_n < a_{n+1}$$

Proof   If n is even, $S_n < S$, and

$$0 < S - S_n = a_{n+1} - (a_{n+2} - a_{n+3}) - \cdots < a_{n+1}$$

If n is odd, $S_n > S$, and

$$0 < S_n - S = a_{n+1} - (a_{n+2} - a_{n+3}) - \cdots < a_{n+1}$$

In either case,

$$E_n = |S - S_n| < a_{n+1} \qquad \blacksquare$$

We have seen in Example 1 that the alternating harmonic series converges. Table 2 provides estimates of the sum of this series and the maximum error associated with the estimate using $n = 3, 9, 99,$ and 999 terms. Observe the slow rate of convergence, as evidenced by the rather small change in the maximum error. For example, using an additional 900 terms reduces the maximum error by merely $|E_{999} - E_{99}| = 0.009$.

Table 2

Number of Terms	Estimate of the Sum	Maximum Error
$n = 3$	$1 - \dfrac{1}{2} + \dfrac{1}{3} \approx 0.83333$	$E_3 < \dfrac{1}{4} = 0.25$
$n = 9$	$\displaystyle\sum_{k=1}^{9} \dfrac{(-1)^{k+1}}{k} \approx 0.74563$	$E_9 < \dfrac{1}{10} = 0.1$
$n = 99$	$\displaystyle\sum_{k=1}^{99} \dfrac{(-1)^{k+1}}{k} \approx 0.7$	$E_{99} < \dfrac{1}{100} = 0.01$
$n = 999$	$\displaystyle\sum_{k=1}^{999} \dfrac{(-1)^{k+1}}{k} \approx 0.694$	$E_{999} < \dfrac{1}{1000} = 0.001$

EXAMPLE 2   Approximate the sum of the alternating series below to within 0.0001.

$$\sum_{k=0}^{+\infty} \frac{(-1)^k}{(2k)!} = 1 - \frac{1}{2!} + \frac{1}{4!} - \frac{1}{6!} + \cdots$$

Solution   We leave it to you to verify that the series actually converges. The fifth term of the series, $1/8! \approx 0.000025$, is the first term less than 0.0001, and it represents an upper estimate to the error due to using the sum of the first four terms as an approximation to the sum. Hence,

$$\sum_{k=0}^{+\infty} \frac{(-1)^k}{(2k)!} \approx 1 - \frac{1}{2!} + \frac{1}{4!} - \frac{1}{6!} = 1 - \frac{1}{2} + \frac{1}{24} - \frac{1}{720} \approx 0.54028$$

accurate to within 0.0001.*   ∎

ABSOLUTE CONVERGENCE

In order to discuss ways of determining the convergence or divergence of a series $\sum_{k=1}^{+\infty} a_k$ in which the terms a_k are sometimes positive and sometimes negative (not necessarily alternating), we require the following definition:

A series $\sum_{k=1}^{+\infty} a_k$ is *absolutely convergent* if

$$\sum_{k=1}^{+\infty} |a_k| = |a_1| + |a_2| + \cdots + |a_n| + \cdots \quad \text{is convergent}$$

The next result gives a test for convergence of a series $\sum_{k=1}^{+\infty} a_k$, in which the terms a_k are sometimes positive and sometimes negative.

(11.35)   THEOREM   **If a series $\sum_{k=1}^{+\infty} a_k$ is absolutely convergent, then it is convergent.**

Proof   For each n,

$$-|a_n| \le a_n \le |a_n|$$

Hence, by adding $|a_n|$,

$$0 \le a_n + |a_n| \le 2|a_n|$$

If $\sum_{k=1}^{+\infty} |a_k|$ converges, then $\sum_{k=1}^{+\infty} 2|a_k| = 2\sum_{k=1}^{+\infty} |a_k|$ converges. Thus, by comparison test I (11.23), $\sum_{k=1}^{+\infty}(a_k + |a_k|)$ converges. But $a_n = (a_n + |a_n|) - |a_n|$. Since $\sum_{k=1}^{+\infty} a_k$ is the difference of two convergent series, it also converges.   ∎

EXAMPLE 3   The series

$$1 - \frac{1}{2} - \frac{1}{4} + \frac{1}{8} - \frac{1}{16} - \frac{1}{32} + \frac{1}{64} + \cdots$$

---

* You will see in Section 8 [see (11.57)] that the actual sum of this series is cos 1. Since it turns out that other values of cos x can be similarly computed, this example illustrates one way to obtain a table of trigonometric values.

converges, since $1 + \frac{1}{2} + \frac{1}{4} + \cdots = \sum_{k=1}^{+\infty} (\frac{1}{2})^{k-1}$, a geometric series with $r = \frac{1}{2}$, converges. ∎

EXAMPLE 4 Test the series below for convergence or divergence.

$$\sum_{k=1}^{+\infty} \frac{\sin k}{k^2} = \frac{\sin 1}{1^2} + \frac{\sin 2}{2^2} + \frac{\sin 3}{3^2} + \cdots$$

Solution This series is not a series of positive terms, nor is it an alternating series. To use theorem (11.35), we investigate the series $\sum_{k=1}^{+\infty} |(\sin k)/k^2|$. Since

$$\left| \frac{\sin n}{n^2} \right| \leq \frac{1}{n^2}$$

for all n, and since $\sum_{k=1}^{+\infty}(1/k^2)$ is a convergent p-series, we conclude by comparison test I (11.23) that $\sum_{k=1}^{+\infty} |(\sin k)/k^2|$ is convergent. Hence, $\sum_{k=1}^{+\infty}[(\sin k)/k^2]$ is absolutely convergent (by definition) and, by (11.35), it follows that $\sum_{k=1}^{+\infty}[(\sin k)/k^2]$ is convergent. ∎

The converse of theorem (11.35) is not necessarily true. That is, there are convergent series that are not absolutely convergent. For instance, we have shown that the alternating harmonic series

$$\sum_{k=1}^{+\infty} \frac{(-1)^{k+1}}{k} = 1 - \frac{1}{2} + \frac{1}{3} - \frac{1}{4} + \frac{1}{5} - \cdots$$

converges. The series of absolute values yields the harmonic series $\sum_{k=1}^{+\infty}(1/k)$, which is divergent. A series that is convergent without being absolutely convergent, is called *conditionally convergent*.

EXAMPLE 5 Determine the numbers p for which the series $\sum_{k=1}^{+\infty}[(-1)^{k+1}/k^p]$ is absolutely convergent, conditionally convergent, and divergent.

Solution We begin by testing the series for absolute convergence. Therefore, we consider the series of absolute values, namely, $\sum_{k=1}^{+\infty}(1/k^p)$. This is the p-series, which converges if $p > 1$ and diverges if $p \leq 1$. Hence, $\sum_{k=1}^{+\infty}[(-1)^{k+1}/k^p]$ is absolutely convergent if $p > 1$.

It remains to determine what happens when $p \leq 1$. By using the alternating series test (11.30), we find:

$$\lim_{n \to +\infty} \frac{1}{n^p} = 0 \text{ if } 0 < p \leq 1; \quad \lim_{n \to +\infty} \frac{1}{n^p} = 1 \text{ if } p = 0; \quad \lim_{n \to +\infty} \frac{1}{n^p} = +\infty \text{ if } p < 0$$

Consequently, $\sum_{k=1}^{+\infty}[(-1)^{k+1}/k^p]$ diverges if $p \leq 0$. To verify part (b) of the alternating series test, when $0 < p \leq 1$, we observe that

$$\frac{d}{dn} \frac{1}{n^p} = \frac{-p}{n^{p+1}} < 0 \qquad \text{for all } n, \text{ when } 0 < p \leq 1$$

Thus, $\sum_{k=1}^{+\infty}[(-1)^{k+1}/k^p]$ is conditionally convergent if $0 < p \leq 1$. ∎

Figure 16 summarizes the preceding discussion. As illustrated, a series $\sum_{k=1}^{+\infty} a_k$ is either convergent or divergent; and, if it is convergent, it is either absolutely convergent or conditionally convergent.

The flowchart in Figure 17 may be helpful in determining whether a given series is absolutely convergent, conditionally convergent, or divergent.

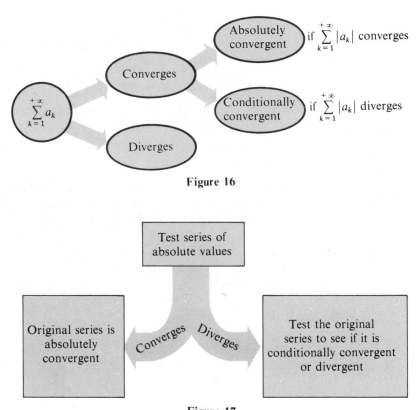

Figure 16

Figure 17

SOME PROPERTIES

Absolute convergence of a series is stronger than conditional convergence, and absolutely convergent series possess properties that conditionally convergent series do not. We state some of these properties without a proof:

1. If a series is absolutely convergent, any rearrangement of its terms will result in a series that is also absolutely convergent to the same sum.

2. If two series $\sum_{k=1}^{+\infty} a_k$ and $\sum_{k=1}^{+\infty} b_k$ are absolutely convergent, the series $a_1b_1 + a_1b_2 + a_2b_1 + a_1b_3 + a_2b_2 + a_3b_1 + \cdots$ formed by multiplying the terms of one series by the terms of the other series will converge absolutely, and its sum will be the product of the sums of the original series.

3. If a series converges absolutely, then the series consisting of just the positive terms converges, as does the series consisting of just the negative terms.

4. If a series is conditionally convergent, its terms may be rearranged to form a new series that converges to any sum we like. In fact, they may be rearranged to form a divergent series.

5. If a series is conditionally convergent, then the series consisting of just the positive terms diverges, as does the series consisting of just the negative terms.

EXERCISE 4

In Problems 1–8 test each alternating series for convergence or divergence by using the alternating series test (11.30).

1. $\displaystyle\sum_{k=1}^{+\infty} (-1)^{k+1} \frac{1}{k^2}$ **2.** $\displaystyle\sum_{k=1}^{+\infty} (-1)^{k+1} \frac{1}{2\sqrt{k}}$ **3.** $\displaystyle\sum_{k=2}^{+\infty} (-1)^k \frac{1}{k \ln k}$ **4.** $\displaystyle\sum_{k=1}^{+\infty} (-1)^{k+1} \frac{k+1}{k^2}$

5. $\displaystyle\sum_{k=1}^{+\infty} (-1)^{k+1} \frac{k+1}{k}$ **6.** $\displaystyle\sum_{k=1}^{+\infty} (-1)^{k+1} \frac{k^2}{5k^2+2}$ **7.** $\displaystyle\sum_{k=1}^{+\infty} \frac{(-1)^{k+1}}{(k+1)2^k}$ **8.** $\displaystyle\sum_{k=0}^{+\infty} (-1)^k \frac{1}{k!}$

In Problems 9–12 approximate the sum of each series accurate to within 0.001.

9. $\displaystyle\sum_{k=0}^{+\infty} (-1)^k \frac{1}{k!}$ **10.** $\displaystyle\sum_{k=0}^{+\infty} (-1)^k \frac{1}{2k+1} \left(\frac{1}{3}\right)^{2k+1}$ **11.** $\displaystyle\sum_{k=1}^{+\infty} (-1)^{k+1} \frac{1}{k^4}$

12. $\displaystyle\sum_{k=0}^{+\infty} (-1)^k \frac{1}{k!} \left(\frac{1}{2}\right)^k$

In Problems 13–18 find the least positive integer n such that S_n approximates the sum of the given convergent alternating series to within 0.0001; that is, find n such that $|S - S_n| < 10^{-4}$.

13. $\displaystyle\sum_{k=1}^{+\infty} (-1)^{k+1} \frac{1}{k^2}$ **14.** $\displaystyle\sum_{k=0}^{+\infty} (-1)^k \frac{1}{k!}$ **15.** $\displaystyle\sum_{k=0}^{+\infty} (-1)^k \frac{1}{k!} \left(\frac{1}{3}\right)^k$ **16.** $\displaystyle\sum_{k=1}^{+\infty} (-1)^{k+1} \frac{1}{k^k}$

17. $\displaystyle\sum_{k=1}^{+\infty} (-1)^{k+1} \frac{1}{k}$ **18.** $\displaystyle\sum_{k=1}^{+\infty} (-1)^{k+1} \frac{1}{\sqrt{k}}$

In Problems 19–38 determine whether each series is conditionally convergent, absolutely convergent, or divergent.

19. $\displaystyle\sum_{k=1}^{+\infty} \frac{(-1)^{k+1}}{2k}$ **20.** $\displaystyle\sum_{k=1}^{+\infty} \frac{(-1)^{k+1}}{3k-4}$ **21.** $\displaystyle\sum_{k=1}^{+\infty} \frac{(-1)^{k+1}}{k(k+1)}$ **22.** $\displaystyle\sum_{k=1}^{+\infty} \frac{(-1)^{k+1}k}{2^k}$

23. $\displaystyle\sum_{k=1}^{+\infty} \frac{(-1)^{k+1}}{k\sqrt{k+3}}$ **24.** $\displaystyle\sum_{k=0}^{+\infty} \frac{(-1)^{k+1}}{k!}$ **25.** $\displaystyle\sum_{k=1}^{+\infty} \frac{(-1)^{k+1}\sqrt{k}}{k^2+1}$ **26.** $\displaystyle\sum_{k=1}^{+\infty} \frac{(-1)^{k+1}\sqrt{k}}{k+1}$

27. $\displaystyle\sum_{k=1}^{+\infty} (-1)^{k+1} \left(\frac{1}{5}\right)^k$ **28.** $\displaystyle\sum_{k=1}^{+\infty} \frac{(-1)^{k+1}}{e^k}$ **29.** $\displaystyle\sum_{k=2}^{+\infty} \frac{(-1)^k}{k \ln k}$ **30.** $\displaystyle\sum_{k=2}^{+\infty} \frac{(-1)^k}{k(\ln k)^2}$

31. $\displaystyle\sum_{k=1}^{+\infty} (-1)^{k+1}\frac{\sin k}{k^2+1}$ **32.** $\displaystyle\sum_{k=1}^{+\infty} (-1)^{k+1}\frac{\cos k}{k^2}$ **33.** $\displaystyle\sum_{k=1}^{+\infty} (-1)^{k+1}\frac{e^k}{k}$ **34.** $\displaystyle\sum_{k=1}^{+\infty} (-1)^{k+1}\frac{k}{e^k}$

35. $\displaystyle\sum_{k=1}^{+\infty} \frac{(-1)^{k+1}k!}{k^2 2^k}$ **36.** $\displaystyle\sum_{k=1}^{+\infty} \frac{(-1)^{k+1}(k!)^2}{(2k)!}$ **37.** $\displaystyle\sum_{k=1}^{+\infty} (-1)^{k+1}\frac{\ln k}{k}$ **38.** $\displaystyle\sum_{k=1}^{+\infty} \frac{(-1)^{k+1}}{k^{1/k}}$

39. Show that the series $\sum_{k=1}^{+\infty}[(-1)^{k+1}/k^{1/3}]$ is conditionally convergent.

40. Show that the positive terms of $\sum_{k=1}^{+\infty}[(-1)^{k+1}/k]$ diverge.

41. Show that the negative terms of $\sum_{k=1}^{+\infty}[(-1)^{k+1}/k]$ diverge.

42. Show that the series $\sum_{k=1}^{+\infty}[(-1)^{k+1}/k]$ can be rearranged so that the resulting series converges to 0.

43. Show that the series $\sum_{k=1}^{+\infty}[(-1)^{k+1}/k]$ can be rearranged so that the resulting series converges to 2.

44. Show that the series $\sum_{k=1}^{+\infty}[(-1)^{k+1}/k]$ can be rearranged so that the resulting series diverges.

45. Show that the series $e^{-x}\cos x + e^{-2x}\cos 2x + e^{-3x}\cos 3x + \cdots$ is absolutely convergent for all positive values of x.   [*Hint:* Use the fact that $|\cos\theta|\le 1$.]

46. What can you say about the convergence of the series below?

$$1 + r\cos\theta + r^2\cos 2\theta + r^3\cos 3\theta + \cdots$$

47. What is wrong with the following argument?

$$A = 1 - \tfrac{1}{2} + \tfrac{1}{3} - \tfrac{1}{4} + \tfrac{1}{5} - \tfrac{1}{6} + \tfrac{1}{7} - \tfrac{1}{8} + \cdots$$
$$(\tfrac{1}{2})A = \tfrac{1}{2} - \tfrac{1}{4} + \tfrac{1}{6} - \tfrac{1}{8} + \cdots$$

Thus,

$$A + (\tfrac{1}{2})A = 1 + \tfrac{1}{3} - \tfrac{1}{2} + \tfrac{1}{5} + \tfrac{1}{7} - \tfrac{1}{4} + \cdots$$

The series on the right is a rearrangement of the terms of the series A. Hence, its sum is A, so that

$$A + (\tfrac{1}{2})A = A$$
$$A = 0$$

But

$$A = (1 - \tfrac{1}{2}) + (\tfrac{1}{3} - \tfrac{1}{4}) + (\tfrac{1}{5} - \tfrac{1}{6}) + \cdots > 0???$$

5. Summary of Tests

The preceding sections contain many tests for convergence or divergence of a series. In the exercises following these sections, each time a series was to be tested, the test to be applied was also cited or the section in which the question appeared virtually gave the test away. Such advantages, unfortunately, are not of the real world. Therefore, in this section we summarize the tests already encountered and give some clues as to what test, in general, gives the best chance of working.

We begin with Table 3, a list of the tests we've encountered, and Table 4 (p. 618), a list of the important series we've analyzed.

Table 3   Tests for Convergence and Divergence of Series

Name	*Description*	*Comment*
Divergence test (11.18)	$\sum_{k=1}^{+\infty} a_k$ diverges if $\lim_{n \to +\infty} a_n \neq 0$	Test gives no information about convergence of $\sum_{k=1}^{+\infty} a_k$
Comparison test I for series of positive terms (11.23)	$\sum_{k=1}^{+\infty} a_k$ converges if $0 \leq a_k \leq b_k$ and if $\sum_{k=1}^{+\infty} b_k$ converges	You must choose a series $\sum_{k=1}^{+\infty} b_k$ of positive terms whose convergence is known
Comparison test II for series of positive terms (11.24)	$\sum_{k=1}^{+\infty} a_k$ diverges if $a_k \geq c_k \geq 0$ and if $\sum_{k=1}^{+\infty} c_k$ diverges	You must choose a series $\sum_{k=1}^{+\infty} c_k$ of positive terms whose divergence is known
Limit comparison test III for series of positive terms (11.26)	$\sum_{k=1}^{+\infty} a_k$ converges (diverges) if $\sum_{k=1}^{+\infty} b_k$ converges (diverges) and if $\lim_{n \to +\infty}(a_n/b_n) = L > 0$	You must choose a series $\sum_{k=1}^{+\infty} b_k$ of positive terms whose convergence (divergence) is known
Integral test for series of positive terms (11.27)	$\sum_{k=1}^{+\infty} a_k$ converges (diverges) if for $f(k) = a_k \geq 0,$ f continuous and non-increasing, then $\int_1^{+\infty} f(x)\,dx$ converges (diverges)	Good to use if f is easy to integrate
Ratio test for series of positive terms (11.29)	$\sum_{k=1}^{+\infty} a_k$ converges if $\lim_{n \to +\infty}(a_{n+1}/a_n) < 1$ $\sum_{k=1}^{+\infty} a_k$ diverges if $\lim_{n \to +\infty}(a_{n+1}/a_n) > 1$	Good to use if a_k consists of factorials or kth powers The test gives no information if $\lim_{n \to +\infty}(a_{n+1}/a_n) = 1$
Alternating series test (11.30)	$\sum_{k=1}^{+\infty}(-1)^{k+1}a_k$ converges if a_k are nonincreasing and if $\lim_{n \to +\infty} a_n = 0$ $(a_k > 0)$	Refer to (11.34) for the procedure to estimate the error
Absolute convergence test (11.35)	If $\sum_{k=1}^{+\infty} |a_k|$ converges, then $\sum_{k=1}^{+\infty} a_k$ converges	

The following outline offers one method of attack for determining the convergence or divergence of a series. Obviously, there are other methods, equally good, that you can devise for yourself.

1. Check to see if the given series is either a geometric series or a *p*-series.

2. Take the limit of the *n*th term of the series to see if the divergence test applies.

Table 4   Important Series

Name	Description	Comment
Geometric series (11.16)	$\sum\limits_{k=1}^{+\infty} ar^{k-1} = a + ar + ar^2 + \cdots, \quad a \neq 0$	Converges to $a/(1-r)$ if $|r| < 1$; diverges otherwise
Harmonic series (p. 588)	$\sum\limits_{k=1}^{+\infty} \dfrac{1}{k} = 1 + \dfrac{1}{2} + \dfrac{1}{3} + \cdots$	Diverges
p-series (11.25)	$\sum\limits_{k=1}^{+\infty} \dfrac{1}{k^p} = 1 + \dfrac{1}{2^p} + \dfrac{1}{3^p} + \cdots$	Converges if $p > 1$; diverges if $p \leq 1$
Factorial series (p. 604)	$\sum\limits_{k=0}^{+\infty} \dfrac{1}{k!} = 1 + 1 + \dfrac{1}{2} + \dfrac{1}{6} + \dfrac{1}{24} + \cdots$	Converges
k-to-the-k series (p. 595)	$\sum\limits_{k=1}^{+\infty} \dfrac{1}{k^k} = 1 + \dfrac{1}{2^2} + \dfrac{1}{3^3} + \dfrac{1}{4^4} + \cdots$	Converges

3. If the series is positive and has terms involving products, factorials, or powers, the ratio test is a good first choice.

4. If the series is positive and meets the conditions of the integral test, the integral test may be tried.

5. If the series is positive and the nth term is a quotient of sums or differences of powers of n, the limit comparison test with an appropriate p-series will usually do the job.

6. If the series is positive and the preceding attempts fail to provide a conclusion, then try comparison test I or II.

7. *Series with negative terms.* For alternating series that meet the proper conditions, use the alternating series test (it is sometimes better to apply the test for absolute convergence first). For other series containing negative terms, always test for absolute convergence first.

Here are some series to test your skill.

EXERCISE 5

Test each series for convergence (absolute or conditional) or divergence. Use any test you wish.

1. $\sum\limits_{k=1}^{+\infty} \dfrac{9k^3 + 5k^2}{k^{5/2} + 4}$

2. $\sum\limits_{k=1}^{+\infty} \dfrac{(-1)^{k+1}}{\sqrt{2k+1}}$

3. $6 + 2 + \dfrac{2}{3} + \dfrac{2}{9} + \dfrac{2}{27} + \cdots$

4. $1 + \dfrac{1 \cdot 2}{1 \cdot 3} + \dfrac{1 \cdot 2 \cdot 3}{1 \cdot 3 \cdot 5} + \dfrac{1 \cdot 2 \cdot 3 \cdot 4}{1 \cdot 3 \cdot 5 \cdot 7} + \cdots$

5. $\sum\limits_{k=1}^{+\infty} \dfrac{1}{k^2} \sin \dfrac{\pi}{k}$

6. $\sum\limits_{k=1}^{+\infty} \dfrac{3k+2}{k^3+1}$

7. $1 + \dfrac{2^2+1}{2^3+1} + \dfrac{3^2+1}{3^3+1} + \dfrac{4^2+1}{4^3+1} + \cdots$

8. $2 + \dfrac{3}{2} \cdot \dfrac{1}{4} + \dfrac{4}{3} \cdot \dfrac{1}{4^2} + \dfrac{5}{4} \cdot \dfrac{1}{4^3} + \cdots$

9. $\displaystyle\sum_{k=1}^{+\infty} \dfrac{k+4}{k\sqrt{3k-2}}$

10. $\displaystyle\sum_{k=1}^{+\infty} \dfrac{\sin k}{k^3}$

11. $\displaystyle\sum_{k=1}^{+\infty} \dfrac{3^{2k-1}}{k^2+2k}$

12. $\displaystyle\sum_{k=1}^{+\infty} \dfrac{5^k}{k!}$

13. $\displaystyle\sum_{k=1}^{+\infty} \left(1 + \dfrac{2}{k}\right)^k$

14. $\displaystyle\sum_{k=1}^{+\infty} \dfrac{k^2+4}{e^k}$

15. $\dfrac{2}{3} - \dfrac{3}{4} \cdot \dfrac{1}{2} + \dfrac{4}{5} \cdot \dfrac{1}{3} - \dfrac{5}{6} \cdot \dfrac{1}{4} + \cdots$

16. $\dfrac{1}{2} - \dfrac{4}{2^3+1} + \dfrac{9}{3^3+1} - \dfrac{16}{4^3+1} + \cdots$

17. $\displaystyle\sum_{k=1}^{+\infty} \dfrac{k!}{(2k)!}$

18. $\displaystyle\sum_{k=1}^{+\infty} k^3 e^{-k^4}$

19. $1 + \dfrac{1 \cdot 3}{2!} + \dfrac{1 \cdot 3 \cdot 5}{3!} + \dfrac{1 \cdot 3 \cdot 5 \cdot 7}{4!} + \cdots$

20. $\dfrac{1}{\sqrt{1 \cdot 2 \cdot 3}} + \dfrac{1}{\sqrt{2 \cdot 3 \cdot 4}} + \dfrac{1}{\sqrt{3 \cdot 4 \cdot 5}} + \cdots$

21. $\displaystyle\sum_{k=4}^{+\infty} \left(\dfrac{1}{k-3} - \dfrac{1}{k}\right)$

22. $\displaystyle\sum_{k=1}^{+\infty} \dfrac{1}{\sqrt{k}+100}$

23. $\displaystyle\sum_{k=1}^{+\infty} \dfrac{k^2+5k}{3+5k^2}$

24. $\displaystyle\sum_{k=1}^{+\infty} \dfrac{1}{\sqrt[3]{k^4+4}}$

25. $\displaystyle\sum_{k=1}^{+\infty} \dfrac{1}{11}\left(\dfrac{-3}{2}\right)^k$

26. $\displaystyle\sum_{k=1}^{+\infty} \dfrac{1}{\sqrt{k^3+1}}$

27. $1 - \dfrac{2!}{1 \cdot 3} + \dfrac{3!}{1 \cdot 3 \cdot 5} - \dfrac{4!}{1 \cdot 3 \cdot 5 \cdot 7} + \cdots$

28. $\dfrac{1}{3} - \dfrac{2}{4} + \dfrac{3}{5} - \dfrac{4}{6} + \cdots$

29. $\displaystyle\sum_{k=1}^{+\infty} (-1)^k \dfrac{k^3}{k^3+e^k}$

30. $\displaystyle\sum_{k=1}^{+\infty} \dfrac{k(-4)^{3k}}{5^k}$

31. $\displaystyle\sum_{k=1}^{+\infty} \left(-\dfrac{1}{k}\right)^k$

32. $\displaystyle\sum_{k=1}^{+\infty} \dfrac{5}{2^k+1}$

33. $\displaystyle\sum_{k=1}^{+\infty} \dfrac{\ln k}{2k^3-1}$

34. $\displaystyle\sum_{k=1}^{+\infty} (\sqrt{k+1} - \sqrt{k})$

35. $\displaystyle\sum_{k=1}^{+\infty} \sin^3\left(\dfrac{1}{k}\right)$

36. $\displaystyle\sum_{k=1}^{+\infty} e^{-k^2}$

37. $\dfrac{\sin\sqrt{1}}{1^{3/2}} - \dfrac{\sin\sqrt{2}}{2^{3/2}} + \dfrac{\sin\sqrt{3}}{3^{3/2}} - \cdots$

38. $\displaystyle\sum_{k=2}^{+\infty} \dfrac{(-1)^{k-1}}{k(\ln k)^3}$

6. Power Series

In this section we study series whose terms are variable. In particular, if x is a variable, then a series of the form

$$\sum_{k=0}^{+\infty} a_k x^k = a_0 + a_1 x + a_2 x^2 + \cdots$$

where the coefficients a_0, a_1, a_2 are constants, is called a *power series in x.**

---

* For power series, we agree that when $x = 0$, the series equals a_0.

A series of the form

$$\sum_{k=0}^{+\infty} a_k(x - c)^k = a_0 + a_1(x - c) + a_2(x - c)^2 + \cdots \qquad c \text{ a constant}$$

is called a *power series in* $(x - c)$.

For a particular x, a power series in x reduces to a series of real numbers like the series studied thus far. For example, $\sum_{k=1}^{+\infty}(x^k/k)$ is a power series in x. It certainly converges (to 0) if $x = 0$. If $x = 1$, it becomes the harmonic series $\sum_{k=1}^{+\infty}(1/k)$, which is divergent. If $x = -1$, it becomes the alternating harmonic series $\sum_{k=1}^{+\infty}[(-1)^k/k]$, which is convergent.

To find all numbers x for which a power series in x is convergent, we rely on the ratio test (11.29), as the next examples illustrate.

EXAMPLE 1   Find all x for which the following power series are convergent.

(a) $\displaystyle\sum_{k=0}^{+\infty} \frac{x^k}{k!} = 1 + x + \frac{x^2}{2!} + \frac{x^3}{3!} + \cdots$ (b) $\displaystyle\sum_{k=0}^{+\infty} \frac{kx^k}{4^k} = \frac{x}{4} + \frac{2x^2}{4^2} + \frac{3x^3}{4^3} + \cdots$

(c) $\displaystyle\sum_{k=0}^{+\infty} k!x^k = 1 + x + 2!x^2 + 3!x^3 + \cdots$

Solution   (a) To test the series $\sum_{k=0}^{+\infty}(x^k/k!)$ for convergence, we look at the series of absolute values, namely, $\sum_{k=0}^{+\infty}(|x|^k/k!)$, and apply the ratio test. For the series under investigation, we take

$$b_n = \frac{|x|^n}{n!} \qquad \text{and} \qquad b_{n+1} = \frac{|x|^{n+1}}{(n+1)!}$$

By forming the ratio of b_{n+1} to b_n and taking the limit as $n \to +\infty$, we find

$$\lim_{n \to +\infty} \frac{b_{n+1}}{b_n} = \lim_{n \to +\infty} \frac{|x|^{n+1}n!}{(n+1)!|x|^n} = \lim_{n \to +\infty} \frac{|x|}{n+1} = 0$$

Since the limit is less than 1 for every x, it follows from the ratio test that the given series is absolutely convergent for all real numbers.*

(b) For $\sum_{k=0}^{+\infty}(kx^k/4^k)$, $b_n = n|x|^n/4^n$ and $b_{n+1} = (n+1)|x|^{n+1}/4^{n+1}$. Then

$$\lim_{n \to +\infty} \frac{b_{n+1}}{b_n} = \lim_{n \to +\infty} \frac{(n+1)|x|^{n+1}(4^n)}{4^{n+1}(n|x|^n)} = \lim_{n \to +\infty} \frac{(n+1)|x|}{4n} = \frac{1}{4}|x|$$

By the ratio test, the series will converge if $|x|/4 < 1$ or $|x| < 4$. It will diverge if $|x|/4 > 1$ or $|x| > 4$. The ratio test gives no information when $|x|/4 = 1$. However, we may check these values directly by replacing x by 4

---

* This suggests the following interesting result: Since $\sum_{k=0}^{+\infty}(x^k/k!)$ converges absolutely for every x, the limit of the nth term tends to 0. Thus, we have $\lim_{n \to +\infty}(x^n/n!) = 0$ for every x.

and -4. For $x = 4$, the series becomes

$$\sum_{k=0}^{+\infty} \frac{k4^k}{4^k} = \sum_{k=1}^{+\infty} k = 1 + 2 + \cdots$$

which diverges. For $x = -4$, we obtain

$$\sum_{k=1}^{+\infty} \frac{k(-4)^k}{4^k} = \sum_{k=1}^{+\infty} \frac{(-1)^k k(4^k)}{4^k} = \sum_{k=1}^{+\infty} (-1)^k k = -1 + 2 - 3 + \cdots$$

which also diverges. Thus, the series $\sum_{k=0}^{+\infty}(kx^k/4^k)$ converges absolutely for $-4 < x < 4$ and diverges for $|x| \geq 4$.

(c)   For $\sum_{k=0}^{+\infty} k!x^k$, $b_n = n!|x|^n$ and $b_{n+1} = (n+1)!|x|^{n+1}$. Thus,

$$\lim_{n \to +\infty} \frac{b_{n+1}}{b_n} = \lim_{n \to +\infty} \frac{(n+1)!|x|^{n+1}}{n!|x|^n} = \lim_{n \to +\infty} (n+1)|x| = \begin{cases} 0 & \text{if } x = 0 \\ +\infty & \text{if } x \neq 0 \end{cases}$$

Thus, this power series converges only for $x = 0$. For any other x, it diverges. ∎

The results obtained in Example 1 are typical of all power series. In order to prove this, we need the following preliminary result:

(11.36)   **THEOREM**

(a)   **If the power series $\sum_{k=0}^{+\infty} a_k x^k$ converges for a number $x_0 \neq 0$, then it converges absolutely for all numbers x such that $|x| < |x_0|$.**

(b)   **If the power series $\sum_{k=0}^{+\infty} a_k x^k$ diverges for a number x_1, then it diverges for all numbers x such that $|x| > |x_1|$.**

Proof of Part (a)   Assume that $\sum_{k=0}^{+\infty} a_k x_0^k$ converges. Then $\lim_{n \to +\infty} a_n x_0^n = 0$. Thus, there is a positive integer N such that $|a_n x_0^n| < 1$ for all $n \geq N$. For any number x such that $|x| < |x_0|$,

$$|a_n x^n| = \left| \frac{a_n x^n x_0^n}{x_0^n} \right| = |a_n x_0^n| \left| \frac{x}{x_0} \right|^n < \left| \frac{x}{x_0} \right|^n \quad \text{for } n \geq N$$

But the series $\sum_{k=0}^{+\infty} |x/x_0|^k$ is a convergent geometric series because $|x/x_0| < 1$. Therefore, by comparison test I (11.23), the series $\sum_{k=0}^{+\infty} |a_k x^k|$ converges, and so the power series $\sum_{k=0}^{+\infty} a_k x^k$ converges absolutely for all numbers x such that $|x| < |x_0|$.   ∎

Proof of Part (b)   If the series converges for some number x such that $|x| > |x_1|$, it must converge for x_1 [by part (a)], which is contrary to the hypothesis. Therefore, the series diverges for all x such that $|x| > |x_1|$.   ∎

From theorem (11.36), we conclude that:

For a power series $\sum_{k=0}^{+\infty} a_k x^k$, exactly one of the following is true:

1. The series converges only for $x = 0$.

2. The series converges absolutely for all x.

3. There is a positive number R such that the series converges absolutely for all x for which $|x| < R$ and diverges for all x for which $|x| > R$.

The set of all numbers x for which a power series converges is called the *interval of convergence* of the power series. The number R in Case 3 above is called its *radius of convergence*. If Case 1 holds, then $R = 0$; if Case 2 holds, we write $R = +\infty$.

As Example 1 illustrates, the ratio test is the most useful method for obtaining the interval of convergence and the radius of convergence of a power series. However, no conclusion may be drawn about convergence or divergence at the endpoints of the interval of convergence. At an endpoint, a power series may be absolutely convergent, conditionally convergent, or divergent.

Let's return to Example 1.

In Example 1(a) the series converges absolutely for all x. In this case, the radius of convergence is $+\infty$, and the interval of convergence is $(-\infty, +\infty)$.

In Example 1(b) the series converges absolutely for $|x| < 4$ and diverges for $|x| \geq 4$. In this case, 4 is the radius of convergence, and the open interval $(-4, 4)$ is the interval of convergence (see Fig. 18).

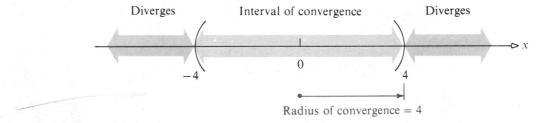

Figure 18

In Example 1(c) the series converges absolutely only for $x = 0$. In this case the radius of convergence is 0.

EXAMPLE 2 Find the interval of convergence of: $\displaystyle\sum_{k=1}^{+\infty} \frac{x^{2k}}{k}$

Solution By applying the ratio test, we obtain

$$\left| \frac{x^{2n+2}}{n+1} \left(\frac{n}{x^{2n}} \right) \right| = \left| \frac{n}{n+1} x^2 \right| = \frac{n}{n+1} x^2$$

$$\lim_{n \to +\infty} \frac{n}{n+1} x^2 = x^2 \lim_{n \to +\infty} \frac{n}{n+1} = x^2$$

For the series to converge, we require $x^2 < 1$; that is, $-1 < x < 1$. At the endpoints, we find that for $x = -1$ or for $x = 1$, the series reduces to the harmonic series $\sum_{k=1}^{+\infty}(1/k)$, which diverges. Consequently, the interval of convergence is $-1 < x < 1$ (see Fig. 19).

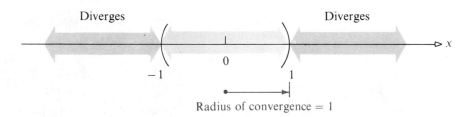

Figure 19   ∎

To find the interval of convergence of a power series in $(x - c)$, namely, $\sum_{k=0}^{+\infty} a_k(x - c)^k$, we let $z = x - c$. The result is a power series in z: $\sum_{k=0}^{+\infty} a_k z^k$. The interval of convergence of the latter series may be used to find the interval of convergence of the former. For example if $\sum_{k=0}^{+\infty} a_k z^k$ has $(-R, R]$ for its interval of convergence, then $-R < z \le R$, so that when we replace z by $(x - c)$, we get $-R < (x - c) \le R$ or $c - R < x \le c + R$. Thus, the interval of convergence is $(c - R, c + R]$ and R is the radius of convergence.

EXAMPLE 3   Find the interval of convergence of: $\displaystyle\sum_{k=0}^{+\infty} (-1)^k \frac{(x - 2)^k}{k + 1}$

Solution   By the ratio test, we obtain

$$\left|\frac{(x - 2)^{n+1}(n + 1)}{(n + 2)(x - 2)^n}\right| = \left|\frac{n + 1}{n + 2}(x - 2)\right|$$

$$\lim_{n \to +\infty} \left|\frac{(n + 1)(x - 2)}{n + 2}\right| = |x - 2|$$

For the series to converge, we require that $|x - 2| < 1$; that is, $-1 < x - 2 < 1$ or $1 < x < 3$. The endpoints $x = 1$ and $x = 3$ must be checked separately. If $x = 1$, the series becomes

$$1 + \frac{1}{2} + \frac{1}{3} + \cdots + \frac{1}{n + 1} + \cdots$$

which is the divergent harmonic series. If $x = 3$, we get

$$1 - \frac{1}{2} + \frac{1}{3} - \frac{1}{4} + \cdots + \frac{(-1)^n}{n + 1} + \cdots$$

the alternating harmonic series, which converges. Hence, the interval of convergence is $(1, 3]$, as shown in Figure 20.

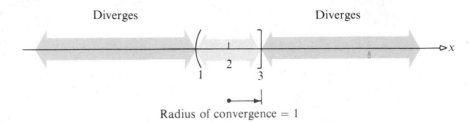

Figure 20   ∎

EXERCISE 6

In Problems 1–20 find the interval of convergence.

1. $\displaystyle\sum_{k=0}^{+\infty} \frac{x^k}{k+5}$

2. $\displaystyle\sum_{k=0}^{+\infty} \frac{x^k}{1+k^2}$

3. $\displaystyle\sum_{k=0}^{+\infty} \frac{k^2 x^k}{3^k}$

4. $\displaystyle\sum_{k=0}^{+\infty} \frac{2^k x^k}{3^k}$

5. $\displaystyle\sum_{k=0}^{+\infty} \frac{kx^k}{2k+1}$

6. $\displaystyle\sum_{k=0}^{+\infty} (6x)^k$

7. $\displaystyle\sum_{k=0}^{+\infty} \frac{k(2x)^k}{3^k}$

8. $\displaystyle\sum_{k=0}^{+\infty} (-1)^k \frac{(2x)^k}{k!}$

9. $\displaystyle\sum_{k=0}^{+\infty} \frac{(-1)^k}{(2k+1)!} x^{2k+1}$

10. $\displaystyle\sum_{k=1}^{+\infty} (kx)^k$

11. $\displaystyle\sum_{k=1}^{+\infty} \frac{kx^k}{\ln(k+1)}$

12. $\displaystyle\sum_{k=0}^{+\infty} (-1)^k \frac{(x-3)^{2k}}{9^k}$

13. $\displaystyle\sum_{k=1}^{+\infty} \frac{k(k+1)x^k}{4^k}$

14. $\displaystyle\sum_{k=1}^{+\infty} \frac{(-1)^k (x-5)^k}{k(k+1)}$

15. $\displaystyle\sum_{k=1}^{+\infty} \frac{x^k}{\ln(k+1)}$

16. $\displaystyle\sum_{k=0}^{+\infty} \frac{x^k}{e^k}$

17. $\displaystyle\sum_{k=0}^{+\infty} (x-3)^k$

18. $\displaystyle\sum_{k=0}^{+\infty} \frac{(x+1)^k}{k!}$

19. $\displaystyle\sum_{k=0}^{+\infty} (-1)^k \frac{(x-1)^{4k}}{k!}$

20. $\displaystyle\sum_{k=1}^{+\infty} \frac{(x+1)^k}{k(k+1)(k+2)}$

In Problems 21 and 22 find the radius of convergence.

21. $\displaystyle\sum_{k=1}^{+\infty} \frac{k^k x^k}{k!}$

22. $\displaystyle\sum_{k=0}^{+\infty} \frac{3^k (x-2)^k}{k!}$

In Problems 23–28 find all x for which the series converges.

23. $\displaystyle\sum_{k=1}^{+\infty} \frac{1}{kx^k}$

24. $\displaystyle\sum_{k=1}^{+\infty} \frac{(\sin x)^k}{2k+1}$

25. $\displaystyle\sum_{k=1}^{+\infty} \frac{\sin k\pi x}{k^2}$

26. $\displaystyle\sum_{k=1}^{+\infty} \frac{1}{k(x-4)^k}$

27. $\displaystyle\sum_{k=0}^{+\infty} (-1)^k 2^k (\sin x)^k, \quad -\pi/2 \le x \le \pi/2$

28. $\displaystyle\sum_{k=1}^{+\infty} (-1)^k (\tan x)^k, \quad -\pi/2 < x < \pi/2$

29. If $\sum_{k=1}^{+\infty} a_k x^k$ converges for $x = 3$, what, if anything, can be said about the convergence at $x = 2$? Can anything be said about the convergence at $x = 5$?

30. If $\sum_{k=1}^{+\infty} a_k (x-2)^k$ converges for $x = 6$, at what other numbers x must the series necessarily converge?

31. If the series $\sum_{k=1}^{+\infty} a_k x^k$ converges for $x = 6$ and diverges for $x = -8$, what, if anything, can be said about the truth of the following statements?
    (a)   The series converges for $x = 2$.   (b)   The series diverges for $x = 7$.

(c) The series is absolutely convergent for $x = 6$.   (d) The series converges for $x = -6$.

(e) The series diverges for $x = 10$.   (f) The series is absolutely convergent for $x = 4$.

32. If the series $\sum_{k=1}^{+\infty} a_k 3^k$ converges, show that the series $\sum_{k=1}^{+\infty} k a_k 2^k$ also converges.

33. If $R > 0$ is the radius of convergence of $\sum_{k=1}^{+\infty} a_k x^k$, show that $\lim_{n \to +\infty} |a_{n+1}/a_n| = 1/R$, provided this limit exists.

34. If R is the radius of convergence of $\sum_{k=1}^{+\infty} a_k x^k$, show that the radius of convergence of $\sum_{k=1}^{+\infty} a_k x^{2k}$ is $\sqrt{R}$.

35. Prove that if a power series converges absolutely at one endpoint of its interval of convergence, then the power series is absolutely convergent at the other endpoint.

7. Functions Represented by Power Series

A power series $\sum_{k=0}^{+\infty} a_k x^k$ defines a function f whose domain is the interval of convergence of the power series. If I is the interval of convergence of $\sum_{k=0}^{+\infty} a_k x^k$, we say that f is *represented by* the power series $\sum_{k=0}^{+\infty} a_k x^k$, and we write

$$f(x) = a_0 + a_1 x + a_2 x^2 + \cdots + a_n x^n + \cdots$$

for all x in I.*

For example, if f is represented by the power series $\sum_{k=0}^{+\infty} a_k x^k$ whose interval of convergence is I and if x_0 is a number in I, then $f(x_0)$ may be found by finding the sum of the series

$$f(x_0) = a_0 + a_1 x_0 + a_2 x_0^2 + \cdots + a_n x_0^n + \cdots$$

Thus, a power series representation for a function f may be used to calculate functional values.

The function f represented by a power series has properties similar to a polynomial. We state three of these properties without proof.

(11.37)   **Let $\sum_{k=0}^{+\infty} a_k x^k$ be a power series in x having a nonzero radius of convergence R. Define the function f to be**

$$f(x) = a_0 + a_1 x + a_2 x^2 + \cdots + a_n x^n + \cdots$$

Then at every number x within the interval of convergence of the given series:

(a) f is continuous:

$$\lim_{x \to x_0} \left(\sum_{k=0}^{+\infty} a_k x^k \right) = \sum_{k=0}^{+\infty} \left(\lim_{x \to x_0} a_k x^k \right) = \sum_{k=0}^{+\infty} a_k x_0^k$$

(b) f is differentiable:

$$\frac{d}{dx} \left(\sum_{k=0}^{+\infty} a_k x^k \right) = \sum_{k=0}^{+\infty} \left(\frac{d}{dx} a_k x^k \right) = \sum_{k=1}^{+\infty} k a_k x^{k-1}$$

---

* As before, we define $f(0) = a_0$.

(c) f **can be integrated:**

$$\int_0^x \left(\sum_{k=0}^{+\infty} a_k t^k \right) dt = \sum_{k=0}^{+\infty} \left(\int_0^x a_k t^k \, dt \right) = \sum_{k=0}^{+\infty} \frac{a_k x^{k+1}}{k+1}$$

Parts (b) and (c) state that a power series can be differentiated or integrated term-by-term and that the resulting series represents the derivative and integral, respectively, of the function represented by the power series at every number x within the interval of convergence of the original power series. Moreover, it can be shown that the power series obtained by differentiating or integrating a power series whose radius of convergence is R, both converge and have the same radius of convergence R.*

As the following example illustrates, (11.37) provides a tool for obtaining new power series representations from existing ones.

EXAMPLE 1   The geometric series $1 + x + x^2 + \cdots$ converges for $-1 < x < 1$ and has the sum $1/(1-x)$.   That is,

$$\frac{1}{1-x} = 1 + x + x^2 + \cdots + x^n + \cdots \qquad -1 < x < 1$$

An application of part (b) of (11.37) reveals that

$$\frac{1}{(1-x)^2} = 1 + 2x + 3x^2 + \cdots + nx^{n-1} + \cdots$$

whose interval of convergence is $-1 < x < 1$.   ∎

The next example is a little more subtle.

EXAMPLE 2   Use the geometric series $1 + x + x^2 + \cdots$ to find a power series representation for $\ln[1/(1-x)]$.

Solution   The geometric series $1 + x + x^2 + \cdots$ converges for $-1 < x < 1$ and has the sum $1/(1-x)$.   Thus,

(11.38)
$$\frac{1}{1-x} = 1 + x + x^2 + \cdots + x^n + \cdots \qquad -1 < x < 1$$

By part (c) of (11.37) we may integrate (11.38), obtaining

$$\int_0^x \frac{1}{1-t} \, dt = x + \frac{x^2}{2} + \frac{x^3}{3} + \cdots + \frac{x^{n+1}}{n+1} + \cdots$$

---

* You are cautioned as to the meaning of this statement. It states that the *radii of convergence of* $\sum_{k=0}^{+\infty} a_k x^k$, $\sum_{k=1}^{+\infty} k a_k x^{k-1}$, and $\sum_{k=0}^{+\infty} [a_k x^{k+1}/(k+1)]$ are the same; it does not imply that the *intervals of convergence* are the same. For example, the interval of convergence of $\sum_{k=1}^{+\infty} (x^k/k)$ is $[-1, 1)$, whereas the interval of convergence of its derivative $\sum_{k=1}^{+\infty} x^{k-1}$ is $(-1, 1)$.

But

$$\int_0^x \frac{1}{1-t}\, dt = -\ln(1-t)\Big|_0^x = -\ln(1-x) = \ln \frac{1}{1-x}$$

Hence,

(11.39)
$$\ln \frac{1}{1-x} = x + \frac{x^2}{2} + \frac{x^3}{3} + \cdots + \frac{x^{n+1}}{n+1} + \cdots \quad \blacksquare$$

It is easy to verify that the interval of convergence of the power series in (11.39) is $-1 \le x < 1$. As a consequence, we may set $x = -1$ to obtain

$$\ln \tfrac{1}{2} = -1 + \tfrac{1}{2} - \tfrac{1}{3} + \tfrac{1}{4} - \cdots$$

Since $\ln \tfrac{1}{2} = -\ln 2$, we have found that the sum of the alternating harmonic series is $\ln 2$. That is,

(11.40)
$$1 - \tfrac{1}{2} + \tfrac{1}{3} - \tfrac{1}{4} + \cdots = \ln 2$$

EXAMPLE 3   Develop the series

(11.41)
$$\tan^{-1} x = x - \frac{x^3}{3} + \frac{x^5}{5} - \frac{x^7}{7} + \cdots + (-1)^n \frac{x^{2n+1}}{2n+1} + \cdots \quad -1 \le x \le 1$$

This is called *Gregory's series*.

Solution   In the geometric series (11.38), replace x by $-x^2$. The result is

(11.42)
$$\frac{1}{1+x^2} = 1 - x^2 + x^4 - x^6 + \cdots \quad -1 < x < 1$$

Hence, by part (c) of (11.37), we may integrate (11.42), obtaining

$$\int_0^x \frac{dt}{1+t^2} = \int_0^x (1 - t^2 + t^4 - \cdots)\, dt$$

Thus, we have

$$\tan^{-1} x = x - \frac{x^3}{3} + \frac{x^5}{5} - \frac{x^7}{7} + \cdots + (-1)^n \frac{x^{2n+1}}{2n+1} + \cdots$$

The radius of convergence of (11.41) is $R = 1$. To find the interval of convergence, we test $x = -1$ and $x = 1$. Each of these gives a form of the alternating harmonic series, which converges. Hence, the interval of convergence of (11.41) is $-1 \le x \le 1$. $\blacksquare$

As the above examples illustrate, it is often possible to obtain a power series representation for a given function by starting with a known series and differentiating, integrating, or substituting in it. We now give a more direct method for finding the power series representation for a large class of functions.

TAYLOR AND MACLAURIN SERIES

Consider the power series

(11.43)
$$\sum_{k=0}^{+\infty} a_k(x - c)^k = a_0 + a_1(x - c) + a_2(x - c)^2 + \cdots + a_n(x - c)^n + \cdots$$

and suppose its interval of convergence is $(c - R, c + R)$, $0 < R \le +\infty$. For a given function f, we show how to obtain its power series representation. We begin by assuming* that for $c - R < x < c + R$, the function f is actually represented by the power series (11.43). That is,

(11.44)
$$f(x) = \sum_{k=0}^{+\infty} a_k(x - c)^k = a_0 + a_1(x - c) + a_2(x - c)^2 + \cdots + a_n(x - c)^n + \cdots$$

We may obtain the coefficients $a_0, a_1, \ldots$ in terms of f and its derivatives by using part (b) of (11.37):

$$f'(x) = \sum_{k=1}^{+\infty} k a_k(x - c)^{k-1} = a_1 + 2a_2(x - c) + 3a_3(x - c)^2 + \cdots$$

$$f''(x) = \sum_{k=2}^{+\infty} k(k - 1)a_k(x - c)^{k-2} = 2a_2 + 6a_3(x - c) + 12a_4(x - c)^2 + \cdots$$

$$f'''(x) = \sum_{k=3}^{+\infty} k(k - 1)(k - 2)a_k(x - c)^{k-3} = 6a_3 + 24a_4(x - c) + \cdots$$

and for any positive integer n,

$$f^{(n)}(x) = \sum_{k=n}^{+\infty} k(k - 1)(k - 2) \cdots (k - n + 1)a_k(x - c)^{k-n}$$

$$= n!a_n + (n + 1)!a_{n+1}(x - c) + \cdots$$

Now we set $x = c$ in each of these equations:

$$f(c) = a_0 \qquad a_0 = f(c)$$
$$f'(c) = a_1 \qquad a_1 = f'(c)$$
$$f''(c) = 2a_2 \qquad a_2 = \frac{f''(c)}{2!}$$
$$f'''(c) = 3!a_3 \qquad a_3 = \frac{f'''(c)}{3!}$$
$$\vdots \qquad\qquad \vdots$$
$$f^{(n)}(c) = n!a_n \qquad a_n = \frac{f^{(n)}(c)}{n!}$$

---

* The conditions under which this assumption is justified are considered later.

Hence, by substituting in (11.44), we obtain

$$f(x) = f(c) + f'(c)(x - c) + \frac{f''(c)}{2!}(x - c)^2 + \cdots + \frac{f^{(n)}(c)}{n!}(x - c)^n + \cdots$$

And we have proved the following result:

(11.45)   **THEOREM**   **If a function f can be represented by the power series $\sum_{k=0}^{+\infty} a_k(x - c)^k$ whose radius of convergence is R, then**

(11.46) $$f(x) = f(c) + f'(c)(x - c) + \frac{f''(c)(x - c)^2}{2!} + \cdots + \frac{f^{(n)}(c)(x - c)^n}{n!} + \cdots$$

for all x such that $c - R < x < c + R$.

Expression (11.46) is called a *Taylor series in* $(x - c)$ of the function f.

We have shown that if f has a series representation of the form (11.44), then it must be the Taylor series (11.46). That is, there is only one power series in $(x - c)$ that defines a function f.

When $c = 0$,   the Taylor series becomes

(11.47) $$f(x) = f(0) + f'(0)x + \frac{f''(0)x^2}{2!} + \cdots + \frac{f^{(n)}(0)x^n}{n!} + \cdots$$

This series representation of f is called a *Maclaurin series*. Note that in a Maclaurin series all the derivatives that appear are evaluated at 0, and the interval of convergence of this series has its center at 0.

The Taylor series in $(x - c)$ of a function f is usually referred to as the *Taylor expansion of f about c;* the Maclaurin series of a function f is the *Maclaurin expansion of f about 0.*

EXAMPLE 4   Assuming that $f(x) = e^x$ can be represented by a power series in x, find its Maclaurin series.

Solution   To use (11.47), we must evaluate f and its derivatives at 0:

$$f(x) = e^x \qquad f(0) = 1$$
$$f'(x) = e^x \qquad f'(0) = 1$$
$$f''(x) = e^x \qquad f''(0) = 1$$
$$\vdots \qquad\qquad \vdots$$

Hence, by (11.47), we find

$$f(x) = e^x = 1 + x + \frac{x^2}{2!} + \frac{x^3}{3!} + \cdots + \frac{x^n}{n!} + \cdots$$

$$= \sum_{k=0}^{+\infty} \frac{x^k}{k!} \quad \blacksquare$$

We have shown in Example 4 that if e^x can be represented by a power series, then the power series takes the form $\sum_{k=0}^{+\infty}(x^k/k!)$. Earlier (see p. 620), we showed that this series converges absolutely for all x. What we have not demonstrated yet, but soon will, is that this series actually converges to e^x for all x.

The conditions on a function f that guarantee that its power series representation actually converges to f are obtained as follows: We begin with the observation that the $(n + 1)$st partial sum of the Taylor series in $(x - c)$ of f is the Taylor polynomial $P_n(x)$ of degree n for f at c [see (10.19)]. As a result,

$$P_n(x) = f(x) - R_{n+1}(x)$$

where the remainder R_{n+1} is

(11.48)
$$R_{n+1}(x) = \frac{f^{(n+1)}(u)(x - c)^{n+1}}{(n + 1)!}$$

for some u between c and x. (The number u depends on n.)

The next result uses this remainder R_{n+1} to establish sufficient conditions for the power series representation of a function f to actually converge to f.

(11.49)   **THEOREM** **If a function f has derivatives of all orders in some interval $(c - R, c + R)$ and if**

$$\lim_{n \to +\infty} R_{n+1}(x) = 0$$

for all x in $(c - R, c + R)$,   then

$$f(x) = \sum_{k=0}^{+\infty} \frac{f^{(k)}(c)}{k!}(x - c)^k$$

$$= f(c) + f'(c)(x - c) + \frac{f''(c)}{2!}(x - c)^2 + \cdots + \frac{f^{(n)}(c)}{n!}(x - c)^n + \cdots$$

for all x in $(c - R, c + R)$.

Proof If f has derivatives of all orders in the interval $(c - R, c + R)$,   then $P_n(x) = f(x) - R_{n+1}(x)$ is   the $(n + 1)$st   partial   sum   of   the   infinite   series $\sum_{k=0}^{+\infty}[f^{(k)}(c)/k!](x - c)^k$.   If we also have $\lim_{n \to +\infty} R_{n+1}(x) = 0$,   then

$$\lim_{n \to +\infty} P_n(x) = \lim_{n \to +\infty} [f(x) - R_{n+1}(x)] = f(x)$$

and the series $\sum_{k=0}^{+\infty}[f^{(k)}(c)/k!](x - c)^k$ converges to f.   ∎

Although at first glance it may appear that the theorem is easy to use, in practice it is not always easy to show that $\lim_{n \to +\infty} R_{n+1}(x) = 0$.   One of the reasons for this is that the term $f^{(n+1)}(u)$ that appears in R_{n+1} depends on n, making the limit difficult to evaluate. Let's look at an example that illustrates this.

EXAMPLE 5 Prove that the series developed in Example 4 converges to e^x for every number x. That is, prove that

(11.50)
$$e^x = 1 + \frac{x}{1!} + \frac{x^2}{2!} + \frac{x^3}{3!} + \cdots + \frac{x^n}{n!} + \cdots \qquad \text{for all } x$$

Solution In order to prove that (11.50) is actually true for every x, we have to verify that $\lim_{n \to +\infty} R_{n+1}(x) = 0$. Since $f^{(n+1)}(x) = e^x$,

$$R_{n+1}(x) = \frac{x^{n+1}}{(n+1)!} e^u \qquad \text{where } u \text{ is between } 0 \text{ and } x$$

To show that $\lim_{n \to +\infty} R_{n+1}(x) = 0$, we consider two cases: (a) $x > 0$ and (b) $x < 0$.

(a) When $x > 0$, we have $0 < u < x$, so that $e^u < e^x$ and, hence, for every positive integer n,

$$0 < R_{n+1}(x) = \frac{x^{n+1} e^u}{(n+1)!} < \frac{e^x x^{n+1}}{(n+1)!}$$

Now, by the ratio test (11.29), the series $\sum_{k=0}^{+\infty} [x^{k+1}/(k+1)!]$ converges for all x. Therefore, from theorem (11.17), it follows that $\lim_{n \to +\infty} [x^{n+1}/(n+1)!] = 0$. Hence,

$$\lim_{n \to +\infty} \frac{e^x x^{n+1}}{(n+1)!} = e^x \lim_{n \to +\infty} \frac{x^{n+1}}{(n+1)!} = 0$$

so that by the squeezing theorem (2.20), $\lim_{n \to +\infty} R_{n+1}(x) = 0$.

(b) When $x < 0$, we have $x < u < 0$ and $0 < e^u < 1$, so that

$$|R_{n+1}(x)| = \frac{|x|^{n+1}}{(n+1)!} e^u < \frac{|x|^{n+1}}{(n+1)!}$$

The right member approaches 0 as $n \to +\infty$, so that by (2.20), $\lim_{n \to +\infty} R_{n+1}(x) = 0$. Thus, by theorem (11.49), the Maclaurin series representation (11.50) for e^x is true for all x, and we can write

(11.51)
$$e^x = 1 + x + \frac{x^2}{2!} + \frac{x^3}{3!} + \cdots + \frac{x^n}{n!} + \cdots = \sum_{k=0}^{+\infty} \frac{x^k}{k!} \qquad \text{for all } x \qquad \blacksquare$$

In particular, if $x = 1$, we find

(11.52)
$$e = 1 + 1 + \frac{1}{2!} + \frac{1}{3!} + \cdots + \frac{1}{n!} + \cdots = \sum_{k=0}^{+\infty} \frac{1}{k!}$$

There are functions whose Taylor series converge for all x in some interval—but to some other function. An example is

(11.53)
$$f(x) = \begin{cases} e^{-1/x^2} & \text{if } x \neq 0 \\ 0 & \text{if } x = 0 \end{cases}$$

With the aid of L'Hospital's rule (10.4), we can show that $f'(0) = 0$ (see Problem 43 on p. 542). Although it is much more difficult, it also can be shown that f has derivatives of all orders at 0 and that $f^{(n)}(0) = 0$ for all $n \geq 1$. Hence, the Maclaurin series for f is

$$0 + 0(x) + 0(x^2) + \cdots + 0(x^n) + \cdots$$

This series converges for every x and its sum is obviously the zero function—not f. Consequently, the series has $f(x)$ as its sum only when $x = 0$ and for no other x. This illustrates that convergence of the Taylor series of a function is not sufficient to ensure that the series converges to the function—*the remainder R_{n+1} must tend to 0 as $n \to +\infty$*. In fact, for the function f in (11.53), it can be shown that R_{n+1} does not approach 0 as $n \to +\infty$.

EXERCISE 7

In Problems 1–18 use (11.46) to find the Taylor expansion of each function about the given number c, under the assumption that such an expansion exists.

1. $f(x) = \sin x$; $c = 0$

2. $f(x) = \cos x$; $c = 0$

3. $f(x) = e^x$; $c = 1$

4. $f(x) = e^{2x}$; $c = -1$

5. $f(x) = \sin x$; $c = \pi/4$

6. $f(x) = \cos x$; $c = \pi/3$

7. $f(x) = \dfrac{1}{x}$; $c = 1$

8. $f(x) = \dfrac{1}{\sqrt{x}}$; $c = 4$

9. $f(x) = \dfrac{1}{(1 + x)^2}$; $c = 0$

10. $f(x) = \dfrac{1}{1 + x^2}$; $c = 0$

11. $f(x) = \dfrac{1}{1 - 3x}$; $c = 0$

12. $f(x) = \dfrac{1}{1 + 2x^3}$; $c = 0$

13. $f(x) = \ln(1 + x)$; $c = 0$

14. $f(x) = \ln x$; $c = 1$

15. $f(x) = 3x^3 + 2x^2 + 5x - 6$; $c = 0$

16. $f(x) = 4x^4 - 2x^3 - x$; $c = 0$

17. $f(x) = 3x^3 + 2x^2 + 5x - 6$; $c = 1$

18. $f(x) = 4x^4 - 2x^3 + x$; $c = 1$

19. In the geometric series (11.38), replace x by x^2 to obtain the power series representation for $1/(1 - x^2)$. What is its interval of convergence?

20. Integrate the power series of Problem 19 to obtain the power series for $\frac{1}{2} \ln[(1 + x)/(1 - x)]$. What is its interval of convergence?

21. Use the power series found in Problem 20 to get an approximation for $\ln 2$ accurate to within 0.001.

22. Use Gregory's series (11.41) to obtain a series for $\pi/4$. How many terms of this series are required to obtain an approximation for π accurate to within 0.0001?

23. By letting $x = 1$ in Gregory's series (11.41),

$$\tan^{-1} x = x - \frac{x^3}{3} + \frac{x^5}{5} - \cdots \qquad -1 \leq x \leq 1$$

it seems possible to use it to obtain an approximation to π. However, since this series converges very slowly (see Problem 22), it is next to useless for this purpose. A more rapidly convergent series is obtained by using the identity

$$\tan^{-1}(1) = \tan^{-1}(\tfrac{1}{2}) + \tan^{-1}(\tfrac{1}{3})$$

Use $x = \frac{1}{2}$ and $x = \frac{1}{3}$ in Gregory's series, together with this identity, to get an approximation for π accurate to within 0.0001.

8. Other Important Series

In Section 7 we saw that the Maclaurin series representation for $f(x) = e^x$,

(11.54)
$$e^x = 1 + x + \frac{x^2}{2!} + \frac{x^3}{3!} + \cdots + \frac{x^n}{n!} + \cdots$$

converges and equals e^x for all real numbers x.

In this section we develop the Maclaurin series representation for $\sin x$, $\cos x$, $\cosh x$, and some other functions.

SERIES FOR sin x

For $f(x) = \sin x$, the value of f and its derivatives at 0 are

$$
\begin{array}{ll}
f(x) = \sin x & f(0) = 0 \\
f'(x) = \cos x & f'(0) = 1 \\
f''(x) = -\sin x & f''(0) = 0 \\
f'''(x) = -\cos x & f'''(0) = -1
\end{array}
$$

Successive derivatives follow this same pattern, so that for $n = 0, 1, 2, 3, \ldots$, we have

$$
\begin{array}{ll}
f^{(2n)}(x) = (-1)^n \sin x & f^{(2n)}(0) = 0 \\
f^{(2n+1)}(x) = (-1)^n \cos x & f^{(2n+1)}(0) = (-1)^n
\end{array}
$$

By Taylor's formula with remainder (10.20), we find that

(11.55)
$$\sin x = x - \frac{x^3}{3!} + \frac{x^5}{5!} - \frac{x^7}{7!} + \cdots + (-1)^n \frac{x^{2n+1}}{(2n+1)!} + R_{n+1}(x)$$

At this point, we know that if $\sin x$ can be represented by a power series in x, then it is given by (11.55). To prove that the series on the right of (11.55) actually converges to $\sin x$ for all x, we need to show that

$$\lim_{n \to +\infty} R_{n+1}(x) = \lim_{n \to +\infty} (-1)^{n+1} \frac{f^{(2n+3)}(u) x^{2n+3}}{(2n+3)!} = 0 \qquad \text{for all } x$$

where u is between 0 and x. Since $|f^{(2n+3)}(u)| = |\cos u| \le 1$ for every number u, we see that

$$|R_{n+1}(x)| = \frac{|f^{(2n+3)}(u)|}{(2n+3)!} |x|^{2n+3} \le \frac{|x|^{2n+3}}{(2n+3)!}$$

But by the ratio test (11.29), the series $\sum_{k=0}^{+\infty}[|x|^{2k+3}/(2k+3)!]$ converges for all x. Hence, it follows by theorem (11.17) that $\lim_{n \to +\infty}[|x|^{2n+3}/(2n+3)!] = 0$, so that by the squeezing theorem (2.20), $\lim_{n \to +\infty}|R_{n+1}(x)| = 0$ and, therefore, $\lim_{n \to +\infty} R_{n+1}(x) = 0$. Thus, by theorem (11.49), the series converges to $\sin x$ for all x and we may write

(11.56) $$\sin x = x - \frac{x^3}{3!} + \frac{x^5}{5!} - \cdots + (-1)^n \frac{x^{2n+1}}{(2n+1)!} + \cdots \qquad \text{for all } x$$

From the discussion so far, it is clear that the task of finding the Taylor series representation of a function by taking successive derivatives and then showing that $\lim_{n \to +\infty} R_{n+1}(x) = 0$ is a difficult one. Consequently, to minimize this difficulty, we make use of the series developed thus far and properties (11.37). For example, we can obtain the Taylor series for $f(x) = \cos x$ quite easily by merely differentiating the series for $\sin x$.

SERIES FOR cos x

By using part (b) of (11.37), we can obtain the series for $\cos x$ by differentiating the series (11.56) term-by-term. The result is

(11.57) $$\cos x = 1 - \frac{x^2}{2!} + \frac{x^4}{4!} - \cdots + (-1)^n \frac{x^{2n}}{(2n)!} + \cdots \qquad \text{for all } x$$

We could have derived (11.57) in the same way we obtained (11.56). It could also have been derived by integrating (11.56) term-by-term.

CALCULATION OF LOGARITHMS

In Example 2 of Section 7, we discovered that

(11.58) $$\ln \frac{1}{1-x} = x + \frac{x^2}{2} + \frac{x^3}{3} + \cdots + \frac{x^n}{n} + \cdots \qquad -1 \le x < 1$$

It might appear that we could use (11.58) to compute logarithms of numbers. Unfortunately, though, this series only converges for $-1 \le x < 1$. In addition, unless x is close to 0, the series converges so slowly that too many terms would be required for practical use. (Recall the rate of convergence of the alternating harmonic series, p. 611.)

We now develop a useful formula for computing logarithms. If we multiply (11.58) by (-1) and use the fact that $-\ln(1/A) = \ln(1/A)^{-1} = \ln A$, we find that

(11.59) $$\ln(1 - x) = -x - \frac{x^2}{2} - \frac{x^3}{3} - \cdots$$

In (11.59), we replace x by $-x$. The result is

(11.60) $$\ln(1 + x) = x - \frac{x^2}{2} + \frac{x^3}{3} - \cdots$$

Now, we subtract (11.59) from (11.60) to get

(11.61)
$$\ln \frac{1+x}{1-x} = 2\left(x + \frac{x^3}{3} + \frac{x^5}{5} + \cdots \right)$$

This series converges for $-1 < x < 1$.

If N is a positive integer, we set $x = 1/(2N+1)$. Then $0 < x < 1$. A little computation shows that

$$\frac{1+x}{1-x} = \frac{N+1}{N}$$

By substituting in (11.61) and using some properties of logarithms, we get

(11.62)
$$\ln(N+1) = \ln N + 2\left[\frac{1}{2N+1} + \frac{1}{3}\left(\frac{1}{2N+1}\right)^3 + \frac{1}{5}\left(\frac{1}{2N+1}\right)^5 + \cdots \right]$$

This series converges quite rapidly for all positive integers N.

For example, if $N = 1$, we find

(11.63)
$$\ln 2 = 2\left[\frac{1}{3} + \frac{1}{3}\left(\frac{1}{3}\right)^3 + \frac{1}{5}\left(\frac{1}{3}\right)^5 + \cdots \right]$$

$$\approx 2\left[\frac{1}{3} + \frac{1}{3}\left(\frac{1}{3}\right)^3 + \frac{1}{5}\left(\frac{1}{3}\right)^5 \right] \approx 0.693004$$

The use of three terms of this series achieves accuracy to within 0.0002.

To get $\ln 3$, we set $N = 2$ in (11.62) and use (11.63). By continuing in this way, a table of natural logarithms may be written.

OTHER SERIES

We can use the series arrived at in this section to obtain other power series representations. For example, if in (11.54), we replace x by $-x$, we obtain

(11.64)
$$e^{-x} = 1 - x + \frac{x^2}{2!} - \frac{x^3}{3!} + \cdots + (-1)^n \frac{x^n}{n!} + \cdots \qquad \text{for all } x$$

Since

$$\cosh x = \frac{e^x + e^{-x}}{2}$$

its power series representation can be found by adding corresponding terms in (11.54) and (11.64) and then dividing by 2. The result is

(11.65)
$$\cosh x = 1 + \frac{x^2}{2!} + \frac{x^4}{4!} + \frac{x^6}{6!} + \cdots + \frac{x^{2n}}{(2n)!} + \cdots \qquad \text{for all } x$$

We can obtain the series representation for $e^x \cos x$ by multiplying the series (11.54) by the series (11.57). That is,

$$e^x \cos x = \left(1 + x + \frac{x^2}{2!} + \frac{x^3}{3!} + \frac{x^4}{4!} + \cdots \right)\left(1 - \frac{x^2}{2!} + \frac{x^4}{4!} - \cdots \right)$$

For the sake of illustration, we list the terms up to and including the fifth power:

$$e^x \cos x = 1\left(1 - \frac{x^2}{2!} + \frac{x^4}{4!}\right) + x\left(1 - \frac{x^2}{2!} + \frac{x^4}{4!}\right) + \frac{x^2}{2!}\left(1 - \frac{x^2}{2!}\right)$$

$$+ \frac{x^3}{3!}\left(1 - \frac{x^2}{2!}\right) + \frac{x^4}{4!}(1) + \frac{x^5}{5!}(1) + \cdots$$

$$= \left(1 - \frac{x^2}{2} + \frac{x^4}{24}\right) + \left(x - \frac{x^3}{2} + \frac{x^5}{24}\right) + \left(\frac{x^2}{2} - \frac{x^4}{4}\right)$$

$$+ \left(\frac{x^3}{6} - \frac{x^5}{12}\right) + \frac{x^4}{24} + \frac{x^5}{120} + \cdots$$

$$= 1 + x + \left(-\frac{1}{2} + \frac{1}{2}\right)x^2 + \left(-\frac{1}{2} + \frac{1}{6}\right)x^3$$

$$+ \left(\frac{1}{24} - \frac{1}{4} + \frac{1}{24}\right)x^4 + \left(\frac{1}{24} - \frac{1}{12} + \frac{1}{120}\right)x^5 + \cdots$$

$$= 1 + x - \frac{1}{3}x^3 - \frac{1}{6}x^4 - \frac{1}{30}x^5 + \cdots$$

NUMERICAL TECHNIQUES

As mentioned in Section 11 of Chapter 9, power series are sometimes useful for approximating definite integrals. Let's look at an illustration.

EXAMPLE 1   Using power series, approximate $\int_0^{1/2} e^{-x^2}\, dx$ accurate to within 0.001.

Solution   In (11.64), we replace x by x^2. The power series for e^{-x^2} is, therefore,

$$e^{-x^2} = 1 - x^2 + \frac{x^4}{2!} - \frac{x^6}{3!} + \frac{x^8}{4!} - \cdots$$

Then, by part (c) of (11.37),

$$\int_0^{1/2} e^{-x^2}\, dx = \int_0^{1/2}\left(1 - x^2 + \frac{x^4}{2!} - \frac{x^6}{3!} + \frac{x^8}{4!} - \cdots\right) dx$$

$$= \left(x - \frac{x^3}{3} + \frac{x^5}{2!5} - \frac{x^7}{3!7} + \cdots\right)\Bigg|_0^{1/2}$$

$$= \frac{1}{2} - \frac{1}{3(2)^3} + \frac{1}{2!5(2)^5} - \frac{1}{3!7(2)^7} + \cdots$$

$$= 0.5 - 0.041666 + 0.003125 - 0.00019 + \cdots$$

Since this is an alternating series, the error due to using the first three terms as an approximation is less than 0.00019 [see (11.34)]. Thus, by summing the first three terms, we have

$$\int_0^{1/2} e^{-x^2}\, dx \approx 0.46146$$

to within 0.001.   ∎

Note that only three terms were needed to obtain the desired accuracy. To get this same accuracy by Simpson's rule or the trapezoidal rule would have required a partition of $[0, \frac{1}{2}]$ so fine that even the use of a computer would be more time-consuming, and, of course, more costly than merely adding three terms. Thus, efficiency of technique is an important consideration for solving numerical problems.

We close this section with an illustration of one procedure for actually finding the sum of an infinite series.

EXAMPLE 2   Find the exact sum of the series below.

$$\frac{1}{1!3} + \frac{1}{2!4} + \frac{1}{3!5} + \cdots + \frac{1}{n!(n+2)} + \cdots$$

Solution   We start with

$$e^x = 1 + x + \frac{x^2}{2!} + \frac{x^3}{3!} + \cdots$$

We multiply this by x and integrate from 0 to 1 to get

$$\int_0^1 xe^x\, dx = \int_0^1 \left(x + x^2 + \frac{x^3}{2!} + \frac{x^4}{3!} + \cdots \right) dx$$

$$= \frac{1}{2} + \left(\frac{1}{1!3} + \frac{1}{2!4} + \frac{1}{3!5} + \cdots \right)$$

Since $\int_0^1 xe^x\, dx = 1$ (use integration by parts), we find that

$$\frac{1}{1!3} + \frac{1}{2!4} + \frac{1}{3!5} + \cdots + \frac{1}{n!(n+2)} + \cdots = 1 - \frac{1}{2} = \frac{1}{2} \qquad \blacksquare$$

EXERCISE 8

In Problems 1–8 find a Maclaurin series that converges to each function.

1. e^{2x} **2.** $\sin^3 x$ **3.** $\ln(1 + x^2)$ **4.** e^{-x^2}

5. $\sinh x$ **6.** $\sin x^2$ **7.** $e^x \sin x$ **8.** $e^{-x} \cos x$

In Problems 9 and 10 use a Maclaurin series for f to get one for g. The first four nonzero terms will be sufficient.

9. $f(x) = \dfrac{1}{\sqrt{1 - x^2}}$; $g(x) = \sin^{-1} x$ **10.** $f(x) = \tan x$; $g(x) = \ln(\cos x)$

In Problems 11–16 approximate each expression accurate to within 0.001 by using power series.

11. $\int_0^1 \sin x^2 \, dx$ **12.** $\int_0^1 e^{-x^2} \, dx$ **13.** $\cos 0.1$

14. $\sin 0.1$ **15.** $\int_0^1 \dfrac{\sin x}{x} \, dx$ **16.** $\int_0^{0.1} \dfrac{\ln(1 + x)}{x} \, dx$

In Problems 17–20 find each limit by means of series.

17. $\lim\limits_{x \to 0} \dfrac{\sin x - x}{x^2}$ **18.** $\lim\limits_{x \to 0} \dfrac{\sin x}{x}$ **19.** $\lim\limits_{x \to 0} \dfrac{x \sin x}{1 - \cos x}$ **20.** $\lim\limits_{h \to 0} \dfrac{e^h - 1}{h}$

21. Find the exact sum of the infinite series below.

$$\frac{x^3}{1(3)} - \frac{x^5}{3(5)} + \frac{x^7}{5(7)} - \frac{x^9}{7(9)} + \cdots \qquad \text{for} \quad x = 1$$

22. Obtain the series (11.57) for $\cos x$ by integrating the series (11.56) for $\sin x$.

23. Obtain the series (11.57) for $\cos x$ in the same way as that of the series (11.56) for $\sin x$.

24. Use the result (11.62) to show that $\ln 3 \approx 1.09861$.

9. The Binomial Series

In algebra we study the binomial theorem, which states that if m is a positive integer and a, b are real numbers, then

$$(a + b)^m = a^m + m a^{m-1} b + \cdots + b^m = \sum_{k=0}^{m} \binom{m}{k} a^{m-k} b^k$$

where

$$\binom{m}{k} = \frac{m(m - 1) \cdots (m - k + 1)}{k!}$$

In particular, for $a = 1$ and $b = x$,   this becomes

(11.66) $(1 + x)^m = 1 + mx + \dfrac{m(m - 1)}{2} x^2 + \cdots + mx^{m-1} + x^m$ m a positive integer

In this section we obtain a Maclaurin series for the function $f(x) = (1 + x)^m$,   where m *is any real number.*

We begin by investigating the derivatives of f at 0:

$$f(x) = (1 + x)^m \qquad\qquad f(0) = 1$$
$$f'(x) = m(1 + x)^{m-1} \qquad\qquad f'(0) = m$$
$$f''(x) = m(m - 1)(1 + x)^{m-2} \qquad\qquad f''(0) = m(m - 1)$$
$$\vdots \qquad\qquad\qquad\qquad\qquad \vdots$$
$$f^{(n)}(x) = m(m - 1)(m - 2) \cdots \qquad\qquad f^{(n)}(0) = m(m - 1)(m - 2) \cdots$$
$$(m - n + 1)(1 + x)^{m-n} \qquad\qquad (m - n + 1)$$

The Maclaurin series is

(11.67)
$$(1 + x)^m = 1 + mx + \frac{m(m-1)}{2!} x^2 + \cdots + \binom{m}{n} x^n + \cdots = \sum_{k=0}^{+\infty} \binom{m}{k} x^k$$

where

(11.68)
$$\binom{m}{k} = \frac{m(m-1)\cdots(m-k+1)}{k!} \qquad k \neq 0*$$

Prompted by (11.66), the series given by (11.67) is called the *binomial series*. The expression in (11.68) is called the *binomial coefficient of* x^k.

The following result, which we state without proof, gives the conditions under which the infinite series (11.67) actually converges to $(1 + x)^m$:

(11.69)
$$(1 + x)^m = \sum_{k=0}^{+\infty} \binom{m}{k} x^k$$

(a)   for all x if m is a nonnegative integer

(b)   for $-1 < x < 1$ if $m \leq -1$

(c)   for $-1 < x \leq 1$ if $-1 < m < 0$

(d)   for $-1 \leq x \leq 1$ if $m > 0$, m not an integer

EXAMPLE 1   Represent $\sqrt{1 + x}$ in a Maclaurin series and find the interval of convergence.

Solution   We use the binomial series (11.69) with $m = \frac{1}{2}$.   The result is

$$(1 + x)^{1/2} = 1 + \frac{1}{2} x + \frac{(\frac{1}{2})(-\frac{1}{2})}{2!} x^2 + \frac{\frac{1}{2}(-\frac{1}{2})(-\frac{3}{2})}{3!} x^3 + \cdots$$

By part (d) of (11.69), the series converges for $-1 \leq x \leq 1$.   ■

EXAMPLE 2   Represent the function $f(x) = \sin^{-1} x$ by a Maclaurin series and find the interval of convergence.

Solution   We observe that

$$\sin^{-1} x = \int_0^x \frac{dt}{\sqrt{1 - t^2}} = \int_0^x (1 - t^2)^{-1/2} \, dt$$

By expanding the integrand in a binomial series, we find

$$(1 - t^2)^{-1/2} = 1 + \frac{t^2}{2} + \left(\frac{1}{2}\right)\left(\frac{3}{2}\right)\left(\frac{t^4}{2!}\right) + \left(\frac{1}{2}\right)\left(\frac{3}{2}\right)\left(\frac{5}{2}\right)\left(\frac{t^6}{3!}\right) + \cdots$$

---

* We define $\binom{m}{0} = 1$ for any m.

which is valid for $-1 < t \le 1$.   Integrating the right side term-by-term, we obtain

$$\sin^{-1}x = x + \left(\frac{1}{2}\right)\left(\frac{x^3}{3}\right) + \left(\frac{1}{2}\right)\left(\frac{3}{4}\right)\left(\frac{x^5}{5}\right) + \left(\frac{1}{2}\right)\left(\frac{3}{4}\right)\left(\frac{5}{6}\right)\left(\frac{x^7}{7}\right) + \cdots \qquad -1 < x \le 1$$

∎

Table 5   Important Series

Name	Description	Comment
Taylor series (11.49)	$f(x) = \sum\limits_{k=0}^{n} \dfrac{f^{(k)}(c)(x-c)^k}{k!} + R_{n+1}(x)$	Converges to $f(x)$ if $\lim_{n \to +\infty} R_{n+1}(x) = 0$
Remainder formula (11.50)	$R_{n+1}(x) = \dfrac{f^{(n+1)}(u)(x-c)^{n+1}}{(n+1)!}$	u between c and x
$\dfrac{1}{1-x}$ (11.38)	$\dfrac{1}{1-x} = \sum\limits_{k=0}^{+\infty} x^k$	Converges to $1/(1-x)$ for $-1 < x < 1$
e^x (11.51)	$e^x = \sum\limits_{k=0}^{+\infty} \dfrac{x^k}{k!}$	Converges to e^x for all x
$\sin x$ (11.56)	$\sin x = \sum\limits_{k=0}^{+\infty} \dfrac{(-1)^k x^{2k+1}}{(2k+1)!}$	Converges to $\sin x$ for all x
$\cos x$ (11.57)	$\cos x = \sum\limits_{k=0}^{+\infty} \dfrac{(-1)^k x^{2k}}{(2k)!}$	Converges to $\cos x$ for all x
$\ln(1+x)$ (11.60)	$\ln(1+x) = \sum\limits_{k=0}^{+\infty} \dfrac{(-1)^k x^{k+1}}{k+1}$	Converges to $\ln(1+x)$ for $-1 < x \le 1$
$\ln(N+1)$ (11.62)	$\ln(N+1) = \ln N + 2 \sum\limits_{k=1}^{+\infty} \dfrac{1}{2k-1}\left(\dfrac{1}{2N+1}\right)^{2k-1}$	N is a positive integer; converges for all N
$(1+x)^m$ (11.69)	$(1+x)^m = \sum\limits_{k=0}^{+\infty} \binom{m}{k} x^k$	For convergence information, see (11.69)

EXERCISE 9

In Problems 1–6 use a binomial series to represent each function. Determine the interval of convergence.

1. $\sqrt{1+x^2}$ **2.** $\dfrac{1}{\sqrt{1-x}}$ **3.** $(1+x)^{1/5}$ **4.** $\dfrac{1}{1+x^2}$

5. $\dfrac{1}{(1+x)^{3/4}}$ **6.** $\dfrac{2x}{\sqrt{1-x}}$

In Problems 7–10 approximate each definite integral accurate to within 0.001.

7. $\displaystyle\int_0^{0.2} \sqrt[3]{1+x^4}\,dx$ **8.** $\displaystyle\int_0^{1/2} \sqrt[3]{1+x}\,dx$ **9.** $\displaystyle\int_0^{1/2} \dfrac{1}{\sqrt[3]{1+x^2}}\,dx$ **10.** $\displaystyle\int_0^{1/2} \dfrac{1}{\sqrt{1+x^3}}\,dx$

11. Show that $[1 + (1/n)]^n < e$ for all $n > 0$.

12. Prove that $(1 + x)^m = \sum_{k=0}^{+\infty} \binom{m}{k} x^k$ when m is a nonnegative integer by showing that $R_{n+1}(x) \to 0$ as $n \to +\infty$.

13. Show that the series $\sum_{k=0}^{+\infty} \binom{m}{k} x^k$ converges absolutely for $|x| < 1$ and diverges for $|x| > 1$. [*Hint:* Use the ratio test (11.29).]

Miscellaneous Exercises

1. One of the earliest uses of the comparison test was made by Jakob Bernoulli to comment on the series $1 + 1/\sqrt{2} + 1/\sqrt{3} + \cdots$. Reproduce his conclusion!

In Problems 2–4 you are asked to consider an incorrect argument given by Jakob Bernoulli to prove that $\frac{1}{2} + \frac{1}{6} + \frac{1}{12} + \cdots = 1$. Bernoulli's argument went as follows: Let $N = 1 + \frac{1}{2} + \frac{1}{3} + \frac{1}{4} + \cdots$. Then $N - 1 = \frac{1}{2} + \frac{1}{3} + \frac{1}{4} + \frac{1}{5} + \cdots$. Now, subtract term-by-term to get $N - (N - 1) = (1 - \frac{1}{2}) + (\frac{1}{2} - \frac{1}{3}) + (\frac{1}{3} - \frac{1}{4}) + \cdots$, or $1 = \frac{1}{2} + \frac{1}{6} + \frac{1}{12} + \cdots$.

2. What is wrong with Bernoulli's argument?

3. In general, what can be said about the convergence or divergence of a series formed by taking the term-by-term difference (or sum) of two divergent series? Support your answer with examples.

4. While the method is wrong, Bernoulli's conclusion is correct. That is, it is true that $1 = \frac{1}{2} + \frac{1}{6} + \frac{1}{12} + \cdots$. Prove it! [*Hint:* Look at the partial sums using the form that the series was in just after our term-by-term subtraction above.]

5. Euler believed that $\frac{1}{2} = 1 - 1 + 1 - 1 + 1 - 1 + \cdots$. He based his argument to support this equation on his belief in the identification of a series and the values of the function from which it was derived.
(a) Write the Maclaurin series expansion for $1/(1 + x)$. Do this without calculating any derivatives.
(b) Evaluate both sides of the equation you derived in part (a) at $x = 1$ to arrive at the formula above.
(c) Criticize the procedure in part (b).

An interesting relationship between the nth partial sum of the harmonic series and $\ln n$ was discovered by Euler. In particular, he showed that

$$\gamma = \lim_{n \to +\infty} \left(1 + \frac{1}{2} + \frac{1}{3} + \cdots + \frac{1}{n} - \ln n \right)$$

exists and is approximately equal to 0.5772. *Euler's number*, as γ is called, appears in many interesting areas of mathematics. For example, it is involved in the evaluation of the exponential integral, $\int_x^{+\infty} (e^{-t}/t) \, dt$, which is important in applied mathematics. It is also related to two special functions—the *gamma function* and *Riemann's zeta function* (see Problem 9). Surprisingly, it is still unknown whether Euler's number is rational or irrational. Problems 6–8 concern Euler's number.

6. The harmonic series diverges quite slowly. For example, the partial sums S_{10}, S_{20}, S_{50}, and S_{100} have approximate values 2.92897, 3.59774, 4.49921, and 5.18738, respectively. In fact, the sum of the first million terms of the harmonic series is about 14.4. With this in mind, what would you conjecture about the rate of convergence of the limit defining γ? Test your conjecture by calculating approximate values for γ by using the partial sums given above.

*7. Let $C_n = 1 + \frac{1}{2} + \cdots + (1/n) - \ln n$.

(a)   By consulting the figure, show that if $f(x) = 1/x$,   then

$$\int_1^n f(x)\, dx \le f(1) + \cdots + f(n-1)$$

Show further that this reduces to $0 < 1/n \le C_n$.

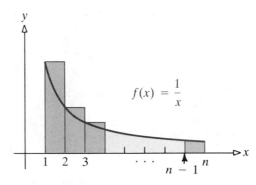

(b)   By consulting the figure, show that if $f(x) = 1/x$,   then

$$\int_n^{n+1} f(x)\, dx > f(n+1)$$

Compute $C_n - C_{n+1}$ and deduce that $\{C_n\}$ is a decreasing sequence bounded from below by 0, and, thus, that it converges to Euler's number γ.

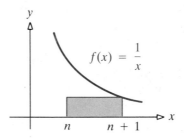

(c)   Since $\lim_{n \to +\infty} C_n = \gamma$,   we have the approximation

$$S_n = 1 + \frac{1}{2} + \frac{1}{3} + \cdots + \frac{1}{n} \approx \gamma + \ln n$$

In fact, this approximation is very good. It can be used to estimate the growth of S_n. Suppose we want to find the smallest n for which $S_n \ge M$.   We can solve $M = \gamma + \ln n$ for n and round the result to an integer. Solve this for $M = 3$ and $M = 4$.   Repeat this for $M = 10$ and $M = 20$.

8. Use the value of Euler's number given above and your calculator to estimate

$$1 + \frac{1}{2} + \frac{1}{3} + \cdots + \frac{1}{1{,}000{,}000{,}000}$$

---

* Based on an article by R. Boas, *American Mathematics Monthly* (April 1977).

9. *Riemann's zeta function* is defined as

$$\zeta(s) = 1 + \frac{1}{2^s} + \frac{1}{3^s} + \cdots \qquad \text{for } s > 1$$

As mentioned in the footnote on page 602, Euler showed that $\zeta(2) = \pi^2/6$. He also found the value of the zeta function for many other even values of s. As of now, no one knows the value of the zeta function for odd values of s. However, it is not too difficult to approximate these values, as this problem demonstrates.
(a)   Use your calculator to find $\sum_{k=1}^{10}(1/k^3)$.
(b)   Use integrals, in a way analogous to their use in the integral test, to set upper and lower bounds on $\sum_{k=1}^{+\infty}(1/k^3)$.
(c)   What can you conclude about $\zeta(3)$?

10. Leibniz derived the following formula for $\pi/4$: $\pi/4 = 1 - \frac{1}{3} + \frac{1}{5} - \frac{1}{7} + \frac{1}{9} - \cdots$.

(a)   Evaluate: $\int_0^1 \frac{1}{1 + x^2}\, dx$
(b)   Expand the integrand in part (a) into a power series and integrate it term-by-term to get Leibniz's formula.
(c)   Find the sum of the first 10 terms in the above series. Does it appear that Leibniz's formula is a useful one for approximating π?

11. Show that: $\displaystyle\sum_{k=1}^{+\infty} \frac{1}{k(k+2)} = \frac{3}{4}$

12. Show that: $\displaystyle\sum_{k=1}^{+\infty} \frac{1}{k(k+1)(k+2)} = \frac{1}{4}$

13. Show that: $\displaystyle\sum_{k=1}^{+\infty} \frac{1}{k(k+1)(k+2)(k+3)} = \frac{1}{18}$

14. Show that: $\displaystyle\sum_{k=1}^{+\infty} \frac{1}{k(k+1)(k+2)\cdots(k+\alpha)} = \frac{1}{\alpha}\left(\frac{1}{\alpha!}\right)$ $\alpha \geq 1$, an integer

15. Show that: $\displaystyle\sum_{k=1}^{+\infty} \frac{k}{(k+1)!} = 1$

16. Find the interval of convergence of $\sum_{k=1}^{+\infty} k^x$.

17. Determine whether the following series is convergent: $\dfrac{1}{3} - \dfrac{2^3}{3^2} + \dfrac{3^3}{3^3} - \dfrac{4^3}{3^4} + \cdots + \dfrac{(-1)^{n-1}n^3}{3^n} + \cdots$

18. Find the interval of convergence of the series $\sum_{k=1}^{+\infty}[(x-2)^k/k(3^k)]$.

19. Show that the following series converges: $1 + \dfrac{2}{2^2} + \dfrac{3}{3^3} + \dfrac{1}{4^4} + \dfrac{2}{5^5} + \dfrac{3}{6^6} + \cdots$

20. Show that the series $\sum_{k=1}^{+\infty}(1/k^{1+1/k})$ diverges.

In Problems 21–29 determine whether the given series converges or diverges.

21. $\displaystyle\sum_{k=1}^{+\infty} \frac{k^2 - 3k - 2}{k^2(k+1)^2}$;   find the sum of the series if it converges.

22. $1 - 1 - \frac{1}{2} + \frac{1}{3} + \frac{1}{3} - \frac{1}{9} - \frac{1}{4} + \frac{1}{27} + \frac{1}{5} - \frac{1}{81} - \cdots$; find the sum of the series if it converges.

23. $\displaystyle\sum_{k=2}^{+\infty} \ln \frac{k}{k+1}$

24. $\displaystyle\sum_{k=2}^{+\infty} \frac{\ln(2k+1)}{\sqrt{k^2-2}\sqrt{k^3-2k-3}}$

25. $\displaystyle\sum_{k=1}^{+\infty} \frac{\sqrt{k}}{\sqrt{k^3-k+1}\,\ln(2k+1)}$

26. $\displaystyle\sum_{k=1}^{+\infty} \frac{1}{\cosh^2 k}$

27. $\displaystyle\sum_{k=1}^{+\infty} \frac{\tan^{-1}k}{k^2}$

28. $\displaystyle\sum_{k=2}^{+\infty} \frac{1}{\sqrt{k}(\ln k)^4}$

29. $\displaystyle\sum_{k=2}^{+\infty} \frac{(\ln k)^2}{k^{5/2}}$

30. Let $a_k = (-1)^{k+1} \int_0^{\pi/k} \sin kx \, dx$.
   (a) Evaluate a_k.
   (b) Show that the infinite series $\sum_{k=1}^{+\infty} a_k$ converges.
   (c) Show that $1 \le \sum_{k=1}^{+\infty} a_k \le \frac{3}{2}$.

31. Consider the finite sum $S_n = \sum_{k=1}^{n}[1/(1+k^2)]$.
   (a) By comparing S_n with an appropriate integral, prove that $S_n \le \tan^{-1}n$ for $n \ge 1$.
   (b) Use part (a) to deduce that $\sum_{k=1}^{+\infty}[1/(1+k^2)]$ exists.
   (c) Prove that $\pi/4 \le \sum_{k=1}^{+\infty}[1/(1+k^2)] \le \pi/2$.

32. Let $s_n = 1/1! + 1/2! + \cdots + 1/n!,\quad n = 1, 2, 3, \ldots$.
   (a) Show that $n! \ge 2^{n-1}$.
   (b) Show that $0 < s_n \le 1 + \frac{1}{2} + (\frac{1}{2})^2 + \cdots + (\frac{1}{2})^{n-1}$.
   (c) Show that $0 < s_n < s_{n+1} < 2$. Hence, conclude that there exists $S = \lim_{n \to +\infty} s_n \le 2$.
   (d) Let $t_n = [1 + (1/n)]^n$. Show that $t_n = 1 + 1 + (1/2!)[1 - (1/n)] + (1/3!)[1 - (1/n)][1 - (2/n)] + \cdots$
        $+ (1/n!)[1 - (1/n)][1 - (2/n)] \cdots [1 - (n-1)/n] < s_n + 1$.
   (e) Show that $0 < t_n < t_{n+1} < 3$. Hence, conclude that there exists $e = \lim_{n \to +\infty} t_n \le 3$.

33. Let $\sum_{n=1}^{+\infty}[P(n)/Q(n)]$ be a series where P and Q are polynomials of degree r and s, respectively. Prove that the series converges if $r < s - 1$ and diverges if $r \ge s - 1$.

34. From the fact that $\sin t \le t$ for all $t \ge 0$, use integration repeatedly to prove

$$1 - \frac{x^2}{2!} \le \cos x \le 1 - \frac{x^2}{2!} + \frac{x^4}{4!} \qquad \text{for all} \quad x \ge 0$$

35. Find an elementary expression for $\sum_{k=1}^{+\infty}[x^{k+1}/k(k+1)]$. [*Hint:* Integrate the series for $\ln[1/(1-x)]$.]

36. Let x be a fixed positive number and define a sequence by $a_{n+1} = (\frac{1}{2})[a_n + (x/a_n)]$, where a_1 is chosen an arbitrary positive number.
   (a) Show that this sequence converges to a limit $\sqrt{x}$.
   (b) *Calculator problem.* Use this sequence to approximate $\sqrt{28}$. How accurate is a_3? a_6?

37. Use series to approximate $\int_0^{0.16}[(\sin x)/\sqrt{x}]\,dx$ to within 10^{-3}.

38. Prove that $\{(3^n + 5^n)^{1/n}\}$ converges.

39. (a) Write the first three nonzero terms and the general term of the Taylor series expansion about $x = 0$ of $f(x) = 3 \sin(x/2)$.
   (b) What is the interval of convergence for the series found in part (a)?
   (c) What is the minimum number of terms of the series in part (a) that are *necessary* to approximate f on the interval $(-2, 2)$ with an error not exceeding 0.1?

40. The integral $\int [(\sin x)/x]\,dx$ cannot be expressed in terms of functions known to us. Show that

$$\int_0^x \frac{\sin t}{t}\,dt = x - \frac{x^3}{3(3!)} + \frac{x^5}{5(5!)} - \cdots$$

For which x is this valid? Find the value of the integral accurate to within 0.0001 when $x = 0.01$.

*41. (a) By considering graphs like those shown earlier in Problem 7, show that if f is decreasing, positive, and continuous, then

$$f(n + 1) + \cdots + f(m) \le \int_n^m f(x)\, dx \le f(n) + \cdots + f(m - 1)$$

(b) Under the assumption of part (a), prove that if $\sum_{k=1}^{+\infty} f(k)$ converges, then

$$\sum_{k=n+1}^{+\infty} f(k) \le \int_n^{+\infty} f(x)\, dx \le \sum_{k=n}^{+\infty} f(k)$$

(c) Let $f(x) = 1/x^2$. Use the inequality in part (b) to determine *exactly* how many terms of the series $\sum_{k=1}^{+\infty}(1/k^2)$ one must take in order to have $|\text{Error}| < (\frac{1}{2})10^{-2}$. To have $|\text{Error}| < (\frac{1}{2})10^{-10}$.

†42. Let $\{a_n\}$ be a sequence that decreases to 0. Define

$$R_n = \sum_{k=n+1}^{+\infty} (-1)^{k+1} a_k \qquad \Delta a_k = a_k - a_{k+1}$$

Suppose that the sequence $\{\Delta a_k\}$ decreases.

(a) Show that the series $\sum_{k=1}^{+\infty}(-1)^{k+1}\Delta a_k$ is a convergent alternating series.

(b) Derive

$$|R_n| = \frac{a_n}{2} + \frac{1}{2}\sum_{k=1}^{+\infty}(-1)^k \Delta a_{k+n-1}$$

Show that $\sum_{k=1}^{+\infty}(-1)^k \Delta a_{k+n-1} < 0$. Deduce $|R_n| < \frac{1}{2}a_n$.

(c) Derive

$$|R_n| = \frac{a_{n+1}}{2} + \frac{1}{2}\sum_{k=1}^{+\infty}(-1)^{k+1}\Delta a_{k+n}$$

Deduce $a_{n+1}/2 < |R_n|$.

---

* Based on an article by R. Boas, *American Mathematics Monthly* (April 1977).

† Based on R. Johnsonbaugh, "Summing an Alternating Series," *American Mathematics Monthly* (1979).

12

Conics

1. Introduction

The word *conic* is derived from the word *cone*. A cone can be generated as follows: Let *a* and *g* be two distinct lines that intersect at a point *V*. Rotate the line *g* around the line *a*, maintaining the angle between *a* and *g*. The resulting surface is called a *cone*. The line *a* is called the *axis* of the cone, the line *g* is called a *generator* of the cone. The point *V* where the axis and generator intersect is called the *vertex* of the cone. See Figure 1.

The curves that result by intersecting a cone with a plane are called *conic sections*, or, simply, *conics*. The most interesting conics occur when the intersecting plane does *not* pass through the vertex. These conics are either *circles* (the intersecting plane is perpendicular to the axis of the cone), *ellipses* (the plane is tipped from the perpendicular position), *hyperbolas* (the plane is parallel to the axis), or *parabolas* (the plane is parallel to a generator). Figure 2 shows all these possibilities.

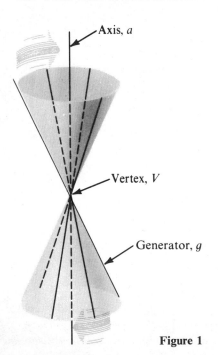

Axis, *a*

Vertex, *V*

Generator, *g*

Figure 1

646

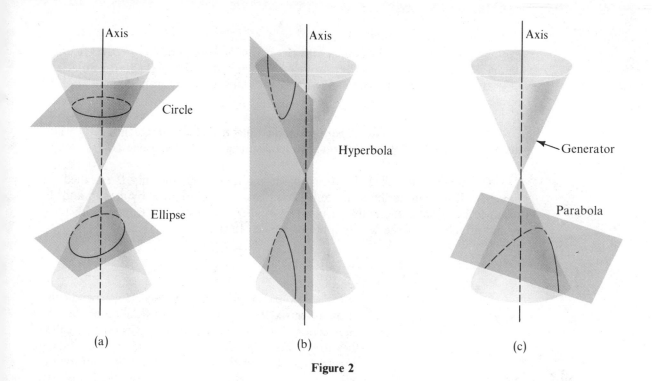

Figure 2

Conics of less interest occur when the intersecting plane passes through the vertex of the cone. These conics, which may be either a point, a line, or a pair of intersecting lines, are called *degenerate conics*.

Historically, many of the interesting properties of conics were discovered and studied by Apollonius around 200 BC. Because they have many current uses, conics are still studied. For example, parabolic reflectors are used in radar, solar energy devices, and reflecting telescopes. Artificial satellites circle the earth in approximately elliptical orbits. Elliptic surfaces are useful for reflecting light beams and sound waves from one place to another. Hyperbolic curves are sometimes used to locate the positions of ships at sea.

Apollonius used the methods of *euclidean geometry* to study conics, but we shall use the more powerful methods of *analytic geometry*. First, we define the conics in terms of the distances from certain fixed lines and points. Then we use cartesian coordinates and algebra to derive the equations of these conics.

We could use many words to describe the shape, size, and position of a conic in the xy-plane. But, to paraphrase an old saying, "An equation is worth a thousand words." As this chapter shows, the equation of a conic contains *complete* information concerning the shape and position of that conic in the xy-plane.

2. The Parabola

We begin our study of the conics with the *parabola*. The parabola occurs naturally as the path of a projectile (such as a thrown ball or an artillery shell). It also occurs in technical applications such as the parabolic reflector, the parabolic arch, and the parabolic suspension cable.

(12.1)   **DEFINITION**   *Parabola.*   **A *parabola* is the collection of all points *P* in a plane whose distance from a fixed point *F* is equal to its distance from a fixed line *D*.**

The point *F* is called the *focus* of the parabola, and the line *D* is called the *directrix*. If we denote the distance between two points *F* and *P* by $d(F, P)$ and the distance between a line *D* and a point *P* by $d(D, P)$, then the parabola with focus *F* and directrix *D* is the set of all points *P* such that

(12.2) $$d(F, P) = d(D, P)$$

See Figure 3.

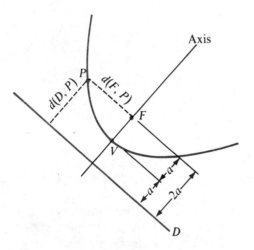

Figure 3

The line passing through the focus *F* perpendicular to the directrix *D* is called the *axis* of the parabola. The intersection *V* of the axis with the parabola is called the *vertex*. This point is midway between the focus and the directrix and lies on the parabola. Usually (and for the purpose of this section), the distance from the focus *F* to the directrix *D* is denoted by $2a$, where *a* is a positive real number. As a result, the distance from the focus to the vertex of a parabola is *a* (refer to Fig. 3).

To obtain the equation of a parabola, we position the axes of our coordinate systems so that the vertex *V*, focus *F*, and directrix *D* are conveniently located. A convenient position for the vertex *V* is $(0, 0)$. Then the focus *F* can be placed on the *x*-axis or the *y*-axis.

Suppose we begin by placing *F* on the *positive x*-axis. Since the distance from the focus to the vertex is *a*, the focus *F* is located at $(a, 0)$, $a > 0$.   Also, because the vertex *V* is equidistant from *F* and the directrix, the directrix must be given by $x = -a$. If $P = (x, y)$ is any point on the parabola, then by equation (12.2), $d(F, P) = d(D, P)$,   and by using the distance formula, we find

$$\sqrt{(x - a)^2 + y^2} = |x + a|$$

By squaring both sides, we obtain

$$(x - a)^2 + y^2 = (x + a)^2$$
$$x^2 - 2ax + a^2 + y^2 = x^2 + 2ax + a^2$$
$$y^2 = 4ax$$

Equation of a Parabola.   **The equation of a parabola with vertex at (0, 0), focus at (a, 0), and directrix   x = −a,   a > 0,   is**

(12.3) $$y^2 = 4ax$$

Conversely, any equation of the form (12.3) is a parabola with vertex at (0, 0), focus at $(a, 0)$, and directrix $x = -a, \quad a > 0$.

EXAMPLE 1   (a)   The parabola with vertex at (0, 0) and focus at (2, 0) has the equation $y^2 = 8x$.
(b)   The equation $y^2 = 16x$ represents a parabola with vertex (0, 0), focus (4, 0), and directrix $x = -4$.   ∎

EXAMPLE 2   Find the equation of a parabola with vertex at the origin, focus at (3, 0), and directrix the line $x = -3$.   Graph the parabola.

Solution   *Method I [by definition (12.1)]:*

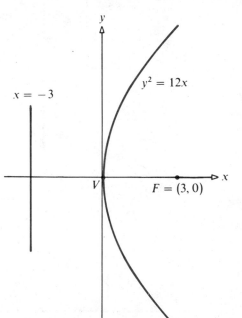

$$d(F, P) = d(D, P)$$
$$\sqrt{(x - 3)^2 + y^2} = |x + 3|$$
$$(x - 3)^2 + y^2 = (x + 3)^2$$
$$y^2 = 12x$$

Method II (by formula):   Here, $a = 3$ and the parabola is of the type described by (12.3). Thus, the form of the equation is

$$y^2 = 4ax$$

With $a = 3$,

$$y^2 = 12x$$

See Figure 4 for the graph.   ∎

Figure 4

The parabola $y^2 = 4ax$ is symmetric with respect to the x-axis since, when y is replaced by $-y$, the relationship between x and y is unchanged.

If the focus F is located on the negative x-axis, negative y-axis, or positive y-axis, a different parabola and, hence, a different equation results. The four possible parabolas with vertex at (0, 0) and focus on a coordinate axis are illustrated in Figure 5. In each case, the distance from the focus to the vertex is denoted by a, where a is positive. The four equations are given in Table 1.

Table 1

Equation	Vertex	Focus	Directrix	Graph	Symmetry
$y^2 = 4ax$	$(0, 0)$	$(a, 0)$	$x = -a$	Figure 5(a)	x-axis
$y^2 = -4ax$	$(0, 0)$	$(-a, 0)$	$x = a$	Figure 5(b)	x-axis
$x^2 = 4ay$	$(0, 0)$	$(0, a)$	$y = -a$	Figure 5(c)	y-axis
$x^2 = -4ay$	$(0, 0)$	$(0, -a)$	$y = a$	Figure 5(d)	y-axis

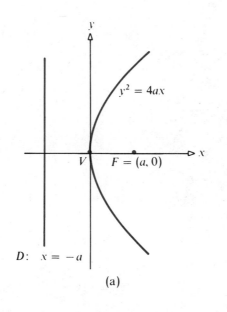

(a)

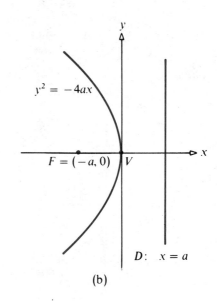

(b)

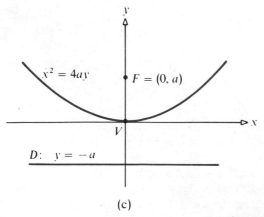

(c)

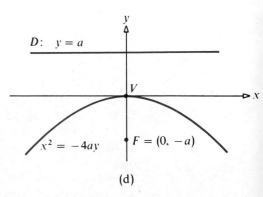

(d)

Figure 5

EXAMPLE 3   Find the vertex, focus, and directrix of the parabola $x^2 = -16y$.

Solution   First, we observe that this type of parabola has vertex at $(0, 0)$, focus on the negative y-axis, and directix parallel to the x-axis. By comparing the given equation $x^2 = -16y$ to $x^2 = -4ay$, we find that $a = 4$. Thus, the focus is at $(0, -4)$ and the directix is $y = 4$. See Figure 6 for a graph.

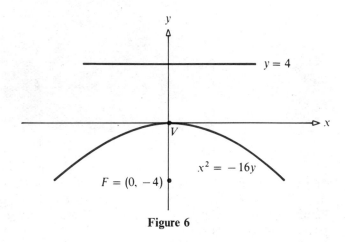

Figure 6   ■

EXAMPLE 4   Find an equation for the tangent line to the parabola $y^2 = -8x$ at the point $(-2, 4)$.

Solution   The slope of the tangent line is obtained by computing the derivative y'. By implicit differentiation, we find

$$2yy' = -8$$

$$y' = \frac{-4}{y}$$

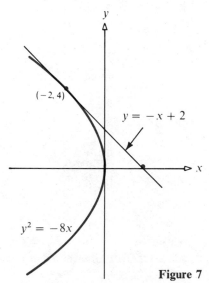

$y = -x + 2$

$y^2 = -8x$

Figure 7

At $y = 4$, we find $y' = -1$. The slope of the tangent line to $y^2 = -8x$ at $(-2, 4)$ is therefore -1 and an equation of this tangent line is

$$y - 4 = (-1)(x + 2)$$

$$y = -x + 2$$

See Figure 7.   ■

SUSPENSION PROPERTY

In a suspension bridge the main cables are of parabolic shape. The reason for this is that if the total weight of a bridge is uniformly distributed along its length, the only cable shape that will bear the load evenly is that of a parabola.

EXAMPLE 5   A suspension bridge with weight uniformly distributed along its length has twin towers that extend 100 meters above the road surface and are 400 meters apart. The cables are parabolic in shape and are tangent to the road surface at the center of the bridge. Find the height of the cables at a point 100 meters from the center. (The road is assumed horizontal.)

Solution   We begin by drawing Figure 8. The cable has the shape of a parabola with vertex at $(0, 0)$ and focus along the positive y-axis. The equation of this parabola is $x^2 = 4ay$.   Since the cable is 100 meters high when $x = 200$,   we find

$$(200)(200) = 4a(100)$$
$$a = 100$$

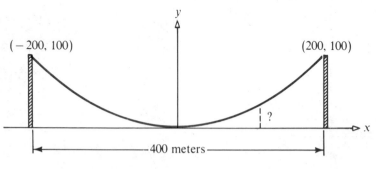

Figure 8

To find the height of the cable when $x = 100$,   we solve for y in $x^2 = 400y$, obtaining $y = 25$.   The cable is 25 meters high at 100 meters from the center of the bridge.   ■

REFLECTING PROPERTY

Another important property of a parabola is its application to reflecting. Suppose a mirror has the shape of a *paraboloid of revolution*, that is, a surface formed by rotating a parabola about its axis. If a light is placed at the focus of the mirror, all the rays emanating from it will be reflected off the mirror in lines parallel to the axis. This property is the principle behind the design of automobile headlights, where the bulb is placed at the focus (see Fig. 9).

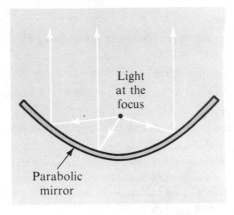

Figure 9

Conversely, when rays of light emanating from a distant source (like the sun) strike a parabolic mirror, they will be reflected to a single point—the focus. This latter property is the principle behind some solar energy devices. You are asked to prove this property in Problem 38 of Exercise 2.

EXERCISE 2

In Problems 1–16 find the equation of the parabola(s) having the stated properties.

1. Focus at $(1, 0)$;   vertex at $(0, 0)$

2. Focus at $(0, 2)$;   vertex at $(0, 0)$

3. Focus at $(-2, 0)$;   vertex at $(0, 0)$

4. Focus at $(-3, 0)$;   vertex at $(0, 0)$

5. Focus at $(2, 0)$;   directrix $x = -2$

6. Focus at $(0, -3)$;   directrix $y = 3$

7. Focus at $(0, 3)$;   directrix $y = -3$

8. Focus at $(-4, 0)$;   directrix $x = 4$

9. Vertex at $(0, 0)$;   directrix $x = 2$

10. Vertex at $(0, 0)$;   directrix $y = 3$

11. Vertex at $(0, 0)$; $a = 3$;   directrix to the right of $(0, 0)$ and parallel to the y-axis

12. Vertex at $(0, 0)$; $a = 4$;   directrix below $(0, 0)$ and parallel to the x-axis

13. Vertex at $(0, 0)$;   axis coinciding with the x-axis;   containing the point $(2, 8)$

14. Vertex at $(0, 0)$;   axis coinciding with the y-axis;   containing the point $(4, -4)$

15. Vertex at $(0, 0)$; $a = 3$;   axis coinciding with the negative x-axis

16. Vertex at $(0, 0)$; $a = 2$;   axis coinciding with the negative y-axis

In Problems 17–24 find the vertex, focus, and directrix of each parabola. Graph each parabola.

17. $y^2 = 8x$ **18.** $x^2 = 4y$ **19.** $x^2 = -12y$ **20.** $y^2 = -4x$

21. $y^2 = -16x$ **22.** $x^2 = 16y$ **23.** $x^2 = 8y$ **24.** $y^2 = 8x$

25. Find an equation of the tangent line to the parabola $x^2 = -8y$ at the point $(4, -2)$.

26. Find an equation of the tangent line to the parabola $y^2 = -12x$ at the point $(-3, 6)$.

27. Find the area of the region in the first quadrant enclosed by the parabola $x^2 = 8y$,   the y-axis, and the line $y = 2$.

28. Find the area of the region enclosed by the parabola $y^2 = 4x$,   the x-axis, and the line $x = 4$.

29. Find the volume of the solid of revolution obtained by rotating the area described in Problem 27 about the x-axis.

30. Find the volume of the solid of revolution obtained by rotating the area described in Problem 28 about the y-axis.

31. A suspension bridge with weight uniformly distributed along its length has twin towers that extend 75 meters above the road surface and are 300 meters apart. The cables are parabolic in shape and are suspended from the tops of the towers. The cables are tangent to the road surface at the center of the bridge. Find the height of the cables at a point 80 meters from the center. Assume the road is horizontal.

32. A parabolic arch has a span of 120 feet and a maximum height of 25 feet. Choose suitable rectangular axes and find the equation of the parabola. Then calculate the height of the arch at points 10 feet, 20 feet, and 40 feet from the center.

33. A radar antenna is constructed so that any cross section through its axis is a parabola. Suppose the receiver is located at the focus. Find the location of this receiver if the antenna is 5 feet across at the opening and is 1 foot deep.

34. If a parabolic reflector has a diameter of 8 inches and is 6 inches deep, how far from the vertex of the parabola should a light be placed so that the rays may be reflected parallel to the axis?

35. Show that the area of a parabolic segment cut off by a chord perpendicular to the axis of the parabola is equal to two-thirds of the area of the circumscribed rectangle.

36. A trough 10 feet long with a vertical parabolic cross section 4 feet deep and 4 feet across the top is filled with water. Find the work done in pumping out the trough.

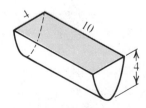

37. A plate in the form of a parabolic segment cut off by a chord perpendicular to the axis is immersed vertically in water. The vertex is at the surface and the axis is vertical. The plate is 20 feet deep and 12 feet broad. Find the force on one face of the segment in tons. (The mass density ρ of water is 1.94 slugs per cubic foot;   1 ton = 2000 pounds.)

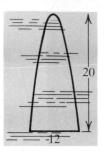

38. In the figure show that the angle of incidence α equals the angle of reflection β.

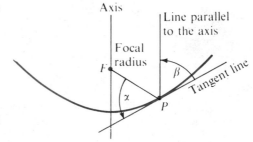

The tangent line makes equal angles with the focal radius and the line parallel to the axis.

3. The Ellipse

An *ellipse* is defined as follows:

(12.4)   **DEFINITION**   *Ellipse.*   **An *ellipse* is the collection of all points in the plane, the sum of whose distances from two fixed points, called the *foci*, is a constant.**

An ellipse can be drawn with the help of two thumbtacks, a pencil, and some string. With the thumbtacks, fix two points (the foci) on a piece of paper and attach the end of a piece of string to each thumbtack. The length of the string is the constant referred to in the definition. Loop the string around a pencil and make the string taut. Keeping the string taut, move the pencil about the two thumbtacks. The pencil will then trace out an ellipse as shown in Figure 10 (p. 656).

In Figure 10 the foci are F and F'. The line joining the foci is called the *major axis*. The midpoint of the foci is called the *center* of the ellipse. The line through the center and perpendicular to the major axis is called the *minor axis*. The points of intersection V and V' of the ellipse with the major axis are the *vertices* of the ellipse.

For convenience in obtaining the standard equation of an ellipse, we place the center at $(0, 0)$ and use the x-axis as the major axis (see Fig. 11). If one focus is at $F = (c, 0)$, $c > 0$,   the other focus must be at $F' = (-c, 0)$.   Now we let the constant sum of the distances of a point $P = (x, y)$ on the ellipse from the foci F and

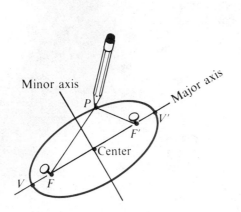

Minor axis
Major axis
P
V'
F'
Center
V
F

Figure 10

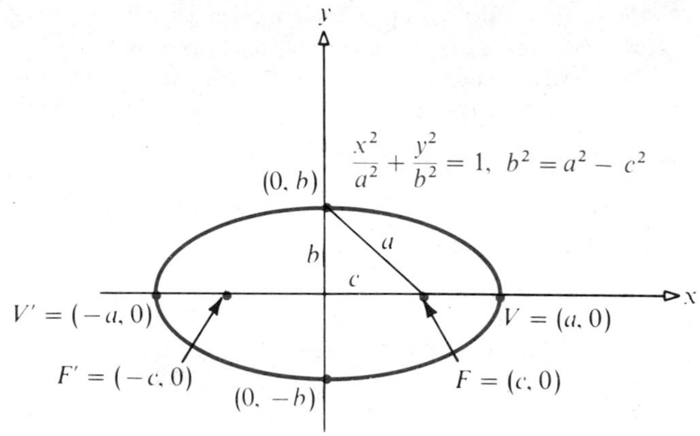

y
$\frac{x^2}{a^2} + \frac{y^2}{b^2} = 1, \; b^2 = a^2 - c^2$
(0, b)
b
a
c
x
V' = (−a, 0)
V = (a, 0)
F' = (−c, 0)
(0, −b)
F = (c, 0)

Figure 11

F' be denoted by $2a$. Then, by (12.4), we have

$$d(F, P) + d(F', P) = 2a$$

By the distance formula,

$$\begin{aligned}
\sqrt{(x-c)^2 + y^2} + \sqrt{(x+c)^2 + y^2} &= 2a \\
\sqrt{(x-c)^2 + y^2} &= 2a - \sqrt{(x+c)^2 + y^2} \\
(x-c)^2 + y^2 &= 4a^2 - 4a\sqrt{(x+c)^2 + y^2} + (x+c)^2 + y^2 \\
x^2 - 2cx + c^2 + y^2 &= 4a^2 - 4a\sqrt{(x+c)^2 + y^2} + x^2 + 2cx + c^2 + y^2 \\
4a\sqrt{(x+c)^2 + y^2} &= 4a^2 + 4cx \\
a\sqrt{(x+c)^2 + y^2} &= a^2 + cx
\end{aligned}$$

Square both sides again and simplify:

$$\begin{aligned}
a^2[(x+c)^2 + y^2] &= (a^2 + cx)^2 \\
a^2[x^2 + 2cx + c^2 + y^2] &= a^4 + 2a^2cx + c^2x^2 \\
a^2x^2 + a^2c^2 + a^2y^2 &= a^4 + c^2x^2 \\
(a^2 - c^2)x^2 + a^2y^2 &= a^4 - a^2c^2 \\
(a^2 - c^2)x^2 + a^2y^2 &= a^2(a^2 - c^2)
\end{aligned}$$

To get points on the ellipse off the x-axis, we must have $a > c$ (see Fig. 11). Since $a > c$, it follows that $a^2 > c^2$, so that $a^2 - c^2 > 0$. As a result, we may set $b^2 = a^2 - c^2$, $b > 0$, to get

$$b^2x^2 + a^2y^2 = a^2b^2$$

or

$$\frac{x^2}{a^2} + \frac{y^2}{b^2} = 1 \qquad \text{where} \quad b^2 = a^2 - c^2, \quad a > b$$

Equation of an Ellipse.   **The standard equation of an ellipse with center at (0, 0) and foci at $(c, 0)$ and $(-c, 0)$ is**

(12.5)
$$\frac{x^2}{a^2} + \frac{y^2}{b^2} = 1 \quad \text{where} \quad b^2 = a^2 - c^2, \quad a > b$$

Conversely, any equation of the form (12.5) is an ellipse with center at $(0, 0)$, foci at $(c, 0)$ and $(-c, 0)$, and major axis along the x-axis.

The graph of the ellipse (12.5) is symmetric with respect to the x-axis, the y-axis, and the origin.

By setting $y = 0$, we find that the vertices of the ellipse obey the equation $x^2/a^2 = 1$, and, hence, the vertices are at $V = (a, 0)$ and $V' = (-a, 0)$, as shown in Figure 11. If we set $x = 0$, we obtain the y-intercepts of the ellipse, $(0, b)$ and $(0, -b)$, which are often called the *covertices* of the ellipse.

We agree that the length of the major axis is the distance from V to V', and the length of the minor axis is the distance between the two y-intercepts. Thus,

Length of major axis is $2a$;   length of minor axis is $2b$

Because $b^2 + c^2 = a^2$, the length of the major axis is greater than the length of the minor axis.

If we position the ellipse so that its center is at (0, 0) and the y-axis is the major axis, then its foci are at $(0, c)$ and $(0, -c)$ and its equation is

(12.6)
$$\frac{x^2}{b^2} + \frac{y^2}{a^2} = 1 \quad \text{where} \quad b^2 = a^2 - c^2, \quad a > b$$

See Figure 12.

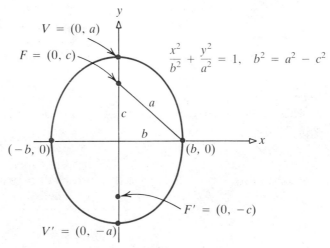

Figure 12

The next example illustrates the difference between equations (12.5) and (12.6).

EXAMPLE 1   Discuss and graph the equation $9x^2 + 16y^2 = 144$.

Solution   To get the equation in standard form, we divide both sides by 144:

$$\frac{x^2}{16} + \frac{y^2}{9} = 1$$

Since the denominator of the x^2 term is larger than the denominator of the y^2 term, this equation is of the form (12.5). Thus, this is the equation of an ellipse whose major axis is the x-axis, with $a^2 = 16$, $b^2 = 9$,   and $c^2 = a^2 - b^2 = 7$.   The foci are at $(\sqrt{7}, 0)$ and $(-\sqrt{7}, 0)$; the vertices are at $(4, 0)$ and $(-4, 0)$. The graph is given in Figure 13.

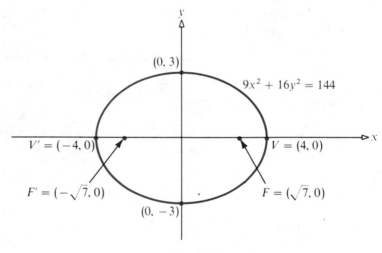

Figure 13   ■

EXAMPLE 2   Find the equation of the ellipse with center at $(0, 0)$, a focus at $(2, 0)$, and a vertex at $(-3, 0)$.

Solution   The second focus is at $(-2, 0)$ and the other vertex is at $(3, 0)$. Since the major axis is along the x-axis, the form of the equation of the ellipse is

$$\frac{x^2}{a^2} + \frac{y^2}{b^2} = 1 \qquad \text{where} \quad b^2 = a^2 - c^2, \quad a > b$$

Since a focus is at $(2, 0)$ and a vertex is at $(3, 0)$, we know that $c = 2$ and $a = 3$. As a result, $b^2 = a^2 - c^2 = 9 - 4 = 5$.   The equation of the ellipse is

$$\frac{x^2}{9} + \frac{y^2}{5} = 1$$

See Figure 14.

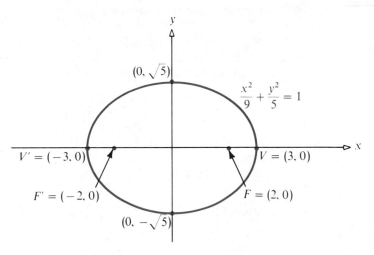

Figure 14   ■

EXAMPLE 3   Find the equation of the ellipse with center at $(0, 0)$, length of major axis 8, and a focus at $(0, 2)$.

Solution   The second focus is at $(0, -2)$, and the major axis is the y-axis. The form of the equation of this ellipse is

$$\frac{x^2}{b^2} + \frac{y^2}{a^2} = 1 \qquad \text{where} \quad b^2 = a^2 - c^2, \quad a > b$$

Since the length of the major axis is $2a = 8$, we have $a = 4$. In addition, $c = 2$. Thus, $b^2 = 16 - 4 = 12$ and the equation is

$$\frac{x^2}{12} + \frac{y^2}{16} = 1$$

See Figure 15.   ■

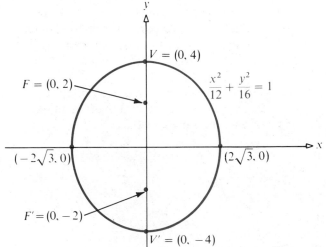

Figure 15

A circle can be thought of as a special kind of ellipse. Indeed, if the two foci of an ellipse move toward the center, then $c \to 0$. Consequently, $b \to a$ (since $b = \sqrt{a^2 - c^2}$). Thus, the length of the major axis becomes equal to the length of the minor axis, and the equation becomes

$$\frac{x^2}{a^2} + \frac{y^2}{a^2} = 1 \qquad \text{or} \qquad x^2 + y^2 = a^2$$

a circle of radius a and center at $(0, 0)$.

EXAMPLE 4   Find an equation for the tangent line to the ellipse $4x^2 + y^2 = 5$ at the point $(1, -1)$.

Solution   We need to compute the derivative y' at $(1, -1)$. By implicit differentiation, we have

$$8x + 2yy' = 0$$

$$y' = \frac{-4x}{y}$$

At $(1, -1)$, the slope of the tangent line is $-4/(-1) = 4$. An equation of the tangent line is

$$y + 1 = 4(x - 1)$$
$$y = 4x - 5 \qquad \blacksquare$$

The ellipse has many uses in science and engineering. For example, the orbits of the planets and some comets about the sun are elliptical, with the sun positioned at a focus. If a source of light (or sound) is placed at one focus of an ellipse, the rays reflect off the ellipse and meet at the other focus (see Fig. 16). This is because the tangent lines to an ellipse make equal angles with the focal radii (see Problem 26 in Exercise 3). This property of an ellipse is used in so-called "whispering galleries," which are rooms designed with ceilings that are elliptical arcs. A person standing at a focus of the ellipse can whisper and be heard by another person standing at the other focus because all the sound rays emanating from the first person reflect off the ceiling to the other person.

Semielliptical arches are often used in stone and concrete bridges. Elliptical gears are used in some types of machinery when variable rates of motion are required. And springs are often formed in the shape of an ellipse or semiellipse.

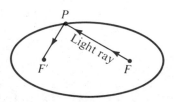

Figure 16

EXERCISE 3

In Problems 1–8 find the equation of the ellipse having the stated properties.

  1. Center at $(0, 0)$;   focus at $(4, 0)$;   vertex at $(6, 0)$
  2. Center at $(0, 0)$;   focus at $(3, 0)$;   vertex at $(5, 0)$
  3. Center at $(0, 0)$;   focus at $(0, 2)$;   vertex at $(0, -3)$
  4. Center at $(0, 0)$;   focus at $(0, -3)$;   vertex at $(0, 4)$
  5. Foci at $(3, 0)$ and $(-3, 0)$;   vertex at $(5, 0)$
  6. Focus at $(0, -2)$;   vertices at $(0, 3)$ and $(0, -3)$
  7. Foci at $(4, 0)$ and $(-4, 0)$;   length of minor axis is 2
  8. Vertices at $(0, 5)$ and $(0, -5)$;   length of minor axis is 4

In Problems 9–16 discuss and graph each equation.

  9. $\dfrac{x^2}{4} + \dfrac{y^2}{9} = 1$ 10. $\dfrac{x^2}{9} + \dfrac{y^2}{4} = 1$ 11. $x^2 + 4y^2 = 16$ 12. $9x^2 + y^2 = 18$

  13. $4x^2 + y^2 = 4$ 14. $9x^2 + 4y^2 = 36$ 15. $x^2 + y^2 = 16$ 16. $x^2 + y^2 = 4$

  17. Find an equation for the tangent line to the ellipse $\dfrac{x^2}{4} + \dfrac{y^2}{3} = 1$ at the point $(1, \tfrac{3}{2})$.

  18. Find an equation for the tangent line to the ellipse $x^2 + \dfrac{y^2}{4} = 1$ at the point $(\tfrac{1}{2}, \sqrt{3})$.

  19. The *eccentricity e* of an ellipse is defined as the number c/a. Since $c < a$,   it follows that $e < 1$.   Describe the general shape of an ellipse whose eccentricity is:
  (a)   Close to 0   (b)   Equal to $\tfrac{1}{2}$ (c)   Close to 1

  20. The orbit of the earth is an ellipse with the sun at one focus. If the *semimajor axis* (half the major axis) is approximately 92 million miles and the eccentricity is $\tfrac{1}{60}$, find the greatest and least distances of the earth from the sun.

  21. In order to support a bridge, an arch in the shape of the upper half of an ellipse is built. Write an equation of the ellipse if the bridge is to span a river 20 meters wide and the center of the arch is 8 meters above the center of the river. Let the *x*-axis coincide with the water level and the *y*-axis pass through the center of the arch.

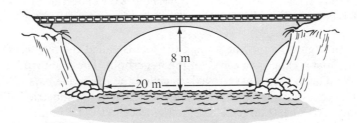

  22. The arch of a bridge is a semiellipse with horizontal major axis. The span is 30 feet, and the top of the arch is 10 feet above the major axis. The roadway is horizontal and is 2 feet above the top of the arch. Find the vertical distance from the roadway to the arch at 5 foot intervals along the roadway.

23. An arch in the form of half an ellipse is 40 feet wide and 15 feet high at the center. Find the height of the arch at intervals of 10 feet along its width.

24. The Colosseum in Rome is in the form of an ellipse 615 feet long and 510 feet wide. Find the area enclosed by the Colosseum.

25. Show that the area enclosed by the ellipse $\dfrac{x^2}{a^2} + \dfrac{y^2}{b^2} = 1$ is πab.

26. Show that the tangent line to an ellipse makes equal angles with the focal radii (see the figure).

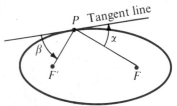

The tangent line makes equal angles with the focal radii.

27. Prove that an equation for the tangent line at any point (x_0, y_0) on the ellipse $\dfrac{x^2}{a^2} + \dfrac{y^2}{b^2} = 1$ is $\dfrac{xx_0}{a^2} + \dfrac{yy_0}{b^2} = 1$.

4. The Hyperbola

A hyperbola is defined as follows:

(12.7)   DEFINITION   *Hyperbola*.   A *hyperbola* is the collection of all points in the plane, the difference of whose distances from two fixed points, called the *foci*, is a constant.

In Figure 17 the line joining the foci is called the *transverse axis*. The midpoint of the foci is called the *center* of the hyperbola. The line through the center and perpendicular to the transverse axis is called the *conjugate axis*. The hyperbola consists of two separate curves, called *branches*, which are symmetric with respect to the conjugate axis, the transverse axis, and the center. The points of intersection of these branches with the transverse axis are the *vertices* of the hyperbola.

For convenience in obtaining the standard equation of a hyperbola, we place the center at $(0, 0)$ and use the x-axis as the transverse axis. If one focus is at

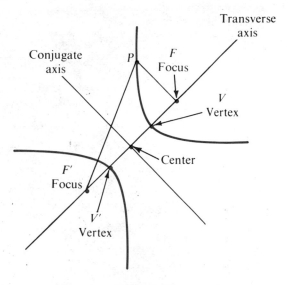

Figure 17

$F = (c, 0)$, $c > 0$,   the other focus must be at $F' = (-c, 0)$,   as shown in Figure 18. Then, if we let the constant difference of the distances from a point $P = (x, y)$ on the hyperbola to the foci F and F' be denoted by $\pm 2a$,* by the definition, we have

$$d(F', P) - d(F, P) = \pm 2a$$

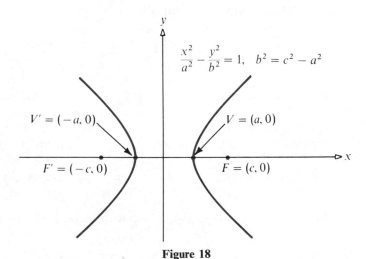

Figure 18

---

* If P is on the right branch, the plus sign is used; if P is on the left branch, the minus sign is used.

By the distance formula,

$$\sqrt{(x + c)^2 + y^2} - \sqrt{(x - c)^2 + y^2} = \pm 2a$$
$$\sqrt{(x + c)^2 + y^2} = \pm 2a + \sqrt{(x - c)^2 + y^2}$$
$$(x + c)^2 + y^2 = 4a^2 \pm 4a\sqrt{(x - c)^2 + y^2} + (x - c)^2 + y^2$$
$$x^2 + 2cx + c^2 + y^2 = 4a^2 \pm 4a\sqrt{(x - c)^2 + y^2} + x^2 - 2cx + c^2 + y^2$$
$$4cx - 4a^2 = \pm 4a\sqrt{(x - c)^2 + y^2}$$
$$cx - a^2 = \pm a\sqrt{(x - c)^2 + y^2}$$

We square both sides again, to get

$$(cx - a^2)^2 = a^2[(x - c)^2 + y^2]$$
$$c^2x^2 - 2ca^2x + a^4 = a^2[x^2 - 2cx + c^2 + y^2]$$
$$c^2x^2 + a^4 = a^2x^2 + a^2c^2 + a^2y^2$$
$$(c^2 - a^2)x^2 - a^2y^2 = a^2c^2 - a^4$$
$$(c^2 - a^2)x^2 - a^2y^2 = a^2(c^2 - a^2)$$

To get points on the hyperbola off the x-axis, we must have $a < c$ (refer to Fig. 18). Since $0 < a < c$, it follows that $a^2 < c^2$, so that $c^2 - a^2 > 0$. By setting $b^2 = c^2 - a^2$, $b > 0$, we get

$$b^2x^2 - a^2y^2 = a^2(b^2)$$

$$\frac{x^2}{a^2} - \frac{y^2}{b^2} = 1 \qquad \text{where} \quad b^2 = c^2 - a^2$$

Equation of a Hyperbola. **The standard equation of a hyperbola with center at (0, 0) and foci at (c, 0) and (−c, 0) is**

(12.8)
$$\frac{x^2}{a^2} - \frac{y^2}{b^2} = 1 \qquad \text{where} \quad b^2 = c^2 - a^2$$

Conversely, any equation of the form (12.8) is a hyperbola with center at (0, 0), foci at $(c, 0)$ and $(-c, 0)$, and transverse axis along the x-axis.

The graph is symmetric with respect to the x-axis, the y-axis, and the origin. By setting $y = 0$, we find that the vertices of the hyperbola obey the equation $x^2/a^2 = 1$ or $x^2 = a^2$, and, hence, the vertices are at $V = (a, 0)$ and $V' = (-a, 0)$. By setting $x = 0$, we obtain the equation $y^2/b^2 = -1$, which has no solution. Thus, this hyperbola never crosses the y-axis; and, in fact, there are no points on this hyperbola for which $|x| < a$.

EXAMPLE 1   Find the equation of the hyperbola with center at (0, 0), a focus at (3, 0), and a vertex at (2, 0).

Solution   The second focus is at $(-3, 0)$ and the second vertex is at $(-2, 0)$. Since the transverse axis is the x-axis, the form of the equation of the hyperbola is

$$\frac{x^2}{a^2} - \frac{y^2}{b^2} = 1$$

Since $a = 2$ and $c = 3$,   we have $b^2 = c^2 - a^2 = 9 - 4 = 5$.   The equation of this hyperbola is

$$\frac{x^2}{4} - \frac{y^2}{5} = 1$$

See Figure 19.

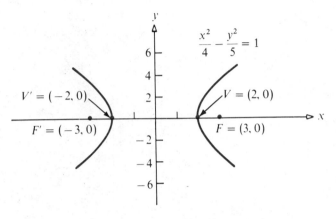

Figure 19

If we position the hyperbola so that $(0, 0)$ is its center and the y-axis is the transverse axis (see Fig. 20), then its foci are at $(0, c)$ and $(0, -c)$ and its equation is

(12.9)
$$\frac{y^2}{a^2} - \frac{x^2}{b^2} = 1 \qquad \text{where} \quad b^2 = c^2 - a^2$$

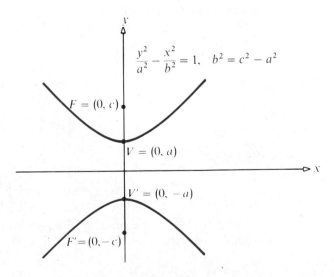

Figure 20

EXAMPLE 2   Discuss and graph the equation $9x^2 - 16y^2 = 144$.

Solution   To get the equation in standard form, we divide both sides by 144:

$$\frac{x^2}{16} - \frac{y^2}{9} = 1$$

This is the equation of a hyperbola with center at $(0, 0)$, whose transverse axis is the x-axis. Also, $a^2 = 16$, $b^2 = 9$, and $c^2 = a^2 + b^2 = 25$. The foci are at $(5, 0)$ and $(-5, 0)$; the vertices are at $(4, 0)$ and $(-4, 0)$; see Figure 21.

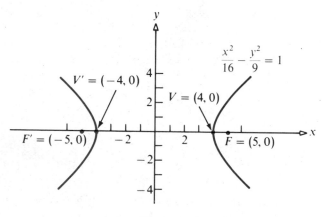

Figure 21   ∎

ASYMPTOTES

Hyperbolas have one feature that parabolas and ellipses do not possess—hyperbolas have *asymptotes*. Recall that an asymptote is a line with the property that the distance from an asymptote to points on the graph gets arbitrarily close to 0 as the points on the graph recede indefinitely far from the origin.

The hyperbola

$$\frac{x^2}{a^2} - \frac{y^2}{b^2} = 1$$

has the two asymptotes

$$y = \frac{b}{a}x \qquad \text{and} \qquad y = -\frac{b}{a}x$$

See Figure 22. A portion of the proof of this statement follows.
   Consider the hyperbola

$$\frac{x^2}{a^2} - \frac{y^2}{b^2} = 1$$

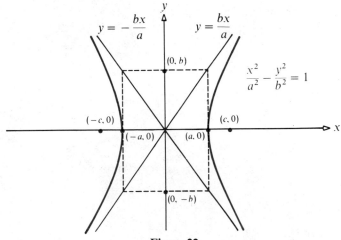

Figure 22

If we solve for y, we get

$$y^2 = \frac{b^2}{a^2}(x^2 - a^2)$$

$$y = \pm\frac{b}{a}\sqrt{x^2 - a^2}$$

Hence, the two branches of the hyperbola are given by

$$y_1 = \frac{b}{a}\sqrt{x^2 - a^2} \quad \text{and} \quad y_2 = -\frac{b}{a}\sqrt{x^2 - a^2}$$

To prove that the line $y = (b/a)x$ is an asymptote to $y_1 = (b/a)\sqrt{x^2 - a^2}$, we must show that the vertical distance between the line $y = (b/a)x$ and y_1 tends to 0 as x gets large. That is,

$$\lim_{x \to +\infty}(y_1 - y) = \lim_{x \to +\infty}\left(\frac{b}{a}\sqrt{x^2 - a^2} - \frac{b}{a}x\right) = 0$$

To show this, we first note that

$$\frac{b}{a}\sqrt{x^2 - a^2} - \frac{b}{a}x = \frac{b}{a}(\sqrt{x^2 - a^2} - x)$$

$$= \frac{b}{a}\left(\frac{\sqrt{x^2 - a^2} + x}{\sqrt{x^2 - a^2} + x}\right)(\sqrt{x^2 - a^2} - x)$$

$$= \frac{b}{a}\left(\frac{-a^2}{\sqrt{x^2 - a^2} + x}\right)$$

$$= \frac{-ab}{\sqrt{x^2 - a^2} + x}$$

By taking the limit, we get

$$\lim_{x \to +\infty} \left(\frac{b}{a}\sqrt{x^2 - a^2} - \frac{b}{a}x \right) = \lim_{x \to +\infty} \left(\frac{-ab}{\sqrt{x^2 - a^2} + x} \right) \underset{\underset{\sqrt{x^2 - a^2} \,\approx\, x}{\uparrow}}{=} \lim_{x \to +\infty} \left(\frac{-ab}{2x} \right) = 0$$

Similarly, we can show that $y = (b/a)x$ is also an asymptote to

$$y_2 = -\left(\frac{b}{a}\right)\sqrt{x^2 - a^2} \qquad \text{as} \qquad x \to -\infty$$

and that $y = -(b/a)x$ is an asymptote to

$$y_1 = \left(\frac{b}{a}\right)\sqrt{x^2 - a^2} \qquad \text{as} \qquad x \to -\infty$$

and to

$$y_2 = -\left(\frac{b}{a}\right)\sqrt{x^2 - a^2} \qquad \text{as} \qquad x \to +\infty$$

(See Problems 28 and 29 in Exercise 4.)

The asymptotes can be used as a guide for approximating the graph of a hyperbola. This is accomplished by locating the vertices $(a, 0)$ and $(-a, 0)$ of the hyperbola and the two points $(0, b)$ and $(0, -b)$. These four points are used to form a rectangle (see Fig. 22). The diagonals of this rectangle have slope b/a and $-b/a$, and their extensions are the asymptotes $y = (b/a)x$ and $y = -(b/a)x$. The graph of the hyperbola can now be obtained by using the rectangle and the asymptotes as guides.

For the hyperbola

$$\frac{y^2}{a^2} - \frac{x^2}{b^2} = 1$$

the asymptotes are the lines

$$y = \frac{a}{b}x \qquad \text{and} \qquad y = -\frac{a}{b}x$$

EXAMPLE 3   Discuss the equation $4y^2 - x^2 = 4$, and graph it by using the asymptotes as guides.

Solution   To get the equation in standard form, we divide both sides by 4:

$$y^2 - \frac{x^2}{4} = 1$$

This is the equation of a hyperbola with center at $(0, 0)$ and y-axis as transverse axis. Furthermore, $a^2 = 1$, $b^2 = 4$, and $c^2 = a^2 + b^2 = 5$. The foci are at $(0, \sqrt{5})$ and $(0, -\sqrt{5})$; the vertices are at $(0, 1)$ and $(0, -1)$. We form the rectangle containing the vertices $(0, 1)$ and $(0, -1)$ and the points $(2, 0)$ and $(-2, 0)$. The extension of the diagonals of this rectangle are the asymptotes $y = x/2$ and $y = -x/2$ (see Fig. 23).

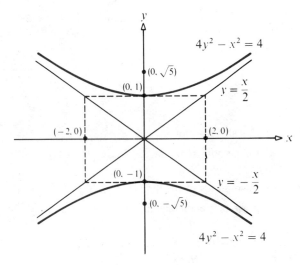

Figure 23   ■

EXERCISE 4

In Problems 1–8 find the equation of the hyperbola that has the stated properties.

1. Center at $(0, 0)$;   focus at $(6, 0)$;   vertex at $(4, 0)$

2. Center at $(0, 0)$;   focus at $(5, 0)$;   vertex at $(3, 0)$

3. Center at $(0, 0)$;   focus at $(0, -3)$;   vertex at $(0, 2)$

4. Center at $(0, 0)$;   focus at $(0, 4)$;   vertex at $(0, -3)$

5. Foci at $(5, 0)$ and $(-5, 0)$;   vertex at $(3, 0)$

6. Focus at $(0, -3)$;   vertices at $(0, 2)$ and $(0, -2)$

7. Foci at $(4, 0)$ and $(-4, 0)$;   asymptote $y = 2x$

8. Vertices at $(0, 5)$ and $(0, -5)$;   asymptote $y = 3x$

In Problems 9–16 discuss and graph each equation.

9. $\dfrac{x^2}{4} - \dfrac{y^2}{9} = 1$ **10.** $\dfrac{y^2}{9} - \dfrac{x^2}{4} = 1$ **11.** $x^2 - 4y^2 = 16$ **12.** $y^2 - 9x^2 = 18$

13. $y^2 - 4x^2 = 4$ **14.** $9x^2 - y^2 = 36$ **15.** $x^2 - y^2 = 16$ **16.** $y^2 - x^2 = 16$

17. The *eccentricity* e of a hyperbola is defined as the number c/a. Since $c > a$, it follows that $e > 1$. Describe the general shape of a hyperbola whose eccentricity is close to 1. What is the shape if e is very large?

18. Use the result of Problem 17 to calculate the eccentricity of the hyperbolas in Problems 9–16.

19. A hyperbola for which $a = b$ is called an *equilateral hyperbola.* Find the eccentricity of an equilateral hyperbola.

20. Two hyperbolas that have the same set of asymptotes are called *conjugate.* Show that the hyperbolas

$$\frac{x^2}{4} - y^2 = 1 \quad \text{and} \quad y^2 - \frac{x^2}{4} = 1$$

are conjugate. Graph each hyperbola.

21. Show that all of the hyperbolas

$$\frac{x^2}{\cos^2\alpha} - \frac{y^2}{\sin^2\alpha} = 1$$

have their foci at $(\pm 1, 0)$ for all values of α.

22. Find the equation of the hyperbola that has the foci of the ellipse $4x^2 + 9y^2 = 36$ for vertices and the vertices of the ellipse as foci.

23. Find the equation of the asymptotes of the hyperbolas: (a) $16x^2 - 25y^2 = 400$ (b) $9y^2 - x^2 = 18$

24. Prove that the product of the perpendicular distances of any point on a hyperbola from its asymptotes is constant.

25. Find equations for the tangent and the normal to each of the following hyperbolas at the point indicated:
   (a) $4x^2 - 16y^2 = 48$; $(4, -1)$ (b) $x^2 - y^2 = 5$; $(3, -2)$

26. Prove that an ellipse and a hyperbola that have the same foci intersect at right angles.

27. *Cooling tower* (see Chap. 6, p. 341). Find the equation of the hyperbola with center $(0, 0)$, one vertex at $(147, 0)$, and passing through the point $(155, 123)$. (A calculator will be very helpful for this exercise.)

28. Show that $y = (b/a)x$ is an asymptote to $y_2 = -(b/a)\sqrt{x^2 - a^2}$ as $x \to -\infty$.

29. Show that $y = -(b/a)x$ is an asymptote to $y_1 = (b/a)\sqrt{x^2 - a^2}$ as $x \to -\infty$ and $y_2 = -(b/a)\sqrt{x^2 - a^2}$ as $x \to +\infty$.

5. Translation and Rotation of Axes

In this section we show that the graph of a general second-degree polynomial in two variables, that is,

(12.10)
$$Ax^2 + Bxy + Cy^2 + Dx + Ey + F = 0$$

is a conic. We assume that A, B, and C are not all simultaneously 0, since, if they were, (12.10) would represent a line,* provided D and E are not both 0.

We begin with the situation for which $B = 0$ in (12.10). In this case, the xy term is absent, so that (12.10) has the form

(12.11)
$$Ax^2 + Cy^2 + Dx + Ey + F = 0$$

---

* It may happen that (12.10) represents a degenerate conic. For example, $x^2 - y^2 = 0$ is the equation of two lines, $x - y = 0$ and $x + y = 0$. The equation $x^2 + y^2 + 1 = 0$ has no graph. We shall not consider such equations here.

The approach is to complete the squares by grouping together the terms involving x and those involving y.

EXAMPLE 1   Complete the squares involving x and y in the polynomial below, and identify the conic represented by this equation.

$$x^2 + 2y^2 - 4x - 12y + 20 = 0$$

Solution   First, we group the terms:

$$(x^2 - 4x) + 2(y^2 - 6y) = -20$$

By adding 4 to $(x^2 - 4x)$ and 9 to $(y^2 - 6y)$,   we get

$$(x^2 - 4x + 4) + 2(y^2 - 6y + 9) = -20 + 4 + 2(9)$$

or

$$(x - 2)^2 + 2(y - 3)^2 = 2$$

By dividing by 2, we find

$$\frac{(x - 2)^2}{2} + (y - 3)^2 = 1$$

Next, we replace $(x - 2)$ by x' and $(y - 3)$ by y'. Then

$$\frac{x'^2}{2} + y'^2 = 1$$

This is an ellipse, with center at $(x', y') = (0, 0)$ and vertices at $x' = \pm\sqrt{2}$, $y' = 0$. As Figure 24 illustrates, the substitutions $x' = x - 2$ and $y' = y - 3$ enable us to identify the point $(2, 3)$ in the xy-coordinate system as the origin $(0, 0)$ in the $x'y'$-coordinate system.

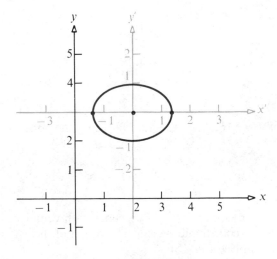

Figure 24   ∎

In general, if we wish to transform the point (h, k) in the xy-coordinate system to become identified as the origin of the $x'y'$-coordinate system, we use the substitutions

$$x' = x - h \quad \text{and} \quad y' = y - k$$

This procedure is referred to as a *translation of axes*. The result of a translation of axes is that new axes are obtained that are parallel to the old ones (see Fig. 25).

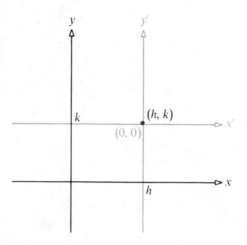

Figure 25

EXAMPLE 2   By a translation of axes, determine the nature of the conic represented by the equation

$$2x^2 - y^2 + 8x + 4y + 3 = 0$$

Graph the equation in the xy-coordinate system.

Solution   First, we group terms:

$$2(x^2 + 4x) - (y^2 - 4y) = -3$$

Next, we complete the squares:

$$2(x^2 + 4x + 4) - (y^2 - 4y + 4) = -3 + 8 - 4$$
$$2(x + 2)^2 - (y - 2)^2 = 1$$

Finally, we translate the axes by using the substitutions $x' = x + 2$ and $y' = y - 2$:

$$2x'^2 - y'^2 = 1$$

$$\frac{x'^2}{\frac{1}{2}} - y'^2 = 1$$

This is the equation of a hyperbola with center at $(0, 0)$ and vertices at $(\pm 1/\sqrt{2}, 0)$ of the $x'y'$-coordinate system. The origin of the $x'y'$ system is the point $(-2, 2)$ of the xy system (see Fig. 26).

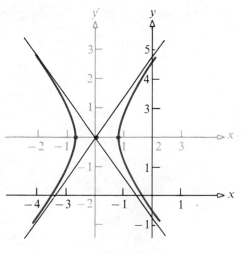

Figure 26   ∎

ROTATION OF AXES (IF TIME PERMITS)

Next, we consider the more difficult case where $B \neq 0$ in

(12.12) $$Ax^2 + Bxy + Cy^2 + Dx + Ey + F = 0$$

Determination of the nature of this conic requires a *rotation of axes*. In a rotation of axes, the origin remains fixed while the x-axis and y-axis are rotated through a positive acute angle θ, becoming the x''-axis and y''-axis, respectively.

Using Figure 27, we obtain the relationship between the coordinates of a point $P = (x'', y'')$ in the new coordinate system and the coordinates of this same point $P = (x, y)$ in the old coordinate system. Let r denote the distance from the origin

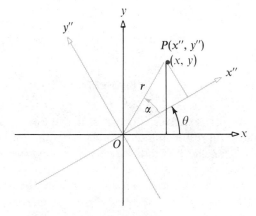

Figure 27

O to the point P and let α denote the angle between the positive x''-axis and the line OP. Then

$$x'' = r \cos \alpha \qquad x = r \cos(\alpha + \theta)$$
$$y'' = r \sin \alpha \qquad y = r \sin(\alpha + \theta)$$

We want to express x and y in terms of x'', y'', and θ. To accomplish this, we use some identities from trigonometry:

(12.13) $$x = r \cos(\alpha + \theta) = r(\cos \alpha \cos \theta - \sin \alpha \sin \theta) = x'' \cos \theta - y'' \sin \theta$$
$$y = r \sin(\alpha + \theta) = r(\sin \alpha \cos \theta + \cos \alpha \sin \theta) = y'' \cos \theta + x'' \sin \theta$$

Although these equations may appear to be cumbersome, they are not difficult to use in practice.

EXAMPLE 3   Transform the equation $xy = 1$ by rotating the axes through an angle of $\pi/4$.

Solution   Replace θ by $\pi/4$ in (12.13). Then

$$x = \frac{1}{\sqrt{2}} x'' - \frac{1}{\sqrt{2}} y'' \qquad \text{and} \qquad y = \frac{1}{\sqrt{2}} y'' + \frac{1}{\sqrt{2}} x''$$

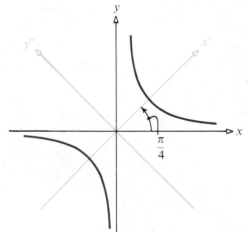

Substitute these into $xy = 1$:

$$\frac{1}{\sqrt{2}}(x'' - y'')\frac{1}{\sqrt{2}}(y'' + x'') = 1$$

$$\frac{1}{2}(x''^2 - y''^2) = 1$$

$$\frac{x''^2}{2} - \frac{y''^2}{2} = 1$$

This is the equation of a hyperbola with center at the origin of the $x''y''$-coordinate system and transverse axis along the x''-axis (see Fig. 28).   ∎

Figure 28

As this example illustrates, a rotation through an appropriate positive acute angle θ will result in a second-degree equation involving x'' and y'', but with no $x''y''$ term. Then the methods already discussed (including possibly a translation of axes) may be used to determine the nature of the conic and its graph.

To derive a formula for determining the appropriate angle θ, we begin with the equation

(12.14) $$Ax^2 + Bxy + Cy^2 + Dx + Ey + F = 0$$

We rotate the axes through an angle θ by using (12.13):

$$A(x'' \cos \theta - y'' \sin \theta)^2 + B(x'' \cos \theta - y'' \sin \theta)(y'' \cos \theta + x'' \sin \theta)$$
$$+ C(y'' \cos \theta + x'' \sin \theta)^2 + D(x'' \cos \theta - y'' \sin \theta) + E(y'' \cos \theta + x'' \sin \theta) + F = 0$$

By expanding the expression and collecting terms, we obtain

$$A''x''^2 + B''x''y'' + C''y''^2 + D''x'' + E''y'' + F = 0$$

where

$$A'' = A \cos^2\theta + B \sin \theta \cos \theta + C \sin^2\theta$$
$$B'' = 2(C - A)\sin \theta \cos \theta + B(\cos^2\theta - \sin^2\theta)$$
$$C'' = A \sin^2\theta - B \sin \theta \cos \theta + C \cos^2\theta$$
$$D'' = D \cos \theta + E \sin \theta$$
$$E'' = -D \sin \theta + E \cos \theta$$

Since we want the $x''y''$ term to be absent, we set $B'' = 0$. By using the double-angle formulas, we get

$$(C - A)\sin 2\theta + B \cos 2\theta = 0$$

Since we are assuming that $B \neq 0$, we find

(12.15)
$$\cot 2\theta = \frac{A - C}{B}$$

Therefore, a rotation of axes through the angle θ, where $\cot 2\theta = (A - C)/B$, $0 < \theta < \pi/2$, will cause $B'' = 0$ and eliminate the xy term from (12.14).

For example, to eliminate the xy term from the equation $xy = 1$, we note that $A = 0$, $B = 1$, and $C = 0$, so that

$$\cot 2\theta = 0$$

$$2\theta = \frac{\pi}{2}$$

$$\theta = \frac{\pi}{4}$$

This was the angle that worked successfully in Example 3.

EXAMPLE 4   By a rotation of axes, determine the nature of the conic with the equation

(12.16)
$$x^2 + 10\sqrt{3}xy + 11y^2 = 4$$

Solution   Here $A = 1$, $B = 10\sqrt{3}$, and $C = 11$, so that the angle of rotation needed to eliminate the xy term is given by

$$\cot 2\theta = \frac{A - C}{B} = \frac{1 - 11}{10\sqrt{3}} = \frac{-1}{\sqrt{3}}$$

This implies that

$$2\theta = \frac{2\pi}{3}$$

$$\theta = \frac{\pi}{3}$$

By using this angle in the rotation formulas (12.13), we obtain

$$x = \frac{1}{2} x'' - \frac{\sqrt{3}}{2} y'' \qquad \text{and} \qquad y = \frac{1}{2} y'' + \frac{\sqrt{3}}{2} x''$$

By substituting for x and y in (12.16), we find that

$$\left(\frac{1}{2} x'' - \frac{\sqrt{3}}{2} y'' \right)^2 + 10\sqrt{3} \left(\frac{1}{2} x'' - \frac{\sqrt{3}}{2} y'' \right) \left(\frac{1}{2} y'' + \frac{\sqrt{3}}{2} x'' \right) + 11 \left(\frac{1}{2} y'' + \frac{\sqrt{3}}{2} x'' \right)^2 = 4$$

By expanding and collecting terms, we obtain

$$16x''^2 - 4y''^2 = 4$$

$$\frac{x''^2}{\frac{1}{4}} - y''^2 = 1$$

This is the equation of a hyperbola with center at the origin of the $x''y''$-coordinate system and transverse axis along the x''-axis (see Fig. 29).

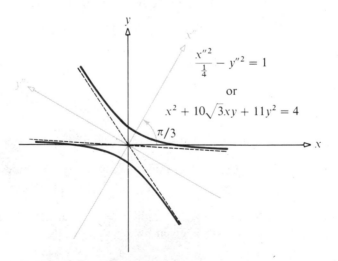

$$\frac{x''^2}{\frac{1}{4}} - y''^2 = 1$$

or

$$x^2 + 10\sqrt{3}xy + 11y^2 = 4$$

Figure 29

It may happen that an equation of the form (12.14) will require a rotation of axes to eliminate the xy term, and will then require a translation of axes. In any case, an equation of the form (12.14) is a conic.

We conclude this section with the following remark: For the second-degree equation

(12.17) $$Ax^2 + Bxy + Cy^2 + Dx + Ey + F = 0$$

the quantity $B^2 - 4AC$, called the *discriminant* of (12.17), may be used to identify the type of conic (12.17) represents, provided the conic is nondegenerate.

1. If $B^2 - 4AC > 0$, then the graph of (12.17) is a hyperbola.

(12.18) 2. If $B^2 - 4AC < 0$, then the graph of (12.17) is an ellipse.

3. If $B^2 - 4AC = 0$, then the graph of (12.17) is a parabola.

EXERCISE 5

In Problems 1–12 use a translation of axes to determine the nature of each conic. Graph each equation in the xy-coordinate system.

1. $x^2 + 4y^2 - 2x = 0$
2. $4x^2 + y^2 - 4y = 0$
3. $4x^2 - 4y^2 + 8y - 5 = 0$
4. $x^2 - y^2 - 2x + 2y - 1 = 0$
5. $4x^2 - y - 16x + 13 = 0$
6. $y^2 + 4x - 4y = 0$
7. $9x^2 + 16y^2 + 54x - 32y - 47 = 0$
8. $4x^2 + 9y^2 + 24x + 18y + 9 = 0$
9. $25y^2 - 9x^2 - 54x - 100y + 10 = 0$
10. $4y^2 - 8x - x^2 + 32y + 49 = 0$
11. $4x^2 - 3y^2 + 8x + 12y - 20 = 0$
12. $3x^2 + 4y^2 - 12x + 8y + 4 = 0$

In Problems 13–18 use a rotation of axes to determine the nature of each conic. Graph each equation in the xy-coordinate system.

13. $xy = 4$
14. $xy = -4$
15. $3x^2 + 5\sqrt{3}xy + 8y^2 = 10$
16. $x^2 + 4\sqrt{3}xy + 5y^2 = 10$
17. $2x^2 - 2\sqrt{3}xy = 5$
18. $y^2 - \sqrt{3}xy = 4$

In Problems 19–22 use (12.18) to determine whether the given second-degree equation is a parabola, ellipse, or hyperbola.

19. $9x^2 - 6xy + y^2 - 3x + y - 2 = 0$
20. $6x^2 + 11xy + 3y^2 + 11x - y - 10 = 0$
21. $10x^2 + 12xy + 4y^2 - x - y = 0$
22. $9x^2 + 12xy + 4y^2 - x - y = 0$

23. Show that under a rotation of axes the sum $A + C$ is invariant. That is, show that $A'' + C'' = A + C$ for any angle θ.

24. Show that under a rotation of axes the expression $B^2 - 4AC$ is invariant.

25. Determine the conic given by the equation $17x^2 - 12xy + 8y^2 - 80 = 0$ by rotating the coordinate axes through an angle θ such that $\cos \theta = \sqrt{5}/5$ and $\sin \theta = 2\sqrt{5}/5$.

Miscellaneous Exercises

In Problems 1–4 determine the nature of the graph described.

1. Each point is such that its distance from the point $(3, 0)$ is $\frac{3}{5}$ of its distance from the line $x = \frac{25}{3}$.

2. Each point is such that its distance from the point $(4, 0)$ is $\frac{4}{5}$ of its distance from the line $x = \frac{25}{4}$.

3. Each point is such that its distance from the point $(5, 0)$ is $\frac{5}{3}$ of its distance from the line $x = \frac{9}{5}$.

4. Each point is such that its distance from the point $(5, 0)$ is $\frac{5}{2}$ of its distance from the line $x = \frac{9}{5}$.

5. Prove that an equation of the tangent line at any point (x_0, y_0) of the parabola $y^2 = 4ax$ is $y_0 y = 2a(x + x_0)$.

6. Prove that an equation of the tangent line at any point (x_0, y_0) of the hyperbola $(x^2/a^2) - (y^2/b^2) = 1$ is $(x_0 x/a^2) - (y_0 y/b^2) = 1$.

7. Find an equation for the tangent lines to $x^2 + 4y^2 = 4$ that contain the point $(4, 1)$.

8. Find the arc length of the parabola $x^2 = 4ay$ from $x = 0$ to $x = b$.

9. An arched window with base width $2b$ and height h is to be set into a wall. The arch is to be either an arc of a parabola or a half-cycle of a cosine curve. Of these two window designs, which has the greater area? Justify your answer.

10. Let A be the area in the first quadrant enclosed by the graphs of $(x^2/9) + (y^2/81) = 1$ and $3x + y = 9$.
    (a) Set up—but do not evaluate—an integral representing the area A. Express the integrand as a function of a single variable.
    (b) Set up—but do not evaluate—an integral representing the volume of the solid generated when A is rotated about the x-axis. Express the integrand as a function of a single variable.
    (c) Set up—but do not evaluate—an integral representing the volume of the solid generated when A is rotated about the y-axis. Express the integrand as a function of a single variable.

11. By examining the standard forms presented in this chapter, explain why (except for degenerate cases) the equation $Ax^2 + Cy^2 + Dx + Ey + F = 0$ (A and C not both 0) represents an ellipse if $AC > 0$ (a circle if $A = C$), a parabola if $AC = 0$, and a hyperbola if $AC < 0$.

12. When a rotation of axes eliminates the xy term from the equation $Ax^2 + Bxy + Cy^2 + Dx + Ey + F = 0$, a new equation $A''x''^2 + B''x''y'' + C''y''^2 + D''x'' + E''y'' + F = 0$ is obtained, with $B'' = 0$. According to Problem 24 in Exercise 5, $B''^2 - 4A''C'' = B^2 - 4AC$. Use this fact, together with Problem 11, to explain why (except for degenerate cases) the graph is a hyperbola, parabola, or ellipse, depending on whether $B^2 - 4AC$ is positive, 0, or negative, respectively.

13. Show that the graph of $x^{1/2} + y^{1/2} = a^{1/2}$ is a parabola.

14. The points A and B on the line segment AP are free to move on the x-axis and y-axis, respectively (see the figure). If $\overline{AP} = a$ and $\overline{BP} = b$, show that P traces out an ellipse as A and B move on their respective axes.

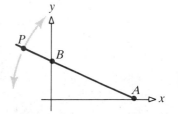

15. The base of a solid coincides with the ellipse $4x^2 + 9y^2 = 36$. Find the volume of the solid if cross sections taken perpendicular to the x-axis are:
    (a) Semicircles　　(b) Squares

16. The base of a solid is the sector of the parabola $y = 4x^2$ below $y = 1$. Find the volume of the solid if cross sections taken perpendicular to the y-axis are:
(a) Rectangles of height 3   (b) Equilateral triangles with two vertices on the parabola

17. Given an ellipse, we may choose a coordinate system so that an equation of the ellipse is given by $(x^2/a^2) + (y^2/b^2) = 1$, where $a > b > 0$. Let F be the focus $(c, 0)$, where $c > 0$. The corresponding *directrix* D is the line $x = a/e$, where e is the eccentricity. Show that the ellipse may be described as the set of points $P = (x, y)$ satisfying the condition $d(F, P)/d(D, P) = e$. [The same condition characterizes the ellipse if F is $(-c, 0)$ and D is the line $x = -a/e$, as you can verify. Hence, the ellipse has a directrix corresponding to each focus.]

18. Rework Problem 17 for an arbitrary hyperbola placed in a coordinate system so that its equation is $(x^2/a^2) - (y^2/b^2) = 1$, where $a > 0$ and $b > 0$.

19. In view of Problems 17 and 18 (and the definition of a parabola in terms of its focus F and directrix D), explain why the condition $d(F, P)/d(D, P) = e$ characterizes every conic (except for degenerate cases), where the conic is an ellipse, parabola, or hyperbola depending on whether $e < 1$, $e = 1$, or $e > 1$, respectively.

20. Given the parabola $y^2 = 4ax$, $a > 0$, show that the length of the chord through the focus perpendicular to the axis is $4a$. (This chord is called the *latus rectum* of the parabola.)

21. Show that any chord through the focus of the parabola $y^2 = 4ax$, $a > 0$, has length $2a + x_1 + x_2$, where (x_1, y_1) and (x_2, y_2) are the endpoints of the chord.

22. At each endpoint of the latus rectum of the parabola $y^2 = 4ax$, a tangent line is drawn. Show that these tangent lines intersect at right angles on the directrix of the parabola.

23. Given the ellipse $(x^2/a^2) + (y^2/b^2) = 1$, show that the length of the chord through either focus perpendicular to the line joining the vertices is $2b^2/a$. (Each of these chords is a *latus rectum* of the ellipse.)

24. Rework Problem 23 for the hyperbola $(x^2/a^2) - (y^2/b^2) = 1$.

25. A tangent line is drawn at each endpoint of the *latus rectum* through the focus $(c, 0)$ of the ellipse $(x^2/a^2) + (y^2/b^2) = 1$. Show that these tangent lines intersect on the directrix $x = a/e$ and have slopes $\pm e$.

26. Rework Problem 25 for the hyperbola $(x^2/a^2) - (y^2/b^2) = 1$.

27. Two lines perpendicular to the asymptotes are drawn through the focus $(c, 0)$ of the hyperbola given by $(x^2/a^2) - (y^2/b^2) = 1$. Prove that each line intersects an asymptote on the directrix of the hyperbola.

28. Find the two parabolas with axes parallel to the coordinate axes having the common tangent line $y = x$ at $P = (1, 1)$; the focus of one parabola is at $(0, 1)$ and the focus of the other parabola is at $(1, 0)$.

13

Polar Coordinates; Parametric Equations

1. Polar Coordinates

Until now we have used a system of rectangular coordinates to locate a point in the plane. In this section we describe another system called *polar coordinates*. As we shall soon discover, polar coordinates offer certain advantages over rectangular coordinates.

In a rectangular coordinate system a point in the plane is represented by a pair of numbers (x, y), where x and y represent the signed distance of the point from the y- and x-axis, respectively. In a polar coordinate system a point is represented by a pair of numbers (r, θ), where r is the distance from the origin (called the *pole*) to the point, and θ is an angle between the positive x-axis (called the *polar axis*) and a ray from the origin through the point (see Fig. 1).

For example, in polar coordinates the point $P = (3, 2\pi/3)$ is located by first drawing the angle with radian measure $2\pi/3$, with vertex at the pole and initial side along the polar axis. Then the point on the terminal side that is 3 units from the pole is the point P, as shown in Figure 2(a). We could also have located this point P by using the polar coordinates $(3, -4\pi/3)$, as shown in Figure 2(b). In fact, the polar coordinates $(3, 8\pi/3)$ also define this same point P, as shown in Figure 2(c).

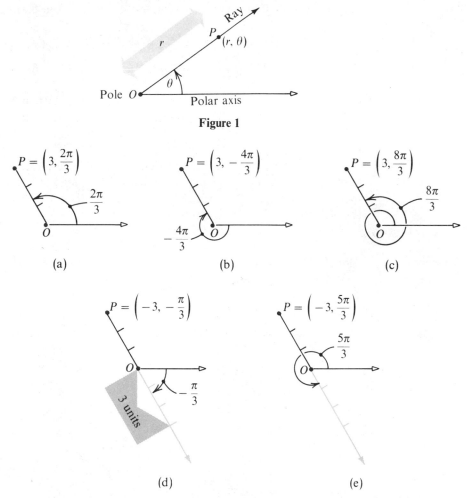

Figure 1

Figure 2   Different pairs of polar coordinates can represent the same point

This point P may also be located by the polar coordinates $(-3, -\pi/3)$ because when the first member of the ordered pair of polar coordinates is negative, we follow the convention that this means *to reflect the point about the origin*. In Figure 2(d) we first measure out 3 units along the ray that makes an angle of $-\pi/3$ radians with the polar axis. Then, following this convention, we obtain point P. In Figure 2(e) we follow the same procedure, using the polar coordinates $(-3, 5\pi/3)$ to obtain point P.

We can now see a major difference between a rectangular coordinate system and a polar coordinate system. In the former, each point has exactly one pair of rectangular coordinates; in the latter, a point may have infinitely many polar coordinates.

In general, if (r, θ) are polar coordinates of a point, then that same point can be represented by

$$(r, \theta + 2n\pi) \qquad \text{or by} \qquad (-r, \theta + (2n + 1)\pi) \qquad n \text{ an integer}$$

The pole itself has polar coordinates of the form $(0, \theta)$, where θ can be any number.

Points (r, θ), where r is negative, deserve special attention. The location of the point, instead of being on the terminal side of the angle θ, is on the ray from the pole extending in the direction opposite to the terminal side and at a distance $|r|$ from the pole. Again, refer to Figures 2(d) and 2(e).

To summarize, every pair of numbers (r, θ) determines a unique point P such that $|OP| = |r|$, and θ is the radian measure of an angle having its initial side along the polar axis and its terminal side along OP if $r > 0$, and along OP extended through the origin in the opposite direction if $r < 0$.

Let's look at a specific example.

EXAMPLE 1   Plot the point whose polar coordinates are $(-2, -\pi/4)$, and find three other polar coordinates of the same point such that:
(a) $r > 0$ and $0 < \theta < 2\pi$ (b) $r > 0$ and $-2\pi < \theta < 0$
(c) $r < 0$ and $0 < \theta < 2\pi$

Solution   The point $(-2, -\pi/4)$ is located by first drawing the direction indicated by the angle $-\pi/4$. Then we locate P on the extension of the terminal side of θ, 2 units from the pole (see Fig. 3). Figure 4 illustrates the answers to parts (a), (b), and (c).

$$P = \left(-2, -\frac{\pi}{4}\right)$$

Figure 3

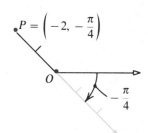

(a)   (b)   (c)

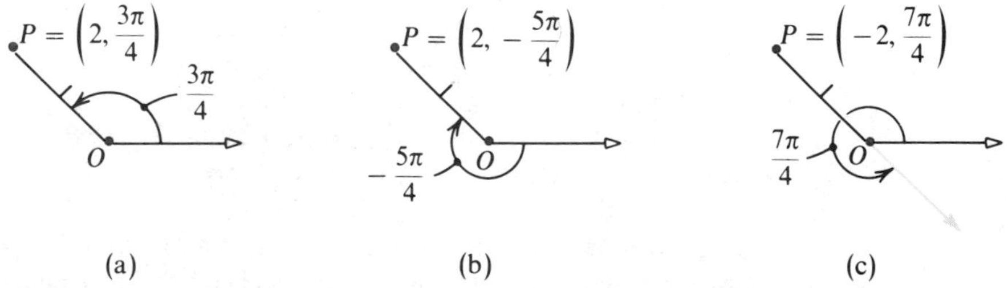

Figure 4

RELATIONSHIP BETWEEN POLAR AND
RECTANGULAR COORDINATES OF A POINT

It is frequently convenient and of great advantage to be able to transform coordinates or equations in rectangular form into polar form, or vice versa. To do this, we recall that the origin in rectangular coordinates is the pole in polar coordinates and that the positive x-axis in rectangular coordinates is the polar axis in polar coordinates. The positive y-axis is represented by the ray $\theta = \pi/2$.

If P is any point in the plane with polar coordinates (r, θ) and rectangular coordinates (x, y), then r, θ and x, y are related by the equations

(13.1)
$$x = r \cos \theta \qquad y = r \sin \theta$$

If $r = 0$, then regardless of θ, the point P under consideration is the pole that has rectangular coordinates $(0, 0)$. Thus, (13.1) holds in this case.

If r is positive, the point P is on the terminal side of θ and $r = |OP|$ (see Fig. 5):

$$\cos \theta = \frac{x}{|OP|} = \frac{x}{r} \qquad \sin \theta = \frac{y}{|OP|} = \frac{y}{r}$$

or

$$x = r \cos \theta \qquad\qquad y = r \sin \theta$$

If r is negative, P is on the extension of the terminal side of θ and $r = -|OP|$ (see Fig. 6). If P' is the point $(-x, -y)$ shown in the figure, then by symmetry with respect to the origin, we have

$$\cos \theta = \frac{-x}{|OP'|} = \frac{-x}{|OP|} = \frac{-x}{-r} = \frac{x}{r} \qquad \sin \theta = \frac{-y}{|OP'|} = \frac{-y}{|OP|} = \frac{-y}{-r} = \frac{y}{r}$$

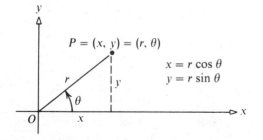

Figure 5

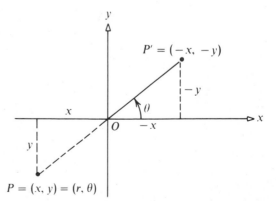

Figure 6

Thus, in this case also, we have

$$x = r \cos \theta \qquad y = r \sin \theta$$

EXAMPLE 2   Find the rectangular coordinates of a point whose polar coordinates are:
(a) $(4, \pi/3)$ (b) $(-2, 3\pi/4)$ (c) $(-3, -5\pi/6)$

Solution   We use equations (13.1):

(a)
$$x = 4 \cos \frac{\pi}{3} = 4\left(\frac{1}{2}\right) = 2$$

$$y = 4 \sin \frac{\pi}{3} = 4\left(\frac{\sqrt{3}}{2}\right) = 2\sqrt{3}$$

The rectangular coordinates are $(2, 2\sqrt{3})$.

(b)
$$x = -2 \cos \frac{3\pi}{4} = -2\left(-\frac{\sqrt{2}}{2}\right) = \sqrt{2}$$

$$y = -2 \sin \frac{3\pi}{4} = -2\left(\frac{\sqrt{2}}{2}\right) = -\sqrt{2}$$

The rectangular coordinates are $(\sqrt{2}, -\sqrt{2})$.

(c)
$$x = -3 \cos\left(-\frac{5\pi}{6}\right) = -3\left(-\frac{\sqrt{3}}{2}\right) = \frac{3\sqrt{3}}{2}$$

$$y = -3 \sin\left(-\frac{5\pi}{6}\right) = -3\left(-\frac{1}{2}\right) = \frac{3}{2}$$

The rectangular coordinates are $(3\sqrt{3}/2, 3/2)$.   ■

Equations (13.1) may be manipulated to obtain a pair of equations that can be used to transform rectangular coordinates into polar coordinates. First, square each equation and then add them to obtain the equation on the left in (13.2). Second, in (13.1), divide y by x; the result is the equation on the right in (13.2).

(13.2) $$r^2 = x^2 + y^2 \qquad \tan \theta = \frac{y}{x} \qquad x \neq 0$$

Care must be taken when using (13.2). The solution for θ in $\tan \theta = y/x$, $x \neq 0$, will yield two values of θ, $0 \leq \theta < 2\pi$. If the value of θ that corresponds to the quadrant where P is located is chosen, then r must equal the positive square root of $x^2 + y^2$. If the other value of θ is chosen, then r must equal the negative of the square root of $x^2 + y^2$.

EXAMPLE 3   Find polar coordinates of the point P whose rectangular coordinates are $(4, -4)$.

Solution   From (13.2), we find

$$r^2 = (4)^2 + (-4)^2 = 32$$
$$r = \pm\sqrt{32} = \pm 4\sqrt{2}$$

$$\tan \theta = -\frac{4}{4} = -1 \quad \text{and, hence,} \quad \theta = \frac{3\pi}{4} \quad \text{or} \quad \frac{7\pi}{4}$$

Since $(4, -4)$ is located in the fourth quadrant, it follows that

$$r = -4\sqrt{2} \quad \text{when} \quad \theta = \frac{3\pi}{4}$$

$$r = 4\sqrt{2} \quad \text{when} \quad \theta = \frac{7\pi}{4}$$

We have thus found two (equivalent) polar coordinates for the point P, namely, $(-4\sqrt{2}, 3\pi/4)$ and $(4\sqrt{2}, 7\pi/4)$. See Figures 7 and 8.

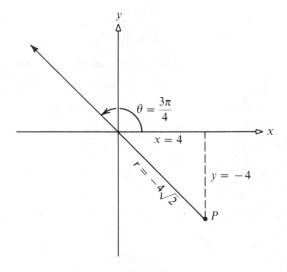

Figure 7

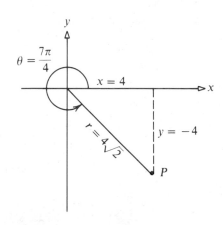

Figure 8   ∎

EXAMPLE 4   Transform the equation $r = 2 \cos \theta$ from polar coordinates to rectangular coordinates and identify the graph.

Solution   Multiply both sides of the equation by r to obtain

$$r^2 = 2(r \cos \theta)$$

By replacing r^2 by $x^2 + y^2$ and $r \cos \theta$ by x, we get

(13.3) $$x^2 + y^2 = 2x$$

Note that (13.3) can be written as $x^2 - 2x + 1 + y^2 = 1$ or as $(x - 1)^2 + y^2 = 1$, which is the equation of a circle with $(1, 0)$ as its center and radius 1. ∎

EXAMPLE 5   Transform the equation $2xy = a^2$ from rectangular coordinates to polar coordinates.

Solution   By substituting $x = r \cos \theta$ and $y = r \sin \theta$ in the given equation, we get

$$2r^2 \sin \theta \cos \theta = a^2$$

Since $2 \sin \theta \cos \theta = \sin 2\theta$, the equation becomes

$$r^2 \sin 2\theta = a^2 \quad ∎$$

EXERCISE 1

In Problems 1–8 plot each point (given in polar coordinates) and find two other polar coordinates of each point—one where $r > 0$ and another where $r < 0$.

1. $(4, \pi/3)$ **2.** $(-4, \pi/3)$ **3.** $(-4, -\pi/3)$ **4.** $(4, -\pi/3)$
5. $(\sqrt{2}, \pi/4)$ **6.** $(7, 7\pi/4)$ **7.** $(6, 4\pi/3)$ **8.** $(5, \pi/2)$

In Problems 9–16 the polar coordinates of a point are given. Find the rectangular coordinates of each point.

9. $(6, \pi/6)$ **10.** $(-6, \pi/6)$ **11.** $(-6, -\pi/6)$ **12.** $(6, -\pi/6)$
13. $(5, \pi/2)$ **14.** $(8, \pi/4)$ **15.** $(2\sqrt{2}, -\pi/4)$ **16.** $(-5, -\pi/3)$

In Problems 17–22 the rectangular coordinates of a point are given. Find polar coordinates of each point for which $r > 0$ and $0 \le \theta < 2\pi$. Plot the point.

17. $(2, -2)$ **18.** $(-2, 2)$ **19.** $(-2, -2\sqrt{3})$ **20.** $(0, -3)$
21. $(-\sqrt{3}, 1)$ **22.** $(3\sqrt{2}, -3\sqrt{2})$

In Problems 23–30 the letters x, y represent rectangular coordinates. Write each equation in terms of the polar coordinates r, θ.

23. $\dfrac{x^2}{4} + \dfrac{y^2}{9} = 1$ **24.** $x - 4y + 4 = 0$ **25.** $x^2 + y^2 - 4x = 0$
26. $y = -6$ **27.** $x^2 = 1 - 4y$ **28.** $y^2 = 1 - 4x$
29. $xy = 1$ **30.** $x^2 + y^2 - 2x + 4y = 0$

In Problems 31–40 the letters r, θ represent polar coordinates. Write each equation in terms of the rectangular coordinates x, y.

31. $r = \cos \theta$ **32.** $r = 2 + \cos \theta$ **33.** $r^2 = \sin \theta$ **34.** $r^2 = 1 - \sin \theta$
35. $r = \dfrac{4}{1 - \cos \theta}$ **36.** $r = \dfrac{4}{4 - \cos \theta}$ **37.** $r^2 = \theta$ **38.** $\theta = -\dfrac{\pi}{4}$
39. $r = 2$ **40.** $r = -5$

41. Show that the formula for the distance d between two points $P_1 = (r_1, \theta_1)$ and $P_2 = (r_2, \theta_2)$ is

$$d = \sqrt{r_1^2 + r_2^2 - 2r_1 r_2 \cos(\theta_2 - \theta_1)}$$

2. Polar Equations and Graphs

The *graph of an equation in polar coordinates* is the set of all points whose polar coordinates satisfy the given equation. The equation is then termed a *polar equation* of the graph.

The basic method for graphing a polar equation is to calculate a table of values of r and θ from the given equation, plot the point representing each such pair of coordinates, and then connect these points. If the equation can be solved for r in terms of θ, convenient values are assigned to θ, and the corresponding values of r are calculated.

A preliminary analysis or discussion of an equation in polar coordinates, in addition to plotting particular points, will generally simplify the construction of the graph. In this analysis we use the familiar notions of intercepts and symmetry, as well as a few new ideas.

INTERCEPTS

Intercepts of the graph on the polar axis or its extension in the opposite direction through the pole are found by putting $\theta = 0, \pi, 2\pi$, and so forth,   and solving for r. Similarly, intercepts on the $\pi/2$ axis or its extension in the opposite direction are obtained by putting $\theta = \pi/2, 3\pi/2$, and so forth,   and solving for r.

EXAMPLE 1   The intercepts of the graph of $r = 2 \cos \theta$ are $r = 2$ (when $\theta = 0$); $r = 0$ (when $\theta = \pi/2$); $r = -2$ (when $\theta = \pi$); and $r = 0$ (when $\theta = 3\pi/2$).   See Figure 9.   ∎

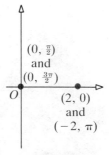

$(0, \frac{\pi}{2})$ and $(0, \frac{3\pi}{2})$

$(2, 0)$ and $(-2, \pi)$

Figure 9

SYMMETRY

Certain types of *symmetry* are readily detected. For example, a graph is:

1. Symmetric about the pole (origin) if the equation is unchanged when r is replaced by $-r$; see Figure 10(a)

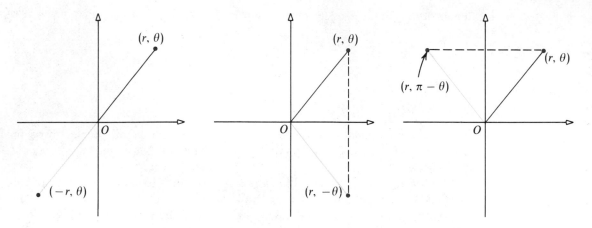

(a) Symmetry about the origin

(b) Symmetry with respect to the polar axis (*x*-axis)

(c) Symmetry with respect to the $\pi/2$ axis (*y*-axis)

Figure 10

2. Symmetric with respect to the polar axis (*x*-axis) if the equation is unchanged when θ is replaced by $-\theta$; see Figure 10(b)

3. Symmetric with respect to the $\pi/2$ axis (*y*-axis) if the equation is unchanged when θ is replaced by $\pi - \theta$; see Figure 10(c)

EXAMPLE 2   The graph of the equation $r = 2 \cos \theta$ is symmetric with respect to the polar axis (*x*-axis) since $\cos(-\theta) = \cos \theta$.   Since the equation $r = 2 \cos \theta$ changes if r is replaced by $-r$, the graph may not be symmetric with respect to the pole (origin). Finally, since $\cos(\pi - \theta) = -\cos \theta$,   the graph of $r = 2 \cos \theta$ may not be symmetric with respect to the $\pi/2$ axis (*y*-axis).   ■

Note the use of the words *may not* in Example 2. The three tests for symmetry given above are *sufficient* conditions for symmetry, but they are not *necessary* conditions; that is, an equation may fail the above tests and yet its graph may be symmetric with respect to the polar axis, pole, or $\pi/2$ axis. A case in point is the graph of $r = \sin 2\theta$,   which is symmetric with respect to the polar axis, pole, and $\pi/2$ axis, yet all the tests given above fail.

TANGENTS AT THE POLE

If the graph passes through the pole, the direction of any tangents to the graph at the pole are found by putting $r = 0$ in the equation and solving for θ.

EXAMPLE 3   The direction of the tangent lines to the graph of $r = 2 \cos \theta$ obey the equation $\cos \theta = 0$,   from which we get

$$\theta = \frac{\pi}{2} \qquad \theta = \frac{3\pi}{2} \qquad \blacksquare$$

We saw in Example 4 of Section 1 that the equation $r = 2 \cos \theta$ is a circle whose rectangular equation is $(x - 1)^2 + y^2 = 1$.   Figure 11 shows the graph of $r = 2 \cos \theta$ and illustrates the results found in Examples 1, 2, and 3.

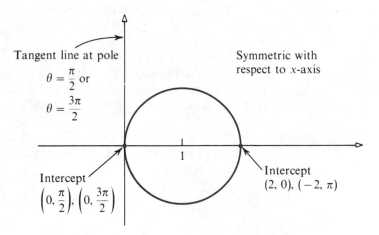

Figure 11   Graph of $r = 2 \cos \theta$

GRAPHING POLAR EQUATIONS

When the given equation expresses r as a trigonometric function of θ, as occurs in many polar equations, a knowledge of the variation of the trigonometric functions in the various quadrants and of the periodicity of these functions can be used to trace the variation of r as θ varies.

The following examples will help clarify these ideas.

EXAMPLE 4   Sketch the graph of $r = 2 \cos 2\theta$.

Solution   To determine the intercepts, we check the value of r at

$$\theta = 0 \qquad \text{and} \qquad \theta = \pi \qquad \text{for which} \qquad r = 2$$

and at

$$\theta = \frac{\pi}{2} \qquad \text{and} \qquad \theta = \frac{3\pi}{2} \qquad \text{for which} \qquad r = -2$$

Hence, the intercepts are $(2, 0)$, $(-2, \pi/2)$, $(2, \pi)$, and $(-2, 3\pi/2)$.

To check for symmetry, we notice that $\cos 2(-\theta) = \cos 2(\theta)$. Hence, the equation is unchanged when θ is replaced by $-\theta$, so that the graph is symmetric about the polar axis (x-axis). Since

$$2 \cos 2(\pi - \theta) = 2 \cos(2\pi - 2\theta) = 2 \cos(-2\theta) = 2 \cos 2\theta$$

the equation $r = 2 \cos 2\theta$ is unchanged when θ is replaced by $\pi - \theta$. Thus, the graph is symmetric with respect to the $\pi/2$ axis. Although our test for symmetry with respect to the pole fails, the graph is symmetric with respect to the pole since it is symmetric with respect to both the polar axis and the $\pi/2$ axis.

To find the tangents to the graph at the pole, we put $r = 0$ in the given equation and obtain $\cos 2\theta = 0$. Therefore, the tangents at the pole are $\theta = \pi/4$, $3\pi/4, 5\pi/4$, and $7\pi/4$, $0 \le \theta \le 2\pi$.

If θ starts at 0 and increases to $\pi/4$, we see that $r = 2 \cos 2\theta$ starts at $r = 2$ and decreases to 0. By symmetry with respect to the polar axis, this part of the graph is repeated below, forming a loop [see Fig. 12(a)]. As θ increases from $\pi/4$ to $\pi/2$ to $3\pi/4$, we see that $r = 2 \cos 2\theta$ is negative and goes from $r = 0$ to $r = -2$ and back to 0. Thus, we get another loop, as shown in Figure 12(b). For θ going from $3\pi/4$ to $5\pi/4$, we get a third loop plotted forward, and from $5\pi/4$ to $7\pi/4$, a fourth loop plotted backward. The final graph, given in Figure 12(c), is called a *four-leaved rose*.*

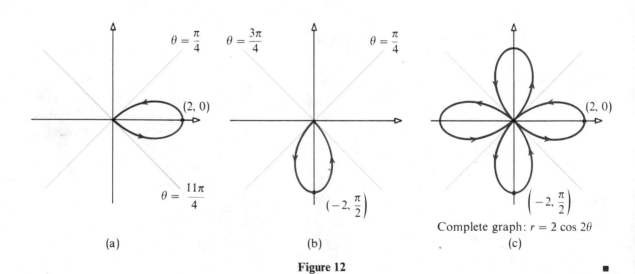

(a)   (b)   Complete graph: $r = 2 \cos 2\theta$
                                              (c)

Figure 12

---

* In general, the graph of a polar equation of the form $r = a \cos n\theta$ or $r = a \sin n\theta$, $a > 0$, n an integer, is called an *n-leaved rose* if n is odd, and a *2n-leaved rose* if n is even.

EXAMPLE 5   Sketch the graph of $r = 1 + 2 \cos \theta$.

Solution   The intercepts are $(3, 0)$, $(1, \pi/2)$, $(-1, \pi)$, $(1, 3\pi/2)$. We check for symmetry and find that the graph is symmetric with respect to the polar axis since

$$1 + 2 \cos(-\theta) = 1 + 2 \cos \theta$$

The tests for symmetry with respect to the pole and the $\pi/2$ axis fail.

The tangents at the pole are $\theta = 2\pi/3$ and $\theta = 4\pi/3$.   To sketch the graph, we limit θ to values less than or equal to 2π, since the graph will repeat itself for other values (due to the periodicity of the cosine function). With the aid of the values given in Table 1 and the rest of the information obtained, we get the graph of $r = 1 + 2 \cos \theta$ shown in Figure 13. This graph is called a *limaçon.** The type of graph paper shown in the figure is called *polar coordinate graph paper*; it has lines through the pole at various angles and circles with centers at the pole.

Table 1

θ	0	$\pi/6$	$\pi/3$	$\pi/2$	$2\pi/3$	$5\pi/6$	π	$7\pi/6$	$4\pi/3$	$3\pi/2$	$5\pi/3$	$11\pi/6$	2π
$r = 1 + 2 \cos \theta$	3	$1 + \sqrt{3}$	2	1	0	$1 - \sqrt{3}$	-1	$1 - \sqrt{3}$	0	1	2	$1 + \sqrt{3}$	3

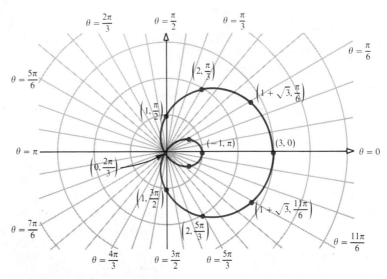

Figure 13 $r = 1 + 2 \cos \theta$ ∎

---

* In general, the graph of any polar equation of the form $r = a \pm b \cos \theta$ or $r = a \pm b \sin \theta$, $a > 0$, $b > 0$,   is a limaçon. If $b > a$,   as in Example 5 the limaçon has a loop. If $a = b$,   the limaçon is a *cardioid*, which is heart-shaped (see Fig. 23, in Section 4).

EXAMPLE 6   Sketch the graph of $r^2 = \sin \theta$.

Solution   The intercepts are $(0, 0)$, $(1, \pi/2)$, and $(-1, \pi/2)$. Since by replacing r by $-r$ we obtain the same equation, the graph is symmetric with respect to the pole; that is, $(-r, \theta)$ is on the graph whenever (r, θ) is. Also, since $\sin(\pi - \theta) = \sin \theta$,   the graph is symmetric with respect to the $\pi/2$ axis. The tangents at the pole are $\theta = 0$ and $\theta = \pi$. As θ increases from 0 to $\pi/2$, r^2 increases from 0 to 1. This gives the portion of the graph shown in Figure 14(a). As θ increases from $\pi/2$ to π, r^2 decreases from 1 to 0. This gives the portion of the graph shown in Figure 14(b). For $\pi < \theta < 2\pi$, $r^2 = \sin \theta < 0$,   so that there is no graph for such angles θ. By combining the graphs in Figure 14(a) and 14(b), we get the complete graph of $r^2 = \sin \theta$ shown in Figure 14(c); this is referred to as a *lemniscate*.

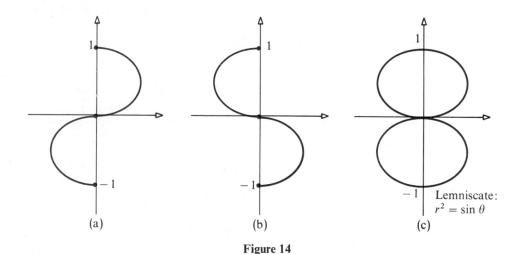

(a)   (b)   (c)

Figure 14

■

EXAMPLE 7   Sketch the graph of $r = e^{\theta/5}$.

Solution   The tests for symmetry with respect to the pole, the polar axis, and the $\pi/2$ axis fail. Furthermore, there is no number θ for which $r = 0$.   Hence, the graph does not pass through the pole. We observe that r is positive for all θ, that r increases as θ increases, that $r \to 0$ as $\theta \to -\infty$,   and that $r \to +\infty$ as $\theta \to +\infty$.   With

Table 2

	A	B	C	D	E	F	G	H	I	J
θ	0	$\pi/4$	$\pi/2$	π	$3\pi/2$	2π	$-\pi/4$	$-\pi/2$	$-\pi$	$-3\pi/2$
$r = e^{\theta/5}$	1	1.17	1.37	1.87	2.57	3.51	0.85	0.73	0.53	0.39

the help of a calculator, we get the values in Table 2. See Figure 15 for the graph. This is called a *logarithmic spiral* since its equation may be written as $\theta = 5 \ln r$ and since it spirals infinitely often both in approaching the pole and in receding from it.

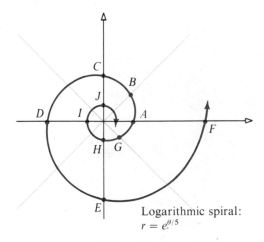

Logarithmic spiral:
$r = e^{\theta/5}$

Figure 15   ∎

Some equations in polar coordinates that are of interest to us are:

1. $\theta = \alpha$;   the line containing the origin with inclination α

2. $r \cos \theta = a$;   the vertical line $x = a$

3. $r \sin \theta = b$;   the horizontal line $y = b$

4. $r(A \cos \theta + B \sin \theta) + C = 0$;   the line $Ax + By + C = 0$

The equations of some circles in polar coordinates and their corresponding equations in rectangular coordinates are given in Table 3 (p. 694). Also included are the graphs of a few of the more frequently encountered polar equations.

We close this section with a summary of some procedures that may be helpful in graphing a polar equation:

1. Change over to a rectangular equation—it may be one familiar to you.

2. Locate the intercepts.

3. Examine for symmetry.

4. Determine the tangent line(s) at the pole.

5. Examine the equation to see if it repeats (periodicity).

6. Make a table of suitably chosen points on the graph.

Table 3

Form	Equation		Graph
Rectangular Polar	$x^2 + y^2 = a^2$ $r = a$	$a > 0$	
Rectangular Polar	$x^2 + y^2 = ax$ $r = a \cos \theta$	$a > 0$	
Rectangular Polar	$x^2 + y^2 = ay$ $r = a \sin \theta$	$a > 0$	
Rectangular Polar	$x^2 + y^2 = ax$ $r = a \cos \theta$	$a < 0$	
Rectangular Polar	$x^2 + y^2 = ay$ $r = a \sin \theta$	$a < 0$	
Cardioid	$r = a \pm a \cos \theta$ $r = a \pm a \sin \theta$	$a > 0$	
Limaçon **with loop**	$r = a \pm b \cos \theta$ $r = a \pm b \sin \theta$	$b > a > 0$	
Lemniscate	$r^2 = a^2 \cos b\theta$ $r^2 = a^2 \sin b\theta$	$b > 0$	
Three-leaved rose	$r = a \sin 3\theta$ $r = a \cos 3\theta$	$a > 0$	

EXERCISE 2

In Problems 1–26 sketch the graph of each polar equation. Locate the intercepts, test for symmetry, and find the tangents at the pole.

1. $r = 2 \sin \theta$ **2.** $r = 2 \cos \theta$ **3.** $\theta = 11\pi/6$ **4.** $r = 2$

5. $r = 4$ **6.** $\theta = \pi/4$

7. $r = 1 - \cos \theta$ (cardioid)   **8.** $r = 4 - 4 \sin \theta$ (cardioid)

9. $r = 4 + 4 \cos \theta$ (cardioid)   **10.** $r = 2 - 6 \cos \theta$ (limaçon)

11. $r = 4 - 3 \sin \theta$ (limaçon)   **12.** $r = 5 \cos 5\theta$ (five-leaved rose)

13. $r = 4 \sin 2\theta$ (four-leaved rose)   **14.** $r = 2 \cos 4\theta$ (eight-leaved rose)

15. $r = 3 \cos 3\theta$ (three-leaved rose)   **16.** $r = e^{\theta}$ (logarithmic spiral)

17. $r = \theta, \quad \theta \geq 0$ (spiral of Archimedes)   **18.** $r = 3/\theta$ (reciprocal spiral)

19. $r^2 = 16 \cos 2\theta$ (lemniscate)   **20.** $r^2 = 1 - \sin \theta$

21. $r = \csc \theta - 2, \quad 0 < \theta < \pi$ (conchoid)   **22.** $r = \sec \theta - \cos \theta = \sin \theta \tan \theta$ (cissoid)

23. $r^2 = 4 \sin \theta$ **24.** $r^2 = 1 + \sin 2\theta$ **25.** $r = \dfrac{2}{1 - \cos \theta}$ **26.** $r = \dfrac{2}{1 - 2 \cos \theta}$

27. Let e denote a positive number. This number e is called the *eccentricity*. (Do not confuse this e with the number e that is the base of the natural logarithm function.) Let d denote the distance from a focus to the directrix of a conic. Position the conic so that its focus is at the pole and its directrix is the line $r \cos \theta = -d, \quad d > 0$. Show that the polar equation of this conic is

$$r = \frac{ed}{1 - e \cos \theta}$$

where the conic is:
(a)   An ellipse if $e < 1$ (b)   A parabola if $e = 1$ (c)   A hyperbola if $e > 1$

28. Graph $r = \sec \theta - 2 \cos \theta$ (a *strophoid*). For what subset of θ values is the entire graph traced out?

3. Intersection of Graphs of Polar Equations

When two rectangular equations in x and y are given, we obtain all the points of intersection of the graphs of the two equations by simply solving the two equations simultaneously. However, this technique does not necessarily yield all the points of intersection of two graphs described by polar equations. The next example illustrates this fact.

EXAMPLE 1   Find the points of intersection of $r = \sin \theta$ and $r = \cos \theta$ for $0 \leq \theta < 2\pi$.

Solution   Certain points of intersection of these two graphs may be found by solving the equations $\sin \theta = \cos \theta$.   By dividing by $\cos \theta$, we get

(13.4) $$\frac{\sin \theta}{\cos \theta} = 1 \quad \text{or} \quad \tan \theta = 1 \quad \text{so that} \quad \theta = \frac{\pi}{4}, \frac{5\pi}{4}$$

From $r = \sin \theta$ or $r = \cos \theta,$ we see that a point of intersection is $(\sqrt{2}/2, \pi/4)$. The point $(-\sqrt{2}/2, 5\pi/4)$ also satisfies each equation, but it is another representation of $(\sqrt{2}/2, \pi/4)$, so we drop it. As Figure 16 illustrates, the graphs of $r = \sin \theta$ and $r = \cos \theta$ are circles that intersect at the point $(\sqrt{2}/2, \pi/4)$ and at the pole. The reason we failed to detect the pole as a point of intersection is that on $r = \sin \theta$ it has coordinates $(0, 0)$, while on $r = \cos \theta,$ it has coordinates $(0, \pi/2)$. In fact, there is no single pair of coordinates of the pole that satisfies both equations.

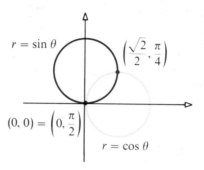

Figure 16   ∎

We see, then, that it is important to make rough sketches of the graphs when we are looking for points of intersection of two polar graphs. This is the only way to find out whether they have points in common other than those found by solving their equations simultaneously. Since the pole presents particular difficulties, always set $r = 0$ in both equations to determine whether the graphs go through the pole.

Let's look at some more examples.

EXAMPLE 2   Find the points of intersection of $r = \cos 2\theta$ and $r = \cos \theta,$ for $0 \le \theta < 2\pi.$

Solution   We set $\cos 2\theta = \cos \theta.$ By using the identity $\cos 2\theta = 2\cos^2\theta - 1,$ we get

$$2\cos^2\theta - 1 = \cos \theta$$
$$2\cos^2\theta - \cos \theta - 1 = 0$$
$$(\cos \theta - 1)(2\cos \theta + 1) = 0$$

Thus,

$$\cos \theta = 1 \qquad \text{and} \qquad \cos \theta = -\tfrac{1}{2}$$

Hence, $\theta = 0, 2\pi/3, 4\pi/3,$ so that the points of intersection are $(1, 0), (-\tfrac{1}{2}, 2\pi/3),$ and $(-\tfrac{1}{2}, 4\pi/3)$. To check the pole, we set $r = 0$ in each equation:

For $r = \cos 2\theta$ with $r = 0,$

$$\cos 2\theta = 0, \qquad \text{so} \qquad \theta = \frac{\pi}{4} \qquad \text{Pole is a point on the graph}$$

For $r = \cos \theta$ with $r = 0$,

$$\cos \theta = 0, \quad \text{so} \quad \theta = \frac{\pi}{2} \qquad \text{Pole is a point on the graph}$$

Figure 17 illustrates the two graphs.

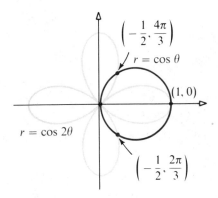

Figure 17   ■

EXAMPLE 3   Find the points of intersection of $r = 2 + 2 \cos \theta$ (cardioid) and $r = 3$ (circle), for $0 \le \theta < 2\pi$

Solution
$$2 + 2 \cos \theta = 3$$
$$2 \cos \theta = 1$$
$$\cos \theta = \tfrac{1}{2}$$

Hence, $\theta = \pi/3$ and $\theta = 5\pi/3$.   The two points of intersection are $(3, \pi/3)$ and $(3, 5\pi/3)$. The pole is not a point of intersection. See Figure 18.

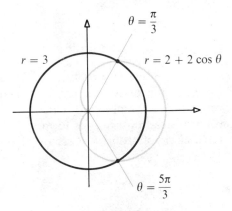

Figure 18   ■

EXERCISE 3

In Problems 1–14 find the points of intersection of the graph of each pair of polar equations for $0 \leq \theta < 2\pi$. Graph both equations.

1. $r = 4\theta; \quad r = \pi/2$

2. $r = \theta/2; \quad r = \pi$

3. $r = 1; \quad r = \sin^2\theta$

4. $r = 1; \quad r = 2 \sin 2\theta$

5. $r = 1, \quad r = 2 \sin \theta$

6. $r = 4; \quad r = 8 \cos \theta$

7. $r = 2(1 - \cos \theta); \quad r = -4 \cos \theta$

8. $r = 2(1 - \cos \theta); \quad r = -6 \cos \theta$

9. $r = 2(1 + \cos \theta); \quad r = 2(1 - \cos \theta)$

10. $r = (1 + \sin \theta); \quad r = (1 - \sin \theta)$

11. $r = \sin 2\theta; \quad r = \cos 2\theta$

12. $r = 2(1 + \cos \theta); \quad r = 2(1 - \sin \theta)$

13. $r = 2 - 2 \cos \theta; \quad r = \dfrac{1}{1 + \cos \theta}$

14. $r = 3(1 - \cos \theta); \quad r = \dfrac{6}{1 - \cos \theta}$

15. The planet Mercury travels around the sun in an elliptical orbit given approximately by

$$r = \frac{(3.442)10^7}{1 - 0.206 \cos \theta}$$

where r is measured in miles and the sun is at the pole. Find the distance from Mercury to the sun at *aphelion* (greatest distance from the sun) and at *perihelion* (shortest distance from the sun).

16. Kepler showed that a line joining a planet to the sun swept out equal areas in space in equal intervals of time (see the figure). Use this information to determine whether a planet travels faster or slower at aphelion than at perihelion (refer to Problem 15).

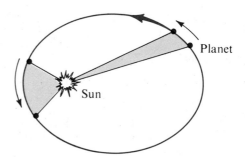

4. Area in Polar Coordinates

In this section we consider the problem of finding the area of a region in the plane enclosed by the graph of a polar equation and two rays that have the pole as a common vertex. The technique used is analogous to the one introduced in Chapter 5. Here, however, instead of using the formula for the area of a rectangle, we use the fact that for a circle of radius r, a sector of central angle θ (measured in radians) has area

$$A = \tfrac{1}{2}r^2\theta$$

(see Fig. 19).

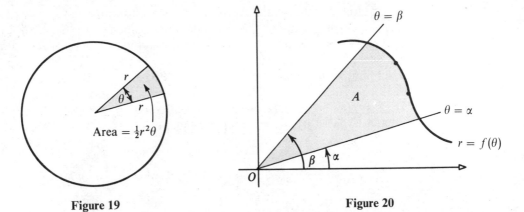

Figure 19   **Figure 20**

The formula for area in polar coordinates is developed as follows: In Figure 20, $r = f(\theta)$ denotes a continuous, nonnegative function defined on $\alpha \le \theta \le \beta$, and A is the area enclosed by the graph of the equation $r = f(\theta)$ and by the rays $\theta = \alpha$ and $\theta = \beta$, where $0 \le \alpha < \beta \le 2\pi$.

Now refer to Figure 21, where we have partitioned $[\alpha, \beta]$ into n subintervals

$$[\alpha, \theta_1], \quad [\theta_1, \theta_2], \quad \ldots, \quad [\theta_{i-1}, \theta_i], \quad \ldots, \quad [\theta_{n-1}, \beta]$$

We denote the length of the ith subinterval by $\Delta\theta_i = \theta_i - \theta_{i-1}$, and we select an angle θ_i^* in $[\theta_{i-1}, \theta_i]$. The quantity $\frac{1}{2}[f(\theta_i^*)]^2\Delta\theta_i$ is the area of the circular sector with radius $r = f(\theta_i^*)$ and central angle $\Delta\theta_i = \theta_i - \theta_{i-1}$ and gives an approximation to the area OPQ. The sum of the areas of these sectors

$$\sum_{i=1}^{n} \frac{1}{2}[f(\theta_i^*)]^2\Delta\theta_i$$

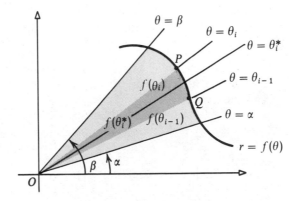

Figure 21

is an estimate of the area A. As the length of each subinterval gets smaller and smaller, that is, as the norm of the partition approaches 0, the sum $\sum_{i=1}^{n} \frac{1}{2}[f(\theta_i^*)]^2 \Delta\theta_i$ becomes a better and better approximation to the area we seek. However, this sum is a Riemann sum. Hence, since f is continuous, as the norm of the partition approaches 0, its limit is the definite integral $\int_{\alpha}^{\beta} \frac{1}{2}[f(\theta)]^2 \, d\theta$.   This suggests the following:

(13.5)   If $r = f(\theta)$ is continuous and nonnegative on $\alpha \leq \theta \leq \beta, \; 0 < \beta - \alpha \leq 2\pi,$ the area A enclosed by the graph of the equation $r = f(\theta)$ and the rays $\theta = \alpha$ and $\theta = \beta,$ is given by

$$A = \int_{\alpha}^{\beta} \frac{1}{2} [f(\theta)]^2 \, d\theta = \int_{\alpha}^{\beta} \frac{1}{2} r^2 \, d\theta$$

EXAMPLE 1   Find the area enclosed by the graph of $r = 2 \cos 3\theta,$ a three-leaved rose.

Solution   Figure 22 depicts the area we seek. We concentrate on the upper portion of the right leaf, which encloses $\frac{1}{6}$ of the area. The upper portion of the right leaf is obtained as θ varies from 0 to $\pi/6$. Hence, the area is

$$A = 6 \int_0^{\pi/6} \frac{1}{2}(4 \cos^2 3\theta) \, d\theta = 12 \int_0^{\pi/6} \frac{1 + \cos 6\theta}{2} \, d\theta$$

$$= 12 \left(\frac{\theta}{2} + \frac{\sin 6\theta}{12} \right)\Big|_0^{\pi/6} = 12 \left(\frac{\pi}{12} \right) = \pi \text{ square units} \qquad \blacksquare$$

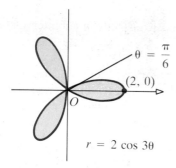

$\theta = \dfrac{\pi}{6}$

$(2, 0)$

O

$r = 2 \cos 3\theta$

Figure 22

EXAMPLE 2   Find the area enclosed by the graph of the limaçon $r = 2 + \cos \theta.$

Solution   The area A we wish to find is shown in Figure 23. Since the graph is symmetric with respect to the polar axis, it is only necessary to calculate the area enclosed by $r = 2 + \cos \theta$ and the rays $\theta = 0$ and $\theta = \pi,$ and then multiply this result by 2. Thus, the area A is

$$A = 2 \int_0^{\pi} \frac{1}{2} (2 + \cos \theta)^2 \, d\theta = \int_0^{\pi} (4 + 4 \cos \theta + \cos^2\theta) \, d\theta$$

$$= \int_0^{\pi} \left[4 + 4 \cos \theta + \frac{1}{2}(1 + \cos 2\theta) \right] d\theta$$

$$= \left(4\theta + 4 \sin \theta + \frac{\theta}{2} + \frac{1}{4} \sin 2\theta \right)\Big|_0^{\pi} = \frac{9\pi}{2} \text{ square units}$$

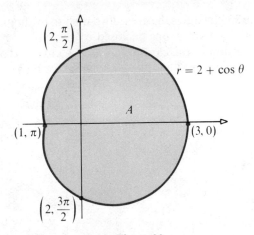

Figure 23   ■

To calculate the area A enclosed by the graphs of two polar equations

$$r = f(\theta) \quad \text{and} \quad r = g(\theta) \qquad 0 \le g(\theta) \le f(\theta)$$

and the rays

$$\theta = \alpha \qquad \theta = \beta \qquad 0 \le \alpha < \beta \le 2\pi$$

as shown in Figure 24, we first calculate the area up to $r = f(\theta)$ and then subtract from it the area up to $r = g(\theta)$. This gives us formula (13.6):

(13.6)
$$A = \int_{\alpha}^{\beta} \frac{1}{2} \left[f(\theta)^2 - g(\theta)^2 \right] d\theta$$

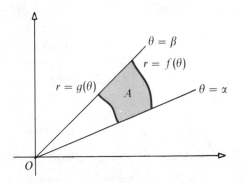

Figure 24

EXAMPLE 3   Find the area that lies outside the cardioid $r = 1 + \cos \theta$ and inside the circle $r = 3 \cos \theta$.

Solution   Figure 25 illustrates the area A we wish to evaluate. Note that the area is the difference
between two areas, one enclosed by the circle $r = 3 \cos \theta$ and the other enclosed
by a portion of the cardioid $r = 1 + \cos \theta$. To find the points of intersection, we
set $1 + \cos \theta = 3 \cos \theta$. By solving, we find

$$\cos \theta = \frac{1}{2} \quad \text{so that} \quad \theta = \pm \frac{\pi}{3}$$

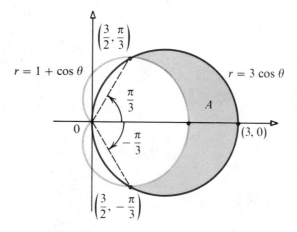

Figure 25

Making use of symmetry, we calculate the area above the polar axis and multiply this
result by 2. Thus, the desired area A is

$$A = 2 \int_0^{\pi/3} \frac{1}{2} [(3 \cos \theta)^2 - (1 + \cos \theta)^2] \, d\theta$$

$$= \int_0^{\pi/3} (9 \cos^2\theta - 1 - 2 \cos \theta - \cos^2\theta) \, d\theta$$

$$= \int_0^{\pi/3} (8 \cos^2\theta - 2 \cos \theta - 1) \, d\theta = \int_0^{\pi/3} \left[8\left(\frac{1 + \cos 2\theta}{2} \right) - 2 \cos \theta - 1 \right] d\theta$$

$$= (4\theta + 2 \sin 2\theta - 2 \sin \theta - \theta)\Big|_0^{\pi/3} = \pi \text{ square units} \quad \blacksquare$$

EXERCISE 4

In each of the following problems draw a figure and indicate the area to be found.

In Problems 1–4 find the area enclosed by the given polar equation and the given rays.

1. $r = 3 \cos \theta, \quad \theta = 0, \quad \theta = \pi/3$ **2.** $r = 3 \sin \theta, \quad \theta = 0, \quad \theta = \pi/4$

3. $r = a\theta, \quad \theta = 0, \quad \theta = 2\pi$ **4.** $r = e^{a\theta}, \quad \theta = 0, \quad \theta = \pi/2$

In Problems 5–8 find the area enclosed by each polar equation.

5. $r = 1 + \cos \theta$ **6.** $r = 2 - \sec \theta$ **7.** $r = 2 \sin^2(\theta/2)$ **8.** $r = 3(1 - \sin \theta)$

In Problems 9–12 find the area of one loop of the given polar equation.

9. $r = 4 \cos 2\theta$ **10.** $r = 2 \sin 3\theta$ **11.** $r^2 = 4 \cos 2\theta$ **12.** $r = a^2 \cos 2\theta$

In Problems 13–16 find the indicated area.

13. Inside $r = 2 \sin \theta$;   outside $r = 1$ **14.** Inside $r = 4 \cos \theta$;   outside $r = 2$

15. Inside $r = \sin \theta$;   outside $r = 1 - \cos \theta$ **16.** Inside $r^2 = 4 \cos 2\theta$;   outside $r = \sqrt{2}$

17. Find the area of the small loop of the limaçon $r = 1 + 2 \cos \theta$.

18. Find the area of the small loop of the graph of $r = 1 + 2 \sin 2\theta$.

19. Find the area inside the circle $r = 8 \cos \theta$ and to the right of the line $r = 2 \sec \theta$.

20. Find the area inside the circle $r = 10 \sin \theta$ and above the line $r = 2 \csc \theta$.

21. Find the area outside the circle $r = 3$ and inside the cardioid $r = 2 + 2 \cos \theta$.

22. Find the area inside the circle $r = \sin \theta$ and outside the cardioid $r = 1 + \cos \theta$.

23. Find the area common to the circle $r = \cos \theta$ and the cardioid $r = 1 - \cos \theta$.

24. Find the area common to the circles $r = \cos \theta$ and $r = \sin \theta$.

25. Find the area common to the inside of the cardioid $r = 1 + \sin \theta$ and the outside of the cardioid $r = 1 + \cos \theta$.

26. Find the area enclosed by the loop of the strophoid $r = \sec \theta - 2 \cos \theta$, $-\pi/2 < \theta < \pi/2$.

5. The Angle ψ (If Time Permits)

In rectangular coordinates, the inclination ϕ of the tangent line to the graph of a differentiable function $y = f(x)$ is obtained by using the derivative $dy/dx = \tan \phi$.   In polar coordinates we find the inclination of the tangent line to a polar graph $r = f(\theta)$ at a point P by making use of the angle ψ* between the ray from O through a point P on the graph of $r = f(\theta)$ and the tangent line to the graph at P (see Fig. 26).

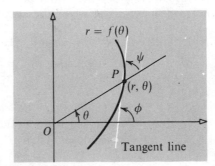

Figure 26

---

* The Greek letter psi, pronounced "see."

The relationship among the angles θ, ϕ, and ψ can be determined from Figure 26 to be

(13.7)
$$\phi = \theta + \psi$$

This relation is used to calculate the angle ϕ, once θ and ψ are known.

Suppose $r = f(\theta)$ is differentiable. As it turns out, there is a relationship between $dr/d\theta$ and the angle ψ. We now proceed to find that relationship.

From (13.7), we see that

$$\tan \psi = \tan(\phi - \theta) = \frac{\tan \phi - \tan \theta}{1 + \tan \phi \tan \theta}$$

In terms of rectangular coordinates, $\tan \phi = dy/dx$ and $\tan \theta = y/x$. Hence,

(13.8)
$$\tan \psi = \frac{(dy/dx) - (y/x)}{1 + (dy/dx)(y/x)} = \frac{x\, dy - y\, dx}{x\, dx + y\, dy}$$

To express this in polar coordinates, we use the relationships

(13.9)
$$x^2 + y^2 = r^2 \qquad x = r \cos \theta \qquad y = r \sin \theta$$

The differential of the first equation is

$$x\, dx + y\, dy = r\, dr$$

From the other two, we calculate

$$x\, dy - y\, dx = r \cos \theta(\sin \theta\, dr + r \cos \theta\, d\theta) - r \sin \theta(\cos \theta\, dr - r \sin \theta\, d\theta)$$
$$= r^2(\cos^2\theta + \sin^2\theta)\, d\theta = r^2\, d\theta$$

Thus, if $r \neq 0$ and $dr/d\theta \neq 0$, then (13.8) reduces to $\tan \psi = (r^2\, d\theta)/(r\, dr)$, which we write as

(13.10)
$$\tan \psi = \frac{r}{dr/d\theta}$$

EXAMPLE 1 Find the angle ψ for the graph of the polar equation $r = 2(1 - \cos \theta)$. Find all points on the graph at which the tangent line is horizontal or vertical.

Solution The derivative is $dr/d\theta = 2 \sin \theta$. Hence, from (13.10),

(13.11)
$$\tan \psi = \frac{r}{dr/d\theta} = \frac{2(1 - \cos \theta)}{2 \sin \theta} = \frac{1 - \cos 2(\theta/2)}{\sin 2(\theta/2)} = \frac{2 \sin^2(\theta/2)}{2 \sin(\theta/2) \cos(\theta/2)} = \tan \frac{\theta}{2}$$

From (13.11), we conclude that $\psi = \theta/2$.

To find points at which the tangent line is horizontal, we use equation (13.7) with $\phi = k\pi$, k an integer. Then

$$k\pi = \psi + \theta = \frac{\theta}{2} + \theta = \frac{3\theta}{2}$$

Consequently, $\theta = 2k\pi/3$. By setting $k = 0, \pm 1, \pm 2, \ldots,$ we find that horizontal tangent lines occur at the three distinct points $(0, 0)$, $(3, 2\pi/3)$, and $(3, 4\pi/3)$.

The tangent line is vertical when $\phi = \pi/2 + k\pi$. Again, by equation (13.7), $\pi/2 + k\pi = 3\theta/2$, so that $\theta = \pi/3 + 2k\pi/3$. By setting $k = 0, \pm 1, \pm 2, \ldots,$ we find that vertical tangent lines occur at the three distinct points $(1, \pi/3)$, $(4, \pi)$, and $(1, -\pi/3)$. The graph is shown in Figure 27.

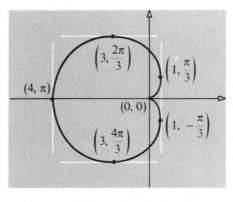

Figure 27   ∎

APPLICATION TO ASTROPHYSICS

Magnetic field lines are stretched out into spiral shapes as the solar wind carries them outward from the sun, much like streams of water from a rotating lawn sprinkler (see Fig. 28). These field lines are approximately described by

$$r = f(\theta) = \frac{VT\theta}{2\pi}$$

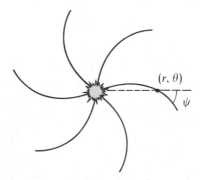

Figure 28

where V is the solar wind velocity (radially outward) and T is the rotation period of the sun. Suppose we want to determine how the angle ψ varies with r and what ψ is near the earth   ($r = 1.5 \times 10^8$ kilometers)   if the average velocity V is 300 kilometers per second and T is 25 days. From (13.10),

$$\tan \psi = \frac{r}{dr/d\theta} = \frac{2\pi r}{VT} \quad \text{and} \quad \psi = \tan^{-1}\left(\frac{2\pi r}{VT}\right)$$

The angle ψ increases smoothly from 0 at $r = 0$ toward 90° as $r \rightarrow +\infty$.

$$\psi_{\text{earth}} \approx \tan^{-1} \frac{(6.28)(1.5 \times 10^8 \text{ km})}{(300 \text{ km/sec})(2.16 \times 10^6 \text{ sec})} = \tan^{-1} 1.45 \approx 55°$$

EXERCISE 5

In Problems 1–10 use (13.10) to find $\tan \psi$ for the graph of each polar equation at the given value of θ.

1. $r = 3(1 - \cos \theta)$; $\theta = 3\pi/4$ **2.** $r = 2(1 - \cos \theta)$; $\theta = \pi/3$ **3.** $r = 4 \cos \theta$; $\theta = \pi/4$

4. $r = 2 \cos 2\theta$; $\theta = \pi/4$ **5.** $r = 2 \sin 2\theta$; $\theta = \pi/6$ **6.** $r = \dfrac{4}{1 + \sin \theta}$; $\theta = -\pi/6$

7. $r = a\theta$; $\theta = -\pi/2$ **8.** $r = \dfrac{a}{\sin \theta}$; $\theta = 3\pi/4$ **9.** $r = 2 + \cos \theta$; $\theta = \pi/3$

10. $r = 2 \sin 3\theta$; $\theta = \pi/4$

In Problems 11–18 find all points on the graph of each polar equation at which the tangent line is horizontal or vertical.

11. $r = 2$ **12.** $r = 3$ **13.** $r = 2 \cos \theta$ **14.** $r = \sin \theta$

15. $r = 3(1 + \cos \theta)$ **16.** $r = 2 + 3 \sin \theta$ **17.** $r = 1 + \sin \theta$ **18.** $r = 2 \sin 3\theta$

19. Find the angle at which the graphs of $r = 3 \cos \theta$ and $r = 1 + \cos \theta$ intersect.

20. Find the angle at which the graphs of $r = \sin 2\theta$ and $r = \cos \theta$ intersect.

21. Show that the graphs of $r = a \cos \theta$ and $r = b \sin \theta$ intersect at right angles.

22. Show that the graph of the cardioid $r = a(1 - \cos \theta)$ and the graph of the parabola $r = a/(1 - \cos \theta)$ intersect at right angles.

23. Refer to Figure 15 (p. 693). Consider the logarithmic spiral $r = ke^{\alpha\theta}$, k and α positive constants. Show that the angle ψ between the tangent line and the line joining the pole and the point of tangency is a constant.

6. Parametric Equations

Equations of the form $y = f(x)$, f a function, have graphs that are cut no more than once by any vertical line. To study more complicated graphs for which the above rule may not hold, we need a different method. One method is to employ a pair of equations and a third variable.

Let $x(t)$ and $y(t)$ be two functions defined on some interval I and set

(13.12)
$$x = x(t) \quad \text{and} \quad y = y(t)$$

To each number t in I, there corresponds a value of x and a value of y that in turn determine the rectangular coordinates of a point in the xy-plane. The collection of all such points is called a *curve in the plane*, or, simply, a *plane curve*. The equations in (13.12) are called *parametric equations* of the curve, and the variable t is called a *parameter* of the curve.

For example, consider the parametric equations

$$x = 4t \qquad y = t^2$$

Here the parameter is t. For each number t, there corresponds a value for x and a value for y. When, say, $t = 3$, then $x = 12$ and $y = 9$. When $t = -1$, then $x = -4$ and $y = 1$. We can set up a table listing various choices for t and the corresponding values of x and y, as shown in Table 4. Figure 29 illustrates the curve whose parametric equations are $x = 4t$, $y = t^2$. The arrows on the graph indicate the direction, or *orientation*, of the curve for increasing values of the parameter t.

Table 4

t	-2	-1	0	1	2	3
$x = 4t$	-8	-4	0	4	8	12
$y = t^2$	4	1	0	1	4	9
(x, y)	$(-8, 4)$	$(-4, 1)$	$(0, 0)$	$(4, 1)$	$(8, 4)$	$(12, 9)$

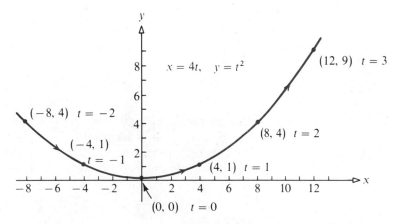

Figure 29

This curve is familiar. In fact, we can identify it readily, once we eliminate the parameter t from the two equations $x = 4t$ and $y = t^2$. We do this as follows: In the equation $x = 4t$, we solve for t to get $t = x/4$. Then we replace t in the other equation, $y = t^2$, by $t = x/4$. The result is the single equation $y = x^2/16$, which is a parabola. We shall refer to such an equation as the *rectangular equation* of the curve in order to distinguish it from the parametric equations.

EXAMPLE 1   Find the rectangular equation of the curve whose parametric equations are

$$x = R \cos t \qquad y = R \sin t \qquad R > 0$$

Graph this curve, indicating its orientation.

Solution   We can eliminate the parameter t by using a familiar trigonometric identity:

$$\cos^2 t + \sin^2 t = \left(\frac{x}{R}\right)^2 + \left(\frac{y}{R}\right)^2 = 1$$

$$x^2 + y^2 = R^2$$

The curve is a circle with center at the origin and radius R. As the parameter t increases, the points (x, y) on the circle are traced out in the counterclockwise direction. See Figure 30.

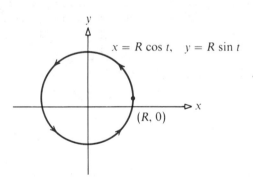

Figure 30   ∎

Some observations should be made concerning Example 1. The parameter t has no restrictions placed on it, so it is assumed to vary from $-\infty$ to $+\infty$. Thus, the graph in Figure 30 is actually being repeated each time t increases by 2π. If we wanted the curve to consist of exactly one revolution in the counterclockwise direction, we could write

$$x = R \cos t \qquad y = R \sin t \qquad 0 \le t \le 2\pi$$

In this case, the curve starts at $(R, 0)$, $t = 0$, and ends at $(R, 0)$, $t = 2\pi$.

If we wanted the curve to consist of exactly three revolutions in the counter-clockwise direction, we could write

$$x = R \cos t \qquad y = R \sin t \qquad -2\pi \le t \le 4\pi$$

Here the choice of the t interval is completely arbitrary. We could just as well have used $0 \le t \le 6\pi$ or $2\pi \le t \le 8\pi$ or any interval of length 6π.

If we wanted the curve to consist of the upper semicircle of radius R with a counterclockwise orientation, we could write

$$x = R \cos t \qquad y = R \sin t \qquad 0 \le t \le \pi$$

See Figure 31.

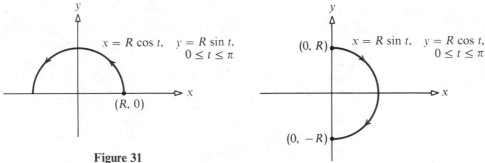

Figure 31

Figure 32

If we wanted the curve to consist of a right semicircle of radius R with a *clockwise* orientation, we could use the parametric equation

$$x = R \sin t \qquad y = R \cos t \qquad 0 \le t \le \pi$$

See Figure 32. Note that the parametric equations are not unique. Other choices that also give the graph in Figure 32 are

$$x = R \sin t \qquad\qquad y = -R \cos(\pi - t) \qquad 0 \le t \le \pi$$

or

$$x = R \cos\left(t - \frac{\pi}{2}\right) \qquad y = R \sin\left(t + \frac{\pi}{2}\right) \qquad 0 \le t \le \pi$$

and infinitely many others.

These observations should convince you of the flexibility and potential usefulness of using parametric equations to define curves. However, in the examples we chose, it was easy to eliminate the parameter t to get the rectangular equation. This may not always be the case. Futhermore, in our examples, the graph of the rectangular equation (obtained by the elimination of the parameter) and the graph of the curve defined by the parametric equations (obtained by locating the values of x and y

corresponding to each t) were identical. But it sometimes happens that the graph of the rectangular equation obtained by the elimination of the parameter contains more points than the graph of the curve defined by the parametric equations. The next example illustrates this.

EXAMPLE 2   Find the rectangular equation of the curve whose parametic equations are

$$x = \cos 2t \qquad y = \sin t$$

Solution   We can eliminate the parameter t from these equations by using the trigonometric identity $\sin^2 t = \frac{1}{2}(1 - \cos 2t)$.   Since $y = \sin t$ and $x = \cos 2t$, it follows that

$$y^2 = \frac{1}{2}(1 - x)$$

so the curve seems to be the parabola $y^2 = \frac{1}{2}(1 - x)$, as shown in Figure 33(a). However, the curve described by the parametric equations does not consist of all points on this parabola. Since $x = \cos 2t$, then $-1 \le x \le 1$, and also, since $y = \sin t$, then $-1 \le y \le 1$. Thus, the curve described by the given parametric equations is only part of the parabola, as shown in Figure 33(b).

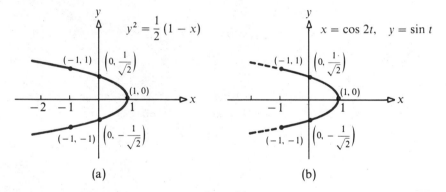

(a)   (b)

Figure 33   ∎

TIME AS A PARAMETER

If we use the parameter t to define time, then the parametric equations $x = x(t)$ and $y = y(t)$ specify how the x- and y-coordinates of a moving point vary with time.

One advantage in using parametric equations, rather than a rectangular equation, to describe the motion of an object, is that with the help of the parametric equations we are able to specify not only *where* the object travels but also *when* it gets to any given place. The rectangular equation tells us only the path along which the object travels.

EXAMPLE 3   Describe the motion of an object that moves so that at time t it has coordinates

$$x = 3 \cos t \qquad y = 4 \sin t \qquad 0 \le t \le 2\pi$$

Solution   We eliminate the parameter t by using the familiar identity $\sin^2 t + \cos^2 t = 1$.   The result is

$$\frac{x^2}{9} + \frac{y^2}{16} = 1$$

so the path is an ellipse. When $t = 0$,   the object is at $(3, 0)$. As t increases, the object moves around the ellipse in a counterclockwise direction, reaching $(0, 4)$ when $t = \pi/2$; $(-3, 0)$ when $t = \pi$; $(0, -4)$ when $t = 3\pi/2$;   and getting back to the starting point when $t = 2\pi$.   See Figure 34.

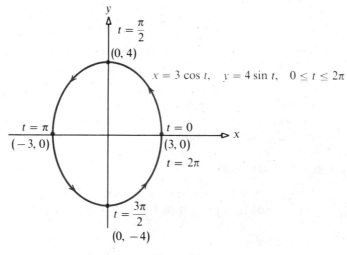

Figure 34   ■

TANGENTS

Next, we develop a formula for finding the slope of the tangent line to a curve directly from its parametric equations.

Let $x = x(t)$, $y = y(t)$ be the parametric equations of a curve, and suppose that the curve can also be represented by the rectangular equation $y = f(x)$.*   Also, assume that $x(t)$, $y(t)$, and f are each differentiable. Then, if $(x(t), y(t))$ is some point on the curve, it will satisfy the equation $y = f(x)$.   Therefore, we can write

$$y = y(t) = f(x(t))$$

---

* Sufficient conditions under which this will happen are twofold: (*1*)   the curve must be *smooth;* that is, the derivatives dx/dt, dy/dt are continuous and never simultaneously zero; and   (*2*) dx/dt is never 0 for any t.

By the chain rule, we have

(13.13) $$\frac{dy}{dt} = \left(\frac{dy}{dx}\right)\left(\frac{dx}{dt}\right) \quad \text{or} \quad \frac{dy}{dx} = \frac{dy/dt}{dx/dt} \quad \text{provided} \quad \frac{dx}{dt} \neq 0$$

Formula (13.13) does not apply if $dx/dt = 0$. However, at a number t where $dx/dt = 0$ and $dy/dt \neq 0$, there is generally a *vertical tangent line*. The behavior of the graph at points for which dy/dt and dx/dt are both 0 can be determined only by more detailed analysis.

EXAMPLE 4   Find an equation of the tangent line to the curve with parametric equations $x = 3t^2$, $y = 2t$ at $t = 1$.

Solution   By equation (13.13),

$$\frac{dy}{dx} = \frac{dy/dt}{dx/dt} = \frac{2}{6t} = \frac{1}{3t}$$

At $t = 1$, the slope of the tangent line is $\frac{1}{3}$ and $x = 3$ and $y = 2$. Therefore, an equation of the tangent line is $y - 2 = \frac{1}{3}(x - 3)$ or $y = x/3 + 1$.   ∎

For the parametric equations in Example 4, $dx/dt = 0$ at $t = 0$. Since $dy/dt = 2 \neq 0$ at $t = 0$, we expect a vertical tangent line at $(0, 0)$. In fact, if we eliminate the parameter, we obtain the parabola $x = \frac{3}{4}y^2$, which, indeed, has a vertical tangent line at $(0, 0)$.

FINDING PARAMETRIC EQUATIONS

If the rectangular equation of a curve is known, a variety of parametric equations can be used to represent it.

EXAMPLE 5   Consider the ellipse

$$\frac{x^2}{4} + y^2 = 1$$

Let the parameter t denote time (in seconds). The position of an object, whose motion around this ellipse begins at $(2, 0)$, is counterclockwise, and requires 1 second for a complete revolution, is given by the parametric equations

$$x = 2 \cos 2\pi t \qquad y = \sin 2\pi t \qquad 0 \leq t \leq 1$$

The position of an object, whose motion around the ellipse begins at $(0, 1)$, is clockwise, and requires 2 seconds for a complete revolution, is given by

$$x = 2 \sin \pi t \qquad y = \cos \pi t \qquad 0 \leq t \leq 2 \qquad ∎$$

The cycloid, which is discussed next, provides an example of how the parametric equations of a curve can be developed. Observe that the parametric equations provide a "natural way" to write the equation of this particular curve.

THE CYCLOID

Suppose that a circle rolls along a horizontal line without slipping. As the circle rolls along the line, a point P on the circle will trace out a curve called a *cycloid* (see Fig. 35). An attempt to derive the equation of the cycloid in rectangular coordinates will soon demonstrate how complicated the task is. However, the derivation in terms of parametric equations is relatively easy.

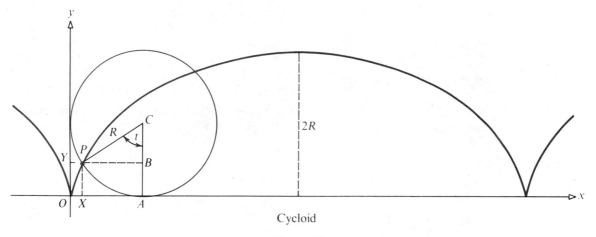

Cycloid

Figure 35

We begin with a circle of radius R and take the fixed line on which the circle rolls as the x-axis. Let the origin be one of the points at which the point P comes in contact with the x-axis. Figure 35 illustrates the position of this point P after the circle has rolled somewhat. The angle t (in radians) measures the angle through which the circle has rolled.

Since we require no slippage, it follows that

$$\text{Arc } AP = |OA|$$

Therefore,

$$Rt = |OA|$$

The x-coordinate of the point P is

$$|OX| = |OA| - |XA| = Rt - R \sin t = R(t - \sin t)$$

The y-coordinate of the point P is equal to

$$|OY| = |AC| - |BC| = R - R \cos t = R(1 - \cos t)$$

Thus, the parametric equations of the cycloid are

(13.14) $$x = R(t - \sin t) \qquad y = R(1 - \cos t)$$

EXAMPLE 6   Show that the slope of the tangent line to the cycloid

$$x = R(t - \sin t) \qquad y = R(1 - \cos t) \qquad 0 \le t \le 2\pi$$

is cot $(t/2)$. Find the numbers t, $\;0 \le t \le 2\pi$, for which the tangent line is horizontal or vertical.

Solution   The derivatives dx/dt, dy/dt are

$$\frac{dx}{dt} = R - R \cos t \qquad \frac{dy}{dt} = R \sin t$$

Hence, the slope of the tangent line is given by

$$\frac{dy}{dx} = \frac{dy/dt}{dx/dt} = \frac{R \sin t}{R(1 - \cos t)} = \frac{\sin t}{1 - \cos t} = \frac{\sin 2(t/2)}{1 - \cos 2(t/2)} = \frac{2 \sin(t/2) \cos (t/2)}{2 \sin^2(t/2)} = \cot \frac{t}{2}$$

The tangent line is horizontal when $\;\sin t = 0$, that is, when $\;t = \pi$. The tangent line is vertical when $\;1 - \cos t = 0$, that is, when $\;t = 0\;$ or $\;t = 2\pi$.   ■

APPLICATIONS TO MECHANICS

If, in the equations (13.14) of the cycloid, we let $\;R < 0$, we obtain an inverted cycloid [see Fig. 36(a)]. The inverted cycloid occurs as a result of some remarkable applications in the field of mechanics. We shall mention two of them—the *brachistochrone* and the *tautochrone*.*

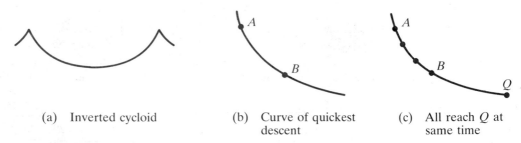

(a) Inverted cycloid   (b) Curve of quickest   (c) All reach Q at
                                    descent   same time

Figure 36

Brachistochrone.   This is the curve of quickest descent. If a particle is constrained to follow some path from one point A to a lower point B (not on the same vertical line) and is acted upon only by gravity, the time needed to make the descent is least if the path is an inverted cycloid [see Fig. 36(b)]. This remarkable discovery, which is attributed to many famous mathematicians (including Johann Bernoulli and Pascal), was a significant step in creating the branch of mathematics known as the *calculus of variations.*

---

* In Greek, *brachistochrone* means "the shortest time," and *tautochrone* means "equal time."

Tautochrone.   Let Q be the lowest point on an inverted cycloid. If several particles placed at various positions on an inverted cycloid simultaneously begin to slide down the cycloid, they will reach the point Q at the same time, as indicated in Figure 36(c). The tautochrone property of the cycloid was used by the Dutch mathematician, phys-icist, and astronomer Christian Huygens (1629–1695) to construct a pendulum clock with a bob that swings along a cycloid (see Fig. 37). In Huygens' clock, the bob was made to swing along a cycloid by suspending the bob on a thin wire constrained by two plates shaped like cycloids. In a clock of this design, the period of the pendulum is independent of its amplitude.

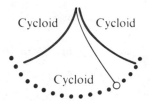

A flexible pendulum constrained by cycloids swings in a cycloid

Figure 37

EXERCISE 6

In Problems 1–14 sketch the graph of the curve defined by the given parametric equations and show its orientation. Find the rectangular equation of each curve.

1. $x = 3, \quad y = 2t, \quad -\infty < t < +\infty$

2. $x = 4t + 1, \quad y = 2t, \quad -\infty < t < +\infty$

3. $x = 2, \quad y = t^2 + 4, \quad 0 \le t < +\infty$

4. $x = t + 3, \quad y = t^3, \quad -4 \le t \le 4$

5. $x = t + 5, \quad y = \sqrt{t}, \quad t \ge 0$

6. $x = 2t^2, \quad y = 2t^3, \quad 0 \le t \le 3$

7. $x = t^{1/2} + 1, \quad y = t^{3/2}, \quad t \ge 1$

8. $x = 2e^t, \quad y = 1 - e^t, \quad t \ge 0$

9. $x = e^t, \quad y = t, \quad -\infty < t < +\infty$

10. $x = t, \quad y = 1/t, \quad -\infty < t < +\infty, \quad t \ne 0$

11. $x = \sec t, \quad y = \tan t, \quad -\pi/2 < t < \pi/2$

12. $x = 3 \sinh t, \quad y = 2 \cosh t, \quad -\infty < t < +\infty$

13. $x = 3 \sin t - 2, \quad y = 2 \cos t, \quad 0 \le t \le 2\pi$

14. $x = 1 + 2 \sin t, \quad y = 2 - \cos t, \quad 0 \le t \le 2\pi$

In Problems 15–18 find two different pairs of parametric equations for each rectangular equation.

15. $y = 4x^3$

16. $y = 2x^2$

17. $x = \frac{1}{3}\sqrt{y} - 3$

18. $x = 3y^3 - 2\sqrt{y} + 5y + 2$

In Problems 19 and 20 use the rectangular equation $(x^2/9) + (y^2/4) = 1$. Find its parametric equations under the given conditions on the motion of a particle along this curve.

19. The motion begins at (3, 0), is counterclockwise, and requires 3 seconds for a complete revolution.

20. The motion begins at (3, 0), is clockwise, and requies 3 seconds for a complete revolution.

In Problems 21–26 find the slope of the tangent line to each curve at the given number t.

21. $x = 3t, \quad y = 2t^2 - 1, \quad$ at $\quad t = 1$

22. $x = 2t, \quad y = t^2 - 2, \quad$ at $\quad t = 2$

23. $x = t^2 - 1, \quad y = 2e^t, \quad$ at $\quad t = 2$

24. $x = 6t + 1, \quad y = 5t + 1, \quad$ at $\quad t = 1$

25. $x = 3 \cos t, \quad y = 4 \sin t, \quad$ at $\quad t = \pi/4$

26. $x = \tan t, \quad y = \sec t, \quad$ at $\quad t = 3\pi/4$

In Problems 27–32 find:
(a) All points on the curve where the curve has a horizontal tangent line
(b) All points on the curve where the curve has a vertical tangent line

27. $x = 3t^2, \quad y = t^3 - 4t$

28. $x = 2 - t^2, \quad y = 1 + 2t^2$

29. $x = 1/t, \quad y = t^2 + 3$

30. $x = 1/t, \quad y = 2t$

31. $x = 2 \cos t, \quad y = 1 + \cos 2t$

32. $x = 2 + 3 \sin t, \quad y = 3 - 2 \cos t$

In Problems 33–36 find a rectangular equation of the tangent line to each curve at the given number t.

33. $x = 2t + 1, \quad y = t^2 - 2, \quad$ at $\quad t = 2$

34. $x = t^2 + 1, \quad y = t^3 + 1, \quad$ at $\quad t = 1$

35. $x = \cos 2t, \quad y = \sin t, \quad$ at $\quad t = \pi/4$

36. $x = \cos^3 t, \quad y = \sin^3 t, \quad$ at $\quad t = \pi/4$

In Problems 37–42 find dy/dx and d^2y/dx^2. [*Hint:* Use the chain rule to calculate d^2y/dx^2.]

37. $x = e^t \cos t, \quad y = e^t \sin t$

38. $x = 2 \sin t, \quad y = \cos 2t$

39. $x = t + 1/t, \quad y = 4 + t$

40. $x = t^2 + 1, \quad y = \sqrt{t}$

41. $x = a(\cos \theta + \theta \sin \theta), \quad y = a(\sin \theta - \theta \cos \theta)$

42. $x = \cos^3 \theta, \quad y = \sin^3 \theta$

43. The position of a projectile (air resistance neglected) at the end of t seconds, fired with an initial velocity v_0 feet per second and at angle θ from the horizontal is given by the parametric equations

$$x = (v_0 \cos \theta)t \qquad y = (v_0 \sin \theta)t - 16t^2$$

(a) By eliminating the parameter t, show that the trajectory is the arc of a parabola from the initial point to the point of impact.
(b) The speed v at any time t is given by

$$v = \left[\left(\frac{dx}{dt} \right)^2 + \left(\frac{dy}{dt} \right)^2 \right]^{1/2}$$

Find v when $\quad t = 1 \quad$ and $\quad t = 2$.
(c) Show that the projectile hits the ground when $\quad t = \frac{1}{16} v_0 \sin \theta$. Find how far the projectile has traveled horizontally at that time.

44. Let a circle C of radius b roll, without slipping, inside a fixed circle with radius $a, \quad a > b$. A fixed point P on the circle C traces out a curve called a *hypocycloid*. If $\quad A = (a, 0) \quad$ is the initial position of the tracing point P and if t denotes the angle from the positive x-axis to the line segment from the origin to the center of C, show that the parametric equations of the hypocycloid are

$$x = (a - b)\cos t + b \cos \frac{a - b}{b} t$$

$$y = (a - b)\sin t - b \sin \frac{a - b}{b} t \qquad 0 \le t \le 2\pi$$

45. The figure shows a hypocycloid with $\quad a = 4b$. Show that the rectangular equation of the hypocycloid with $a = 4b$ is $\quad x^{2/3} + y^{2/3} = a^{2/3}$.

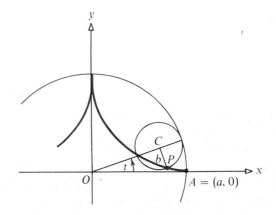

46. If the circle C of Problem 44 rolls on the outside of the second circle, find the parametric equations of the curve traced by P. This curve is called an *epicycloid*.

47. The figure shows an epicycloid with $a = 4b$. Given that $x = a \cos g(t)$, $y = b \sin g(t)$, prove that

$$xy^2 \frac{d^2y}{dx^2} = b^2 \frac{dy}{dx}$$

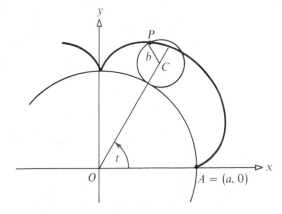

7. Arc Length

In Chapter 6 we developed the formula (6.22) for the arc length of the graph of a function $y = f(x)$.   For some graphs, such as the one in Figure 38, this formula cannot be used since such a graph is not the graph of a function. In this section we develop a formula for the arc length of a graph defined by parametric equations. In addition, we shall show that this formula reduces to the formula given in Chapter 6 when the parametric equations define a graph of a function $y = f(x)$.

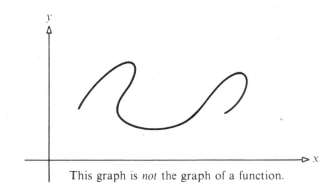

This graph is *not* the graph of a function.

Figure 38

We begin with a curve C defined by the parametric equations

$$x = x(t) \qquad y = y(t) \qquad a \le t \le b$$

where $x(t)$ and $y(t)$ are each continuous and differentiable on the closed interval $[a, b]$. We further assume that dx/dt and dy/dt are each continuous and never simultaneously zero on $[a, b]$. Such a curve C is referred to as *smooth* since under these conditions its graph will have no corners.

Our procedure is much like the one we used in Chapter 6. First, we partition the closed interval $[a, b]$ into n subintervals,

$$[a, t_1], \quad [t_1, t_2], \quad \ldots, \quad [t_{i-1}, t_i], \quad \ldots, \quad [t_{n-1}, b]$$

and we let $\Delta t_i = t_i - t_{i-1}$. Corresponding to each number $a, t_1, t_2, \ldots, t_{n-1}, b$, there is a succession of points $P_0, P_1, P_2, \ldots, P_n$, on the curve (see Fig. 39). Join each point to its successor by a line segment. The sum of the lengths of these line segments provides an approximation to the length of the curve from $t = a$ to $t = b$. This sum may be written as

(13.15)
$$d(P_0, P_1) + d(P_1, P_2) + \cdots + d(P_{n-1}, P_n) = \sum_{i=1}^{n} d(P_{i-1}, P_i)$$

where $d(P_{i-1}, P_i)$ is the length of the line segment joining P_{i-1} and P_i. By the formula for the distance between two points, it follows that the length of each line segment is

(13.16)
$$d(P_{i-1}, P_i) = \sqrt{[x(t_i) - x(t_{i-1})]^2 + [y(t_i) - y(t_{i-1})]^2}$$

By using (13.15), the sum of the lengths of the line segments can be written as

(13.17)
$$\sum_{i=1}^{n} \sqrt{[x(t_i) - x(t_{i-1})]^2 + [y(t_i) - y(t_{i-1})]^2}$$

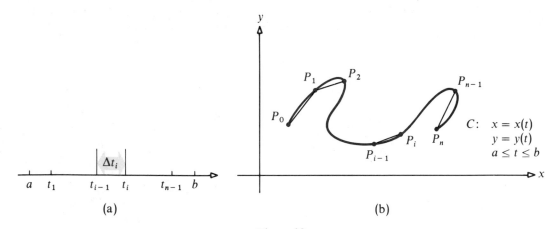

Figure 39

Since we assume that each of the functions $x(t)$ and $y(t)$ has a continuous derivative on $[a, b]$, it follows that $x(t)$ and $y(t)$ have derivatives on each subinterval $[t_{i-1}, t_i]$. When we apply the mean value theorem (4.26) to $x(t)$ and $y(t)$ on $[t_{i-1}, t_i]$, the result is that there exist numbers u_i and v_i in each open interval (t_{i-1}, t_i) such that

(13.18)
$$x(t_i) - x(t_{i-1}) = \left[\frac{dx}{dt}(u_i)\right]\Delta t_i \qquad y(t_i) - y(t_{i-1}) = \left[\frac{dy}{dt}(v_i)\right]\Delta t_i$$

Thus, the sum of the lengths of the line segments in (13.17) becomes

(13.19)
$$\sum_{i=1}^{n} \sqrt{\left[\frac{dx}{dt}(u_i)\right]^2 + \left[\frac{dy}{dt}(v_i)\right]^2}\; \Delta t_i$$

This sum is not a Riemann sum since the numbers u_i and v_i are not necessarily equal. However, there is a result (usually given in advanced calculus), that states that the limit of the sum in (13.19) as the norm of the partition approaches 0 is a definite integral.* But as the length of each subinterval gets smaller and smaller—that is, as the norm of the partition approaches 0—it is plausible to define the limit of this sum as the length of the curve from $t = a$ to $t = b$. This suggests the following definition:

(13.20)   **Let $x(t)$ and $y(t)$ be functions having continuous derivatives on a closed interval $a \le t \le b$.   Then the length s of the curve**

$$x = x(t) \qquad y = y(t)$$

from $t = a$ to $t = b$ is given by the formula

(13.21)
$$s = \int_a^b \sqrt{\left(\frac{dx}{dt}\right)^2 + \left(\frac{dy}{dt}\right)^2}\; dt$$

EXAMPLE 1   Find the length of the curve defined by the parametric equations
$$x = t^3 + 2 \qquad y = 2t^{9/2}$$
from the point where $t = 0$ to the point where $t = 3$.

Solution   We have $dx/dt = 3t^2$ and $dy/dt = 9t^{7/2}$. Therefore, by (13.21), the length s of the curve from $t = 0$ to $t = 3$ is

$$s = \int_0^3 \sqrt{(3t^2)^2 + (9t^{7/2})^2}\; dt = \int_0^3 \sqrt{9t^4 + 81t^7}\; dt = \int_0^3 3t^2 \sqrt{1 + 9t^3}\; dt$$

Make the substitution $u = 1 + 9t^3$. Then $du = 27t^2\, dt$, and

$$s = \int_0^3 3t^2 \sqrt{1 + 9t^3}\; dt = \int_1^{244} \sqrt{u}\left(\frac{du}{9}\right) = \frac{1}{9}\left(\frac{u^{3/2}}{\frac{3}{2}}\right)\Bigg|_1^{244} = \frac{2}{27}(244\sqrt{244} - 1) \approx 282.3$$

∎

EXAMPLE 2   Use (13.21) to verify the familiar formula $s = R\theta$ for the length s of arc of a circle of radius R subtended by a central angle of θ radians.

---

* This is where the continuity of the derivatives is used—to guarantee the existence of the definite integral.

Solution

We set up our coordinate system so that the circle has its center at the origin and the central angle θ has its initial side along the positive x-axis (see Fig. 40). We can represent the circle by the parametric equations

$$x = R \cos t \qquad y = R \sin t \qquad 0 \leq t \leq 2\pi$$

in which the parameter t is a central angle. We seek the length s of arc from $t = 0$ to $t = \theta$.   By (13.21), we find

$$s = \int_0^\theta \sqrt{\left(\frac{dx}{dt}\right)^2 + \left(\frac{dy}{dt}\right)^2} \, dt = \int_0^\theta \sqrt{R^2 \sin^2 t + R^2 \cos^2 t} \, dt = \int_0^\theta R \, dt = R\theta$$

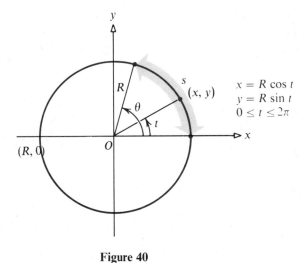

Figure 40   ∎

In case the circular arc is a full circle, that is, if $\theta = 2\pi$, then we get $s = 2\pi R$, the formula for the circumference of the circle.

EXAMPLE 3

Find the length of one arch of the cycloid

$$x = R(t - \sin t) \qquad y = R(1 - \cos t) \qquad R > 0$$

Solution

We obtain one arch of the cycloid when t changes from 0 to 2π. By using equation (13.21) with $dx/dt = R - R \cos t$ and $dy/dt = R \sin t$, we get

$$s = \int_0^{2\pi} \sqrt{R^2(1 - \cos t)^2 + R^2 \sin^2 t} \, dt$$

We simplify the radical before going any further. Since $\sin^2 t + \cos^2 t = 1$,

$$\sqrt{R^2(1 - \cos t)^2 + R^2 \sin^2 t} = \sqrt{R^2(2 - 2 \cos t)} = R\sqrt{2}\sqrt{1 - \cos t}$$

By using the half-angle identity $1 - \cos t = 2 \sin^2(t/2)$ and remembering that $\sin(t/2) \geq 0$ if $0 \leq t \leq 2\pi$, we get

$$R\sqrt{2}\sqrt{1 - \cos t} = 2R \sin \frac{t}{2}$$

Thus, the arc length s from $t = 0$ to $t = 2\pi$ is

$$s = \int_0^{2\pi} 2R \sin \frac{t}{2} \, dt = -4R \cos \frac{t}{2} \Big|_0^{2\pi} = 8R \qquad \blacksquare$$

ARC LENGTH OF A RECTANGULAR EQUATION $y = f(x)$

We verify next that formula (13.21) reduces to formula (6.22) in the case where the curve C is the graph of a function. If the curve C, defined by the parametric equations

$$x = x(t) \qquad y = y(t) \qquad a \leq t \leq b$$

is the graph of a function $y = f(x)$, then we can use x as the parameter t and write the parametric equations for C as

$$x = t \qquad y = f(t) \qquad a \leq t \leq b$$

Since $dx/dt = 1$ and $dy/dt = dy/dx = f'(x)$, equation (13.21) takes the form

$$s = \int_a^b \sqrt{\left(\frac{dx}{dt}\right)^2 + \left(\frac{dy}{dt}\right)^2} \, dt = \int_a^b \sqrt{1 + [f'(x)]^2} \, dx$$

which is formula (6.22).

ARC LENGTH OF A POLAR EQUATION $r = f(\theta)$

Suppose a curve C is given by the polar equation $r = f(\theta)$, where f is continuous and has a continuous derivative on $\alpha \leq \theta \leq \beta$. We can obtain a pair of parametric equations for C by using the rules for changing over to rectangular coordinates, namely,

(13.22) $$x = r \cos \theta \qquad y = r \sin \theta$$

We may replace r by $f(\theta)$ to obtain

(13.23) $$x = f(\theta) \cos \theta \qquad y = f(\theta) \sin \theta \qquad \alpha \leq \theta \leq \beta$$

These are parametric equations for the curve whose polar equation is $r = f(\theta)$.
Now, in preparation for using (13.21), we calculate, from (13.23),

$$\frac{dx}{d\theta} = -f(\theta) \sin \theta + f'(\theta) \cos \theta \qquad \frac{dy}{d\theta} = f(\theta) \cos \theta + f'(\theta) \sin \theta$$

After simplification, we obtain

$$\left(\frac{dx}{d\theta}\right)^2 + \left(\frac{dy}{d\theta}\right)^2 = [f(\theta)]^2 + [f'(\theta)]^2$$

Thus, from equation (13.21), the arc length s of the curve $r = f(\theta)$ from $\theta = \alpha$ to $\theta = \beta$ is

(13.24)
$$s = \int_\alpha^\beta \sqrt{[f(\theta)]^2 + [f'(\theta)]^2}\, d\theta$$

EXAMPLE 4   Find the arc length of the logarithmic spiral

$$r = f(\theta) = e^{3\theta}   \quad 0 \le \theta \le 2$$

Solution   By applying (13.24) with $f(\theta) = e^{3\theta}$ and $f'(\theta) = 3e^{3\theta}$, we get

$$s = \int_0^2 \sqrt{(e^{3\theta})^2 + (3e^{3\theta})^2}\, d\theta = \int_0^2 \sqrt{10e^{6\theta}}\, d\theta$$

$$= \int_0^2 \sqrt{10}\, e^{3\theta}\, d\theta = \frac{\sqrt{10}}{3} e^{3\theta}\bigg|_0^2 = \frac{\sqrt{10}}{3}(e^6 - 1) \approx 424.197 \quad \blacksquare$$

THE DIFFERENTIAL OF ARC LENGTH

We have shown that for a smooth curve C defined by the parametric equations

$$x = x(t)   \quad y = y(t)   \quad a \le t \le b$$

the arc length s along C from a to b is given by (13.21). The arc length s along C from a to some variable t will, in general, be a function of t given by

(13.25)
$$s = s(t) = \int_a^t \sqrt{\left(\frac{dx}{du}\right)^2 + \left(\frac{dy}{du}\right)^2}\, du$$

where, for convenience, we use u as an artificial variable.

Differentiating (13.25) with respect to t gives us

$$\frac{ds}{dt} = \sqrt{\left(\frac{dx}{dt}\right)^2 + \left(\frac{dy}{dt}\right)^2}$$

so that

(13.26)
$$\left(\frac{ds}{dt}\right)^2 = \left(\frac{dx}{dt}\right)^2 + \left(\frac{dy}{dt}\right)^2$$

In terms of differentials, we may write (13.26) as

(13.27)
$$(ds)^2 = (dx)^2 + (dy)^2$$
$$ds = \sqrt{(dx)^2 + (dy)^2}$$

Let's interpret (13.27) geometrically. As Figure 41 illustrates, the differential $ds = \sqrt{(dx)^2 + (dy)^2}$ is the length of the hypotenuse of a right triangle with sides of

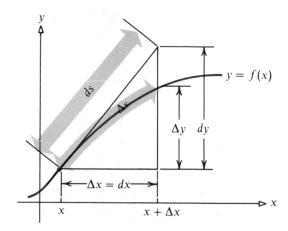

Figure 41

lengths dx and dy. The hypotenuse is along the tangent line to the curve because dy/dx is the slope of the tangent line.

The differential ds may be used to approximate the arc length Δs between two nearby points. This approximating tool is particularly useful when the definite integral (13.21) is difficult, or perhaps impossible, to evaluate.

EXAMPLE 5   Use the differential ds to approximate the arc length along the curve $\quad x = t^2, \quad y = t^3$ from $\quad t = 1 \quad$ to $\quad t = 1.1$.

Solution   In preparation for using (13.27), we calculate the differentials

$$dx = 2t \, dt \qquad \text{and} \qquad dy = 3t^2 \, dt$$

The differential $\quad dt = 0.1, \quad$ so that the values of dx and dy at $\quad t = 1 \quad$ are

$$dx = 2(1)(0.1) = 0.2 \qquad dy = 3(1)^2(0.1) = 0.3$$

Hence, an approximation to the arc length s is

$$ds = \sqrt{(dx)^2 + (dy)^2} = \sqrt{(0.2)^2 + (0.3)^2} = \sqrt{0.13} \approx 0.36 \qquad \blacksquare$$

EXERCISE 7

In Problems 1–12 find the arc length of each curve on the given interval.

1. $x(t) = t^3, \quad y(t) = t^2, \quad 0 \le t \le 2$

2. $x(t) = 3t^2 + 1, \quad y(t) = t^3 - 1, \quad 0 \le t \le 2$

3. $x(t) = t - 1, \quad y(t) = \frac{1}{2}t^2, \quad 0 \le t \le 2$

4. $x(t) = t^2, \quad y(t) = 2t, \quad 1 \le t \le 3$

5. $x(t) = 4 \sin t, \quad y(t) = 4 \cos t, \quad -\pi/2 \le t \le \pi/2$

6. $x(t) = 6 \sin t, \quad y(t) = 6 \cos t, \quad -\pi/2 \le t \le \pi/2$

7. $x(t) = 2 \sin t - 1, \quad y(t) = 2 \cos t + 1, \quad 0 \le t \le 2\pi$

8. $x(t) = e^t \sin t, \quad y(t) = e^t \cos t, \quad 0 \le t \le \pi$

9. $r = f(\theta) = e^{\theta/2}, \quad 0 \le \theta \le 2$

10. $r = f(\theta) = e^{2\theta}, \quad 0 \le \theta \le 2$

11. $r = f(\theta) = \cos^2(\theta/2), \quad 0 \le \theta \le \pi$ **12.** $r = f(\theta) = \sin^2(\theta/2), \quad 0 \le \theta \le \pi$

13. Find the arc length of the spiral $r = \theta, \quad 0 \le \theta \le 2\pi$.

14. Find the perimeter of the cardioid $r = f(\theta) = 1 - \cos\theta, \quad -\pi \le \theta \le \pi$.

15. Find the arc length of one arch of the four-cusped hypocycloid $x = b \sin^3 t, \quad y = b \cos^3 t, \quad 0 \le t \le \pi/2$.

16. Find the entire arc length of the curve $r = a \sin^3(\theta/3)$.

In Problems 17–22 find the distance a particle travels along the given path during the indicated time.

17. $x(t) = 3t, \quad y(t) = t^2 - 3, \quad 0 \le t \le 2$ **18.** $x(t) = t^2, \quad y(t) = 3t, \quad 0 \le t \le 2$

19. $x(t) = (t^2/2) + 1, \quad y(t) = \frac{1}{3}(2t + 3)^{3/2}, \quad 0 \le t \le 2$ **20.** $x(t) = a \cos t, \quad y(t) = a \sin t, \quad 0 \le t \le \pi$

21. $x(t) = \cos 2t, \quad y(t) = \sin^2 t, \quad 0 \le t \le \pi/2$ **22.** $x(t) = 1/t, \quad y(t) = \ln t, \quad 1 \le t \le 2$

In Problems 23–26 use the differential ds to approximate the indicated arc length.

23. $x = t^{1/3}, \quad y = t^2;$ from $t = 1$ to $t = 1.1$ **24.** $x = \sqrt{t}, \quad y = t^3;$ from $t = 1$ to $t = 1.2$

25. $x = a \sin t, \quad y = b \cos t;$ from $t = 0$ to $t = 0.1$ **26.** $x = e^{at}, \quad y = e^{bt};$ from $t = 0$ to $t = 0.2$

27. Find the length of the circumference of a circle of radius 1 by inscribing regular polygons in the circle, finding their perimeters, and then allowing the number of sides of the polygons to tend to infinity.

8. Curvature

Figure 42

Consider the curve given in Figure 42. It is evident that at the point P it is "more curved" than at the point Q. In this section we derive a formula to measure this property of "being curved"; the formula gives us a number called the *curvature of the curve at a point*.

Let P be any point on a smooth curve C, and let ϕ be the inclination to the positive x-axis of the tangent line to this curve at P, as shown in Figure 43. Let Q be

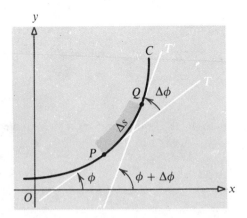

Figure 43

another point on the curve and let $\phi + \Delta\phi$ be the inclination to the positive x-axis of the tangent line at Q. Finally, let Δs be the arc length along C from P to Q. The amount by which the curve C bends at P can be described in terms of the rate at which the tangent line is turning as we travel along the curve at P, and this, in turn, can be measured by the rate at which the angle ϕ is changing at P. But the rate of change with respect to what? Surely not with respect to time since this does not make the notion of curvature intrinsic to the curve. To ensure that we are measuring a property of the curve itself rather than properties involving the way we move along the curve, we use the only "natural" parameter of the curve—its arc length. So, we study the rate of change of the angle ϕ with respect to the rate of change of the arc length. The ratio $\Delta\phi/\Delta s$ represents the average rate of change in the angle ϕ per unit of arc length along C. In general, this ratio, $\Delta\phi/\Delta s$, will approach a limit as $\Delta s \to 0$, namely, the derivative $d\phi/ds$.

(13.28) **DEFINITION** The *curvature κ of a curve C at a point (x, y) on C is*

$$\kappa = \left|\frac{d\phi}{ds}\right|$$

where ϕ is the inclination to the positive x-axis of the tangent line to C at (x, y) and s is the arc length as measured along the curve C.*

Thus, κ is the absolute value of the rate of change of ϕ with respect to s.

We will use (13.28) to calculate the curvature at any point on a straight line, but before we do this, let's guess the answer. A straight line does not bend at all. Hence, its curvature must be 0.

The curvature of a straight line is everywhere 0.

Proof In the case of a straight line, $\Delta\phi = 0$ everywhere on the line since the tangent line coincides with the line itself. Then $d\phi/ds = 0$, and, therefore, by (13.28), $\kappa = |0| = 0$ at all points of the line. ∎

Now let's use (13.28) to calculate the curvature at any point on a circle. Again, let's see if we can guess the answer. A circle is a curve that bends uniformly. Hence, we suspect that the curvature of a circle is a constant. What the constant is may surprise you!

The curvature of a circle is the same at every point on it, and it is equal to the reciprocal of the radius. That is, if R is the radius of a circle, then the curvature κ at any point on the circle is

(13.29)
$$\kappa = \frac{1}{R}$$

---

* The symbol κ is the Greek letter kappa.

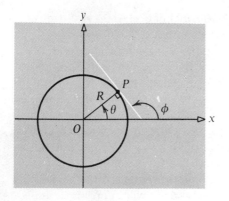

Figure 44

Proof We position the circle so that its center is at the origin, as shown in Figure 44. At any point P on the circle, the inclination ϕ to the positive x-axis of the tangent line is

$$\phi = \frac{\pi}{2} + \theta$$

where θ is a central angle. The arc length s is $s = R\theta$. Hence,

$$\phi = \frac{\pi}{2} + \frac{s}{R}$$

By differentiating ϕ with respect to s, we find that

$$\kappa = \left|\frac{d\phi}{ds}\right| = \frac{1}{R} \qquad \blacksquare$$

A FORMULA FOR THE CURVATURE OF THE GRAPH OF A FUNCTION $y = f(x)$

Let's now derive a formula for the curvature of the graph of a twice differentiable function $y = f(x)$ at a given point $P = (x, y)$ on the graph. The slope of the tangent line to the graph of $y = f(x)$ is $\tan \phi = dy/dx$,   so that

(13.30)
$$\phi = \tan^{-1}\left(\frac{dy}{dx}\right) = \tan^{-1} y'$$

By using (13.28) and the chain rule, we have

(13.31)
$$\kappa = \left|\frac{d\phi}{ds}\right| = \left|\left(\frac{d\phi}{dx}\right)\left(\frac{dx}{ds}\right)\right| = \left|\frac{d\phi/dx}{ds/dx}\right|$$

We first find $d\phi/dx$ from (13.30):

(13.32)
$$\frac{d\phi}{dx} = \frac{d}{dx}(\tan^{-1} y') = \frac{1}{1 + (y')^2}\left[\frac{d}{dx}(y')\right] = \frac{y''}{1 + (y')^2}$$

To obtain ds/dx, we use (13.27), namely, $(ds)^2 = (dx)^2 + (dy)^2$,   from which we conclude that

$$\left(\frac{ds}{dx}\right)^2 = 1 + \left(\frac{dy}{dx}\right)^2$$

(13.33)
$$\frac{ds}{dx} = \sqrt{1 + (y')^2}$$

By substituting the results (13.32) and (13.33) into (13.31), we obtain

$$\kappa = \left|\frac{y''/[1 + (y')^2]}{\sqrt{1 + (y')^2}}\right| = \frac{|y''|}{[1 + (y')^2]^{3/2}}$$

(13.34)   **The curvature κ of the graph of a twice differentiable function $y = f(x)$ at a point (x, y) on its graph is given by**

$$\kappa = \frac{|y''|}{[1 + (y')^2]^{3/2}}$$

where y' and y'' are to be evaluated at (x, y).

EXAMPLE 1   Find the curvature of the parabola $y = \frac{1}{4}x^2$ at the point $(2, 1)$. Compare this to the curvature of the graph of $y = (x^3 + 4)/12$ at $(2, 1)$.

Solution   From the equation $y = \frac{1}{4}x^2$,   we find that $y' = \frac{1}{2}x$ and $y'' = \frac{1}{2}$.   At the point $(2, 1)$, $y' = 1$, $y'' = \frac{1}{2}$.   By using this information in (13.34), we obtain

$$\kappa = \frac{\frac{1}{2}}{(1 + 1)^{3/2}} = \frac{1}{4\sqrt{2}} \approx 0.177$$

For $y = (x^3 + 4)/12$,   we find $y' = \frac{1}{4}x^2$ and $y'' = \frac{1}{2}x$.   At $(2, 1)$ these become $y' = 1$ and $y'' = 1$.   The curvature κ of this graph at $(2, 1)$ is

$$\kappa = \frac{1}{(1 + 1)^{3/2}} = \frac{1}{2\sqrt{2}} \approx 0.354$$

Figure 45 illustrates the graphs of these two functions. Note that they have the same tangent line at $(2, 1)$. The tangent line to the graph of $y = \frac{1}{4}x^2$ at $(2, 1)$ turns more slowly $(\kappa \approx 0.177)$ than does the tangent line to the graph of $y = (x^3 + 4)/12$ at $(2, 1)$ $(\kappa \approx 0.354)$.

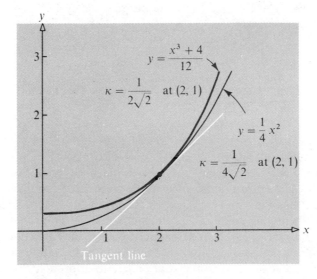

Figure 45

CURVATURE FOR PARAMETRIC EQUATIONS

The curvature κ for a smooth curve defined by the parametric equations $x = x(t)$, $y = y(t)$, $a \leq t \leq b$, is given by the formula

(13.35)
$$\kappa(t) = \frac{|x'(t)y''(t) - x''(t)y'(t)|}{[(x')^2 + (y')^2]^{3/2}}$$

The derivation of (13.35) is left as an exercise (Problem 36 in Exercise 8).

EXAMPLE 2   Find the curvature of the cycloid

$$x(t) = t - \sin t \qquad y(t) = 1 - \cos t \qquad \text{at any point}$$

Solution
$$x'(t) = 1 - \cos t \qquad x''(t) = \sin t$$
$$y'(t) = \sin t \qquad y''(t) = \cos t$$

Then, by (13.35),

$$\kappa = \frac{|(1 - \cos t)\cos t - (\sin t)(\sin t)|}{[(1 - \cos t)^2 + (\sin t)^2]^{3/2}} = \frac{|\cos t - (\cos^2 t + \sin^2 t)|}{(1 - 2\cos t + \cos^2 t + \sin^2 t)^{3/2}}$$

$$= \frac{|\cos t - 1|}{(2 - 2\cos t)^{3/2}} = \frac{|\cos t - 1|}{2^{3/2}(1 - \cos t)^{3/2}}$$

$$= \left(\frac{1}{2\sqrt{2}}\right)\left[\frac{1 - \cos t}{(1 - \cos t)^{3/2}}\right] = \frac{1}{2\sqrt{2}(1 - \cos t)^{1/2}}$$

Since $y = 1 - \cos t$, the curvature κ of a cycloid can be written compactly in terms of y as $\kappa = 1/\sqrt{8y}$.   ∎

RADIUS OF CURVATURE

If the curvature κ at a point P on a smooth curve C is not 0, we use the expression

(13.36)
$$\rho = \frac{1}{\kappa}$$

to represent the *radius of curvature* of C at P. This name was chosen because if the curve is a circle, then $\rho = 1/\kappa$ is indeed its radius.

The circle that is tangent to a given curve at a point P on its concave side and that has the same curvature as the given curve at P is called the *circle of curvature* or *osculating circle* of the given curve at P. The center (h, k) of this circle is called the *center of curvature*. The circle of curvature at a point P on a smooth curve C is, in fact, the limiting circle that results from the circle passing through three points P, Q, and R on C as Q and R are made to approach P. See Figure 46.

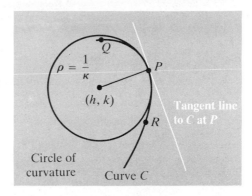

Figure 46

EXERCISE 8

In Problems 1–14 find the curvature of each curve.

1. $y = x^2$, at $(1, 1)$

2. $y = 2x - x^2$, at $(1, 1)$

3. $y = x^2 - x^3$, at $x = 1$

4. $y = x^3$, at $x = 2$

5. $y = \sqrt{x}$, at $x = 2$

6. $y = 1/x$, at $(1, 1)$

7. $y = x^{-3/2}$, at $x = 1$

8. $y = 1/\sqrt{x}$, at $(1, 1)$

9. $x^2 - y^2 = a^2$, at $x = 2a$

10. $4x^2 + 9y^2 = 36$, at $(0, 2)$

11. $y = \cos x$, at $x = 0$

12. $y = \sec x - 1$, at $x = \frac{1}{4}\pi$

13. $y = e^x$, at $(0, 1)$

14. $y = \ln(x + 1)$, at $x = 2$

15. Find the curvature of the curve $x = t^2$, $y = 2/t$ at the point where $t = t_1$.

16. Find the curvature of the curve $x = 2t$, $y = t^3$ at the point where $t = t_1$.

In Problems 17–22 find the radius of curvature of each curve.

17. $y = x^3 - 6x$, at $(1, -5)$

18. $y = 1/x^2$, at $(-1, 1)$

19. $y = \sin x$, at $(\pi/2, 1)$

20. $y = e^{-x}$, at $(0, 1)$

21. $x^2 + xy + y^2 = 3$, at $(1, 1)$

22. $y^2 - y + x = 0$, at $(0, 0)$

23. Find the radius of curvature of the curve $x = 3t^2$, $y = 3t - t^3$ at the point where $t = 1$.

24. Find the radius of curvature of the curve $x = \sin t$, $y = \cos 2t$ at the point where $t = \frac{1}{4}\pi$.

25. Show that the radius of curvature of the parabola $y = ax^2 + bx + c$ is a minimum at its vertex.

26. Show that the radii of curvature at the ends of the axes of the ellipse $b^2x^2 + a^2y^2 = a^2b^2$ are b^2/a and a^2/b.

27. Find the point at which the curvature is maximum on the curve $y = \ln x$.

28. Find the point at which the curvature is maximum on the curve $y = e^x$.

29. Find the point at which the curvature is maximum on the curve $y = \frac{1}{3}x^3$.

30. Show that the formula for the curvature of a polar curve $r = f(\theta)$ is

$$\kappa = \frac{|r^2 + 2(dr/d\theta)^2 - r(d^2r/d\theta^2)|}{[r^2 + (dr/d\theta)^2]^{3/2}}$$

In Problems 31–34 use the result of Problem 30 to find the curvature of each polar curve.

31. $r = 2 \cos 2\theta$, at $\theta = \pi/12$ **32.** $r = e^{a\theta}$, at $\theta = \pi/2$ **33.** $r = a\theta$, at $\theta = 1$, $a > 0$

34. $r = 1 - \cos \theta$, at $\theta = 0$

35. Show that the curvature of the catenary $y = a \cosh(x/a)$ at any point is a/y^2.

36. Derive formula (13.35) for κ using the same approach we used in deriving (13.34).

37. What is the curvature at a point of inflection of a plane curve?

38. Find the curvature of the cissoid $y^2(2 - x) = x^3$ at the point $(1, 1)$.

39. Find the curvature of the cycloid $x = \theta - \sin \theta$, $y = 1 - \cos \theta$ at the highest point of an arch.

9. Surface Area of a Solid of Revolution

We begin this discussion with the simple situation illustrated in Figure 47(a). If the line segment of length L is revolved about the axis of revolution, we obtain a frustum of a right circular cone whose surface area is

(13.37)
$$\frac{2\pi(R + r)L}{2} = \pi(R + r)L$$

where r and R are the base radii and L is the slant height, as shown in Figure 47(b). We shall use this equation to obtain a general formula for the surface area of a solid of revolution.

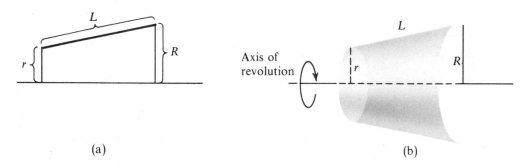

(a)   (b)

Figure 47

Let C denote a smooth curve defined by the parametric equations $x = x(t)$, $y = y(t)$, $a \le t \le b$, where $y = y(t) \ge 0$ on $[a, b]$. Revolve C about the x-axis to obtain a solid of revolution. We wish to find the surface area of this solid of revolution.

We begin by partitioning the interval $a \le t \le b$ into n subintervals $[a, t_1]$, $[t_1, t_2], \ldots, [t_{i-1}, t_i], \ldots, [t_{n-1}, b]$. We denote the length of the ith subinterval by Δt_i. Corresponding to each number $a, t_1, t_2, \ldots, t_{i-1}, t_i, \ldots, t_{n-1}, b$, there is a succession of points $P_0, P_1, \ldots, P_{i-1}, P_i, \ldots, P_{n-1}, P_n$ on the curve C. Join each

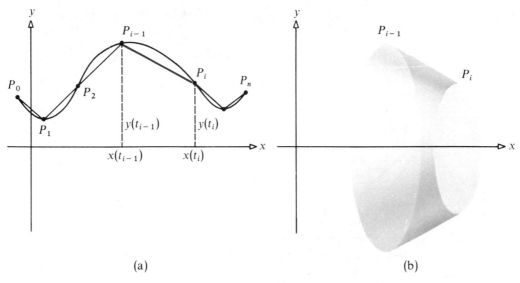

(a)   (b)

Figure 48

point to its successor by a line segment and concentrate on the line segment joining the points P_{i-1} and P_i. See Figure 48(a). When this segment, of length $d(P_{i-1}, P_i)$, is revolved about the x-axis, it generates a frustum of a right circular cone whose surface area is

(13.38)
$$\Delta S_i = \pi[y(t_{i-1}) + y(t_i)][d(P_{i-1}, P_i)]$$

By (13.37)

Looking back at the development for arc length in Section 7, we can combine (13.16) and (13.18) to get

(13.39)
$$d(P_{i-1}, P_i) = \sqrt{\left[\frac{dx}{dt}(u_i)\right]^2 + \left[\frac{dy}{dt}(v_i)\right]^2}\, \Delta t_i$$

where u_i and v_i are numbers in the ith subinterval. By using (13.39) in (13.38), the area of the surface generated by the sum of the line segments is

(13.40)
$$\sum_{i=1}^{n} \Delta S_i = \sum_{i=1}^{n} \pi[y(t_{i-1}) + y(t_i)] \sqrt{\left[\frac{dx}{dt}(u_i)\right]^2 + \left[\frac{dy}{dt}(v_i)\right]^2}\, \Delta t_i$$

This sum provides an approximation to the surface area we seek. By using an argument similar to the one used in developing formula (13.21) for arc length, it is plausible to define the surface area of the solid of revolution generated by revolving the curve C about the x-axis as

(13.41)
$$S = \int_a^b 2\pi y \sqrt{\left(\frac{dx}{dt}\right)^2 + \left(\frac{dy}{dt}\right)^2}\, dt$$

EXAMPLE 1 Find the surface area that is generated by revolving the curve $x = 2t^3$, $y = 3t^2$, $0 \le t \le 1$ about the x-axis.

Solution By using formula (13.41) with $dx/dt = 6t^2$, $dy/dt = 6t$, we get

$$S = 2\pi \int_0^1 3t^2 \sqrt{36t^4 + 36t^2} \, dt = 36\pi \int_0^1 t^3 \sqrt{t^2 + 1} \, dt$$

$$\underset{\underset{\text{Set } u = t^2 + 1}{\uparrow}}{=} \frac{36\pi}{2} \int_1^2 (u - 1)\sqrt{u} \, du = 18\pi \left(\frac{2}{5} u^{5/2} - \frac{2}{3} u^{3/2} \right) \Big|_1^2$$

$$= \frac{24\pi}{5} (\sqrt{2} + 1) \qquad \blacksquare$$

As an aid in memorizing formula (13.41), regard the integrand as the product of the slant height $\sqrt{(dx/dt)^2 + (dy/dt)^2}$ and the circumference $2\pi y$ of the circle traced by a point (x, y) on the corresponding subarc. Remember, too, that the limits of integration are parameter values, not x values.

From symmetry, we see that the surface area of a solid of revolution obtained by revolving the curve C: $x = x(t)$, $y = y(t)$, $a \le t \le b$, where $x = x(t) \ge 0$ on $[a, b]$, about the y-axis is

(13.42) $$S = \int_a^b 2\pi x \sqrt{\left(\frac{dx}{dt} \right)^2 + \left(\frac{dy}{dt} \right)^2} \, dt$$

**SURFACE AREA OF A SOLID OF REVOLUTION
USING A RECTANGULAR EQUATION $y = f(x)$**

Suppose we have a smooth curve C that is defined by the rectangular equation $y = f(x)$, $a \le x \le b$ where $f(x) \ge 0$ on $[a, b]$. A set of parametric equations for this curve is $x = t$ and $y = f(t)$. Then $dx/dt = 1$, $dy/dt = f'(t) = f'(x)$, and $dt = dx$. The surface area S of the solid of revolution obtained by revolving C about the x-axis is

(13.43) $$S = \int_a^b 2\pi y \sqrt{1 + [f'(x)]^2} \, dx$$

EXAMPLE 2 Find the surface area of the solid generated by revolving $y = \sqrt{x}$ about the x-axis from $x = 0$ to $x = 1$.

Solution Here we use equation (13.43):

$$S = \int_0^1 2\pi \sqrt{x} \sqrt{1 + \frac{1}{4x}} \, dx = 2\pi \int_0^1 \frac{1}{2} \sqrt{4x + 1} \, dx$$

$$\underset{\underset{\text{Set } u = 4x + 1}{\uparrow}}{=} \pi \int_1^5 \sqrt{u} \left(\frac{du}{4} \right) = \frac{\pi}{4} \left(\frac{u^{3/2}}{\frac{3}{2}} \right) \Big|_1^5 = \frac{\pi}{6} (5\sqrt{5} - 1) \qquad \blacksquare$$

SURFACE AREA OF A SOLID OF REVOLUTION
USING A POLAR EQUATION $r = f(\theta)$

Suppose a smooth curve C is given by the polar equation $r = f(\theta)$, $\alpha \le \theta \le \beta$. One set of parametric equations for this curve is

$$x(\theta) = f(\theta) \cos \theta \qquad\qquad y(\theta) = f(\theta) \sin \theta$$

Then

$$\frac{dx}{d\theta} = f'(\theta) \cos \theta - f(\theta) \sin \theta \qquad \frac{dy}{d\theta} = f'(\theta) \sin \theta + f(\theta) \cos \theta$$

$$\left(\frac{dx}{d\theta}\right)^2 + \left(\frac{dy}{d\theta}\right)^2 = f(\theta)^2 + f'(\theta)^2$$

So, the surface area S of the solid of revolution obtained by revolving C about the polar axis is

(13.44) $$S = 2\pi \int_\alpha^\beta f(\theta) \sin \theta \sqrt{[f(\theta)]^2 + [f'(\theta)]^2} \, d\theta$$

EXAMPLE 3 Find the surface area of the solid generated by revolving the arc of the circle $r = a$, $0 \le \theta \le \pi/4$,   about the polar axis.

Solution By using (13.44) with $r = f(\theta) = a$,   we find

$$S = 2\pi \int_0^{\pi/4} a \sin \theta \sqrt{a^2} \, d\theta = 2\pi a^2 \int_0^{\pi/4} \sin \theta \, d\theta$$

$$= 2\pi a^2 (-\cos \theta) \Big|_0^{\pi/4} = 2\pi a^2 \left(\frac{-\sqrt{2}}{2} + 1\right) = \pi a^2 (2 - \sqrt{2}) \qquad \blacksquare$$

EXERCISE 9

In Problems 1–14 find the surface area of the solid generated by revolving the given curve about the x-axis.

1. $x = 3t^2$, $y = 6t$, $0 \le t \le 1$

2. $x = t^2$, $y = 2t$, $0 \le t \le 3$

3. $x = a \cos^3\theta$, $y = a \sin^3 \theta$, $0 \le \theta \le \pi/2$

4. $x = a(t - \sin t)$, $y = a(1 - \cos t)$, $0 \le t \le \pi$

5. $y = x^3$, $0 \le x \le 1$

6. $y = \dfrac{x^4}{8} + \dfrac{1}{4x^2}$, $1 \le x \le 2$

7. $y = \dfrac{a}{2}(e^{x/a} + e^{-x/a})$, $0 \le x \le a$

8. $x = \dfrac{1}{4}y^2$, $0 \le y \le 2$

9. $y = \sqrt{a^2 - x^2}$, $-a \le x \le a$

10. $y = e^{-x}$, $0 \le x \le 1$

11. $y = e^x$, $0 \le x \le 1$

12. $y = e^{-x}$, $0 \le x < +\infty$

13. $x^{2/3} + y^{2/3} = a^{2/3}$, $0 \le x \le a$, $y \ge 0$

14. $y = \dfrac{1}{x}$, $1 \le x \le 2$

In Problems 15 and 16 find the surface area of the solid generated by revolving the given curve about the *y*-axis.

15. $x = 3t^2$, $y = 2t^3$, $0 \le t \le 1$ **16.** $x = 2t + 1$, $y = t^2 + 3$, $0 \le t \le 3$

In Problems 17–20 find the surface area of the solid generated by revolving the given curve about the polar axis.

17. $r = \sin \theta$, $0 \le \theta \le \pi/2$ **18.** $r = 1 + \cos \theta$, $0 \le \theta \le \pi$ **19.** $r = e^\theta$, $0 \le \theta \le \pi$

20. $r = 2a \cos \theta$, $0 \le \theta \le \pi/2$

21. Show that if the curve $y = 1/x$, $x > 1$, is revolved about the *x*-axis, then the volume of the solid generated is finite, but its surface area is infinite.

22. Prove that the surface area of a right circular cone of altitude *h* and base *b* is $\pi b \sqrt{h^2 + b^2}$.

23. When an arc of a catenary $y = \cosh x$, $a \le x \le b$, is rotated about the *x*-axis, it generates a surface called a *catenoid*, which has the least area of all surfaces generated by rotating curves having the same endpoints. What is its area?

24. Develop a formula for the surface area of a sphere of radius *R*.

Miscellaneous Exercises

In Problems 1–4 find the area.

1. Enclosed by the lines $\theta = 0$ and $\theta = 1$ and $r = e^{-\theta}$, $0 \le \theta \le 1$

2. Enclosed by the lines $\theta = 0$ and $\theta = 1$ and $r = e^\theta$, $0 \le \theta \le 1$

3. Enclosed by the lines $\theta = 1$ and $\theta = \pi$ and $r = 1/\theta$, $1 \le \theta \le \pi$

4. Inside the lemniscate $r^2 = 8 \cos 2\theta$ and outside the circle $r = 2$

In Problems 5 and 6 find the arc length of each graph.

5. $r = 4 \cos \theta$ **6.** $r = 3\theta^2$ from $\theta = 0$ to $\theta = 3$

In Problems 7–12 eliminate the parameter to obtain the rectangular equation.

7. $x = \dfrac{1}{t^2}$, $y = \dfrac{2}{t^2 + 1}$ **8.** $x = \dfrac{3t}{\sqrt{t^2 + 1}}$, $y = \dfrac{3}{\sqrt{t^2 + 1}}$

9. $x = \dfrac{4}{\sqrt{4 - t^2}}$, $y = \dfrac{4t}{\sqrt{4 - t^2}}$ **10.** $x = \sqrt{t - 3}$, $y = \sqrt{t + 1}$

11. $x = \sin \theta - 2$, $y = 4 - 2 \cos \theta$ **12.** $x = 2 + \tan \theta$, $y = 3 - 2 \sec \theta$

In Problems 13–18 find dy/dx and d^2y/dx^2 in terms of the parameter. [*Hint:* Use the chain rule when calculating d^2y/dx^2.]

13. $x = 3t^2$, $y = 2t$ **14.** $x = t + \dfrac{1}{t}$, $y = t - \dfrac{1}{t}$ **15.** $x = 2 \cos 2\theta$, $y = \sin \theta$

16. $x = 1 + e^{-t}$, $y = e^{3t}$ **17.** $x = \cot^2 \theta$, $y = \cot \theta$ **18.** $x = \sin t$, $y = \sec^2 t$

19. A sphere of radius *R* has a hole of radius $a < R$ drilled through it. The axis of the hole coincides with a diameter of the sphere. Find the surface area of that part of the sphere that remains.

20. A plug is made to repair the hole of the sphere in Problem 19. What is its surface area?

21. Sketch and compare the graphs of the following parametric equations.
   (a) $x = t$, $y = t^2$ (b) $x = \sqrt{t}$, $y = t$ (c) $x = e^t$, $y = e^{2t}$ (d) $x = \cos t$, $y = 1 - \sin^2 t$

22. Sketch and compare the graphs of the following parametric equations.
   (a) $x = \sec t$, $y = \tan t$ (b) $x = t$, $y = \sqrt{t^2 - 1}$ (c) $x = \sqrt{t+1}$, $y = \sqrt{t}$

23. Find an expression for d^2y/dx^2 if $x = f(t)$, $y = g(t)$, where f and g have second derivatives.

24. Find d^2y/dx^2 if $x = a\cos^3\theta$, $y = a\sin^3\theta$.

25. Use the figure to show that the coordinates (h, k) of the center of curvature of $y = f(x)$ are

$$h = x - \rho\sin\phi \qquad k = y + \rho\cos\phi$$

where ρ is the radius of curvature. Show that

$$\sin\phi = \frac{y'}{\sqrt{1 + (y')^2}} \qquad \cos\phi = \frac{1}{\sqrt{1 + (y')^2}}$$

so

$$h = x - \frac{y'[1 + (y')^2]}{y''} \qquad k = y + \frac{1 + (y')^2}{y''}$$

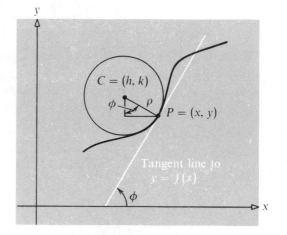

In Problems 26–30 use the result obtained in Problem 25.

26. Find the center of curvature of $y = x^2$ at $x = 1$.

27. Find the center of curvature of $y = \sin x$ at $x = \pi/2$.

28. Find the center of curvature of $y = x/(x + 1)$ at $(0, 0)$.

29. Find the center of curvature of $x^3 + y^3 = 4xy$ at $(2, 2)$.

30. Find the center of curvature of $xy = 4$ at $x = 2$.

31. Sketch the graph of $x = t^2 + 2$, $y = t^3 - 4t$. Find the points where the tangent line is horizontal and where it is vertical. Find an equation of the tangent line(s) at the point $(6, 0)$.

32. Express $r^2 = \cos 2\theta$ in rectangular coordinates free of radicals.

33. Find the area inside the outer loop but outside the inner loop of $r = 1 + 2\sin\theta$.

34. Find parametric equations for the circle $x^2 + y^2 = R^2$, using as parameter the slope m of the line through $(-R, 0)$ and the general point $P = (x, y)$ on the circle.

35. Find parametric equations for the parabola $y = x^2$ using as parameter the slope m of the line joining the point $(1, 1)$ to the general point $P = (x, y)$ on the parabola.

36. Show that the area enclosed by the graph of $r\theta = a$ and any two radii r_1 and r_2 is proportional to the difference of the radii, $r_1 - r_2$.

37. Find the point on the curve $x = \frac{4}{3}t^3 + 3t^2$, $y = t^3 - 4t^2$ such that the length of the curve from $(0, 0)$ to (x, y) is $(80\sqrt{2} - 40)/3$.

38. Show that the surface area of the solid generated by revolving the first-quadrant arc of $(x^2/a^2) + (y^2/b^2) = 1$ about the x-axis is

$$S = \pi b^2 + \frac{\pi ab}{e} \sin^{-1} e$$

where e is the eccentricity of the ellipse. (See p. 661, Problem 19.)

39. Show that $r = a \cos \theta + b \sin \theta$ is the equation of a circle. Find the center and radius of the circle.

40. Prove that the area of the triangle with vertices $(0, 0)$, (r_1, θ_1), (r_2, θ_2) is

$$A = \tfrac{1}{2} r_1 r_2 \sin(\theta_2 - \theta_1) \qquad 0 \le \theta_1 < \theta_2 \le \pi$$

41. Find the surface area of a parabolic reflector of a searchlight, generated by revolving an arc of a parabola about its axis, if the searchlight is 1 meter across and $\tfrac{1}{4}$ meter deep.

42. As a point P moves along a curve C, the center of curvature corresponding to P traces out a curve C_1 called the *evolute* of C; conversely, C is the *involute* of C_1. Show that parametric equations of the evolute of $y = \tfrac{1}{2}x^2$ are $h = -x^3$, $k = \tfrac{3}{2}x^2 + 1$. Then, eliminate the parameter x to obtain

$$h^2 = \frac{8}{27}(k - 1)^3$$

43. Refer to Problems 42 and 25. Find parametric equations and a rectangular equation for the evolute of $x = a \cos \theta$, $y = b \sin \theta$.

14

Vectors; Analytic Geometry in Space

1. Introduction to Vectors

Roughly speaking, a *vector** is a quantity that has both *magnitude and direction*. A vector is usually portrayed by an arrow of length equal to the magnitude of the vector and pointing in the appropriate direction.

We've already encountered vectors earlier in this book. For example, *velocity* is a vector. The velocity of an airplane can be represented by an arrow that points in the direction of movement (see Fig. 1, p. 738). The length of the arrow represents the speed of the airplane. If the airplane speeds up, the length of the arrow increases. If the direction of the airplane changes, the direction of the arrow changes.

Force and *acceleration* are also vectors.[†] If a force **F** is exerted on an object of mass *m*, the force causes the object to accelerate *in the direction of the force*. Thus,

---

* From the Latin *vehere:* "to carry."

[†] Boldface letters will be used to denote vectors, in order to distinguish them from numbers. For handwritten work, an arrow placed over a letter may be used to denote a vector.

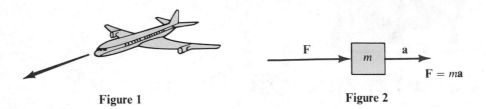

Figure 1   Figure 2

force and acceleration are quantities that have both magnitude and direction, and, hence, they are vectors. On the other hand, mass has magnitude but no direction, and, hence, it is not a vector. Figure 2 illustrates the situation just described. The relationship among force **F**, mass m, and acceleration **a** is, of course, given by Newton's second law of motion,

$$\mathbf{F} = m\mathbf{a}$$

which you may remember from Chapter 6.

Vectors are closely related to the geometric idea of a *directed line segment*. Consider two points P and Q. If P and Q are distinct points, there is exactly one line containing both P and Q. The points that are on the part of the line that joins P and Q—including P and Q—make up the *line segment from P to Q*. If we order the points P and Q, say, as PQ, then we have a *directed line segment from P to Q*, which we denote by $\overline{PQ}$. For the directed line segment $\overline{PQ}$, we call P the *initial point* and Q the *terminal point* (see Fig. 3).

If P and Q coincide, the directed line segment $\overline{PQ}$, which consists of a single point, is said to be *degenerate*.

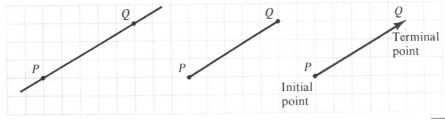

(a) Line containing P and Q (b) Line segment   (c) Directed line segment $\overrightarrow{PQ}$

Figure 3

Directed line segments that are not degenerate have both magnitude and direction. The magnitude of the directed line segment $\overline{PQ}$ is the distance from the point P to the point Q. The direction of $\overline{PQ}$ is from P to Q. The directed line segment $\overline{QP}$ is equal in magnitude to the directed line segment $\overline{PQ}$, but is opposite in direction since it is directed from Q to P (see Fig. 4).

Two nondegenerate directed line segments are said to be *equivalent* if they have the same magnitude and the same direction. In Figure 5 we see that the directed line

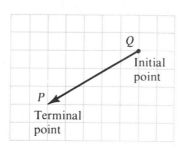

Directed line segment $\overrightarrow{QP}$

Figure 4

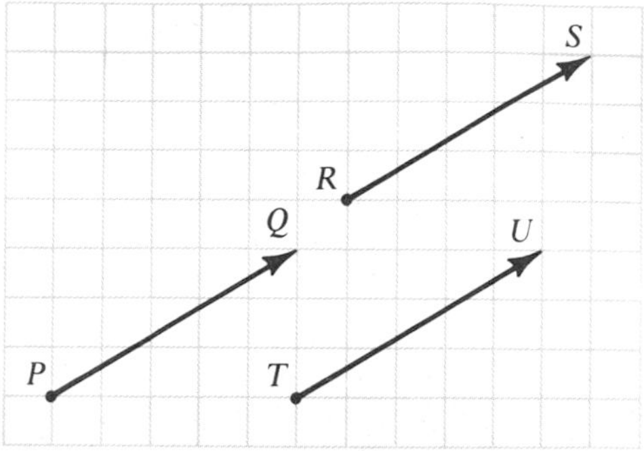

Figure 5

segments $\overrightarrow{PQ}$, $\overrightarrow{RS}$, and $\overrightarrow{TU}$ are all equivalent. For completeness, we agree that all degenerate directed line segments are equivalent.

We will find it convenient to define a vector in terms of equivalent directed line segments.

A *vector* is defined as a collection of equivalent directed line segments.

We may represent a vector by any one of the equivalent directed line segments in the collection. For example, if **v** is the vector defined by the collection of equivalent directed line segments $\overrightarrow{PQ}$, $\overrightarrow{RS}$, $\overrightarrow{TU}$,..., then we may represent **v** by $\overrightarrow{PQ}$ or by $\overrightarrow{RS}$ or by $\overrightarrow{TU}$ or .... Thus, even though the directed line segments $\overrightarrow{PQ}$, $\overrightarrow{RS}$, and $\overrightarrow{TU}$ are not equal, since they are equivalent, any one of them can be used to represent the vector **v**.

In our illustrations, we shall usually select a single directed line segment from the collection defining the vector to represent it. The vector defined by the collection of all degenerate directed line segments is called the *zero vector* **0**.

We shall find it useful to think of a vector as simply an arrow, keeping in mind that two arrows (vectors) are equal if they have the same direction and the same magnitude.

ADDING VECTORS

We define the *sum* **v** + **w** of two vectors as follows: We position the vectors **v** and **w** so that the terminal point of **v** coincides with the initial point of **w**. The vector **v** + **w** is then represented by the arrow directed from the initial point of **v** to the terminal point of **w**. Figure 6 illustrates this idea.

Vector addition is *commutative*. That is, if **v** and **w** are any two vectors, then

(14.1) $$\mathbf{v} + \mathbf{w} = \mathbf{w} + \mathbf{v}$$

Figure 6

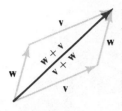

Figure 7

Figure 7 illustrates the validity of this fact. [Observe that statement (14.1) is another way of saying that opposite sides of a parallelogram are equal and parallel.]

We have said earlier that forces are vectors. But how do we know that forces "add" the same way vectors do? Well, physicists tell us they do, and laboratory experiments bear it out. Thus, if $\mathbf{F}_1$ and $\mathbf{F}_2$ are two forces simultaneously acting on an object, the vector sum $\mathbf{F}_1 + \mathbf{F}_2$ is equal to the force that produces the same effect as that obtained when the force $\mathbf{F}_1$ is applied followed by the force $\mathbf{F}_2$. The force $\mathbf{F}_1 + \mathbf{F}_2$ is sometimes called the *resultant* of $\mathbf{F}_1$ and $\mathbf{F}_2$ (see Fig. 8).

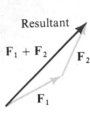

Resultant

$\mathbf{F}_1 + \mathbf{F}_2$

Figure 8

Another physical example of a vector is velocity. For example, if $\mathbf{w}$ is a vector describing the velocity of the wind relative to the earth and $\mathbf{v}$ is a vector describing the velocity of an airplane in the air, then $\mathbf{w} + \mathbf{v}$ is the vector describing the velocity of the airplane relative to the earth (see Fig. 9).

(a) Velocity of wind relative to earth

(b) Velocity of airplane relative to air

(c) Resultant equals velocity of airplane relative to earth

Figure 9

Vector addition is also *associative*; that is, if $\mathbf{u}$, $\mathbf{v}$, and $\mathbf{w}$ are vectors, then

$$(\mathbf{u} + \mathbf{v}) + \mathbf{w} = \mathbf{u} + (\mathbf{v} + \mathbf{w})$$

See Figure 10.

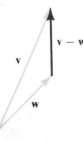

Figure 12

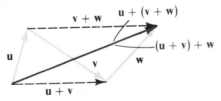

Figure 10

Figure 11

If $\mathbf{w}$ is a vector, then $-\mathbf{w}$ is defined as the vector having the same magnitude as $\mathbf{w}$, but with direction opposite to $\mathbf{w}$ (Fig. 11). We define the *difference* $\mathbf{v} - \mathbf{w}$ as

$$\mathbf{v} - \mathbf{w} = \mathbf{v} + (-\mathbf{w})$$

As Figure 12 illustrates, to subtract $\mathbf{w}$ from $\mathbf{v}$, that is, to find $\mathbf{v} - \mathbf{w}$, we need to locate the vector that, when added to $\mathbf{w}$, produces $\mathbf{v}$.

Each definition given above applies to the zero vector **0** when it is represented by a degenerate line segment. We list two properties of the zero vector:

$$\mathbf{v} + \mathbf{0} = \mathbf{v} \qquad \mathbf{v} - \mathbf{v} = \mathbf{0} \qquad \textbf{for any vector v}$$

MULTIPLYING VECTORS BY NUMBERS

We use the word *scalar* to mean a real number. Scalars are quantities that have only *magnitude*. In physics, the quantities *mass*, *time*, *density*, *temperature*, and *speed* are examples of scalars.

If $\alpha \neq 0$ is a scalar and **v** is a vector, the *product* $\alpha\mathbf{v}$ is defined as:

1. The vector having magnitude $|\alpha|$ times the magnitude of **v** and having the same direction as **v** if $\alpha > 0$

2. The vector having magnitude $|\alpha|$ times the magnitude of **v** and having the opposite direction as **v** if $\alpha < 0$

3. If $\alpha = 0$, then $\alpha\mathbf{v} = \mathbf{0}$. See Figure 13.

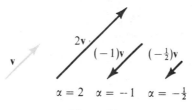

Figure 13

The product of a scalar and a vector has the following properties:

$$(0)\mathbf{v} = \mathbf{0} \qquad (1)\mathbf{v} = \mathbf{v} \qquad (-1)\mathbf{v} = -\mathbf{v}$$

$$(\alpha + \beta)\mathbf{v} = \alpha\mathbf{v} + \beta\mathbf{v} \qquad \alpha(\mathbf{v} + \mathbf{w}) = \alpha\mathbf{v} + \alpha\mathbf{w} \qquad \alpha(\beta\mathbf{v}) = (\alpha\beta)\mathbf{v}$$

EXAMPLE 1   Given the vectors **v**, **w**, and **u** in Figure 14, construct:

(a) $\mathbf{v} - \mathbf{w}$ (b) $2\mathbf{v} - \mathbf{w} + \mathbf{u}$ (c) $\frac{2}{3}\mathbf{u} + \frac{1}{2}(\mathbf{v} - \mathbf{w})$

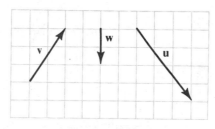

Figure 14

Solution   The solutions are illustrated in Figure 15.

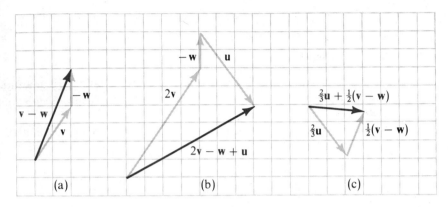

Figure 15   ∎

MAGNITUDES OF VECTORS

If $\mathbf{v}$ is a vector, we use the symbol $\|\mathbf{v}\|$ to denote the magnitude of $\mathbf{v}$. Since $\|\mathbf{v}\|$ equals the length of the directed line segment that represents $\mathbf{v}$, it is not surprising that $\|\mathbf{v}\|$ has many of the properties of absolute value. For example,

$$\|\mathbf{v}\| \geq 0$$
$$\|\mathbf{v}\| = 0 \qquad \text{if and only if} \qquad \mathbf{v} = \mathbf{0}$$
$$\|-\mathbf{v}\| = \|\mathbf{v}\|$$

Other important properties of the magnitude of a vector are:

(14.2)   **If α is a scalar and $\mathbf{v}$ is a vector, then $\|\alpha\mathbf{v}\| = |\alpha|\,\|\mathbf{v}\|$.**

(14.3)   **If $\mathbf{v}$ and $\mathbf{w}$ are any two vectors, then $\|\mathbf{v} + \mathbf{w}\| \leq \|\mathbf{v}\| + \|\mathbf{w}\|$.**

Property (14.2) is a direct consequence of the definition. Property (14.3), usually referred to as the *triangle inequality*, is the vector expression of the fact that the length of any side of a triangle does not exceed the sum of the lengths of the other two sides. (Refer back to Fig. 7 for an illustration.) An analytic proof of this result will be given in Section 5.

A vector $\mathbf{v}$ whose magnitude is 1 is called a *unit vector*. If $\mathbf{v} \neq \mathbf{0}$,   we can obtain a unit vector in the same direction as $\mathbf{v}$ by forming the vector $\mathbf{v}/\|\mathbf{v}\|$. This is a unit vector because

$$\left\|\frac{\mathbf{v}}{\|\mathbf{v}\|}\right\| = \frac{\|\mathbf{v}\|}{\|\mathbf{v}\|} = 1$$

In the above discussion we did not concern ourselves with *dimensionality*. In fact, everything we have said thus far is true in two-, three-, or n-dimensional space.

However, there are some vector concepts that relate directly to the dimension. For example, the cross product, which will be taken up in Section 6, is used only in three-dimensional space.

Vectors in the plane (two dimensions) are discussed next; vectors in space (three dimensions) are discussed in Section 4.

EXERCISE 1

1. State which of the following are scalars and which are vectors.
   (a) Volume   (b) Speed   (c) Weight
   (d) Work   (e) Mass   (f) Momentum
   (g) Distance   (h) Magnetic field intensity (i) Age

2. Given the vectors **v**, **w**, and **u** in the figure, construct:
   (a) $2\mathbf{v}$ (b) $-2\mathbf{v}$ (c) $\mathbf{v} + \mathbf{w}$ (d) $\mathbf{v} - \mathbf{w}$
   (e) $\mathbf{w} - \mathbf{v}$ (f) $\mathbf{v} - 2\mathbf{w}$ (g) $\mathbf{v} - \mathbf{w} + 3\mathbf{u}$ (h) $2\mathbf{u} - \frac{1}{3}(\mathbf{v} - \mathbf{w})$

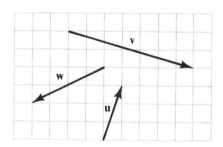

In Problems 3–10 use the figure at the right.

3. Find the vector **x** if $\mathbf{x} + \mathbf{B} = \mathbf{F}$.

4. Find the vector **x** if $\mathbf{x} + \mathbf{D} = \mathbf{E}$.

5. Write **C** in terms of **E**, **D**, and **F**.

6. Write **G** in terms of **C**, **D**, **E**, and **K**.

7. Write **E** in terms of **G**, **H**, and **D**.

8. Write **E** in terms of **A**, **B**, **C**, and **D**.

9. What is $\mathbf{A} + \mathbf{B} + \mathbf{K} + \mathbf{G}$?

10. What is $\mathbf{A} + \mathbf{B} + \mathbf{C} + \mathbf{H} + \mathbf{G}$?

11. If $\|\mathbf{v}\| = 3$,   find $\|4\mathbf{v}\|$.

12. If $\|\mathbf{v}\| = 6$,   find $\|-2\mathbf{v}\|$.

13. Let **v** and **w** be nonzero vectors represented by arrows with the same initial point. Let the terminal points of **v** and **w** be P and Q, respectively. Let **u** denote the vector represented by an arrow from the initial point of **v** to the midpoint of the directed line segment $\overrightarrow{PQ}$. Write **u** in terms of **v** and **w**.

14. Find nonzero scalars α and β so that

$$\alpha\mathbf{v} + \beta(\mathbf{v} - \mathbf{w}) + 4(\mathbf{v} + \mathbf{w}) = \mathbf{0}$$

for every pair of vectors $\mathbf{v}$ and $\mathbf{w}$.

2. Vectors in the Plane

The advantage of using a coordinate system to provide analytic justification for geometric results has been evident since Chapter 1. We use the same idea to advantage with vectors in the plane.

Consider a rectangular coordinate system in which each point P is given by an ordered pair (x, y) of real numbers. Let $\mathbf{i}$ denote a unit vector (that is, $\|\mathbf{i}\| = 1$) directed along the positive x-axis, and let $\mathbf{j}$ denote a unit vector directed along the positive y-axis. Then every vector $\mathbf{v}$ in the plane can be written uniquely in terms of the vectors $\mathbf{i}$ and $\mathbf{j}$ as

$$\mathbf{v} = a\mathbf{i} + b\mathbf{j}$$

for some choice of numbers a and b. The numbers a and b are called the *components* of $\mathbf{v}$ in the x direction and y direction, respectively. Figure 16(a) illustrates the idea, where $b = 1$ and $a = 2$. In Figure 16(b) the vector $\mathbf{v} = 3\mathbf{i} - 5\mathbf{j}$ has the component 3 in the x direction and the component -5 in the y direction.

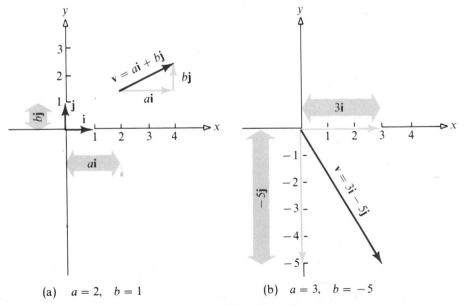

(a) $a = 2$, $b = 1$ (b) $a = 3$, $b = -5$

Figure 16

To determine the components of a vector **v**, we select any one of the directed line segments from the collection that defines **v**. Thus, if

$$P_1 = (x_1, y_1) \qquad \text{and} \qquad P_2 = (x_2, y_2)$$

are the initial and terminal points, respectively, of a directed line segment from the collection that defines **v**, then

$$\mathbf{v} = (x_2 - x_1)\mathbf{i} + (y_2 - y_1)\mathbf{j}$$

See Figure 17.

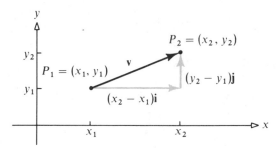

Figure 17

The use of any of the equivalent directed line segments will give us the same components. For example, the directed line segment from the point (3, 2) to the point (8, 6) is equivalent to the directed line segment from the point $(-1, 2)$ to the point (4, 6). Each of these directed line segments represent the vector $5\mathbf{i} + 4\mathbf{j}$ (see Fig. 18).

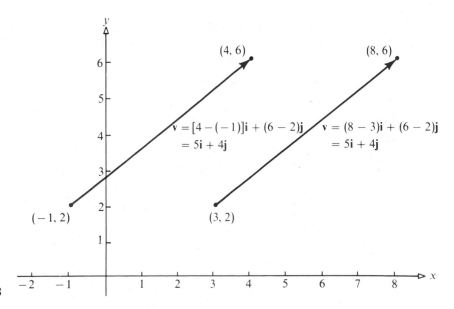

Figure 18

Two vectors v and w are equal if and only if their corresponding components are equal. That is:

$$\text{If}\quad \mathbf{v} = a_1\mathbf{i} + b_1\mathbf{j}\quad \text{and}\quad \mathbf{w} = a_2\mathbf{i} + b_2\mathbf{j}$$
$$\text{then}\quad \mathbf{v} = \mathbf{w}\quad \text{if and only if}\quad a_1 = a_2,\ \ b_1 = b_2$$

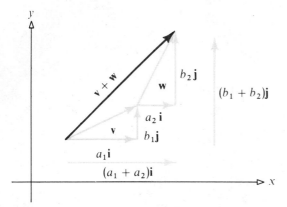

The magnitude of the vector $\mathbf{v} = a\mathbf{i} + b\mathbf{j}$ is found by using the Pythagorean theorem. Thus,

$$\|\mathbf{v}\| = \sqrt{a^2 + b^2}$$

To add two vectors, we merely add the respective components. That is, if $\mathbf{v} = a_1\mathbf{i} + b_1\mathbf{j}$ and $\mathbf{w} = a_2\mathbf{i} + b_2\mathbf{j}$, then

$$\mathbf{v} + \mathbf{w} = (a_1 + a_2)\mathbf{i} + (b_1 + b_2)\mathbf{j}$$

Figure 19 illustrates how this works.

Figure 19

To subtract two vectors, we subtract the respective components. For example, if

$$\mathbf{v} = 2\mathbf{i} - 3\mathbf{j}\quad \text{and}\quad \mathbf{w} = -\mathbf{i} + 8\mathbf{j}$$

then

$$\mathbf{v} - \mathbf{w} = (2\mathbf{i} - 3\mathbf{j}) - (-\mathbf{i} + 8\mathbf{j}) = 3\mathbf{i} - 11\mathbf{j}$$

To multiply a vector by a scalar, we multiply each component by the scalar. That is, if α is a scalar and if the vector is $\mathbf{v} = a\mathbf{i} + b\mathbf{j}$, then

$$\alpha\mathbf{v} = \alpha(a\mathbf{i} + b\mathbf{j}) = (\alpha a)\mathbf{i} + (\alpha b)\mathbf{j}$$

Figure 20 illustrates the idea when $\alpha > 0$.

As a result of the above properties, we may treat vectors according to the usual rules of algebra.

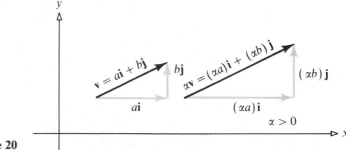

Figure 20

EXAMPLE 1   If $\mathbf{v} = 2\mathbf{i} - 3\mathbf{j}$ and $\mathbf{w} = -\mathbf{i} + 2\mathbf{j}$,   find:
(a) $2\mathbf{v} - \mathbf{w}$ (b) $\|\mathbf{v}\|$ (c) $\|-3\mathbf{v} + 2\mathbf{w}\|$

Solution   (a) $2\mathbf{v} - \mathbf{w} = 2(2\mathbf{i} - 3\mathbf{j}) - (-\mathbf{i} + 2\mathbf{j}) = (4\mathbf{i} - 6\mathbf{j}) - (-\mathbf{i} + 2\mathbf{j}) = 5\mathbf{i} - 8\mathbf{j}$

(b) $\|\mathbf{v}\| = \sqrt{2^2 + (-3)^2} = \sqrt{4 + 9} = \sqrt{13}$

(c) $-3\mathbf{v} + 2\mathbf{w} = -3(2\mathbf{i} - 3\mathbf{j}) + 2(-\mathbf{i} + 2\mathbf{j}) = (-6\mathbf{i} + 9\mathbf{j}) + (-2\mathbf{i} + 4\mathbf{j})$
$$= -8\mathbf{i} + 13\mathbf{j}$$
$$\|-3\mathbf{v} + 2\mathbf{w}\| = \sqrt{(-8)^2 + (13)^2} = \sqrt{64 + 169} = \sqrt{233} \quad \blacksquare$$

EXAMPLE 2   Find a unit vector in the same direction as $\mathbf{v} = 3\mathbf{i} - 4\mathbf{j}$.   What is the unit vector in the opposite direction?

Solution   Recall that a unit vector in the direction of $\mathbf{v}$ is the vector $\mathbf{v}/\|\mathbf{v}\|$.   Since $\|\mathbf{v}\| = \sqrt{3^2 + (-4)^2} = \sqrt{25} = 5$,   the unit vector in the direction of $3\mathbf{i} - 4\mathbf{j}$ is $\frac{3}{5}\mathbf{i} - \frac{4}{5}\mathbf{j}$.   The unit vector in the opposite direction is $(-1)(\frac{3}{5}\mathbf{i} - \frac{4}{5}\mathbf{j}) = -\frac{3}{5}\mathbf{i} + \frac{4}{5}\mathbf{j}$. $\blacksquare$

The next example illustrates one use of vectors in navigation.

EXAMPLE 3   An airplane has an air speed of 400 kilometers per hour in an easterly direction. The wind velocity is 80 kilometers per hour in a southeasterly direction.
(a)   Find a unit vector having southeast as direction.
(b)   Find a vector 80 units long having the same direction as the vector found in part (a).
(c)   Use parts (a) and (b) to find the actual speed of the airplane relative to the ground.

Solution   We set up a coordinate system in which the direction north is along the positive y-axis. Then the direction east is along the positive x-axis (see Fig. 21). Using a scale of   1 unit = 1 kilometer per hour,   we set

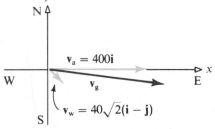

$\mathbf{v}_a$ = Velocity of the airplane in the air
       = $400\mathbf{i}$

$\mathbf{v}_g$ = Velocity of the airplane relative to the ground

$\mathbf{v}_w$ = Velocity of the wind

Figure 21

(a)   A vector having southeast as direction is $\mathbf{i} - \mathbf{j}$.   The corresponding unit vector in this direction is
$$\frac{\mathbf{i} - \mathbf{j}}{\|\mathbf{i} - \mathbf{j}\|} = \frac{\mathbf{i} - \mathbf{j}}{\sqrt{1 + 1}} = \frac{1}{\sqrt{2}}(\mathbf{i} - \mathbf{j})$$

(b) The velocity of the wind $\mathbf{v}_w$ is a vector in the direction of $(1/\sqrt{2})(\mathbf{i} - \mathbf{j})$ with magnitude 80. Hence,

$$\mathbf{v}_w = 80 \frac{1}{\sqrt{2}}(\mathbf{i} - \mathbf{j}) = 40\sqrt{2}(\mathbf{i} - \mathbf{j})$$

(c) The velocity $\mathbf{v}_g$ of the airplane relative to the ground is

$$\mathbf{v}_g = \mathbf{v}_a + \mathbf{v}_w = 400\mathbf{i} + 40\sqrt{2}(\mathbf{i} - \mathbf{j}) = (400 + 40\sqrt{2})\mathbf{i} - 40\sqrt{2}\mathbf{j}$$

The actual speed of the airplane relative to the ground is

$$\|\mathbf{v}_g\| = \sqrt{(400 + 40\sqrt{2})^2 + (40\sqrt{2})^2} \approx 460.06 \text{ kilometers per hour} \qquad \blacksquare$$

$$\uparrow$$
$$\text{Use a calculator}$$

In Section 5 we find the direction of the airplane relative to the ground. Those of you who are well versed in trigonometry can find the direction now, but as you will see, the method we use in Section 5 is much easier.

EXERCISE 2

In Problems 1–6 write the vector represented by the directed line segment $\overrightarrow{P_1P_2}$.

1. $P_1 = (2, 3);\quad P_2 = (6, -2)$ **2.** $P_1 = (4, -1);\quad P_2 = (-3, 2)$ **3.** $P_1 = (0, 5);\quad P_2 = (-1, 6)$

4. $P_1 = (-2, 0);\quad P_2 = (3, 2)$ **5.** $P_1 = (0, 0);\quad P_2 = (x, y)$ **6.** $P_1 = (0, 0);\quad P_2 = (0, y)$

In Problems 7–12 find the magnitude of $\mathbf{v}$.

7. $\mathbf{v} = 4\mathbf{i} - 3\mathbf{j}$ **8.** $\mathbf{v} = \mathbf{i} - 12\mathbf{j}$ **9.** $\mathbf{v} = \mathbf{i} + \mathbf{j}$

10. $\mathbf{v} = \mathbf{i} - \mathbf{j}$ **11.** $\mathbf{v} = a\mathbf{i} - a\mathbf{j}$ **12.** $\mathbf{v} = (\cos \theta)\mathbf{i} + (\sin \theta)\mathbf{j}$

In Problems 13–18 use $\mathbf{v} = 2\mathbf{i} - 3\mathbf{j}$ and $\mathbf{w} = \mathbf{i} + 2\mathbf{j}$ to find the indicated quantity.

13. $2\mathbf{v} - \mathbf{w}$ **14.** $\mathbf{v} + 5\mathbf{w}$ **15.** $\frac{1}{3}\mathbf{v} + \frac{1}{2}\mathbf{w}$

16. $\frac{2}{3}\mathbf{v} - \frac{1}{2}\mathbf{w}$ **17.** $\|\mathbf{v} - \mathbf{w}\|$ **18.** $\|\mathbf{v} + \mathbf{w}\|$

In Problems 19–24 find a unit vector in the same direction as $\mathbf{v}$. What is the unit vector in the opposite direction?

19. $\mathbf{v} = 5\mathbf{i} - 12\mathbf{j}$ **20.** $\mathbf{v} = 3\mathbf{i} + 4\mathbf{j}$ **21.** $\mathbf{v} = 2\mathbf{i} + \mathbf{j}$

22. $\mathbf{v} = \mathbf{i} - \mathbf{j}$ **23.** $\mathbf{v} = \frac{1}{2}\mathbf{i} + \frac{\sqrt{3}}{2}\mathbf{j}$ **24.** $\mathbf{v} = \frac{\sqrt{2}}{2}(\mathbf{i} - \mathbf{j})$

In Problems 25–30 find all vectors $\mathbf{v} = a\mathbf{i} + b\mathbf{j}$ that have the described properties.

25. Magnitude is 4; makes an angle of 30° with the positive x-axis

26. Magnitude is 2; makes an angle of 45° with the positive x-axis

27. Magnitude is 5; $\mathbf{i}$ component is twice the $\mathbf{j}$ component

28. Magnitude is 2; **i** component equals **j** component

29. **j** component is 1; makes an angle of 135° with the positive *x*-axis

30. **i** component is −3; makes an angle of 210° with the positive *x*-axis

31. An airplane has an air speed of 500 kilometers per hour in an easterly direction. If the wind velocity is 60 kilometers per hour in a northwesterly direction, find the speed of the airplane relative to the ground.

32. An airplane, after 1 hour of flying, arrives at a point 200 miles due south of the departure point. If, during the flight, there was a steady wind of 20 miles per hour from the northwest, what was the airplane's average air speed?

33. Forces of 2 and 5 pounds act at an angle of 30° on an object. Determine graphically what force is necessary to balance them.

34. Forces of 1 and 2 pounds act at an angle of 135° on an object. Determine graphically what force is necessary to balance them.

35. Represent graphically:
   (a)   A velocity of 20 kilometers per hour in a direction 60° west of north
   (b)   A velocity of 30 kilometers per hour in a direction 30° north of east

36. Forces $\mathbf{F}_1$, $\mathbf{F}_2$, $\mathbf{F}_3$, and $\mathbf{F}_4$ act as shown in the figure on an object Q. What force is needed to prevent Q from moving?

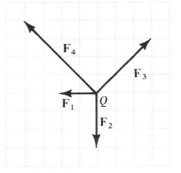

37. An airplane travels in a northeasterly direction at 250 kilometers per hour relative to the ground, due to the fact that there is an easterly wind of 50 kilometers per hour relative to the ground. How fast would the plane be going if there were no wind?   [*Hint:*   Use   1 unit = 50 kilometers per hour.]

38. A woman traveling east at 8 miles per hour finds that the wind appears to be coming from the north. Upon doubling her speed, she finds that the wind appears to be coming from the northeast. Find the velocity of the wind.

39. Find all numbers x so that $\|\mathbf{v} + \mathbf{w}\| = 5$,   where $\mathbf{v} = 2\mathbf{i} - \mathbf{j}$ and $\mathbf{w} = x\mathbf{i} + 3\mathbf{j}$.

40. Find all numbers x so that the vector represented by $\overrightarrow{P_1P_2}$ with $P_1 = (-3, 1)$ and $P_2 = (x, 4)$ has length 5.

3. Rectangular Coordinates in Space

In Chapter 1 we established a correspondence between the points on a line and the real numbers. Then we showed that each point in a plane can be associated with an ordered pair of real numbers. Here we show that each point in three-dimensional space can be associated with an *ordered triple* of real numbers.

We begin by selecting a fixed point called the *origin*. Through the origin, we draw three mutually perpendicular lines; these are called the *coordinate axes*, and they are usually labeled the *x-axis, y-axis, and z-axis.* On each of the three lines we choose one direction as positive and select an appropriate scale, as indicated in Figure 22.

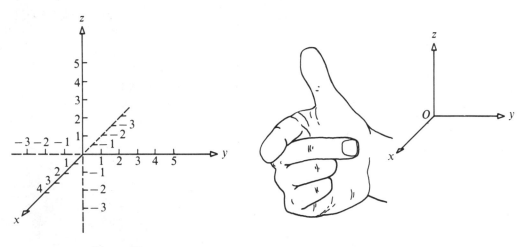

Figure 22   **Figure 23**

As indicated in Figure 23, we position the positive *z*-axis so that the system is *right-handed*. This conforms to the so-called *right-hand rule*, which asserts that if the index finger of the right hand points in the direction of the positive *x*-axis and the middle finger points in the direction of the positive *y*-axis, then the thumb will point in the direction of the positive *z*-axis.*

Just as we did in one and two dimensions, we assign coordinates to each point P in three dimensions. Specifically, we identify a point P with an ordered triple of real numbers (x, y, z), and we refer to it as "the point (x, y, z)." Thus, "the point $(3, 5, 7)$" is the point for which $x = 3$, $y = 5$, $z = 7$; that is, starting from the origin, we reach P by moving 3 units along the positive *x*-axis, then 5 units in the direction of the positive *y*-axis, and finally, 7 units in the direction of the positive *z*-axis. Figure 24 illustrates the location of the point $(3, 5, 7)$, as well as the points $(3, 5, 0)$ and $(0, 5, 0)$. Observe that any point on the *x*-axis will have the form $(x, 0, 0)$. Similarly, $(0, y, 0)$ and $(0, 0, z)$ represent points on the *y*-axis and *z*-axis, respectively.

In addition, all points of the form $(x, y, 0)$ constitute a plane called the *xy*-plane. This plane is perpendicular to the *z*-axis. Similarly, the points $(0, y, z)$ form the *yz*-plane, which is perpendicular to the *x*-axis; and the points $(x, 0, z)$ form the *xz*-plane, which is perpendicular to the *y*-axis (see Fig. 25). Figure 25 also illustrates

---

* Although there are left-handed systems and left-handed rules, we shall adopt the usual convention and only use a right-handed system.

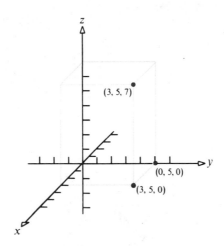

Figure 24

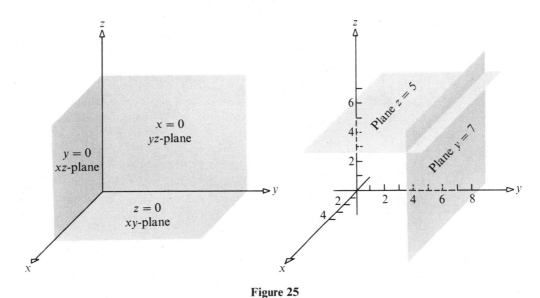

Figure 25

that points of the form (x, y, z), where $z = 5$, lie in a plane parallel to the xy-plane. Similarly, points (x, y, z), where $y = 7$, lie in a plane parallel to the xz-plane.

DISTANCE IN SPACE

To derive a formula for the distance $|P_1 P_2|$ between two points $P_1 = (x_1, y_1, z_1)$ and $P_2 = (x_2, y_2, z_2)$, we apply the Pythagorean theorem twice. As Figure 26 illustrates, we utilize the point $A = (x_2, y_2, z_1)$. The first application of the

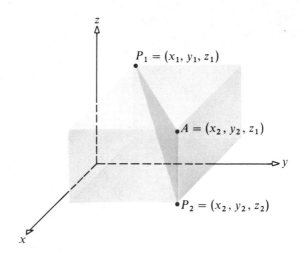

Figure 26

Pythagorean theorem involves observing that the triangle P_1AP_2 is a right triangle in which the side of length $|P_1P_2|$ is the hypotenuse. As a result,

$$(14.4) \qquad |P_1P_2| = \sqrt{|P_1A|^2 + |AP_2|^2}$$

The points P_1 and A lie in a plane parallel to the xy-plane. Thus, we can use the formula for distance in two dimensions and obtain

$$(14.5) \qquad |P_1A| = \sqrt{(x_2 - x_1)^2 + (y_2 - y_1)^2}$$

The points P_2 and A lie along a line parallel to the z-axis, so that $|AP_2| = |z_2 - z_1|$ and $|AP_2|^2 = (z_2 - z_1)^2$. This fact, together with (14.4) and (14.5), gives a formula for the distance between two points in three-dimensional space, namely,

$$(14.6) \qquad |P_1P_2| = \sqrt{(x_2 - x_1)^2 + (y_2 - y_1)^2 + (z_2 - z_1)^2}$$

For example, if $P_1 = (1, 3, -2)$ and $P_2 = (2, -1, -3)$, the distance $|P_1P_2|$ is

$$|P_1P_2| = \sqrt{(2 - 1)^2 + (-1 - 3)^2 + (-3 + 2)^2}$$
$$= \sqrt{1 + 16 + 1} = \sqrt{18} = 3\sqrt{2}$$

THE SPHERE

In three-dimensional space the collection of all points that are the same distance from some fixed point is called a *sphere* (see Fig. 27). The constant distance is called the *radius* and the fixed point is the *center* of the sphere. Any point $P = (x, y, z)$ on a sphere of radius R and center at the point $P_0 = (x_0, y_0, z_0)$ obeys $|PP_0| = R$. By (14.6), the equation of this sphere is

$$\sqrt{(x - x_0)^2 + (y - y_0)^2 + (z - z_0)^2} = R$$

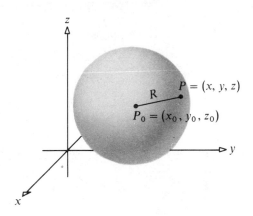

Figure 27

or

(14.7) $$(x - x_0)^2 + (y - y_0)^2 + (z - z_0)^2 = R^2$$

EXAMPLE 1   The equation of the sphere illustrated in Figure 28, with radius 2 and center $(-1, 2, 0)$, is

$$(x + 1)^2 + (y - 2)^2 + z^2 = 4$$

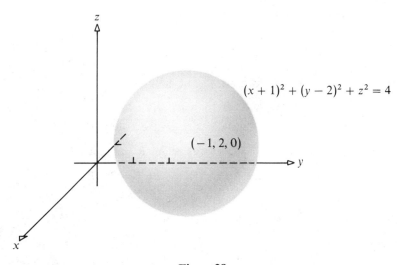

Figure 28   ■

EXAMPLE 2   Complete the squares in the expression

$$x^2 + y^2 + z^2 + 2x + 4y - 2z = 10$$

and show that it is the equation of a sphere. Find its center and radius.

Solution   Rewrite the expression as

$$(x^2 + 2x) + (y^2 + 4y) + (z^2 - 2z) = 10$$

and complete the squares. The result is

$$(x^2 + 2x + 1) + (y^2 + 4y + 4) + (z^2 - 2z + 1) = 10 + 1 + 4 + 1$$
$$(x + 1)^2 + (y + 2)^2 + (z - 1)^2 = 16$$

This is the equation of a sphere with radius 4 and center at $(-1, -2, 1)$.   ∎

EXERCISE 3

In Problems 1–6 plot each point in a three-dimensional coordinate system.

1. $(1, 1, 1)$ **2.** $(0, 0, 1)$ **3.** $(0, 2, 5)$
4. $(-1, 5, 0)$ **5.** $(-3, 1, 0)$ **6.** $(4, -1, -3)$

In Problems 7–12 opposite vertices of a rectangular box whose edges are parallel to the coordinate axes are given. List the coordinates of the other six vertices of the box.

7. $(0, 0, 0)$; $(2, 1, 3)$ **8.** $(0, 0, 0)$; $(4, 2, 2)$ **9.** $(1, 2, 3)$; $(3, 4, 5)$
10. $(5, 6, 1)$; $(3, 8, 2)$ **11.** $(-1, 0, 2)$; $(4, 2, 5)$ **12.** $(-2, -3, 0)$; $(-6, 7, 1)$

In Problems 13–18 describe in words the set of all points (x, y, z) that satisfy the given conditions.

13. $y = 3$ **14.** $z = -3$ **15.** $x = 0$
16. $x = 1$ and $y = 0$ **17.** $z = 5$ **18.** $x = y$ and $z = 0$

In Problems 19–24 find the distance between each pair of points.

19. $(1, 3, 0)$ and $(4, 1, 2)$ **20.** $(3, 2, 1)$ and $(1, 2, 3)$ **21.** $(-1, 2, -3)$ and $(4, -2, 1)$
22. $(-2, 1, 3)$ and $(4, 0, -3)$ **23.** $(4, -2, -2)$ and $(3, 2, 1)$ **24.** $(2, -3, -3)$ and $(4, 1, -1)$

In Problems 25–28 find the equation of a sphere with radius R and center P_0.

25. $R = 1$; $P_0 = (3, 1, 1)$ **26.** $R = 2$; $P_0 = (1, 2, 2)$
27. $R = 3$; $P_0 = (-1, 1, 2)$ **28.** $R = 1$; $P_0 = (-3, 1, -1)$

In Problems 29–34 find the radius and center of each sphere.

29. $x^2 + y^2 + z^2 + 2x - 2y = 2$ **30.** $x^2 + y^2 + z^2 + 2x - 2z = -1$
31. $x^2 + y^2 + z^2 - 4x + 4y + 2z = 0$ **32.** $x^2 + y^2 + z^2 - 4x = 0$
33. $2x^2 + 2y^2 + 2z^2 - 8x + 4z = -1$ **34.** $3x^2 + 3y^2 + 3z^2 + 6x - 6y = 3$

In Problems 35–38 write the equation of the sphere described.

35. The endpoints of a diameter are $(-2, 0, 4)$ and $(2, 6, 8)$.
36. The endpoints of a diameter are $(1, 3, 6)$ and $(-3, 1, 4)$.

37. The center is at $(-3, 2, 1)$, and the sphere passes through the point $(4, -1, 3)$.

38. The center is at $(0, -3, 4)$, and the sphere passes through the point $(2, 1, 1)$.

39. Show that the points $(-2, 6, 0)$, $(4, 9, 1)$, and $(-3, 2, 18)$ are the vertices of a right triangle.

40. Show that the points $(2, 2, 2)$, $(0, 1, 2)$, $(-1, 3, 3)$, and $(3, 0, 1)$ are the vertices of a parallelogram.

41. Show that the points $(2, 2, 2)$, $(2, 0, 1)$, $(4, 1, -1)$, and $(4, 3, 0)$ are the vertices of a rectangle.

42. Find an equation of the sphere that passes through the vertices of the right triangle cited in Problem 39 and that has a diameter along the hypotenuse of this triangle.

43. Show that the points $(2, 4, 2)$, $(2, 1, 5)$, and $(5, 1, 2)$ are the vertices of an equilateral triangle.

4. Vectors in Space

Consider a rectangular coordinate system in which each point P is given by an ordered triple (x, y, z) of real numbers. Let $\mathbf{i}, \mathbf{j}$, and $\mathbf{k}$ denote unit vectors directed along the positive x-axis, positive y-axis, and positive z-axis, respectively. Every vector $\mathbf{v}$ in space can then be written uniquely in terms of the vectors $\mathbf{i}, \mathbf{j}$, and $\mathbf{k}$ as

$$\mathbf{v} = a\mathbf{i} + b\mathbf{j} + c\mathbf{k}$$

for some choice of numbers a, b, and c. The numbers a, b, and c are called the *components of* $\mathbf{v}$ in the x, y, and z directions, respectively. See Figure 29.

   We determine the components of a vector $\mathbf{v}$ in space in a manner similar to vectors in the plane. If $P_1 = (x_1, y_1, z_1)$ and $P_2 = (x_2, y_2, z_2)$ are the initial and

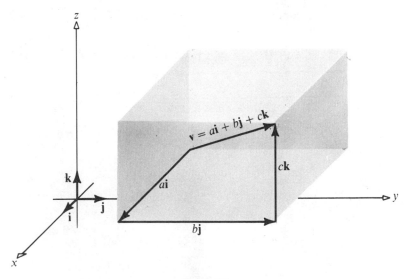

Figure 29

terminal points, respectively, of a directed line segment from the collection that defines the vector **v**, then

(14.8) $$\mathbf{v} = (x_2 - x_1)\mathbf{i} + (y_2 - y_1)\mathbf{j} + (z_2 - z_1)\mathbf{k}$$

Any of the equivalent directed line segments will give the same components. See Figure 30.

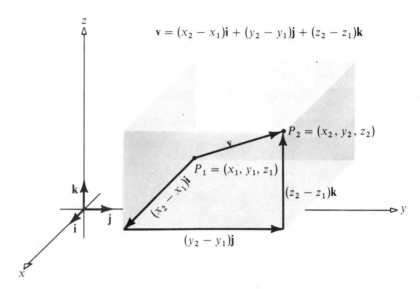

Figure 30

Two vectors v and w are equal if and only if their corresponding components are equal. That is:

$$\text{If} \quad \mathbf{v} = a_1\mathbf{i} + b_1\mathbf{j} + c_1\mathbf{k} \quad \text{and} \quad \mathbf{w} = a_2\mathbf{i} + b_2\mathbf{j} + c_2\mathbf{k}$$
$$\text{then} \quad \mathbf{v} = \mathbf{w} \quad \text{if and only if} \quad a_1 = a_2, \quad b_1 = b_2, \quad c_1 = c_2$$

The magnitude of the vector $\mathbf{v} = a\mathbf{i} + b\mathbf{j} + c\mathbf{k}$ is

$$\|\mathbf{v}\| = \sqrt{a^2 + b^2 + c^2}$$

EXAMPLE 1   Let $P_1 = (3, 4, -2)$ and $P_2 = (7, -8, 5)$.
(a)   Find the vector **v** from P_1 to P_2.
(b)   What is **v** if P_2 is replaced by $(5, -1, 4)$? By $(0, 0, 0)$?

Solution   (a)   By (14.8), the vector **v** is

$$\mathbf{v} = (7 - 3)\mathbf{i} + (-8 - 4)\mathbf{j} + (5 + 2)\mathbf{k} = 4\mathbf{i} - 12\mathbf{j} + 7\mathbf{k}$$

(b) $$\mathbf{v} = (5 - 3)\mathbf{i} + (-1 - 4)\mathbf{j} + (4 + 2)\mathbf{k} = 2\mathbf{i} - 5\mathbf{j} + 6\mathbf{k}$$
$$\mathbf{v} = (0 - 3)\mathbf{i} + (0 - 4)\mathbf{j} + (0 + 2)\mathbf{k} = -3\mathbf{i} - 4\mathbf{j} + 2\mathbf{k} \quad \blacksquare$$

To add two vectors in space, we add respective components. Thus, if $\mathbf{v} = a_1\mathbf{i} + b_1\mathbf{j} + c_1\mathbf{k}$ and $\mathbf{w} = a_2\mathbf{i} + b_2\mathbf{j} + c_2\mathbf{k}$,   then

$$\mathbf{v} + \mathbf{w} = (a_1 + a_2)\mathbf{i} + (b_1 + b_2)\mathbf{j} + (c_1 + c_2)\mathbf{k}$$

To subtract vectors in space, we subtract respective components. Thus, if $\mathbf{v} = a_1\mathbf{i} + b_1\mathbf{j} + c_1\mathbf{k}$ and $\mathbf{w} = a_2\mathbf{i} + b_2\mathbf{j} + c_2\mathbf{k}$,   then

$$\mathbf{v} - \mathbf{w} = (a_1 - a_2)\mathbf{i} + (b_1 - b_2)\mathbf{j} + (c_1 - c_2)\mathbf{k}$$

To multiply a scalar times a vector in space, we multiply each component by the scalar. Thus, if α is a scalar and if $\mathbf{v} = a\mathbf{i} + b\mathbf{j} + c\mathbf{k}$ is a vector, then

$$\alpha\mathbf{v} = \alpha(a\mathbf{i} + b\mathbf{j} + c\mathbf{k}) = (\alpha a)\mathbf{i} + (\alpha b)\mathbf{j} + (\alpha c)\mathbf{k}$$

EXAMPLE 2   If $\mathbf{v} = 2\mathbf{i} + 3\mathbf{j} - \mathbf{k}$ and $\mathbf{w} = -\mathbf{i} - 2\mathbf{j} + 4\mathbf{k}$,   find:

(a) $\|\mathbf{v}\|$ (b) $2\mathbf{v} - 3\mathbf{w}$ (c) $\dfrac{\mathbf{v}}{\|\mathbf{v}\|}$

Solution   (a) $$\|\mathbf{v}\| = \sqrt{2^2 + 3^2 + (-1)^2} = \sqrt{4 + 9 + 1} = \sqrt{14}$$

(b) $$2\mathbf{v} - 3\mathbf{w} = 2(2\mathbf{i} + 3\mathbf{j} - \mathbf{k}) - 3(-\mathbf{i} - 2\mathbf{j} + 4\mathbf{k})$$
$$= (4\mathbf{i} + 6\mathbf{j} - 2\mathbf{k}) - (-3\mathbf{i} - 6\mathbf{j} + 12\mathbf{k}) = 7\mathbf{i} + 12\mathbf{j} - 14\mathbf{k}$$

(c) $$\frac{\mathbf{v}}{\|\mathbf{v}\|} = \frac{2\mathbf{i} + 3\mathbf{j} - \mathbf{k}}{\sqrt{14}} = \frac{2}{\sqrt{14}}\mathbf{i} + \frac{3}{\sqrt{14}}\mathbf{j} - \frac{1}{\sqrt{14}}\mathbf{k} \qquad \blacksquare$$

Note that the vector found in part (c) of Example 2 is a unit vector, since

$$\left\|\frac{2}{\sqrt{14}}\mathbf{i} + \frac{3}{\sqrt{14}}\mathbf{j} - \frac{1}{\sqrt{14}}\mathbf{k}\right\| = \sqrt{\left(\frac{2}{\sqrt{14}}\right)^2 + \left(\frac{3}{\sqrt{14}}\right)^2 + \left(\frac{-1}{\sqrt{14}}\right)^2}$$

$$= \sqrt{\frac{4}{14} + \frac{9}{14} + \frac{1}{14}} = \sqrt{\frac{14}{14}} = 1$$

PARALLEL VECTORS

Let $\mathbf{v}$ and $\mathbf{w}$ be two vectors. If there is a scalar α so that $\mathbf{v} = \alpha\mathbf{w}$,   then $\mathbf{v}$ and $\mathbf{w}$ are called *parallel* vectors.

For example, the vectors

$$\mathbf{v} = 2\mathbf{i} + 3\mathbf{j} - \mathbf{k} \qquad \text{and} \qquad \mathbf{w} = 4\mathbf{i} + 6\mathbf{j} - 2\mathbf{k}$$

are parallel since $\mathbf{v} = \frac{1}{2}\mathbf{w}$.   In this case, $\mathbf{v}$ and $\mathbf{w}$ have the same direction.
The vectors

$$\mathbf{v} = \mathbf{i} + 2\mathbf{j} + 3\mathbf{k} \qquad \text{and} \qquad \mathbf{w} = -3\mathbf{i} - 6\mathbf{j} - 9\mathbf{k}$$

are parallel since $\mathbf{v} = -\frac{1}{3}\mathbf{w}$.   In this case, $\mathbf{v}$ and $\mathbf{w}$ have opposite direction.

DIRECTION ANGLES OF VECTORS

A vector **v** in space can be described by specifying its three *direction angles* α, β, γ and its magnitude. These direction angles are defined as

α = Angle between **v** and the positive x-axis, $0 \leq \alpha \leq \pi$

β = Angle between **v** and the positive y-axis, $0 \leq \beta \leq \pi$

γ = Angle between **v** and the positive z-axis, $0 \leq \gamma \leq \pi$

See Figure 31.

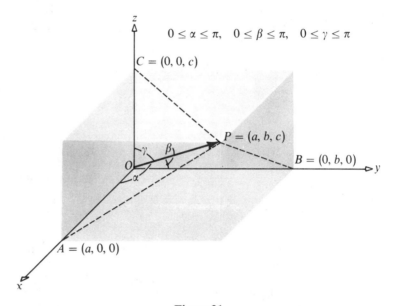

$$0 \leq \alpha \leq \pi, \quad 0 \leq \beta \leq \pi, \quad 0 \leq \gamma \leq \pi$$

Figure 31

Our first problem will be to find an expression for α, β, and γ in terms of the components of a vector **v**. Let $\mathbf{v} = a\mathbf{i} + b\mathbf{j} + c\mathbf{k}$.   Refer again to Figure 31. There we see that triangle OAP is a right triangle with right angle at A. Thus,

$$\cos \alpha = \frac{a}{\|\mathbf{v}\|}$$

Similarly, from the right triangles OBP and OCP, we get

$$\cos \beta = \frac{b}{\|\mathbf{v}\|} \qquad \cos \gamma = \frac{c}{\|\mathbf{v}\|}$$

Since $\|\mathbf{v}\| = \sqrt{a^2 + b^2 + c^2}$,   we conclude that

(14.9) $$\cos \alpha = \frac{a}{\sqrt{a^2 + b^2 + c^2}} \qquad \cos \beta = \frac{b}{\sqrt{a^2 + b^2 + c^2}} \qquad \cos \gamma = \frac{c}{\sqrt{a^2 + b^2 + c^2}}$$

The numbers $\cos \alpha$, $\cos \beta$, and $\cos \gamma$ are called the *direction cosines* of the vector **v**. They play the same role in three dimensions as slope does in two dimensions.

EXAMPLE 3   Find the direction cosines of $\mathbf{v} = -3\mathbf{i} + 2\mathbf{j} - 6\mathbf{k}$.

Solution
$$\|\mathbf{v}\| = \sqrt{(-3)^2 + 2^2 + (-6)^2} = \sqrt{49} = 7$$

By using formula (14.9), we get

$$\cos \alpha = \frac{-3}{7} \qquad \cos \beta = \frac{2}{7} \qquad \cos \gamma = \frac{-6}{7} \qquad \blacksquare$$

From (14.9), it follows that

(14.10)
$$\cos^2\alpha + \cos^2\beta + \cos^2\gamma = 1$$

Based on (14.10), once two direction cosines are known, the third is determined up to its sign.

EXAMPLE 4   The vector **v** makes an angle of $\alpha = \pi/3$ with the positive x-axis, $\beta = \pi/3$ with the positive y-axis, and an acute angle γ with the positive z-axis. Find γ.

Solution   By (14.10), we have

$$\cos^2\left(\frac{\pi}{3}\right) + \cos^2\left(\frac{\pi}{3}\right) + \cos^2\gamma = 1$$

$$(\tfrac{1}{2})^2 + (\tfrac{1}{2})^2 + \cos^2\gamma = 1$$

$$\cos^2\gamma = \tfrac{1}{2}$$

$$\cos \gamma = \frac{\sqrt{2}}{2} \qquad \text{or} \qquad \cos \gamma = -\frac{\sqrt{2}}{2}$$

$$\gamma = \frac{\pi}{4} \qquad \text{or} \qquad \gamma = \frac{3\pi}{4}$$

Since we are requiring that γ be acute, the answer is $\gamma = \pi/4$. $\blacksquare$

The direction cosines of a vector only give information about the direction of the vector; they provide no information about its magnitude. For example, *any* vector parallel to the xy-plane and making an angle of $\pi/4$ radian with the positive x- and y-axes has direction cosines

$$\cos \alpha = \frac{\sqrt{2}}{2} \qquad \cos \beta = \frac{\sqrt{2}}{2} \qquad \cos \gamma = 0$$

However, if the direction angles *and* the magnitude of a vector are known, then the vector is uniquely determined.

EXAMPLE 5   Show that any non-zero vector **v** can be written in terms of its magnitude and direction cosines as

(14.11)
$$\mathbf{v} = \|\mathbf{v}\|[(\cos \alpha)\mathbf{i} + (\cos \beta)\mathbf{j} + (\cos \gamma)\mathbf{k}]$$

Solution   Let $\mathbf{v} = a\mathbf{i} + b\mathbf{j} + c\mathbf{k}$.   From (14.9), we see that

$$a = \|\mathbf{v}\|\cos \alpha \qquad b = \|\mathbf{v}\|\cos \beta \qquad c = \|\mathbf{v}\|\cos \gamma$$

Thus,

$$\mathbf{v} = a\mathbf{i} + b\mathbf{j} + c\mathbf{k} = \|\mathbf{v}\|(\cos \alpha)\mathbf{i} + \|\mathbf{v}\|(\cos \beta)\mathbf{j} + \|\mathbf{v}\|(\cos \gamma)\mathbf{k}$$
$$= \|\mathbf{v}\|[(\cos \alpha)\mathbf{i} + (\cos \beta)\mathbf{j} + (\cos \gamma)\mathbf{k}] \quad \blacksquare$$

VECTORS IN n DIMENSIONS

In a space of n dimensions, the rectangular coordinates of a point P are given by an ordered n-tuple $(x_1, x_2, \ldots, x_n)$ of real numbers. Corresponding to each of these numbers, there is a positive axis along each of which we may define a unit vector $\mathbf{u}_1, \mathbf{u}_2, \ldots, \mathbf{u}_n$. A vector $\mathbf{v}$ may then be written uniquely in terms of the vectors $\mathbf{u}_1, \mathbf{u}_2, \ldots, \mathbf{u}_n$ as

$$\mathbf{v} = a_1\mathbf{u}_1 + a_2\mathbf{u}_2 + \cdots + a_n\mathbf{u}_n$$

for some choice of numbers $a_1, a_2, \ldots, a_n$.   These numbers are called the *components* of $\mathbf{v}$ in the $\mathbf{u}_1, \mathbf{u}_2, \ldots, \mathbf{u}_n$ directions, respectively.

We add two vectors in n-space by adding respective components. To multiply a scalar times a vector, we multiply each component by the scalar.

The magnitude of the vector $\mathbf{v} = a_1\mathbf{u}_1 + a_2\mathbf{u}_2 + \cdots + a_n\mathbf{u}_n$ is

$$\|\mathbf{v}\| = \sqrt{a_1^2 + a_2^2 + \cdots + a_n^2}$$

A non-zero vector $\mathbf{v}$ in n-space will have n *direction angles* $\alpha_1, \alpha_2, \ldots, \alpha_n$, each defined as the angle α_i, $0 \le \alpha_i \le \pi$, between $\mathbf{v}$ and the corresponding positive axis.

If $\mathbf{v} = a_1\mathbf{u}_1 + a_2\mathbf{u}_2 + \cdots + a_n\mathbf{u}_n$ is a non-zero vector and $\alpha_1, \alpha_2, \ldots, \alpha_n$ are its direction angles, then the direction cosines of $\mathbf{v}$ are

$$\cos \alpha_1 = \frac{a_1}{\|\mathbf{v}\|}, \quad \cos \alpha_2 = \frac{a_2}{\|\mathbf{v}\|}, \quad \ldots, \quad \cos \alpha_n = \frac{a_n}{\|\mathbf{v}\|}$$

and

$$\cos^2\alpha_1 + \cos^2\alpha_2 + \cdots + \cos^2\alpha_n = 1$$

EXERCISE 4

In Problems 1–6 write the vector represented by the directed line segment $\overrightarrow{P_1P_2}$.

1. $P_1 = (6, 2, 1)$; $P_2 = (3, 0, 2)$ **2.** $P_1 = (4, 7, 0)$; $P_2 = (0, 5, 6)$

3. $P_1 = (-1, 0, 1)$; $P_2 = (2, 0, 0)$ **4.** $P_1 = (6, 2, 2)$; $P_2 = (2, 6, 2)$

5. $P_1 = (0, 0, 0)$; $P_2 = (x, y, z)$ **6.** $P_1 = (0, 0, 0)$; $P_2 = (0, y, 0)$

In Problems 7–16 find the indicated quantities if $\mathbf{u} = \mathbf{i} - 2\mathbf{j} + 3\mathbf{k}$, $\mathbf{v} = 3\mathbf{i} + \mathbf{j} - \mathbf{k}$,   and $\mathbf{w} = 6\mathbf{i} + \mathbf{j} + \mathbf{k}$.

7. $3\mathbf{v} - 2\mathbf{w}$ **8.** $-2\mathbf{v} + \mathbf{w}$ **9.** $\|\mathbf{v} - \mathbf{w}\|$ **10.** $\|2\mathbf{v} + \mathbf{w}\|$

11. $2\mathbf{u} - 3\mathbf{v} + 4\mathbf{w}$ **12.** $\dfrac{\mathbf{u} + \mathbf{v}}{\|\mathbf{w}\|}$ **13.** $\|5\mathbf{u} - \mathbf{v} + \mathbf{w}\|$ **14.** $\|\mathbf{u}\| + \|\mathbf{v}\| + \|\mathbf{w}\|$

15. $\|\mathbf{u} + \mathbf{v} + \mathbf{w}\|$ **16.** $\dfrac{\mathbf{u}}{\|\mathbf{v} + \mathbf{w}\|}$

17. Find a vector whose magnitude is 4 that is parallel to $2\mathbf{i} - 3\mathbf{j} + 4\mathbf{k}$.

18. Find a vector whose magnitude is 2 that is parallel to $-\mathbf{i} + \mathbf{j} + 2\mathbf{k}$.

In Problems 19–24 find the magnitude of $\mathbf{v}$ and write the direction cosines of $\mathbf{v}$.

19. $\mathbf{v} = 4\mathbf{i} + 2\mathbf{j} - \mathbf{k}$ **20.** $\mathbf{v} = \mathbf{i} - \mathbf{j} + \mathbf{k}$ **21.** $\mathbf{v} = \mathbf{i} + \mathbf{j} + \mathbf{k}$

22. $\mathbf{v} = 2\mathbf{i} - \mathbf{k}$ **23.** $\mathbf{v} = a\mathbf{i} + a\mathbf{j} + a\mathbf{k}, \quad a > 0$ **24.** $\mathbf{v} = (\cos\theta)\mathbf{i} + (\sin\theta)\mathbf{j} + \mathbf{k}$

In Problems 25–28 find a vector $\mathbf{v}$ that has the given magnitude and direction angles.

25. $\|\mathbf{v}\| = 3, \quad \alpha = \pi/3, \quad \beta = \pi/4, \quad 0 < \gamma < \pi/2$

26. $\|\mathbf{v}\| = 3,$ direction angles are equal, $\mathbf{v}$ has positive components

27. $\|\mathbf{v}\| = 2, \quad \alpha = \pi/4, \quad \pi/2 < \beta < \pi, \quad \gamma = \pi/3$

28. $\|\mathbf{v}\| = \frac{1}{2}, \quad \cos\alpha > 0, \quad \cos\beta = \frac{1}{4}, \quad \cos\gamma = \sqrt{\frac{7}{8}}$

29. Let A, B, C, and D be the vertices of a tetrahedron (triangular pyramid) in space. Let $\mathbf{b} = \overrightarrow{AB}, \quad \mathbf{c} = \overrightarrow{AC}, \quad$ and $\quad \mathbf{d} = \overrightarrow{AD}.$ Express the directed edges $\overrightarrow{BC}$, $\overrightarrow{BD}$, and $\overrightarrow{CD}$ of the tetrahedron in terms of the vectors $\mathbf{b}$, $\mathbf{c}$, and $\mathbf{d}$.

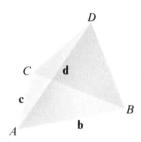

30. Let A, B, C, D, E, F, G, and H be the vertices of a parallelepiped in space, whose faces are the parallelograms $ABCD$, $ABFE$, $AEHD$, and so forth. Let $\mathbf{b} = \overrightarrow{AB}, \quad \mathbf{e} = \overrightarrow{AE}, \quad$ and $\quad \mathbf{d} = \overrightarrow{AD}.$ Express the vectors $\overrightarrow{AC}$, $\overrightarrow{AF}$, $\overrightarrow{AG}$, $\overrightarrow{FG}$, and $\overrightarrow{EG}$ in terms of $\mathbf{b}$, $\mathbf{e}$, and $\mathbf{d}$.

31. Find α so that the vectors $2\mathbf{i} + \mathbf{j} - \mathbf{k}$ and $2\alpha\mathbf{i} + \mathbf{j} + \mathbf{k}$ have the same magnitude.

32. Find α so that $\|\alpha\mathbf{i} + (\alpha + 1)\mathbf{j} + 2\mathbf{k}\| = 3$.

33. Use (14.10) to show that:
  (a)   There is no vector for which $\alpha = \pi/6$ and $\beta = \pi/4$.
  (b)   If α and β are positive acute direction angles of a vector, then $\alpha + \beta \geq \pi/2$.

5. The Dot Product

In this section we discuss a product of two vectors called the *dot product*. As it turns out, we will also define another product of two vectors, which is called the *cross product*. We take up the dot product here, and we will discuss cross products in the next section.

(14.12)   **DEFINITION**   If $\mathbf{v} = a_1\mathbf{i} + b_1\mathbf{j} + c_1\mathbf{k}$ and $\mathbf{w} = a_2\mathbf{i} + b_2\mathbf{j} + c_2\mathbf{k}$ **are two vectors, the *dot product* $\mathbf{v} \cdot \mathbf{w}$ is defined as**

$$\mathbf{v} \cdot \mathbf{w} = a_1 a_2 + b_1 b_2 + c_1 c_2$$

For example, if $\mathbf{v} = 2\mathbf{i} - \mathbf{j} + \mathbf{k}$ and $\mathbf{w} = 4\mathbf{i} + 2\mathbf{j} - \mathbf{k}$, then

$$\mathbf{v} \cdot \mathbf{w} = (2)(4) + (-1)(2) + (1)(-1) = 8 - 2 - 1 = 5$$
$$\mathbf{v} \cdot \mathbf{v} = 4 + 1 + 1 = 6$$
$$\mathbf{w} \cdot \mathbf{w} = 16 + 4 + 1 = 21$$

Note that the dot product $\mathbf{v} \cdot \mathbf{w}$ of two vectors is a real number. Because of this, the dot product is sometimes referred to as the *scalar product*.

We now list some algebraic properties of the dot product.

(14.13)   **THEOREM**   **If u, v, and w are vectors and α is any scalar, then**

(a) $\mathbf{u} \cdot \mathbf{v} = \mathbf{v} \cdot \mathbf{u}$ (b) $\mathbf{u} \cdot (\mathbf{v} + \mathbf{w}) = \mathbf{u} \cdot \mathbf{v} + \mathbf{u} \cdot \mathbf{w}$

(c) $\alpha(\mathbf{u} \cdot \mathbf{v}) = (\alpha\mathbf{u}) \cdot \mathbf{v}$ (d) $\mathbf{0} \cdot \mathbf{v} = 0$ (e) $\mathbf{v} \cdot \mathbf{v} = \|\mathbf{v}\|^2$

Proof

(a) If $\mathbf{u} = a_1\mathbf{i} + b_1\mathbf{j} + c_1\mathbf{k}$ and $\mathbf{v} = a_2\mathbf{i} + b_2\mathbf{j} + c_2\mathbf{k}$, then

$$\mathbf{u} \cdot \mathbf{v} = a_1a_2 + b_1b_2 + c_1c_2 = a_2a_1 + b_2b_1 + c_2c_1 = \mathbf{v} \cdot \mathbf{u}$$

(e) If $\mathbf{v} = a\mathbf{i} + b\mathbf{j} + c\mathbf{k}$, then

$$\mathbf{v} \cdot \mathbf{v} = (a)(a) + (b)(b) + (c)(c) = a^2 + b^2 + c^2 = \|\mathbf{v}\|^2$$

The proofs of parts (b), (c), and (d) are left as exercises.   ∎

As a result of part (e) of (14.13), we conclude that

(14.14) $\mathbf{v} \cdot \mathbf{v} \geq 0$ and $\mathbf{v} \cdot \mathbf{v} = 0$ if and only if $\mathbf{v} = \mathbf{0}$

ANGLE BETWEEN VECTORS

The dot product may be used to calculate the angle between two non-zero vectors. Consider triangle ABC in Figure 32. Let θ denote the angle at vertex A, and let the vectors $\mathbf{v}$ and $\mathbf{w}$ be represented by the directed line segments $\overrightarrow{AB}$ and $\overrightarrow{AC}$, respectively. Then the vector $\mathbf{w} - \mathbf{v}$ can be represented by the directed line segment $\overrightarrow{BC}$. The sides of the triangle have lengths $\|\mathbf{v}\|$, $\|\mathbf{w}\|$, and $\|\mathbf{w} - \mathbf{v}\|$. If we use the law of cosines,* we find

(14.15) $\|\mathbf{w} - \mathbf{v}\|^2 = \|\mathbf{v}\|^2 + \|\mathbf{w}\|^2 - 2\|\mathbf{v}\|\,\|\mathbf{w}\|\cos\theta$

Now we use part (e) of (14.13) to rewrite (14.15) as

(14.16) $(\mathbf{w} - \mathbf{v}) \cdot (\mathbf{w} - \mathbf{v}) = \mathbf{v} \cdot \mathbf{v} + \mathbf{w} \cdot \mathbf{w} - 2\|\mathbf{v}\|\,\|\mathbf{w}\|\cos\theta$

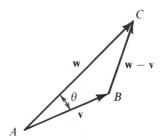

Figure 32

---

* The law of cosines states that $c^2 = a^2 + b^2 - 2ab\cos\theta$, where a, b, and c are the lengths of the sides of a triangle and θ is the angle opposite the side of length c.

A double application of part (b) of (14.13) to the term $(\mathbf{w} - \mathbf{v}) \cdot (\mathbf{w} - \mathbf{v})$ yields

(14.17)
$$(\mathbf{w} - \mathbf{v}) \cdot (\mathbf{w} - \mathbf{v}) = \mathbf{w} \cdot (\mathbf{w} - \mathbf{v}) - \mathbf{v} \cdot (\mathbf{w} - \mathbf{v}) = \mathbf{w} \cdot \mathbf{w} - \mathbf{w} \cdot \mathbf{v} - \mathbf{v} \cdot \mathbf{w} + \mathbf{v} \cdot \mathbf{v}$$
$$= \underset{\underset{(14.13(a))}{\uparrow}}{\mathbf{w} \cdot \mathbf{w} - \mathbf{v} \cdot \mathbf{w} - \mathbf{v} \cdot \mathbf{w} + \mathbf{v} \cdot \mathbf{v}} = \underset{\underset{(14.13(e))}{\uparrow}}{\|\mathbf{w}\|^2 - 2\mathbf{v} \cdot \mathbf{w} + \|\mathbf{v}\|^2}$$

Since $\|\mathbf{w} - \mathbf{v}\|^2 = (\mathbf{w} - \mathbf{v}) \cdot (\mathbf{w} - \mathbf{v})$, we may combine (14.15) and (14.17) to get

$$-2\|\mathbf{v}\| \|\mathbf{w}\| \cos \theta = -2(\mathbf{v} \cdot \mathbf{w})$$
$$\mathbf{v} \cdot \mathbf{w} = \|\mathbf{v}\| \|\mathbf{w}\| \cos \theta$$

So, we have proved the following result:

(14.18)   **The angle θ, $0 \le \theta \le \pi$, between two nonzero vectors $\mathbf{v}$ and $\mathbf{w}$ is obtained by the formula**

$$\cos \theta = \frac{\mathbf{v} \cdot \mathbf{w}}{\|\mathbf{v}\| \|\mathbf{w}\|}$$

EXAMPLE 1   Find the angle θ between the vectors $\mathbf{v} = 2\mathbf{i} - \mathbf{j} + \mathbf{k}$ and $\mathbf{w} = -\mathbf{i} + \mathbf{j}$.

Solution $\mathbf{v} \cdot \mathbf{w} = -2 - 1 = -3 \qquad \|\mathbf{v}\| = \sqrt{4 + 1 + 1} = \sqrt{6} \qquad \|\mathbf{w}\| = \sqrt{1 + 1} = \sqrt{2}$

Thus,

$$\cos \theta = \frac{\mathbf{v} \cdot \mathbf{w}}{\|\mathbf{v}\| \|\mathbf{w}\|} = \frac{-3}{\sqrt{6}\sqrt{2}} = \frac{-3}{\sqrt{12}} = \frac{-\sqrt{3}}{2}$$

Since $0 \le \theta \le \pi$, the angle θ between $\mathbf{v}$ and $\mathbf{w}$ is $5\pi/6$ radians (see Fig. 33).

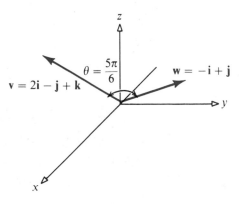

Figure 33   ■

The next example illustrates a use for (14.18).

EXAMPLE 2   An airplane has an air speed of 400 kilometers per hour in an easterly direction. The wind velocity is 80 kilometers per hour in a southeasterly direction. Find the direction of the airplane relative to the ground.

Solution   This is the information supplied in Example 3 of Section 2. There, we found that the velocity of the airplane relative to the ground was

$$\mathbf{v}_g = (400 + 40\sqrt{2})\mathbf{i} - 40\sqrt{2}\mathbf{j}$$

The angle θ between $\mathbf{v}_g$ and the vector $\mathbf{v}_a = 400\mathbf{i}$ (the velocity of the airplane in the air) obeys

$$\cos \theta = \frac{\mathbf{v}_g \cdot \mathbf{v}_a}{\|\mathbf{v}_g\| \, \|\mathbf{v}_a\|} = \frac{(400 + 40\sqrt{2})(400)}{(460.06)(400)} \underset{\underset{\text{Use a calculator}}{\uparrow}}{\approx} 0.9924$$

$$\theta \simeq 7.06°$$

The direction of the plane relative to the ground is 7.06° south of east.   ∎

PERPENDICULAR VECTORS

If the angle θ between two nonzero vectors $\mathbf{v}$ and $\mathbf{w}$ is 0 or π, the vectors $\mathbf{v}$ and $\mathbf{w}$ are parallel. If the angle θ between two nonzero vectors $\mathbf{v}$ and $\mathbf{w}$ is $\pi/2$, the vectors $\mathbf{v}$ and $\mathbf{w}$ are *perpendicular* (or *orthogonal*).

It follows from (14.18) that if $\mathbf{v}$ and $\mathbf{w}$ are perpendicular, then $\mathbf{v} \cdot \mathbf{w} = 0$ since $\cos(\pi/2) = 0$.

On the other hand, if $\mathbf{v} \cdot \mathbf{w} = 0$, then either $\mathbf{v} = \mathbf{0}$ or $\mathbf{w} = \mathbf{0}$ or $\cos \theta = 0$. In the latter case, $\theta = \pi/2$ and $\mathbf{v}$ and $\mathbf{w}$ are perpendicular. If $\mathbf{v}$ or $\mathbf{w}$ is the zero vector, then since the zero vector has no specific direction, we adopt the convention that the zero vector is perpendicular to every vector. Thus,

(14.19)   **$\mathbf{v}$ is perpendicular to $\mathbf{w}$ if and only if $\mathbf{v} \cdot \mathbf{w} = 0$**

See Figure 34.

For example, the vectors

$$\mathbf{v} = 2\mathbf{i} - \mathbf{j} + 5\mathbf{k} \qquad \text{and} \qquad \mathbf{w} = 3\mathbf{i} + \mathbf{j} - \mathbf{k}$$

are perpendicular since

$$\mathbf{v} \cdot \mathbf{w} = 6 - 1 - 5 = 0$$

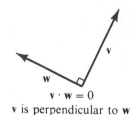

$\mathbf{v} \cdot \mathbf{w} = 0$
$\mathbf{v}$ is perpendicular to $\mathbf{w}$

Figure 34

By applying the definition of dot product to the unit vectors

$$\mathbf{i} = 1\mathbf{i} + 0\mathbf{j} + 0\mathbf{k} \qquad \mathbf{j} = 0\mathbf{i} + 1\mathbf{j} + 0\mathbf{k} \qquad \mathbf{k} = 0\mathbf{i} + 0\mathbf{j} + 1\mathbf{k}$$

we find that

$$\mathbf{i} \cdot \mathbf{j} = 0 \qquad \mathbf{i} \cdot \mathbf{k} = 0 \qquad \mathbf{j} \cdot \mathbf{k} = 0$$

Hence, we conclude that the unit vectors $\mathbf{i}$, $\mathbf{j}$, and $\mathbf{k}$ are mutually perpendicular.

EXAMPLE 3   Find a scalar α so that the vectors $\mathbf{v} = 2\alpha\mathbf{i} + \mathbf{j} - \mathbf{k}$ and $\mathbf{w} = \mathbf{i} - \alpha\mathbf{j} + \mathbf{k}$ are perpendicular.

Solution   It is required that $\mathbf{v} \cdot \mathbf{w} = 0$,   so

$$\mathbf{v} \cdot \mathbf{w} = 2\alpha - \alpha - 1 = 0$$

$$\alpha = 1 \quad \blacksquare$$

CAUCHY–SCHWARZ INEQUALITY

For any angle θ we have $|\cos\theta| \le 1$.   Thus, for nonzero vectors $\mathbf{v}$ and $\mathbf{w}$, it follows from (14.18) that

$$\frac{|\mathbf{v} \cdot \mathbf{w}|}{\|\mathbf{v}\| \, \|\mathbf{w}\|} \le 1$$

Hence, for any pair of nonzero vectors $\mathbf{v}$ and $\mathbf{w}$ (the result is trivially true for $\mathbf{v} = \mathbf{0}$ or $\mathbf{w} = \mathbf{0}$),

(14.20) $$|\mathbf{v} \cdot \mathbf{w}| \le \|\mathbf{v}\| \, \|\mathbf{w}\|$$

This result, referred to as the *Cauchy–Schwarz inequality*, is used to prove the triangle inequality (14.3), which we restate below:

(14.21)   *Triangle Inequality.*   **If v and w are vectors, then**

$$\|\mathbf{v} + \mathbf{w}\| \le \|\mathbf{v}\| + \|\mathbf{w}\|$$

Proof   We use the Cauchy–Schwarz inequality (14.20) and several properties of the dot product:

$$\|\mathbf{v} + \mathbf{w}\|^2 \underset{(14.13(e))}{=} (\mathbf{v} + \mathbf{w}) \cdot (\mathbf{v} + \mathbf{w}) \underset{(14.13(b))}{=} \mathbf{v} \cdot (\mathbf{v} + \mathbf{w}) + \mathbf{w} \cdot (\mathbf{v} + \mathbf{w})$$

$$\underset{(14.13(b))}{=} (\mathbf{v} \cdot \mathbf{v}) + (\mathbf{v} \cdot \mathbf{w}) + (\mathbf{w} \cdot \mathbf{v}) + (\mathbf{w} \cdot \mathbf{w}) \underset{(14.13(a)),\,(14.13(e))}{=} \|\mathbf{v}\|^2 + 2(\mathbf{v} \cdot \mathbf{w}) + \|\mathbf{w}\|^2$$

$$\le \|\mathbf{v}\|^2 + 2|\mathbf{v} \cdot \mathbf{w}| + \|\mathbf{w}\|^2 \underset{(14.20)}{\le} \|\mathbf{v}\|^2 + 2\|\mathbf{v}\| \, \|\mathbf{w}\| + \|\mathbf{w}\|^2 = (\|\mathbf{v}\| + \|\mathbf{w}\|)^2$$

By taking square roots, we find

$$\|\mathbf{v} + \mathbf{w}\| \le \|\mathbf{v}\| + \|\mathbf{w}\| \quad \blacksquare$$

PROJECTION OF VECTORS

In many physical applications it is often necessary to find "how much" of a vector is applied along a given direction. This requires finding the *projection* of a vector **v** along a vector **w**. Based on Figure 35, if θ is the angle between **v** and **w**, we define

(14.22)　　　　　　　**Projection of v along w** $= \|\mathbf{v}\| \cos \theta$

$$= \|\mathbf{v}\| \frac{\mathbf{v} \cdot \mathbf{w}}{\|\mathbf{v}\| \|\mathbf{w}\|} = \frac{\mathbf{v} \cdot \mathbf{w}}{\|\mathbf{w}\|}$$

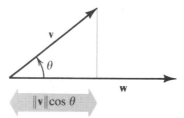

Figure 35

As (14.22) indicates, the projection of **v** along **w** is a scalar. Furthermore, if $0 \le \theta < \pi/2$, the projection of **v** along **w** is positive; see Figure 36(a). If $\theta = \pi/2$, the projection of **v** along **w** is 0; see Figure 36(b). If $\pi/2 < \theta \le \pi$, the projection of **v** along **w** is negative; see Figure 36(c).

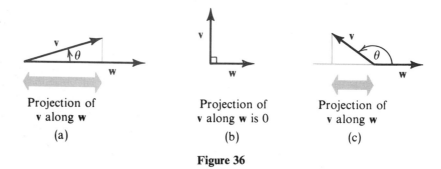

Figure 36

EXAMPLE 4　　Find the projection of $\mathbf{v} = 2\mathbf{i} - \mathbf{j} + \mathbf{k}$ along $\mathbf{w} = \mathbf{i} + \mathbf{j} + \mathbf{k}$.

Solution　　The projection of **v** along **w** is

$$\|\mathbf{v}\| \cos \theta = \frac{\mathbf{v} \cdot \mathbf{w}}{\|\mathbf{w}\|} = \frac{2 - 1 + 1}{\sqrt{3}} = \frac{2}{\sqrt{3}}$$　■

The components of a vector are the projections of the vector along $\mathbf{i}$, $\mathbf{j}$, and $\mathbf{k}$, respectively. If $\mathbf{v} = a\mathbf{i} + b\mathbf{j} + c\mathbf{k}$, then

$$\text{Projection of } \mathbf{v} \text{ along } \mathbf{i} = \|\mathbf{v}\|\cos\alpha = \|\mathbf{v}\|\frac{\mathbf{v}\cdot\mathbf{i}}{\|\mathbf{v}\|\,\|\mathbf{i}\|} = a$$

$$\text{Projection of } \mathbf{v} \text{ along } \mathbf{j} = \|\mathbf{v}\|\cos\beta = \|\mathbf{v}\|\frac{\mathbf{v}\cdot\mathbf{j}}{\|\mathbf{v}\|\,\|\mathbf{j}\|} = b$$

$$\text{Projection of } \mathbf{v} \text{ along } \mathbf{k} = \|\mathbf{v}\|\cos\gamma = \|\mathbf{v}\|\frac{\mathbf{v}\cdot\mathbf{k}}{\|\mathbf{v}\|\,\|\mathbf{k}\|} = c$$

Since the angle θ between two vectors $\mathbf{v}$ and $\mathbf{w}$ obeys the formula

$$\cos\theta = \frac{\mathbf{v}\cdot\mathbf{w}}{\|\mathbf{v}\|\,\|\mathbf{w}\|}$$

$$\mathbf{v}\cdot\mathbf{w} = \|\mathbf{w}\|(\|\mathbf{v}\|\cos\theta)$$

we have the following interpretation of the dot product:

(14.23)   **$\mathbf{v} \cdot \mathbf{w} = $ (Length of w)(Projection of v along w)**

We shall find this interpretation useful for defining work.

WORK

In Chapter 6 we defined the work W done by a constant force F in moving an object a distance x by the equation $W = Fx$. We assumed that the force F was applied *along the line of motion*. Now suppose that $\mathbf{F}$ is a vector and is *not* applied along the line of motion (see Fig. 37). Then the work W done by $\mathbf{F}$ in moving an object along a straight line from point A to point B is

$$W = (\text{Projection of } \mathbf{F} \text{ along } \overrightarrow{AB})(\text{Length of } AB) = (\|\mathbf{F}\|\cos\theta)(\|\mathbf{D}\|)$$

where $\mathbf{D}$ is the vector represented by the directed line segment $\overrightarrow{AB}$ and θ is the angle between $\mathbf{D}$ and $\mathbf{F}$. By (14.23), we have

(14.24) $$W = \mathbf{F} \cdot \mathbf{D}$$

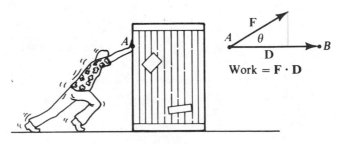

Figure 37

EXAMPLE 5   Find the work done by a force of 2 newtons acting in the direction $\mathbf{i} + \mathbf{j} + \mathbf{k}$ in
moving an object 1 meter from $(0, 0, 0)$ to $(1, 0, 0)$.

Solution   First, we must express the force $\mathbf{F}$ as a vector. From (14.9), the direction cosines of
the force are

$$\cos \alpha = \frac{1}{\sqrt{3}} \qquad \cos \beta = \frac{1}{\sqrt{3}} \qquad \cos \gamma = \frac{1}{\sqrt{3}}$$

Since the force has magnitude 2, by (14.11), it can be written as

$$\mathbf{F} = \frac{2}{\sqrt{3}} (\mathbf{i} + \mathbf{j} + \mathbf{k})$$

The line of motion of the object is along $\mathbf{D} = \mathbf{i}.$ The work W is therefore

$$W = \mathbf{F} \cdot \mathbf{D} = \frac{2}{\sqrt{3}} \text{ joules} \qquad \blacksquare$$

EXAMPLE 6   Figure 38(a) shows a person pushing on a lawnmower handle with a force of 30
pounds, moving the lawnmower a distance of 75 feet. How much work is done if the
handle makes an angle of 60° with the ground?

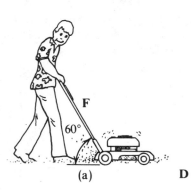

Figure 38

Solution   We choose to set up our coordinates in such a way that the lawnmower is moved
from $(0, 0)$ to $(75, 0)$. Thus, the motion is along $\mathbf{D} = 75\mathbf{i}.$ As a result of (14.11),
the force vector $\mathbf{F}$, as shown in Figure 38(b), is

$$\mathbf{F} = 30(\cos 60°)\mathbf{i} - 30(\sin 60°)\mathbf{j} = 15\mathbf{i} - 15\sqrt{3}\mathbf{j}$$

By (14.24), the work W done is

$$W = \mathbf{F} \cdot \mathbf{D} = (15\mathbf{i} - 15\sqrt{3}\mathbf{j}) \cdot 75\mathbf{i} = 1125 \text{ foot-pounds} \qquad \blacksquare$$

EXERCISE 5

In Problems 1–6 find the dot product $\mathbf{v} \cdot \mathbf{w}$,   and find the cosine of the angle θ between $\mathbf{v}$ and $\mathbf{w}$.

1. $\mathbf{v} = 2\mathbf{i} - 3\mathbf{j} + \mathbf{k}, \quad \mathbf{w} = \mathbf{i} - \mathbf{j} + \mathbf{k}$ **2.** $\mathbf{v} = -3\mathbf{i} + 2\mathbf{j} - \mathbf{k}, \quad \mathbf{w} = 2\mathbf{i} + \mathbf{j} - \mathbf{k}$

3. $\mathbf{v} = \mathbf{i} - \mathbf{j}, \quad \mathbf{w} = \mathbf{j} + \mathbf{k}$ **4.** $\mathbf{v} = \mathbf{j} - \mathbf{k}, \quad \mathbf{w} = \mathbf{i} + \mathbf{k}$

5. $\mathbf{v} = 3\mathbf{i} + \mathbf{j} - \mathbf{k}, \quad \mathbf{w} = -2\mathbf{i} - \mathbf{j} + \mathbf{k}$ **6.** $\mathbf{v} = \mathbf{i} - 3\mathbf{j} + 4\mathbf{k}, \quad \mathbf{w} = 4\mathbf{i} - \mathbf{j} + 3\mathbf{k}$

In Problems 7–12 calculate the projection of $\mathbf{v}$ along $\mathbf{w}$.

7. $\mathbf{v} = 2\mathbf{i} - 3\mathbf{j} + \mathbf{k}, \quad \mathbf{w} = \mathbf{i} - \mathbf{j} + \mathbf{k}$ **8.** $\mathbf{v} = -3\mathbf{i} + 2\mathbf{j} - \mathbf{k}, \quad \mathbf{w} = 2\mathbf{i} + \mathbf{j} - \mathbf{k}$

9. $\mathbf{v} = \mathbf{i} - \mathbf{j}, \quad \mathbf{w} = \mathbf{j} + \mathbf{k}$ **10.** $\mathbf{v} = \mathbf{j} - \mathbf{k}, \quad \mathbf{w} = \mathbf{i} + \mathbf{k}$

11. $\mathbf{v} = 3\mathbf{i} + \mathbf{j} - \mathbf{k}, \quad \mathbf{w} = -2\mathbf{i} - \mathbf{j} + \mathbf{k}$ **12.** $\mathbf{v} = \mathbf{i} - 3\mathbf{j} + 4\mathbf{k}, \quad \mathbf{w} = 4\mathbf{i} - \mathbf{j} + 3\mathbf{k}$

In Problems 13–16 find a real number α so that the vectors $\mathbf{v}$ and $\mathbf{w}$ are perpendicular.

13. $\mathbf{v} = 2\alpha\mathbf{i} + \mathbf{j} - \mathbf{k}, \quad \mathbf{w} = \mathbf{i} - \mathbf{j} + \mathbf{k}$ **14.** $\mathbf{v} = \mathbf{i} + 2\alpha\mathbf{j} - \mathbf{k}, \quad \mathbf{w} = \mathbf{i} - \mathbf{j} + \mathbf{k}$

15. $\mathbf{v} = \alpha\mathbf{i} + \mathbf{j} + \mathbf{k}, \quad \mathbf{w} = \mathbf{i} + \alpha\mathbf{j} + 4\mathbf{k}$ **16.** $\mathbf{v} = \mathbf{i} - \alpha\mathbf{j} + 2\mathbf{k}, \quad \mathbf{w} = 2\alpha\mathbf{i} + \mathbf{j} + \mathbf{k}$

17. Find a real number α so that the angle θ between the vectors $\mathbf{v} = \alpha\mathbf{i} + \mathbf{j} + \mathbf{k}$ and $\mathbf{w} = \mathbf{i} + \alpha\mathbf{j} + \mathbf{k}$ is $\pi/3$.

18. Find a real number α so that the angle θ between the vectors $\mathbf{v} = \alpha\mathbf{i} - \mathbf{j} + \mathbf{k}$ and $\mathbf{w} = \mathbf{i} - \mathbf{j} + \alpha\mathbf{k}$ is $\pi/3$.

19. Show that the projection of $\mathbf{v}$ along $\mathbf{i}$ is $\mathbf{v} \cdot \mathbf{i}$.   In fact, show that we can always write a vector $\mathbf{v}$ as

$$\mathbf{v} = (\mathbf{v} \cdot \mathbf{i})\mathbf{i} + (\mathbf{v} \cdot \mathbf{j})\mathbf{j} + (\mathbf{v} \cdot \mathbf{k})\mathbf{k}$$

20. Find all numbers α and β so that the vectors $2\alpha\mathbf{i} - 2\mathbf{j} + \mathbf{k}$ and $\beta\mathbf{i} + 2\mathbf{j} + 2\mathbf{k}$ are perpendicular and have the same magnitude.

21. If $\|\mathbf{v}\| = 2$, $\|\mathbf{w}\| = 6$,   and the angle between $\mathbf{v}$ and $\mathbf{w}$ is $\pi/6$, find $\|\mathbf{v} + \mathbf{w}\|$ and $\|\mathbf{v} - \mathbf{w}\|$.

22. A wagon is pulled horizontally by exerting a force of 20 pounds on the handle at an angle of $30°$ with the horizontal. How much work is done in moving the wagon 100 feet?

23. Find the work done in moving an object along a vector $\mathbf{u} = 3\mathbf{i} + 2\mathbf{j} - 5\mathbf{k}$ if the applied force is $\mathbf{F} = 2\mathbf{i} - \mathbf{j} - \mathbf{k}$.

24. Find the work done by a force of 3 newtons acting in the direction $2\mathbf{i} + \mathbf{j} + 2\mathbf{k}$ in moving an object 2 meters from $(0, 0, 0)$ to $(0, 2, 0)$.

25. Find the work done by a force of 1 newton acting in the direction $2\mathbf{i} + 2\mathbf{j} + \mathbf{k}$ in moving an object 3 meters from $(0, 0, 0)$ to $(1, 2, 2)$.

26. Find the acute angle that a constant unit force vector makes with the positive x-axis if the work done by the force in moving a particle from $(0, 0)$ to $(4, 0)$ equals 2.

27. Find all numbers z so that the triangle with vertices $A = (1, -1, 0)$, $B = (-2, 2, 1)$,   and $C = (0, 2, z)$ is a right triangle with right angle at C.

28. An airplane has an air speed of 500 kilometers per hour in a northerly direction. The wind velocity is 60 kilometers per hour in a southeasterly direction. Find the actual speed and direction of the plane relative to the ground.

29. A stream 1 kilometer wide has a constant current of 5 kilometers per hour. At what angle to the shore should a person head a boat, which is capable of maintaining a constant speed of 15 kilometers per hour, in order to reach a point directly opposite?

30. A river is 500 meters wide and has a current of 1 kilometer per hour. If a person can swim at a rate of 2 kilometers per hour, at what angle to the shore should she swim if she wishes to cross the river to a point directly opposite? How long will it take to swim across the river?

31. An airplane travels 200 miles due west and then 150 miles 60° north of west. Determine the resultant displacement.

32. Prove the *polarization identity:* $\|\mathbf{u} + \mathbf{v}\|^2 - \|\mathbf{u} - \mathbf{v}\|^2 = 4(\mathbf{u} \cdot \mathbf{v})$

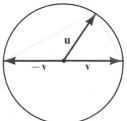

33. (a) If $\mathbf{u}$ and $\mathbf{v}$ have the same length, then show that $\mathbf{u} + \mathbf{v}$ and $\mathbf{u} - \mathbf{v}$ are perpendicular.
    (b) Use this to prove that an angle inscribed in a semicircle is a right angle (see the figure).

34. Let $\mathbf{r}_1$ and $\mathbf{r}_2$ be unit vectors in the xy-plane making angles α and β with the positive x-axis.
    (a) Prove that $\mathbf{r}_1 = (\cos \alpha)\mathbf{i} + (\sin \alpha)\mathbf{j}$ and $\mathbf{r}_2 = (\cos \beta)\mathbf{i} + (\sin \beta)\mathbf{j}$.
    (b) By considering $\mathbf{r}_1 \cdot \mathbf{r}_2$, prove the trigonometric formulas below:

$$\cos(\alpha - \beta) = \cos \alpha \cos \beta + \sin \alpha \sin \beta \qquad \cos(\alpha + \beta) = \cos \alpha \cos \beta - \sin \alpha \sin \beta$$

35. Let $\mathbf{v}$ and $\mathbf{w}$ denote nonzero vectors. Show that the vector $\mathbf{v} - \alpha\mathbf{w}$ is perpendicular to $\mathbf{w}$ if $\alpha = \dfrac{\mathbf{v} \cdot \mathbf{w}}{\|\mathbf{w}\|^2}$.

36. Let $\mathbf{v}$ and $\mathbf{w}$ denote nonzero vectors. Show that the vectors $\|\mathbf{w}\|\mathbf{v} + \|\mathbf{v}\|\mathbf{w}$ and $\|\mathbf{w}\|\mathbf{v} - \|\mathbf{v}\|\mathbf{w}$ are perpendicular.

37. In the definition of work given in (14.24), what is the work done if $\mathbf{F}$ is perpendicular to $\mathbf{D}$?

38. Let $\mathbf{v} = a_1\mathbf{i} + b_1\mathbf{j} + c_1\mathbf{k}$ and $\mathbf{w} = a_2\mathbf{i} + b_2\mathbf{j} + c_2\mathbf{k}$ be two nonzero vectors that are not parallel. Find a vector $\mathbf{u} \neq \mathbf{0}$ that is perpendicular to both $\mathbf{v}$ and $\mathbf{w}$.

39. Let $\mathbf{v}$ and $\mathbf{w}$ denote nonzero vectors, and let $\mathbf{u} = \|\mathbf{w}\|\mathbf{v} + \|\mathbf{v}\|\mathbf{w}$. Show that the angle between $\mathbf{v}$ and $\mathbf{u}$ equals the angle between $\mathbf{w}$ and $\mathbf{u}$.

40. Let $\mathbf{w}$ be a nonzero vector, and let $\mathbf{u}$ denote a unit vector. Prove that the unit vector $\mathbf{u}$ making $\mathbf{w} \cdot \mathbf{u}$ a maximum is the unit vector pointing in the same direction as $\mathbf{w}$.

41. Let $\mathbf{u}$ and $\mathbf{w}$ be two unit vectors, and let θ be the angle between them. Find $\frac{1}{2}\|\mathbf{u} - \mathbf{w}\|$ in terms of θ.

42. Under what conditions is the Cauchy–Schwarz inequality an equality?

43. Prove part (b) of theorem (14.13).

44. Prove part (c) of theorem (14.13).

45. Prove part (d) of theorem (14.13).

46. Use the fact that $\|\mathbf{v}\|^2 = \mathbf{v} \cdot \mathbf{v}$ to prove (14.2) algebraically; that is, prove

$$\|\alpha\mathbf{v}\| = |\alpha|\,\|\mathbf{v}\| \qquad \alpha \text{ a scalar,} \quad \mathbf{v} \text{ a vector}$$

47. If $\mathbf{v}$ is a vector for which $\mathbf{v} \cdot \mathbf{i} = 0$, $\mathbf{v} \cdot \mathbf{j} = 0$, and $\mathbf{v} \cdot \mathbf{k} = 0$, find $\mathbf{v}$.

48. Solve for $\mathbf{x}$ in terms of α, $\mathbf{a}$, $\mathbf{b}$, and $\mathbf{c}$, if

$$\alpha\mathbf{x} + (\mathbf{x} \cdot \mathbf{b})\mathbf{a} = \mathbf{c} \qquad \alpha \neq 0, \quad \alpha + \mathbf{a} \cdot \mathbf{b} \neq 0$$

[*Hint:* First find $\mathbf{x} \cdot \mathbf{b}$, then $\mathbf{x}$.]

49. Provide the details of the following outline of an alternate proof of the Cauchy–Schwarz inequality: If $\mathbf{u}$ and $\mathbf{v}$ denote nonzero vectors, first show that

$$0 \le \|\alpha\mathbf{u} + \beta\mathbf{v}\|^2 = \alpha^2\|\mathbf{u}\|^2 + 2\alpha\beta\mathbf{u}\cdot\mathbf{v} + \beta^2\|\mathbf{v}\|^2$$

for any scalars α and β. Then set $\alpha = \|\mathbf{v}\|$ and $\beta = -\|\mathbf{u}\|$ to get the desired result.

50. The dot product of two vectors $\mathbf{v} = a_1\mathbf{u}_1 + a_2\mathbf{u}_2 + \cdots + a_n\mathbf{u}_n$ and $\mathbf{w} = b_1\mathbf{u}_1 + b_2\mathbf{u}_2 + \cdots + b_n\mathbf{u}_n$ in n-space is defined as

$$\mathbf{v}\cdot\mathbf{w} = a_1 b_1 + a_2 b_2 + \cdots + a_n b_n$$

Show that the five properties listed in (14.13) hold.

6. The Cross Product

The cross product of two vectors is of special interest for those studying physics, particularly mechanics and electricity. It is used, for example, to describe angular velocity, torque, and angular momentum.

We begin with an algebraic definition of the cross product and list some algebraic properties that follow from that definition. Later in the section we look at some of the geometric and physical properties of the cross product. As mentioned earlier, the cross product is only defined for vectors in 3-space.

(14.25)   **DEFINITION** If $\mathbf{v} = a_1\mathbf{i} + b_1\mathbf{j} + c_1\mathbf{k}$ and $\mathbf{w} = a_2\mathbf{i} + b_2\mathbf{j} + c_2\mathbf{k}$ are vectors, the *cross product* $\mathbf{v} \times \mathbf{w}$ is defined as the vector

$$\mathbf{v} \times \mathbf{w} = (b_1 c_2 - b_2 c_1)\mathbf{i} - (a_1 c_2 - a_2 c_1)\mathbf{j} + (a_1 b_2 - a_2 b_1)\mathbf{k}$$

Note that the cross product $\mathbf{v} \times \mathbf{w}$ of two vectors is itself a vector. Because of this, the cross product is sometimes referred to as the *vector product*.

EXAMPLE 1   Let $\mathbf{v} = 2\mathbf{i} - \mathbf{j} + \mathbf{k}$ and $\mathbf{w} = 4\mathbf{i} + 2\mathbf{j} - \mathbf{k}$. Find:
(a) $\mathbf{v} \times \mathbf{w}$ (b) $\mathbf{w} \times \mathbf{v}$ (c) $\mathbf{v} \times \mathbf{v}$ (d) $\mathbf{w} \times \mathbf{w}$

Solution   (a) $\mathbf{v} \times \mathbf{w} = [(-1)(-1) - (2)(1)]\mathbf{i} - [(2)(-1) - (4)(1)]\mathbf{j} + [(2)(2) - (4)(-1)]\mathbf{k}$
$= -\mathbf{i} + 6\mathbf{j} + 8\mathbf{k}$

(b) $\mathbf{w} \times \mathbf{v} = [(2)(1) - (-1)(-1)]\mathbf{i} - [(4)(1) - (2)(-1)]\mathbf{j} + [(4)(-1) - (2)(2)]\mathbf{k}$
$= \mathbf{i} - 6\mathbf{j} - 8\mathbf{k}$

(c) $\mathbf{v} \times \mathbf{v} = [(-1)(1) - (-1)(1)]\mathbf{i} - [(2)(1) - (2)(1)]\mathbf{j} + [(2)(-1) - (2)(-1)]\mathbf{k} = \mathbf{0}$

(d) $\mathbf{w} \times \mathbf{w} = \mathbf{0}$ ∎

Determinants may be used as an aid in remembering (14.25). A *second-order determinant* is defined as

$$\begin{vmatrix} a_1 & b_1 \\ a_2 & b_2 \end{vmatrix} = a_1 b_2 - a_2 b_1$$

A *third-order determinant* is defined as

$$\begin{vmatrix} A & B & C \\ a_1 & b_1 & c_1 \\ a_2 & b_2 & c_2 \end{vmatrix} = \begin{vmatrix} b_1 & c_1 \\ b_2 & c_2 \end{vmatrix} A - \begin{vmatrix} a_1 & c_1 \\ a_2 & c_2 \end{vmatrix} B + \begin{vmatrix} a_1 & b_1 \\ a_2 & b_2 \end{vmatrix} C$$

For example,

$$\begin{vmatrix} 2 & 3 \\ -1 & 2 \end{vmatrix} = (2)(2) - (-1)(3) = 4 + 3 = 7$$

and

$$\begin{vmatrix} A & B & C \\ 2 & 3 & 1 \\ -1 & 2 & -1 \end{vmatrix} = \begin{vmatrix} 3 & 1 \\ 2 & -1 \end{vmatrix} A - \begin{vmatrix} 2 & 1 \\ -1 & -1 \end{vmatrix} B + \begin{vmatrix} 2 & 3 \\ -1 & 2 \end{vmatrix} C$$

$$= (-3 - 2)A - (-2 + 1)B + (4 + 3)C = -5A + B + 7C$$

By using determinants, the cross product of the vectors $\mathbf{v} = a_1\mathbf{i} + b_1\mathbf{j} + c_1\mathbf{k}$ and $\mathbf{w} = a_2\mathbf{i} + b_2\mathbf{j} + c_2\mathbf{k}$ may be written as

(14.26) $$\mathbf{v} \times \mathbf{w} = \begin{vmatrix} \mathbf{i} & \mathbf{j} & \mathbf{k} \\ a_1 & b_1 & c_1 \\ a_2 & b_2 & c_2 \end{vmatrix} = \begin{vmatrix} b_1 & c_1 \\ b_2 & c_2 \end{vmatrix} \mathbf{i} - \begin{vmatrix} a_1 & c_1 \\ a_2 & c_2 \end{vmatrix} \mathbf{j} + \begin{vmatrix} a_1 & b_1 \\ a_2 & b_2 \end{vmatrix} \mathbf{k}$$

$$= (b_1 c_2 - b_2 c_1)\mathbf{i} - (a_1 c_2 - a_2 c_1)\mathbf{j} + (a_1 b_2 - a_2 b_1)\mathbf{k}$$

EXAMPLE 2   If $\mathbf{v} = 2\mathbf{i} - \mathbf{j} + \mathbf{k}$ and $\mathbf{w} = 4\mathbf{i} + 2\mathbf{j} - \mathbf{k}$, find: (a) $\mathbf{v} \times \mathbf{w}$ (b) $\mathbf{w} \times \mathbf{v}$

Solution   (a) $$\mathbf{v} \times \mathbf{w} = \begin{vmatrix} \mathbf{i} & \mathbf{j} & \mathbf{k} \\ 2 & -1 & 1 \\ 4 & 2 & -1 \end{vmatrix} = (1 - 2)\mathbf{i} - (-2 - 4)\mathbf{j} + (4 + 4)\mathbf{k}$$

$$= -\mathbf{i} + 6\mathbf{j} + 8\mathbf{k}$$

(b) $$\mathbf{w} \times \mathbf{v} = \begin{vmatrix} \mathbf{i} & \mathbf{j} & \mathbf{k} \\ 4 & 2 & -1 \\ 2 & -1 & 1 \end{vmatrix} = \mathbf{i} - 6\mathbf{j} - 8\mathbf{k}$$ ∎

The cross product has some interesting algebraic and geometric properties. We first list the algebraic properties.

(14.27)   **THEOREM**   **If v, w, and u are vectors and if α is a scalar, then:**

(a) $\mathbf{v} \times \mathbf{v} = 0$ (b) $\alpha(\mathbf{v} \times \mathbf{w}) = (\alpha\mathbf{v}) \times \mathbf{w} = \mathbf{v} \times (\alpha\mathbf{w})$

(c) $\mathbf{v} \times \mathbf{w} = -(\mathbf{w} \times \mathbf{v})$ (d) $\mathbf{v} \times (\mathbf{w} + \mathbf{u}) = (\mathbf{v} \times \mathbf{w}) + (\mathbf{v} \times \mathbf{u})$

(e) $\|\mathbf{v} \times \mathbf{w}\|^2 = \|\mathbf{v}\|^2\|\mathbf{w}\|^2 - (\mathbf{v} \cdot \mathbf{w})^2$

We derive parts (a), (c), and (e) below and leave the proofs of parts (b) and (d) as exercises.

Proof

(a) If $v = ai + bj + ck,$ then

$$v \times v = (bc - bc)i - (ac - ac)j + (ab - ab)k = 0$$

(c) If $v = a_1i + b_1j + c_1k$ and $w = a_2i + b_2j + c_2k,$ then

$$v \times w = (b_1c_2 - b_2c_1)i - (a_1c_2 - a_2c_1)j + (a_1b_2 - a_2b_1)k$$

and

$$w \times v = (b_2c_1 - b_1c_2)i - (a_2c_1 - a_1c_2)j + (a_2b_1 - a_1b_2)k$$

Consequently, $v \times w = -(w \times v).$

(e) If $v = a_1i + b_1j + c_1k$ and $w = a_2i + b_2j + c_2k,$ then

$$\|v \times w\|^2 = (b_1c_2 - b_2c_1)^2 + (a_1c_2 - a_2c_1)^2 + (a_1b_2 - a_2b_1)^2$$
$$= (b_1c_2)^2 - 2b_1b_2c_1c_2 + (b_2c_1)^2 + (a_1c_2)^2$$
$$- 2a_1a_2c_1c_2 + (a_2c_1)^2 + (a_1b_2)^2 - 2a_1a_2b_1b_2 + (a_2b_1)^2$$

$$\|v\|^2\|w\|^2 - (v \cdot w)^2 = (a_1^2 + b_1^2 + c_1^2)(a_2^2 + b_2^2 + c_2^2) - (a_1a_2 + b_1b_2 + c_1c_2)^2$$
$$= (a_1b_2)^2 + (a_1c_2)^2 + (b_1a_2)^2 + (b_1c_2)^2 + (c_1a_2)^2$$
$$+ (c_1b_2)^2 - 2a_1b_1a_2b_2 - 2a_1c_1a_2c_2 - 2b_1c_1b_2c_2$$

Reordering the terms on the right in both equations gives the same result.   ■

We shall find the cross products of the unit vectors **i**, **j**, and **k** to be particularly useful:

$$i \times j = \begin{vmatrix} i & j & k \\ 1 & 0 & 0 \\ 0 & 1 & 0 \end{vmatrix} = 0i + 0j + k = k$$

Similarly,

$$j \times k = \begin{vmatrix} i & j & k \\ 0 & 1 & 0 \\ 0 & 0 & 1 \end{vmatrix} = i \qquad k \times i = \begin{vmatrix} i & j & k \\ 0 & 0 & 1 \\ 1 & 0 & 0 \end{vmatrix} = j$$

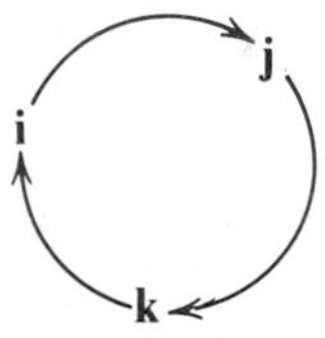

Note the cyclic pattern for the cross products of **i**, **j**, and **k**:

$$i \times j = k \qquad j \times k = i \qquad k \times i = j$$

Figure 39   Figure 39 may be helpful in remembering this pattern.

We list the remaining cross products involving **i**, **j**, and **k**; they are a consequence of properties (a) and (c) of (14.27):

$$j \times i = -k \qquad k \times j = -i \qquad i \times k = -j$$
$$i \times i = 0 \qquad j \times j = 0 \qquad k \times k = 0$$

GEOMETRIC PROPERTIES OF THE CROSS PRODUCT

The geometric properties of the cross product are summarized in the following theorem:

(14.28) **THEOREM** **If v and w are vectors, then:**

(a) $\mathbf{v} \times \mathbf{w}$ **is perpendicular to both v and w**

(b) $\|\mathbf{v} \times \mathbf{w}\| = \|\mathbf{v}\| \|\mathbf{w}\| \sin \theta$, **where θ is the angle between $\mathbf{v} \neq 0$ and $\mathbf{w} \neq 0$**

(c) $\|\mathbf{v} \times \mathbf{w}\|$ **is the area of the parallelogram having $\mathbf{v} \neq 0$ and $\mathbf{w} \neq 0$ as adjacent sides**

(d) $\mathbf{v} \times \mathbf{w} = 0$ **if and only if v and w are parallel**

Proof of Part (a)   Let $\mathbf{v} = a_1\mathbf{i} + b_1\mathbf{j} + c_1\mathbf{k}$ and $\mathbf{w} = a_2\mathbf{i} + b_2\mathbf{j} + c_2\mathbf{k}$.   Then

$$\mathbf{v} \times \mathbf{w} = (b_1c_2 - b_2c_1)\mathbf{i} - (a_1c_2 - a_2c_1)\mathbf{j} + (a_1b_2 - a_2b_1)\mathbf{k}$$

so that

$$\mathbf{v} \cdot (\mathbf{v} \times \mathbf{w}) = a_1b_1c_2 - a_1b_2c_1 - b_1a_1c_2 + b_1a_2c_1 + c_1a_1b_2 - c_1a_2b_1 = 0$$

and

$$\mathbf{w} \cdot (\mathbf{v} \times \mathbf{w}) = a_2b_1c_2 - a_2b_2c_1 - b_2a_1c_2 + b_2a_2c_1 + c_2a_1b_2 - c_2a_2b_1 = 0$$

Thus, $\mathbf{v} \times \mathbf{w}$ is perpendicular to v, and $\mathbf{v} \times \mathbf{w}$ is perpendicular to w.   ∎

The vector $\mathbf{v} \times \mathbf{w}$ is perpendicular to the plane containing v and w. As Figure 40 illustrates, there are two vectors that are perpendicular to the plane containing v and w. If the plane is represented by this page, one such vector is directed up and the other is directed down. Which of these is $\mathbf{v} \times \mathbf{w}$?   It can be shown that the direction of $\mathbf{v} \times \mathbf{w}$ is determined by the thumb of the right hand when the other fingers of the right hand are cupped so that they point in the direction of the angle θ from v to w (see Fig. 41). As in Section 3, this is usually referred to as the *right-hand rule* and the coordinate system we have been using is called *right-handed*.

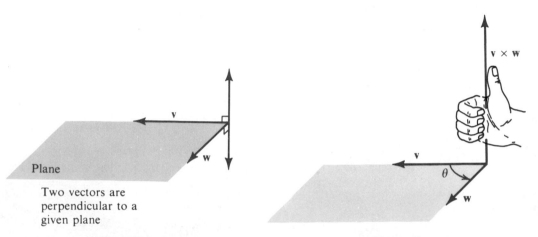

Plane

Two vectors are
perpendicular to a
given plane

Figure 40   **Figure 41**

EXAMPLE 3 Find a vector perpendicular to each of the vectors $\mathbf{v} = 2\mathbf{i} - \mathbf{j} + \mathbf{k}$ and $\mathbf{w} = 4\mathbf{i} + 2\mathbf{j} - \mathbf{k}$.

Solution Based on the preceding discussion, such a vector is

$$\mathbf{v} \times \mathbf{w} = -\mathbf{i} + 6\mathbf{j} + 8\mathbf{k}$$

$$\uparrow$$

Example 2

This may be verified as follows:

$$\mathbf{v} \cdot (\mathbf{v} \times \mathbf{w}) = (2\mathbf{i} - \mathbf{j} + \mathbf{k}) \cdot (-\mathbf{i} + 6\mathbf{j} + 8\mathbf{k}) = -2 - 6 + 8 = 0$$

and

$$\mathbf{w} \cdot (\mathbf{v} \times \mathbf{w}) = (4\mathbf{i} + 2\mathbf{j} - \mathbf{k}) \cdot (-\mathbf{i} + 6\mathbf{j} + 8\mathbf{k}) = -4 + 12 - 8 = 0$$

Hence, $\mathbf{v} \times \mathbf{w}$ is perpendicular to both $\mathbf{v}$ and $\mathbf{w}$. ∎

Proof of Part (b) of (14.28) Let θ be the angle between the nonzero vectors $\mathbf{v}$ and $\mathbf{w}$. Then, by part (e) of (14.27), we have

$$\|\mathbf{v} \times \mathbf{w}\|^2 = \|\mathbf{v}\|^2\|\mathbf{w}\|^2 - (\mathbf{v} \cdot \mathbf{w})^2$$

Since $\mathbf{v} \cdot \mathbf{w} = \|\mathbf{v}\|\,\|\mathbf{w}\|\cos\theta$, we find

$$\begin{aligned}\|\mathbf{v} \times \mathbf{w}\|^2 &= \|\mathbf{v}\|^2\|\mathbf{w}\|^2 - \|\mathbf{v}\|^2\|\mathbf{w}\|^2\cos^2\theta \\ &= \|\mathbf{v}\|^2\|\mathbf{w}\|^2(1 - \cos^2\theta) \\ &= \|\mathbf{v}\|^2\|\mathbf{w}\|^2\sin^2\theta\end{aligned}$$

By taking square roots, we have the result

$$\|\mathbf{v} \times \mathbf{w}\| = \|\mathbf{v}\|\,\|\mathbf{w}\|\sin\theta \quad\blacksquare$$

Proof of Part (c) of (14.28) Part (b) of (14.28) is used to prove part (c). If $\mathbf{v}$ and $\mathbf{w}$ are nonzero vectors and θ is the angle between $\mathbf{v}$ and $\mathbf{w}$, then $\|\mathbf{v}\|$ and $\|\mathbf{w}\|$ represent the lengths of the sides of a parallelogram (see Fig. 42) and the quantity $\|\mathbf{v}\|\,\|\mathbf{w}\|\sin\theta$ is the area of the parallelogram. Thus, the magnitude of $\mathbf{v} \times \mathbf{w}$ is the area of the parallelogram whose sides are the vectors $\mathbf{v}$ and $\mathbf{w}$.

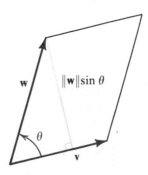

Figure 42 Area − Base × Height = $\|\mathbf{v}\|\,\|\mathbf{w}\|\sin\theta$ ∎

As the next example illustrates, some care has to be taken when using property (c) of (14.28).

EXAMPLE 4   Find the area of the parallelogram whose vertices are $P_1 = (0, 0, 0)$, $P_2 = (-1, 2, 4)$, $P_3 = (2, -1, 4)$,   and $P_4 = (1, 1, 8)$.

Solution   The area of the parallelogram is $\|\mathbf{v} \times \mathbf{w}\|$,   where $\mathbf{v}$ and $\mathbf{w}$ are two *adjacent sides* of the parallelogram. We may pick

$$\mathbf{v} = \overrightarrow{P_1 P_2} = -\mathbf{i} + 2\mathbf{j} + 4\mathbf{k} \qquad \text{and} \qquad \mathbf{w} = \overrightarrow{P_1 P_3} = 2\mathbf{i} - \mathbf{j} + 4\mathbf{k}$$

Be careful! Not all pairs of vertices give a side. For example, $\overrightarrow{P_1 P_4}$ is not a side; it is a diagonal of the parallelogram since $\overrightarrow{P_1 P_2} + \overrightarrow{P_1 P_3} = \overrightarrow{P_1 P_4}$.   Now

$$\mathbf{v} \times \mathbf{w} = (-\mathbf{i} + 2\mathbf{j} + 4\mathbf{k}) \times (2\mathbf{i} - \mathbf{j} + 4\mathbf{k}) = 12\mathbf{i} + 12\mathbf{j} - 3\mathbf{k}$$

So, the area of the parallelogram is

$$\|\mathbf{v} \times \mathbf{w}\| = \sqrt{144 + 144 + 9} = \sqrt{297} = 3\sqrt{33} \approx 17.23 \text{ square units} \qquad \blacksquare$$

Proof of Part (d) of (14.28)   If $\mathbf{v}$ and $\mathbf{w}$ are parallel vectors $(\sin \theta = 0)$,   it follows from part (b) of (14.28) that $\|\mathbf{v} \times \mathbf{w}\| = 0$ or $\mathbf{v} \times \mathbf{w} = \mathbf{0}$.
On the other hand, if $\mathbf{v} \times \mathbf{w} = \mathbf{0}$,   then

$$\|\mathbf{v}\| = 0 \qquad \text{or} \qquad \|\mathbf{w}\| = 0 \qquad \text{or} \qquad \sin \theta = 0$$

In the latter case, $\theta = 0$ or $\theta = \pi$ and the vectors $\mathbf{v}$ and $\mathbf{w}$ are parallel. If $\mathbf{v}$ or $\mathbf{w}$ is the zero vector, then, since the zero vector has no specific direction, we adopt the convention that the zero vector is parallel to every vector.* Thus, we have the result we seek. $\blacksquare$

We use part (d) of (14.28) as a criterion for parallel vectors. For example, the vectors $\mathbf{v} = 2\mathbf{i} + \mathbf{j} - \mathbf{k}$ and $\mathbf{w} = -4\mathbf{i} - 2\mathbf{j} + 2\mathbf{k}$ are parallel since

$$\mathbf{v} \times \mathbf{w} = \begin{vmatrix} \mathbf{i} & \mathbf{j} & \mathbf{k} \\ 2 & 1 & -1 \\ -4 & -2 & 2 \end{vmatrix} = \begin{vmatrix} 1 & -1 \\ -2 & 2 \end{vmatrix} \mathbf{i} - \begin{vmatrix} 2 & -1 \\ -4 & 2 \end{vmatrix} \mathbf{j} + \begin{vmatrix} 2 & 1 \\ -4 & -2 \end{vmatrix} \mathbf{k}$$

$$= 0\mathbf{i} + 0\mathbf{j} + 0\mathbf{k} = \mathbf{0}$$

For these vectors $\mathbf{v}$ and $\mathbf{w}$, we note that $\mathbf{v} = -2\mathbf{w}$.   Thus, the magnitude of $\mathbf{v}$ is twice that of $\mathbf{w}$, and $\mathbf{v}$ and $\mathbf{w}$ are opposite in direction.

---

* With this convention, the zero vector is both parallel and perpendicular to every vector. This apparent contradiction will cause us no trouble, however, since the angle between two vectors is never applied when one of the vectors is the zero vector.

TRIPLE PRODUCTS

There are two types of *triple products*. The triple product $\mathbf{u} \cdot (\mathbf{v} \times \mathbf{w})$ is called the *triple scalar product* of $\mathbf{u}$, $\mathbf{v}$, and $\mathbf{w}$. The triple product $\mathbf{u} \times (\mathbf{v} \times \mathbf{w})$ is called the *triple vector product* of $\mathbf{u}$, $\mathbf{v}$, and $\mathbf{w}$. Some properties of triple products are given below (their proofs are left as exercises).

(14.29)   (a) $\mathbf{u} \cdot (\mathbf{v} \times \mathbf{w}) = (\mathbf{u} \times \mathbf{v}) \cdot \mathbf{w}$

(b) $|\mathbf{u} \cdot (\mathbf{v} \times \mathbf{w})|$ **is the volume of the parallelepiped with sides u, v, and w**

(c) $\mathbf{u} \times (\mathbf{v} \times \mathbf{w}) = (\mathbf{u} \cdot \mathbf{w})\mathbf{v} - (\mathbf{u} \cdot \mathbf{v})\mathbf{w}$

EXERCISE 6

In Problems 1–10 compute the cross product $\mathbf{v} \times \mathbf{w}$. Check your answer by showing that $\mathbf{v}$ and $\mathbf{w}$ are each perpendicular to $\mathbf{v} \times \mathbf{w}$.

1. $\mathbf{v} = 2\mathbf{i} + \mathbf{j} - \mathbf{k}$, $\mathbf{w} = \mathbf{i} - \mathbf{j} + \mathbf{k}$ **2.** $\mathbf{v} = 4\mathbf{i} - \mathbf{j} + 2\mathbf{k}$, $\mathbf{w} = 2\mathbf{i} + \mathbf{j} + \mathbf{k}$

3. $\mathbf{v} = \mathbf{i} + \mathbf{j}$, $\mathbf{w} = \mathbf{i} - \mathbf{j}$ **4.** $\mathbf{v} = \mathbf{j} - \mathbf{k}$, $\mathbf{w} = \mathbf{i} - \mathbf{j}$

5. $\mathbf{v} = 3\mathbf{i} - 2\mathbf{j} + \mathbf{k}$, $\mathbf{w} = \mathbf{i} + \mathbf{j}$ **6.** $\mathbf{v} = 2\mathbf{i} - \mathbf{j}$, $\mathbf{w} = \mathbf{i} + \mathbf{j} - 3\mathbf{k}$

7. $\mathbf{v} = -\mathbf{i} + 8\mathbf{j} + 3\mathbf{k}$, $\mathbf{w} = 7\mathbf{i} + 2\mathbf{j}$ **8.** $\mathbf{v} = 2\mathbf{j} - \mathbf{k}$, $\mathbf{w} = -3\mathbf{i} + \mathbf{j} + \mathbf{k}$

9. $\mathbf{v} = 2\mathbf{i} + 3\mathbf{j} - 4\mathbf{k}$, $\mathbf{w} = -\mathbf{i} + \mathbf{j} - 4\mathbf{k}$ **10.** $\mathbf{v} = (\cos \theta)\mathbf{i} + (\sin \theta)\mathbf{j}$, $\mathbf{w} = (\sin \theta)\mathbf{i} + (\cos \theta)\mathbf{j}$

In Problems 11–16 find the area of the parallelogram with one corner at P and sides PQ and PR.

11. $P = (1, -3, 7)$; $Q = (2, 1, 1)$; $R = (6, -1, 2)$ **12.** $P = (0, 1, 1)$; $Q = (2, 0, -4)$; $R = (-3, -2, 1)$

13. $P = (-2, 1, 6)$; $Q = (2, 1, -7)$; $R = (4, 1, 1)$ **14.** $P = (0, 0, 3)$; $Q = (2, -5, 3)$; $R = (1, 1, -2)$

15. $P = (1, 1, -6)$; $Q = (5, -3, 0)$; $R = (-2, 4, 1)$ **16.** $P = (-4, 6, 3)$; $Q = (1, 1, -5)$; $R = (2, 2, 2)$

In Problems 17–20 find the area of the parallelogram whose vertices are P_1, P_2, P_3, and P_4.

17. $P_1 = (0, 0, 0)$; $P_2 = (1, 2, 3)$; $P_3 = (2, -1, 1)$; $P_4 = (3, 1, 4)$

18. $P_1 = (0, 0, 0)$; $P_2 = (-1, 2, 0)$; $P_3 = (2, 3, -4)$; $P_4 = (1, 5, -4)$

19. $P_1 = (1, 2, -1)$; $P_2 = (4, 2, -3)$; $P_3 = (6, -5, 2)$; $P_4 = (9, -5, 0)$

20. $P_1 = (-1, 1, 1)$; $P_2 = (-1, 2, 2)$; $P_3 = (-3, 4, -5)$; $P_4 = (-3, 5, -4)$

21. Show that the area of the triangle whose vertices are the endpoints of the vectors $\mathbf{u}$, $\mathbf{v}$, and $\mathbf{w}$ is

$$A = \tfrac{1}{2}\|(\mathbf{v} - \mathbf{u}) \times (\mathbf{w} - \mathbf{u})\|$$

22. Use the result of Problem 21 to find the area of the triangle with vertices $(0, 0, 0)$, $(2, 3, -2)$, and $(-1, 1, 4)$.

23. Derive part (b) of (14.27).

24. Derive part (d) of (14.27).

25. Give an example to show that the cross product is not associative; that is, find vectors $\mathbf{u}$, $\mathbf{v}$, and $\mathbf{w}$ so that
$\mathbf{u} \times (\mathbf{v} \times \mathbf{w}) \neq (\mathbf{u} \times \mathbf{v}) \times \mathbf{w}$.

26. If $\mathbf{v} \times \mathbf{w} = \mathbf{0}$ and $\mathbf{v} \cdot \mathbf{w} = 0,$ can you draw any conclusions about $\mathbf{v}$ or $\mathbf{w}$?

27. Show that $\mathbf{u} \cdot (\mathbf{v} \times \mathbf{w})$ is given by the determinant

$$\mathbf{u} \cdot (\mathbf{v} \times \mathbf{w}) = \begin{vmatrix} a_1 & b_1 & c_1 \\ a_2 & b_2 & c_2 \\ a_3 & b_3 & c_3 \end{vmatrix}$$

where $\mathbf{u} = a_1\mathbf{i} + b_1\mathbf{j} + c_1\mathbf{k},$ $\mathbf{v} = a_2\mathbf{i} + b_2\mathbf{j} + c_2\mathbf{k},$ and $\mathbf{w} = a_3\mathbf{i} + b_3\mathbf{j} + c_3\mathbf{k}.$ Use this to prove part (a) of (14.29).

28. Prove part (b) of (14.29).

29. Prove part (c) of (14.29).

30. Use part (c) of (14.29) to show that $\mathbf{u} \times (\mathbf{v} \times \mathbf{w}) + \mathbf{v} \times (\mathbf{w} \times \mathbf{u}) + \mathbf{w} \times (\mathbf{u} \times \mathbf{v}) = \mathbf{0}.$

31. Determine a unit vector perpendicular to the plane containing $\mathbf{v} = 2\mathbf{i} - 6\mathbf{j} - 3\mathbf{k}$ and $\mathbf{w} = 4\mathbf{i} + 3\mathbf{j} - \mathbf{k}.$

32. If $\mathbf{v}, \mathbf{w}, \mathbf{u}$ and $\mathbf{v}', \mathbf{w}', \mathbf{u}'$ are such that

$$\mathbf{v}' \cdot \mathbf{v} = \mathbf{w}' \cdot \mathbf{w} = \mathbf{u}' \cdot \mathbf{u} = 1$$

$$\mathbf{v}' \cdot \mathbf{w} = \mathbf{v}' \cdot \mathbf{u} = \mathbf{w}' \cdot \mathbf{v} = \mathbf{w}' \cdot \mathbf{u} = \mathbf{u}' \cdot \mathbf{v} = \mathbf{u}' \cdot \mathbf{w} = 0$$

prove that

$$\mathbf{v}' = \frac{\mathbf{w} \times \mathbf{u}}{\mathbf{v} \cdot \mathbf{w} \times \mathbf{u}} \qquad \mathbf{w}' = \frac{\mathbf{u} \times \mathbf{v}}{\mathbf{v} \cdot \mathbf{w} \times \mathbf{u}} \qquad \mathbf{u}' = \frac{\mathbf{v} \times \mathbf{w}}{\mathbf{v} \cdot \mathbf{w} \times \mathbf{u}}$$

33. Prove Lagrange's identity: $(\mathbf{a} \times \mathbf{b}) \cdot (\mathbf{c} \times \mathbf{d}) = (\mathbf{a} \cdot \mathbf{c})(\mathbf{b} \cdot \mathbf{d}) - (\mathbf{a} \cdot \mathbf{d})(\mathbf{b} \cdot \mathbf{c})$

34. Prove that: $(\mathbf{a} \times \mathbf{b}) \times (\mathbf{c} \times \mathbf{d}) = [\mathbf{a} \cdot (\mathbf{b} \times \mathbf{d})]\mathbf{c} - [\mathbf{a} \cdot (\mathbf{b} \times \mathbf{c})]\mathbf{d}$

In Problems 35 and 36 use vector methods to prove each statement.

35. The diagonals of a parallelogram are perpendicular if and only if the parallelogram is a rhombus.

36. The altitudes of a triangle meet at one point; the medians of a triangle meet at one point.

37. Solve for $\mathbf{x}$ in terms of $\alpha, \mathbf{a},$ and $\mathbf{b}$ if

$$\alpha\mathbf{x} + \mathbf{x} \times \mathbf{a} = \mathbf{b} \qquad \alpha \neq 0$$

[*Hint:* First find $\mathbf{x} \cdot \mathbf{a},$ then $\mathbf{x} \times \mathbf{a}.$]

7. Lines

When a vector $\mathbf{v} = x\mathbf{i} + y\mathbf{j} + z\mathbf{k}$ has its initial point at the origin and terminal point at $P = (x, y, z),$ the vector $\mathbf{v}$ is called the *position vector* at the point $P.$ *Thus, every point has a position vector associated with it.*

With this in mind, we take up the question of finding the vector equation of a line in three space. Let $P_1 = (x_1, y_1, z_1)$ and $P_2 = (x_2, y_2, z_2)$ be two distinct points on a line $L.$ The vector $\mathbf{D}$ represented by the directed line segment $\overrightarrow{P_1P_2}$ is a nonzero vector parallel to L (see Fig. 43).

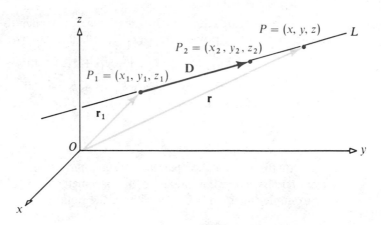

Figure 43

Let $\mathbf{r}_1$ denote the position vector of P_1 and let $\mathbf{r}$ denote the position vector of any point $P = (x, y, z)$. If we insist that P is on L, then the vector $\mathbf{r} - \mathbf{r}_1$ must be parallel to the vector $\mathbf{D}$, so that

$$\mathbf{r} - \mathbf{r}_1 = t\mathbf{D}$$
$$\mathbf{r} = \mathbf{r}_1 + t\mathbf{D} \qquad \text{for some scalar } t$$

Thus, the vector equation of a line L parallel to $\mathbf{D} \neq 0$ and passing through the point $P_1 = (x_1, y_1, z_1)$ is

(14.30) $$\mathbf{r} = \mathbf{r}_1 + t\mathbf{D}$$

where $\mathbf{r}_1$ is the position vector of P_1 and $\mathbf{r}$ is the position vector of any point P on L.

In the vector equation (14.30), the vector $\mathbf{r}_1$ is known (it is the position vector of the point P_1 on L) and the vector $\mathbf{D}$ is known (it is parallel to L). To locate any point P on L, we simply assign an appropriate number to the scalar t. For each number t, we obtain the position vector of a point on L.

EXAMPLE 1   Find the vector equation of a line L passing through the two points $(1, 2, -1)$ and $(4, 3, -2)$.

Solution   The vector $\mathbf{D}$ in the direction from $(1, 2, -1)$ to $(4, 3, -2)$ is

$$\mathbf{D} = 3\mathbf{i} + \mathbf{j} - \mathbf{k}$$

The vector equation of L is

$$\mathbf{r} = \mathbf{r}_1 + t\mathbf{D} = \mathbf{i} + 2\mathbf{j} - \mathbf{k} + t(3\mathbf{i} + \mathbf{j} - \mathbf{k})$$
$$= (1 + 3t)\mathbf{i} + (2 + t)\mathbf{j} + (-1 - t)\mathbf{k} \qquad \blacksquare$$

The vector equation (14.30) of a line L parallel to $\mathbf{D} = a\mathbf{i} + b\mathbf{j} + c\mathbf{k}$ and passing through the point $P_1 = (x_1, y_1, z_1)$ may be written as

(14.31)
$$x\mathbf{i} + y\mathbf{j} + z\mathbf{k} = (x_1\mathbf{i} + y_1\mathbf{j} + z_1\mathbf{k}) + t(a\mathbf{i} + b\mathbf{j} + c\mathbf{k})$$

By using the fact that two vectors $a_1\mathbf{i} + b_1\mathbf{j} + c_1\mathbf{k}$ and $a_2\mathbf{i} + b_2\mathbf{j} + c_2\mathbf{k}$ are equal if and only if $a_1 = a_2$, $b_1 = b_2$, and $c_1 = c_2$, we equate coefficients in (14.31) to find

(14.32)
$$x = x_1 + at \qquad y = y_1 + bt \qquad z = z_1 + ct$$

These equations are called *parametric equations of the line L*, and the variable t is a *parameter*. By assigning values to the parameter t, we obtain points (x, y, z) on L. For example, when $t = 0$, we obtain the point (x_1, y_1, z_1) on L.

EXAMPLE 2　Find the parametric equations of a line L containing the point $(2, -3, 1)$ and parallel to $4\mathbf{i} + \frac{3}{4}\mathbf{j} - \mathbf{k}$.

Solution　Set $(x_1, y_1, z_1) = (2, -3, 1)$ and $a\mathbf{i} + b\mathbf{j} + c\mathbf{k} = 4\mathbf{i} + \frac{3}{4}\mathbf{j} - \mathbf{k}$. Then, by substituting in (14.32), the parametric equations of L are

$$x = 2 + 4t \qquad y = -3 + \tfrac{3}{4}t \qquad z = 1 - t \qquad ■$$

In the parametric equations (14.32), if the numbers a, b, and c (the components of the vector $\mathbf{D}$) are each nonzero, we may solve for t, obtaining

$$t = \frac{x - x_1}{a} = \frac{y - y_1}{b} = \frac{z - z_1}{c}$$

By dropping the t, we have

(14.33)
$$\frac{x - x_1}{a} = \frac{y - y_1}{b} = \frac{z - z_1}{c}$$

These equations are referred to as *symmetric equations of the line L*.

EXAMPLE 3　Find symmetric equations of the line L passing through the point $(1, -1, 2)$ and parallel to $5\mathbf{i} - 2\mathbf{j} + 3\mathbf{k}$.

Solution　By using (14.33) with $a = 5$, $b = -2$, $c = 3$, $x_1 = 1$, $y_1 = -1$, and $z_1 = 2$, we get

(14.34)
$$\frac{x - 1}{5} = \frac{y + 1}{-2} = \frac{z - 2}{3} \qquad ■$$

Symmetric equations of a line are not unique. For example, since the vector $-10\mathbf{i} + 4\mathbf{j} - 6\mathbf{k}$ is parallel to $5\mathbf{i} - 2\mathbf{j} + 3\mathbf{k}$, we can write symmetric equations of the line given by (14.34) as

$$\frac{x - 1}{-10} = \frac{y + 1}{4} = \frac{z - 2}{-6}$$

EXAMPLE 4   Find symmetric equations for the line L containing the points $P_1 = (3, -2, 1)$ and $P_2 = (1, -5, 2)$.

Solution   In order to use (14.33) we must find a vector $\mathbf{D}$ parallel to L. Since P_1 and P_2 are distinct points lying on L, the directed line segment $\overrightarrow{P_1 P_2}$ can be used for $\mathbf{D}$. Therefore, $\mathbf{D} = -2\mathbf{i} - 3\mathbf{j} + \mathbf{k}$. If we use $P_1 = (3, -2, 1)$ in (14.33), we obtain the symmetric equations

$$\frac{x - 3}{-2} = \frac{y + 2}{-3} = \frac{z - 1}{1}$$

If we use $P_2 = (1, -5, 2)$, instead of P_1, we get another form of the symmetric equations, namely,

$$\frac{x - 1}{-2} = \frac{y + 5}{-3} = \frac{z - 2}{1}$$

Either representation is correct.   ∎

In (14.32), if one of the numbers a, b, or c equals 0, we may still write symmetric equations for L. For example, if $a = 0$, but $b \neq 0$ and $c \neq 0$. we rearrange (14.32) as

$$x = x_1 \qquad \frac{y - y_1}{b} = \frac{z - z_1}{c}$$

and call these *symmetric equations of the line L*. This particular line lies in the plane $x = x_1$.

EXAMPLE 5   Find the symmetric equations of the line L that contains the point $(5, -2, 3)$ and is parallel to $3\mathbf{j} - 2\mathbf{k}$.

Solution   In this case, $a = 0$, $b = 3$, and $c = -2$ in (14.32). Hence, the symmetric equations of L are

$$x = 5 \qquad \frac{y + 2}{3} = \frac{z - 3}{-2} \qquad ∎$$

EXAMPLE 6   Let

$$\frac{x - 6}{3} = \frac{y + 2}{1} = \frac{z + 3}{-2}$$

be symmetric equations of the line L. Find a vector parallel to L and find two points on L.

Solution   The denominators give the numbers $a = 3$, $b = 1$, and $c = -2$ in (14.33). Therefore, a vector parallel to L is $3\mathbf{i} + \mathbf{j} - 2\mathbf{k}$. We also know from (14.33) that $P_1 = (6, -2, -3)$ lies on L. To obtain a second point, we arbitrarily assign x a number, say, $x = 0$, to obtain

$$\frac{0 - 6}{3} = \frac{y + 2}{1} = \frac{z + 3}{-2}$$

from which we find $y = -4$, $z = 1$. Hence, another point on L is $(0, -4, 1)$. ∎

EXAMPLE 7   Find symmetric equations of the line passing through the point $(1, 0, -1)$ that is perpendicular to each of the lines

$$\frac{x + 1}{-1} = \frac{y - 2}{2} = \frac{z + 3}{1} \quad \text{and} \quad \frac{x}{3} = \frac{y - 1}{1} = \frac{z + 2}{-2}$$

Solution   The two lines are parallel, respectively, to the vectors

$$\mathbf{v} = -\mathbf{i} + 2\mathbf{j} + \mathbf{k} \quad \text{and} \quad \mathbf{w} = 3\mathbf{i} + \mathbf{j} - 2\mathbf{k}$$

A vector perpendicular to both $\mathbf{v}$ and $\mathbf{w}$ is

$$\mathbf{v} \times \mathbf{w} = \begin{vmatrix} \mathbf{i} & \mathbf{j} & \mathbf{k} \\ -1 & 2 & 1 \\ 3 & 1 & -2 \end{vmatrix} = -5\mathbf{i} + \mathbf{j} - 7\mathbf{k}$$

The line parallel to this vector and passing through the point $(1, 0, -1)$ is given by the symmetric equations

$$\frac{x - 1}{-5} = \frac{y}{1} = \frac{z + 1}{-7} \qquad ∎$$

SKEW LINES

In three-dimensional space, two distinct lines intersect, are parallel, or are *skew*. They *intersect* when they have exactly one point in common. They are *parallel* when they do not intersect and are parallel to the *same* vector. They are *skew* if they do not intersect and are not parallel. See Figure 44.

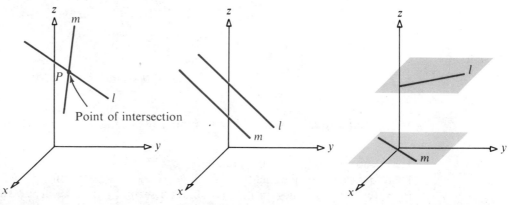

Figure 44   (a) l and m intersect at P (b) l and m are parallel   (c) l and m are skew

As an example of skew lines, think of the path of two airplanes—one at an altitude of 1000 meters traveling north and the other at an altitude of 3000 meters traveling east. The next example illustrates a method for determining whether a given pair of lines are skew, parallel, or intersecting.

EXAMPLE 8   Determine whether the lines given below intersect, are parallel, or are skew.

$$l: \quad \mathbf{r} = (3\mathbf{i} + 2\mathbf{j} + \mathbf{k}) + t(\mathbf{i} - \mathbf{j} + \mathbf{k}) = (3 + t)\mathbf{i} + (2 - t)\mathbf{j} + (1 + t)\mathbf{k}$$
$$m: \quad \mathbf{R} = (5\mathbf{i} + 6\mathbf{j} + \mathbf{k}) + T(\mathbf{i} - 4\mathbf{j} + 2\mathbf{k}) = (5 + T)\mathbf{i} + (6 - 4T)\mathbf{j} + (1 + 2T)\mathbf{k}$$

Solution   The line l is parallel to the vector $\mathbf{i} - \mathbf{j} + \mathbf{k}$ and the line m is parallel to the vector $\mathbf{i} - 4\mathbf{j} + 2\mathbf{k}$. Since these vectors are not parallel, l and m either intersect or are skew. To see which is the case, we suppose they intersect. In order for this to occur, some value of the parameter t and some value of the parameter T must give a point (position vector $\mathbf{r}$) on l that is the same as a point (position vector $\mathbf{R}$) on m. We set $\mathbf{r} = \mathbf{R}$, obtaining

$$(3 + t)\mathbf{i} + (2 - t)\mathbf{j} + (1 + t)\mathbf{k} = (5 + T)\mathbf{i} + (6 - 4T)\mathbf{j} + (1 + 2T)\mathbf{k}$$

Then, by equating the components of each vector, we obtain

$$3 + t = 5 + T \qquad 2 - t = 6 - 4T \qquad 1 + t = 1 + 2T$$

From the last of these conditions, $t = 2T$. Using this in the first two conditions, we find

$$3 + 2T = 5 + T \qquad 2 - 2T = 6 - 4T$$
$$T = 2 \qquad\qquad T = 2$$

Thus, $t = 4$ and $T = 2$. The point of intersection can be found by setting $t = 4$ in l or $T = 2$ in m. The result is $\mathbf{r} = \mathbf{R} = 7\mathbf{i} - 2\mathbf{j} + 5\mathbf{k}$. The point of intersection is $(7, -2, 5)$.   ∎

The next example illustrates the situation when two lines are skew.

EXAMPLE 9   Show that the lines given below are skew.

$$l: \quad \mathbf{r} = (3\mathbf{i} + 2\mathbf{j} + \mathbf{k}) + t(\mathbf{i} - \mathbf{j} + \mathbf{k})$$
$$m: \quad \mathbf{R} = (5\mathbf{i} + 6\mathbf{j} + \mathbf{k}) + T(\mathbf{i} - \mathbf{j} + 2\mathbf{k})$$

Solution   The line l is parallel to the vector $\mathbf{i} - \mathbf{j} + \mathbf{k}$ and the line m is parallel to the vector $\mathbf{i} - \mathbf{j} + 2\mathbf{k}$. Since these vectors are not parallel, l and m either intersect or are skew. As in Example 8, we set $\mathbf{r} = \mathbf{R}$, obtaining

$$3 + t = 5 + T \qquad 2 - t = 6 - T \qquad 1 + t = 1 + 2T$$

From the last of these, $t = 2T$. Then

$$3 + 2T = 5 + T \qquad 2 - 2T = 6 - T$$
$$T = 2 \qquad\qquad T = -4$$

Thus, l and m have at least two points in common. This is impossible since l and m are distinct lines. (Why?) Consequently, l and m are skew.   ■

ANGULAR VELOCITY

Figure 45 illustrates a rigid body that is rotating about a fixed axis l with a constant angular speed ω.* The *angular velocity* ω is defined as the vector of magnitude ω whose direction is parallel to the axis l, so that if the fingers of the right hand are wrapped about l in the direction of the rotation, the thumb will point in the direction of ω.

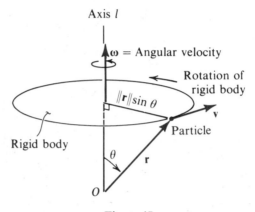

Figure 45

We are interested in obtaining a formula for the velocity $\mathbf{v}$ of a particle on the rigid body. To fix our ideas, let the origin O be on the axis l of rotation and let $\mathbf{r}$ be the position vector of the particle. Since the motion is circular and $\|\mathbf{r}\| \sin \theta$ equals the distance of the particle from the axis l, it follows that†

(14.35)
$$\|\mathbf{v}\| = \omega \|\mathbf{r}\| \sin \theta$$

Next, since the velocity $\mathbf{v}$ is necessarily perpendicular to both $\mathbf{r}$ and ω, it follows that $\mathbf{v}$ is parallel to $\omega \times \mathbf{r}$. By the right-hand rule used to define ω, $\mathbf{v}$ is in the direction of $\omega \times \mathbf{r}$, rather than $\mathbf{r} \times \omega$. So, by part (b) of (14.28),

$$\|\omega \times \mathbf{r}\| = \|\omega\| \, \|\mathbf{r}\| \sin \theta = \|\mathbf{v}\|$$

It follows that

(14.36)
$$\mathbf{v} = \omega \times \mathbf{r}$$

---

* Think of a phonograph record (rigid body) on a spindle (axis) rotating at $\omega = 45$ rpm. Its angular speed ω is $\omega = (45 \text{ revolutions/minute})(2\pi \text{ radians/revolution}) = 90\pi$ radians/minute.

† See page A-2, equation (3).

EXAMPLE 10   A rigid body rotates with constant angular speed ω radians per second about the line $x/2 = y = z/2$. Find the speed of a particle on this body at the instant it passes through the point $(1, 3, 5)$. Assume the distance is in meters.

Solution   The vector $2\mathbf{i} + \mathbf{j} + 2\mathbf{k}$ is parallel to the axis l of rotation of the rigid body and hence is parallel to ω. Since $\frac{2}{3}\mathbf{i} + \frac{1}{3}\mathbf{j} + \frac{2}{3}\mathbf{k}$ is a unit vector parallel to l, we have $\omega = \pm\omega(\frac{2}{3}\mathbf{i} + \frac{1}{3}\mathbf{j} + \frac{2}{3}\mathbf{k})$. (We use a $\pm$ sign since the problem gives no information as to the direction of ω.)

The position vector of the particle is $\mathbf{r} = \mathbf{i} + 3\mathbf{j} + 5\mathbf{k}$. Hence, from (14.36), the velocity $\mathbf{v}$ of the particle is

$$\mathbf{v} = \omega \times \mathbf{r} = \pm\omega \begin{vmatrix} \mathbf{i} & \mathbf{j} & \mathbf{k} \\ \frac{2}{3} & \frac{1}{3} & \frac{2}{3} \\ 1 & 3 & 5 \end{vmatrix}$$

$$= \pm\omega(-\tfrac{1}{3}\mathbf{i} - \tfrac{8}{3}\mathbf{j} + \tfrac{5}{3}\mathbf{k})$$

The speed v of the particle is

$$v = \|\mathbf{v}\| = \omega\sqrt{\tfrac{1}{9} + \tfrac{64}{9} + \tfrac{25}{9}} = \sqrt{10}\,\omega \text{ meters per second} \quad \blacksquare$$

EXERCISE 7

In Problems 1–8 find the vector equation, parametric equations, and symmetric equations of the line described.

1. Passing through the point $(1, 2, 3)$ and parallel to $2\mathbf{i} - \mathbf{j} + \mathbf{k}$
2. Passing through the point $(-1, 1, 5)$ and parallel to $\mathbf{i} + \mathbf{j} - \mathbf{k}$
3. Passing through the point $(4, -1, 6)$ and parallel to $\mathbf{i} + \mathbf{j}$
4. Passing through the point $(3, 2, -1)$ and parallel to $\mathbf{j} - \mathbf{k}$
5. Passing through the points $(1, -1, 3)$ and $(4, 2, 1)$
6. Passing through the points $(-2, 3, 0)$ and $(1, -1, 2)$
7. Passing through the points $(0, 0, 1)$ and $(2, 3, 1)$
8. Passing through the points $(0, 0, 0)$ and $(1, 2, -1)$

In Problems 9–12 determine the point of intersection and the cosine of the angle between the lines l and m.

9. l: $\mathbf{r} = (2 - t)\mathbf{i} + (4 + 2t)\mathbf{j} + (-5 + t)\mathbf{k}$
   m: $\mathbf{R} = (4 - T)\mathbf{i} + (3 + T)\mathbf{j} + (-13 + 3T)\mathbf{k}$

10. l: $\mathbf{r} = (5 + 2t)\mathbf{i} + (6 - t)\mathbf{j} + 2t\mathbf{k}$
    m: $\mathbf{R} = (7 + 3T)\mathbf{i} + (5 - 2T)\mathbf{j} + (2 - T)\mathbf{k}$

11. l: $\mathbf{r} = t\mathbf{i} + (1 + 2t)\mathbf{j} + (-3 + t)\mathbf{k}$
    m: $\mathbf{R} = (-3 + T)\mathbf{i} + (1 - 4T)\mathbf{j} + (2 - 7T)\mathbf{k}$

12. l: $\mathbf{r} = (2 - 3t)\mathbf{i} + 6t\mathbf{j} + (-2 + 5t)\mathbf{k}$
    m: $\mathbf{R} = -2T\mathbf{i} + (1 + T)\mathbf{j} + 2T\mathbf{k}$

In Problems 13–16 find symmetric equations of the line passing through the point P that is perpendicular to the two directions given.

13. $P = (0, 0, 0)$; $\mathbf{a} = 2\mathbf{i} + 3\mathbf{j} - 2\mathbf{k}$; $\mathbf{b} = 3\mathbf{i} + \mathbf{j} + 2\mathbf{k}$
14. $P = (0, 0, 0)$; $\mathbf{a} = 4\mathbf{i} + 2\mathbf{j} + \mathbf{k}$; $\mathbf{b} = 5\mathbf{i} + 3\mathbf{j} + \mathbf{k}$

15. $P = (1, 2, -1);$ $\mathbf{a} = 2\mathbf{i} + 4\mathbf{j} - 2\mathbf{k};$ $\mathbf{b} = -3\mathbf{i} - 2\mathbf{j} + \mathbf{k}$

16. $P = (-1, 3, 2);$ $\mathbf{a} = -2\mathbf{i} + 2\mathbf{j} - 3\mathbf{k};$ $\mathbf{b} = 4\mathbf{i} + 2\mathbf{j} + \mathbf{k}$

In Problems 17–27 determine whether the lines l and m intersect, are parallel, or are skew.

17. l: $\mathbf{r} = (2 - 3t)\mathbf{i} + 6t\mathbf{j} + (-2 + 6t)\mathbf{k}$
  m: $\mathbf{R} = (6 - T)\mathbf{i} + (2 + 2T)\mathbf{j} + (5 + 2T)\mathbf{k}$

18. l: $\mathbf{r} = (3 + 3t)\mathbf{i} + 6t\mathbf{j} + (3 - 2t)\mathbf{k}$
  m: $\mathbf{R} = (3 - 2T)\mathbf{i} + 4T\mathbf{j} + (3 + 7T)\mathbf{k}$

19. l: $\mathbf{r} = (4 + t)\mathbf{i} + (3 - t)\mathbf{j} + 6t\mathbf{k}$
  m: $\mathbf{R} = (4 + T)\mathbf{i} + (3 - T)\mathbf{j} + (2 - 2T)\mathbf{k}$

20. l: $\mathbf{r} = (2 - 2t)\mathbf{i} + (7 + 8t)\mathbf{j} - 6t\mathbf{k}$
  m: $\mathbf{R} = (6 + T)\mathbf{i} + (-5 - 4T)\mathbf{j} + 3T\mathbf{k}$

21. l: $\mathbf{r} = (5 + 2t)\mathbf{i} + (6 - t)\mathbf{j} + (8 - t)\mathbf{k}$
  m: $\mathbf{R} = (4 + T)\mathbf{i} + T\mathbf{j} + (2 + T)\mathbf{k}$

22. l: $\mathbf{r} = (2 - t)\mathbf{i} + t\mathbf{j} + (1 - t)\mathbf{k}$
  m: $\mathbf{R} = (6 + T)\mathbf{i} + (-4 + T)\mathbf{j} + T\mathbf{k}$

23. l: $\dfrac{x - 3}{2} = \dfrac{y + 2}{3} = \dfrac{z - 1}{4}$

  m: $\dfrac{x + 4}{-4} = \dfrac{y - 3}{-6} = \dfrac{z + 4}{-8}$

24. l: $\dfrac{x}{3} = \dfrac{y - 2}{4} = \dfrac{z + 4}{1}$

  m: $\dfrac{x - 6}{3} = \dfrac{y + 2}{4} = \dfrac{z - 3}{2}$

25. l: $\dfrac{x + 1}{5} = \dfrac{y - 2}{4} = \dfrac{z - 3}{-3}$

  m: $\dfrac{x + 1}{6} = \dfrac{y - 2}{3} = \dfrac{z + 3}{2}$

26. l: $\dfrac{x + 5}{6} = \dfrac{y - 2}{3} = \dfrac{z + 4}{-1}$

  m: $\dfrac{x}{1} = \dfrac{y - 2}{3} = \dfrac{z - 8}{2}$

27. l: $\dfrac{2x + 5}{6} = \dfrac{y - 2}{3} = \dfrac{z + 4}{7}$

  m: $\dfrac{x + 5}{6} = \dfrac{y + 1}{3} = \dfrac{z - 2}{7}$

28. Write the parametric equations of a line passing through the point $(-1, 5, 6)$ and parallel to the line

$$\frac{x + 1}{5} = \frac{y - 2}{4} = \frac{z - 3}{-3}$$

29. Write the parametric equations of a line passing through the point $(1, -2, -3)$ and parallel to the line

$$\frac{x + 1}{6} = \frac{y + 2}{2} = \frac{z}{-1}$$

30. What property can you assign to the lines l and m, given below, if $a_1 b_1 + a_2 b_2 + a_3 b_3 = 0$?

$$l: \quad \frac{x - x_1}{a_1} = \frac{y - y_1}{a_2} = \frac{z - z_1}{a_3} \qquad m: \quad \frac{x - x_1}{b_1} = \frac{y - y_1}{b_2} = \frac{z - z_1}{b_3}$$

31. What property can you assign to the lines l and m, given below, if $a_1/b_1 = a_2/b_2 = a_3/b_3$?

$$l: \quad \frac{x - x_1}{a_1} = \frac{y - y_1}{a_2} = \frac{z - z_1}{a_3} \qquad m: \quad \frac{x - x_1}{b_1} = \frac{y - y_1}{b_2} = \frac{z - z_1}{b_3}$$

32. Let

$$\frac{x - 4}{2} = \frac{y + 1}{-1} = \frac{z - 2}{2}$$

be symmetric equations of the line L. Find a vector parallel to L, and find two points on L.

33. Let

$$x + 1 = y + 3 = \frac{z + 4}{2}$$

be symmetric equations of the line L. Find a vector parallel to L, and find two points on L.

34. Find symmetric equations of the line passing through the centers of the spheres

$$x^2 + y^2 + z^2 - 2x - 4y + 4z = 8 \quad \text{and} \quad x^2 + y^2 + z^2 + 2x + 6y + 4z = 20$$

35. A rigid body rotates about an axis through the origin with a constant angular speed of 30 radians per second. If the angular velocity ω points in the direction of $\mathbf{i} + \mathbf{j} + \mathbf{k}$, find the speed of a particle at the instant it passes through the point $(-1, 2, 3)$. Assume the distance scale is in meters.

36. For the rigid body in Problem 35, at what points will the speed be 60 meters per second?

37. A rigid body rotates with constant angular speed ω about the line $x/3 = y = z/-2$. Find the speed of a particle at the instant that it passes through the point $(4, 4, 0)$. Assume that distance is measured in meters. Find ω if the speed of the particle at $(4, 4, 0)$ is $8\sqrt{14}$ meters per second.

8. Planes

Recall that the vector equation of a line is determined once a point on the line and a vector parallel to the line are known. To determine the vector equation of a plane, we use a point on the plane and a vector perpendicular to the plane.

Suppose the point $P_1 = (x_1, y_1, z_1)$ is on a plane and suppose the nonzero vector $\mathbf{N}$ is perpendicular to the plane (see Fig. 46). Let $P = (x, y, z)$ be any point on the plane. Denote by $\mathbf{r}_1$ and $\mathbf{r}$ the position of the points P_1 and P, respectively. Since we are insisting that P is a point on the plane, the vector $\mathbf{r} - \mathbf{r}_1$ will always be perpendicular to the vector $\mathbf{N}$. That is, $(\mathbf{r} - \mathbf{r}_1) \cdot \mathbf{N} = 0$.

The vector equation of the plane containing the point P_1 and perpendicular to the vector $\mathbf{N}$ is

(14.37)
$$(\mathbf{r} - \mathbf{r}_1) \cdot \mathbf{N} = 0$$

where $\mathbf{r}_1$ denotes the position vector of P_1.

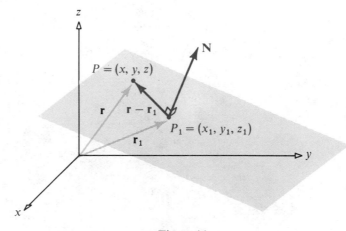

Figure 46

If, in the above, the coordinates of P_1 are (x_1, y_1, z_1) and the vector $\mathbf{N} = A\mathbf{i} + B\mathbf{j} + C\mathbf{k}$, then

(14.38)
$$[(x - x_1)\mathbf{i} + (y - y_1)\mathbf{j} + (z - z_1)\mathbf{k}] \cdot (A\mathbf{i} + B\mathbf{j} + C\mathbf{k}) = 0$$
$$A(x - x_1) + B(y - y_1) + C(z - z_1) = 0$$

or

(14.39)
$$Ax + By + Cz = D$$

where $D = Ax_1 + By_1 + Cz_1$. Equation (14.39) is called the *general equation of a plane*. Notice that the coefficients of x, y, and z in the general equation are the components of the perpendicular vector $\mathbf{N}$.

Conversely, any equation of the form $Ax + By + Cz = D$, with at least one of the numbers A, B, C not 0, is the equation of a plane and the vector $A\mathbf{i} + B\mathbf{j} + C\mathbf{k}$ is perpendicular to this plane.

EXAMPLE 1   Find the general equation of the plane passing through the point $(1, 2, -1)$ and perpendicular to the vector $2\mathbf{i} + 3\mathbf{j} - 4\mathbf{k}$.

Solution   Here a point in the plane is $(x_1, y_1, z_1) = (1, 2, -1)$ and a vector perpendicular to the plane is $\mathbf{N} = 2\mathbf{i} + 3\mathbf{j} - 4\mathbf{k}$. Using (14.38), the equation of the plane is

$$2(x - 1) + 3(y - 2) - 4(z + 1) = 0$$

which simplifies to $2x + 3y - 4z = 12$. Figure 47 depicts the plane given by $2x + 3y - 4z = 12$. Note the use of the intercepts to locate three points on the plane; these are useful for graphing it.

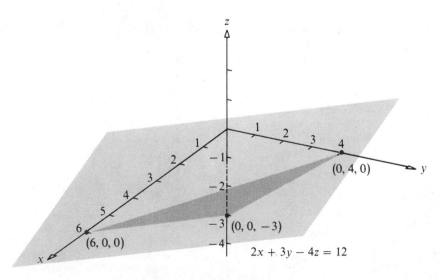

Figure 47   ■

A vector that is perpendicular to a plane is called a *normal* to that plane. For example, the vector $N = Ai + Bj + Ck$ is a normal to the plane

$$Ax + By + Cz = D$$

The next example illustrates the use of (14.38) to find the equation of the plane through three noncollinear points.

EXAMPLE 2   Find the general equation of the plane determined by the points $P_1 = (1, -1, 2)$, $P_2 = (3, 0, 0)$,   and $P_3 = (4, 2, 1)$.

Solution   The vectors $v = \overrightarrow{P_1 P_2} = 2i + j - 2k$ and $w = \overrightarrow{P_1 P_3} = 3i + 3j - k$ lie in the plane. The vector

$$N = v \times w = \begin{vmatrix} i & j & k \\ 2 & 1 & -2 \\ 3 & 3 & -1 \end{vmatrix} = \begin{vmatrix} 1 & -2 \\ 3 & -1 \end{vmatrix} i - \begin{vmatrix} 2 & -2 \\ 3 & -1 \end{vmatrix} j + \begin{vmatrix} 2 & 1 \\ 3 & 3 \end{vmatrix} k$$

$$= 5i - 4j + 3k$$

is perpendicular to both **v** and **w** and, thus, is normal to the plane. By using this normal vector and the point $P_1 = (1, -1, 2)$ (we could have used P_2 or P_3), we obtain the equation

$$5(x - 1) - 4(y + 1) + 3(z - 2) = 0$$
$$5x - 4y + 3z = 15 \quad \blacksquare$$

PARALLEL PLANES; ANGLE BETWEEN PLANES

Two distinct planes are *parallel* if they have parallel normals. If two planes have parallel normals and a point in common, they are *identical*. For example, the planes

$$p_1: \quad 2x + 3y - z = 3 \qquad p_2: \quad 2x + 3y - z = 4$$

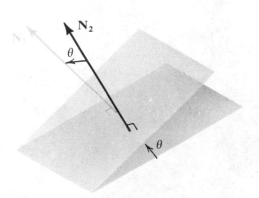

have parallel normals. Since these planes cannot have any points in common (Why?), the planes p_1 and p_2 are parallel. The plane $p_3: \quad 4x + 6y - 2z = 6$ is identical to the plane p_1. (Why?)

If two planes have nonparallel normals, they *intersect*. In this case, the *angle θ between the planes* is defined as the nonobtuse angle between the normals (see Fig. 48). Depending on the choice of normals, there are actually two angles between the normals. Since these angles are always supplementary, there is no confusion in agreeing to name the nonobtuse angle as the angle between the planes. This is the angle whose cosine is not negative. Thus, if N_1 and N_2 are normals of two intersecting planes, the angle θ between these planes is

Figure 48

(14.40) $$\cos \theta = \frac{|\mathbf{N}_1 \cdot \mathbf{N}_2|}{\|\mathbf{N}_1\|\,\|\mathbf{N}_2\|} \qquad 0 \le \theta \le \frac{\pi}{2}$$

EXAMPLE 3   Find the angle between the planes

$$p_1: \quad 2x - 3y + 6z = 5 \qquad p_2: \quad 3x + y + 2z = 6$$

Solution   First, we note that the normals $\mathbf{N}_1$ and $\mathbf{N}_2$ to p_1 and p_2, respectively, can be chosen to be

$$\mathbf{N}_1 = 2\mathbf{i} - 3\mathbf{j} + 6\mathbf{k} \qquad \mathbf{N}_2 = 3\mathbf{i} + \mathbf{j} + 2\mathbf{k}$$

Since the normals are not parallel, the planes intersect. The cosine of the angle θ between p_1 and p_2 is

$$\cos \theta = \frac{|\mathbf{N}_1 \cdot \mathbf{N}_2|}{\|\mathbf{N}_1\|\,\|\mathbf{N}_2\|} = \frac{|6 - 3 + 12|}{\sqrt{4 + 9 + 36}\,\sqrt{9 + 1 + 4}} = \frac{15}{7\sqrt{14}} \approx 0.5727$$

Thus, $\theta \approx 0.96$ radian.   ∎

In the above example, notice that we take the absolute value of $\mathbf{N}_1 \cdot \mathbf{N}_2$ to ensure that the nonobtuse angle between p_1 and p_2 is being calculated.

DISTANCE FROM A POINT TO A PLANE

To find the distance from a point P_0 to a plane, we use projections. As Figure 49 illustrates, if P_1 is any point on the plane itself, the distance from P_0 to the plane may be found by calculating the absolute value of the projection of $\overrightarrow{P_1 P_0}$ on the normal $\mathbf{N}$ of the plane.

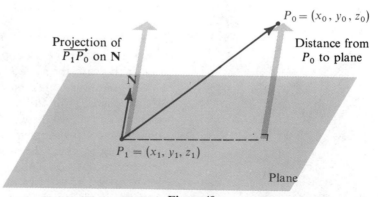

Figure 49

Let the plane be given by the equation

$$Ax + By + Cz = D$$

The distance from the point $P_0 = (x_0, y_0, z_0)$ to the plane is given by

$$\begin{bmatrix} \text{Distance from} \\ P_0 \text{ to plane} \end{bmatrix} = |\text{Projection of } \overrightarrow{P_1 P_0} \text{ on } \mathbf{N}| = \frac{|\overrightarrow{P_1 P_0} \cdot \mathbf{N}|}{\|\mathbf{N}\|}$$

(14.22)

where $P_1 = (x_1, y_1, z_1)$ is a point on the plane. But $\mathbf{N} = A\mathbf{i} + B\mathbf{j} + C\mathbf{k}$ and
$\overrightarrow{P_1 P_0} = (x_0 - x_1)\mathbf{i} + (y_0 - y_1)\mathbf{j} + (z_0 - z_1)\mathbf{k}$.   Hence,

$$\begin{bmatrix} \text{Distance from} \\ P_0 \text{ to plane} \end{bmatrix} = \frac{|A(x_0 - x_1) + B(y_0 - y_1) + C(z_0 - z_1)|}{\sqrt{A^2 + B^2 + C^2}}$$

$$= \frac{|(Ax_0 + By_0 + Cz_0) - (Ax_1 + By_1 + Cz_1)|}{\sqrt{A^2 + B^2 + C^2}}$$

$$= \frac{|Ax_0 + By_0 + Cz_0 - D|}{\sqrt{A^2 + B^2 + C^2}}$$

P_1 is on the plane, so
$Ax_1 + By_1 + Cz_1 = D$

(14.41)   **The distance from the point** $P_0 = (x_0, y_0, z_0)$ **to the plane** $Ax + By + Cz = D$ **is**

$$\frac{|Ax_0 + By_0 + Cz_0 - D|}{\sqrt{A^2 + B^2 + C^2}}$$

EXAMPLE 4   Find the distance from the point $(2, 3, -1)$ to the plane $x + 4y + z = 5$.

Solution   The normal to the plane is $\mathbf{N} = \mathbf{i} + 4\mathbf{j} + \mathbf{k}$,   so that $A = 1$, $B = 4$, $C = 1$,
and $D = 5$;   we are given $P_0 = (2, 3, -1)$.   Thus, the distance is

$$\frac{|(1)(2) + (4)(3) + (1)(-1) - 5|}{\sqrt{1 + 16 + 1}} = \frac{8}{3\sqrt{2}} \qquad \blacksquare$$

EXERCISE 8

In Problems 1–16 find the general equation of the plane whose properties are stated.

1. Parallel to the xy-plane and 4 units above it
2. Parallel to the yz-plane and 2 units to the right of it
3. Parallel to the xz-plane and containing the point $(1, -2, 3)$
4. Parallel to the xy-plane and containing the point $(2, -3, 4)$
5. Containing the point $(1, -1, 2)$ and perpendicular to $2\mathbf{i} - \mathbf{j} + \mathbf{k}$

6. Containing the point $(-3, 2, 1)$ and perpendicular to $\mathbf{i} + \mathbf{j} - 2\mathbf{k}$

7. Containing the point $(0, 1, 5)$ and perpendicular to $2\mathbf{i} + \mathbf{j} + 3\mathbf{k}$

8. Containing the point $(1, 0, -2)$ and perpendicular to $-3\mathbf{i} + \mathbf{j} + \mathbf{k}$

9. Containing the point $(0, 5, -2)$ and parallel to the plane $x + 2y - z = 6$

10. Containing the point $(1, -2, 0)$ and parallel to the plane $2x - y + 3z = 10$

11. Containing the point $(10, 3, -4)$ and parallel to the plane $x - y + 3z = 5$

12. Containing the point $(-1, 2, 3)$ and parallel to the plane $2x + 3y - 4z = 8$

13. Containing the point $(2, 3, -1)$ and perpendicular to the line $\dfrac{x - 1}{2} = \dfrac{y - 3}{5} = \dfrac{z + 1}{-2}$

14. Containing the point $(-1, 2, 3)$ and perpendicular to the line $\dfrac{x + 5}{3} = \dfrac{y + 2}{4} = \dfrac{z - 4}{4}$

15. Containing the point $(0, 1, -2)$ and perpendicular to the line $\dfrac{x - 1}{2} = \dfrac{y + 3}{3} = \dfrac{z - 2}{2}$

16. Containing the point $(-1, 0, 4)$ and perpendicular to the line $\dfrac{x + 2}{3} = \dfrac{y + 4}{2} = \dfrac{z - 1}{1}$

17. Determine which of the points below lie on the plane $3x - 2y = 0$.

$$(2, 3, 0) \qquad (5, -1, 2) \qquad (6, 0, 3) \qquad (0, 0, 0)$$

18. Determine which of the points below lie on the plane $2x + y - z = 2$.

$$(0, 1, -1) \qquad (1, 1, 1) \qquad (3, 2, 1) \qquad (-1, 1, 1)$$

19. Find parametric equations of the line passing through the point $(1, 2, -1)$ and perpendicular to the plane $2x - y + z = 6$.

20. Find parametric equations of the line passing through the point $(2, 3, -1)$ and perpendicular to the plane $x + y - z = 3$.

21. Find the point of intersection of the plane $2x + y - z = 5$ and the line $\dfrac{x - 1}{2} = \dfrac{y + 3}{4} = \dfrac{z - 1}{1}$.

22. Find the point of intersection of the plane $x + y - 2z = 8$ and the line $\dfrac{x + 1}{2} = \dfrac{y - 3}{1} = \dfrac{z - 4}{-2}$.

23. Find the point of intersection of the plane $2x + 3y + z = 5$ and the line $\dfrac{x - 3}{1} = \dfrac{y + 4}{2} = \dfrac{z - 1}{2}$.

24. Find the point of intersection of the plane $x + y - z = 3$ and the line $\dfrac{x + 2}{2} = \dfrac{y - 3}{1} = \dfrac{z}{2}$.

In Problems 25–30 find the general equation of the plane determined by the points P_1, P_2, and P_3.

25. $P_1 = (0, 0, 0); \quad P_2 = (1, 2, -1); \quad P_3 = (-1, 1, 0)$

26. $P_1 = (0, 0, 0); \quad P_2 = (3, -1, 2); \quad P_3 = (-3, 1, 0)$

27. $P_1 = (1, 2, 1); \quad P_2 = (3, 2, 2); \quad P_3 = (4, -1, -1)$

28. $P_1 = (-1, 2, 0); \quad P_2 = (3, 4, -1); \quad P_3 = (-2, -1, 0)$

29. $P_1 = (6, 8, -2);$ $P_2 = (4, -1, 0);$ $P_3 = (1, 0, 0)$

30. $P_1 = (-3, -4, 0);$ $P_2 = (6, -7, 2);$ $P_3 = (0, 0, 1)$

In Problems 31–38 determine whether the two planes intersect, are parallel, or are identical. If they intersect, find the cosine of the angle between them.

31. p_1: $2x - y + z = 2;$ p_2: $x + y + z = 3$

32. p_1: $x + y - z = 5;$ p_2: $2x + 3y - 4z = 1$

33. p_1: $2x - 3y + z = 1;$ p_2: $2x - 3y + 4z = 2$

34. p_1: $x - y = 2;$ p_2: $y - z = 2$

35. p_1: $2x - y + z = 3;$ p_2: $4x - y + 6z = 7$

36. p_1: $x + y - z = 1;$ p_2: $-2x - 2y + 2z = -2$

37. p_1: $x - 2y + z = 1;$ p_2: $3x - 6y + 3z = 3$

38. p_1: $x + y - z = 1;$ p_2: $x + y - z = 2$

In Problems 39–42 find the distance from the point to the plane.

39. $(1, 2, -1);$ $2x - y + z = 1$

40. $(-1, 3, -2);$ $x + 2y - 3z = 4$

41. $(2, -1, 1);$ $-x + y - 3z = 6$

42. $(-2, 1, 1);$ $-3x + 2y + z = 1$

9. Quadric Surfaces

Earlier, we showed that the graph of a second-degree equation in two variables,

$$Ax^2 + Bxy + Cy^2 + Dx + Ey + F = 0$$

is, except for degenerate cases, a conic section; that is, a parabola, ellipse, circle, or hyperbola. (You may want to review Chap. 12.)

The graph of a second-degree equation in three variables,

$$Ax^2 + By^2 + Cz^2 + Dxy + Exz + Fyz + Gx + Hy + Iz + J = 0$$

is called a *quadric surface*, and, except for degenerate cases,* there are nine distinct types. As usual, the equations we use to identify the nine quadric surfaces are the so-called "standard equations." For those that have centers, the center will be at the origin; for those that are symmetric, the axis of symmetry will be a coordinate axis.

In the equations that follow we characterize each quadric surface by citing its *intercepts* (points at which the coordinate axes are crossed), its *traces* (the intersection of the surface with the coordinate planes), and its *sections* (the intersection of the surface with other planes, usually taken parallel to the coordinate planes). The letters a, b, and c denote positive numbers.

(14.42)   **1.**   *Ellipsoid:* $\dfrac{x^2}{a^2} + \dfrac{y^2}{b^2} + \dfrac{z^2}{c^2} = 1$ (Fig. 50, p. 794)

The intercepts are the points $(\pm a, 0, 0)$, $(0, \pm b, 0)$, and $(0, 0, \pm c)$.

---

* For example, we exclude the equation $x^2 + y^2 + z^2 = 0$, whose graph is the point $(0, 0, 0)$. We also exclude the equation $x^2 + y^2 + 3z^2 = -2$, whose graph consists of no points.

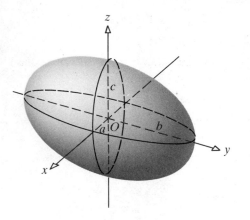
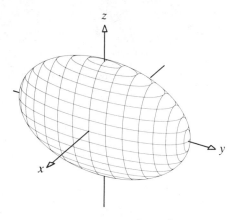

(a)   Ellipsoid: $\dfrac{x^2}{a^2} + \dfrac{y^2}{b^2} + \dfrac{z^2}{c^2} = 1$ (b)   Ellipsoid: $x^2 + \dfrac{y^2}{9} + \dfrac{z^2}{4} = 1$

Figure 50

The traces are all ellipses. For example, the trace in the xy-plane, obtained by setting $z = 0$ in (14.42), is the ellipse $(x^2/a^2) + (y^2/b^2) = 1$. All sections parallel to the coordinate planes are ellipses.

A computer graph of the ellipsoid $x^2 + (y^2/9) + (z^2/4) = 1$ is provided in Figure 50(b).*

(14.43)   **2.**   *Elliptic Cone:* $z^2 = \dfrac{x^2}{a^2} + \dfrac{y^2}{b^2}$ (Fig. 51)

The lone intercept is the origin $(0, 0, 0)$, and it is referred to as the *vertex* of the cone.

The trace in the xy-plane is the origin; the trace in the yz-plane consists of the pair of intersecting lines $z = \pm y/b$; the trace in the xz-plane consists of the pair of intersecting lines $z = \pm x/a$.

Sections parallel to the xy-plane are ellipses.

If $a = b$, the elliptic cone (14.43) becomes a *circular cone*. A computer graph of the circular cone $z^2 = 3x^2 + 3y^2$ is provided in Figure 51(b).

(14.44)   **3.**   *Elliptic Paraboloid:* $z = \dfrac{x^2}{a^2} + \dfrac{y^2}{b^2}$ (Fig. 52)

The lone intercept is at the origin $(0, 0, 0)$, and it is referred to as the *vertex* of the elliptic paraboloid.

---

* The computer-generated graphs were prepared by Douglas Dunham and Robert Hapy, Department of Mathematical Sciences, University of Minnesota, Duluth. See the Preface for a description of their techniques.

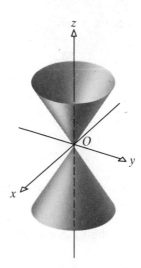
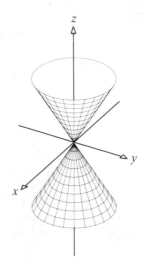

Figure 51   (a)   Elliptic cone: $z^2 = \dfrac{x^2}{a^2} + \dfrac{y^2}{b^2}$ (b)   Circular cone: $z^2 = 3x^2 + 3y^2$

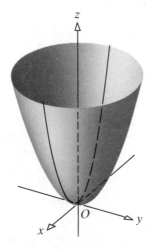

Figure 52   (a)   Elliptic paraboloid: $z = \dfrac{x^2}{a^2} + \dfrac{y^2}{b^2}$ (b)   Paraboloid of revolution: $z = \tfrac{1}{2}(x^2 + y^2)$

The trace in the xy-plane is the origin; the trace in the yz-plane is the parabola $z = y^2/b^2$;   the trace in the xz-plane is the parabola $z = x^2/a^2$.

Sections parallel to the xy-plane are ellipses; sections parallel to the other coordinate planes are parabolas.

Observe that $z \geq 0$,   so that this surface (except for the origin) lies above the xy-plane.

If $a = b$,   the surface is called a *paraboloid of revolution*. A computer graph of the paraboloid of revolution $z = \frac{1}{2}(x^2 + y^2)$ is provided in Figure 52(b).

(14.45)   4.   Hyperbolic Paraboloid: $z = \dfrac{y^2}{b^2} - \dfrac{x^2}{a^2}$ (Fig. 53)

The lone intercept is at the origin.

The trace in the xy-plane is the pair of lines $y/b = \pm x/a$ that intersect at the origin; the trace in the yz-plane is the parabola $z = y^2/b^2$;   the trace in the xz-plane is the parabola $z = -x^2/a^2$.

Sections parallel to the xy-plane are hyperbolas; sections parallel to the other coordinate planes are parabolas.

Note that the origin is a minimum point for the trace in the yz-plane and is a maximum point for the trace in the xy-plane. Such a point is called a *saddle point* of the surface.

A computer graph of the hyperbolic paraboloid $z = y^2 - x^2$ is provided in Figure 53(b).

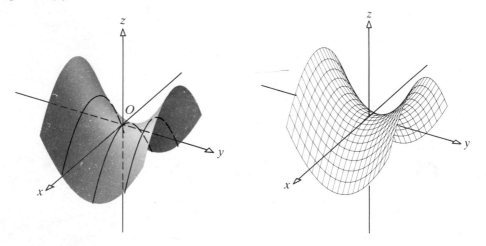

(a)   Hyperbolic paraboloid: $z = \dfrac{y^2}{b^2} - \dfrac{x^2}{a^2}$ (b)   Hyperbolic paraboloid: $z = y^2 - x^2$

Figure 53

(14.46)   5.   Hyperboloid of One Sheet: $\dfrac{x^2}{a^2} + \dfrac{y^2}{b^2} - \dfrac{z^2}{c^2} = 1$ (Fig. 54)

The intercepts are at $(\pm a, 0, 0)$ and $(0, \pm b, 0)$.

The trace in the xy-plane is the ellipse $(x^2/a^2) + (y^2/b^2) = 1$;   the trace in the yz-plane is the hyperbola $(y^2/b^2) - (z^2/c^2) = 1$;   the trace in the xz-plane is the hyperbola $(x^2/a^2) - (z^2/c^2) = 1$.

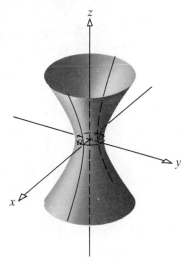

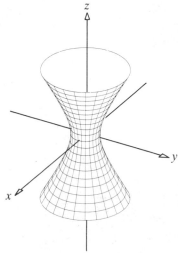

(a)   Hyperboloid of one sheet:

$$\frac{x^2}{a^2} + \frac{y^2}{b^2} - \frac{z^2}{c^2} = 1$$

(b)   Hyperboloid of revolution:

$$x^2 + y^2 - \frac{z^2}{3} = 1$$

Figure 54

Sections parallel to the xy-plane are ellipses; sections parallel to the other coordinate planes are hyperbolas.

If $a = b$,   the surface (14.46) is called a *hyperboloid of revolution*. A computer graph of the hyperboloid of revolution $x^2 + y^2 - (z^2/3) = 1$ is provided in Figure 54(b).

(14.47)   6.   *Hyperboloid of Two Sheets:* $\dfrac{x^2}{a^2} + \dfrac{y^2}{b^2} - \dfrac{z^2}{c^2} = -1$ (Fig. 55, p. 798)

The intercepts are the two points $(0, 0, \pm c)$.

There is no trace in the xy-plane; the trace in the yz-plane is the hyperbola $(z^2/c^2) - (y^2/b^2) = 1$;   and in the xz-plane the trace is $(z^2/c^2) - (x^2/a^2) = 1$,   a hyperbola .

Sections parallel to the xy-plane are ellipses; sections parallel to the other coordinate axes are hyperbolas.

Note that this surface consists of two parts, one for which $z \geq c$,   and the other for which $z \leq -c$.

A computer graph of the hyperboloid of two sheets $x^2 + y^2 - (z^2/3) = -1$ is provided in Figure 55(b).

The three remaining quadric surfaces are termed *cylinders*. Their standard equations are characterized by the fact that one of the variables is missing from the equation. In naming these surfaces, we have chosen z as the missing variable. As a result, z is unrestricted, and the cylinder will be unbounded in the z direction.

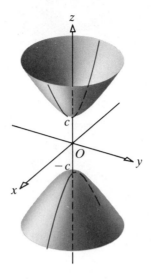
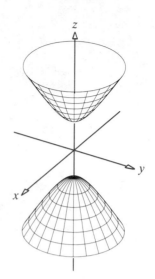

(a)   Hyperboloid of two sheets:

$$\frac{x^2}{a^2} + \frac{y^2}{b^2} - \frac{z^2}{c^2} = -1$$

(b)   Hyperboloid of two sheets:

$$x^2 + y^2 - \frac{z^2}{3} = -1$$

Figure 55

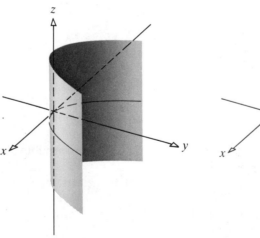
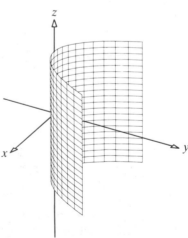

(a)   Parabolic cylinder: $x^2 = 4ay$ (b)   Parabolic cylinder: $x^2 = 4y$

Figure 56

Cylinders are surfaces that are generated by a line moving along a curve while remaining parallel to a fixed line. We therefore describe cylinders in terms of how they are generated.

(14.48)   7.   *Parabolic Cylinder:* $x^2 = 4ay$ (Fig. 56)

This surface is generated by a line parallel to the z-axis moving along the parabola $x^2 = 4ay$.   This surface is easily visualized by taking a piece of paper and folding it so that two opposite edges trace out a portion of a parabola.

A computer graph of the parabolic cylinder $x^2 = 4y$ is provided in Figure 56(b).

(14.49)   8.   *Elliptic Cylinder:* $\dfrac{x^2}{a^2} + \dfrac{y^2}{b^2} = 1$ (Fig. 57)

This surface is generated by a line parallel to the z-axis moving along the ellipse $(x^2/a^2) + (y^2/b^2) = 1$.   This surface can be visualized by taking a piece of paper and attaching the two opposite edges in the shape of an ellipse.

A computer graph of the elliptic cylinder $(x^2/4) + (y^2/9) = 1$ is provided in Figure 57(b).

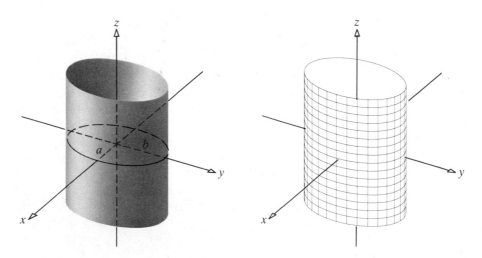

(a)   Elliptic cylinder: $\dfrac{x^2}{a^2} + \dfrac{y^2}{b^2} = 1$ (b)   Elliptic cylinder: $\dfrac{x^2}{4} + \dfrac{y^2}{9} = 1$

Figure 57

(14.50)   **9.**   *Hyperbolic Cylinder:* $\dfrac{x^2}{a^2} - \dfrac{y^2}{b^2} = 1$ (Fig. 58)

This surface is generated by a line parallel to the z-axis moving along the hyperbola $(x^2/a^2) - (y^2/b^2) = 1$. This surface consists of two parts, each of which may be visualized by folding a piece of paper so that two opposite edges trace out a portion of a hyperbola.

A computer graph of the hyperbolic cylinder $(x^2/9) - (y^2/4) = 1$ is provided in Figure 58(b).

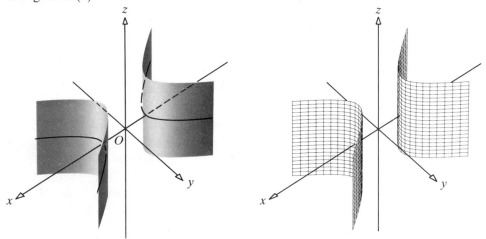

(a) Hyperbolic cylinder: $\dfrac{x^2}{a^2} - \dfrac{y^2}{b^2} = 1$ (b) Hyperbolic cylinder: $\dfrac{x^2}{9} - \dfrac{y^2}{4} = 1$

Figure 58

The quadric surfaces classified here are in standard position, with centers (or vertex) at the origin and symmetries with respect to the coordinate axes. When the center (or vertex) of a quadric surface is not located at the origin, but there is symmetry with respect to lines parallel to the coordinate axes, a simple translation of axes may be applied to place the equation in standard form. The correct translation is obtained by completing squares. If the symmetry of the quadric surface is with respect to lines that are not parallel to the coordinate axes, a rotation of axes, perhaps followed by a translation of axes, is required to place the equation in standard form. Although rotations in three-space are somewhat involved, the idea is quite similar to that of rotations in the plane (see Section 5, Chap. 12).

IDENTIFYING AND SKETCHING QUADRIC SURFACES

To identify and sketch a quadric surface, compare the given equation to the ones discussed in this section. Once you have matched it, the nature of the graph is apparent. Here are some suggestions to help you make the comparison:

1. Remember that the nine types of quadric surfaces listed are given in standard positions. An interchange of variables does not affect the classification. For example, based on (14.44), the equation $z = (x^2/4) + (y^2/9)$ is an elliptic paraboloid and its graph lies above the xy-plane. The equation $y = (x^2/9) + (z^2/4)$ is also an elliptic paraboloid. (Do you see why?) Its graph lies to the right of the xz-plane.

2. Cylinders are characterized by a variable missing from the equation. For example, the equation $(y^2/4) + (z^2/9) = 1$ is an elliptic cylinder whose graph is perpendicular to the yz-plane.

3. The technique of completing the square is sometimes required to identify the given quadric surface. In this case, a translation of axes will help to get the correct location of the graph.

EXAMPLE 1   Identify the surface given by each equation and sketch its graph.

(a) $4x^2 - 18y^2 + 9z^2 = 36$ (b) $4y^2 + 9z^2 - 36x = 0$

(c) $x^2 + z^2 - 4x = 0$ (d) $4y^2 + 9z^2 - 36x - 8y + 36z + 148 = 0$

Solution   (a)   Divide both sides of the equation by 36 to get

$$\frac{x^2}{9} - \frac{y^2}{2} + \frac{z^2}{4} = 1$$

This equation resembles (14.46). We conclude that the surface is a hyperboloid of one sheet whose axis is the y-axis (see Fig. 59).

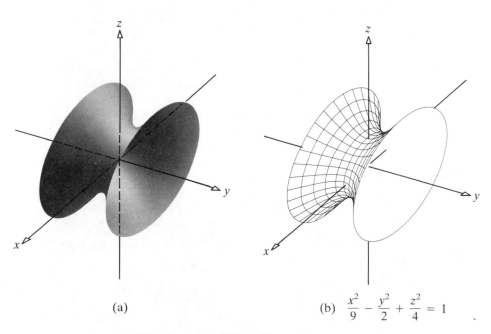

(a)   (b) $\dfrac{x^2}{9} - \dfrac{y^2}{2} + \dfrac{z^2}{4} = 1$

Figure 59

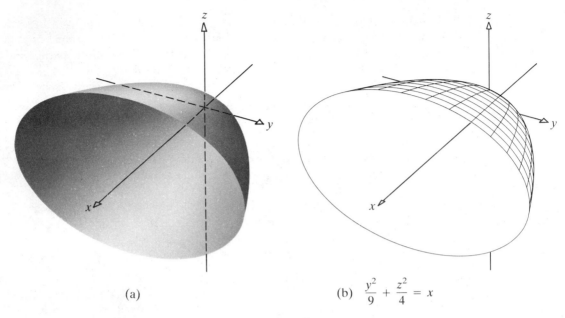

(a)   (b) $\dfrac{y^2}{9} + \dfrac{z^2}{4} = x$

Figure 60

(b) The equation can be written as

$$\frac{y^2}{9} + \frac{z^2}{4} = x$$

which resembles (14.44). Its graph is therefore an elliptic paraboloid with vertex at the origin whose axis is the x-axis (see Fig. 60).

(c) Since the y variable is missing, this surface is a cylinder. By completing the square in x, the equation can be written as $(x - 2)^2 + z^2 = 4$. This is the equation of a circular cylinder with center axis parallel to the y-axis passing through $(2, 0, 0)$ and of radius 2. Its graph is perpendicular to the xz-plane (see Fig. 61).

(d) We rewrite the equation and complete squares:

$$4y^2 - 8y + 9z^2 + 36z = 36x - 148$$
$$4(y^2 - 2y) + 9(z^2 + 4z) = 36x - 148$$
$$4(y - 1)^2 + 9(z + 2)^2 = 36x - 148 + 40$$
$$4(y - 1)^2 + 9(z + 2)^2 = 36(x - 3)$$

$$\frac{(y - 1)^2}{9} + \frac{(z + 2)^2}{4} = x - 3$$

We recognize this [see (14.44) or compare with part (b) above] as an elliptic

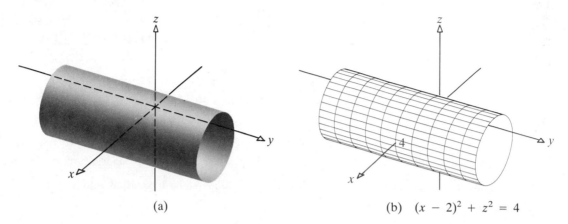

(a)   (b) $(x - 2)^2 + z^2 = 4$

Figure 61

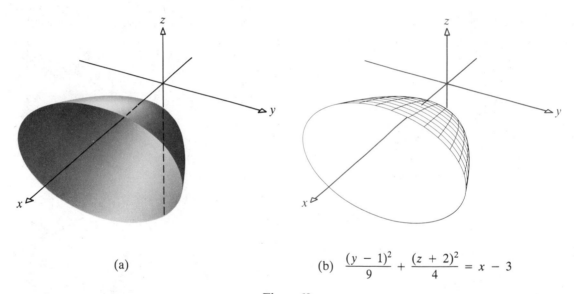

(a)   (b) $\dfrac{(y - 1)^2}{9} + \dfrac{(z + 2)^2}{4} = x - 3$

Figure 62

paraboloid with vertex at $(3, 1, -2)$ whose axis is the x-axis (see Fig. 62 and compare it to Fig. 60).   ∎

EXERCISE 9

In Problems 1–22 identify and sketch each quadric surface.

1. $z = x^2 + y^2$ **2.** $z = x^2 - y^2$ **3.** $4x^2 + y^2 + 4z^2 = 4$

4. $2x^2 + y^2 + z^2 = 1$ **5.** $z^2 = x^2 + 2y^2$ **6.** $x^2 + 2y^2 - z^2 = 1$

7. $x = 4z^2$ **8.** $x^2 + y^2 = 1$ **9.** $x^2 + 2y^2 - z^2 = -4$

10. $z = x^2 - 2y^2$ **11.** $y^2 - x^2 = 4$ **12.** $2x = y^2$

13. $z = x^2 + 2y^2$ **14.** $z^2 = 3x^2 + 4y^2$ **15.** $4y^2 - x^2 = 1$

16. $4x^2 + y^2 - z^2 = 1$ **17.** $x^2 + 4x + y^2 = z^2 + 4z$ **18.** $x^2 + y^2 + 2y = z - 1$

19. $y = 4x^2 + 9z^2$ **20.** $x + 4 = 2y^2$

21. $x^2 - y^2 - 4z^2 + 4y + 8z - 9 = 0$ **22.** $4x^2 + y^2 + 4z^2 - 16x - 4y - 8z + 20 = 0$

23. Explain why the graph of $xy = 1$ is a cylinder. Sketch it. Is the graph a quadric surface?

24. Explain why the graph of $z = \sin y$ is a cylinder. Sketch it. Is the graph a quadric surface?

Figures A–J are graphs of quadric surfaces. In Problems 25–30 match each equation with its graph.

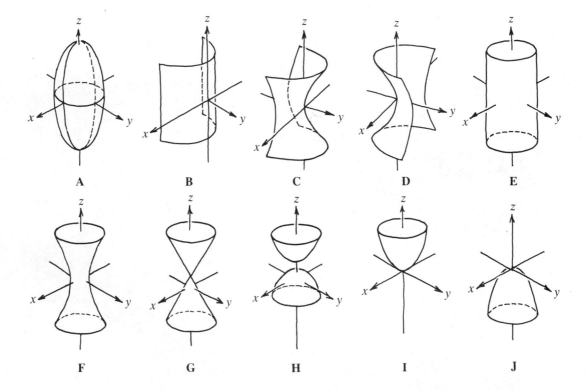

A   B   C   D   E

F   G   H   I   J

25. $3x^2 + 4y^2 + z = 0$ **26.** $3x^2 + 4y^2 + 4y = 0$

27. $3x^2 + 2y^2 - (z - 2)^2 + 1 = 0$ **28.** $4z^2 - x^2 = 3y$

29. $x^2 + 2y^2 - z^2 + 4z = 4$ **30.** $3x^2 + 3y^2 + z^2 = 1$

31. Figures K–O are computer graphs of quadric surfaces. Match each graph to one of the following:
   (a) $z = 4y^2 - x^2$ (b) $2z = x^2 + 4y^2$ (c) $2x^2 + y^2 - z^2 = 1$
   (d) $2x^2 + y^2 + 3z^2 = 1$ (e) $y^2 = 4x$

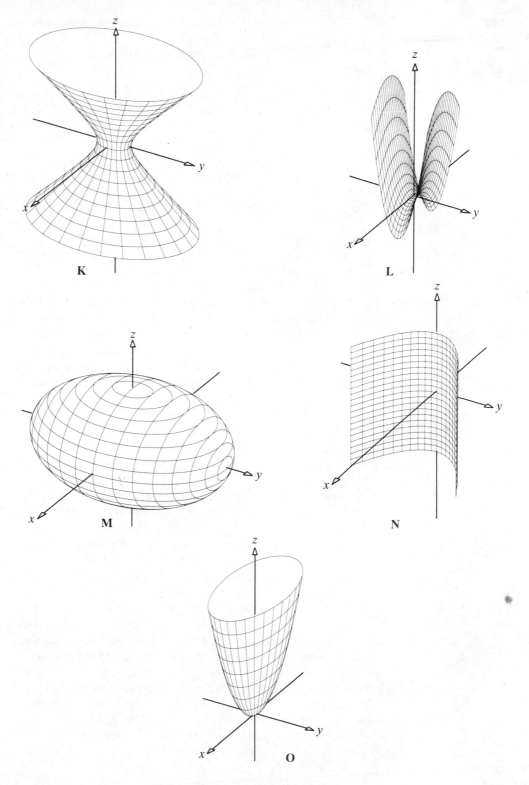

K

L

M

N

O

Historical Perspectives

Most mathematical concepts have undergone many changes and interpretations after their introduction into mathematics, but this process is especially apparent in the development of the concept of vector. Vector analysis is a fairly young branch of calculus since it was developed in the late nineteenth century. However, the concept of a vector quantity—that is, a quantity with both magnitude and direction—was undoubtedly present in much mathematical thought long before it became formalized. The *parallelogram law for addition* was known to Aristotle and was stated explicitly by Galileo, but the first extensive use of vectors came about in attempts to visualize *complex numbers*. Two-dimensional vectors and vector algebra form an essential part of the framework of the geometric representation of complex numbers introduced by the French mathematician Argand in 1813 and later by the great German mathematician and physicist Karl Friedrich Gauss (1777–1855).

Gauss was born in Brunswick, Germany, to working-class parents. He received his education through the generosity of Duke Wilhelm, after showing great promise in elementary school. He worked at the University of Göttingen from 1807 until his death, rarely leaving the city of Göttingen even for short periods of time. Gauss's accomplishments were extraordinary in both mathematics and physics, where he did first-rate work in the study of optics, electricity and magnetism, and astronomy. In mathematics he made fundamental discoveries in algebra, differential geometry, complex analysis, potential theory, and noneuclidean geometry. Even though he published only a fraction of his work, Gauss was widely known and acclaimed to be the greatest mathematician since Newton. The awe-inspiring scope of his life's work has led many to consider him the last of the "universal geniuses." His work on complex numbers did much to popularize the geometric interpretation of complex numbers discussed in Problems 1–3.

The success of the two-dimensional interpretation of complex numbers motivated the Irish mathematician Rowan Hamilton (1805–1865) to search for an analogous three-dimensional "complex" algebra. Hamilton was born in Dublin, Ireland, and ranks second only to Newton among British mathematicians. Unlike Newton, Hamilton's genius was apparent at an early age. When he was 8 years old, he is said to have been literate in six languages: Latin, Hebrew, Greek, Italian, French, and, of course, English. As a 23-year-old undergraduate, he was appointed Professor of Astronomy at Trinity College. Today, he is perhaps most famous for *Hamilton's principle* in physics, but his discovery of four-dimensional quantities, which he called *quaternions*, was an accomplishment with equally important implications in the motivation that it provided for the introduction of vectors into calculus later on. Hamilton's quaternions consisted of a sum of a scalar (real number) part and a vector part, although he did not treat the vector part separately (see Problems 4 and 5).

The German mathematician Grassmann and others also worked on the problem of extending or generalizing the complex numbers to higher dimensions, but it was Hamilton's quaternions that found the most immediate application and eventually led to vector analysis. James Clerk Maxwell (1831–1879), the Scottish mathematical physicist, used and developed Hamilton's theory of quaternions, distinguishing the scalar and vector parts of the quantities. His famous work on electromagnetic theory, based on what came to be known as *Maxwell's equations*, was very influential in making vectors and vector methods in analysis well known and in motivating their further development.

However, it was in the independent work of the American physicist and mathematician Josiah Willard Gibbs (1839–1903) and the Englishman Oliver Heaviside (1850–1925) that the distinction between scalar and vector was drawn once and for all, and that the concept of vector was finally divorced from the theory of quaternions. Both men introduced the vector cross product and placed the scalar product in proper perspective. Modern vector analysis is usually considered to have begun with Gibbs and Heaviside; their developments of the subject proceed from a starting point not too dissimilar from the treatment in this chapter. In the early 1880's, while Professor of Mathematical Physics at Yale University, Gibbs produced some notes on vector analysis for his students' use. These notes formed the basis for a textbook in vector analysis that was published in 1901 by one of his former students. Gibb's work on vector analysis was among the first major American contributions to mathematics.

Throughout the early development of the theory of vectors, the geometric interpretation, as opposed to the algebraic interpretation, was the prevalent viewpoint. The basic vector operations were defined geometrically and the **i, j, k** notation was not used except when attempts were made to evade vector arguments with the aid of a rectangular coordinate system. Of course, today we recognize that the real advantages of vector analysis are realized through the combination of geometry and algebra inherent in the uses of vector algebra, or linear algebra, to simplify and clarify many of its applications.

HISTORICAL EXERCISES

1. A *complex number z* can be thought of as a number of the form $z = a + bi$, where a and b are real numbers and $i^2 = -1$. (This use of the symbol i should not be confused with the vector **i**.) The number a is called the *real part of z*, and b is called the *imaginary part of z*. The *Argand diagram* of $z = a + bi$ is shown in the figure. The quantity $r = \sqrt{a^2 + b^2}$ is called the *modulus* (length) of the complex number z. To add two complex numbers $a + bi$ and $c + di$, we simply add the corresponding real and imaginary parts.
   (a) Use the above definition to add z_1 to z_2, where $z_1 = 3 + 2i$ and $z_2 = -1 + 2i$.
   (b) Sketch the Argand diagrams for z_1, z_2, and the sum $z_1 + z_2$.

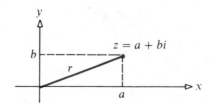

2. To multiply two complex numbers $a + bi$ and $c + di$, we multiply as though we were multiplying two binomials, use the fact that $i^2 = -1$, and collect the real and imaginary parts.
   (a) Use the above definition to multiply z_1 and z_2, as given in Problem 1(a).
   (b) Find the product of $2 + 3i$ and $2 - 3i$.
   (c) What can you say, in general, about the product of a complex number $z = a + bi$ and its *complex conjugate* $\bar{z} = a - bi$? Interpret this result geometrically.
   (d) What vector operation is suggested by your geometric interpretation in part (c)?

3. Recall from Problem 2(c) that the conjugate of $z = a + bi$ is $\bar{z} = a - bi$.
   (a) Define a new operation on complex numbers, say, $\Diamond$, as follows: $z \Diamond w = (z)(\bar{w})$. Let $z = a + bi$ and $w = c + di$, and write the expression for $z \Diamond w$. What kind of number is $z \Diamond w$?
   (b) Show that the operation $\Diamond$ satisfies all the properties of the dot product (inner product) of two vectors listed in theorem (14.13) except (a). (This is the reason that $z \Diamond w$ is called the *inner product* of the complex numbers z and w.)

4. Hamilton was the first person to think of complex numbers as ordered pairs of real numbers (a, b), where a is the real part and b is the imaginary part of the complex number. Since $(a + bi) + (c + di) = (a + c) + (b + d)i$, we can write this in ordered-pair notation as

$$(a, b) + (c, d) = (a + c, \quad b + d)$$

   (a) Write the definition of complex multiplication in ordered-pair notation.
   (b) Show that complex multiplication is a commutative operation. That is, show that $(a, b) \times (c, d) = (c, d) \times (a, b)$ for all complex numbers (a, b) and (c, d).

5. Hamilton's *quaternions* can be represented as ordered four-tuples, or quadruples. Multiplication of quaternions can then be defined (in a way analogous to complex multiplication, as in Problem 4) as

$$(a, b, c, d) \times (e, f, g, h) = (ae - bf - cg - dh, \quad af + be + ch - dg, \quad ag - bh + ce + df, \quad ah + bg - cf + de)$$

   (a) Verify that quaternion multiplication is *not* commutative by providing a counterexample.
   (b) Is there an identity for quaternion multiplication? That is, is there a quaternion that when multiplied by any other quaternion Q will produce Q as the result? If so, what is the identity?

Miscellaneous Exercises

1. Let (r, θ) be any point of the polar coordinate curve $r = f(\theta)$, where $r \geq 0$, and let

$$\mathbf{u}_r = (\cos \theta)\mathbf{i} + (\sin \theta)\mathbf{j} \qquad \mathbf{u}_\theta = (-\sin \theta)\mathbf{i} + (\cos \theta)\mathbf{j}$$

   (a) Show that $\mathbf{u}_r$ and $\mathbf{u}_\theta$ are unit vectors.
   (b) Explain why $\mathbf{u}_r$ and $\mathbf{u}_\theta$ are perpendicular.
   (c) Show that $\mathbf{u}_r$ has the same direction as the ray from the origin to (r, θ) and that $\mathbf{u}_\theta$ is 90° counterclockwise from $\mathbf{u}_r$.

2. Prove that the vector $A\mathbf{i} + B\mathbf{j}$ is perpendicular to the line $Ax + By + C = 0$ in the plane.

3. Let θ be the angle between the nonzero vectors $\mathbf{v}$ and $\mathbf{w}$, $0 \leq \theta \leq \pi$.
   (a) Show that

$$\sin^2\theta = \frac{\|\mathbf{v}\|^2\|\mathbf{w}\|^2 - (\mathbf{v} \cdot \mathbf{w})^2}{\|\mathbf{v}\|^2\|\mathbf{w}\|^2}$$

(b) Use part (a) to prove that **v** and **w** are parallel if and only if $|\mathbf{v} \cdot \mathbf{w}| = \|\mathbf{v}\| \, \|\mathbf{w}\|$. What connection does this have with the Cauchy–Schwarz inequality?

4. Two unidentified flying objects are at $(t, -t, 1 - t)$ and $(t - 3, 2t, 4t - 1)$ at time t.
   (a) Describe the paths of the objects.
   (b) Find where the paths intersect (or determine that they don't). Will the objects collide?
   (c) Find the acute angle between the paths.

5. Find the line perpendicular to the lines

$$l: \quad x = 1 - t, \quad y = t, \quad z = 2t - 1 \qquad \text{and} \qquad m: \quad x = t + 1, \quad y = -t, \quad z = t - 1$$

at their point of intersection. Why is this line parallel to the xy-plane?

6. Find all points (x, y, z) satisfying the system of equations $5x + 2y - z = 3$, $2x + y + z = 1$. Give a geometric interpretation of the set of all such points.

7. Find an equation of the plane containing the lines

$$l: \quad x = t + 1, \quad y = 3t, \quad z = t - 1 \qquad \text{and} \qquad m: \quad \frac{x + 3}{2} = \frac{y - 3}{3} = z$$

8. Find an equation of the plane with nonzero intercepts $(a, 0, 0)$, $(0, b, 0)$, and $(0, 0, c)$. Then show that your equation can be written in the form $(x/a) + (y/b) + (z/c) = 1$.

9. Find an equation of the plane tangent to the sphere $(x - 1)^2 + (y + 2)^2 + (z - 2)^2 = 6$ at the point $(2, -1, 0)$.

10. Explain why the set of points (x, y, z) equidistant from the points $(1, 3, 0)$ and $(-1, 1, 2)$ is a plane. Then find its equation in two ways, as follows:
    (a) Use the distance formula to equate the distances between (x, y, z) and the given points, simplifying the result to obtain an equation of the plane.
    (b) Name a point of the plane and a vector perpendicular to the plane, and use the answers to find an equation of the plane.

11. We proved in the text that the distance between the point $P_0 = (x_0, y_0, z_0)$ and the plane $Ax + By + Cz = D$ is

$$d = \frac{|Ax_0 + By_0 + Cz_0 - D|}{\sqrt{A^2 + B^2 + C^2}}$$

Derive this formula differently, as follows:
   (a) Show that the line through P_0 perpendicular to the plane has the parametric equations $x = x_0 + At$, $y = y_0 + Bt$, $z = z_0 + Ct$.
   (b) If $P = (x, y, z)$ is the point of intersection of the line in part (a) with the plane, show that $x - x_0 = At$, $y - y_0 = Bt$, $z - z_0 = Ct$, where

$$t = \frac{-(Ax_0 + By_0 + Cz_0) + D}{A^2 + B^2 + C^2}$$

   (c) Explain why d is the distance between P_0 and P, and use the distance formula to finish the proof.

12. Prove that the distance between the parallel planes $Ax + By + Cz = D_1$ and $Ax + By + Cz = D_2$ is

$$d = \frac{|D_2 - D_1|}{\sqrt{A^2 + B^2 + C^2}}$$

13. Suppose that the cross product of $\mathbf{v} = a_1\mathbf{i} + b_1\mathbf{j} + c_1\mathbf{k}$ and $\mathbf{w} = a_2\mathbf{i} + b_2\mathbf{j} + c_2\mathbf{k}$ has not yet been defined, and we seek a vector $\mathbf{u} = x\mathbf{i} + y\mathbf{j} + z\mathbf{k}$ perpendicular to both $\mathbf{v}$ and $\mathbf{w}$.
    (a) Explain why (x, y, z) satisfies the system of equations $a_1x + b_1y + c_1z = 0$, $a_2x + b_2y + c_2z = 0$.
    (b) By eliminating first y and then x from the system, show that

$$(a_1b_2 - a_2b_1)x = (b_1c_2 - b_2c_1)z \qquad \text{and} \qquad (a_1b_2 - a_2b_1)y = -(a_1c_2 - a_2c_1)z$$

    (c) Use the result in part (b) to show that one solution of the system is

$$x = b_1c_2 - b_2c_1 \qquad y = -(a_1c_2 - a_2c_1) \qquad z = a_1b_2 - a_2b_1$$

    Compare $\mathbf{u} = x\mathbf{i} + y\mathbf{j} + z\mathbf{k}$ (for this choice of x, y, z) with the definition of $\mathbf{v} \times \mathbf{w}$ in Section 6.

14. Let points $P_1 = (x_1, y_1, z_1)$ and $P_2 = (x_2, y_2, z_2)$ have respective position vectors $\mathbf{r}_1$ and $\mathbf{r}_2$. Show that the point $P = (x, y, z)$ that divides the segment P_1P_2 in the ratio $t = \overline{P_1P}/\overline{P_1P_2}$ has position vector $\mathbf{r} = \mathbf{r}_1 + t(\mathbf{r}_2 - \mathbf{r}_1)$. Thus, conclude that

$$x = x_1 + t(x_2 - x_1) \qquad y = y_1 + t(y_2 - y_1) \qquad z = z_1 + t(z_2 - z_1)$$

15. Let $A_1, A_2, \ldots, A_n$ be consecutive vertices of a regular polygon (n sides of equal length) with center at the origin. Show that $\sum_{i=1}^{n} \overrightarrow{OA_i} = \mathbf{0}$.

16. Given $\mathbf{u} = \mathbf{i} - 2\mathbf{j}$ and $\mathbf{v} = -3\mathbf{i} + \mathbf{j}$, find vectors $\mathbf{w}_1$ and $\mathbf{w}_2$ so that $\mathbf{u} = \mathbf{w}_1 - \mathbf{w}_2$, $\mathbf{v}$ is parallel to $\mathbf{w}_1$, and $\mathbf{u}$ is perpendicular to $\mathbf{w}_2$.

17. Let $\mathbf{u}_1$, $\mathbf{u}_2$, and $\mathbf{u}_3$ be noncoplanar vectors. Let $\mathbf{w}_1 = \mathbf{u}_1/\|\mathbf{u}_1\|$, $\mathbf{v}_2 = \mathbf{u}_2 - (\mathbf{u}_2 \cdot \mathbf{w}_1)\mathbf{w}_1$, $\mathbf{w}_2 = \mathbf{v}_2/\|\mathbf{v}_2\|$, $\mathbf{v}_3 = \mathbf{u}_3 - (\mathbf{u}_3 \cdot \mathbf{w}_1)\mathbf{w}_1 - (\mathbf{u}_3 \cdot \mathbf{w}_2)\mathbf{w}_2$, and $\mathbf{w}_3 = \mathbf{v}_3/\|\mathbf{v}_3\|$. Show that $\mathbf{w}_1$, $\mathbf{w}_2$, and $\mathbf{w}_3$ are mutually orthogonal unit vectors. (This is the *Gram–Schmidt orthogonalization process*.)

18. Use the procedure of Problem 17 to transform $-\mathbf{i} + \mathbf{j}$, $2\mathbf{i} + \mathbf{k}$, and $3\mathbf{i} - \mathbf{j} + 2\mathbf{k}$ into a set of mutually orthogonal unit vectors.

19. Let $\mathbf{u}$ and $\mathbf{v}$ be nonparallel vectors in the plane. Show that for each vector $\mathbf{w}$ there exist unique constants s and t so that $\mathbf{w} = s\mathbf{u} + t\mathbf{v}$. Interpret your result geometrically.

20. Find an equation of the sphere with center at $(-2, 1, 5)$ and tangent to $x + 4y - z = 7$.

21. Find an equation of the sphere having radius $\sqrt{26}$ and tangent to $3x + y - 4z = 20$ at the point $(1, 1, -4)$.

22. Show that the graph of $Ax^2 + Ay^2 + Az^2 + Dx + Ey + Fz = G$, $A \neq 0$, is a sphere, a point, or the empty set. Give a condition involving the constants A, D, E, F, G so that the graph is a sphere, and find the center and radius of the sphere.

23. Find an equation of the sphere passing through the points $(3, 0, 0)$, $(0, 0, 1)$, $(-1, -3, 1)$, and $(2, 0, 1)$.

24. Find an equation of the sphere with $(2, 2, -3)$ and $(-2, 6, 5)$ as endpoints of a diameter.

25. Find an equation of the plane containing the origin that is perpendicular to $x - 2y - z = 0$ and makes an angle of $60°$ with the positive y-axis.

26. Show that the volume of a tetrahedron with adjacent sides $\mathbf{u}$, $\mathbf{v}$, and $\mathbf{w}$ is $\frac{1}{6}|\mathbf{u} \cdot (\mathbf{v} \times \mathbf{w})|$.

27. Find the volume of the parallelepiped determined by $\overrightarrow{AB}$, $\overrightarrow{AC}$, and $\overrightarrow{AD}$, where $A = (0, 1, 0)$, $B = (1, 1, 4)$, $C = (2, 0, 1)$, and $D = (3, 5, 8)$.

28. Find symmetric equations of the line perpendicular to the plane containing the lines

$$x - 2 = \frac{y + 1}{2} = \frac{z - 1}{2} \qquad \text{and} \qquad x + 1 = \frac{y + 8}{3} = \frac{z}{-3}$$

    at their point of intersection.

29. Find the total work done by gravity when a particle of mass m moves once around a vertical triangle. [*Hint:* The force of gravity is $m\mathbf{g}$, where $g \approx 9.80$ meters per second per second is the acceleration of gravity.]

30. Let points A, B, and C determine a plane p. If $\mathbf{u} = \overrightarrow{OA}$, $\mathbf{v} = \overrightarrow{OB}$, and $\mathbf{w} = \overrightarrow{OC}$, prove that the vector $\mathbf{u} \times \mathbf{v} + \mathbf{v} \times \mathbf{w} + \mathbf{w} \times \mathbf{u}$ is perpendicular to p.

31. Find symmetric equations of the line of intersection of the planes $2x + y - z = 6$ and $x - y + 3z = 4$.

32. Let $\mathbf{u}$ and $\mathbf{v}$ be fixed vectors and define $g(t) = \|\mathbf{u} - t\mathbf{v}\|^2$. Find the minimum value of g and deduce the Cauchy–Schwarz inequality $|\mathbf{u} \cdot \mathbf{v}| \le \|\mathbf{u}\| \, \|\mathbf{v}\|$.

33. Let two skew lines have respective direction vectors $\mathbf{M}$ and $\mathbf{N}$. Let A and B be points on the respective lines, and let $\mathbf{w} = \overrightarrow{AB}$. Show that the distance between the two lines is the absolute value of the projection of $\mathbf{w}$ on $\mathbf{M} \times \mathbf{N}$:

$$|p| = \frac{|\mathbf{w} \cdot (\mathbf{M} \times \mathbf{N})|}{\|\mathbf{M} \times \mathbf{N}\|}$$

(The shortest distance is measured along the common perpendicular to the two lines.)

34. Find the distance between the lines

$$\frac{x-3}{2} = \frac{y}{3} = z \quad \text{and} \quad x = \frac{y+1}{-2} = \frac{z-2}{-1}$$

35. Show that $\mathbf{w} = \|\mathbf{v}\|\mathbf{u} + \|\mathbf{u}\|\mathbf{v}$ bisects the angle between $\mathbf{u}$ and $\mathbf{v}$.

36. Suppose that $\mathbf{v} = c_1\mathbf{u}_1 + c_2\mathbf{u}_2$ is in the plane. Find c_1 and c_2 in terms of $\mathbf{u}_1$, $\mathbf{u}_2$, and $\mathbf{v}$. What are c_1 and c_2 in case $\mathbf{u}_1$ and $\mathbf{u}_2$ are unit orthogonal vectors?

37. Find symmetric equations of the line that contains $(2, 0, -3)$, is perpendicular to $\mathbf{i} + 2\mathbf{j} - \mathbf{k}$, and is parallel to $2x + 3y - z = 1$.

38. Show that two vectors in the plane are parallel if and only if their projections on two fixed mutually perpendicular vectors are proportional.

39. Show that $(\mathbf{r} - \mathbf{b}) \cdot (\mathbf{r} + \mathbf{b}) = 0$ is a vector equation of the sphere, with $\mathbf{b}$ and $-\mathbf{b}$ as endpoints of a diameter. Show that $(\mathbf{r} - \mathbf{r}_0 - \mathbf{b}) \cdot (\mathbf{r} - \mathbf{r}_0 + \mathbf{b}) = 0$ is a vector equation of the sphere with center $\mathbf{r}_0$ and radius $\|\mathbf{b}\|$.

40. Sketch the set of points in the plane with position vector $\mathbf{r}$ satisfying $\mathbf{r} \cdot (\mathbf{i} + \mathbf{j}) \ge 0$.

41. Solve for vectors $\mathbf{u}$ and $\mathbf{v}$ if

$$\mathbf{u} - 3\mathbf{v} = 2\mathbf{i} + \mathbf{j} - \mathbf{k}$$
$$2\mathbf{u} + \mathbf{v} = \mathbf{i} + \mathbf{j}$$

42. Find the point of intersection of the line through the points $(0, 2, -2)$ and $(2, 1, -3)$ and the plane through $(0, 4, -2)$, $(1, 3, -2)$, and $(2, 2, -3)$.

43. Find an equation of the plane parallel to the line $\mathbf{r} = 2\mathbf{i} + t(-\mathbf{i} + \mathbf{j} + 2\mathbf{k})$ and containing the points $(2, 2, -1)$ and $(1, 0, 1)$.

44. Let A, B, C, and D be the vertices of a square in the plane with center at Q. Show that

$$\overrightarrow{OQ} = \frac{\overrightarrow{OA} + \overrightarrow{OB} + \overrightarrow{OC} + \overrightarrow{OD}}{4}$$

45. If the points P, Q, and R, not all lying on the same straight line, have position vectors $\mathbf{v}$, $\mathbf{w}$, and $\mathbf{u}$ relative to a given origin, show that $\mathbf{v} \times \mathbf{w} + \mathbf{w} \times \mathbf{u} + \mathbf{u} \times \mathbf{v}$ is a vector perpendicular to the plane containing P, Q, and R.

46. (a) Prove that the vectors $\mathbf{u} = 3\mathbf{i} + \mathbf{j} - 2\mathbf{k}$, $\mathbf{v} = -\mathbf{i} + 3\mathbf{j} + 4\mathbf{k}$, and $\mathbf{w} = 4\mathbf{i} - 2\mathbf{j} - 6\mathbf{k}$ can form the sides of a triangle.

(b) Find the lengths of the medians of the triangle.

47. Let θ be the angle between the nonzero vectors $\mathbf{u} = a_1\mathbf{i} + a_2\mathbf{j} + a_3\mathbf{k}$ and $\mathbf{v} = b_1\mathbf{i} + b_2\mathbf{j} + b_3\mathbf{k}$. Show that

$$\sin^2\theta = \frac{(a_2b_3 - a_3b_2)^2 + (a_1b_3 - a_3b_1)^2 + (a_1b_2 - a_2b_1)^2}{\|\mathbf{u}\|^2\|\mathbf{v}\|^2}$$

[*Hint:* Begin with $\sin^2\theta = 1 - \cos^2\theta$.]

48. Sketch $z = xy$. (This surface is a hyperbolic paraboloid rotated $45°$ about the z-axis.)

49. Show that through each point on the hyperboloid of one sheet

$$\frac{x^2}{a^2} + \frac{y^2}{b^2} - \frac{z^2}{c^2} = 1$$

there are two lines lying entirely on the surface. [*Hint:* Write the equation as $\dfrac{x^2}{a^2} - \dfrac{z^2}{c^2} = 1 - \dfrac{y^2}{b^2}$ and factor.]

15

Vector Functions

1. Vector Functions and Their Derivatives

Until now, we have only dealt with functions whose domain and range are both sets of real numbers. Such functions may be appropriately described as *real functions*. A function whose domain is a set of real numbers, but whose range is a collection of vectors, is called a *vector function*. We shall denote a vector function by $\mathbf{r} = \mathbf{r}(t)$, where t is a real number defined on some interval $a \leq t \leq b$ and $\mathbf{r}$ is a position vector.

A vector function in two dimensions has as components two real functions $x = x(t)$ and $y = y(t)$, each defined on $a \leq t \leq b$, and takes the form

$$\mathbf{r} = \mathbf{r}(t) = x(t)\mathbf{i} + y(t)\mathbf{j}$$

A vector function in three dimensions has as components three real functions $x = x(t)$, $y = y(t)$, and $z = z(t)$, each defined on $a \leq t \leq b$, and takes the form

$$\mathbf{r} = \mathbf{r}(t) = x(t)\mathbf{i} + y(t)\mathbf{j} + z(t)\mathbf{k}$$

In general, a vector function $\mathbf{r} = \mathbf{r}(t)$ with domain $a \leq t \leq b$ is a curve in the sense that, as t varies over the interval $a \leq t \leq b$, the tip of the vector $\mathbf{r} = \mathbf{r}(t)$, whose initial point is the origin, traces out the curve. The components of the vector function are the *parametric equations* of the curve. We shall refer to the curve C: $\mathbf{r} = \mathbf{r}(t) = x(t)\mathbf{i} + y(t)\mathbf{j}$, $a \leq t \leq b$, as a *curve in the plane* or as a *plane curve*. The curve C: $\mathbf{r} = \mathbf{r}(t) = x(t)\mathbf{i} + y(t)\mathbf{j} + z(t)\mathbf{k}$, $a \leq t \leq b$, will be referred to as a *curve in three-space* or as a *space curve* (see Fig. 1, p. 814).

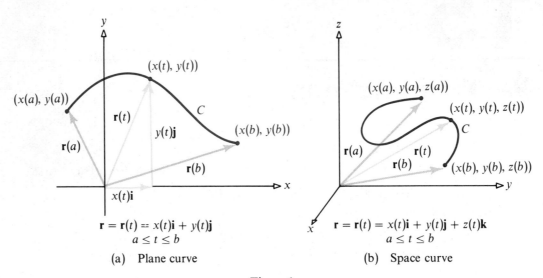

Figure 1

The curve traced out by a vector function $\mathbf{r} = \mathbf{r}(t)$, $a \le t \le b$, has a *direction*, or *orientation*, at each point. We shall take as the *positive direction* along the curve, the direction in which the vector $\mathbf{r} = \mathbf{r}(t)$ moves as t increases.

EXAMPLE 1   The vector function

$$\mathbf{r} = \mathbf{r}(t) = \cos t\,\mathbf{i} + \sin t\,\mathbf{j} \qquad 0 \le t \le 2\pi$$

traces out a circle, with center at the origin and radius equal to 1 unit. The positive direction along the circle is counterclockwise, beginning at $\mathbf{r}(0) = \mathbf{i}$. See Figure 2.

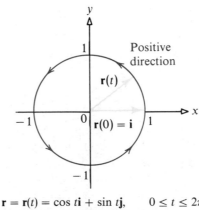

$$\mathbf{r} = \mathbf{r}(t) = \cos t\,\mathbf{i} + \sin t\,\mathbf{j}, \qquad 0 \le t \le 2\pi$$
$$\|\mathbf{r}(t)\| = \sqrt{\cos^2 t + \sin^2 t} = 1 \quad \text{for all } t$$

Figure 2   ∎

EXAMPLE 2   The vector function

$$\mathbf{r} = \mathbf{r}(t) = \sin t\mathbf{i} + \cos t\mathbf{j} \qquad 0 \le t \le \pi$$

traces out a semicircle, with center at the origin
and radius equal to 1 unit. The positive direction along the semicircle is clockwise, begin-
ning at $\mathbf{r}(0) = \mathbf{j}$ and ending at $\mathbf{r}(\pi) = -\mathbf{j}$.
See Figure 3.   ■

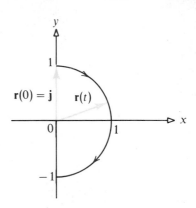

Figure 3 $r(t) = \sin t\mathbf{i} + \cos t\mathbf{j}, \quad 0 \le t \le \pi$

EXAMPLE 3   Discuss the curve traced out by the vector function

$$\mathbf{r} = \mathbf{r}(t) = (2 + 3t)\mathbf{i} + (3 - t)\mathbf{j} + 2t\mathbf{k}$$

Solution   The components of this vector function,

$$x(t) = 2 + 3t \qquad y(t) = 3 - t \qquad z(t) = 2t$$

are the parametric equations of a line in space that passes through the point $(2, 3, 0)$.
Symmetric equations of this line are

$$\frac{x - 2}{3} = \frac{y - 3}{-1} = \frac{z}{2}$$

Since $\mathbf{r}(1) = 5\mathbf{i} + 2\mathbf{j} + 2\mathbf{k}$ and $\mathbf{r}(0) = 2\mathbf{i} + 3\mathbf{j}$, the positive direction of this line
is $\mathbf{r}(1) - \mathbf{r}(0) = 3\mathbf{i} - \mathbf{j} + 2\mathbf{k}$.   ■

EXAMPLE 4   Discuss the curve traced out by

$$\mathbf{r}(t) = a \cos t\mathbf{i} + a \sin t\mathbf{j} + bt\mathbf{k} \qquad t \ge 0$$

where a and b are positive constants.

Solution   The parametric equations of this curve are

$$x = a \cos t \qquad y = a \sin t \qquad z = bt$$

For any t, a point (x, y, z) on this curve lies
on the right circular cylinder

$$x^2 + y^2 = a^2$$

For $t = 0$, the point $(a, 0, 0)$ is on the
curve. As t increases, the vector $\mathbf{r} = \mathbf{r}(t)$
starts at $(a, 0, 0)$ and winds around the cir-
cular cylinder, one revolution for every
increase of 2π in t. See Figure 4.   ■

Circular
helix

$(a, 0, 0)$

Figure 4 x

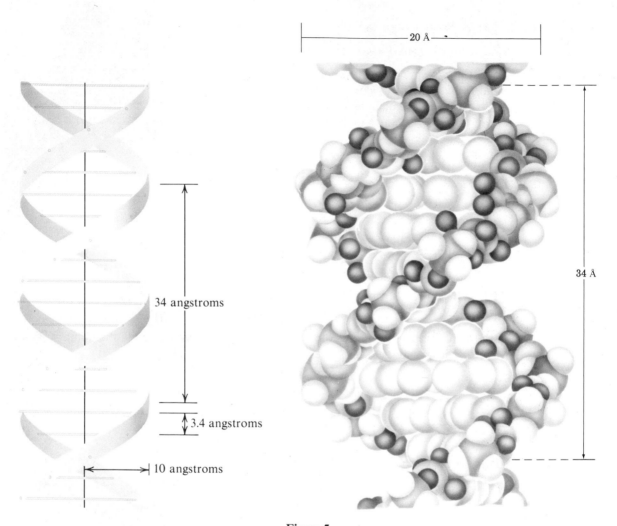

Figure 5

The space curve shown in Figure 4 is called a *circular helix* and may be viewed as a coiled spring, much like a Slinky® toy. One of the many occurrences in nature of a helix is in the model of deoxyribonucleic acid (DNA), which consists of a *double helix*—two intertwined helices (see Fig. 5).

LIMIT, CONTINUITY, AND DERIVATIVE

The notions of limit, continuity, and derivative for vector functions are extensions of these same ideas for real functions. Although the discussion that follows deals specifically with vector functions in three dimensions, the ideas are the same for vector functions in two dimensions.

Consider the vector function

$$\mathbf{r} = \mathbf{r}(t) = x(t)\mathbf{i} + y(t)\mathbf{j} + z(t)\mathbf{k}$$

where x, y, and z are three real functions defined on some interval I, and suppose t_0 is in I.

The *limit of* $\mathbf{r}(t)$ *as* t *approaches* t_0 is the vector $\mathbf{L} = l_1\mathbf{i} + l_2\mathbf{j} + l_3\mathbf{k}$,   provided

$$\lim_{t \to t_0} x(t) = l_1 \qquad \lim_{t \to t_0} y(t) = l_2 \qquad \lim_{t \to t_0} z(t) = l_3$$

In this event, we write

$$\lim_{t \to t_0} \mathbf{r}(t) = \mathbf{L}$$

A vector function

$$\mathbf{r} = \mathbf{r}(t) = x(t)\mathbf{i} + y(t)\mathbf{j} + z(t)\mathbf{k}$$

is continuous at t_0, provided each of its components $x(t)$, $y(t)$, and $z(t)$ is continuous at t_0. Thus, $\mathbf{r}(t)$ is continuous at t_0 means

$$\lim_{t \to t_0} \mathbf{r}(t) = \mathbf{r}(t_0)$$

EXAMPLE 5   For the vector function

$$\mathbf{r} = \mathbf{r}(t) = t^2\mathbf{i} + (1 + t)\mathbf{j} + \sin t\mathbf{k}$$

find $\lim_{t \to 0} \mathbf{r}(t)$.   Determine whether $\mathbf{r}$ is continuous at 0.

Solution
$$\lim_{t \to 0} \mathbf{r}(t) = \lim_{t \to 0}[t^2\mathbf{i} + (1 + t)\mathbf{j} + \sin t\mathbf{k}]$$

$$= \lim_{t \to 0} t^2\mathbf{i} + \lim_{t \to 0}(1 + t)\mathbf{j} + \lim_{t \to 0} \sin t\mathbf{k}$$

$$= (0)\mathbf{i} + (1)\mathbf{j} + (0)\mathbf{k} = \mathbf{j}$$

This vector function is continuous at 0 because $\mathbf{r}(0) = \mathbf{j}$.   ∎

The definition of continuity on an interval for vector functions follows the pattern developed in Chapter 2 for real functions.

Our main interest, however, is with the derivative. Just as the derivative of a real function is defined as the limit of a difference quotient, we define the derivative of a vector function as the limit of a vector difference quotient.

(15.1)   **DEFINITION**   The *derivative* of the vector function $\mathbf{r} = \mathbf{r}(t)$,   denoted by $\mathbf{r}'(t)$, is defined by

$$\mathbf{r}'(t) = \lim_{h \to 0} \frac{\mathbf{r}(t + h) - \mathbf{r}(t)}{h}$$

provided this limit exists. In this event, we say $\mathbf{r}(t)$ is *differentiable*.

The derivative of a vector function can be found by differentiating its components.

(15.2) **THEOREM**   Let $\mathbf{r}(t) = x(t)\mathbf{i} + y(t)\mathbf{j} + z(t)\mathbf{k}$ denote a differentiable vector function. The derivative $\mathbf{r}'(t)$ is given by

$$\mathbf{r}'(t) = x'(t)\mathbf{i} + y'(t)\mathbf{j} + z'(t)\mathbf{k}$$

Proof

$$\mathbf{r}'(t) = \lim_{h \to 0} \frac{\mathbf{r}(t + h) - \mathbf{r}(t)}{h}$$

$$= \lim_{h \to 0} \frac{[x(t + h) - x(t)]\mathbf{i} + [y(t + h) - y(t)]\mathbf{j} + [z(t + h) - z(t)]\mathbf{k}}{h}$$

$$= \lim_{h \to 0} \left[\frac{x(t + h) - x(t)}{h} \right]\mathbf{i} + \lim_{h \to 0} \left[\frac{y(t + h) - y(t)}{h} \right]\mathbf{j} + \lim_{h \to 0} \left[\frac{z(t + h) - z(t)}{h} \right]\mathbf{k}$$

$$= x'(t)\mathbf{i} + y'(t)\mathbf{j} + z'(t)\mathbf{k} \quad \blacksquare$$

In terms of the Leibniz notation, the derivative of the vector function $\mathbf{r} = \mathbf{r}(t) = x(t)\mathbf{i} + y(t)\mathbf{j} + z(t)\mathbf{k}$ is

$$\frac{d\mathbf{r}}{dt} = \frac{dx}{dt}\mathbf{i} + \frac{dy}{dt}\mathbf{j} + \frac{dz}{dt}\mathbf{k}$$

EXAMPLE 6   Find the derivative $\mathbf{r}'(t)$ of each vector function:
(a) $\mathbf{r}(t) = 2 \sin t\mathbf{i} + 3 \cos t\mathbf{j}$ (b) $\mathbf{r}(t) = t^2\mathbf{i} + (1 + t)\mathbf{j} + \sin t\mathbf{k}$

Solution   (a) $\dfrac{d\mathbf{r}}{dt} = \mathbf{r}'(t) = 2 \cos t\mathbf{i} - 3 \sin t\mathbf{j}$

(b) $\dfrac{d\mathbf{r}}{dt} = \mathbf{r}'(t) = 2t\mathbf{i} + \mathbf{j} + \cos t\mathbf{k}$ $\blacksquare$

GEOMETRIC INTERPRETATION

We have already seen that a vector function $\mathbf{r} = \mathbf{r}(t)$ that is defined and differentiable on some interval traces out a curve as t ranges over the interval. For a given number t, $\mathbf{r} = \mathbf{r}(t)$ is the position vector of a point P on the curve. If $h \neq 0$ is an increment of t, the vector $\mathbf{r}(t + h)$ is also the position vector of a point on the curve (see Fig. 6).
The vector

$$\frac{\mathbf{r}(t + h) - \mathbf{r}(t)}{h} = \frac{1}{h} [\mathbf{r}(t + h) - \mathbf{r}(t)]$$

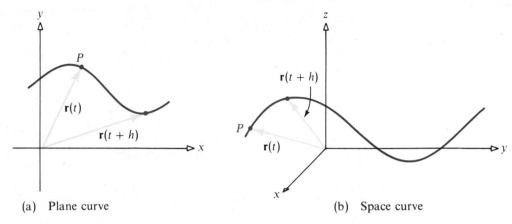

(a)   Plane curve   (b)   Space curve

Figure 6

represents a vector parallel to $\mathbf{r}(t + h) - \mathbf{r}(t)$.   If we allow h to approach 0, the limiting position of this vector is a vector tangent to the curve at P (see Fig. 7). But

$$\mathbf{r}'(t) = \lim_{h \to 0} \frac{\mathbf{r}(t + h) - \mathbf{r}(t)}{h}$$

If this limit is not $\mathbf{0}$ at P, then the direction of $\mathbf{r}'(t)$ defines the *direction of the tangent line* to the curve at P.

A curve C defined by the vector function $\mathbf{r} = \mathbf{r}(t)$, $a \le t \le b$,   is called *smooth* if its derivative $\mathbf{r}'(t)$ is continuous on $[a, b]$ and if $\mathbf{r}'(t) \ne \mathbf{0}$ throughout $[a, b]$. Thus, smooth curves have tangent lines at every point. The vector equation of the tangent

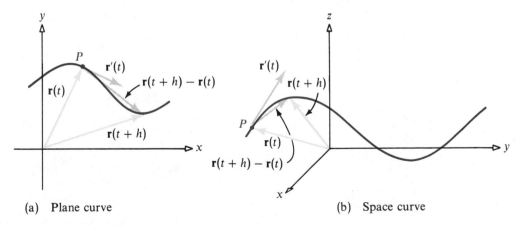

(a)   Plane curve   (b)   Space curve

Figure 7

line to a smooth curve $\mathbf{r} = \mathbf{r}(t)$ at t_0 is given by

(15.3) $$R(u) = \mathbf{r}(t_0) + u\mathbf{r}'(t_0)$$

where, to avoid confusion, we use u as the parameter for the line.

EXAMPLE 7   Find the direction of the tangent line:
(a)   To the plane curve $\mathbf{r} = \mathbf{r}(t) = 2t^2\mathbf{i} + 4t\mathbf{j}$ at $t = 1$
(b)   To the helix $\mathbf{r}(t) = \cos t\mathbf{i} + \sin t\mathbf{j} + t\mathbf{k}$ at $t = \pi$
What is the vector equation of these tangent lines?

Solution   (a) $\mathbf{r}'(t) = 4t\mathbf{i} + 4\mathbf{j}$
The direction of the tangent line at 1 is $\mathbf{r}'(1) = 4\mathbf{i} + 4\mathbf{j}$.   Since

$$\mathbf{r}(1) = 2\mathbf{i} + 4\mathbf{j}$$

the vector equation of the tangent line is

$$R(u) = (2\mathbf{i} + 4\mathbf{j}) + u(4\mathbf{i} + 4\mathbf{j})$$
$$= (2 + 4u)\mathbf{i} + (4 + 4u)\mathbf{j}$$

See Figure 8.

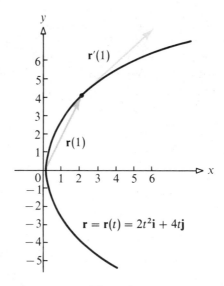

Figure 8

(b) $\mathbf{r}'(t) = -\sin t\mathbf{i} + \cos t\mathbf{j} + \mathbf{k}$
The direction of the tangent line at π is $\mathbf{r}'(\pi) = -\mathbf{j} + \mathbf{k}$.   Since

$$\mathbf{r}(\pi) = -\mathbf{i} + \pi\mathbf{k}$$

the vector equation of the tangent line is

$$\mathbf{R}(u) = (-\mathbf{i} + \pi\mathbf{k}) + u(-\mathbf{j} + \mathbf{k})$$
$$= -\mathbf{i} - u\mathbf{j} + (\pi + u)\mathbf{k}$$

See Figure 9.

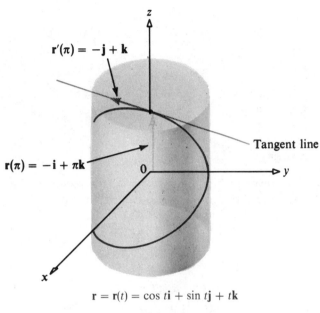

$$\mathbf{r} = \mathbf{r}(t) = \cos t\mathbf{i} + \sin t\mathbf{j} + t\mathbf{k}$$

Figure 9   ■

EXAMPLE 8   For the helix

$$\mathbf{r}(t) = \cos t\mathbf{i} + \sin t\mathbf{j} + t\mathbf{k} \qquad 0 \le t \le 2\pi$$

show that the acute angle between the tangent line and the z-axis is $\pi/4$ radian.

Solution   The direction of the tangent line at any point P on the helix is given by

$$\mathbf{r}'(t) = -\sin t\mathbf{i} + \cos t\mathbf{j} + \mathbf{k}$$

The direction of the z-axis is $\mathbf{k}$. The cosine of the acute angle θ between $\mathbf{r}'(t)$ and $\mathbf{k}$
is

$$\cos \theta = \frac{|\mathbf{r}'(t) \cdot \mathbf{k}|}{\|\mathbf{r}'(t)\| \, \|\mathbf{k}\|} \underset{\substack{\big\uparrow \\ \|\mathbf{k}\| = 1}}{=} \frac{1}{\sqrt{\sin^2 t + \cos^2 t + 1}} = \frac{1}{\sqrt{2}}$$

Hence, $\theta = \pi/4$ radian.   ■

HIGHER-ORDER DERIVATIVES

If it is possible to take the derivative of $\mathbf{r}'(t)$, we obtain the second derivative $\mathbf{r}''(t)$. Continuing in this fashion, we may compute $\mathbf{r}'''(t)$, $\mathbf{r}^{(4)}(t)$, and so on, provided these derivatives exist.

EXAMPLE 9   Find $\mathbf{r}'(t)$, $\mathbf{r}''(t)$, and $\mathbf{r}'''(t)$ for $\mathbf{r}(t) = e^t\mathbf{i} + \ln t\mathbf{j} - \cos t\mathbf{k}$.

Solution

$$\mathbf{r}'(t) = e^t\mathbf{i} + \frac{1}{t}\mathbf{j} + \sin t\mathbf{k}$$

$$\mathbf{r}''(t) = e^t\mathbf{i} - \frac{1}{t^2}\mathbf{j} + \cos t\mathbf{k}$$

$$\mathbf{r}'''(t) = e^t\mathbf{i} + \frac{2}{t^3}\mathbf{j} - \sin t\mathbf{k} \qquad \blacksquare$$

DIFFERENTIATION FORMULAS

The manner in which the derivative of a vector function is given is quite similar to the way the derivative of a real-valued function is defined. This similarity extends to the formulas for vector differentiation.

If $u(t)$ is a differentiable scalar function and if $\mathbf{f}(t)$ and $\mathbf{g}(t)$ are differentiable vector functions, we have the following rules:

(15.4)   (a) $[\mathbf{f}(t) + \mathbf{g}(t)]' = \mathbf{f}'(t) + \mathbf{g}'(t)$

         (b) $[u(t)\mathbf{f}(t)]' = u'(t)\mathbf{f}(t) + u(t)\mathbf{f}'(t)$

         (c) $[\mathbf{f}(t) \cdot \mathbf{g}(t)]' = \mathbf{f}'(t) \cdot \mathbf{g}(t) + \mathbf{f}(t) \cdot \mathbf{g}'(t)$

         (d) $[\mathbf{f}(t) \times \mathbf{g}(t)]' = \mathbf{f}'(t) \times \mathbf{g}(t) + \mathbf{f}(t) \times \mathbf{g}'(t)$

We verify part (b) here, and leave the proofs of parts (a), (c), and (d) as exercises.

Proof of Part (b)   Let $\mathbf{f}(t) = f_1(t)\mathbf{i} + f_2(t)\mathbf{j} + f_3(t)\mathbf{k}$.   Then

$$u(t)\mathbf{f}(t) = u(t)f_1(t)\mathbf{i} + u(t)f_2(t)\mathbf{j} + u(t)f_3(t)\mathbf{k}$$

By using the rule for differentiating the product of two real functions, we obtain

$$
\begin{aligned}
[u(t)\mathbf{f}(t)]' &= [u(t)f_1(t)]'\mathbf{i} + [u(t)f_2(t)]'\mathbf{j} + [u(t)f_3(t)]'\mathbf{k} \\
&= [u'(t)f_1(t) + u(t)f_1'(t)]\mathbf{i} + [u'(t)f_2(t) + u(t)f_2'(t)]\mathbf{j} + [u'(t)f_3(t) + u(t)f_3'(t)]\mathbf{k} \\
&= u'(t)[f_1(t)\mathbf{i} + f_2(t)\mathbf{j} + f_3(t)\mathbf{k}] + u(t)[f_1'(t)\mathbf{i} + f_2'(t)\mathbf{j} + f_3'(t)\mathbf{k}] \\
&= u'(t)\mathbf{f}(t) + u(t)\mathbf{f}'(t) \qquad \blacksquare
\end{aligned}
$$

EXAMPLE 10   Find the derivative of $\mathbf{f}(t) \times \mathbf{g}(t)$ for

$$\mathbf{f}(t) = \cos t\mathbf{i} + \sin t\mathbf{j} + t\mathbf{k} \qquad \mathbf{g}(t) = t\mathbf{i} + \ln t\mathbf{j} + \mathbf{k}$$

Solution
$$\mathbf{f}'(t) = -\sin t\mathbf{i} + \cos t\mathbf{j} + \mathbf{k} \qquad \mathbf{g}'(t) = \mathbf{i} + \frac{1}{t}\mathbf{j}$$

$$[\mathbf{f}(t) \times \mathbf{g}(t)]' = \mathbf{f}'(t) \times \mathbf{g}(t) + \mathbf{f}(t) \times \mathbf{g}'(t)$$
$$= (-\sin t\mathbf{i} + \cos t\mathbf{j} + \mathbf{k}) \times (t\mathbf{i} + \ln t\mathbf{j} + \mathbf{k})$$
$$+ (\cos t\mathbf{i} + \sin t\mathbf{j} + t\mathbf{k}) \times \left(\mathbf{i} + \frac{1}{t}\mathbf{j}\right)$$

$$= \begin{vmatrix} \mathbf{i} & \mathbf{j} & \mathbf{k} \\ -\sin t & \cos t & 1 \\ t & \ln t & 1 \end{vmatrix} + \begin{vmatrix} \mathbf{i} & \mathbf{j} & \mathbf{k} \\ \cos t & \sin t & t \\ 1 & \frac{1}{t} & 0 \end{vmatrix}$$

$$= (\cos t - \ln t)\mathbf{i} - (-\sin t - t)\mathbf{j} + [(-\sin t)(\ln t) - t \cos t]\mathbf{k}$$
$$+ (-1)\mathbf{i} - (-t)\mathbf{j} + \left(\frac{1}{t}\cos t - \sin t\right)\mathbf{k}$$

$$= (\cos t - \ln t - 1)\mathbf{i} + (\sin t + 2t)\mathbf{j}$$
$$+ \left[\frac{1}{t}\cos t - \sin t - (\sin t)(\ln t) - t \cos t\right]\mathbf{k} \qquad ■$$

In using part (d) of (15.4), care must be taken. Remember that the cross product is not commutative, so order is important.

The solution to Example 10 could have been obtained by first taking the cross product and then differentiating. Which operation to perform first is determined by convenience, ease of computation, and, most often, the particular situation.

EXERCISE 1

In Problems 1–8 sketch the graph of the given vector function, indicating its orientation.

1. $\mathbf{r}(t) = t\mathbf{i}, \quad -1 \le t \le 1$ **2.** $\mathbf{r}(t) = t\mathbf{j}, \quad -1 \le t \le 1$ **3.** $\mathbf{r}(t) = t\mathbf{i} + t\mathbf{j}$

4. $\mathbf{r}(t) = 3t\mathbf{i} + 2t\mathbf{j}$ **5.** $\mathbf{r}(t) = 3t\mathbf{i} - 2t\mathbf{j}$ **6.** $\mathbf{r}(t) = t\mathbf{i} + t^2\mathbf{j}$

7. $\mathbf{r}(t) = \cos t\mathbf{i} - \sin t\mathbf{j}, \quad 0 \le t \le \pi/2$ **8.** $\mathbf{r}(t) = \cos t\mathbf{i} + \sin t\mathbf{j}, \quad 0 \le t \le \pi/2$

In Problems 9–14 find $\mathbf{r}'(t)$ and $\mathbf{r}''(t)$.

9. $\mathbf{r}(t) = t^2\mathbf{i} + t^3\mathbf{j} - t\mathbf{k}$ **10.** $\mathbf{r}(t) = (1 + t)\mathbf{i} - 3t^2\mathbf{j} + t\mathbf{k}$

11. $\mathbf{r}(t) = e^t \cos t\mathbf{i} + e^t \sin t\mathbf{j} + t\mathbf{k}$ **12.** $\mathbf{r}(t) = e^{-t} \cos t\mathbf{i} + e^{-t} \sin t\mathbf{j} - t\mathbf{k}$

13. $\mathbf{r}(t) = (t - t^3)\mathbf{i} + (t + t^3)\mathbf{j} - t\mathbf{k}$ **14.** $\mathbf{r}(t) = (t^2 - t)\mathbf{i} + (t^2 + t)\mathbf{j} + t\mathbf{k}$

In Problems 15–22 sketch the graph of the given vector function, indicating its orientation. Include in the sketch the vectors $\mathbf{r}(0)$ and $\mathbf{r}'(0)$.

15. $\mathbf{r}(t) = t\mathbf{i} + t^2\mathbf{j}$ **16.** $\mathbf{r}(t) = 2t^2\mathbf{i} - t\mathbf{j}$

17. $\mathbf{r}(t) = t\mathbf{i} + e^t\mathbf{j}$

18. $\mathbf{r}(t) = t\mathbf{i} + \ln(1 + t)\mathbf{j}$

19. $\mathbf{r}(t) = 3 \sin t\mathbf{i} - 3 \cos t\mathbf{j}$

20. $\mathbf{r}(t) = 4 \sin t\mathbf{i} + 4 \cos t\mathbf{j}$

21. $\mathbf{r}(t) = 2 \cos t\mathbf{i} - 3 \sin t\mathbf{j}$

22. $\mathbf{r}(t) = -\cos t\mathbf{i} + 2 \sin t\mathbf{j}$

In Problems 23–30 find the direction of the tangent line to $\mathbf{r}(t)$ at $t = 0$. Write the vector equation of this tangent line.

23. $\mathbf{r}(t) = (1 - 3t)\mathbf{i} + 2t\mathbf{j} - (5 + t)\mathbf{k}$

24. $\mathbf{r}(t) = (2 + t)\mathbf{i} + (2 - t)\mathbf{j} + 3t\mathbf{k}$

25. $\mathbf{r}(t) = \cos 2t\mathbf{i} - \sin 2t\mathbf{j} - 5\mathbf{k}$

26. $\mathbf{r}(t) = 3\mathbf{i} + \cos t\mathbf{j} + \sin t\mathbf{k}$

27. $\mathbf{r}(t) = 2 \cos t\mathbf{i} + \mathbf{j} + 2 \sin t\mathbf{k}$

28. $\mathbf{r}(t) = 2 \cos 2t\mathbf{i} + 2 \sin 2t\mathbf{j} + 5\mathbf{k}$

29. $\mathbf{r}(t) = e^t \cos t\mathbf{i} + e^t \sin t\mathbf{j} + e^t\mathbf{k}$

30. $\mathbf{r}(t) = e^{-t} \cos t\mathbf{i} + e^{-t} \sin t\mathbf{j} - t\mathbf{k}$

In Problems 31–36 find $[\mathbf{f}(t) \cdot \mathbf{g}(t)]'$ and $[\mathbf{f}(t) \times \mathbf{g}(t)]'$.

31. $\mathbf{f}(t) = 2t\mathbf{i} + t^2\mathbf{j} - 5\mathbf{k}, \quad \mathbf{g}(t) = t^2\mathbf{i} + 2t\mathbf{j} + \mathbf{k}$

32. $\mathbf{f}(t) = t^3\mathbf{i} - t^2\mathbf{j} + t\mathbf{k}, \quad \mathbf{g}(t) = t\mathbf{i} - t^2\mathbf{j} + t^3\mathbf{k}$

33. $\mathbf{f}(t) = \cos 2t\mathbf{i} + \sin 2t\mathbf{j} + \mathbf{k}, \quad \mathbf{g}(t) = \cos t\mathbf{i} + \sin t\mathbf{j} + \mathbf{k}$

34. $\mathbf{f}(t) = \tan t\mathbf{i} + t\mathbf{j} + \mathbf{k}, \quad \mathbf{g}(t) = \sec t\mathbf{i} + 3t\mathbf{j} - \mathbf{k}$

35. $\mathbf{f}(t) = e^{2t}\mathbf{i} + e^{-2t}\mathbf{j} + t\mathbf{k}, \quad \mathbf{g}(t) = e^{-t}\mathbf{i} + e^{-2t}\mathbf{j} - t\mathbf{k}$

36. $\mathbf{f}(t) = \ln t\mathbf{i} + t\mathbf{j} - \ln(t + 1)\mathbf{k}, \quad \mathbf{g}(t) = \dfrac{1}{t}\mathbf{i} + t^2\mathbf{j} - t\mathbf{k}$

37. For $\mathbf{f}(t) = \sin t\mathbf{i} - \cos t\mathbf{j}$, show that $\mathbf{f}(t)$ and $\mathbf{f}''(t)$ are parallel.

38. For $\mathbf{f}(t) = e^{3t}\mathbf{i} + e^{-3t}\mathbf{j}$, show that $\mathbf{f}(t)$ and $\mathbf{f}''(t)$ are parallel.

39. Find all points on the curve traced out by

$$\mathbf{r}(t) = t^2\mathbf{i} + (t^2 - 1)\mathbf{j} - t\mathbf{k}$$

at which $\mathbf{r}(t)$ and its tangent line are orthogonal.

40. Show that the helix

$$\mathbf{r}(t) = \alpha \cos t\mathbf{i} + \alpha \sin t\mathbf{j} + \beta t\mathbf{k} \qquad \alpha, \beta \text{ real numbers}$$

intersects the direction of the z-axis at a constant angle.

41. Given the vectors $\mathbf{u} = \cos \omega t\mathbf{i} + \sin \omega t\mathbf{j}$ and $\mathbf{v} = t\mathbf{i} + t^2\mathbf{j} + t^3\mathbf{k}$, find:

(a) $\dfrac{d\mathbf{u}}{dt}$ and $\left\| \dfrac{d\mathbf{u}}{dt} \right\|$ (b) $\dfrac{d^2\mathbf{v}}{dt^2}$ and $\left\| \dfrac{d^2\mathbf{v}}{dt^2} \right\|$

If two curves intersect, the angle between their tangent lines at the point of intersection is called the *angle between the curves*. In Problems 42 and 43 find the angle between the curves $\mathbf{r}_1(t)$ and $\mathbf{r}_2(T)$ at the point of intersection indicated.

42. $\mathbf{r}_1(t) = (e^t - 1)\mathbf{i} - \cos \pi t\mathbf{j} + t\mathbf{k} \qquad$ at $(0, -1, 0)$

$\mathbf{r}_2(T) = (1 - T)\mathbf{i} - \mathbf{j} + (T - 1)\mathbf{k}$

43. $\mathbf{r}_1(t) = t^2\mathbf{i} + \sin \pi t\mathbf{j} + \mathbf{k} \qquad$ at $(1, 0, 1)$

$\mathbf{r}_2(T) = \mathbf{i} + T\mathbf{j} + (1 + T)\mathbf{k}$

44. If $\mathbf{f}(t)$ and $\mathbf{g}(t)$ are differentiable vector functions, show that

$$[\mathbf{f}(t) + \mathbf{g}(t)]' = \mathbf{f}'(t) + \mathbf{g}'(t) \qquad \text{Part (a) of (15.4)}$$

45. If $\mathbf{f}(t)$ and $\mathbf{g}(t)$ are differentiable vector functions, show that

$$[\mathbf{f}(t) \cdot \mathbf{g}(t)]' = \mathbf{f}'(t) \cdot \mathbf{g}(t) + \mathbf{f}(t) \cdot \mathbf{g}'(t) \qquad \text{Part (c) of (15.4)}$$

46. If $\mathbf{f}(t)$ and $\mathbf{g}(t)$ are differentiable vector functions, show that

$$[\mathbf{f}(t) \times \mathbf{g}(t)]' = \mathbf{f}'(t) \times \mathbf{g}(t) + \mathbf{f}(t) \times \mathbf{g}'(t) \qquad \text{Part (d) of (15.4)}$$

47. If $\mathbf{f}(t)$ and $\mathbf{g}(t)$ are differentiable vector functions, show that

$$[\mathbf{f}(t) \times \mathbf{g}(t)]' = -[\mathbf{g}(t) \times \mathbf{f}(t)]'$$

48. If $\mathbf{r}(t)$ is a twice differentiable vector function, show that

$$[\mathbf{r}(t) \times \mathbf{r}'(t)]' = \mathbf{r}(t) \times \mathbf{r}''(t)$$

49. If $\mathbf{r}(t)$ is a differentiable vector function, show that $\|\mathbf{r}(t)\|$ is constant if and only if $\mathbf{r}(t) \cdot \mathbf{r}'(t) = 0$ for all t.

50. Show that

$$\lim_{t \to t_0} \mathbf{r}(t) = \mathbf{L} \qquad \text{if and only if} \qquad \lim_{t \to t_0} \|\mathbf{r}(t) - \mathbf{L}\| = 0$$

2. Velocity and Acceleration

An important physical application of vector functions and their derivatives is found in the study of the motion of an object. If we think of the mass of the object as being concentrated at the object's center of gravity, then the object may be represented as a point. In this case, we refer to the object as a *particle*. As the particle moves through space, its coordinates x, y, and z are each functions of time t; and its position vector at time t is given by the vector function:

$$\text{Position:} \quad \mathbf{r}(t) = x(t)\mathbf{i} + y(t)\mathbf{j} + z(t)\mathbf{k}$$

If we assume that $\mathbf{r}(t)$ is twice differentiable, we define the *velocity*, *acceleration*, and *speed* of the particle as

$$\text{Velocity:} \quad \mathbf{v}(t) = \mathbf{r}'(t) = \frac{dx}{dt}\mathbf{i} + \frac{dy}{dt}\mathbf{j} + \frac{dz}{dt}\mathbf{k}$$

$$\text{Acceleration:} \quad \mathbf{a}(t) = \mathbf{r}''(t) = \frac{d^2x}{dt^2}\mathbf{i} + \frac{d^2y}{dt^2}\mathbf{j} + \frac{d^2z}{dt^2}\mathbf{k}$$

$$\text{Speed:} \quad v = \|\mathbf{v}(t)\| = \|\mathbf{r}'(t)\| = \sqrt{\left(\frac{dx}{dt}\right)^2 + \left(\frac{dy}{dt}\right)^2 + \left(\frac{dz}{dt}\right)^2}$$

These definitions have analogous counterparts for particles whose position vector at time t is given by a vector function in two dimensions. Figure 10 illustrates the position vector and velocity vector of a particle moving along a plane curve.

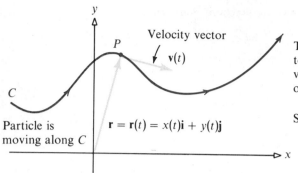

Figure 10

The velocity vector **v** is tangent to C at P. The magnitude of the velocity vector equals the speed of the object:

$$\text{Speed} = \sqrt{\left(\frac{dx}{dt}\right)^2 + \left(\frac{dy}{dt}\right)^2}$$

EXAMPLE 1   Find the velocity, acceleration, and speed of a particle whose motion is along:

(a)   The plane curve $\mathbf{r}(t) = (\tfrac{1}{2}t^2 + t)\mathbf{i} + t^3\mathbf{j}, \quad 0 \le t \le 2$

(b)   The space curve $\mathbf{r}(t) = t\mathbf{i} + t^2\mathbf{j} + t^3\mathbf{k}, \quad 0 \le t \le 2$

For each curve, sketch the path of motion of the particle and illustrate $\mathbf{v}(1)$ and $\mathbf{a}(1)$.

Solution   (a)   The velocity, acceleration, and speed are

$$\mathbf{v}(t) = \mathbf{r}'(t) = (t + 1)\mathbf{i} + 3t^2\mathbf{j}$$
$$\mathbf{a}(t) = \mathbf{r}''(t) = \mathbf{i} + 6t\mathbf{j}$$
$$v = \|\mathbf{v}(t)\| = \sqrt{9t^4 + t^2 + 2t + 1}$$

For $t = 1$,   we have

$$\mathbf{v}(1) = \mathbf{r}'(1) = 2\mathbf{i} + 3\mathbf{j} \qquad \text{and} \qquad \mathbf{a}(1) = \mathbf{r}''(1) = \mathbf{i} + 6\mathbf{j}$$

Figure 11 illustrates the path of motion, $\mathbf{v}(1)$, and $\mathbf{a}(1)$.

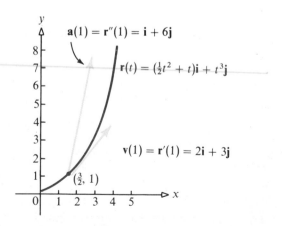

Figure 11

(b)   The velocity, acceleration, and speed are

$$\mathbf{v}(t) = \mathbf{r}'(t) = \mathbf{i} + 2t\mathbf{j} + 3t^2\mathbf{k}$$
$$\mathbf{a}(t) = \mathbf{r}''(t) = 2\mathbf{j} + 6t\mathbf{k}$$
$$v = \|\mathbf{v}(t)\| = \sqrt{1 + 4t^2 + 9t^4}$$

When $t = 1$,   we have

$$\mathbf{v}(1) = \mathbf{i} + 2\mathbf{j} + 3\mathbf{k} \qquad \text{and} \qquad \mathbf{a}(1) = 2\mathbf{j} + 6\mathbf{k}$$

Figure 12 illustrates the graph of $\mathbf{r}(t)$, $\mathbf{v}(1)$, and $\mathbf{a}(1)$.

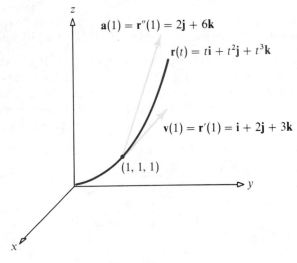

Figure 12   ■

EXAMPLE 2   Find the force acting on an object of mass m whose motion is along the elliptic path

$$\mathbf{r}(t) = \alpha \cos \omega t\mathbf{i} + \beta \sin \omega t\mathbf{j} \qquad 0 \le t \le 2\pi$$

Solution   To find the force, we must first find the acceleration of the particle and then apply Newton's law, $\mathbf{F} = m\mathbf{a}$.

$$\mathbf{v}(t) = -\alpha\omega \sin \omega t\mathbf{i} + \beta\omega \cos \omega t\mathbf{j}$$
$$\mathbf{a}(t) = -\alpha\omega^2 \cos \omega t\mathbf{i} - \beta\omega^2 \sin \omega t\mathbf{j} = -\omega^2\mathbf{r}(t)$$

Hence, by Newton's law,

$$\mathbf{F}(t) = m\mathbf{a} = -m\omega^2\mathbf{r}(t)$$

Thus, the direction of the force vector $\mathbf{F}$ is opposite to that of the position vector (see Fig. 13, p. 828).

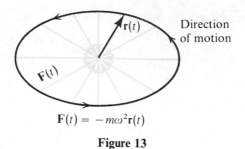

$$\mathbf{F}(t) = -m\omega^2\mathbf{r}(t)$$

Figure 13   ∎

The next example will be referred to frequently, so study it carefully.

EXAMPLE 3   Find the position vector of a particle that moves counterclockwise along a circle of radius R with a constant speed v_0. Find the velocity and acceleration of this particle.

Solution   For convenience, we place the circle so that it lies in the xy-plane, with its center at the origin. Furthermore, at time $t = 0$, we assume the particle is on the positive x-axis. After an arbitrary time t, the particle has moved counterclockwise along the circle. As Figure 14 illustrates, if $\mathbf{r}(t)$ is the position vector of the particle at time t and if $\theta(t)$ is the angle between $\mathbf{r}(t)$ and the positive x-axis, it follows that

(15.5) $$\mathbf{r}(t) = R \cos \theta(t)\mathbf{i} + R \sin \theta(t)\mathbf{j}$$

The function $\theta(t)$ can be found by using the fact that the speed of the particle is constant. Then

$$\mathbf{v}(t) = \frac{d\mathbf{r}}{dt} = -R \sin \theta(t) \frac{d\theta}{dt}\mathbf{i} + R \cos \theta(t) \frac{d\theta}{dt}\mathbf{j}$$

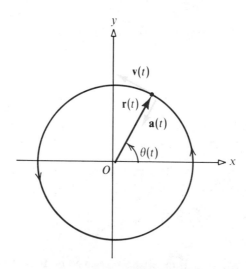

Figure 14

so that

$$v_0 = \|\mathbf{v}(t)\| = \sqrt{R^2 \sin^2\theta \left(\frac{d\theta}{dt}\right)^2 + R^2 \cos^2\theta \left(\frac{d\theta}{dt}\right)^2}$$

$$= \sqrt{R^2 \left(\frac{d\theta}{dt}\right)^2} = R \left|\frac{d\theta}{dt}\right|$$

Since θ is increasing, $d\theta/dt > 0$.   Hence,

$$v_0 = R \frac{d\theta}{dt}$$

$$\frac{d\theta}{dt} = \frac{v_0}{R}$$

But the angular speed ω equals v_0/R [compare to equation (3) on p. A-2], so we have

$$\frac{d\theta}{dt} = \omega$$

Therefore, $\theta(t) = \omega t + \theta_0$.   But the initial conditions given are that at $t = 0$, $\theta = 0$.   Hence, $\theta(t) = \omega t$.*   The position vector $\mathbf{r}(t)$ of the particle is given by

$$\mathbf{r}(t) = R \cos \omega t \mathbf{i} + R \sin \omega t \mathbf{j}$$

The velocity $\mathbf{v}(t)$ and acceleration $\mathbf{a}(t)$ are

(15.6)
$$\mathbf{v}(t) = - R\omega \sin \omega t \mathbf{i} + R\omega \cos \omega t \mathbf{j}$$
$$\mathbf{a}(t) = - R\omega^2 \cos \omega t \mathbf{i} - R\omega^2 \sin \omega t \mathbf{j} = - \omega^2 \mathbf{r}(t) \quad \blacksquare$$

We observe that although the speed of the particle in Example 3 is constant, its velocity is not. Furthermore, the direction of the acceleration is opposite that of the position vector $\mathbf{r}$ and, hence, is directed toward the center of the circle. By Newton's law $\mathbf{F} = m\mathbf{a}$,   the force vector $\mathbf{F}$ is also directed toward the center of the circle. Such a force is termed a *centripetal force*.

The results obtained in Example 3 may be used to calculate the speed required of a satellite to maintain a near-earth circular orbit[†] (the gravitational attractions of other bodies are ignored). Let R be the distance of a satellite from the center of the earth. From (15.6), the magnitude of the acceleration of the satellite whose motion

---

* The function $\theta(t)$ may be found another way. Since the speed of the particle about the circle is constant, the rate at which the angle θ changes per unit time ($d\theta/dt$), the *angular speed*, is a constant ω. That is, $d\theta/dt = \omega$.   Solving this differential equation with the initial condition that $\theta = 0$ when $t = 0$,   yields $\theta = \omega t$.

† Near-earth orbits are above 100 miles (out of the earth's atmosphere) up to an altitude of approximately 15,000 miles.

is circular is

(15.7)
$$\|\mathbf{a}(t)\| = \omega^2 \|\mathbf{r}(t)\| = \omega^2 R = \frac{v_0^2}{R}$$

But the magnitude of the acceleration of the satellite must equal g, the acceleration of gravity for earth.* Hence,

(15.8)
$$\frac{v_0^2}{R} = g$$

$$v_0 = \sqrt{gR}$$

For example, the speed required of a communications satellite whose circular orbit is to be 4500 miles from the center of the earth, is

$$v_0 = \sqrt{(79{,}036)(4500)} \approx 18{,}859 \text{ miles per hour}$$

Let's look at another use of formula (15.7).

EXAMPLE 4   A motorcycle that weighs 150 kilograms is driven at a constant speed of 120 kilometers per hour on a circular track whose radius is 100 meters. To keep the motorcycle from skidding, what frictional force must be exerted by the tires on the track?

Solution   By Newton's law, the force $\mathbf{F}$ required to keep an object of mass m traveling along a curve traced out by $\mathbf{r} = \mathbf{r}(t)$ is $\mathbf{F} = m\mathbf{a}$. The magnitude of the frictional force exerted by the tires must therefore equal

$$\|\mathbf{F}\| = m\|\mathbf{a}\|$$

But the motion is circular. Hence, from (15.7), we find

$$\|\mathbf{F}\| = m\left(\frac{v_0^2}{R}\right) = 150 \text{ kg}\left(\frac{120^2 \text{ km}^2/\text{hr}^2}{100 \text{ m}}\right) = (150)(144)\left(\frac{1000^2}{3600^2}\right)\frac{\text{kg m}}{\text{sec}^2} \approx 1667 \text{ newtons}$$

$\blacksquare$

EXERCISE 2

In Problem 1–8 sketch the motion of a particle traveling along the curve traced out by $\mathbf{r} = \mathbf{r}(t)$. Include in the sketch the vectors $\mathbf{v}(0)$ and $\mathbf{a}(0)$.

1. $\mathbf{r}(t) = t\mathbf{i} + t^2\mathbf{j}$

2. $\mathbf{r}(t) = 2t^2\mathbf{i} - t\mathbf{j}$

3. $\mathbf{r}(t) = t\mathbf{i} + e^t\mathbf{j}$

4. $\mathbf{r}(t) = t\mathbf{i} + \ln(1 + t)\mathbf{j}$

5. $\mathbf{r}(t) = 3 \sin t\mathbf{i} - 3 \cos t\mathbf{j}$

6. $\mathbf{r}(t) = 4 \sin t\mathbf{i} + 4 \cos t\mathbf{j}$

7. $\mathbf{r}(t) = 2 \cos t\mathbf{i} - 3 \sin t\mathbf{j}$

8. $\mathbf{r}(t) = -\cos t\mathbf{i} + 2 \sin t\mathbf{j}$

---

* Although the acceleration of gravity at such altitudes is somewhat less than $g \approx 32.2$ feet per second per second $\approx 79{,}036$ miles per hour per hour, we shall ignore this discrepancy in our calculations.

In Problems 9–20 find the velocity, acceleration, and speed of a particle whose motion is along the curve traced out by $\mathbf{r} = \mathbf{r}(t)$.

9. $\mathbf{r}(t) = 2t\mathbf{i} + (t + 1)\mathbf{j}$

10. $\mathbf{r}(t) = e^t\mathbf{i} + e^{2t}\mathbf{j}$

11. $\mathbf{r}(t) = 2 \cos t\mathbf{i} + 3 \sin t\mathbf{j}$

12. $\mathbf{r}(t) = e^{-t}\mathbf{i} + e^{-2t}\mathbf{j}$

13. $\mathbf{r}(t) = (1 + t)\mathbf{i} - (3 - t)\mathbf{j} + t\mathbf{k}$

14. $\mathbf{r}(t) = (2 - t)\mathbf{i} + t\mathbf{j} + (1 + t)\mathbf{k}$

15. $\mathbf{r}(t) = 3 \cos t\mathbf{i} + 3 \sin t\mathbf{j} + 5\mathbf{k}$

16. $\mathbf{r}(t) = 4 \cos t\mathbf{i} + 5 \sin t\mathbf{j} + 6\mathbf{k}$

17. $\mathbf{r}(t) = (t^2 - t)\mathbf{i} + t\mathbf{j} + (t^2 + t)\mathbf{k}$

18. $\mathbf{r}(t) = e^t \cos t\mathbf{i} + e^t \sin t\mathbf{j} - t\mathbf{k}$

19. $\mathbf{r}(t) = \ln t\mathbf{i} + \sqrt{t}\mathbf{j} + t^{3/2}\mathbf{k}, \quad t > 0$

20. $\mathbf{r}(t) = e^{2t}\mathbf{i} + \ln\sqrt{t}\mathbf{j} + \cosh t\mathbf{k}, \quad t > 0$

21. Show that if the speed of a particle along a curve is constant, then the velocity and acceleration vectors are orthogonal.

22. If the motion of a particle is along the curve traced out by $\mathbf{r}(t) = \alpha \cosh t\mathbf{i} + \beta \sinh t\mathbf{j}, \quad \alpha > 0, \quad \beta > 0,$ show that the force acting on the particle is parallel to $\mathbf{r}(t)$.

In Problems 23–31 use the idea expressed by formulas (15.7) and (15.8).

23. A girl is spinning a bucket of water in a horizontal plane at the end of a rope of length L.
    (a) If she triples the speed of the bucket, how many times as hard must she pull on the rope?
    (b) If, instead, she doubles the length of the rope but keeps the speed of the bucket the same, will she have to pull more or less? How much?

24. Find the distance of a satellite from the surface of the earth if it moves around the earth in a circular orbit at a constant speed of 18,630 miles per hour. (The radius of the earth is approximately 4000 miles.)

25. Find the speed of a satellite that moves in a circular orbit around the earth at a height of 100 miles.

26. At the Fermi National Accelerator Laboratory in Batavia, Illinois, protons are accelerated along a circular route of radius 1 kilometer. Find the magnitude of the force necessary to give a proton of mass m a constant speed of 280,000 kilometers per second.

27. A weather satellite orbits the earth in a circle every $1\frac{1}{2}$ hours. How high is it above the earth?

28. Some communications satellites remain stationary above a fixed point on the equator of the earth's surface. How high are such satellites and what is their common velocity? (Assume the earth turns once every 24 hours.)

29. If, in Example 4 in this section, the motorcycle is driven at a speed that is 10% faster, by how much is the frictional force of the tires increased?

30. If, in Example 4 in this section, the radius of the circular track is halved, how much slower should the motorcycle be driven so as not to increase the frictional force?

31. A race car of mass 1000 kilograms is driven at a constant speed of 200 kilometers per hour around a circular track whose radius is 75 meters. What frictional force must be exerted by the tires on the track to keep the car from skidding?

32. A particle moves on the circle $x^2 + y^2 = 1$ so that at time $t \geq 0$ the position is given by the vector

$$\mathbf{r}(t) = \frac{1 - t^2}{1 + t^2}\mathbf{i} + \frac{2t}{1 + t^2}\mathbf{j}$$

(a) Find the velocity vector.
(b) Is the particle ever at rest? Justify your answer.
(c) Give the coordinates of the point that the particle approaches as t increases without bound.

33. The *power* expended by a force $\mathbf{F}(t)$ acting on an object of mass m with velocity $\mathbf{v}(t)$ is the dot product

$$\text{Power} = \mathbf{F}(t) \cdot \mathbf{v}(t)$$

The *kinetic energy* of an object of mass m and velocity $\mathbf{v}(t)$ equals one-half the product of its mass times the square of its speed:

$$\text{Kinetic energy} = \tfrac{1}{2}m\|\mathbf{v}(t)\|^2$$

Show that the power of an object equals the instantaneous rate of change of its kinetic energy with respect to time.

34. In classical mechanics, the *momentum* $\mathbf{p}(t)$ of an object of mass m at time t is defined as $\mathbf{p}(t) = m\mathbf{v}(t)$, where $\mathbf{v}(t)$ is the velocity of the object at time t. Show that force equals the instantaneous rate of change of momentum with respect to time.

35. The *torque* τ^* produced by a force $\mathbf{F}(t)$ acting on an object whose position at time t is $\mathbf{r}(t)$ is defined as $\tau(t) = \mathbf{r}(t) \times \mathbf{F}(t)$. The torque measures the twist imparted on the object by the force. The *angular momentum* $\mathbf{L}$ of an object of mass m and velocity $\mathbf{v}(t)$ whose position at time t is $\mathbf{r}(t)$ is $\mathbf{L}(t) = \mathbf{r}(t) \times m\mathbf{v}(t)$. Show that the instantaneous rate of change of angular momentum with respect to time equals the torque.

36. A central force $\mathbf{F}(t)$ is one whose direction is proportional to the position vector $\mathbf{r}(t)$ of the object it acts upon; that is, $\mathbf{F}(t) = u(t)\mathbf{r}(t)$, where $u(t)$ is a scalar function. Show that a central force produces no torque.

37. A particle moves on a disk from the center directly toward the edge. If the disk is revolving in the counterclockwise direction at a constant angular speed ω, the position of the particle at time t is $\mathbf{r}(t) = t \cos \omega t \mathbf{i} + t \sin \omega t \mathbf{j}$.
(a)   Show that the velocity $\mathbf{v}$ of the particle is

$$\mathbf{v} = \cos \omega t \mathbf{i} + \sin \omega t \mathbf{j} + t\mathbf{v}_d$$

where $\mathbf{v}_d$ is the velocity of the rotating disk.
(b)   Also show that the acceleration $\mathbf{a}$ of the particle is

$$\mathbf{a} = 2\mathbf{v}_d + t\mathbf{a}_d$$

where $\mathbf{a}_d$ is the acceleration of the rotating disk. The extra term $2\mathbf{v}_d$ is the so-called *Coriolis acceleration*, which results from the interaction of the rotation of the disk and the motion of the particles on the disk.

38. Refer to Problem 37.
(a)   Find the velocity and acceleration of a particle revolving on a rotating disk according to

$$\mathbf{r}(t) = t^2 \cos \omega t \mathbf{i} + t^2 \sin \omega t \mathbf{j}$$

(b)   What is the Coriolis acceleration?

3. Tangential and Normal Components of Acceleration

We have already seen that the derivative $\mathbf{r}'(t)$ of a vector function $\mathbf{r} = \mathbf{r}(t)$ is a vector in the direction of the tangent line. We define the *unit tangent vector* $\mathbf{T}$ of the curve traced out by a differentiable vector function $\mathbf{r} = \mathbf{r}(t)$, $a \leq t \leq b$, for which

---

* Greek letter tau.

$\mathbf{r}'(t) \neq \mathbf{0}$,   to be

(15.9)
$$\mathbf{T} = \frac{\mathbf{r}'(t)}{\|\mathbf{r}'(t)\|}$$

Observe that the vector $\mathbf{T}$ defined above is indeed a unit vector parallel to the direction of the tangent line to $\mathbf{r} = \mathbf{r}(t)$.

EXAMPLE 1   Show that the unit tangent vector $\mathbf{T}$ of the circle of radius R,

$$\mathbf{r} = \mathbf{r}(t) = R \cos t\mathbf{i} + R \sin t\mathbf{j}   \qquad 0 \leq t \leq 2\pi$$

is everywhere perpendicular to $\mathbf{r}$.

Solution   We calculate $\mathbf{r}'(t)$ and $\|\mathbf{r}'(t)\|$:

$$\mathbf{r}'(t) = -R \sin t\mathbf{i} + R \cos t\mathbf{j}$$
$$\|\mathbf{r}'(t)\| = \sqrt{(-R \sin t)^2 + (R \cos t)^2} = R$$

From (15.9),

$$\mathbf{T} = \frac{\mathbf{r}'(t)}{\|\mathbf{r}'(t)\|} = -\sin t\mathbf{i} + \cos t\mathbf{j}$$

It follows that

$$\mathbf{T} \cdot \mathbf{r} = (-\sin t\mathbf{i} + \cos t\mathbf{j}) \cdot (R \cos t\mathbf{i} + R \sin t\mathbf{j})$$
$$= -R \sin t \cos t + R \sin t \cos t = 0$$

for all t. Hence, $\mathbf{T}$ is everywhere perpendicular to $\mathbf{r}$ (see Fig. 15).

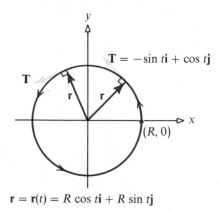

$$\mathbf{r} = \mathbf{r}(t) = R \cos t\mathbf{i} + R \sin t\mathbf{j}$$

Figure 15   ∎

Suppose a curve is traced out by a twice differentiable vector function $\mathbf{r} = \mathbf{r}(t)$. Then the unit tangent vector $\mathbf{T}$ is itself differentiable. Furthermore, since $\mathbf{T}$ is a unit vector, we know that

$$\mathbf{T} \cdot \mathbf{T} = 1$$

If we differentiate this expression, we find that

$$\frac{d}{dt}(\mathbf{T}\cdot\mathbf{T}) = \frac{d}{dt}(1)$$

$$\mathbf{T}'\cdot\mathbf{T} + \mathbf{T}\cdot\mathbf{T}' = 0$$

$$\mathbf{T}\cdot\mathbf{T}' = 0$$

That is, $\mathbf{T}'$ is a vector that is everywhere perpendicular to $\mathbf{T}$. By virtue of this fact, we define the *principal* unit normal vector* $\mathbf{N}$ of a curve traced out by a twice differentiable vector function $\mathbf{r} = \mathbf{r}(t)$, $a \le t \le b$, for which $\mathbf{r}''(t) \ne \mathbf{0}$, to be

(15.10)
$$\mathbf{N} = \frac{\mathbf{T}'}{\|\mathbf{T}'\|}$$

To summarize, the unit tangent vector $\mathbf{T}$ points in the direction of the tangent line to $\mathbf{r} = \mathbf{r}(t)$; that is, it points in the direction in which the particle is moving. The principal unit normal vector $\mathbf{N}$ points in a direction perpendicular to the motion of the particle.

For example, the principal unit normal vector $\mathbf{N}$ of the circle discussed in Example 1 is

$$\mathbf{N} = \frac{\mathbf{T}'}{\|\mathbf{T}'\|} = \frac{-\cos t\mathbf{i} - \sin t\mathbf{j}}{\sqrt{(-\cos t)^2 + (-\sin t)^2}} = -\cos t\mathbf{i} - \sin t\mathbf{j}$$

We observe that, for a circle, $\mathbf{N} = -\mathbf{r}(t)/\|\mathbf{r}(t)\|$, so that the principal unit normal vector of a circle is opposite in direction to the vector $\mathbf{r}$ and, hence, is directed toward the center of the circle (see Fig. 16).

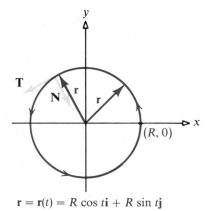

$$\mathbf{r} = \mathbf{r}(t) = R\cos t\mathbf{i} + R\sin t\mathbf{j}$$

Figure 16

---

* The vector opposite in direction to $\mathbf{T}'$ is also perpendicular to $\mathbf{T}$, and, hence, it is also normal to the curve.

EXAMPLE 2   Show that the principal unit normal vector $\mathbf{N}$ of the helix

$$\mathbf{r} = \mathbf{r}(t) = \cos t\mathbf{i} + \sin t\mathbf{j} + t\mathbf{k}$$

is perpendicular to and directed toward the z-axis.

Solution

$$\mathbf{r}'(t) = -\sin t\mathbf{i} + \cos t\mathbf{j} + \mathbf{k}$$

$$\|\mathbf{r}'(t)\| = \sqrt{(-\sin t)^2 + (\cos t)^2 + 1} = \sqrt{2}$$

$$\mathbf{T} = \frac{\mathbf{r}'(t)}{\|\mathbf{r}'(t)\|} = \frac{-\sin t\mathbf{i} + \cos t\mathbf{j} + \mathbf{k}}{\sqrt{2}}$$

$$\mathbf{T}' = \frac{-\cos t\mathbf{i} - \sin t\mathbf{j}}{\sqrt{2}}$$

$$\|\mathbf{T}'\| = \frac{1}{\sqrt{2}}$$

$$\mathbf{N} = \frac{\mathbf{T}'}{\|\mathbf{T}'\|} = -\cos t\mathbf{i} - \sin t\mathbf{j}$$

Since the direction of the z-axis is $\mathbf{k}$, it follows that $\mathbf{N} \cdot \mathbf{k} = 0$,   and, hence, $\mathbf{N}$ is perpendicular to the direction of the z-axis. It is also clear that $\mathbf{N}$ points toward the z-axis (see Fig. 17).

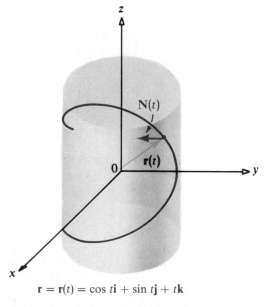

$$\mathbf{r} = \mathbf{r}(t) = \cos t\mathbf{i} + \sin t\mathbf{j} + t\mathbf{k}$$

Figure 17

The acceleration vector **a** of a particle whose motion is along the curve traced out by a twice differentiable vector function $\mathbf{r} = \mathbf{r}(t)$ has been defined as $\mathbf{a} = \mathbf{r}''(t)$. We show now that the acceleration vector **a** lies in the plane determined by the tangent vector **T** and the normal vector **N**, and we express **a** in terms of **T** and **N**.

We begin by expressing the velocity vector **v** in terms of **T** as

(15.11)
$$\mathbf{v} = \mathbf{r}'(t) = \|\mathbf{r}'(t)\|\mathbf{T} = \|\mathbf{v}\|\mathbf{T} = v\mathbf{T}$$

$$\uparrow$$
$$\text{Use (15.9)}$$

where $v = \|\mathbf{v}\|$ is the speed of the particle. Thus, we have shown that

The velocity vector v of a particle whose motion is along a curve traced out by a twice differentiable vector function $\mathbf{r} = \mathbf{r}(t)$ is directed along the tangent to the curve.

By differentiating (15.11), we can get an expression for the acceleration vector:

$$\mathbf{a} = \frac{d\mathbf{v}}{dt} = \frac{d}{dt}(v\mathbf{T}) = \frac{dv}{dt}\mathbf{T} + v\frac{d\mathbf{T}}{dt}$$

From (15.10), $d\mathbf{T}/dt = \mathbf{T}' = \|\mathbf{T}'\|\mathbf{N}$, so that

(15.12)
$$\mathbf{a} = \frac{dv}{dt}\mathbf{T} + v\|\mathbf{T}'\|\mathbf{N}$$

Thus, the acceleration vector a of a particle whose motion is along a curve traced out by a twice differentiable vector function $\mathbf{r} = \mathbf{r}(t)$ lies in the plane of the tangent vector T and the normal vector N of the curve.

The *tangential component*, denoted by $a_\mathbf{T}$, and the *normal component*, denoted by $a_\mathbf{N}$, of the acceleration vector **a** are

(15.13)
$$a_\mathbf{T} = \frac{dv}{dt} \qquad a_\mathbf{N} = v\|\mathbf{T}'\|$$

Although the formula for $a_\mathbf{T}$ is easy to use, the one for $a_\mathbf{N}$ is usually not easy to calculate. However, by combining (15.12) and (15.13), we may write

$$\mathbf{a} = a_\mathbf{T}\mathbf{T} + a_\mathbf{N}\mathbf{N}$$

Then

$$\mathbf{a} \cdot \mathbf{a} = (a_\mathbf{T}\mathbf{T} + a_\mathbf{N}\mathbf{N}) \cdot (a_\mathbf{T}\mathbf{T} + a_\mathbf{N}\mathbf{N})$$
$$\|\mathbf{a}\|^2 = a_\mathbf{T}^2\|\mathbf{T}\|^2 + 2a_\mathbf{T}a_\mathbf{N}\mathbf{T} \cdot \mathbf{N} + a_\mathbf{N}^2\|\mathbf{N}\|$$

Since $\mathbf{T} \cdot \mathbf{N} = 0$ and $\|\mathbf{N}\| = \|\mathbf{T}\| = 1$, we get

$$\|\mathbf{a}\|^2 = a_\mathbf{T}^2 + a_\mathbf{N}^2$$

Hence, we find that

(15.14)
$$a_\mathbf{N} = \sqrt{\|\mathbf{a}\|^2 - a_\mathbf{T}^2}$$

Formula (15.14) is usually easier to use in practice than the formula for $a_\mathbf{N}$ in (15.13).

EXAMPLE 3   Find the tangential and normal components of the acceleration of a particle whose motion is along the ellipse

$$\mathbf{r} = \mathbf{r}(t) = 3 \sin t\mathbf{i} + 4 \cos t\mathbf{j}$$

Graph the ellipse, showing $\mathbf{a}$, a_T, and a_N when $t = \pi/4$.

Solution   First,

$$\mathbf{v} = \mathbf{r}'(t) = 3 \cos t\mathbf{i} - 4 \sin t\mathbf{j}$$
$$v = \|\mathbf{v}\| = \sqrt{(3 \cos t)^2 + (-4 \sin t)^2} = \sqrt{9 \cos^2 t + 16 \sin^2 t}$$
$$\mathbf{a} = \mathbf{r}''(t) = -3 \sin t\mathbf{i} - 4 \cos t\mathbf{j}$$

The tangential component of the acceleration is

$$a_T = \frac{dv}{dt} = \frac{-18 \cos t \sin t + 32 \sin t \cos t}{2\sqrt{9 \cos^2 t + 16 \sin^2 t}}$$

$$= \frac{7 \sin t \cos t}{\sqrt{9 \cos^2 t + 16 \sin^2 t}}$$

To find the normal component of the acceleration, we find $\|\mathbf{a}\|$ and then use (15.14):

$$\|\mathbf{a}\| = \sqrt{(-3 \sin t)^2 + (-4 \cos t)^2} = \sqrt{9 \sin^2 t + 16 \cos^2 t}$$

$$a_N = \sqrt{\|\mathbf{a}\|^2 - a_T^2} = \sqrt{9 \sin^2 t + 16 \cos^2 t - \frac{49 \sin^2 t \cos^2 t}{9 \cos^2 t + 16 \sin^2 t}}$$

$$= \sqrt{\frac{144(\sin^4 t + 2 \sin^2 t \cos^2 t + \cos^4 t)}{9 \cos^2 t + 16 \sin^2 t}} = \frac{12}{\sqrt{9 \cos^2 t + 16 \sin^2 t}}$$

At $t = \pi/4$,   we have

$$\mathbf{a} = \frac{-3}{\sqrt{2}}\mathbf{i} - \frac{4}{\sqrt{2}}\mathbf{j} \qquad a_T = \frac{7\sqrt{2}}{10} \qquad a_N = \frac{12\sqrt{2}}{5}$$

The motion of the particle on the ellipse is clockwise, starting at $\mathbf{r}(0) = 4\mathbf{j}$.   When $t = \pi/4$,   the particle is at the position $\mathbf{r}(\pi/4) = (3/\sqrt{2})\mathbf{i} + (4/\sqrt{2})\mathbf{j}$.   See Figure 18.   ■

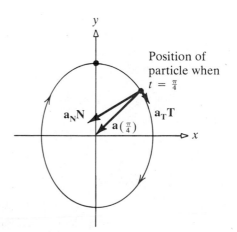

Position of particle when $t = \frac{\pi}{4}$

Figure 18

EXAMPLE 4   Find the tangential and normal components of the acceleration of a particle whose motion is along the helix

$$\mathbf{r} = \mathbf{r}(t) = \cos t\mathbf{i} + \sin t\mathbf{j} + t\mathbf{k}$$

Solution $$\mathbf{v} = \mathbf{r}'(t) = -\sin t\mathbf{i} + \cos t\mathbf{j} + \mathbf{k}$$
$$v = \|\mathbf{v}\| = \sqrt{(-\sin t)^2 + (\cos t)^2 + 1} = \sqrt{2}$$

The tangential component of the acceleration

$$a_{\mathbf{T}} = \frac{dv}{dt} = 0$$

To find the normal component, we first find $\mathbf{a}$ and then use (15.14):

$$\mathbf{a} = \frac{d\mathbf{v}}{dt} = -\cos t\mathbf{i} - \sin t\mathbf{j}$$

$$\|\mathbf{a}\| = \sqrt{(-\cos t)^2 + (-\sin t)^2} = 1$$

$$a_{\mathbf{N}} = \sqrt{\|\mathbf{a}\|^2 - a_{\mathbf{T}}^2} = 1 \qquad \blacksquare$$

APPLICATION: THE UVW-AXES IN ORBITAL MECHANICS
(If Time Permits)

A spacecraft in orbit about the earth follows a curve in three-space, which is modeled as the graph of a vector function of one variable—time, t. The spacecraft carries a gyroscope whose three axes remain fixed in direction throughout time and are aligned with the $\mathbf{i}$, $\mathbf{j}$, $\mathbf{k}$ unit vectors of the xyz-coordinate system (see Fig. 19). The xyz-coordinate system has its origin at the center of the earth and provides the same basic frame of reference for both the earthbound observers and the spacecraft.

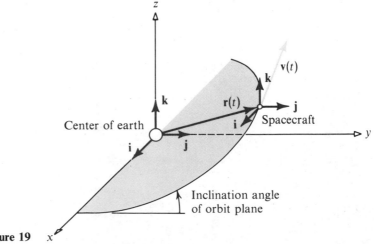

Figure 19

The position vector from the origin is given by the function

$$\mathbf{r}(t) = x(t)\mathbf{i} + y(t)\mathbf{j} + z(t)\mathbf{k}$$

The velocity is the vector tangent to the orbit expressed by $\mathbf{v}(t) = \mathbf{r}'(t)$.

If the earth is taken to have all its mass concentrated at the center, then the laws of Kepler and Newton tell us that any orbit remains in a *fixed plane* containing that center. Furthermore, a closed orbit describes an ellipse (or circle) in that plane with the earth's center placed at *one of the foci;* the other focus remains vacant.

The UVW system of axes is an orthogonal set of unit vectors $\mathbf{i}_U(t), \mathbf{i}_V(t), \mathbf{i}_W(t)$, which are "attached" to the spacecraft's orbit at each point. This set of vectors is particularly useful in manned spacecraft since it corresponds to the axes of *yaw*, *roll*, and *pitch* customarily used in airplanes flying over the surface of the earth. We define these time-dependent unit vectors as follows:

(15.15)
$$\mathbf{i}_U = \frac{\mathbf{r}}{\|\mathbf{r}\|} \qquad \mathbf{i}_W = \frac{\mathbf{r} \times \mathbf{v}}{\|\mathbf{r} \times \mathbf{v}\|} \qquad \mathbf{i}_V = \mathbf{i}_W \times \mathbf{i}_U$$

Note that $\mathbf{i}_U(t)$ lies in the orbit plane and points radially *outward* (see Fig. 20). The vector $\mathbf{i}_W(t)$ is *normal* to the orbit plane since both $\mathbf{r}(t)$ and $\mathbf{v}(t)$ are vectors in that plane. The vector $\mathbf{i}_V(t)$ completes the set, and points, roughly, in the direction of the velocity; it lies in the orbit plane.

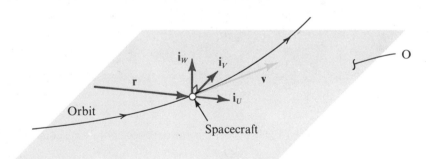

Figure 20

Now, consider a circular orbit intersecting the x-axis, inclined to the xy-plane at $30°$. Suppose that its radius is a and the motion has angular speed ω. Then its position vector is

$$\mathbf{r}(t) = a\left(\cos \omega t \mathbf{i} + \frac{\sqrt{3}}{2} \sin \omega t \mathbf{j} + \frac{1}{2} \sin \omega t \mathbf{k} \right)$$

Therefore,

$$\mathbf{v}(t) = \mathbf{r}'(t) = a\omega\left(-\sin \omega t \mathbf{i} + \frac{\sqrt{3}}{2} \cos \omega t \mathbf{j} + \frac{1}{2} \cos \omega t \mathbf{k} \right)$$

In this case, using equation (15.15), we obtain

$$i_U = \cos \omega t\, i + \frac{\sqrt{3}}{2} \sin \omega t\, j + \frac{1}{2} \sin \omega t\, k$$

(15.16) $$i_W = -\frac{1}{2} j + \frac{\sqrt{3}}{2} k$$

$$i_V = -\sin \omega t\, i + \frac{\sqrt{3}}{2} \cos \omega t\, j + \frac{1}{2} \cos \omega t\, k$$

Thus, by (15.16), we have the directions of the UVW-axes expressed in terms of the onboard gyroscope axes.

EXERCISE 3

In Problems 1–8 sketch the motion of a particle traveling along the curve traced out by $r = r(t)$. Include in the sketch a, a_T, and a_N when $t = 0$.

1. $r(t) = ti + t^2 j$ **2.** $r(t) = 2t^2 i - tj$ **3.** $r(t) = ti + e^t j$

4. $r(t) = ti + \ln(1 + t)j$ **5.** $r(t) = 3 \sin ti - 3 \cos tj$ **6.** $r(t) = 4 \sin ti + 4 \cos tj$

7. $r(t) = 2 \cos ti - 3 \sin tj$ **8.** $r(t) = -\cos ti + 2 \sin tj$

In Problems 9–20 find the tangential and normal components of the acceleration of a particle traveling along the curve traced out by $r = r(t)$.

9. $r(t) = 2ti + (t + 1)j$ **10.** $r(t) = e^t i + e^{2t} j$ **11.** $r(t) = 2 \cos ti + 3 \sin tj$

12. $r(t) = e^{-t} i + e^{-2t} j$ **13.** $r(t) = (1 - 3t)i + 2tj - (5 + t)k$ **14.** $r(t) = (2 + t)i + (2 - t)j + 3tk$

15. $r(t) = \cos 2ti + \sin 2tj - 5k$ **16.** $r(t) = 3i + \cos tj + \sin tk$ **17.** $r(t) = t^2 i + (t - t^3)j + tk$

18. $r(t) = ti + \sin tj + \cos tk$ **19.** $r(t) = ti + t^2 j + t^3 k$

20. $r(t) = a \cos ti + b \sin tj + ctk$, $a > 0$, $b > 0$, $c > 0$

21. Show that if a particle moves along a path at constant speed, the tangential component of the acceleration is 0.

22. Suppose that the function $r(t) = e^t i + e^{-t} j$ gives the position vector of a particle at time t.
   (a) Show that the force on the particle is directed away from the origin.
   (b) What is the minimum speed of the particle and where does it occur?
   (c) Find the tangential and normal components of acceleration at the point found in part (b).
   (d) The answers in part (c) are $a_T = 0$ and $a_N = \|a\| = \sqrt{2}$. How could these have been predicted?

4. Integrals of Vector Functions

The integration of a vector function $r = r(t)$ is performed by integrating each component. That is, if $r(t) = x(t)i + y(t)j + z(t)k$, then

(15.17) $$\int r(t)\, dt = \left[\int x(t)\, dt \right] i + \left[\int y(t)\, dt \right] j + \left[\int z(t)\, dt \right] k$$

In actually doing integration problems, remember to insert the vector constant of integration. For example,

$$\int \left(\sin t\mathbf{i} + \cos t\mathbf{j} + 2\mathbf{k} \right) dt = \left(\int \sin t\, dt \right) \mathbf{i} + \left(\int \cos t\, dt \right) \mathbf{j} + \left(\int 2\, dt \right) \mathbf{k}$$

$$= (-\cos t + c_1)\mathbf{i} + (\sin t + c_2)\mathbf{j} + (2t + c_3)\mathbf{k}$$

The next example illustrates how to determine a vector function when its derivative and a boundary condition are given.

EXAMPLE 1   Find $\mathbf{r}(t)$ if $\mathbf{r}'(t) = 2t\mathbf{i} + e^t\mathbf{j} + e^{-t}\mathbf{k}$ and $\mathbf{r}(0) = \mathbf{i} - \mathbf{j} + \mathbf{k}$.

Solution
$$\mathbf{r}(t) = \int \mathbf{r}'(t)\, dt = \int (2t\mathbf{i} + e^t\mathbf{j} + e^{-t}\mathbf{k})\, dt$$

$$= \left(\int 2t\, dt \right)\mathbf{i} + \left(\int e^t\, dt \right)\mathbf{j} + \left(\int e^{-t}\, dt \right)\mathbf{k}$$

$$= (t^2 + c_1)\mathbf{i} + (e^t + c_2)\mathbf{j} + (c_3 - e^{-t})\mathbf{k}$$

By applying the condition that $\mathbf{r}(0) = \mathbf{i} - \mathbf{j} + \mathbf{k}$,   we find

$$\mathbf{i} - \mathbf{j} + \mathbf{k} = c_1\mathbf{i} + (1 + c_2)\mathbf{j} + (c_3 - 1)\mathbf{k}$$

so

$$1 = c_1 \qquad\qquad -1 = 1 + c_2 \qquad\qquad 1 = c_3 - 1$$
$$c_1 = 1 \qquad\qquad c_2 = -2 \qquad\qquad c_3 = 2$$

Hence, the vector function $\mathbf{r}$ is

$$\mathbf{r}(t) = (t^2 + 1)\mathbf{i} + (e^t - 2)\mathbf{j} + (2 - e^{-t})\mathbf{k} \qquad \blacksquare$$

NEWTON'S SECOND LAW OF MOTION

From Newton's second law of motion, $\mathbf{F}(t) = m\mathbf{a}(t)$,   once the force $\mathbf{F}(t)$ acting on a particle of known constant mass m is given, the acceleration vector $\mathbf{a}(t)$ is determined. Since the velocity vector $\mathbf{v}(t)$ is the integral of $\mathbf{a}(t)$, that is,

$$\mathbf{v}(t) = \int \mathbf{a}(t)\, dt$$

we can calculate the velocity and speed of the particle, provided a boundary condition on $\mathbf{v}(t)$ is known. Furthermore, since

$$\mathbf{r}(t) = \int \mathbf{v}(t)\, dt$$

we can determine the path of the particle when a boundary condition on $\mathbf{r}(t)$ is known.

PROJECTILE PROBLEM

We use these facts to find the path of a projectile fired at an inclination θ to the horizontal and with initial speed v_0. For convenience, we select the starting point as the origin, the x-axis as horizontal, and the y-axis as positive upward (see Fig. 21).

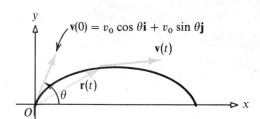

Figure 21

When time $t = 0$, the position of the projectile is the origin, and the initial velocity vector $\mathbf{v}(0)$ has magnitude v_0 and inclination θ. Consequently,

(15.18) $\mathbf{r}(0) = \mathbf{0}$ and $\mathbf{v}(0) = v_0 \cos \theta \mathbf{i} + v_0 \sin \theta \mathbf{j}$

We let $\mathbf{r}(t)$ represent the position vector of the projectile after time t. We wish to find an expression for $\mathbf{r}(t)$ in terms of v_0 and θ. We assume that the only force acting on the projectile is the force $\mathbf{F} = -mg\mathbf{j}$,* where g is the acceleration of gravity and m is the mass of the projectile.

From Newton's second law of motion, the acceleration $\mathbf{a}(t)$ of the projectile obeys

$$m\mathbf{a}(t) = -mg\mathbf{j} \qquad \text{or} \qquad \mathbf{r}''(t) = -g\mathbf{j}$$

Integrating both sides with respect to t gives

$$\mathbf{v}(t) = -gt\mathbf{j} + \mathbf{c}$$

where $\mathbf{c}$ is a constant vector. When $t = 0$, $\mathbf{v}(0) = \mathbf{c}$. Hence, from (15.18), the velocity vector of the projectile is

$$\mathbf{v}(t) = v_0 \cos \theta \mathbf{i} + v_0 \sin \theta \mathbf{j} - gt\mathbf{j}$$

Integrating both sides with respect to t gives

$$\mathbf{r}(t) = (v_0 \cos \theta)t\mathbf{i} + (v_0 \sin \theta)t\mathbf{j} - \tfrac{1}{2}gt^2\mathbf{j} + \mathbf{d}$$

where $\mathbf{d}$ is a constant vector. When $t = 0$, the position of the projectile is at the origin. Hence, $\mathbf{d} = \mathbf{0}$. Thus, the position vector of the projectile at time t is

$$\mathbf{r}(t) = (v_0 \cos \theta)t\mathbf{i} + (v_0 \sin \theta)t\mathbf{j} - \tfrac{1}{2}gt^2\mathbf{j}$$

The parametric equations of the motion of the projectile are

(15.19) $x = (v_0 \cos \theta)t \qquad y = -\tfrac{1}{2}gt^2 + (v_0 \sin \theta)t$

By eliminating t from these equations, we find

(15.20) $$y = \frac{-g}{2v_0^2 \cos^2\theta} x^2 + x \tan \theta$$

which is the equation of a parabola.

---

* This is called the *flat earth approximation*. It is a good approximation for short-range missiles.

The x-intercepts of this parabola are found by setting $y = 0$. Hence, we find that

(15.21) $$x = 0 \quad \text{and} \quad x = \frac{2v_0^2 \cos^2\theta \tan\theta}{g} = \frac{v_0^2 \sin 2\theta}{g}$$

The number $(v_0^2 \sin 2\theta)/g$ is called the *range* of the projectile. In Problem 15 of Exercise 4 you are asked to show that an inclination of $\theta = \pi/4$ gives the maximum range.

The projectile hits the ground when $x = (v_0^2 \sin 2\theta)/g$. Therefore, by substituting into equation (15.19) and solving for t, we find that the projectile is in the air for

(15.22) $$t = \frac{(v_0^2 \sin 2\theta)/g}{v_0 \cos\theta} = \frac{2v_0 \sin\theta}{g} \text{ seconds}$$

The projectile reaches its maximum height when $dy/dt = 0$, or $t = (v_0 \sin\theta)/g$, and, therefore, the maximum height is

(15.23) $$y = -\tfrac{1}{2}g\left(\frac{v_0^2 \sin^2\theta}{g^2}\right) + (v_0 \sin\theta)\left(\frac{v_0 \sin\theta}{g}\right) = \frac{v_0^2 \sin^2\theta}{2g}$$

The formula for the range of a projectile and its maximum height as given by (15.21) and (15.23) are valid only for the initial conditions specified. For other initial conditions, the same pattern of solution must be followed.

EXERCISE 4

In Problems 1–4 find each integral.

1. $\int (\sin t\mathbf{i} - \cos t\mathbf{j} + t\mathbf{k})\, dt$

2. $\int (t^2\mathbf{i} - t\mathbf{j} + e^t\mathbf{k})\, dt$

3. $\int (\cos t\mathbf{i} + \sin t\mathbf{j} - \mathbf{k})\, dt$

4. $\int (e^t\mathbf{i} - \sqrt{t}\,\mathbf{j} + t^2\mathbf{k})\, dt$

In Problems 5–8 find the velocity, speed, and position of a particle having the given acceleration, initial velocity, and initial position.

5. $\mathbf{a} = -32\mathbf{k}, \quad \mathbf{v}(0) = 0, \quad \mathbf{r}(0) = 0$

6. $\mathbf{a} = -32\mathbf{k}, \quad \mathbf{v}(0) = \mathbf{i} + \mathbf{j}, \quad r(0) = 0$

7. $\mathbf{a} = \cos t\mathbf{i} + \sin t\mathbf{j}, \quad \mathbf{v}(0) = \mathbf{i}, \quad \mathbf{r}(0) = \mathbf{j}$

8. $\mathbf{a} = \cos t\mathbf{i} + \sin t\mathbf{j}, \quad \mathbf{v}(0) = \mathbf{j}, \quad \mathbf{r}(0) = \mathbf{i}$

9. Find $\mathbf{r}(t)$ if $\mathbf{r}'(t) = e^t\mathbf{i} - \ln t\mathbf{j} + 2t\mathbf{k}$ and $\mathbf{r}(1) = \mathbf{j} + \mathbf{k}$.

10. Find $\mathbf{r}(t)$ if $\mathbf{r}'(t) = t\mathbf{i} + e^{-t}\mathbf{j} - (1/t)\mathbf{k}$ and $\mathbf{r}(1) = \mathbf{i} - \mathbf{j} + 2\mathbf{k}$.

In Problems 11–17 assume $g = 9.8$ meters per second per second, and neglect air resistance.

11. A projectile was fired at an angle of $30°$ with the horizontal with an initial speed of 520 meters per second. What was its range, the time of flight, and the greatest height reached?

12. A projectile was fired with an initial speed of 200 meters per second at an inclination of $60°$ with the horizontal. What was its range, the time of flight, and the greatest height reached?

13. A projectile was fired with an initial speed of 100 meters per second at an angle of inclination of $\tan^{-1}(\frac{5}{12})$. Find the equation of the path and the range.

14. A projectile was fired with an initial speed of 120 meters per second at an angle of inclination of $\tan^{-1}(\frac{3}{4})$. Find the equation of the path and the range.

15. Show that the maximum range of the projectile of (15.21) occurs when $\theta = \pi/4$ and has the value v_0^2/g.

16. Show that the speed of the projectile of (15.19) is least when the projectile is at its highest point.

17. A projectile is fired with a speed of 500 meters per second at an inclination of 45° from a point 30 meters above level ground. Find the point where the projectile will strike the ground.

18. A force whose magnitude is 5 and whose direction is along the positive x-axis is continuously applied to an object of mass $m = 1$ kilogram. If at $t = 0$, the position of the object is the origin and if its velocity is 3 meters per second in the direction of the positive y-axis, find:
   (a)   The speed and velocity after t seconds
   (b)   The position after force has been applied for t seconds
   (c)   The path of the object

19. An object of mass m leaves the point $(1, 2)$ with initial velocity $v_0 = 3i + 4j$. Thereafter, it is subjected only to the force $F = m(-i - j)/\sqrt{2}$. Find the formula for the position of the object at any time $t > 0$.

20. A projectile is propelled horizontally at a height of 3 meters above the ground in order to hit a target 1 meter high that is 30 meters away (see the figure). What should the initial velocity of the projectile be?

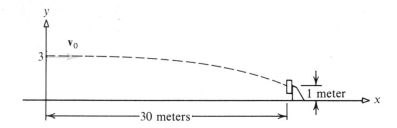

21. *Calculator problem.* In a Metro Conference basketball game on January 21, 1980, between Florida State University and Virginia Tech, a record was set. Les Henson, who is 6 feet 6 inches tall, made a basket from $89\frac{1}{4}$ feet downcourt, to win the game for Virginia Tech by a score of 79 to 77. Assuming he released the ball at a height of 6 feet 6 inches and threw it at an angle of 45° (to maximize distance), with what initial velocity was the ball tossed? See the accompanying figure. (Assume $g = 32$ feet per second per second.)

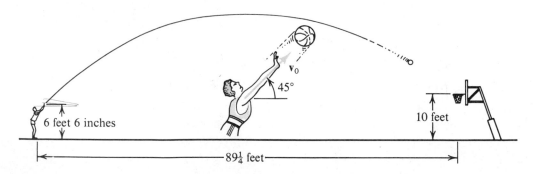

22. *Calculator problem.*   A baseball is hit at an angle of 45° to the horizontal from an initial height of 3 feet. If the ball just clears the vines in front of the bleachers in Wrigley Field, which are 10 feet high and a distance of 400 feet from the plate, what was the initial speed of the ball? How long did it take the ball to reach the vines?

23. *Calculator problem.*   A certain outfielder throws a baseball at an angle of 45° to the horizontal from an initial height of 6 feet. If he can throw the ball with an initial velocity of 100 feet per second, what is the maximum distance he can be from home plate to ensure that the ball reaches home plate on the fly? How long is the ball in flight? (The answers reveal how fast a runner must be to get from third base to home plate on a fly ball and beat the throw to score a run.)

24. A gun, lifted at an angle θ_0 to the horizontal, is aimed at an elevated target, which is released at the moment the gun is fired. No matter what the initial speed v_0 of the bullet, show that it will always hit the falling target. Use the accompanying figure as a guide.

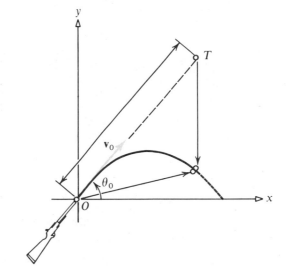

25. A plane is flying at an elevation of 4.0 kilometers with a constant horizontal speed of 400 kilometers per hour toward a point directly above its target. At what angle of sight α should a package be released in order to strike the target? Use the accompanying figure as a guide.   [*Hint:* $g \approx 127{,}008$ kilometers per hour per hour.]

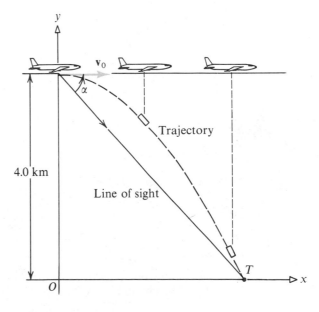

26. Show that a particle subject to no outside forces is either stationary or moves with constant speed along a straight line.

27. Show that if $\mathbf{r}'(t) = \mathbf{0}$ for all t on some interval I, then $\mathbf{r}(t) = \mathbf{c}$, for all t in I.

28. Show that if $\mathbf{f}'(t) = \mathbf{g}'(t)$ for all t in some interval I, then $\mathbf{f}(t) = \mathbf{g}(t) + \mathbf{c}$ for all t in I.

29. If $\mathbf{c}$ is a constant vector, show that

$$\int_a^b [\mathbf{c} \cdot \mathbf{r}(t)] \, dt = \mathbf{c} \cdot \int_a^b \mathbf{r}(t) \, dt$$

30. Use the result of Problem 29 to show that

$$\left\| \int_a^b \mathbf{r}(t) \, dt \right\| \leq \int_a^b \|\mathbf{r}(t)\| \, dt$$

[*Hint:* Set $\mathbf{c} = \int_a^b \mathbf{r}(t) \, dt$, and evaluate $\|\mathbf{c}\|^2$.]

5. Arc Length; Curvature

In Chapter 13 we derived a formula for the arc length s of a smooth plane curve. There, we showed that if a smooth plane curve C is defined by the parametric equations

$$x = x(t) \qquad y = y(t) \qquad a \leq t \leq b$$

then the arc length s along C from $t = a$ to $t = b$ is

(15.24)
$$s = \int_a^b \sqrt{\left(\frac{dx}{dt}\right)^2 + \left(\frac{dy}{dt}\right)^2} \, dt$$

We state here, without proof, the extension of this result for a smooth curve in space. In particular, if a smooth space curve C is defined by the parametric equations

$$x = x(t) \qquad y = y(t) \qquad z = z(t) \qquad a \leq t \leq b$$

the arc length s along C from $t = a$ to $t = b$ is

(15.25)
$$s = \int_a^b \sqrt{\left(\frac{dx}{dt}\right)^2 + \left(\frac{dy}{dt}\right)^2 + \left(\frac{dz}{dt}\right)^2} \, dt$$

In vector notation, if a curve is defined by the vector function $\mathbf{r} = \mathbf{r}(t)$, $a \leq t \leq b$, then formulas (15.24) and (15.25) become

(15.26)
$$s = \int_a^b \left\| \frac{d\mathbf{r}}{dt} \right\| \, dt$$

EXAMPLE 1   Find the arc length of:
(a)   The circle $\mathbf{r} = \mathbf{r}(t) = R \cos t \mathbf{i} + R \sin t \mathbf{j}$ from $t = 0$ to $t = 2\pi$
(b)   The circular helix $\mathbf{r} = \mathbf{r}(t) = R \cos t \mathbf{i} + R \sin t \mathbf{j} + t \mathbf{k}$ from $t = 0$ to $t = 2\pi$

Solution   (a)
$$\frac{d\mathbf{r}}{dt} = -R \sin t\mathbf{i} + R \cos t\mathbf{j}$$

$$\left\|\frac{d\mathbf{r}}{dt}\right\| = \sqrt{(-R \sin t)^2 + (R \cos t)^2} = R$$

By using (15.26), we find the arc length s to be

$$s = \int_0^{2\pi} R \, dt = 2\pi R$$

which is the familiar formula for the circumference of a circle.

(b)
$$\frac{d\mathbf{r}}{dt} = -R \sin t\mathbf{i} + R \cos t\mathbf{j} + \mathbf{k}$$

$$\left\|\frac{d\mathbf{r}}{dt}\right\| = \sqrt{(-R \sin t)^2 + (R \cos t)^2 + 1^2} = \sqrt{R^2 + 1}$$

Hence, the desired arc length s is

$$s = \int_0^{2\pi} \sqrt{R^2 + 1} \, dt = 2\pi\sqrt{R^2 + 1} \qquad \blacksquare$$

ARC LENGTH AS A PARAMETER

Suppose a curve C is defined by the vector equation

$$\mathbf{r} = \mathbf{r}(s) \qquad 0 \le s \le L$$

where the parameter s is the arc length as measured along C, and L is the length of C (see Fig. 22). Then, by (15.26), the arc length s of C from 0 to an arbitrary number s is given by

(15.27)
$$s = \int_0^s \left\|\frac{d\mathbf{r}}{dt}\right\| dt$$

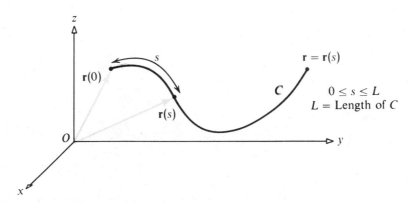

Figure 22

Differentiating both sides of (15.27) with respect to s gives

$$\frac{d}{ds}(s) \underset{\uparrow}{=} \frac{d}{ds} \int_0^s \left\| \frac{d\mathbf{r}}{dt} \right\| dt = \left\| \frac{d\mathbf{r}}{ds} \right\|$$

Thus

(5.35)

(15.28)
$$1 = \left\| \frac{d\mathbf{r}}{ds} \right\|$$

(15.29) **Thus, when the parameter of a curve C: $\mathbf{r} = \mathbf{r}(s)$ is its arc length s, the derivative $\mathbf{r}'(s)$ is a unit vector; in fact, it is the unit tangent vector.**

We shall have occasion to use this fact shortly.

CURVATURE IN THREE-SPACE

In Chapter 13 we defined the curvature κ of a curve C in the plane to be

$$\kappa = \left| \frac{d\phi}{ds} \right|$$

where ϕ is the inclination to the positive x-axis of the tangent line at some point on the curve, and s is the arc length as measured along C. In three-space this approach to curvature does not work.

In space a curve can "bend" in infinitely many directions, whereas in the plane it can only bend away from its tangent. Our purpose in this section is to obtain a satisfactory definition for the curvature of a space curve, namely, a definition that, as a special case, will agree with the one given for a plane curve.

As with plane curves, the most natural approach to the curvature of space curves is to use the arc length s as the parameter. Thus, we choose to define the curve C by the twice differentiable vector function

$$\mathbf{r} = \mathbf{r}(s) \qquad 0 \le s \le L$$

where the parameter s is the arc length as measured along C, and L is the length of C.

Based on (15.29), the unit tangent vector $\mathbf{T} = \mathbf{T}(s)$ to C at a point on C is

(15.30)
$$\mathbf{T} = \mathbf{T}(s) = \frac{d\mathbf{r}}{ds}$$

From (15.10), the principal unit normal vector $\mathbf{N} = \mathbf{N}(s)$ to C at a point on C is

(15.31)
$$\mathbf{N} = \mathbf{N}(s) = \frac{d\mathbf{T}/ds}{\|d\mathbf{T}/ds\|}$$

The number $\|d\mathbf{T}/ds\|$ is defined as the *curvature* κ of the curve C.

Based on (15.31), we have the formulas

(15.32)
$$\kappa = \left\|\frac{d\mathbf{T}}{ds}\right\| \qquad \frac{d\mathbf{T}}{ds} = \kappa \mathbf{N}(s)$$

PARAMETERS OTHER THAN s

To use the definition $\kappa = \|d\mathbf{T}/ds\|$ effectively to compute the curvature κ of a space curve C requires that the curve be expressed with arc length s as a parameter. Since this is usually not the case, it is helpful to have a formula for the curvature κ when the curve is expressed by a parameter other than arc length. We now give such a formula.

(15.33) **If a space curve C is defined by the twice differentiable vector function $\mathbf{r} = \mathbf{r}(t)$, its curvature κ is given by the formula**

$$\kappa = \frac{\|\mathbf{r}'(t) \times \mathbf{r}''(t)\|}{\|\mathbf{r}'(t)\|^3}$$

Proof   We use the chain rule. The unit tangent vector $\mathbf{T}$ is

(15.34)
$$\mathbf{T} = \frac{d\mathbf{r}}{ds} = \frac{d\mathbf{r}/dt}{ds/dt} \qquad \text{or} \qquad \mathbf{r}'(t) = \frac{d\mathbf{r}}{dt} = \frac{ds}{dt}\mathbf{T}$$

We differentiate $\mathbf{r}'(t) = d\mathbf{r}/dt$ in (15.34) with respect to t, remembering to use the chain rule:

(15.35)
$$\mathbf{r}''(t) = \frac{d^2\mathbf{r}}{dt^2} = \frac{d}{dt}\left(\frac{ds}{dt}\mathbf{T}\right) = \frac{d^2s}{dt^2}\mathbf{T} + \left(\frac{ds}{dt}\right)\left(\frac{d\mathbf{T}}{dt}\right)$$

$$= \frac{d^2s}{dt^2}\mathbf{T} + \left(\frac{ds}{dt}\right)\left(\frac{ds}{dt}\right)\left(\frac{d\mathbf{T}}{ds}\right) = \frac{d^2s}{dt^2}\mathbf{T} + \left(\frac{ds}{dt}\right)^2\left(\frac{d\mathbf{T}}{ds}\right)$$

From (15.32), we may write (15.35) as

(15.36)
$$\mathbf{r}''(t) = \frac{d^2s}{dt^2}\mathbf{T} + \left(\frac{ds}{dt}\right)^2 \kappa \mathbf{N}$$

From (15.34) and (15.36), we get

$$\mathbf{r}'(t) \times \mathbf{r}''(t) = \frac{ds}{dt}\mathbf{T} \times \left[\frac{d^2s}{dt^2}\mathbf{T} + \left(\frac{ds}{dt}\right)^2 \kappa \mathbf{N}\right]$$

$$= \frac{ds}{dt}\left[\frac{d^2s}{dt^2}(\mathbf{T} \times \mathbf{T}) + \left(\frac{ds}{dt}\right)^2 \kappa(\mathbf{T} \times \mathbf{N})\right] = \left(\frac{ds}{dt}\right)^3 \kappa(\mathbf{T} \times \mathbf{N})$$

$$\underset{\mathbf{T} \times \mathbf{T} = 0}{\uparrow}$$

Hence,

(15.37)
$$\|\mathbf{r}'(t) \times \mathbf{r}''(t)\| = \left(\frac{ds}{dt}\right)^3 \kappa \|\mathbf{T} \times \mathbf{N}\|$$

Since $\mathbf{T}$ and $\mathbf{N}$ are perpendicular unit vectors, $\mathbf{T} \times \mathbf{N}$ is also a unit vector. From (15.34), it follows that $\|\mathbf{r}'(t)\| = ds/dt$. Using these facts and solving (15.37) for κ gives the required formula. ∎

EXAMPLE 2   Find the curvature κ of the curve $\mathbf{r}(t) = t\mathbf{i} + t^2\mathbf{j} + t^3\mathbf{k}$.

Solution
$$\mathbf{r}'(t) = \mathbf{i} + 2t\mathbf{j} + 3t^2\mathbf{k}$$
$$\|\mathbf{r}'(t)\| = \sqrt{1 + 4t^2 + 9t^4}$$

Since $\|\mathbf{r}'(t)\| \neq 1$, we know that the parameter t is not arc length [refer to (15.29)]. As a result, we proceed to calculate $\mathbf{r}''(t)$ in order to use (15.33):

$$\mathbf{r}''(t) = 2\mathbf{j} + 6t\mathbf{k}$$

$$\mathbf{r}'(t) \times \mathbf{r}''(t) = \begin{vmatrix} \mathbf{i} & \mathbf{j} & \mathbf{k} \\ 1 & 2t & 3t^2 \\ 0 & 2 & 6t \end{vmatrix} = 6t^2\mathbf{i} - 6t\mathbf{j} + 2\mathbf{k}$$

$$\|\mathbf{r}'(t) \times \mathbf{r}''(t)\| = \sqrt{36t^4 + 36t^2 + 4} = 2\sqrt{9t^4 + 9t^2 + 1}$$

Thus,

$$\kappa = \frac{2\sqrt{9t^4 + 9t^2 + 1}}{(1 + 4t^2 + 9t^4)^{3/2}} \qquad \blacksquare$$

We now show that the definition of curvature as $\kappa = \|d\mathbf{T}/ds\|$ reduces to $|d\phi/ds|$ if the curve C lies in a plane. In the event the curve C is a plane curve, its unit tangent vector $\mathbf{T} = \mathbf{T}(s)$ is

$$\mathbf{T} = \mathbf{T}(s) = \cos\phi\,\mathbf{i} + \sin\phi\,\mathbf{j}$$

where $\phi = \phi(s)$ is the inclination to the positive x-axis of the tangent line. Differentiating $\mathbf{T}$ with respect to s yields

$$\frac{d\mathbf{T}}{ds} = -\sin\phi\,\frac{d\phi}{ds}\mathbf{i} + \cos\phi\,\frac{d\phi}{ds}\mathbf{j} = \frac{d\phi}{ds}(-\sin\phi\,\mathbf{i} + \cos\phi\,\mathbf{j})$$

Hence,

$$\left\|\frac{d\mathbf{T}}{ds}\right\| = \left|\frac{d\phi}{ds}\right| \|-\sin\phi\,\mathbf{i} + \cos\phi\,\mathbf{j}\| = \left|\frac{d\phi}{ds}\right|$$

TANGENTIAL AND NORMAL COMPONENTS OF ACCELERATION

From (15.12), the acceleration vector $\mathbf{a}$ of a particle whose motion is along the curve C traced out by a twice differentiable vector function $\mathbf{r} = \mathbf{r}(t)$ is

(15.38)
$$\mathbf{a} = \frac{dv}{dt}\mathbf{T} + v\left\|\frac{d\mathbf{T}}{dt}\right\|\mathbf{N}$$

where v is the particle's speed along C. Since

$$\left\|\frac{d\mathbf{T}}{dt}\right\| = \left\|\frac{ds}{dt}\frac{d\mathbf{T}}{ds}\right\| = v\left\|\frac{d\mathbf{T}}{ds}\right\|$$

it follows from (15.32) that

(15.39)
$$\mathbf{a} = \frac{dv}{dt}\mathbf{T} + v^2\kappa\mathbf{N}$$

Thus, the tangential component $a_\mathbf{T}$ and normal component $a_\mathbf{N}$ of the acceleration vector $\mathbf{a}$ are given by

(15.40)
$$a_\mathbf{T} = \frac{dv}{dt} \quad \text{and} \quad a_\mathbf{N} = v^2\kappa$$

Formula (15.40) provides an alternative to (15.13) and (15.14) for calculating $a_\mathbf{N}$.

EXERCISE 5

In Problems 1–6 find the arc length s of each curve.

1. $\mathbf{r}(t) = t\mathbf{i} + 2t\mathbf{j} + t\mathbf{k}$ from $t = 0$ to $t = 1$
2. $\mathbf{r}(t) = 2t\mathbf{i} + t\mathbf{j} + 3t\mathbf{k}$ from $t = 1$ to $t = 3$
3. $\mathbf{r}(t) = \sin 2t\mathbf{i} + \cos 2t\mathbf{j} + t\mathbf{k}$ from $t = 0$ to $t = \pi$
4. $\mathbf{r}(t) = \sin t\mathbf{i} + \cos t\mathbf{j} + bt\mathbf{k}$ from $t = 0$ to $t = 2\pi$
5. $\mathbf{r}(t) = e^t\mathbf{i} + e^{-t}\mathbf{j} + \sqrt{2}\,t\mathbf{k}$ from $t = 0$ to $t = 1$
6. $\mathbf{r}(t) = \cos^3 t\mathbf{i} + \sin^3 t\mathbf{j} + \mathbf{k}$ from $t = 0$ to $t = \pi/2$

In Problems 7–14 find the curvature κ of each curve.

7. $\mathbf{r}(t) = t\mathbf{i} + 2t\mathbf{j} + t\mathbf{k}$

8. $\mathbf{r}(t) = 2t\mathbf{i} + t\mathbf{j} + 3t\mathbf{k}$

9. $\mathbf{r}(t) = \sin 2t\mathbf{i} + \cos 2t\mathbf{j} + t\mathbf{k}$

10. $\mathbf{r}(t) = \sin t\mathbf{i} + \cos t\mathbf{j} + bt\mathbf{k}$

11. $\mathbf{r}(t) = e^t\mathbf{i} + e^{-t}\mathbf{j} + \sqrt{2}\,t\mathbf{k}$

12. $\mathbf{r}(t) = \cos^3 t\mathbf{i} + \sin^3 t\mathbf{j} + \mathbf{k}$

13. $\mathbf{r}(t) = a(3t - t^3)\mathbf{i} + 3at^2\mathbf{j} + a(3t + t^3)\mathbf{k}$

14. $\mathbf{r}(t) = 4a\cos^3\theta\mathbf{i} + 4a\sin^3\theta\mathbf{j} + 3a\cos 2\theta\mathbf{k}$

15. Find the curvature of the curve $\mathbf{r}(t) = (1 - t^3)\mathbf{i} + t^2\mathbf{j}$.   Where is the curvature undefined? Do you see any geometrical reason for this?

16. Find the curvature of the spiral $\mathbf{r}(t) = e^{-t}\cos t\mathbf{i} + e^{-t}\sin t\mathbf{j}$.   How does it behave when $t \to +\infty$?   Do you see any geometrical reason for this?

17. Find the curvature κ of the curve $\mathbf{r}(t) = 2a\cos\theta\mathbf{i} + 2a\sin\theta\mathbf{j} + b\theta^2\mathbf{k}$.

18. Suppose $\mathbf{r} = \mathbf{r}(t)$ is the position vector of a particle at time t. If the normal component of acceleration equals zero at any time t, explain why the motion of the particle must be in a straight line.

19. We have shown that if $\mathbf{r}(t) = x(t)\mathbf{i} + y(t)\mathbf{j} + z(t)\mathbf{k}$ is the position vector of a particle at time t, the normal component of acceleration is $a_N = \kappa v^2$,   where $v = ds/dt$ is speed. Use this fact to find a_N when the motion of a particle is given by $\mathbf{r}(t) = \cos t\mathbf{i} + \sin t\mathbf{j} + t\mathbf{k}$.

20. Use the formula

$$\mathbf{F} = m\mathbf{a} = m\frac{dv}{dt}\mathbf{T} + m\kappa v^2\mathbf{N}$$

to discuss the forces on a passenger in a car. What force corresponds to the push against the seat experienced by the passenger when the accelerator is depressed? What force corresponds to the push against the door experienced by the passenger when the car is going around a curve? How does this latter force vary with the curvature of the road? How does it vary with the speed of the car?

In Problems 21–24 use the fact that the graph of a twice differentiable function $y = f(x)$ can be described as the space curve $\mathbf{r} = \mathbf{r}(t) = t\mathbf{i} + f(t)\mathbf{j} + 0 \cdot \mathbf{k}$ to prove each result.

21. $\dfrac{dt}{ds} = \dfrac{1}{\sqrt{1 + f'^2}}$

22. $\mathbf{T} = \dfrac{\mathbf{i} + f'\mathbf{j}}{\sqrt{1 + f'^2}}$

23. $\dfrac{d\mathbf{T}}{ds} = \dfrac{f''}{(1 + f'^2)^2}(-f'\mathbf{i} + \mathbf{j})$

24. $\kappa = \dfrac{|f''|}{(1 + f'^2)^{3/2}}$

25. Find $a > 0$ so that $\mathbf{r}(t) = a \cos t\mathbf{i} + a \sin t\mathbf{j} + t\mathbf{k}$ has maximum curvature.

6. Kepler's Laws of Planetary Motion (If Time Permits)

In the sixteenth century, Copernicus (1473–1543) conjectured the *heliocentric theory of planetary motion*—that is, that the planets travel in circular orbits with the sun as center. This theory agreed better with observation than the earlier *geocentric theory*—that the sun, moon, and planets travel in circular paths around the earth.

However, Copernicus' theory still did not explain some observable facts. Early in the seventeenth century, Kepler, working with the observations of the Danish astronomer Tycho Brahe (1546–1601), stated three laws to explain the orbits of planets:

1. The orbit of each planet is an ellipse with the sun at a focus (1609).

2. The speed of a planet is such that a line joining the sun to the planet sweeps out an area at a constant rate (1609).

3. The square of the period of revolution of the planet is proportional to the cube of the length of the semimajor axis of the planet's elliptical orbit (1619).

Although Kepler's laws of planetary motion agreed well with observable facts, a sound footing for this theory was not obtained until Newton used calculus to organize a theory that explained the motion of both planetary bodies and the earth. Here, we develop the first and second of Kepler's laws with the aid of Newton's ideas and vector calculus.

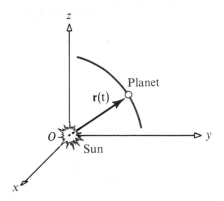

Figure 23

We position our coordinate system so that the sun is at the origin and let $\mathbf{r}(t)$ denote the position vector of a planet in orbit about the sun (see Fig. 23). Then $\mathbf{r}$ is subject to two laws:

(15.41)   Newton's second law of motion: $\mathbf{F} = m\mathbf{r}''(t)$

(15.42)   Newton's inverse square law: $\mathbf{F} = \dfrac{-GmM}{\|\mathbf{r}\|^2} \dfrac{\mathbf{r}}{\|\mathbf{r}\|}$

The first of these is the familiar law stating that the force acting on an object is proportional to the acceleration of the object, its mass being the constant of proportionality. Equation (15.42) states that the force $\mathbf{F}$ of attraction between two objects of mass m and M is inversely proportional to the square of the distance between them and is directed along a line joining them, G ($\approx 6.67 \times 10^{-11}$ newton-meter^2 per kilogram^2)   being the constant of proportionality.

From (15.41) and (15.42), we may write

$$\mathbf{r}''(t) = \frac{-GM}{\|\mathbf{r}\|^3} \mathbf{r}$$

If $\mathbf{v}(t)$ is the velocity of the planet, it follows that

(15.43)
$$\frac{d\mathbf{v}}{dt} = \frac{-GM}{\|\mathbf{r}\|^3} \mathbf{r}$$

Then we can conclude that $d\mathbf{v}/dt$ and $\mathbf{r}$ are parallel, so that

$$\mathbf{r} \times \frac{d\mathbf{v}}{dt} = \mathbf{0}$$

Because of this,

$$\frac{d}{dt}(\mathbf{r} \times \mathbf{v}) = \frac{d\mathbf{r}}{dt} \times \mathbf{v} + \mathbf{r} \times \frac{d\mathbf{v}}{dt} = \mathbf{v} \times \mathbf{v} + \mathbf{0} = \mathbf{0}$$

Hence, it follows by integrating that

(15.44) $$\mathbf{r} \times \mathbf{v} = \mathbf{D}$$

where $\mathbf{D}$ is a constant vector.

Kepler's second law may now be verified. As Figure 24 illustrates, if A is the area swept out by the vector $\mathbf{r}$ in time t, then

$$\Delta A = \text{Area swept out by } \mathbf{r} \text{ in time } \Delta t \approx \frac{1}{2} \|\mathbf{r} \times \Delta \mathbf{r}\|$$

$$\frac{\Delta A}{\Delta t} \approx \frac{1}{2} \left\| \mathbf{r} \times \frac{\Delta \mathbf{r}}{\Delta t} \right\|$$

$$\frac{dA}{dt} = \frac{1}{2} \left\| \mathbf{r} \times \frac{d\mathbf{r}}{dt} \right\| = \frac{1}{2} \|\mathbf{r} \times \mathbf{v}\| = \frac{1}{2} \|\mathbf{D}\| = \text{Constant}$$

By (15.44)

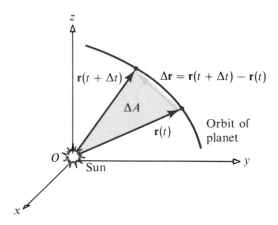

Figure 24

That is:

(15.45)   The speed of the planet is such that vectorial area is swept out at a constant rate.

To put it another way, in equal amounts of time, equal vectorial areas are swept out (see Fig. 25).

To obtain Kepler's first law, we return to equations (15.43) and (15.44). Then, using the formula for triple cross products, $\mathbf{r} \times (\mathbf{r} \times \mathbf{v}) = (\mathbf{r} \cdot \mathbf{v})\mathbf{r} - (\mathbf{r} \cdot \mathbf{r})\mathbf{v}$,   we find that

$$\frac{d\mathbf{v}}{dt} \times \mathbf{D} = \frac{-GM}{\|\mathbf{r}\|^3} \mathbf{r} \times (\mathbf{r} \times \mathbf{v}) = \frac{-GM}{\|\mathbf{r}\|^3} \left[(\mathbf{r} \cdot \mathbf{v})\mathbf{r} - (\mathbf{r} \cdot \mathbf{r})\mathbf{v} \right]$$

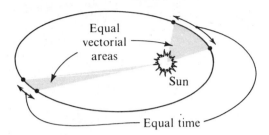

Figure 25

But

$$\frac{d}{dt}(\mathbf{r} \cdot \mathbf{r}) = \frac{d\mathbf{r}}{dt} \cdot \mathbf{r} + \mathbf{r} \cdot \frac{d\mathbf{r}}{dt} = 2\mathbf{r} \cdot \frac{d\mathbf{r}}{dt} = 2\mathbf{r} \cdot \mathbf{v}$$

and

$$\frac{d}{dt}(\mathbf{r} \cdot \mathbf{r}) = \frac{d}{dt}\|\mathbf{r}\|^2 = 2\|\mathbf{r}\|\frac{d}{dt}\|\mathbf{r}\|$$

Hence,

$$\|\mathbf{r}\|\frac{d}{dt}\|\mathbf{r}\| = \mathbf{r} \cdot \mathbf{v}$$

so that

$$\frac{d\mathbf{v}}{dt} \times \mathbf{D} = \frac{-GM}{\|\mathbf{r}\|^3}\left(\|\mathbf{r}\|\frac{d}{dt}\|\mathbf{r}\|\mathbf{r} - \|\mathbf{r}\|^2\mathbf{v}\right) = GM\left(\frac{\mathbf{v}}{\|\mathbf{r}\|} - \frac{\mathbf{r}}{\|\mathbf{r}\|^2}\frac{d}{dt}\|\mathbf{r}\|\right)$$

But

$$\frac{d}{dt}\frac{\mathbf{r}}{\|\mathbf{r}\|} = \frac{d\mathbf{r}/dt}{\|\mathbf{r}\|} - \frac{\mathbf{r}}{\|\mathbf{r}\|^2}\frac{d}{dt}\|\mathbf{r}\| = \frac{\mathbf{v}}{\|\mathbf{r}\|} - \frac{\mathbf{r}}{\|\mathbf{r}\|^2}\frac{d}{dt}\|\mathbf{r}\|$$

Hence,

$$\frac{d\mathbf{v}}{dt} \times \mathbf{D} = GM\frac{d}{dt}\frac{\mathbf{r}}{\|\mathbf{r}\|}$$

Integrating yields

(15.46)
$$\mathbf{v} \times \mathbf{D} = GM\frac{\mathbf{r}}{\|\mathbf{r}\|} + \mathbf{H}$$

where **H** is a constant vector.

If we then find the dot product of both sides of (15.46) by **r**, the result is

$$\mathbf{r} \cdot (\mathbf{v} \times \mathbf{D}) = GM\|\mathbf{r}\| + \mathbf{r} \cdot \mathbf{H}$$

But $A \cdot (B \times C) = (A \times B) \cdot C$;   hence,

$$(r \times v) \cdot D = GM\|r\| + r \cdot H$$

By (15.44), we have

$$\|D\|^2 = GM\|r\| + r \cdot H$$

If θ is the angle between r and H, it follows that

$$\|D\|^2 = GM\|r\| + \|r\| \|H\| \cos \theta$$

By solving for $\|r\|$, we obtain

(15.47) $$\|r\| = \frac{\|D\|^2}{GM + \|H\| \cos \theta} = \frac{\|D\|^2}{GM} \left(\frac{1}{1 + e \cos \theta} \right) \qquad 0 \le e < 1$$

where $e = \|H\|/GM$.   Equation (15.47) is the polar coordinate equation of an ellipse with eccentricity e. Thus, we have Kepler's first law of planetary motion:

(15.48)   **The orbit of each planet is an ellipse with the sun at a focus.**

EXERCISE 6

1. If the period of Jupiter is $5\sqrt{5}$ years, what is the distance of Jupiter from the sun? Assume that orbits are circles and that the average distance of the earth from the sun is 9.3×10^7 miles.

2. Mercury has a "year" of 88 days. How far is Mercury from the sun?

3. The mean distance of Pluto from the sun is 39.5 times that of the earth. What is the "year" of Pluto?

4. From (15.47), deduce that:

$$\text{Length of semimajor axis} = \frac{\|D\|^2}{GM} \left(\frac{1}{1 - e^2} \right)$$

$$\text{Length of semiminor axis} = \frac{\|D\|^2}{GM} \left(\frac{1}{\sqrt{1 - e^2}} \right)$$

$$\text{Area of ellipse} = \frac{\pi \|D\|^4}{G^2 M^2} \left[\frac{1}{(1 - e^2)^{3/2}} \right]$$

Miscellaneous Exercises

1. The position vector of a particle at time t is $r = (1 - t^3)i + t^2 j$.   Eliminate the parameter to find y as a function of x, and sketch the path of the particle. Indicate the direction of travel.

2. Rework Problem 1 if $r = t^2 i + (4t^2 - t^4)j$.

3. Find an equation in x and y of the graph of the vector function $r = \sin^2 t\, i + \tan t\, j$, $-\pi/2 < t < \pi/2$, and sketch the graph. Indicate the positive direction.

4. The position vector of a particle at time $t \geq 0$ is $\mathbf{r} = e^{-t} \cos t \mathbf{i} + e^{-t} \sin t \mathbf{j}$. Show that the path of the particle is part of the spiral $r = e^{-\theta}$ (in polar coordinates). Sketch the path and indicate the direction of travel.

5. In Problem 4 show that the angle between the position vector and velocity vector is always $135°$.

6. Find an equation in x and y of the graph of the vector function $\mathbf{r} = e^t \mathbf{i} + e^{-t} \mathbf{j}$, and sketch the graph. Indicate the direction of travel.

7. Suppose a particle moves along the curve $\mathbf{r}(t) = 4 \cos t \mathbf{i} - 2 \cos 2t \mathbf{j}$.
   (a) Show that the particle oscillates on an arc of a parabola.
   (b) Sketch the path.
   (c) Find the acceleration $\mathbf{a}$ at points of zero velocity.

8. Find the maximum magnitude of the force acting on a particle of mass m whose motion is along the curve $\mathbf{r}(t) = 4 \cos t \mathbf{i} - 2 \cos 2t \mathbf{j}$.

9. If $\mathbf{r}'(t) = \mathbf{b} \times \mathbf{r}(t)$ for all t, where $\mathbf{b}$ is a constant vector, show that the acceleration $\mathbf{a}(t)$ is perpendicular to $\mathbf{b}$ and that the speed is constant.

10. Show that if $\mathbf{r}'(t) \cdot \mathbf{r}(t) = 0$ for all t, then the motion $\mathbf{r}(t)$ is on the surface of a sphere centered at the origin. What can be said about a motion for which $\mathbf{r}''(t) \cdot \mathbf{r}'(t) = 0$ for all t? What can be said about a motion for which $\mathbf{r}'(t) \times \mathbf{r}(t) = \mathbf{0}$ for all t?

11. A particle moves along the path $y = 3x^2 - x^3$ with the horizontal component of the velocity identically equal to $\frac{1}{3}$. Find the acceleration at the points where the velocity $\mathbf{v}$ is horizontal. Sketch the path and indicate $\mathbf{v}$ and $\mathbf{a}$ at these points.

12. If a particle moves along the graph of $y = f(x)$, show that $a_N = 0$ at a point of inflection of the graph.

13. A particle moves along the graph of $y = \frac{1}{2}x^2$ with constant speed $v(t) = 2$. Find a_T and a_N in terms of x alone at a general point $(x, \frac{1}{2}x^2)$ on the graph.

14. Find the length of $\mathbf{r}(t) = e^t \cos t \mathbf{i} + e^t \sin t \mathbf{j} + e^t \mathbf{k}$, $0 \leq t \leq 2\pi$.

15. Find the length of $\mathbf{r}(t) = t^2 \mathbf{i} - 2\sqrt{2}t \mathbf{j} + (t^2 - 1)\mathbf{k}$, $0 \leq t \leq 1$.

16. Use Simpson's rule (9.28) with $n = 4$ to approximate the length of the curve $\mathbf{r}(t) = t^2 \mathbf{i} + t^3 \mathbf{j} + (2t + 3)\mathbf{k}$, $0 \leq t \leq 2$.

17. Find $\int (te^t \mathbf{i} + t^2 \cos t \mathbf{j} + \ln t \mathbf{k}) \, dt$.

18. Evaluate $\int_0^3 \| t^2 \mathbf{i} - 2t \mathbf{j} + 2\mathbf{k} \| \, dt$.

19. Write a vector equation for the curve $\mathbf{r}(t) = 2t \mathbf{i} + (2t - 1)\mathbf{j} + t\mathbf{k}$, $0 \leq t \leq 2$, using arc length s as a parameter. [*Hint:* For each t, $0 \leq t \leq 2$, calculate the length $s(t)$ of the curve from 0 to t.]

20. Rework Problem 19 for a coil of the helix $\mathbf{r}(t) = \cos t \mathbf{i} + \sin t \mathbf{j} + t\mathbf{k}$, $0 \leq t \leq 2\pi$.

21. For a curve $\mathbf{r} = \mathbf{r}(\theta)$, let $\mathbf{B} = \mathbf{T} \times \mathbf{N}$, be the *binormal* to the curve. Then $\mathbf{T}$, $\mathbf{N}$, and $\mathbf{B}$ form a right-hand system of unit vectors at each point of $\mathbf{r}(\theta)$, the *moving trihedral* on the curve. Calculate κ, $\mathbf{T}$, $\mathbf{N}$, and $\mathbf{B}$ for $\mathbf{r}(\theta) = \sin \theta \mathbf{i} + \cos \theta \mathbf{j} + \theta \mathbf{k}$.

22. Let $\mathbf{u}_r = \cos \theta \mathbf{i} + \sin \theta \mathbf{j}$ and $\mathbf{u}_\theta = -\sin \theta \mathbf{i} + \cos \theta \mathbf{j}$.
   (a) Show that $\mathbf{u}_r$ and $\mathbf{u}_\theta$ are unit orthogonal vectors, and that $d\mathbf{u}_r/d\theta = \mathbf{u}_\theta$ and $d\mathbf{u}_\theta/d\theta = -\mathbf{u}_r$.
   (b) Suppose that $\mathbf{r}(t)$ is the position vector of a plane curve and that the tip of $\mathbf{r}(t)$ has polar coordinates $(r(t), \theta(t))$. Show that $\mathbf{r}(t) = \|\mathbf{r}(t)\| \mathbf{u}_r = r(t)\mathbf{u}_r(t)$.
   (c) Use the chain rule to show that $\mathbf{v}(t) = \dfrac{dr}{dt} \mathbf{u}_r + r \dfrac{d\theta}{dt} \mathbf{u}_\theta = r'\mathbf{u}_r + r\theta'\mathbf{u}_\theta$.
   (d) Show that $\mathbf{a}(t) = [r'' - r(\theta')^2]\mathbf{u}_r + [r\theta'' + 2r'\theta']\mathbf{u}_\theta$.

23. Express $\mathbf{v}$ and $\mathbf{a}$ in terms of $\mathbf{u}_r$ and $\mathbf{u}_\theta$ for a motion along the polar coordinate curve $r = 2 + \cos t$, $\theta = 2t$.

24. Find polar coordinate equations for $\mathbf{r}(t) = e^{2t} \cos t\, \mathbf{i} + e^{2t} \sin t\, \mathbf{j}$ and express $\mathbf{v}$ and $\mathbf{a}$ in terms of $\mathbf{u}_r$ and $\mathbf{u}_\theta$.

25. (a) For a particle moving under the influence of a central force directed toward the origin, show that
    $$r\theta'' + 2r'\theta' = 0.$$
    (b) Multiply this equation by r and integrate to obtain $r^2\theta' = c$, c a constant.
    (c) Using the expression for area in polar coordinates, deduce Kepler's second law for *any* central force.

26. If $\mathbf{c}$ is a constant vector and $\mathbf{r}(t)$ is continuous on a closed interval $[a, b]$, prove that

$$\int_a^b [\mathbf{c} \times \mathbf{r}(t)]\, dt = \mathbf{c} \times \int_a^b \mathbf{r}(t)\, dt$$

27. Estimate, correct to five decimal places, the length of the curve $\mathbf{r}(t) = \frac{1}{3}t^3\mathbf{i} + (t - 1)\mathbf{j} + 2\mathbf{k}$, $0 \le t \le \frac{1}{2}$, by expanding the integrand in a power series.

16

Partial Derivatives

1. Preliminaries

Many important problems in physics, engineering, and other sciences require functions of more than one variable. For example, to calculate the volume V of water stored in a cylindrical tank, we need to know both the height h of the tank and its radius R. The volume $V = \pi R^2 h$ depends on both h and R, and we say that V is a *function of two variables*, R and h. In function notation, we write

$$V = f(R, h) = \pi R^2 h$$

For example, if $R = 10$ centimeters and $h = 3$ centimeters, then

$$V = f(10, 3) = \pi(10^2)(3) = 300\pi \text{ cubic centimeters}$$

(16.1) **DEFINITION** **A *real function of two variables* is a rule that assigns a unique real number to each ordered pair (x, y) of real numbers in a certain set D of the xy-plane.**

In the equation $z = f(x, y)$ we refer to z as the *dependent* variable and we refer to x and y as the *independent* variables. The set D in the definition is called the *domain* of the function f. It is the set of points in the xy-plane for which the function is defined. The *range* of f consists of all real numbers $f(x, y)$ where (x, y) is in D. Figure 1 (p. 860) illustrates one way of depicting $z = f(x, y)$. We shall have occasion to use this kind of "picture" of f later on.

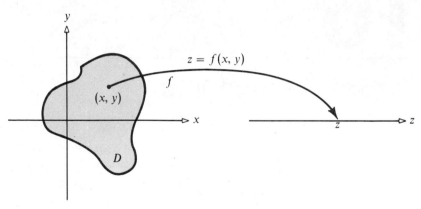

Figure 1 D is the domain of $z = f(x, y)$ Value of f

EXAMPLE 1   Let $f(x, y) = \sqrt{x} + x\sqrt{y}$.   Find:
(a) $f(0, 0)$ (b) $f(1, 4)$ (c) $f(a^2, 9b^2)$, $a > 0$, $b > 0$
(d) $f(x + \Delta x, y)$ (e) $f(x, y + \Delta y)$

Solution   (a) $f(0, 0) = \sqrt{0} + 0\sqrt{0} = 0$ (b) $f(1, 4) = \sqrt{1} + 1\sqrt{4} = 1 + 2 = 3$
(c) $f(a^2, 9b^2) = \sqrt{a^2} + a^2\sqrt{9b^2} = a + 3a^2b$
(d) $f(x + \Delta x, y) = \sqrt{x + \Delta x} + (x + \Delta x)\sqrt{y}$
(e) $f(x, y + \Delta y) = \sqrt{x} + x\sqrt{y + \Delta y}$ ∎

The function f defined by $w = f(x, y, z)$ is a function of the three independent
variables x, y, and z. For each ordered triple (x, y, z) in the domain, the rule f assigns
a value of w, the dependent variable. In this case, the domain is a collection of points
in three-space. Figure 2 illustrates a way of depicting $w = f(x, y, z)$.

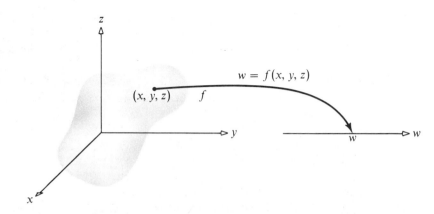

Figure 2   Domain of $w = f(x, y, z)$ Value of f

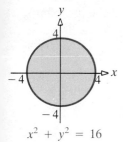

$x^2 + y^2 = 16$

Figure 3
Domain of $f(x, y)$

As with functions of a single variable, a function of several variables is usually given by a formula, and, unless otherwise stated, the domain is taken to be the largest set of points (in the plane or in space) for which this rule or formula makes sense. For example, the domain of the function f defined by $z = f(x, y) = \sqrt{16 - x^2 - y^2}$ consists of all points in the plane for which $16 - x^2 - y^2 \geq 0$.   That is, the domain consists of all points inside and on the circle $x^2 + y^2 = 16$.   See the shaded portion of Figure 3.

Similarly, the domain of the function

$$w = f(x, y, z) = \frac{1}{x^2 + y^2 + z^2}$$

consists of all points in space, except $(0, 0, 0)$.

In practical problems the domain often is determined by physical considerations. Table 1 lists a few examples.

Table 1

Description	Formula	Domain
Volume V of a cylindrical tank of height h and radius R	$V(R, h) = \pi R^2 h$	$R > 0, \quad h > 0$
Area of a rectangle with sides x and y	$A(x, y) = xy$	$x > 0, \quad y > 0$
Surface area A of a closed box with sides of length x, y, and z	$A(x, y, z) = 2xy + 2yz + 2xz$	$x > 0, \quad y > 0, \quad z > 0$
Volume of a rectangular parallelepiped with sides x, y, and z	$V(x, y, z) = xyz$	$x > 0, \quad y > 0, \quad z > 0$
Magnitude of force of attraction between two bodies, one located at the origin and the other at (x, y, z), of masses m and M; G is a constant	$F(x, y, z) = \dfrac{GmM}{x^2 + y^2 + z^2}$	$(x, y, z) \neq (0, 0, 0)$

EXAMPLE 2   Find the domain of the function $f(x, y) = \ln(y^2 - 4x)$ and graph the domain.

Solution   The logarithm function is defined only for positive numbers. Therefore, the domain of $f(x, y) = \ln(y^2 - 4x)$ is limited to those points for which $y^2 - 4x > 0$ or $y^2 > 4x$.   We graph this set of points by first graphing the parabola $y^2 = 4x$.   We indicate that points on the parabola itself are not part of the domain by using a dashed curve to graph the parabola. The parabola $y^2 = 4x$ divides the plane into two sets of points: those for which $y^2 < 4x$ and those for which $y^2 > 4x$. To determine which points are "inside" and which are "outside" the parabola, we merely choose a point not on the curve $y^2 = 4x$ and test it. For example, the point $(2, 0)$ belongs to the set $y^2 < 4x$ since $0^2 < (4)(2)$. The set of points for which $y^2 > 4x$, the domain of f, is shaded in Figure 4.

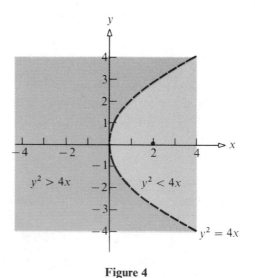

Figure 4   ■

EXAMPLE 3   Find the domain of the function $w = f(x, y, z) = \sqrt{x^2 + y^2 + z^2 - 1}$.

Solution   Since square roots of negative numbers are not permitted, the domain of this function consists of all points for which $x^2 + y^2 + z^2 - 1 \geq 0$. Therefore, the domain consists of all points outside and on the unit sphere (see Figure 5).

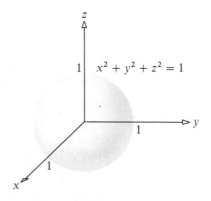

Figure 5   ■

EXERCISE 1

1. Let $f(x, y) = 3x + 2y + xy$.   Find:
   (a) $f(1, 0)$ (b) $f(0, 1)$ (c) $f(2, 1)$ (d) $f(x + \Delta x, y)$ (e) $f(x, y + \Delta y)$

2. Let $f(x, y) = x^2 y + x + 1$.   Find:
   (a) $f(0, 0)$ (b) $f(0, 1)$ (c) $f(2, 1)$ (d) $f(x + \Delta x, y)$ (e) $f(x, y + \Delta y)$

3. Let $f(x, y) = \sqrt{xy} + x$.   Find:
   (a) $f(0, 0)$ (b) $f(0, 1)$ (c) $f(a^2, t^2)$, $a > 0$, $t > 0$ (d) $f(x + \Delta x, y)$ (e) $f(x, y + \Delta y)$

4. Let $f(x, y) = e^{x+y}$.   Find:
   (a) $f(0, 0)$ (b) $f(1, -1)$ (c) $f(x + \Delta x, y)$ (d) $f(x, y + \Delta y)$

5. Let $f(x, y, z) = \dfrac{3xy + z}{x^2 + y^2 - z^2}$.   Find:

   (a) $f(1, 1, 1)$ (b) $f(0, 0, 1)$ (c) $f(0, 1, 0)$ (d) $f(\sin t, \cos t, 0)$

6. Let $f(x, y) = x^2 + xy + y^2$, $x(t) = t$, $y(t) = t^2$.   Find:
   (a) $f(x(0), y(0))$ (b) $f(x(1), y(1))$ (c) $f(x(2), y(2))$

7. Let $F(x, y) = e^{x^2+y^2}$, $g(x) = x^2$, $h(y) = 2y - 1$.   Find $F(g(x), h(y))$.

8. Let $g(x, y) = x \cos(xy)$, $u(x, y) = x^2 y$, $v(x, y) = y^2 x$.   Find $g(u(x, y), v(x, y))$.

In Problems 9–20 find the domain of each function.

9. $z = f(x, y) = \dfrac{\sqrt{x}}{\sqrt{y}}$

10. $z = f(x, y) = \sqrt{x}\sqrt{y}$

11. $z = f(x, y) = \sqrt{xy}$

12. $z = f(x, y) = \dfrac{xy}{x^2 + y^2}$

13. $z = f(x, y) = e^x \sin y$

14. $z = f(x, y) = \dfrac{\ln x}{\ln y}$

15. $z = f(x, y) = \dfrac{x}{\sqrt{x^2 + y^2 - 4}}$

16. $z = f(x, y) = \dfrac{y}{\sqrt{9 - x^2 - y^2}}$

17. $w = f(x, y, z) = \dfrac{x^2 + y^2}{z^2}$

18. $w = f(x, y, z) = e^z \ln(x^2 + y^2)$

19. $w = f(x, y, z) = \dfrac{z \sin x}{\cos y}$

20. $w = f(x, y, z) = \dfrac{xyz}{\sqrt{x^2 + y^2 + z^2}}$

21. Write the equation for the surface area of an open box as a function of its length x, width y, and depth z.

22. The cost of the bottom and top of a cylindrical tank is \$300 per square meter and the cost of the sides is \$500 per square meter. Write the total cost of constructing such a tank as a function of the radius R and height h (both in meters).

23. Write the total cost of constructing an open rectangular box if the cost per square centimeter of the material to be used for the bottom is \$4, for two of the opposite sides is \$3, and for the remaining pair of opposite sides is \$2.

2. Graphing Functions of Two or More Variables

The *graph* of a function $z = f(x, y)$ of two variables consists of all points (x, y, z) for which $z = f(x, y)$ and (x, y) is in the domain of f.

EXAMPLE 1   Describe the graph of the function $z = f(x, y) = 1 - x - y$.

Solution   In Chapter 14 we showed that the equation $z = 1 - x - y$,   or $x + y + z = 1$,   is the equation of a plane whose intercepts are the points $(1, 0, 0)$, $(0, 1, 0)$, and $(0, 0, 1)$. See Figure 6.

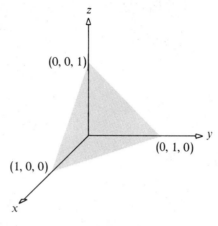

Figure 6   ■

EXAMPLE 2   Describe the graph of the function $z = f(x, y) = x^2 + 4y^2$.

Solution   The equation $z = x^2 + 4y^2$ is the equation of an elliptic paraboloid (a quadric surface*) whose vertex is at the origin (see Fig. 7).

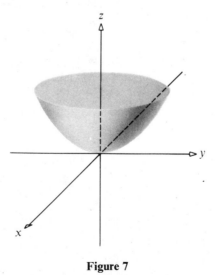

Figure 7   ■

---

* These surfaces are discussed in detail in Section 9, Chapter 14.

EXAMPLE 3   Describe the graph of the function $z = f(x, y) = \sqrt{x^2 + y^2}$.

Solution   The equation $z = \sqrt{x^2 + y^2}$,   or equivalently, $z^2 = x^2 + y^2$, $z \geq 0$,   is the equation of part of a circular cone whose vertex is at the origin. Since z must be nonnegative, the graph of the function consists only of the upper portion of the cone (see Fig. 8).   ■

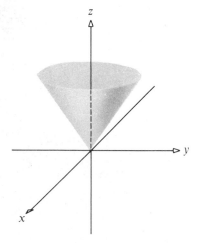

Figure 8

Ordinarily, the graph of a function of two variables is difficult to draw. In practice, such as in *topography* (map-making), the idea of the graph of $z = f(x, y)$ is conveyed by drawing properly labeled curves on which z is fixed. Such curves are of two types: level curves and contour lines. A *level curve* is a curve in the xy-plane for which $f(x, y) = c$,   where c is a number in the range of f (see Fig. 9). A *contour line* is the curve resulting from the intersection of the graph of f with a plane parallel to the xy-plane. The corresponding level curve is the projection of the contour line on the xy-plane. If several level curves of a surface are drawn, the surface

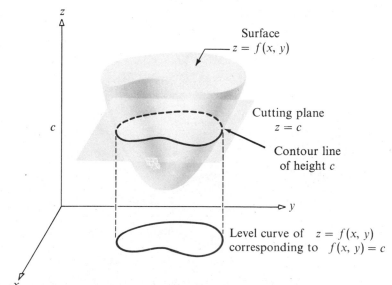

Surface
$z = f(x, y)$

Cutting plane
$z = c$

Contour line
of height c

Level curve of $z = f(x, y)$
corresponding to $f(x, y) = c$

Figure 9

Figure 10 Menan Buttes, Idaho. Contour intervals are shown in color for each 100 foot change in altitude. The crater on the left is about 100 feet deep, and the crater on the right is about 150 feet deep. (Courtesy U.S. Geological Survey.)

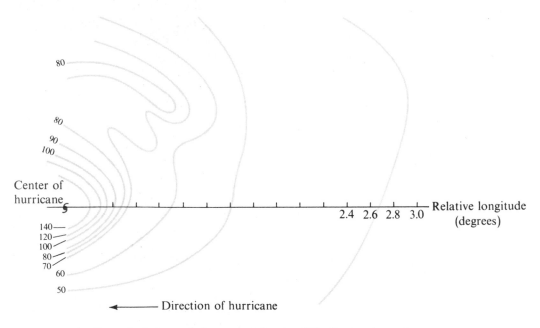

Isolines of wind speeds in rear quadrants of Pacific hurricane Ava

Figure 11 Pacific hurricane Ava, June 6, 1973. (Courtesy U.S. Department of Commerce, National Oceanic and Atmospheric Administration.)

may be visualized by viewing them together, in much the same way a movie film results from a collection of still pictures.

For example, to represent a hilly terrain, a topographer draws level curves corresponding to contour lines for various heights at standard intervals. See Figure 10.

Level curves are also used by meteorologists to indicate points at which barometric pressure is fixed (*isobars*) and to illustrate places at which wind speed remains constant (*isolines*).* See Figure 11.

Figure 12 illustrates the use of level curves to show the annual average summer rainfall (in inches) in the Chicago area. This diagram was once used to argue that cities caused increased rainfall.

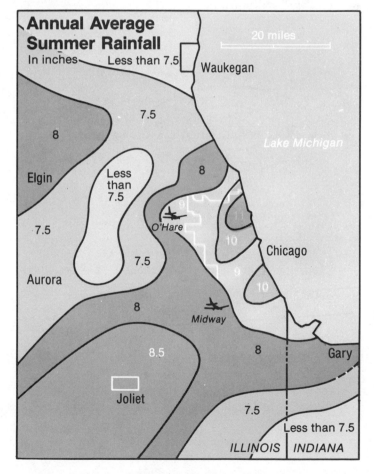

Figure 12   Chicago rainfall. (Adapted from *Chicago Tribune* graphic by Terry Volpp. Copyright 1980. Used with permission.)

---

* The prefix *iso* is Greek, meaning "same."

We shall use the technique of level curves to try to visualize the graphs of functions of two variables.

EXAMPLE 4   Sketch the level curves of the function $z = f(x, y) = x^2 + 4y^2$.

Solution   Since $z \geq 0$,   the level curves of f obey

$$x^2 + 4y^2 = c \qquad c \geq 0$$

For $c = 0$,   this is a point—the origin. For $c >$ 0,   the level curves are concentric ellipses. Figure 13 shows the level curves for $c = 0$, $c = 1$, $c = 4$,   and $c = 16$. A computer graph of $z = x^2 + 4y^2$ is shown in Figure 14.

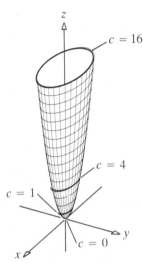

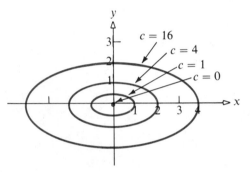

Level curves are concentric
ellipses: $x^2 + 4y^2 = c$

Figure 13

Elliptic paraboloid:
$$z = x^2 + 4y^2$$

Figure 14   ∎

EXAMPLE 5   Sketch the level curves of the function $z = f(x, y) = e^{x^2 + y^2}$.

Solution   Since $z \geq 1$,   the level curves obey

$$e^{x^2 + y^2} = c \qquad \text{or} \qquad x^2 + y^2 = \ln c$$
$$c \geq 1$$

The level curves are concentric circles if $c > 1$. For $c = 1$, the level curve reduces to a single point at the origin. Figure 15 illustrates the level curves of f.

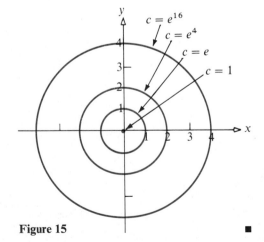

Figure 15   ∎

LEVEL SURFACES

The graph of a function $w = f(x, y, z)$ of three variables consists of all points (x, y, z, w) for which $w = f(x, y, z)$ and (x, y, z) is in the domain of f. We cannot draw the graph of such a function since it requires four dimensions. However, we may try to visualize such graphs by examining *level surfaces*—that is, the surfaces obtained by setting w equal to a constant. Although level surfaces are usually difficult to sketch, they do provide useful information. We shall limit ourselves to some simple examples.

EXAMPLE 6   Sketch the level surfaces of the function $w = f(x, y, z) = x^2 + y^2 + z^2$.

Solution   Since $w \geq 0$, the level surfaces obey

$$x^2 + y^2 + z^2 = c \qquad c \geq 0$$

These consist of concentric spheres if $c > 0$ and the origin if $c = 0$ (Fig. 16).
∎

EXAMPLE 7   Sketch the level surfaces of the function $w = f(x, y, z) = 2x + 3y + z$.

Solution   The level surfaces obey

$$2x + 3y + z = c$$

This is a collection of parallel planes, each having the vector $\mathbf{N} = 2\mathbf{i} + 3\mathbf{j} + \mathbf{k}$ as normal (see Fig. 17).   ∎

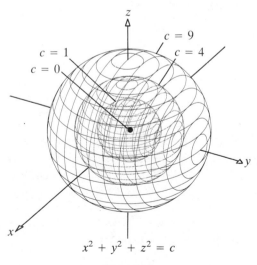

$$x^2 + y^2 + z^2 = c$$

Figure 16

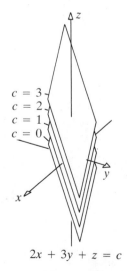

$$2x + 3y + z = c$$

Figure 17

EXERCISE 2

For the functions given in Problems 1–8, sketch the level curves corresponding to the given values of c.

1. $z = f(x, y) = x^2 - y^2$; $c = 0, 1, 4, 9$

2. $z = f(x, y) = 2x^2 + y^2$; $c = 0, 1, 4, 9$

3. $z = f(x, y) = \sqrt{1 - x^2 - y^2}$; $c = 0, \frac{1}{4}, \frac{1}{9}$

4. $z = f(x, y) = \sqrt{x^2 + y^2 - 4}$; $c = 0, 1, 4, 9$

5. $z = f(x, y) = x^2 - 2y$; $c = 0, 1, 4, 9$

6. $z = f(x, y) = y^2 - x$; $c = 0, 1, 4$

7. $z = f(x, y) = x + \sin y$; $c = 0, 2, 4, 8$

8. $z = f(x, y) = y - \ln x$; $c = 1, 2, 4$

In Problems 9–14 describe the level surfaces associated with f.

9. $f(x, y, z) = x^2 + y^2 + z^2$

10. $f(x, y, z) = x + y - z$

11. $f(x, y, z) = z - 2x - 2y$

12. $f(x, y, z) = 4 - x^2 - y^2$

13. $f(x, y, z) = x^2 + y^2$

14. $f(x, y, z) = z$

15. The formula

$$V(x, y) = \frac{9}{\sqrt{4 - (x^2 + y^2)}}$$

gives the electrical potential V (in volts) at a point (x, y) in the xy-plane. Draw the equipotential curves (level curves) for $V = 36, 18, 9$.   Describe the surface $z = V(x, y)$.

16. The temperature T (in °C) at any point (x, y) of a flat plate situated in the xy-plane is $T = 60 - 2x^2 - 3y^2$. Draw the isothermal curves (level curves) for $T = 60, 54, 48, 6, 0$.   Describe the surface $z = T(x, y)$.

17. The strength E of an electric field at a point (x, y, z) due to an infinitely long charged wire lying along the x-axis is

$$E(x, y, z) = \frac{3}{\sqrt{y^2 + z^2}}$$

Describe the level surfaces of E.

18. The magnitude F of the force of attraction between two bodies, one located at the origin and the other at the point $(x, y, z) \neq (0, 0, 0)$,   of masses m and M is given by

$$F = \frac{GmM}{x^2 + y^2 + z^2}$$

where $G = 6.67 \times 10^{-11}$ in SI units   is a positive constant. Describe the level surfaces of F.

3. Partial Derivatives

When a function is expressed in terms of several variables rather than in terms of only one variable, the concept of a *partial derivative* is usually applicable. If, for example, z is a function of x and y—that is, if $z = f(x, y)$—then the function f_x *is the derivative of f with respect to x, with y treated as a constant*, and the function f_y *is the derivative of f with respect to y, with x treated as a constant*.

Suppose $z = f(x, y) = x^2 - 2xy + e^{x^3 + y^3}$.   By differentiating with respect to x, with y treated as a constant, we obtain the partial derivative of f with respect to x at (x, y), namely,

$$f_x(x, y) = 2x - 2y + e^{x^3 + y^3}(3x^2)$$

Similarly, the partial derivative of f with respect to y at (x, y) is found by treating x as a constant and differentiating with respect to y:

$$f_y(x, y) = -2x + e^{x^3 + y^3}(3y^2)$$

We define these partial derivatives as the limits of certain difference quotients.

(16.2)   **DEFINITION**   **Let f be a function of two variables. Then the *partial derivatives of f with respect to x and y* are functions f_x and f_y defined as follows:**

$$f_x(x, y) = \lim_{\Delta x \to 0} \frac{f(x + \Delta x, y) - f(x, y)}{\Delta x}$$

$$f_y(x, y) = \lim_{\Delta y \to 0} \frac{f(x, y + \Delta y) - f(x, y)}{\Delta y}$$

provided these limits exist.

Observe the similarity between the above definitions and the definition of a derivative given in Chapter 3. Observe also that in $f_x(x, y)$, an increment Δx is given to x, while y is fixed; in $f_y(x, y)$, an increment Δy is given to y, while x is fixed.

The next example illustrates how (16.2) is used to find partial derivatives.

EXAMPLE 1   Use (16.2) to find $f_x(1, 2)$, $f_x(x, y)$, and $f_y(x, y)$ for $z = f(x, y) = x^2y + 3xy^2$.

Solution

$$f_x(1, 2) = \lim_{\Delta x \to 0} \frac{f(1 + \Delta x, 2) - f(1, 2)}{\Delta x}$$

$$= \lim_{\Delta x \to 0} \frac{(1 + \Delta x)^2(2) + 3(1 + \Delta x)(4) - [1^2(2) + (3)(1)(4)]}{\Delta x}$$

$$= \lim_{\Delta x \to 0} \frac{2 + 4\Delta x + 2(\Delta x)^2 + 12 + 12\Delta x - 14}{\Delta x}$$

$$= \lim_{\Delta x \to 0} \frac{2(\Delta x)^2 + 16\Delta x}{\Delta x} = 16$$

$$f_x(x, y) = \lim_{\Delta x \to 0} \frac{f(x + \Delta x, y) - f(x, y)}{\Delta x}$$

$$= \lim_{\Delta x \to 0} \frac{(x + \Delta x)^2(y) + 3(x + \Delta x)y^2 - (x^2y + 3xy^2)}{\Delta x}$$

$$= \lim_{\Delta x \to 0} \frac{x^2y + 2x(\Delta x)y + (\Delta x)^2 y + 3xy^2 + 3y^2\Delta x - x^2y - 3xy^2}{\Delta x}$$

$$= \lim_{\Delta x \to 0} \frac{\Delta x(2xy + y\Delta x + 3y^2)}{\Delta x} = 2xy + 3y^2$$

$$f_y(x, y) = \lim_{\Delta y \to 0} \frac{f(x, y + \Delta y) - f(x, y)}{\Delta y}$$

$$= \lim_{\Delta y \to 0} \frac{x^2(y + \Delta y) + 3x(y + \Delta y)^2 - (x^2 y + 3xy^2)}{\Delta y}$$

$$= \lim_{\Delta y \to 0} \frac{x^2 y + x^2 \Delta y + 3xy^2 + 6xy\Delta y + 3x(\Delta y)^2 - x^2 y - 3xy^2}{\Delta y}$$

$$= \lim_{\Delta y \to 0} \frac{\Delta y(x^2 + 6xy + 3x\Delta y)}{\Delta y} = x^2 + 6xy \quad \blacksquare$$

The usual rules for finding derivatives may be used to find partial derivatives—just remember that to find $f_x(x, y)$, y is treated as a constant while differentiating f with respect to x; and to find $f_y(x, y)$, x is treated as a constant while differentiating f with respect to y.

EXAMPLE 2   Find $f_x(x, y)$ and $f_y(x, y)$ for:
(a) $f(x, y) = 3x^2 y + 2x - 3y$ (b) $f(x, y) = x \sin y + y \sin x$
(c) $f(x, y) = e^{y/x}$

Solution   (a) $f_x(x, y) = 6xy + 2$; $f_y(x, y) = 3x^2 - 3$
(b) $f_x(x, y) = \sin y + y \cos x$; $f_y(x, y) = x \cos y + \sin x$

(c) $f_x(x, y) = \left(\frac{-y}{x^2}\right) e^{y/x}$; $f_y(x, y) = \frac{1}{x} e^{y/x}$ $\blacksquare$

ANOTHER NOTATION

There is another notation used for the partial derivatives $f_x(x, y)$ and $f_y(x, y)$ of a function $z = f(x, y)$, which we introduce here:

$$f_x(x, y) = \frac{\partial f}{\partial x} = \frac{\partial z}{\partial x} \qquad f_y(x, y) = \frac{\partial f}{\partial y} = \frac{\partial z}{\partial y}$$

The symbols $\partial/\partial x$ and $\partial/\partial y$ denote operations on a function to obtain the partial derivatives with respect to x in the case of $\partial/\partial x$ and with respect to y in the case of $\partial/\partial y$. For example,

$$\frac{\partial}{\partial x}(e^x \cos y) = e^x \cos y \qquad \frac{\partial}{\partial y}(e^x \cos y) = -e^x \sin y$$

GEOMETRIC INTERPRETATION

For a geometric interpretation of the partial derivatives of $z = f(x, y)$, look at the graph of the surface $z = f(x, y)$ shown in Figure 18. In computing $f_x(x, y)$, we hold y fixed, say, at $y = y_0$, and then differentiate with respect to x. But holding y fixed at y_0 is equivalent to intersecting the surface $z = f(x, y)$ with the plane $y = y_0$,

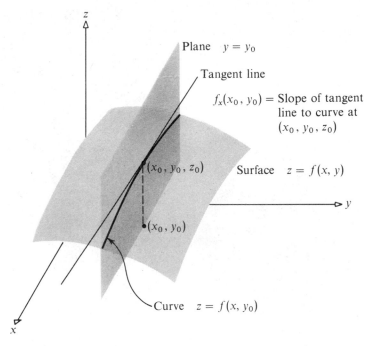

Figure 18

and the result is the curve $z = f(x, y_0)$. Thus, the partial derivative f_x is the slope of the tangent line to this curve. In particular:

The partial derivative $f_x(x_0, y_0)$ equals the slope of the tangent line to the curve of intersection of the surface $z = f(x, y)$ and the plane $y = y_0$ at the point (x_0, y_0, z_0) on the surface.

The partial derivative $f_y(x_0, y_0)$ equals the slope of the tangent line to the curve of intersection of the surface $z = f(x, y)$ and the plane $x = x_0$ at the point (x_0, y_0, z_0) on the surface.

See Figure 19 on p. 874.

EXAMPLE 3   Find an equation of the tangent line to the curve of intersection of the surface $z = f(x, y) = 16 - x^2 - y^2$:
(a) With the plane $y = 2$ at the point $(1, 2, 11)$
(b) With the plane $x = 1$ at the point $(1, 2, 11)$

Solution   (a) The slope of the tangent line to the curve of intersection of $z = 16 - x^2 - y^2$ and the plane $y = 2$ at any point is $f_x(x, y) = -2x$. At the point $(1, 2, 11)$, the slope is -2. An equation of this tangent line is

$$z - 11 = -2(x - 1) \qquad y = 2$$

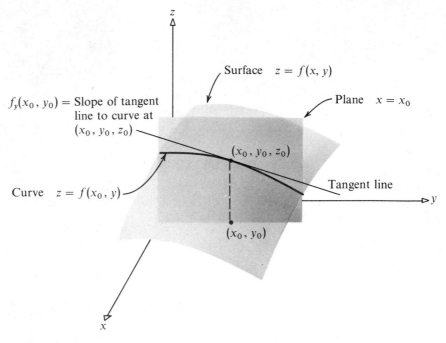

$f_y(x_0, y_0)$ = Slope of tangent line to curve at (x_0, y_0, z_0)

Surface $z = f(x, y)$

Plane $x = x_0$

(x_0, y_0, z_0)

Tangent line

Curve $z = f(x_0, y)$

(x_0, y_0)

Figure 19

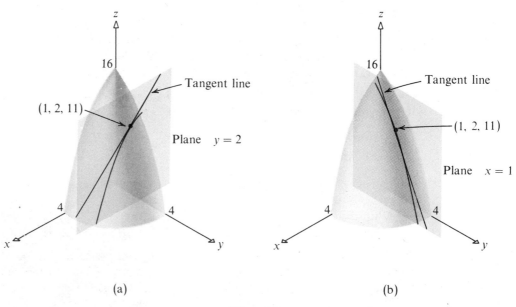

16

Tangent line

$(1, 2, 11)$

Plane $y = 2$

4

4

(a)

16

Tangent line

$(1, 2, 11)$

Plane $x = 1$

4

4

(b)

Figure 20

(b) The slope of the tangent line to the curve of intersection of $z = 16 - x^2 - y^2$
and the plane $x = 1$ at any point is $f_y(x, y) = -2y$. At the point $(1, 2, 11)$,
the slope is -4. An equation of this tangent line is

$$z - 11 = -4(y - 2) \qquad x = 1$$

Figure 20 illustrates these solutions.   ∎

RATE OF CHANGE

As we noted above, the definition of the partial derivatives of $z = f(x, y)$ is a
generalization of the definition of the derivative of a function of one variable. It is
not surprising, therefore, to see similarities in the interpretations. For example, $f_x(x, y)$
equals the rate of change of f in a direction parallel to the x-axis (while y is held
fixed), as shown in Figure 21.

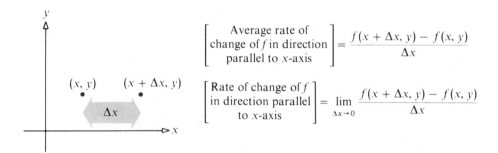

$$\left[\begin{array}{c} \text{Average rate of} \\ \text{change of } f \text{ in direction} \\ \text{parallel to } x\text{-axis} \end{array} \right] = \frac{f(x + \Delta x, y) - f(x, y)}{\Delta x}$$

$$\left[\begin{array}{c} \text{Rate of change of } f \\ \text{in direction parallel} \\ \text{to } x\text{-axis} \end{array} \right] = \lim_{\Delta x \to 0} \frac{f(x + \Delta x, y) - f(x, y)}{\Delta x}$$

Figure 21

Similarly, $f_y(x, y)$ equals the rate of change of f in a direction parallel to the
y-axis (x is held fixed), as shown in Figure 22.

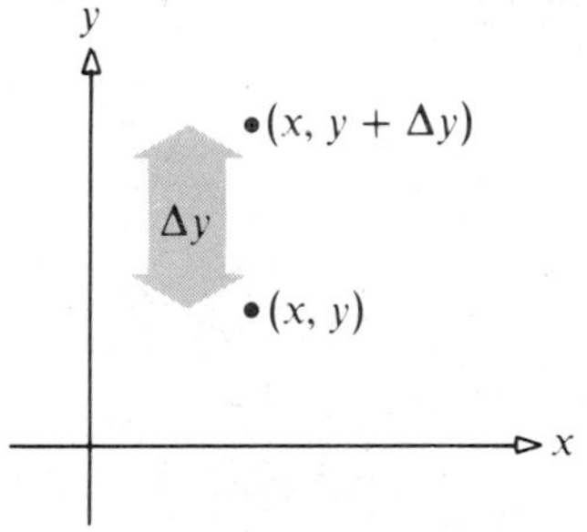

$$\left[\begin{array}{c} \text{Average rate of} \\ \text{change of } f \text{ in direction} \\ \text{parallel to } y\text{-axis} \end{array} \right] = \frac{f(x, y + \Delta y) - f(x, y)}{\Delta y}$$

$$\left[\begin{array}{c} \text{Rate of change of } f \\ \text{in direction parallel} \\ \text{to } y\text{-axis} \end{array} \right] = \lim_{\Delta y \to 0} \frac{f(x, y + \Delta y) - f(x, y)}{\Delta y}$$

Figure 22

EXAMPLE 4   A particle moves along the path obtained from the intersection of the sphere
$x^2 + y^2 + z^2 = 14$ and the plane $y = 1$. At what rate is z changing with respect
to x when the particle is at $P = (2, 1, 3)$?

Solution   The path of intersection of the sphere and the plane $y = 1$ is a circle (see Fig. 23). The point P lies on the upper hemisphere since the z-coordinate is positive. Therefore,

$$z = f(x, y) = \sqrt{14 - x^2 - y^2}$$

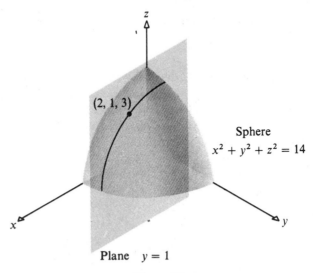

Plane $y = 1$

Figure 23

The rate at which z changes with respect to x is $f_x(x, y)$. From the above equation, we obtain

$$f_x(x, y) = \frac{\partial}{\partial x} (14 - x^2 - y^2)^{1/2} = \frac{1}{2}(14 - x^2 - y^2)^{-1/2}(-2x) = \frac{-x}{\sqrt{14 - x^2 - y^2}}$$

At the point $(2, 1, 3)$, the rate at which z is changing with respect to x is

$$f_x(2, 1) = \frac{-2}{\sqrt{14 - 4 - 1}} = -\frac{2}{3} \quad \blacksquare$$

EXAMPLE 5   The temperature (in °C) of a metal plate, located in the xy-plane, at any point (x, y) is given by the formula $T = 24(x^2 + y^2)^2$. Find the rate of change of T in a direction parallel to the x-axis at the point $(1, -2)$.

Solution   The rate of change of temperature in a direction parallel to the x-axis is given by

$$T_x(x, y) = 2(24)(x^2 + y^2)(2x)$$

At $(1, -2)$, we find $T_x(1, -2) = 96(5) = 480°C$.

    We interpret $T_x(1, -2) = 480°C$ to mean that as one moves in a horizontal direction (y fixed) away from the point $(1, -2)$, the temperature of the plate is increasing at the rate of 480°C per unit of distance moved.

HIGHER-ORDER PARTIAL DERIVATIVES

For a function $z = f(x, y)$ of two variables for which the limits (16.2) exist, there are two *first-order partial derivatives*: $f_x(x, y)$ and $f_y(x, y)$. If it is possible to differentiate each of these partially with respect to x or y, there will result four *second-order partial derivatives*, namely,

$$f_{xx}(x, y) = \frac{\partial}{\partial x} f_x(x, y) = \frac{\partial}{\partial x} \frac{\partial z}{\partial x} = \frac{\partial^2 z}{\partial x^2} \qquad f_{xy}(x, y) = \frac{\partial}{\partial y} f_x(x, y) = \frac{\partial}{\partial y} \frac{\partial z}{\partial x} = \frac{\partial^2 z}{\partial y\, \partial x}$$

$$f_{yx}(x, y) = \frac{\partial}{\partial x} f_y(x, y) = \frac{\partial}{\partial x} \frac{\partial z}{\partial y} = \frac{\partial^2 z}{\partial x\, \partial y} \qquad f_{yy}(x, y) = \frac{\partial}{\partial y} f_y(x, y) = \frac{\partial}{\partial y} \frac{\partial z}{\partial y} = \frac{\partial^2 z}{\partial y^2}$$

The two second-order partial derivatives

$$\frac{\partial^2 z}{\partial x\, \partial y} = f_{yx}(x, y) \qquad \text{and} \qquad \frac{\partial^2 z}{\partial y\, \partial x} = f_{xy}(x, y)$$

are called *mixed partials*. *Be careful!* Observe the differences in these two equations. The notation f_{yx} means that first we should differentiate f partially with respect to y and then differentiate the result partially with respect to x—in that order! On the other hand, f_{xy} means we should differentiate with respect to x and then with respect to y.

EXAMPLE 6   Find all second-order partial derivatives of $z = f(x, y) = x \ln y + ye^x$.

Solution
$$f_x = \ln y + ye^x \qquad f_y = \frac{x}{y} + e^x$$

Therefore,

$$f_{xx} = \frac{\partial}{\partial x}(f_x) = \frac{\partial}{\partial x}(\ln y + ye^x) = ye^x \qquad f_{xy} = \frac{\partial}{\partial y}(f_x) = \frac{\partial}{\partial y}(\ln y + ye^x) = \frac{1}{y} + e^x$$

$$f_{yx} = \frac{\partial}{\partial x}(f_y) = \frac{\partial}{\partial x}\left(\frac{x}{y} + e^x\right) = \frac{1}{y} + e^x \qquad f_{yy} = \frac{\partial}{\partial y}(f_y) = \frac{\partial}{\partial y}\left(\frac{x}{y} + e^x\right) = \frac{-x}{y^2} \qquad \blacksquare$$

Note in the above example that $f_{xy} = f_{yx}$ for all (x, y). As it turns out, this will be the case for most functions we encounter. The conditions under which the equality of the mixed partials holds is given in the next section.

FUNCTIONS OF THREE VARIABLES

The idea of partial differentiation may be extended to a function of three variables. That is, if $w = f(x, y, z)$ is a function of three variables, there will be three partial derivatives: the partial derivative with respect to x is f_x; the partial derivative with respect to y is f_y; and the partial derivative with respect to z is f_z. Each of these is calculated by differentiating with respect to the indicated variable, while holding the other two fixed. The function f_x equals the rate of change of w with respect to x (y and z are fixed); the function f_y equals the rate of change of w with respect to

y (x and z are fixed); and the function f_z equals the rate of change of w with respect to z (x and y are fixed). For example, if $w = f(x, y, z) = 5x^2yz^3$, then

$$f_x = 10xyz^3 \qquad f_y = 5x^2z^3 \qquad f_z = 15x^2yz^2$$

We may also write the partial derivatives of $w = f(x, y, z)$ as

$$f_x = \frac{\partial w}{\partial x} \qquad f_y = \frac{\partial w}{\partial y} \qquad f_z = \frac{\partial w}{\partial z}$$

EXAMPLE 7   The area of a triangle is given by $A = \frac{1}{2}ab \sin \gamma$. See Figure 24.

(a)   Find the rate of change of A with respect to a when b and γ are held fixed.
(b)   Find the rate of change of A with respect to γ when a and b are held fixed.
(c)   Find the rate of change of b with respect to a when A and γ are held fixed.
(d)   Calculate each of the above if $a = 20$, $b = 30$, and $\gamma = \pi/6$.

Solution   (a) $\dfrac{\partial A}{\partial a} = \dfrac{1}{2} b \sin \gamma$ (b) $\dfrac{\partial A}{\partial \gamma} = \dfrac{1}{2} ab \cos \gamma$

(c) $b = \dfrac{2A}{a \sin \gamma}; \quad \dfrac{\partial b}{\partial a} = -\dfrac{2A}{a^2 \sin \gamma} = -\dfrac{2(\frac{1}{2}ab \sin \gamma)}{a^2 \sin \gamma} = -\dfrac{b}{a}$

Figure 24

(d) $\dfrac{\partial A}{\partial a} = \dfrac{1}{2}(30)\left(\sin \dfrac{\pi}{6}\right) = \dfrac{15}{2}; \quad \dfrac{\partial A}{\partial \gamma} = \dfrac{1}{2}(20)(30)\left(\cos \dfrac{\pi}{6}\right) = 150\sqrt{3}; \quad \dfrac{\partial b}{\partial a} = -\dfrac{3}{2}$ ∎

Higher-order partial derivatives of $w = f(x, y, z)$ are defined in the usual way. For example,

$$f_{zy} = \frac{\partial}{\partial y}\left(\frac{\partial w}{\partial z}\right) = \frac{\partial^2 w}{\partial y\, \partial z} \qquad f_{xz} = \frac{\partial}{\partial z}\left(\frac{\partial w}{\partial x}\right) = \frac{\partial^2 w}{\partial z\, \partial x}$$

As in the case of functions of two variables, for most functions $w = f(x, y, z)$, the mixed partials have the property that $f_{xy} = f_{yx}$, $f_{xz} = f_{zx}$, and $f_{yz} = f_{zy}$.

EXERCISE 3

In Problems 1–14 find f_x and f_y.

1. $f(x, y) = x^2y + 6y^2$ 2. $f(x, y) = 3x^2 + 6xy^3$ 3. $f(x, y) = x^3/y^3$

4. $f(x, y) = \dfrac{x + y}{y}$ 5. $f(x, y) = e^y \cos x + e^x \sin y$ 6. $f(x, y) = x^2 \cos y + y^2 \sin x$

7. $f(x, y) = e^{2x + 3y}$ 8. $f(x, y) = \cos(x^2y^3)$ 9. $f(x, y) = \ln\sqrt{x^2 + y^2}$

10. $f(x, y) = \tan^{-1}(y/x)$ 11. $f(x, y) = \sin^2(2xy)$ 12. $f(x, y) = e^{(x^2 + y^2)^{1/2}}$

13. $f(x, y) = x^y, \quad x > 0$ 14. $f(x, y) = \sin[\ln(x^2 + y^2)]$

In Problems 15–22 compute f_{xx}, f_{xy}, f_{yx}, and f_{yy}. Check to verify that $f_{xy} = f_{yx}$.

15. $f(x, y) = 6x^2 - 8xy + 9y^2$ 16. $f(x, y) = x^3/y^3$ 17. $f(x, y) = \ln(x^3 + y^2)$

18. $f(x, y) = \ln(y/x)$ **19.** $f(x, y) = e^{2x+3y}$ **20.** $f(x, y) = \tan^{-1}(y/x)$

21. $f(x, y) = \cos(x^2 y^3)$ **22.** $f(x, y) = \sin^2(xy)$

In Problems 23–32 compute f_x, f_y, and f_z.

23. $f(x, y, z) = xy + yz + xz$ **24.** $f(x, y, z) = xe^y + ye^z + ze^x$ **25.** $f(x, y, z) = xy \sin z - yz \sin x$

26. $f(x, y, z) = 1/\sqrt{x^2 + y^2 + z^2}$ **27.** $f(x, y, z) = z \tan^{-1}(y/x)$ **28.** $f(x, y, z) = e^{xyz}$

29. $f(x, y, z) = (x + y)^z$ **30.** $f(x, y, z) = x^{y+z}$ **31.** $f(x, y, z) = x^{yz}$

32. $f(x, y, z) = \sin[\ln(x^2 + y^2 + z^2)]$

In Problems 33–36 use (16.2) to calculate the required partial derivatives.

33. Find $f_x(1, 2)$ and $f_y(1, 2)$ if $f(x, y) = x^2 y^3$.

34. Find $f_x(2, -1)$ and $f_y(2, -1)$ if $f(x, y) = 2x^2 + 3xy$.

35. Find $f_x(0, 0)$ and $f_y(0, 0)$ if

$$f(x, y) = \begin{cases} \dfrac{x^3 + y^3}{x^2 + y^2} & \text{if} \ (x, y) \neq (0, 0) \\ 0 & \text{if} \ (x, y) = (0, 0) \end{cases}$$

36. Find $f_x(0, 0)$ and $f_y(0, 0)$ if

$$f(x, y) = \begin{cases} \dfrac{x^2 y^3}{x^2 + 4y^3} & \text{if} \ (x, y) \neq (0, 0) \\ 0 & \text{if} \ (x, y) = (0, 0) \end{cases}$$

In Problems 37–42 find an equation of the tangent line to the curve of intersection of the surface with the plane at the point indicated.

37. $z = x^2 + y^2$, $y = 2$,   at $(1, 2, 5)$ **38.** $z = x^2 - y^2$, $x = 3$,   at $(3, 1, 8)$

39. $z = \sqrt{1 - x^2 - y^2}$, $x = 0$,   at $(0, \frac{1}{2}, \sqrt{3}/2)$ **40.** $z = \sqrt{16 - x^2 - y^2}$, $y = 2$,   at $(\sqrt{3}, 2, 3)$

41. $z = \sqrt{x^2 - y^2}$, $x = 4$,   at $(4, 1, \sqrt{15})$ **42.** $z = e^x \ln y$, $y = e$,   at $(0, e, 1)$

43. Find the rate of change of $z = \ln\sqrt{x^2 + y^2}$ at $(3, 4, \ln 5)$:
   (a) In a direction parallel to the x-axis   (b) In a direction parallel to the y-axis

44. Find the rate of change of $z = e^y \sin x$ at $(\pi/3, 0, \sqrt{3}/2)$:
   (a) In a direction parallel to the x-axis   (b) In a direction parallel to the y-axis

45. The temperature distribution T of a heated plate located in the xy-plane is given by

$$T = T(x, y) = \left(\frac{100}{\ln 2}\right) \ln(x^2 + y^2) \qquad \text{for} \ \ 1 \leq x^2 + y^2 \leq 9$$

   (a) Show that $T = 0$ if $x^2 + y^2 = 1$;   and $T = 200$ if $x^2 + y^2 = 4$.
   (b) Find the rate of change of T in a direction parallel to the x-axis at the point $(1, 0)$ and at the point $(0, 1)$.
   (c) Find the rate of change of T in a direction parallel to the y-axis at the point $(2, 0)$ and at the point $(0, 2)$.

46. Rework parts (b) and (c) of Problem 45 if the temperature distribution is $T = T(x, y) = 100/\sqrt{x^2 + y^2}$.

47. Find $\partial x/\partial r$, $\partial x/\partial \theta$, $\partial y/\partial r$, and $\partial y/\partial \theta$ if $x = r \cos \theta$, $y = r \sin \theta$.

48. Find $\partial r/\partial x$, $\partial \theta/\partial x$, $\partial r/\partial y$, and $\partial \theta/\partial y$ by solving for r and θ in $x = r \cos \theta$, $y = r \sin \theta$.

49. Show that $\partial u/\partial x = \partial v/\partial y$ and $\partial u/\partial y = -\partial v/\partial x$ for $u = e^x \cos y$, $v = e^x \sin y$.

50. Rework Problem 49 for $u = \ln\sqrt{x^2 + y^2}$, $v = \tan^{-1}(y/x)$.

51. If $u = x^2 + 4y^2$,   show that $x(\partial u/\partial x) + y(\partial u/\partial y) = 2u$.

52. If $u = xy^2$, show that $x(\partial u/\partial x) + y(\partial u/\partial y) = 3u$.

53. If $w = x^2 + y^2 - 3yz$, show that $x(\partial w/\partial x) + y(\partial w/\partial y) + z(\partial w/\partial z) = 2w$.

54. If $w = (xz + y^2)/yz$, show that $x(\partial w/\partial x) + y(\partial w/\partial y) + z(\partial w/\partial z) = 0$.

55. If $z = \cos(x + y) + \cos(x - y)$, show that $\partial^2 z/\partial x^2 - \partial^2 z/\partial y^2 = 0$.

56. If $z = \sin(x - y) + \ln(x + y)$, show that $\partial^2 z/\partial x^2 = \partial^2 z/\partial y^2$.

A function $z = f(x, y)$ that obeys the partial differential equation $\partial^2 z/\partial x^2 + \partial^2 z/\partial y^2 = 0$ is called *harmonic*.*

57. Show that $z = \ln\sqrt{x^2 + y^2}$ is harmonic.

58. Show that $z = \tan^{-1}(y/x)$ is harmonic.

59. Show that $z = e^{x^2+y^2}/(x^2 + y^2)$ obeys the partial differential equation $y(\partial z/\partial x) = x(\partial z/\partial y)$.

60. Show that $z = x^2y^2/(x + y)$ obeys the partial differential equation $x(\partial z/\partial x) + y(\partial z/\partial y) = 3z$.

61. The ideal gas law $PV = nrT$ is used to describe the relationship between pressure P, volume V, and temperature T of a confined gas, where n is the number of moles of the gas and r is the universal gas constant. Show that $(\partial V/\partial T)(\partial T/\partial P)(\partial P/\partial V) = -1$.

62. The function

$$z = f(x, y, r) = \frac{1 + (1 - x)y}{1 + r} - 1$$

describes the net gain or loss of money invested, where x = annual marginal tax rate, y = annual effective yield on an investment, and r = annual inflation rate.
  (a) Find the relative annual net gain or loss if money is invested at an effective yield of 6% when the marginal tax rate is 25% and the inflation rate is 10%; that is, find $f(0.25, 0.06, 0.1)$.
  (b) Given $r = 10\%$, find f_x.
  (c) Given $r = 10\%$, find f_y.
  (d) Evaluate f_x and f_y when $x = 0.25$ and $y = 0.06$.

63. The velocity of sound v in a gas depends upon the pressure p and density d of the gas according to the formula $v(p, d) = k\sqrt{p/d}$, where k is some constant. Find the rate of change of velocity with respect to p and with respect to d.

64. Suppose a vibrating string is governed by the equation $f(x, t) = 2\cos 5t \sin x$, where x is the horizontal distance of a point on the string, t is time, and $f(x, t)$ is the amplitude of the point. Verify that $\partial^2 f/\partial t^2 = 25(\partial^2 f/\partial x^2)$ at all points (x, t).

4. Limit and Continuity

In Chapter 2 we defined $\lim_{x \to x_0} f(x) = L$ to mean that given any positive number ε, there can be found a positive number δ, so that:

Whenever $\quad 0 < |x - x_0| < \delta, \quad$ then $\quad |f(x) - L| < \varepsilon$

Limits of functions of two or more variables are defined in much the same way.

---

* This equation is referred to as *Laplace's equation*. It is of great importance in many branches of mathematical physics, such as flow of fluids, heat, elasticity, electricity, and so forth.

Intuitively, if (x_0, y_0) is some point in the plane and L is a number, the statement $\lim_{(x, y) \to (x_0, y_0)} f(x, y) = L$ means that $f(x, y)$ can be as close to L as we wish if (x, y) is close enough to (x_0, y_0). For a function of three variables,

$$\lim_{(x, y, z) \to (x_0, y_0, z_0)} f(x, y, z) = L$$

has a similar meaning. To be precise, we use the following definition:

(16.3)   **DEFINITION**   **Let f be a function of two variables and let x_0, y_0, and L be numbers. Then L is the *limit of f at* (x_0, y_0) if, given any $\varepsilon > 0$, there exists a corresponding number $\delta > 0$, such that:**

Whenever $0 < \sqrt{(x - x_0)^2 + (y - y_0)^2} < \delta,$ then $|f(x, y) - L| < \varepsilon$

In this case we say that the limit exists and write

$$\lim_{(x, y) \to (x_0, y_0)} f(x, y) = L$$

See Figure 25.

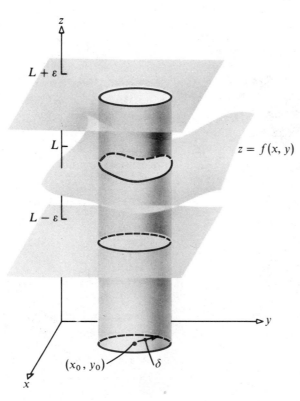

$\lim_{(x, y) \to (x_0, y_0)} f(x, y) = L$
means that for any $\varepsilon > 0$
there is a $\delta > 0$ so that
whenever (x, y) is within δ
of (x_0, y_0), then $z = f(x, y)$
is within ε of L

Figure 25

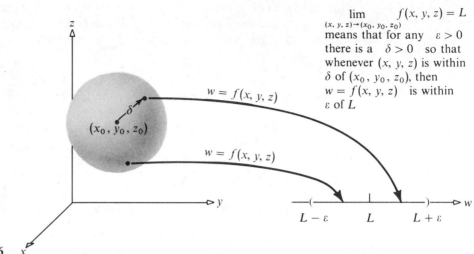

$$\lim_{(x,\,y,\,z)\to(x_0,\,y_0,\,z_0)} f(x, y, z) = L$$
means that for any $\varepsilon > 0$ there is a $\delta > 0$ so that whenever (x, y, z) is within δ of (x_0, y_0, z_0), then $w = f(x, y, z)$ is within ε of L

$w = f(x, y, z)$

$w = f(x, y, z)$

Figure 26

A similar definition is used for the statement

$$\lim_{(x,\,y,\,z)\to(x_0,\,y_0,\,z_0)} f(x, y, z) = L$$

where f is a function of three variables, and x_0, y_0, z_0, and L are numbers. See Figure 26.

Although limits of functions of more than three variables can also be defined, we shall have no need to discuss such functions.

A function $f(x, y)$ does not have to be defined at (x_0, y_0) in order for the limit at (x_0, y_0) to exist. Furthermore, even if $f(x_0, y_0)$ exists, the limit does not necessarily exist; and even if both $\lim_{(x,\,y)\to(x_0,\,y_0)} f(x, y)$ and $f(x_0, y_0)$ exist, they do not have to be equal. The next example illustrates this idea.

EXAMPLE 1　　The function,

$$f(x, y) = \frac{\sin(x^2 + y^2)}{x^2 + y^2}$$

is not defined at $(x, y) = (0, 0)$. Show, however, that

$$\lim_{(x,\,y)\to(0,\,0)} \frac{\sin(x^2 + y^2)}{x^2 + y^2} = 1$$

Solution　　Let $t = x^2 + y^2$ and note that t is close to 0 if (x, y) is close to $(0, 0)$. In fact, $\sqrt{x^2 + y^2} < \sqrt{\varepsilon}$ implies that $t = x^2 + y^2 < \varepsilon$. Thus,

$$\lim_{(x,\,y)\to(0,\,0)} \frac{\sin(x^2 + y^2)}{x^2 + y^2} = \lim_{t\to 0} \frac{\sin t}{t} = 1 \qquad \blacksquare$$

From Section 7, Chapter 3

As with limits of functions of one variable, the limit of a function of more than one variable, if it exists, is unique.

The algebra of limits for functions of two variables is analogous to that for functions of a single variable.

THEOREM   **Let f and g be functions of two variables for which**

$$\lim_{(x, y) \to (x_0, y_0)} f(x, y) = L \qquad \text{and} \qquad \lim_{(x, y) \to (x_0, y_0)} g(x, y) = M$$

where L and M are two real numbers.

Limit of a Sum:

$$\lim_{(x, y) \to (x_0, y_0)} [f(x, y) + g(x, y)] = \lim_{(x, y) \to (x_0, y_0)} f(x, y) + \lim_{(x, y) \to (x_0, y_0)} g(x, y) = L + M$$

Limit of a Difference:

$$\lim_{(x, y) \to (x_0, y_0)} [f(x, y) - g(x, y)] = \lim_{(x, y) \to (x_0, y_0)} f(x, y) - \lim_{(x, y) \to (x_0, y_0)} g(x, y) = L - M$$

If k is any real number, then

$$\lim_{(x, y) \to (x_0, y_0)} k[f(x, y)] = kL$$

Limit of a Product:

$$\lim_{(x, y) \to (x_0, y_0)} f(x, y)g(x, y) = \left[\lim_{(x, y) \to (x_0, y_0)} f(x, y) \right]\left[\lim_{(x, y) \to (x_0, y_0)} g(x, y) \right] = LM$$

Limit of a Quotient: **If** $\displaystyle\lim_{(x, y) \to (x_0, y_0)} g(x, y) = M \neq 0,$

$$\lim_{(x, y) \to (x_0, y_0)} \frac{f(x, y)}{g(x, y)} = \frac{\lim_{(x, y) \to (x_0, y_0)} f(x, y)}{\lim_{(x, y) \to (x_0, y_0)} g(x, y)} = \frac{L}{M}$$

A criterion for the existence of the limit of a function of one variable was that the left-hand limit equals the right-hand limit. More simply stated, this says that the limit must be the same no matter how the point x_0 is approached. Of course, there are only two ways to approach x_0—from the left or from the right. For $P_0 = (x_0, y_0)$ or for $P_0 = (x_0, y_0, z_0)$, however, there are infinitely many ways to approach P_0, namely, along any curve that passes through P_0. Thus, in order for the limit to exist, it is necessary that the function approach this limit if we approach P_0 along any curve that contains P_0. To put it another way:

(16.4)   **If $\lim_{P \to P_0} f(P)$ is computed along two different curves of approach to P_0 and if two different answers are obtained, then $\lim_{P \to P_0} f(P)$ does not exist.**

Let's look at an example that illustrates how (16.4) can be used.

EXAMPLE 2

Show that the following limit does not exist: $\displaystyle\lim_{(x,\,y)\to(0,\,0)} \frac{xy}{x^2 + y^2}$

Solution

We seek two curves of approach to $(0, 0)$ that result in different limits. Suppose we evaluate the limit using the curve $y = 2x.$ Then

$$\lim_{(x,\,y)\to(0,\,0)} \frac{xy}{x^2 + y^2} \underset{\substack{\uparrow \\ y = 2x}}{=} \lim_{x\to0} \frac{2x^2}{x^2 + 4x^2} = \lim_{x\to0} \frac{2x^2}{5x^2} = \frac{2}{5}$$

Using the curve $y = -x,$ we find

$$\lim_{(x,\,y)\to(0,\,0)} \frac{xy}{x^2 + y^2} \underset{\substack{\uparrow \\ y = -x}}{=} \lim_{x\to0} \frac{-x^2}{x^2 + x^2} = \lim_{x\to0} -\frac{x^2}{2x^2} = -\frac{1}{2}$$

Since $\frac{2}{5} \neq -\frac{1}{2},$ we conclude that the limit does not exist. See Figure 27.

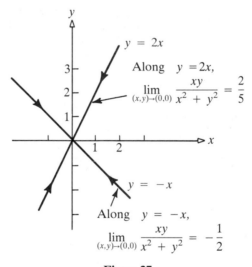

Figure 27   ∎

Be careful! Even if we obtain the same limit along two curves of approach (or three, or even several hundred), we would not be able to draw any conclusion about the existence of the limit of the function. For instance, in the example above, using the two curves of approach $y = 2x$ and $x = 2y$ would have resulted in the same value for the limit; and yet the limit does not exist.

If a limit does not exist, it is often (but not always) possible to find two curves of approach that yield different values. On the other hand, if a limit is known to exist, its value can be found by computing along any curve of approach. To show that the limit exists, however, is more difficult (we may even have to use an ε, δ argument).

CONTINUITY

The definition of continuity for a function of two or more variables is analogous to the definition for functions of one variable.

A function f of two variables is said to be *continuous* at (x_0, y_0) if all three of the following statements are true:

1. $f(x_0, y_0)$ is defined

2. $\lim_{(x, y) \to (x_0, y_0)} f(x, y)$ exists

3. $\lim_{(x, y) \to (x_0, y_0)} f(x, y) = f(x_0, y_0)$

We state without proof four results that are helpful in identifying continuous functions.

1. Just as for functions of one variable, the sum, difference, and product of two continuous functions are continuous functions; and the quotient of two continuous functions is continuous provided the denominator is not zero.

2. If f is a function of one variable that is continuous at b in its domain, then the function $g(x, y) = f(x)$ is continuous at (b, y) for all y, and the function $h(x, y) = f(y)$ is continuous at (x, b) for all x.

3. If h is a continuous function of one variable and g is a continuous function of two variables, then their composite $f(x, y) = h(g(x, y))$ is a continuous function of two variables.

4. If $f(x, y)$ is continuous at (x_0, y_0), then f is continuous in each variable separately;* that is,

$$\lim_{x \to x_0} \left[\lim_{y \to y_0} f(x, y) \right] = \lim_{y \to y_0} \left[\lim_{x \to x_0} f(x, y) \right] = f(x_0, y_0)$$

EXAMPLE 3   Discuss the continuity of: (a) $f(x, y) = \dfrac{x^2 + y^2}{x^2 - y^2}$ (b) $g(x, y) = \dfrac{x^2 + 2xy - y^2}{x^2 + y^2 + 2}$

Solution   (a) As a consequence of 1 and 2 above, any rational function in two variables is continuous wherever its denominator is nonzero. Thus,

$$f(x, y) = \frac{x^2 + y^2}{x^2 - y^2}$$

is continuous at all (x_0, y_0) where $x_0 \neq y_0$ and $x_0 \neq -y_0$.

---

* The converse is false. It is possible for a function of two variables to be continuous in each variable separately and yet fail to be continuous in both variables simultaneously.

(b)   The function

$$g(x, y) = \frac{x^2 + 2xy - y^2}{x^2 + y^2 + 2}$$

is continuous everywhere since its denominator is never zero.   ∎

EXAMPLE 4   The function

$$f(x, y) = \tan^{-1}\left(\frac{x^2 + 2xy - y^2}{x^2 + y^2 + 2}\right)$$

is continuous everywhere since it is the composite of the arctangent function, which is continuous, and the rational function discussed in part (b) of Example 3.   ∎

We can obtain the limits of continuous functions by direct substitution. For example,

$$\lim_{(x, y)\to(0, 0)} \tan^{-1}\left(\frac{x^2 + 2xy - y^2}{x^2 + y^2 + 2}\right) = \tan^{-1}(0) = 0$$

In Chapter 3 we proved the important result that if a function f of one variable has a derivative at x_0, then f is continuous at x_0. However, if a function f of more than one variable has partial derivatives at a point P_0, it may fail to be continuous at P_0.

In the next section we see that the function

$$f(x, y) = \begin{cases} \dfrac{xy}{x^2 + y^2} & \text{if}\quad (x, y) \neq (0, 0) \\ 0 & \text{if}\quad (x, y) = (0, 0) \end{cases}$$

has partial derivatives at $(0, 0)$ but is not continuous at that point.

INTERIOR POINTS; BOUNDARY POINTS

Some of the following concepts will be used immediately; others will be needed in the remaining chapters. We group them together here for your convenience.

If S is a set of points, then a point P_0 is called an *interior point* of S if all points P lying within some distance $\delta > 0$ of P_0 are also in S. See Figure 28(a). A point P_0 is a *boundary point* of S if the set of all points P within any distance $\delta > 0$ of P_0 contains points both in S and not in S. See Figure 28(b). (Note that every interior point of S is in S, while a boundary point may either be in S or not. In fact, a boundary point of a set is also a boundary point of its complement.)

A set of points is *open* if all its points are interior points. A set is *closed* if it includes all its own boundary points. A set is a *region* if all its boundary points are also boundary points of its interior. A *bounded* region in the plane is a region that can be enclosed in a circle of sufficiently large radius.

For example, the set of points satisfying the inequality

$$(x - h)^2 + (y - k)^2 \leq R^2$$

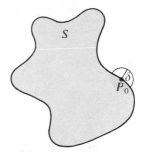

(a)   Interior point P_0
For some $\delta > 0$,   all points P
for which $d(P, P_0) < \delta$ lie in
the set S

(b)   Boundary point P_0
For any $\delta > 0$,   look at the
points P for which $d(P, P_0) < \delta$;
some will lie in the set S and
some will lie outside the set S

Figure 28

is a closed bounded region whose boundary is the circle of radius R centered at (h, k).
The set of points (x, y) such that $a < x < b$ and $c < y < d$ is the open rectangular region whose boundary is the rectangle with vertices at (a, c), (a, d), (b, c), and
(b, d).

EQUALITY OF MIXED PARTIALS

We observed earlier that the equality $f_{xy} = f_{yx}$ is ordinarily true for most functions
$z = f(x, y)$ we encounter in calculus. However, since for some particular functions,
the mixed partials are not equal, it is advisable to know conditions that guarantee
the equality of the mixed partials.

(16.5)   **THEOREM   Let $z = f(x, y)$ denote a function of two variables, and let P_0 be an
interior point of its domain. If the partial derivatives f_x, f_y, f_{xy}, and f_{yx} exist on the
domain of f and if f_{xy} and f_{yx} are continuous at P_0, then $f_{xy} = f_{yx}$ at P_0.**

A proof of this result may be found in most advanced calculus texts.

EXERCISE 4

In Problems 1–11 find each limit.

1. $\lim\limits_{(x, y)\to(1, 2)} (x^2 + xy - y^2)$

2. $\lim\limits_{(x, y)\to(-1, 3)} (x^2 + y^2 - 3x)$

3. $\lim\limits_{(x, y)\to(0, 0)} \dfrac{x^2 - y^2}{x - y}$

4. $\lim\limits_{(x, y)\to(1, 2)} \dfrac{4x^2 - y^2}{2x - y}$

5. $\lim\limits_{(x, y)\to(0, 4)} \dfrac{4x - xy}{x(4y - y^2)}$

6. $\lim\limits_{(x, v)\to(1,2)} \dfrac{x^2 + 2xy + y^2 - 9}{x + y - 3}$

7. $\displaystyle\lim_{(x,\,y,\,z)\to(0,\,0,\,1)} \frac{x^2z - y^2z}{x - y}$

8. $\displaystyle\lim_{(x,\,y,\,z)\to(1,\,0,\,1)} \frac{x^2 + y^2 + z^2}{x + y + z}$

9. $\displaystyle\lim_{(x,\,y)\to(1,\,0)} \frac{x \sin y}{x^2 + 1}$

10. $\displaystyle\lim_{(x,\,y)\to(0,\,0)} x \sin \frac{1}{y}$

11. $\displaystyle\lim_{(x,\,y)\to(0,\,0)} \cos \frac{x^2 + y^2}{x + y + 1}$

In Problems 12 and 13 calculate the given limit along each curve.

12. $\displaystyle\lim_{(x,\,y)\to(0,\,0)} \frac{xy}{x^2 + y^2}$
    (a) Along the x-axis   (b) Along the y-axis   (c) Along the line $y = 2x$
    (d) Along the parabola $y = x^2$

13. $\displaystyle\lim_{(x,\,y)\to(0,\,0)} \frac{2xy}{x^2 + 3y^2}$
    (a) Along the x-axis   (b) Along the y-axis   (c) Along the line $y = 3x$
    (d) Along the parabola $y = x^2$

In Problems 14–17 show that each limit does not exist.

14. $\displaystyle\lim_{(x,\,y)\to(0,\,0)} \frac{2xy}{x^2 + y^2}$

15. $\displaystyle\lim_{(x,\,y)\to(0,\,0)} \frac{2x^2 + y^2}{x^2 + y^2}$

16. $\displaystyle\lim_{(x,\,y)\to(0,\,0)} \frac{x^2 + y^4}{x^2 + y^2}$

17. $\displaystyle\lim_{(x,\,y)\to(0,\,0)} \frac{x^4 - y^2}{x^2 + y^2}$

In Problems 18–23 determine the points of discontinuity of the given function f, if any.

18. $f(x, y) = \dfrac{1}{x^2 + y^2 - 1}$

19. $f(x, y) = \dfrac{x^2y}{x^2 - y^2}$

20. $f(x, y) = y^2 e^{xy}$

21. $f(x, y) = \ln(x + y + 1)$

22. $f(x, y, z) = \sqrt{xy} \cot z$

23. $f(x, y, z) = \dfrac{xyz}{1 + x^2 + y^2 + z^2}$

In Problems 24 27 find f_{xy} and f_{yx} and show that $f_{xy} = f_{yx}$.

24. $f(x, y) = \tan^{-1}\left(\dfrac{y}{x}\right)$

25. $f(x, y) = \dfrac{x}{x^2 + y^2}$

26. $f(x, y) = x \sin x \cos y$

27. $f(x, y) = e^{x \ln y}$

28. If you are told that f is a function of two variables whose partial derivatives are $f_x(x, y) = 3x - y$ and $f_y(x, y) = x - 3y$, should you believe it? Explain.

29. (a) Consider two coordinate systems as given in the figure. Let $f(x, y) = 3x^2 + 4y^3$.
        Evaluate $f_x(1, 6)$ and $f_y(1, 6)$.
    (b) Let (a, b) be the $x'y'$-coordinates of $(1, 6)$. Let $\bar{f}(x', y') = f(x, y)$. Evaluate $\bar{f}_{x'}(a, b)$
        and $f_{y'}(a, b)$.

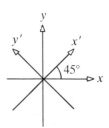

30. Define the function

$$f(x, y) = \begin{cases} \dfrac{xy^3}{x^2 + y^2} & \text{if } (x, y) \neq (0, 0) \\ 0 & \text{if } (x, y) = (0, 0) \end{cases}$$

(a)   Find f_x and f_y.   [*Hint:* $f_x(x, y)$, $(x, y) \neq (0, 0)$,   can be found by differentiating. To find $f_x(0, 0)$, use the definition of the partial derivative.]

(b)   Show that $f_{xy}(0, 0) \neq f_{yx}(0, 0)$.

5. Differentials

In the case of a differentiable function of one variable, the differential is a good approximation to the change in that function; that is, the change in $y = f(x)$ from x to $(x + \Delta x)$,   namely,

$$\Delta y = f(x + \Delta x) - f(x)$$

may be approximated by the differential

$$dy = f'(x)\Delta x$$

for $\Delta x \approx 0$ (see Chap. 4).

The idea behind approximating Δy by dy can also be introduced in the following manner: Define η (Greek letter eta) as

(16.6)
$$\eta = \frac{\Delta y}{\Delta x} - f'(x)$$

By multiplying by Δx and solving for Δy, we find that

(16.7)
$$\Delta y = f'(x)\Delta x + \eta \Delta x$$

where η is a function depending on Δx, for which $\lim_{\Delta x \to 0} \eta = 0$.

Figure 29 gives a geometric interpretation of equation (16.7). The term $\eta \Delta x$ represents the difference between Δy and $dy = f'(x)\Delta x$.   It is evident from Figure 29 that $\eta \Delta x \to 0$ as $\Delta x \to 0$.   But Figure 29 does not show something that equation (16.6) does show—that $\eta \to 0$ as $\Delta x \to 0$.   (Do you see why? Since f is differentiable, it follows that $\Delta y/\Delta x \to f'(x)$ as $\Delta x \to 0$.)

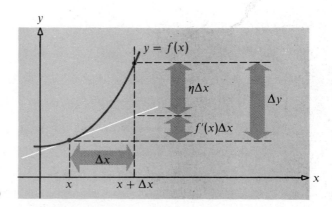

Figure 29

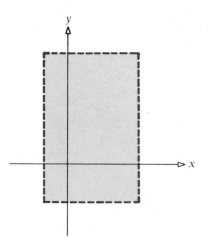

Figure 30   (a)   Open rectangular region   (b)   Open rectangular region
                            in the plane   in space

In this section, we extend (16.7) to a function f of two or more variables. For convenience, we take the domain of f to be a *rectangular region*. In particular, an *open rectangular region* is the collection of points that are inside a rectangle whose sides are parallel to the coordinate axes. Figure 30(a) illustrates an open rectangular region in the plane; Figure 30(b) shows an open rectangular region in space.

The points in an open rectangular region are interior points. The boundary of an open rectangular region consists of the points on the rectangle itself. A *closed*

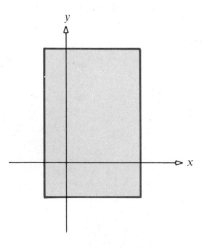
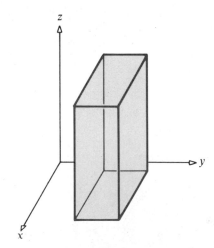

Figure 31   (a)   Closed rectangular region   (b)   Closed rectangular region
                            in the plane   in space

rectangular region consists of an open rectangular region and its boundary (see Fig. 31).

CHANGE IN $z = f(x, y)$

Let $z = f(x, y)$ denote a function of two variables defined on some open rectangular region H. Let (x, y) be a point in H, and let Δx and Δy (which represent changes in x and in y, respectively) be chosen so that the point $(x + \Delta x, y + \Delta y)$ is also in H. The *change in z*, denoted by Δz, is defined as

(16.8)
$$\Delta z = f(x + \Delta x, y + \Delta y) - f(x, y)$$

Figure 32(a) on p. 892 illustrates these concepts in the plane; Figure 32(b) illustrates them in space.

EXAMPLE 1   Find the change Δz in $z = f(x, y) = x^2 y - 1$ from (x, y) to $(x + \Delta x, y + \Delta y)$. Use the answer to calculate the change in z from $(1, 2)$ to $(1.1, 1.9)$.

Solution   From (16.8), we find

$$\Delta z = f(x + \Delta x, y + \Delta y) - f(x, y)$$
$$= (x + \Delta x)^2 (y + \Delta y) - 1 - (x^2 y - 1)$$
$$= x^2 \Delta y + 2xy\Delta x + 2x\Delta x\Delta y + y(\Delta x)^2 + (\Delta x)^2 \Delta y$$

The change in z from $(1, 2)$ to $(1.1, 1.9)$ may be computed by substituting $x = 1$, $y = 2$, $\Delta x = 0.1$, and $\Delta y = -0.1$ in the above expression. The change is found to be $\Delta z = 0.299$. This answer could also have been found by computing $f(1.1, 1.9) - f(1, 2)$. ∎

DIFFERENTIABILITY

We use an equation similar to (16.7) to define *differentiability* of a function of two variables:

(16.9)   **DEFINITION**   **Let $z = f(x, y)$ denote a function of two variables defined on some open rectangular region H. Let (x, y) be a point in H and let Δx and Δy be chosen so that the point $(x + \Delta x, y + \Delta y)$ is also in H. Suppose also that $f_x(x, y)$ and $f_y(x, y)$ exist. If the change Δz from (x, y) to $(x + \Delta x, y + \Delta y)$ can be expressed in the form**

(16.10)
$$\Delta z = f_x(x, y)\Delta x + f_y(x, y)\Delta y + \eta_1 \Delta x + \eta_2 \Delta y$$

where η_1 and η_2 are each functions of Δx and Δy such that

$$\lim_{(\Delta x, \Delta y) \to (0, 0)} \eta_1 = 0 \quad \text{and} \quad \lim_{(\Delta x, \Delta y) \to (0, 0)} \eta_2 = 0$$

then $z = f(x, y)$ is said to be *differentiable at (x, y)*.

Let's use the definition to show that the function given in Example 1 is differentiable at any point (x, y).

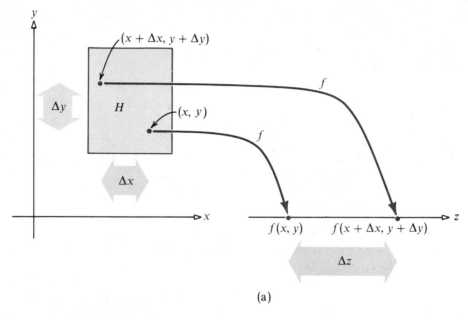

(a)

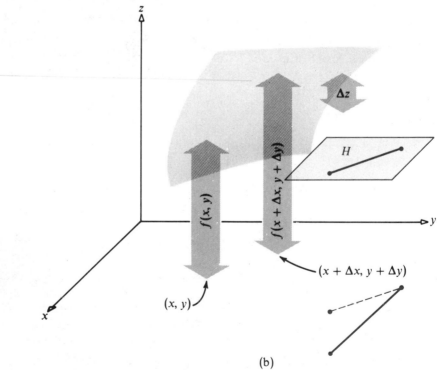

(b)

Figure 32

EXAMPLE 2   Show that $z = f(x, y) = x^2y - 1$ is differentiable at any point (x, y).

Solution   We need to find η_1 and η_2 so that

(16.11)
$$\Delta z - f_x(x, y)\Delta x - f_y(x, y)\Delta y = \eta_1 \Delta x + \eta_2 \Delta y$$

where

$$\lim_{(\Delta x, \Delta y) \to (0, 0)} \eta_1 = 0 \qquad \text{and} \qquad \lim_{(\Delta x, \Delta y) \to (0, 0)} \eta_2 = 0$$

Since $f(x, y) = x^2y - 1$,   it follows that

$$f_x(x, y) = 2xy \qquad \text{and} \qquad f_y(x, y) = x^2$$

By using these facts and the calculation of Δz obtained in Example 1, we find that

$$\Delta z - f_x(x, y)\Delta x - f_y(x, y)\Delta y = x^2\Delta y + 2xy\Delta x + 2x\Delta x\Delta y + y(\Delta x)^2$$
$$+ (\Delta x)^2\Delta y - 2xy\Delta x - x^2\Delta y$$
$$= 2x\Delta x\Delta y + y(\Delta x)^2 + (\Delta x)^2\Delta y$$

There are several ways in which we can write the right-hand side of the above equation in such a way that it conforms to (16.11). One way is the following:

$$\Delta z - f_x(x, y)\Delta x - f_y(x, y)\Delta y = (2x\Delta y + y\Delta x)\Delta x + (\Delta x)^2\Delta y$$

where $\eta_1 = 2x\Delta y + y\Delta x$ and $\eta_2 = (\Delta x)^2$.   It is now clear that η_1 and η_2 both approach 0 as $(\Delta x, \Delta y)$ approaches $(0, 0)$; thus, the function is differentiable everywhere.   ■

We emphasize that *the choice of η_1 and η_2 in (16.10) is not necessarily unique.* In Example 2, there are several other convenient and natural ways to write the equation, including the following:

1. $\eta_1\Delta x + \eta_2\Delta y = (y\Delta x)\Delta x + [2x\Delta x + (\Delta x)^2]\Delta y$
2. $\eta_1\Delta x + \eta_2\Delta y = (y\Delta x + \Delta x\Delta y)\Delta x + (2x\Delta x)\Delta y$
3. $\eta_1\Delta x + \eta_2\Delta y = (2x\Delta y + y\Delta x + \Delta x\Delta y)\Delta x + (0)\Delta y$

In each case (as you may easily verify), η_1 and η_2 are functions that approach 0 as Δx and Δy approach 0.

Later in this section we give an example of a function $z = f(x, y)$ that is *not* differentiable. For now, we proceed to give conditions on a function $z = f(x, y)$ that will guarantee that it is differentiable.

(16.12)   **THEOREM   Let $z = f(x, y)$ denote a function of two variables defined on some open rectangular region H. Let (x_0, y_0) be a point in H. If the partial derivatives f_x and f_y exist in H and if f_x and f_y are each continuous at (x_0, y_0), then f is differentiable at (x_0, y_0).**

The proof of (16.12) depends on the mean value theorem for derivatives.

Proof Let Δx and Δy be changes, not both 0, in x and in y, respectively, so that the point $(x_0 + \Delta x, y_0 + \Delta y)$ lies in H. The change in z is

$$(16.13) \quad \Delta z = f(x_0 + \Delta x, y_0 + \Delta y) - f(x_0, y_0)$$
$$= f(x_0 + \Delta x, y_0 + \Delta y) - f(x_0, y_0 + \Delta y) + f(x_0, y_0 + \Delta y) - f(x_0, y_0)$$

The expression $f(x, y_0 + \Delta y)$ is a function of x alone, and its (partial) derivative $f_x(x, y_0 + \Delta y)$ exists in H. Hence, by the mean value theorem, there is a u between x_0 and $(x_0 + \Delta x)$, so that

$$(16.14) \quad f(x_0 + \Delta x, y_0 + \Delta y) - f(x_0, y_0 + \Delta y) = f_x(u, y_0 + \Delta y)\Delta x$$

Similarly, the expression $f(x_0, y)$ is a function of y alone, and its derivative $f_y(x_0, y)$ exists in H. Hence, by the mean value theorem, there is a v between y_0 and $(y_0 + \Delta y)$, so that

$$(16.15) \quad f(x_0, y_0 + \Delta y) - f(x_0, y_0) = f_y(x_0, v)\Delta y$$

By substituting (16.14) and (16.15) in (16.13), we obtain

$$(16.16) \quad \Delta z = f_x(u, y_0 + \Delta y)\Delta x + f_y(x_0, v)\Delta y$$

We introduce the functions η_1 and η_2 defined by

$$(16.17) \quad \eta_1 = f_x(u, y_0 + \Delta y) - f_x(x_0, y_0) \quad \text{and} \quad \eta_2 = f_y(x_0, v) - f_y(x_0, y_0)$$

and observe that η_1 and η_2 have the desired property that

$$\lim_{(\Delta x, \Delta y)\to(0,0)} \eta_1 = \lim_{(\Delta x, \Delta y)\to(0,0)} [f_x(u, y_0 + \Delta y) - f_x(x_0, y_0)]$$
$$= f_x(x_0, y_0) - f_x(x_0, y_0) = 0$$
$$\lim_{(\Delta x, \Delta y)\to(0,0)} \eta_2 = \lim_{(\Delta x, \Delta y)\to(0,0)} [f_y(x_0, v) - f_y(x_0, y_0)]$$
$$= f_y(x_0, y_0) - f_y(x_0, y_0) = 0$$

since f_x and f_y are continuous at (x_0, y_0) and $u \to x_0$, $v \to y_0$ as $(\Delta x, \Delta y) \to (0, 0)$. As a result of (16.17), we may write (16.16) as

$$\Delta z = f_x(x_0, y_0)\Delta x + f_y(x_0, y_0)\Delta y + \eta_1\Delta x + \eta_2\Delta y$$

If we compare the above expression to (16.10), we conclude that f is differentiable at (x_0, y_0). ∎

DIFFERENTIALS

(16.18) **DEFINITION** Let $z = f(x, y)$ denote a function of two variables defined on some open rectangular region H. Let (x, y) be a point in H, and let Δx and Δy be chosen so that the point $(x + \Delta x, y + \Delta y)$ is also in H. If f is differentiable at the point (x, y), we define the differentials dx and dy as

$$dx = \Delta x \qquad dy = \Delta y$$

The *differential dz*, also called the *total differential of* $z = f(x, y)$, is defined as

(16.19)
$$dz = f_x(x, y)\,dx + f_y(x, y)\,dy$$

If $z = f(x, y)$ is a differentiable function, a comparison of (16.19) with (16.10) yields the expression

(16.20)
$$\Delta z = dz + \eta_1 \Delta x + \eta_2 \Delta y$$

where $\lim_{(\Delta x, \Delta y) \to (0, 0)} \eta_1 = 0$ and $\lim_{(\Delta x, \Delta y) \to (0, 0)} \eta_2 = 0$. When $dx\,(=\Delta x)$ and $dy\,(=\Delta y)$ are close to 0—and hence η_1 and η_2 are also close to 0—we see from (16.20) that dz is, therefore, approximately equal to Δz. Since dz is usually easier to calculate than Δz, we make use of the fact that when Δx and Δy are close to 0, then dz is an approximation to Δz. Of course, in using dz as an approximation to Δz, the error that results equals the expression $\eta_1 \Delta x + \eta_2 \Delta y$.

EXAMPLE 3   For the function $z = f(x, y) = x^2 y - 1$ use the total differential dz to approximate the change in z from $(1, 2)$ to $(1.1, 1.9)$.

Solution   We showed in Example 2 that f is differentiable. We use (16.19), with $f_x(x, y) = 2xy$ and $f_y(x, y) = x^2$, and find the total differential dz at any point (x, y) to be

$$dz = 2xy\,dx + x^2\,dy$$

At $(1, 2)$, $dz = 4\,dx + dy$. By using $dx = 0.1$ and $dy = -0.1$, we estimate the change in z to be

$$\Delta z \approx dz = 4(0.1) + (-0.1) = 0.3 \qquad \blacksquare$$

The actual change in z was computed in Example 1 to be 0.299, so the use of differentials resulted in an error of 0.001. Hence, by (16.20), we see that, in this case, $\eta_1 \Delta x + \eta_2 \Delta y = -0.001$.

EXAMPLE 4   A cola company requires cans in the shape of a right circular cylinder of height 10 centimeters and radius 3 centimeters. If the manufacturer of the cans claims a percentage error of no more than 0.2% in the height and no more than 0.1% in the radius, what is the maximum error in the volume?

Solution   The volume V of a right circular cylinder of height h and radius R is $V = \pi R^2 h$. By (16.19), the total differential dV is

$$dV = \frac{\partial V}{\partial R}\,dR + \frac{\partial V}{\partial h}\,dh = 2\pi R h\,dR + \pi R^2\,dh$$

The relative error in R is $|\Delta R|/R = |dR|/R = 0.001$, and the relative error in h is $|\Delta h|/h = |dh|/h = 0.002$. The relative error in the volume is

$$\frac{|\Delta V|}{V} \approx \frac{|dV|}{V} = \frac{|2\pi R h\,dR + \pi R^2\,dh|}{\pi R^2 h} \leq 2\,\frac{|dR|}{R} + \frac{|dh|}{h} = 2(0.001) + 0.002 = 0.004$$

The maximum percentage error in the volume is therefore 0.4%, so the actual volume of the container is $90\pi \pm (0.004)(90\pi)$ and lies between $89.64\pi \approx 281.612$ and $90.36\pi \approx 283.874$ cubic centimeters. ■

DIFFERENTIABILITY AND CONTINUITY

We showed in Chapter 3 that differentiable functions of a single variable are necessarily continuous. We now extend this result to functions of two variables.

(16.21)   **THEOREM**   Let $z = f(x, y)$ denote a function of two variables defined on some open rectangular region H. Let (x_0, y_0) be a point in H. If f is differentiable at (x_0, y_0), then f is continuous at (x_0, y_0).

Proof   Since $z = f(x, y)$ is differentiable at (x_0, y_0), we can express Δz by

$$\Delta z = f_x(x_0, y_0)\Delta x + f_y(x_0, y_0)\Delta y + \eta_1 \Delta x + \eta_2 \Delta y$$

where $\lim_{(\Delta x, \Delta y) \to (0, 0)} \eta_1 = 0$ and $\lim_{(\Delta x, \Delta y) \to (0, 0)} \eta_2 = 0$. We write Δz as

$$\Delta z = [f_x(x_0, y_0) + \eta_1]\Delta x + [f_y(x_0, y_0) + \eta_2]\Delta y$$

and set $\Delta x = x - x_0$ and $\Delta y = y - y_0$. Then $(\Delta x, \Delta y) \to (0, 0)$ is equivalent to $(x, y) \to (x_0, y_0)$, so that

$$\lim_{(x, y) \to (x_0, y_0)} \Delta z = \lim_{(x, y) \to (x_0, y_0)} \{[f_x(x_0, y_0) + \eta_1](x - x_0) + [f_y(x_0, y_0) + \eta_2](y - y_0)\} = 0$$

Since $\Delta z = f(x, y) - f(x_0, y_0)$, then $\lim_{(x, y) \to (x_0, y_0)} \Delta z = 0$ is equivalent to $\lim_{(x, y) \to (x_0, y_0)} f(x, y) = f(x_0, y_0)$. Hence, it follows that f is continuous at (x_0, y_0). ■

Theorem (16.21) states that differentiability implies continuity. We use the equivalent statement, "a function that is not continuous is not differentiable" to show that the *mere existence of partial derivatives at a point does not imply differentiability at that point*.

EXAMPLE 5   Show that the function below has partial derivatives at $(0, 0)$ but is not continuous at $(0, 0)$ and, therefore, is not differentiable at that point.

$$f(x, y) = \begin{cases} \dfrac{xy}{x^2 + y^2} & \text{if } (x, y) \neq (0, 0) \\ 0 & \text{if } (x, y) = (0, 0) \end{cases}$$

Solution   We use (16.2) to compute $f_x(0, 0)$ and $f_y(0, 0)$. Then, since

$$f(0 + \Delta x, 0) = f(\Delta x, 0) = \frac{0}{(\Delta x)^2 + 0} = 0$$

and

$$f(0, 0 + \Delta y) = f(0, \Delta y) = \frac{0}{0 + (\Delta y)^2} = 0$$

we have

$$f_x(0, 0) = \lim_{\Delta x \to 0} \frac{f(0 + \Delta x, 0) - f(0, 0)}{\Delta x} = \lim_{\Delta x \to 0} \frac{0}{\Delta x} = 0$$

$$f_y(0, 0) = \lim_{\Delta y \to 0} \frac{f(0, 0 + \Delta y) - f(0, 0)}{\Delta y} = \lim_{\Delta y \to 0} \frac{0}{\Delta y} = 0$$

Thus, $f_x(0, 0)$ and $f_y(0, 0)$ both exist. However, as was shown in Example 2 (p. 884), $\lim_{(x, y) \to (0, 0)} f(x, y)$ does not exist, and, hence, f is not continuous at $(0, 0)$. See Figure 33.

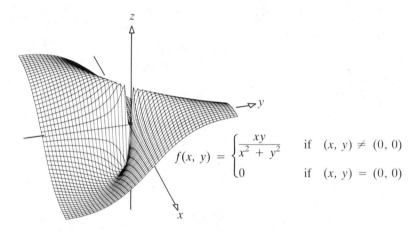

$$f(x, y) = \begin{cases} \dfrac{xy}{x^2 + y^2} & \text{if } (x, y) \neq (0, 0) \\ 0 & \text{if } (x, y) = (0, 0) \end{cases}$$

Figure 33

Since f is not continuous at $(0, 0)$, it follows from (16.21) that f cannot be differentiable at $(0, 0)$. Thus, we have exhibited a function f that has partial derivatives at a point, but is not differentiable at that point. ∎

Based on (16.12) and (16.21), the following result is evident:

(16.22) **Let $z = f(x, y)$ denote a function of two variables defined on some open rectangular region H. Let (x_0, y_0) be a point in H. If the partial derivatives f_x and f_y exist in H and if f_x and f_y are each continuous at (x_0, y_0), then f is continuous at (x_0, y_0).**

Although the precise formulations are given as theorems, the following summary may be helpful:

1. Continuity of f_x and f_y $\Rightarrow$ Differentiability of f.
2. Differentiability of f $\Rightarrow$ Continuity of f.

3. Existence of f_x and f_y does not necessarily mean f is differentiable.

4. Existence of f_x and f_y does not necessarily mean f is continuous.

DIFFERENTIALS FOR FUNCTIONS OF MORE THAN TWO VARIABLES

Under suitable conditions the definitions and theorems given for functions of two variables extend to functions of three or more variables. Thus, if $w = f(x, y, z)$ is a function of three variables, we define f to be differentiable at a point (x, y, z) if the change Δw in w can be expressed in the form

$$\Delta w = f_x(x, y, z)\Delta x + f_y(x, y, z)\Delta y + f_z(x, y, z)\Delta z + \eta_1\Delta x + \eta_2\Delta y + \eta_3\Delta z$$

where η_1, η_2, and η_3 are each functions of Δx, Δy, and Δz such that

$$\lim_{(\Delta x, \Delta y, \Delta z)\to(0, 0, 0)} \eta_1 = 0 \qquad \lim_{(\Delta x, \Delta y, \Delta z)\to(0, 0, 0)} \eta_2 = 0 \qquad \lim_{(\Delta x, \Delta y, \Delta z)\to(0, 0, 0)} \eta_3 = 0$$

If $w = f(x, y, z)$ is differentiable at a point (x, y, z), the *total differential dw* is defined as

$$dw = f_x(x, y, z)\, dx + f_y(x, y, z)\, dy + f_z(x, y, z)\, dz$$

where $dx = \Delta x$, $dy = \Delta y$, and $dz = \Delta z$.

EXAMPLE 6   If $w = f(x, y, z) = 3x^2 \sin^2 y \cos z$, find the total differential dw.

Solution $dw = 6x \sin^2 y \cos z\, dx + 6x^2 \sin y \cos y \cos z\, dy - 3x^2 \sin^2 y \sin z\, dz$ ∎

It can be shown that if $w = f(x, y, z)$ is defined in an open rectangular region containing (x_0, y_0, z_0) and if f_x, f_y, and f_z exist in this region and are continuous at (x_0, y_0, z_0), then f is differentiable at (x_0, y_0, z_0).

The remarks above extend to functions of more than three variables in a completely analogous way.

APPLICATION IN ASTROPHYSICS

The luminosity L (total power output in watts) of a star is given by the formula

$$L = 4\pi R^2 \sigma T^4$$

where R is its radius (in meters), T is its effective surface temperature (in degrees Kelvin),* and σ is the Stefan–Boltzmann constant. Our sun presently has $L_0 = 3.90 \times 10^{26}$ watts, $R_0 = 6.94 \times 10^8$ meters, and $T_0 = 4800$ K. Suppose another billion years of evolution is expected to result in the changes $\Delta R = +0.08 \times 10^8$ meters and $\Delta T = +100$ K. What will be the resulting percent increase in luminosity?

---

* 0 K $= -273°$C; 100 K $= -173°$C

The change in L is

$$\Delta L \approx dL = \frac{\partial L}{\partial R} \, dR + \frac{\partial L}{\partial T} \, dT = 4\pi\sigma(2RT^4 \, dR + 4R^2T^3 \, dT)$$

The relative error in luminosity is, therefore,

$$\frac{\Delta L}{L} \approx \frac{dL}{L} = \frac{4\pi\sigma(2RT^4 \, dR + 4R^2T^3 \, dT)}{4\pi R^2\sigma T^4}$$

$$= 2\frac{dR}{R} + 4\frac{dT}{T} \approx \frac{2(0.08) \times 10^8}{6.94 \times 10^8} + \frac{4(100)}{4800}$$

$$= 2(0.0115) + 4(0.0208) = 0.106$$

The percent increase in luminosity is approximately 10.6%.

Incidentally, a reasonable, though rough, guess at how this would affect the earth's temperature is

$$\Delta T_e \approx (\tfrac{1}{4})(0.106)T_e = (0.0265)(290 \text{ K}) = +7.69 \text{ K}$$

Such a change in temperature would be enough to modify the earth's climate.

EXERCISE 5

In Problems 1–14 find the total differential of each function.

1. $z = x^2 + y^2$ **2.** $z = 2x^2 + xy - y^2$ **3.** $z = x \sin y + y \sin x$

4. $z = \tan^{-1}(y/x)$ **5.** $z = e^x \cos y + e^{-x} \sin y$ **6.** $z = \ln(y/x)$

7. $z = \ln(x^2 + y^2)$ **8.** $z = e^{xy}$ **9.** $w = x^2y + y^2z + z^2x$

10. $w = xyz$ **11.** $w = xe^{yz} + ye^{xz} + ze^{xy}$ **12.** $w = \ln(x^2 + y^2 + z^2)$

13. $w = e^t(\ln xy + \ln xz + \ln yz)$ **14.** $w = \dfrac{xyzt}{x + y + z}$

In Problems 15–18 show that the function $z = f(x, y)$ is differentiable at any point (x, y) in its domain by: (1) Finding Δz; (2) finding η_1 and η_2 so that (16.10) holds; and (3) showing that $\lim_{(\Delta x, \Delta y) \to (0, 0)} \eta_1 = 0$ and $\lim_{(\Delta x, \Delta y) \to (0, 0)} \eta_2 = 0$.

15. $z = f(x, y) = xy^2 - 2xy$ **16.** $z = f(x, y) = 3x^2 + y^2$ **17.** $z = f(x, y) = y^2/x$

18. $z = f(x, y) = 2x/y$

19. Use differentials to estimate the change in $z = x^2 + y^2$ from $(1, 3)$ to $(1.1, 3.2)$.

20. Use differentials to estimate the change in $z = 2x^2 + xy - y^2$ from $(2, -1)$ to $(2.1, -1.1)$.

21. Use differentials to estimate the change in $z = e^x \ln(xy)$ from $(1, 2)$ to $(0.9, 2.1)$.

22. Use differentials to estimate the change in $z = xy/(x + y)$ from $(-1, 2)$ to $(-0.9, 1.9)$.

23. Use differentials to estimate $(\sqrt[4]{16.01})(\sqrt[5]{32.1})$.

24. Use differentials to estimate $(2.01)^6/\sqrt{3.89}$.

25. Using differentials, estimate the change in the volume of a right circular cylinder if the height changes from 2 to 2.1 centimeters and the radius changes from 0.5 to 0.51 centimeter.

26. For the data in Problem 25, what is the estimated change in surface area? Assume that the cylinder is closed at both the top and the bottom.

27. Estimate the increase in area of a triangle if its base is increased from 2 to 2.05 centimeters and its altitude is increased from 5 to 5.1 centimeters.

28. If the base of a triangle is increased from 5 to 5.1 centimeters and its altitude is decreased from 10 to 9.8 centimeters, how is the area affected?

29. If $x = r \cos \theta$, $y = r \sin \theta$, show that $x \, dy - y \, dx = r^2 \, d\theta$.

30. The specific gravity of an object is defined as $s = a/(a - w)$, where a is the weight of the object in air and w is its weight in water. If a is found to be 6 pounds with a possible error of 1% and w is 5 pounds with a possible error of 2%, what is the maximum error in the specific gravity?

31. In a parallel circuit the total resistance R due to two resistances R_1 and R_2 obeys $1/R = (1/R_1) + (1/R_2)$. If $R_1 = 50$ ohms with a possible error of 1.2% and $R_2 = 75$ ohms with a possible error of 1%, what is the maximum error in the total resistance?

32. A tank consists of a hemisphere mounted on a cylinder of the same radius (see the figure). The height and radius of the cylinder were measured as 14 meters and 5 meters, respectively. However, the device used to make this measurement was found to be in error by 1%. What is the maximum error in the volume of the tank?

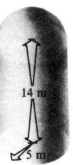

14 m

5 m

33. The index of refraction is defined as $\mu = (\sin i)/(\sin r)$, where i is the angle of incidence and r is the angle of refraction. If $i = 30°$ and $r = 60°$, and each is subject to a possible error of 2%, what is the maximum relative error for μ?

34. The equation $PV = kT$, where k is a constant, relates the pressure P, volume V, and temperature T of a confined ideal gas. If $P = 0.1$ gram per square millimeter, $V = 12$ cubic millimeters, and $T = 32°C$, approximate the change in P if V and T change to 15 cubic millimeters and 29°C, respectively.

35. Two sides and the included angle of a triangle are measured by a ruler and protractor, which are subject to errors of 2% and 3%, respectively. The area of the triangle is then computed from the formula $A = \frac{1}{2}bc \sin \alpha$.

(a) Show that: $\dfrac{dA}{A} = \dfrac{db}{b} + \dfrac{dc}{c} + \cot \alpha \, d\alpha$

(b) If $\alpha = \pi/4$, what is the maximum error in the computation of A?

36. Show that the function below has partial derivatives at $(0, 0)$, but that f is not continuous at $(0, 0)$ and, hence, is not differentiable at that point.

$$f(x, y) = \begin{cases} \dfrac{2xy}{x^2 + y^2} & \text{if } (x, y) \neq (0, 0) \\ 0 & \text{if } (x, y) = (0, 0) \end{cases}$$

37. Show that the function below has partial derivatives at $(0, 0)$, but that f is not continuous at $(0, 0)$ and, hence, is not differentiable at that point.

$$f(x, y) = \begin{cases} \dfrac{xy(1 + y^2)}{x^2 + y^2} & \text{if} \quad (x, y) \neq (0, 0) \\ 0 & \text{if} \quad (x, y) = (0, 0) \end{cases}$$

38. For the function below, show that $f_x(1, 1)$ and $f_y(1, 1)$ each exist, but that f is not differentiable at $(1, 1)$.

$$f(x, y) = \begin{cases} \dfrac{xy - 1}{x^2 + y^2 - 2} & \text{if} \quad x^2 + y^2 \neq 2 \\ \frac{1}{2} & \text{if} \quad x^2 + y^2 = 2 \end{cases}$$

39. For the function below, show that $f_x(0, 0)$ and $f_y(0, 0)$ each exist, but that f is not differentiable at $(0, 0)$.

$$f(x, y) = \begin{cases} \dfrac{x^2 y^2}{x^4 + y^4} & \text{if} \quad (x, y) \neq (0, 0) \\ 0 & \text{if} \quad (x, y) = (0, 0) \end{cases}$$

6. The Chain Rule

The notion of a composite function may be extended to functions of several variables. For example, if $z = u \ln v$ and if $u = x^2 + y^2$, $v = xy$, then the composite function is

$$z = (x^2 + y^2) \ln(xy)$$

In general, if $z = f(u, v)$ and $u = g(x, y)$, $v = h(x, y)$, then the composite function $z = f(g(x, y), h(x, y))$ is a function of the two variables x and y, provided the points (x, y) from the intersection of the domains of g and h result in values of (u, v) found in the domain of f.

The chain rule gives a formula for finding the partial derivatives $\partial z/\partial x$, $\partial z/\partial y$ of a composite function $z = f(u, v) = f(g(x, y), h(x, y))$ in terms of $\partial z/\partial u$, $\partial z/\partial v$, and $\partial u/\partial x$, $\partial u/\partial y$, $\partial v/\partial x$, $\partial v/\partial y$.

(16.23)   **THEOREM** *Chain Rule.* **Let** $z = f(g(x, y), h(x, y))$ **be the composite of** $z = f(u, v)$ **and** $u = g(x, y)$, $v = h(x, y)$. **If** g **and** h **are each continuous and have continuous first-order partial derivatives at a point** (x, y), **which is an interior point of the domains of both** g **and** h, **and if** f **is differentiable in some open rectangular region that contains the point** $(u, v) = (g(x, y), h(x, y))$, **then**

$$\frac{\partial z}{\partial x} = \left(\frac{\partial z}{\partial u}\right)\left(\frac{\partial u}{\partial x}\right) + \left(\frac{\partial z}{\partial v}\right)\left(\frac{\partial v}{\partial x}\right) \qquad \frac{\partial z}{\partial y} = \left(\frac{\partial z}{\partial u}\right)\left(\frac{\partial u}{\partial y}\right) + \left(\frac{\partial z}{\partial v}\right)\left(\frac{\partial v}{\partial y}\right)$$

Proof We shall prove the formula for $\partial z/\partial x$; the formula for $\partial z/\partial y$ is left as an exercise.

We begin by computing the changes Δz, Δu, Δv corresponding to a change Δx in x and a change $\Delta y = 0$ in y (that is, y is held constant). Then

(16.24) $$\Delta z = f(g(x + \Delta x, y), h(x + \Delta x, y)) - f(g(x, y), h(x, y))$$

(16.25) $$\Delta u = g(x + \Delta x, y) - g(x, y) \qquad \Delta v = h(x + \Delta x, y) - h(x, y)$$

Since $u = g(x, y)$ and $v = h(x, y)$, we see from (16.25) that

(16.26) $$g(x + \Delta x, y) = u + \Delta u \qquad h(x + \Delta x, y) = v + \Delta v$$

By substituting into (16.24), we find that

$$\Delta z = f(u + \Delta u, v + \Delta v) - f(u, v)$$

But the function f is differentiable, so Δz can be written in the form

(16.27) $$\Delta z = f_u(u, v)\Delta u + f_v(u, v)\Delta v + \eta_1 \Delta u + \eta_2 \Delta v$$

$$= \frac{\partial z}{\partial u} \Delta u + \frac{\partial z}{\partial v} \Delta v + \eta_1 \Delta u + \eta_2 \Delta v$$

where η_1 and η_2 are functions of Δu and Δv, and $\lim_{(\Delta u, \Delta v) \to (0, 0)} \eta_1 = 0$, $\lim_{(\Delta u, \Delta v) \to (0, 0)} \eta_2 = 0$. When we divide both sides of (16.27) by Δx, we obtain

(16.28) $$\frac{\Delta z}{\Delta x} = \frac{\partial z}{\partial u} \frac{\Delta u}{\Delta x} + \frac{\partial z}{\partial v} \frac{\Delta v}{\Delta x} + \eta_1 \frac{\Delta u}{\Delta x} + \eta_2 \frac{\Delta v}{\Delta x}$$

Now, it follows from the definition of a partial derivative that, as $\Delta x \to 0$,

$$\lim_{\Delta x \to 0} \frac{\Delta z}{\Delta x} = \frac{\partial z}{\partial x} \qquad \lim_{\Delta x \to 0} \frac{\Delta u}{\Delta x} = \frac{\partial u}{\partial x} \qquad \lim_{\Delta x \to 0} \frac{\Delta v}{\Delta x} = \frac{\partial v}{\partial x}$$

In addition, as $\Delta x \to 0$, Δu and Δv also approach 0.* As a result, η_1 and η_2 approach 0. With these facts in mind, we take the limit of both sides of (16.28) as $\Delta x \to 0$. The result is

$$\frac{\partial z}{\partial x} = \left(\frac{\partial z}{\partial u}\right)\left(\frac{\partial u}{\partial x}\right) + \left(\frac{\partial z}{\partial v}\right)\left(\frac{\partial v}{\partial x}\right) \qquad \blacksquare$$

EXAMPLE 1   If $z = u^2 + uv - v^2$ and $u = e^{2x+y}$, $v = \ln(y/x)$, find $\partial z/\partial x$ and $\partial z/\partial y$.

Solution

$$\frac{\partial z}{\partial x} = \left(\frac{\partial z}{\partial u}\right)\left(\frac{\partial u}{\partial x}\right) + \left(\frac{\partial z}{\partial v}\right)\left(\frac{\partial v}{\partial x}\right) = (2u + v)(2e^{2x+y}) + (u - 2v)\left(\frac{-y/x^2}{y/x}\right)$$

$$= \left[4e^{2x+y} + 2\ln\left(\frac{y}{x}\right)\right]e^{2x+y} + \left[e^{2x+y} - 2\ln\left(\frac{y}{x}\right)\right]\left(\frac{-1}{x}\right)$$

---

* If Δu and Δv both equal 0, we take $\eta_1 = 0$ and $\eta_2 = 0$, because if $\eta_1 \neq 0$ and $\eta_2 \neq 0$ in this case, they may be replaced by functions η_1', η_2' that equal 0 when $\Delta u = 0$, $\Delta v = 0$ and equal η_1, η_2 everywhere else.

$$\frac{\partial z}{\partial y} = \left(\frac{\partial z}{\partial u}\right)\left(\frac{\partial u}{\partial y}\right) + \left(\frac{\partial z}{\partial v}\right)\left(\frac{\partial v}{\partial y}\right) = (2u + v)(e^{2x+y}) + (u - 2v)\left(\frac{1/x}{y/x}\right)$$

$$= \left[2e^{2x+y} + \ln\left(\frac{y}{x}\right)\right]e^{2x+y} + \left[e^{2x+y} - 2\ln\left(\frac{y}{x}\right)\right]\frac{1}{y} \quad \blacksquare$$

As the above example illustrates, when $\partial z/\partial x$ and $\partial z/\partial y$ are found, they should be expressed as functions of x and y alone.

The symmetry of the chain rule formula remains even when the number of variables is increased. Thus, if $z = f(u_1, u_2, \ldots, u_m)$ is a differentiable function and if $u_1 = g_1(x_1, x_2, \ldots, x_n)$, $u_2 = g_2(x_1, x_2, \ldots, x_n)$, $\ldots$, $u_m = g_m(x_1, x_2, \ldots, x_n)$ each possess continuous first-order partial derivatives, then the composite function $z = f(g_1, g_2, \ldots, g_m)$ is a function of $x_1, x_2, \ldots, x_n$, and

(16.29)

$$\frac{\partial z}{\partial x_1} = \left(\frac{\partial z}{\partial u_1}\right)\left(\frac{\partial u_1}{\partial x_1}\right) + \left(\frac{\partial z}{\partial u_2}\right)\left(\frac{\partial u_2}{\partial x_1}\right) + \cdots + \left(\frac{\partial z}{\partial u_m}\right)\left(\frac{\partial u_m}{\partial x_1}\right)$$

$$\frac{\partial z}{\partial x_2} = \left(\frac{\partial z}{\partial u_1}\right)\left(\frac{\partial u_1}{\partial x_2}\right) + \left(\frac{\partial z}{\partial u_2}\right)\left(\frac{\partial u_2}{\partial x_2}\right) + \cdots + \left(\frac{\partial z}{\partial u_m}\right)\left(\frac{\partial u_m}{\partial x_2}\right)$$

$$\vdots$$

$$\frac{\partial z}{\partial x_n} = \left(\frac{\partial z}{\partial u_1}\right)\left(\frac{\partial u_1}{\partial x_n}\right) + \left(\frac{\partial z}{\partial u_2}\right)\left(\frac{\partial u_2}{\partial x_n}\right) + \cdots + \left(\frac{\partial z}{\partial u_m}\right)\left(\frac{\partial u_m}{\partial x_n}\right)$$

More compactly, we may write

$$\frac{\partial z}{\partial x_i} = \sum_{j=1}^{m} \left(\frac{\partial z}{\partial u_j}\right)\left(\frac{\partial u_j}{\partial x_i}\right) \quad i = 1, 2, \ldots, n$$

Let's look at an example to see how (16.29) is used in practice.

EXAMPLE 2   If $f(u, v, w) = u^2 + v^2 + w^2$ and $u = xyzt$, $v = e^{x+y+z+t}$, $w = x + 2y + 3z + 4t$, find $\partial f/\partial x$, $\partial f/\partial y$, $\partial f/\partial z$, and $\partial f/\partial t$.

Solution   We use (16.29):

$$\frac{\partial f}{\partial x} = \left(\frac{\partial f}{\partial u}\right)\left(\frac{\partial u}{\partial x}\right) + \left(\frac{\partial f}{\partial v}\right)\left(\frac{\partial v}{\partial x}\right) + \left(\frac{\partial f}{\partial w}\right)\left(\frac{\partial w}{\partial x}\right)$$

$$= (2u)(yzt) + (2v)(e^{x+y+z+t}) + (2w)(1)$$

$$= 2xy^2z^2t^2 + 2e^{2(x+y+z+t)} + 2(x + 2y + 3z + 4t)$$

$$\frac{\partial f}{\partial y} = \left(\frac{\partial f}{\partial u}\right)\left(\frac{\partial u}{\partial y}\right) + \left(\frac{\partial f}{\partial v}\right)\left(\frac{\partial v}{\partial y}\right) + \left(\frac{\partial f}{\partial w}\right)\left(\frac{\partial w}{\partial y}\right)$$

$$= (2u)(xzt) + (2v)(e^{x+y+z+t}) + (2w)(2)$$

$$= 2x^2yz^2t^2 + 2e^{2(x+y+z+t)} + 4(x + 2y + 3z + 4t)$$

$$\frac{\partial f}{\partial z} = \left(\frac{\partial f}{\partial u}\right)\left(\frac{\partial u}{\partial z}\right) + \left(\frac{\partial f}{\partial v}\right)\left(\frac{\partial v}{\partial z}\right) + \left(\frac{\partial f}{\partial w}\right)\left(\frac{\partial w}{\partial z}\right)$$

$$= (2u)(xyt) + (2v)(e^{x+y+z+t}) + (2w)(3)$$

$$= 2x^2y^2zt^2 + 2e^{2(x+y+z+t)} + 6(x + 2y + 3z + 4t)$$

$$\frac{\partial f}{\partial t} = \left(\frac{\partial f}{\partial u}\right)\left(\frac{\partial u}{\partial t}\right) + \left(\frac{\partial f}{\partial v}\right)\left(\frac{\partial v}{\partial t}\right) + \left(\frac{\partial f}{\partial w}\right)\left(\frac{\partial w}{\partial t}\right)$$

$$= (2u)(xyz) + (2v)(e^{x+y+z+t}) + (2w)(4)$$

$$= 2x^2y^2z^2t + 2e^{2(x+y+z+t)} + 8(x + 2y + 3z + 4t)   \quad\blacksquare$$

Again, note that the partial derivatives of f in Example 2 are expressed in terms of x, y, z, and t alone.

EXAMPLE 3   Let $z = f(v - w, \; v - u, \; u - w)$.   Show that

$$\frac{\partial z}{\partial u} + \frac{\partial z}{\partial v} + \frac{\partial z}{\partial w} = 0$$

Solution   If we let $x = v - w$, $y = v - u$,   and $t = u - w$,   then $z = f(x, y, t)$ and we can use (16.29):

$$\frac{\partial z}{\partial u} = \left(\frac{\partial z}{\partial x}\right)\left(\frac{\partial x}{\partial u}\right) + \left(\frac{\partial z}{\partial y}\right)\left(\frac{\partial y}{\partial u}\right) + \left(\frac{\partial z}{\partial t}\right)\left(\frac{\partial t}{\partial u}\right)$$

$$= \frac{\partial z}{\partial x}(0) + \frac{\partial z}{\partial y}(-1) + \frac{\partial z}{\partial t}(1) = -\frac{\partial z}{\partial y} + \frac{\partial z}{\partial t}$$

$$\frac{\partial z}{\partial v} = \left(\frac{\partial z}{\partial x}\right)\left(\frac{\partial x}{\partial v}\right) + \left(\frac{\partial z}{\partial y}\right)\left(\frac{\partial y}{\partial v}\right) + \left(\frac{\partial z}{\partial t}\right)\left(\frac{\partial t}{\partial v}\right)$$

$$= \frac{\partial z}{\partial x}(1) + \frac{\partial z}{\partial y}(1) + \frac{\partial z}{\partial t}(0) = \frac{\partial z}{\partial x} + \frac{\partial z}{\partial y}$$

$$\frac{\partial z}{\partial w} = \left(\frac{\partial z}{\partial x}\right)\left(\frac{\partial x}{\partial w}\right) + \left(\frac{\partial z}{\partial y}\right)\left(\frac{\partial y}{\partial w}\right) + \left(\frac{\partial z}{\partial t}\right)\left(\frac{\partial t}{\partial w}\right)$$

$$= \frac{\partial z}{\partial x}(-1) + \frac{\partial z}{\partial y}(0) + \frac{\partial z}{\partial t}(-1) = -\frac{\partial z}{\partial x} - \frac{\partial z}{\partial t}$$

But

$$\left(-\frac{\partial z}{\partial y} + \frac{\partial z}{\partial t}\right) + \left(\frac{\partial z}{\partial x} + \frac{\partial z}{\partial y}\right) + \left(-\frac{\partial z}{\partial x} - \frac{\partial z}{\partial t}\right) = 0$$

Therefore,

$$\frac{\partial z}{\partial u} + \frac{\partial z}{\partial v} + \frac{\partial z}{\partial w} = 0   \quad\blacksquare$$

A special case of the chain rule occurs when each of the functions $u_1, u_2, \ldots,$ u_m in (16.29) is a function of a single variable, say, t. Then, after composition, the function $z = f(u_1, u_2, \ldots, u_m)$ is a function of the one variable t. In this case, formula (16.29) becomes

(16.30)
$$\frac{dz}{dt} = \left(\frac{\partial z}{\partial u_1}\right)\left(\frac{du_1}{dt}\right) + \left(\frac{\partial z}{\partial u_2}\right)\left(\frac{du_2}{dt}\right) + \cdots + \left(\frac{\partial z}{\partial u_m}\right)\left(\frac{du_m}{dt}\right)$$

where each of the partial derivatives $\partial z/\partial u_1, \ldots, \partial z/\partial u_m$ is expressed in terms of t.

EXAMPLE 4   If $z = u^2 v + v^2 w$ and $u = t$, $v = t^2$, $w = t^3$, then z is a function of t and

$$\frac{dz}{dt} = \left(\frac{\partial z}{\partial u}\right)\left(\frac{du}{dt}\right) + \left(\frac{\partial z}{\partial v}\right)\left(\frac{dv}{dt}\right) + \left(\frac{\partial z}{\partial w}\right)\left(\frac{dw}{dt}\right)$$

$$= (2uv)(1) + (u^2 + 2vw)(2t) + (v^2)(3t^2)$$

$$= 2t^3 + (t^2 + 2t^5)(2t) + (t^4)(3t^2) = 7t^6 + 4t^3 \qquad \blacksquare$$

In Example 4, z can easily be found in terms of t, namely,

$$z = u^2 v + v^2 w$$
$$= (t)^2 t^2 + (t^2)^2 t^3 = t^4 + t^7 = t^7 + t^4$$

Therefore,

$$\frac{dz}{dt} = 7t^6 + 4t^3$$

Of course, (16.30) was introduced so that we can deal with cases where z cannot be so easily expressed in terms of t.

IMPLICIT DIFFERENTIATION

If a differentiable function $y = f(x)$ is defined implicitly by the equation $F(x, y) = 0$, we can find the derivative dy/dx by applying the chain rule. (Recall that in Chap. 3 we used implicit differentiation to get dy/dx.) If $y = f(x)$ is a function defined by the equation $F(x, y) = 0$, it follows that $F(x, f(x)) \equiv 0$. For convenience, we set

$$z = F(u, y) \qquad u = x \qquad \text{and} \qquad y = f(x)$$

Then

(16.31)
$$\frac{dz}{dx} = \left(\frac{\partial F}{\partial u}\right)\left(\frac{du}{dx}\right) + \left(\frac{\partial F}{\partial y}\right)\left(\frac{dy}{dx}\right)$$

But the composite $z = F(u, y) = F(x, f(x)) \equiv 0$, so $dz/dx = 0$. Since $u = x$, then $du/dx = 1$, and (16.31) becomes

$$\left(\frac{\partial F}{\partial x}\right)(1) + \left(\frac{\partial F}{\partial y}\right)\left(\frac{dy}{dx}\right) = 0$$

And, if $\partial F/\partial y \neq 0$,

$$(16.32) \qquad \frac{dy}{dx} = -\frac{F_x}{F_y}$$

EXAMPLE 5   For $F(x, y) = x^2y + y^2 - 2x = 0$,   find dy/dx by using (16.32).

Solution
$$F_x = \frac{\partial F}{\partial x} = 2xy - 2 \qquad F_y = \frac{\partial F}{\partial y} = x^2 + 2y$$

Hence, by (16.32), if $x^2 + 2y \neq 0$,

$$\frac{dy}{dx} = \frac{-F_x}{F_y} = -\frac{2xy - 2}{x^2 + 2y} = \frac{2(1 - xy)}{x^2 + 2y} \qquad \blacksquare$$

You may wish to compare the above method with the method of implicit differentiation used in Chapter 3.

If a differentiable function $z = f(x, y)$ is defined implicitly by the equation $F(x, y, z) = 0$,   we can find the partial derivatives $\partial z/\partial x$ and $\partial z/\partial y$ by applying the chain rule. For convenience, we set $w = F(u, v, z)$ and $u = x$, $v = y$,   and $z = f(x, y)$.   Since the composite $w = F(x, y, f(x, y)) \equiv 0$,   it follows that $\partial w/\partial x = 0$, $\partial w/\partial y = 0$.   For $\partial w/\partial x$, we obtain

$$\frac{\partial w}{\partial x} = \left(\frac{\partial F}{\partial u}\right)\left(\frac{\partial u}{\partial x}\right) + \left(\frac{\partial F}{\partial v}\right)\left(\frac{\partial v}{\partial x}\right) + \left(\frac{\partial F}{\partial z}\right)\left(\frac{\partial z}{\partial x}\right) = 0$$

Since $u = x$ and $v = y$,   we have

$$\left(\frac{\partial F}{\partial x}\right)(1) + \left(\frac{\partial F}{\partial y}\right)(0) + \left(\frac{\partial F}{\partial z}\right)\left(\frac{\partial z}{\partial x}\right) = 0$$

and if $\partial F/\partial z \neq 0$,   it follows that

$$(16.33) \qquad \frac{\partial z}{\partial x} = -\frac{\partial F/\partial x}{\partial F/\partial z} = -\frac{F_x(x, y, z)}{F_z(x, y, z)}$$

In a similar way we can show that

$$(16.34) \qquad \frac{\partial z}{\partial y} = -\frac{F_y(x, y, z)}{F_z(x, y, z)}$$

EXAMPLE 6   For $F(x, y, z) = x^2z^2 + y^2 - z^2 + 6yz - 10 = 0$ find $\partial z/\partial x$ and $\partial z/\partial y$ by using (16.33) and (16.34).

Solution
$$F_x = \frac{\partial F}{\partial x} = 2xz^2 \qquad F_y = 2y + 6z \qquad F_z = 2zx^2 - 2z + 6y$$

Thus, if $F_z = 2zx^2 - 2z + 6y \neq 0$,

$$\frac{\partial z}{\partial x} = -\frac{2xz^2}{2zx^2 - 2z + 6y} \qquad \frac{\partial z}{\partial y} = -\frac{2y + 6z}{2zx^2 - 2z + 6y}$$

$$= \frac{-xz^2}{zx^2 - z + 3y} \qquad\qquad = -\frac{y + 3z}{zx^2 - z + 3y} \qquad \blacksquare$$

EXERCISE 6

In Problems 1–12 find $\partial z/\partial x$ and $\partial z/\partial y$ by using the chain rule.

1. $z = u^2 + v^2,\quad u = xe^y,\quad v = ye^x$

2. $z = u^2 - v^2,\quad u = x \ln y,\quad v = y \ln x$

3. $z = e^u \sin v,\quad u = x^2 y,\quad v = \ln(xy)$

4. $z = \dfrac{1}{v} \ln u,\quad u = \sqrt{xy},\quad v = \dfrac{y}{x}$

5. $z = se^r,\quad r = x^2 + y^2,\quad s = \dfrac{y}{x}$

6. $z = \sqrt{s^2 + r^2},\quad s = \ln(xy),\quad r = \sqrt{xy}$

7. $z = uv^2w^3,\quad u = 2x + y,\quad v = 5x - 3y,\quad w = 2x + 3y$

8. $z = u^2 - v^2 + w,\quad u = e^{x+y},\quad v = xy,\quad w = \dfrac{y}{x}$

9. $z = \ln(u^2 + v^2),\quad u = \dfrac{y^2}{x},\quad v = \dfrac{x}{y^2}$

10. $z = u \sin v - v \sin u,\quad u = x^2 y,\quad v = yx^2$

11. $z = u^2 + v^2,\quad u = \sin(x - y),\quad v = \cos(x + y)$

12. $z = e^u + v,\quad u = \tan^{-1}\left(\dfrac{x}{y}\right),\quad v = \ln(x + y)$

In Problems 13–22 find dz/dt.

13. $z = u^2 + v^2,\quad u = te^t,\quad v = te^{-t}$

14. $z = u^2 - v^2,\quad u = te^{-t},\quad v = t^2 e^{-t}$

15. $z = e^u \sin v,\quad u = \sqrt{t},\quad v = \pi t$

16. $z = \ln(uv),\quad u = t^5,\quad v = \sqrt{t + 1}$

17. $z = \ln(u/v),\quad u = te^t,\quad v = e^{t^2}$

18. $z = e^{u/v},\quad u = \sqrt{t},\quad v = t^3 + 1$

19. $z = u^2 vw^3,\quad u = \sin t,\quad v = \cos t,\quad w = e^t$

20. $z = \sqrt{uvw},\quad u = e^t,\quad v = te^t,\quad w = t^2 e^t$

21. $z = \dfrac{uv}{u^2 + v^2},\quad u = \sin t,\quad v = \cos t$

22. $z = v \ln u + uv + \tan v,\quad u = \dfrac{t}{t + 1},\quad v = t^3 - t$

In Problems 23–28 find dy/dx by using (16.32).

23. $F(x, y) = x^2 y - y^2 x + xy - 5 = 0$

24. $F(x, y) = x^3 y^2 - xy + x^2 y - 10 = 0$

25. $F(x, y) = x \sin y + y \sin x - 2 = 0$

26. $F(x, y) = xe^y + ye^x - xy = 0$

27. $F(x, y) = x^{1/3} + y^{1/3} - 1 = 0$

28. $F(x, y) = x^{2/3} + y^{2/3} - 1 = 0$

In Problems 29–32 find $\partial z/\partial x$ and $\partial z/\partial y$ by using (16.33) and (16.34).

29. $F(x, y, z) = xz + 3yz^2 + x^2 y^3 - 5z = 0$

30. $F(x, y, z) = x^2 z + y^2 z + x^3 y - 10z = 0$

31. $F(x, y, z) = \sin z + y \cos z + xyz - 10 = 0$

32. $F(x, y, z) = xe^{yz} + ye^{xz} + xyz = 0$

33. If $z = f(x, y),\quad x = r \cos \theta,\quad y = r \sin \theta,\quad$ show that

$$\left(\frac{\partial z}{\partial r}\right)^2 + \frac{1}{r^2}\left(\frac{\partial z}{\partial \theta}\right)^2 = \left(\frac{\partial z}{\partial x}\right)^2 + \left(\frac{\partial z}{\partial y}\right)^2$$

34. If $z = f(x, y)$ and $x = u \cos \theta - v \sin \theta,\ y = u \sin \theta + v \cos \theta,$ with θ a constant, show that $(\partial f/\partial u)^2 + (\partial f/\partial v)^2 = (\partial f/\partial x)^2 + (\partial f/\partial y)^2.$

35. If $z = f(x - y, y - x)$, show that $(\partial z/\partial x) + (\partial z/\partial y) = 0.$ [*Hint:* Let $u = x - y,\ v = y - x.$]

36. If $z = yf(x^2 - y^2)$, show that $y(\partial z/\partial x) + x(\partial z/\partial y) = xz/y.$

37. If $w = f(u)$ and $u = \sqrt{x^2 + y^2 + z^2}$, show that $(\partial w/\partial x)^2 + (\partial w/\partial y)^2 + (\partial w/\partial z)^2 = (dw/du)^2.$

38. If $z = f(y/x)$, show that $x(\partial z/\partial x) + y(\partial z/\partial y) = 0$.

39. Show that if $z = f(u/v, v/w)$, then $u(\partial z/\partial u) + v(\partial z/\partial v) + w(\partial z/\partial w) = 0$.

40. Let $z = f(y + ax) + g(y - ax)$, $a \neq 0$. Show that z satisfies the wave equation $\partial^2 z/\partial x^2 = a^2(\partial^2 z/\partial y^2)$.

41. If $z = f(u, v)$ and $u = g(x, y)$, $v = h(x, y)$, find expressions for $\partial^2 z/\partial x^2$, $\partial^2 z/\partial x\,\partial y$, and $\partial^2 z/\partial y^2$.

42. Prove the formula for $\partial z/\partial y$ in the chain rule (16.23).

43. Prove the formula for $\partial z/\partial y$ in (16.34).

44. A certain confined gas obeys the ideal gas law $PV = 20T$. If the temperature of the gas is increasing at the rate of 5°C per second and if, when the temperature is 80°C, the pressure is 10 newtons per square meter and is decreasing at the rate of 2 newtons per square meter per second, find the rate of change of the volume.

45. Prove that if $F(x, y, z) = 0$ is differentiable, then $(\partial z/\partial x)(\partial x/\partial y)(\partial y/\partial z) = -1$.

In Problems 46 and 47 find $\partial w/\partial x$, $\partial w/\partial y$, and $\partial w/\partial z$.

46. $w = (2x + 3y)^{4z}$ **47.** $w = (2x)^{3y+4z}$

48. Let $y = f(a, b)$, $a = h(s, t)$, $b = k(s, t)$. When $s = 1$ and $t = 3$ we know that

$$\frac{\partial h}{\partial s} = 4 \qquad \frac{\partial k}{\partial s} = -3 \qquad \frac{\partial h}{\partial t} = 1 \qquad \frac{\partial k}{\partial t} = -5$$

Also,

$$h(1, 3) = 6 \qquad k(1, 3) = 2 \qquad f_a(6, 2) = 7 \qquad f_b(6, 2) = 2$$

What are $\partial y/\partial s$ and $\partial y/\partial t$ at $(1, 3)$?

49. Suppose we denote the expression $(\partial^2/\partial x^2) + (\partial^2/\partial y^2)$ by Δ. If $z = f(x, y)$, show that

$$\Delta f = \frac{\partial^2 f}{\partial r^2} + \frac{1}{r}\left(\frac{\partial f}{\partial r}\right) + \frac{1}{r^2}\left(\frac{\partial^2 f}{\partial \theta^2}\right)$$

where $x = r \cos \theta$ and $y = r \sin \theta$.

Miscellaneous Exercises

In Problems 1–4 find all the second partial derivatives of the given functions.

1. $w = e^{xyz}$ **2.** $w = ze^{xy}$

3. $F(x, y, z) = e^x \sin y + e^y \sin z$ **4.** $u = z \tan^{-1}(y/x)$

5. If $f(x, y) = \sqrt{x^2 - y^2}$, find $f_x(2, 1)$ and $f_y(2, -1)$.

6. If $F(x, y) = e^x \sin y$, find $F_x(0, \pi/6)$ and $F_y(0, \pi/6)$.

7. Find an equation of the tangent line to the curve at $(1, -2, 1)$ cut from the surface $z = 4x^2 - y^2 + 1$ by the plane $x = 1$; by the plane $y = -2$.

8. Interpret $f_x(1, -\frac{1}{3})$ and $f_y(1, -\frac{1}{3})$ geometrically if $z = f(x, y) = 4x^2 + 9y^2 - 12$.

9. If $u = z \tan^{-1}(x/y)$, show that $(\partial^2 u/\partial x^2) + (\partial^2 u/\partial y^2) + (\partial^2 u/\partial z^2) = 0$.

10. Show that $u = e^{-\alpha^2 t} \sin \alpha x$ satisfies the equation $\partial u/\partial t = \partial^2 u/\partial x^2$ for all values of the constant α.

11. Describe the set of points (x, y, z) satisfying the conditions $x^2 + y^2 + z^2 < 1$, $x^2 + y^2 < z^2$, and $z > 0$.

12. Sketch the surfaces $x^2 + y^2 + z^2 = 4$ and $z = \frac{1}{3}(x^2 + y^2)$ and their curve of intersection above the xy-plane. What is the length of this curve?

13. Use (16.2) to show that the function $z = \sqrt{x^2 + y^2}$ does not have partial derivatives at $(0, 0)$. By discussing the graph of the function, give a geometric reason why this should be so.

14. Find the first partial derivatives of $w = x^{yz}$.

15. If $x = r \cos \theta$ and $y = r \sin \theta$, show that

$$\begin{vmatrix} \dfrac{\partial x}{\partial r} & \dfrac{\partial x}{\partial \theta} \\[2mm] \dfrac{\partial y}{\partial r} & \dfrac{\partial y}{\partial \theta} \end{vmatrix} = r \quad \text{and} \quad \begin{vmatrix} \dfrac{\partial r}{\partial x} & \dfrac{\partial r}{\partial y} \\[2mm] \dfrac{\partial \theta}{\partial x} & \dfrac{\partial \theta}{\partial y} \end{vmatrix} = \frac{1}{r}$$

16. Suppose that $F(x, y)$ has continuous second-order partial derivatives and $F(x, y) = 0$ defines y as a function of x. Show that

$$\frac{d^2 y}{dx^2} = -\frac{F_y^2 F_{xx} - 2F_x F_y F_{xy} + F_x^2 F_{yy}}{F_y^3} \quad \text{where} \quad F_y \neq 0$$

17. Use the result of Problem 16 to find $d^2 y/dx^2$ if $x^3 + 3xy - y^3 = 6$.

18. Let

$$f(x, y) = \begin{cases} \dfrac{xy(x^2 - y^2)}{x^2 + y^2} & \text{if } (x, y) \neq (0, 0) \\[3mm] 0 & \text{if } (x, y) = (0, 0) \end{cases}$$

Use the definition of partial derivative as the limit of a difference quotient to show that:
(a) $f_x(0, y) = -y$ (b) $f_y(x, 0) = x$ (c) $f_{xy}(0, 0) = -1$ (d) $f_{yx}(0, 0) = 1$
(This provides an example of a function whose mixed second partial derivatives f_{xy} and f_{yx} are different.)

19. If you are told that $f(x, y)$ is a function whose partial derivatives are $f_x(x, y) = 2x - y$ and $f_y(x, y) = x - 2y$, should you believe it? Explain.

20. Let

$$f(x, y) = \begin{cases} \dfrac{\sin(x^2 + y^2)}{x^2 + y^2} & \text{if } (x, y) \neq (0, 0) \\[3mm] 1 & \text{if } (x, y) = (0, 0) \end{cases}$$

Is f continuous at $(0, 0)$? Explain why or why not.

21. Let

$$f(x, y) = \begin{cases} \dfrac{\sin(x^2 - y^2)}{x^2 + y^2} & \text{if } (x, y) \neq (0, 0) \\[3mm] 1 & \text{if } (x, y) = (0, 0) \end{cases}$$

Is f continuous at $(0, 0)$? Explain why or why not.

In Problems 22–24 sketch the cylindrical surface.

22. $y = \sin x$ 23. $y = \ln z$ 24. $z = e^x$

25. Sketch on the same set of axes the level curves of $f(x, y) = x^2 - y^2$ and $g(x, y) = xy$ for $c = \pm 1$, $\pm 2, \pm 3$.

26. Refer to Problem 25. Show that at each point $P_0 \neq (0, 0)$, the level curve of $f(x, y) = x^2 - y^2$ through P_0 is perpendicular to the level curve of $g(x, y) = xy$ through P_0. The two families of level curves are said to be *orthogonal*.

27. Find f_x and f_y at $(0, 0)$ if

$$f(x, y) = \begin{cases} e^{-1/(x^2 + y^2)} & \text{if } (x, y) \neq (0, 0) \\ 0 & \text{if } (x, y) = (0, 0) \end{cases}$$

28. Calculate $f_x, f_y, f_{xx}, f_{yy},$ and f_{xy} for $f(x, y) = (xy)^{xy}$. What is the domain of f?

29. Show that $xf_x + yf_y = 0$ for $f(x, y) = \sin^{-1}(y/x)$.

30. Show that $xf_x + yf_y + zf_z = 0$ for $f(x, y, z) = e^{x/y} + e^{y/z} + e^{z/x}$.

31. Show that $f(x, y) = xy/(x + y)$ satisfies $x^2 f_{xx} + 2xy f_{xy} + y^2 f_{yy} = 0$.

32. Show that $f(r, \theta) = r^n \sin n\theta$ satisfies $f_{rr} + (1/r)f_r + (1/r^2)f_{\theta\theta} = 0$. (This is Laplace's equation in polar coordinates.)

33. Find a in terms of b and c so that $f(t, x, y) = e^{at} \sin bx \cos cy$ satisfies $f_t = f_{xx} + f_{yy}$.

34. Suppose $u(x, y)$ and $v(x, y)$ have continuous second partial derivatives, and $u_x = v_y$, $u_y = -v_x$. Show that u and v are harmonic functions. (See Problems 57 and 58 in Exercise 3.)

35. Let $f(x, y) = 3xy^2/(x^2 + y^4)$. Show that $\lim_{(x, y) \to (0, 0)} f(x, y) = 0$ along each line of approach $y = mx$ to the origin. Find $\lim_{(x, y) \to (0, 0)} f(x, y)$ along $x = y^2$. Does $\lim_{(x, y) \to (0, 0)} f(x, y)$ exist? Explain.

36. Show that $f(x, y, z) = (x^2 + y^2 + z^2)^{-1/2}$ satisfies the three-dimensional Laplace equation $f_{xx} + f_{yy} + f_{zz} = 0$.

37. Show that $f(x, t) = \cos(x + ct)$ satisfies the one-dimensional wave equation $f_{tt} = c^2 f_{xx}$, where c is a constant.

38. Find symmetric equations of the tangent lines at $(x_0, y_0, f(x_0, y_0))$ to the curve of intersection of $z = f(x, y)$ and $y = y_0$, and the curve of intersection of $z = f(x, y)$ and $x = x_0$. Write an equation of the plane determined by these two lines. What is the geometric relationship of this plane to the surface $z = f(x, y)$?

39. Find f_x and f_y if $f(x, y) = \sqrt{(x + 2y)/(3x - y)}$. [*Hint:* Use logarithmic differentiation.]

40. Find f_x and f_y if $f(x, y) = (x + y)(x - y)/[(2x + 3y)(3x - 2y)]$.

41. Suppose a thin metal rod extends along the x-axis from $x = 0$ to $x = 20$, and for each x, $0 \leq x \leq 20$, the temperature of the rod at time $t \geq 0$ is $T(t, x) = 40e^{-\lambda t} \sin(\pi x/20)$, where $\lambda > 0$ is a constant.
    (a) Show that $T_t = -\lambda T$, $T_{xx} = -(\pi^2/400)T$, and $T_t = (1/k^2)T_{xx}$ for some k.
    (b) Graph the initial temperature distribution, $y = T(0, x)$, $0 \leq x \leq 20$.
    (c) At what point(s) on the rod is the rate of cooling one-half the maximum rate of cooling?

42. The law of cosines is $a^2 = b^2 + c^2 - 2bc \cos A$. Find $\partial a/\partial b$, $\partial a/\partial c$, $\partial a/\partial A$, and $\partial c/\partial A$.

43. Find f_x and f_y if $f(x, y) = \int_x^y \ln(\cos \sqrt{t}) \, dt$.

44. Let

$$f(t, x) = \int_0^{x/2\sqrt{\lambda t}} e^{-u^2} \, du$$

Show that $f_t = \lambda f_{xx}$. [*Hint:* Use the chain rule to show that $(d/dx) \int_a^{g(x)} h(u) \, du = h(g(x))g'(x)$.]

45. Let $u = r^m \cos m\theta$. Show that

$$\frac{\partial^2 u}{\partial r^2} + \frac{1}{r^2}\left(\frac{\partial^2 u}{\partial \theta^2}\right) + \frac{1}{r}\left(\frac{\partial u}{\partial r}\right) = 0 \qquad \text{for all } m$$

46. Show that $u = e^{ax} \sin ay$ is harmonic.

47. Show that $u = e^{ax} \cos ay$ is harmonic.

48. Show that the function below has first partial derivatives at all points.

$$f(x, y) = \begin{cases} \dfrac{x^3 - y^3}{x^2 + y^2} & \text{if } (x, y) \neq (0, 0) \\ 0 & \text{if } (x, y) = (0, 0) \end{cases}$$

In Problems 49–52 find the total differential of each function.

49. $u = x\sqrt{1 - y^2}$ **50.** $u = \sin^{-1}(x/y)$ **51.** $u = ze^{xy}$ **52.** $u = \ln(xyz)$

In Problem 53 and 54 find the derivative du/dt of each function.

53. $u = \sin(xy) - x \sin y$;   if $x = e^t$, $y = te^t$ **54.** $u = \dfrac{x}{y} + \dfrac{y}{z} + \dfrac{z}{x}$;   if $x = \dfrac{1}{t}$, $y = \dfrac{1}{t^2}$, $z = \dfrac{1}{t^3}$

In Problems 55 and 56 find $\partial u/\partial r$ and $\partial u/\partial s$ in terms of r and s.

55. $u = xy + yz - zx$; $x = r + s$, $y = rs$, $z = s$

56. $u = \sqrt{x^2 + y^2 + z^2}$; $x = r \cos s$, $y = r \sin s$, $z = \sqrt{r^2 + s^2}$

57. Use the differential of $f(x, y) = y^2 \cos x$ to find an approximate value of $f(0.05, 1.98)$. Compare the result with the value 3.9155 found from a calculator.

58. The electrical resistance R of a wire is proportional to the length L of the wire and inversely proportional to the square of its diameter D; that is, $R = kL/D^2$,   where k is a constant. Measurements of L and D are subject to errors of 1% and 2%, respectively.

    (a)   Show that: $\dfrac{dR}{R} = \dfrac{dL}{L} - 2\left(\dfrac{dD}{D}\right)$

    (b) What is the maximum error in the computation of R?

59. Find the total differential of $f(x, y, z) = x^{y^z}$ and $g(x, y, z) = (x^y)^z$.

17

Directional Derivative, Gradient, and Extrema

1. Directional Derivative; Gradient
2. Tangent Planes
3. Extrema of Functions of Two Variables
4. The Method of Lagrange Multipliers
   Miscellaneous Exercises

1. Directional Derivative; Gradient

Recall that the partial derivatives of a function $z = f(x, y)$ at (x_0, y_0) have been defined by

(17.1)

$$f_x(x_0, y_0) = \lim_{\Delta x \to 0} \frac{f(x_0 + \Delta x, y_0) - f(x_0, y_0)}{\Delta x}$$

$$f_y(x_0, y_0) = \lim_{\Delta y \to 0} \frac{f(x_0, y_0 + \Delta y) - f(x_0, y_0)}{\Delta y}$$

The partial derivative $f_x(x_0, y_0)$ equals the rate of change of f at (x_0, y_0) in a *direction parallel to the x-axis*; similarly, $f_y(x_0, y_0)$ equals the rate of change of f at (x_0, y_0) in a *direction parallel to the y-axis*. The *directional derivative*, a generalization of the partial derivative, equals the rate of change of f at (x_0, y_0) in *any chosen direction in the xy-plane.*

To lay the groundwork for the definition of the directional derivative, we first let $P_0 = (x_0, y_0)$ denote a point in the domain of a function $z = f(x, y)$. Then we let $\mathbf{u} = (\cos \theta)\mathbf{i} + (\sin \theta)\mathbf{j}$ denote a unit vector with initial point at P_0 so that $\mathbf{u}$ makes an angle θ with the positive x-axis (see Fig. 1). Finally, we let L denote a directed line segment with initial point P_0 in the direction of $\mathbf{u}$. We choose the point $P = (x, y)$ on L, different from P_0, so that the directed line segment $\overrightarrow{P_0P}$ is in the domain of f. If $t = |P_0P|$, then, as shown in Figure 2, the coordinates of P are $(x_0 + t \cos \theta, y_0 + t \sin \theta)$.

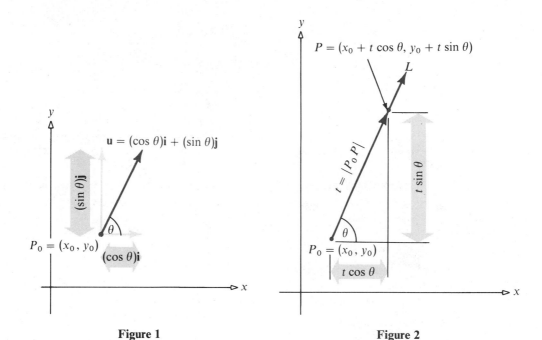

Figure 1   **Figure 2**

The average rate of change of f from P_0 to P is

$$\frac{f(x_0 + t \cos \theta, y_0 + t \sin \theta) - f(x_0, y_0)}{t}$$

By taking the limit as $t \to 0$, we obtain the rate of change of f at (x_0, y_0) in the direction of $\mathbf{u}$, provided this limit exists. This limit, denoted by $D_\mathbf{u} f(x_0, y_0)$, is the *directional derivative of f at (x_0, y_0) in the direction of $\mathbf{u}$.*

(17.2)   **DEFINITION** *Directional Derivative.* **We define the derivative of f at (x_0, y_0) in the direction of the unit vector** $\mathbf{u} = (\cos \theta)\mathbf{i} + (\sin \theta)\mathbf{j}$ **as**

(17.3) $$D_\mathbf{u} f(x_0, y_0) = \lim_{t \to 0} \frac{f(x_0 + t \cos \theta, y_0 + t \sin \theta) - f(x_0, y_0)}{t}$$

provided this limit exists. We shall refer to $D_\mathbf{u} f$ as the *directional derivative of f at (x_0, y_0) in the direction of* u.

As (17.3) implies, the directional derivative of f in the direction of $\mathbf{u}$ is a number. In fact, it is easy to show that the partial derivatives of f at (x_0, y_0) are special cases of the directional derivative $D_\mathbf{u} f(x_0, y_0)$. If we set $\theta = 0$, then $\mathbf{u} = (\cos 0)\mathbf{i} + (\sin 0)\mathbf{j} = \mathbf{i}$ is a vector parallel to the x-axis. From (17.3), the directional derivative of f at (x_0, y_0) in this direction is

$$D_\mathbf{u} f(x_0, y_0) = D_\mathbf{i} f(x_0, y_0) = \lim_{t \to 0} \frac{f(x_0 + t, y_0) - f(x_0, y_0)}{t} = f_x(x_0, y_0)$$

(17.1)

Similarly, if $\theta = \pi/2$, then $\mathbf{u} = \mathbf{j}$ and $D_\mathbf{j} f(x_0, y_0) = f_y(x_0, y_0)$.

When $z = f(x, y)$ is differentiable, a simple formula can be used to calculate directional derivatives:

(17.4)   **THEOREM** If $z = f(x, y)$ **is differentiable, then the directional derivative of f at** (x_0, y_0) **in the direction of** $\mathbf{u} = (\cos \theta)\mathbf{i} + (\sin \theta)\mathbf{j}$ **is given by**

(17.5) $$D_\mathbf{u} f(x_0, y_0) = f_x(x_0, y_0) \cos \theta + f_y(x_0, y_0) \sin \theta$$

Proof We define the function g as

$$g(t) = f(x_0 + t \cos \theta, y_0 + t \sin \theta)$$

Then the derivative of g at $t = 0$ is $D_\mathbf{u} f(x_0, y_0)$ since

$$g'(0) = \lim_{t \to 0} \frac{g(t) - g(0)}{t} = \lim_{t \to 0} \frac{f(x_0 + t \cos \theta, y_0 + t \sin \theta) - f(x_0, y_0)}{t}$$

$$= D_\mathbf{u} f(x_0, y_0)$$

(17.3)

We apply the chain rule (16.30) to

$$g(t) = f(u, v) \qquad u = x_0 + t \cos \theta \qquad \text{and} \qquad v = y_0 + t \sin \theta$$

to get

$$g'(t) = \left(\frac{\partial f}{\partial u}\right)\left(\frac{du}{dt}\right) + \left(\frac{\partial f}{\partial v}\right)\left(\frac{dv}{dt}\right) = \left(\frac{\partial f}{\partial u}\right) \cos \theta + \left(\frac{\partial f}{\partial v}\right) \sin \theta$$

At $t = 0$, $u = x_0$ and $v = y_0$, so

$$D_\mathbf{u} f(x_0, y_0) = g'(0) = f_x(x_0, y_0) \cos \theta + f_y(x_0, y_0) \sin \theta \qquad \blacksquare$$

EXAMPLE 1   Find the directional derivative $D_\mathbf{u} f(x, y)$ of $f(x, y) = x^2 y + y^2$ in the direction of

$$\mathbf{u} = \left(\cos \frac{\pi}{4}\right)\mathbf{i} + \left(\sin \frac{\pi}{4}\right)\mathbf{j}$$

What is $D_\mathbf{u} f(1, 2)$?

Solution   The partial derivatives of f are

$$f_x(x, y) = 2xy \qquad f_y(x, y) = x^2 + 2y$$

Hence, by (17.5),

$$D_{\mathbf{u}}f(x, y) = (2xy) \cos \frac{\pi}{4} + (x^2 + 2y) \sin \frac{\pi}{4} = \sqrt{2}xy + \frac{\sqrt{2}}{2}(x^2 + 2y)$$

$$D_{\mathbf{u}}f(1, 2) = 2\sqrt{2} + \frac{\sqrt{2}}{2}(1 + 4) = \frac{9\sqrt{2}}{2} \qquad ■$$

The directional derivative of a function f in the direction of $\mathbf{a}$, where $\mathbf{a}$ is any nonzero vector, is defined as the directional derivative of f in the direction of $\mathbf{u}$, where $\mathbf{u} = \mathbf{a}/\|\mathbf{a}\|$, the unit vector having the same direction as $\mathbf{a}$.

EXAMPLE 2   Find the directional derivative of $f(x, y) = x \sin y$ at $(2, \pi/3)$ in the direction of $\mathbf{a} = 3\mathbf{i} + 4\mathbf{j}$.

Solution   The vector $\mathbf{a}$ is not a unit vector. However, the unit vector $\mathbf{u}$ in the direction of $\mathbf{a}$ is

$$\mathbf{u} = \frac{\mathbf{a}}{\|\mathbf{a}\|} = \frac{3}{5}\mathbf{i} + \frac{4}{5}\mathbf{j}$$

The partial derivatives of f are

$$f_x(x, y) = \sin y \qquad f_y(x, y) = x \cos y$$

At $(2, \pi/3)$,

$$f_x\left(2, \frac{\pi}{3}\right) = \frac{\sqrt{3}}{2} \qquad f_y\left(2, \frac{\pi}{3}\right) = 1$$

The directional derivative of f at $(2, \pi/3)$ in the direction of $\mathbf{a} = 3\mathbf{i} + 4\mathbf{j}$ is

$$D_{\mathbf{a}}f\left(2, \frac{\pi}{3}\right) = \left(\frac{\sqrt{3}}{2}\right)\left(\frac{3}{5}\right) + (1)\left(\frac{4}{5}\right) = \frac{3\sqrt{3} + 8}{10} \qquad ■$$

GEOMETRIC INTERPRETATION

The partial derivative $f_x(x_0, y_0)$ of $z = f(x, y)$ equals the slope of the tangent line to the curve of intersection of the surface and the plane $y = y_0$. A similar interpretation has been given for $f_y(x_0, y_0)$. The directional derivative $D_{\mathbf{u}}f(x_0, y_0)$ of f at (x_0, y_0) in the direction of $\mathbf{u}$ may be interpreted as follows: Let L denote the line through (x_0, y_0) in the direction of $\mathbf{u}$. The plane through L perpendicular to the xy-plane will intersect the surface $z = f(x, y)$ in a curve C (see Fig. 3, p. 916).

The directional derivative $D_{\mathbf{u}}f(x_0, y_0)$ equals the slope of the tangent line to C at the point $(x_0, y_0, f(x_0, y_0))$ on the surface.

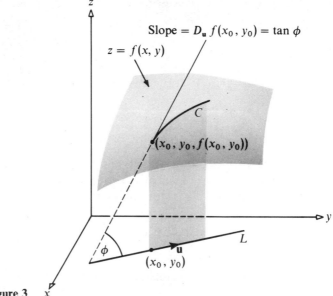

Figure 3

GRADIENT

The directional derivative of f in the direction of $\mathbf{u}$ may be compactly written in terms of vector notation. We define the vector $\nabla f(x, y)$, read "del f" and called the *gradient of f*,[*] as

(17.6)
$$\nabla f(x, y) = f_x(x, y)\mathbf{i} + f_y(x, y)\mathbf{j}$$

The dot product of the vector $\nabla f(x, y)$ and the unit vector $\mathbf{u} = (\cos \theta)\mathbf{i} + (\sin \theta)\mathbf{j}$ is a scalar that is equal to the directional derivative $D_\mathbf{u} f(x, y)$. That is,

(17.7)
$$\nabla f(x, y) \cdot \mathbf{u} = D_\mathbf{u} f(x, y)$$

Just as the differential operator $D = d/dx$ is used to denote the operation of differentiation, the *vector differential operator* ∇ may be used to symbolize the operation

$$\nabla = \mathbf{i}\frac{\partial}{\partial x} + \mathbf{j}\frac{\partial}{\partial y}$$

It is important to remember that ∇ is an operator and has no meaning except when it operates on a function $z = f(x, y)$.

MAXIMIZING THE RATE OF CHANGE OF $z = f(x, y)$

We let $z = f(x, y)$ denote a differentiable function. The directional derivative $D_\mathbf{u} f(x_0, y_0)$ of $z = f(x, y)$ at the point (x_0, y_0) in the direction of $\mathbf{u}$ equals the

---

[*] The abbreviation "del f" is short for "delta f," the symbol ∇ being the inverted Greek letter delta. In some books, ∇f is written as "grad f."

rate of change of f at (x_0, y_0) in the direction of **u**. We seek the direction of **u** that maximizes $D_{\mathbf{u}}f(x_0, y_0)$.   This direction is the direction along which the rate of change of f at (x_0, y_0) is greatest (see Fig. 4).

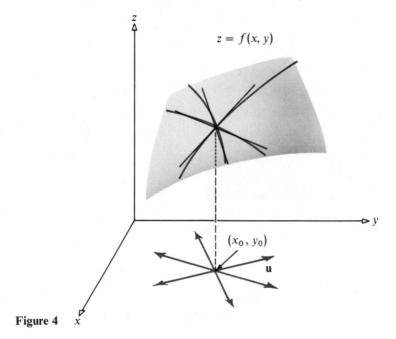

Figure 4

We proceed as follows: We let ϕ denote the angle between the unit vector **u** and the gradient $\nabla f(x_0, y_0)$, as shown in Figure 5. Then

$$D_{\mathbf{u}}f(x_0, y_0) = \nabla f(x_0, y_0) \cdot \mathbf{u} = \|\nabla f(x_0, y_0)\| \, \|\mathbf{u}\| \cos \phi$$

(17.8)   (14.19)

$$= \|\nabla f(x_0, y_0)\| \cos \phi$$

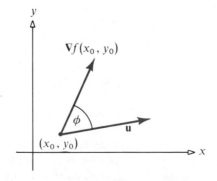

Figure 5

The value of $D_{\mathbf{u}}f(x_0, y_0)$ is a maximum when $\cos \phi = 1$, that is, when the angle ϕ between $\mathbf{u}$ and ∇f is 0. Hence, the directional derivative $D_{\mathbf{u}}f(x_0, y_0)$ at (x_0, y_0) is a maximum when the direction of $\mathbf{u}$ is the same as that of ∇f at (x_0, y_0). It follows from (17.8) that if $\mathbf{u}$ has the same direction as $\nabla f(x_0, y_0)$, the maximum value of the directional derivative $D_{\mathbf{u}}f(x_0, y_0)$ is $\|\nabla f(x_0, y_0)\|$. Thus, we have the following result:

(17.9)   **THEOREM** Let $z = f(x, y)$ **have continuous partial derivatives** f_x **and** f_y, **and let** (x_0, y_0) **be a point in the plane for which** ∇f **is not 0. Then:**

(a)   **The directional derivative** $D_{\mathbf{u}}f(x_0, y_0)$ **of** f **at** (x_0, y_0) **is maximum in the direction of** $\nabla f(x_0, y_0)$; **that is,** $D_{\mathbf{u}}f(x_0, y_0)$ **is maximum when**

$$\mathbf{u} = \frac{\nabla f(x_0, y_0)}{\|\nabla f(x_0, y_0)\|}$$

(b)   **The maximum value of** $D_{\mathbf{u}}f(x_0, y_0)$ **is** $\|\nabla f(x_0, y_0)\|$.

(c)   **The directional derivative** $D_{\mathbf{u}}f(x_0, y_0)$ **is minimum in the direction opposite to** $\nabla f(x_0, y_0)$, **that is, when**

$$\mathbf{u} = -\frac{\nabla f(x_0, y_0)}{\|\nabla f(x_0, y_0)\|}$$

EXAMPLE 3   Find the direction for which the directional derivative of $f(x, y) = x^2 - xy + y^2$ at $(1, -2)$ is a maximum, and find that maximum value.

Solution   From theorem (17.9), we need to calculate the gradient of f at $(1, -2)$:

$$\nabla f(x, y) = (2x - y)\mathbf{i} + (2y - x)\mathbf{j} \qquad \nabla f(1, -2) = 4\mathbf{i} - 5\mathbf{j}$$

The direction for which $D_{\mathbf{u}}f(1, -2)$ is a maximum is the same as that of the unit vector $(4/\sqrt{41})\mathbf{i} - (5/\sqrt{41})\mathbf{j}$.

The maximum value of the directional derivative at $(1, -2)$ equals the magnitude of the gradient, namely, $\|\nabla f(1, -2)\| = \sqrt{16 + 25} = \sqrt{41}$.   ■

The directional derivative $D_{\mathbf{u}}f(x_0, y_0)$ equals the rate of change of f at (x_0, y_0) in the direction of $\mathbf{u}$. When $\mathbf{u}$ has the same direction as $\nabla f(x_0, y_0)$, the rate at which f changes at (x_0, y_0) is a maximum. It follows that $f(x, y)$ will increase most rapidly in this direction. Thus, we may restate part of (17.9) in the following way:

(17.10)   **The value of** $z = f(x, y)$ **at** (x_0, y_0) **increases most rapidly in the direction of** $\nabla f(x_0, y_0)$ **and decreases most rapidly in the direction of** $-\nabla f(x_0, y_0)$.

The next example illustrates another use of (17.10).

EXAMPLE 4   A metal plate is situated on the xy-plane in such a way that the temperature T at any point $P = (x, y)$ is inversely proportional to the distance of P from $(0, 0)$. If the temperature at $(-3, 4)$ equals 50°C, in what direction will the temperature at $(-3, 4)$ increase the fastest?

Solution   The temperature T at any point (x, y) is given by

$$T(x, y) = \frac{k}{\sqrt{x^2 + y^2}}$$

where k is the constant of proportionality. Since $T = 50$ when $(x, y) = (-3, 4)$, we find that $k = T\sqrt{x^2 + y^2} = 50(5) = 250$. The gradient of T is

$$\nabla T(x, y) = T_x(x, y)\mathbf{i} + T_y(x, y)\mathbf{j} = \frac{-250x}{(x^2 + y^2)^{3/2}}\,\mathbf{i} + \frac{-250y}{(x^2 + y^2)^{3/2}}\,\mathbf{j}$$

$$\nabla T(-3, 4) = 6\mathbf{i} - 8\mathbf{j}$$

By (17.10), the direction along which the temperature increases the fastest is $6\mathbf{i} - 8\mathbf{j}$.   ∎

In Example 4 the temperature increases the fastest in the direction of $6\mathbf{i} - 8\mathbf{j}$. The temperature decreases the fastest in the opposite direction, that is, in the direction of $-6\mathbf{i} + 8\mathbf{j}$. Along any direction perpendicular to $6\mathbf{i} - 8\mathbf{j}$, the temperature neither increases nor decreases. Such a direction is therefore along an isothermal curve (a curve of constant temperature).

Let's look at another interpretation of (17.10): Suppose the surface $z = f(x, y)$ represents a mountain whose elevation is z (see Fig. 6). The level curves of f correspond to contours along which the elevation remains fixed; that is, along which

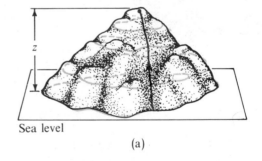

Sea level

(a)

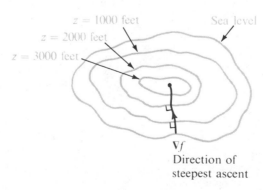

$z = 1000$ feet   Sea level
$z = 2000$ feet
$z = 3000$ feet

∇f
Direction of
steepest ascent

Figure 6   (b)

there is no increase or decrease in the value of z. From (17.8), it follows that the direction of the level curves of $z = f(x, y)$ are those for which $\cos \phi = 0$, or $\phi = \pi/2$. Therefore, the direction along which the elevation increases most rapidly is perpendicular to the level curves. In other words, the most direct route to the summit is in the direction of ∇f, which is perpendicular to the level curves.* Based on (17.8), the shortest route down the mountain $(\cos \phi = -1)$ is in the direction of $-\nabla f$, which is also perpendicular to the level curves (see Fig. 7). A stream of water would flow this way down the mountain.

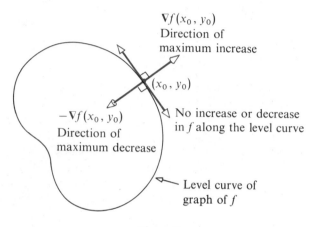

Figure 7

We formalize this discussion in the following theorem:

(17.11)   **THEOREM**   **Let f denote a function of two variables. If f and its first partial derivatives are continuous at a point $P_0 = (x_0, y_0)$ and $\nabla f(x_0, y_0) \neq 0$, then $\nabla f(x_0, y_0)$ is perpendicular to the level curve of f at P_0.**

Proof   Let $f(x, y) = k$ be the level curve through P_0. Suppose this level curve is represented parametrically by $x = x(t)$ and $y = y(t)$, with $x(t_0) = x_0$ and $y(t_0) = y_0$. Then

(17.12) $$f(x(t), y(t)) = k$$

By differentiating with respect to t, we have

$$f_x(x(t), y(t))x'(t) + f_y(x(t), y(t))y'(t) = 0$$

or, equivalently,

$$\nabla f(x, y) \cdot [x'(t)\mathbf{i} + y'(t)\mathbf{j}] = 0$$

---

* The *grade* of a mountain (or hill) is a measure of its steepness; hence, the name *gradient* given to ∇f—the direction of steepest ascent.

In particular, when $t = t_0$,   this equation says that $\nabla f(x_0, y_0)$ is normal to the level curve of f at (x_0, y_0).   ∎

EXAMPLE 5   For the function $f(x, y) = \sqrt{x^2 + y^2}$,   sketch the level curve passing through the point (3, 4) and sketch the gradient at this point.

Solution   Since $z = \sqrt{x^2 + y^2}$ is a cone, the level curves of this surface are circles. Because $f(3, 4) = \sqrt{9 + 16} = 5$,   the level curve through (3, 4) is the circle $x^2 + y^2 = 25$. Since

$$\nabla f(x, y) = \frac{x}{\sqrt{x^2 + y^2}}\,\mathbf{i} + \frac{y}{\sqrt{x^2 + y^2}}\,\mathbf{j}$$

the gradient at (3, 4) is

$$\nabla f(3, 4) = \tfrac{3}{5}\mathbf{i} + \tfrac{4}{5}\mathbf{j}$$

See Figure 8.

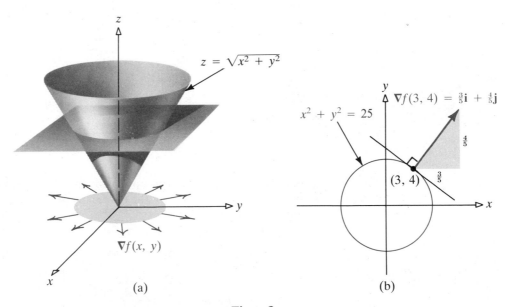

(a)   (b)

Figure 8   ∎

FUNCTIONS OF THREE VARIABLES

We conclude this section with an informal extension of the directional derivative and the gradient to functions of three variables.

We begin with the directional derivative $D_{\mathbf{u}}f(x_0, y_0, z_0)$ of a function $w = f(x, y, z)$ at (x_0, y_0, z_0) in the direction of $\mathbf{u}$ in which the unit vector $\mathbf{u}$ is given by

(17.13) $\mathbf{u} = (\cos \alpha)\mathbf{i} + (\cos \beta)\mathbf{j} + (\cos \gamma)\mathbf{k}$

where α, β, and γ are the direction angles of $\mathbf{u}$ and $\cos^2\alpha + \cos^2\beta + \cos^2\gamma = 1$ [refer to equation (14.10)]. Then

(17.14) $$D_{\mathbf{u}}f(x_0, y_0, z_0) = f_x(x_0, y_0, z_0)\cos\alpha + f_y(x_0, y_0, z_0)\cos\beta + f_z(x_0, y_0, z_0)\cos\gamma$$

is the rate of change of f at (x_0, y_0, z_0) in the direction of $\mathbf{u}$.

The gradient of f at (x_0, y_0, z_0) is the vector

(17.15) $$\nabla f(x_0, y_0, z_0) = f_x(x_0, y_0, z_0)\mathbf{i} + f_y(x_0, y_0, z_0)\mathbf{j} + f_z(x_0, y_0, z_0)\mathbf{k}$$

From (17.13)–(17.15), we have the formula

$$D_{\mathbf{u}}f(x_0, y_0, z_0) = \nabla f(x_0, y_0, z_0) \cdot \mathbf{u} = \|\nabla f(x_0, y_0, z_0)\| \cos\phi$$

where ϕ is the angle between $\nabla f(x_0, y_0, z_0)$ and $\mathbf{u}$.

As we saw with functions of two variables, the gradient $\nabla f(x_0, y_0, z_0)$ is the direction along which the value of f increases most rapidly, and $-\nabla f(x_0, y_0, z_0)$ is the direction along which the value of f decreases most rapidly. For example, suppose $T = T(x, y, z)$ is the temperature of a homogeneous body at the point (x, y, z). At the point (x_0, y_0, z_0) on the body, heat will flow in the direction of greatest decrease in temperature, namely, $-\nabla T(x_0, y_0, z_0)$, and this direction is perpendicular to the level surface through (x_0, y_0, z_0).*

We now state the analog of theorem (17.11) to functions of three variables.

(17.16) **THEOREM** **Let f denote a function of three variables. If f and its first partial derivatives are continuous, then at each point $P_0 = (x_0, y_0, z_0)$ in the domain of f at which $\nabla f(x_0, y_0, z_0) \neq 0$, the gradient $\nabla f(x_0, y_0, z_0)$ is perpendicular to the level surface S of f through P_0.**

Proof Suppose that $f(x, y, z) = k$ is the equation of the level surface. Suppose C is any curve that lies on S. We let C be defined by the parametric equations $x = x(t)$, $y = y(t)$, $z = z(t)$. Suppose that $x_0 = x(t_0)$, $y_0 = y(t_0)$, $z_0 = z(t_0)$. Since C lies on S, we have $f(x(t), y(t), z(t)) = k$ for all t. Thus, we have

$$f_x(x(t), y(t), z(t))x'(t) + f_y(x(t), y(t), z(t))y'(t) + f_z(x(t), y(t), z(t))z'(t) = 0$$

or, equivalently,

$$\nabla f(x, y, z) \cdot [x'(t)\mathbf{i} + y'(t)\mathbf{j} + z'(t)\mathbf{k}] = 0$$

In particular,

$$\nabla f(x_0, y_0, z_0) \cdot [x'(t_0)\mathbf{i} + y'(t_0)\mathbf{j} + z'(t_0)\mathbf{k}] = 0$$

Since C is an arbitrary curve on S, it follows that $\nabla f(x_0, y_0, z_0)$ is perpendicular to the tangent vector of every curve lying on S that passes through (x_0, y_0, z_0). Thus, ∇f is perpendicular to S at P_0. ∎

---

* Recall that level surfaces were defined in Section 2, Chapter 16.

EXERCISE 1

In Problems 1–12 find the directional derivative of each function at the indicated point and in the indicated direction.

1. $f(x, y) = xy^2 + x^2$ at $(-1, 2)$ in the direction $\theta = \pi/3$

2. $f(x, y) = 3xy + y^2$ at $(2, 1)$ in the direction $\theta = \pi/4$

3. $f(x, y) = 2xy - y^2$ at $(-1, 3)$ in the direction $\theta = 2\pi/3$

4. $f(x, y) = 2xy + x^2$ at $(0, 3)$ in the direction $\theta = 4\pi/3$

5. $f(x, y) = xe^y + ye^x$ at $(0, 0)$ in the direction $\theta = \pi/6$

6. $f(x, y) = x \ln y$ at $(5, 1)$ in the direction $\theta = \pi/4$

7. $f(x, y) = \tan^{-1}(y/x)$ at $(1, 1)$ in the direction of $\mathbf{u} = (3\mathbf{i} - 4\mathbf{j})/5$

8. $f(x, y) = \ln\sqrt{x^2 + y^2}$ at $(3, 4)$ in the direction of $\mathbf{u} = (5\mathbf{i} + 12\mathbf{j})/13$

9. $f(x, y, z) = z \tan^{-1}(y/x)$ at $(1, 1, 3)$ in the direction of $\mathbf{a} = \mathbf{i} + \mathbf{j} - \mathbf{k}$

10. $f(x, y, z) = \sqrt{x^2 + y^2 + z^2}$ at $(3, 4, 0)$ in the direction of $\mathbf{a} = \mathbf{i} - \mathbf{j} + \mathbf{k}$

11. $f(x, y, z) = xe^{yz}$ at $(1, 0, 1)$ in the direction of $\mathbf{a} = 2\mathbf{i} + \mathbf{j}$

12. $f(x, y, z) = z \ln(x/y)$ at $(1, 1, 2)$ in the direction of $\mathbf{a} = \mathbf{j} + \mathbf{k}$

In Problems 13–16 find ∇f.

13. $f(x, y) = xy^2 + x^2$

14. $f(x, y) = 2xy + x^2$

15. $f(x, y) = \tan^{-1}(y/x)$

16. $f(x, y) = \ln\sqrt{x^2 + y^2}$

In Problems 17–20 find ∇f at the indicated point.

17. $f(x, y) = 2xy + x^2$; $P = (0, 3)$

18. $f(x, y) = x \ln y$; $P = (5, 1)$

19. $f(x, y, z) = xe^{yz}$; $P = (1, 0, 1)$

20. $f(x, y, z) = z \ln(x/y)$; $P = (1, 1, 2)$

In Problems 21–26 find the direction at P along which each function increases most rapidly. Find the rate of increase in this direction.

21. $z = xy^2 + x^2$; $P = (-1, 2)$

22. $z = 3xy + y^2$; $P = (2, 1)$

23. $z = xe^y + ye^x$; $P = (0, 0)$

24. $z = x \ln y$; $P = (5, 1)$

25. $w = z \tan^{-1}(y/x)$; $P = (1, 1, 3)$

26. $w = \sqrt{x^2 + y^2 + z^2}$; $P = (3, 4, 0)$

In Problems 27–32 sketch the level curve of f that passes through the point P, and sketch ∇f at P.

27. $f(x, y) = x^2 + y^2$; $P = (3, 4)$

28. $f(x, y) = x^2 - y^2$; $P = (2, -1)$

29. $f(x, y) = x^2 - 4y^2$; $P = (3, \sqrt{5}/2)$

30. $f(x, y) = x^2 + 4y^2$; $P = (-2, 0)$

31. $f(x, y) = x^2 y$; $P = (3, \frac{1}{9})$

32. $f(x, y) = xy$; $P = (1, 1)$

33. A metal plate is situated on the xy-plane in such a way that the temperature T at any point (x, y) is given by $T = e^x \sin y + e^y \sin x$. What is the rate of change in temperature at $(0, 0)$ in the direction of $3\mathbf{i} - 4\mathbf{j}$? At $(0, 0)$, in what direction is the rate of change of temperature the greatest? In what direction is it the least? In what direction is it 0?

34. Rework Problem 33 if $\;T = \ln\sqrt{1 - (x^2 + y^2)}$.

35. The electrical potential V at any point (x, y) is given by $\;V = \ln\sqrt{x^2 + y^2}$. Find the rate of change of potential V at any point $\;(x, y) \neq (0, 0)$:
    (a) In a direction toward $(0, 0)$
    (b) In the two directions perpendicular to a direction toward $(0, 0)$
    (c) In what direction is the rate of change in potential V maximum?
    (d) In what direction is the rate of change least?

36. The surface of a hill may be represented by the equation $\;z = 8 - 2x^2 - y^2$. If a freshwater spring is located at the point $(1, 2, 2)$, in what direction will the water flow?

37. Suppose that you are climbing a mountain in the shape of the surface $\;x^2 + y^2 - 5x + z = 0$. (The x-axis points east, the y-axis north, and the z-axis up.)
    (a) Your route is such that as you pass through the point $(1, 1, 3)$ you are heading northeast. At what rate (with respect to distance) is your altitude changing at that point?
    (b) Another climber at $(1, 2, 0)$ wants to move upward as quickly as possible. In what direction should he start? At what rate is his altitude changing when he starts?
    (c) A third climber at $(2, 1, 5)$ wants to remain at the same altitude. In what direction(s) may she go?

38. Suppose that the temperature at each point of the coordinate plane is $\;T = 3x^2 + 4y^2 + 5$ (in degrees Fahrenheit).
    (a) If we leave the point $(3, 4)$ heading for the point $(4, 3)$, how fast (in degrees per unit of distance) is the temperature changing as we leave?
    (b) After reaching $(4, 3)$, in what direction should we go if we want to cool off as fast as possible? How fast is the temperature changing as we leave $(4, 3)$?
    (c) Suppose that we want to move away from $(4, 3)$ along a path of constant temperature. What is an equation of our path and what is the temperature?
    (d) A person at the origin may go in *any* direction and will experience the same rate of change of temperature as he leaves. Why? What is the rate?

39. Suppose that $\;z = xy^2$. In what direction(s) may we go from the point $(-1, 1)$ if we want the rate of change of z to be 2?

40. An Eskimo whose igloo is in the shape of the surface $\;z = 4 - x^2 - y^2$ wants to drill a hole perpendicular to the igloo at the point $(1, 1, 2)$ so that a flat solar panel can be bolted to the igloo. In what direction should the Eskimo drill?

41. Show that the level curves of $\;f(x, y) = x^2 - y^2$ are perpendicular to the level curves of $\;h(x, y) = xy$ for all $\;(x_0, y_0) \neq (0, 0)$.

42. Find a unit vector $\mathbf{u}$ that is perpendicular to the level curve of $\;f(x, y) = 4x^2y$ through $\;P = (1, -2)$ at P.

43. Find a unit vector $\mathbf{u}$ that is perpendicular to the level curve of $\;f(x, y) = 2x^2 + y^2 + 1$ through $P = (1, 1)$ at P.

44. Under the hypotheses of theorem (17.11), show that $\;D_{\mathbf{u}}f(x_0, y_0) = 0\;$ in the directions perpendicular to that of $\;\nabla f(x_0, y_0)$.

45. If $\;u = f(x, y)\;$ and $\;v = g(x, y)\;$ are differentiable, show that:
    (a) $\mathbf{V}(ku) = k\mathbf{V}u,\;\;k$ a constant   (b) $\mathbf{V}(u + v) = \mathbf{V}u + \mathbf{V}v$ (c) $\mathbf{V}(uv) = u\mathbf{V}v + v\mathbf{V}u$

    (d) $\mathbf{V}\left(\dfrac{u}{v}\right) = \dfrac{v\mathbf{V}u - u\mathbf{V}v}{v^2}$ (e) $\mathbf{V}u^\alpha = \alpha u^{\alpha-1}\mathbf{V}u,\;\;\alpha$ a real number

46. Show that for a vector $\mathbf{a} = a_1\mathbf{i} + a_2\mathbf{j}$ and a function f,

$$D_{\mathbf{u}}f(x, y) = \frac{a_1(\partial f/\partial x) + a_2(\partial f \, \partial y)}{\sqrt{a_1^2 + a_2^2}}$$

where $\mathbf{u} = \mathbf{a}/\|\mathbf{a}\|$.

2. Tangent Planes

Consider a surface S defined by an equation of the form $F(x, y, z) = 0$.   Let $P_0 = (x_0, y_0, z_0)$ denote a point on S such that F is differentiable at P_0 and $\nabla F(x_0, y_0, z_0)$ is nonzero. A line L that passes through P_0 is said to be *tangent to S at P_0* if, for some curve C lying in S that also passes through P_0, the tangent vector to C at P_0 has the same direction as L. Since all tangent lines to S at P_0 are perpendicular to the vector $\nabla F(x_0, y_0, z_0)$ [see (17.11)] all such tangent lines lie in the same plane, which we call a *tangent plane*. See Figure 9.

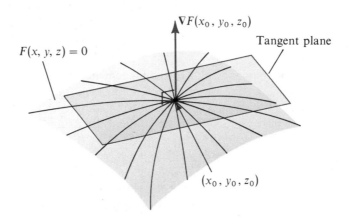

Figure 9

(17.17)   **DEFINITION**   Let $F(x, y, z) = 0$ **be the equation of a surface. If at a point** $P_0 = (x_0, y_0, z_0)$ **on this surface, the gradient vector** $\nabla F(x_0, y_0, z_0)$ **is nonzero, then the plane through P_0 and perpendicular to** $\nabla F(x_0, y_0, z_0)$ **is called the *tangent plane to the surface at P_0.*** *

Since $\nabla F(x_0, y_0, z_0)$ is normal to the tangent plane at $P_0 = (x_0, y_0, z_0)$,   a vector equation for the tangent plane is given by

$$\nabla F(x_0, y_0, z_0) \cdot [(x - x_0)\mathbf{i} + (y - y_0)\mathbf{j} + (z - z_0)\mathbf{k}] = 0$$

---

* If $\nabla F(x_0, y_0, z_0) = \mathbf{0}$ then (x_0, y_0, z_0) is a *singular point* of the surface and no tangent plane is defined there.

Since

$$\nabla F(x_0, y_0, z_0) = F_x(x_0, y_0, z_0)\mathbf{i} + F_y(x_0, y_0, z_0)\mathbf{j} + F_z(x_0, y_0, z_0)\mathbf{k}$$

the equivalent rectangular equation is

(17.18)
$$F_x(x_0, y_0, z_0)(x - x_0) + F_y(x_0, y_0, z_0)(y - y_0) + F_z(x_0, y_0, z_0)(z - z_0) = 0$$

EXAMPLE 1　　Find an equation of the tangent plane to the surface $x^2 + y^2 - z^2 - 24 = 0$ at the point $(3, -4, 1)$.

Solution　　Here, $F(x, y, z) = x^2 + y^2 - z^2 - 24$, so that

$$F_x(x, y, z) = 2x \qquad F_y(x, y, z) = 2y \qquad F_z(x, y, z) = -2z$$

At the point $(3, -4, 1)$,

$$F_x(3, -4, 1) = 6 \qquad F_y(3, -4, 1) = -8 \qquad F_z(3, -4, 1) = -2$$

An equation of the tangent plane is, therefore,

$$6(x - 3) - 8(y + 4) - 2(z - 1) = 0$$
$$3x - 4y - z - 24 = 0 \qquad \blacksquare$$

NORMAL LINE

The line perpendicular to the tangent plane of a surface $F(x, y, z) = 0$ at a point $P_0 = (x_0, y_0, z_0)$ is called the *normal line to the surface at* P_0. Since the gradient $\nabla F(x_0, y_0, z_0)$ is perpendicular to the tangent plane at P_0, it follows that the normal line is in the direction of the gradient. By using the notation of Section 7 in Chapter 14, we can write the vector equation for the normal line as

$$\mathbf{r}(t) = \mathbf{r}_0 + t\nabla F(x_0, y_0, z_0)$$

where $\mathbf{r}_0 = x_0\mathbf{i} + y_0\mathbf{j} + z_0\mathbf{k}$ is the position vector of P_0.

The corresponding set of parametric equations is

$$x = x_0 + at \qquad y = y_0 + bt \qquad z = z_0 + ct$$

where $a = F_x(x_0, y_0, z_0)$, $b = F_y(x_0, y_0, z_0)$, $c = F_z(x_0, y_0, z_0)$. If $abc \neq 0$, the symmetric equations

$$\frac{x - x_0}{a} = \frac{y - y_0}{b} = \frac{z - z_0}{c}$$

can also be used.

EXAMPLE 2　　Find symmetric equations of the normal line to the surface $x^2 + y^2 - z^2 - 24 = 0$ at the point $(3, -4, 1)$.

Solution　　The normal line is perpendicular to the plane

$$6(x - 3) - 8(y + 4) - 2(z - 1) = 0$$

found in Example 1. Hence, the symmetric equations of the normal line are

$$\frac{x-3}{6} = \frac{y+4}{-8} = \frac{z-1}{-2} \qquad \blacksquare$$

SURFACES GIVEN BY $z = f(x, y)$

Frequently, the equation of a surface is given by an equation of the form $z = f(x, y)$. To obtain an equation of the tangent plane to the surface $z = f(x, y)$ at (x_0, y_0, z_0), we write the equation of the surface as

$$F(x, y, z) = f(x, y) - z = 0$$

Then

$$F_x = f_x \qquad F_y = f_y \qquad F_z = -1$$

so that by (17.18), an equation of the tangent plane to $z = f(x, y)$ at (x_0, y_0, z_0) is

$$f_x(x_0, y_0)(x - x_0) + f_y(x_0, y_0)(y - y_0) - (z - z_0) = 0$$

or

(17.19) $$z - z_0 = f_x(x_0, y_0)(x - x_0) + f_y(x_0, y_0)(y - y_0)$$

The parametric equations for the normal line are

$$x = x(t) = x_0 + t\nabla f(x_0, y_0) \cdot \mathbf{i} = x_0 + tf_x(x_0, y_0)$$
$$y = y(t) = y_0 + t\nabla f(x_0, y_0) \cdot \mathbf{j} = y_0 + tf_y(x_0, y_0)$$
$$z = z(t) = z_0 - t$$

In Examples 1 and 2, equivalent results could have been obtained by solving the equation $x^2 + y^2 - z^2 - 24 = 0$ for z and using the function

$$z = f(x, y) = \sqrt{x^2 + y^2 - 24}$$

whose graph is the part of the surface lying above the xy-plane. This would yield $f_x(3, -4) = 3$, $f_y(3, -4) = -4$, and

$$z = 3x - 4y - 24$$

as an equation for the tangent plane at $(3, -4, 1)$. The parametric equations for the normal line would be

$$x = 3 + 3t \qquad y = -4 - 4t \qquad z = 1 - t$$

EXERCISE 2

In Problems 1–14 find an equation of the tangent plane and an equation of the normal line to each surface at the point indicated.

1. $x^2 + y^2 + z^2 = 14$ at $(1, -2, 3)$ **2.** $x^2 - y^2 + z^2 = 4$ at $(-1, 1, 2)$

3. $2x^2 + 3y^2 - z^2 = 10$ at $(2, 1, -1)$ **4.** $4x^2 + y^2 + 2z^2 = 7$ at $(1, 1, -1)$

5. $z^2 = x^2 + 3y^2$ at $(1, -1, -2)$

6. $x^2 + y^2 - 2z^2 = -13$ at $(2, 1, 3)$

7. $z = x^2 + y^2$ at $(-2, 1, 5)$

8. $z = 2x^2 - 3y^2$ at $(2, 1, 5)$

9. $2x^2 + y^2 = z$ at $(1, 0, 2)$

10. $2x^2 - y^2 = z$ at $(1, 0, 2)$

11. $z = e^x \cos y$ at $(0, \pi/2, 0)$

12. $z = \ln(x^2 + y^2)$ at $(1, -1, \ln 2)$

13. $x^{2/3} + y^{2/3} + z^{2/3} = 9$ at $(1, 8, -8)$

14. $x^{1/2} + y^{1/2} + z^{1/2} = 6$ at $(1, 4, 9)$

In Problems 15–20 determine those point(s) on the surface at which the tangent plane is parallel to the xy-plane.

15. $z = 6x - 4y - x^2 - 2y^2$ **16.** $z = 4x - 2y + x^2 + y^2$ **17.** $z = x^2 + 2xy + y^2$

18. $z = x^2 - 3xy + y^2$ **19.** $z = 2x^4 - y^2 - x^2 - 2y$ **20.** $z = x^2 + y^4 - 4y^2 - 2x$

21. Two surfaces are said to be *tangent at a common point* P_0 if each has the same tangent plane at P_0. Show that the surfaces $x^2 + z^2 + 4y = 0$ and $x^2 + y^2 + z^2 - 6z + 7 = 0$ are tangent at the point $(0, -1, 2)$.

22. Two surfaces are *orthogonal at a common point* P_0 if their normal lines at P_0 are orthogonal. Show that the surfaces $x^2 + y^2 + z^2 = 4$ and $x^2 + y^2 - z^2 = 0$ are orthogonal at $(0, \sqrt{2}, \sqrt{2})$. In fact, show that the surfaces are orthogonal at every point of intersection.

23. Prove that the normal lines of a sphere given by $x^2 + y^2 + z^2 = a^2$ pass through the center of the sphere.

24. Write equations of the tangent plane and normal line to $x^2 - 3y^2 - 4z^2 = 2$ at $(3, 1, 1)$.

25. Write equations of the tangent plane and normal line to $x^2 + 2y^2 - 5z^2 = 4$ at $(1, 2, 1)$.

26. Let $P_0 = (x_0, y_0, z_0)$ be a point on the sphere $x^2 + y^2 + z^2 = R^2$.   Show that the normal line to the sphere at P_0 passes through the point $(-x_0, -y_0, -z_0)$.

27. Find equations for the tangent line to the curve of intersection of the surfaces $x \sin yz = 1$, $z e^{y^2 - x^2} = \pi/2$ at the point $(1, 1, \pi/2)$.

28. Find the points on the surface $(x^2/9) + (y^2/4) + (z^2/36) = 1$ where the tangent plane is parallel to the plane $2x - 3y + z = 4$.

3. Extrema of Functions of Two Variables

We saw in Chapter 4 that an important application of the derivative is to find the local extrema (maxima and minima) of a function of a single variable. In this section we find that partial derivatives are used in a similar way to find the local extrema of a function of two variables.

Let $z = f(x, y)$ denote a function of two variables. We say that *f has a local maximum at the point* (x_0, y_0) if there is some open rectangular region H containing (x_0, y_0) so that

(17.20)
$$f(x_0, y_0) \geq f(x, y)$$

for all (x, y) in H. Similarly, *f has a local minimum at* (x_1, y_1) if

(17.21)
$$f(x_1, y_1) \leq f(x, y)$$

for all (x, y) in H. We shall refer to the local maxima and local minima of f as *local extrema.*

Figure 10 illustrates the fact that a function of two variables, like a function of one variable, can have many local extrema.

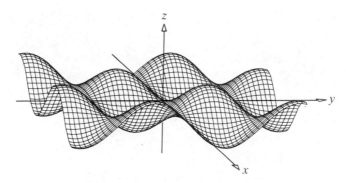

Figure 10   The function $f(x, y) = \sin x \sin y$ has many local maxima and minima

In contrast to the definition of local maxima and minima, we now define absolute maximum and absolute minimum. If the inequality

$$f(x_0, y_0) \geq f(x, y)$$

holds for all points (x, y) in the domain of f, then $f(x_0, y_0)$ is the *absolute maximum of f.* Similarly, $f(x_1, y_1)$ is the *absolute minimum of f* if

$$f(x_1, y_1) \leq f(x, y)$$

holds for all points (x, y) in the domain of f.

It can be shown that if a function f of two variables is continuous on a closed and bounded region R, then f attains its absolute maximum $f(x_0, y_0)$ and its absolute minimum $f(x_1, y_1)$ for some (x_0, y_0) and (x_1, y_1) in R. Just as in the case of one variable, the absolute extrema will occur either on the boundary of the region R (analogous to the endpoints of a closed interval) or else at an interior point of R. In the latter case, a local extremum also occurs.

To develop a procedure for locating local extrema, we first let $z = f(x, y)$ denote a function of two variables that is defined and continuous on an open rectangular region H. Suppose f has a local maximum at the point (x_0, y_0) in H. Then the curve that results from the intersection of the surface $z = f(x, y)$ and any plane through $(x_0, y_0, 0)$ that is perpendicular to the xy-plane will have a local maximum

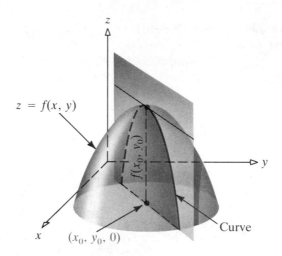

Figure 11 f has a local maximum at (x_0, y_0)

at (x_0, y_0), as shown in Figure 11. In particular, the curve that results from intersecting $z = f(x, y)$ with the plane $x = x_0$ has this property. This means that if $f_y(x_0, y_0)$ exists, then $f_y(x_0, y_0) = 0$. By a similar argument, we must also have $f_x(x_0, y_0) = 0$. This leads us to formulate the following necessary condition for local extrema:

(17.22)   THEOREM *A Necessary Condition for Local Extrema.* **Let $z = f(x, y)$ denote a function of two variables defined and continuous on an open rectangular region H containing the point (x_0, y_0). Suppose $f_x(x_0, y_0)$ and $f_y(x_0, y_0)$ each exist. If f has a local extremum at (x_0, y_0), then**

$$f_x(x_0, y_0) = 0 \qquad \text{and} \qquad f_y(x_0, y_0) = 0$$

From this theorem we see that local extrema of a function occur at those points at which the partial derivatives exist and are zero simultaneously. It can also happen that the local extrema of a function occur at points at which one or both of the partial derivatives fail to exist. Thus, we say that f has a *critical point* at (x_0, y_0) if $f_x(x_0, y_0) = f_y(x_0, y_0) = 0$, or if one of the partial derivatives does not exist at (x_0, y_0). We can now restate theorem (17.22):

A function of two variables has a local maximum or minimum only at critical points in its domain.

EXAMPLE 1   Locate all local maxima and local minima for

$$z = f(x, y) = x^2 + y^2 - 2x + 4y$$

Solution   The partial derivatives of f are

$$f_x = 2x - 2 \qquad \text{and} \qquad f_y = 2y + 4$$

By setting each of these equal to 0 and solving, we find that $(1, -2)$ is the only critical point of f.

Whether the function actually has a local maximum or a local minimum or neither at the critical point $(1, -2)$ can be verified by algebraic means. For this particular function, we may complete the squares, obtaining $z = (x - 1)^2 + (y + 2)^2 - 5$, from which we conclude that at $(1, -2)$ there is a local minimum and that $z = -5$ is in fact the absolute minimum. This conclusion is evident from a graph of the surface (an elliptic paraboloid, as seen in Fig. 12).

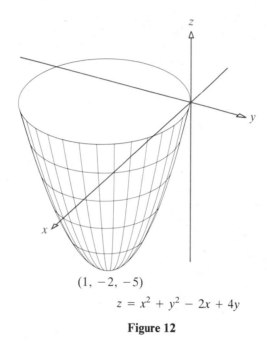

$$(1, -2, -5)$$

$$z = x^2 + y^2 - 2x + 4y$$

Figure 12   ■

The next example illustrates that it is possible for a function to have a critical point without having either a local maximum or a local minimum.

EXAMPLE 2   Consider the hyperbolic paraboloid $z = f(x, y) = y^2 - x^2$.   Show that $(0, 0)$ is the only critical point and that neither a local maximum nor a local minimum occurs at $(0, 0)$.

Solution   We have

$$f_x(x, y) = -2x \qquad \text{and} \qquad f_y(x, y) = 2y$$

Both equal 0 at $(0, 0)$; this is the only critical point. Next, we observe that $f(0, 0) = 0$, $f(x, 0) = -x^2 < 0$ for all $x \neq 0$,   and $f(0, y) = y^2 > 0$ for all $y \neq 0$.

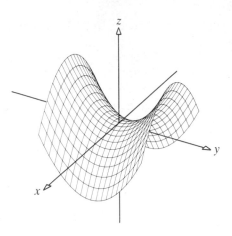

Figure 13

See Figure 13.

If we consider the values of $f(x, 0)$ along the x-axis, we have a function of x that attains a maximum value of 0 at the origin. However, if we consider the values of $f(0, y)$ along the y-axis, we have a function that attains a minimum value at the origin. In other words, at the critical point $(0, 0)$, the function appears to have a maximum when viewed in one direction and to have a minimum when viewed in another direction and, thus, has neither a maximum nor a minimum.   ∎

We say that the function f has a *saddle point* on its graph at $(x_0, y_0, f(x_0, y_0))$ if (x_0, y_0) is a critical point and f does not have a local extremum at (x_0, y_0). Thus, the function $z = f(x, y) = y^2 - x^2$ of Example 2 has a saddle point at $(0, 0, 0)$.

The following test, which we state without proof, is often useful in classifying critical points. This test is analogous to the second derivative test for functions of one variable [see equation (4.35)]. A proof may be found in texts on advanced calculus.

(17.23)   **Second Derivative Test.   Let $z = f(x, y)$ denote a function of two variables for which the first- and second-order partial derivatives are continuous in some open rectangular region H containing the point (x_0, y_0). Suppose that $f_x(x_0, y_0) = 0$ and $f_y(x_0, y_0) = 0.$ Set $A = f_{xx}(x_0, y_0)$, $B = f_{xy}(x_0, y_0)$, $C = f_{yy}(x_0, y_0)$,   and $D = AC - B^2$.**

(a)   If $D > 0$ and $A > 0$,   then f has a local minimum at (x_0, y_0).
(b)   If $D > 0$ and $A < 0$,   then f has a local maximum at (x_0, y_0).
(c)   If $D < 0$,   then f has a saddle point at $(x_0, y_0, f(x_0, y_0))$.
(d)   If $D = 0$,   then the test gives no information.

We can express the quantity D by a 2×2 determinant:

$$D = \begin{vmatrix} f_{xx} & f_{xy} \\ f_{xy} & f_{yy} \end{vmatrix} = f_{xx}f_{yy} - (f_{xy})^2$$

EXAMPLE 3   Find all local maxima and local minima for

$$z = f(x, y) = x^2 + xy + y^2 - 6x + 6$$

Solution   The critical points of f obey the equations

$$f_x = 2x + y - 6 = 0 \qquad f_y = x + 2y = 0$$

Solving these equations, we find that $(4, -2)$ is the only critical point. The second-order partial derivatives of f are

$$A = f_{xx} = 2 \qquad C = f_{yy} = 2 \qquad B = f_{xy} = f_{yx} = 1$$

At the critical point $(4, -2)$, we have

$$D = (2)(2) - 1 = 3 > 0 \qquad A > 0$$

It follows from (17.23) that f has a local minimum at $(4, -2)$. The value of f at this local minimum is $z = f(4, -2) = -6$.   ∎

To find the critical points of a function $z = f(x, y)$ requires that we be able to solve the system of equations $f_x = 0$, $f_y = 0$.   If this results in a system of linear equations, techniques for finding the solution are abundant. However, if the system is nonlinear, no general method is available. In this event, various manipulations such as substitution, addition/subtraction, and so forth, are used. Care must be taken because extraneous roots are sometimes introduced.

EXAMPLE 4   Find all local maxima and local minima for

$$z = f(x, y) = x^3 + y^2 + 2xy - 4x - 3y + 5$$

Solution   The critical points obey the equations

$$f_x = 3x^2 + 2y - 4 = 0 \qquad f_y = 2y + 2x - 3 = 0$$

By solving for $2y$ in $f_y = 0$ and substituting this into $f_x = 0$,   we find

$$3x^2 - 2x - 1 = 0$$

$$x = -\tfrac{1}{3} \quad \text{or} \quad x = 1$$

so the critical points are $(-\tfrac{1}{3}, \tfrac{11}{6})$ and $(1, \tfrac{1}{2})$.* The second-order partial derivatives of f are

$$f_{xx} = 6x \qquad f_{xy} = f_{yx} = 2 \qquad f_{yy} = 2$$

At $(-\tfrac{1}{3}, \tfrac{11}{6})$, $A = -2$, $B = 2$,   and $C = 2$,   so

$$D = -4 - 4 = -8 < 0$$

It follows from (17.23) that f has a saddle point at $(-\tfrac{1}{3}, \tfrac{11}{6}, \tfrac{317}{108})$.
At $(1, \tfrac{1}{2})$, $A = 6$, $B = 2$,   and $C = 2$,   so

$$D = 12 - 4 = 8 > 0 \qquad A > 0$$

Hence, f has a local minimum at $(1, \tfrac{1}{2})$. A graph of

$$z = f(x, y) = x^3 + y^2 + 2xy - 4x - 3y + 5$$

is provided in Figure 14 on p. 934.

---

* You should check to make certain that neither of these roots is extraneous.

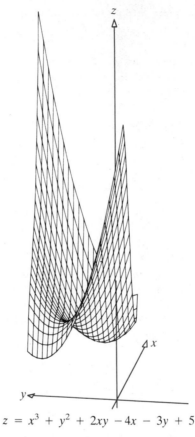

$$z = x^3 + y^2 + 2xy - 4x - 3y + 5$$

Figure 14   ∎

EXAMPLE 5   A manufacturer wishes to make an open rectangular box of volume $V = 500$ cubic centimeters,   using the least possible amount of material. Find the dimensions of the box. See Figure 15.

Solution   Let x and y be the dimensions of the base of the box, and let z be the height of the box. Then

(17.24) $V = 500 = xyz \qquad x > 0, \quad y > 0, \quad z > 0$

The surface area S—that is, the amount of material used—is

(17.25) $S = xy + 2yz + 2zx$

Now solve for z in (17.24) and substitute into (17.25):

(17.26) $S = xy + \dfrac{1000}{x} + \dfrac{1000}{y}$

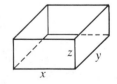

Figure 15

This is the function to be minimized. Differentiation yields

$$S_x = y - \dfrac{1000}{x^2} \qquad S_y = x - \dfrac{1000}{y^2}$$

To determine the possible extrema, we set these derivatives equal to 0:

(17.27)
$$S_x = y - \frac{1000}{x^2} = 0 \qquad S_y = x - \frac{1000}{y^2} = 0$$

and we find

$$y = \frac{1000}{x^2} \qquad \text{and} \qquad x = \frac{1000}{y^2}$$

By solving these simultaneously (using substitution), we get

$$x = y = (1000)^{1/3} = 10$$

From (17.27), it follows that

$$S_{xx} = \frac{2000}{x^3} \qquad S_{xy} = 1 \qquad S_{yy} = \frac{2000}{y^3}$$

and, hence, at $(x, y) = (10, 10)$,

$$S_{yy} = S_{xx} = 2 \qquad D = S_{xx}S_{yy} - (S_{xy})^2 = 3$$

The conditions from (17.23) for a local minimum have been met. Therefore, S has a local minimum at $(10, 10)$. What we are seeking, though, is the point where S attains its absolute minimum. Since the domain of S is not bounded, we have no guarantee that S has an absolute minimum. But the physical considerations of the problem require that the absolute minimum exists. Therefore, it must be the case that the local minimum is also the absolute minimum. Consequently, we know that the dimensions (in centimeters) of the open box of fixed volume that uses the least amount of material are

$$x = 10 \qquad y = 10 \qquad z = \frac{500}{100} = 5 \quad \blacksquare$$

Note: To show that an absolute minimum occurs where a local minimum has been found (and that an absolute maximum occurs where a local maximum has been found) can be extremely difficult. Fortunately, in applied problems such as Example 5, it is often possible to argue from physical or geometrical considerations either that an absolute minimum must exist and that it must occur at a local minimum, or, similarly, that an absolute maximum must occur where one of the local maxima has been found. If an absolute extremum occurs on the boundary, the method of Lagrange multipliers can sometimes be used to find it. (See Example 4 in the next section.)

The theory just discussed can be generalized to functions of more than two variables. If, for example, $w = f(x, y, z)$, the critical points of f occur when $f_x, f_y,$ and f_z are simultaneously 0, but it is more difficult to determine their nature. Fortunately, in practice, the correct answer can often be anticipated on physical grounds.

EXERCISE 3

In Problems 1–6 find all the critical points for each function.

1. $f(x, y) = x^4 - 2x^2 + y^2 + 5$

2. $f(x, y) = x^2 - y^2 + 6x - 2y + 4$

3. $f(x, y) = 4xy - x^4 - y^4 + 2$

4. $f(x, y) = x^3 + 6xy + 3y^2 + 3$

5. $f(x, y) = x^4 + y^4$ **6.** $f(x, y) = xy + \dfrac{2}{x} + \dfrac{4}{y}$

In Problems 7–26 find all local maxima and local minima for each function.

7. $z = x^2 + y^2 - 2x + 4y + 2$ **8.** $z = x^2 + y^2 - 4x + 2y - 4$

9. $z = x^2 + 4y^2 - 4x + 8y - 1$ **10.** $z = 2x^2 - y^2 + 4x - 4y + 8$

11. $z = x^3 - 6xy + y^3$ **12.** $z = x^2 - 3xy - y^2$

13. $z = x^2 + 3xy - y^2 + 4y - 6x$ **14.** $z = 2x^2 + xy + y^2 - 2x + 3y + 6$

15. $z = x^3 + 3xy + y^3$ **16.** $z = x^3 - 3xy - y^3$ **17.** $z = x^3 + x^2y + y^2$

18. $z = 3y^3 - x^2y + x$ **19.** $z = \dfrac{y}{x + y}$ **20.** $z = \dfrac{x}{x + y}$

21. $z = \cos x + \cos y$ **22.** $z = x^2 + 4 - 4x \cos y$ **23.** $z = y^2 - 6y \cos x + 6$

24. $z = x^2 - 6x \sin y + 2$ **25.** $z = e^{xy}$ **26.** $z = xye^{-(x+y)}$

27. Find the highest point of the paraboloid $x^2 + y^2 - 6x + 4y + z + 12 = 0$ as follows:
(a) Find the vertex of the paraboloid without using calculus.
(b) Express z as a function of x and y and use calculus to find its maximum value.

28. A certain mountain is in the shape of the surface $z = 2xy - 2x^2 - y^2 - 8x + 6y + 4$. (The unit of distance is 1000 feet.) If sea level is the xy-plane, how high is the mountain?

29. An open rectangular box has a fixed surface area. Find the dimensions that make its volume a maximum.

30. Rework Example 5 if the box is closed.

31. Rework Problem 29 if the box is closed.

32. Find the dimensions of an open rectangular box having a volume of 72 cubic meters if the cost per square meter of the material to be used is \$4 for the bottom, \$3 for one of the sides, and \$2 for the three remaining sides, and the total cost is to be minimized.

33. The U.S. Post Office regulations state that the combined (sum) length and girth of a parcel post package being sent to a first-class post office in the United States may not exceed 84 inches. If this combined length and girth exceeds 84 inches, extra postage will be charged according to weight. Find:
(a) The length, width and height of the rectangular box of maximum volume that can be mailed, subject to the 84 inch restriction.
(b) The dimensions of a circular tube of maximum volume that can be mailed, subject to the 84 inch restriction.

34. The reaction to an injection of x units of a certain drug, t hours after the injection, is given by

$$y = x^2(a - x)te^{\cdot} \qquad a \text{ constant}$$

Find the values of x and t, if any, which will maximize y.

35. An open irrigation channel is to be made into a symmetric form with three straight sides, as indicated in the illustration. If the perimeter is of length L, find a channel design that will allow the maximum possible flow. Is this design preferable to an open semicircular channel?

36. The cost of material for the sides of a rectangular shipping container is a cents per square foot; the cost of the top and bottom material is $\frac{3}{2}a$ cents per square foot. If the volume is to be $\frac{3}{2}$ cubic feet, what dimensions of the container will minimize its cost?

37. A fuel reservoir at D (see the figure) is to service plants located at A, B, and C, as shown. The cost, in thousands of dollars, of connecting the plants to D is determined by the formula

$$F = 6x^2 + 6y^2 - 4x - 6y + 5$$

Find the location of D that will minimize this cost.

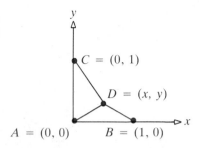

38. A metal detector is used to locate an underground pipe. When several meter readings of the detector are compared, it is found that the reading at an arbitrary point (x, y) is given by the formula

$$M = y(x - x^2) - y^2 \text{ volts} \qquad x \ge 0, \quad y \ge 0$$

Find the point (x, y) where the reading is largest.

39. A steel manufacturer produces two grades of steel, x tons of grade A and y tons of grade B. His cost C and revenue R are given in dollars by the formulas

$$C = \tfrac{1}{20}x^2 + 700x + y^2 - 150y - \tfrac{1}{2}xy$$
$$R = 2700x - \tfrac{3}{20}x^2 + 1000y - y^2 + \tfrac{1}{2}xy + 10,000$$

If $P = \text{Profit} = R - C$,   find the production (in tons) of grades A and B that maximizes the manufacturer's profit.

40. Find the point in the plane $3x + 2y + z = 14$ that is nearest the origin.

41. Let $g(x, y) = Ax^2 + 2Bxy + Cy^2$,   where A, B, and C are constants. Show that:
    (a) If $AC - B^2 > 0$, $A \ne 0$,   then $g(x, y)$ has a maximum at $(0, 0)$ if $A < 0$,   or a minimum if $A > 0$.
    (b) If $AC - B^2 < 0$,   then $g(x, y)$ has both positive and negative values at points (x, y) near $(0, 0)$. Thus, $(0, 0, 0)$ is a saddle point for g.

4. The Method of Lagrange Multipliers

We begin with an example.

EXAMPLE 1   At what point in the first quadrant on the hyperbola $xy = 4$ is the value of $z = 12x + 3y$ a minimum? What is this minimum?

Solution   In this problem we are asked to find the minimum of $z = 12x + 3y$ subject to the condition that x and y obey the equation $xy = 4$.   We can express this as a minimum problem in one variable by setting $y = 4/x$ in the expression $z = 12x + 3y$, obtaining

$$z = 12x + \frac{12}{x}$$

The critical numbers obey

$$\frac{dz}{dx} = 12 - \frac{12}{x^2} = 0$$

$$\frac{12}{x^2} = 12$$

$$x = 1 \quad \text{or} \quad x = -1$$

We ignore $x = -1$ since the point $(-1, -4)$ is not in the first quadrant. Since $d^2z/dx^2 = 24/x^3 > 0$ for $x = 1$, we conclude that when $x = 1$ (and, therefore, $y = 4/x = 4$), the value of z is a minimum. Thus, $z = 12x + 3y$ is a minimum at the point $(1, 4)$ on the hyperbola, and this minimum value is $z = 24$.   ∎

In this example we found the minimum value of $z = 12x + 3y$ subject to the condition that $xy = 4$. We solved the problem by eliminating the variable y and treating the problem as a typical minimum problem in one variable. Sometimes, though, it is not easy—and perhaps even impossible—to eliminate a variable. In such cases, the *method of Lagrange multipliers* may be used.

The purpose of the method of *Lagrange multipliers* is to maximize (or minimize) a function, subject to an auxiliary condition (or constraint) on the domain of the function. Specifically, for two variables: $z = f(x, y)$ may be maximized (or minimized) subject to the constraint $g(x, y) = 0$. Similarly, for three variables: $w = f(x, y, z)$ may be maximized (or minimized) subject to the constraint $g(x, y, z) = 0$.

GEOMETRIC ARGUMENT

We begin by looking at the level curves of $z = 12x + 3y$. On a given level curve, the value of z is fixed, and we seek the smallest such value when x and y are restricted to the hyperbola $xy = 4$. We observe in Figure 16 that the minimum value of z occurs precisely at the point of tangency of the line $12x + 3y = 24$ with the hyperbola $xy = 4$. At this point of tangency, the two curves have a common tangent and, hence, the same normal line. Since the gradient gives the direction of the normal line to a curve, it follows that ∇f is parallel to ∇g. That is, there is a scalar $\lambda \neq 0$ so that

$$\nabla f = \lambda \nabla g \quad \text{where} \quad g(x, y) = 0$$

We now state and prove a theorem that suggests a method for solving problems of the type under discussion.

(17.28)   **THEOREM**   **Let f and g be functions of two variables with continuous partial derivatives. Suppose that the function f, when restricted to the level curve $g(x, y) = 0$ of the function g, has a local extremum at the point (x_0, y_0). Suppose also that $\nabla g(x_0, y_0) \neq 0$. Then there is a number λ such that**

$$\nabla f(x_0, y_0) = \lambda \nabla g(x_0, y_0)$$

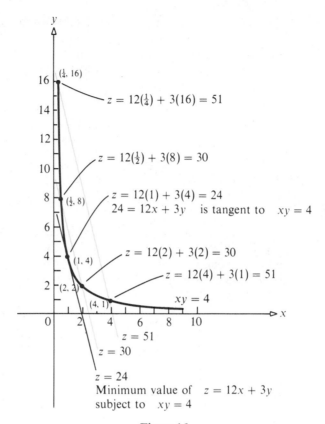

Figure 16

Note: If $\lambda = 0$, this equation says that $f_x(x_0, y_0) = f_y(x_0, y_0) = 0$. If $\lambda \neq 0$, it says that the level curves of f and g passing through (x_0, y_0) have the same tangent line.

Proof Let C be the curve $g(x, y) = 0$. Suppose that $\mathbf{r}(t) = x(t)\mathbf{i} + y(t)\mathbf{j}$ is a parametric equation in vector form for C. Define $s(t) = f(x(t), y(t))$ and choose t_0 so that $x(t_0) = x_0$, $y(t_0) = y_0$. We are assuming that $s(t)$ has either a local maximum or a local minimum at t_0. Thus, $s'(t_0) = 0$. Then, by the chain rule,

$$0 = s'(t_0) = f_x(x_0, y_0)x'(t_0) + f_y(x_0, y_0)y'(t_0)$$
$$= \nabla f(x_0, y_0) \cdot \mathbf{r}'(t_0)$$

If $\nabla f(x_0, y_0)$ is nonzero, it is perpendicular to $\mathbf{r}'(t_0)$, which is a tangent vector to C at (x_0, y_0). By theorem (17.11), the vector $\nabla g(x_0, y_0)$ is also perpendicular to $\mathbf{r}'(t_0)$. Thus, there is a number λ such that $\nabla f(x_0, y_0) = \lambda \nabla g(x_0, y_0)$. ∎

As a consequence of theorem (17.28), we have the following:

The extreme values of $z = f(x, y)$, **subject to the condition that** $g(x, y) = 0$, **are found at points** (x_0, y_0) **which are solutions of the equations**

(17.29)
$$\nabla f(x, y) = \lambda \nabla g(x, y) \qquad g(x, y) = 0$$

where λ, **called a** *Lagrange multiplier*, **represents one or more numbers to be determined.**

The equations (17.29) are, in fact, a system of three equations in the three unknowns, x, y, and λ, namely,

(17.30)
$$f_x(x, y) = \lambda g_x(x, y) \qquad f_y(x, y) = \lambda g_y(x, y) \qquad g(x, y) = 0$$

We call a solution (x_0, y_0) of (17.29) or (17.30) a *test point* since it is a candidate for the desired extrema. The maximum and minimum values, if they exist, may be found by choosing the largest and smallest values of $f(x, y)$ at the test points. Another approach is to define an *auxiliary function* of three variables:

$$L(x, y, \lambda) = f(x, y) - \lambda g(x, y)$$

Note that if (x, y) obeys the constraint $g(x, y) = 0$, then $L(x, y, \lambda) = f(x, y)$ for all λ.

If we set all three partial derivatives of L equal to zero, we have

$$
\begin{aligned}
L_x = f_x - \lambda g_x = 0 &\qquad \text{or} \qquad& f_x = \lambda g_x \\
L_y = f_y - \lambda g_y = 0 &\qquad \text{or} \qquad& f_y = \lambda g_y \\
L_\lambda = -g(x, y) = 0 &\qquad \text{or} \qquad& g(x, y) = 0
\end{aligned}
$$

The equations in the right-hand column above show that the process of solving the equations in (17.30) to find test points is equivalent to finding the critical points of the auxiliary function.

Let's rework Example 1, this time using the method of Lagrange multipliers.

EXAMPLE 2 Find the minimum value of $z = f(x, y) = 12x + 3y$ subject to the condition $xy = 4$, $x > 0$, $y > 0$.

Solution The test points obey the equations

(17.31)
$$\nabla f(x, y) = \lambda \nabla g(x, y) \qquad g(x, y) = xy - 4 = 0$$

where λ is a number that is not known. The gradients of f and g are

$$\nabla f = 12\mathbf{i} + 3\mathbf{j} \qquad \nabla g = y\mathbf{i} + x\mathbf{j}$$

so that the equations (17.29) are

$$12 = \lambda y \qquad 3 = \lambda x \qquad xy - 4 = 0$$

Eliminating λ from the first two equations results in $3y = 12x$, or $y = 4x$, so the third equation becomes

$$4x^2 - 4 = 0$$
$$x^2 = 1$$
$$x = 1 \quad \text{or} \quad x = -1$$

We ignore $x = -1$ since $x > 0$ in the domain of the function. When $x = 1$, we have $y = 4$, so the only test point is $(1, 4)$. The corresponding minimum value of $z = 12x + 3y$ is $z = 24$. ∎

EXAMPLE 3   Find the maximum and minimum value of $z = f(x, y) = 3x - y + 1$ subject to the condition $3x^2 + y^2 = 9$.

Solution   The test points satisfy the equations

$$\nabla f = \lambda \nabla g \qquad g(x, y) = 3x^2 + y^2 - 9 = 0$$

where λ is to be determined. These equations are

$$3 = \lambda 6x \qquad -1 = \lambda 2y \qquad 3x^2 + y^2 - 9 = 0$$

From the first two equations, we find that $x = 1/2\lambda$ and $y = -1/2\lambda$, so the third equation may be written as

$$\frac{3}{4\lambda^2} + \frac{1}{4\lambda^2} = 9$$

$$\frac{1}{\lambda^2} = 9$$

$$\lambda = \pm\tfrac{1}{3}$$

Hence, $x = \pm\tfrac{3}{2}$, $y = \mp\tfrac{3}{2}$, and the test points are $(\tfrac{3}{2}, -\tfrac{3}{2})$ and $(-\tfrac{3}{2}, \tfrac{3}{2})$. The corresponding values of z are $z = \tfrac{9}{2} + \tfrac{3}{2} + 1 = 7$ and $z = -\tfrac{9}{2} - \tfrac{3}{2} + 1 = -5$. Thus, the maximum value of z on the ellipse is 7 and the minimum value is -5. ∎

The next example shows how a Lagrange multiplier may be used in combination with the derivative tests of Section 3 to find the extrema of a function whose domain is a closed region.

EXAMPLE 4   Find the absolute maximum and minimum of the function

$$f(x, y) = x^2 + y^2 + 4x - 4y + 3 \qquad \text{on} \quad x^2 + y^2 \le 4$$

Solution   We first look for critical points:

$$f_x = 2x + 4 = 0 \qquad x = -2$$
$$f_y = 2y - 4 = 0 \qquad y = 2$$

But the inequality $x^2 + y^2 \le 4$ is not satisfied at $x = -2$, $y = 2$; the point $(-2, 2)$ lies outside the domain of the function. Thus, the function has no critical points. The extrema must occur on the boundary of the domain; that is, on the curve $x^2 + y^2 = 4$. We apply the Lagrange method with the constraint $g(x, y) = x^2 + y^2 - 4 = 0$. The test point conditions are

$$f_x = 2x + 4 = 2x\lambda \qquad f_y = 2y - 4 = 2y\lambda \qquad x^2 + y^2 = 4$$

Eliminating λ from the first two equations yields $y = -x$. By substituting in the third equation and solving, we get two test points: $(\sqrt{2}, -\sqrt{2})$ and $(-\sqrt{2}, \sqrt{2})$. Thus, the extreme values are

$$f(\sqrt{2}, -\sqrt{2}) = 7 + 8\sqrt{2} \qquad \text{Maximum value}$$
$$f(-\sqrt{2}, \sqrt{2}) = 7 - 8\sqrt{2} \qquad \text{Minimum value} \qquad \blacksquare$$

FUNCTIONS OF THREE VARIABLES

For functions of three variables, the method of Lagrange multipliers works in the same way. The extreme values of $w = f(x, y, z)$ subject to the condition $g(x, y, z) = 0$ occur at the solutions (x, y, z) of the system of equations

$$\mathbf{V}f = \lambda \mathbf{V}g \qquad g(x, y, z) = 0$$

where λ is some number to be determined.

EXAMPLE 5   A box that is open on top is to have a fixed volume of 12 cubic meters. The material used to make the bottom of the box costs \$3 per square meter, while the material used for the sides costs \$1 per square meter. What should the dimensions be so that the cost is a minimum?

Solution   Let x be the width of the box, y be the depth of the box, and z be the height of the box. Then xy is the area of the bottom of the box. We seek to minimize

$$C = 3xy + 2xz + 2yz$$

subject to the condition $g(x, y, z) = xyz - 12 = 0$. The test points satisfy

$$\mathbf{V}C = \lambda \mathbf{V}g \qquad g(x, y, z) = xyz - 12 = 0$$

That is,

(17.32) $3y + 2z = \lambda yz \qquad 3x + 2z = \lambda xz \qquad 2x + 2y = \lambda xy \qquad xyz - 12 = 0$

Since $x > 0$, $y > 0$, $z > 0$, we can solve the first three equations for λ, obtaining

$$\lambda = \frac{3}{z} + \frac{2}{y} \qquad \lambda = \frac{3}{z} + \frac{2}{x} \qquad \lambda = \frac{2}{y} + \frac{2}{x}$$

From these, we find that

$$y = x \qquad \text{and} \qquad z = \tfrac{3}{2}x$$

Hence, from the last equation in (17.32), we get

$$\tfrac{3}{2}x^3 = 12$$
$$x = 2$$

The test point is therefore $(2, 2, 3)$. The dimensions of the box are 2 meters $\times$ 2 meters $\times$ 3 meters, and the minimum cost is \$36. ■

SEVERAL CONSTRAINTS

The method of Lagrange multipliers may also be used for problems involving several constraints. The next example illustrates the case of two constraints.

EXAMPLE 6 Find the points on the intersection of the ellipsoid $x^2 + y^2 + 9z^2 = 25$ and the plane $x + 3y - 2z = 0$ that are the farthest from the origin. Also, find the points that are the closest to the origin.

Solution Let C be the curve resulting from the intersection of the ellipsoid and the plane, and let $P = (x, y, z)$ be a point on C. The function we wish to maximize and minimize is the distance from P to the origin, namely, $d = f(x, y, z) = \sqrt{x^2 + y^2 + z^2}$. The constraints under which this is to be done are

$$g(x, y, z) = x^2 + y^2 + 9z^2 - 25 = 0 \quad \text{and} \quad h(x, y, z) = x + 3y - 2z = 0$$

Set

$$\nabla f(x, y, z) = \lambda_1 \nabla g(x, y, z) + \lambda_2 \nabla h(x, y, z)$$

where λ_1 and λ_2 are to be determined. The desired values of x, y, and z are solutions of

(17.33)
$$\frac{x}{\sqrt{x^2 + y^2 + z^2}} = \lambda_1(2x) + \lambda_2(1)$$

$$\frac{y}{\sqrt{x^2 + y^2 + z^2}} = \lambda_1(2y) + \lambda_2(3)$$

$$\frac{z}{\sqrt{x^2 + y^2 + z^2}} = \lambda_1(18z) + \lambda_2(-2)$$

Elimination of λ_1 and λ_2 from (17.33) gives $3xz - yz = 0$. By using this equation together with the equations of the plane and the ellipsoid, we obtain the following four points:

$$\left(\frac{-15}{\sqrt{10}}, \frac{5}{\sqrt{10}}, 0 \right), \qquad \left(\frac{15}{\sqrt{10}}, \frac{-5}{\sqrt{10}}, 0 \right),$$

$$\left(\frac{5}{\sqrt{235}}, \frac{15}{\sqrt{235}}, \frac{25}{\sqrt{235}} \right), \qquad \left(\frac{-5}{\sqrt{235}}, \frac{-15}{\sqrt{235}}, \frac{-25}{\sqrt{235}} \right)$$

The distance of each of these from $(0, 0, 0)$ is

$$d = 5 \qquad d = 5 \qquad d = \sqrt{\tfrac{175}{47}} \qquad d = \sqrt{\tfrac{175}{47}}$$

respectively. Thus, the points $(-15/\sqrt{10}, 5/\sqrt{10}, 0)$ and $(15/\sqrt{10}, -5\sqrt{10}, 0)$ are farthest from the origin, while the points $(5/\sqrt{235}, 15/\sqrt{235}, 25/\sqrt{235})$ and $(-5/\sqrt{235}, -15/\sqrt{235}, -25/\sqrt{235})$ are closest. Since the intersection of the ellipsoid and the plane is an ellipse, the points we have found are the ends of the axes of this ellipse (see Fig. 17).

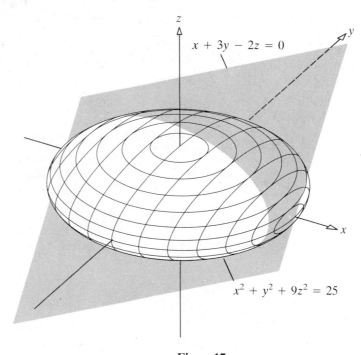

$x + 3y - 2z = 0$

$x^2 + y^2 + 9z^2 = 25$

Figure 17　　　　　　　　　　　　　　■

EXERCISE 4

In Problems 1–8 use the method of Lagrange multipliers to find the maximum and minimum value of f subject to the given condition. Tell where the extreme values occur.

1. $f(x, y) = 3xy;$ $g(x, y) = x^2 + 2y^2 - 4 = 0$

2. $f(x, y) = x - 2y^2;$ $g(x, y) = x^2 + y^2 - 1 = 0$

3. $f(x, y) = x^2 + 4y^3;$ $g(x, y) = x^2 + 2y^2 - 2 = 0$

4. $f(x, y) = 2xy;$ $g(x, y) = x^2 + y^2 - 2 = 0$

5. $f(x, y) = xy;$ $g(x, y) = 9x^2 + 4y^2 - 36 = 0$

6. $f(x, y) = x^2 - 4xy + 4y^2;$ $g(x, y) = x^2 + y^2 - 4 = 0$

7. $f(x, y, z) = 4x - 3y + 2z;$ $g(x, y, z) = x^2 + y^2 - 6z = 0$

8. $f(x, y, z) = x^2 + 2y^2 + z^2;$ $g(x, y, z) = 2x - 3y + z - 6 = 0$

9. Find the point on the line $x - 3y = 6$ that is closest to the origin.

10. Find the point on the plane $2x + y - 3z = 6$ that is closest to the origin.

11. At which points on the ellipse $x^2 + 2y^2 = 2$ is the product xy a maximum?

12. Find the dimensions of a box that is open at the top, so that the volume is a maximum when the surface area is fixed at 24 square centimeters.

13. A box that is open at the top and has a volume of 12 cubic meters is to be made from material costing $1 per square meter. What dimensions minimize the cost?

14. Suppose that $T = T(x, y, z) = 100x^2yz$ is the temperature (in degrees Celsius) at any point (x, y, z) on the sphere given by $x^2 + y^2 + z^2 = 1.$ Find the points on the sphere where the temperature is greatest and least. What is the temperature at these points?

15. A farmer has 340 meters of fencing for enclosing two separate fields, one of which is to be a rectangle twice as long as it is wide and the other a square. The square field must contain at least 100 square meters, and the rectangular one must contain at least 800 square meters.
(a)   If x is the width of the rectangular field, what are the maximum and minimum possible values of x?
(b)   What is the greatest number of square meters that can be enclosed in the two fields? Justify your answer.

16. The surface $xyz = -1$ is cut by the plane $x + y + z = 1,$ resulting in a curve C. Find the points on C that are nearest to the origin and farthest from it.

17. Find the maximum and minimum values of the function $f(x, y, z) = x^2 + y^2 + z^2$ subject to $z^2 = x^2 + y^2$ and $x + y - z + 1 = 0.$

Miscellaneous Exercises

1. Find the direction through the point $(2, 1)$ in which the function $u = 4x^2 + 9y^2$ has the maximum rate of change. Show that this direction is that of the normal to the curve $4x^2 + 9y^2 = 25$ at the point $(2, 1)$. Also, find the value of this maximum rate of change.

2. The temperature at any point (x, y) of a rectangular plate lying in the xy-plane is given by $T = x \sin 2y.$ Find the rate of change of temperature at the point $(1, \pi/4)$ in the direction making an angle of $\pi/6$ with the x-axis.

3. Suppose that the electric potential (voltage V) at each point in space is $V = e^{xyz}$ and that electric charges move in the direction of greatest *potential drop* (most rapid decrease of potential). In what direction does a charge at the point $(1, -1, 2)$ move? How fast does the potential change as the charge leaves this point?

4. Show that the tangent plane to the cylinder $x^2 + y^2 = a^2$ at the point (x_0, y_0, z_0) is given by the equation $x_0x + y_0y = a^2.$

5. Show that the tangent plane to the cone $(z^2/c^2) = (x^2/a^2) + (y^2/b^2)$ at the point $(x_0, y_0, z_0) \neq (0, 0, 0)$ is given by the equation $(z_0z/c^2) = (x_0x/a^2) + (y_0y/b^2).$ Why do we restrict the point of tangency to be different from the origin?

6. Let $F(x, y) = 0$ be the equation of a curve in the xy-plane, where F is differentiable. If (x_0, y_0) is a point on the curve, show that $\nabla F(x_0, y_0)$ is normal to the curve at (x_0, y_0).

7. Use Problem 6 to show that the tangent to the curve $F(x, y) = 0$ at (x_0, y_0) is given by $a(x - x_0) + b(y - y_0) = 0$, where $a = F_x(x_0, y_0)$ and $b = F_y(x_0, y_0)$. (Assume that a and b are not both 0.)

8. Use Problem 7 to find the following tangents to the hyperbola $x^2 - y^2 = 16$:
   (a) The tangent at $(5, 3)$; check by the methods of single variable calculus, finding dy/dx at $(5, 3)$ by implicit differentiation
   (b) The tangent at $(4, 0)$; note that this tangent is vertical, which requires special treatment in single variable calculus

9. Assuming that $b \neq 0$ in Problem 7 show that the slope of the tangent to the curve $F(x, y) = 0$ at (x_0, y_0) is $m = -F_x(x_0, y_0)/F_y(x_0, y_0)$. [This is a proof of (16.32).]

10. What points of the surface $xy - z^2 - 6y + 36 = 0$ are closest to the origin?

11. Find the distance between the skew lines

$$L_1: \quad x = t - 6, \quad y = t, \quad z = 2t \qquad \text{and} \qquad L_2: \quad x = t, \quad y = t, \quad z = -t$$

12. Show that if the sum of the sines of the angles of a triangle is a maximum, then the triangle must be equilateral.

13. Find the point of the paraboloid $z = 2 - x^2 - y^2$ that is closest to the point $(1, 1, 2)$.

14. It is shown in single variable calculus that a cylindrical can of fixed surface area and maximum volume has altitude equal to the diameter of its base. Use the method of Lagrange multipliers to confirm this fact.

15. You found in Problem 31 of Exercise 3 that the closed rectangular box of fixed surface area and maximum volume is a cube. Use the method of Lagrange multipliers to confirm this fact.

16. A scientist plots the points $(x_1, y_1), \ldots, (x_n, y_n)$ from her experimental data. Her theory tells her that the points should lie on a straight line, but they do not—experimental error, perhaps. She is looking for the line $y = mx + b$ that "best" fits the data. The most often used criterion is the least squares fit, for which m and b are chosen to minimize

$$\sum_{k=1}^{m} (mx_k + b - y_k)^2$$

Show that the minimizing values of m and b are

$$m = \frac{n \sum_{k=1}^{n} x_k y_k - \sum_{k=1}^{n} x_k \sum_{k=1}^{n} y_k}{n \sum_{k=1}^{n} x_k^2 - (\sum_{k=1}^{n} x_k)^2} \qquad b = \frac{1}{n} \left(\sum_{k=1}^{n} y_k - m \sum_{k=1}^{n} x_k \right)$$

17. Find the least squares estimate $y = mx + b$ for the points $(0, 2)$, $(1, 1)$, $(2, 2)$, $(3, 4)$, and $(4, 4)$. Plot the points, and draw the line.

18. Show that at any point, the sum of the intercepts on the coordinate axes of the tangent plane to the surface $x^{1/2} + y^{1/2} + z^{1/2} = a^{1/2}$ is a constant.

19. Find equations of the tangent plane and normal line to $\sin(x + y) + \sin(y + z) = 1$ at $(0, \pi/2, \pi/2)$.

20. Let $\mathbf{r} = x\mathbf{i} + y\mathbf{j}$ and $r = \|\mathbf{r}\| = \sqrt{x^2 + y^2}$.
   (a) Show that: $\nabla r^n = nr^{n-2}\mathbf{r}$ (b) Show that: $\nabla g(r) = g'(r) \dfrac{\mathbf{r}}{r}$

   (c) Show that: $D_\mathbf{u}[g(r)] = \dfrac{g'(r)}{r}(\mathbf{u} \cdot \mathbf{r})$ (d) Show that: $\nabla \left(\dfrac{x}{r} \right) = \dfrac{\mathbf{i}}{r} - \left(\dfrac{x}{r} \right) \left(\dfrac{\mathbf{r}}{r^2} \right)$

21. Show that if $\nabla f(x, y) = c(x\mathbf{i} + y\mathbf{j})$, where c is a constant, then $f(x, y)$ is constant on any circle of radius k, centered at $(0, 0)$. [Hint: Set $x = k \cos \theta$, $y = k \sin \theta$ and calculate $df/d\theta$.]

22. A hill is in the shape of the surface $z = 5 - x^2 - 3y^2$. Standing at the point $(1, 1, 1)$, we may ski in the direction **i** or **j**. Use partial derivatives to determine which direction is best if we want the descent to start out as slowly as possible.

23. The temperature at each point of region $x^2 + y^2 + z^2 \leq 9$ is $T = \sqrt{9 - x^2 - y^2 - z^2}$. Use a partial derivative to find the (instantaneous) rate of change of T if we start at the point $(0, 1, 2)$ and move across the region in a straight path ending at $(2, 1, 2)$.

24. Find equations of the tangent plane and normal line to $(yz)^{xz} = 16$ at $(2, 1, 2)$.

25. Find the acute angle between the normal lines to the surfaces $z = xy$ and $y^2 = xz$ at $(1, 1, 1)$.

26. Suppose that $z^2 + xz - y + w^2 = 0$ and $x^2 + y^2 + z^2 + w^2 = 18$. Determine z and w as functions of x and y. Find $D_{\mathbf{u}}z$ at $x = -2$, $y = 3$, $z = 1$, $w = 2$ in the direction from $(0, 1)$ to $(2, -3)$.

27. If $f(x, y, z) = z^3 + 3xz - y^2$, find the directional derivatives of f at $(1, 2, 1)$ in the direction of the line $x - 1 = y - 2 = z - 1$.

28. Find the directional derivatives of $f(x, y, z) = 2xy + xz - z^2$ at $(1, -1, 0)$ in the direction of the line of intersection of the planes $2x + y - z = 0$ and $x - 2z = 0$.

29. Find the minimum value of $w = x^2 + y^2 + z^2$ subject to the contraints $2x + y + 2z = 9$, $5x + 5y + 7z = 29$.

18

Multiple Integrals

The definite integral of a function of a single variable can be extended to functions of two or more variables. Integrals of a function of two or more variables are called *multiple integrals*. Specifically, the integral of a function of two variables is called a *double integral*, and integrals of a function of three variables are called *triple integrals*. For consistency, we shall call the integral of a function of one variable a *single integral*. Since the development of multiple integration parallels that of a single integral, we recommend that you review Chapter 5 before going ahead in this chapter.

1. The Double Integral

For $\int_a^b f(x)\,dx$, we require that f be defined on a closed interval $a \le x \le b$. For the double integral of a function of two variables, we require that the function be defined over a *closed bounded region* R of the xy-plane. By a *closed* region, we mean that the boundary of R is included. By a *bounded* region, we mean that R can be enclosed in a circle of sufficiently large radius.*

---

* If R is not bounded, we have an *improper* double integral. We do not discuss such integrals in this book.

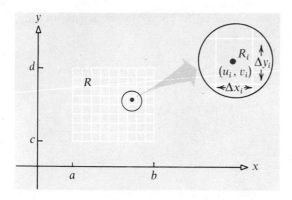

Figure 1

We begin by considering a function f of two variables defined over a rectangular region R given by

$$R: \quad a \le x \le b, \quad c \le y \le d$$

We divide the rectangular region R into subrectangles by drawing lines parallel to the coordinate axes and let n be the total number of subrectangles constructed. We now have a *partition* P of R into n subrectangles $R_1, R_2, \ldots, R_n$ (see Fig. 1).

The *norm* of the partition P, denoted by $\|P\|$, is defined as the length of the largest diagonal of the rectangles R_i, $i = 1, 2, \ldots, n$. If Δx_i denotes the length of the rectangle R_i and Δy_i is its width, then its area is $\Delta A_i = \Delta x_i \Delta y_i$. In each rectangle R_i, we arbitrarily select a point (u_i, v_i) and evaluate the function f there. Then we can form the sum

(18.1)
$$\sum_{i=1}^{n} f(u_i, v_i) \Delta A_i$$

This sum is called a *Riemann sum* of f for the partition P.

EXAMPLE 1   Let $f(x, y) = y/x$ be a function defined over the unit square having its lower left corner at $(1, 0)$ and its upper right corner at $(2, 1)$. Evaluate the Riemann sum by partitioning the unit square into four congruent subsquares and using the lower left corner of each as (u_i, v_i).

Solution   Figure 2 illustrates the situation. By using (18.1), in which $\Delta A_i = (\frac{1}{2})(\frac{1}{2})$ for $i = 1, 2, 3, 4$, the Riemann sum is

$$\sum_{i=1}^{4} f(u_i, v_i) \Delta A_i = f(1, 0)\Delta A_1 + f\left(\frac{3}{2}, 0\right)\Delta A_2$$

$$+ f\left(1, \frac{1}{2}\right)\Delta A_3 + f\left(\frac{3}{2}, \frac{1}{2}\right)\Delta A_4$$

$$= \left(\frac{0}{1}\right)\left(\frac{1}{4}\right) + \left(\frac{0}{\frac{3}{2}}\right)\left(\frac{1}{4}\right)$$

$$+ \left(\frac{\frac{1}{2}}{1}\right)\left(\frac{1}{4}\right) + \left(\frac{\frac{1}{2}}{\frac{3}{2}}\right)\left(\frac{1}{4}\right)$$

$$= \frac{1}{4}\left(0 + 0 + \frac{1}{2} + \frac{1}{3}\right)$$

$$= \left(\frac{1}{4}\right)\left(\frac{5}{6}\right) = \frac{5}{24} \quad \blacksquare$$

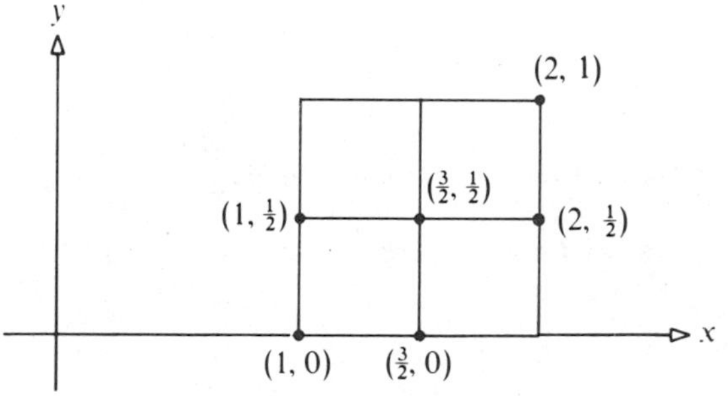

Figure 2

EXAMPLE 2   In Example 1 use the upper right corners as (u_i, v_i) to compute the Riemann sum.

Solution   By using (18.1), the Riemann sum is

$$\sum_{i=1}^{4} f(u_i, v_i)\Delta A_i = f\left(\frac{3}{2}, \frac{1}{2}\right)\Delta A_1 + f\left(2, \frac{1}{2}\right)\Delta A_2 + f\left(\frac{3}{2}, 1\right)\Delta A_3 + f(2, 1)\Delta A_4$$

$$= \left(\frac{\frac{1}{2}}{\frac{3}{2}}\right)\left(\frac{1}{4}\right) + \left(\frac{\frac{1}{2}}{2}\right)\left(\frac{1}{4}\right) + \left(\frac{1}{\frac{3}{2}}\right)\left(\frac{1}{4}\right) + \left(\frac{1}{2}\right)\left(\frac{1}{4}\right)$$

$$= \frac{1}{4}\left(\frac{1}{3} + \frac{1}{4} + \frac{2}{3} + \frac{1}{2}\right) = \frac{7}{16}   \blacksquare$$

The sum (18.1) depends on both the choice of a partition and on the choice of the points (u_i, v_i). However, if all such sums can be made as close as we please to a number I by choosing partitions whose norm is sufficiently close to 0, then I is defined as the limit of these sums as $\|P\|$ approaches 0.

(18.2)   **DEFINITION**   **Let f be a function defined on a closed bounded rectangular region R. If there is a number I with the property that for any $\varepsilon > 0$,   there is a $\delta > 0$ such that**

$$\left|\sum_{i=1}^{n} f(u_i, v_i)\Delta A_i - I\right| < \varepsilon$$

for every partition P for which $\|P\| < \delta$ and for any choice of (u_i, v_i) in R_i,   then I is called the limit* of the sums of the form $\sum_{i=1}^{n} f(u_i, v_i)\Delta A_i$,   and we write

$$I = \lim_{\|P\| \to 0} \sum_{i=1}^{n} f(u_i, v_i)\Delta A_i$$

The above statement asserts that the Riemann sums of f for the partition P can be made as close as we please to I if the norm of the partition P is sufficiently close to 0. We now give a name to this number I, provided it exists.

(18.3)   **DEFINITION**   *Double Integral.*   **Let f be a function defined on a closed bounded rectangular region R. Then the *double integral of f over R*, denoted by $\iint_R f(x, y)\, dA$,   is defined by**

$$\iint_R f(x, y)\, dA = \lim_{\|P\| \to 0} \sum_{i=1}^{n} f(u_i, v_i)\Delta A_i$$

provided this limit exists. In this case, f is said to be *integrable on R*.

Other symbols for the double integral of f over R are

$$\iint_R f(x, y)\, dx\, dy \qquad \text{and} \qquad \iint_R f(x, y)\, dy\, dx$$

---

* The proof that such a number, if it exists, is unique is similar to the proof given in Appendix II.

THE DOUBLE INTEGRAL OVER A MORE GENERAL REGION

Consider a closed bounded region R whose boundary consists of a finite number of arcs of smooth curves that do not intersect themselves, but are joined together to form a closed curve, as illustrated in Figure 3(a). Such curves are referred to as *piecewise smooth*.* In this book we shall deal only with such curves. We draw lines parallel to the coordinate axes to obtain a rectangular partition of R, and we discard any rectangle thus formed that does not lie entirely in R [see Fig. 3(b)]. Let n be the number of rectangles that remain. Now we may proceed as we did in the development of definitions (18.2) and (18.3) to define the double integral of a function f defined on the region R as

(18.4)
$$\iint\limits_{R} f(x, y)\, dA = \lim_{\|P\| \to 0} \sum_{i=1}^{n} f(u_i, v_i)\Delta x_i \Delta y_i$$

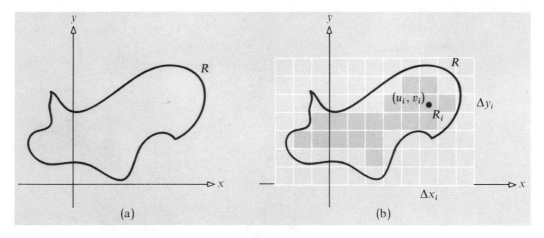

Figure 3

Two comments are in order: First, it is apparent (at least intuitively) that the areas of the discarded rectangles will approach 0 as the norm of the partition approaches 0, so that having discarded them will not affect the limit of the sum. Second, if the limit of the sum exists for rectangular partitions, it can be shown that the manner in which R is partitioned does not matter, as long as each subregion obtained has an area.

We state, without proof, a condition under which the limit in (18.4) exists. The similarity of this result and (5.23) for a single integral is apparent.

(18.5)   **THEOREM**   **If $z = f(x, y)$ is a continuous function defined on a closed bounded region R, whose boundary is piecewise smooth, then f is integrable on R.**

---

* Such curves are sometimes referred to as *simple closed curves* or *Jordan curves*.

GEOMETRIC INTERPRETATION

If a function f of a single variable is continuous and nonnegative on a closed interval $[a, b]$, then the single integral $\int_a^b f(x)\,dx$ equals the area under the graph of f from $x = a$ to $x = b$. Similarly, suppose $z = f(x, y)$ is a continuous and nonnegative function defined on a closed bounded region R. The graph of f is then a surface lying above the xy-plane. Figure 4 shows a typical rectangle that results from partitioning R. The area of this rectangle is $\Delta x_i \Delta y_i$. The value $f(u_i, v_i)$ is the height of a rectangular solid whose base has the area $\Delta x_i \Delta y_i$. The volume of this rectangular solid is the product $f(u_i, v_i)\Delta x_i \Delta y_i$. The sum $\sum_{i=1}^{n} f(u_i, v_i)\Delta x_i \Delta y_i$ over all the rectangles of the partition will approximate the volume of the solid under the graph of the surface $z = f(x, y)$ and above R. But this sum is also an approximation to the double integral $\iint_R f(x, y)\,dA$. As a result, we define the volume of the solid as equal to the value of the double integral.*

(18.6) **Let f denote a continuous, nonnegative function defined on a closed bounded region R, whose boundary is a piecewise smooth curve. The volume V under the graph of $z = f(x, y)$ and over the region R is given by**

$$V = \iint_R f(x, y)\,dA$$

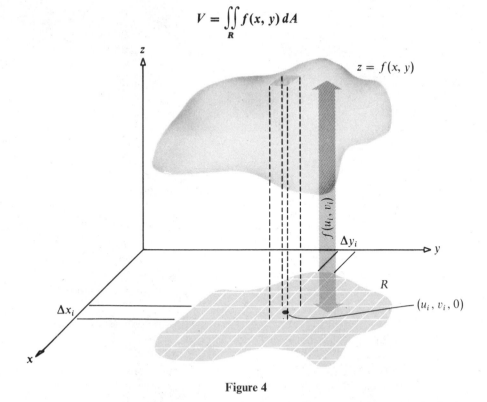

Figure 4

---

* It can be proved that this formula for volume is consistent with the formulas for volume given in Chapter 6.

In particular, if $f(x, y) = 1$,   the double integral $\iint_R dA$ represents the *area* of the region R since, when the height of the solid is 1, the area and the volume are numerically equal.

We postpone problems on evaluating area and volume using double integrals until Section 3. The use of double integrals to calculate mass and moments is taken up in Section 5.

PROPERTIES OF DOUBLE INTEGRALS

We close this section by listing, without proof, some properties of the double integral. The similarity between these properties and those of a single integral is apparent. It is assumed that the functions and the regions are suitably chosen so that the indicated integrals exist.

$$(18.7) \qquad \iint_R [f(x, y) + g(x, y)]\, dA = \iint_R f(x, y)\, dA + \iint_R g(x, y)\, dA$$

$$(18.8) \qquad \iint_R cf(x, y)\, dA = c \iint_R f(x, y)\, dA \qquad c \text{ constant}$$

If R consists of two subregions R_1 and R_2, which have no points in common except for points lying on portions of their common boundary, then

$$(18.9) \qquad \iint_R f(x, y)\, dA = \iint_{R_1} f(x, y)\, dA + \iint_{R_2} f(x, y)\, dA$$

(18.10) If $f(x, y) \geq 0$ throughout R, then $\iint_R f(x, y)\, dA \geq 0.$

EXERCISE 1

In Problems 1–4 calculate the Riemann sum of each function over the given rectangular region R, using the indicated partition of R and choosing (u_i, v_i) as the lower left corner of each subregion.

1. $f(x, y) = x^2 + y^2$; $1 \leq x \leq 3$, $2 \leq y \leq 4$;
$x_0 = 1$, $x_1 = 2$, $x_2 = 3$, $y_0 = 2$, $y_1 = 3$, $y_2 = 4$

2. $f(x, y) = x^2 - y^2$; $0 \leq x \leq 4$, $1 \leq y \leq 3$;
$x_0 = 0$, $x_1 = 1$, $x_2 = 2$, $x_3 = 3$, $x_4 = 4$, $y_0 = 1$, $y_1 = 2$, $y_2 = 3$

3. $f(x, y) = 2xy - y^2$; $0 \leq x \leq 4$, $0 \leq y \leq 2$;
$x_0 = 0$, $x_1 = 2$, $x_2 = 4$, $y_0 = 0$, $y_1 = 1$, $y_2 = 2$

4. $f(x, y) = 2x^2y + x$; $-1 \leq x \leq 1$, $-2 \leq y \leq 0$;
$x_0 = -1$, $x_1 = 0$, $x_2 = 1$, $y_0 = -2$, $y_1 = -1$, $y_2 = 0$

5. Find the Riemann sum for Example 1 if R is divided into nine equal squares and the lower left corners are taken as the (u_i, v_i).

6. Rework Problem 5 using the upper right corners for the (u_i, v_i).

7. Let R be the region enclosed by $y = 1$, $y = -1$, $x = 0$, and $x = 1$; and let R_1 and R_2 be the subregions of R in the first and fourth quadrants, respectively. Suppose $f(x, y)$ is continuous on R and

$$\iint_R 3f(x, y)\, dA - 2 \iint_{R_1} f(x, y)\, dA = \iint_{R_2} f(x, y)\, dA \qquad \iint_{R_2} 5f(x, y)\, dA - 2 \iint_{R_1} f(x, y)\, dA = 18$$

Find $\iint_R f(x, y)\, dA$.

8. Using the properties of double integrals, prove that if the functions f and g are integrable on R and $f(x, y) \le g(x, y)$ for all (x, y) in R, then $\iint_R f(x, y)\, dA \le \iint_R g(x, y)\, dA$.

2. Evaluation of Double Integrals; Iterated Integrals

The evaluation of double integrals would be limited to a few very elementary cases if it were necessary to find the value by a direct application of definition (18.3). Fortunately, there is a method of evaluation for double integrals, called *iteration*, which can be used under certain conditions. We need some preliminary remarks, though, before getting to the formula.

Let f be a function defined and continuous on a closed bounded region R. Under certain conditions on the boundary of R we can evaluate $\iint_R f(x, y)\, dA$ as two single integrals, called *repeated* or *iterated integrals*. We proceed to classify two types of boundaries for R.

Type I. **The region R lies between $x = a$ and $x = b$, and every vertical line between $x = a$ and $x = b$ intersects the boundary of R in at most two points. In addition, Type I regions have a boundary that consists of two smooth curves, $y = g_1(x)$ and $y = g_2(x)$, $g_1(x) \le g_2(x)$, $a \le x \le b$ (see Fig. 5).**

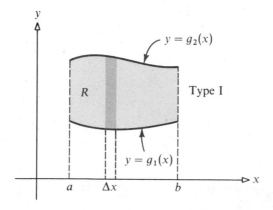

Figure 5

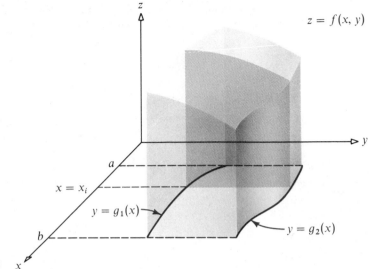

Figure 6 x

We give an intuitive (geometric) argument to develop a method for evaluating double integrals of nonnegative functions $z = f(x, y)$ over Type I regions.

At a number x_i in the closed interval $[a, b]$, let $A(x_i)$ denote the area of the intersection of the plane $x = x_i$ and the solid under the graph of $z = f(x, y)$ and over R (see Fig. 6). By the method developed in Chapter 6, the volume V of this solid is

$$V = \int_a^b A(x)\, dx$$

But this volume V is also given by the double integral of f over R. Thus,

(18.11)
$$\iint\limits_R f(x, y)\, dA = \int_a^b A(x)\, dx$$

But $A(x_i)$ is the area of the plane region under the graph of $z = f(x_i, y)$ from $y = g_1(x_i)$ to $y = g_2(x_i)$. That is,

(18.12)
$$A(x) = \int_{g_1(x)}^{g_2(x)} f(x, y)\, dy$$

where x is held fixed. By substituting expression (18.12) for $A(x)$ in (18.11), we find

(18.13)
$$\iint\limits_R f(x, y)\, dA = \int_a^b \left[\int_{g_1(x)}^{g_2(x)} f(x, y)\, dy \right] dx$$

It can be shown that (18.13) is valid for *any* continuous function f defined on a closed bounded region R of Type I, even if $f(x, y) < 0$ over all or part of R.*

---

* An analytic proof is given in most books in advanced calculus.

If f is a continuous function defined on a closed bounded region R of Type I, then the double integral of f on R is given by

(18.14)
$$\iint\limits_{R} f(x, y) \, dA = \int_a^b \left[\int_{g_1(x)}^{g_2(x)} f(x, y) \, dy \right] dx$$

The integrals on the right of (18.14) are called *iterated integrals*. In evaluating the inside integral, $\int_{g_1(x)}^{g_2(x)} f(x, y) \, dy$, we treat x as a constant and integrate with respect to y. This is analogous to reversing the operation of partial differentiation of f with respect to y.

Let's look at an example.

EXAMPLE 1   Evaluate the iterated integral

$$\int_0^1 \left[\int_{x^2}^{\sqrt{x}} xy \, dy \right] dx$$

Solution   First, we evaluate the inside integral:

$$\int_{x^2}^{\sqrt{x}} xy \, dy = x \int_{x^2}^{\sqrt{x}} y \, dy = x \left(\frac{y^2}{2} \Big|_{x^2}^{\sqrt{x}} \right) = \frac{x}{2} \left[(\sqrt{x})^2 - (x^2)^2 \right]$$

Treat x as
a constant
$$= \frac{x}{2} (x - x^4) = \frac{1}{2} (x^2 - x^5)$$

Thus,

$$\int_0^1 \left[\int_{x^2}^{\sqrt{x}} xy \, dy \right] dx = \int_0^1 \frac{1}{2} (x^2 - x^5) \, dx = \frac{1}{2} \left(\frac{x^3}{3} - \frac{x^6}{6} \right) \Big|_0^1 = \frac{1}{12} \qquad \blacksquare$$

EXAMPLE 2   Evaluate the double integral $\iint_R x^2 y \, dA$,   if the region R is the rectangle given by $-1 \le x \le 1, \quad 0 \le y \le 2$.

Solution   Figure 7 illustrates that R is a Type I region, with $g_1(x) = 0$ and $g_2(x) = 2$. Thus,

$$\iint\limits_{R} x^2 y \, dA = \int_{-1}^1 \left[\int_0^2 x^2 y \, dy \right] dx = \int_{-1}^1 \left[x^2 \left(\frac{y^2}{2} \right) \Big|_0^2 \right] dx$$

$$= \int_{-1}^1 2x^2 \, dx = \frac{2x^3}{3} \Big|_{-1}^1 = \frac{4}{3} \qquad \blacksquare$$

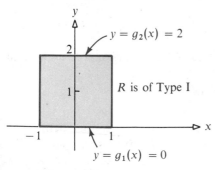

R is of Type I

Figure 7

Regions of Type II are defined as follows:

Type II. **The region R lies between $y = c$ and $y = d$,   and every horizontal line between $y = c$ and $y = d$ intersects the boundary of R in at most two points. In addition, Type II regions have a boundary that consists of two smooth curves, $x = h_1(y)$ and $x = h_2(y)$, $h_1(y) \leq h_2(y)$, $c \leq y \leq d$ (see Fig. 8).**

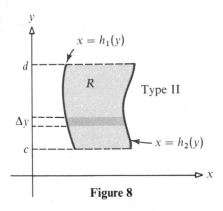

Figure 8

Double integrals of continuous function f defined over regions of Type II have a geometric interpretation analogous to those of Type I. They are evaluated by using iterated integrals as follows:

If f is a continuous function defined on a closed bounded region R of Type II, the double integral of f on R is given by

(18.15)
$$\iint_R f(x, y)\, dA = \int_c^d \left[\int_{h_1(y)}^{h_2(y)} f(x, y)\, dx \right] dy$$

In evaluating the inside integral, $\int_{h_1(y)}^{h_2(y)} f(x, y)\, dx$,   we treat y as a constant and integrate with respect to x. This is analogous to reversing the operation of partial differentiation of f with respect to x.

EXAMPLE 3   Evaluate $\iint_R 3xy^2\, dA$,   if the region R is enclosed by $y = x^2$ and $y = 2x$.

Solution   Figure 9 illustrates the region R. Observe that R is both of Type I and of Type II. If we treat R as Type I, the bottom curve is $y = x^2$,   the top curve is $y = 2x$, and $0 \leq x \leq 2$.

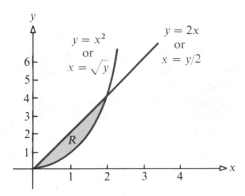

Figure 9

Type I: $\iint\limits_{R} 3xy^2 \, dA = 3 \int_0^2 \left[\int_{x^2}^{2x} xy^2 \, dy \right] dx$

$$\int_{x^2}^{2x} xy^2 \, dy = x\left(\frac{y^3}{3} \Big|_{x^2}^{2x}\right) = \frac{x}{3}(8x^3 - x^6) = \frac{1}{3}(8x^4 - x^7)$$

$$\iint\limits_{R} 3xy^2 \, dA = 3\left(\frac{1}{3}\right) \int_0^2 (8x^4 - x^7) \, dx = \left(\frac{8x^5}{5} - \frac{x^8}{8}\right)\Big|_0^2 = \frac{256}{5} - 32 = \frac{96}{5}$$

If we treat R as Type II, the left curve is $x = y/2$, the right curve is $x = \sqrt{y}$, and $0 \le y \le 4$.

Type II: $\iint\limits_{R} 3xy^2 \, dA = 3 \int_0^4 \left[\int_{y/2}^{\sqrt{y}} xy^2 \, dx \right] dy$

$$\int_{y/2}^{\sqrt{y}} xy^2 \, dx = y^2 \left(\frac{x^2}{2} \Big|_{y/2}^{\sqrt{y}}\right) = \frac{y^2}{2}\left(y - \frac{y^2}{4}\right) = \frac{1}{2}\left(y^3 - \frac{y^4}{4}\right)$$

$$\iint\limits_{R} 3xy^2 \, dA = 3\left(\frac{1}{2}\right) \int_0^4 \left(y^3 - \frac{y^4}{4}\right) dy$$

$$= \frac{3}{2}\left(\frac{y^4}{4} - \frac{y^5}{20}\right)\Big|_0^4 = \frac{3}{2}\left(64 - \frac{256}{5}\right) = \frac{96}{5} \quad \blacksquare$$

As the above example illustrates, a *choice* of the order of integration is available *when the region R is both of Type I and of Type II.* In some cases, integration in one order requires simpler techniques than is required by the opposite order. If you decide to change the order of integration, be careful to determine the new limits of integration.

The next example illustrates that one order of integration may be impossible while the opposite is straightforward.

EXAMPLE 4   By changing the order of integration evaluate

$$\int_0^1 \left[\int_{2x}^2 e^{y^2} \, dy \right] dx$$

Solution   We observe that the integration cannot be done in the order indicated since e^{y^2} has no elementary antiderivative. Thus, we are forced to change the order of the integration. To accomplish this, we need to determine the region R on which the integration is performed. We infer from the given iterated integrals that R is of Type I and R is enclosed by $y = 2x$ and $y = 2$, $0 \le x \le 1$. In Figure 10 we see that region R is also of Type II and is enclosed by $x = 0$ and $x = y/2$, $0 \le y \le 2$. Thus,

$$\int_0^1 \left[\int_{2x}^2 e^{y^2} \, dy \right] dx = \iint\limits_{R} e^{y^2} \, dA = \int_0^2 \left[\int_0^{y/2} e^{y^2} \, dx \right] dy = \int_0^2 e^{y^2} \left[\int_0^{y/2} dx \right] dy$$

$$= \int_0^2 \frac{ye^{y^2}}{2} \, dy = \frac{1}{4} e^{y^2} \Big|_0^2 = \frac{1}{4}(e^4 - 1)$$

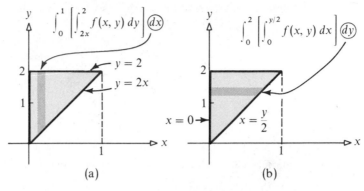

(a)   (b)

Figure 10   ∎

If a region R does not have smooth curves for a boundary or if it is neither of Type I nor of Type II, it may be possible to partition R into subregions, each of which is of Type I or Type II. In this situation, we use property (18.9) to evaluate the double integral. Figure 11 illustrates how this may be done.

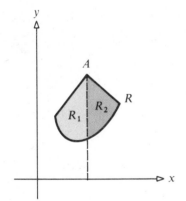

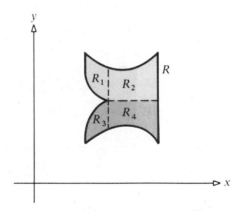

(a) R is Type I, but has a corner at A; draw a vertical line through A, decomposing R into two subregions R_1 and R_2, each of Type I with smooth boundaries

(b) R is neither Type I nor Type II
$$R = R_1 \cup R_2 \cup R_3 \cup R_4$$
R_1, R_2, R_3, R_4 are Type I

$$\iint_R f(x, y)\, dA = \iint_{R_1} f(x, y)\, dA + \iint_{R_2} f(x, y)\, dA$$

$$+ \iint_{R_3} f(x, y)\, dA + \iint_{R_4} f(x, y)\, dA$$

Figure 11

We conclude this section with a capsule summary.

Type I Region: The inside integration is from a bottom curve $y = g_1(x)$ to a top curve $y = g_2(x)$, where $a \le x \le b$:

$$\iint_R f(x, y)\, dA = \int_a^b \left[\int_{g_1(x)}^{g_2(x)} f(x, y)\, dy \right] dx$$

Type II Region: The inside integration is from a left curve $x = h_1(y)$ to a right curve $x = h_2(y)$,   where $c \le y \le d$:

$$\iint\limits_R f(x, y)\, dA = \int_c^d \left[\int_{h_1(y)}^{h_2(y)} f(x, y)\, dx \right] dy$$

EXERCISE 2

In working the following problems, start by drawing an accurate figure illustrating the region R used in the integration. Pay particular attention to the boundary since it determines the correct limits of integration.

In Problems 1–18 evaluate each iterated integral.

1. $\int_0^1 \left[\int_{x^2}^{\sqrt{x}} dy \right] dx$

2. $\int_0^1 \left[\int_{y^2}^{\sqrt{y}} x\, dx \right] dy$

3. $\int_{-1}^2 \left[\int_{y^2}^{y+2} dx \right] dy$

4. $\int_0^1 \left[\int_x^{2x} y\, dy \right] dx$

5. $\int_0^1 \left[\int_1^{e^y} \frac{y}{x}\, dx \right] dy$

6. $\int_0^1 \left[\int_0^{x^2} xe^y\, dy \right] dx$

7. $\int_0^2 \left[\int_y^{2y} xy\, dx \right] dy$

8. $\int_2^4 \left[\int_1^{y^2} \frac{y}{x^2}\, dx \right] dy$

9. $\int_0^1 \left[\int_{x^2}^x \sqrt{xy}\, dy \right] dx$

10. $\int_0^1 \left[\int_{x^2}^x \sqrt{x}\, dy \right] dx$

11. $\int_0^a \left[\int_{a-x}^{\sqrt{a^2-x^2}} y\, dy \right] dx$

12. $\int_0^1 \left[\int_x^1 \frac{dy}{x^2+1} \right] dx$

13. $\int_0^{\sqrt{\pi}} \left[\int_0^{x^2} x \sin y\, dy \right] dx$

14. $\int_0^1 \left[\int_y^{\sqrt{y}} (x^2 + y^2)\, dx \right] dy$

15. $\int_1^2 \left[\int_0^{\ln x} xe^y\, dy \right] dx$

16. $\int_0^1 \left[\int_{x^5}^{x^2} (x + y)\, dy \right] dx$

17. $\int_2^3 \left[\int_0^{1/y} \ln y\, dx \right] dy$

18. $\int_0^\pi \left[\int_{\pi-y}^{\pi+y} \sin(x + y)\, dx \right] dy$

In Problems 19–22 evaluate each double integral.

19. $\iint_R (x + y)\, dA$; R is the region enclosed by $y = x^2$ and $y^2 = 8x$
20. $\iint_R (x^2 - y^2)\, dA$; R is the region enclosed by $y = x$ and $y = x^2$
21. $\iint_R y^2\, dA$; R is the region enclosed by $y = 2 - x$ and $y = x^2$
22. $\iint_R xy\, dA$; R is the region enclosed by $y^2 = x + 1$ and $y = 1 - x$

In Problems 23–28 change the order of integration of each iterated integral. Do not integrate.

23. $\int_0^1 \left[\int_0^x f(x, y)\, dy \right] dx$

24. $\int_0^2 \left[\int_{y^2}^{2y} f(x, y)\, dx \right] dy$

25. $\int_0^a \left[\int_0^{\sqrt{a^2-y^2}} f(x, y)\, dx \right] dy$

26. $\int_0^{2\sqrt[3]{2}} \left[\int_{x^2/4}^{\sqrt{x}} f(x, y)\, dy \right] dx$

27. $\int_0^4 \left[\int_{y^2-2y}^{2y} f(x, y)\, dx \right] dy$

28. $\int_2^5 \left[\int_{x^2-6x+9}^{x-1} f(x, y)\, dy \right] dx$

In Problems 29–32 change the order of integration of the given iterated integral and then evaluate it.

29. $\int_0^1 \left[\int_x^{3x} xy\, dy \right] dx$

30. $\int_0^1 \left[\int_y^{\sqrt{y}} x^2 y\, dx \right] dy$

31. $\int_0^1 \left[\int_{1-x}^{2-x} (x + y)\, dy \right] dx$

32. $\int_0^{\sqrt{\pi/2}} \left[\int_y^{\sqrt{\pi/2}} \sin x^2 \, dx \right] dy$

(Notice the difficulty in
evaluating this integral directly.)

33. Evaluate: $\int_0^{1/2} \left[\int_{2x}^1 e^{y^2} \, dy \right] dx$ _ (see Example 4)

34. Evaluate: $\int_0^1 \left[\int_y^1 \frac{\sin x}{x} \, dx \right] dy$

35. Evaluate: $\int_0^1 \left[\int_0^{\sqrt{1+x^2}} \frac{dy}{x^2 + y^2 + 1} \right] dx$

36. Evaluate:
(a) $\iint_R xy^2 \, dA$; R is the square enclosed by the lines $x = -1$, $x = 1$, $y = -1$,   and $y = 1$
(b) $4 \iint_{R_1} xy^2 \, dA$; R_1 is the square enclosed by the lines $x = 0$, $x = 1$, $y = 0$,   and $y = 1$
(This shows that although the region we are integrating over is symmetric about the x-axis and the y-axis,
we must also take into consideration the behavior of the function on this region before we use symmetry.)

37. Evaluate:
(a) $\iint_R x \, dA$; R is the region enclosed by the circle of radius 1 centered at the origin
(b) $4 \iint_{R_1} x \, dA$; R_1 is the region in the first quadrant enclosed by the circle of radius 1 centered at the origin
(See the comment to Problem 36.)

3. Area and Volume by Double Integrals

AREA

If we restrict our attention to a region R in the xy-plane that is of Type I or Type
II, or to a region that can be decomposed into a finite number of subregions of these
types, then the area A of such a region is given by either of the integrals

(18.16)
$$A = \iint_R dA = \iint_R dx \, dy \qquad A = \iint_R dA = \iint_R dy \, dx$$

The choice of which double integral to use is, of course, dependent on the region R.
It is easy to verify (18.16) if R is a Type I region,
by using (18.14):

$$\iint_R dA = \int_a^b \left[\int_{g_1(x)}^{g_2(x)} dy \right] dx = \int_a^b \left[y \Big|_{g_1(x)}^{g_2(x)} \right] dx$$

$$= \int_a^b \left[g_2(x) - g_1(x) \right] dx$$

As Figure 12 illustrates, this is the area of the region R.
You are asked to verify (18.16) if R is a Type II region
in Problem 28 of Exercise 3.

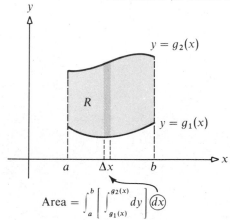

$$\text{Area} = \int_a^b \left[\int_{g_1(x)}^{g_2(x)} dy \right] \widehat{dx}$$

Figure 12

EXAMPLE 1   Find the area enclosed by the curves $y^2 = 4x$ and $x^2 = y/2$.

Solution   The desired area equals

$$\iint_R dA = \iint_R dy\,dx$$

where R is the region indicated in Figure 13(a). The points of intersection of the two curves are $(0, 0)$ and $(1, 2)$. If x is kept fixed first, then y ranges from $2x^2$ to $\sqrt{4x}$ [these are the $g_1(x)$ and $g_2(x)$ of Type I] and x ranges from 0 to 1. Therefore,

$$A = \int_0^1 \left[\int_{2x^2}^{\sqrt{4x}} dy \right] dx$$

and on performing the integration in that order, we obtain

$$A = \iint_R dy\,dx = \int_0^1 \left[\int_{2x^2}^{\sqrt{4x}} dy \right] dx = \int_0^1 \left[y \Big|_{2x^2}^{\sqrt{4x}} \right] dx$$

$$= \int_0^1 (\sqrt{4x} - 2x^2)\,dx = \left(\frac{4x^{3/2}}{3} - \frac{2x^3}{3} \right) \Big|_0^1 = \frac{2}{3} \text{ square units}$$

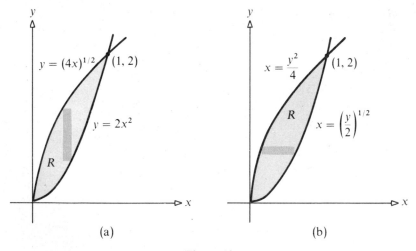

Figure 13

Alternatively, if we first keep y fixed, then x varies from $y^2/4$ to $(y/2)^{1/2}$ [these are the $h_1(y)$ and $h_2(y)$ of Type II] and y varies from 0 to 2, as shown in Figure 13(b). The corresponding iterated integral is

$$A = \iint_R dx\,dy = \int_0^2 \left[\int_{y^2/4}^{\sqrt{y/2}} dx \right] dy$$

Upon evaluation, we obtain the same result, that is, $A = \frac{2}{3}$.   ∎

Sometimes the region R has to be split up in order to integrate.

EXAMPLE 2   Find the area of the region R enclosed by the parabola $y = 6x - x^2$,   the x-axis, and the line $y = 4x - 8$.

Solution   Figure 14 illustrates the region R, which is both of Type I and Type II. However, for Type I, the lower boundary curve has a corner at $(2, 0)$; and for Type II, the right boundary curve has a corner at $(4, 8)$. Because of this, we need to decompose R into regions that have smooth boundaries. We choose to split R with a vertical line through the point $(2, 0)$, thus obtaining two regions R_1 and R_2, each of which we may treat as Type I.* The desired area is

$$\iint_R dA \underset{\uparrow}{=} \iint_{R_1} dA + \iint_{R_2} dA = \int_0^2 \left[\int_0^{6x-x^2} dy \right] dx + \int_2^4 \left[\int_{4x-8}^{6x-x^2} dy \right] dx$$

$$\text{(18.9)}$$

$$= \int_0^2 (6x - x^2)\,dx + \int_2^4 (-x^2 + 2x + 8)\,dx$$

$$= \frac{28}{3} + \frac{28}{3} = \frac{56}{3}$$

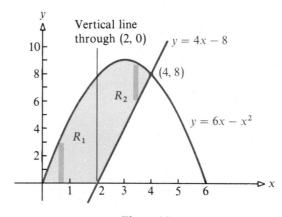

Figure 14   ■

VOLUME

Because of (18.6), the volumes of certain solids can be expressed as double integrals that may then be evaluated as iterated integrals. The next example illustrates such a use.

EXAMPLE 3   Find the volume of the solid enclosed in the first octant by the plane $x + y + z = 1$.

---

* In Problem 19 of Exercise 3 you are asked to do this example by splitting R with a horizontal line through the point $(4, 8)$.

Solution   Figure 15 illustrates the solid and the region R in the xy-plane. We see that R is enclosed by the coordinate axes $x = 0$ and $y = 0$ and by the line $x + y = 1$ (or $y = 1 - x$). We shall treat R as a Type I region so that y will vary according to $0 \le y \le 1 - x$ and x will vary according to $0 \le x \le 1$. Thus, the volume of the solid is

$$\iint_R z \, dA = \int_0^1 \left[\int_0^{1-x} (1 - x - y) \, dy \right] dx = \int_0^1 \left[\left(y - xy - \frac{y^2}{2} \right) \Big|_0^{1-x} \right] dx$$

$$= \int_0^1 \left[(1 - x) - x(1 - x) - \frac{(1 - x)^2}{2} \right] dx = \int_0^1 \left(\frac{1}{2} - x + \frac{x^2}{2} \right) dx$$

$$= \left(\frac{x}{2} - \frac{x^2}{2} + \frac{x^3}{6} \right) \Big|_0^1 = \frac{1}{6} \text{ cubic unit}$$

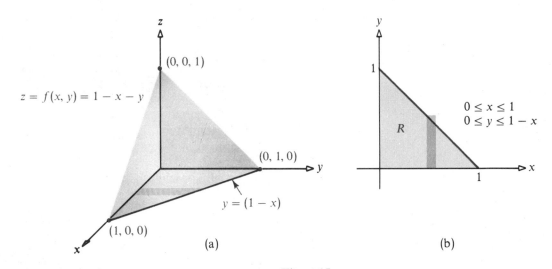

(a)   (b)

Figure 15   ■

EXAMPLE 4   Find the volume of the solid under the graph of the elliptic paraboloid $z = 8 - 2x^2 - y^2$ and above the xy-plane.

Solution   We notice that the solid $z = 8 - 2x^2 - y^2$, $z \ge 0$, can be divided into four parts, each with the same volume. One of these parts is illustrated in Figure 16(a). We treat the region R as being of Type I. Then

$$V = \iint_R (8 - 2x^2 - y^2) \, dA = \int_{-2}^2 \left[\int_{-\sqrt{8-2x^2}}^{\sqrt{8-2x^2}} (8 - 2x^2 - y^2) \, dy \right] dx$$

$$\underset{\underset{\text{By symmetry}}{\uparrow}}{=} 4 \int_0^2 \left[\int_0^{\sqrt{8-2x^2}} (8 - 2x^2 - y^2) \, dy \right] dx$$

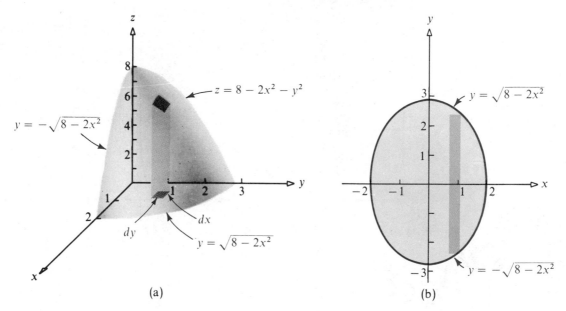

Figure 16

The inside integral is

$$\int_0^{\sqrt{8-2x^2}} (8 - 2x^2 - y^2)\, dy = \left[(8 - 2x^2)y - \frac{y^3}{3}\right]\Big|_0^{\sqrt{8-2x^2}} = \frac{2}{3}(8 - 2x^2)^{3/2}$$

The required volume is, therefore,

$$V = \frac{8}{3} \int_0^2 (8 - 2x^2)^{3/2}\, dx$$

If we use the substitution $x = 2 \sin \theta$, $dx = 2 \cos \theta\, d\theta$, then we obtain

$$V = \frac{8}{3} \int_0^{\pi/2} 16\sqrt{2}(\cos^3\theta)(2 \cos \theta\, d\theta) = \frac{256\sqrt{2}}{3} \int_0^{\pi/2} \cos^4\theta\, d\theta$$

$$\underset{\underset{\text{Wallis' formula (9.7)}}{\uparrow}}{=} \frac{256\sqrt{2}}{3} \left[\frac{(3)(1)}{(4)(2)} \frac{\pi}{2}\right] = 16\sqrt{2}\pi \text{ cubic units} \qquad \blacksquare$$

EXAMPLE 5   Compute the volume common to the two cylinders $x^2 + y^2 = 1$ and $x^2 + z^2 = 1$.

Solution   Each cylinder has radius 1; their axes are perpendicular to each other and lie on the z-axis and y-axis, respectively. Figure 17(a) shows the portion of the solid lying in the first octant. We compute the volume of this portion of the solid. Then, since the volume in each octant is the same, the required volume will be eight times as much

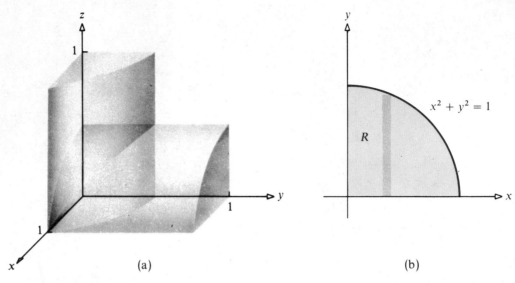

Figure 17

as the volume in the first octant; that is,

$$V = 8 \iint_R z \, dx \, dy = 8 \iint_R (1 - x^2)^{1/2} \, dx \, dy$$

where R is the region in the first quadrant inside the circle $x^2 + y^2 = 1$ [see Fig. 17(b)]. A quick look at the integrand will convince us that it is simpler to integrate with respect to y first, thus treating R as a Type I region. In this event, y varies from 0 to $(1 - x^2)^{1/2}$ and x varies from 0 to 1, so that the required volume is

$$V = 8 \int_0^1 \left[\int_0^{\sqrt{1-x^2}} (1 - x^2)^{1/2} \, dy \right] dx$$

$$= 8 \int_0^1 \left[(1 - x^2)^{1/2}(y) \Big|_0^{\sqrt{1-x^2}} \right] dx$$

$$= 8 \int_0^1 (1 - x^2) \, dx = 8 \left(x - \frac{x^3}{3} \right) \Big|_0^1 = \frac{16}{3} \text{ cubic units} \quad \blacksquare$$

EXERCISE 3

In Problems 1–10 use double integration to find the area of each region.

1. Enclosed by $y = x^3$ and $y = x^2$
2. Enclosed by $y = 2\sqrt{x}$ and $y = x^2/4$
3. Enclosed by $y = x^2 - 9$ and $y = 9 - x^2$
4. Enclosed by $x^2 + y^2 = 16$ and $y^2 = 6x$

5. Enclosed by $y = x^2$ and $x + y - 2 = 0$

6. Enclosed by the triangle with vertices $(0, 0)$, $(5, 0)$, and $(2, 3)$

7. Enclosed by $y = 1/\sqrt{x - 1}$, $y = 0$, $x = 2$, and $x = 5$

8. Enclosed by $y = x^{3/2}$ and $y = x$

9. Enclosed by the line $x + y = 3$ and the hyperbola $xy = 2$

10. Enclosed by the hyperbola $xy = \sqrt{3}$ and the circle $x^2 + y^2 = 4$, in the first quadrant only

In Problems 11–18 find the volume of each solid.

11. The tetrahedron enclosed by the plane $x + 2y + z = 2$ and the coordinate planes

12. The elliptic paraboloid $4x^2 + 9y^2 = 36z$ between the planes $z = 0$ and $z = 1$

13. Below the paraboloid $z = x^2 + y^2$ and above the square in the xy-plane enclosed by the lines $x = \pm 1$ and $y = \pm 1$

14. Enclosed by the cylinder $z = 9 - y^2$ and the planes $y = x$, $x = 0$, and $z = 0$

15. Enclosed by the surfaces $z = 4 - x^2 - y^2$, $z = 0$, and $x = 0$ (front half only)

16. Enclosed by the surfaces $x^2 + y^2 = 1$, $z = x$, and $z = x^2$

17. In the first octant below $y = z^2$, above $z = 0$, and enclosed on the sides by $y = x^2$ and $y = 4$

18. Enclosed by $y = e^x$ and the planes $x = 0$, $x = 1$, $z = 0$, and $z = y$, in the first octant only

19. Find the area of the region R enclosed by $y = 6x - x^2$ and the x-axis above the line $y = 4x - 8$ by splitting R with a horizontal line through the point $(4, 8)$. (Refer to Fig. 14.)

20. Use double integration to find the area in the first quadrant enclosed by the parabola $x^2 = 9y$, the y-axis, and the circle $x^2 + y^2 = 10$.

21. Find the volume of the solid below the graph of $z = x^2 + y^2$ and inside $x^2 + y^2 = 1$.

In Problems 22–25 each of the iterated integrals represents the volume of a solid. Describe the solid.

22. $\int_0^2 \left[\int_{y/2}^{\sqrt{y}} (x + y)\,dx \right] dy$

23. $\int_0^1 \left[\int_0^x \sqrt{1 - x^2}\,dy \right] dx$

24. $\int_0^2 \left[\int_{-2}^2 2\,dy \right] dx$

25. $\int_0^\pi \left[\int_0^{\sin x} 3\,dy \right] dx$

26. Interchange the order of integration of the expression below, and calculate the integral.

$$\int_1^5 \left[\int_{-\sqrt{y-1}}^{\sqrt{y-1}} y\,dx \right] dy$$

27. Evaluate $\iint_R xy\,dA$, where R is the region enclosed by $y = 4 - x^2$ and $y = x^2$ to the right of the y-axis.

28. Verify (18.16) if R is a Type II region.

29. Find the volume of the solid enclosed by a portion of the parabolic cylinder $z = x^2/2$ and the four planes $y = 0$, $y = x$, $x = 2$, and $z = 0$.

30. Find the volume of the solid enclosed by the coordinate planes, the plane $y = 3$, and the surface $z = y + 1 - x^2$.

31. Find the volume of the solid enclosed by the surfaces $z = xe^y$, $z = 0$, $y = 0$, $x = 1$, and $y = x^2$.

4. Double Integrals in Polar Coordinates

Some regions (such as circles, cardioids, and so forth) over which a double integral is to be evaluated are more readily described in polar coordinates than in rectangular coordinates. In such cases, it may be easier to evaluate the double integral by utilizing polar coordinates.

We start with a region R enclosed by the rays $\theta = \alpha$, $\theta = \beta$ $(0 < \beta - \alpha \leq 2\pi)$ and the smooth curves $r = r_1(\theta)$, $r = r_2(\theta)$ $(r_1 \leq r_2)$, as shown in Figure 18. We proceed to partition the two intervals $[\alpha, \beta]$ and $[r_1, r_2]$ as depicted in Figure 19. The result is a collection of circular segments. Let n denote the number of such circular segments that lie entirely in R. We now have a partition P of the region whose norm $\|P\|$ is taken as the length of the longest diagonal of the circular segments.

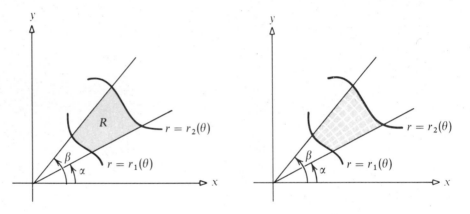

Figure 18   Figure 19

As has been the usual practice, we calculate the area ΔA_i of a typical circular segment R_i (see Fig. 20). Because this area ΔA_i is the difference of the areas of two circular sectors, we have

$$\Delta A_i = \tfrac{1}{2}r_i^2 \Delta\theta_i - \tfrac{1}{2}r_{i-1}^2 \Delta\theta_i = \tfrac{1}{2}(r_i^2 - r_{i-1}^2)\Delta\theta_i$$
$$= \tfrac{1}{2}(r_i + r_{i-1})(r_i - r_{i-1})\Delta\theta_i = \tfrac{1}{2}(r_i + r_{i-1})\Delta r_i \Delta\theta_i$$
$$\uparrow$$
$$\Delta r_i = r_i - r_{i-1}$$

Now we pick a point $(\bar{r}_i, \bar{\theta}_i)$ in the circular segment R_i in such a way that $\bar{r}_i = \tfrac{1}{2}(r_i + r_{i-1})$. Then

(18.17) $$\Delta A_i = \bar{r}_i \Delta r_i \Delta\theta_i$$

Let f be a function of the polar coordinates r and θ, and form the sum

(18.18) $$\sum_{i=1}^{n} f(\bar{r}_i, \bar{\theta}_i)\Delta A_i = \sum_{i=1}^{n} f(\bar{r}_i, \bar{\theta}_i)\bar{r}_i \Delta r_i \Delta\theta_i$$

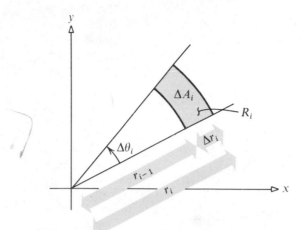

Figure 20

It can be shown that if f is continuous on the region R, then the limit of the sum in (18.18) as the norm $\|P\|$ approaches 0 exists and equals the double integral of f over R. That is,

$$\iint\limits_{R} f(r, \theta)\, dA = \lim_{\|P\|\to 0} \sum_{i=1}^{n} f(\bar{r}_i, \bar{\theta}_i)\Delta A_i$$

Alternatively, we may write

(18.19)
$$\iint\limits_{R} f(r, \theta) r\, dr\, d\theta = \lim_{\|P\|\to 0} \sum_{i=1}^{n} f(\bar{r}_i, \bar{\theta}_i)\bar{r}_i \Delta r_i \Delta \theta_i$$

It is important to remember that the integrand contains a factor of r. This is due to the fact that in polar coordinates the differential dA of area A [see (18.17)] is given by

$$dA = r\, dr\, d\theta$$

The evaluation of the double integral in (18.19) may be accomplished by using iterated integrals:

(18.20)
$$\iint\limits_{R} f(r, \theta) r\, dr\, d\theta = \int_{\alpha}^{\beta} \left[\int_{r_1(\theta)}^{r_2(\theta)} f(r, \theta) r\, dr \right] d\theta$$

As before, in evaluating the inside integral $\int_{r_1(\theta)}^{r_2(\theta)} f(r, \theta) r\, dr$, we treat θ as a constant and integrate (partially) with respect to r.

EXAMPLE 1   Evaluate the iterated integral: $\int_{0}^{2} \left[\int_{0}^{\sqrt{4-x^2}} (x^2 + y^2)\, dy \right] dx$

Solution   The region R is enclosed by the graphs of $y = \sqrt{4 - x^2}$ (a portion of a circle of radius 2) and $0 \le x \le 2$ (see Fig. 21). In polar coordinates this region R is simply $0 \le r \le 2$, $0 \le \theta \le \pi/2$. The integrand $x^2 + y^2$ becomes r^2 and the differential of area $dy\,dx$ becomes $r\,dr\,d\theta$. By changing the limits of integration, we find

$$\int_0^2 \left[\int_0^{\sqrt{4-x^2}} (x^2 + y^2)\,dy \right] dx = \int_0^{\pi/2} \left[\int_0^2 r^2 r\,dr \right] d\theta$$

$$= \int_0^{\pi/2} \left[\frac{r^4}{4} \Big|_0^2 \right] d\theta = 4\left(\frac{\pi}{2} \right) = 2\pi$$

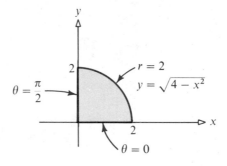

Figure 21   ■

When the integrand of a double integral is expressed as a function of $x^2 + y^2$, it is sometimes easier to perform the integration in polar coordinates than in rectangular coordinates. Our next example illustrates that sometimes turning to polar coordinates is the only available choice.

EXAMPLE 2   Evaluate $\iint_R e^{x^2+y^2}\,dx\,dy$, where R is the region in the first quadrant inside the circle $x^2 + y^2 = a^2$.

Solution   Without using polar coordinates, we can approximate this integral only by numerical techniques. However, a change to polar coordinates results in

$$\iint_R e^{x^2+y^2}\,dx\,dy = \iint_R e^{r^2} r\,dr\,d\theta = \int_0^{\pi/2} \left[\int_0^a e^{r^2} r\,dr \right] d\theta$$

$$= \frac{1}{2} \int_0^{\pi/2} (e^{a^2} - 1)\,d\theta = \left(\frac{\pi}{4} \right)(e^{a^2} - 1)$$ ■

There is another type of region R that lends itself to iteration in polar coordinates. This type of region R is the area enclosed by the two circles $r = a$, $r = b$, $0 \le a < b$, and by the graphs of two smooth curves $\theta = \theta_1(r)$, $\theta = \theta_2(r)$, where $0 \le \theta_1 \le \theta_2 \le 2\pi$ (see Fig. 22).

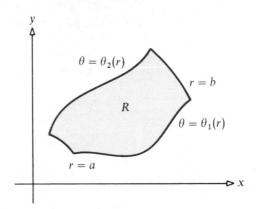

$$\theta = \theta_2(r)$$

$$r = b$$

$$R$$

$$\theta = \theta_1(r)$$

$$r = a$$

Figure 22

If f is a continuous function of the polar coordinates r and θ on R, the double integral of f over R may be evaluated by

(18.21)
$$\iint_R f(r, \theta) r\, dr\, d\theta = \int_a^b \left[\int_{\theta_1(r)}^{\theta_2(r)} f(r, \theta)\, d\theta \right] r\, dr$$

In evaluating the inside integral $\int_{\theta_1(r)}^{\theta_2(r)} f(r, \theta)\, d\theta$, we treat r as a constant and integrate (partially) with respect to θ.

GEOMETRIC INTERPRETATION

Provided $f(r, \theta) \geq 0$ and f is continuous over a region R, the double integral $\iint_R f(r, \theta)\, dA$ represents the volume of the solid under the graph of the surface $z = f(r, \theta)$ and above the region R in the xy-plane. Figure 23 illustrates the situation. An intuitive argument follows.

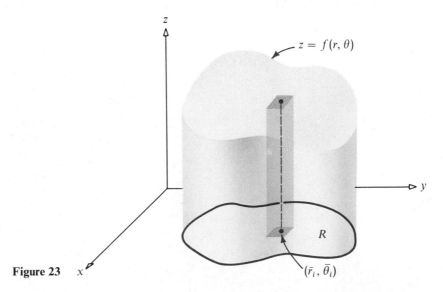

$$z = f(r, \theta)$$

$$R$$

$$(\bar{r}_i, \bar{\theta}_i)$$

Figure 23 x

Suppose R is partitioned into n circular segments. As we did earlier, select a point $(\bar{r}_i, \bar{\theta}_i)$, $\bar{r}_i = \frac{1}{2}(r_i + r_{i-1})$, in the ith circular segment R_i. The quantity $f(\bar{r}_i, \bar{\theta}_i)$ measures the altitude of a solid whose base is R_i. The volume of this solid is $f(\bar{r}_i, \bar{\theta}_i)\Delta A_i$. The sum of all such solids corresponding to the n circular segments is

$$\sum_{i=1}^{n} f(\bar{r}_i, \bar{\theta}_i)\Delta A_i = \sum_{i=1}^{n} f(\bar{r}_i, \bar{\theta}_i)\bar{r}_i\Delta r_i\Delta\theta_i$$

The volume V of the solid under the graph of $z = f(r, \theta)$ and above the region R in the xy-plane is

(18.22)
$$V = \lim_{\|P\|\to 0} \sum_{i=1}^{n} f(\bar{r}_i, \bar{\theta}_i)\bar{r}_i\Delta r_i\Delta\theta_i = \iint_{R} f(r, \theta)r\,dr\,d\theta$$

EXAMPLE 3   Find the volume of the solid under the graph of $z = x^2 + y^2$, above the xy-plane, and inside the cylinder $x^2 + y^2 = 2y$.

Solution   The plane $x = 0$ divides the solid into two parts with equal volume. Hence, we may evaluate the volume for $x \geq 0$ and double the result. Let $x = r\cos\theta$, $y = r\sin\theta$. Then $x^2 + y^2 = 2y$ becomes $r = 2\sin\theta$ and the restrictions on r and θ become $0 \leq \theta \leq \pi/2$ and $0 \leq r \leq 2\sin\theta$ (see Fig. 24). The required volume V is

$$V = 2\int_0^{\pi/2}\left[\int_0^{2\sin\theta} f(r, \theta)r\,dr\right]d\theta$$

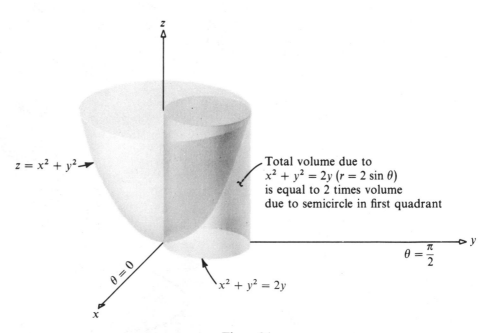

Figure 24

The surface $z = f(r, \theta)$ under which the volume lies is the paraboloid of revolution $z = x^2 + y^2 = r^2$. Thus,

$$V = 2 \int_0^{\pi/2} \left[\int_0^{2\sin\theta} r^2 r \, dr \right] d\theta = 8 \int_0^{\pi/2} \sin^4\theta \, d\theta \underset{\underset{(9.7)}{\uparrow}}{=} \frac{3\pi}{2} \qquad \blacksquare$$

If $f(r, \theta) = 1$, then (18.22) represents the volume of a solid whose height is constantly 1. As a result, the value of (18.22) in this case gives the area A of the region R.

EXAMPLE 4   Use double integration to find the area enclosed by the cardioid $r = a(1 - \cos\theta)$.

Solution   We see that the area we wish to calculate is twice the shaded area in Figure 25; hence, the required area is

$$A = 2 \int_0^\pi \left[\int_0^{a(1-\cos\theta)} (1)r \, dr \right] d\theta = 2 \int_0^\pi \frac{a^2}{2} (1 - 2\cos\theta + \cos^2\theta) \, d\theta = 3\pi \frac{a^2}{2} \qquad \blacksquare$$

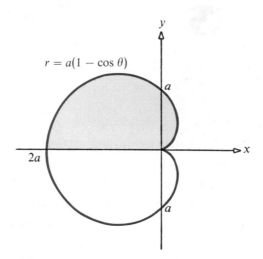

Figure 25

EXERCISE 4

In Problems 1–14 evaluate each double integral by changing to polar coordinates.

1. $\int_0^1 \left[\int_0^{\sqrt{1-x^2}} dy \right] dx$

2. $\int_0^3 \left[\int_0^y \sqrt{x^2 + y^2} \, dx \right] dy$

3. $\int_0^4 \left[\int_0^{\sqrt{4y-y^2}} (x^2 + y^2) \, dx \right] dy$

4. $\int_0^2 \left[\int_0^{\sqrt{4-x^2}} \sqrt{4 - x^2 - y^2} \, dy \right] dx$

5. $\int_0^1 \left[\int_0^{\sqrt{1-x^2}} \cos(x^2 + y^2)\, dy \right] dx$

6. $\int_0^1 \left[\int_0^{\sqrt{1-y^2}} e^{\sqrt{x^2+y^2}}\, dx \right] dy$

7. $\iint_R e^{-(x^2+y^2)}\, dx\, dy;$ R is the region in the first quadrant enclosed by the circles $x^2 + y^2 = 1$ and $x^2 + y^2 = 4$

8. $\iint_R (y/\sqrt{x^2 + y^2})\, dx\, dy;$ R is the region in the first quadrant inside the circle $x^2 + y^2 = a^2$

9. $\iint_R x\, dx\, dy;$ R is the region enclosed by the circle $x^2 + y^2 = x$

10. $\iint_R y^2\, dx\, dy;$ R is the region enclosed by the circle $x^2 + y^2 = 2y$

11. $\iint_R \sqrt{x^2 + y^2}\, dx\, dy;$ R is the region enclosed by $r = 3 + \cos\theta$

12. $\iint_R (x^2 + y^2)\, dx\, dy;$ R is the region enclosed by $r = 2(1 + \sin\theta)$

13. $\int_{-2}^2 \left[\int_{-\sqrt{4-y^2}}^{\sqrt{4-y^2}} (x^2 + y^2)^2\, dx \right] dy$

14. $\int_0^2 \left[\int_{\sqrt{2x-x^2}}^{\sqrt{4x-x^2}} dy \right] dx + \int_2^4 \left[\int_0^{\sqrt{4x-x^2}} dy \right] dx$

In Problems 15–18 evaluate the iterated integral in polar coordinates, and sketch the region of integration.

15. $\int_{\pi/2}^\pi \left[\int_0^1 r \cos\theta\, dr \right] d\theta$

16. $\int_1^2 \left[\int_0^\pi re^r\, d\theta \right] dr$

17. $\int_{1/2\pi}^{1/\pi} \left[\int_{1/r}^{2\pi} d\theta \right] dr$

18. $\int_0^1 \left[\int_0^{\cos^{-1}(r/2)} r \sin\theta\, d\theta \right] dr$

In Problems 19–28 use double integrals in polar coordinates.

19. Find the volume of the solid cut from the sphere $x^2 + y^2 + z^2 = a^2$ by the cylinder $x^2 + y^2 = ay$.

20. Find the volume of the solid enclosed by the paraboloid $x^2 + y^2 = az$,   the xy-plane, and the cylinder $x^2 + y^2 = a^2$.

21. Find the volume of the solid enclosed by the ellipsoid $x^2 + y^2 + 4z^2 = 4$.

22. Find the volume of the solid cut from the ellipsoid $x^2 + y^2 + 4z^2 = 4$ by the cylinder $x^2 + y^2 = 1$.

23. Find the area enclosed by one leaf of the rose $r = \sin 3\theta$.

24. Find the area enclosed by one loop of $r^2 = 9 \sin 2\theta$.

25. Find the area that lies inside the circle $r = 4 \cos\theta$ but outside the circle $r = \cos\theta$.

26. Find the area that lies inside the circle $r = 1$ but outside the cardioid $r = 1 + \cos\theta$.

27. Find the area that lies inside the cardioid $r = 1 + \cos\theta$ but outside the circle $r = \frac{1}{2}$.

28. Find the area that lies inside the limaçon $r = 3 - \cos\theta$ but outside the circle $r = 5 \cos\theta$.

In Problems 29 and 30 replace the given iterated integral(s) in polar coordinates with iterated integral(s) in cartesian coordinates. Do not evaluate the integrals.

29. $\int_0^{1/\sqrt{2}} \left[\int_0^{\sin^{-1} r} r\, d\theta \right] dr$

30. $\int_0^1 \left[\int_{\cos^{-1} r}^{\pi/2} r^2 \sin\theta\, d\theta \right] dr$

In Problems 31 and 32 reverse the order of integration in the indicated iterated integrals in polar coordinates.

31. $\int_0^{\sqrt{2}/2} \left[\int_{\sin^{-1} r}^{\cos^{-1} r} f(r, \theta)\, d\theta \right] dr$

32. $\int_{\sqrt{2}/2}^1 \left[\int_{\cos^{-1} r}^{\sin^{-1} r} f(r, \theta)\, d\theta \right] dr$

5. Center of Mass; Moment of Inertia

CENTER OF MASS

In many practical situations it is convenient to regard thin sheets of material, such as copper stripping, as two-dimensional. A *lamina* is a plane area that represents a two-dimensional distribution of matter. If the material is of constant mass density,* the lamina is called *homogeneous*. The mass m of a homogeneous lamina is ρA, where A is the area of the lamina and ρ is its constant mass density.

In general, however, substances are not homogeneous and so the mass density is variable. Suppose a lamina is represented by a certain region R of the xy-plane and its mass density $\rho = \rho(x, y)$ varies continuously over R. To find the total mass m of such a lamina, we may use double integration.

First, partition the region R into n rectangles. Then, in a representative rectangle R_i of area ΔA_i, choose a point (u_i, v_i). An approximation to the mass due to the ith rectangle is

$$[\text{Mass density}] \times [\text{Area}] = \rho(u_i, v_i)\Delta A_i$$

By adding up all the masses, the total mass of the lamina may be approximated by

$$\sum_{i=1}^{n} \rho(u_i, v_i)\Delta A_i$$

If the norm $\|P\|$ of the partition is allowed to approach 0, the total mass m of the lamina is given by

(18.23)
$$m = \lim_{\|P\| \to 0} \sum_{i=1}^{n} \rho(u_i, v_i)\Delta A_i = \iint_R \rho(x, y)\, dA$$

Observe that if the mass density ρ is constant, we have the familiar formula

$$m = \iint_R \rho\, dA = \rho \iint_R dA = \rho A$$

If a particle of mass m is located a distance l from a fixed axis a, the *moment of mass* M_a[†] with respect to this axis is defined as

$$M_a = ml$$

Consider a lamina of variable mass density ρ represented by a certain region R of the xy-plane. Partition R into n rectangles and concentrate on the ith rectangle

---

* The mass density of a two-dimensional material is defined as the mass per unit area of the material. In SI units, mass density is measured in kilograms per square meter (kg/m^2); in U.S. customary units, it is measured in slugs per square foot ($slug/ft^2$).

† Sometimes, M_a is referred to as the *first moment* of the particle about the axis a.

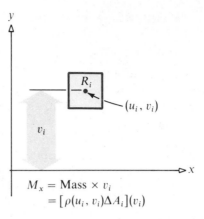

Figure 26

R_i, whose area is ΔA_i. Pick a point (u_i, v_i) in R_i (see Fig. 26). The moment of mass M_x with respect to the x-axis of the ith rectangle may be approximated by

$$[\text{Mass}] \times [\text{Distance from } x\text{-axis}] = [\rho(u_i, v_i)\Delta A_i](v_i)$$

By adding up all these moments and taking the limit as the norm $\|P\|$ approaches 0, the moment of mass M_x with respect to the x-axis of the entire lamina is given by

(18.24)
$$M_x = \lim_{\|P\| \to 0} \sum_{i=1}^{n} v_i \rho(u_i, v_i)\Delta A_i = \iint_R y\rho(x, y)\,dA$$

Similarly, the moment of mass M_y with respect to the y-axis of the entire lamina is given by

(18.25)
$$M_y = \lim_{\|P\| \to 0} \sum_{i=1}^{n} u_i \rho(u_i, v_i)\Delta A_i = \iint_R x\rho(x, y)\,dA$$

The *center of mass** of a lamina is defined as the point $(\bar{x}, \bar{y})$ whose coordinates satisfy the equations

(18.26)
$$m\bar{x} = M_y \qquad m\bar{y} = M_x$$

From (18.23), (18.24), (18.25), and (18.26), we conclude that

(18.27)
$$\bar{x} = \frac{\iint_R x\rho(x, y)\,dA}{\iint_R \rho(x, y)\,dA} \qquad \bar{y} = \frac{\iint_R y\rho(x, y)\,dA}{\iint_R \rho(x, y)\,dA}$$

---

* We distinguish here between the *center of mass* (or *center of gravity*) of a lamina, which is defined in (18.26), and the *centroid* of a lamina. The latter is a purely geometric property of the lamina and coincides with the center of mass in the case of a homogeneous body. To see the distinction, a square lamina always has its centroid at the (geometric) center, but the same square with variable density will almost always have its center of mass off-center.

For a physical interpretation of the center of mass of a lamina, consider a flat piece of corrugated cardboard cut in the shape of the lamina and weighted appropriately to reflect the mass density of the lamina. If a piece of string is attached to the cardboard at the center of mass $(\bar{x}, \bar{y})$, then the cardboard, when suspended from the string, will lie in a horizontal position.

EXAMPLE 1   Find the center of mass of a lamina in the shape of an isosceles right triangle, if the mass density ρ is directly proportional to the square of the distance from the vertex opposite the hypotenuse.

Solution   We situate the lamina in the xy-plane in such a way that the vertex opposite the hypotenuse is at the origin and the two equal sides, say, of length a, lie along the positive coordinate axes (see Fig. 27).

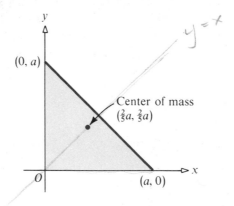

Figure 27

The mass density of the lamina is   — Given!

$$\rho(x, y) = k(x^2 + y^2)$$

where k is a constant. From (18.23), the mass m of this lamina is

$$m = \iint_R \rho(x, y)\, dA = \iint_R k(x^2 + y^2)\, dx\, dy = k \int_0^a \left[\int_0^{a-x} (x^2 + y^2)\, dy \right] dx$$

$$= k \int_0^a \left[\left(x^2 y + \frac{y^3}{3} \right) \Big|_0^{a-x} \right] dx = k \int_0^a \frac{1}{3}(a^3 - 3a^2 x + 6ax^2 - 4x^3)\, dx$$

$$= \frac{k}{3} \left(a^4 - \frac{3a^4}{2} + 2a^4 - a^4 \right) = \frac{ka^4}{6}$$

Due to the symmetric character of the region R and the mass density, the center of mass $(\bar{x}, \bar{y})$ must lie on the line $y = x$. Consequently, if we find $\bar{x}$, we also know

$\bar{y}$. Using (18.25), we find

$$M_y = \iint\limits_R x\rho(x, y)\,dA = \iint\limits_R xk(x^2 + y^2)\,dx\,dy = k \int_0^a \left[\int_0^{a-x} (x^3 + xy^2)\,dy \right] dx$$

$$= \frac{k}{3} \int_0^a (a^3x - 3a^2x^2 + 6ax^3 - 4x^4)\,dx = \frac{ka^5}{15}$$

From (18.27), we conclude that

$$\bar{x} = \frac{M_y}{m} = \frac{ka^5/15}{ka^4/6} = \frac{2}{5}a$$

The center of mass of the lamina is $(\tfrac{2}{5}a, \tfrac{2}{5}a)$.   ∎

EXAMPLE 2   Find the center of mass of a lamina in the shape of a region R in the xy-plane that lies outside the circle $x^2 + y^2 = a^2$ and inside the circle $x^2 + y^2 = 2ax$, if the mass density ρ is inversely proportional to the distance from the origin.

Solution   Figure 28 illustrates the region R. Since the distance of a point (x, y) from the origin is $\sqrt{x^2 + y^2}$, the mass density of the lamina at any point (x, y) in R is

$$\rho(x, y) = \frac{k}{\sqrt{x^2 + y^2}}$$

where k is a constant. From (18.23), the mass m of the lamina is given by

$$m = \iint\limits_R \rho(x, y)\,dA = k \iint\limits_R \frac{dx\,dy}{\sqrt{x^2 + y^2}}$$

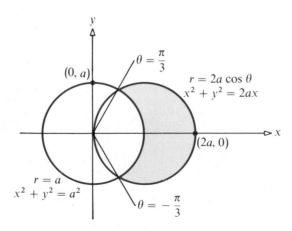

Figure 28

Due to the character of the region R and the integrand, it appears that iteration in polar coordinates is preferable. By transforming to polar coordinates, we have

$$m = k \iint\limits_{R} \frac{1}{r} r \, dr \, d\theta = k \int_{-\pi/3}^{\pi/3} \left[\int_{a}^{2a \cos \theta} dr \right] d\theta = k \int_{-\pi/3}^{\pi/3} (2a \cos \theta - a) \, d\theta$$

$$= k(2a \sin \theta - a\theta) \Big|_{-\pi/3}^{\pi/3} = k \left(2a\sqrt{3} - a \frac{2\pi}{3} \right) = \frac{2ka}{3} (3\sqrt{3} - \pi)$$

Due to the symmetric character of the mass density and the region R, it follows that the center of mass lies on the x-axis. Consequently, $\bar{y} = 0$. To find $\bar{x}$, we first need M_y. From (18.25),

$$M_y = \iint\limits_{R} x\rho(x, y) \, dx \, dy = k \iint\limits_{R} r \cos \theta \, dr \, d\theta = k \int_{-\pi/3}^{\pi/3} \left[\int_{a}^{2a \cos \theta} r \cos \theta \, dr \right] d\theta$$

$$= \frac{k}{2} \int_{-\pi/3}^{\pi/3} \cos \theta (4a^2 \cos^2\theta - a^2) \, d\theta = \frac{k}{2} \int_{-\pi/3}^{\pi/3} (3a^2 \cos \theta - 4a^2 \sin^2\theta \cos \theta) \, d\theta$$

$$= \frac{k}{2} \left(3a^2 \sin \theta - \frac{4a^2 \sin^3\theta}{3} \right) \Big|_{-\pi/3}^{\pi/3} = ka^2 \sqrt{3}$$

Thus,

$$\bar{x} = \frac{M_y}{m} = \frac{ka^2\sqrt{3}}{(2ka/3)(3\sqrt{3} - \pi)} = \frac{3a\sqrt{3}}{2(3\sqrt{3} - \pi)} \approx 1.26a$$

The center of mass is approximately $(1.26a, 0)$.   ∎

MOMENT OF INERTIA

If a particle of mass m is located a distance l from a fixed axis a, its *moment of inertia* $I_a{}^*$ about this axis is defined as

$$I_a = ml^2$$

Proceeding in the same fashion as we did for the moment of mass, the moment of inertia I_x about the x-axis of a lamina of variable mass density $\rho(x, y)$ represented by a region R of the xy-plane is given by

$$
\overset{\text{Mass}}{} \quad \overset{\substack{\text{Square of} \\ \text{distance}}}{}
$$

(18.28)
$$I_x = \lim_{\|P\| \to 0} \sum_{i=1}^{n} [\overbrace{\rho(u_i, v_i)\Delta A_i}][\overbrace{(v_i^2)}] = \iint\limits_{R} y^2 \rho(x, y) \, dA$$

---

* Sometimes I_a is referred to as the *second moment* of the particle about the axis a.

Similarly, the moment of inertia I_y about the y-axis is given by

$$\text{(18.29)} \qquad I_y = \lim_{||P|| \to 0} \sum_{i=1}^{n} [\overbrace{\rho(u_i, v_i)\Delta A_i}^{\text{Mass}}](\overbrace{u_i^2}^{\substack{\text{Square of}\\\text{distance}}}) = \iint_R x^2 \rho(x, y)\, dA$$

and the moment of inertia I_O about the origin (z-axis) is given by

$$\text{(18.30)} \qquad I_O = \lim_{||P|| \to 0} \sum_{i=1}^{n} [\overbrace{\rho(u_i, v_i)\Delta A_i}^{\text{Mass}}](\overbrace{u_i^2 + v_i^2}^{\substack{\text{Square of}\\\text{distance}}}) = \iint_R (x^2 + y^2)\rho(x, y)\, dA$$

The moment of inertia I_O about the origin is sometimes referred to as the *polar moment of inertia* or the *polar second moment*.

A consequence of formulas (18.28), (18.29), and (18.30) is that

$$\text{(18.31)} \qquad I_O = I_x + I_y$$

The polar moment is frequently used to find the moments I_x and I_y when symmetry is present, that is, when $I_x = I_y$.

EXAMPLE 3   Find the polar moment of inertia of a homogeneous lamina of mass density ρ in the shape of a region R in the xy-plane enclosed by the circle $x^2 + y^2 = a^2$. Use the polar moment to find the moments of inertia I_x and I_y of this lamina.

Solution   We shall use polar coordinates. From (18.30), we have

$$I_O = \iint_R (x^2 + y^2)\rho\, dA = \rho \iint_R r^2 r\, dr\, d\theta = \rho \int_0^{2\pi} \left[\int_0^a r^3\, dr \right] d\theta$$

$$= \frac{\rho}{4} \int_0^{2\pi} a^4\, d\theta = \frac{\pi a^4 \rho}{2}$$

By symmetry, it is evident that $I_x = I_y$. From (18.31), we conclude that

$$I_x = I_y = \frac{\pi a^4 \rho}{4} \qquad \blacksquare$$

In dynamics, the moment of inertia of a lamina occurs in connection with the study of rotational motion. If a rigid body (lamina) is rotated about an axis a with angular velocity ω, its kinetic energy K is given by

$$\text{(18.32)} \qquad K = \tfrac{1}{2}I_a\omega^2$$

For example, suppose the lamina in Example 3 has mass density $\rho = 20$ kilograms per square meter and is of radius $a = 1$ meter. If it is rotated about the origin at a constant angular velocity of 2π radians per second, its kinetic energy is

$$K = \left(\frac{1}{2}\right)\left(\frac{\pi a^4 \rho}{2}\right)(4\pi^2) = 20\pi^3 \text{ kg-m}^2/\text{sec}^2 \approx 620 \text{ joules}$$

Observe the similarity between formula (18.32) and the formula for the kinetic energy K of a particle of mass m moving in a straight line with speed v, namely,

$$K = \tfrac{1}{2}mv^2$$

EXERCISE 5

In Problems 1–12 use double integration to find the mass and center of mass of each lamina for the given mass density ρ.

1. Lamina in the shape of a rectangle enclosed by the lines $x = 2$, $y = 4$, and the coordinate axes; $\rho = 3x^2 y$

2. Lamina in the shape of a rectangle enclosed by the lines $x = 1$, $y = 2$, and the coordinate axes; $\rho = 2x^2 y^2$

3. Lamina in the shape of a region in the first quadrant enclosed by $y^2 = x$, $x = 1$, and the x-axis; $\rho = 2x + 3y$

4. Lamina in the shape of a region in the first quadrant enclosed by $y^2 = 4x$, $x = 1$, and the x-axis; $\rho = x + 1$

5. Lamina in the shape of a region enclosed by $y^2 = x$ and $y = x$; ρ is proportional to the distance from the y-axis

6. Lamina in the shape of a region enclosed by $y = \sin x$, $x = 0$, $x = \pi$, and $y = 0$; ρ is proportional to the distance from the x-axis

7. Lamina in the shape of a triangle enclosed by the lines $2x + 3y = 6$, $x = 0$, and $y = 0$; ρ is proportional to the sum of the distances from the coordinate axes

8. Lamina in the shape of a triangle enclosed by the lines $3x + 4y = 12$, $x = 0$, and $y = 0$; ρ is proportional to the product of the distances from the coordinate axes

9. Lamina in the shape of the region inside the cardioid $r = 1 + \sin\theta$; ρ is proportional to the distance from the pole

10. Lamina in the shape of the region enclosed by one leaf of the rose $r = \cos 2\theta$; ρ is proportional to the distance from the pole

11. Lamina in the shape of the region inside the graph of $r = 2a\sin\theta$ and outside the graph of $r = a$; ρ is inversely proportional to the distance from the pole

12. Lamina in the shape of the region outside the limaçon $r = 2 - \cos\theta$ and inside the circle $r = 4\cos\theta$; ρ is inversely proportional to the distance from the pole

In Problems 13–20 use double integration to find the moment of inertia about the indicated axis for each homogeneous lamina of mass density ρ.

13. Lamina in the shape of a triangle enclosed by the lines $2x + 3y = 6$, $x = 0$, $y = 0$; about the x-axis

14. Rework Problem 13 for the moment of inertia about the y-axis.

15. Lamina in the shape of a rectangle enclosed by the lines $x = a$, $y = b$, $x = 0$, $y = 0$; about the x-axis; $a > 0$, $b > 0$

16. Rework Problem 15 for the moment of inertia about the y-axis.

17. Lamina in the shape of the region enclosed by $y = x^2$ and $y = 2 - x^2$; about the y-axis

18. Lamina in the shape of the region enclosed by the loop of $y^2 = x^2(4 - x)$; about the y-axis

19. Lamina in the shape of the ellipse $b^2x^2 + a^2y^2 = a^2b^2$; about the x-axis

20. Lamina in the shape of the region enclosed by $x^{2/3} + y^{2/3} = a^{2/3}$; about the x-axis

21. Find the mass of a circular washer with inner radius a and outer radius b, if its mass density is inversely proportional to the square of the distance from the center.

22. Rework Problem 21 if the mass density is inversely proportional to the distance from the center.

23. Find the mass and center of mass of a lamina in the shape of the region enclosed on the left by the line $x = a$ ($a > 0$) and on the right by the circle $r = 2a \cos \theta$, if its mass density is inversely proportional to the distance from the y-axis.

24. Find the mass and center of mass of a lamina in the shape of the smaller region cut from the circle $r = 6$ by the line $r \cos \theta = 3$, if its mass density is $\rho = \cos^2\theta$.

25. Show that the center of mass of a rectangular homogeneous lamina lies at the intersection of its diagonals.

26. Show that the center of mass of a triangular homogeneous lamina lies at the point of intersection of its medians.

27. Find the center of mass of the lamina inside $r = 4 \cos \theta$ and outside $r = 2\sqrt{3}$, if the mass density is inversely proportional to the distance from the origin.

28. Find the moment of inertia about the x-axis of the lamina inside the limaçon $r = 3 + 2 \cos \theta$ and outside $r = 2$, if the mass density is inversely proportional to the square of the distance from the origin.

29. A homogeneous lamina is in the shape of a right triangle of base b and altitude h. Show that its moment of inertia about the base is $\frac{1}{6}mh^2$, where m is the mass.

30. Find the mass of the lamina enclosed by $y = x^2$ and $y = x^3$; $\rho = \sqrt{xy}$ at any point (x, y).

31. Find the center of mass and the mass of the lamina enclosed by $x = y - 2$ and $x = -y^2$; $\rho = x^2$ at any point (x, y).

32. Find the center of mass of the lamina enclosed by $x = 0$, $y = x^2$, and $x - 2y + 1 = 0$; $\rho = 2x + 8y + 2$ at any point (x, y).

6. Area of a Surface

In Chapter 13 we used a single integral to calculate the surface area of a solid of revolution—a surface obtained by revolving the graph of a continuous function $y = f(x)$ about an axis. In this section we show how double integration can be used to calculate the area of the part of a surface $z = f(x, y)$ (not necessarily a solid of revolution) that lies above a closed bounded region R of the xy-plane.

We begin with the simplest case in which the surface is a plane $z = ax + by + c$ and R is a rectangle with sides of length Δx and Δy. See Figure 29. We let S be the area of that part of the plane that lies over the rectangle.

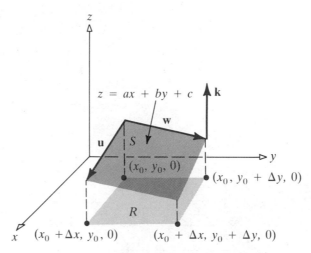

Figure 29

THEOREM   Let R be a closed rectangular region in the xy-plane. If R has sides of length Δx and Δy, then the surface area S of that part of the plane $z = ax + by + c$ that projects onto the region R is given by

(18.33)
$$S = \sqrt{a^2 + b^2 + 1}\; \Delta x \Delta y$$

Proof   As suggested by Figure 29, the surface is a parallelogram. What we want then is the area of this parallelogram.

The vector marked **u** in the figure has $(x_0, y_0, ax_0 + by_0 + c)$ as the initial point and $(x_0 + \Delta x, y_0, a(x_0 + \Delta x) + by_0 + c)$ as the terminal point. Hence,

$$\mathbf{u} = \Delta x \mathbf{i} + a\Delta x \mathbf{k} = \Delta x(\mathbf{i} + a\mathbf{k})$$

The vector marked **w** has $(x_0, y_0, ax_0 + by_0 + c)$ as the initial point and the point $(x_0, y_0 + \Delta y, ax_0 + b(y_0 + \Delta y) + c)$ as the terminal point. Hence,

$$\mathbf{w} = \Delta y \mathbf{j} + b\Delta y \mathbf{k} = \Delta y(\mathbf{j} + b\mathbf{k})$$

The area of the parallelogram is

$$\|\mathbf{u} \times \mathbf{w}\| = \|\Delta x(\mathbf{i} + a\mathbf{k}) \times \Delta y(\mathbf{j} + b\mathbf{k})\|$$

$$\uparrow$$
$$\text{(14.28(c))}$$

$$= \|(\mathbf{i} + a\mathbf{k}) \times (\mathbf{j} + b\mathbf{k})\|\Delta x \Delta y$$
$$= \|-a\mathbf{i} - b\mathbf{j} + \mathbf{k}\|\Delta x \Delta y$$
$$= \sqrt{a^2 + b^2 + 1}\; \Delta x \Delta y \quad \blacksquare$$

We are now prepared to derive a formula for the surface area of the graph of a nonnegative function $z = f(x, y)$ whose graph is not a plane.

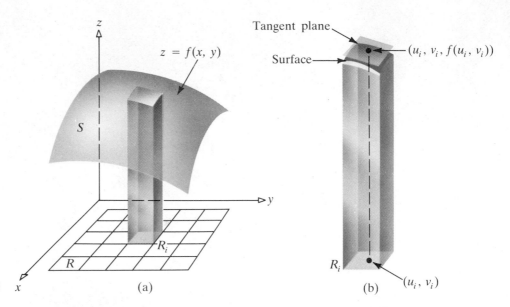

Figure 30   (a)   (b)

Figure 30 shows a surface which projects onto a region R of the xy-plane. We state the following result:

If S is the area of the part of the surface $z = f(x, y)$ that lies over the closed and bounded region R (whose boundary is a piecewise smooth curve) and if f_x and f_y are continuous on R, then

(18.34)
$$S = \iint_R \sqrt{f_x^2 + f_y^2 + 1}\, dx\, dy$$

Formula (18.34) can be developed as follows: We enclose R within a rectangle whose sides are parallel to the x-axis and y-axis. Using lines parallel to the x- and y-axes, we partition R into n rectangles and concentrate on the ith rectangle R_i having sides of length Δx_i and Δy_i [see Fig. 30(a)]. We pick a point (u_i, v_i) in R_i and at the corresponding point $P_i = (u_i, v_i, f(u_i, v_i))$ on the surface, we construct the tangent plane to the surface [see Fig. 30(b)]. The equation of this tangent plane can be written

(18.35)
$$z = f_x(u_i, v_i)(x - u_i) + f_y(u_i, v_i)(y - v_i) + f(u_i, v_i)$$

If the rectangle R_i is projected upward parallel to the z-axis, the result is a cylinder. Let ΔS_i denote the area cut from the tangent plane at P_i by the cylinder. The number ΔS_i is an approximation to the area of the part of the surface that lies above the ith rectangle R_i. By adding up all these parts (there are n of them) and taking the limit as the norm $\|P\|$ of the partition approaches 0, we arrive at a definition for the area S of the part of the surface $z = f(x, y)$ that lies above the region R, namely,

(18.36)
$$S = \lim_{\|P\| \to 0} \sum_{i=1}^{n} \Delta S_i$$

provided the limit exists.

Now we require a formula for ΔS_i. By (18.33) and (18.35), this area above R_i is given by

(18.37)
$$\Delta S_i = \sqrt{[f_x(u_i, v_i)]^2 + [f_y(u_i, v_i)]^2 + 1}\, \Delta x_i \Delta y_i$$

In (18.36) we replace ΔS_i by the expression for ΔS_i in (18.37):

(18.38)
$$S = \lim_{\|P\| \to 0} \sum_{i=1}^{n} \sqrt{[f_x(u_i, v_i)]^2 + [f_y(u_i, v_i)]^2 + 1}\, \Delta x_i \Delta y_i$$

Since the first-order partial derivatives of f are continuous on R, expression (18.38) leads to the following double integral for the surface area:

(18.39)
$$S = \iint_R \sqrt{[f_x(x, y)]^2 + [f_y(x, y)]^2 + 1}\, dx\, dy$$

EXAMPLE 1   Find the surface area of the graph of $z = f(x, y) = \frac{2}{3}(x^{3/2} + y^{3/2})$ that lies above the rectangle enclosed by $x = 0$, $x = 1$, $y = 0$, and $y = 2$.

Solution   Figure 31 illustrates the situation. Since $f_x(x, y) = x^{1/2}$ and $f_y(x, y) = y^{1/2}$, it follows from (18.39) that

$$S = \iint_R \sqrt{[f_x(x, y)]^2 + [f_y(x, y)]^2 + 1}\, dx\, dy = \iint_R \sqrt{x + y + 1}\, dx\, dy$$

$$= \int_0^1 \left[\int_0^2 \sqrt{x + y + 1}\, dy \right] dx = \int_0^1 \left[\tfrac{2}{3}(x + y + 1)^{3/2} \Big|_0^2 \right] dx$$

$$= \tfrac{2}{3} \int_0^1 [(x + 3)^{3/2} - (x + 1)^{3/2}]\, dx = (\tfrac{2}{3})(\tfrac{2}{5})[(x + 3)^{5/2} - (x + 1)^{5/2}] \Big|_0^1$$

$$= \tfrac{4}{15}(32 - 9\sqrt{3} - 4\sqrt{2} + 1) = \tfrac{4}{15}(33 - 9\sqrt{3} - 4\sqrt{2}) \approx 3.135 \text{ square units}$$

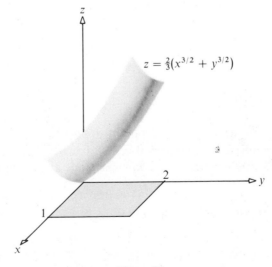

$$z = \tfrac{2}{3}(x^{3/2} + y^{3/2})$$

Figure 31   ■

EXAMPLE 2   Find the surface area of the part of the paraboloid $z = f(x, y) = 1 - x^2 - y^2$ that lies above the xy-plane.

Solution   Figure 32 shows the graph of the part of the surface $z = 1 - x^2 - y^2$ that lies above the xy-plane and its projection onto the xy-plane. The region R is enclosed by the circle $x^2 + y^2 = 1$.   Since $f_x(x, y) = -2x$ and $f_y(x, y) = -2y$,   it follows from (18.39) that the desired surface area S is given by

$$S = \iint_R \sqrt{(-2x)^2 + (-2y)^2 + 1}\, dx\, dy = \iint_R \sqrt{4(x^2 + y^2) + 1}\, dx\, dy$$

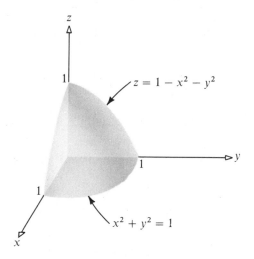

Figure 32

Given the character of both the region R (a circle) and the integrand (it involves $x^2 + y^2$),   we turn to polar coordinates. Then

$$S = \iint_R \sqrt{4r^2 + 1}\, r\, dr\, d\theta = \int_0^{2\pi} \left[\int_0^1 \sqrt{4r^2 + 1}\, r\, dr \right] d\theta = \int_0^{2\pi} \left[\frac{1}{12}(4r^2 + 1)^{3/2} \Big|_0^1 \right] d\theta$$

$$= \frac{1}{12}(5\sqrt{5} - 1) \int_0^{2\pi} d\theta = \frac{\pi}{6}(5\sqrt{5} - 1) \approx 5.33 \text{ square units} \quad \blacksquare$$

If the surface is defined implicitly by the equation $F(x, y, z) = 0$,   and if F has continuous first-order partial derivatives, then the vector $\nabla F = F_x \mathbf{i} + F_y \mathbf{j} + F_z \mathbf{k}$ is normal to the surface. The tangent plane at any point (x^*, y^*, z^*) is given by

$$F_x(x^*, y^*, z^*)x + F_y(x^*, y^*, z^*)y + F_z(x^*, y^*, z^*)z = D$$

where D is an appropriate constant. If $F_z(x^*, y^*, z^*) \neq 0$,   we solve for z:

$$z = -\frac{F_x(x^*, y^*, z^*)}{F_z(x^*, y^*, z^*)}x - \frac{F_y(x^*, y^*, z^*)}{F_z(x^*, y^*, z^*)}y + \frac{D}{F_z(x^*, y^*, z^*)}$$

Using this expression in place of (18.35), the formula for the area S of the part of the surface $F(x, y, z) = 0$ that lies above a region R is

(18.40)
$$S = \iint\limits_{R} \frac{\sqrt{[F_x(x, y, z)]^2 + [F_y(x, y, z)]^2 + [F_z(x, y, z)]^2}}{|F_z(x, y, z)|} \, dx \, dy$$

We use this formula in the next example.

EXAMPLE 3   Find the surface area of the part of the sphere $x^2 + y^2 + z^2 = a^2$ that is contained within the cylinder $x^2 + y^2 = ax$ and lies above the xy-plane.

Solution   Figure 33 illustrates the situation. Since the surface is given by $F(x, y, z) = x^2 + y^2 + z^2 - a^2 = 0$, we have

$$F_x(x, y, z) = 2x \qquad F_y(x, y, z) = 2y \qquad F_z(x, y, z) = 2z$$

Thus,

$$\sqrt{[F_x(x, y, z)]^2 + [F_y(x, y, z)]^2 + [F_z(x, y, z)]^2} = \sqrt{4x^2 + 4y^2 + 4z^2}$$
$$= 2(x^2 + y^2 + z^2)^{1/2} = 2a$$

Since (x, y, z) is on the surface,
$x^2 + y^2 + z^2 = a^2$

and

$$\frac{\sqrt{[F_x(x, y, z)]^2 + [F_y(x, y, z)]^2 + [F_z(x, y, z)]^2}}{|F_z(x, y, z)|} = \frac{2a}{2z} = \frac{a}{\sqrt{a^2 - (x^2 + y^2)}}$$

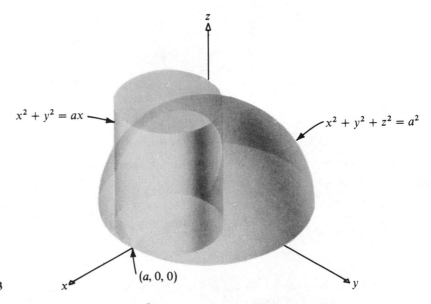

Figure 33

By (18.40), the area S is

$$S = \iint_R \frac{a}{\sqrt{a^2 - (x^2 + y^2)}} \, dx \, dy$$

Since the integrand involves the expression $(x^2 + y^2)$ and the region R is enclosed by the circle $x^2 + y^2 = ax$ $(r = a \cos \theta)$, we employ polar coordinates in evaluating it. Also, half the required area is to the right of the xz-plane and half is to the left of it. The required area S is, therefore,

$$S = 2 \int_0^{\pi/2} \left[\int_0^{a \cos \theta} \frac{a}{\sqrt{a^2 - r^2}} \, r \, dr \right] d\theta = -2a \int_0^{\pi/2} \left[\sqrt{a^2 - r^2} \, \Big|_0^{a \cos \theta} \right] d\theta$$

$$= -2a \int_0^{\pi/2} (a \sin \theta - a) \, d\theta = -2a^2 (-\cos \theta - \theta) \Big|_0^{\pi/2} = a^2 (\pi - 2) \text{square units} \qquad \blacksquare$$

EXERCISE 6

1. Find the surface area of the graph of $z = \frac{2}{3}(x^{3/2} + y^{3/2})$ that lies above the triangle enclosed by the lines $x = 0$, $y = 0$, and $2x + 3y = 6$.

2. Find the surface area of the graph of $z = \frac{2}{3}(x^{3/2} + y^{3/2})$ that lies above the triangle enclosed by $x = 0$, $y = 0$, and $3x + y = 3$.

3. Find the surface area of the paraboloid $z = 4 - x^2 - y^2$ that lies above the xy-plane.

4. Find the surface area of the part of the cone $z = \sqrt{x^2 + y^2}$ that lies inside the cylinder $x^2 + y^2 = 2x$.

5. Find the surface area of the part of the cylinder $z = \sqrt{a^2 - x^2}$ that lies above the square $-\frac{1}{2}a \le x \le \frac{1}{2}a$, $-\frac{1}{2}a \le y \le \frac{1}{2}a$.

6. Find the surface area of the part of the sphere $x^2 + y^2 + z^2 = 4z$ that lies within the paraboloid $x^2 + y^2 = 2z$.

7. Find the surface area of the part of $z = xy$ that lies in the first octant and within the cylinder $x^2 + y^2 = a^2$.

8. Find the surface area of the part of $z = x^2 - y^2$ in the first octant that lies within the cylinder $x^2 + y^2 = 4$.

9. Find the surface area of the part of the cylinder $y^2 + z^2 = 2z$ that is cut off by the cone $x^2 = y^2 + z^2$.

10. Find the surface area of the part of the cylinder $x^2 + y^2 = 2ax$ that lies inside the sphere $x^2 + y^2 + z^2 = 4a^2$.

11. Derive a formula for the surface area of a sphere of radius a.

12. Find the surface area cut from the hyperbolic paraboloid $y^2 - x^2 = 6z$ by the cylinder $x^2 + y^2 = 36$.

13. Find the surface area of the part of $z = 4 - y^2$ that lies in the first octant and is enclosed by $z = 0$, $x = 0$, $x = y$, and $y = 2$.

14. Find the surface area of the part of $x^2 = y$ that lies in the first octant and is enclosed by $x + z = 3$. [*Hint:* Project the surface onto the xz-plane.]

15. Find the surface area of the part of the paraboloid $z = 9 - x^2 - y^2$ that lies between the planes $z = 0$ and $z = 8$.

16. Set up, but do not evaluate, the integral to find the surface area of the part of the surface $z = 1 - x^4 - y^2$ that lies above the xy-plane.

17. Show that, for the plane surface $F(x, y, z) = Ax + By + Cz - D = 0,$ formula (18.40) may be written as

$$S = \iint_R \sec \gamma \, dx \, dy$$

where γ is the positive acute angle between the normal $\mathbf{n} = A\mathbf{i} + B\mathbf{j} + C\mathbf{k}$ to the plane and $\mathbf{k}$.

7. The Triple Integral

The triple integral of a function of three variables is defined in a way analogous to the definition of the double integral.

We begin by considering a function f of three variables defined over a box-shaped region S of three-dimensional space. We divide the region S into rectangular boxes by drawing planes parallel to the coordinate planes. Let n be the total number of boxes thus constructed. We now have a partition P of S into n boxes $S_1, S_2, \ldots, S_n$ (see Fig. 34).

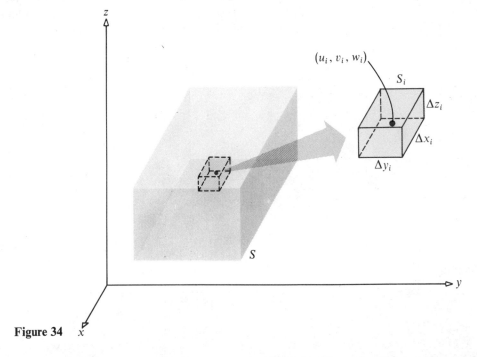

Figure 34

The norm of the partition P, denoted by $\|P\|$, is defined as the length of the longest diagonal of the boxes S_i. If Δx_i, Δy_i, and Δz_i denote the length, width, and height of the ith box, respectively, then its volume is $\Delta V_i = \Delta x_i \Delta y_i \Delta z_i$.   In each box S_i we arbitrarily select a point (u_i, v_i, w_i) and evaluate the function f there. Then we form the sum

$$\sum_{i=1}^{n} f(u_i, v_i, w_i) \Delta V_i$$

All sums of this form are referred to as *Riemann sums* of f for the partition P. If f is continuous on S, all such sums will approach a limit as the norm of the partition approaches 0. This limit is called the *triple integral of f over S*, and we write

$$\lim_{\|P\| \to 0} \sum_{i=1}^{n} f(u_i, v_i, w_i) \Delta V_i = \iiint_S f(x, y, z) \, dV$$

Other symbols for the triple integral of f over S are

$$\iiint_S f(x, y, z) \, dx \, dy \, dz \qquad \iiint_S f(x, y, z) \, dx \, dz \, dy$$

and so on.

THE TRIPLE INTEGRAL OVER A MORE GENERAL REGION

Consider a closed bounded region S of three-dimensional space whose boundary consists of the planes $x = a$ and $x = b$, $a < b$; the cylinders $y = y_1(x)$ and $y = y_2(x)$, $y_1 \leq y_2$; and the surfaces $z = z_1(x, y)$ and $z = z_2(x, y)$, $z_1 \leq z_2$,

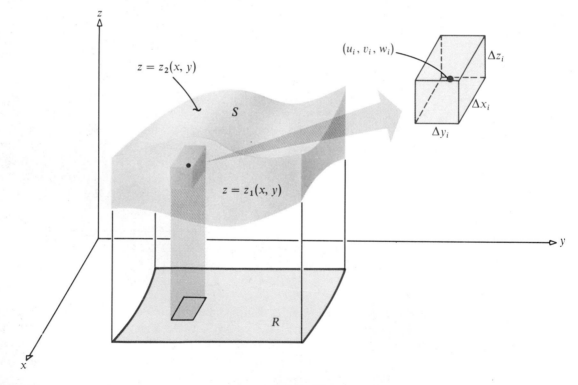

Figure 35

where the functions $y_1, y_2, z_1,$ and z_2 each have continuous derivatives or partial derivatives (see Fig. 35).

Draw planes parallel to the coordinate planes to obtain a partition of S into boxes and discard any box thus formed that does not lie entirely in S. Let n be the number of boxes that remain. We now proceed as before to define the triple integral of a function f defined on S as

(18.41)
$$\iiint\limits_{S} f(x, y, z)\, dV = \lim_{||P|| \to 0} \sum_{i=1}^{n} f(u_i, v_i, w_i)\Delta V_i$$

provided this limit exists.

We state, without proof, a condition under which the limit in (18.41) exists:

(18.42)   If $w = f(x, y, z)$ **is a continuous function defined on a closed bounded region S, then the limit in (18.41) exists.**

ITERATED TRIPLE INTEGRALS

It can be proved that the triple integral (18.41) can be evaluated by an iterated integral of the form

(18.43)
$$\iiint\limits_{S} f(x, y, z)\, dV = \iint\limits_{R} \left[\int_{z_1(x, y)}^{z_2(x, y)} f(x, y, z)\, dz \right] dA$$

where R denotes the region in the xy-plane obtained by projecting the points of S perpendicularly onto that plane (refer again to Fig. 35). In evaluating the inside integral $\int_{z_1(x, y)}^{z_2(x, y)} f(x, y, z)\, dz$,   we treat x and y as constants and integrate with respect to z. When the limits of integration $z_1(x, y)$ and $z_2(x, y)$ are substituted for z, the result is a function of x and y alone to be (doubly) integrated over the region R.

Depending on the order of integration used to evaluate this double integral, there are two iterated forms of the triple integral (18.43). If the region R is of Type I, then

(18.44)
$$\iiint\limits_{S} f(x, y, z)\, dV = \iint\limits_{R} \left[\int_{z_1(x, y)}^{z_2(x, y)} f(x, y, z)\, dz \right] dA$$
$$= \int_{a}^{b} \left\{ \int_{y_1(x)}^{y_2(x)} \left[\int_{z_1(x, y)}^{z_2(x, y)} f(x, y, z)\, dz \right] dy \right\} dx$$

If the region R is of Type II, then

(18.45)
$$\iiint\limits_{S} f(x, y, z)\, dV = \iint\limits_{R} \left[\int_{z_1(x, y)}^{z_2(x, y)} f(x, y, z)\, dz \right] dA$$
$$= \int_{c}^{d} \left\{ \int_{x_1(y)}^{x_2(y)} \left[\int_{z_1(x, y)}^{z_2(x, y)} f(x, y, z)\, dz \right] dx \right\} dy$$

Since the symbols x, y, and z can be permuted in six different ways, there are four additional formulas similar to (18.44) and (18.45), one for each of the six possible orders of integration. Which one to use will depend on the nature of the region S and the character of the function f.

The determination of the limits of integration in (18.44) may be obtained by noting that the first two integrations, with respect to z and y, are carried out on a typical plane on which x is constant. This plane is perpendicular to the x-axis and intersects the region S in a certain plane region. The limits of integration for z and y are exactly those that would be found by evaluating a double integral by iteration, first with respect to z and then y, over this plane section of S. Similar remarks hold for (18.45).

EXAMPLE 1   Evaluate the triple integral $\iiint_S 4x\,dV$ over the tetrahedron formed by the coordinate planes and the plane $2x + 3y + 4z = 12$.

Solution   The region S is pictured in Figure 36. A typical plane section of S by a plane perpendicular to the x-axis reveals that the upper surface is the plane $z = z_2(x, y) = \frac{1}{4}(12 - 2x - 3y)$ and the lower surface is the plane $z = z_1(x, y) = 0$. The region R in the xy-plane is the triangle enclosed by the x-axis, the y-axis, and the line $2x + 3y = 12$ and is of Type I. Formula (18.44) can therefore be used. [Note that region R is also of Type II, so (18.45) could have been used.]

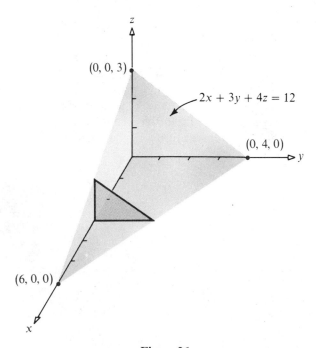

Figure 36

$$\iiint\limits_{S} 4x \, dV = \int_0^6 \left\{ \int_0^{(1/3)(12-2x)} \left[\int_0^{(1/4)(12-2x-3y)} 4x \, dz \right] dy \right\} dx$$

$$= \int_0^6 \left[\int_0^{(1/3)(12-2x)} (12x - 2x^2 - 3xy) \, dy \right] dx$$

$$= \int_0^6 (24x - 8x^2 + \tfrac{2}{3}x^3) \, dx = 72 \qquad \blacksquare$$

EXAMPLE 2   Express $\iiint_S f(x, y, z) \, dV$ as an iterated integral if S is the region in the first octant that is enclosed by the paraboloid $z = 16 - 4x^2 - y^2$ and the xy-plane.

Solution   Figure 37 illustrates the region S. The upper surface is the paraboloid $z = z_2(x, y) = 16 - 4x^2 - y^2$ and the lower surface is the plane $z = z_1(x, y) = 0$.   The region R in the xy-plane is enclosed by the x-axis, the y-axis, and one-fourth of the ellipse $4x^2 + y^2 = 16$ and is of Type I. We use formula (18.44):

$$\iiint\limits_{S} f(x, y, z) \, dV = \int_0^2 \left\{ \int_0^{\sqrt{16-4x^2}} \left[\int_0^{16-4x^2-y^2} f(x, y, z) \, dz \right] dy \right\} dx$$

[We could also have used (18.45) here since R is also a Type II region.]

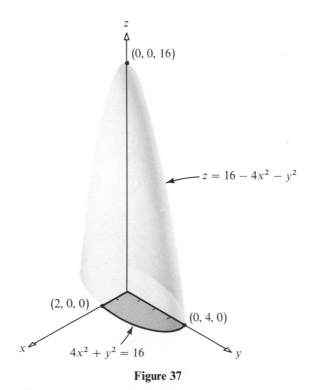

Figure 37   ■

If $f(x, y, z) = 1$ on S, then formula (18.44) or (18.45) may be used to calculate the volume of the region S. For example, the volume V of the tetrahedron of Example 1 is

$$V = \iiint_S dV = \int_0^6 \left\{ \int_0^{(1/3)(12 - 2x)} \left[\int_0^{(1/4)(12 - 2x - 3y)} dz \right] dy \right\} dx$$

$$= \frac{1}{4} \int_0^6 \left[\int_0^{(1/3)(12 - 2x)} (12 - 2x - 3y) \, dy \right] dx$$

$$= \frac{1}{6} \int_0^6 (36 - 12x + x^2) \, dx = 12 \text{ cubic units}$$

MASS; CENTER OF MASS; MOMENT OF INERTIA

Another useful application of triple integrals is in finding mass, centers of mass, and moments of inertia. The formulas are obtained in a manner much like those involving double integrals, so we merely state them here.

The mass m of a body of continuous mass density $\rho = \rho(x, y \, z)$ in a region S of volume V is

$$m = \iiint_S \rho \, dV$$

The center of mass $(\bar{x}, \bar{y}, \bar{z})$ of this body obeys

$$m\bar{x} = \iiint_S x\rho \, dV \qquad m\bar{y} = \iiint_S y\rho \, dV \qquad m\bar{z} = \iiint_S z\rho \, dV$$

The moment of inertia is

$$I = \iiint_S r^2 \rho \, dV$$

where r is the distance from the point (x, y, z) of the body to the axis about which the moment is to be calculated.

EXAMPLE 3   Find the mass of the body in the shape of a tetrahedron cut from the first octant by the plane $x + y + z = 1$, if the mass density is proportional to the distance from the yz-plane. Locate its center of mass.

Solution   The mass density ρ is given by $\rho = kx$, where k is a constant. The mass m of the body is therefore given by

$$m = \iiint_S kx \, dV = k \int_0^1 \left\{ \int_0^{1-x} \left[\int_0^{1-x-y} x \, dz \right] dy \right\} dx = \frac{k}{24}$$

The center of mass $(\bar{x}, \bar{y}, \bar{z})$ is

$$\bar{x} = \frac{\iiint_S kx^2 \, dV}{m} = 24 \int_0^1 \left\{ \int_0^{1-x} \left[\int_0^{1-x-y} x^2 \, dz \right] dy \right\} dx = \frac{2}{5}$$

$$\bar{y} = \frac{\iiint_S kxy\, dV}{m} = 24 \int_0^1 \left\{ \int_0^{1-x} \left[\int_0^{1-x-y} xy\, dz \right] dy \right\} dx = \frac{1}{5}$$

$$\bar{z} = \frac{\iiint_S kxz\, dV}{m} = 24 \int_0^1 \left\{ \int_0^{1-x} \left[\int_0^{1-x-y} xz\, dz \right] dy \right\} dx = \frac{1}{5} \quad \blacksquare$$

EXAMPLE 4   Find the moment of inertia about the z-axis of the homogeneous body of mass density ρ in the first octant enclosed by the surface $z = 4xy$ and the planes $z = 0$, $x = 3$, $y = 2$.

Solution   The moment of inertia I_z about the z-axis is

$$I_z = \iiint_S \rho(x^2 + y^2)\, dV = \rho \int_0^3 \left\{ \int_0^2 \left[\int_0^{4xy} (x^2 + y^2)\, dz \right] dy \right\} dx = 234\rho \quad \blacksquare$$

EXERCISE 7

In Problems 1–4 evaluate each iterated triple integral.

1. $\int_0^2 \left\{ \int_0^{2-3x} \left[\int_0^{x+y} x\, dz \right] dy \right\} dx$

2. $\int_0^1 \left\{ \int_0^{4-x} \left[\int_0^{2x+y} z\, dz \right] dy \right\} dx$

3. $\int_0^3 \left\{ \int_z^{z+2} \left[\int_y^{y+z} 2x\, dx \right] dy \right\} dz$

4. $\int_0^{\pi/2} \left\{ \int_y^{\pi/2} \left[\int_0^{xy} \sin \frac{z}{y}\, dz \right] dx \right\} dy$

5. Evaluate $\iiint_S x\, dV$ using six different iterated integrals, if S is the region enclosed by $x = 0$, $x = 1$, $y = 2$, $y = 3$, $z = 0$, and $z = 2$.

6. Evaluate $\iiint_S y\, dV$ using six different iterated integrals, if S is the region enclosed by $x = 0$, $x = 2$, $y = -1$, $y = 1$, $z = 0$, and $z = 1$.

In Problems 7 and 8 express $\iiint_S f(x, y, z)\, dV$ as an iterated integral in six different ways for each region S.

7. S is the region enclosed by the coordinate planes and the plane $x + 2y + 3z = 6$

8. S is the region enclosed by the coordinate planes and the plane $x + y + z = 3$

In Problems 9–12 evaluate each triple integral.

9. $\iiint_S x\, dV$,   if S is the region enclosed by the tetrahedron having vertices at $(0, 0, 0)$, $(1, 1, 0)$, $(1, 0, 0)$, $(1, 0, 1)$

10. $\iiint_S (x^2 + z^2)\, dV$,   if S is the same region as in Problem 9

11. $\iiint_S (xy + 3y)\, dV$,   if S is the region enclosed by the cylinder $x^2 + y^2 = 9$ and the planes $x + z = 3$, $y = 0$, and $z = 0$

12. $\iiint_S xyz\, dV$,   if S is the region enclosed by the cylinders $x^2 + y^2 = 1$ and $x^2 + z^2 = 1$

In Problems 13–18 use triple integration to find the volume of the indicated region.

13. Enclosed by $z = 4 - y^2$, $z = 9 - x$, $x = 0$, and $z = 0$

14. Enclosed by $z = 0$, $z = 1 - x^2$, and $z = 1 - y^2$

15. Enclosed by $y^2 = z$, $x = 0$, and $x = y - z$

16. Enclosed by $z = x^2 + y^2$ and $z = 16 - x^2 - y^2$

17. Enclosed by $z = x^2 + y^2$ and $z = 2 - x$

18. Enclosed by $z^2 = 4x$ and $x^2 + y^2 = 2x$

19. Find the mass of a body in the shape of a right circular cylinder of height h and radius a, if its mass density is proportional to the square of the distance from the axis of the cylinder.

20. Find the mass of a body in the shape of a cube of edge a if its mass density is proportional to the square of the distance from one corner.

21. Find the mass of a body in the shape of a tetrahedron cut from the first octant by the plane $x + y + z = 1$, if the mass density is proportional to the product of the distances from the three coordinate planes.

22. Set up, but do not evaluate, the integral in rectangular coordinates to integrate the function $f(x, y, z) = x^2yz$ over the region enclosed by the cone $3x^2 + 3y^2 = z^2$, $z \geq 0$, and the plane $z = 3$.

8. Triple Integrals in Cylindrical Coordinates

We have seen that in many instances the evaluation of a double integral is more easily accomplished using iteration in polar coordinates rather than in rectangular coordinates. For triple integrals, at least two alternatives to integration in rectangular coordinates are available: one utilizes *cylindrical coordinates* and the other uses *spherical coordinates* (introduced in Section 9).

If the rectangular coordinates of a point P in three-dimensional space are (x, y, z) and if the polar coordinates for the projection of P onto the xy-plane are (r, θ), then P may be located by the ordered triple (r, θ, z), called the *cylindrical coordinates of P*.*

Figure 38 illustrates the role of r, θ, and z in locating a point P whose cylindrical coordinates are (r, θ, z). The algebraic relationship of the cylindrical coordinates

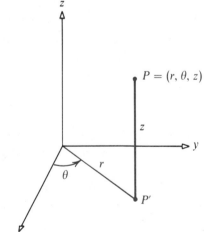

Figure 38 x

---

* The reason for the name *cylindrical* coordinates is that the surfaces $r = $ a constant   are cylinders.

(r, θ, z) and the rectangular coordinates (x, y, z) of a point P are given by the formulas

(18.46)
$$x = r \cos \theta \qquad y = r \sin \theta \qquad z = z$$

EXAMPLE 1   If the cylindrical coordinates of a point P are $(6, \pi/3, -2)$, find the rectangular coordinates for P.

Solution   From (18.46), we find that

$$x = 6 \cos \frac{\pi}{3} = 3 \qquad y = 6 \sin \frac{\pi}{3} = 3\sqrt{3} \qquad z = -2 \quad \blacksquare$$

Table 1 is a list of several equations in rectangular coordinates and their respective equations in cylindrical coordinates. Figure 39 (p. 998) illustrates the graph of each equation.

Table 1

	Surface	Rectangular	Cylindrical
(a)	Half plane	$y = x \tan k$	$\theta = k$
(b)	Plane	$z = k$	$z = k$
(c)	Cylinder	$x^2 + y^2 = a^2$	$r = a$
(d)	Sphere	$x^2 + y^2 + z^2 = R^2$	$r^2 + z^2 = R^2$
(e)	Circular cone	$x^2 + y^2 = a^2 z^2$	$r = az$
(f)	Circular paraboloid	$x^2 + y^2 = az$	$r^2 = az$

For triple integral problems in which symmetry about an axis occurs, particularly problems concerned with cylinders and cones, it may be easier to evaluate the triple integral by utilizing cylindrical coordinates.

The solid regions most easily described by cylindrical coordinates are cylindrical wedges of the form

$$a_1 \leq r \leq a_2 \qquad b_1 \leq \theta \leq b_2 \qquad c_1 \leq z \leq c_2$$

Such a region is enclosed by two half planes ($\theta = b_1$ and $\theta = b_2$) passing through the z-axis, two concentric circular cylinders ($r = a_1$ and $r = a_2$) centered about the z-axis, and two parallel planes ($z = c_1$ and $z = c_2$) perpendicular to the z-axis. The volume is given by

$$V = \bar{r} \Delta r \Delta \theta \Delta z$$

where

$$\bar{r} = \frac{a_1 + a_2}{2} \qquad \Delta r = a_2 - a_1 \qquad \Delta \theta = b_2 - b_1 \qquad \Delta z = c_2 - c_1$$

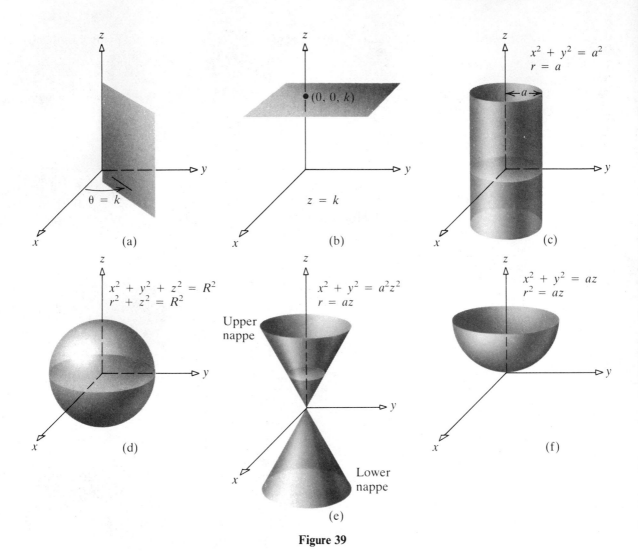

Figure 39

To evaluate a triple integral over a solid region S, we divide S into a number of wedge-shaped subregions by passing planes and cylinders through S. The shaded portion of Figure 40 illustrates a typical subregion, denoted S_i. We let n denote the number of such subregions that lie entirely in S. We now have a partition P of the region whose norm $\|P\|$ is taken as the length of the longest diagonal of these subregions.

We proceed to calculate the volume ΔV_i of a typical subregion S_i. Using an argument similar to the one employed for double integrals in polar coordinates, the volume ΔV_i is given by

$$\Delta V_i = \bar{r}_i \Delta r_i \Delta \theta_i \Delta z_i$$

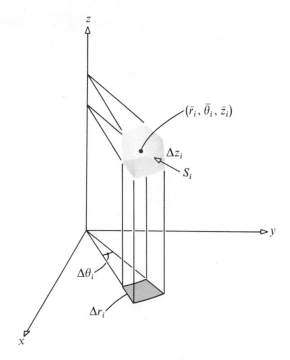

Figure 40

where Δr_i, $\Delta \theta_i$, and Δz_i are the dimensions of the subregion S_i and $\bar{r}_i$ is the r-coordinate of a point in its interior. (Again, refer to Fig. 40.)

Now we let f be a function of the cylindrical coordinates r, θ, and z. We pick a point $(\bar{r}_i, \bar{\theta}_i, \bar{z}_i)$ in S_i and evaluate f there. Then we form the sum

(18.47)
$$\sum_{i=1}^{n} f(\bar{r}_i, \bar{\theta}_i, \bar{z}_i) \bar{r}_i \Delta r_i \Delta \theta_i \Delta z_i$$

It can be shown that if f is continuous on S, then the limit of the sum in (18.47), as the norm $\|P\|$ approaches 0, exists and equals the triple integral of f over S. That is,

$$\lim_{\|P\| \to 0} \sum_{i=1}^{n} f(\bar{r}_i, \bar{\theta}_i, \bar{z}_i) \bar{r}_i \Delta r_i \Delta \theta_i \Delta z_i = \iiint_S f(r, \theta, z)\, dV$$

or

$$\lim_{\|P\| \to 0} \sum_{i=1}^{n} f(\bar{r}_i, \bar{\theta}_i, \bar{z}_i) \bar{r}_i \Delta r_i \Delta \theta_i \Delta z_i = \iiint_S f(r, \theta, z) r\, dr\, d\theta\, dz$$

It is important to remember that the integrand contains a factor of r. In other words, in cylindrical coordinates,

$$dV = r\, dr\, d\theta\, dz$$

We can evaluate a triple integral by using iterated integrals. Suppose the region S is enclosed by the half planes $\theta = \alpha$, $\theta = \beta$, $0 < \beta - \alpha \leq 2\pi$; the cylinders $r = r_1(\theta)$, $r = r_2(\theta)$, $r_1 \leq r_2$; and the surfaces $z = z_1(r, \theta)$, $z = z_2(r, \theta)$, $z_1 \leq z_2$. If the functions r_1, r_2, z_1, and z_2 have continuous derivatives or partial derivatives, the triple integral of f over S is given by an iterated integral of the form

(18.48)
$$\iiint\limits_{S} f(r, \theta, z) r \, dr \, d\theta \, dz = \int_{\alpha}^{\beta} \left\{ \int_{r_1(\theta)}^{r_2(\theta)} \left[\int_{z_1(r, \theta)}^{z_2(r, \theta)} f(r, \theta, z) r \, dz \right] dr \right\} d\theta$$

Five other formulations for iteration are also possible; the choice of which to use depends on convenience.

EXAMPLE 2   Find the volume for the region enclosed by the sphere $x^2 + y^2 + z^2 = 4$ and the cylinder $(x - 1)^2 + y^2 = 1$. The portion of S in the first octant is pictured in Figure 33 (p. 987). (Let $a = 2$.)

Solution   We use cylindrical coordinates because the variables x and y appear in the equations for the sphere and the cylinder only in combinations of $x^2 + y^2$. The equation of the sphere in cylindrical coordinates is $r^2 + z^2 = 4$ and that of the cylinder is $r = 2 \cos \theta$. Using (18.48) gives

$$V = \int_{-\pi/2}^{\pi/2} \left\{ \int_0^{2 \cos \theta} \left[\int_{-\sqrt{4 - r^2}}^{\sqrt{4 - r^2}} r \, dz \right] dr \right\} d\theta$$

Since both the region and the integral are symmetric about the half plane $\theta = 0$ and the plane $z = 0$, we have

$$V = 4 \int_0^{\pi/2} \left\{ \int_0^{2 \cos \theta} \left[\int_0^{\sqrt{4 - r^2}} r \, dz \right] dr \right\} d\theta = 4 \int_0^{\pi/2} \left[\int_0^{2 \cos \theta} (4 - r^2)^{1/2} r \, dr \right] d\theta$$

$$= -2 \int_0^{\pi/2} \tfrac{2}{3} (4 - r^2)^{3/2} \Big|_0^{2 \cos \theta} d\theta = \tfrac{4}{3} \int_0^{\pi/2} \left[8 - (4 - 4 \cos^2 \theta)^{3/2} \right] d\theta$$

$$= \tfrac{32}{3} \int_0^{\pi/2} (1 - \sin^3 \theta) \, d\theta = \tfrac{16}{9} (3\pi - 4) \quad \blacksquare$$

EXAMPLE 3   Find the moment of inertia about the z-axis of a homogeneous body of mass density ρ in the shape of a region enclosed by the paraboloid $z = 1 - x^2 - y^2$ and the xy-plane.

Solution   In cylindrical coordinates, the region S is $0 \leq \theta \leq 2\pi$, $0 \leq r \leq 1$, and $z = 0$, $z = 1 - r^2$. The moment of inertia I_z about the z-axis is

$$I_z = \iiint\limits_{S} r^2 \rho \, dV = \rho \iiint\limits_{S} r^2 r \, dr \, d\theta \, dz = \rho \int_0^{2\pi} \left\{ \int_0^1 \left[\int_0^{1 - r^2} r^3 \, dz \right] dr \right\} d\theta$$

$$= \rho \int_0^{2\pi} \left[\int_0^1 r^3 (1 - r^2) \, dr \right] d\theta = \rho \int_0^{2\pi} \frac{1}{12} \, d\theta = \rho \frac{\pi}{6} \quad \blacksquare$$

EXAMPLE 4   Find the moment of inertia of a homogeneous body in the shape of a sphere about a diameter.

Solution   Let ρ denote the constant mass density of the body and let a be the radius of the sphere. Situate the sphere so that its center is at the origin. The equation of the sphere is then $x^2 + y^2 + z^2 = a^2$,   or $r^2 + z^2 = a^2$.   Since the moment of inertia about a diameter is given by the moment of inertia about the z-axis, we have

$$I_z = \iiint_S \rho r^2 \, dV = \rho \int_0^{2\pi} \left\{ \int_0^a \left[\int_{-\sqrt{a^2-r^2}}^{\sqrt{a^2-r^2}} r^3 \, dz \right] dr \right\} d\theta$$

$$= \rho \int_0^{2\pi} \left[\int_0^a (r^3) 2\sqrt{a^2-r^2} \, dr \right] d\theta = 2\rho \int_0^{2\pi} \frac{2a^5}{15} \, d\theta = \frac{8\pi\rho a^5}{15} \quad \blacksquare$$

The next example illustrates a triple integral stated in rectangular coordinates that is more easily evaluated using cylindrical coordinates.

EXAMPLE 5   Use cylindrical coordinates to evaluate

$$\int_{-1}^1 \left\{ \int_{-\sqrt{1-x^2}}^{\sqrt{1-x^2}} \left[\int_0^{2\sqrt{1-x^2-y^2}} 1 \, dz \right] dy \right\} dx$$

Solution   From the limits of integration on z, we see that S is the upper half of the ellipsoid $4x^2 + 4y^2 + z^2 = 4$, $z \geq 0$.   From the x and y limits of integration, the projection R in the xy-plane is enclosed by the circle $x^2 + y^2 = 1$.   See Figure 41. The region

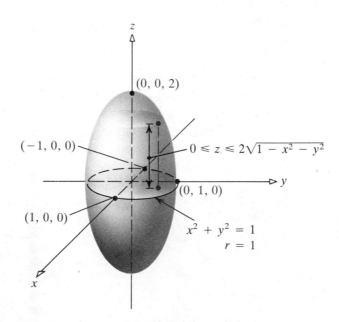

Figure 41

S of integration and its projection on the xy-plane can be described by the inequalities

$$0 \le z \le 2\sqrt{1 - x^2 - y^2} \qquad -\sqrt{1 - x^2} \le y \le \sqrt{1 - x^2} \qquad -1 \le x \le 1$$

In cylindrical coordinates the region S is described by the inequalities

$$0 \le r \le 1 \qquad 0 \le \theta \le 2\pi \qquad 0 \le z \le 2\sqrt{1 - r^2}$$

Thus,

$$V = \int_0^{2\pi} \left\{ \int_0^1 \left[\int_0^{2\sqrt{1-r^2}} r \, dz \right] dr \right\} d\theta = \int_0^{2\pi} \left[\int_0^1 \left(zr \Big|_{z=0}^{z=2\sqrt{1-r^2}} \right) dr \right] d\theta$$

$$= \int_0^{2\pi} \left[\int_0^1 2r\sqrt{1 - r^2} \, dr \right] d\theta = \int_0^{2\pi} \left[-\tfrac{2}{3}(1 - r^2)^{3/2} \Big|_{r=0}^{r=1} \right] d\theta = \frac{4\pi}{3} \qquad \blacksquare$$

EXERCISE 8

In Problems 1–4 find the cylindrical coordinates of the point with the given rectangular coordinates.

1. $(-\sqrt{3}, -1, -5)$ **2.** $(-1, \sqrt{3}, 4)$ **3.** $(1, 1, \sqrt{2})$ **4.** $(2, -2, 4)$

In Problems 5–8 find the rectangular coordinates of the point with the given cylindrical coordinates.

5. $(2, \pi/6, -5)$ **6.** $(4, \pi/3, 3)$ **7.** $(4, \pi/6, 2)$ **8.** $(2, \pi/2, 0)$

In Problems 9 and 10 evaluate each iterated integral.

9. $\int_{\pi/6}^{\pi/2} \left\{ \int_0^3 \left[\int_0^{r\sin\theta} r \csc^3\theta \, dz \right] dr \right\} d\theta$ **10.** $\int_0^{\pi/3} \left\{ \int_0^{\sin\theta} \left[\int_0^{r\sin\theta} r \, dz \right] dr \right\} d\theta$

In Problems 11–16 use triple integration in cylindrical coordinates.

11. Find the mass of a homogeneous body in the shape of a sphere of radius a.

12. Find the mass of a body in the shape of a sphere of radius a, if the mass density is proportional to the square of the distance from the center.

13. Find the center of mass of a homogeneous body in the shape of a region enclosed by the surface $x^2 + y^2 = 4z$ and the plane $z = 2$.

14. Find the center of mass of a homogeneous body in the shape of a region in the first octant enclosed by the surface $z = xy$ and the cylinder $x^2 + y^2 = 4$.

15. Find the center of mass of a homogeneous body in the shape of a region enclosed by the inside of the sphere $x^2 + y^2 + z^2 = 12$ and above the paraboloid $z = x^2 + y^2$.

16. Find the center of mass of a homogeneous body in the shape of a region enclosed by the paraboloid $z = x^2 + y^2$ and the plane $z = 4$.

In Problems 17 and 18 each integral is given in cylindrical coordinates. Express each integral in rectangular coordinates. Do not evaluate.

17. $\int_0^{2\pi} \left\{ \int_0^4 \left[\int_{-r}^{\sqrt{16-r^2}} z^2 r^5 \cos^4\theta \, dz \right] dr \right\} d\theta$ **18.** $\int_0^{\pi/2} \left\{ \int_0^2 \left[\int_{-r^2}^9 z^2 r^4 \sin\theta \, dz \right] dr \right\} d\theta$

9. Triple Integrals in Spherical Coordinates

Suppose (x, y, z) are the rectangular coordinates of a point P (different from the origin) in three-dimensional space. We define the numbers ρ, θ, and ϕ by

$$\rho = |OP|, \quad \text{distance from } O \text{ to } P$$

$$\theta = \text{Angle between positive } x\text{-axis and } OP', \text{ where } P'$$
$$\text{is the projection of } P \text{ onto the } xy\text{-plane}$$

$$\phi = \text{Angle between positive } z\text{-axis and } OP, \quad 0 \le \phi \le \pi$$

As Figure 42(a) illustrates, the point P may be located by the ordered triple (ρ, θ, ϕ), called the *spherical* coordinates of P. The algebraic relationship of the spherical coordinates (ρ, θ, ϕ) and the rectangular coordinates (x, y, z) of a point P are obtained through the use of Figure 42(b). There, we conclude that

$$x = |OP'| \cos \theta \qquad y = |OP'| \sin \theta$$

Since $|OP'| = |QP| = \rho \sin \phi$ and $|OQ| = z = \rho \cos \phi$, it follows that

(18.49)
$$x = \rho \sin \phi \cos \theta \qquad y = \rho \sin \phi \sin \theta \qquad z = \rho \cos \phi$$

Based on (18.49), we have

(18.50)
$$\rho = \sqrt{x^2 + y^2 + z^2} \qquad \tan \theta = \frac{y}{x} \qquad \cos \phi = \frac{z}{\rho}$$

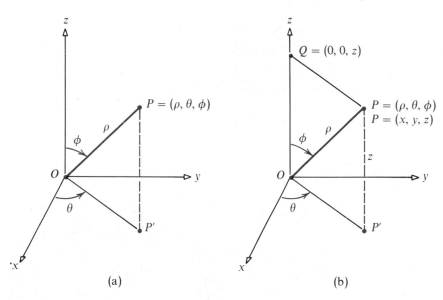

(a)   (b)

Figure 42

---

* The reason for the name *spherical* coordinates is that the surfaces $\rho = $ a constant   are spheres.

EXAMPLE 1   If the spherical coordinates of a point P are $(4, \pi/6, 2\pi/3)$, find the rectangular and cylindrical coordinates for P.

Solution   From (18.49), we find that

$$x = 4 \sin \frac{2\pi}{3} \cos \frac{\pi}{6} = 4\left(\frac{\sqrt{3}}{2}\right)\left(\frac{\sqrt{3}}{2}\right) = 3$$

$$y = 4 \sin \frac{2\pi}{3} \sin \frac{\pi}{6} = 4\left(\frac{\sqrt{3}}{2}\right)\left(\frac{1}{2}\right) = \sqrt{3}$$

$$z = 4 \cos \frac{2\pi}{3} = 4\left(-\frac{1}{2}\right) = -2$$

Thus, the rectangular coordinates of P are $(3, \sqrt{3}, -2)$. Since $r^2 = x^2 + y^2 = 9 + 3 = 12$, the cylindrical coordinates of P are $(2\sqrt{3}, \pi/6, -2)$.   ∎

EXAMPLE 2   If the rectangular coordinates of a point P are $(1, \sqrt{3}, -2)$, find the spherical coordinates of P.

Solution   From (18.50), we find that

$$\rho = \sqrt{x^2 + y^2 + z^2} = \sqrt{1 + 3 + 4} = \sqrt{8} = 2\sqrt{2}$$

$$\tan \theta = \sqrt{3}, \qquad \theta = \frac{\pi}{3}$$

$$\cos \phi = \frac{-1}{\sqrt{2}}, \qquad \phi = \frac{3\pi}{4}$$

Thus, the spherical coordinates of P are $(2\sqrt{2}, \pi/3, 3\pi/4)$.   ∎

Spherical coordinates are frequently useful in problems concerned with spheres, planes, and cones because their spherical coordinate equations are relatively simple, as shown by Table 2.

Table 2

Surface	Equation
(a) Sphere	$\rho = a$
(b) Cone	$\phi = a$
(c) Half plane	$\theta = a$

The surface $\phi = a$, $0 < a < \pi$, $a \neq \pi/2$, requires some explanation. The surface is a half cone, which is generated by revolving any ray emanating from the

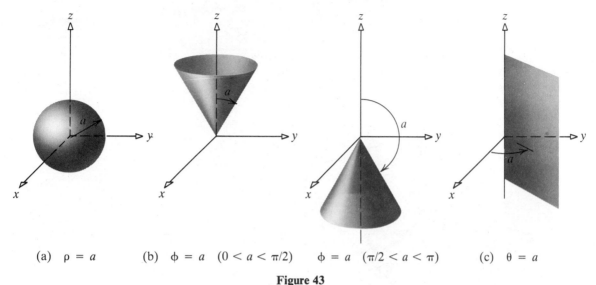

(a) $\rho = a$ (b) $\phi = a$ $(0 < a < \pi/2)$ $\phi = a$ $(\pi/2 < a < \pi)$ (c) $\theta = a$

Figure 43

origin at an angle a about the positive z-axis. The various surfaces are illustrated in Figure 43.

Let (ρ, θ, ϕ) be spherical coordinates of a point. We partition a region S by drawing concentric spheres (ρ a constant), planes through the z-axis (θ a constant), and circular cones with vertex at the origin and axis along the z-axis (ϕ a constant). We let n denote the number of such subregions that lie entirely in S. We now have a partition of S whose norm $\|P\|$ is taken as the length of the longest diagonal of these subregions.

We proceed to calculate the volume ΔV_i of a typical subregion S_i (see Fig. 44, p. 1006). By treating the subregion as though it were a rectangular parallelepiped, we obtain the following approximation to ΔV_i:

$$\Delta V_i = (\bar{\rho}_i \Delta \phi_i)(\Delta \rho_i)(\bar{\rho}_i \sin \bar{\phi}_i \Delta \theta_i) = \bar{\rho}_i^2 \sin \bar{\phi}_i \Delta \rho_i \Delta \theta_i \Delta \phi_i$$

where $(\bar{\rho}_i, \bar{\theta}_i, \bar{\phi}_i)$ is some point in S_i. Now we let f be a function in spherical coordinates ρ, θ, and ϕ. We pick a point $(\bar{\rho}_i, \bar{\theta}_i, \bar{\phi}_i)$ in each S_i and evaluate f there. Then we form the sum

(18.51)
$$\sum_{i=1}^{n} f(\bar{\rho}_i, \bar{\theta}_i, \bar{\phi}_i) \bar{\rho}_i^2 \sin \bar{\phi}_i \Delta \rho_i \Delta \theta_i \Delta \phi_i$$

It can be shown that if f is continuous on S, then the limit of the sum in (18.51), as the norm $\|P\|$ approaches 0, exists and equals the triple integral of f over S. That is,

$$\lim_{\|P\| \to 0} \sum_{i=1}^{n} f(\bar{\rho}_i, \bar{\theta}_i, \bar{\phi}_i) \rho_i^2 \sin \bar{\phi}_i \Delta \rho_i \Delta \theta_i \Delta \phi_i = \iiint\limits_{S} f(\rho, \theta, \phi)\, dV$$

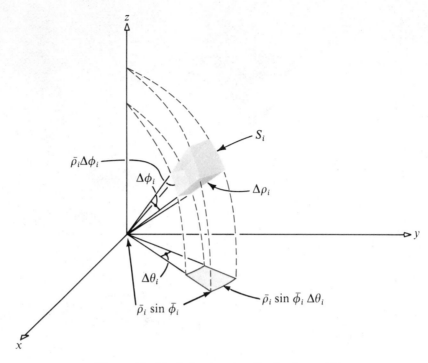

Figure 44   Typical subregion S_i of volume ΔV_i

or

(18.52)
$$\lim_{\|P\| \to 0} \sum_{i=1}^{n} f(\bar{\rho}_i, \bar{\theta}_i, \bar{\phi}_i)\bar{\rho}_i^2 \sin \bar{\phi}_i \Delta\rho_i \Delta\theta_i \Delta\phi_i = \iiint\limits_{S} f(\rho, \theta, \phi)\rho^2 \sin \phi \, d\rho \, d\theta \, d\phi$$

It is important to remember that the integrand contains the factor $\rho^2 \sin \phi$. This is due to the fact that in spherical coordinates,

$$dV = \rho^2 \sin \phi \, d\rho \, d\theta \, d\phi$$

EXAMPLE 3   Find the volume that is cut from a sphere whose radius is 1 by a cone that makes an angle of 30° with the positive z-axis.

Solution   From Figure 45, we see that the limits on the variables of integration ρ, θ, and ϕ are

$$0 \le \rho \le 1 \qquad 0 \le \theta \le 2\pi \qquad 0 \le \phi \le \pi/6$$

Thus,

$$V = \int_0^{\pi/6} \left\{ \int_0^{2\pi} \left[\int_0^1 \rho^2 \sin \phi \, d\rho \right] d\theta \right\} d\phi = \frac{2\pi}{3}\left(1 - \frac{\sqrt{3}}{2} \right) \text{ cubic units}$$

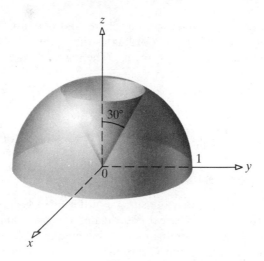

Figure 45   ■

EXAMPLE 4   Use spherical coordinates to evaluate the integral

$$\iiint_S z \, dx \, dy \, dz$$

where S is defined by the inequalities

$$\sqrt{x^2 + y^2} \le z \qquad x^2 + y^2 + z^2 \le 1 \qquad z \ge 0$$

Solution S can be described in spherical coordinates as follows: $z \ge 0$ corresponds to $\phi < \pi/2$, and $x^2 + y^2 + z^2 \le 1$ corresponds to $\rho \le 1$ (it is the upper half of the hemispherical solid $x^2 + y^2 + z^2 \le 1$). For the surface $z = \sqrt{x^2 + y^2}$ (which is a half cone), we get $x^2 + y^2 = z^2$. The cross section of this cone with the xz-plane is the pair of lines $z^2 = x^2$ or $z = \pm x$, which form an angle $\phi = \pi/4$ with the positive z-axis. Since $z = \rho \cos \phi$, our integral becomes

$$\iiint_S z \, dx \, dy \, dz = \iiint_S \rho \cos \phi \, \rho^2 \sin \phi \, d\rho \, d\theta \, d\phi$$

$$= \int_0^{2\pi} \left\{ \int_0^{\pi/4} \sin \phi \cos \phi \left[\int_0^1 \rho^3 \, d\rho \right] d\phi \right\} d\theta$$

$$= \frac{1}{4} \int_0^{2\pi} \left[\int_0^{\pi/4} \frac{1}{2} \sin 2\phi \, d\phi \right] d\theta = \frac{1}{16} \int_0^{2\pi} d\theta = \frac{\pi}{8} \qquad ■$$

EXAMPLE 5   Find the mass of a body in the shape of a sphere if the mass density δ^* is proportional to the distance from the center.

---

* Since we are using ρ as a spherical coordinate here, we will use δ as the symbol for mass density in order to avoid confusion.

Solution   Let a be the radius of the sphere, and position the sphere so that its center is at the origin. In spherical coordinates the equation of this sphere is $\rho = a.$ The density δ of the sphere is $\delta = k\rho.$ The mass m is given by the formula

$$m = \iiint_S \delta \, dV = k \iiint_S \rho\rho^2 \sin\phi \, d\rho \, d\phi \, d\theta = k \int_0^{2\pi} \left\{ \int_0^\pi \left[\int_0^a \rho^3 \sin\phi \, d\rho \right] d\phi \right\} d\theta$$

$$= \frac{ka^4}{4} \int_0^{2\pi} \left[\int_0^\pi \sin\phi \, d\phi \right] d\theta = \frac{ka^4}{4} \int_0^{2\pi} 2 \, d\theta = k\pi a^4 \quad \blacksquare$$

SPHERICAL COORDINATES IN NAVIGATION

If we assume that the surface of the earth is spherical, there is a simple relationship between the spherical coordinates we have defined and the system of latitude and longitude measurements used in geography. The origin is placed at the center of the earth and the z-axis is chosen to be the diameter through the North and South Poles. See Figure 46(a). The equator is then the great circle in the xy-plane. The x-axis is chosen so that the xz-plane will pass through the Greenwich Observatory in London. The longitude for a point P on the surface of the earth is then the angle we have called θ, except that degree measure is used and east and west are measured from the great circle through the poles and Greenwich. London itself thus has longitude 0°.

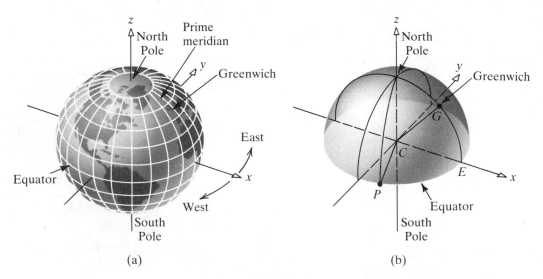

(a)   (b)

Figure 46

The latitude for a point on the surface of the earth north of the equator is $90° - \phi;$ thus, latitude is measured from the equator rather than from the north pole. For a point on the earth south of the equator, the latitude is $\phi - 90°.$ Points on the equator have $\phi = 90°$ and latitude $0°$. For the north pole, we write $\phi = 0°$ and north latitude $90°$; for the south pole, we write $\phi = 180°$ and south latitude $90°$.

EXAMPLE 6   Greenwich has a longitude of $0°$ and an approximate latitude of $51\frac{1}{2}°$ N. If we use $R_e = 4000$ miles for the radius of the earth and use degrees instead of radians, the spherical coordinates for Greenwich are $(4000, 0°, 38\frac{1}{2}°)$. See Figure 46(b), where $ECG = 51\frac{1}{2}°$.   ∎

EXERCISE 9

In Problems 1–8 find the spherical coordinates of the point with the given rectangular coordinates.

1. $(-\sqrt{2}, -\sqrt{2}, 2\sqrt{3})$ **2.** $(-1, \sqrt{3}, 2)$ **3.** $(1, 1, \sqrt{2})$ **4.** $(1, -\sqrt{3}, -2)$

5. $(1, 2, 3)$ (Use a calculator to find the answer to the nearest 0.01.)

6. $(1, -1, \sqrt{2})$ **7.** $(0, 3\sqrt{3}, 3)$ **8.** $(-5\sqrt{3}, 5, 0)$

In Problems 9 and 10 fill in the missing coordinates. Assume that the three ordered triples represent the rectangular, cylindrical, and spherical coordinates of some point in space.

9. (x, y, z) (r, θ, z) (ρ, θ, ϕ)
     $(-1, \quad , \quad)$ $(\quad , \quad , -2\sqrt{3}/3)$ $(\quad , 4\pi/3, \quad)$

10. (x, y, z) (r, θ, z) (ρ, θ, ϕ)
     $(\quad , \quad , 2\sqrt{2})$ $(\sqrt{8}, \quad , \quad)$ $(\quad , \pi/6, \quad)$

In Problems 11–14 evaluate each iterated integral.

11. $\int_0^{\pi/2} \left\{ \int_0^{\sin \phi} \left[\int_0^{\pi/4} \rho^2 \sin \phi \, d\theta \right] d\rho \right\} d\phi$

12. $\int_0^{\pi/2} \left\{ \int_0^{\pi/2} \left[\int_0^{\sin \phi} \rho^2 \sin \phi \cos \phi \, d\rho \right] d\theta \right\} d\phi$

13. $\int_0^{2\pi} \left\{ \int_0^{\pi/4} \left[\int_0^{\sec \phi} \rho^2 \sin^2 \phi \, d\rho \right] d\phi \right\} d\theta$

14. $\int_0^{\pi/4} \left\{ \int_0^{\cos \phi} \left[\int_0^{2\pi} \rho^2 \sin \phi \, d\theta \right] d\rho \right\} d\phi$

In Problems 15–18 use triple integration in spherical coordinates.

15. Find the mass of a homogeneous body in the shape of a sphere of radius a.

16. Find the mass of a body in the shape of a sphere of radius a, if the mass density is proportional to the square of the distance from the center.

17. Find the center of mass of a body in the shape of a hemisphere of radius a, if the mass density is proportional to the distance from the center.

18. Find the center of mass of a body in the shape of a hemisphere, if the mass density is proportional to the distance from the axis of symmetry.

In Problems 19 and 20 use either cylindrical or spherical coordinates to evaluate each triple integral.

19. $\int_0^2 \left\{ \int_0^2 \left[\int_0^{\sqrt{4-x^2}} \sqrt{x^2 + y^2} \, dy \right] dx \right\} dz$

20. $\int_0^2 \left\{ \int_0^{\sqrt{4-y^2}} \left[\int_0^{\sqrt{4-x^2-y^2}} \frac{2z}{\sqrt{x^2 + y^2}} \, dz \right] dx \right\} dy$

21. The magnitude of the resultant gravitational force of a solid hemisphere of radius a and constant mass density δ on a unit mass particle situated at the center of the base of the hemisphere is given by the triple integral

$$F = k\delta \iiint\limits_{V} \frac{\cos \phi}{\rho^2} \, dV$$

where the center of the sphere is at the origin and spherical coordinates are used. Evaluate the integral.

22. Use spherical coordinates to integrate $f(x, y, z) = \sqrt{x^2 + y^2 + z^2}$ over the region above the cone $z = -\sqrt{3x^2 + 3y^2}$ and inside the sphere $x^2 + y^2 + z^2 = 4$.

In Problems 23 and 24 set up the triple integral $\iiint_S f(x, y, z) \, dV$ for S in rectangular, cylindrical, and spherical coordinates.

23. S is the solid sphere of radius a with center at the origin.

24. S is the region inside the cylinder $x^2 + y^2 = 4$ and inside the sphere $x^2 + y^2 + z^2 = 9$.

The integrals in Problems 25 and 26 are given in spherical coordinates. Express each in rectangular coordinates. Do not evaluate.

25. $\int_0^\pi \left\{ \int_{3\pi/4}^\pi \left[\int_0^4 \rho^5 \cos \theta \sin^2 \phi \, d\rho \right] d\phi \right\} d\theta$

26. $\int_{\pi/2}^{3\pi/2} \left\{ \int_{\pi/2}^\pi \left[\int_0^2 \frac{\rho^4 \sin \phi \cos \phi}{\rho^2 + 3} \, d\rho \right] d\phi \right\} d\theta$

27. The integral below is given in cylindrical coordinates. Express it in spherical coordinates. Do not evaluate.

$$\int_0^\pi \left\{ \int_0^3 \left[\int_0^{\sqrt{9-z^2}} r^2 \sin \theta \, dr \right] dz \right\} d\theta$$

28. A solid occupies the region $\sqrt{x^2 + y^2} \le z \le 1$ and has density $\delta(x, y, z) = z\sqrt{x^2 + y^2 + z^2}$. Determine its mass.

29. Find the volume of the solid enclosed on the outside by the sphere $\rho = 2$ and on the inside by the surface $\rho = 1 + \cos \phi$.

30. Use spherical coordinates to find the centroid of the hemisphere of radius a, whose base is on the xy-plane.

Miscellaneous Exercises

In Problems 1–4 evaluate each integral.

1. $\int_0^1 \left[\int_y^1 ye^{-x^3} \, dx \right] dy$

2. $\int_0^1 \left[\int_0^{\sqrt{1+x^2}} \frac{dy}{x^2 + y^2 + 1} \right] dx$

3. $\int_0^{\pi/4} \left[\int_0^{\tan x} \sec x \, dy \right] dx$

4. $\int_0^1 \left[\int_{-x}^x e^{x+y} \, dy \right] dx$

5. Use cylindrical coordinates to evaluate: $\int_0^a \{ \int_0^{\sqrt{a^2-x^2}} [\int_0^{\sqrt{a^2-x^2-y^2}} dz] \, dy \} \, dx$

6. Rework Problem 5 using spherical coordinates.

7. Let R be the square region $-2 \le x \le 2$, $-2 \le y \le 2$ in the xy-plane, and suppose that $f(x, y) = xy$.
   (a) Using a partition of R into four squares of equal area and evaluating f at the midpoint of each square, compute the corresponding Riemann sum of f over R.
   (b) Using the same partition as in part (a), but evaluating f at an arbitrary point of each square, determine the largest possible Riemann sum and the smallest. What is the average of these values?
   (c) What is the actual value of $\iint_R f(x, y) \, dA$?

8. Suppose that $f(x, y)$ is integrable over a rectangular region R in the xy-plane, and let P be a partition of R into n subrectangles of equal area ΔA. Evaluating f at the midpoint (u_i, v_i) of the ith subrectangle $(i = 1, 2, \ldots, n)$, let $M(P)$ be the average of these n functional values.
   (a) Show that $M(P) = (1/A) \sum_{i=1}^{n} f(u_i, v_i)\Delta A$, where A is the area of R.
   (b) Explain why $\lim_{\|P\| \to 0} M(P) = (1/A) \iint_R f(x, y)\, dA$. (This is called the *average value of f over R*.)

9. By analogy with Problem 8 we define the average value of f over a region R that is not necessarily rectangular to be the number $(1/A) \iint_R f(x, y)\, dA$.
   (a) In single variable calculus the average value of a function f over the interval $a \le x \le b$ is defined to be the number $[1/(b - a)] \int_a^b f(x)\, dx$. In what sense is this a special case of the above definition of the average value of f over R?
   (b) Let $f(x, y, z)$ be integrable over the region S in space. What definition would you give for the average value of f over S?

10. Suppose that a homogeneous lamina of mass density 1 occupies a region R in the xy-plane. Show that the average value of $f(x, y) = x$ over R (as defined in Problem 9) is $\bar{x}$, the first coordinate of the center of mass of the lamina. What is the average value of $g(x, y) = y$ over R?

11. Evaluate $\iint_R x \sin y^3\, dA$, where R is the triangle with vertices at $(0, 0)$, $(0, 2)$, and $(2, 2)$.

12. Reverse the order of integration in the iterated integral $\int_1^e \left[\int_0^{\ln x} (1/x)\, dy \right] dx$ and evaluate it both ways.

13. Reverse the order of integration in the iterated integral $\int_0^{\pi/2} \left[\int_0^{\cos x} \sin x\, dy \right] dx$ and evaluate it both ways.

14. Find the volume of the tetrahedron enclosed by the coordinate planes and the plane $(x/a) + (y/b) + (z/c) = 1$. (Assume that a, b, and c are positive.)

15. Use multiple integration to show that the volume of the ellipsoid $(x^2/a^2) + (y^2/b^2) + (z^2/c^2) = 1$ is $\frac{4}{3}\pi abc$. (Assume that a, b, and c are positive.) What does this formula reduce to if $a = b = c$?

16. Find the volume of the solid in the first octant enclosed by the coordinate planes, $a^2 y = b(a^2 - x^2)$, and $a^2 z = c(a^2 - x^2)$.

17. The mass density at each point of a circular washer of inner radius a and outer radius b is inversely proportional to the square of the distance from the center.
   (a) Find the mass. Then discuss its behavior as $a \to 0$. (Note that the mass density becomes infinite as we approach the center of the washer.)
   (b) Find the moment of inertia of the washer about its center, and show that (unlike the mass) it remains finite as $a \to 0$.

18. Find the mass of a square lamina of side a, if the mass density is proportional to the distance from a fixed vertex to the square.

19. Find the center of mass of the homogeneous body enclosed by $bx^2 = a^2 y$ and $ay = bx$.

20. Find the center of mass of the homogeneous hemispherical shell $a \le r \le b$, $0 \le \phi \le \pi/2$.

21. Derive the formula $S = \pi a \sqrt{a^2 + h^2}$ for the lateral surface area of a right circular cone of base radius a and altitude h.

22. Show that the area of the first-octant portion of the plane $(x/a) + (y/b) + (z/c) = 1$ (where a, b, and c are positive) is $S = \frac{1}{2}\sqrt{a^2 b^2 + b^2 c^2 + c^2 a^2}$.

23. Describe the solid S whose volume is given by the iterated integral $\int_0^1 \{\int_{y^2}^1 [\int_0^{1-x} dz]\, dx\}\, dy$. Set up the other five iterated integrals for the volume of S.

24. Describe the solid S whose volume is given by the iterated integral $\int_0^1 \{\int_0^{x^2} [\int_0^y dz]\, dy\}\, dx$. Set up the other five iterated integrals for the volume of S.

25. Show that $\int_a^b \left[\int_a^y f(x) \, dx \right] dy = \int_a^b (b - x) f(x) \, dx.$

26. Show that $\int_a^b \left\{ \int_a^z \left[\int_a^y f(x) \, dx \right] dy \right\} dz = \int_a^b \left[(b - x)^2 / 2 \right] f(x) \, dx.$

27. To calculate $\int_{-\infty}^{+\infty} e^{-x^2} \, dx,$ let $I_a = \int_0^a e^{-x^2} \, dx.$

    (a) Show that $I_a^2 = \int_0^a e^{-x^2} \, dx \int_0^a e^{-y^2} \, dy = \int_0^a \int_0^a e^{-(x^2 + y^2)} \, dx \, dy.$

    (b) Let $J_a = \iint_R e^{-(x^2 + y^2)} \, dA,$ where R is the quarter circle $0 \le \theta \le \pi/2,$ $0 \le r \le a.$ Show that

$$\left| I_a^2 - J_a \right| < \frac{(4 - \pi) a^2}{4} e^{-a^2}$$

    (c) Evaluate $J_a.$

    (d) Show that $\lim_{a \to +\infty} J_a = \pi/4$ and $\lim_{a \to +\infty} \left| I_a^2 - J_a \right| = 0.$ Thus, show that $\int_{-\infty}^{+\infty} e^{-x^2} \, dx = \sqrt{\pi}.$
This integral is of special importance in statistics.

28.*(a) Prove that the moment of inertia of the thin flat plate in the figure, about the z-axis, equals the sum of its moments of inertia about the x- and y-axes.

    (b) Given that the moment of inertia of a disk about an axis through its center and perpendicular to its plane is $mR^2/2$ (where m is mass and R is the radius), use the relation from part (a) to find its moment of inertia about a diameter.

    (c) Derive the result of part (b) by direct integration of the defining equation $I = \int r^2 \, dm.$

    (d) What is the moment of inertia of a disk about an axis tangent to its edge?

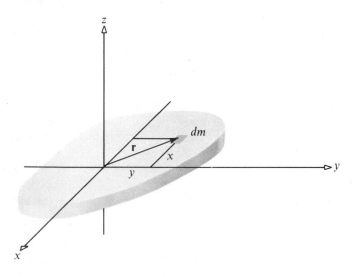

29. Evaluate $\int_1^2 \left\{ \int_0^1 \left[1/(x^2 \sqrt{x^2 + y^2}) \right] dy \right\} dx.$ Use polar coordinates.

30. Evaluate $\iint_R e^{\sqrt{x}} \, dA,$ where R is the region enclosed by $y = x,$ $y = 0,$ and $x = 1.$

---

* Adapted from F. W. Sears, M. W. Zemansky, and H. D. Young, *University Physics* (Reading, Mass.: Addison-Wesley Publishing Co., 1976) p. 177. Reprinted by permission.

19

Topics in
Vector Calculus

1. Line Integrals

In Chapter 5 we defined the definite integral $\int_a^b f(x)\,dx$ whose value is determined by the values of the function f on the closed interval $[a, b]$. In this section we define the *line integral** in the plane, denoted by $\int_C f(x, y)\,ds$, whose value is determined by the values of a function of two variables along a curve C. Similarly, the value of a line integral in three-dimensional space is determined by the values of a function of three variables along a curve in space.

We now proceed to develop the definition of the line integral in the plane. Let C be a smooth curve whose parametric equations are given by

$$x = x(t) \qquad y = y(t) \qquad a \le t \le b$$

See Figure 1. As t increases from a to b, the corresponding points $(x(t), y(t))$ trace out the curve C from the point $A = (x(a), y(a))$ to the point $B = (x(b), y(b))$,

---

* Such an integral might better be called a *curve integral*, but we shall follow the usual practice and use the term *line integral*.

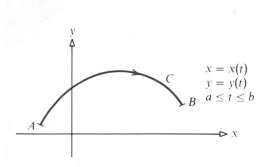

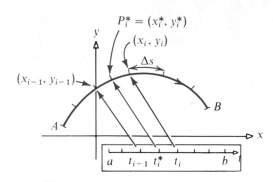

Figure 1   **Figure 2**

that is, the orientation of C is from A to B. Let $f(x, y)$ be a function that is defined and continuous at each point on the curve C.

Partition the closed interval $[a, b]$ into n subintervals,

$$[a, t_1], \quad [t_1, t_2], \quad \ldots, \quad [t_{i-1}, t_i], \quad \ldots, \quad [t_{n-1}, b]$$

and denote the length of each subinterval by $\Delta t_1, \Delta t_2, \ldots, \Delta t_n$. Corresponding to each number $a = t_0, t_1, \ldots, t_n = b$, of the partition, there is a succession of points $P_0, P_1, \ldots, P_n$ on the curve C. These points divide the curve into n subarcs of lengths $\Delta s_1, \Delta s_2, \Delta s_3, \ldots, \Delta s_n$ (see Fig. 2).

Let $P_i^* = (x_i^*, y_i^*)$ be an arbitrary point on the ith subarc, corresponding to the number t_i^*, $t_{i-1} \leq t_i^* \leq t_i$. We define the *norm* $\|\Delta\|$ of the partition to be the largest arc length Δs_i. Now form the sum

(19.1)
$$\sum_{i=1}^{n} f(x_i^*, y_i^*)\Delta s_i$$

If this sum has a limit L as $\|\Delta\| \to 0$, and if this limit is the same for all partitions of $[a, b]$ and all choices of P_i^*, then the number L is called the *line integral of f along C from A to B* and is denoted by

(19.2)
$$\int_C f(x, y)\, ds = \lim_{\|\Delta\| \to 0} \sum_{i=1}^{n} f(x_i^*, y_i^*)\Delta s_i$$

An easily understood physical application of the line integral (19.2) is the problem of finding the mass of a long thin piece of wire of variable density, whose shape is described by the curve C in Figure 2. Suppose the linear density (mass per unit length) at the point (x, y) is $f(x, y)$. Within each subinterval $[t_{i-1}, t_i]$ of $[a, b]$ we arbitrarily choose a number t_i^*. Then the mass of the corresponding short piece of wire between (x_{i-1}, y_{i-1}) and (x_i, y_i) can be approximated by $\Delta m_i = f(x_i^*, y_i^*)\Delta s_i$, where Δs_i is the length of the short piece and (x_i^*, y_i^*) is the point corresponding to t_i^*. The sum

$$\sum_{i=1}^{n} f(x_i^*, y_i^*)\Delta s_i$$

over all the subintervals is an approximation of the mass m of the wire. This approximation can be made as close as we please to m by taking the subintervals to be sufficiently small and sufficiently numerous (that is, by taking the limit as $\|\Delta\|$ approaches 0). Thus,

$$\int_C f(x, y)\, ds = m$$

is the exact value of the mass of the wire.

EVALUATION OF LINE INTEGRALS

We state, without proof, a method for calculating the value of a line integral when the curve C and the function f are specifically given.

(19.3)   **Let C be a smooth curve whose parametric equations are given by**

$$x = x(t) \qquad y = y(t) \qquad a \le t \le b$$

Let $f(x, y)$ be a function that is defined and continuous on C. Then the line integral of f along C from A to B is given by the formula

(19.4)

$$\int_C f(x, y)\, ds = \int_a^b f(x(t), y(t)) \sqrt{\left(\frac{dx}{dt}\right)^2 + \left(\frac{dy}{dt}\right)^2}\, dt$$

This result expresses the fact that under certain conditions a line integral is equal to a definite integral. When this is the case, the definite integral may then be evaluated by techniques already studied.

Although equation (19.4) seems complicated, it is simply the result of substituting the parametric equations $x = x(t)$, $y = y(t)$ of C for x and y in f and using the arc length formula for ds developed in Chapter 13. Here is a concrete example to illustrate the technique.

EXAMPLE 1   Evaluate $\int_C y\, ds$, if C is the curve given by the parametric equations $x = t$, $y = \sqrt{t}$, $2 \le t \le 6$.

Solution   The curve C is part of a parabola. The element ds of arc length along C is given by

$$ds = \sqrt{\left(\frac{dx}{dt}\right)^2 + \left(\frac{dy}{dt}\right)^2}\, dt$$

where

$$\frac{dx}{dt} = 1 \qquad \text{and} \qquad \frac{dy}{dt} = \frac{1}{2\sqrt{t}}$$

Thus,

$$ds = \sqrt{1 + \frac{1}{4t}}\, dt = \sqrt{\frac{4t + 1}{4t}}\, dt = \frac{\sqrt{4t + 1}}{2\sqrt{t}}\, dt$$

By applying (19.4), we obtain

$$\underset{\underset{y=\sqrt{t}}{\uparrow}}{\int_C y\, ds} = \int_2^6 \sqrt{t}\, \frac{\sqrt{4t+1}}{2\sqrt{t}}\, dt = \frac{1}{2} \int_2^6 \sqrt{4t+1}\; dt$$

$$= \left(\frac{1}{8}\right) \frac{(4t+1)^{3/2}}{\frac{3}{2}} \bigg|_2^6 = \frac{49}{6} \qquad \blacksquare$$

It can be shown that the value of the line integral along the curve C *does not* depend on the parametric representation of the curve; all parameterizations satisfying the assumptions made on $x = x(t)$ and $y = y(t)$ in (19.3) will give the same value. The next example illustrates this phenomenon.

EXAMPLE 2   Evaluate $\int_C (x^2 + y)\, ds$, if C is the line segment from $(0, 0)$ to $(1, 2)$ and C is parameterized as

(a) $x = t,\quad y = 2t,\quad 0 \le t \le 1$ (b) $x = \sin t,\quad y = 2 \sin t,\quad 0 \le t \le \pi/2$

Solution   We note that the two parameterizations of C are merely different ways of representing a part of the line $y = 2x$ from $(0, 0)$ to $(1, 2)$.

(a)   Since $dx/dt = 1$, $dy/dt = 2$, the differential ds of arc length is given by $ds = \sqrt{1^2 + 2^2}\, dt = \sqrt{5}\, dt$, so that, by (19.4),

$$\underset{\underset{\substack{x=t\\y=2t}}{\uparrow}}{\int_C (x^2 + y)\, ds} = \int_0^1 (t^2 + 2t)\sqrt{5}\, dt = \sqrt{5}\left(\frac{t^3}{3} + t^2\right)\bigg|_0^1 = \frac{4\sqrt{5}}{3}$$

(b)   With $dx/dt = \cos t$, $dy/dt = 2 \cos t$, the differential ds of arc length is $ds = \sqrt{\cos^2 t + 4 \cos^2 t} = \sqrt{5 \cos^2 t} = \sqrt{5} \cos t$. By using (19.4), we have

$$\underset{\underset{\substack{x=\sin t\\y=2\sin t}}{\uparrow}}{\int_C (x^2 + y)\, ds} = \int_0^{\pi/2} (\sin^2 t + 2 \sin t)\sqrt{5} \cos t\, dt$$

$$= \sqrt{5} \int_0^{\pi/2} (\sin^2 t + 2 \sin t) \cos t\, dt$$

$$= \sqrt{5}\left(\frac{\sin^3 t}{3} + \sin^2 t\right)\bigg|_0^{\pi/2} = \frac{4\sqrt{5}}{3} \qquad \blacksquare$$

OTHER TYPES OF LINE INTEGRALS

If we replace Δs_i by Δx_i or Δy_i in (19.2), we obtain two other types of line integrals:

(19.5) $$\int_C f(x, y)\, dx = \lim_{\|\Delta\| \to 0} \sum_{i=1}^n f(x_i^*, y_i^*)\Delta x_i$$

$$(19.6) \qquad \int_C f(x, y)\, dy = \lim_{||\Delta|| \to 0} \sum_{i=1}^{n} f(x_i^*, y_i^*)\Delta y_i$$

We refer to (19.5) as the *line integral of f along C with respect to x;* and (19.6) is called the *line integral of f along C with respect to y.* In general, (19.5) and (19.6) have different values.

EXAMPLE 3   Evaluate $\int_C (x - 3y)\, dx$ and $\int_C (x - 3y)\, dy$, if C is the part of the parabola $x = y^2$ that joins the points $(1, 1)$ and $(4, 2)$.

Solution   We can write the parametric equations of the curve C as $x = t^2, \quad y = t, \quad 1 \le t \le 2$. Hence, $dx = 2t\, dt, \quad dy = dt, \quad$ and

$$\int_C (x - 3y)\, dx = \int_1^2 (t^2 - 3t)2t\, dt = -\frac{13}{2}$$

$$\int_C (x - 3y)\, dy = \int_1^2 (t^2 - 3t)\, dt = -\frac{13}{6} \qquad \blacksquare$$

It is not always necessary to use the parametric equations for C to evaluate the line integral of some function f. Let's see how (19.5) and (19.6) look if the curve C is given by a rectangular equation.

If C is given by the rectangular equation $y = g(x), \quad a \le x \le b$, a parameterization of C is

$$x = t \qquad y = g(t) \qquad a \le t \le b$$

With this parameterization, (19.5) and (19.6) may be written as

$$(19.7) \qquad \int_C f(x, y)\, dx = \int_a^b f(t, g(t))\, dt = \int_a^b f(x, g(x))\, dx$$

$$(19.8) \qquad \int_C f(x, y)\, dy = \int_a^b f(t, g(t))g'(t)\, dt = \int_a^b f(x, g(x))g'(x)\, dx$$

For example, the curve C in Example 3 has the rectangular equation $y = \sqrt{x}$, $1 \le x \le 4$. If we use (19.7) and (19.8) to evaluate $\int_C (x - 3y)\, dx$ and $\int_C (x - 3y)\, dy$, the computation is a little easier than the solution we gave above:

$$\int_C (x - 3y)\, dx = \int_1^4 (x - 3\sqrt{x})\, dx = -\frac{13}{2}$$

$$\int_C (x - 3y)\, dy = \int_1^4 (x - 3\sqrt{x})\left(\frac{1}{2\sqrt{x}}\right) dx = -\frac{13}{6}$$

LINE INTEGRALS OF THE FORM $\int_C (P\, dx + Q\, dy)$

Let C denote a smooth curve, and let P and Q each be a function of two variables obeying the conditions on f stated in (19.3). We define a line integral of the form

$\int_C (P\,dx + Q\,dy)$ as

(19.9)
$$\int_C (P\,dx + Q\,dy) = \int_C P(x, y)\,dx + \int_C Q(x, y)\,dy$$

and refer to it as the *line integral of $P\,dx + Q\,dy$ along C.*

EXAMPLE 4   Evaluate the line integral $\int_C (y^2\,dx - x^2\,dy)$ along the two curves with common endpoints given below.

C_1:   The parabola $x = t, \quad y = t^2$ joining the two points $(0, 0)$ and $(2, 4)$

C_2:   The line $x = t, \quad y = 2t$ joining the two points $(0, 0)$ and $(2, 4)$

Solution   Along C_1, we have

$$\int_{C_1} (y^2\,dx - x^2\,dy) = \int_0^2 [t^4\,dt - t^2(2t\,dt)] = \int_0^2 (t^4 - 2t^3)\,dt = -\frac{8}{5}$$

Along C_2, we have

$$\int_{C_2} (y^2\,dx - x^2\,dy) = \int_0^2 [4t^2\,dt - t^2(2\,dt)] = \int_0^2 2t^2\,dt = \frac{16}{3} \quad \blacksquare$$

We observe that the value of the line integral in Example 4 depends on the curve C over which the integration takes place. In the next section we investigate conditions under which the value of the integral is independent of path, that is, where the value of the integral depends only on the endpoints of the curve.

ORIENTATION

In evaluating a line integral over a curve C, the orientation of C plays a role. If C is a smooth curve, let $-C$ denote the same curve, but with reverse orientation. Then

$$\int_C (P\,dx + Q\,dy) = -\int_{-C} (P\,dx + Q\,dy)$$

Thus, a reversal of orientation alters the line integral by a factor of -1 (see Fig. 3).

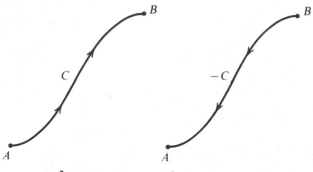

Figure 3 $\int_C (P\,dx + Q\,dy) = -\int_{-C} (P\,dx + Q\,dy)$

PIECEWISE SMOOTH CURVES

In the definition of a line integral (19.3), it is assumed that the curve C is a smooth curve. That is, the two parametric equations $x(t)$ and $y(t)$ defining C have continuous derivatives dx/dt and dy/dt, not simultaneously zero on $a \le t \le b$, and the curve does not intersect itself, except possibly at the endpoints. If a curve C consists of a finite number of arcs of smooth curves, say, $C_1, C_2, \ldots, C_n$ that do not intersect themselves but are joined together to form a curve (see Fig. 4), then we say that C is *piecewise smooth*.

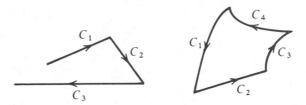

Figure 4   Two piecewise smooth curves

If P and Q are two functions defined and continuous on C and if C is a *piecewise smooth curve*, then

$$(19.10) \quad \int_C (P\,dx + Q\,dy) = \int_{C_1} (P\,dx + Q\,dy) + \int_{C_2} (P\,dx + Q\,dy) + \cdots + \int_{C_n} (P\,dx + Q\,dy)$$

EXAMPLE 5   Evaluate $\int_C (xy\,dx + x^2\,dy)$ along the piecewise smooth curve C illustrated in Figure 5.

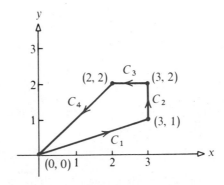

Figure 5

Solution   The values of the line integral along each of the smooth curves C_1, C_2, C_3, and C_4 are

$$C_1: \quad y = \frac{x}{3}, \qquad dy = \frac{dx}{3}, \qquad 0 \le x \le 3$$

$$\int_{C_1} (xy\,dx + x^2\,dy) = \int_0^3 \left[x\left(\frac{x}{3}\right) dx + x^2\,\frac{dx}{3} \right] = 6$$

$$C_2: \quad x = 3, \qquad dx = 0, \qquad 1 \le y \le 2$$

$$\int_{C_2} (xy\,dx + x^2\,dy) = \int_1^2 9\,dy = 9$$

$$C_3: \quad y = 2, \qquad dy = 0 \qquad \text{Watch the orientation here: } x \text{ varies from 3 to 2}$$

$$\int_{C_3} (xy\,dx + x^2\,dy) = \int_3^2 2x\,dx = -5$$

$$C_4: \quad y = x, \qquad dy = dx \qquad \text{Watch the orientation here: } x \text{ varies from 2 to 0}$$

$$\int_{C_4} (xy\,dx + x^2\,dy) = \int_2^0 (x^2\,dx + x^2\,dx) = -\frac{16}{3}$$

Thus,

$$\int_C (xy\,dx + x^2 dy) = 6 + 9 - 5 - \frac{16}{3} = \frac{14}{3} \qquad \blacksquare$$

LINE INTEGRALS IN SPACE

We now consider line integrals in space of the form

(19.11) $$\int_C [P(x, y, z)\,dx + Q(x, y, z)dy + R(x, y, z)dz]$$

where C is a smooth curve defined by the parametric equations

$$x = x(t) \qquad y = y(t) \qquad z = z(t) \qquad a \le t \le b$$

and the functions P, Q, and R are each continuous along C.* Since the definition of a line integral in space is analogous to that in the plane, we shall not give the details here. Instead, we look at an example that illustrates a method for evaluating line integrals in this setting.

EXAMPLE 6   Evaluate $\int_C [xy^2\,dx + x^2z\,dy - (y - x)\,dz]$, where C is the twisted cubic

$$x = t \qquad y = t^2 \qquad z = t^3 \qquad 0 \le t \le 1$$

---

* We will not consider the more general form of the line integral $\int_C f(x, y, z)\,ds$, where f is continuous on C and $ds = \sqrt{[x'(t)]^2 + [y'(t)]^2 + [z'(t)]^2}\,dt$.

Solution   For $x = t$, $y = t^2$, $z = t^3$,   we   find   that $dx = dt$, $dy = 2t\,dt$, $dz = 3t^2\,dt$, so that

$$\int_C [xy^2\,dx + x^2z\,dy - (y - x)\,dz] = \int_0^1 [t^5 + 2t^6 - 3t^2(t^2 - t)]\,dt$$

$$= \int_0^1 [2t^6 + t^5 - 3t^4 + 3t^3]\,dt$$

$$= \left(\frac{2t^7}{7} + \frac{t^6}{6} - \frac{3t^5}{5} + \frac{3t^4}{4}\right)\Bigg|_0^1 = \frac{253}{420} \quad\blacksquare$$

VECTOR NOTATION FOR LINE INTEGRALS

The line integral (19.11) can be expressed in vector notation, as follows: If

$$\mathbf{F}(x, y, z) = P(x, y, z)\mathbf{i} + Q(x, y, z)\mathbf{j} + R(x, y, z)\mathbf{k}$$

and the curve C is defined by the vector function

$$\mathbf{r}(t) = x(t)\mathbf{i} + y(t)\mathbf{j} + z(t)\mathbf{k} \qquad a \le t \le b$$

then

(19.12)
$$\int_C P\,dx + Q\,dy + R\,dz = \int_a^b [P(x(t), y(t), z(t))x'(t) + Q(x(t), y(t), z(t))y'(t)$$

$$+ R(x(t), y(t), z(t))z'(t)]\,dt$$

$$= \int_a^b \mathbf{F}(\mathbf{r}(t)) \cdot \mathbf{r}'(t)\,dt$$

$$= \int_C \mathbf{F} \cdot d\mathbf{r}$$

The second integral expression in (19.12) is the *parametric form* of the line integral; the last expression is the *vector form*.

Similarly, when the vector function *F* is defined in the plane by

$$\mathbf{F}(x, y) = P(x, y)\mathbf{i} + Q(x, y)\mathbf{j}$$

and the curve C is defined by the vector function

$$\mathbf{r}(t) = x(t)\mathbf{i} + y(t)\mathbf{j} \qquad a \le t \le b$$

then

(19.13)
$$\int_C P\,dx + Q\,dy = \int_a^b [P(x(t), y(t))x'(t) + Q(x(t), y(t))y'(t)]\,dt$$

$$= \int_C \mathbf{F} \cdot d\mathbf{r}$$

We shall have occasion a little later to utilize this compact way of writing a line integral; for the time being, an example will demonstrate the mechanics of its use.

EXAMPLE 7   Evaluate $\int_C \mathbf{F} \cdot d\mathbf{r}$ if $\mathbf{F}(x, y, z) = xy^2\mathbf{i} + x^2z\mathbf{j} - (y - x)\mathbf{k}$ and the curve C is $\mathbf{r}(t) = t\mathbf{i} + t^2\mathbf{j} + t^3\mathbf{k}, \quad 0 \le t \le 1$.

Solution   The parametric equations of the curve C are

$$x = t \qquad y = t^2 \qquad z = t^3$$

Hence,

$$\mathbf{F} = xy^2\mathbf{i} + x^2z\mathbf{j} - (y - x)\mathbf{k} = t^5\mathbf{i} + t^5\mathbf{j} - (t^2 - t)\mathbf{k}$$

and,

$$d\mathbf{r} = \frac{d\mathbf{r}}{dt}\,dt = (\mathbf{i} + 2t\mathbf{j} + 3t^2\mathbf{k})\,dt$$

Therefore,

$$\mathbf{F} \cdot d\mathbf{r} = [t^5\mathbf{i} + t^5\mathbf{j} - (t^2 - t)\mathbf{k}] \cdot (\mathbf{i} + 2t\mathbf{j} + 3t^2\mathbf{k})\,dt$$
$$= [t^5 + 2t^6 - 3t^2(t^2 - t)]\,dt = (2t^6 + t^5 - 3t^4 + 3t^3)\,dt$$

so that

$$\int_C \mathbf{F} \cdot d\mathbf{r} = \int_0^1 (2t^6 + t^5 - 3t^4 + 3t^3)\,dt = \frac{253}{420} \qquad \blacksquare$$

EXERCISE 1

In Problems 1–4 evaluate each line integral for the given function f and curve C.

1. $\int_C (x + y^2)\,ds$; C: $x = t, \quad y = t, \quad 0 \le t \le 1$

2. $\int_C (x + y^2)\,ds$; C: $x = \sin t, \quad y = \sin t, \quad 0 \le t \le \pi$

3. $\int_C x^2y\,ds$; C: $x = \cos t, \quad y = \cos t, \quad 0 \le t \le \pi/2$

4. $\int_C x\,ds$; C: $x = t, \quad y = 3t^2, \quad 0 \le t \le 1$

5. Evaluate $\int_C (x^2y\,dx + xy\,dy)$ along the curve C: $x^2 + y^2 = 1$ from $(1, 0)$ to $(0, 1)$, using the following parameterizations:
   (a) $x = \cos t, \quad y = \sin t, \quad 0 \le t \le \pi/2$ (b) $y = \sqrt{1 - x^2}, \quad 0 \le x \le 1$

6. Evaluate $\int_C [y\,dx + (x - 16y)\,dy]$ along each of the given curves from $(2, 0)$ to $(0, 4)$.
   (a) The straight line joining the two points
   (b) The parabola $y = 4 - x^2$
   (c) The straight line from $(2, 0)$ to $(2, 2)$, followed by the straight line from $(2, 2)$ to $(0, 4)$

7. Evaluate $\int_C [(x + 2y)\,dx + (2x + y)\,dy]$, where:
   (a) C is the curve $y = x^2$ from $(0, 0)$ to $(1, 1)$
   (b) C is the curve $y = x^3$ from $(0, 0)$ to $(1, 1)$
   (c) C is the curve $x = \cos t, \quad y = \sin t, \quad 0 \le t \le \pi/2$ from $(1, 0)$ to $(0, 1)$
   (d) C is the curve $x = 1 - t, \quad y = t, \quad 0 \le t \le 1$

8. Evaluate $\int_C (yz\,dx + xz\,dy + xy\,dz)$,   where C consists of line segments connecting the points $(0, 0, 0)$, $(1, 0, 0)$, $(1, 1, 0)$, and $(1, 1, 1)$, in that order.

9. Evaluate $\int_C [y^2\,dx + (xy - x^2)\,dy]$ from $(0, 0)$ to $(1, 3)$, where:
   (a) C is the line $y = 3x$ (b) C is the parabola $y^2 = 9x$

10. Evaluate $\int_C [(x^2 + y^2)\,dx + 3x^2y\,dy]$,   where C is the parabola $y = x^2$ from $(-2, 4)$ to $(2, 4)$.

11. Evaluate $\int_C [(x \cos y)\,dx - (y \sin x)\,dy]$,   where C consists of line segments connecting the points $(0, 0)$, $(1, 0)$, $(1, 1)$, and $(0, 1)$, in that order.

12. Evaluate $\int_C (z\,dx + x\,dy + y\,dz)$,   where C is the circular helix $x = a \cos t$, $y = a \sin t$, $z = t$, $0 \le t \le 2\pi$.

13. Let C_1 be the curve $x = \cos \theta$, $y = \sin \theta$, $0 \le \theta \le 4\pi$,   and let C_2 be the curve $x = \cos t^2$, $y = \sin t^2$, $0 \le t \le 2\sqrt{\pi}$.   For $P = x^3y$, $Q = y^2x$,   show that $\int_{C_1} (P\,dx + Q\,dy) = \int_{C_2} (P\,dx + Q\,dy)$.   Explain why this is so.

14. Evaluate $\int_C [ye^{xy}\,dx + \sin x\,dy + (xy/z)\,dz]$,   where C is the curve $x = t$, $y = t^2$, $z = t^3$, $1 \le t \le 3$.

15. Evaluate $\int_C [(yz/x)\,dx + e^y\,dy + \sin z\,dz]$,   where C is the curve $x = t^3$, $y = t$, $z = t^2$, $2 \le t \le 3$.

16. Evaluate $\int_C (xy\,dx + x^2z\,dy + xyz\,dz)$,   where C is the curve $x = e^t$, $y = e^{-t}$, $z = t^2$, $0 \le t \le 1$.

17. Evaluate the line integral below, where C is the curve $x = \cos t$, $y = \sin t$, $-\pi \le t \le \pi$.

$$\int_C \left(\frac{y\,dx}{\sqrt{x^2 + y^2}} + \frac{x\,dy}{\sqrt{x^2 + y^2}} \right)$$

18. Evaluate the line integral below, where C is the curve $x^2 - y^2 = 16$ from $(4, 0)$ to $(5, 3)$.

$$\int_C \left(\frac{-y}{x\sqrt{x^2 - y^2}}\,dx + \frac{1}{\sqrt{x^2 - y^2}}\,dy \right)$$

19. Evaluate $\int_C [yz\,dx + (y + zx)\,dy + xz\,dz]$,   where C consists of line segments connecting the points $(0, 0, 0)$, $(1, 0, 0)$, $(1, 2, 0)$, $(1, 2, 1)$, and $(0, 0, 1)$, in that order.

20. Evaluate $\int_C 2y\,ds$,   where C is the curve $y = \frac{1}{2}x^3$ joining $(0, 0)$ to $(2, 4)$.

21. Evaluate $\int_C [xz\,dx + (y + z)\,dy + x\,dz]$,   where C is the curve $x = e^t$, $y = e^{-t}$, $z = e^{2t}$ from $t = 0$ to $t = 1$.

22. Evaluate $\int_C [1/(x^2 + y^2 + z^2)]\,ds$,   where C is the helix $x = a \cos t$, $y = a \sin t$, $z = bt$, $0 \le t \le 1$.

23. Evaluate $\int_C (y^n\,dx + x^n\,dy)$,   where C is the ellipse $x = a \sin t$, $y = b \cos t$, $0 \le t \le 2\pi$.

24. Evaluate $\int_C y^2\,ds$,   where C is the first arch of the cycloid $x = a(t - \sin t)$, $y = a(1 - \cos t)$.

25. Evaluate $\int_C \sqrt{x^2 + y^2}\,ds$,   where C is the curve $x = a(\cos t + t \sin t)$, $y = a(\sin t - t \cos t)$, $0 \le t \le 2\pi$.

26. Evaluate $\int_C (x + y)\,ds$,   where C is the curve $x = t$, $y = 3t^2/\sqrt{2}$, $0 \le t \le 1$.

27. Evaluate $\int_C \mathbf{F} \cdot d\mathbf{r}$ if $\mathbf{F}(x, y) = (x + 2y)\mathbf{i} + (2x + y)\mathbf{j}$ and C is $\mathbf{r}(t) = t\mathbf{i} + t^2\mathbf{j}$, $0 \le t \le 1$.

28. Evaluate $\int_C \mathbf{F} \cdot d\mathbf{r}$ if $\mathbf{F}(x, y) = x^2\mathbf{i} + xy\mathbf{j}$ and C is $\mathbf{r}(t) = (\cos t)\mathbf{i} + (\sin t)\mathbf{j}$, $0 \le t \le \pi$.

29. Evaluate $\int_C \mathbf{F} \cdot d\mathbf{r}$ if $\mathbf{F}(x, y, z) = xy\mathbf{i} + x^2\mathbf{j} + xyz\mathbf{k}$ and C is $\mathbf{r}(t) = e^t\mathbf{i} + e^{-t}\mathbf{j} + t^2\mathbf{k}$, $0 \le t \le 1$.

30. Evaluate $\int_C \mathbf{F} \cdot d\mathbf{r}$ if $\mathbf{F}(x, y, z) = xy\mathbf{i} - y\mathbf{j} + z\mathbf{k}$ and:
   (a) C is the line segment from $(0, 0, 0)$ to $(1, 0, 0)$ (b) C is the line segment from $(1, 0, 0)$ to $(1, 2, 0)$
   (c) C is the line segment from $(1, 2, 0)$ to $(1, 2, 3)$ (d) C is the line segment from $(0, 0, 0)$ to $(1, 2, 3)$

2. Independence of Path

Recall Example 4 of Section 1, in which we evaluated a line integral along two different curves with common endpoints and obtained two different values for the integral. By way of contrast, consider the following example, in which $\int_C (P\,dx + Q\,dy)$ turns out to be the same for several different curves connecting two given points.

EXAMPLE 1   Evaluate $\int_C (2xy\,dx + x^2\,dy)$ if:
(a) C consists of line segments from $(3, 1)$ to $(5, 1)$ and from $(5, 1)$ to $(5, 6)$; see Figure 6(a)
(b) C is the line segment from $(3, 1)$ to $(5, 6)$; see Figure 6(b)
(c) C is a part of the parabola $x = 2t + 1$, $y = 2t^2 - t$, $1 \le t \le 2$; see Figure 6(c)

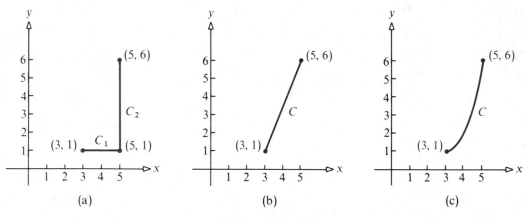

(a)   (b)   (c)

Figure 6

Solution   (a)   The curve C consists of two line segments whose equations are

$$C_1:\ \ 3 \le x \le 5; \qquad y = 1$$
$$C_2:\ \ x = 5, \qquad 1 \le y \le 6$$

The line integral along C is the sum of the line integrals along C_1 and along C_2. Along C_1 we have $y = 1$,   so that $dy = 0$. Hence,

$$\int_{C_1} (2xy\,dx + x^2\,dy) = \int_3^5 (2x\,dx + 0) = x^2\Big|_3^5 = 16$$

Along C_2 we have $x = 5$,   so that $dx = 0$.   Hence,

$$\int_{C_2} (2xy\,dx + x^2\,dy) = \int_1^6 (0 + 25\,dy) = 25y\Big|_1^6 = 125$$

Consequently, the line integral along C equals $16 + 125 = 141$.
(b)   The curve C is a line segment whose equation is $y = (\tfrac{5}{2})x - \tfrac{13}{2}$, $3 \le x \le 5$.

By using the fact that $dy = (\frac{5}{2})\,dx$,   we find

$$\int_C (2xy\,dx + x^2\,dy) = \int_3^5 \left[2x\left(\frac{5x}{2} - \frac{13}{2}\right)dx + x^2\left(\frac{5}{2}\right)dx \right]$$

$$= \int_3^5 \left[\left(\frac{15}{2}\right)x^2 - 13x\right]dx = \left[\left(\frac{5}{2}\right)x^3 - \left(\frac{13}{2}\right)x^2\right]\Big|_3^5 = 141$$

(c)   The curve C is part of a parabola. Since $dx = 2\,dt$ and $dy = (4t - 1)\,dt$, we find

$$\int_C (2xy\,dx + x^2\,dy) = \int_1^2 [2(2t + 1)(2t^2 - t)2\,dt + (2t + 1)^2(4t - 1)\,dt]$$

$$= (8t^4 + 4t^3 - 2t^2 - t)\Big|_1^2 = 141 \qquad \blacksquare$$

Note that in Example 1, $2xy\,dx + x^2\,dy$ is the total differential of $f(x, y) = x^2y$;   that is, the vector function $\mathbf{F} = 2xy\mathbf{i} + x^2\mathbf{j}$ is the gradient of f, or $\nabla f = \mathbf{F}$.   In fact, based on theorem (19.14) below, the value of the line integral $\int_C (2xy\,dx + x^2\,dy)$ depends only on $f(x, y) = x^2y$ and the endpoints (3, 1) and (5, 6) of the curve C. Thus,

$$\int_C (2xy\,dx + x^2\,dy) = f(x, y)\Big|_{(3,\,1)}^{(5,\,6)} = x^2y\Big|_{(3,\,1)}^{(5,\,6)} = f(5, 6) - f(3, 1) = 150 - 9 = 141$$

In a case like this, where the line integral for a given integrand depends only on the endpoints of the curve and not on the path taken by the curve between the two points, we say that the integral is *independent of path*.

The next result, theorem (19.14), and its converse, theorem (19.17), are concerned with independence of path on an open connected set. By a *connected set S* we mean that any pair of points in S can be joined by a piecewise smooth curve that lies entirely in S.

(19.14)   **THEOREM**   *The Fundamental Theorem of Line Integrals.**   **Let $\mathbf{F} = \mathbf{F}(x, y) = P(x, y)\mathbf{i} + Q(x, y)\mathbf{j}$ denote a vector function defined on an open connected set S. We assume that P and Q are continuous on S. Let C denote a piecewise smooth curve in S beginning at the point (x_0, y_0) and ending at the point (x_1, y_1). If $\mathbf{F}$ is the gradient of some function f, that is, if**

$$\mathbf{F} = \nabla f = \frac{\partial f}{\partial x}\mathbf{i} + \frac{\partial f}{\partial y}\mathbf{j} \qquad \textbf{throughout } S$$

then

(19.15) $$\int_C (P\,dx + Q\,dy) = \int_{(x_0,\,y_0)}^{(x_1,\,y_1)} (P\,dx + Q\,dy) = f(x, y)\Big|_{(x_0,\,y_0)}^{(x_1,\,y_1)} = f(x_1, y_1) - f(x_0, y_0)$$

In vector notation, we write

$$\int_C \mathbf{F} \cdot d\mathbf{r} = \int_C \nabla f \cdot d\mathbf{r} = \int_{(x_0,\,y_0)}^{(x_1,\,y_1)} \mathbf{F} \cdot d\mathbf{r} = f(x, y)\Big|_{(x_0,\,y_0)}^{(x_1,\,y_1)} = f(x_1, y_1) - f(x_0, y_0)$$

---

* Note the similarity of (19.14) and the fundamental theorem of calculus.

Proof We give a proof for a smooth curve C. (The proof for a piecewise smooth curve may be obtained in the same manner by considering one piece at a time.) Let $x = x(t)$, $y = y(t)$, $a \le t \le b$, be the parametric equations of the curve C. The initial point and endpoint of C may be written as

$$(x_0, y_0) = (x(a), y(a)) \qquad \text{and} \qquad (x_1, y_1) = (x(b), y(b))$$

Since $\mathbf{F} = P\mathbf{i} + Q\mathbf{j} = \nabla f = (\partial f / \partial x)\mathbf{i} + (\partial f / \partial y)\mathbf{j}$, we find that

$$\int_C (P\,dx + Q\,dy) = \int_C \left(\frac{\partial f}{\partial x}\,dx + \frac{\partial f}{\partial y}\,dy \right) = \int_a^b \left(\frac{\partial f}{\partial x}\frac{dx}{dt} + \frac{\partial f}{\partial y}\frac{dy}{dt} \right) dt$$

$$\underset{\uparrow}{=} \int_a^b \frac{d}{dt}\left[f(x(t), y(t)) \right] dt = f(x(t), y(t))\Big|_{t=a}^{t=b}$$

(16.30)

$$= f(x(b), y(b)) - f(x(a), y(a)) = f(x_1, y_1) - f(x_0, y_0) \qquad \blacksquare$$

Some observations about (19.15) are in order. First, the right-hand side of (19.15) depends only on the endpoints (x_0, y_0) and (x_1, y_1) of the curve C. As a result, the value of the line integral will not change if C is replaced by any other piecewise smooth curve in S, so long as the new curve also connects (x_0, y_0) to (x_1, y_1) (see Fig. 7). Second, a special case of some interest occurs when C is a *closed curve*— that is, when $(x_0, y_0) = (x_1, y_1)$. Then $f(x_1, y_1) = f(x_0, y_0)$ and the value of the line integral is zero.

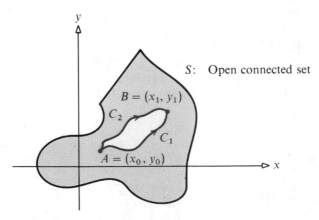

Figure 7 If $\mathbf{F} = P\mathbf{i} + Q\mathbf{j} = \nabla f$, then
(1) $\int_C (P\,dx + Q\,dy) = f(x_1, y_1) - f(x_0, y_0)$
for any curve C joining A to B
(2) $\int_{C_1} (P\,dx + Q\,dy) = \int_{C_2} (P\,dx + Q\,dy)$

If $\nabla f = P\mathbf{i} + Q\mathbf{j}$ for some function f, then independence of path implies that the line integral of $P\,dx + Q\,dy$ over a closed curve C is zero:

(19.16)
$$\oint_C (P\,dx + Q\,dy) = 0$$

EXAMPLE 2   The vector function $\mathbf{F} = \mathbf{F}(x, y) = (2xy + 24x)\mathbf{i} + (x^2 + 16)\mathbf{j}$ is the gradient of $f(x, y) = x^2y + 12x^2 + 16y$. Use this fact to evaluate

$$\int_C [(2xy + 24x)\,dx + (x^2 + 16)\,dy]$$

where C is any path joining the point $(1, 1)$ to $(2, 4)$.

Solution   We use two methods to evaluate the given line integral.

Method I.   This method utilizes the function $f(x, y) = x^2y + 12x^2 + 16y$ whose gradient is

$$\nabla f = (2xy + 24x)\mathbf{i} + (x^2 + 16)\mathbf{j} = \mathbf{F}(x, y)$$

Thus, by (19.15),

$$\int_C [(2xy + 24x)\,dx + (x^2 + 16)\,dy] = f(x, y)\Big|_{(1, 1)}^{(2, 4)} = f(2, 4) - f(1, 1) = 99$$

Method II.   This method uses the fact that the given line integral is independent of path, so that we can evaluate the line integral along *any* path joining $(1, 1)$ and $(2, 4)$.

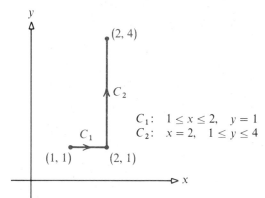

C_1: $1 \le x \le 2$, $y = 1$
C_2: $x = 2$, $1 \le y \le 4$

Figure 8

We choose the path shown in Figure 8 since it makes the integration easy. Then

$$\int_C [(2xy + 24x)\,dx + (x^2 + 16)\,dy] = \int_{C_1} [(2xy + 24x)\,dx + (x^2 + 16)\,dy]$$
$$+ \int_{C_2} [(2xy + 24x)\,dx + (x^2 + 16)\,dy]$$
$$= \int_1^2 (2x + 24x)\,dx + \int_1^4 (4 + 16)\,dy$$
$$= 39 + 60 = 99 \quad \blacksquare$$

In general, if it is required to evaluate the line integral $\int_C (P\,dx + Q\,dy)$, where $\mathbf{F} = P\mathbf{i} + Q\mathbf{j}$ is the gradient of some function f,* then either of the methods outlined in the above example may be used:

Method I.   Use (19.15) directly by evaluating f at the endpoints (x_0, y_0) and (x_1, y_1) of C. That is,

$$\int_C (P\,dx + Q\,dy) = f(x_1, y_1) - f(x_0, y_0)$$

Method II.   Use the fact that $\int_C (P\,dx + Q\,dy)$ is independent of path and select some suitable path joining the endpoints of C to evaluate the line integral.

Some care must be exercised when Method II is used. For example, suppose we are asked to evaluate the line integral

$$\int_C \frac{-y\,dx + x\,dy}{x^2}$$

along the curve C joining the points $(1, -1)$ and $(4, 2)$. As you can verify, $(-y\mathbf{i} + x\mathbf{j})/x^2$ is the gradient of the function $f(x, y) = y/x$. In using Method II to evaluate this line integral, you may not, for example, choose a path (such as $x = y^2$) that crosses the y-axis $(x = 0)$ since the open connected set S over which $P = -y/x^2$ and $Q = 1/x$ are continuous does not contain points on the y-axis $(x = 0)$ and the path $x = y^2$, which passes through the origin, does not lie entirely in S. However, any path that lies entirely in the first and fourth quadrants can be used (see Fig. 9).

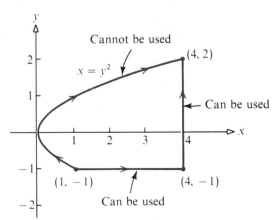

Figure 9

RECONSTRUCTING A FUNCTION FROM ITS GRADIENT

We have seen, by applying theorem (19.14) to Examples 1 and 2, that the principle of independence of path can be used to evaluate a line integral of the form

---

* We give a criterion for determining when $\mathbf{F}$ is the gradient of some function f shortly.

$\int_C (P\,dx + Q\,dy)$ or $\int_C \mathbf{F} \cdot d\mathbf{r}$ if there happens to be a function f such that $\nabla f = \mathbf{F}$.   Theorem (19.17) shows, conversely, that if independence of path applies, then $\mathbf{F}$ is necessarily the gradient of some function. The proof of this result is of importance since it provides a way of finding the function f when its gradient $\nabla f = \mathbf{F}$ is known. This is known as *reconstructing a function from its gradient.*

(19.17)   **THEOREM**   **Let $P = P(x, y)$ and $Q = Q(x, y)$ denote functions that are continuous on an open connected set S, and let (x_0, y_0) be a fixed point in S. If, for every point (x, y) in S, the line integral $\int_C (P\,dx + Q\,dy)$ has the same value for every piecewise smooth curve C joining (x_0, y_0) to (x, y), then the function $\mathbf{F} = P\mathbf{i} + Q\mathbf{j}$ is the gradient of some function f.**

Proof   The line integral $\int_C (P\,dx + Q\,dy)$ is independent of path. Hence, if C is any piecewise smooth curve in S joining a fixed point (x_0, y_0) to an arbitrary point (x, y), then the line integral $\int_C (P\,dx + Q\,dy)$ will define a function f that depends only on (x, y) and not on C. Thus, we may define

(19.18)
$$f(x, y) = \int_{(x_0, y_0)}^{(x, y)} (P\,dx + Q\,dy)$$

Any path joining (x_0, y_0) and (x, y) is allowed. We shall use two paths, one in which a horizontal line segment is used, and another in which a vertical line segment is used.* As Figure 10 illustrates,

C_1:   Consists of a piecewise smooth curve joining (x_0, y_0) to (x_1, y)
plus a horizontal line segment joining (x_1, y) to (x, y)

C_2:   Consists of a piecewise smooth curve joining (x_0, y_0) to (x, y_1)
plus a vertical segment joining (x, y_1) to (x, y)

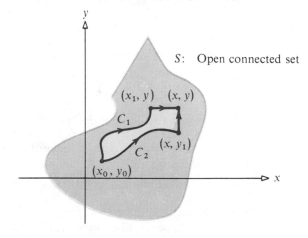

Figure 10

---

* Such line segments must exist since S is open and connected.

For C_1, (19.18) takes the form

$$f(x, y) = \int_{(x_0, y_0)}^{(x_1, y)} (P\,dx + Q\,dy) + \int_{(x_1, y)}^{(x, y)} (P\,dx + Q\,dy)$$

In the first integral on the right, only y is treated as a variable. Hence, the partial derivative of this integral with respect to x is zero. That is,

$$\frac{\partial}{\partial x} f(x, y) = \frac{\partial}{\partial x} \int_{(x_1, y)}^{(x, y)} (P\,dx + Q\,dy)$$

In the second integral on the right, y is constant, so that $dy = 0$. Hence,

(19.19)
$$\frac{\partial}{\partial x} f(x, y) = \frac{\partial}{\partial x} \int_{x_1}^{x} P\,dx \underset{\uparrow}{=} P(x, y)$$

$$\text{By (5.35)}$$

For C_2, (19.18) takes the form

$$f(x, y) = \int_{(x_0, y_0)}^{(x, y_1)} (P\,dx + Q\,dy) + \int_{(x, y_1)}^{(x, y)} (P\,dx + Q\,dy)$$

In the first integral on the right, only x is treated as a variable. Hence, the partial derivative of this integral with respect to y is zero. That is,

$$\frac{\partial}{\partial y} f(x, y) = \frac{\partial}{\partial y} \int_{(x, y_1)}^{(x, y)} (P\,dx + Q\,dy)$$

In the second integral on the right, x is constant, so that $dx = 0$. Hence,

(19.20)
$$\frac{\partial}{\partial y} f(x, y) = \frac{\partial}{\partial y} \int_{y_1}^{y} Q\,dy \underset{\uparrow}{=} Q(x, y)$$

$$\text{By (5.35)}$$

By combining (19.19) and (19.20), we find that

$$\nabla f = \frac{\partial f}{\partial x}\mathbf{i} + \frac{\partial f}{\partial y}\mathbf{j} = P(x, y)\mathbf{i} + Q(x, y)\mathbf{j} = \mathbf{F} \quad \blacksquare$$

We are now ready to discuss the following two fundamental questions:

1. How can we tell whether a vector function $\mathbf{F}(x, y) = P(x, y)\mathbf{i} + Q(x, y)\mathbf{j}$ is the gradient of some function $f(x, y)$?
2. If $\mathbf{F}$ is known to be the gradient of some function f, then how can f be reconstructed?

We have just shown that these two questions are equivalent to asking, "When can independence of path be applied to the evaluation of a line integral and, in particular, how can formula (19.15) be used?"

The forthcoming theorem (19.21) gives a nice answer to the first question; an answer to the second question (but not an easy one) has been suggested by the proof of theorem (19.17). However, the discussion following theorem (19.21) illustrates a much more practical method of reconstruction.

But, before we state theorem (19.21), we need the following definitions: Recall that a piecewise smooth curve defined by $\mathbf{r}(t) = x(t)\mathbf{i} + y(t)\mathbf{j}$, $a \le t \le b$, is said to be *closed* if $\mathbf{r}(a) = \mathbf{r}(b)$, that is, if it intersects itself only at its initial point and its terminal point and nowhere else. See Figure 11. A set S is said to be *simply connected* if, for every piecewise smooth closed curve C that lies in S, every point of the interior of C also lies in S. Intuitively, S is simply connected if it has no "holes," as shown in Figure 12.

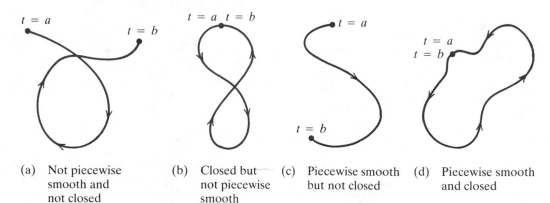

(a) Not piecewise smooth and not closed   (b) Closed but not piecewise smooth   (c) Piecewise smooth but not closed   (d) Piecewise smooth and closed

Figure 11

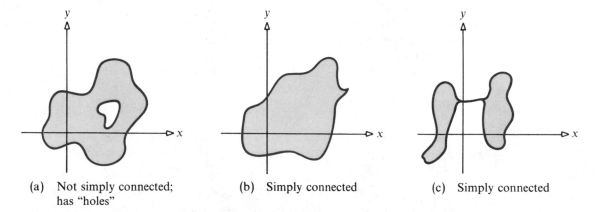

(a) Not simply connected; has "holes"   (b) Simply connected   (c) Simply connected

Figure 12

We now state theorem (19.21):

(19.21)   **THEOREM** **Let** $P = P(x, y)$ **and** $Q = Q(x, y)$ **denote functions that are continuous on a simply connected open set S. Suppose that $\partial P/\partial y$ and $\partial Q/\partial x$ are continuous**

on S. Then $\mathbf{F}(x, y) = P(x, y)\mathbf{i} + Q(x, y)\mathbf{j}$ **is the gradient of some function f defined on S if and only if**

(19.22)

$$\frac{\partial P}{\partial y} = \frac{\partial Q}{\partial x}$$

throughout S.

Proof Suppose that $\mathbf{F} = P\mathbf{i} + Q\mathbf{j}$ is the gradient of some function f. Then

$$\frac{\partial f}{\partial x}\mathbf{i} + \frac{\partial f}{\partial y}\mathbf{j} = P\mathbf{i} + Q\mathbf{j}$$

or

$$\frac{\partial f}{\partial x} = P \quad \text{and} \quad \frac{\partial f}{\partial y} = Q$$

Thus,

$$\frac{\partial^2 f}{\partial y\, \partial x} = \frac{\partial P}{\partial y} \quad \text{and} \quad \frac{\partial^2 f}{\partial x\, \partial y} = \frac{\partial Q}{\partial x}$$

Since we assume that $\partial P/\partial y$ and $\partial Q/\partial x$ are continuous in S, we apply (16.5), which states that under these conditions, the mixed partials are equal. As a result,

$$\frac{\partial P}{\partial y} = \frac{\partial Q}{\partial x} \quad \blacksquare$$

The proof of the converse—that condition (19.22) implies the existence of a function f such that $\nabla f = \mathbf{F}$—requires Green's theorem, which is given in Section 4. For the moment, however, we assume that the converse is true.

The next example uses the converse of (19.21) and shows a method for reconstructing a function f when its gradient ∇f is known.

EXAMPLE 3   (a) Show that the line integral $\int_C [(2xy + 24x)\, dx + (x^2 + 16)\, dy]$ is independent of path.

(b) Find a function f such that $\nabla f = (2xy + 24x)\mathbf{i} + (x^2 + 16)\mathbf{j}$.

(c) Evaluate $\int_C [(2xy + 24x)\, dx + (x^2 + 16)\, dy]$ where C is any piecewise smooth curve joining $(0, 1)$ to $(1, 2)$.

Solution   (a) Let $P(x, y) = 2xy + 24x$ and $Q(x, y) = x^2 + 16$. Then $\partial P/\partial y = 2x$ and $\partial Q/\partial x = 2x$. Since P, Q, $\partial P/\partial y$, and $\partial Q/\partial x$ are continuous in the entire plane and since $\partial P/\partial y = \partial Q/\partial x$, according to theorem (19.21) there is a function f such that $\nabla f = P\mathbf{i} + Q\mathbf{j}$. By theorem (19.14), the integral is independent of path.

(b) Since $\nabla f = P\mathbf{i} + Q\mathbf{j}$, we have $\partial f/\partial x = 2xy + 24x$ and $\partial f/\partial y = x^2 + 16$. By integrating the first of these partial derivatives with respect to x while holding

y fixed, we obtain

$$f(x, y) = \int (2xy + 24x)\,dx = x^2y + 12x^2 + h(y)$$

Here we integrate $\;2xy + 24x\;$ as though y were a constant, realizing, however, that the "constant of integration" may actually be a function of y. If we now differentiate f with respect to y, we get

$$\frac{\partial f}{\partial y} = x^2 + h'(y)$$

But $\;\partial f/\partial y = Q = x^2 + 16;\;$ thus, $\;h'(y) = 16.\;$ By now integrating with respect to y, we obtain

$$h(y) = 16y + K$$

where K is a constant. Thus,

$$f(x, y) = x^2y + 12x^2 + 16y + K$$

We may verify the correctness of this formula by differentiating:

$$\frac{\partial f}{\partial x} = 2xy + 24x = P \qquad \frac{\partial f}{\partial y} = x^2 + 16 = Q$$

(c)   By applying (19.15), we find

$$\int_C [(2xy + 24x)\,dx + (x^2 + 16)\,dy] = \int_{(0,\,1)}^{(1,\,2)} [(2xy + 24x)\,dx + (x^2 + 16)\,dy]$$

$$= (x^2y + 12x^2 + 16y + K)\Big|_{(0,\,1)}^{(1,\,2)}$$

$$= (2 + 12 + 32 + K) - (16 + K)$$

$$= 30 \quad \blacksquare$$

Note that in part (b) above, we could have begun by integrating Q with respect to y instead of P with respect to x:

$$f(x, y) = \int (x^2 + 16)\,dy = x^2y + 16y + k(x)$$

Then

$$\frac{\partial f}{\partial x} = 2xy + k'(x) = P = 2xy + 24x$$

Thus,

$$k(x) = \int 24x\,dx = 12x^2 + K$$

and

$$f(x, y) = x^2y + 16y + 12x^2 + K$$
$$= x^2y + 12x^2 + 16y + K$$

as before.

EXAMPLE 4   (a)   Show that the line integral $\int_C [(y \cos x + 2xe^y)\, dx + (\sin x + x^2e^y + 4)\, dy]$ is independent of path.

(b)   Find a function f such that $\nabla f = (y \cos x + 2xe^y)\mathbf{i} + (\sin x + x^2e^y + 4)\mathbf{j}$.

Solution   (a)   Let $P(x, y) = y \cos x + 2xe^y$ and $Q(x, y) = \sin x + x^2e^y + 4$.   Then

$$\frac{\partial P}{\partial y} = \cos x + 2xe^y = \frac{\partial Q}{\partial x}$$

Since P, Q, $\partial P/\partial y$, and $\partial Q/\partial x$ are continuous in the entire plane, and since $\partial P/\partial y = \partial Q/\partial x$,   according to theorem (19.21), there is a function f such that $\nabla f = P\mathbf{i} + Q\mathbf{j}$.   Thus, according to theorem (19.14), the integral is independent of path.

(b)   Since $\nabla f = P\mathbf{i} + Q\mathbf{j}$,   we have

(19.23)
$$\frac{\partial f}{\partial x} = y \cos x + 2xe^y \qquad \frac{\partial f}{\partial y} = \sin x + x^2e^y + 4$$

We integrate the second of these functions with respect to y, while holding x fixed,* to obtain

(19.24)
$$f(x, y) = y \sin x + x^2e^y + 4y + k(x)$$

in which the constant of integration is denoted by the function $k(x)$. Now we differentiate (19.24) with respect to x, to get

$$\frac{\partial f}{\partial x} = y \cos x + 2xe^y + k'(x)$$

We equate this with the expression for $\partial f/\partial x$ found in (19.23):

$$y \cos x + 2xe^y + k'(x) = y \cos x + 2xe^y$$
$$k'(x) = 0$$

So

$$k(x) = K \qquad \text{where } K \text{ is a constant}$$

By substituting for $k(x)$ in (19.24), we obtain

$$f(x, y) = y \sin x + x^2e^y + 4y + K \qquad \blacksquare$$

---

* Again, it makes no difference which equation we begin with. Start with the first one and see for yourself.

EXAMPLE 5   Determine whether $\int_C (x^2 y\, dx + xy^2\, dy)$ is independent of path anywhere in the plane.

Solution   Let $P = x^2 y$ and $Q = xy^2$. Then $\partial P / \partial y = x^2$ and $\partial Q / \partial x = y^2$. Since these two functions are not equal (except on the graph of the equation $x^2 = y^2$, which is not an open set), the line integral is not independent of path anywhere in the plane.

∎

SUMMARY

Let P and Q be two functions that are continuous on an open connected set S containing a piecewise smooth curve C.

(19.14)*   **If $\mathbf{F} = P\mathbf{i} + Q\mathbf{j}$ is the gradient of some function f, then $\int_C (P\, dx + Q\, dy) = f(B) - f(A)$ is independent of the path joining $A = (x_0, y_0)$ to $B = (x_1, y_1)$.**

(19.16)   **If $\int_C (P\, dx + Q\, dy)$ is independent of path and if C is a closed curve in S, then $\int_C (P\, dx + Q\, dy) = 0$.**

(19.17)   **If $\int_C (P\, dx + Q\, dy)$ is independent of path, then $\mathbf{F} = P\mathbf{i} + Q\mathbf{j}$ is the gradient of some function f.**

(19.21)   **If $\partial P / \partial y$ and $\partial Q / \partial x$ are also continuous on S and if S is simply connected, then $\mathbf{F} = P\mathbf{i} + Q\mathbf{j}$ is the gradient of some function f if and only if $\partial P / \partial y = \partial Q / \partial x$.**

EXERCISE 2

In Problems 1–4 use theorem (19.21) to determine whether there exists a function f for which $\mathbf{F} = \nabla f$.

1. $\mathbf{F}(x, y) = x^2 \mathbf{i} + y^2 \mathbf{j}$

2. $\mathbf{F}(x, y) = xy \mathbf{i} + xy \mathbf{j}$

3. $\mathbf{F}(x, y) = xe^y \mathbf{i} + \frac{1}{2} x^2 e^y \mathbf{j}$

4. $\mathbf{F}(x, y) = (x^2 + y^2) \mathbf{i} + (2xy - \sin y) \mathbf{j}$

In Problems 5–10:
(a) Show that the line integral $\int_C (P\, dx + Q\, dy)$ is independent of path.
(b) Find a function f such that $\nabla f = P(x, y)\mathbf{i} + Q(x, y)\mathbf{j}$.
(c) Evaluate the integral $\int_C (P\, dx + Q\, dy)$.

5. $\int_C (x\, dx + y\, dy)$; C is any curve joining the points $(1, 3)$ and $(2, 5)$

6. $\int_C [2xy\, dx + (x^2 + 1)\, dy]$; C is any curve joining the points $(1, -4)$ and $(-2, 3)$

7. $\int_C [(x^2 + 3y)\, dx + 3x\, dy]$; C is any curve joining the points $(1, 2)$ and $(-3, 5)$

8. $\int_C [(2x + y + 1)\, dx + (x + 3y + 2)\, dy]$; C is any curve joining the points $(0, 0)$ and $(1, 2)$

9. $\int_C [(4x^3 + 20xy^3 - 3y^4)\, dx + (30x^2 y^2 - 12xy^3 + 5y^4)\, dy]$; C is any curve joining the points $(0, 0)$ and $(1, 1)$

10. $\int_C [(2yx^{-1}\, dx + (\ln x^2)\, dy]$; C is any curve in the first quadrant joining the points $(1, 1)$ and $(5, 5)$

---

* The numbers are the numbers of the original statements in this section.

In Problems 11–22 show that each line integral $\int_C (P\,dx + Q\,dy)$ is independent of path in the entire plane by finding a function f such that $\nabla f = P\mathbf{i} + Q\mathbf{j}$.

11. $\int_C (3x^2y^2\,dx + 2x^3y\,dy)$

12. $\int_C [(2x + y)\,dx + (x - 2y)\,dy]$

13. $\int_C [(x + 3y)\,dx + 3x\,dy]$

14. $\int_C [(2x + y)\,dx + (2y + x)\,dy]$

15. $\int_C [(2xy - y^2)\,dx + (x^2 - 2xy)\,dy]$

16. $\int_C [y^2\,dx + (2yx - e^y)\,dy]$

17. $\int_C [(x^2 - x + y^2)\,dx - (ye^y - 2xy)\,dy]$

18. $\int_C [(3x^2y + xy^2 + e^x)\,dx + (x^3 + x^2y + \sin y)\,dy]$

19. $\int_C [(y \cos x - 2 \sin y)\,dx - (2x \cos y - \sin x)\,dy]$

20. $\int_C [(e^x \sin y + 2y \sin x)\,dx + (e^x \cos y - 2 \cos x)\,dy]$

21. $\int_C [(2x + y \cos x)\,dx + \sin x\,dy]$

22. $\int_C [(\cos y - \cos x)\,dx + (e^y - x \sin y)\,dy]$

23. Show that $\int_C \left(\dfrac{-y}{x^2 + y^2}\,dx + \dfrac{x}{x^2 + y^2}\,dy \right)$ is independent of the path in the rectangle R whose vertices are $(1/a, -a)$, $(a, -a)$, $(1/a, -1/a)$, and $(a, -1/a)$, $a > 1$. Find f so that $\nabla f = [-y/(x^2 + y^2)]\mathbf{i} + [x/(x^2 + y^2)]\mathbf{j}$.

3. Work

One important physical application of line integrals is to the concept of work. Here is a review of our development to date.

If an object is pushed along a straight line segment of length s by a constant force of magnitude F acting in the direction of the motion, then the work W done is

(19.25)
$$W = Fs$$

If the force acts in the opposite direction, then the work is $-Fs$.

In Chapter 6 we generalized (19.25) to the case where the force F is variable. There we showed that the work W done by a variable force $F = F(x)$ acting in the direction of the motion of an object as that object moves along a straight line from $x = a$ to $x = b$ is

(19.26)
$$W = \int_a^b F(x)\,dx$$

Next, in Chapter 14, we took up the question of the work W done by a constant force vector $\mathbf{F}$ acting on an object as that object moves in the direction of the vector $\mathbf{r}$, and we found that

(19.27)
$$W = \mathbf{F} \cdot \mathbf{r}$$

The purpose of this section is to generalize (19.26) and (19.27) in order to give a definition for the work done by a variable vector force $\mathbf{F}$ acting on an object as that object moves along a smooth curve C from the point A on C to the point B on C.

Let

$$\mathbf{F} = \mathbf{F}(x, y) = P(x, y)\mathbf{i} + Q(x, y)\mathbf{j}$$

denote the force vector exerted on an object at the point (x, y) in some open connected set S. We assume P and Q have continuous first-order partial derivatives at each point in S. Let C be a smooth curve lying entirely in S and defined by the vector function $\mathbf{r} = \mathbf{r}(t) = x(t)\mathbf{i} + y(t)\mathbf{j}$, $a \leq t \leq b$.

We proceed as usual to partition the closed interval $[a, b]$ into n subintervals,

$$[a, t_1], \quad [t_1, t_2], \quad \dots, \quad [t_{i-1}, t_i], \quad \dots, \quad [t_{n-1}, b]$$

and denote the length of each subinterval by

$$\Delta t_1, \Delta t_2, \dots, \Delta t_n$$

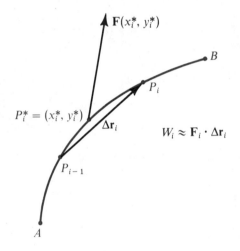

Figure 13

$\mathbf{F}(x_i^*, y_i^*)$

B

P_i

$P_i^* = (x_i^*, y_i^*)$

$\Delta \mathbf{r}_i$

$W_i \approx \mathbf{F}_i \cdot \Delta \mathbf{r}_i$

P_{i-1}

A

Corresponding to each number $a = t_0, t_1, t_2, \dots, t_n = b$ of the partition, there is a sequence of points

$$P_0, P_1, \dots, P_n$$

on the curve C. These points subdivide the curve into n subarcs of lengths $\Delta s_1, \Delta s_2, \dots, \Delta s_n$.   The norm $\|\Delta\|$ of this partition is the largest of the arc lengths Δs_i, $i = 1, 2, \dots, n$. Let $P_i^* = (x_i^*, y_i^*)$ be some point on the ith subarc $\widehat{P_{i-1}P_i}$ of C. If the norm $\|\Delta\|$ is small enough, then the work W_i done by the variable force $\mathbf{F}$ in moving an object along the arc $\widehat{P_{i-1}P_i}$ can be approximated by the work done by the constant force

$$\mathbf{F}_i^* = \mathbf{F}(x_i^*, y_i^*)$$

in moving the object along the directed line segment $\Delta \mathbf{r}_i = \overrightarrow{P_{i-1}P_i}$ (see Fig. 13). That is, by (19.27),

(19.28) $$W_i \approx \mathbf{F}(x_i^*, y_i^*) \cdot \Delta \mathbf{r}_i$$

The total work W done by $\mathbf{F}$ along C from the point A: $\mathbf{r} = \mathbf{r}(a)$ to the point B: $\mathbf{r} = \mathbf{r}(b)$ is $W = \sum_{i=1}^{n} W_i$.   By virtue of (19.28), we define the work W done by the force $\mathbf{F}$ in moving an object along the curve C from A to B as

(19.29) $$W = \lim_{\|\Delta\| \to 0} \sum_{i=1}^{n} \mathbf{F}(x_i^*, y_i^*) \cdot \Delta \mathbf{r}_i = \int_C \mathbf{F} \cdot d\mathbf{r}$$

EXAMPLE 1   Find the work done by the force $\mathbf{F}(x, y) = -y\mathbf{i} + x\mathbf{j}$ in moving an object along the half circle

$$C: \quad \mathbf{r}(t) = (\cos t)\mathbf{i} + (\sin t)\mathbf{j} \qquad 0 \leq t \leq \pi$$

Solution   On C we have $x(t) = \cos t,$ $y(t) = \sin t.$ Hence,

$$\mathbf{F}(x(t),\, y(t)) = (-\sin t)\mathbf{i} + (\cos t)\mathbf{j}$$
$$d\mathbf{r}(t) = [(-\sin t)\mathbf{i} + (\cos t)\mathbf{j}]\, dt$$
$$\mathbf{F} \cdot d\mathbf{r} = (\sin^2 t + \cos^2 t)\, dt = dt$$

Therefore,

$$W = \int_C \mathbf{F} \cdot d\mathbf{r} = \int_0^\pi dt = \pi \qquad \blacksquare$$

APPLICATION TO KINETIC ENERGY

The work W done by a force $\mathbf{F}$ in moving an object along the smooth curve C defined by $\mathbf{r} = \mathbf{r}(t),$ $a \leq t \leq b,$ is

(19.30)
$$W = \int_C \mathbf{F} \cdot d\mathbf{r} = \int_a^b \mathbf{F} \cdot \mathbf{r}'(t)\, dt$$

By utilizing Newton's second law of motion, $\mathbf{F} = m\mathbf{r}''(t),$ where $\mathbf{r}''(t)$ is the acceleration of the object, we find that (19.30) takes the form

$$W = \int_a^b m\mathbf{r}''(t) \cdot \mathbf{r}'(t)\, dt = \frac{m}{2} \int_a^b \frac{d}{dt} [\mathbf{r}'(t) \cdot \mathbf{r}'(t)]\, dt$$

(19.31)
$$= \frac{m}{2} \int_a^b \frac{d}{dt} \|\mathbf{r}'(t)\|^2\, dt = \frac{m}{2} \|\mathbf{r}'(t)\|^2 \Big|_a^b$$

Recall that the *kinetic energy* K of an object of mass m that moves along a curve with velocity $\mathbf{v}$ is

(19.32)
$$K = \frac{1}{2} m\|\mathbf{v}\|^2 = \frac{m}{2} \|\mathbf{r}'(t)\|^2$$
$$\underset{\mathbf{v}\,=\,\mathbf{r}'(t)}{\uparrow}$$

By combining (19.31) and (19.32), we obtain the result

$$W = K(b) - K(a)$$

In other words,

(19.33)
$$\begin{bmatrix} \text{Work done by} \\ \text{a force } \mathbf{F} \text{ in} \\ \text{moving an object} \\ \text{from } A \text{ to } B \end{bmatrix} = \begin{bmatrix} \text{Change in} \\ \text{kinetic energy} \\ \text{from } A \text{ to } B \end{bmatrix}$$

CONSERVATIVE FIELDS OF FORCE

Let $P = P(x, y)$ and $Q = Q(x, y)$ denote two functions that are continuous on an open connected region S. If $\mathbf{F} = \mathbf{F}(x, y) = P(x, y)\mathbf{i} + Q(x, y)\mathbf{j}$ is a vector (force) function and if **F** equals the gradient of some function f—that is, if $\nabla f = \mathbf{F}$—then **F** is called a *conservative field of force.**

Because of (19.14), it follows that the work W done by a conservative field of force $\mathbf{F} = P\mathbf{i} + Q\mathbf{j}$ along a curve C from the point A to the point B in S is independent of path. Furthermore,

(19.34)
$$W = \int_C \mathbf{F} \cdot d\mathbf{r} = \int_C \nabla f \cdot d\mathbf{r} = \int_C (P\,dx + Q\,dy) = f(B) - f(A)$$

If the curve C is closed, so that $A = B$ and, hence, $f(A) = f(B)$, the work done by a conservative force **F** is 0.

(19.35)   **The work done by a conservative force field in moving an object along a closed path is 0. By virtue of (19.33), it follows that the object returns to its initial position with the same kinetic energy it started with.**

Physicists paraphrase this by saying that work is a function of position and not path.

CONSERVATION OF ENERGY

For a conservative field of force, the principle of conservation of energy holds.

Suppose an object of mass m moves along a smooth curve C defined by $\mathbf{r} = \mathbf{r}(t)$ in a conservative field of force $\mathbf{F} = \mathbf{F}(x, y)$. The *potential energy* $U = U(x, y)$ of the object due to **F** is that function U for which[†]

(19.36)
$$\nabla U = -\mathbf{F}$$

The principle of conservation of energy may be stated as follows:

(19.37)   **In a conservative field of force the sum of the potential and kinetic energies of an object is constant.**

The proof of this statement is relatively straightforward.

Proof   The potential energy $U = U(x, y)$ of an object moving along a curve C in a conservative field of force **F** obeys

(19.38)
$$\nabla U = -\mathbf{F}(x, y)$$

If the object moves along the curve $C: \mathbf{r} = \mathbf{r}(t)$, its kinetic energy is

$$\tfrac{1}{2}m\|\mathbf{r}'(t)\|^2 = \tfrac{1}{2}m[\mathbf{r}'(t) \cdot \mathbf{r}'(t)]$$

---

* In physics a *vector field* consists of a set of points and a vector-valued function **F** that associates to each point P a vector $\mathbf{F}(P)$ located at P. In a *field of force*, the vectors represent forces due to fluid pressure, electromagnetic fields, gravity, and so forth.

† The existence of $U = U(x, y)$ is guaranteed since **F** is conservative.

Let $E = E(t)$ equal the sum of the potential and kinetic energies of the object at time t. Then

$$E(t) = U(x, y) + \frac{1}{2} m\|\mathbf{r}'(t)\|^2 = U(x, y) + \left(\frac{m}{2}\right)[\mathbf{r}'(t) \cdot \mathbf{r}'(t)]$$

We show that E is constant by showing that $E'(t) = 0$:

(19.39)
$$E'(t) = \left(\frac{\partial U}{\partial x}\right)\left(\frac{dx}{dt}\right) + \left(\frac{\partial U}{\partial y}\right)\left(\frac{dy}{dt}\right) + m[\mathbf{r}'(t) \cdot \mathbf{r}''(t)]$$
$$\uparrow$$
$$\text{Chain rule}$$

But

(19.40)
$$\mathbf{r}'(t) = \frac{dx}{dt}\mathbf{i} + \frac{dy}{dt}\mathbf{j} \quad \text{and} \quad \nabla U = \frac{\partial U}{\partial x}\mathbf{i} + \frac{\partial U}{\partial y}\mathbf{j} = -\mathbf{F} = -m\mathbf{r}''(t)$$
$$\uparrow \qquad\qquad \uparrow$$
$$\text{(19.36)} \quad \text{Newton's law}$$

By combining (19.39) and (19.40), we find

$$E'(t) = -m[\mathbf{r}''(t) \cdot \mathbf{r}'(t)] + m[\mathbf{r}'(t) \cdot \mathbf{r}''(t)] = 0 \quad \blacksquare$$

We close this section by rephrasing some important points.

(19.41)   **Let $\mathbf{F} = \mathbf{F}(x, y)$ denote a field of force defined on some open simply connected set S. The following statements are equivalent:**

1. **$\mathbf{F}$ is the gradient of some function.**
2. **$\mathbf{F}$ is conservative.**
3. **The work done by $\mathbf{F}$ in moving an object of mass m from a point A to a point B in S is independent of the path chosen from A to B.**
4. **The work done by $\mathbf{F}$ in moving an object of mass m along any piecewise smooth closed curve C in S is 0.**

The equivalence of statements 1 and 2 is true by definition. The equivalence of 1 and 3 was proved in (19.14) and (19.16), and implication $3 \Rightarrow 4$ follows from (19.15). We shall not prove that 4 implies 3.

Some words of caution: First, statement 4 above asserts that the integral around *every* closed curve in S is 0 if and only if $\mathbf{F}$ is the gradient of some function f. It is possible to have $\int_C \mathbf{F} \cdot d\mathbf{r} = 0$ for a particular closed curve or for an infinite number of closed curves without $\mathbf{F}$ being the gradient of any function. Second, (19.41) asserts the *equivalency* of statements 1, 2, 3, and 4. If any one of them is true, so are the others. If any one of them is not true, then neither are the others. Finally, the equivalence of the statements applies only for simply connected sets; the difficulty of integrating around "holes" in a nonsimply connected set is demonstrated at the end of Section 4.

EXERCISE 3

In Problems 1–10 calculate the work done by the field of force **F** in moving an object along each curve between the indicated points.

1. $\mathbf{F} = y\mathbf{i} + x\mathbf{j}$ along $\mathbf{r}(t) = t\mathbf{i} + t^2\mathbf{j}$ from $t = 0$ to $t = 1$

2. $\mathbf{F} = xy\mathbf{i} + y^2\mathbf{j}$ along $\mathbf{r}(t) = t\mathbf{i} + t^2\mathbf{j}$ from $t = 0$ to $t = 1$

3. $\mathbf{F} = (x - 2y)\mathbf{i} + xy\mathbf{j}$ along $\mathbf{r}(t) = (3 \cos t)\mathbf{i} + (2 \sin t)\mathbf{j}$ from $t = 0$ to $t = \pi/2$

4. $\mathbf{F} = x\mathbf{i} - y\mathbf{j}$ along $\mathbf{r}(t) = (\cos t)\mathbf{i} + (\sin t)\mathbf{j}$ from $t = 0$ to $t = 2\pi$

5. $\mathbf{F} = x^2 y\mathbf{i} + (x^2 - y^2)\mathbf{j}$ along $y = 2x^2$ from $(0, 0)$ to $(1, 2)$

6. $\mathbf{F} = (y \sin x)\mathbf{i} - (x \cos y)\mathbf{j}$ along $y = x$ from $(0, 0)$ to $(1, 1)$

7. $\mathbf{F} = (y - x^2)\mathbf{i} + x\mathbf{j}$ along the upper half of the circle $x^2 + y^2 = 1$ from $(1, 0)$ to $(-1, 0)$

8. $\mathbf{F} = (y - x^2)\mathbf{i} + x\mathbf{j}$ along the line segments joining $(1, 0)$ to $(1, 1)$ to $(-1, 1)$ to $(-1, 0)$

9. $\mathbf{F} = y^3\mathbf{i} + x^3\mathbf{j}$ along the ellipse $\mathbf{r}(t) = (a \cos t)\mathbf{i} + (b \sin t)\mathbf{j}$ from $t = 0$ to $t = 2\pi$

10. $\mathbf{F} = -y^2\mathbf{i} + x^2\mathbf{j}$ along the upper half of the ellipse $(x^2/a^2) + (y^2/b^2) = 1$ from $(a, 0)$ to $(-a, 0)$

11. Verify that the force function $\mathbf{F} = y\mathbf{i} - x\mathbf{j}$ is nonconservative by showing that the integral $\int \mathbf{F} \cdot d\mathbf{r}$ is dependent on the path of integration, by taking two paths in which the starting point is the origin $(0, 0)$ and the endpoint is $(1, 1)$. For one path, take the line $x = y$.   For the other path, take the x-axis out to the point $(1, 0)$ and then the line $x = 1$ up to the point $(1, 1)$.

12. A particle moves in a clockwise direction from the origin to the point $(2a, 0)$ along the upper half of the circle $(x - a)^2 + y^2 = a^2$ and is acted upon by a force with constant magnitude 3 and with direction $\mathbf{i} + \mathbf{j}$.   Find the work done by this force.

13. A particle moves along the curve $\mathbf{r}(t) = 64\sqrt{3}\,t\mathbf{i} + (64t - 16t^2)\mathbf{j}$ from $t = 0$ to $t = 4$ and is acted upon by a force whose magnitude is directly proportional to the **speed** of the particle and whose direction is opposite to that of the velocity. Find the work done by this force.

14. A particle moves from the point $(0, 0)$ to the point $(1, 0)$ along the curve $y = ax(1 - x)$ and is acted upon by a force given by $\mathbf{F}(x, y) = (y^2 + 1)\mathbf{i} + (x + y)\mathbf{j}$.   Find a so that the work done is a minimum.

15. The repelling force between a charged particle P at the origin and an oppositely charged particle Q at (x, y) is

$$\mathbf{F}(x, y) = \frac{x}{(x^2 + y^2)^{3/2}}\mathbf{i} + \frac{y}{(x^2 + y^2)^{3/2}}\mathbf{j}$$

Find the work done by **F** as Q moves along the line segment from $(1, 0)$ to $(-1, 2)$.

4. Green's Theorem in the Plane

*Green's theorem** relates the value of a line integral along a closed curve C to a certain double integral over the region R enclosed by C. Green's theorem may be stated in

---

* Named in honor of the Englishman George Green (1793–1841), who first used it in 1828. Although the theorem was actually discovered earlier by Gauss and Lagrange, it became widely known as *Green's theorem*, because Green used it in applications to electricity and magnetism, fluid flow, and other areas of physics.

several ways, but the most common form is the following:

(19.42)
$$\oint_C (P\ dx + Q\ dy) = \iint_R \left(\frac{\partial Q}{\partial x} - \frac{\partial P}{\partial y}\right) dx\ dy$$

The curve C is the boundary of the region R, and the symbol $\oint$ indicates that the closed curve C is to be traversed in a counterclockwise direction; that is, the region R will always lie to the left as one moves along the curve (see Fig. 14).

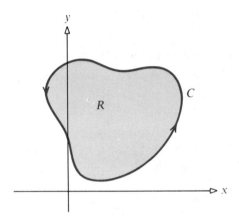

Figure 14

The result (19.42) requires two assumptions:

1. The functions P and Q possess continuous first-order partial derivatives at each point in the region R.*

2. The curve C may be any closed piecewise smooth curve, and the region R is a simply connected closed region—that is, R is the union of C and its interior.

We now give a rather general formulation of Green's theorem, but we will confine our proof to regions that are of both Type I and Type II (see Section 2, Chap. 18).

(19.43)   *Green's Theorem.*   **Let R be a closed simply connected region whose boundary is a closed piecewise smooth curve C. Let P and Q denote functions that possess continuous first-order partial derivatives on some open rectangular region that contains R. Then**

$$\oint_C (P\ dx + Q\ dy)) = \iint_R \left(\frac{\partial Q}{\partial x} - \frac{\partial P}{\partial y}\right) dx\ dy$$

where the line integral is taken around C in the counterclockwise direction.

---

* Somewhat less stringent conditions will still work.

Proof for Regions Which Are Both Type I and Type II   It is sufficient to prove that

(19.44)
$$\oint_C P(x, y)\, dx = -\iint_R \frac{\partial P}{\partial y}\, dx\, dy$$

and

(19.45)
$$\oint_C Q(x, y)\, dy = \iint_R \frac{\partial Q}{\partial x}\, dx\, dy$$

since the result will then follow from adding these two equations. Suppose the region R is both of Type I and Type II. Then, since it is of Type I (see Fig. 15), the boundary of R consists of two smooth curves, C_1: $y = g_1(x)$ and C_2: $y = g_2(x)$, $a \le x \le b$.

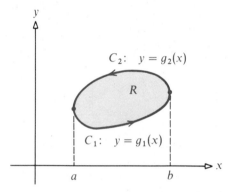

Figure 15

To establish (19.44) we begin with the double integral on the right side of (19.44). From (18.14), the double integral of $-\partial P/\partial y$ over R is given by

$$-\iint_R \frac{\partial P}{\partial y}\, dx\, dy = -\int_a^b \left[\int_{g_1(x)}^{g_2(x)} \frac{\partial P}{\partial y}\, dy \right] dx$$

$$= -\int_a^b \left[P(x, y)\Big|_{g_1(x)}^{g_2(x)} \right] dx = -\int_a^b \left[P(x, g_2(x)) - P(x, g_1(x)) \right] dx$$

$$= \int_a^b \left[P(x, g_1(x)) - P(x, g_2(x)) \right] dx$$

$$= \int_a^b P(x, g_1(x))\, dx - \int_a^b P(x, g_2(x))\, dx$$

$$= \int_{C_1} P(x, y)\, dx - \int_{C_2} P(x, y)\, dx = \int_{C_1} P(x, y)\, dx + \int_{-C_2} P(x, y)\, dx$$

$$= \oint_C P(x, y)\, dx$$

Thus, we have established (19.44). The proof of (19.45) uses the fact that the region R is also of Type II. It is obtained in a similar manner and is left as an exercise. ∎

EXAMPLE 1   Use Green's theorem to evaluate the line integral

(19.46)
$$\oint_C [(-2xy + y^2)\,dx + x^2\,dy]$$

where C is the boundary of the region R enclosed by $y = 4x$ and $y = 2x^2$.

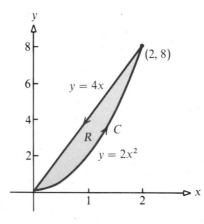

Figure 16

Solution   Figure 16 illustrates the curve C and the region R. By applying Green's theorem to (19.46), with

$$P(x, y) = -2xy + y^2 \qquad \text{and} \qquad Q(x, y) = x^2$$

we find

$$\oint_C [(-2xy + y^2)\,dx + x^2\,dy] = \iint_R \left[\frac{\partial}{\partial x}(x^2) - \frac{\partial}{\partial y}(-2xy + y^2) \right] dx\,dy$$

$$= \iint_R (4x - 2y)\,dx\,dy$$

$$= \underset{\substack{\uparrow \\ R \text{ is Type I}}}{\int_0^2} \left[\int_{2x^2}^{4x} (4x - 2y)\,dy \right] dx = \int_0^2 \left[(4xy - y^2) \Big|_{2x^2}^{4x} \right] dx$$

$$= \int_0^2 [(16x^2 - 16x^2) - (8x^3 - 4x^4)]\,dx$$

$$= \left(-2x^4 + \frac{4}{5}x^5 \right) \Big|_0^2 = -32 + \frac{128}{5} = \frac{-32}{5} \qquad ∎$$

You should verify that the same answer may be obtained by evaluating the line integral using the method developed in Section 1. The next example illustrates how Green's theorem is used to evaluate line integrals that may not be easily evaluated by the techniques of Section 1.

EXAMPLE 2   Use Green's theorem to evaluate the line integral

(19.47)
$$\oint_C [(e^{-x^2} + y^2)\,dx + (\ln y - x^2)\,dy]$$

where C is the square illustrated in Figure 17.

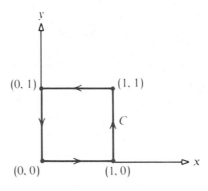

Figure 17

Solution   Because e^{-x^2} appears in the integrand of (19.47), the use of prior techniques may be useless. However, an application of Green's theorem produces

$$\oint_C [(e^{-x^2} + y^2)\,dx + (\ln y - x^2)]\,dy = \iint_R (-2x - 2y)\,dx\,dy$$

$$= -2 \int_0^1 \left[\int_0^1 (x + y)\,dy \right] dx$$

$$= -2 \int_0^1 \left[\left(xy + \frac{y^2}{2} \right)\Big|_0^1 \right] dx = -2 \int_0^1 \left(x + \frac{1}{2} \right) dx$$

$$= -2 \left(\frac{x^2}{2} + \frac{x}{2} \right)\Big|_0^1 = -2 \quad ■$$

APPLICATION OF GREEN'S THEOREM TO AREA

Green's theorem can be used to express the area A of a region R enclosed by a closed piecewise smooth curve C as a line integral.

(19.48) **If C denotes a closed piecewise smooth curve C, the area A of the region R enclosed by C is given by**

$$A = \frac{1}{2} \oint_C (-y\,dx + x\,dy)$$

Proof   In Green's theorem, put $P = P(x, y) = -y$ and $Q = Q(x, y) = x$.   Then

$$\oint_C (-y\,dx + x\,dy) = \iint_R \left[\frac{\partial}{\partial x}(x) - \frac{\partial}{\partial y}(-y) \right] dx\,dy = \iint_R 2\,dx\,dy = 2A \quad \blacksquare$$

In Problem 30 of Exercise 4 you are asked to show that the area A in (19.48) can also be expressed by the formulas

(19.49) $$A = \oint_C x\,dy \quad \text{and} \quad A = \oint_C (-y)\,dx$$

EXAMPLE 3   Use (19.48) to find the area of the region enclosed by the ellipse $(x^2/a^2) + (y^2/b^2) = 1$.

Solution   To evaluate the integral (19.48), we use the parametric equations of the ellipse: $x = a\cos t$, $y = b\sin t$, $0 \le t \le 2\pi$.   We   thus   have $dx = -a\sin t\,dt$ and $dy = b\cos t\,dt$.   Then

$$A = \frac{1}{2} \oint_C (-y\,dx + x\,dy)$$

$$= \frac{1}{2} \int_0^{2\pi} [-b\sin t(-a\sin t\,dt) + a\cos t(b\cos t\,dt)]$$

$$= \frac{1}{2} \int_0^{2\pi} ab(\sin^2 t + \cos^2 t)\,dt$$

$$= \frac{1}{2}(ab) \int_0^{2\pi} dt = \frac{1}{2}ab(2\pi) = \pi ab \quad \blacksquare$$

APPLICATION OF GREEN'S THEOREM TO CONSERVATIVE FIELDS OF FORCE

Another application of Green's theorem is to conservative fields of force. Suppose that $\mathbf{F} = \mathbf{F}(x, y) = P(x, y)\mathbf{i} + Q(x, y)\mathbf{j}$ denotes a conservative field of force for which the components P and Q possess continuous first-order partial derivatives on an open connected region S. It follows from (19.21) that $\partial P/\partial y = \partial Q/\partial x$ throughout S. We shall now prove that the converse is true, provided S is simply connected.

(19.50) **THEOREM**   **Let $\mathbf{F} = P(x, y)\mathbf{i} + Q(x, y)\mathbf{j}$ denote a field of force whose components P and Q possess continuous first-order partial derivatives throughout an open simply connected region S.**

If $\partial P/\partial y = \partial Q/\partial x$ throughout S, then $\mathbf{F}$ is conservative.

To show that **F** is conservative, we need only show that for any curve C in S, $\int_C \mathbf{F} \cdot d\mathbf{r} = \int_C (P\,dx + Q\,dy)$ is independent of path; that is, that it has the same value for any two piecewise smooth curves joining a fixed point A in S to an arbitrary point (x, y) in S [see (19.41)].

Proof for a Special Case   Let C_1 and C_2 be two piecewise smooth curves in S joining A to (x, y). We assume these curves do not intersect or coincide except at A and (x, y).* The curves C_1 and C_2 will therefore form the boundary of a region R that is a subset of S since S is simply connected (see Fig. 18).

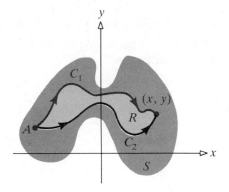

Figure 18

We apply Green's theorem on this region R to get

$$\iint\limits_R \left(\frac{\partial Q}{\partial x} - \frac{\partial P}{\partial y} \right) dx\,dy \underset{\underset{(19.43)}{\uparrow}}{=} \int_{C_1 \cup -C_2} (P\,dx + Q\,dy)$$

$$= \int_{C_1} (P\,dx + Q\,dy) + \int_{-C_2} (P\,dx + Q\,dy)$$

$$= \int_{C_1} (P\,dx + Q\,dy) - \int_{C_2} (P\,dx + Q\,dy)$$

But $\partial P/\partial y = \partial Q/\partial x$ throughout S. So,

$$\iint\limits_R \left(\frac{\partial Q}{\partial x} - \frac{\partial P}{\partial y} \right) dx\,dy = 0$$

Hence, we conclude that $\int_{C_1} (P\,dx + Q\,dy) = \int_{C_2} (P\,dx + Q\,dy)$.   Thus, the line integral is independent of path, and **F** is conservative. This completes the proof of theorem (19.21).   ∎

---

* This assumption may be relaxed so that the curves C_1 and C_2 may sometimes intersect or even coincide in S.

Because of (19.41) and (19.50), we may summarize much of the discussion in this chapter as follows:

(19.51)   Let $F = P(x, y)\mathbf{i} + Q(x, y)\mathbf{j}$ denote a field of force whose components P and Q possess continuous first-order partial derivatives throughout an open connected region S. If S is simply connected, the following statements are equivalent:

1. **F is conservative.**
2. **F is the gradient of some function.**
3. **The work done by F in moving an object of mass m from a point A to a point B in S is independent of the path chosen from A to B.**
4. **The work done by F in moving an object of mass m along any closed piecewise smooth curve C in S is 0.**
5. $\partial P/\partial y = \partial Q/\partial x$ **throughout** S.

You may recall that (19.41) gave equivalent statements for a force F to be conservative on an open connected set S. If S is simply connected, we now have the additional equivalence $\partial P/\partial y = \partial Q/\partial x$.   This is the condition that is easiest to use in practice.

EXAMPLE 4   Each of the following fields of force is conservative on the indicated region:
(a) $F = 2xy\mathbf{i} + (x^2 + 1)\mathbf{j}$ Entire plane

Since $\dfrac{\partial}{\partial x}(x^2 + 1) = 2x$ and $\dfrac{\partial}{\partial y}(2xy) = 2x$

(b) $F = \dfrac{x}{y^2}\mathbf{i} - \dfrac{x^2}{y^3}\mathbf{j}$ Any region not containing the x-axis

Since $\dfrac{\partial}{\partial x}\left(-\dfrac{x^2}{y^3}\right) = \dfrac{-2x}{y^3}$ and $\dfrac{\partial}{\partial y}\left(\dfrac{x}{y^2}\right) = \dfrac{-2x}{y^3}$ ■

EXAMPLE 5   (a)   For the line integral

$$\int_C \left(\frac{x}{y^2}\,dx - \frac{x^2}{y^3}\,dy\right)$$

we have

$$\frac{\partial}{\partial x}\left(\frac{-x^2}{y^3}\right) = \frac{-2x}{y^3} \qquad \text{and} \qquad \frac{\partial}{\partial y}\left(\frac{x}{y^2}\right) = \frac{-2x}{y^3}$$

Thus, the integral is independent of path in any simply connected region for which $y \neq 0$.

(b)   For the line integral

$$\int_C \frac{-y\,dx + x\,dy}{x^2 + y^2}$$

we have

$$\frac{\partial}{\partial x}\left(\frac{x}{x^2 + y^2}\right) = \frac{y^2 - x^2}{(x^2 + y^2)^2} \quad \text{and} \quad \frac{\partial}{\partial y}\left(\frac{-y}{x^2 + y^2}\right) = \frac{y^2 - x^2}{(x^2 + y^2)^2}$$

Thus, the integral will be independent of path, provided the region does not contain the origin in its interior.   ■

GREEN'S THEOREM FOR MULTIPLY
CONNECTED REGIONS (IF TIME PERMITS)

For Green's theorem to hold, we assume that the closed piecewise smooth curve C and its interior form the region R over which the double integral is evaluated. We now state the formulation of Green's theorem for *multiply connected regions*.

(19.52)   **THEOREM**   Let $C_1, C_2, \ldots, C_n$ be n closed piecewise smooth curves for which:

1. No two of the curves intersect.

2. Each of the curves lies in the interior of a closed piecewise smooth curve C.

3. Any curve C_i is exterior to every curve C_j, $i \neq j$, $i, j = 1, 2, \ldots, n$.

Let R denote the region consisting of the interior of C and its boundary, less the interior of each of the curves $C_1, C_2, \ldots, C_n$.   Let P and Q denote functions that possess continuous first-order partial derivatives throughout R. Then

$$\iint_R \left(\frac{\partial Q}{\partial x} - \frac{\partial P}{\partial y}\right) dx\,dy = \oint_C (P\,dx + Q\,dy) - \sum_{i=1}^{n} \oint_{C_i} (P\,dx + Q\,dy)$$

Figure 19(a) illustrates an example of a multiply connected region.

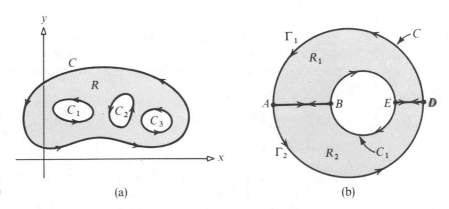

Figure 19   (a)   (b)

We give the idea of the proof for $n = 1$: In Figure 19(b), curve C_1 is a circle contained within the larger circle C. We construct line segments $\overline{AB}$ and $\overline{ED}$ that join the curves C and C_1. This decomposes R into two regions R_1 and R_2, with boundaries Γ_1 and Γ_2, respectively. Since R_1 and R_2 each satisfies the conditions of Green's theorem, we find

$$\iint_R \left(\frac{\partial Q}{\partial x} - \frac{\partial P}{\partial y}\right) dx\, dy = \iint_{R_1} \left(\frac{\partial Q}{\partial x} - \frac{\partial P}{\partial y}\right) dx\, dy + \iint_{R_2} \left(\frac{\partial Q}{\partial x} - \frac{\partial P}{\partial y}\right) dx\, dy$$

$$= \oint_{\Gamma_1} (P\, dx + Q\, dy) + \oint_{\Gamma_2} (P\, dx + Q\, dy)$$

$$= \int_{\widehat{DA}} (P\, dx + Q\, dy) + \int_{\overline{AB}} (P\, dx + Q\, dy) + \int_{\widehat{BE}} (P\, dx + Q\, dy)$$

$$+ \int_{\overline{ED}} (P\, dx + Q\, dy) + \int_{\overline{DE}} (P\, dx + Q\, dy) + \int_{\widehat{EB}} (P\, dx + Q\, dy)$$

$$+ \int_{\overline{BA}} (P\, dx + Q\, dy) + \int_{\widehat{AD}} (P\, dx + Q\, dy)$$

Along the line segments $\overline{AB}$ and $\overline{ED}$,

$$\int_{\overline{AB}} (P\, dx + Q\, dy) = -\int_{\overline{BA}} (P\, dx + Q\, dy)$$

and

$$\int_{\overline{ED}} (P\, dx + Q\, dy) = -\int_{\overline{DE}} (P\, dx + Q\, dy)$$

Hence,

$$\iint_R \left(\frac{\partial Q}{\partial x} - \frac{\partial P}{\partial y}\right) dx\, dy = \oint_C (P\, dx + Q\, dy) + \oint_{-C_1} (P\, dx + Q\, dy)$$

$$= \oint_C (P\, dx + Q\, dy) - \oint_{C_1} (P\, dx + Q\, dy)$$

The next result is a consequence of theorem (19.52).

(19.53)   **THEOREM   Let P and Q denote functions that possess continuous first-order partial derivatives throughout an open connected set S. Let C_1 and C_2 denote two closed piecewise smooth curves, each lying entirely in S, for which:**

1. Curve C_2 lies in the interior C_1.

2. All points inside C_1 that are also outside C_2 form a simply connected subset of S.

If $\partial P/\partial y = \partial Q/\partial x$ throughout S, then

$$\int_{C_1} (P\, dx + Q\, dy) = \int_{C_2} (P\, dx + Q\, dy)$$

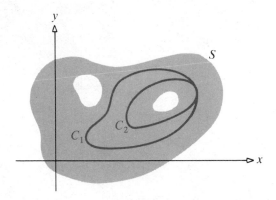

Figure 20

Figure 20 illustrates a set S satisfying the conditions of (19.53). This theorem may be restated as: If $\partial P/\partial y = \partial Q/\partial x$ throughout S, then the value of a line integral along a closed piecewise smooth curve equals the value of the same line integral along any other closed piecewise smooth curve that can be obtained from the original curve by a continuous deformation with all of the intermediate curves of the deformation being entirely in S.

Let's look at an example.

EXAMPLE 6   Evaluate: $\displaystyle\oint_C \left(\frac{-y}{x^2 + y^2}\, dx + \frac{x}{x^2 + y^2}\, dy \right)$

(a)   If C is the curve $x^{2/3} + y^{2/3} = 1$
(b)   If C is any closed piecewise smooth curve not containing the origin in its interior

Solution   We first note that, for

$$P = \frac{-y}{x^2 + y^2} \qquad \text{and} \qquad Q = \frac{x}{x^2 + y^2}$$

we have

$$\frac{\partial P}{\partial y} = \frac{y^2 - x^2}{(x^2 + y^2)^2} \qquad \text{and} \qquad \frac{\partial Q}{\partial x} = \frac{y^2 - x^2}{(x^2 + y^2)^2}$$

Next, we note that, except for $(0, 0)$, $\partial P/\partial y$ and $\partial Q/\partial x$ are continuous.

(a)   The line integral is difficult to evaluate. We cannot change the problem as was done in Example 2 because $x^{2/3} + y^{2/3} = 1$ contains $(0, 0)$ in its interior, so Green's theorem does not apply. However, we can change the curve and evaluate a different line integral and obtain the desired result. We choose to replace

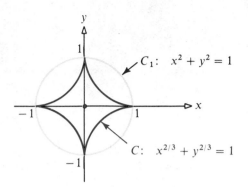

Figure 21

$x^{2/3} + y^{2/3} = 1$ by the unit circle C_1: $x^2 + y^2 = 1$, noting that the conditions required by (19.53) are met (see Fig. 21). Thus,

$$\oint_C \left(\frac{-y}{x^2 + y^2}\, dx + \frac{x}{x^2 + y^2}\, dy \right) = \oint_{C_1} \left(\frac{-y}{x^2 + y^2}\, dx + \frac{x}{x^2 + y^2}\, dy \right)$$

$$= \int_0^{2\pi} \left[\frac{-\sin\theta(-\sin\theta)\, d\theta}{1} + \frac{\cos\theta(\cos\theta)\, d\theta}{1} \right]$$

Use $x = \cos\theta$
$y = \sin\theta$

$$= \int_0^{2\pi} (\sin^2\theta + \cos^2\theta)\, d\theta = 2\pi$$

(b) In this case, the interior of the curve C does not contain the origin. Since $\partial P/\partial y = \partial Q/\partial x$ in the simply connected region R enclosed by C, we have, by (19.51),

$$\oint_C \left(\frac{-y}{x^2 + y^2}\, dx + \frac{x}{x^2 + y^2}\, dy \right) = 0 \quad \blacksquare$$

EXERCISE 4

In Problems 1–6 use Green's theorem to evaluate the given line integral. In each case, assume that C is the perimeter of the rectangle with vertices $(0, 0)$, $(4, 0)$, $(4, 3)$, $(0, 3)$, and that the orientation of C is counterclockwise.

1. $\oint_C y\, dx$ 2. $\oint_C x\, dy$ 3. $\oint_C [xy\, dx + (x + y)\, dy]$

4. $\oint_C (3y\, dx - 2x\, dy)$ 5. $\oint_C (xy^2\, dx + x^2 y\, dy)$ 6. $\oint_C (y \sin xy\, dx + x \sin xy\, dy)$

In Problems 7–10 use Green's theorem to evaluate the given line integral. In each case, assume that C is the unit circle $x^2 + y^2 = 1$, and that the orientation of C is counterclockwise.

7. $\oint_C (-x^2 y\, dx + y^2 x\, dy)$ 8. $\oint_C [y(x^2 + y^2)\, dx - x(x^2 + y^2)\, dy]$

9. $\oint_C [(x^2 - y^3) dx + (x^2 + y^2) dy]$ **10.** $\oint_C [(xy^3 + \sin x) dx + (x^2 y^2 + 4x) dy]$

In Problems 11–16 use Green's theorem to evaluate the indicated integral. In each case, C is to be traversed in the counterclockwise direction.

11. $\oint_C [(x^2 + y) dx + (x - y^2) dy]$,   where C is the boundary of the region R enclosed by $y = x^{3/2}$, the x-axis, and $x = 1$

12. $\oint_C [(4x^2 - 8y^2) dx + (y - 6xy) dy]$,   where C is the boundary of the region R enclosed by $y = \sqrt{x}$ and $y = x^2$

13. $\oint_C [(x^3 - x^2 y) dx + xy^2 \, dy]$,   where C is the boundary of the region R enclosed by $y = x^2$ and $x = y^2$

14. $\oint_C [(x^2 - y^2) dx + xy \, dy]$,   where C is the boundary of the region R enclosed by $x = y^2$ and $x = 1$

15. $\oint_C [(1/y) dx + (1/x) dy]$,   where C is the boundary of the region R enclosed by $y = 1$, $x = 9$, and $y = \sqrt{x}$

16. $\oint_C (x^2 y \, dx - y^2 x \, dy)$,   where C is the boundary of the region R enclosed by $y = \sqrt{a^2 - x^2}$ and $y = 0$

In Problems 17–20 use Green's theorem to find the area of the region described.

17. Enclosed by $y = x^2$ and $y = x + 2$ **18.** Enclosed by $y = x^2 - 1$ and $y = 0$

19. Under one arch of the cycloid $\mathbf{r}(t) = (2\pi t - \sin 2\pi t)\mathbf{i} + (1 - \cos 2\pi t)\mathbf{j}$, $0 \le t \le 1$

20. Enclosed by the hypocycloid $x^{2/3} + y^{2/3} = 1$ [*Hint:* Use the parametric equations $x = \cos^3 t$, $y = \sin^3 t$.]

State why each of the line integrals in Problems 21–24 is equal to 0, where C is the given piecewise smooth closed curve.

21. $\oint_C (xe^{x^2 + y^2} dx + ye^{x^2 + y^2} dy)$,   where C is any piecewise smooth closed curve in the plane

22. $\oint_C [e^{xy}(xy + 1) dx + x^2 e^{xy} dy]$,   where C is any piecewise smooth closed curve in the plane

23. $\oint_C (e^x \sin y \, dx + e^x \cos y \, dy)$,   where C is any piecewise smooth closed curve in the plane

24. $\oint_C \left[\dfrac{x \, dx}{(x^2 + y^2)^{1/2}} + \dfrac{y \, dy}{(x^2 + y^2)^{1/2}} \right]$,   where C is any piecewise smooth closed curve not containing the origin in its interior

In Problems 25–28 evaluate $\oint_C \dfrac{y \, dx - x \, dy}{x^2 + y^2}$ about the indicated curve C.

25. C: $x^2 + y^2 = 4$ **26.** C: $(x - 2)^2 + (y + 3)^2 = 1$

27. C: $x^{2/3} + y^{2/3} = 1$ **28.** C: $3x^2 + 6y^2 = 4$

29. Complete the proof of Green's theorem by establishing (19.45).

30. Use (19.48) to obtain the formulas (19.49).

31. Let $r = f(\theta)$ be continuous and nonnegative on $\alpha \leq \theta \leq \beta$ and let R be the region as depicted in the figure. Show that the area of R is $A = \frac{1}{2} \int_{\alpha}^{\beta} [f(\theta)]^2 \, d\theta$.

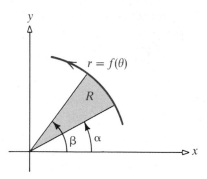

5. Surface Integrals

Surface integrals are similar to line integrals, but the integration takes place on a surface instead of taking place along a curve.

Let Σ be a surface of the type discussed in Section 6 of Chapter 18, having projection R on the xy-plane, as illustrated in Figure 22. Assume that the surface Σ is

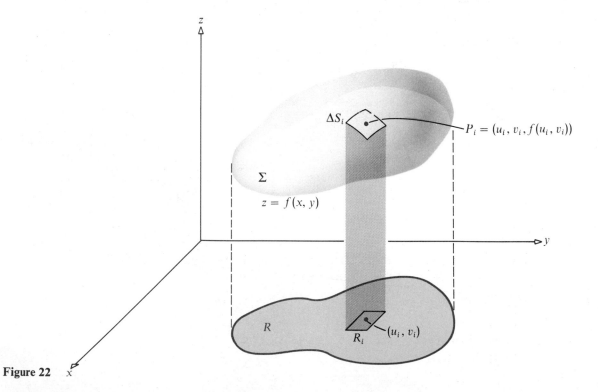

Figure 22

given by $z = f(x, y)$,   where f is continuous and has continuous first-order partial derivatives in R. Following our usual practice, we partition R into n rectangles and concentrate on the ith rectangle R_i, whose area is ΔA_i. Then we pick a point (u_i, v_i) in each rectangle R_i and erect a vertical column on each of these rectangles to intersect the tangent plane at $(u_i, v_i, f(u_i, v_i))$ on Σ in an area ΔS_i. The number ΔS_i is an approximation to the area of the part of the surface that lies above the ith rectangle R_i.

Let $\phi(x, y, z)$ be a function that is continuous at all points of Σ. Evaluate ϕ at the point $(u_i, v_i, f(u_i, v_i))$ on Σ, and form the product $\phi(u_i, v_i, f(u_i, v_i))\Delta S_i$.   Then form the sum

(19.54)
$$\sum_{i=1}^{n} \phi(u_i, v_i, f(u_i, v_i))\Delta S_i$$

We take the limit of this sum as the norm $\|P\|$ of the partition approaches 0. If this limit exists, we refer to it as the *surface integral of ϕ over Σ*, and we write

(19.55)
$$\iint_{\Sigma} \phi(x, y, z)\,dS = \lim_{\|P\| \to 0} \sum_{i=1}^{n} \phi(u_i, v_i, f(u_i, v_i))\Delta S_i$$

By (18.37), $\Delta S_i = \sqrt{[f_x(u_i, v_i)]^2 + [f_y(u_i, v_i)]^2 + 1}\ \Delta x_i \Delta y_i$ so that (19.55) can be written as

(19.56) $\displaystyle\iint_{\Sigma} \phi(x, y, z)\,dS = \lim_{\|P\| \to 0} \sum_{i=1}^{n} \phi(u_i, v_i, f(u_i, v_i))\sqrt{[f_x(u_i, v_i)]^2 + [f_y(u_i, v_i)]^2 + 1}\ \Delta x_i \Delta y_i$

Thus,

(19.57) $\displaystyle\iint_{\Sigma} \phi(x, y, z)\,dS = \iint_{R} \phi(x, y, f(x, y))\sqrt{[f_x(x, y)]^2 + [f_y(x, y)]^2 + 1}\ dx\,dy$

Let's see how (19.57) is used to calculate surface integrals.

EXAMPLE 1   Evaluate $\iint_{\Sigma} (x^2 + y^2)\,dS$,   where Σ is the part of the surface of the paraboloid $z = f(x, y) = 1 - x^2 - y^2$ that lies above the xy-plane.

Solution   For $f(x, y) = 1 - x^2 - y^2$,   it is easy to calculate that $\sqrt{f_x^2 + f_y^2 + 1} = \sqrt{4(x^2 + y^2) + 1}$. Using this fact in (19.57), with $\phi(x, y, z) = x^2 + y^2$,   we obtain

$$\iint_{\Sigma} (x^2 + y^2)\,dS = \iint_{R} (x^2 + y^2)\sqrt{4(x^2 + y^2) + 1}\ dx\,dy$$

Because the integrand is a function of $x^2 + y^2$,   we choose to use polar coordinates to evaluate the double integral. The result is

$$\iint_{\Sigma} (x^2 + y^2)\,dS = \int_0^{2\pi} \left[\int_0^1 r^2\sqrt{4r^2 + 1}\ r\,dr \right] d\theta$$

$$= \int_0^{2\pi} \left[\int_1^{\sqrt{5}} \frac{u^4 - u^2}{16}\,du \right] d\theta = \frac{1}{16} \int_0^{2\pi} \frac{50\sqrt{5} + 2}{15}\,d\theta \approx 0.948\pi \qquad \blacksquare$$

$$\uparrow$$
$$\text{Set}\quad u = \sqrt{4r^2 + 1}$$

If $\phi(x, y, z) = 1$,   then equation (19.57) takes the form

$$\iint_{\Sigma} dS = \iint_{R} \sqrt{[f_x(x, y)]^2 + [f_y(x, y)]^2 + 1} \, dx \, dy$$

which is the same as formula (18.39) defining surface area.

OTHER TYPES OF SURFACE INTEGRALS

We have been evaluating surface integrals over surfaces that have projections on the xy-plane. If the surface Σ has a projection R on the xz-plane, so that Σ is given by the equation $y = h(x, z)$ with (x, z) in R, then

(19.58) $$\iint_{\Sigma} \phi(x, y, z) \, dS = \iint_{R} \phi(x, h(x, z), z) \sqrt{[h_x(x, z)]^2 + [h_z(x, z)]^2 + 1} \, dx \, dz$$

Similarly, if Σ is a surface that has a projection R on the yz-plane, so that Σ is given by the equation $x = g(y, z)$ with (y, z) in R, then

(19.59) $$\iint_{\Sigma} \phi(x, y, z) \, dS = \iint_{R} \phi(g(y, z), y, z) \sqrt{[g_y(y, z)]^2 + [g_z(y, z)]^2 + 1} \, dy \, dz$$

We omit the proof.

EXERCISE 5

In Problems 1–8 evaluate $\iint_{\Sigma} \phi(x, y, z) \, dS$.

1. $\phi(x, y, z) = x^2 + y^2 + z$; Σ: $z = x + y + 1$, $0 \le x \le 1$, $0 \le y \le 1$

2. $\phi(x, y, z) = x + y + z$; Σ: portion of the plane $z = x - 3y$ above the region enclosed by $y = 0$, $x = 1$,   and $x = 3y$ in the xy-plane

3. $\phi(x, y, z) = x$; Σ: portion of the plane $x + y + 2z = 4$ above the region enclosed by $x = 1$, $y = 1$, $x = 0$,   and $y = 0$

4. $\phi(x, y, z) = y$; Σ: portion of the plane $x + y + z = 2$ inside the cylinder $x^2 + y^2 = 1$

5. $\phi(x, y, z) = x^2 z$; Σ: $x^2 + y^2 = 1$, $0 \le z \le 1$

6. $\phi(x, y, z) = yz$; Σ: first octant part of the plane $x + 2y + 3z = 6$

7. $\phi(x, y, z) = x^2$; Σ: piece of the cylinder $x^2 + y^2 = 1$ that is in the first octant and between the planes $z = 0$ and $z = 1$

8. $\phi(x, y, z) = (x + y)z$; Σ: surface of the unit cube

6. Outer Unit Normals

We begin with the definition:

Let T denote a closed, bounded region in three-space and let Σ be the surface that forms the boundary of T. At any point on the surface Σ the *outer unit normal* n is the

unit vector parallel to the normal line to Σ that points away from the region T enclosed by Σ.

Although at first this definition seems easy enough, it contains many hidden difficulties. The region T enclosed by Σ may be a familiar solid such as a cube, a solid sphere, a right circular cylinder, and so on. See Figure 23(a). But T might be the region between two concentric spheres or the region left when a sphere is removed from the inside of a large ellipsoid. See Figure 23(b). We shall limit our discussion here and in the next section to simple types of regions T and leave the general discussion to courses such as metric topology.

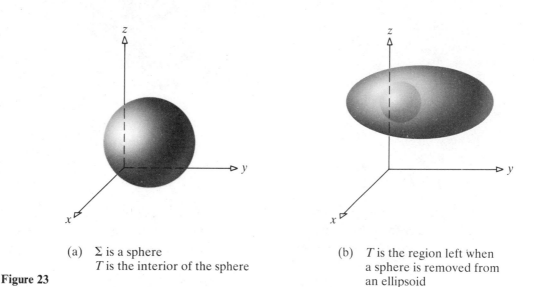

(a) Σ is a sphere
      T is the interior of the sphere

(b) T is the region left when
      a sphere is removed from
      an ellipsoid

Figure 23

REGIONS ENCLOSED BY SURFACES GIVEN BY $z = f(x, y)$

Consider the surface Σ that forms the boundary of the region T depicted in Figure 24. There we see that Σ actually consists of three surfaces, Σ_1, Σ_2, and Σ_3. Note that the outer unit normals $\mathbf{n}_1$, $\mathbf{n}_2$, and $\mathbf{n}_3$ of the three surfaces each point away from the region T enclosed by $\Sigma = \Sigma_1 \cup \Sigma_2 \cup \Sigma_3$.

We assume the surfaces Σ_1 and Σ_2 are defined by the equations

$$\Sigma_2: \quad z = f_2(x, y) \qquad \Sigma_1: \quad z = f_1(x, y)$$

where $f_1(x, y) < f_2(x, y)$ on D, f_1 and f_2 are continuous on D, and f_1, f_2 possess continuous partial derivatives on D. The surface Σ_3 is the portion of a cylindrical surface between Σ_1 and Σ_2 formed by lines parallel to the z-axis and along the boundary of D. The region T is enclosed by Σ_2 on the top, Σ_1 on the bottom, and the lateral surface Σ_3 on the sides (see Fig. 24, p. 1058).

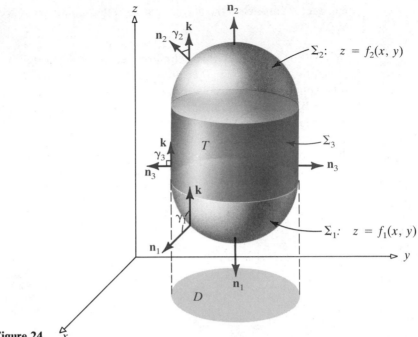

Figure 24

The direction angle γ_3 between the outer unit normal $\mathbf{n}_3$ to Σ_3 and the direction of the positive z-axis, $\mathbf{k}$, is a right angle so that $\cos \gamma_3 = 0$.   We seek expressions for the outer unit normals $\mathbf{n}_2$ and $\mathbf{n}_1$, to Σ_2 and Σ_1, respectively.

For the top surface Σ_2: $z = f_2(x, y)$,   the direction angle γ_2 between $\mathbf{n}_2$ and the positive z-axis, $\mathbf{k}$, is necessarily acute. Hence, the direction cosine, $\cos \gamma_2$, is positive. Recall from Chapter 17 (p. 927) that the normal line to Σ_2 is parallel to the vector $f_{2_x}\mathbf{i} + f_{2_y}\mathbf{j} - \mathbf{k}$.   The unit vector $\mathbf{n}_2$ in this direction for which the direction cosine, $\cos \gamma_2$, is positive is the outer unit normal $\mathbf{n}_2$ we seek. Based on this, we find $\mathbf{n}_2$ and $\cos \gamma_2$ to be

$$\mathbf{n}_2 = \frac{-f_{2_x}(x, y)\mathbf{i} - f_{2_y}(x, y)\mathbf{j} + \mathbf{k}}{\sqrt{[f_{2_x}(x, y)]^2 + [f_{2_y}(x, y)]^2 + 1}}$$

(19.60)

$$\cos \gamma_2 = \mathbf{n}_2 \cdot \mathbf{k} = \frac{1}{\sqrt{[f_{2_x}(x, y)]^2 + [f_{2_y}(x, y)]^2 + 1}}$$

For the bottom surface Σ_1: $z = f_1(x, y)$,   the direction angle γ_1 between the outer unit normal $\mathbf{n}_1$ and the positive z-axis, $\mathbf{k}$, is necessarily obtuse. Hence, the direction cosine, $\cos \gamma_1$, is negative. The normal line to Σ_1 is parallel to the vector $f_{1_x}\mathbf{i} + f_{1_y}\mathbf{j} - \mathbf{k}$. The unit vector in this direction for which the direction cosine, $\cos \gamma_1$, is negative is the outer unit normal $\mathbf{n}_1$ we seek. Based on this, we find $\mathbf{n}_1$ and cos γ_1 to be

$$\mathbf{n}_1 = \frac{f_{1_x}(x, y)\mathbf{i} + f_{1_y}(x, y)\mathbf{j} - \mathbf{k}}{\sqrt{[f_{1_x}(x, y)]^2 + [f_{1_y}(x, y)]^2 + 1}}$$

(19.61)

$$\cos \gamma_1 = \mathbf{n}_1 \cdot \mathbf{k} = \frac{-1}{\sqrt{[f_{1_x}(x, y)]^2 + [f_{1_y}(x, y)]^2 + 1}}$$

EXAMPLE 1 Find the outer unit normal to the region T enclosed by

$$z = \sqrt{R^2 - x^2 - y^2} \qquad 0 \le x^2 + y^2 \le R^2$$

Solution The region T is the interior of a hemisphere with center at $(0, 0, 0)$ and radius R (see Fig. 25). The top surface Σ_2 and the bottom surface Σ_1 are defined by

$$\Sigma_2: \quad z = \sqrt{R^2 - x^2 - y^2} \qquad \Sigma_1: \quad z = 0 \qquad 0 \le x^2 + y^2 \le R^2$$

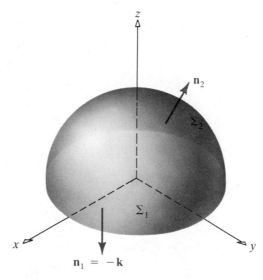

Figure 25

Since

$$\frac{\partial z}{\partial x} = \frac{-x}{\sqrt{R^2 - x^2 - y^2}} = \frac{-x}{z} \qquad \text{and} \qquad \frac{\partial z}{\partial y} = \frac{-y}{\sqrt{R^2 - x^2 - y^2}} = \frac{-y}{z}$$

the outer unit normal $\mathbf{n}_2$ of Σ_2 is given by

$$\mathbf{n}_2 = \frac{x\mathbf{i} + y\mathbf{i} + z\mathbf{k}}{R}$$

The outer unit normal $\mathbf{n}_1$ of Σ_1 is $-\mathbf{k}$. ∎

APPLICATION TO FLUID FLOW

Suppose $\mathbf{F}$ denotes the velocity of a fluid whose variable mass density is $\rho = \rho(x, y, z)$. Then $\rho\mathbf{F}$ equals the *rate of flow* (mass per unit time per unit area). We wish to obtain a formula for the total mass of fluid that flows across a given surface Σ per unit time. Let $\mathbf{n}$ denote the outer unit normal of Σ.

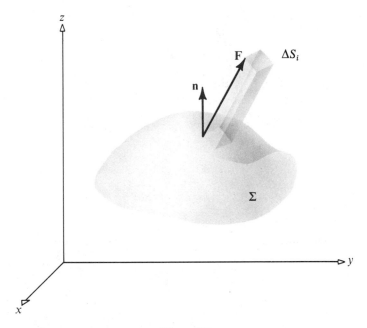

Figure 26

As Figure 26 illustrates, the mass of fluid crossing a patch of surface with area ΔS_i per unit time in the direction of the outer unit normal $\mathbf{n}_i$ is approximately*

$$\rho\mathbf{F} \cdot \mathbf{n}_i\Delta S_i$$

where $\mathbf{F}$ is evaluated at some point on the patch of surface. The total mass of fluid flowing across the surface Σ per unit time in the direction of $\mathbf{n}$ is

$$\lim_{\|\Delta\| \to 0} \sum_{i=1}^{n} \rho\mathbf{F} \cdot \mathbf{n}_i\Delta S_i = \iint_{\Sigma} \rho\mathbf{F} \cdot \mathbf{n} \, dS$$

EXAMPLE 2   Find the total mass of fluid of constant mass density ρ flowing across the paraboloid $z = 4 - x^2 - y^2$, $z \geq 0$,   per unit time in the direction of the outer unit nor-

---

* The velocity $\mathbf{F}$ is decomposed into a component parallel to the patch and a component normal to it. The parallel component carries no fluid across the surface. Hence, the mass carried across per unit time is the normal component of $\rho\mathbf{F}$.

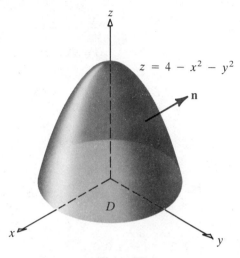

$z = 4 - x^2 - y^2$

Figure 27

mal, if the velocity of the fluid at any point on the paraboloid is $\mathbf{F} = \mathbf{F}(x, y, z) = x\mathbf{i} + y\mathbf{j} + 2z\mathbf{k}$ (see Fig. 27).

Solution

The outer unit normal $\mathbf{n}$ to the surface $z = 4 - x^2 - y^2$ is

$$\mathbf{n} = \frac{2x\mathbf{i} + 2y\mathbf{i} + \mathbf{k}}{\sqrt{4x^2 + 4y^2 + 1}}$$

Thus,

$$\mathbf{F} \cdot \mathbf{n} = \frac{2x^2 + 2y^2 + 2z}{\sqrt{4x^2 + 4y^2 + 1}}$$

The total mass of fluid flowing across $z = 4 - x^2 - y^2$ per unit time in the direction of $\mathbf{n}$ is, therefore,

$$\iint_{\Sigma} \rho \mathbf{F} \cdot \mathbf{n}\, dS = \rho \iint_{D} \frac{2x^2 + 2y^2 + 2z}{\sqrt{4x^2 + 4y^2 + 1}} \sqrt{4x^2 + 4y^2 + 1}\; dx\, dy$$

$$= 8\rho \iint_{D} dx\, dy = 8\rho\pi(2)^2 = 32\,\rho\pi \quad\blacksquare$$

$$2z = 8 - 2x^2 - 2y^2 \quad D \text{ is a disk of radius 2}$$

EXAMPLE 3

Find the total mass of fluid of constant mass density ρ flowing across the cube enclosed by the planes $x = 0$, $x = 1$, $y = 0$, $y = 1$, $z = 0$,   and $z = 1$ (see Fig. 28) per unit time, in the direction of the outer unit normal if the velocity of the fluid at any point on the cube is $\mathbf{F} = \mathbf{F}(x, y, z) = 4xz\mathbf{i} - y^2\mathbf{j} + yz\mathbf{k}$.

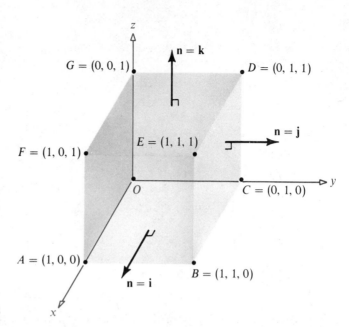

Figure 28

Solution

The total mass of fluid flowing across the cube per unit time in the direction of the outer unit normal **n** is

$$\iint_{\Sigma} \rho \mathbf{F} \cdot \mathbf{n}\, dS = \rho \iint_{\Sigma} \mathbf{F} \cdot \mathbf{n}\, dS$$

We decompose Σ into its six faces and calculate the surface integral over each one in Table 1. From the table, we can see that the total mass of fluid is $0 + 0 + 0 + 2\rho - \rho + \frac{1}{2}\rho = \frac{3}{2}\rho$.

Table 1

	Face	**n**	**F**	$\mathbf{F} \cdot \mathbf{n}$	$\rho \iint \mathbf{F} \cdot \mathbf{n}\, dS$
Σ_1	*ABCO*	$z = 0$	$-\mathbf{k}$	$-y^2\mathbf{j}$	0	0
Σ_2	*OAFG*	$y = 0$	$-\mathbf{j}$	$4xz\mathbf{i}$	0	0
Σ_3	*OCDG*	$x = 0$	$-\mathbf{i}$	$-y^2\mathbf{j} + yz\mathbf{k}$	0	0
Σ_4	*ABEF*	$x = 1$	$\mathbf{i}$	$4z\mathbf{i} - y^2\mathbf{j} + yz\mathbf{k}$	$4z$	$\rho \int_0^1 \int_0^1 4z\, dy\, dz = 2\rho$
Σ_5	*BCDE*	$y = 1$	$\mathbf{j}$	$4xz\mathbf{i} - \mathbf{j} + z\mathbf{k}$	-1	$-\rho \int_0^1 \int_0^1 dx\, dz = -\rho$
Σ_6	*DEFG*	$z = 1$	$\mathbf{k}$	$4x\mathbf{i} - y^2\mathbf{j} + y\mathbf{k}$	y	$\rho \int_0^1 \int_0^1 y\, dx\, dy = \frac{1}{2}\rho$

∎

EXERCISE 6

In Problems 1–6 find the total mass of fluid of constant mass density ρ flowing out of the cube in Figure 28 per unit time in the direction of the outer unit normal, if the velocity of the fluid at any point on the cube is given by **F**.

1. $\mathbf{F} = x\mathbf{i}$ **2.** $\mathbf{F} = y\mathbf{i}$ **3.** $\mathbf{F} = z\mathbf{i}$ **4.** $\mathbf{F} = x\mathbf{i} + y\mathbf{j}$

5. $\mathbf{F} = x\mathbf{i} + y\mathbf{j} + z\mathbf{k}$ **6.** $\mathbf{F} = z^2\mathbf{i}$

In Problems 7–12 evaluate $\iint_{\Sigma} \mathbf{F} \cdot \mathbf{n}\, dS$ where **n** is the outer unit normal of Σ.

7. $\mathbf{F} = x\mathbf{i} + y\mathbf{j} + z\mathbf{k};$ Σ: the upper half of the hemispherical region $x^2 + y^2 + z^2 = 1, \quad z \ge 0$

8. $\mathbf{F} = x\mathbf{i} + y\mathbf{j} + z\mathbf{k};$ Σ: the lower half of the hemispherical region $x^2 + y^2 + z^2 = 1, \quad z \le 0$

9. $\mathbf{F} = -y\mathbf{i} + x\mathbf{j} + z\mathbf{k};$ Σ: same as in Problem 7

10. $\mathbf{F} = x\mathbf{i} + y\mathbf{j} + z\mathbf{k};$ Σ: the sphere $x^2 + y^2 + z^2 = 1$

11. $\mathbf{F} = (x + y)\mathbf{i} + (2x - z)\mathbf{j} + y\mathbf{k};$ Σ: the tetrahedron formed by the coordinate planes and the plane $z + 2x + 2y = 8$

12. $\mathbf{F} = 2x\mathbf{i} - x^2\mathbf{j} + (z - 2x + 2y)\mathbf{k};$ Σ: the tetrahedron formed by the coordinate planes and the plane $2x + 2y + z = 6$

13. Refer to Figure 24, p. 1058. If **n** is the outer unit normal to Σ_2 and $\mathbf{F} = P\mathbf{i} + Q\mathbf{j} + R\mathbf{k},$ show that

$$\iint_{\Sigma_2} \mathbf{F} \cdot \mathbf{n}\, dS = \iint_{D} \left(-P\frac{\partial f}{\partial x} - Q\frac{\partial f}{\partial y} + R \right) dx\, dy$$

7. The Divergence Theorem

Recall that Green's theorem expresses a relationship between a certain double integral extended over a plane region and a line integral taken around its boundary. There are two ways to generalize this result to three-space. One of these, known as the *divergence theorem* (or *Gauss' theorem*), is the subject of this section, and the other, known as *Stokes' theorem*, is the subject of the next section.

DIVERGENCE

Let $\mathbf{F} = P\mathbf{i} + Q\mathbf{j} + R\mathbf{k}$ be a vector function such that $\partial P/\partial x$, $\partial Q/\partial y$, and $\partial R/\partial z$ all exist. The *divergence of* **F**, denoted by div **F** or $\nabla \cdot \mathbf{F}$, is the function

(19.62)
$$\nabla \cdot \mathbf{F} = \operatorname{div} \mathbf{F} = \frac{\partial P}{\partial x} + \frac{\partial Q}{\partial y} + \frac{\partial R}{\partial z}$$

The definition (19.62) makes it clear that the divergence of a vector function is a real-valued, or scalar, function.

EXAMPLE 1   The divergence of

$$\mathbf{F}(x, y, z) = x^2 y z^2 \mathbf{i} + (2xz + y^3)\mathbf{j} + x^2 y^3 z\mathbf{k}$$

is

$$\text{div } \mathbf{F} = \frac{\partial}{\partial x}(x^2 y z^2) + \frac{\partial}{\partial y}(2xz + y^3) + \frac{\partial}{\partial z}(x^2 y^3 z)$$

$$= 2xyz^2 + 3y^2 + x^2 y^3 \qquad \blacksquare$$

(19.63)   *Divergence Theorem.*   **Let P, Q, and R be functions of x, y, and z that are continuous and have continuous first-order partial derivatives in a closed and bounded region T. Let Σ be the surface that forms the boundary of T, and let $\cos \alpha$, $\cos \beta$, and $\cos \gamma$ be the direction cosines of the outer unit normal to Σ. Then,**

(19.64)
$$\iiint_T \left(\frac{\partial P}{\partial x} + \frac{\partial Q}{\partial y} + \frac{\partial R}{\partial z} \right) dV = \iint_\Sigma (P \cos \alpha + Q \cos \beta + R \cos \gamma)\, dS$$

This statement can be written more compactly in vector form. Thus, if $\mathbf{F} = P\mathbf{i} + Q\mathbf{j} + R\mathbf{k}$ is a vector function and if $\mathbf{n} = \cos \alpha \mathbf{i} + \cos \beta \mathbf{j} + \cos \gamma \mathbf{k}$ is the outer unit normal to Σ, then

(19.65)
$$\iiint_T \text{div } \mathbf{F}\, dV = \iiint_T \nabla \cdot \mathbf{F}\, dV = \iint_\Sigma \mathbf{F} \cdot \mathbf{n}\, dS$$

It is beyond the scope of this book to prove (19.63) for the general conditions on the region T. However, we can give the idea behind the proof. The basis of the proof is to verify each of the three equations

(19.66(a))
$$\iiint_T \frac{\partial P}{\partial x}\, dV = \iint_\Sigma P \cos \alpha\, dS$$

(19.66(b))
$$\iiint_T \frac{\partial Q}{\partial y}\, dV = \iint_\Sigma Q \cos \beta\, dS$$

(19.66(c))
$$\iiint_T \frac{\partial R}{\partial z}\, dV = \iint_\Sigma R \cos \gamma\, sS$$

Once these are verified, the result (19.64) follows by adding these three equations. We shall give a proof for (19.66(c)) for a special version of the region T.

To fix our ideas, we reconsider the surface Σ that forms the boundary of the region T depicted in Figure 29. Again, we see that Σ actually consists of three surfaces, Σ_1, Σ_2, and Σ_3, with the outer unit normals $\mathbf{n}_1$, $\mathbf{n}_2$, and $\mathbf{n}_3$, respectively. We assume that surfaces Σ_1 and Σ_2 are defined by the equations

$$\Sigma_2: \quad z = f_2(x, y) \qquad \Sigma_1: \quad z = f_1(x, y)$$

where $f_1(x, y) < f_2(x, y)$ on D, f_1 and f_2 are continuous on D, and f_1, f_2 have continuous partial derivatives on D. The surface Σ_3 is the portion of a cylindrical surface

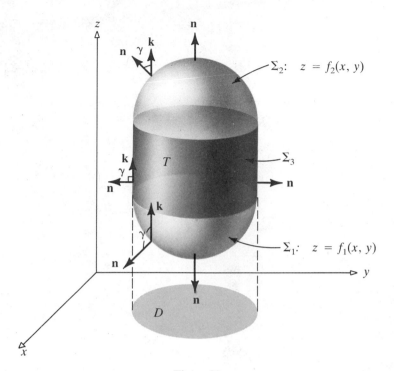

Σ_2: $z = f_2(x, y)$

Σ_3

Σ_1: $z = f_1(x, y)$

Figure 29

between Σ_1 and Σ_2 formed by lines parallel to the z-axis and along the boundary of D. The region T is enclosed by Σ_2 on the top, Σ_1 on the bottom, and the lateral surface Σ_3 on the sides.

We shall now show that

(19.67)
$$\iiint_T \frac{\partial R}{\partial z}\, dV = \iint_\Sigma R \cos \gamma\, dS$$

We start with the right side, using the fact that $\Sigma = \Sigma_1 \cup \Sigma_2 \cup \Sigma_3$:

(19.68)
$$\iint_\Sigma R \cos \gamma\, dS = \iint_{\Sigma_1} R \cos \gamma\, dS + \iint_{\Sigma_2} R \cos \gamma\, dS + \iint_{\Sigma_3} R \cos \gamma\, dS$$

On the lateral surface Σ_3, $\gamma = \pi/2$ so that $\cos \gamma = 0$.   Therefore,

(19.69)
$$\iint_{\Sigma_3} R \cos \gamma\, dS = 0$$

On the top surface Σ_2: $z = f_2(x, y)$,   the direction angle γ is acute so that based on (19.60) and (18.39), we have

$$\iint_{\Sigma_2} R \cos \gamma \, dS \underset{\substack{\uparrow \\ (19.60)}}{=} \iint_{\Sigma_2} R(x, y, f_2(x, y)) \frac{1}{\sqrt{[f_{2_x}(x, y)]^2 + [f_{2_y}(x, y)]^2 + 1}} \, dS$$

(19.70)
$$\underset{\substack{\uparrow \\ (18.39)}}{=} \iint_D R(x, y, f_2(x, y)) \frac{\sqrt{[f_{2_x}(x, y)]^2 + [f_{2_y}(x, y)]^2 + 1} \, dx \, dy}{\sqrt{[f_{2_x}(x, y)]^2 + [f_{2_y}(x, y)]^2 + 1}}$$

$$= \iint_D R(x, y, f_2(x, y)) \, dx \, dy$$

On the bottom surface Σ_1: $z = f_1(x, y)$, the direction angle γ is obtuse so that based on (19.61) and (18.39), we have

$$\iint_{\Sigma_1} R \cos \gamma \, dS \underset{\substack{\uparrow \\ (19.61)}}{=} \iint_{\Sigma_1} R(x, y, f_1(x, y)) \frac{-1}{\sqrt{[f_{1_x}(x, y)]^2 + [f_{1_y}(x, y)]^2 + 1}} \, dS$$

(19.71)
$$\underset{\substack{\uparrow \\ (18.39)}}{=} \iint_D R(x, y, f_1(x, y)) \frac{\sqrt{[f_{1_x}(x, y)]^2 + [f_{1_y}(x, y)]^2 + 1} \, dx \, dy}{\sqrt{[f_{1_x}(x, y)]^2 + [f_{1_y}(x, y)]^2 + 1}}$$

$$= - \iint_D R(x, y, f_1(x, y)) \, dx \, dy$$

Thus, based on (19.68), (19.69), (19.70), and (19.71), we have

$$\iint_{\Sigma} R \cos \gamma \, dS = \iint_D R(x, y, f_2(x, y)) \, dx \, dy - \iint_D R(x, y, f_1(x, y)) \, dx \, dy$$

$$= \iint_D [R(x, y, f_2(x, y)) - R(x, y, f_1(x, y))] \, dx \, dy$$

$$= \iint_D \left[\int_{f_1(x,y)}^{f_2(x,y)} \frac{\partial R}{\partial z} \, dz \right] dx \, dy$$

$$= \iiint_T \frac{\partial R}{\partial z} \, dV$$

EXAMPLE 2　　Redo Example 3 of Section 6 using the divergence theorem.

Solution   Suppose T is the solid cube, Σ is its surface, and $\mathbf{F} = 4xz\mathbf{i} - y^2\mathbf{j} + yz\mathbf{k}$. Since div $\mathbf{F} = 4z - 2y + y = 4z - y$, an application of (19.65) yields

$$\iint_\Sigma \rho \mathbf{F} \cdot \mathbf{n} \, dS = \rho \iint_\Sigma \mathbf{F} \cdot \mathbf{n} \, dS = \rho \iiint_T \text{div } \mathbf{F} \, dV$$

$$= \rho \int_0^1 \left\{ \int_0^1 \left[\int_0^1 (4z - y) \, dz \right] dy \right\} dx$$

$$= \rho \int_0^1 \left[\int_0^1 (2 - y) \, dy \right] dx = \rho \int_0^1 (2 - y) \, dy = \frac{3}{2} \rho \qquad \blacksquare$$

EXAMPLE 3   Let Σ be the surface of a cylindrical solid T whose boundary is $x^2 + y^2 = 4$, $z = 0$, and $z = 1$. Let $\mathbf{F} = x^3\mathbf{i} + y^3\mathbf{j} + z^2\mathbf{k}$ and let $\mathbf{n}$ be the outer unit normal to Σ (see Fig. 30). Use the divergence theorem to evaluate $\iint_\Sigma \mathbf{F} \cdot \mathbf{n} \, dS$.

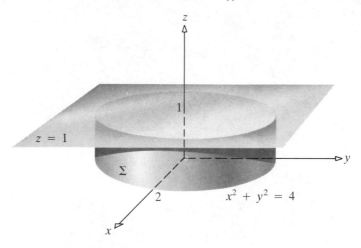

Figure 30

Solution   Since div $\mathbf{F} = 3x^2 + 3y^2 + 2z$, an application of (19.65) yields

$$\iint_\Sigma \mathbf{F} \cdot \mathbf{n} \, dS = \iiint_T (3x^2 + 3y^2 + 2z) \, dV$$

By using cylindrical coordinates to evaluate the triple integral, we obtain

$$\iint_\Sigma \mathbf{F} \cdot \mathbf{n} \, dS = \int_0^{2\pi} \left\{ \int_0^2 \left[\int_0^1 (3r^2 + 2z)r \, dz \right] dr \right\} d\theta$$

$$= \int_0^{2\pi} \left[\int_0^2 (3r^3z + z^2r) \Big|_{z=0}^{z=1} \, dr \right] d\theta$$

$$= \int_0^{2\pi} \left[\int_0^2 (3r^3 + r) \, dr \right] d\theta$$

$$= \int_0^{2\pi} \left(\frac{3r^4}{4} + \frac{r^2}{2} \right) \Big|_{r=0}^{r=2} \, d\theta = \int_0^{2\pi} 14 \, d\theta = 28\pi \qquad \blacksquare$$

EXAMPLE 4   Let $\mathbf{F}(x, y, z) = x^3\mathbf{i} + y^3\mathbf{j} + z^3\mathbf{k}$.   Let Σ be the surface of the hemispherical region T, enclosed by $z = \sqrt{4 - x^2 - y^2}$ and $z = 0$.   Let $\mathbf{n}$ be the outer unit normal to Σ (see Fig. 31). Use the divergence theorem to evaluate $\iint_\Sigma \mathbf{F} \cdot \mathbf{n}\,dS$.

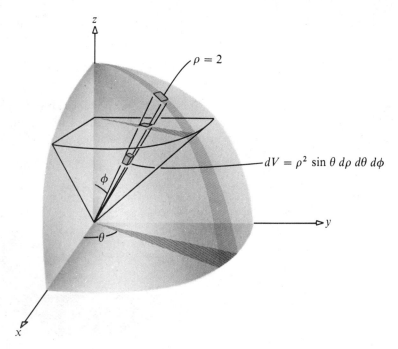

Figure 31   Note that only the first octant is shown

Solution   Since   div $\mathbf{F} = 3x^2 + 3y^2 + 3z^2 = 3(x^2 + y^2 + z^2)$,   an application of (19.65) yields

$$\iint_\Sigma \mathbf{F} \cdot \mathbf{n}\,dS = 3 \iiint_T (x^2 + y^2 + z^2)\,dV$$

We use spherical coordinates to evaluate the triple integral, obtaining

$$\iint_\Sigma \mathbf{F} \cdot \mathbf{n}\,dS = 3 \int_{\phi=0}^{\pi/2} \left\{ \int_{\theta=0}^{2\pi} \left[\int_{\rho=0}^{2} (\rho^2)\rho^2 \sin\phi\,d\rho \right] d\theta \right\} d\phi = \frac{6\pi(2^5)}{5} = \frac{192\pi}{5} \qquad \blacksquare$$

INTERPRETATION OF div F

Recall that the law of the mean for a single integral states that if a function f of one variable is continuous on a closed interval $[a, b]$, then there is a number c in (a, b) such that

$$\int_a^b f(x)\,dx = f(c)L \qquad \text{where} \quad L = b - a$$

In an analogous way, there is a law of the mean for triple integrals that asserts that if f is a continuous function defined on a simply connected closed bounded region T, then there is a point (x^*, y^*, z^*) in T such that

(19.72)
$$\iiint_T f(x, y, z)\, dV = f(x^*, y^*, z^*)V$$

where V is the volume of T.

Let $\mathbf{F}$ denote a continuous vector function. We apply (19.72) to $\iiint_{T_a} \text{div } \mathbf{F}\, dV_a$, where T_a consists of a sphere of radius a with center at (x_1, y_1, z_1), together with its interior. The volume of T_a is denoted by V_a. When we apply (19.72), we find

(19.73)
$$\iiint_{T_a} \text{div } \mathbf{F}\, dV_a = \text{div } \mathbf{F}(x^*, y^*, z^*)V_a$$

But from (19.65),

(19.74)
$$\iiint_{T_a} \text{div } \mathbf{F}\, dV_a = \iint_{\Sigma_a} \mathbf{F} \cdot \mathbf{n}\, dS_a$$

where Σ_a denotes the sphere of radius a with center at (x_1, y_1, z_1). By putting (19.73) and (19.74) together, we find

(19.75)
$$\text{div } \mathbf{F}(x^*, y^*, z^*) = \frac{\iint_{\Sigma_a} \mathbf{F} \cdot \mathbf{n}\, dS_a}{V_a}$$

The ratio on the right side of (19.75) is *the flux of* $\mathbf{F}$ *per unit volume* across the sphere.

If we let the radius $a \to 0$, the point $(x^*, y^*, z^*) \to (x_1, y_1, z_1)$, so that

(19.76)
$$\text{div } \mathbf{F}(x_1, y_1, z_1) = \lim_{a \to 0}\left(\frac{\iint_{\Sigma_a} \mathbf{F} \cdot \mathbf{n}\, dS_a}{V_a}\right)$$

In words, formula (19.76) states that

div $\mathbf{F}(x_1, y_1, z_1) = $ **Limiting value of the flux of F per unit volume across** Σ_a

If $\mathbf{F}$ is the velocity of a steady fluid flow and if div $\mathbf{F}(x_1, y_1, z_1) > 0$, then the net flow is out of V_a and the point (x_1, y_1, z_1) is called a *source*. If div $\mathbf{F}(x_1, y_1, z_1) < 0$, then the net flow is into V_a and the point (x_1, y_1, z_1) is called a *sink*. In terms of this vocabulary, the divergence theorem states that the net flow out of a volume V equals the sum of the sources and sinks in V.

If there are no sources or sinks in a region, then div $\mathbf{F} = 0$, and we call $\mathbf{F}$ a *solenoidal* vector field. For such fields, the equation

(19.77)
$$\text{div } \mathbf{F} = \frac{\partial P}{\partial x} + \frac{\partial Q}{\partial y} + \frac{\partial R}{\partial z} = 0$$

is referred to as the *equation of continuity*.

Result (19.76) can be taken as a starting point for defining the divergence of $\mathbf{F}$ and all the properties may be derived from it, including the divergence theorem. In fact, this was the original formulation of the theorem as given by Gauss. Using (19.76) as a definition has the advantage of being coordinate-free.

We close this section with an example that is important in the study of electrostatic fields. It is *Coulomb's law of electrostatic attraction (gravitation)*.

EXAMPLE 5

Let $\mathbf{F} = \mathbf{r}/a^3$, where $a = \sqrt{x^2 + y^2 + z^2}$ and $\mathbf{r} = x\mathbf{i} + y\mathbf{j} + z\mathbf{k}$. Show that:
(a) $\iint_\Sigma \mathbf{F} \cdot \mathbf{n} \, dS = 0$ if neither Σ nor its interior contain the point $(0, 0, 0)$
(b) $\iint_\Sigma \mathbf{F} \cdot \mathbf{n} \, dS = 4\pi$ if the interior of Σ contains the point $(0, 0, 0)$

Solution

(a)   Here, Σ is a surface that forms the boundary of a region T satisfying the assumptions of the divergence theorem. Furthermore,

$$\text{div } \mathbf{F} = \frac{\partial}{\partial x}\left[\frac{x}{(x^2 + y^2 + z^2)^{3/2}}\right] + \frac{\partial}{\partial y}\left[\frac{y}{(x^2 + y^2 + z^2)^{3/2}}\right] + \frac{\partial}{\partial z}\left[\frac{z}{(x^2 + y^2 + z^2)^{3/2}}\right]$$

$$= \frac{(x^2 + y^2 + z^2)^{3/2} - x(\frac{3}{2})(2x)\sqrt{x^2 + y^2 + z^2}}{(x^2 + y^2 + z^2)^3}$$

$$+ \frac{(x^2 + y^2 + z^2)^{3/2} - y(\frac{3}{2})(2y)\sqrt{x^2 + y^2 + z^2}}{(x^2 + y^2 + z^2)^3}$$

$$+ \frac{(x^2 + y^2 + z^2)^{3/2} - z(\frac{3}{2})(2z)\sqrt{x^2 + y^2 + z^2}}{(x^2 + y^2 + z^2)^3}$$

$$= \frac{3(x^2 + y^2 + z^2)^{3/2} - 3\sqrt{x^2 + y^2 + z^2}(x^2 + y^2 + z^2)}{(x^2 + y^2 + z^2)^3} = 0$$

Therefore, by the divergence theorem,

$$\iint_\Sigma \mathbf{F} \cdot \mathbf{n} \, dS = \iiint_T \text{div } \mathbf{F} \, dV = 0$$

(b)   We note that since $\mathbf{F}$ is not continuous at the origin, the divergence theorem cannot be applied. However, the following argument will help us prove (b). Let T be the closed region enclosed by two separate surfaces: the surface Σ and a sphere Σ_1 of radius a, with center at $(0, 0, 0)$, as shown in Figure 32. The outer surface is Σ and the inner surface is Σ_1. Now $\mathbf{F}$ is continuous throughout T. Therefore, we can apply the divergence theorem to get

$$\iiint_T \text{div } \mathbf{F} \, dV = \iint_\Sigma \mathbf{F} \cdot \mathbf{n} \, dS + \iint_{\Sigma_1} \mathbf{F} \cdot \mathbf{n} \, dS$$

From part (a), we know that $\iiint_T \text{div } \mathbf{F} \, dV = 0$.   Thus,

$$\iint_\Sigma \mathbf{F} \cdot \mathbf{n} \, dS = -\iint_{\Sigma_1} \mathbf{F} \cdot \mathbf{n} \, dS$$

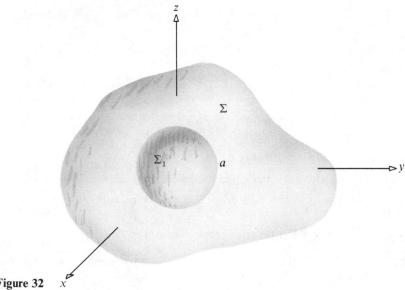

Figure 32

On the inner surface Σ_1, the outer unit normal is

$$\mathbf{n} = -\frac{x\mathbf{i} + y\mathbf{j} + z\mathbf{k}}{a}$$

Hence,

$$\mathbf{F} \cdot \mathbf{n} = -\frac{x^2 + y^2 + z^2}{a^4} = -\frac{1}{a^2}$$

Thus,

$$\iint_{\Sigma} \mathbf{F} \cdot \mathbf{n}\, dS = -\iint_{\Sigma_1} \left(-\frac{1}{a^2}\right) dS = \frac{1}{a^2} \iint_{\Sigma_1} dS = \frac{1}{a^2}(4\pi a^2) = 4\pi \quad \blacksquare$$

Surface area
of the sphere Σ_1
is $4\pi a^2$

EXERCISE 7

In Problems 1–4 find the divergence of **F**.

1. $\mathbf{F}(x, y, z) = x^2\mathbf{i} + y^2\mathbf{j} + z^2\mathbf{k}$

2. $\mathbf{F}(x, y, z) = x\mathbf{i} + xy\mathbf{j} + xyz\mathbf{k}$

3. $\mathbf{F}(x, y, z) = (x + \cos x)\mathbf{i} + (y + y\sin x)\mathbf{j} + 2z\mathbf{k}$

4. $\mathbf{F}(x, y, z) = xye^z\mathbf{i} + x^2e^z\mathbf{j} + x^2ye^z\mathbf{k}$

In Problems 5–14 use the divergence theorem to evaluate $\iint_{\Sigma} \mathbf{F} \cdot \mathbf{n}\, dS$.   Here **n** denotes the outer unit normal to Σ.

5. $\mathbf{F} = (2xy + 2z)\mathbf{i} + (y^2 + 1)\mathbf{j} - (x + y)\mathbf{k}$; Σ is the surface of the region enclosed by $x + y + z = 4$, $x = 0$, $y = 0$,   and $z = 0$

6. $\mathbf{F} = (2xy + z)\mathbf{i} + y^2\mathbf{j} - (x + 4y)\mathbf{k}$; Σ is the surface of the region enclosed by $2x + 2y + z = 6$, $x = 0$, $y = 0$,   and $z = 0$

7. $\mathbf{F} = x^2\mathbf{i} + y^2\mathbf{j} + z^2\mathbf{k}$; Σ is the surface of the region enclosed by $x = 0$, $x = 1$, $y = 0$, $y = 1$, $z = 0$, and $z = 1$

8. $\mathbf{F} = (x - y)\mathbf{i} + (y - z)\mathbf{j} + (x - y)\mathbf{k}$; Σ is the surface of a cube with center at the origin and faces in the planes $x = \pm 1$, $y = \pm 1$,   and $z = \pm 1$

9. $\mathbf{F} = x\mathbf{i} + y\mathbf{j} + z\mathbf{k}$; Σ is the sphere $x^2 + y^2 + z^2 = 1$

10. $\mathbf{F} = 2x\mathbf{i} + 2y\mathbf{j} + 2z\mathbf{k}$; Σ is the sphere $x^2 + y^2 + z^2 = 2$

11. $\mathbf{F} = x^2\mathbf{i} + 2y\mathbf{j} + 4z^2\mathbf{k}$; Σ is the surface of the cylinder $x^2 + y^2 \leq 4$, $0 \leq z \leq 2$

12. $\mathbf{F} = x\mathbf{i} + 2y^2\mathbf{j} + 3z^2\mathbf{k}$; Σ is the surface of the cylinder $x^2 + y^2 \leq 9$, $0 \leq z \leq 1$

13. $\mathbf{F} = (x + \cos x)\mathbf{i} + (y + y \sin x)\mathbf{j} + 2z\mathbf{k}$; Σ is the tetrahedron with vertices $(0, 0, 0)$, $(1, 0, 0)$, $(0, 1, 0)$, and $(0, 0, 1)$

14. $\mathbf{F} = yz\mathbf{i} + xz\mathbf{j} + xy\mathbf{k}$; Σ is the tetrahedron described in Problem 13

15. Let $\mathbf{F} = x^2\mathbf{i} + y^2\mathbf{j} + z^2\mathbf{k}$ and $\mathbf{n} = (\cos \alpha)\mathbf{i} + (\cos \beta)\mathbf{j} + (\cos \gamma)\mathbf{k}$.   Use the divergence theorem to show that $\iint_\Sigma \mathbf{F} \cdot \mathbf{n} \, dS = 8\pi q^4/3$,   if Σ is the surface $x^2 + y^2 + z^2 = 2qz$.

16. What is the value of the integral in Problem 15, if Σ is the surface of the cube enclosed by $x = 0$, $x = q$. $y = 0$, $y = q$, $z = 0$,   and $z = q$?

17. Let $\mathbf{F} = x\mathbf{i} + y\mathbf{j} + z\mathbf{k}$;   let Σ be the surface of a region T obeying the divergence theorem; and let $\mathbf{n}$ be the outer unit normal to Σ. Show that the volume V of T is given by the formula $V = \frac{1}{3} \iint_\Sigma \mathbf{F} \cdot \mathbf{n} \, dS$.

In Problems 18–20 use the expression given in Problem 17 to find each volume.

18. A rectangular parallelepiped with sides of length a, b, and c

19. A right circular cone with height h and base radius R
   [*Hint:*   The calculation is simplified with the cone oriented as shown in the figure.]

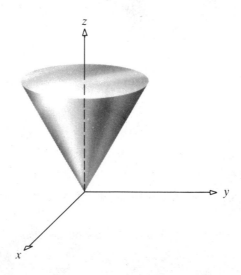

20. A sphere of radius R

21. Let $\mathbf{F} = 3x\mathbf{i} + 4y\mathbf{j} + (7z + 2x)\mathbf{k}$ and $\mathbf{G} = 2x\mathbf{i} + 3y\mathbf{j} + (9z + 6y)\mathbf{k}$. Let Σ be the surface of a region T obeying the assumption of the divergence theorem. Prove that $\iint_\Sigma \mathbf{F} \cdot \mathbf{n}\, dS = \iint_\Sigma \mathbf{G} \cdot \mathbf{n}\, dS$.

22. Let f and g be two scalar functions. Let Σ be the surface of a region T obeying the assumption of the divergence theorem. Prove that $\iint_\Sigma (f\nabla g \cdot \mathbf{n})\, dS = \iiint_T (f\nabla^2 g + \nabla f \cdot \nabla g)\, dV$. [*Hint:* Let $\mathbf{F} = f\nabla g$ in (19.65) and use the fact that $\nabla^2 g = \nabla \cdot \nabla g = g_{xx} + g_{yy} + g_{zz}$.

In Problems 23 and 24 verify the divergence theorem.

23. $\mathbf{F}(x, y, z) = x\mathbf{i} + y\mathbf{j} + z\mathbf{k}$; Σ is the surface of the sphere $x^2 + y^2 + z^2 = 100$

24. $\mathbf{F}(x, y, z) = x\mathbf{i} + y\mathbf{i} + z\mathbf{k}$; Σ is the closed cylindrical surface $x^2 + y^2 = 1$ between $z = 0$ and $z = 2$

25. Prove that if $\mathbf{F}$ is a constant vector field and Σ is the surface of a region T obeying the assumption of the divergence theorem, then $\iint_\Sigma \mathbf{F} \cdot \mathbf{n}\, dS = 0$.

8. Stokes' Theorem

CURL OF F

Let $\mathbf{F} = P(x, y, z)\mathbf{i} + Q(x, y, z)\mathbf{j} + R(x, y, z)\mathbf{k}$ be a vector function such that the partial derivatives of P, Q, and R exist. Then the *curl of* $\mathbf{F}$, denoted by curl $\mathbf{F}$ or $\nabla \times \mathbf{F}$, is the vector function defined by

(19.78)
$$\operatorname{curl} \mathbf{F} = \nabla \times \mathbf{F} = \left(\frac{\partial R}{\partial y} - \frac{\partial Q}{\partial z}\right)\mathbf{i} - \left(\frac{\partial R}{\partial x} - \frac{\partial P}{\partial z}\right)\mathbf{j} + \left(\frac{\partial Q}{\partial x} - \frac{\partial P}{\partial y}\right)\mathbf{k}$$

Rather than memorize (19.78), the curl can be written in the form of the symbolic determinant:

(19.79)
$$\operatorname{curl} \mathbf{F} = \begin{vmatrix} \mathbf{i} & \mathbf{j} & \mathbf{k} \\ \dfrac{\partial}{\partial x} & \dfrac{\partial}{\partial y} & \dfrac{\partial}{\partial z} \\ P & Q & R \end{vmatrix}$$

EXAMPLE 1   Find curl $\mathbf{F}$ if $\mathbf{F} = x^2 y\mathbf{i} - 2xz\mathbf{j} + 2yz\mathbf{k}$

Solution
$$\operatorname{curl} \mathbf{F} = \begin{vmatrix} \mathbf{i} & \mathbf{j} & \mathbf{k} \\ \dfrac{\partial}{\partial x} & \dfrac{\partial}{\partial y} & \dfrac{\partial}{\partial z} \\ x^2 y & -2xz & 2yz \end{vmatrix}$$

$$= \left[\frac{\partial}{\partial y}(2yz) - \frac{\partial}{\partial z}(-2xz)\right]\mathbf{i} - \left[\frac{\partial}{\partial x}(2yz) - \frac{\partial}{\partial z}(x^2 y)\right]\mathbf{j} + \left[\frac{\partial}{\partial x}(-2xz) - \frac{\partial}{\partial y}(x^2 y)\right]\mathbf{k}$$

$$= (2z + 2x)\mathbf{i} - (0 - 0)\mathbf{j} + (-2z - x^2)\mathbf{k}$$

$$= (2z + 2x)\mathbf{i} + (-2z - x^2)\mathbf{k} \qquad \blacksquare$$

Now we are ready to introduce the generalization of (19.43) to surfaces and curves in three dimensions. This is called *Stokes' theorem*,* and it may be stated in vector form as

(19.80)
$$\oint_C \mathbf{F} \cdot d\mathbf{r} = \iint_\Sigma \text{curl } \mathbf{F} \cdot \mathbf{n} \, dS \qquad \text{under certain conditions on } \Sigma$$

We briefly discuss the conditions on Σ. The surface Σ must be *smooth;* that is, if the surface is defined by $z = f(x, y)$, then f has continuous first-order partial derivatives. The surface Σ must also be *simply connected.* By this we mean that there are no "holes" in Σ. Finally, we require the surface Σ to be *orientable.* This means that it is possible to assign a direction to the unit normal $\mathbf{n}$ at each point of Σ, which we take as the positive direction of $\mathbf{n}$. The positively oriented vector $\mathbf{n}$ must vary continuously, and, as it moves around any closed curve on the surface, $\mathbf{n}$ must return to its original direction. An example of a nonorientable surface is the Möbius strip (see Fig. 33).

The curve C in (19.80) is oriented, and its orientation is such that if you stand with your feet on the surface and your head in the direction of the positively oriented vector $\mathbf{n}$, you see the curve C_1 oriented counterclockwise (see Fig. 34).

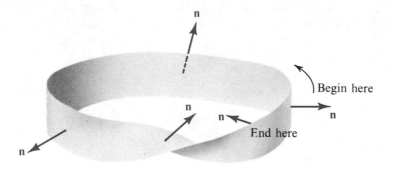

Figure 33 Möbius strip

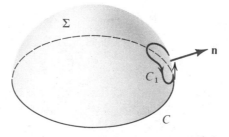

Figure 34

---

* Named after the English scientist George Gabriel Stokes (1819–1903).

We now give a precise formulation of Stokes' theorem.

(19.81)   *Stokes' Theorem.*   **Let Σ denote a smooth simply connected orientable surface bounded by a piecewise smooth closed curve C. Let $\mathbf{F} = P(x, y, z)\mathbf{i} + Q(x, y, z)\mathbf{j} + R(x, y, z)\mathbf{k}$ denote a vector function, where P, Q, and R have continuous first-order partial derivatives throughout a region D containing Σ and C. Let $\mathbf{n}$ denote the positive unit normal to Σ, and let C be positively oriented, as described above. Then**

$$\oint_C \mathbf{F} \cdot d\mathbf{r} = \iint_\Sigma \operatorname{curl} \mathbf{F} \cdot \mathbf{n}\, dS$$

This statement can also be written as

$$\oint_C (P\, dx + Q\, dy + R\, dz) = \iint_\Sigma \left[\left(\frac{\partial R}{\partial y} - \frac{\partial Q}{\partial z} \right) dy\, dz - \left(\frac{\partial R}{\partial x} - \frac{\partial P}{\partial z} \right) dz\, dx + \left(\frac{\partial Q}{\partial x} - \frac{\partial P}{\partial y} \right) dx\, dy \right]$$

The statement of Stokes' theorem as given in (19.81) is quite general, and its proof is beyond the scope of this book.

EXAMPLE 2   Verify Stokes' theorem for $\mathbf{F} = y\mathbf{i} - x\mathbf{j}$,   where Σ is the paraboloid $z = x^2 + y^2$, with the circle $x^2 + y^2 = 1$, $z = 1$,   as its boundary (see Fig. 35).

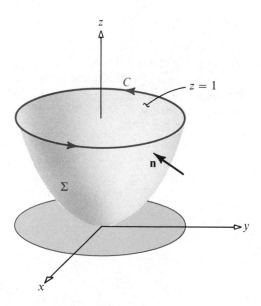

Figure 35

Solution   A set of parametric equations for C are $x = \cos t$, $y = \sin t$, $z = 1$.   We use these to evaluate the line integral $\oint_C \mathbf{F} \cdot d\mathbf{r}$.

$$\oint_C \mathbf{F} \cdot d\mathbf{r} = \oint_C (y\, dx - x\, dy) = \int_0^{2\pi} [\sin t(-\sin t)\, dt - \cos t \cos t\, dt]$$

$$= -\int_0^{2\pi} [\sin^2 t + \cos^2 t]\, dt = -2\pi$$

To evaluate the surface integral, $\iint_\Sigma \text{curl } \mathbf{F} \cdot \mathbf{n} \, dS,$ we compute curl $\mathbf{F}$ and $\mathbf{n}$:

$$\text{curl } \mathbf{F} = \begin{vmatrix} \mathbf{i} & \mathbf{j} & \mathbf{k} \\ \dfrac{\partial}{\partial x} & \dfrac{\partial}{\partial y} & \dfrac{\partial}{\partial z} \\ y & -x & 0 \end{vmatrix} = 0\mathbf{i} - 0\mathbf{j} - 2\mathbf{k} = -2\mathbf{k}$$

$$\mathbf{n} = \frac{-2x\mathbf{i} - 2y\mathbf{j} + \mathbf{k}}{\sqrt{4x^2 + 4y^2 + 1}}$$

Then

$$\iint_\Sigma \text{curl } \mathbf{F} \cdot \mathbf{n} \, dS = \iint_\Sigma \frac{-2}{\sqrt{4x^2 + 4y^2 + 1}} \, dS$$

By using equation (19.57) to evaluate the surface integral, we get

$$\iint_\Sigma \text{curl } \mathbf{F} \cdot \mathbf{n} \, dS = -\iint_\Gamma 2 \, dx \, dy = -2\pi$$

where Γ is the interior of the circle with radius 1.   ∎

INTERPRETATION OF CURL F

Just as the divergence theorem gives a new interpretation for the divergence of $\mathbf{F}$, we may use Stokes' theorem to give an interpretation for the curl of a vector. Let $Q = (x_1, y_1, z_1)$ be the center of a circular disk Σ_ρ of radius ρ and let C_ρ be the boundary of Σ_ρ. (See Fig. 36.) There is a law of the mean for double integrals which, when combined with Stokes' theorem, gives us

(19.82)
$$\oint_{C_\rho} \mathbf{F} \cdot d\mathbf{r} = \iint_{\Sigma_\rho} \text{curl } \mathbf{F} \cdot \mathbf{n} \, dS = (\text{curl } \mathbf{F} \cdot \mathbf{n})_{Q*}(\pi\rho^2)$$

where $(\text{curl } \mathbf{F} \cdot \mathbf{n})_{Q*}$ denotes the value of curl $\mathbf{F} \cdot \mathbf{n}$ evaluated at a suitably chosen point $Q* = (x*, y*, z*)$ in Σ_ρ, and $\pi\rho^2$ is the area of Σ_ρ. Thus,

$$(\text{curl } \mathbf{F} \cdot \mathbf{n})_{Q*} = \frac{1}{\pi\rho^2} \oint_{C_\rho} \mathbf{F} \cdot d\mathbf{r}$$

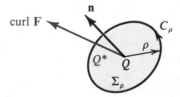

Figure 36

Now, if ρ is allowed to approach 0, then $Q^* \to Q$, and hence,

(19.83)
$$(\text{curl } \mathbf{F} \cdot \mathbf{n})_Q = \lim_{\rho \to 0} \frac{1}{\pi \rho^2} \oint_{C_\rho} \mathbf{F} \cdot d\mathbf{r}$$

In the case where $\mathbf{F}$ is the velocity of a fluid motion, the integral $\oint_{C_\rho} \mathbf{F} \cdot d\mathbf{r}$ in (19.83) is referred to as the *circulation*, or *whirling tendency*, around C_ρ, and it measures the extent to which the corresponding fluid motion is a rotation around the circle C_ρ in the given direction. Equation (19.83) therefore states that the component of curl $\mathbf{F}$ at (x_1, y_1, z_1) in the direction of $\mathbf{n}$ is the limiting ratio of circulation to area for a circle about (x_1, y_1, z_1) with $\mathbf{n}$ as a normal. Thus,

(19.84)   **Circulation per unit of area at** $(x_1, y_1, z_1) = (\text{curl } \mathbf{F} \cdot \mathbf{n})_Q$

The left-hand side of (19.84) is a maximum at Q when $\mathbf{n}$ has the same direction as curl $\mathbf{F}$. Suppose that a small paddle wheel of radius ρ is introduced into the fluid at Q with its axle directed along $\mathbf{n}$. The rate of spin of the paddle wheel will be affected by the circulation of the fluid around C_ρ. The wheel will spin fastest when the circulation integral is maximized; that is, it will spin fastest when the axle of the paddle wheel is in the direction of curl $\mathbf{F}$ (see Fig. 37).

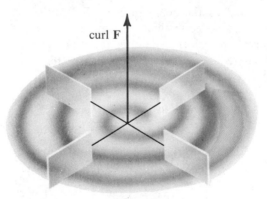

curl F

Figure 37

We now give another interpretation of the curl based on motion. Knowledge of angular velocity (discussed in Chap. 14) will be helpful here.

(19.85)   **Let v be the velocity vector of a fluid and let ω be the angular velocity. Then**

$$\text{curl } \mathbf{v} = 2\omega$$

Proof Suppose that the motion of a fluid is simply a rotation about a given axis fixed in space. Then we may represent the angular velocity by a constant vector

$$\omega = \omega_1 \mathbf{i} + \omega_2 \mathbf{j} + \omega_3 \mathbf{k}$$

where the length of ω is the angular speed and the direction of ω is along the direction of the axis of rotation, in accordance with the right-hand rule. If we take the origin of a rectangular coordinate system on the axis and denote the position vector to a point (x, y, z) by $\mathbf{r} = x\mathbf{i} + y\mathbf{j} + z\mathbf{k}$, then

$$\mathbf{v} = \omega \times \mathbf{r}$$

and

$$\operatorname{curl} \mathbf{v} = \begin{vmatrix} \mathbf{i} & \mathbf{j} & \mathbf{k} \\ \dfrac{\partial}{\partial x} & \dfrac{\partial}{\partial y} & \dfrac{\partial}{\partial z} \\ \omega_2 z - \omega_3 y & \omega_3 x - \omega_1 z & \omega_1 y - \omega_2 x \end{vmatrix}$$

Since ω is constant, straightforward computation shows that

$$\operatorname{curl} \mathbf{v} = 2\omega_1 \mathbf{i} + 2\omega_2 \mathbf{j} + 2\omega_3 \mathbf{k} = 2\omega \quad \blacksquare$$

Thus, if a fluid experiences a rotation, the curl of the velocity vector is a constant vector equal to twice the angular velocity vector.

Stokes' theorem can be used to avoid calculation of a surface integral by substituting an equivalent line integral. However, it is probably more frequently employed in the opposite direction, that is, in cases when the line integral is not easily evaluated but the quantity $\operatorname{curl} \mathbf{F} \cdot \mathbf{n}$ is of relatively simple form on some open surface with boundary C. This fact is illustrated in the next example.

EXAMPLE 3   Evaluate $I = \oint_C [(e^{-x^2/2} - yz)\,dx + (e^{-y^2/2} + xz + 2x)\,dy + (e^{-z^2/2} + 5)\,dz]$, where C is the circle $x = \cos t$, $y = \sin t$, $z = 2$, $0 \le t \le 2\pi$.

Solution   If we let $\mathbf{F} = P\mathbf{i} + Q\mathbf{j} + R\mathbf{k}$, with $P = e^{-x^2/2} - yz$, $Q = e^{-y^2/2} + xz + 2x$, and $R = e^{-z^2/2} + 5$, it can be verified that

$$\operatorname{curl} \mathbf{F} = -x\mathbf{i} - y\mathbf{j} + (2 + 2z)\mathbf{k}$$

To apply Stokes' theorem, we take Σ to be a plane region enclosed by the circle C in the plane $z = 2$ so that $\mathbf{n} = \mathbf{k}$. Hence, $\operatorname{curl} \mathbf{F} \cdot \mathbf{n} = 6$ on Σ, and we may write

$$I = \iint_\Sigma \operatorname{curl} \mathbf{F} \cdot \mathbf{n}\,dS = 6 \iint_\Sigma dx\,dy = 6\pi \quad \blacksquare$$

APPLICATION OF STOKES' THEOREM TO CONSERVATIVE FIELDS OF FORCE

The notion of a conservative field of force extends to space. A field of force $\mathbf{F} = P(x, y, z)\mathbf{i} + Q(x, y, z)\mathbf{j} + R(x, y, z)\mathbf{k}$ is *conservative* if $\mathbf{F}$ is the gradient of some function $f = f(x, y, z)$. It is an extension of earlier proofs to show that $\mathbf{F}$ is conservative if and only if $\oint_C \mathbf{F} \cdot d\mathbf{r} = 0$ for every piecewise smooth closed curve C. We now provide an equivalent statement for saying $\mathbf{F}$ is conservative.

(19.86)   **THEOREM**   Let $\mathbf{F} = P(x, y, z)\mathbf{i} + Q(x, y, z)\mathbf{j} + R(x, y, z)\mathbf{k}$ **denote a field of force whose components P, Q, and R possess continuous first-order partial derivatives throughout a region D whose boundary is a surface Σ satisfying the conditions of Stokes' theorem. Suppose the surface Σ is simply connected. Then $\oint_C \mathbf{F} \cdot d\mathbf{r} = 0$ for every piecewise smooth closed curve C in D if and only if   curl $\mathbf{F} = 0$ throughout D.**

Proof   Suppose   curl $\mathbf{F} = \mathbf{0}$.   Then, by Stokes' theorem,

$$\oint_C \mathbf{F} \cdot d\mathbf{r} = \iint_\Sigma \text{curl } \mathbf{F} \cdot \mathbf{n}\, dS = 0$$

Conversely, suppose $\oint_C \mathbf{F} \cdot d\mathbf{r} = 0$ around every closed path C and assume that curl $\mathbf{F} \neq \mathbf{0}$ at some point P. Then, assuming that curl $\mathbf{F}$ is continuous, that is, that $\mathbf{F}$ has continuous partial derivatives, there will be a region with P as an interior point where   curl $\mathbf{F} \neq \mathbf{0}$.   Let Σ be a surface contained in this region whose normal $\mathbf{n}$ at each point has the same direction as curl $\mathbf{F}$, that is,   curl $\mathbf{F} = a\mathbf{n}$,   where a is a positive constant. If C is a boundary of Σ, then, by Stokes' theorem,

$$\oint_C \mathbf{F} \cdot d\mathbf{r} = \iint_\Sigma \text{curl } \mathbf{F} \cdot \mathbf{n}\, dS = a \iint_\Sigma \mathbf{n} \cdot \mathbf{n}\, dS > 0$$

which contradicts the hypothesis that $\oint_C \mathbf{F} \cdot d\mathbf{r} = 0$. Hence, we conclude that curl $\mathbf{F} = \mathbf{0}$.   ■

Under the hypothesis of (19.86), we may now list the following equivalent statements for a field of force $\mathbf{F} = P\mathbf{i} + Q\mathbf{j} + R\mathbf{k}$ in space; these are parallel to those developed in (19.51) for a field of force in the plane.

1. **F is conservative.**
2. **F is the gradient of some function.**
3. **The work done by F in moving an object of mass m from a point A to a point B in Σ is independent of the path chosen from A to B.**
4. **The work done by F in moving an object of mass m along any closed piecewise smooth curve C in Σ is 0.**
5. **curl F = 0.**

Of all these equivalent conditions, the easiest to establish is statement 5, as the next example illustrates.

EXAMPLE 4   Show that $\mathbf{F} = (\frac{3}{5}y^5 + 2z^2)\mathbf{i} + 3xy^4\mathbf{j} + 4xz\mathbf{k}$ is conservative.

Solution

$$\text{curl } \mathbf{F} = \begin{vmatrix} \mathbf{i} & \mathbf{j} & \mathbf{k} \\ \dfrac{\partial}{\partial x} & \dfrac{\partial}{\partial y} & \dfrac{\partial}{\partial z} \\ \tfrac{3}{5}y^5 + 2z^2 & 3xy^4 & 4xz \end{vmatrix}$$

$$= (0 - 0)\mathbf{i} - (4z - 4z)\mathbf{j} + (3y^4 - 3y^4)\mathbf{k} = 0$$

Thus, $\mathbf{F}$ is conservative.   ∎

EXERCISE 8

In Problems 1–12 find the curl of $\mathbf{F}$.

1. $\mathbf{F}(x, y) = x\mathbf{i} + y\mathbf{j}$

2. $\mathbf{F}(x, y) = y\mathbf{i} + x\mathbf{j}$

3. $\mathbf{F}(x, y, z) = xyz\mathbf{i} + xz\mathbf{j} + z\mathbf{k}$

4. $\mathbf{F}(x, y, z) = 4x\mathbf{i} - y\mathbf{j} - 2z\mathbf{k}$

5. $\mathbf{F}(x, y, z) = 3xyz^2\mathbf{i} + (y^2 \sin z)\mathbf{j} + xe^{2z}\mathbf{k}$

6. $\mathbf{F}(x, y, z) = yz\mathbf{i} + z^2x\mathbf{j} + yz\mathbf{k}$

7. $\mathbf{F}(x, y, z) = \dfrac{x\mathbf{i}}{x^2 + y^2 + z^2} + \dfrac{y\mathbf{j}}{x^2 + y^2 + z^2} + \dfrac{z\mathbf{k}}{x^2 + y^2 + z^2}$

8. $\mathbf{F}(x, y, z) = e^x\mathbf{i} + x^2y\mathbf{j} + e^z\mathbf{k}$

9. $\mathbf{F}(x, y, z) = (\cos x)\mathbf{i} + (\sin y)\mathbf{j} + e^{xz}\mathbf{k}$

10. $\mathbf{F}(x, y, z) = (\sin xy)\mathbf{i} + (\cos xy^2)\mathbf{j} + x\mathbf{k}$

11. $\mathbf{F}(x, y, z) = (x + y)\mathbf{i} + (y + z)\mathbf{j} + (z + x)\mathbf{k}$

12. $\mathbf{F}(x, y, z) = (y + z)\mathbf{i} + (z + x)\mathbf{j} + (z + y + x)\mathbf{k}$

13. Show that $\text{div(curl } \mathbf{F}) = 0$, where $\mathbf{F} = P\mathbf{i} + Q\mathbf{j} + R\mathbf{k}$ and P, Q, and R are continuously twice differentiable functions of x, y, and z.

14. Show that $\text{curl}(\mathbf{F} + \mathbf{G}) = \text{curl } \mathbf{F} + \text{curl } \mathbf{G}$ and $\text{curl } c\mathbf{F} = c(\text{curl } \mathbf{F})$, c a constant.

15. Determine which of the following forces in space are conservative.
    (a) $\mathbf{F} = x\mathbf{i} + y\mathbf{j}$
    (b) $\mathbf{F} = y\mathbf{i} + x\mathbf{j}$
    (c) $\mathbf{F} = y\mathbf{i} - x\mathbf{j}$
    (d) $\mathbf{F} = xy\mathbf{i} + yz\mathbf{j} + zx\mathbf{k}$
    (e) $\mathbf{F} = yz\mathbf{i} + zx\mathbf{j} + xy\mathbf{k}$

16. Find the value of the constant c so that the following forces in space are conservative:

    (a) $\mathbf{F} = xy\mathbf{i} + cx^2\mathbf{j}$ (b) $\mathbf{F} = \dfrac{z}{y}\mathbf{i} + c\dfrac{xz}{y^2}\mathbf{j} + \dfrac{x}{y}\mathbf{k}$, $y \neq 0$

In Problems 17–20 verify Stokes' theorem for the given $\mathbf{F}$ and Σ.

17. $\mathbf{F} = y\mathbf{i} - x\mathbf{j}$; Σ is the hemisphere $z = \sqrt{1 - x^2 - y^2}$

18. $\mathbf{F} = z\mathbf{i} + x\mathbf{j} + y\mathbf{k}$; Σ is the hemisphere $z = \sqrt{1 - x^2 - y^2}$

19. $\mathbf{F} = (z - y)\mathbf{i} + (z + x)\mathbf{j} - (x + y)\mathbf{k}$; Σ is the portion of the paraboloid $z = 1 - x^2 - y^2$ that lies above the plane $z = 0$

20. $\mathbf{F} = y\mathbf{i} + z\mathbf{j} + x\mathbf{k}$; Σ is the portion of the paraboloid $z = 1 - x^2 - y^2$, $z \geq 0$

In Problems 21–24 apply Stokes' theorem to evaluate the given line integral, and verify your answer by a direct calculation of the line integral.

21. $\displaystyle\oint_C [(y + z)\,dx + (z + x)\,dy + (x + y)\,dz]$; C is the circle $x^2 + y^2 + z^2 = 1$, $x + y + z = 0$

22. $\oint_C [(y - z)dx + (z - x)dy + (x - y)dz]$; C is the ellipse $x^2 + y^2 = 1$, $x + x = 1$

23. $\oint_C [x\,dx + (x + y)\,dy + (x + y + z)\,dz]$; C is the curve $x = 2\cos t$, $y = 2\sin t$, $z = 2$, $0 \le t \le 2\pi$

24. $\oint_C (y^2 dx + z^2 dy + x^2 dz)$; C is the triangle with vertices $(1, 0, 0)$, $(0, 1, 0)$, and $(0, 0, 1)$

25. Let $\mathbf{F} = 2xy^2 z\mathbf{i} + 2x^2 yz\mathbf{j} + (x^2 y^2 - 2z)\mathbf{k}$.   Show that F is conservative.

26. Show that $\mathbf{F} = y\mathbf{i} - x\mathbf{j} + z\mathbf{k}$ is not a conservative field. Nevertheless, there are certain paths C for which $\oint_C \mathbf{F} \cdot d\mathbf{r} = 0$.   Find one.

27. A particle is moved from the origin to the point (a, b, c) in a field of force $\mathbf{F} = (x + y)\mathbf{i} + (x - z)\mathbf{j} + (z - y)\mathbf{k}$. Show that the work done depends only on a, b, and c, and find this value.

28. Verify Stokes' theorem when $\mathbf{F}(x, y, z) = y^2\mathbf{i} + x\mathbf{j} - xz\mathbf{k}$ and Σ is the surface $z = 1 - x^2 - y^2$, $z \ge 0$.

29. If $\mathbf{F}(x, y, z) = z\mathbf{i} + x\mathbf{j} + y\mathbf{k}$,   calculate $\iint_\Sigma (\text{curl } \mathbf{F} \cdot \mathbf{n})\,dS$ where Σ is the hemisphere $z = \sqrt{1 - x^2 - y^2}$.

30. Rework Problem 29 where Σ is the circular region $x^2 + y^2 \le 1$, $z = 0$.

31. By direct calculation show that $\int_C (z\,dx + x\,dy + y\,dz) = \sqrt{3}\pi$,   where C is the circular region given by $x + y + z = 0$, $x^2 + y^2 + z^2 = 1$.   Obtain the same result by using Stokes' theorem.

Miscellaneous Exercises

1. Evaluate $\int_C \left(\dfrac{dx}{y} + \dfrac{dy}{x}\right)$,   where C is the arc of the parabola $y = x^2$ from $(1, 1)$ to $(2, 4)$.

2. Find $\int_C y\cos xy\,dx + x\cos xy\,dy$,   where C is the curve $x = t^2$, $y = t^3$, $0 \le t \le 1$.

3. Confirm that the integral in Problem 2 is independent of path, and compute it by following the right-angle path from $(0, 0)$ to $(1, 0)$ to $(1, 1)$.

4. Find a function f whose gradient is $(y \cos xy)\mathbf{i} + (x \cos xy)\mathbf{j}$ and explain why Problem 2 can be worked by writing $\int_C y \cos xy\,dx + x \cos xy\,dy = f(1, 1) - f(0, 0)$.

5. Find $\int_C (yz - y - z)\,dx + (xz - x - z)\,dy + (xy - x - y)\,dz$,   where C is the twisted cubic $x = t$, $y = t^2$, $z = t^3$, $0 \le t \le 1$.

6. Find a function $f(x, y, z)$ whose gradient is $(yz - y - z)\mathbf{i} + (xz - x - z)\mathbf{j} + (xy - x - y)\mathbf{k}$,   and use it to confirm the answer to Problem 5.

7. Find $\int_C e^x \sin y\,dx + e^x \cos y\,dy$,   where C is the arc of the parabola $y = x^2$ from $(0, 0)$ to $(1, 1)$.

8. Given the constant vector $\mathbf{c}$, show that $\nabla(\mathbf{c} \cdot \mathbf{r}) = \mathbf{c}$,   where $\mathbf{r} = x\mathbf{i} + y\mathbf{j} + z\mathbf{k}$.   Use the result to prove that $\int_C \mathbf{c} \cdot d\mathbf{r} = 0$,   where C is any circle in space.

9. Suppose an object travels through a force field along a path that is always normal to the force. Show that the work done in moving the object from one point to another is zero. Why does it follow that the gravitational field of the earth does no work on a satellite in circular orbit?

10. Suppose sea level is the xy-plane and the z-axis points upward. Then the gravitational force on an object of mass m near sea level may be taken to be $\mathbf{F} = mg\mathbf{k}$,   where g is the acceleration due to gravity. Show that the work done by gravity on an object moving from (x_1, y_1, z_1) to (x_2, y_2, z_2) along any path is $W = mg(z_2 - z_1)$.

11. Suppose that $F(x, y, z)$ is a force directed toward the origin with magnitude inversely proportional to the square of the distance from the origin. (Such "inverse square law" forces are common in nature.) Show that F is conservative, and find a potential function.

12. Evaluate $\int_C (y^2\, dx - x^2\, dy)$, where C is the square with vertices $(0, 0)$, $(1, 0)$, $(1, 1)$, and $(0, 1)$ traversed counterclockwise.

13. Rework Problem 12 by using Green's theorem.

14. Evaluate $\int_C [(x - y)\, dx + (x + y)\, dy]$, where C is the ellipse $x = 2 \cos t$, $y = 3 \sin t$, $0 \le t \le 2\pi$.

15. Rework Problem 14 by using Green's theorem.

16. Suppose f and g are (differentiable) functions of one variable. Show that $\int_C f(x)\, dx + g(y)\, dy = 0$, where C is any circle in the coordinate plane.

17. A homogeneous lamina occupies a region in the xy-plane with boundary C (oriented counterclockwise). Show that its center of mass has the coordinates given below, where A is the area of the region.

$$\bar{x} = \frac{1}{2A} \int_C x^2\, dy \qquad \bar{y} = -\frac{1}{2A} \int_C y^2\, dx$$

18. In Problem 17 show that the moment of inertia about the origin is

$$I_0 = \frac{1}{3} \int_C (x^3\, dy - y^3\, dx)$$

(Assume that the mass density of the lamina is 1.)

19. Use Problem 17 to find the center of mass of a homogeneous lamina enclosed by the coordinate axes and the hypocycloid $x = a \cos^3 t$, $y = a \sin^3 t$, $0 \le t \le \pi/2$.

20. Find the center of mass of a homogeneous wire in the shape of the hypocycloid $x = a \cos^3 t$, $y = a \sin^3 t$, $0 \le t \le \pi/2$.

21. Prove that no twice differentiable vector function exists whose curl is $x\mathbf{i} + y\mathbf{j} + z\mathbf{k}$.

22. Explain why the area of a smooth surface σ is $S = \iint_\sigma dS$. When the surface is the graph of $z = f(x, y)$ over the region R in the xy-plane, show that this formula reduces to

$$S = \iint_R \sqrt{(\partial z/\partial x)^2 + (\partial z/\partial y)^2 + 1}\, dA$$

as given in Chapter 18.

23. The sphere $x^2 + y^2 + z^2 = a^2$ is covered by a thin material with mass density at each point proportional to the distance from the xy-plane. Find the mass of the material.

24. Find the center of mass of the upper half of the sphere in Problem 23.

25. Let $F(x, y, z) = (1/\rho)(x\mathbf{i} + y\mathbf{j} + z\mathbf{k})$, where $\rho = \sqrt{x^2 + y^2 + z^2}$. Show that $\operatorname{div} F = 2/\rho$.

26. Use Problem 25 and the divergence theorem to compute

$$\iiint_T \frac{dV}{\sqrt{x^2 + y^2 + z^2}}$$

where T is the region inside the sphere $x^2 + y^2 + z^2 = 1$. Check by evaluating the triple integral directly.

27. Suppose the velocity of a fluid flow in space is constant. Show that the flux through any closed surface is zero. What is the physical interpretation?

28. Let $F(x, y, z) = x^3\mathbf{i} + y^3\mathbf{j} + z^3\mathbf{k}$ be the velocity of a fluid flow in space, where the mass density of the fluid is 1.
(a)   Find the flux through the sphere $x^2 + y^2 + z^2 = 1$.
(b)   Find the circulation around the circle $x^2 + y^2 = 1$ in the xy-plane.

29. Suppose $F(x, y, z) = P(x, y, z)\mathbf{i} + Q(x, y, z)\mathbf{j} + R(x, y, z)\mathbf{k}$ is a vector field with continuous and differentiable components in a simply connected region of space. Use Stokes' theorem to show that F is a gradient field if $Q_x = P_y$, $R_y = Q_z$,   and $P_z = R_x$.   Why does it follow that $\int_C F \cdot d\mathbf{r}$ is independent of path when these conditions hold?

30. Use Problem 29 to show that $\int_C (yz - y - z)\,dx + (xz - x - z)\,dy + (xy - x - y)\,dz$ is independent of path. (See Problems 5 and 6.)

31. Find $\int_C [(x - y)\,dx + (y - z)\,dy + (z - x)\,dz]$,   where C is the boundary of the first octant portion of the plane $x + y + z = 1$ (traversed counterclockwise when viewed from above).

32. Rework Problem 31 by using Stokes' theorem.

33. Let $F(x, y, z)$ be a vector function with continuous and differentiable components, and let Σ be a sphere with outer unit normal $\mathbf{n}$. Use Stokes' theorem to show that $\iint_\Sigma \text{curl } F \cdot \mathbf{n}\,dS = 0$.

Appendix I
Review Material

I.1 A Review of Trigonometry
I.2 The Binomial Theorem

I.1. A Review of Trigonometry

In this appendix we review definitions and properties of the six trigonometric functions: sine, cosine, tangent, secant, cosecant, cotangent, including some of the basic identities involving them.

We begin by reviewing angles and their measurement.

ANGLES

An *angle* is formed when two *rays* (*half-lines*) have the same endpoint or *vertex*. One ray is called the *initial side* and the other is called the *terminal side*. The angle may be measured by the amount of rotation needed for the initial side to coincide with the terminal side. We agree that when this rotation is counterclockwise, the angle is measured positively; when the rotation is clockwise, the angle is measured negatively (see Fig. 1).

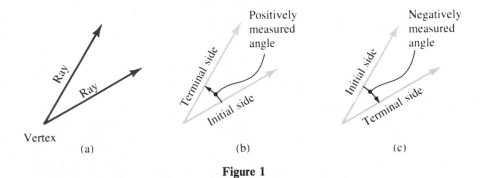

Figure 1

Angles are measured either in *degrees* or in *radians*. By definition, 1 revolution = 360 degrees and 1 revolution = 2π radians. Hence, 1 degree = $\pi/180$ radians and 1 radian = $180/\pi$ degrees. Table 1 lists the degree and radian measures of some frequently encountered angles.

Table 1

Degrees	0	30	45	60	90	180	270	360
Radians	0	$\pi/6$	$\pi/4$	$\pi/3$	$\pi/2$	π	$3\pi/2$	2π

If θ is a central angle of a circle of radius R, the length s of the intercepted arc is given by the formula

(1)
$$s = R\theta$$

where θ is measured in radians. The area A of the sector determined by this central angle θ is

(2)
$$A = (\tfrac{1}{2})R^2\theta$$

See Figure 2.

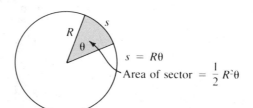

Figure 2

The simplicity of the formula, which is a consequence of measuring the angle θ in radians, is the first evidence of why we prefer radian measure to degree measure when working with calculus. As an example of formulas (1) and (2), for a circle of radius 3 meters, a central angle of $\pi/6$ radian intercepts an arc of length $3(\pi/6) = \pi/2 \approx 1.57$ meters. The area A of the sector determined by this angle is $A = \tfrac{1}{2}(3)^2(\pi/6) = 3\pi/4$ square meters.

Formula (1) is used to define two kinds of speed that involve circular motion. These velocities, called *linear speed on a circle* and *angular speed*, are used in a variety of applications. If a particle is moving along a circular path with constant speed, then a line segment drawn from the center of the circle to the particle sweeps out an angle θ at a constant rate. We define

$$V = \text{Linear speed on a circle} = \frac{\text{Change in length } s \text{ of arc}}{\text{Change in time } t} = \frac{\Delta s}{\Delta t}$$

$$\omega = \text{Angular speed} = \frac{\text{Changed in angle } \theta}{\text{Change in time } t} = \frac{\Delta \theta}{\Delta t}$$

From (1), it follows that

(3)
$$V = R\omega$$

It is important to remember that ω is measured in radians per unit of time. For example, a particle on a drive shaft of radius 2 centimeters rotating at the rate of 1 revolution per minute has the linear speed

$$V = R\omega = (2)(2\pi) = 4\pi \text{ centimeters per minute}$$

An angle is in the *standard position* when its vertex is placed at the origin and its initial side coincides with the positive x-axis of a rectangular coordinate system (see Fig. 3).

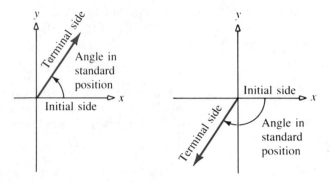

Figure 3

THE TRIGONOMETRIC FUNCTIONS

Figure 4 illustrates an angle θ in the standard position, together with a circle of radius R and center at the origin. If $P = (x, y)$ is a point on the circle, the six trigonometric functions of θ are defined by the equations below:

(4)
$$\sin \theta = y/R \qquad \cos \theta = x/R \qquad \tan \theta = y/x$$
$$\csc \theta = R/y \qquad \sec \theta = R/x \qquad \cot \theta = x/y$$

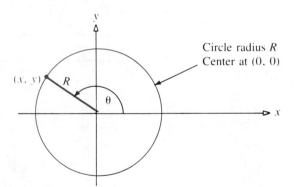

Figure 4

Of course, $\sec \theta$ and $\tan \theta$ are not defined when $x = 0$, and $\csc \theta$ and $\cot \theta$ are not defined when $y = 0$. As a result, the values $\pm\pi/2, \pm 3\pi/2, \ldots$ are not in the domain of either the secant function or the tangent function. Similarly, the values $0, \pm\pi, \pm 2\pi, \ldots$ are not in the domain of either the cosecant function or the cotangent function.

Based on the relationships in (4), we have the identities

$$\sin\theta = 1/\csc\theta \qquad \cos\theta = 1/\sec\theta \qquad \tan\theta = 1/\cot\theta$$

(5)

$$\tan\theta = \sin\theta/\cos\theta \qquad \cot\theta = \cos\theta/\sin\theta$$

which are valid for all θ where the functions are defined. Furthermore, the Pythagorean theorem requires that $x^2 + y^2 = R^2$, so that $(x/R)^2 + (y/R)^2 = 1$. Hence, we have the identity

(6)

$$(\sin\theta)^2 + (\cos\theta)^2 = 1$$

We shall follow the usual convention and write $(\sin\theta)^n$ as $\sin^n\theta$ when n is a positive integer. As a result, formula (6) becomes $\sin^2\theta + \cos^2\theta = 1$. In (6), if we divide each side by $\cos^2\theta$ and use two of the identities in (5), we obtain

(7)

$$1 + \tan^2\theta = \sec^2\theta$$

In (6), if we divide each side by $\sin^2\theta$ and use two of the identities in (5), we obtain

(8)

$$1 + \cot^2\theta = \csc^2\theta$$

If the point $P = (x, y)$ of definition (4) lies on an axis, the value of each trigonometric function is easy to calculate. For example, for an angle of $\pi/2$ radians, $P = (0, R)$, so that

$$\sin\pi/2 = R/R = 1 \qquad \cos\pi/2 = 0/R = 0 \qquad \tan\pi/2 \text{ is not defined}$$
$$\csc\pi/2 = R/R = 1 \qquad \sec\pi/2 \text{ is not defined} \qquad \cot\pi/2 = 0/R = 0$$

Table 2

	θ							
	0	$\pi/6$	$\pi/4$	$\pi/3$	$\pi/2$	π	$3\pi/2$	2π
$\sin\theta$	0	$1/2$	$\sqrt{2}/2$	$\sqrt{3}/2$	1	0	-1	0
$\cos\theta$	1	$\sqrt{3}/2$	$\sqrt{2}/2$	$1/2$	0	-1	0	1
$\tan\theta$	0	$1/\sqrt{3}$	1	$\sqrt{3}$	Not defined	0	Not defined	0
$\sec\theta$	1	$2/\sqrt{3}$	$\sqrt{2}$	2	Not defined	-1	Not defined	1
$\csc\theta$	Not defined	2	$\sqrt{2}$	$2/\sqrt{3}$	1	Not defined	-1	Not defined
$\cot\theta$	Not defined	$\sqrt{3}$	1	$1/\sqrt{3}$	0	Not defined	0	Not defined

For angles of $\pi/6$, $\pi/4$, and $\pi/3$ radians, there exist particularly simple relationships among x, y, and R. For example, for an angle of $\pi/6$ radian, if $R = 2$, then $y = 1$ and, by the Pythagorean theorem, $x = \sqrt{3}$. Hence,

$$\sin \pi/6 = 1/2 \qquad \cos \pi/6 = \sqrt{3}/2 \qquad \tan \pi/6 = 1/\sqrt{3}$$
$$\csc \pi/6 = 2 \qquad \sec \pi/6 = 2/\sqrt{3} \qquad \cot \pi/6 = \sqrt{3}$$

Table 2 lists the value of each trigonometric function for some of the most frequently encountered angles.

The graphs of the trigonometric functions are given in Figure 5.

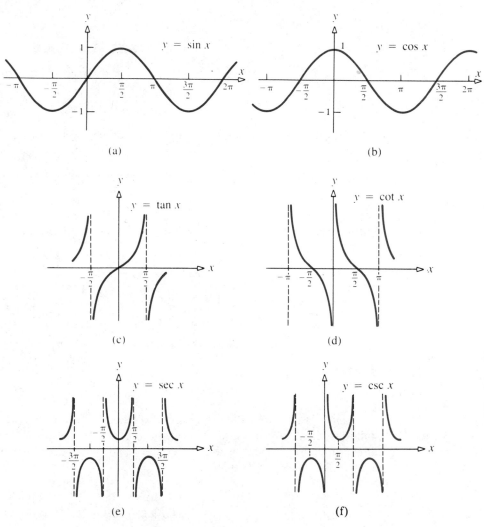

(a)

(b)

(c)

(d)

(e)

(f)

Figure 5

IDENTITIES

We have already established certain identities for the trigonometric functions. Other identities from trigonometry that are useful in calculus are listed below.

(9)   *Sum Formulas*

$$\sin(A + B) = \sin A \cos B + \cos A \sin B$$
$$\cos(A + B) = \cos A \cos B - \sin A \sin B$$
$$\tan(A + B) = \frac{\tan A + \tan B}{1 - \tan A \tan B}$$

(10)   *Difference Formulas*

$$\sin(A - B) = \sin A \cos B - \cos A \sin B$$
$$\cos(A - B) = \cos A \cos B + \sin A \sin B$$
$$\tan(A - B) = \frac{\tan A - \tan B}{1 + \tan A \tan B}$$

Based on the identities in (9), we can establish the following:

(11)   *Double-Angle Formulas*

$$\sin 2A = 2 \sin A \cos A$$
$$\cos 2A = \cos^2 A - \sin^2 A$$

Alternate forms for $\cos 2A$ can be obtained by using (6) in (11):

(12)
$$\cos 2A = 1 - 2 \sin^2 A$$
$$\cos 2A = 2 \cos^2 A - 1$$

Additionally, by using (10), we find that

$$\sin(-x) = \sin(0 - x) = (\sin 0)(\cos x) - (\cos 0)(\sin x) = -\sin x$$
$$\cos(-x) = \cos(0 - x) = (\cos 0)(\cos x) + (\sin 0)(\sin x) = \cos x$$

Hence,

(13)
$$\sin(-x) = -\sin x \qquad \cos(-x) = \cos x$$

Finally, as Figure 5 illustrates, the graphs of the sine, cosine, secant, and cosecant functions repeat over an interval of length 2π, and the graphs of the tangent and co-tangent functions repeat over an interval of length π. These facts may be verified by using (9) and showing that

$$\sin(x + 2\pi) = \sin x \qquad \sec(x + 2\pi) = \sec x \qquad \tan(x + \pi) = \tan x$$
$$\cos(x + 2\pi) = \cos x \qquad \csc(x + 2\pi) = \csc x \qquad \cot(x + \pi) = \cot x$$

Functions having the above property are called *periodic*. More precisely, a function f is periodic with period p, if p is the smallest positive number for which $f(x + p) = f(x)$, for all values of x. Thus, the sine, cosine, secant, and cosecant functions are periodic with period 2π, and the tangent and cotangent functions are periodic with period π.

Some of the important properties of the trigonometric functions are summarized in Table 3.

Table 3

Function	Domain	Range	Period	Symmetry
$y = \sin x$	All reals	$-1 \le y \le 1$	2π	Origin
$y = \cos x$	All reals	$-1 \le y \le 1$	2π	y-axis
$y = \tan x$	$x \ne \ldots, -\pi/2, \pi/2, 3\pi/2, \ldots$	All reals	π	Origin
$y = \sec x$	$x \ne \ldots, -\pi/2, \pi/2, 3\pi/2, \ldots$	$y \le -1, \quad y \ge 1$	2π	y-axis
$y = \csc x$	$x \ne \ldots, -\pi, 0, \pi, 2\pi, \ldots$	$y \le -1, \quad y \ge 1$	2π	Origin
$y = \cot x$	$x \ne \ldots, -\pi, 0, \pi, 2\pi, \ldots$	All reals	π	Origin

In the spirit of review we present a number of exercises to reacquaint you with trigonometry.

EXERCISE I.1

1. Convert the given degree measure of an angle to radian measure. Express the answer as a multiple of π and then obtain an approximation by using 3.14 for π.
(a) $45°$ (b) $-90°$ (c) $135°$ (d) $180°$ (e) $-210°$ (f) $300°$
(g) $-720°$ (h) $450°$ (i) $10°$ (j) $-80°$ (k) $2°$ (l) $120°$

2. Convert the given radian measure of an angle to degree measure. Where an approximation is needed, take 1 radian as $57.3°$.
(a) $\pi/3$ (b) $3\pi/4$ (c) $-5\pi/6$ (d) 2 (e) $-3\pi/2$ (f) π
(g) $\pi/6$ (h) $-\pi/2$ (i) 0.3 (j) $-\sqrt{2}$ (k) $11\pi/3$ (l) $5\pi/4$

3. Draw, in standard position, the angle whose measure is:
(a) $30°$ (b) $-225°$ (c) $5\pi/4$ (d) $-\pi$ (e) $3\pi/2$ (f) $480°$
(g) $-300°$ (h) $\frac{1}{2}$ (i) -2

4. Determine the exact numerical value of:
(a) $\sin \theta, \cos \theta$, and $\tan \theta$; $\theta = \pi/6$ (b) $\sin x, \cos x$, and $\sec x$; $x = 2\pi/3$
(c) $\cos \phi, \tan \phi$, and $\csc \phi$; $\phi = -\pi/3$ (d) $\sin t, \cos t$, and $\cot t$; $t = 3\pi/4$
(e) $\sin \alpha, \tan \alpha$, and $\sec \alpha$; $\alpha = 7\pi/6$ (f) $\cos \beta, \cot \beta$, and $\csc \beta$; $\beta = -3\pi/4$
(g) $\sin y, \tan y$, and $\cot y$; $y = 5\pi/6$ (h) $\sin w, \cos w$, and $\tan w$; $w = 11\pi/6$

5. Determine $\sin x, \cos x, \tan x, \cot x, \sec x$, and $\csc x$ for the given angle.
(a) $x = 0$ (b) $x = \pi/2$ (c) $x = \pi$ (d) $x = 3\pi/2$

6. Determine the "sign" of each of $\sin\theta$, $\cos\theta$, and $\tan\theta$, given that:
   (a) $0 < \theta < \pi/2$ (b) $\pi/2 < \theta < \pi$ (c) $\pi < \theta < 3\pi/2$ (d) $3\pi/2 < \theta < 2\pi$

7. Determine the value (or values) of x in the interval $[0, 2\pi)$ for which:
   (a) $\sin x = \frac{1}{2}$ (b) $\cos x = \sqrt{2}/2$ (c) $\tan x = 0$ (d) $\sin x = -\sqrt{3}/2$
   (e) $\cos x = -\frac{1}{2}$ (f) $\tan x = -1$ (g) $\cot x = -\sqrt{3}$ (h) $\sec x = \frac{2}{3}\sqrt{3}$
   (i) $\csc x = \sqrt{2}$ (j) $\cot x = 1$ (k) $\sec x = 1$ (l) $\csc x = -1$

8. Determine the value (or values) of x in the interval $(-\pi, \pi]$ for which:
   (a) $\sin x = -\frac{1}{2}$ (b) $\cos x = 0$ (c) $\tan x = -1$ (d) $\sin x = -1$
   (e) $\cos x = -\sqrt{3}/2$ (f) $\cot x = 0$ (g) $\sec x = -2$ (h) $\csc x = 2$
   (i) $\sec x = -1$

9. Determine the other five trigonometric functions of θ, given that:
   (a) $\sin\theta = \frac{3}{5}$; $\pi/2 < \theta < \pi$ (b) $\cos\theta = -\frac{5}{13}$; $-\pi < \theta < -\pi/2$
   (c) $\tan\theta = \frac{12}{5}$; $0 < \theta < \pi/2$ (d) $\sin\theta = -\frac{1}{3}$; $-\pi/2 < \theta < 0$
   (e) $\cos\theta = \frac{2}{5}$; $3\pi/2 < \theta < 2\pi$ (f) $\tan\theta = -\frac{1}{10}$; $\pi/2 < \theta < \pi$

10. Determine:
   (a) $\cos x$, given that $\sin x = 1/n$, where $n > 1$ and $0 < x < \pi/2$
   (b) $\cos\theta$, given that $\sin\theta = \sqrt{1 - x^2}$, where $0 \le x \le 1$ and $\pi/2 \le \theta \le \pi$
   (c) $\sin\alpha$, given that $\cos\alpha = 2x$, where $-\frac{1}{2} \le x \le 0$ and $\pi/2 \le \alpha \le \pi$
   (d) $\tan\theta$, given that $\sec\theta = 4x$, where $x \ge \frac{1}{4}$ and $0 \le \theta < \pi/2$
   (e) $\sin\beta$, given that $\tan\beta = 3x$, where $x \le 0$ and $-\pi/2 < \beta \le 0$
   (f) $\tan\theta$, given that $\cos\theta = x - 1$, where $1 \le x \le 2$ and $3\pi/2 < \theta \le 2\pi$

11. Find the length of the arc intercepted on a circle of radius 2 by each central angle.
   (a) $\pi/3$ (b) $2\pi/3$ (c) $60°$ (d) $120°$

12. Find the area of the sector determined by each central angle in Problem 11.

In Problems 13–20 reduce each expression to a trigonometric function of θ.

13. $\sin(\theta + \pi)$ **14.** $\tan\left(\dfrac{3\pi}{2} + \theta\right)$ **15.** $\cos\left(\dfrac{3\pi}{2} - \theta\right)$ **16.** $\sin(2\pi - \theta)$

17. $\cos(\theta + \pi)$ **18.** $\cos(\theta - \pi)$ **19.** $\sin\left(\theta + \dfrac{3\pi}{2}\right)$ **20.** $\tan\left(\theta + \dfrac{\pi}{2}\right)$

In Problems 21–32 prove that the given statement is true by transforming the expression on the left side of the equality into the expression on the right side.

21. $\sec\alpha - \cos\alpha = \tan\alpha \sin\alpha$

22. $(1 - \sin^2\alpha)\csc^2\alpha = \cot^2\alpha$

23. $\cos^4(2x) - \sin^4(2x) = \cos(4x)$

24. $\dfrac{2\tan A}{1 + \tan^2 A} = \sin(2A)$

25. $\tan x + \tan y = \dfrac{\sin(x + y)}{\cos x \cos y}$

26. $\cos\left(x - \dfrac{\pi}{4}\right) = \dfrac{\sqrt{2}}{2}(\sin x + \cos x)$

27. $\cot\theta - \tan\theta = 2\cot(2\theta)$

28. $\dfrac{1 + \sec\beta}{\sec\beta} = 2\cos^2(\frac{1}{2}\beta)$

29. $\cos(7\alpha) + \cos(5\alpha) = 2\cos(6\alpha)\cos\alpha$

30. $\cos(2\theta) - \cos\theta = -2\sin(3\theta/2)\sin(\theta/2)$

31. $\sin^2 x - \sin^2 y = \sin(x + y)\sin(x - y)$

32. $\cos(3\theta) = 4\cos^3\theta - 3\cos\theta$

In Problems 33–48 sketch the graph of the given equation for the indicated value of x. Give the period of the function.

33. $y = 2\sin x;$ x in $[-\pi, 2\pi]$

34. $y = -\frac{1}{2}\cos x;$ x in $[-2\pi, 2\pi]$

35. $y = 3\sin(2x);$ x in $[-\pi, \pi]$

36. $y = 2\cos(2x);$ x in $[0, 2\pi]$

37. $y = 4\cos(x/2);$ x in $[-\pi, 4\pi]$

38. $y = \sin(3x);$ x in $[0, 3\pi]$

39. $y = \tan(2x);$ x in $[-\pi, 2\pi]$

40. $y = \sec(2x);$ x in $[0, 2\pi]$

41. $y = 4\sin(x/2);$ x in $[0, 8\pi]$

42. $y = 2\csc x;$ x in $[-2\pi, 2\pi]$

43. $y = 2\cot(x/2);$ x in $[-\pi, 4\pi]$

44. $y = \sin^2 x;$ x in $[0, 2\pi]$

45. $y = \sin(x/4);$ x in $[-\pi, 2\pi]$

46. $y = |\sin x|;$ x in $[0, 3\pi]$

47. $y = 2\sin(\pi x);$ x in $[-2, 4]$

48. $y = \frac{1}{2}\cos(2\pi x);$ x in $[-1, 2]$

In Problems 49–54 solve the given trigonometric equation. For example, to solve $\sin(2x) = \frac{1}{2}$; either

$$2x = \frac{\pi}{6} + 2\pi n \qquad \text{or} \qquad 2x = \frac{5\pi}{6} + 2\pi n \qquad n \text{ an integer}$$

Therefore,

$$x = \frac{\pi}{12} + \pi n \qquad \text{or} \qquad x = \frac{5\pi}{12} + \pi n \qquad n \text{ an integer}$$

49. $\sin(2x) = 1$

50. $\sin x - \cos x = 0$

51. $\cos(3x) = \sqrt{3}/2$

52. $\tan(2x) = -1$

53. $\sin^2 x = \frac{1}{4}$

54. $\cos(2x + 1) = \frac{1}{2}$

In Problems 55–58 use the following facts to help solve the given equations.

If $\sin\alpha = \sin\beta$,   then either $\alpha = \beta + 2\pi n$ or $\alpha = (\pi - \beta) + 2\pi n$, n an integer

If $\cos\alpha = \cos\beta$,   then either $\alpha = \beta + 2\pi n$ or $\alpha = -\beta + 2\pi n$, n an integer

If $\tan\alpha = \tan\beta$,   then $\alpha = \beta + \pi n$, n an integer

55. $\sin x = \sin(2x)$

56. $\cos x = \cos(3x)$

57. $\tan x = \tan\left(3x + \dfrac{\pi}{3}\right)$

58. $\sin(2x) = \sin(3x - 1)$

59. Let f and g be functions that are periodic with period p. Prove that:
   (a)   The function $f + g$ is periodic with period p.
   (b)   The function $f \cdot g$ is periodic with period p.
   (c)   The function f/g is periodic with period p.

60. Let $f(x) = \dfrac{1}{k}\cos(kx)$.   For what number k does f have period 3?

61. Show that the area A of an isosceles triangle is given by the formula $A = \frac{1}{2}a^2\sin\theta$,   where a is the length of each of the equal sides and θ is the included angle.

62. A pendulum of length L is displaced through an angle θ with the vertical (see the figure). Show that the vertical displacement s of the end of the pendulum is given by the formula $s = 2L \sin^2(\theta/2)$.

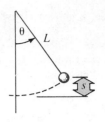

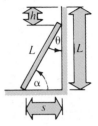

63. A plank leans aganist a wall and forms an angle θ with the wall (see the figure). Show that the distance h that the top of the plank is below its highest point is given by $h = s \tan(\theta/2)$, where s is the distance of the bottom of the plank from the wall. Determine a formula for h if the angle α that the plank forms with the ground is used instead of θ.

64. The minute and hour hands of a clock have lengths a and b, respectively $(a > b)$. Show that the distance s between the tips of the minute and hour hands is given by $s = \sqrt{a^2 + b^2 - 2ab \cos(\frac{11}{6}\pi t)}$ where t is time in hours. Assume that $t = 0$ when both hands point to 12. Determine the times when both hands point in the same direction.

65. Rod OA (see the figure) rotates about the fixed point O so that point A travels on a circle of radius R. Connected to point A is another rod AB of length $L \geq R$. Point B is connected to a piston that slides in a cylinder. Show that the distance x between point O and point B is given by $x = R \cos \theta + \sqrt{R^2 \cos^2\theta + L^2 - R^2}$, where θ is the angle of rotation of rod OA.

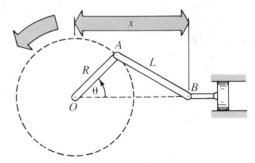

66. Consider a general triangle with angles A, B, C and corresponding opposite sides a, b, c. Show that the following two important properties hold:

$$\textit{Law of sines:} \quad \frac{\sin A}{a} = \frac{\sin B}{b} = \frac{\sin C}{c}$$

$$\textit{Law of cosines:} \quad a^2 + 2bc \cos A = b^2 + c^2$$

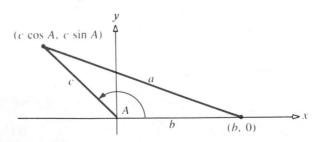

[*Hint:*   To prove the law of sines, drop a perpendicular from one vertex to the opposite side (or possibly its extension) and consider the two right triangles so determined. To prove the law of cosines, compute the square of the distance from $(b, 0)$ to $(c \cos A, c \sin A)$ (see the figure).]

67. *Calculator problem.*   Use "sin x" to stand for the "sine of an angle of x radians," and complete the following table:

x	1.5	1	0.1	0.01	0.001	0.0001
$\sin x$						
$(\sin x)/x$						

Can you conjecture as to what happens to $(\sin x)/x$ as $x \to 0^+$?

I.2. The Binomial Theorem

Expressions such as $(x + y)^2$ and $(x + y)^3$ are not too difficult to expand. For example,

$$(x + y)^2 = x^2 + 2xy + y^2 \qquad (x + y)^3 = x^3 + 3x^2y + 3xy^2 + y^3$$

However, expanding expressions such as $(x + y)^6$ or $(x + y)^8$ can be tedious and time-consuming. In this section we present a procedure for expanding expressions of the form $(x + y)^n$,   where n is a positive integer. This procedure, which is called the *binomial theorem*, requires the expression

$$\binom{n}{r} = \frac{n!}{r!(n - r)!}$$

Then, for example,

$$(x + y)^2 = x^2 + 2xy + y^2$$

can be written as

$$(x + y)^2 = \binom{2}{0}x^2 + \binom{2}{1}xy + \binom{2}{2}y^2$$

The expansion of $(x + y)^3$ can be written as

$$(x + y)^3 = x^3 + 3x^2y + 3xy^2 + y^3 = \binom{3}{0}x^3 + \binom{3}{1}x^2y + \binom{3}{2}xy^2 + \binom{3}{3}y^3$$

In general, we have the following theorem:

BINOMIAL THEOREM   **If n is a positive integer, then**

$$(x + y)^n = \binom{n}{0}x^n + \binom{n}{1}x^{n-1}y + \binom{n}{2}x^{n-2}y^2 + \cdots + \binom{n}{k}x^{n-k}y^k + \cdots + \binom{n}{n}y^n$$

Observe that the powers of x begin at n and decrease by 1, while the powers of y begin with 0 and increase by 1. Also, in $(x + y)^n$, the coefficient of y^k is always $\binom{n}{k}$.

EXAMPLE 1   Expand $(x + y)^6$ by using the binomial theorem.

Solution $(x + y)^6 = \binom{6}{0}x^6 + \binom{6}{1}x^5y + \binom{6}{2}x^4y^2 + \binom{6}{3}x^3y^3 + \binom{6}{4}x^2y^4 + \binom{6}{5}xy^5 + \binom{6}{6}y^6$

$$= x^6 + 6x^5y + 15x^4y^2 + 20x^3y^3 + 15x^2y^4 + 6xy^5 + y^6 \qquad \blacksquare$$

EXAMPLE 2   Find the coefficient of y^4 in the expansion of $(x + y)^7$.

Solution   The coefficient of y^4 is

$$\binom{7}{4} = \frac{7 \cdot 6 \cdot 5}{3 \cdot 2 \cdot 1} = 35 \qquad \blacksquare$$

EXAMPLE 3   Expand $(x + 2y)^4$ by using the binomial theorem.

Solution $(x + 2y)^4 = \binom{4}{0}x^4 + \binom{4}{1}x^3(2y) + \binom{4}{2}x^2(2y)^2 + \binom{4}{3}x(2y)^3 + \binom{4}{4}(2y)^4$

$$= x^4 + 8x^3y + 24x^2y^2 + 32xy^3 + 16y^4 \qquad \blacksquare$$

EXERCISE I.2

In Problems 1–6 use the binomial theorem to expand each expression.

1. $(x + y)^5$ **2.** $(x + y)^4$ **3.** $(x + 3y)^3$

4. $(2x + y)^3$ **5.** $(2x - y)^4$ **6.** $(x - y)^4$

7. What is the coefficient of y^3 in the expansion of $(x + y)^5$?

8. What is the coefficient of y^6 in the expansion of $(x + y)^8$?

9. What is the coefficient of x^8 in the expansion of $(x + 3)^{10}$?

10. What is the coefficient of x^3 in the expansion of $(x + 2)^5$?

Appendix II
Theorems, Proofs, and Definitions

II.1 Limit Theorems, Proofs, and Definitions

II.2 Derivative Theorems and Proofs

II.3 Integral Theorems and Proofs

II.1. Limit Theorems, Proofs, and Definitions

UNIQUENESS OF LIMIT

The limit of a function, if it exists, is unique. That is, a function cannot approach two different limits at the same time.

THEOREM If $\lim_{x \to c} f(x) = L_1$ and $\lim_{x \to c} f(x) = L_2$, then $L_1 = L_2$.

Proof Let us assume that $L_1 \neq L_2$. We will show that this assumption leads to a contradiction. Since $\lim_{x \to c} f(x) = L_1$, by the definition of limit, for any given $\varepsilon > 0$, there is a $\delta_1 > 0$ such that for x in the domain of f,

(1)
$$|f(x) - L_1| < \varepsilon \qquad \text{whenever} \qquad 0 < |x - c| < \delta_1$$

Similarly, since $\lim_{x \to c} f(x) = L_2$, by the definition of limit, for any given $\varepsilon > 0$, there is a $\delta_2 > 0$ such that for x in the domain of f,

(2)
$$|f(x) - L_2| < \varepsilon \qquad \text{whenever} \qquad 0 < |x - c| < \delta_2$$

Now, we know that $L_1 - L_2 = L_1 - f(x) + f(x) - L_2$, so, by applying the triangle inequality, we have

(3)
$$|L_1 - L_2| = |L_1 - f(x) + f(x) - L_2| \leq |L_1 - f(x)| + |f(x) - L_2|$$

From (1), (2), and (3), we conclude that for any given $\varepsilon > 0$, there exist δ_1 and δ_2 such that whenever $0 < |x - c| < \delta_1$ and $0 < |x - c| < \delta_2$, we have

(4)
$$|L_1 - L_2| < \varepsilon + \varepsilon = 2\varepsilon$$

But, if we let $\varepsilon = \frac{1}{2}|L_1 - L_2| > 0$,   then, from (4), we have

$$|L_1 - L_2| < 2\varepsilon = |L_1 - L_2|$$

which is a contradiction. Hence, $L_1 = L_2$.   ∎

ALGEBRA OF LIMITS

THEOREM   Let f and g be two functions for which $\lim_{x \to c} f(x) = L$ and $\lim_{x \to c} g(x) = M$, L and M being two real numbers.

(2.9)   *Limit of a Sum.* $\lim_{x \to c}[f(x) + g(x)] = \lim_{x \to c} f(x) + \lim_{x \to c} g(x) = L + M$

(2.11)   If k is any real number, $\lim_{x \to c}[kf(x)] = k \lim_{x \to c} f(x) = kL$

(2.12)   *Limit of a Product.* $\lim_{x \to c}[f(x)g(x)] = [\lim_{x \to c} f(x)][\lim_{x \to c} g(x)] = LM$

Proof of (2.9)   We wish to show that for any $\varepsilon > 0$ there must exist a $\delta > 0$ such that

$$|f(x) + g(x) - (L + M)| < \varepsilon \qquad \text{whenever} \qquad 0 < |x - c| < \delta$$

Since $\lim_{x \to c} f(x) = L$,   by the definition of limit, given $\varepsilon/2 > 0$,   there is a $\delta_1 > 0$ such that

$$|f(x) - L| < \varepsilon/2 \qquad \text{whenever} \qquad 0 < |x - c| < \delta_1$$

Since $\lim_{x \to c} g(x) = M$,   for this same choice of $\varepsilon/2$ there is a $\delta_2 > 0$ such that

$$|g(x) - M| < \varepsilon/2 \qquad \text{whenever} \qquad 0 < |x - c| < \delta_2$$

Now let δ be the smaller of δ_1 and δ_2. Then $\delta \le \delta_1$ and $\delta \le \delta_2$;   and, by using this δ, we can state the following:

$$|f(x) - L| < \varepsilon/2 \qquad \text{whenever} \qquad 0 < |x - c| < \delta$$
$$|g(x) - M| < \varepsilon/2 \qquad \text{whenever} \qquad 0 < |x - c| < \delta$$

Therefore, by the triangle inequality,

$$|f(x) + g(x) - (L + M)| = |f(x) - L + g(x) - M|$$
$$\le |f(x) - L| + |g(x) - M| \le \varepsilon/2 + \varepsilon/2 = \varepsilon$$

whenever $0 < |x - c| < \delta$.   Hence, $\lim_{x \to c}[f(x) + g(x)] = L + M$.   ∎

Proof of (2.11)   If the constant $k = 0$,   the result is obvious. For $k \ne 0$,   we look at

$$|kf(x) - kL| = |k[f(x) - L]| = |k||f(x) - L|$$

Since $\lim_{x \to c} f(x) = L$,   given $\varepsilon/|k| > 0$,   there is a $\delta > 0$ so that whenever $0 < |x - c| < \delta$,   we have $|f(x) - L| < \varepsilon/|k|$.   Hence, for any $\varepsilon > 0$,   there is a

$\delta > 0$ such that whenever $0 < |x - c| < \delta$,   we have

$$|kf(x) - kL| = |k||f(x) - L| < |k|\frac{\varepsilon}{|k|} = \varepsilon \qquad \blacksquare$$

Proof of (2.12)   We look at $|f(x)g(x) - LM|$ and see that we should add and subtract $f(x)M$ to get terms involving $g(x) - M$ and $f(x) - L$:

$$\begin{aligned}|f(x)g(x) - LM| &= |f(x)g(x) - f(x)M + f(x)M - LM| \\ &= |f(x)[g(x) - M] + [f(x) - L]M| \\ &\leq |f(x)||g(x) - M| + |f(x) - L||M|\end{aligned}$$

Since $\lim_{x \to c} f(x) = L$ and $\lim_{x \to c} g(x) = M$,   we know:

(5)   There is a $\delta_1 > 0$ so that if $0 < |x - c| < \delta_1$,   then

$$|f(x) - L| < 1 \qquad \text{and so} \qquad |f(x)| < 1 + |L|$$

(6)   Given $\varepsilon > 0$,   there is a δ_2 so that if $0 < |x - c| < \delta_2$,   then

$$|g(x) - M| < \frac{\varepsilon}{1 + |L| + |M|}$$

(7)   Given $\varepsilon > 0$,   there is a δ_3 so that if $0 < |x - c| < \delta_3$,   then

$$|f(x) - L| < \frac{\varepsilon}{1 + |L| + |M|}$$

So we choose $\delta = \min(\delta_1, \delta_2, \delta_3)$.   Then, if $0 < |x - c| < \delta$,   we may combine (5), (6), and (7) to get

$$|f(x)g(x) - LM| < [1 + |L|]\frac{\varepsilon}{1 + |L| + |M|} + |M|\frac{\varepsilon}{1 + |L| + |M|} = \varepsilon \qquad \blacksquare$$

(2.20)   *Squeezing Theorem.*   **Let f, g, and h be functions such that $f(x) \leq g(x) \leq h(x)$ for all numbers x in some open interval containing c, except possibly at c. If $\lim_{x \to c} f(x) = L$ and $\lim_{x \to c} h(x) = L$,   then $\lim_{x \to c} g(x) = L$.**

Proof   Let $\varepsilon > 0$.   Since $\lim_{x \to c} f(x) = \lim_{x \to c} h(x) = L$,   there   exist   positive numbers δ_1 and δ_2 such that for x in the open interval,

$$|f(x) - L| < \varepsilon \qquad \text{whenever} \qquad 0 < |x - c| < \delta_1$$
$$|h(x) - L| < \varepsilon \qquad \text{whenever} \qquad 0 < |x - c| < \delta_2$$

We choose δ to be the smaller of δ_1 and δ_2. Then $0 < |x - c| < \delta$ implies that $|f(x) - L| < \varepsilon$ and $|h(x) - L| < \varepsilon$.   In   other   words, $0 < |x - c| < \delta$ implies that $L - \varepsilon < f(x) < L + \varepsilon$ and $L - \varepsilon < h(x) < L + \varepsilon$.   Since $f(x) \leq g(x) \leq h(x)$ for all $x \neq c$ in the open interval, it follows that if $0 < |x - c| < \delta$ and x is in the open interval, we have

$$L - \varepsilon < f(x) \le g(x) \le h(x) < L + \varepsilon$$

Thus, for any given $\varepsilon > 0$, there is some $\delta > 0$ such that if $0 < |x - c| < \delta$ and x is in the open interval, then $L - \varepsilon < g(x) < L + \varepsilon$. Hence, $\lim_{x \to c} g(x) = L$.

∎

THEOREM   Let a function f be defined on some interval containing c and let f be continuous at c. If $f(c) > 0,$ then there is a *subinterval* containing c throughout which $f(x) > 0.$

Proof   Since f is continuous at c, we know that $\lim_{x \to c} f(x) = f(c) > 0$. Then, given any $\varepsilon > 0$, and, in particular, $\varepsilon = \frac{1}{2}f(c)$, there is a $\delta > 0$ so that whenever $|x - c| < \delta$, we have

$$\left| f(x) - f(c) \right| < \varepsilon = \tfrac{1}{2}f(c)$$

For x in the interval $c - \delta < x < c + \delta$, it follows that

$$-\tfrac{1}{2}f(c) < f(x) - f(c) < \tfrac{1}{2}f(c)$$
$$\tfrac{1}{2}f(c) < \quad f(x) \quad < \tfrac{3}{2}f(c)$$

That is, $f(x) > \frac{1}{2}f(c) > 0$ in the interval $c - \delta < x < c + \delta$. ∎

LIMIT DEFINITIONS

DEFINITION   *Left-Hand Limit.*   Let f be a function defined on the open interval (a, c). The *limit of $f(x)$ as x approaches c from the left is L,* written

$$\lim_{x \to c^-} f(x) = L$$

if for any given number $\varepsilon > 0,$ a positive number δ exists such that

$$|f(x) - L| < \varepsilon \qquad \text{whenever} \qquad c - \delta < x < c \quad \text{and } x \text{ is in the domain of } f$$

See Figure 1 for a geometric interpretation.

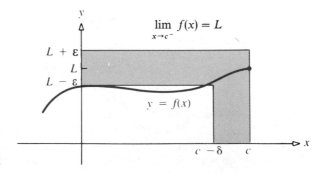

Figure 1

DEFINITION   *Right-Hand Limit.*   **Let f be a function defined on the open interval (c, b). The *limit of $f(x)$ as x approaches c from the right* is L, written**

$$\lim_{x \to c^+} f(x) = L$$

if for any given number $\varepsilon > 0$,   a positive number δ exists such that

$$|f(x) - L| < \varepsilon \qquad \text{whenever} \qquad c < x < c + \delta \quad \text{and } x \text{ is in the domain of } f$$

See Figure 2 for an illustration.

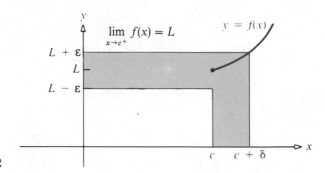

Figure 2

DEFINITION   *Infinite Limits.*   **Let f be a function defined on an open interval containing c. Then $f(x)$ becomes *positively infinite as x approaches c*, written**

$$\lim_{x \to c} f(x) = +\infty$$

if, for every positive number M, a positive number δ exists such that

$$f(x) > M \qquad \text{whenever} \quad 0 < |x - c| < \delta \qquad \text{and } x \text{ is in the domain of } f$$

Similarly, $f(x)$ becomes *negatively infinite*, written

$$\lim_{x \to c} f(x) = -\infty$$

if, for any negative number N, a positive number δ exists such that

$$f(x) < N \qquad \text{whenever} \quad 0 < |x - c| < \delta \qquad \text{and } x \text{ is in the domain of } f$$

See Figure 3 (p. A-18) for an illustration.

DEFINITION   *Limits at Infinity.*   **Let f be a function defined on the open interval $(b, +\infty)$. Then**

$$\lim_{x \to +\infty} f(x) = L$$

if, for any given $\varepsilon > 0$,   there is a positive number M so that

$$|f(x) - L| < \varepsilon \qquad \text{whenever} \qquad x > M \quad \text{and } x \text{ is in the domain of } f$$

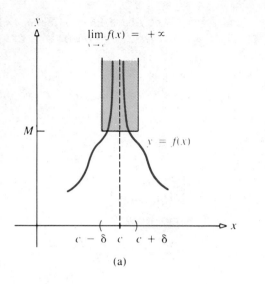

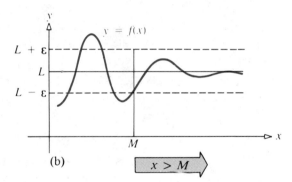

Figure 3   (a)   (b)

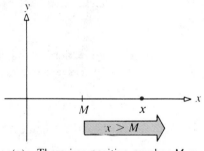

(a) There is a positive number M so   (b)
that whenever $x > M$ and x is in
the domain of f, then
$|f(x) - L| < \varepsilon$ for any $\varepsilon > 0$

Figure 4 $\displaystyle \lim_{x \to +\infty} f(x) = L$

Figure 4 illustrates the idea of limits at infinity.

If f is a function defined on the open interval $(-\infty, a)$ then

$$\lim_{x \to -\infty} f(x) = L$$

if, for any given $\varepsilon > 0$, there is a negative number N so that

$$|f(x) - L| < \varepsilon \qquad \text{whenever} \qquad x < N \quad \text{and x is in the domain of f}$$

Figure 5 illustrates the idea.

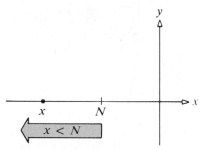

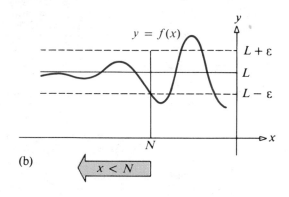

(a) There is a negative number N so that whenever $x < N$ and x is in the domain of f, then $|f(x) - L| < \varepsilon$ for any $\varepsilon > 0$

(b)

Figure 5 $\lim\limits_{x \to -\infty} f(x) = L$

II.2. Derivative Theorems and Proofs

THEOREM *Chain Rule.* **If f and g are differentiable functions, the composite function $f \circ g$ is also differentiable. Moreover, if $y = f(u)$ and $u = g(x)$, then the derivative of $y = (f \circ g)(x)$ is**

$$\frac{dy}{dx} = \left(\frac{dy}{du}\right)\left(\frac{du}{dx}\right)$$

Proof If $y = f(u)$ and $u = g(x)$, the composite function is $y = (f \circ g)(x) = f(g(x))$. First, the function $y = f(u)$ is differentiable, so that

$$\lim_{\Delta u \to 0} \frac{\Delta y}{\Delta u} = \frac{dy}{du}$$

Define a new function t to be

$$t = \begin{cases} 0 & \text{if } \Delta u = 0 \\ \dfrac{\Delta y}{\Delta u} - \dfrac{dy}{du} & \text{if } \Delta u \neq 0 \end{cases}$$

For $\Delta u \neq 0$,

(1)
$$\frac{\Delta y}{\Delta u} = \frac{dy}{du} + t$$

$$\Delta y = \frac{dy}{du} \Delta u + t\Delta u$$

Since for $\Delta u = 0$, we have $\Delta y = 0$ (y is a continuous function of u), formula (1) is valid whether $\Delta u = 0$ or $\Delta u \neq 0$. Division by Δx yields

$$\frac{\Delta y}{\Delta x} = \frac{dy}{du}\left(\frac{\Delta u}{\Delta x}\right) + t\left(\frac{\Delta u}{\Delta x}\right)$$

Take the limit as $\Delta x \to 0$. Then

$$\lim_{\Delta x \to 0} \frac{\Delta y}{\Delta x} = \lim_{\Delta x \to 0} \frac{dy}{du}\left(\frac{\Delta u}{\Delta x}\right) + \lim_{\Delta x \to 0} t\frac{\Delta u}{\Delta x} = \left(\frac{dy}{du}\right)\left(\frac{du}{dx}\right) + \left(\lim_{\Delta x \to 0} t\right)\left(\frac{du}{dx}\right)$$

Now when $\Delta x \to 0$, so does $\Delta u \to 0$ since u is a continuous function of x. Thus,

$$\lim_{\Delta x \to 0} t = \lim_{\Delta u \to 0} t \underset{\underset{\text{By (1)}}{\uparrow}}{=} \lim_{\Delta u \to 0} \left(\frac{\Delta y}{\Delta u} - \frac{dy}{du}\right) = 0$$

Hence,

$$\frac{dy}{dx} = \lim_{\Delta x \to 0} \frac{\Delta y}{\Delta x} = \left(\frac{dy}{du}\right)\left(\frac{du}{dx}\right) + \left(\lim_{\Delta u \to 0} t\right)\left(\frac{du}{dx}\right) = \left(\frac{dy}{du}\right)\left(\frac{du}{dx}\right) + 0\left(\frac{du}{dx}\right) = \left(\frac{dy}{du}\right)\left(\frac{du}{dx}\right)$$

■

INVERSE FUNCTIONS

THEOREM *Continuity of the Inverse Functions.* **If f denotes a continuous one-to-one function defined on the open interval (a, b), then its inverse $g = f^{-1}$ is continuous.**

Proof If f is continuous, then, being one-to-one, f is either increasing on (a, b) or it is decreasing on (a, b). (This proof is left to you.) Suppose f is increasing on (a, b). Set $y_0 = f(x_0)$. We need to show that g is continuous at y_0, given that f is continuous at x_0. The number $g(y_0) = x_0$ is in the open interval (a, b). Choose $\varepsilon > 0$ sufficiently small so that $g(y_0) - \varepsilon$ and $g(y_0) + \varepsilon$ are also in (a, b). We wish to find $\delta > 0$ so that:

$$\text{If} \quad y_0 - \delta < y < y_0 + \delta \quad \text{then} \quad g(y_0) - \varepsilon < g(y) < g(y_0) + \varepsilon$$

This can be done by choosing δ to obey

$$f[g(y_0) - \varepsilon] < y_0 - \delta \quad \text{and} \quad y_0 + \delta < f[g(y_0) + \varepsilon]$$

Then, if $y_0 - \delta < y < y_0 + \delta$, we have

$$f[g(y_0) - \varepsilon] < y < f[g(y_0) + \varepsilon]$$

But g is also increasing since f is. Hence

$$g(y_0) - \varepsilon < g(y) < g(y_0) + \varepsilon$$

We can handle the case where f is decreasing on (a, b) in a similar way. ■

(3.54)   **THEOREM**   Let $y = f(x)$ and $x = g(y)$ be inverse functions. Assume that f is differentiable on an open interval containing x_0 and that $y_0 = f(x_0)$.   If $f'(x_0) \neq 0$,   then g is differentiable at y_0, and

$$g'(y_0) = \frac{1}{f'(x_0)}$$

Proof   Since f and g are inverses of one another, we have $f(x) = y$ if and only if $x = g(y)$.   Hence, we have the following identity, where $g(y_0) = x_0$:

$$\frac{g(y) - g(y_0)}{y - y_0} = \frac{x - x_0}{f(x) - f(x_0)} = \frac{1}{\dfrac{f(x) - f(x_0)}{x - x_0}}$$

By the previous theorem, the continuity of f at x_0 implies the continuity of g at y_0; hence, $y \to y_0$ as $x \to x_0$.   By taking the limits of both sides of the above identity, we have

$$g'(y_0) = \lim_{x \to x_0} \frac{g(y) - g(y_0)}{y - y_0} = \frac{1}{\displaystyle\lim_{x \to x_0} \frac{f(x) - f(x_0)}{x - x_0}} = \frac{1}{f'(x_0)}$$

which completes the proof.   ∎

Theorem (3.54) and its proof are illustrated in Figure 6.

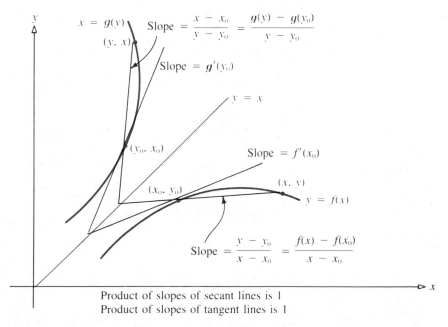

Product of slopes of secant lines is 1
Product of slopes of tangent lines is 1

Figure 6

II.3. Integral Theorems and Proofs

(5.25)   **THEOREM If f is continuous on a closed interval containing the numbers a, b, and c, then**

$$\int_a^b f(x)\,dx + \int_b^c f(x)\,dx = \int_a^c f(x)\,dx$$

no matter what the order of the numbers a, b, and c.

Proof, Part 1   Assume that $a < b < c$. Let $\varepsilon > 0$. Since f is continuous on $[a, b]$ and on $[b, c]$, there is a $\delta_1 > 0$ such that

(1)
$$\left| \sum_{i=1}^k f(u_i)\Delta x_i - \int_a^b f(x)\,dx \right| < \frac{\varepsilon}{2}$$

for every Riemann sum $\sum_{i=1}^k f(u_i)\Delta x_i$ for f on $[a, b]$ whose partition P_1 of $[a, b]$ has norm $\|P_1\| < \delta_1$. There is also a $\delta_2 > 0$ so that

(2)
$$\left| \sum_{i=1}^j f(u_i)\Delta x_i - \int_b^c f(x)\,dx \right| < \frac{\varepsilon}{2}$$

for every Riemann sum $\sum_{i=1}^j f(u_i)\Delta x_i$ for f on $[b, c]$ whose partition P_2 of $[b, c]$ has norm $\|P_2\| < \delta_2$.

Let $\delta = \min\{\delta_1, \delta_2\}$. Then (1) and (2) hold with δ replacing δ_1 and δ_2. If we now add (1) and (2), term-by-term, and if $\|P_1\| < \delta$ and $\|P_2\| < \delta$, we have

$$\left| \sum_{i=1}^k f(u_i)\Delta x_i - \int_a^b f(x)\,dx \right| + \left| \sum_{i=1}^j f(u_i)\Delta x_i - \int_b^c f(x)\,dx \right| < \frac{\varepsilon}{2} + \frac{\varepsilon}{2} = \varepsilon$$

By using the triangle inequality, this result implies that

(3)
$$\left| \sum_{i=1}^k f(u_i)\Delta x_i - \int_a^b f(x)\,dx + \sum_{i=1}^j f(u_i)\Delta x_i - \int_b^c f(x)\,dx \right| < \varepsilon$$

if $\|P_1\| < \delta$ and $\|P_2\| < \delta$.

Denote $P_1 + P_2$ by P^*, a partition of $[a, c]$ having b as a point of division.

$$\sum_{i=1}^k f(u_i)\Delta x_i + \sum_{i=1}^j f(u_i)\Delta x_i = \sum_{i=1}^n f(u_i)\Delta x_i$$

is a Riemann sum for f on P^*. Since $\|P^*\| < \delta$ implies that $\|P_1\| < \delta$ and $\|P_2\| < \delta$, it follows from (3) that

$$\left| \sum_{i=1}^n f(u_i)\Delta x_i - \left[\int_a^b f(x)\,dx + \int_b^c f(x)\,dx \right] \right| < \varepsilon$$

for every Riemann sum $\sum_{i=1}^n f(u_i)\Delta x_i$ for f on $[a, c]$ whose partition P^* of $[a, c]$ has b for a point of division and norm $\|P^*\| < \delta$. Therefore,

$$\int_a^c f(x)\,dx = \int_a^b f(x)\,dx + \int_b^c f(x)\,dx$$

Part 2   There are six possible orderings for the points a, b, and c:

$$a < b < c \qquad a < c < b \qquad b < a < c$$
$$b < c < a \qquad c < a < b \qquad c < b < a$$

In part I we showed that the theorem is true for the order $a < b < c$. Now consider any other order, say, $b < c < a$. From part 1,

(4)
$$\int_b^c f(x)\,dx + \int_c^a f(x)\,dx = \int_b^a f(x)\,dx$$

However,

$$\int_c^a f(x)\,dx = -\int_a^c f(x)\,dx \qquad \int_b^a f(x)\,dx = -\int_a^b f(x)\,dx$$

By substituting this in (4), we obtain

$$\int_b^c f(x)\,dx - \int_a^c f(x)\,dx = -\int_a^b f(x)\,dx$$

$$\int_a^b f(x)\,dx + \int_b^c f(x)\,dx = \int_a^c f(x)\,dx$$

which proves the theorem for the order $b < c < a$.

The proofs for the remaining four orders are similar.   ∎

(5.26)   **THEOREM**   **If a function f is continuous on a closed interval $[a, b]$ and if m and M denote the absolute minimum and absolute maximum of f on $[a, b]$, then**

$$m(b - a) \le \int_a^b f(x)\,dx \le M(b - a)$$

Proof   We shall only prove here that $m(b-a) \le \int_a^b f(x)\,dx$. We assume the contrary, namely, that

(5)
$$m(b - a) > \int_a^b f(x)\,dx$$

We show that this assumption leads to a contradiction, and therefore that the theorem must be true. Since f is continuous on $[a, b]$,

$$\lim_{\|P\| \to 0} \sum_{i=1}^{n} f(u_i)\Delta x_i = \int_a^b f(x)\,dx$$

By (5), $m(b - a) - \int_a^b f(x)\,dx > 0$.   Let

(6)
$$\varepsilon = m(b - a) - \int_a^b f(x)\,dx$$

Then there is a $\delta > 0$ such that for all partitions P of $[a, b]$ with $\|P\| < \delta$, we have

$$\sum_{i=1}^{n} f(u_i)\Delta x_i - \int_a^b f(x)\,dx < \varepsilon$$

which is equivalent to

$$\int_a^b f(x)\,dx - \varepsilon < \sum_{i=1}^n f(u_i)\Delta x_i < \int_a^b f(x)\,dx + \varepsilon$$

By (6), the second inequality can be written as

$$\sum_{i=1}^n f(u_i)\Delta x_i < \int_a^b f(x)\,dx + \varepsilon = \int_a^b f(x)\,dx + m(b-a) - \int_a^b f(x)\,dx = m(b-a)$$

Consequently,

$$\sum_{i=1}^n f(u_i)\Delta x_i < m(b-a) = \sum_{i=1}^n m\Delta x_i$$

which implies that for every partition P of $[a, b]$ with $\|P\| < \delta$,

$$f(u_i) < m$$

for some u_i in $[a, b]$. But this result is impossible because m is the absolute minimum of f on $[a, b]$. Therefore, the assumption that $m(b-a) > \int_a^b f(x)\,dx$ is false, and it follows that

$$m(b-a) \le \int_a^b f(x)\,dx$$

The proof that $\int_a^b f(x)\,dx \le M(b-a)$ is similar.   ∎

Appendix III
Formulas

      III.1 Algebra

      III.2 Geometry

III.1. Algebra

(1) *Quadratic Formula:* The solutions of the quadratic equation $ax^2 + bx + c = 0$ $(a \neq 0)$ are

$$x = \frac{-b \pm \sqrt{b^2 - 4ac}}{2a}$$

provided $b^2 - 4ac \geq 0$. Otherwise, if $b^2 - 4ac < 0$, there are no real solutions.

(2) *Exponents:*

$$a^m \cdot a^n = a^{m+n} \qquad\qquad (ab)^n = a^n b^n$$

$$\frac{a^m}{a^n} = a^{m-n}, \quad a \neq 0 \qquad\qquad a^{-n} = \frac{1}{a^n}, \quad a \neq 0$$

$$(a^m)^n = a^{mn} \qquad\qquad a^{p/q} = \sqrt[q]{a^p} = (\sqrt[q]{a})^p, \quad a > 0$$

(3) *Logarithms:*

$$\log_a x + \log_a y = \log_a xy \qquad\qquad \log_a 1 = 0$$

$$\log_a x - \log_a y = \log_a \frac{x}{y} \qquad\qquad \log_a a = 1$$

$$\qquad\qquad\qquad\qquad\qquad\qquad a^{\log_a x} = x$$

$$r \log_a x = \log_a x^r$$

$$\log_a x = \frac{\log_b x}{\log_b a}$$

(4) *Determinants:*

$$\begin{vmatrix} a & b \\ c & d \end{vmatrix} = ad - bc$$

$$\begin{vmatrix} a_1 & b_1 & c_1 \\ a_2 & b_2 & c_2 \\ a_3 & b_3 & c_3 \end{vmatrix} = a_1 \begin{vmatrix} b_2 & c_2 \\ b_3 & c_3 \end{vmatrix} - b_1 \begin{vmatrix} a_2 & c_2 \\ a_3 & c_3 \end{vmatrix} + c_1 \begin{vmatrix} a_2 & b_2 \\ a_3 & b_3 \end{vmatrix}$$

III.2. Geometry

(1) *Triangle:*

Area $= \frac{1}{2}bh$

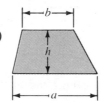

(2) *Parallelogram:*

Area $= bh$

(3) *Trapezoid:*

Area $= \frac{1}{2}h(a + b)$

(4) *Circle:*

Area $= \pi R^2$

Circumference $= 2\pi R$

(5) *Sector of Circle:*

Area $= \frac{1}{2}R^2\theta$

(6) *Ellipse:*

Area $= \pi ab$

(7) *Right Circular Cylinder*

Volume $= \pi R^2 h$

Lateral surface $= 2\pi Rh$

Total surface $= 2\pi R(R + h)$

(8) *Right Circular Cone:*

Volume $= \frac{1}{3}\pi R^2 h$

Lateral surface $= \pi Rs$

Total surface $= \pi R(R + s)$

(9) *Sphere:*

Volume $= \frac{4}{3}\pi R^3$

Surface $= 4\pi R^2$

(10) *Frustum of a Right Circular Cone:*

Volume $= \frac{1}{3}\pi h(R_1^2 + R_1 R_2 + R_2^2)$

Lateral surface $= \pi s(R_1 + R_2)$

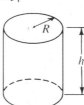

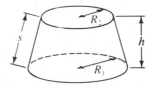

Appendix IV
Tables

Table 1   Squares, Square Roots, and Prime Factors

No.	Sq.	Sq. Rt.	Factors	No.	Sq.	Sq. Rt.	Factors
1	1	1.000		51	2,601	7.141	$3 \cdot 17$
2	4	1.414	2	52	2,704	7.211	$2^2 \cdot 13$
3	9	1.732	3	53	2,809	7.280	53
4	16	2.000	2^2	54	2,916	7.348	$2 \cdot 3^3$
5	25	2.236	5	55	3,025	7.416	$5 \cdot 11$
6	36	2.449	$2 \cdot 3$	56	3,136	7.483	$2^3 \cdot 7$
7	49	2.646	7	57	3,249	7.550	$3 \cdot 19$
8	64	2.828	2^3	58	3,364	7.616	$2 \cdot 29$
9	81	3.000	3^2	59	3,481	7.681	59
10	100	3.162	$2 \cdot 5$	60	3,600	7.746	$2^2 \cdot 3 \cdot 5$
11	121	3.317	11	61	3,721	7.810	61
12	144	3.464	$2^2 \cdot 3$	62	3,844	7.874	$2 \cdot 31$
13	169	3.606	13	63	3,969	7.937	$3^2 \cdot 7$
14	196	3.742	$2 \cdot 7$	64	4,096	8.000	2^6
15	225	3.873	$3 \cdot 5$	65	4,225	8.062	$5 \cdot 13$
16	256	4.000	2^4	66	4,356	8.124	$2 \cdot 3 \cdot 11$
17	289	4.123	17	67	4,489	8.185	67
18	324	4.243	$2 \cdot 3^2$	68	4,624	8.246	$2^2 \cdot 17$
19	361	4.359	19	69	4,761	8.307	$3 \cdot 23$
20	400	4.472	$2^2 \cdot 5$	70	4,900	8.367	$2 \cdot 5 \cdot 7$
21	441	4.583	$3 \cdot 7$	71	5,041	8.426	71
22	484	4.690	$2 \cdot 11$	72	5,184	8.485	$2^3 \cdot 3^2$
23	529	4.796	23	73	5,329	8.544	73
24	576	4.899	$2^3 \cdot 3$	74	5,476	8.602	$2 \cdot 37$
25	625	5.000	5^2	75	5,625	8.660	$3 \cdot 5^2$
26	676	5.099	$2 \cdot 13$	76	5,776	8.718	$2^2 \cdot 19$
27	729	5.196	3^3	77	5,929	8.775	$7 \cdot 11$
28	784	5.292	$2^2 \cdot 7$	78	6,084	8.832	$2 \cdot 3 \cdot 13$
29	841	5.385	29	79	6,241	8.888	79
30	900	5.477	$2 \cdot 3 \cdot 5$	80	6,400	8.944	$2^4 \cdot 5$
31	961	5.568	31	81	6,561	9.000	3^4
32	1,024	5.657	2^5	82	6,724	9.055	$2 \cdot 41$
33	1,089	5.745	$3 \cdot 11$	83	6,889	9.110	83
34	1,156	5.831	$2 \cdot 17$	84	7,056	9.165	$2^2 \cdot 3 \cdot 7$
35	1,225	5.916	$5 \cdot 7$	85	7,225	9.220	$5 \cdot 17$
36	1,296	6.000	$2^2 \cdot 3^2$	86	7,396	9.274	$2 \cdot 43$
37	1,369	6.083	37	87	7,569	9.327	$3 \cdot 29$
38	1,444	6.164	$2 \cdot 19$	88	7,744	9.381	$2^3 \cdot 11$
39	1,521	6.245	$3 \cdot 13$	89	7,921	9.434	89
40	1,600	6.325	$2^3 \cdot 5$	90	8,100	9.487	$2 \cdot 3^2 \cdot 5$
41	1,681	6.403	41	91	8,281	9.539	$7 \cdot 13$
42	1,764	6.481	$2 \cdot 3 \cdot 7$	92	8,464	9.592	$2^2 \cdot 23$
43	1,849	6.557	43	93	8,649	9.644	$3 \cdot 31$
44	1,936	6.633	$2^2 \cdot 11$	94	8,836	9.695	$2 \cdot 47$
45	2,025	6.708	$3^2 \cdot 5$	95	9,025	9.747	$5 \cdot 19$
46	2,116	6.782	$2 \cdot 23$	96	9,216	9.798	$2^5 \cdot 3$
47	2,209	6.856	47	97	9,409	9.849	97
48	2,304	6.928	$2^4 \cdot 3$	98	9,604	9.899	$2 \cdot 7^2$
49	2,401	7.000	7^2	99	9,801	9.950	$3^2 \cdot 11$
50	2,500	7.071	$2 \cdot 5^2$	100	10,000	10.000	$2^2 \cdot 5^2$

Table 2 Common Logarithms

x	0	1	2	3	4	5	6	7	8	9
1.0	.0000	.0043	.0086	.0128	.0170	.0212	.0253	.0294	.0334	.0374
1.1	.0414	.0453	.0492	.0531	.0569	.0607	.0645	.0682	.0719	.0755
1.2	.0792	.0828	.0864	.0899	.0934	.0969	.1004	.1038	.1072	.1106
1.3	.1139	.1173	.1206	.1239	.1271	.1303	.1335	.1367	.1399	.1430
1.4	.1461	.1492	.1523	.1553	.1584	.1614	.1644	.1673	.1703	.1732
1.5	.1761	.1790	.1818	.1847	.1875	.1903	.1931	.1959	.1987	.2014
1.6	.2041	.2068	.2095	.2122	.2148	.2175	.2201	.2227	.2253	.2279
1.7	.2304	.2330	.2355	.2380	.2405	.2430	.2455	.2480	.2504	.2529
1.8	.2553	.2577	.2601	.2625	.2648	.2672	.2695	.2718	.2742	.2765
1.9	.2788	.2810	.2833	.2856	.2878	.2900	.2923	.2945	.2967	.2989
2.0	.3010	.3032	.3054	.3075	.3096	.3118	.3139	.3160	.3181	.3201
2.1	.3222	.3243	.3263	.3284	.3304	.3324	.3345	.3365	.3385	.3404
2.2	.3424	.3444	.3464	.3483	.3502	.3522	.3541	.3560	.3579	.3598
2.3	.3617	.3636	.3655	.3674	.3692	.3711	.3729	.3747	.3766	.3784
2.4	.3802	.3820	.3838	.3856	.3874	.3892	.3909	.3927	.3945	.3962
2.5	.3979	.3997	.4014	.4031	.4048	.4065	.4082	.4099	.4116	.4133
2.6	.4150	.4166	.4183	.4200	.4216	.4232	.4249	.4265	.4281	.4298
2.7	.4314	.4330	.4346	.4362	.4378	.4393	.4409	.4425	.4440	.4456
2.8	.4472	.4487	.4502	.4518	.4533	.4548	.4564	.4579	.4594	.4609
2.9	.4624	.4639	.4654	.4669	.4683	.4698	.4713	.4728	.4742	.4757
3.0	.4771	.4786	.4800	.4814	.4829	.4843	.4857	.4871	.4886	.4900
3.1	.4914	.4928	.4942	.4955	.4969	.4983	.4997	.5011	.5024	.5038
3.2	.5051	.5065	.5079	.5092	.5105	.5119	.5132	.5145	.5159	.5172
3.3	.5185	.5198	.5211	.5224	.5237	.5250	.5263	.5276	.5289	.5302
3.4	.5315	.5328	.5340	.5353	.5366	.5378	.5391	.5403	.5416	.5428
3.5	.5441	.5453	.5465	.5478	.5490	.5502	.5514	.5527	.5539	.5551
3.6	.5563	.5575	.5587	.5599	.5611	.5623	.5635	.5647	.5658	.5670
3.7	.5682	.5694	.5705	.5717	.5729	.5740	.5752	.5763	.5775	.5786
3.8	.5798	.5809	.5821	.5832	.5843	.5855	.5866	.5877	.5888	.5899
3.9	.5911	.5922	.5933	.5944	.5955	.5966	.5977	.5988	.5999	.6010
4.0	.6021	.6031	.6042	.6053	.6064	.6075	.6085	.6096	.6107	.6117
4.1	.6128	.6138	.6149	.6160	.6170	.6180	.6191	.6201	.6212	.6222
4.2	.6232	.6243	.6253	.6263	.6274	.6284	.6294	.6304	.6314	.6325
4.3	.6335	.6345	.6355	.6365	.6375	.6385	.6395	.6405	.6415	.6425
4.4	.6435	.6444	.6454	.6464	.6474	.6484	.6493	.6503	.6513	.6522
4.5	.6532	.6542	.6551	.6561	.6571	.6580	.6590	.6599	.6609	.6618
4.6	.6628	.6637	.6646	.6656	.6665	.6675	.6684	.6693	.6702	.6712
4.7	.6721	.6730	.6739	.6749	.6758	.6767	.6776	.6785	.6794	.6803
4.8	.6812	.6821	.6830	.6839	.6848	.6857	.6866	.6875	.6884	.6893
4.9	.6902	.6911	.6920	.6928	.6937	.6946	.6955	.6964	.6972	.6981
5.0	.6990	.6998	.7007	.7016	.7024	.7033	.7042	.7050	.7059	.7067
5.1	.7076	.7084	.7093	.7101	.7110	.7118	.7126	.7135	.7143	.7152
5.2	.7160	.7168	.7177	.7185	.7193	.7202	.7210	.7218	.7226	.7235
5.3	.7243	.7251	.7259	.7267	.7275	.7284	.7292	.7300	.7308	.7316
5.4	.7324	.7332	.7340	.7348	.7356	.7364	.7372	.7380	.7388	.7396
x	0	1	2	3	4	5	6	7	8	9

Table 2   (*Continued*)

x	0	1	2	3	4	5	6	7	8	9
5.5	.7404	.7412	.7419	.7427	.7435	.7443	.7451	.7459	.7466	.7474
5.6	.7482	.7490	.7497	.7505	.7513	.7520	.7528	.7536	.7543	.7551
5.7	.7559	.7566	.7574	.7582	.7589	.7597	.7604	.7612	.7619	.7627
5.8	.7634	.7642	.7649	.7657	.7664	.7672	.7679	.7686	.7694	.7701
5.9	.7709	.7716	.7723	.7731	.7738	.7745	.7752	.7760	.7767	.7774
6.0	.7782	.7789	.7796	.7803	.7810	.7818	.7825	.7832	.7839	.7846
6.1	.7853	.7860	.7868	.7875	.7882	.7889	.7896	.7903	.7910	.7917
6.2	.7924	.7931	.7938	.7945	.7952	.7959	.7966	.7973	.7980	.7987
6.3	.7993	.8000	.8007	.8014	.8021	.8028	.8035	.8041	.8048	.8055
6.4	.8062	.8069	.8075	.8082	.8089	.8096	.8102	.8109	.8116	.8122
6.5	.8129	.8136	.8142	.8149	.8156	.8162	.8169	.8176	.8182	.8189
6.6	.8195	.8202	.8209	.8215	.8222	.8228	.8235	.8241	.8248	.8254
6.7	.8261	.8267	.8274	.8280	.8287	.8293	.8299	.8306	.8312	.8319
6.8	.8325	.8331	.8338	.8344	.8351	.8357	.8363	.8370	.8376	.8382
6.9	.8388	.8395	.8401	.8407	.8414	.8420	.8426	.8432	.8439	.8445
7.0	.8451	.8457	.8463	.8470	.8476	.8482	.8488	.8494	.8500	.8506
7.1	.8513	.8519	.8525	.8531	.8537	.8543	.8549	.8555	.8561	.8567
7.2	.8573	.8579	.8585	.8591	.8597	.8603	.8609	.8615	.8621	.8627
7.3	.8633	.8639	.8645	.8651	.8657	.8663	.8669	.8675	.8681	.8686
7.4	.8692	.8698	.8704	.8710	.8716	.8722	.8727	.8733	.8739	.8745
7.5	.8751	.8756	.8762	.8768	.8774	.8779	.8785	.8791	.8797	.8802
7.6	.8808	.8814	.8820	.8825	.8831	.8837	.8842	.8848	.8854	.8859
7.7	.8865	.8871	.8876	.8882	.8887	.8893	.8899	.8904	.8910	.8915
7.8	.8921	.8927	.8932	.8938	.8943	.8949	.8954	.8960	.8965	.8971
7.9	.8976	.8982	.8987	.8993	.8998	.9004	.9009	.9015	.9020	.9025
8.0	.9031	.9036	.9042	.9047	.9053	.9058	.9063	.9069	.9074	.9079
8.1	.9085	.9090	.9096	.9101	.9106	.9112	.9117	.9122	.9128	.9133
8.2	.9138	.9143	.9149	.9154	.9159	.9165	.9170	.9175	.9180	.9186
8.3	.9191	.9196	.9201	.9206	.9212	.9217	.9222	.9227	.9232	.9238
8.4	.9243	.9248	.9253	.9258	.9263	.9269	.9274	.9279	.9284	.9289
8.5	.9294	.9299	.9304	.9309	.9315	.9320	.9325	.9330	.9335	.9340
8.6	.9345	.9350	.9355	.9360	.9365	.9370	.9375	.9380	.9385	.9390
8.7	.9395	.9400	.9405	.9410	.9415	.9420	.9425	.9430	.9435	.9440
8.8	.9445	.9450	.9455	.9460	.9465	.9469	.9474	.9479	.9484	.9489
8.9	.9494	.9499	.9504	.9509	.9513	.9518	.9523	.9528	.9533	.9538
9.0	.9542	.9547	.9552	.9557	.9562	.9566	.9571	.9576	.9581	.9586
9.1	.9590	.9595	.9600	.9605	.9609	.9614	.9619	.9624	.9628	.9633
9.2	.9638	.9643	.9647	.9652	.9657	.9661	.9666	.9671	.9675	.9680
9.3	.9685	.9689	.9694	.9699	.9703	.9708	.9713	.9717	.9722	.9727
9.4	.9731	.9736	.9741	.9745	.9750	.9754	.9759	.9763	.9768	.9773
9.5	.9777	.9782	.9786	.9791	.9795	.9800	.9805	.9809	.9814	.9818
9.6	.9823	.9827	.9832	.9836	.9841	.9845	.9850	.9854	.9859	.9863
9.7	.9868	.9872	.9877	.9881	.9886	.9890	.9894	.9899	.9903	.9908
9.8	.9912	.9917	.9921	.9926	.9930	.9934	.9939	.9943	.9948	.9952
9.9	.9956	.9961	.9965	.9969	.9974	.9978	.9983	.9987	.9991	.9996
x	0	1	2	3	4	5	6	7	8	9

Table 3 Natural Logarithms of Numbers

x	$\ln x$	x	$\ln x$	x	$\ln x$
		4.5	1.5041	9.0	2.1972
0.1	− 2.3026	4.6	1.5261	9.1	2.2083
0.2	− 1.6094	4.7	1.5476	9.2	2.2192
0.3	− 1.2040	4.8	1.5686	9.3	2.2300
0.4	− 0.9163	4.9	1.5892	9.4	2.2407
0.5	− 0.6931	5.0	1.6094	9.5	2.2513
0.6	− 0.5108	5.1	1.6292	9.6	2.2618
0.7	− 0.3567	5.2	1.6487	9.7	2.2721
0.8	− 0.2231	5.3	1.6677	9.8	2.2824
0.9	− 0.1054	5.4	1.6864	9.9	2.2925
1.0	0.0000	5.5	1.7047	10	2.3026
1.1	0.0953	5.6	1.7228	11	2.3979
1.2	0.1823	5.7	1.7405	12	2.4849
1.3	0.2624	5.8	1.7579	13	2.5649
1.4	0.3365	5.9	1.7750	14	2.6391
1.5	0.4055	6.0	1.7918	15	2.7081
1.6	0.4700	6.1	1.8083	16	2.7726
1.7	0.5306	6.2	1.8245	17	2.8332
1.8	0.5878	6.3	1.8405	18	2.8904
1.9	0.6419	6.4	1.8563	19	2.9444
2.0	0.6931	6.5	1.8718	20	2.9957
2.1	0.7419	6.6	1.8871	25	3.2189
2.2	0.7885	6.7	1.9021	30	3.4012
2.3	0.8329	6.8	1.9169	35	3.5553
2.4	0.8755	6.9	1.9315	40	3.6889
2.5	0.9163	7.0	1.9459	45	3.8067
2.6	0.9555	7.1	1.9601	50	3.9120
2.7	0.9933	7.2	1.9741	55	4.0073
2.8	1.0296	7.3	1.9879	60	4.0943
2.9	1.0647	7.4	2.0015	65	4.1744
3.0	1.0986	7.5	2.0149	70	4.2485
3.1	1.1314	7.6	2.0281	75	4.3175
3.2	1.1632	7.7	2.0412	80	4.3820
3.3	1.1939	7.8	2.0541	85	4.4427
3.4	1.2238	7.9	2.0669	90	4.4998
3.5	1.2528	8.0	2.0794	100	4.6052
3.6	1.2809	8.1	2.0919	110	4.7005
3.7	1.3083	8.2	2.1041	120	4.7875
3.8	1.3350	8.3	2.1163	130	4.8676
3.9	1.3610	8.4	2.1282	140	4.9416
4.0	1.3863	8.5	2.1401	150	5.0106
4.1	1.4110	8.6	2.1518	160	5.0752
4.2	1.4351	8.7	2.1633	170	5.1358
4.3	1.4586	8.8	2.1748	180	5.1930
4.4	1.4816	8.9	2.1861	190	5.2470

Table 4   Exponential Functions

x	e^x	e^{-x}	x	e^x	e^{-x}
0.00	1.0000	1.0000	1.5	4.4817	0.2231
0.01	1.0101	0.9901	1.6	4.9530	0.2019
0.02	1.0202	0.9802	1.7	5.4739	0.1827
0.03	1.0305	0.9705	1.8	6.0496	0.1653
0.04	1.0408	0.9608	1.9	6.6859	0.1496
0.05	1.0513	0.9512	2.0	7.3891	0.1353
0.06	1.0618	0.9418	2.1	8.1662	0.1225
0.07	1.0725	0.9324	2.2	9.0250	0.1108
0.08	1.0833	0.9331	2.3	9.9742	0.1003
0.09	1.0942	0.9139	2.4	11.023	0.0907
0.10	1.1052	0.9048	2.5	12.182	0.0821
0.11	1.1163	0.8958	2.6	13.464	0.0743
0.12	1.1275	0.8869	2.7	14.880	0.0672
0.13	1.1388	0.8781	2.8	16.445	0.0608
0.14	1.1503	0.8694	2.9	18.174	0.0550
0.15	1.1618	0.8607	3.0	20.086	0.0498
0.16	1.1735	0.8521	3.1	22.198	0.0450
0.17	1.1853	0.8437	3.2	24.533	0.0408
0.18	1.1972	0.8353	3.3	27.113	0.0369
0.19	1.2092	0.8270	3.4	29.964	0.0334
0.20	1.2214	0.8187	3.5	33.115	0.0302
0.21	1.2337	0.8106	3.6	36.598	0.0273
0.22	1.2461	0.8025	3.7	40.447	0.0247
0.23	1.2586	0.7945	3.8	44.701	0.0224
0.24	1.2712	0.7866	3.9	49.402	0.0202
0.25	1.2840	0.7788	4.0	54.598	0.0183
0.30	1.3499	0.7408	4.1	60.340	0.0166
0.35	1.4191	0.7047	4.2	66.686	0.0150
0.40	1.4918	0.6703	4.3	73.700	0.0136
0.45	1.5683	0.6376	4.4	81.451	0.0123
0.50	1.6487	0.6065	4.5	90.017	0.0111
0.55	1.7333	0.5769	4.6	99.484	0.0101
0.60	1.8221	0.5488	4.7	109.95	0.0091
0.65	1.9155	0.5220	4.8	121.51	0.0082
0.70	2.0138	0.4966	4.9	134.29	0.0074
0.75	2.1170	0.4724	5.0	148.41	0.0067
0.80	2.2255	0.4493	5.5	244.69	0.0041
0.85	2.3396	0.4274	6.0	403.43	0.0025
0.90	2.4596	0.4066	6.5	665.14	0.0015
0.95	2.5857	0.3867	7.0	1096.6	0.0009
1.0	2.7183	0.3679	7.5	1808.0	0.0006
1.1	3.0042	0.3329	8.0	2981.0	0.0003
1.2	3.3201	0.3012	8.5	4914.8	0.0002
1.3	3.6693	0.2725	9.0	8103.1	0.0001
1.4	4.0552	0.2466	10.0	22026	0.00005

Table 5   Trigonometric Functions

Degrees	Radians	sin	cos	tan	cot		
0	0.0000	0.0000	1.0000	0.0000		1.5708	90
1	0.0175	0.0175	0.9998	0.0175	57.290	1.5533	89
2	0.0349	0.0349	0.9994	0.0349	28.636	1.5359	88
3	0.0524	0.0523	0.9986	0.0524	19.081	1.5184	87
4	0.0698	0.0698	0.9976	0.0699	14.301	1.5010	86
5	0.0873	0.0872	0.9962	0.0875	11.430	1.4835	85
6	0.1047	0.1045	0.9945	0.1051	9.5144	1.4661	84
7	0.1222	0.1219	0.9925	0.1228	8.1443	1.4486	83
8	0.1396	0.1392	0.9903	0.1405	7.1154	1.4312	82
9	0.1571	0.1564	0.9877	0.1584	6.3138	1.4137	81
10	0.1745	0.1736	0.9848	0.1763	5.6713	1.3963	80
11	0.1920	0.1908	0.9816	0.1944	5.1446	1.3788	79
12	0.2094	0.2079	0.9781	0.2126	4.7046	1.3614	78
13	0.2269	0.2250	0.9744	0.2309	4.3315	1.3439	77
14	0.2443	0.2419	0.9703	0.2493	4.0108	1.3265	76
15	0.2618	0.2588	0.9659	0.2679	3.7321	1.3090	75
16	0.2793	0.2756	0.9613	0.2867	3.4874	1.2915	74
17	0.2967	0.2924	0.9563	0.3057	3.2709	1.2741	73
18	0.3142	0.3090	0.9511	0.3249	3.0777	1.2566	72
19	0.3316	0.3256	0.9455	0.3443	2.9042	1.2392	71
20	0.3491	0.3420	0.9397	0.3640	2.7475	1.2217	70
21	0.3665	0.3584	0.9336	0.3839	2.6051	1.2043	69
22	0.3840	0.3746	0.9272	0.4040	2.4751	1.1868	68
23	0.4014	0.3907	0.9205	0.4245	2.3559	1.1694	67
24	0.4189	0.4067	0.9135	0.4452	2.2460	1.1519	66
25	0.4363	0.4226	0.9063	0.4663	2.1445	1.1345	65
26	0.4538	0.4384	0.8988	0.4877	2.0503	1.1170	64
27	0.4712	0.4540	0.8910	0.5095	1.9626	1.0996	63
28	0.4887	0.4695	0.8829	0.5317	1.8807	1.0821	62
29	0.5061	0.4848	0.8746	0.5543	1.8040	1.0647	61
30	0.5236	0.5000	0.8660	0.5774	1.7321	1.0472	60
31	0.5411	0.5150	0.8572	0.6009	1.6643	1.0297	59
32	0.5585	0.5299	0.8480	0.6249	1.6003	1.0123	58
33	0.5760	0.5446	0.8387	0.6494	1.5399	0.9948	57
34	0.5934	0.5592	0.8290	0.6745	1.4826	0.9774	56
35	0.6109	0.5736	0.8192	0.7002	1.4281	0.9599	55
36	0.6283	0.5878	0.8090	0.7265	1.3764	0.9425	54
37	0.6458	0.6018	0.7986	0.7536	1.3270	0.9250	53
38	0.6632	0.6157	0.7880	0.7813	1.2799	0.9076	52
39	0.6807	0.6293	0.7771	0.8098	1.2349	0.8901	51
40	0.6981	0.6428	0.7660	0.8391	1.1918	0.8727	50
41	0.7156	0.6561	0.7547	0.8693	1.1504	0.8552	49
42	0.7330	0.6691	0.7431	0.9004	1.1106	0.8378	48
43	0.7505	0.6820	0.7314	0.9325	1.0724	0.8203	47
44	0.7679	0.6947	0.7193	0.9657	1.0355	0.8029	46
45	0.7854	0.7071	0.7071	1.0000	1.0000	0.7854	45
		cos	sin	cot	tan	Radians	Degrees

Answers to Odd-Numbered Problems

CHAPTER 1

Exercise 1 (page 13)

1. $>$ **3.** $=$ **5.** $x \leq -1$ **7.** $x \geq -1$ **9.** $x \leq -4$ **11.** $x \geq \frac{9}{5}$ **13.** $x \leq 2$ or $x \geq 3$

15. $-4 < x < -3$ **17.** $x = 5$ or $x = -5$ **19.** $-3 \leq x \leq 3$ **21.** $-1 < x < 7$ **23.** $0 \leq x \leq 4$

25. $x < -1$ or $x > 7$ **27.** $-\infty < x < +\infty$ **29.** $2.99 < x < 3.01$ **31.** $-\frac{7}{4} \leq x \leq \frac{9}{4}$

33. $x < 0$ or $x > \frac{1}{3}$ **35.** $\frac{8}{5} \leq x < 2$ **37.** $-4 < x < 3$ **39.** $x < 1$ or $x > 3$ **41.** $x < 0$ or $1 < x < 3$

43. $x < -\frac{1}{2}$ or $x > \frac{1}{2}$ **45.** $x \geq 0$ **47.** $x \leq 1$ or $x \geq 2$

49. $|x - y| = |(x - a) + (a - y)| \leq |x - a| + |a - y| < \frac{1}{3} + \frac{1}{3} = \frac{2}{3}$

51. $a < b \Rightarrow a + a < a + b \Rightarrow a < \dfrac{a + b}{2}$ **53.** $\dfrac{1}{h} = \left(\dfrac{a + b}{2}\right)\left(\dfrac{1}{ab}\right); \ h = \dfrac{ab}{(a + b)/2}$

$\quad a < b \Rightarrow a + b < b + b \Rightarrow \dfrac{a + b}{2} < b$ $\qquad a < \dfrac{a + b}{2} < b$

$$\dfrac{a}{ab} < \left(\dfrac{a + b}{2}\right)\left(\dfrac{1}{ab}\right) < \dfrac{b}{ab}$$

$$\dfrac{1}{b} < \dfrac{1}{h} < \dfrac{1}{a}$$

$$a < h < b$$

55. $P = $ Perimeter

For a circle, $P = 2\pi R \Rightarrow R = \dfrac{P}{2\pi} \Rightarrow A = \pi R^2 = \dfrac{P^2}{4\pi}$

For a square, $P = 4s \Rightarrow s = \dfrac{P}{4} \Rightarrow A = s^2 = \dfrac{P^2}{16} < \dfrac{P^2}{4\pi}$

57. $b - a \geq 0$ and $-c > 0 \Rightarrow -c(b - a) \geq 0 \Rightarrow ac \geq bc$

59. $b - a \geq 0$ and $b + a > 0 \Rightarrow (b - a)(b + a) \geq 0 \Rightarrow b^2 \geq a^2$

Exercise 2 (page 22)

1. 5   **3.** $\sqrt{4.88} \approx 2.209$

5.

7.
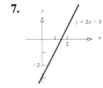

9.
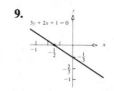

11.
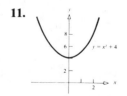

13. $x^2 + y^2 - 4x + 6y - 3 = 0$ **15.** $x^2 + y^2 - 2x + 4y + 4 = 0$ **17.** $(1, 0); 2$ **19.** $(-2, 3); 4$

21. $(4, 4)$ **23.** $(\frac{1}{2}, \frac{1}{2})$ **25.**    **27.**

29. No symmetry   **31.** Origin   **33.** y-axis   **35.** Origin   **37.** $x^2 + y^2 - 2x + 4y + 3 = 0$

39. $5, \sqrt{73}, 2\sqrt{13}$ **41.** $6, 4, 2\sqrt{13}$; right angle triangle   **43.** $\sqrt{68}, \sqrt{34}, \sqrt{34}$; both   **45.** $(240°, 0.16)$

47. $\triangle P_1 R P_2$ similar to $\triangle P_1 Q P$ **49.** P_2 **51.** $(9, 8)$

$$r = \frac{|P_1 P|}{|P_1 P_2|} = \frac{x - x_1}{x_2 - x_1}$$

$$x = (1 - r)x_1 + rx_2$$

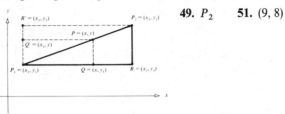

Exercise 3 (page 31)

1. $2x - y + 7 = 0$ **3.** $2x + 3y + 1 = 0$ **5.** $x - 2y + 5 = 0$ **7.** $3x + y - 3 = 0$ **9.** $x - 2y - 2 = 0$

11. $x - 1 = 0$ **13.** $\frac{3}{2}; -3$ **15.** $-\frac{1}{2}; 2$ **17.** Undefined; no y-intercept   **19.** $2x - y = 0$

21. $-\frac{2}{3}$ **23.** 2   **25.** 0   **27.** Parallel   **29.** Intersecting; $(0, -2)$ **31.** Intersecting; $(-\frac{2}{3}, \frac{8}{3})$

33. $x + 3y - 9 = 0$ **35.** \$50,000 at 14\%; \$50,000 at 10\%   **37.** $°C = \frac{5}{9}(°F - 32); °C = 21.11$

39. $\tan[\theta_2 - (\pi/2)] = -1/\tan \theta_2 = -1/m_2 = m_1 = \tan \theta_1 \Rightarrow \theta_2 - \theta_1 = \pi/2$ **41.** -3 **43.** 5

Exercise 4 (page 42)

1. (a) 7   (b) -8 (c) -2 (d) $3x + 4$ (e) $3x + 3\Delta x - 2$ (f) $(3/x) - 2$ **3.** Yes   **5.** Yes   **7.** No

9. Yes   **11.** No   **13.** Yes   **15.** Yes   **17.** $\mathbb{R}$ **19.** $x \geq 1$ **21.** $\mathbb{R}$ **23.** $x \neq 2$ **25.** $x > 0$

27. $-\infty < x < 5$    **29.** $-2 \leq x < +\infty$

31. $\mathbb{R}$ 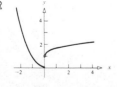   **33.** $\mathbb{R}$ 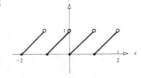   **35.** $\mathbb{R}$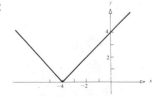

37. (a) $2x + 2\Delta x + 5$ (b) $2\Delta x$ (c) 2

39. (a) $x^2 + 2x\Delta x + (\Delta x)^2 + 3x + 3\Delta x + 4$ (b) $2x\Delta x + (\Delta x)^2 + 3\Delta x$ (c) $2x + (\Delta x) + 3$

41. $\dfrac{\sqrt{x+\Delta x} - \sqrt{x}}{\Delta x} = \dfrac{\sqrt{x+\Delta x} - \sqrt{x}}{\Delta x}\,\dfrac{\sqrt{x+\Delta x} + \sqrt{x}}{\sqrt{x+\Delta x} + \sqrt{x}} = \dfrac{1}{\sqrt{x+\Delta x} + \sqrt{x}}$ **43.** -1

45. (a) 15.1 m; 14.07 m; 12.94 m; 11.72 m (b) 2.02 sec (c) The rock falls faster on the earth

(d) $H(x) = \begin{cases} 20 - 4.9x^2 & \text{if } 0 \le x \le 10\sqrt{2}/7 \approx 2.02 \\ 0 & \text{if } x > 10\sqrt{2}/7 \approx 2.02 \end{cases}$

47. $A(x) = \frac{1}{2}x(3000 - 2x)$; $0 < x < 1500$ **49.** $A = (7 - 2x)(11 - 2x)$; $0 < x < \frac{7}{2}$; $0 < A < 77$

Exercise 5 (page 46)

1. (a) $2x^2 + x - 1$ (b) $-2x^2 + x - 1$ (c) $2x^3 - 2x^2$ (d) $(x - 1)/2x^2$; $\mathbb{R}, \mathbb{R}, \mathbb{R}, \mathbb{R}, x \neq 0$

3. (a) $\sqrt{x+1} + x + 1$ (b) $\sqrt{x+1} - x - 1$ (c) $(x + 1)^{3/2}$
(d) $1/\sqrt{x+1}$; $x \ge -1$, $\mathbb{R}$, $x \ge -1$, $x \ge -1$, $x > -1$

5. (a) $(2/x) + 1$ (b) -1 (c) $(1/x)[(1/x) + 1]$ (d) $1/(1 + x)$; $x \neq 0$, $x \neq 0$, $x \neq 0$, $x \neq 0$, $x \neq 0, -1$

7. Yes **9.** No **11.** No **13.** $x \neq -2$ **15.** $x \neq -2, 2$

17.

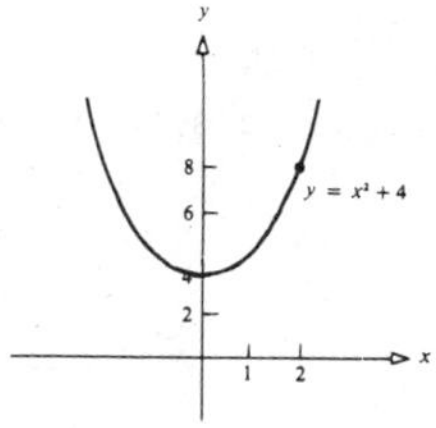

19.

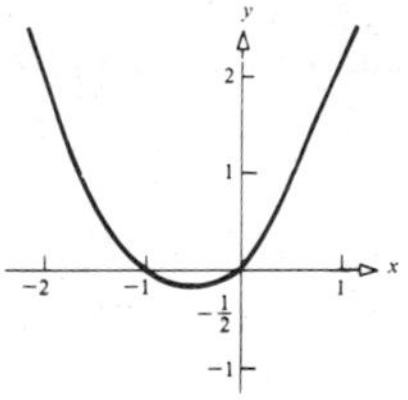

21.

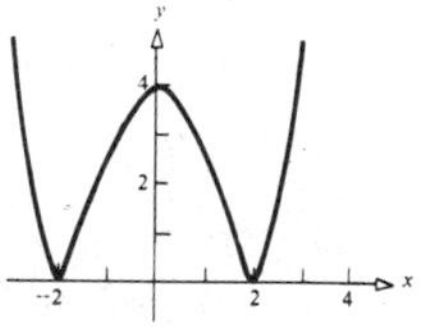

23. Odd **25.** Even **27.** Neither **29.** Use x and x^3 **31.** Use x^2 and x^4 **33.** Use x and x^3

35. $a = 0$, $c = 0$, b any number **37.** $g(x) = 5 - \frac{7}{2}x$

Exercise 6 (page 50)

1. (a) $3x^2 + 1$ (b) $(3x + 1)^2$ (c) x^4 (d) $9x + 4$

3. (a) $\sqrt{x^2 - 1}$ (b) $x - 1$ (c) $(x^2 - 1)^2 - 1 = x^4 - 2x^2$ (d) $\sqrt[4]{x}$

5. (a) $(1 - x)/(1 + x)$ (b) $(x + 1)/(x - 1)$ (c) x (d) $-1/x$

7. (a) $(8/x^2)(6 - x)$ (b) $2/\sqrt{3x^4 - 2x^2}$ (c) $\sqrt{2}\sqrt[4]{x}$ (d) $3(3x^4 - 2x^2)^4 - 2(3x^4 - 2x^2)^2$

9. $f(x) = \sqrt{x}$; $g(x) = x^2 + x - 1$ **11.** $f(x) = x^7$; $g(x) = x^2 - 1$ **13.** $f(x) = 1/x^2$; $g(x) = 3x - 5$

15. $g(x) = \sqrt[3]{x}$ **17.** (a) $4x - 3$ (b) $9 - 12x + 4x^2$ **19.** 2; 41

Exercise 7 (page 54)

1. One-to-one **3.** Not one-to-one **5.** One-to-one **7.** Not one-to-one

9.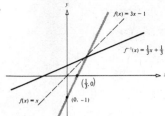

11.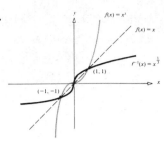

13.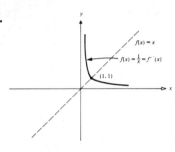

15.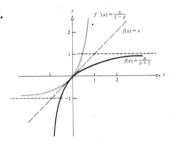

17. First quadrant   **19.** $(f \circ g)(x) = f(g(x)) = \frac{9}{5}[\frac{5}{9}(x - 32)] + 32 = x$ **21.** $f^{-1}(x) = \sqrt[5]{x + 1}$

Miscellaneous Exercises (page 54)

1. $-\frac{14}{3} < x < -\frac{10}{3}, x \neq -4$ **3.** $-2 < x < 2$

5. (a) $\frac{1}{2}, -2, \frac{1}{2}, -2$ (b) AB and CD, BC and AD are parallel; AB is perpendicular to BC and AD, BC is perpendicular to CD and AB (c) Undefined (d) $(2, 4)$ (e) $2\sqrt{5}$ (f) $(3, 2)$

7. $m_{AB} = 2, m_{AC} = -\frac{1}{2}; |AB| = \sqrt{245}, |BC| = \sqrt{325}, |AC| = \sqrt{80} \Rightarrow |AB|^2 + |AC|^2 = |BC|^2$

9. $(-1, \frac{3}{2})$ **11.** $15a + 8b - 102 = 0, 15a + 8b + 34 = 0$ **13.** $(x - 1)^2 + (y + 2)^2 = 16$

15. $x - y + 5 = 0$ **17.** $x^2 + y^2 - 4x - 6y + 9 = 0$ **19.** $x^2 + y^2 + 6x + 4y - 36 = 0$

21. (a) $(-a/2, -b/2), \frac{1}{2}\sqrt{a^2 + b^2}$ (b) $a^2 + b^2 - 4c > 0$ **23.** $x_0 x + y_0 y - R^2 = 0$ **25.** $(1, 4)$

27. (a) Neither (b) Odd (c) Even (d) Odd (e) Even (f) Even   **29.** n is odd   **31.** $p = -1$

33. (a) $(f \circ f)(x_0) = f(f(x_0)) = f(x_0) = x_0$ (b) $-3, 1$
(c) Two fixed points if $(d - a)^2 + 4bc > 0$; one if $(d - a)^2 + 4bc = 0$; none if $(d - a)^2 + 4bc < 0$

35. (a) f is one-to-one by horizontal line test and hence has an inverse
(b) Domain of $f = \mathbb{R}$
Domain of $f^{-1} = $ Range of f
$= -2, -1 < x < 0, 0 < x$

(c)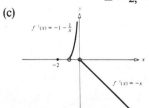

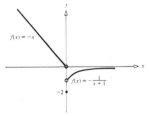

37. $(f \circ g) \circ (f \circ g)^{-1}(x) = x$ and $f(g(g^{-1}(f^{-1}(x)))) = (f \circ g) \circ (g^{-1} \circ f^{-1})(x) = x$; hence $(f \circ g)^{-1} = g^{-1} \circ f^{-1}$

39. $|xy - 6| = |y(x - 2) + 2(y - 3)| \le |y||x - 2| + 2|y - 3| \le (\frac{31}{10})(\frac{1}{5}) + (2)(\frac{1}{10}) = \frac{41}{50}$ **41.** $A = x\sqrt{4R^2 - x^2}$

43. $V = 4x(100 - x)(8 - x)$ **45.** $A = [(100 - 2x)/(\pi + 2)^2](4x + \pi x + 50\pi)$

47. $T - 70 = [-415/2(10^6)][e - 29(10^6)]$ **49.**
where T is temperature and e is elasticity

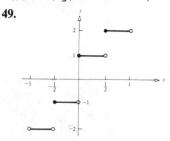

51. Assume a not even; then $a = 2n + 1$ and $a^2 = (2n + 1)^2 = 4n^2 + 4n + 1$, which is odd, contradicting the fact that a^2 is even.

53. (a)   (b)   (c)

 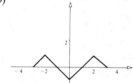 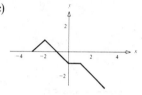

(d)

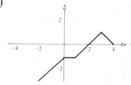

CHAPTER 2

Exercise 1 (page 66)

1.

x	0.9	0.99	0.999	1.001	1.01	1.1
$f(x)$	1.8	1.98	1.998	2.002	2.02	2.2

$\lim\limits_{x \to 1} 2x = 2$

3.

x	0.1	0.01	0.001	-0.001	-0.01	-0.1
$f(x)$	2.01	2.0001	2.000001	2.000001	2.0001	2.01

$\lim\limits_{x \to 0}(x^2 + 2) = 2$

5.

x	-2.5	-2.9	-2.99	-3.01	-3.1	-3.5
$f(x)$	-5.5	-5.9	-5.99	-6.01	-6.1	-6.5

$$\lim_{x \to -3} \frac{x^2 - 9}{x + 3} = -6$$

7.

x	-0.2	-0.1	-0.01	0.01	0.1	0.2
$f(x)$	1.01355	1.00335	1.00003	1.00003	1.00335	1.01355

$$\lim_{x \to 0} \frac{\tan x}{x} = 1$$

9. Exists　　**11.** Exists　　**13.** Does not exist　　**15.** Does not exist

17. $\lim\limits_{x \to 2} f(x)$ exists

$\lim\limits_{x \to 2} f(x) = 9$

19. $\lim\limits_{x \to 1} f(x)$ exists

$\lim\limits_{x \to 1} f(x) = 2$

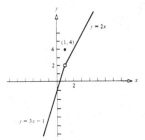

21. $\lim\limits_{x \to 1} f(x)$ exists

$\lim\limits_{x \to 1} f(x) = 2$

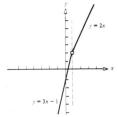

23. Since there is no single number that the values of f are close to when x is close to 0, $\lim_{x \to 0} f(x)$ does not exist.

25. -3　　**27.** -1　　**29.** (a) 15　(b) $y = 12x - 12$　(c)

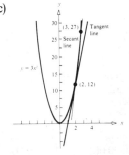

31. (a) $m_{\text{sec}} = 2 + (h/2)$　(b) 1.75; 2.25　(c)

h	-0.5	-0.1	-0.001	0.001	0.1	0.5
m_{sec}	1.75	1.95	1.9995	2.0005	2.05	2.25

31. (d) 2   (e) $y = 2x - 3$ (f)

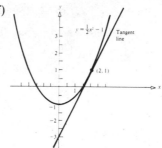

Exercise 2 (page 75)

1. 14   **3.** 5   **5.** 12   **7.** $-\frac{31}{8}$ **9.** 6   **11.** 4   **13.** 4   **15.** 0   **17.** 4   **19.** $\frac{13}{4}$ **21.** 4

23. $\frac{12}{5}$ **25.** -4 **27.** $-\frac{1}{7}$ **29.** $\frac{2}{3}$ **31.** 2   **33.** 12   **35.** $\sqrt{2}/4$ **37.** $-1/x^2$ **39.** $1/2\sqrt{x}$

41. 10   **43.** 8   **45.** 4   **47.** $6x + 4$ **49.** $-2/x^2$ **51.** -6 **53.** na^{n-1} **55.** m/n

57. $(a + b)/2$ **59.** Since $\lim_{x \to 0}(1 - x^2) = 1$ and $\lim_{x \to 0} \cos x = 1$, the result follows.

61. Since $0 \le f(x) \le 1$ for every x and $x^2 > 0$ for $x \ne 0$, then on multiplying the inequality by x^2, we have $0 \le x^2 f(x) \le x^2$. Since $\lim_{x \to 0} x^2 = 0$, then $\lim_{x \to 0} x^2 f(x) = 0$.

63. $x \ne 0, |\sin(1/x)| \le 1$
$x \ne 0, |x^n \sin(1/x)| = |x^n| |\sin(1/x)| \le |x^n|$
$-|x^n| < x^n \sin(1/x) < |x^n|$
Since $\lim_{x \to 0}(-|x^n|) = 0$ and $\lim_{x \to 0}|x^n| = 0$, then $\lim_{x \to 0}[x^n \sin(1/x)] = 0$.

Exercise 3 (page 78)

1. 5   **3.** -19 **5.** $\frac{3}{2}$ **7.** 6   **9.** 3   **11.** 1   **13.** 0   **15.** 1   **17.** 0   **19.** 0

21. $\lim_{x \to 0^-} f(x) = 0$; $\lim_{x \to 0^+} f(x) = 0$; yes, $\lim_{x \to 0} f(x) = 0$

23. $\lim_{x \to 3^-} f(x) = 6$; $\lim_{x \to 3^+} f(x) = 6$; yes, $\lim_{x \to 3} f(x) = 6$ **25.** $\lim_{x \to 1^-} f(x) = -1$; $\lim_{x \to 1^+} f(x) = 2$; no

27. $\lim_{x \to 1^-} f(x) = 2$; $\lim_{x \to 1^+} f(x) = 2$; yes, $\lim_{x \to 1} f(x) = 2$

29. $\lim_{x \to 1^-} f(x) = 2$; $\lim_{x \to 1^+} f(x) = 2$; yes, $\lim_{x \to 1} f(x) = 2$

31. $\lim_{x \to 3^-} f(x) = 0$; $\lim_{x \to 3^+} f(x) = 0$; yes, $\lim_{x \to 3} f(x) = 0$ **33.** $\sqrt{5}$ **35.** 0   **37.** $\sqrt{5}$ **39.** 3

41. Yes, 0   **43.** Yes, 0   **45.** $f(x) = x/(x - c)$, $g(x) = -c/(x - c)$, $\lim_{x \to c}[f(x) + g(x)] = 1$

47. $f(x) = |x|/x$; $c = 0$

Exercise 4 (page 88)

1. Continuous   **3.** Continuous   **5.** Continuous   **7.** Continuous   **9.** Discontinuous

11. Discontinuous   **13.** Continuous   **15.** Continuous   **17.** Discontinuous   **19.** $f(2) = 4$

21. $f(1) = 2$ **23.** All real numbers   **25.** All real numbers   **27.** All real numbers, except $x = 2$

29. All real numbers, except $x = 2$ and $x = -2$ **31.** All real numbers   **33.** All real numbers

35. All real numbers, except $x \geq 2$ **37.** Yes, since $\lim_{x \to 0} f(x) = f(0) = \sqrt{15}$

39. No, since $\lim_{x \to 2^-} f(x) = \sqrt{9} = 3$ and $f(2) = \sqrt{5}$ **41.** f has a zero   **43.** Theorem gives no information

45. f has a zero   **47.** $k = \frac{1}{6}$ **49.** $\lim_{x \to 0^-} |x|/x = -1$, $\lim_{x \to 0^+} |x|/x = 1$

51. Take $f(x) = x$, $g(x) = x^2 - 4$, $c = 2$ or -2

Exercise 5 (page 98)

1. (a) $\delta = 0.025$ (b) $\delta = 0.0025$ (c) $\delta = 0.00025$ (d) $\delta = \varepsilon/4$

3. (a) $\delta = 0.1$ (b) $\delta = 0.01$ (c) $\delta = \varepsilon$ **5.** $\delta = 0.005$ **7.** $\delta = \frac{1}{12}$

9. Take $\delta = \varepsilon/3$. If $0 < |x - 2| < \delta$, then $|3x - 6| < \varepsilon$.

11. Take $\delta = \varepsilon/2$. If $0 < |x| < \delta$, then $|(2x + 5) - 5| < \varepsilon$.

13. Take $\delta = \varepsilon/5$. If $0 < |x + 3| < \delta$, then $|(-5x + 2) - 17| < \varepsilon$.

15. Take $\delta \leq \min(1, \varepsilon/3)$. If $0 < |x - 2| < \delta$, then $|x| < 3$ and $|x^2 - 2x| < \varepsilon$.

17. Take $\delta \leq \min(1, 2\varepsilon/7)$. If $0 < |x - 1| < \delta$, then $|1/(3 - x)| < 1$ and $|(1 + 2x)/(3 - x) - \frac{3}{2}| < \varepsilon$.

19. Take $\delta = \varepsilon^3$. If $0 < |x| < \delta$, then $|\sqrt[3]{x}| < \varepsilon$.

21. Suppose $\varepsilon = 1$. There is a $\delta > 0$ so that whenever $0 < |x - 3| < \delta$, then $|(3x - 1) - 12| < 1 \Rightarrow 4 < x < \frac{14}{3}$. For any $\delta < 1$, say, $\delta = \frac{1}{2}$, we have $0 < |x - 3| < \delta \Rightarrow 2.5 < x < 3.5$. This is impossible.

23. $\left| \dfrac{1}{x^2 + 9} - \dfrac{1}{18} \right| = \dfrac{|18 - (x^2 + 9)|}{18(x^2 + 9)} = \dfrac{|x + 3||x - 3|}{18(x^2 + 9)} \leq \dfrac{7|x - 3|}{18(13)}$

Given any $\varepsilon > 0$, choose δ to be the smaller of 1 or $\frac{234}{7}\varepsilon$.

Historical Exercises (page 100)

1. $y = (-\sqrt{3}/3)x + (4\sqrt{3}/3)$; $m_{L_2} = \sqrt{3}$; $m_{L_1} = -\sqrt{3}/3$ **3.** $y = 4x - 4$ **5.** $y = (2y_0/x_0)x - y_0$

7. $m = \frac{1}{2}$; $y = \frac{1}{2}x$ **9.** $y = -\frac{3}{4}x + \frac{25}{4}$ **11.** $y = 6x - 9$ **13.** $y = 12x - 16$

Miscellaneous Exercises (page 103)

1. $\frac{-1}{16}$ **3.** 8   **5.** Discontinuous at $-2, -1, 0, 1, 2, 3$

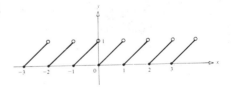

7. x	3.9	3.99	3.999	4.001	4.01	4.1
$f(x)$	0.2516	0.2502	0.2500	0.2499	0.2498	0.2485

$\lim_{x \to 4} f(x) = 0.25$

9. 0   **11.** 1; -1; does not exist

13. Continuous on its domain

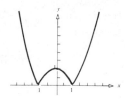

15. Apply the intermediate value theorem to $f(x) = x^3 - (1 - x^2)$ on $[0, 1]$.   **17.** $A = 1$; $B = -1$

19. $K = 12$ **21.** Take $\delta \leq \min(1, \varepsilon/5)$. If $0 < |x| < \delta$, then $|x + 4| < 5$ and $|(2 + x)^2 - 4| < \varepsilon$.

23. Take $\delta \leq \min(1, 26\varepsilon)$. If $0 < |x - 2| < \delta$, then $|(x + 2)/(x^2 + 9)| < \frac{1}{2}$ and $|1/(x^2 + 9) - \frac{1}{13}| < \varepsilon$.

25. Suppose $\varepsilon = 0.1$. There is a $\delta > 0$ so that whenever $0 < |x - 1| < \delta$, then $|x^2 - 1.31| < 0.1 \Rightarrow 1.1 < x < \sqrt{1.41}$. For any $\delta < 0.2$, say, $\delta = 0.01$, we have $0 < |x - 1| < 0.01 \Rightarrow 0.99 < x < 1.01$. This is impossible.

27. Take $\delta = \sqrt{\varepsilon}$. If $0 < |x| < \delta$, then $|(4 - x^2) - 4| < \varepsilon$.   **29.** By the squeezing theorem.

31. $\cos x = \cos[(x - c) + c] = \cos(x - c) \cos c - \sin(x - c) \sin c$; $\therefore \lim_{x \to c} \cos x = \cos c$

CHAPTER 3

Exercise 1 (page 110)

1. (a) 20 (b) 25 (c) 5 (d) 5 (e) 5; average rate of change is always 5 since f is a linear function

3. (a) $-\frac{1}{4}$ (b) $\frac{16}{7}$ (c) $-\frac{1}{8}$ (d) $\frac{4}{7}$ (e) $-\frac{4}{5}$

5. 3

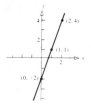

7. 2

9. -2

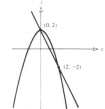

11. (a) 104 ft/sec (b) 97.6 ft/sec   **13.** 3   **15.** $-6 + 3\Delta x$ **17.** $(\Delta x)^2 + \Delta x$ **19.** $\Delta x/(1 + \Delta x)$

21. $1/(\sqrt{4 + \Delta x} + 2)$ **23.** 49.5 mph   **25.** $1, \frac{4}{3}, 4(\sqrt{2} - 1)$ kph

27. 0.046, 0.047, 0.1 bushel/lb   **29.** -0.06 g/hr

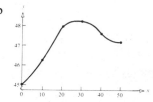

Exercise 2 (page 118)

1. 2   **3.** 0   **5.** -5 **7.** $\frac{1}{4}$ **9.** 2   **11.** $2x$ **13.** $6x + 1$ **15.** 0   **17.** $5/2\sqrt{x}$ **19.** m

21. 4, 16, $6t + 4$ m/sec   **23.** 61, 60.1, 60.01, 60 cm/sec   **25.** 100, 68, -28 ft/sec; 6.25 sec; 3.125 sec

27. $f'(-c) = \lim\limits_{x \to -c} \dfrac{f(x) - f(-c)}{x - (-c)} = \lim\limits_{x \to -c} \dfrac{f(x) - f(c)}{x + c} = \lim\limits_{t \to c} \dfrac{f(-t) - f(c)}{-t + c} = \lim\limits_{t \to c} -\dfrac{f(t) - f(c)}{t - c} = -f'(c)$

29. $f'(x) = \lim\limits_{h \to 0} \dfrac{f(x + h) - f(x)}{h} = \lim\limits_{h \to 0} \dfrac{f(x)f(h) - f(x)}{h} = f(x) \lim\limits_{h \to 0} \dfrac{f(h) - 1}{h} = f(x) \lim\limits_{h \to 0} \dfrac{f(h) - f(0)}{h - 0} = f(x)f'(0)$

Exercise 3 (page 121)

1. $y = 6x - 9$

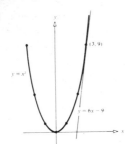

3. $y = 4x$

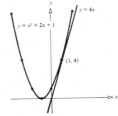

5. $y = -x + 2$

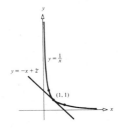

7. No   **9.** (2, 4)

11. $-\frac{1}{2}$

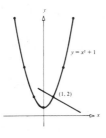

13. $\frac{1}{4}$

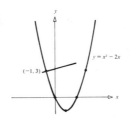

15. 1

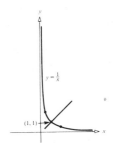

17. P = Pressure; x = Distance from surface; $P'(x) = -kP$, $k > 0$ a constant

19. (a) $2\pi R \Delta R + \pi(\Delta R)^2$ (b) $2\pi \Delta R$ (c) $2\pi R + \pi \Delta R$ (d) 2π (e) 2π **21.** 0.59   **23.** $f'(2) = -3$

Exercise 4 (page 129)

1. $45x^{14}$ **3.** 3   **5.** $2x + 3$ **7.** $40x^4 - 5$ **9.** $\frac{4}{3}x^3 - 3$ **11.** $3\pi x^2 + 3x$ **13.** $\frac{5}{3}x^4$

15. $2ax + b$ **17.** $\sqrt{3}$ **19.** $2\pi R$ **21.** $4\pi R^2$ **23.** 3; $y = 3x - 1$ **25.** 2   **27.** $-1, 1$

29. $y - 2 = 5(x - 1)$ **31.** $y = 3x - 9$; $9x - 3y + 5 = 0$

33. -1, 74 m/sec   **35.** This is the derivative of $5x^8$ at $\frac{1}{2}$, which equals $40(\frac{1}{2})^7 = \frac{5}{16}$.

37. $F(x) = f(x) - g(x) = f(x) + (-1)g(x)$; $F'(x) = \dfrac{d}{dx} f(x) + \dfrac{d}{dx}(-1)g(x) = f'(x) + (-1)g'(x) = f'(x) - g'(x)$

39. $f'(c) = \lim\limits_{x \to c} \dfrac{x^n - c^n}{x - c} = \lim\limits_{x \to c} \dfrac{(x - c)(x^{n-1} + cx^{n-2} + \cdots + c^{n-1})}{x - c} = nc^{n-1}$

Exercise 5 (page 134)

1. $5x^4 + 3x^2 - 2x$ **3.** $18x^2 + 6x - 10$ **5.** $16t^7 + 10t^4 - 4t^3 - 1$ **7.** $-3t^{-4}$ **9.** $\dfrac{-40}{x^5} - \dfrac{6}{x^3}$

11. $\dfrac{2}{(s + 1)^2}$ **13.** $\dfrac{12x^2 + 32x + 6}{(3x + 4)^2}$ **15.** $3 - \dfrac{1}{3t^2}$ **17.** $\dfrac{-4}{(1 + 2u)^2}$ **19.** $9x^2 + \dfrac{2}{3x^3}$

21. $\dfrac{-1}{t^2} + \dfrac{2}{t^3} - \dfrac{3}{t^4}$ **23.** $\dfrac{-3w^2}{(w^3 - 1)^2}$ **25.** $\frac{5}{4}$; $y = \frac{5}{4}x - \frac{3}{4}$ **27.** $0, -2$

29. (a) $y' = 2x(3x - 2) + x^2(3) = 9x^2 - 4x$ (b) $y = 3x^3 - 2x^2$; $y' = 9x^2 - 4x$ (c) They are equal.

31. -2 units/m   **33.** $\dfrac{d}{dx}[f(x)g(x)h(x)] = f(x)\dfrac{d}{dx}[g(x)h(x)] + g(x)h(x)\dfrac{d}{dx}[f(x)]$; now use (3.30)

35. $6x^5 - 5x^4 + 20x^3 - 18x^2 + 2x - 5$ **37.** $9x^2(x^3 + 1)^2$

39. $\dfrac{3}{x^4}\left(1 - \dfrac{1}{x}\right)\left(1 - \dfrac{1}{x^2}\right) + \dfrac{2}{x^3}\left(1 - \dfrac{1}{x}\right)\left(1 - \dfrac{1}{x^3}\right) + \dfrac{1}{x^2}\left(1 - \dfrac{1}{x^2}\right)\left(1 - \dfrac{1}{x^3}\right)$

Exercise 6 (page 139)

1. $2; 0$ **3.** $6x + 1; 6$ **5.** $1 - (1/x^2); 2/x^3$ **7.** $\dfrac{1}{(t + 1)^2}; \dfrac{-2}{(t + 1)^3}$ **9.** $\dfrac{x^2 + 2x}{(x + 1)^2}; \dfrac{2}{(x + 1)^3}$

11. $\dfrac{-12}{(3x + 5)^2}; \dfrac{72}{(3x + 5)^3}$

13. (a) $y' = 20x^3 - 6x^2 + 1$; $y'' = 60x^2 - 12x$ (b) $y' = \dfrac{-1}{x^2}$; $y'' = \dfrac{2}{x^3}$

   (c) $y' = 2(x^3 + 5) + 3x^2(2x + 1) = 8x^3 + 3x^2 + 10$; $y'' = 24x^2 + 6x$

   (d) $y' = \dfrac{(x)(2) - (1)(2x - 5)}{x^2} = \dfrac{5}{x^2}$; $y'' = \dfrac{-10}{x^3}$

15. (a) $y' = 12x^2 - 6x + 1$; $y'' = 24x - 6$; $y''' = 24$ (b) $y' = 3ax^2 + 2bx + c$; $y'' = 6ax + 2b$; $y''' = 6a$

17. 0 **19.** 0 **21.** $7! = 5040$ **23.** $32t + 20; 32$ **25.** $9.8t + 4; 9.8$ **27.** $x^2g''(x) + 4xg'(x) + 2g(x)$

29. (a) $-9.8t + 39.2$ m/sec   (b) 4 sec   (c) 78.4 m   (d) -9.8 m/sec^2 (e) 8 sec   (f) -39.2 m/sec   (g) 156.8 m

31. $F'(x) = f'(x)g(x) + f(x)g'(x)$;
   $F''(x) = f''(x)g(x) + f'(x)g'(x) + f'(x)g'(x) + f(x)g''(x) = f''(x)g(x) + 2f'(x)g'(x) + f(x)g''(x)$

33. $f(x) = \dfrac{1}{x}$; $f'(x) = \dfrac{-1}{x^2}$; $f''(x) = \dfrac{2}{x^3}$; $f'''(x) = \dfrac{-2(3)}{x^4}$; $f^{(n)}(x) = \dfrac{(-1)^n n(n - 1)(n - 2) \cdots (3)(2)(1)}{x^{n+1}}$

Exercise 7 (page 146)

1. $y' = 3\cos x + 2\sin x$ **3.** $y' = \cos^2 x - \sin^2 x$ **5.** $y' = \sec x \tan^2 x + \sec^3 x$

7. $y' = \dfrac{-x\csc^2 x - \cot x}{x^2}$ **9.** $y' = 2x\tan x + x^2 \sec^2 x$ **11.** $y' = x\sec^2 x + \tan x - 3\sec x \tan x$

13. $y' = \dfrac{-1}{1 - \cos x}$ **15.** $y' = \dfrac{x\cos x + \cos x - \sin x}{(1 + x)^2}$ **17.** $y' = \dfrac{-2}{(\sin x - \cos x)^2}$

19. $y' = \dfrac{x\tan^2 x - x + \sec x \tan x - \tan x}{(1 + x\sin x)^2}$ **21.** $y' = -\csc x \cot^2 x - \csc^3 x$ **23.** $y' = \dfrac{2\sec^2 x}{(1 - \tan x)^2}$

25. $y'' = -\sin x$ **27.** $y'' = \sec x \tan^2 x + \sec^3 x$ **29.** $y'' = 2\cos x - x\sin x$ **31.** $y'' = -2\sin x + 3\cos x$

33. $y'' = 2\sin x + 4x\cos x - x^2 \sin x$ **35.** $y'' = -a\sin x - b\cos x$ **37.** -1 **39.** $\dfrac{-2}{2 + \sqrt{3}}$ **41.** 7

43. 1   **45.** 0   **47.** 1   **49.** $y = x$ **51.** $3\sqrt{3}x + 6y = 3 + \sqrt{3}\pi$ **53.** $y = x$ **55.** $y = \sqrt{2}$

57. $(-1)^{n/2}\sin x$ if n is even; $(-1)^{(n-1)/2}\cos x$ if n is odd   **59.** Derivative of $\cos x$ at $\pi/2$, which is -1

61. From (3.37), if $0 < \theta < \pi/2$, then $\sin \theta \le \theta$. Also, $-\sin \theta \ge -\theta$ or $-\theta \le \sin(-\theta)$. Thus, $-|\theta| \le \sin \theta \le |\theta|$ for $-\pi/2 < \theta < \pi/2$ and $|\sin \theta| \le |\theta|$.

63. $y' = A\cos t - B\sin t$; $y'' = -A\sin t - B\cos t = -(A\sin t + B\cos t) = -y$

65. $\dfrac{d}{dx}\sin x = \lim\limits_{h\to 0}\dfrac{\sin(x + h) - \sin x}{h} = \lim\limits_{h\to 0}2\left\{\dfrac{\cos[(2x + h)/2]\,\sin(h/2)}{h}\right\}$

$= 2\lim\limits_{h\to 0}\dfrac{\cos(2x + h)}{2}\cdot\lim\limits_{h\to 0}\dfrac{\sin(h/2)}{2(h/2)} = 2(\cos x)(\tfrac{1}{2}) = \cos x$

67. a/b **69.** $f'(0) = \lim\limits_{x\to 0}\dfrac{f(x) - f(0)}{x - 0} = \lim\limits_{x\to 0}\dfrac{\cos x - 1}{x}$

Exercise 8 (page 154)

1. $6(3x + 5)$ **3.** $-18(6x - 5)^{-4}$ **5.** $8x(x^2 + 5)^3$ **7.** $7(t^5 - t^2 + t)^6(5t^4 - 2t + 1)$

9. $3\left(x - \dfrac{1}{x}\right)^2\left(1 + \dfrac{1}{x^2}\right)$ **11.** $\dfrac{3z^2}{(z + 1)^4}$ **13.** $2\tan x \sec^2 x$ **15.** $4\sin t \cos t$

17. $\dfrac{-6x(3x^2 + 1)(x^3 + x + 2)}{(x^3 - 1)^3}$ **19.** $2x(x^2 + 4)(2x^3 - 1)^2(13x^3 + 36x - 2)$

21. $4[5x + (3x + 6x^2)^3]^3[5 + 9(3x + 6x^2)^2(1 + 4x)]$ **23.** $15x^2(x^3 + 1)^4$ **25.** $\dfrac{2x}{(x^2 + 2)^2}$ **27.** $-\dfrac{2(x + 1)}{x^3}$

29. $\dfrac{-30}{x^7}\left(\dfrac{1}{x^6} - 1\right)^4$ **31.** $4\cos 4x$ **33.** $5\sec^2 5x$ **35.** $6x\cos(3x^2 + 4)$ **37.** $24\sin 3x \cos 3x$

39. $4(x + 1)\cos(x^2 + 2x - 1)$ **41.** $-6\csc^2(1 + 3x)\cot(1 + 3x)$ **43.** $2x(\sin 4x + 2x\cos 4x)$

45. $(-1/x^2)\cos(1/x)$ **47.** $2x\sin x(\sin x + x\cos x)$

49. (a) $y' = (2u)(3x^2) = 6x^2(x^3 + 1)$ (b) $y' = 2(x^3 + 1)(3x^2) = 6x^2(x^3 + 1)$
(c) $y = x^6 + 2x^3 + 1$; $y' = 6x^5 + 6x^2 = 6x^2(x^3 + 1)$ (d) All are equal.

51. $\dfrac{-576(48 + x^4)^2}{x^{13}}$ **53.** $\dfrac{-64}{x^5}$ **55.** $6\tan^2 2x\,\sec^2 2x$ **57.** $-5x^3[5x^5\cos(x^5) + 4\sin(x^5)]$ **59.** $2^n n!$

61. $f(x) = -f(-x); f'(x) = -\dfrac{d}{dx}f(-x) = (-1)(-1)f'(-x) = f'(-x)$ **63.** $2xf'(x^2 + 1)$

65. $\dfrac{-2}{(x-1)^2}f'\left(\dfrac{x+1}{x-1}\right)$ **67.** $\cos x\, f'(\sin x)$ **69.** $\sin^2 x\, f''(\cos x) - \cos x\, f'(\cos x)$ **71.** -12 **73.** 78

75. $3\text{ m/sec}; 6(t-2)\text{ m/sec}^2$ **77.** $\dfrac{-20\pi}{9}\sin\dfrac{\pi}{6}t$

79. $\dfrac{d}{dx}\cos x = \dfrac{d}{dx}\sin(\pi/2 - x) = (-1)\cos(\pi/2 - x) = -\sin x$

Exercise 9 (page 160)

1. $-x/y$ **3.** $-2y/x$ **5.** $\dfrac{2x-y}{x+2y}$ **7.** $\dfrac{4y-2x}{2y-4x-1}$ **9.** $-2x/y^2$ **11.** $\dfrac{1-12x^2}{6y^2}$ **13.** $(y/x)^3$

15. $-(y/x)^2$ **17.** $\dfrac{6x(x^2+y)^2}{1-3(x^2+y)^2}$ **19.** y/x **21.** $\dfrac{-x(y^2-1)^2}{y}$ **23.** $\dfrac{\sin y}{1-x\cos y}$ **25.** $-y/x$

27. $\dfrac{\sec^2(x-y)}{1+\sec^2(x-y)}$ **29.** $\dfrac{\cos(x+y)-\sin(x-y)}{1-\cos(x+y)-\sin(x-y)}$ **31.** $-x/y; \dfrac{-4}{y^3}$ **33.** $\dfrac{-y(1+2x)}{x(1+x)}; \dfrac{2y(3x^2+3x+1)}{x^2(1+x)^2}$

35. $-\frac{1}{2}; y = -\frac{1}{2}x + \frac{5}{2}$ **37.** $m_{\tan} = -x/y$; slope of OP is y/x; product of slopes is -1

39. (a) $\dfrac{-(y+1)}{4y+x}$ (b) $y = -\frac{1}{3}x + \frac{5}{3}$ (c) $(6, -3)$

Exercise 10 (page 166)

1. $\frac{2}{3}x^{-1/3}$ **3.** $\frac{2}{3}x^{-1/3}$ **5.** $\dfrac{-1}{x^{3/2}} - \dfrac{1}{x^{4/3}} + \dfrac{6}{x^{5/2}} - \dfrac{6}{x^{7/4}}$ **7.** $\dfrac{1}{3x^{2/3}} + \dfrac{1}{3x^{4/3}}$ **9.** $\dfrac{-1}{2\sqrt{3x^3}}$ **11.** $\dfrac{3x^2}{2(x^3-1)^{1/2}}$

13. $\dfrac{\cos x}{2\sqrt{\sin x}}$ **15.** $\dfrac{\sec\sqrt{x}\tan\sqrt{x}}{2\sqrt{x}}$ **17.** $\dfrac{2x^2-1}{\sqrt{x^2-1}}$ **19.** $\dfrac{5x^3+2}{2(x^3+1)^{1/2}}$ **21.** $\dfrac{-x}{(3-x^2)^{1/2}} - \dfrac{x}{(4-x^2)^{1/2}}$

23. $\dfrac{x}{\sqrt{x^2+1}}$ **25.** $\dfrac{-1}{(x+1)^{1/2}(x-1)^{3/2}}$ **27.** $\dfrac{|x|(32x+3)}{2\sqrt{x}(8x+1)}$ **29.** $\dfrac{1}{6x^{2/3}\sqrt{4+\sqrt[3]{x}}}$

31. $\dfrac{3x^2}{2}(\cos x)^{1/2}(2\cos x - x\sin x)$ **33.** $\dfrac{-x}{\sqrt{x^2+1}}\sin\sqrt{x^2+1}\cos(\cos\sqrt{x^2+1})$ **35.** $\dfrac{8x^2+x-6}{(x^2-3)^{1/2}(6x+1)^{2/3}}$

37. $\dfrac{18x^3+32x^2+3}{2(3x+4)^{3/2}(2x^3-1)^{1/3}}$ **39.** $\dfrac{18x^4+17x^2+2}{6\sqrt[3]{(2x^3+x)^2}\sqrt[4]{(x^2+1)^3}}$ **41.** $\dfrac{x}{\sqrt{x^2+1}}; \dfrac{1}{(x^2+1)^{3/2}}$ **43.** $y = -\frac{1}{4}x + 1$

45. $\frac{2}{3}x^{-1/3} + \frac{2}{3}y^{-1/3}y' = 0; y' = -y^{1/3}/x^{1/3}$ **47.** $-\frac{1}{2}$ **49.** 2 **51.** $\frac{1}{2}; \frac{1}{5}$ **53.** $\frac{1}{t}$

Exercise 11 (page 170)

1. 1.154 **3.** 0.2128 **5.** 3.13496 **7.** 1.157 **9.** 1.495 **11.** 1.02 ft   **13.** 1.02

Exercise 12 (page 174)

1. Yes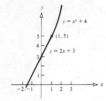

3. No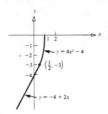

5. No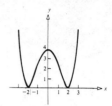

7. No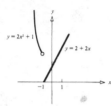

9. (a) Yes (b) Yes (c)

11. 75 ft/sec; 0; **No**

Miscellaneous Exercises (page 177)

1. $na(ax + b)^{n-1}$ **3.** $\dfrac{2 - 3x}{2(1 - x)^{1/2}}$ **5.** $\dfrac{x(x + 2)}{(x + 1)^2}$ **7.** $\dfrac{1}{(a^2 - r^2)^{3/2}}$ **9.** $3x(x^2 + 4)^{1/2}$

11. $-\dfrac{2}{x^2}\left(\dfrac{3}{x} + 1\right)$ **13.** $\dfrac{-2z}{(z^2 + 1)^2}$ **15.** $\dfrac{x(x^2 - 2)}{(x^2 - 1)^{3/2}}$ **17.** $\dfrac{1}{a^2 p^2 (p^2 - 1)^{1/2}}$ **19.** $(a^2 + x^2)^{1/2}(5x^3 + 2xa^2)$

21. $\dfrac{1}{2(x + 2)^{1/2}}$ **23.** $\dfrac{9}{(y^2 + 9)^{3/2}}$ **25.** $\dfrac{-x}{(1 - x^2)^{1/2}}$ **27.** $\dfrac{1 - 2z}{(1 - z + z^2)^2}$ **29.** $\dfrac{-3x^2}{2(1 - x^3)^{1/2}}$

31. $\frac{3}{2}(1 + u)^{1/2}$ **33.** $\dfrac{-2x}{(x - 1)^3}$ **35.** $\dfrac{x^{1/2} - a^{1/2}}{x^{1/2}}$ **37.** $2x \cos 2x + \sin 2x$ **39.** $\sec u(\sec u + \tan u)$

41. $\dfrac{\cos z}{2(1 + \sin z)^{1/2}}$ **43.** $\frac{1}{2}$ **45.** Does not exist **47.** $\dfrac{1 + y \cos xy}{1 - x \cos xy}$

49. (a) $x \geq -\frac{1}{6}; y \geq 0$ (b) $\frac{3}{5}$ (c) $\frac{13}{5}$ (d) $(\frac{4}{3}, 3)$ **51.** $\frac{4}{3}$ **53.** $\dfrac{(-1)^n n!}{2^{n+1}}$

55. $f_1' f_2 f_3 f_4 + f_1 f_2' f_3 f_4 + f_1 f_2 f_3' f_4 + f_1 f_2 f_3 f_4'$ **57.** (a) $f'' g + 2f' g' + f g''$ (b) $\dfrac{2(g')^2 - g g''}{g^3}$

59. $v = (-\pi/2) \sin 4\pi t; a = -2\pi^2 \cos 4\pi t$ **61.** $a = 3, b = 2, c = 0$ **63.** $\dfrac{f^2 g' + f' g^2}{(f + g)^2}$ **65.** $-\frac{4}{9}; \frac{-100}{243}$

67. $y = \sqrt[n]{x}, y' = \frac{1}{n} x^{(1/n)-1} = \frac{1}{n}$ at $(1, 1); y = x^n, y' = nx^{n-1} = -n$ at $(-1, 1)$ **69.** 2

71. $a = 3, b = -6, c = 6, d = -3$

73. $y^2 = 2ax + a^2, 2yy' = 2a, y' = a/y$ } The points of intersection are $(0, a)$ and $(0, -a)$ at which
$y^2 = a^2 - 2ax, 2yy' = -2a, y' = -a/y$ } both the products of the slopes are -1.

75. (a) 4 (b) 12 (c) $4x; 2x^2 + 3$ **77.** (a) 1 (b) $y = 12x - 16; y = 12x + 1$

79. $v = \dfrac{1 + t^2}{(t^2 - 1)^2}; a = \dfrac{2t(t^2 + 3)}{(t^2 - 1)^3}$ **81.** (a) Parallel (b) Perpendicular   **83.** $8x - 16; (4 - 2x)^2$

85. $(-1, -1); 34.7° \approx 0.6055$ radian   **87.** None

CHAPTER 4

Exercise 1 (page 189)

1. $-\frac{8}{3}$ **3.** -1 **5.** 40   **7.** $5\pi^2$ **9.** 900 cm^3/sec   **11.** $-8/\sqrt{2009}$ cm/min   **13.** $-\frac{3}{4}$ m^2/min

15. $\frac{1}{30}$ m/min   **17.** 8500π m/min   **19.** $4/\pi$ m/min   **21.** $3 - (9\pi/32)$ m^3/min

23. (a) $-3/2\sqrt{55}$ m/sec (b) $-1/2\sqrt{3}$ m/sec (c) $-3/2\sqrt{7}$ m/sec   **25.** $4\sqrt{10}/7$ m/sec

27. $2\sqrt{3}\pi/45$ cm^2/min   **29.** 1.75 kg/cm^2/min   **31.** $2\sqrt{5}/5$ m/sec   **33.** $107/\sqrt{241}$ m/sec

35. $3/40$ radian/sec   **37.** -4.8 lb/sec

Exercise 2 (page 198)

1. $(3x^2 - 2)\,dx$ **3.** $\dfrac{-x^2 + 2x - 6}{(x^2 + 2x - 8)^2}\,dx$ **5.** $(6\cos 2x + 1)\,dx$ **7.** $-y/x; -x/y$ **9.** $-x/y; -y/x$

11. $\dfrac{x(x - 2y)}{x^2 - y^2}; \dfrac{x^2 - y^2}{x(x - 2y)}$ **13.** $\dfrac{2}{3\cos 3y}; \dfrac{3\cos 3y}{2}$ **15.** $\dfrac{1}{2\sqrt{x}}\,dx$ **17.** $(3x^2 + 1)\,dx$ **19.** $f(x) \approx 2x - 3$

21. $f(x) \approx \frac{1}{4}x + 1$ **23.** $f(x) \approx \dfrac{\sqrt{3}}{2}x + \dfrac{6 - \pi\sqrt{3}}{12}$

25. (a) $5.916\overline{6}$ (b) 5.12 (c) 0.9, if $x = 1$; 0.9128, if $x = 1.21$ (d) 0.4849   **27.** (a) 0.006 (b) 0.00125

29. 2π cm^2 **31.** 3.6π cm^3 **33.** 6%   **35.** 8π cm^3 **37.** 30 m; 0.9%   **39.** 72 min

41. (a) $dy = f'(x)\,dx = y'\,dx; dy' = f''(x)\,dx = y''\,dx$ (b) $dy'' = f'''(x)\,dx = y'''\,dx; dy'^2 = 2f'(x)f''(x)\,dx$

Exercise 3 (page 208)

1. -1 **3.** 3   **5.** 0, 2   **7.** 0, 1, -1 **9.** 0   **11.** 0   **13.** $-1, 1, -\sqrt{2}/2, \sqrt{2}/2$ **15.** 0, 2

17. $-3, 1, 0$ **19.** 4, 3   **21.** $3, -3, 3\sqrt{3}, -3\sqrt{3}$ **23.** 15, -1 **25.** 1, -8 **27.** 16, -4

29. 9, 0   **31.** 1, 0   **33.** 4, 2   **35.** $\frac{1}{2}, -\frac{1}{2}$ **37.** $0, -\frac{1}{2}$ **39.** $128\sqrt[3]{2}, 0$ **41.** $\frac{1}{3}, -1$

43. $\sqrt[3]{18}/3\sqrt{3} \approx 0.504, 0$ **45.** 9, 1   **47.** 8, 0   **49.** 40 mph   **51.** $A = 6, B = 0, C = 2$

53. Reverse the inequalities given in the proof of (4.18).

Exercise 4 (page 215)

1. $\frac{3}{2}$ **3.** 1   **5.** $-\sqrt{3}/3$ **7.** $-\sqrt{3}/3, \sqrt{3}/3$ **9.** $-1, 0, 1$ **11.** $f(-2) \neq f(1)$

13. $f'(0)$ does not exist   **15.** $\frac{1}{2}$ **17.** 1   **19.** $\frac{7}{3}$ **21.** $\sqrt{3}$ **23.** $(\frac{14}{9})^3$ **25.** $f'(0)$ does not exist

27. $\dfrac{f(b) - f(a)}{b - a}$ = slope of line; f is not differentiable at $x = 0$

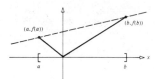

29. By the mean value theorem, $(\sqrt{66} - \sqrt{64})/(66 - 64) = 1/2\sqrt{c}$, $64 < c < 66$. But $8 = \sqrt{64} < \sqrt{c} < \sqrt{66}$ $\Rightarrow 1/\sqrt{66} < 1/\sqrt{c} < \frac{1}{8}$. Since $1/\sqrt{66} > 1/\sqrt{81} = \frac{1}{9}$, then $\frac{1}{9} < 1/\sqrt{c} < \frac{1}{8} \Rightarrow \frac{1}{9} < \sqrt{66} - 8 < \frac{1}{8}$.

31. $f(0) = f(1) = 0$; $f'(x) = (x - 1)\cos x + \sin x = 0$ for some x, $0 < x < 1$. Divide by $\cos x$.

33. Assume the contrary. Then there are at least three roots x_1, x_2, x_3. Set $f(x) = x^n + ax + b$. Apply Rolle's theorem to f on $[x_1, x_2]$ and on $[x_2, x_3]$, obtaining an impossible result.

35. Follow the outline given for the solution to Problem 33.

37. If x_1, x_2, x_3 are the three numbers, apply Rolle's theorem to f on $[x_1, x_2]$ and again on $[x_2, x_3]$.

Exercise 5 (page 224)

1. $f'(x) = 6x^2 - 12x + 6 = 6(x - 1)^2 \geq 0$ **3.** $f'(x) = 1/(x + 1)^2 > 0$

5. $\frac{1}{2}$; decreasing on $(-\infty, \frac{1}{2}]$, increasing on $[\frac{1}{2}, +\infty)$; local minimum at $\frac{1}{2}$

7. 1; increasing on $(-\infty, 1]$, decreasing on $[1, +\infty)$; local maximum at 1

9. $-1, 0$; increasing on $(-\infty, -1]$, decreasing on $[-1, 0]$, increasing on $[0, +\infty)$; local maximum at -1, local minimum at 0

11. -2; increasing for all x; no local extrema

13. 0, 3; decreasing on $(-\infty, 3]$, increasing on $[3, +\infty)$; local minimum at 3

15. 0, 1; decreasing on $(-\infty, 1]$, increasing on $[1, +\infty)$; local minimum at 1

17. $-3, 1$; increasing on $(-\infty, -3]$, decreasing on $[-3, 1]$, increasing on $[1, +\infty)$; local maximum at -3, local minimum at 1

19. $-\frac{2}{3}, 0$; increasing on $(-\infty, -\frac{2}{3}]$, decreasing on $[-\frac{2}{3}, 0]$, increasing on $[0, +\infty)$; local maximum at $-\frac{2}{3}$, local minimum at 0

21. 0; decreasing on $(-\infty, 0]$, increasing on $[0, +\infty)$; local minimum at 0

23. $-\frac{1}{8}, 0$; decreasing on $(-\infty, -\frac{1}{8}]$, increasing on $[-\frac{1}{8}, +\infty)$; local minimum at $-\frac{1}{8}$

25. 0, 4; increasing on $(-\infty, 0]$, decreasing on $[0, 4]$, increasing on $[4, +\infty)$; local maximum at 0, local minimum at 4

27. $-1, 0, 1$; decreasing on $(-\infty, -1]$, increasing on $[-1, 0]$; decreasing on $[0, 1]$, increasing on $[1, +\infty)$; local minima at -1 and 1; local maximum at 0

29. $-1, 0, 1$; decreasing on $(-\infty, -1]$, increasing on $[-1, 0]$, decreasing on $[0, 1]$, increasing on $[1, +\infty)$; local minima at -1 and 1, local maximum at 0

31. $(2k + 1)\pi/2$; decreasing on $[(4k + 1)\pi/2, (4k + 3)\pi/2]$, increasing on $[(4k + 3)\pi/2, (4k + 5)\pi/2]$; local maxima at $(4k + 1)\pi/2$, local minima at $(4k + 3)\pi/2$; k an integer

33. Moves to right if $t > 1$, to left if $t < 1$; reverses direction if $t = 1$; velocity increases for all t

35. Moves to right if $t < -3$ or $t > 1$; moves to left if $-3 < t < 1$; reverses direction if $t = -3$, $t = 1$; velocity increasing if $t > -1$; velocity decreasing if $t < -1$

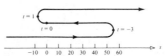

37. Always moves to right; velocity always decreasing

39. Moves to right if $0 < t < \pi/6$ or $\pi/2 < t < 2\pi/3$; moves to left if $\pi/6 < t < \pi/2$; reverses direction at $t = \pi/6$, $t = \pi/2$; velocity decreasing if $0 \le t \le \pi/3$; velocity increasing if $\pi/3 \le t \le 2\pi/3$

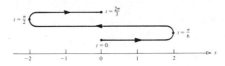

41. $f'(x) = 2ax + b = 0$ if $x = -b/2a$; $a < 0$: $f'(x) = 2a(x + b/2a) > 0$ if $x < -b/2a$; $f'(x) < 0$ if $x > -b/2a$; the case where $a > 0$ is similar

43. $a = -\frac{7}{8}$, $b = \frac{21}{4}$, $c = 0$, $d = 5$

45. f is decreasing on $a \le x \le c$ since $f'(x) < 0$ for $a < x < c$; f is increasing on $c \le x \le b$ since $f'(x) > 0$ for $c < x < b$; for all x in $[a, b]$, $f(c) \le f(x)$; f has local minimum at c

47. $f'(x) = 1/3x^{2/3} > 0$ if $x \ne 0$; 0 is only critical number; f is increasing for all x

49. $\frac{1}{3}$ **51.** $\left(\sin\dfrac{x}{2}\right)^2 \le \left(\dfrac{x}{2}\right)^2$ and $\sin^2\dfrac{x}{2} = \dfrac{1 - \cos x}{2}$; $\therefore$ $\dfrac{1 - \cos x}{2} \le \dfrac{x^2}{4} \Rightarrow \cos x \ge 1 - \dfrac{x^2}{2}$

Exercise 6 (page 230)

1. Concave up for all x; no inflection points

3. Concave up on $[3, +\infty)$; concave down on $(-\infty, 3]$; inflection point at 3

5. Concave up on $(-\infty, 0]$ and $[2, +\infty)$; concave down on $[0, 2]$; inflection points at 0 and 2

7. Concave up on $(0, +\infty)$; concave down on $(-\infty, 0)$; no inflection points

9. Concave up on $(-\infty, 0]$; concave down on $[0, +\infty)$; inflection point at 0

11. Concave up on $(-\infty, 0)$ and $(0, 3]$; concave down on $[3, +\infty)$; inflection point at 3

13. No local extrema;
concave down on $(-\infty, 1]$;
concave up on $[1, +\infty)$;
inflection point at 1

15. Local maximum at 3;
local minimum at 2;
concave up on $(-\infty, \frac{5}{2}]$;
concave down on $[\frac{5}{2}, +\infty)$;
inflection point at $\frac{5}{2}$

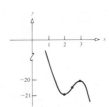

17. Local minimum at 1;
concave up for all x;
no inflection points

19. Local maximum at 4;
local minimum at 0;
concave up on $(-\infty, 3]$;
concave down on $[3, +\infty)$;
inflection point at 3

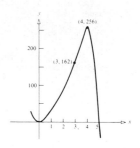

21. Local maximum at -2;
local minimum at 2;
concave down on $(-\infty, -\sqrt{2}]$;
concave up on $[-\sqrt{2}, 0]$;
concave down on $[0, \sqrt{2}]$;
concave up on $[\sqrt{2}, +\infty)$;
inflection points at $-\sqrt{2}, 0, \sqrt{2}$

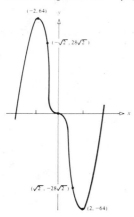

23. Local minimum at $\frac{1}{8}$;
concave up on $(-\infty, -\frac{1}{4}]$;
concave down on $[-\frac{1}{4}, 0]$;
concave up on $[0, +\infty)$;
inflection points at $-\frac{1}{4}, 0$

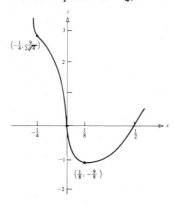

25. Local maximum at 0;
local minimum at 4;
concave down on $(-\infty, -2]$;
concave up on $[-2, +\infty)$;
inflection point at -2

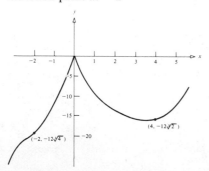

27. Local maximum at 0;
local minima at $-\sqrt{2}$ and $\sqrt{2}$;
concave up for all x;
no inflection points

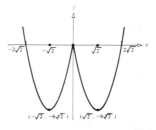

29. Local minimum at 0;
concave down on $(-\infty, -\sqrt{3}/3]$;
concave up on $[-\sqrt{3}/3, \sqrt{3}/3]$;
concave down on $[\sqrt{3}/3, +\infty)$;
inflection points at $-\sqrt{3}/3$ and $\sqrt{3}/3$

31. Local maximum at 1;
concave down on $[0, (3 + 2\sqrt{3})/3]$;
concave up on $[(3 + 2\sqrt{3})/3, +\infty)$;
inflection point at $(3 + 2\sqrt{3})/3$

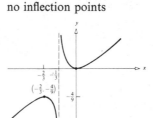

33. Local maximum at $\frac{1}{2}$;
concave down on $[0, 1]$;
no inflection points

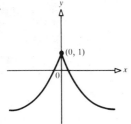

35. Local maximum at $-\frac{2}{3}$,
local minimum at 0;
concave down on $(-\infty, -\frac{1}{3})$;
concave up on $(-\frac{1}{3}, +\infty)$;
no inflection points

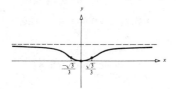

37.

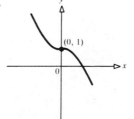

39.

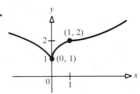

41.

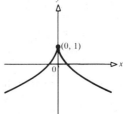

43.

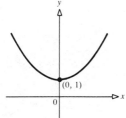

45.

47.

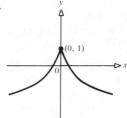

49. $a = -3$; $b = 9$

51. $f(x) = x^2 - 8x + 21$; f has a local minimum at $(4, 5)$; f is concave up for all x; therefore $(4, 5)$ is an absolute minimum and $f(x) > 0$ for all x

53. $f''(x) = 2a \neq 0$ for any x; $a > 0$; $a < 0$

55. Local maximum at $\pi/3$, local minimum at $4\pi/3$; inflection points at $5\pi/6$, $11\pi/6$

57. If P is of degree n, then P' is of degree $n - 1$ and P'' is of degree $n - 2$. A polynomial of degree n has at most n roots.

59. $f''(x) = n(n - 1)(x - a)^{n-2}$; if n is even, so is $n - 2$, and so $f''(x) \geq 0$ for all x

61. Proof is same as part (a) of (4.31) to and including equation (4.33). If f'' is negative throughout (a, b), then from (4.28), f' is decreasing on (a, b). For $x_1 > c$, $f'(x_1) < f'(c)$ so (4.33) may be written as $f(x) < f(c) + f'(c)(x - c)$. For $x_1 < c$, $f'(x_1) > f'(c)$, so (4.33) may be written $f(x) < f(c) + f'(c)(x - c)$. Therefore, $f(x)$ lies below the tangent line to $(c, f(c))$ throughout (a, b) for c arbitrary in (a, b). Thus, f is concave down on $[a, b]$.

Exercise 7 (page 243)

1. 1   **3.** 2   **5.** 3   **7.** $-\infty$ **9.** $+\infty$ **11.** $\frac{1}{3}$ **13.** $-\infty$ **15.** $+\infty$ **17.** $+\infty$ **19.** $\sqrt{5}/5$

21. -1 **23.** 0   **25.** $-\infty$ **27.** 0   **29.** $+\infty$ **31.** $y = 3$; $x = 0$ **33.** $y = 0$; $x = 1$

35. $y = 3$; $x = -1$ **37.** $y = 0$; $x = -1$; $x = 1$ **39.** $y = 1$; no vertical   **41.** No horizontal; $x = 0$

43. No horizontal; no vertical   **45.** $y = a/c$; $x = -d/c$ **47.** $\lim\limits_{x \to 0} |f'(x)| = \lim\limits_{x \to 0} \left| \dfrac{1}{3x^{2/3}} \right| = +\infty$

49. $\lim\limits_{x \to -4^+} \left| \dfrac{1}{2\sqrt{x + 4}} \right| = +\infty$ **51.** $\lim\limits_{x \to 3} \left| \dfrac{2}{3(x - 3)^{1/3}} \right| = +\infty$

53. No x-intercepts; y-intercept $-\frac{1}{2}$;
symmetric about y-axis;
horizontal asymptote $y = 0$; vertical asymptotes $x = \pm 2$;
critical numbers $x = 0$;
local maximum at 0;
concave up on $(-\infty, -2)$ and $(2, +\infty)$;
concave down on $(-2, 2)$;
no inflection points

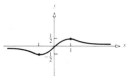

55. x-intercept $\frac{1}{2}$; y-intercept -1;
no symmetry;
horizontal asymptote $y = 2$; vertical asymptote $x = -1$;
no critical numbers;
no local extrema;
concave up on $(-\infty, -1)$;
concave down on $(-1, +\infty)$;
no inflection points

57. x-intercept 0; y-intercept 0;
symmetric about origin;
horizontal asymptote $y = 0$;
critical numbers $x = \pm 1$;
local maximum at 1; local minimum at -1;
concave down on $(-\infty, -\sqrt{3}]$ and $[0, \sqrt{3}]$
concave up on $[-\sqrt{3}, 0]$ and $[\sqrt{3}, +\infty]$;
inflection points at 0, $\pm\sqrt{3}$

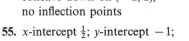

59. No x-intercepts; y-intercept $-\frac{1}{2}$;
symmetric about y-axis;
horizontal asymptote $y = 0$; vertical asymptotes $x = \pm 4$;
critical number $x = 0$;
local maximum at 0;
concave up on $(-\infty, -4)$ and $(4, +\infty)$;
concave down on $(-4, 4)$;
no inflection points

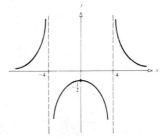

61. No intercepts;
symmetric about origin;
vertical asymptote $x = 0$;
critical numbers $x = -1, 1$;
local maximum at -1; local minimum at 1;
concave up on $(0, +\infty)$;
concave down on $(-\infty, 0)$;
no inflection points

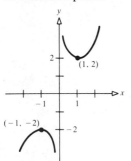

63. x-intercept 0; y-intercept 0;
no symmetry;
vertical asymptote $x = -3$;
critical numbers $x = 0, -6$;
local maximum at -6;
local minimum at 0;
concave down on $(-\infty, -3)$;
concave up on $(-3, +\infty)$;
no inflection points

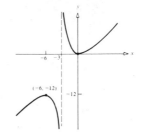

65. x-intercept 0; y-intercept 0;
no symmetry;
horizontal asymptote $y = 0$; vertical asymptote $x = 1$;
critical numbers $x = -2, 0$;
local maximum at 0; local minimum at -2;
concave down on $(-\infty, (-4 - 3\sqrt{2})/2]$ and $[(-4 + 3\sqrt{2})/2, 1)$;
concave up on $[(-4 - 3\sqrt{2})/2, (-4 + 3\sqrt{2})/2]$ and $(1, +\infty)$;
inflection points $x = (-4 \pm 3\sqrt{2})/2$

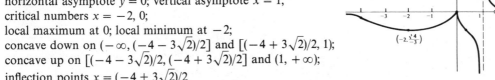

67. No x-intercept; no y-intercept;
no symmetry;
horizontal asymptote $y = 1$; vertical asymptote $x = 0$;
critical number $x = -2$;
local minimum at -2;
concave down on $(-\infty, -3]$;
concave up on $[-3, 0)$ and $(0, +\infty)$;
inflection point at -3

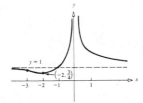

69. For a rational function, discontinuities occur when the denominator becomes 0. If the numerator is not 0 (i.e., the function is irreducible), then the fraction becomes infinitely large ($+$ or $-$) at the discontinuities.

71. For $x^2 + y^2 = 1$, $\dfrac{dy}{dx} = -\dfrac{x}{y}$; $y = 0$ for $x = \pm 1$; $\lim\limits_{x \to \pm 1} |f'(x)| = +\infty$. Tangent lines are vertical.

73. (a) $m > n$; $\lim\limits_{x \to +\infty} \dfrac{a_m x^m + \cdots + a_0}{b_n x^n + \cdots + b_0} = \lim\limits_{x \to +\infty} \left[\dfrac{a_m}{b_n} x^{m-n} + \text{(Lesser powers of } x)\right] = \pm\infty$, according as $\dfrac{a_m}{b_n} > 0$ or

$\dfrac{a_m}{b_n} < 0$

(b) $m = n$; $\lim\limits_{x \to +\infty} \dfrac{a_m x^m + \cdots + a_0}{b_n x^n + \cdots + b_0} = \lim\limits_{x \to +\infty} \dfrac{a_m + \text{(Reciprocals of } x)}{b_n + \text{(Reciprocals of } x)} = \dfrac{a_m + 0}{b_n + 0} = \dfrac{a_m}{b_n}$

(c) $m < n$; $\lim\limits_{x \to +\infty} \dfrac{a_m x^m + \cdots + a_0}{b_n x^n + \cdots + b_0} = \lim\limits_{x \to +\infty} \dfrac{a_m + \text{(Reciprocals of } x)}{b_n x^{n-m} + \text{(Lesser powers of } x)} = 0$ (by $n - m > 0$)

Exercise 8 (page 252)

1. 1,125,000 m²   **3.** $L/4$ by $L/4$ m   **5.** 16,666,667 m²   **7.** Base = $L/3$; legs = $L/3$

9. Height = 4 cm; base = 16 by 16 cm   **11.** $10\sqrt[3]{2}$ by $10\sqrt[3]{2}$ by $10\sqrt[3]{2}$ cm

13. Height = $50/\sqrt[3]{5\pi}$ cm; radius = $4\sqrt[3]{25/\pi}$ cm   **15.** $17.00

17. $dF/d\theta = -mc(c \cos\theta - \sin\theta)/(c \sin\theta + \cos\theta)^2 = 0$ when $\tan\theta = c$

19. (a) 40 mph   (b) 51.64 mph   (c) 53.39 mph   **21.** 1.98 km from box   **23.** $\sqrt{p^2 + 4qr}$

25. Width = $2\sqrt{3}$; depth = $4\sqrt{15}/3$ **27.** $5\sqrt{5}$ m   **29.** $\pi/2$ **31.** Base-to-height ratio is 2 to 1

33. Diameter-to-height ratio is $2/(1 + \pi)$ **35.** $16\sqrt{\pi}$ cm; $16\sqrt{\pi + 4}$ cm

37. Circle of length 35 cm; circle of length $35\pi/(4 + \pi)$ cm

39. Diameter = $20/(\pi + 4)$ m; height = $10/(\pi + 4)$ m

Exercise 9 (page 266)

1. $\frac{2}{3}x^6 + C$ **3.** $2x^{5/2} + C$ **5.** $(-2/x) + C$ **7.** $\frac{2}{3}x^{3/2} + C$ **9.** $x^4 - x^3 + x + C$

11. $-\frac{1}{9}(2 - 3x)^3 + C$ **13.** $2x^{3/2} - 4x^{1/2} + C$ **15.** $\frac{1}{6}x^2 + \frac{7}{3}x + C$ **17.** $x^2 - 3\sin x + C$

19. $y = x^3 - x^2 + x + 1$ **21.** $v = t^3 - t^2 + t + 4$ **23.** $s = \frac{1}{4}t^4 - (1/t) + \frac{11}{4}$

25. $y = \frac{1}{2}x^2 + 2\cos x - 2$ **27.** $s = -16t^2 + 128t$ **29.** $s = \frac{1}{2}t^3 + 18t + 2$

31. $F(x) = x\cos x + \sin x + 1$ **33.** $10\sqrt{3}$ m/sec   **35.** 12.25 m   **37.** 13.13 m/sec   **39.** 8 g-cm/sec²

41. 13.84 m/sec

Exercise 10 (page 273)

1. (a) $8 - 2x$ (b) 2; 0   (c) 2   (d) 5   (e) 8   (f) $x = 1$; $x = 5$ (g) $-x^2 + 6x - 5$ (h) $x = 3$ (i) 4   (j) 2
(k) 15   (l) 6   (m)   (n)

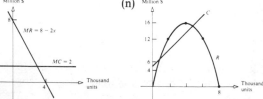

3. 333 units　　**5.** 100 tons　　**7.** (a) $-\frac{1}{100}x + 25$　(b) 985 articles　(c) \$15.15

9. Average cost $= \dfrac{C(x)}{x}$　　$\dfrac{d(\text{Average cost})}{dx} = \dfrac{xC'(x) - C(x)}{x^2} = 0$

So $C'(x) = C(x)/x$ at a critical number or when $MC = $ Average cost.

$\dfrac{d^2(\text{Average cost})}{dx^2} = \dfrac{x^2 C''(x) - 2xC'(x) + 2C(x)}{x^3} = \dfrac{C''(x)}{x} > 0 \, [\text{by } C''(x) > 0]$

Thus, minimum when $C'(x) = C(x)/x$ or minimum when $MC = $ Average cost.

Miscellaneous Exercises (page 274)

1. f is continuous on $[0, 2]$, apply (4.23); absolute maximum at 1; absolute minimum at 0 and at 2. Without calculus, the graph of f is a semicircle with center at $(1, 0)$ and radius $= 1$.

3. Yes; $c = \frac{9}{4}$　　　　　　　　　　　　　　**5.** f is increasing on an open interval

7. x-intercepts 0, 4; y-intercept 0;　　　　　**9.** (d); all others are true　　**11.** (b)

no symmetry;
no asymptotes;
critical numbers 0, 1;
local minimum at 1;
concave up for all $x \geq 0$;
no inflection points

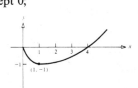

13.

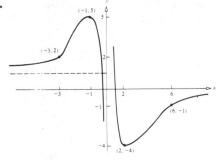

15.

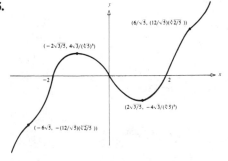

17. $\dfrac{dV}{dt} = kS$ and $V = \frac{4}{3}\pi R^3$

$$\frac{dV}{dt} = 4\pi R^2 \frac{dR}{dt}$$

or　$kS = 4\pi R^2 \dfrac{dR}{dt}$

$$k4\pi R^2 = 4\pi R^2 \frac{dR}{dt}$$

$$k = \frac{dR}{dt}$$

So $\dfrac{dR}{dt}$ is a constant.

19. 0.1 km　　**21.** (2, 2)

23. (a)

x	1	10	100	1000	5000	Guess
$f(x)$	2	2.5937	2.7048	2.7169	2.718	2.718

23. (b)

x	1	10	100	Guess
$f(x)$	2	13,780.6	1.6358×10^{43}	$+\infty$

(c)

x	1	10	100	1000	Guess
$f(x)$	0	0.3487	0.3660	0.36769	0.3678

(d)

x	1	10	100	1000	Guess
$f(x)$	0.84147	-0.054402	-0.0050636	0.00082687	0

(e)

x	1	10	100	Guess
$f(x)$	0.84147	0.0022706	2.7953×10^{-30}	0

25.

Month	Δx	dy	$\sim \sqrt{x}$
2	1	0.5	1.5
3	-1	-0.25	1.75
5	1	0.25	2.25
6	2	0.5	2.5
7	-2	-0.33	2.67
8	-1	-0.166	2.834
10	1	0.166	3.166
11	2	0.33	3.33
12	3	0.5	3.5

27. 49 m/sec; down; 15.7 m

29. $9\sqrt{3}$ m^3/min   **31.** (b) is true

33. (a) $f'(x) = nx^{n-1}(acx^{2n} + 2adx^n - bc)/(cx^n + d)^2$; $f'(x) = 0$ if $x = 0$ or if $acx^{2n} + 2adx^n - bc = 0$ (a disguised quadratic, which can have at most four solutions)   (b) $f(x) = (\frac{1}{2}x^4 - 6)/(2x^2 - 7)$

35. 1.52 sec   **37.** (a) A (b) 0, 1.19   (c) $\frac{2}{3}$ (d) $\frac{2}{3}$

39. (b) $\lim\limits_{n \to +\infty} 2nR \sin(\pi/n) = 2R \lim\limits_{n \to +\infty} \pi(\sin \pi/n)/(\pi/n) = 2\pi R \lim\limits_{x \to 0^+} (\sin \pi x)/(\pi x) = 2\pi R$

CHAPTER 5

Exercise 1 (page 284)

1. $[1, 2], [2, 3], [3, 4]$ **3.** $[-1, -\frac{1}{2}], [-\frac{1}{2}, 0], [0, \frac{1}{2}], [\frac{1}{2}, 1], [1, \frac{3}{2}], [\frac{3}{2}, 2], [2, \frac{5}{2}], [\frac{5}{2}, 3], [3, \frac{7}{2}], [\frac{7}{2}, 4]$

5. $s_4 = 48, S_4 = 56$; $s_8 = 50, S_8 = 54$ **7.** $s_4 = \frac{7}{4}, S_4 = \frac{15}{4}$; $s_8 = \frac{35}{16}, S_8 = \frac{51}{16}$

9. $s_4 = \frac{77}{60}, S_4 = \frac{25}{12}$; $s_8 = \frac{3601}{2520}, S_8 = \frac{4609}{2520}$

Exercise 2 (page 291)

1. $n(n + 1)(n + 2)$ 3. $n^2(n + 3)(n - 1)/4$ 5. $(n + 1)^2/4n^2$ 7. $(1 - n^2)/6n^2$ 9. $\sum_{i=0}^{n-1} (i + 1)^2, \sum_{i=1}^{n} i^2$

11. $\sum_{i=0}^{n} 2^i, \sum_{i=1}^{n+1} 2^{i-1}$ 13. $\sum_{i=0}^{n} \frac{1}{2^i}, \sum_{i=1}^{n-1} \frac{1}{2^{i-1}}$ 15. $\sum_{i=0}^{n-1} (i + 1)(i + 2), \sum_{i=1}^{n} i(i + 1)$ 17. 150   19. 20

21. $\frac{8}{3}$ 23. $\frac{16}{3}$ 25. 4   27. 10   29. 20   31. $\frac{8}{3}$ 33. $\frac{16}{3}$ 35. 4   37. 10

39. $\sum_{i=1}^{n} [(i + 1)^2 - i^2] = \sum_{i=1}^{n} (2i + 1) = 2 \sum_{i=1}^{n} i + \sum_{i=1}^{n} 1 = \frac{2n(n + 1)}{2} + n = n^2 + 2n = (n + 1)^2 - 1$

41. $\sum_{i=1}^{n} \left(\frac{1}{i} - \frac{1}{i + 1}\right) = \left(1 - \frac{1}{2}\right) + \left(\frac{1}{2} - \frac{1}{3}\right) + \cdots + \left(\frac{1}{n} - \frac{1}{n + 1}\right) = 1 - \frac{1}{n + 1}$

43. Let the vertices of the right triangle be $(0, 0)$ $(0, H)$, and $(B, 0)$. The equation of the line determined by $(0, H)$ and $(B, 0)$ is then

$$y = -\frac{H}{B}x + H$$

Choose $\Delta x = B/n$ and right-hand endpoints, that is, $x_i = i\Delta x$. Then,

$$A = \lim_{n \to +\infty} \sum_{i=1}^{n} f(x_i)\Delta x = \lim_{n \to +\infty} \sum_{i=1}^{n} \left[-\frac{H}{B}(i\Delta x) + H\right]\Delta x = \lim_{n \to +\infty} \left[-\frac{BH}{2}\left(\frac{n + 1}{n}\right) + BH\right] = \frac{1}{2}BH$$

45. $m_i = \frac{1}{2}\left[\frac{3(i - 1)}{n} + \frac{3i}{n}\right] = \frac{6i - 3}{2n}$; $\lim_{n \to +\infty} \sum_{i=1}^{n} f(m_i)\Delta x = \lim_{n \to +\infty} \sum_{i=1}^{n} \frac{(6i - 3)^2}{4n^2}\left(\frac{3}{n}\right) = 9$

47. n equal subintervals give $\Delta x = \frac{b - a}{n}$, and the ith subinterval is $\left[a + (i - 1)\left(\frac{b - a}{n}\right), a + i\left(\frac{b - a}{n}\right)\right]$.

$$s_n = \sum_{i=1}^{n}\left[a + (i - 1)\left(\frac{b - a}{n}\right)\right]\left(\frac{b - a}{n}\right) = \frac{b - a}{n}a\sum_{i=1}^{n} 1 + \left(\frac{b - a}{n}\right)^2 \sum_{i=1}^{n}(i - 1) = \frac{b^2 - a^2}{2} - \frac{(b - a)^2}{2n}.$$

Similarly, $S_n = \sum_{i=1}^{n}\left[a + i\left(\frac{b - a}{n}\right)\right]\left(\frac{b - a}{n}\right) = \frac{b^2 - a^2}{2} + \frac{(b - a)^2}{2n}$. Clearly, $s_n < \frac{b^2 - a^2}{2} < S_n$.

49. $2\sum_{i=1}^{n} i = \underbrace{(n + 1) + (n + 1) + \cdots + (n + 1)}_{n \text{ terms}} = n(n + 1)$

Exercise 3 (page 298)

1. $-\frac{7}{2}$ 3. $-\frac{128}{3}$ 5. $\frac{21}{2}$ 7. 2   9. $\frac{11}{4}$ 11. $(1/n^{3/2})\sum_{i=1}^{n} \sqrt{i}$ 13. $\sum_{i=1}^{n} [1/(2n + i)]$

15. $6n \sum_{i=1}^{n} [1/(n + 3i)^2]$

17. $\int_0^4 f(x)\,dx$ equals the sum of four rectangles with bases the intervals $[0, 1]$, $[1, 2]$, $[2, 3]$, $[3, 4]$, namely, $(0 + 1 + 2 + 3) \cdot 1 = 6$.

19. $\int_a^b k\,dx = \lim_{n \to +\infty} \sum_{i=1}^{n} k\frac{b - a}{n} = k(b - a)\lim_{n \to +\infty} \frac{1}{n}\sum_{i=1}^{n} 1 = k(b - a)$

Exercise 4 (page 305)

1. $\frac{7}{2}$ **3.** 1   **5.** $\frac{2}{3}$ **7.** $\frac{1}{2}$ **9.** $-\frac{1}{15}$ **11.** $\frac{5}{3}$ **13.** -3 **15.** $\frac{12}{7}$ **17.** $\frac{20}{3}$ **19.** $(a/5) + b$

21. $(b^3 - a^3)/3 + 2(b^2 - a^2) + 4(b - a)$ **23.** $124/3$ **25.** $\sqrt{x^2 + 1}$ **27.** $(3 + t^2)^{3/2}$ **29.** $f(x)$

31. $6x^2\sqrt{4x^6 + 1}$ **33.** 160   **35.** $\sqrt{3}$ **37.** $2\sqrt{3}/3$ **39.** 12; 32   **41.** $\pi\sqrt{2}/8$; $\pi/4$ **43.** 1; $\sqrt{2}$

45. $F'(x) = G'(x) = f(x)$; $\therefore$ $F(x) = G(x) + C$. If $x = d$, then $F(d) = \int_c^d f(t)\,dt = G(d) + C = \int_d^d f(t)\,dt + C = C$;

$\therefore$ $F(x) - G(x) = \int_c^d f(t)\,dt$

47. $\frac{1}{2}[f(x)]^2$ is an antiderivative of $f(x)f'(x)$. Thus, $\int_a^b f(x)f'(x)\,dx = \frac{1}{2}[f(x)]^2\big|_a^b$

Exercise 5 (page 308)

1. $\int_3^{11} f(x)\,dx - \int_7^{11} f(x)\,dx = \int_3^{11} f(x)\,dx + \int_{11}^7 f(x)\,dx = \int_3^7 f(x)\,dx$

3. $\int_0^4 f(x)\,dx - \int_6^4 f(x)\,dx = \int_0^4 f(x)\,dx + \int_4^6 f(x)\,dx = \int_0^6 f(x)\,dx$

5. 3   **7.** 31   **9.** 17   **11.** On $[0, 1]$, $x \geq x^3$. Apply (5.44).

13. Let $F' = f$. Then kF is an antiderivative of kf. Thus, $\int_a^b kf(x)\,dx = kF(x)\big|_a^b = kF(b) - kF(a) = kF(x)\big|_a^b = k\int_a^b f(x)\,dx$

15. If $f(x) \leq g(x)$ on $[a, b]$, then $g(x) - f(x) \geq 0$ and is continuous on $[a, b]$. By (5.43), $\int_a^b [g(x) - f(x)]\,dx \geq 0$ or $\int_a^b f(x)\,dx \leq \int_a^b g(x)\,dx$.

17. $\frac{11}{6}$

Exercise 6 (page 314)

1. $6x + C$ **3.** $\frac{3}{2}x^2 + C$ **5.** $-\frac{1}{3}t^{-3} + C$ **7.** $\frac{1}{3}x^3 + 2x + C$ **9.** $\frac{1}{8}x^8 + x + C$

11. $x^4 - x^3 + \frac{5}{2}x^2 - 2x + C$ **13.** $2z^{3/2} + \frac{1}{2}z^2 + C$ **15.** $\frac{1}{2}x^2 + \frac{1}{x} + C$ **17.** $\frac{1}{3}u^3 - \frac{1}{2}u^2 + C$

19. $\frac{3}{4}x^4 - \frac{1}{x} + C$ **21.** $\frac{1}{2}t^2 + 2t + C$ **23.** $\frac{1}{12}(2x + 1)^6 + C$ **25.** $\frac{1}{3}(x^2 - 9)^{3/2} + C$

27. $\sqrt{1 + x^2} + C$ **29.** $\frac{2}{5}(x + 3)^{5/2} - 2(x + 3)^{3/2} + C$ **31.** $-\frac{1}{3}\cos 3x + C$ **33.** $-\frac{1}{2}\cos x^2 + C$

35. $\frac{2}{7}(x + 1)^{7/2} - \frac{4}{5}(x + 1)^{5/2} + \frac{2}{3}(x + 1)^{3/2} + C$ **37.** $\frac{2}{3}(x + 1)^{3/2} - 2(x + 1)^{1/2} + C$

39. $\frac{2}{3}(s - 5)^{3/2} + C$ **41.** $\dfrac{-2}{3(1 + \sqrt{x})^3} + C$ **43.** $\frac{2}{5}(x + 1)^{5/2} - 4(x + 1)^{3/2} + C$ **45.** $\frac{4}{9}(4 + t\sqrt{t})^{3/2} + C$

47. $\frac{-2}{21}$ **49.** $\frac{-1}{28}$ **51.** $\frac{-232}{5}$ **53.** $\frac{16}{3}$ **55.** $\frac{300}{7}$ **57.** $2 - \sqrt{3}$ **59.** 1   **61.** $\frac{38}{3}$ **63.** $\frac{2}{3}$

65. (a) $\frac{1}{3}(x + 1)^3 + C_1$ (b) $\frac{1}{3}x^3 + x^2 + x + C_2$; $C_2 = C_1 + \frac{1}{3}$

67. $\int x\sqrt{x}\,dx = \frac{2}{5}x^{5/2} + C_1$; $\int x\,dx = \frac{1}{2}x^2 + C_2$; $\int \sqrt{x}\,dx = \frac{2}{3}x^{3/2} + C_3$

69. $\int \dfrac{x^2 - 1}{x - 1}\,dx = \frac{1}{2}x^2 + x + C_1$; $\int (x^2 - 1)\,dx = \frac{1}{3}x^3 - x + C_2$; $\int (x - 1)\,dx = \frac{1}{2}x^2 - x + C_3$

71. $\frac{1}{2}$ **73.** Since $\sin\theta = \cos(\pi/2 - \theta)$, let $\phi = \pi/2 - \theta$.   **75.** 8

77. $\int_{-a}^a f(x)\,dx = \int_{-a}^0 f(x)\,dx + \int_0^a f(x)\,dx = \int_a^0 f(-u)(-du) + \int_0^a f(x)\,dx = \int_0^a f(u)\,du + \int_0^a f(x)\,dx = 2\int_0^a f(x)\,dx$

$u = -x$

Chapter 6

Historical Exercises (page 317)

1. (a) 20 ft (b)

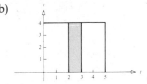

3. (a) Area $ABE = \frac{1}{2}(t_0)(kt_0) = kt_0^2/2$; Area $ABCD = (t_0)(kt_0/2) = kt_0^2/2$ (b) Area $ABE = \int_0^{t_0} kt\,dt = \frac{1}{2}kt^2\Big|_0^{t_0} = \frac{1}{2}kt_0^2$

5. Area $A = \int_a^b [f(x) - g(x)]\,dx = \int_a^b M[h(x) - k(x)]\,dx = M\int_c^b [h(x) - k(x)]\,dx = M(\text{Area } B)$

7. (a) Area $R_1 = ab$; Area $R_2 = AB$; $ka = A$; $kb = B$; Area $R_1/\text{Area } R_2 = ab/AB = ab/A(kb) = a/A(A/a) = a^2/A^2$
(b) Area $T_1 = \frac{1}{2}bh$; Area $T_2 = \frac{1}{2}BH$; $kb = B$; $kh = H$; Area $T_1/\text{Area } T_2 = bh/BH = bh/B(kh) = b/Bk = b/B(B/b) = b^2/B^2$

Miscellaneous Exercises (page 320)

1. $\frac{1}{12}[(z^2 + 1)^4 - 3]^{3/2} + C$ **3.** $\frac{1}{14}[2\sqrt{x^2 + 3} - (4/x) + 9]^7 + C$ **5.** $\frac{8}{3}(3 - \sqrt{2})$ **7.** $b = \sqrt[3]{2}$

9. -2 **11.** $\frac{1}{12}$ **13.** 0.6912 **15.** Because $\int_0^{\pi/2} \cos x\,dx = 1$ **17.** $k = \sqrt{2}$ **19.** All nonnegative n

21. 9   **23.** $\frac{5}{2}$ **25.** (a) 2 (b) $\pi^2/4$ **27.** $-(x - 1)^2$ **29.** $F'(x) = G'(x) = \sqrt{x}; \frac{2}{3}$ **31.** $a = 4$

33. $2\int_0^{\pi/4} \sin^2 y\,dy$ **35.** I and II   **37.** $2(\sqrt{r} - 1); +\infty$ **39.** 0; 84   **41.** $\sqrt[n]{2}$

43. By (5.28) there is u in $[a, b]$ so that $f'(u) = \dfrac{1}{b - a}\int_a^b f'(x)\,dx = \dfrac{f(b) - f(a)}{b - a} = \text{average slope}$

45. No; a different constant of integration may occur.   **47.** Use $f(x) = \begin{cases} 3 & \text{if } 0 \le x \le 1 \\ -3 & \text{if } 1 < x \le 2 \end{cases}$

49. f is odd on $[-a, a]$ implies $f(-x) = -f(x)$;

$$\int_{-a}^a f(x)\,dx = \int_{-a}^0 f(x)\,dx + \int_0^a f(x)\,dx = \int_a^0 f(-u)(-du) + \int_0^a f(x)\,dx = \int_0^a -f(u)\,du + \int_0^a f(x)\,dx = 0$$

51. (a) An extension of (5.39) (b) Mean value theorem for integrals (c) Mean value theorem for derivatives; combine sums (d) Triangle inequality (e) $|f'(x)| \le M$; u_i, t_i in (x_{i-1}, x_i) (f) $\Delta x_i = (b - a)/n$

53. Estimate 0.2; actual 0.048

CHAPTER 6

Exercise 1 (page 331)

1. $A = \frac{1}{2}$

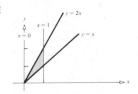

3. $A = \frac{1}{6}$

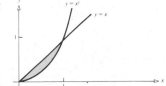

5. $A = \frac{1}{6}$

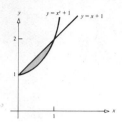

7. $A = \frac{5}{12}$

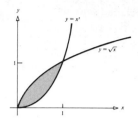

9. $A = \frac{4}{15}$

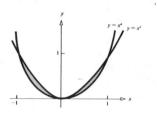

11. $A = \frac{8}{3}$

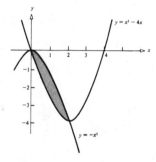

13. $A = \frac{9}{2}$

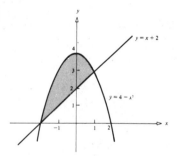

15. $A = 8$

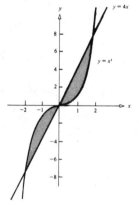

17. $A = \frac{1}{3}$

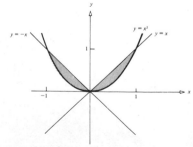

19. $A = 12$

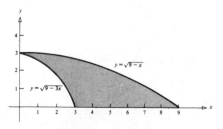

21. $A = \dfrac{125}{6}$

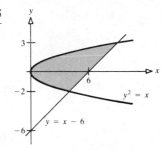

23. $A = \dfrac{\pi}{6} - \dfrac{1}{2}$

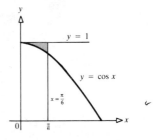

25. $A = 12$

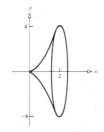

27. $A = \dfrac{9}{2}$　　**29.** $A = \dfrac{125}{24}$　　**31.** $A = \dfrac{1}{6}$　　**33.** $\sqrt[3]{\dfrac{1}{2}}$

35. Shaded area $= \dfrac{9}{2}$; area of parallelogram $= \dfrac{27}{4}$

Exercise 2 (page 343)

1. $V = 32\pi/5$

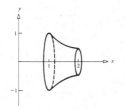

3. $V = 128\pi/7$

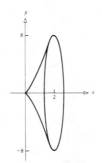

5. $V = 30\pi$

7. $V = \pi/2$

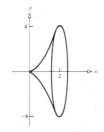

9. $V = 15\pi/2$

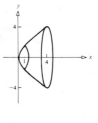

11. $V = 3\pi/4$

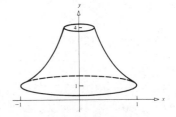

13. $V = 20\pi/7$ **15.** $V = 2\pi/15$ **17.** $V = 2\pi/15$

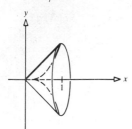   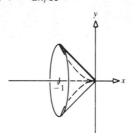   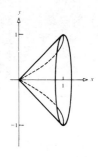

19. $V = 128\pi/5$ **21.** $V = 64\pi/5$ **23.** $V = 32\pi$ **25.** $V = \pi/6$ **27.** $V = 256\pi/15$

Exercise 3 (page 350)

1. 32π **3.** $334\pi/5$ **5.** $\pi/10$ **7.** $768\pi/7$ **9.** $4\pi/5$ **11.** $2\pi/15$ **13.** 128π **15.** $4\pi/15$
17. $308\pi/3$ **19.** 9π **21.** $1177\pi/10$ **23.** $V \approx 53{,}135{,}993 \text{ ft}^3$

Exercise 4 (page 355)

1. 4   **3.** $\frac{16}{3}$ **5.** $\pi/40$ **7.** $\frac{256{,}000}{3} \text{ m}^3$ **9.** $37\pi/1320$ **11.** $V = \pi \int_0^h \left(\frac{R}{h}x\right)^2 dx = \frac{\pi R^2}{h^2} \frac{x^3}{3}\Big|_0^h = \frac{\pi}{3} R^2 h$

13. $\pi h^2 \left(R - \dfrac{h}{3}\right)$ **15.** $\dfrac{250\sqrt{3}}{9}$

Exercise 5 (page 360)

1. $2\sqrt{10}$ **3.** $\sqrt{13}$ **5.** $\frac{1}{27}(80\sqrt{10} - 13\sqrt{13})$ **7.** $\frac{1}{27}(80\sqrt{10} - 8)$ **9.** $\frac{2}{3}(2\sqrt{2} - 1)$ **11.** 45
13. 21   **15.** $\frac{33}{16}$ **17.** $\frac{1}{27}(13\sqrt{13} - 8)$ **19.** $\frac{2}{27}(10\sqrt{10} - 1)$ **21.** $6a$ **23.** $(31\sqrt{31} - 13\sqrt{13})/27$

25. $\frac{8}{243}(82^{3/2} - 1) + 4\sqrt{37}$ **27.** $\frac{17}{12}$ **29.** $\int_0^2 \sqrt{1 + 4x^2}\, dx$ **31.** $5\int_0^4 \dfrac{dx}{\sqrt{25 - x^2}}$

33. $\dfrac{1}{a}\int_0^{a/2} \sqrt{\dfrac{a^4 - (a^2 - b^2)x^2}{a^2 - x^2}}\, dx$

35. (a) $\int_{x_2}^{x_1} \dfrac{1}{\sqrt{1 - x^2}}\, dx$ (b) $2\pi - \int_{x_2}^{x_1} \dfrac{1}{\sqrt{1 - x^2}}\, dx$ (c) $\int_0^{y_1} \dfrac{1}{\sqrt{1 - y^2}}\, dy + \dfrac{\pi}{2} + \int_0^{x_2} \dfrac{1}{\sqrt{1 - x^2}}\, dx$

Exercise 6 (page 368)

1. $\int_0^2 \frac{3}{2}x\, dx = 3 \text{ J}$ **3.** $\frac{10}{3} \text{ J}$ **5.** 10 ft   **7.** $72\pi\rho g \approx 2{,}216{,}707 \text{ J}$ **9.** $\frac{64}{3}\pi\rho g \approx 656{,}802 \text{ J}$
11. $36\pi\rho g \approx 1{,}108{,}354 \text{ J}$ **13.** $10{,}500\rho g \approx 656{,}250 \text{ ft-lb}$ **15.** $1.362 \times 10^8 \text{ J}$ **17.** $-m/2a$
19. 1500 J   **21.** 4950 ft-lb   **23.** 209,210 in.-lb

Exercise 7 (page 373)

1. $36\rho g \approx 352{,}800$ N   **3.** $360\rho g \approx 22{,}488$ lb   **5.** $\frac{14}{3}\rho g \approx 45{,}733$ N   **7.** $\frac{16}{3}\rho g \approx 52{,}267$ N
9. $\rho g/4 \approx 2450$ N   **11.** $\frac{128}{3}\rho g \approx 2{,}555.39$ lb

Exercise 8 (page 377)

1. $\frac{1}{3}$ **3.** $\frac{2}{3}$ **5.** 9   **7.** -26 **9.** $2/\pi$ **11.** $\frac{5}{2}g \approx 24.5$ m/sec; $\frac{10}{3}g \approx 32.7$ m/sec   **13.** $37.5°C$
15. 12 m/sec   **17.** $(15 - 4\sqrt{3})/6$ kg/m³

Miscellaneous Exercises (page 377)

1. $\frac{14}{3}$ **3.** $\frac{4}{3}$ **5.** $\frac{2^{14}}{5} = 3276.8$ **7.** $\pi/2$ **9.** 4   **11.** $a^2/6$
13. (a) $b < \frac{1}{2}$ (b) $A = 9$ (c) $V = 2\pi/3$ **15.** 8π
17. Let A be enclosed by $y = f(x)$, by $y = g(x)$ (assume $f(x) \geq g(x)$ for all x in $[a, b]$), and by $x = a$ and $x = b$.
Then $V = 2\pi \int_a^b x[f(x) - g(x)]\, dx$. Revolved about $x = -k$ $(k > 0)$,

$$V_1 = 2\pi \int_a^b (k + x)[f(x) - g(x)]\, dx = 2\pi k \int_a^b [f(x) - g(x)]\, dx + 2\pi \int_a^b x[f(x) - g(x)]\, dx = 2\pi k A + V$$

19. 1.6 J   **21.** $(8, 32\sqrt{2}/3)$ **23.** (a) 10, 20, 40 N   (b) 0.5, 2, 8 J
25. $\pi = 2^n \sin \dfrac{45}{2^{n-2}} = 2^n \sqrt{\dfrac{1 - \cos(45/2^{n-3})}{2}}$; set $n = m + 2$: $\pi \approx 2^{m+2} \sqrt{\dfrac{1 - \cos(45/2^{m-1})}{2}}$ **27.** $F(x) = x^2/4$

CHAPTER 7

Exercise 2 (page 391)

1. $a + b$ **3.** $\frac{1}{2}b$ **5.** $-2a - b$ **7.** $b - a$ **9.**   **11.**

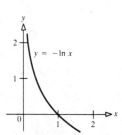

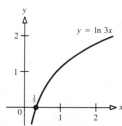

13. $1/x$ **15.** $\dfrac{2x^2}{x^2 + 4} + \ln(x^2 + 4)$ **17.** $\dfrac{x^2}{x^2 + 1} + \dfrac{1}{2}\ln(x^2 + 1)$ **19.** $\dfrac{1}{1 - x^2}$ **21.** $\dfrac{1}{x \ln x}$

23. $\dfrac{1}{x(x^2 + 1)}$ **25.** $\dfrac{2x^4 - 5x^2 + 1}{x(x^4 - 1)}$ **27.** $\cot x$ **29.** $\dfrac{1}{2x(\ln x)^{1/2}}$ **31.** $-x \tan 2x + \frac{1}{2}\ln(\cos 2x)$

33. $\dfrac{1}{\sqrt{x^2 + 4}}$ **35.** $\dfrac{1}{\sqrt{x^2 + a^2}}$ **37.** $4x(x^2 + 1)(2x^3 - 1)^3(8x^3 + 6x - 1)$

39. $(x^3 + 1)(x - 1)(x^4 + 5)\left(\dfrac{3x^2}{x^3 + 1} + \dfrac{1}{x - 1} + \dfrac{4x^3}{x^4 + 5}\right)$ **41.** $\dfrac{x^2(x^3 + 1)}{\sqrt{x^2 + 1}}\left(\dfrac{2}{x} + \dfrac{3x^2}{x^3 + 1} - \dfrac{x}{x^2 + 1}\right)$

43. $\dfrac{x \cos x}{(x^2 + 1)^3 \sin x}\left(\dfrac{1}{x} - \tan x - \dfrac{6x}{x^2 + 1} - \cot x\right)$ **45.** $-\dfrac{y(x \ln y + y)}{x(y \ln x + x)}$ **47.** y/x **49.** $y = 5x - 1$

51. $t > 1 \Rightarrow t > \sqrt{t} \Rightarrow 1/t < 1/\sqrt{t} \Rightarrow \ln x = \int_1^x dt/t < \int_1^x dt/\sqrt{t} = 2(\sqrt{x} - 1),\ x > 1$ **53.** 0

55. $f'(x) > 0$ for $x > 0$, $f'(x) < 0$ for $x < 0$, and $f(0) = 0$. Hence, $-x + \ln(1 + x)^{1+x} \geq 0$ for $x > -1$ and $\ln(1 + x)^{1+x} \geq x$.

57. Local minimum at $(1/e, -1/e)$ **59.** 2   **61.** e **63.** $1/\sqrt{e} \approx 0.607$

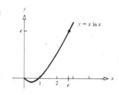

Exercise 3 (page 398)

1. $x - x^2$ **3.** $1/x$ **5.** $-x^2$ **7.** xe^x **9.** $\ln 2$ **11.** $x = \pm e^{-5/2}$ **13.** $x = 0$ or $x = \ln 2$

15. $+\infty$ **17.** 0   **19.** 1   **21.** 1   **23.** $15e^{3x}$ **25.** $\dfrac{e^x - e^{-x}}{2}$ **27.** $e^{-3x}\left(\dfrac{1}{x} - 3 \ln 2x\right)$

29. $e^x(\sin x + \cos x)$ **31.** $e^{ax}(b \cos bx + a \sin bx)$ **33.** $\dfrac{xe^{\sqrt{x^2 - 9}}}{\sqrt{x^2 - 9}}$ **35.** $\dfrac{-e^{1/x}}{x^2}$ **37.** $\dfrac{2ae^{ax}}{(e^{ax} + 1)^2}$

39. $\dfrac{e^{x+y}}{1 - e^{x+y}}$ **41.** $\dfrac{1 + ye^x}{1 - e^x}$ **43.** $\dfrac{9900e^{-x}}{(1 + 99e^{-x})^2}$ **45.** $-2xe^{x^2} \sin e^{x^2}$ **47.** $e^x \cot e^x$ **49.** $ye^x \ln y$

51. $\dfrac{e^y \sin x - e^x \sin y}{e^x \cos y + e^y \cos x}$ **53.** $a^n e^{ax}$ **55.** $y'' = 4e^{2x}$ **57.** $y'' = 4Ae^{2x} + 4Be^{-2x}$

59. $y' = 2Ae^{2x} + 3Be^{3x};\ y'' = 4Ae^{2x} + 9Be^{3x}$ **61.** $e^{-at}[(A\omega - Ba) \cos \omega t - (B\omega + Aa) \sin \omega t]$

63. No local extrema, no points of inflection, and concave up everywhere

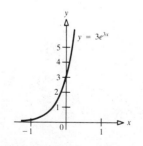

65. Local maximum at $x = 1$, points of inflection at $x = \sqrt{2}/2$ and $x = -\sqrt{2}/2$, concave up on $(-\infty, -\sqrt{2}/2)$, concave down on $(-\sqrt{2}/2, \sqrt{2}/2)$, and concave up on $(\sqrt{2}/2, +\infty)$

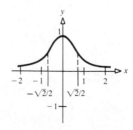

67. No local extrema, an inflection point at $x = -\frac{1}{2}$, concave down on $(-\infty, -\frac{1}{2})$, and concave up on $(-\frac{1}{2}, 0)$ and $(0, +\infty)$

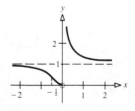

69. Absolute maximum $= 1$ and occurs at $x = 0$. Absolute minimum $= e - 3$ and occurs at $x = 1$.

71. (a) $-2(\sin x)e^{\cos x}$; $-2e^{\cos x}(\cos x - \sin^2 x)$ (b) $-\frac{5}{2}$ **73.** $di/dt = -Ri/L$

75. $f(0) = 0 \; \therefore f(x) > f(0) = 0$ if $x > 0$ **77.** $c = -\ln 3$

79. Tangent line at $x = x_0$; $y - e^{x_0} = e^{x_0}(x - x_0)$. Solving for $y = 0$, $x = x_0 - 1$.

81. Using the squeezing theorem and $1 < x^2 e^{-1/x^2}$ for $x \neq 0$, $f'(x) = \dfrac{2}{x^3} e^{-1/x^2}$; for $x = 0$, $f'(x) = \lim\limits_{h \to 0} \dfrac{1}{he^{1/h^2}}$,

where $0 \geq \dfrac{1}{h^2 e^{1/h^2}} \geq h$. Hence, $f'(0) = 0$.

83. $11e^{10}$ square units per second   **85.** $(\sqrt{2}/2, e^{-1/2})$ and $(-\sqrt{2}/2, e^{-1/2})$

87.

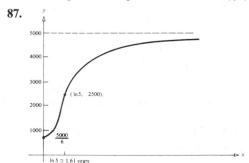

89. $\ln(e^\alpha/e^\beta) = \ln e^\alpha - \ln e^\beta = \alpha - \beta = \ln e^{\alpha - \beta}$

$\ln(e^\alpha/e^\beta) = \ln e^{\alpha - \beta}$

$e^\alpha/e^\beta = e^{\alpha - \beta}$

Exercise 4 (page 404)

1. $\frac{1}{3}e^{3x} + C$ **3.** $\frac{1}{3}e^{3x+1} + C$ **5.** $\frac{1}{2}x^2 + \frac{1}{7}e^{7x} + C$ **7.** $\frac{1}{3}\ln|3x - 1| + C$ **9.** $\frac{1}{2}\ln|x^2 - 1| + C$

11. $\frac{1}{2}[\ln(e^4 + 1) - \ln 2]$ **13.** $\frac{1}{3}(e^2 - e)$ **15.** $-e^{1/x} + C$ **17.** $\ln|e^x - e^{-x}| + C$ **19.** $4(e^x - 1)^{3/4} + C$

21. $\ln(\ln 3) - \ln(\ln 2)$ **23.** $\ln|\ln x| + C$ if $n = 1$; $(\ln x)^{1-n}/(1-n) + C$ if $n \neq 1$ **25.** $\sin e^\pi - \sin 1$

27. $-4\ln(1 + e^{-x}) + C$ **29.** $\frac{1}{3}(\ln 2 + 1)$ **31.** $2\ln(1 + \sqrt{x}) + C$ **33.** $\frac{1}{2}\ln|2 \sin x - 1| + C$

35. $\frac{1}{2}$ **37.** $\ln 2$ **39.** $\frac{1}{3}(e^6 - 3e^2 + 2)$ **41.** $(\pi/2)(1 - e^{-4})$ **43.** $e - 1$

Exercise 5 (page 409)

1. $\sqrt{2}x^{\sqrt{2}-1}$ **3.** $2\sqrt{2}x(1 + x^2)^{\sqrt{2}-1}$ **5.** $(\ln \frac{1}{2})(\frac{1}{2})^x$ **7.** $1/(x \ln 2)$ **9.** $2x/[(\ln 2)(1 + x^2)]$

11. $2^{-x}[\cos x - (\ln 2)\sin x]$ **13.** $(3x)^x[1 + \ln(3x)]$ **15.** $2(\ln x)x^{\ln x - 1}$ **17.** $\dfrac{(3x)^{\sqrt{x}}}{2\sqrt{x}}(\ln 3x + 2)$

19. $e^x x^{e^x}(\ln x + \frac{1}{x})$ **21.** $2^{\sin x}(\cos x)(\ln 2)$ **23.** $x^{\sin x}\left[\dfrac{\sin x}{x} + (\ln x)(\cos x)\right]$

25. $(\sin x)^{\cos x + 1}[\cot^2 x - \ln(\sin x)]$ **27.** $\dfrac{1 - xy \ln 2}{x^2 \ln 2}$ **29.** $\dfrac{2^x}{\ln 2} + C$ **31.** $\dfrac{2^{3x+5}}{3 \ln 2} + C$

33. $\dfrac{1}{3\pi + 1}(2^{3\pi+1} - 1)$ **35.** $(\ln 3)(\ln 2)$ **37.** $\ln 2$ **39.** By (7.34), $\log_b a = \dfrac{\ln a}{\ln b} = \dfrac{1}{(\ln b)/(\ln a)} = \dfrac{1}{\log_a b}$

41. $e^{x_1} < \quad a < e^{x_2}$

$\ln e^{x_1} < \ln a < \ln e^{x_2}$ (ln is increasing)

$x_1 < \ln a < x_2$

43. $y = f(x)^{g(x)}$

$\ln y = g(x) \ln f(x)$

$\dfrac{y'}{y} = g(x)\dfrac{f'(x)}{f(x)} + \ln f(x)g'(x)$

or $y' = g(x)f(x)^{g(x)-1}f'(x) + f(x)^{g(x)}\ln f(x)g'(x)$

45. $(xy)^\alpha = e^{\alpha \ln xy} = e^{\alpha(\ln x + \ln y)} = e^{\alpha \ln x} \cdot e^{\alpha \ln y} = x^\alpha \cdot y^\alpha$

Exercise 6 (page 413)

1. e^2 **3.** $e^{1/3}$ **5.** \$530.92; \$531.99; $6\frac{1}{4}\%$ is better   **7.** \$941.76; \$942.18   **9.** 83.18 months; 84.2 months

Exercise 7 (page 420)

1. $\dfrac{y^3}{3} = \dfrac{x^2}{2} + \dfrac{x^4}{4} + C$ **3.** Not separable   **5.** Not separable   **7.** $(\ln y)^2 = 3x^2 + C$

9. Not separable   **11.** $y + 3\ln|y| = \dfrac{x^3}{3} - x^2 + x + C$ **13.** $y = 10e^{x^3} - 2$ **15.** $\dfrac{1}{2}y^2 = \dfrac{x^3}{3} + 2x - \dfrac{11}{6}$

17. $y^2 + 4y = x^2 - 2x + 57$ **19.** $-e^{-y} = \ln|x| - 1$ **21.** $y^2 + 8y = -3x^2 + 45$

23. 7.678 g   **25.** Approximately 9950 years ago   **27.** 6944   **29.** $7500e^2$

31. (a) $10,000e^{(\ln 5)t/10}$ (b) 250,000   (c) 4.3 min   **33.** 4.6×10^6 years   **35.** 16.5°C

37. (a) 625.3 mb   (b) 303 mb   (c) 429 mb   (d) 348 mb   (e) 57,396 m

Exercise 8 (page 429)

1. $y = \dfrac{3}{2}x - \dfrac{3}{2}\left(\dfrac{1}{x}\right)$ **3.** $y = (x + C)e^{-x^2}$ **5.** $y = C\cos x - \cos x \ln(\cos x)$ **7.** $y = C_1e^{2x} + C_2e^{-6x}$

9. $y = C_1e^{-4x} + C_2xe^{-4x}$ **11.** $y = C_1\cos(\sqrt{2}x) + C_2\sin(\sqrt{2}x)$ **13.** $y = C_1e^{-11x} + C_2xe^{-11x}$

15. $y = C_1e^{-4x} + C_2e^x$ **17.** $y = C_1\cos(5x) + C_2\sin(5x)$ **19.** $y = C_1 + C_2e^{-6x}$ **21.** $y = e^{-2x} + 4xe^{-2x}$

23. $y = -2\cos(3x) + 2\sin(3x)$ **27.** $q = EC(1 - e^{-t/RC})$ **29.** 18.96 lb, 30 lb

Miscellaneous Exercises (page 430)

1. $\dfrac{2y}{y^2 + 1}$ **3.** $\dfrac{1}{a}x^{(1/a)-1} - \dfrac{(\ln a)a^{1/x}}{x^2}$ **5.** $2\cot(2x)$ **7.** $\dfrac{2(x-1)}{x(x-2)}$ **9.** $\dfrac{-4x}{x^4 - 1}$ **11.** $\dfrac{-1}{2\sqrt{x}\sqrt{x+a}}$

13. $-\frac{1}{2}(1 + 3\ln 2)$ **15.** $\ln|e^{x/2} - e^{-x/2}| + C$ or $\ln|e^x - 1| + \ln|1 - e^{-x}| + k$

17. $(\ln 3)\dfrac{(\log_3 15)^2}{2}$

19. (a) $\dfrac{d}{dx}\ln ax = \dfrac{1}{x}$ (b) $f' = g' \Rightarrow f = g + C$; set $x = a$

21. Domain: all $x > 0$;
$\lim_{x \to 0^+} (x^3 - 3 \ln x) = +\infty$ (vertical asymptote at $x = 0$);
local minimum at $(1, 1)$;
concave up for all $x > 0$

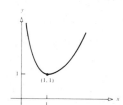

23. $f(x) = e^{1-x}$ **25.** $f(x) = \frac{1}{2}(e^x + e^{-x})$

27. (a) $f(0 + 0) = 1f(0) + 1f(0)$
$f(0) = 2f(0)$
$0 = f(0)$

(b) $2 = f'(0) = \lim_{h \to 0} \dfrac{f(0 + h) - f(0)}{h} = \lim_{h \to 0} \dfrac{e^h f(0) + e^0 f(h) - 0}{h} = \lim_{h \to 0} \dfrac{e^h(0) + f(h)}{h} = \lim_{h \to 0} \dfrac{f(h)}{h}$;

so $\lim_{x \to 0} \dfrac{f(x)}{x} = 2$ (let x replace h)

(c) $f'(x) = \lim_{h \to 0} \dfrac{f(x + h) - f(x)}{h} = \lim_{h \to 0} \dfrac{[e^h f(x) + e^x f(h)] - f(x)}{h} = f(x) \lim_{h \to 0} \dfrac{e^h - 1}{h} + 2e^x$;

$\lim_{h \to 0} \dfrac{e^h - 1}{h} = 1$; so $f'(x) = f(x) + 2e^x$

(d) By inspection of part (c), $p = 2$.

29. (a) $1 + 3 \ln k$ (b) e^2 (c) 25 units per second   **31.** $1/e$ **33.** $\pi(1 - e^{-1/2})$ **35.** (d) is true

37. By the mean value theorem there is c in (a, b) such that $f'(c) = \dfrac{f(b) - f(a)}{b - a}$ or $\dfrac{1}{c} = \dfrac{\ln b - \ln a}{b - a}$.

Now c is in (a, b), so $\dfrac{1}{a} > \dfrac{1}{c} > \dfrac{1}{b}$, and so $\dfrac{1}{b} < \dfrac{1}{c} = \dfrac{\ln b - \ln a}{b - a} < \dfrac{1}{a}$ or $b > \dfrac{b - a}{\ln b - \ln a} > a$

and $b(\ln b - \ln a) > b - a > a(\ln b - \ln a)$

$$a \ln \frac{b}{a} < b - a < b \ln \frac{b}{a}$$

39. (a) Absolute maximum at $x = 1/e$; absolute minimum at $x = 1$ **41.** $\frac{1}{2}$ **43.** e^c **45.** $0; -\infty$
(b) Concave up on $[1/e, 2]$
(c)

47. Local minimum at $(-1, -e^{-1})$;
inflection point at $(-2, -2e^{-2})$

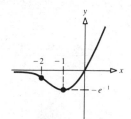

49. $y = 10e^{x-3}$ **51.** 3.57715 and 0.11183 **55.** (a) $nRT \ln(V_2/V_1)$ (b) $[k/(1 - \gamma)](V_2^{1-\gamma} - V_1^{1-\gamma})$

57. $y = 4\sqrt{x} - 2$ **59.** $y = C_1 e^{x/2} + C_2 e^{-2x}$ **61.** $y = C_1 e^{2x} + C_2 e^{-2x}$

63. $y = C_1 e^{(1+\sqrt{6}/2)x} + C_2 e^{(1-\sqrt{6}/2)x}$

CHAPTER 8

Exercise 1 (page 440)

1. $4 \sec 4x \tan 4x$ **3.** $5 \sec^2 5x$ **5.** $e^x(\sec x)(1 + \tan x)$ **7.** $e^{-4x}(3 \cos 3x - 4 \sin 3x)$

9. $\pi e^{\pi x}(\sec^2 \pi x + \tan \pi x)$ **11.** $2x \cos(x^2)$ **13.** $4 \sec^2 x \tan x$

15. $2(x + 1) \sec(x^2 + 2x - 1) \tan(x^2 + 2x - 1)$ **17.** $-4 \csc^2(1 + 2x) \cot(1 + 2x)$

19. $2x(2x \cos 4x + \sin 4x)$ **21.** $-\dfrac{\cos(1/x)}{2x^2 \sqrt{\sin(1/x)}}$ **23.** $x \sec^2 x + \tan x$ **25.** $2x \sin x(x \cos x + \sin x)$

27. $\sec x(\sec^2 x + \tan^2 x)$ **29.** $\dfrac{(1 + x) \sec x \tan x - \sec x}{(1 + x^2)}$ **31.** $\dfrac{1 - \sec y - y \sec^2 x}{x \sec y \tan y + \tan x}$

33. $\dfrac{3 - 2y \sec^2(xy) \tan(xy)}{2x \sec^2(xy) \tan(xy)}$ **35.** $2(3x^2 + 1) \sec^2(x^3 + x) \tan(x^3 + x)$ **37.** $-2 \csc^2 x \cot x e^{\csc^2 x}$

39.

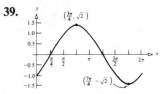

41. One period:

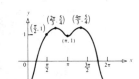

43.

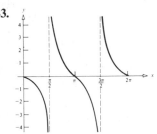

45. $R = \dfrac{4\sqrt{6}}{3}; h = \dfrac{4\sqrt{3}}{3}$ **47.** $\dfrac{\sqrt{3}}{3}$ cm²/min **49.** $\left(\dfrac{11\pi}{120}\right) \cos\left(\dfrac{11\pi}{36}\right) \approx 0.165$ in.²/min

51. $\dfrac{d}{dx}(\csc x) = \dfrac{d}{dx}(\sin x)^{-1} = (-1)(\sin x)^{-2}(\cos x) = -\left(\dfrac{1}{\sin x}\right)\left(\dfrac{\cos x}{\sin x}\right) = -\csc x \cot x$

Exercise 2 (page 444)

1. 1 **3.** $\frac{1}{5} \ln|\sec 5x + \tan 5x| + C$ **5.** $\frac{1}{2} \ln|\sec 2x| + C$ **7.** $\frac{1}{4} \sec 4x + C$ **9.** $\frac{1}{2} \tan 2x + C$

11. $\frac{2}{3}(\tan x)^{3/2} + C$ **13.** $\frac{1}{5} \sin 5x + C$ **15.** $\frac{1}{3} \sec(3x - 1) + C$ **17.** $\sec x + C$ **19.** $(\pi + 4)/4$

21. $-e^{\cos x} + C$ **23.** π **25.** $(2/\pi) \ln 2$

27. $\int \csc x \, dx = \int \dfrac{\csc^2 x - \csc x \cot x}{\csc x - \cot x} \, dx = \ln|\csc x - \cot x| + C$ **29.** $C_2 = \frac{1}{2} + C_1; C_3 = C_1 + \frac{1}{4}$

Exercise 3 (page 450)

1. $\pi/4$ **3.** $-\pi/2$ **5.** 0 **7.** 0 **9.** $\frac{1}{2}$ **11.** 0 **13.** $\frac{4}{5}$ **15.** $2\sqrt{6}$ **17.** $\pi - 2$

19. $(2 + \sqrt{40})/9$ **29.** $\alpha + \beta = \pi/2$; $\cot \alpha = \tan \beta = x$; $\alpha = \cot^{-1} x$, $\beta = \tan^{-1} x$; $\therefore \cot^{-1} x + \tan^{-1} x = \pi/2$

Exercise 4 (page 457)

1. $4/\sqrt{1-16x^2}$ 3. $-2x/\sqrt{1-x^4}$ 5. $-5/(5x+1)\sqrt{(5x+1)^2-1}$ 7. 0   9. $-1/(1+x^2)$

11. $2/(8x^2-4x+1)$ 13. $-1/2x\sqrt{x-1}$ 15. $-2x/\sqrt{2x^2-x^4}$ 17. $2(1+\tan^{-1}x)/(1+x^2)$

19. $-x/\sqrt{x^2-x^4}$ 21. $-2x^2/(x^4+2x^2+2)+\cot^{-1}(1+x^2)$ 23. $1/[\sqrt{x}(x+1)]$

25. $(-\sin x)/\sqrt{\sin^2 x}$ 27. $1/x[1+(\ln x)^2]$ 29. $1/[(1+x^2)\tan^{-1}x]$ 31. $(\cos x)/(1+\sin^2 x)$

33. $\tan^{-1}(x/a)$ 35. $\sqrt{1-y^2}/[\sqrt{1-x^2}(1-\sqrt{1-y^2})]$ 37. $5\pi/4$ 39. $\frac{1}{5}\tan^{-1}(x/5)+C$

41. $\sin^{-1}(x/3)+C$ 43. $\frac{1}{3}\sin^{-1}(3x/4)+C$ 45. $\tan^{-1}(\sin x)+C$ 47. $-\sin^{-1}(\cos x/2)+C$

49. $2\sin^{-1}(2x^2-1)+C$ 51. $\dfrac{2}{\sqrt{5}}\tan^{-1}\dfrac{2x+1}{\sqrt{5}}+C$ 53. $\frac{1}{4}\ln(2x^2+2x+3)-\dfrac{1}{2\sqrt{5}}\tan^{-1}\dfrac{2x+1}{\sqrt{5}}$

55. $\sin^{-1}(x-1)+C$ 57. $\pi/3$ 59. $\dfrac{1}{\sqrt{5}}\tan^{-1}\sqrt{5}$ 61. $\dfrac{d}{dx}\tan^{-1}\left(\dfrac{1}{x}\right)=\dfrac{-1/x^2}{1+(1/x)^2}=\dfrac{-1}{1+x^2}=\dfrac{d}{dx}\cot^{-1}x$

63. $\dfrac{d}{dx}\tan^{-1}(\cot x)=\dfrac{-\csc^2 x}{1+\cot^2 x}=\dfrac{-\csc^2 x}{\csc^2 x}=-1$ 65. 0.0366 radian/sec

67. $y'=\dfrac{1}{\sqrt{1-x^2}}-\dfrac{1}{\sqrt{1-x^2}}=0 \Rightarrow \sin^{-1}x+\cos^{-1}x=C \Rightarrow$ if $x=0$, then $C=\pi/2$

69. $N=\sqrt{1-\dfrac{4}{\pi^2}}\approx 0.77$ 71. $\displaystyle\int \dfrac{dx}{\sqrt{a^2-x^2}}=\dfrac{1}{a}\int\dfrac{dx}{\sqrt{1-(x/a)^2}}\underset{\substack{\uparrow\\u=x/a}}{=}\dfrac{1}{a}\int\dfrac{a\,du}{\sqrt{1-u^2}}=\sin^{-1}\left(\dfrac{x}{a}\right)+C$

Exercise 5 (page 462)

1. $\frac{5}{3}$ 3. $-2\sin(2t)$ 5. $\cos(t/2)$ 7. $\theta(t)=\theta_0\cos[\sqrt{2k/m}(t/R)]$; $T=2\pi R\sqrt{m/2k}$

Exercise 6 (page 468)

1. $\tanh^2 x+\operatorname{sech}^2 x=\dfrac{\sinh^2 x}{\cosh^2 x}+\dfrac{1}{\cosh^2 x}=\dfrac{1+\sinh^2 x}{\cosh^2 x}=\dfrac{\cosh^2 x}{\cosh^2 x}=1$

3. $\sinh(-A)=\dfrac{e^{-A}-e^A}{2}=-\dfrac{e^A-e^{-A}}{2}=-\sinh A$

5. $\sinh A\cosh B+\cosh A\sinh B=\left(\dfrac{e^A-e^{-A}}{2}\right)\left(\dfrac{e^B+e^{-B}}{2}\right)+\left(\dfrac{e^A+e^{-A}}{2}\right)\left(\dfrac{e^B-e^{-B}}{2}\right)$

$$=\dfrac{e^{A+B}-e^{-(A+B)}}{2}=\sinh(A+B)$$

7. $\cosh 3x=\cosh(2x+x)=\cosh 2x\cosh x+\sinh 2x\sinh x$
$$=\cosh x(\cosh^2 x+\sinh^2 x)+\sinh x(2\sinh x\cosh x)$$
$$=2\cosh^3 x-\cosh x+2\cosh^3 x-2\cosh x=4\cosh^3 x-3\cosh x$$

9. $3\cosh 3x$ 11. $2x\sinh(x^2+1)$ 13. $\cosh x\cosh 4x+4\sinh x\sinh 4x$ 15. $2\tanh x\operatorname{sech}^2 x$

17. $[\operatorname{csch}^2(1/x)]/x^2$ 19. $2x\operatorname{sech}^2 x^2$ 21. $2\cosh\sqrt{x}+C$ 23. $\frac{1}{4}\sinh 2x-\frac{1}{2}x+C$

25. $\frac{1}{2}\sinh^2 x+C$ 27. $2\sqrt{\cosh x-1}+C$ 29. $e-(1/e)$ 31. $\dfrac{L^2}{8(d-b)}-\dfrac{d-b}{2}$ 33. $a\sinh\left(\dfrac{x}{a}\right)$

35. $\pi\left[\dfrac{a^3}{4}\sinh\left(\dfrac{2}{a}\right) + \dfrac{a^2}{2} + 2(b-a)\sinh\left(\dfrac{1}{a}\right) + (b-a)^2\right]$

37. $(\cosh x + \sinh x)^n = \left(\dfrac{e^x + e^{-x} + e^x - e^{-x}}{2}\right)^n = e^{nx} = \cosh(nx) + \sinh(nx)$

Exercise 7 (page 470)

1. $3/\sqrt{9x^2 + 1}$ **3.** $2/(2x - x^3)$ **5.** $1/\sqrt{x^2 + x}$ **7.** $-2\csc 2x$ **9.** $(\sec^2 x)/(1 - \tan^2 x)$

11. Let $y = \cosh^{-1}x$ for $x \geq 1$; then $x = \cosh y = (e^y + e^{-y})/2$ or

$$2x = e^y + e^{-y}$$
$$e^{2y} - 2xe^y + 1 = 0$$
$$e^y = \frac{2x \pm \sqrt{4x^2 - 4(1)}}{2(1)}$$
$$= x \pm \sqrt{x^2 - 1}$$

The only possible solution is $e^y = x + \sqrt{x^2 - 1}$, so $\cosh^{-1}x = y = \ln(x + \sqrt{x^2 - 1})$ for $x \geq 1$.

13. $y = \sinh^{-1}x$
  $x = \sinh y$
  $1 = (\cosh y)(y')$

  $y' = \dfrac{1}{\cosh y} = \dfrac{1}{\sqrt{\sinh^2 y + 1}} = \dfrac{1}{\sqrt{x^2 + 1}}$

15. $y = \tanh^{-1}x$
  $x = \tanh y$
  $1 = (\mathrm{sech}^2 y)(y')$

  $y' = \dfrac{1}{1 - \tanh^2 y} = \dfrac{1}{1 - x^2}$

17.    **19.** $\ln\left(\dfrac{3 + \sqrt{8}}{2 + \sqrt{3}}\right)$

Miscellaneous Exercises (page 471)

1. $z\cosh z + \sinh z$ **3.** $-\dfrac{1}{x^2}\mathrm{sech}^2\left(\dfrac{1}{x}\right)$ **5.** $1 - \dfrac{z}{\sqrt{1 - z^2}}\sin^{-1}z$ **7.** $2\sqrt{2x - x^2}$ **9.** $\dfrac{\sqrt{x}}{x + 1}$

11. $\left(\dfrac{x^2 - 4}{x^2 + 4}\right)^2$ **13.** $\dfrac{z^2}{\sqrt{1 - z^2}} + 2z\sin^{-1}z$ **15.** $-\dfrac{\sqrt{a^2 - x^2}}{x^2}$ **17.** $\dfrac{1}{\sqrt{2ax - x^2}}$ **19.** $\tan^{-1}x$

21. $\dfrac{2\sec^2 v}{\tan v}$ **23.** $\dfrac{e^x}{\sqrt{1 - e^{2x}}}$ **25.** $\frac{1}{2}x - \frac{1}{12}\sin 6x + C$ **27.** 1 **29.** $-\sin y$

31. $\sqrt{2} - \cosh^{-1}\sqrt{2}$ **33.** $\dfrac{d}{dx}(x\sin^{-1}x + \sqrt{1 - x^2} + C) = \sin^{-1}x + \dfrac{x}{\sqrt{1 - x^2}} - \dfrac{x}{\sqrt{1 - x^2}} = \sin^{-1}x$

35. $\pi a^2/2$ **37.** $\pi/6$ **39.** $\pi/6$ **41.** $\dfrac{1}{4}\sin^{-1}(e^{4x}/4) + C$ **43.** $2\tan^{-1}r; \pi$ **45.** $\sin x$

47. $\int_a^b \sqrt{1 + \sec^4 x}\,dx$ **49.** $(1, \pi/2)$ **51.** $\pi/6$

53. We get the square root of a negative number. The domain of $\sin^{-1}$ is $[-1, 1]$ and the range of cosh is ≥ 1.

55. (a)

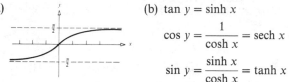

(b) $\tan y = \sinh x$

$$\cos y = \frac{1}{\cosh x} = \operatorname{sech} x$$

$$\sin y = \frac{\sinh x}{\cosh x} = \tanh x$$

57. $B = A \cos \theta_0$, $C = A \sin \theta_0$ **59.** (a) $12\sqrt{2}$ cm (b) $-30\sqrt{2}\pi^2$ dynes (c) $\frac{4}{3}$ sec (d) $-6\sqrt{3}\pi$ cm/sec

CHAPTER 9

Exercise 1 (page 477)

1. $\frac{2}{9}(3x + 1)^{3/2} + C$ **3.** $\frac{1}{36}(3x^2 - 5)^6 + C$ **5.** $-\ln|1 + \cos x| + C$ **7.** $-\frac{1}{9}(1 - 9x^2)^{1/2} + C$

9. $\sin^{-1}(x/2) + C$ **11.** $\frac{1}{2}\ln(1 + 2e^x) + C$ **13.** $2e^{\sqrt{x}} + C$ **15.** $\frac{1}{2}e^{x^2} + C$ **17.** $e^{\sin x} + C$

19. $\frac{1}{2}\sin^2 x + C$ **21.** $\frac{1}{4}\tan^{-1}(x/4) + C$ **23.** $-\frac{1}{2}(\cot 2x + \csc 2x) + C$ **25.** $-\frac{1}{2}\csc x^2 + C$

27. $\sec t + C$ **29.** $\frac{1}{3}\sec^{-1}(x/3) + C$ **31.** $x - \frac{1}{3}\sin(3x) + C$ **33.** $(5^x/\ln 5) + C$

35. $-\ln|1 - \cosh x| + C$

Exercise 2 (page 484)

1. $(e^{2x}/4)(2x - 1) + C$ **3.** $x \sin x + \cos x + C$ **5.** $(2x^{3/2}/3)(\ln x - \frac{2}{3}) + C$

7. $x \cot^{-1}x + \frac{1}{2}\ln(1 + x^2) + C$ **9.** $x(\ln x)^2 - 2x \ln x + 2x + C$ **11.** $(e^x/2)(\sin x - \cos x) + C$

13. $2 - 5e^{-1}$ **15.** $2x \sin x + (2 - x^2) \cos x + C$ **17.** $(x^2/4) + [(x \sin 2x)/4] + [(\cos 2x)/8] + C$

19. $\frac{2}{27}(1 - 25e^{-6})$ **21.** $x \cosh x - \sinh x + C$ **23.** $x^{n+1}[(\ln x)/(n + 1) - 1/(n + 1)^2] + C$

25. $x[(\ln x)^3 - 3(\ln x)^2 + 6 \ln x - 6] + C$ **27.** $(e^x/2)[x(\cos x + \sin x) - \sin x] + C$

29. $(x^3/3)[(\ln x)^2 - \frac{2}{3}\ln x + \frac{2}{9}] + C$ **31.** $(x^3/3)\tan^{-1}x - (x^2/6) + \frac{1}{6}\ln(1 + x^2) + C$

33. Let $u = \ln(x + \sqrt{x^2 + a^2})$, $dv = dx$ **35.** Let $u = \sin^{-1}x$, $dv = x^n \, dx$ **37.** Let $u = x^n$, $dv = (ax + b)^{1/2} \, dx$

39. Let $u = \sin^{n-1}x$, $dv = \sin x \, dx$ **41.** 1   **43.** $\pi(\pi - 2)$ **45.** $\pi(\pi - 2)$

47. (a) $5\pi/32$ (b) $\frac{8}{15}$ (c) $35\pi/256$ (d) $5\pi/32$

Exercise 3 (page 490)

1. $\frac{1}{3}\sin^3 x + C$ **3.** $\sec x + C$ **5.** $-\cos x + \frac{1}{3}\cos^3 x + C$ **7.** $\frac{1}{2}[x - (\sin 6x)/6] + C$

9. $\sin x - \frac{2}{3}\sin^3 x + \frac{1}{5}\sin^5 x + C$ **11.** $-\frac{1}{3}\cos 3x + \frac{2}{9}\cos^3 3x - \frac{1}{15}\cos^5 3x + C$

13. $-\frac{1}{3}\cos^3 x + \frac{1}{5}\cos^5 x + C$ **15.** $-\frac{1}{6}\cos^6 x + \frac{1}{8}\cos^8 x + C$ **17.** $\frac{3}{4}\sin^{4/3}x + C$

19. $\frac{2}{3}\sin^{3/2}x - \frac{2}{7}\sin^{7/2}x + C$ **21.** $\frac{1}{8}[x - (\sin 4x)/4] + C$ **23.** $(\tan^2 x)/2 + \ln|\cos x| + C$

25. $(\tan^3 x)/3 + C$ **27.** $(-\csc^7 x)/7 + (2 \csc^5 x)/5 - (\csc^3 x)/3 + C$

29. $\frac{1}{4}\tan x \sec^3 x - \frac{1}{8}\sec x \tan x - \frac{1}{8}\ln|\sec x + \tan x| + C$ **31.** $-\frac{1}{4}\cos 2x - \frac{1}{8}\cos 4x + C$

33. $\frac{1}{4}\sin 2x - \frac{1}{12}\sin 6x + C$ **35.** $\frac{1}{2}\sin x + \frac{1}{6}\sin 3x + C$ **37.** $\frac{1}{2}\cos x - \frac{1}{4}\cos 2x + C$

39. $(\tan x \sec^{n-2}x)/(n-1) + (n-2)/(n-1)\int \sec^{n-2}x\,dx$ **41.** $\pi^2/2$ **43.** $s(t) = \frac{16}{3}t - \frac{1}{3}\sin t + \frac{1}{9}\sin^3 t$

Exercise 4 (page 495)

1. $\sin^{-1}(x/2) + C$ **3.** $(-\sqrt{9-x^2}/x) - \sin^{-1}(x/3) + C$ **5.** $(x/2)\sqrt{x^2-1} + \frac{1}{2}\ln|x + \sqrt{x^2-1}| + C$

7. $(x/4\sqrt{x^2+4}) + C$ **9.** $(-\sqrt{x^2+4}/4x) + C$ **11.** $\frac{1}{2}[x\sqrt{16-x^2} + 16\sin^{-1}(x/4)] + C$

13. $(x/4\sqrt{4-x^2}) + C$ **15.** $(-x/2)\sqrt{16-x^2} + 8\sin^{-1}(x/4) + C$ **17.** $\sqrt{x^2-1} - \sec^{-1}x + C$

19. $(-x/9\sqrt{x^2-9}) + C$ **21.** $\sin^{-1}(x-2) + C$ **23.** $\ln|(x-1) + \sqrt{(x-1)^2-4}| + C$

25. $[(x-1)/9\sqrt{x^2-2x+10}] + C$ **27.** $\ln|(x+1) + \sqrt{(x+1)^2-4}| + C$ **29.** $(\sqrt{3}/3) - (\pi/6)$

31. $(7\sqrt{17} - \sqrt{5})/24$ **33.** πab **35.** Let $x = a\sec\theta$ **37.** $A = 4\int_0^R \sqrt{R^2 - x^2}\,dx = \pi R^2$ **39.** 13.9

41. $\frac{1}{4}[(2x^2-1)\sin^{-1}x + x\sqrt{1-x^2}] + C$

Exercise 5 (page 498)

1. $\tan^{-1}(x+2) + C$ **3.** $\sin^{-1}\left(\dfrac{x-1}{3}\right) + C$ **5.** $\sin^{-1}\left(\dfrac{x-2}{2}\right) + C$ **7.** $\sin^{-1}\left(\dfrac{x+1}{5}\right) + C$

9. $\sqrt{x^2-2x+5} + \ln\left|\dfrac{\sqrt{x^2-2x+5}}{2} + \dfrac{x-1}{2}\right| + C$ **11.** $\ln|\sqrt{2}+1|$ **13.** $\ln(\sqrt{e^{2x}+e^x+1} + e^x + \frac{1}{2}) + C$

Exercise 6 (page 504)

1. $(x^2/2) - x + 2\ln|x+1| + C$ **3.** $(x^3/3) + x^2 + 7x + 10\ln|x-2| + C$ **5.** $\frac{1}{3}\ln|(x-2)/(x+1)| + C$

7. $\ln[(x-2)^2/|x-1|] + C$ **9.** $\frac{1}{4}\ln|(x+1)(x-1)^3| - [1/2(x-1)] + C$

11. $5\ln|(x+1)/(x+2)| + [4/(x+1)] + C$ **13.** $\frac{2}{21}\ln|3x-2| + \frac{1}{14}\ln|2x+1| + C$

15. $\frac{1}{4}\ln|(x+3)^3(x-1)| + C$ **17.** $\ln|(x-3)^2/x(x+1)| + C$ **19.** $1/(x-1) + \ln|(x-2)^4/(x-1)^3| + C$

21. Let $u = \sin\theta$; $\frac{1}{5}\ln|(\sin\theta-2)/(\sin\theta+3)| + C$ **23.** $(-\ln 2)/6$ **25.** $(\ln 21)/8$ **27.** $\ln\frac{15}{7}$

29. $\pi(\frac{1}{8}\ln\frac{15}{7} + \frac{19}{105})$

Exercise 7 (page 508)

1. $\ln|x/\sqrt{x^2+1}| + C$ **3.** $\ln(|x-1|/\sqrt{x^2+x+1}) + (1/\sqrt{3})\tan^{-1}[(2x+1)/\sqrt{3}] + C$

5. $\frac{1}{8}\ln|x^2/(x^2+4)| - (1/x) - \frac{1}{2}\tan^{-1}(x/2) + C$ **7.** $\frac{1}{3}\ln|(x+1)^2\sqrt{x^2+2x+4}| + C$

9. $\frac{7}{16}\tan^{-1}(x/2) - 1/(x^2+4) - x/8(x^2+4) + C$ **11.** $-(x^2+8)/2(x^2+16)^2 + C$

13. $\frac{1}{2}\ln|(1+\cos^2\theta)/(\cos^2\theta)| + C$ **15.** 8.72

Exercise 8 (page 510)

1. $\dfrac{2}{1-\tan(x/2)} + C$ **3.** $-\cot(x/2) + C$ **5.** $\sqrt{2}\ln\left|\dfrac{\sqrt{2}-1+\tan(x/2)}{\sqrt{2}+1-\tan(x/2)}\right| + C$ **7.** $-\ln|3+\cos x| + C$

9. $-\dfrac{1}{4}\left(\tan\dfrac{x}{2}\right)^{-2} - \dfrac{1}{2}\ln\left|\tan\dfrac{x}{2}\right| + C$ **11.** $\ln\left|\dfrac{\tan(x/2)}{[1+\tan(x/2)]^2}\right| + C$ **13.** 1   **15.** $\pi\left(\dfrac{4}{3\sqrt{3}} - \dfrac{1}{2}\right)$

17. $\sec x + \tan x = \dfrac{1 + \sin x}{\cos x} = \dfrac{1 + 2\sin(x/2)\cos(x/2)}{\cos^2(x/2) - \sin^2(x/2)} = \dfrac{\sec^2(x/2) + 2\tan(x/2)}{1 - \tan^2(x/2)}$

$= \dfrac{1 + \tan^2(x/2) + 2\tan(x/2)}{[1 + \tan(x/2)][1 - \tan(x/2)]} = \dfrac{1 + \tan(x/2)}{1 - \tan(x/2)}$

19. $\csc x - \cot x = \dfrac{1 - \cos x}{\sin x} = \dfrac{1 - \cos x}{\sqrt{(1 - \cos x)(1 + \cos x)}} = \sqrt{\dfrac{1 - \cos x}{1 + \cos x}}$

Exercise 9 (page 512)

1. $\frac{2}{3}x^{3/2} - 3x + 18\sqrt{x} - 54\ln|\sqrt{x} + 3| + C$ **3.** $\frac{3}{2}\ln|x^{2/3} - 1| + C$

5. $2\sqrt{x} - 3\sqrt[3]{x} + 6\sqrt[6]{x} - 6\ln|\sqrt[6]{x} + 1| + C$ **7.** $\frac{1}{2}(2 + 3x)^{2/3} + C$ **9.** $\frac{4}{5}(1 + x)^{5/4} - 4(1 + x)^{1/4} + C$

11. $2[\sqrt{x} - \ln|\sqrt{x} + 1|] + C$ **13.** $x + 3x^{2/3} + 6x^{1/3} + 6\ln|x^{1/3} - 1| + C$ **15.** $959\pi/30$

Exercise 11 (page 520)

1. 21.5   **3.** 0.6970   **5.** 1.9541   **7.** 0.4832   **9.** 0.74298   **11.** 21.333   **13.** 0.5004

15. 1.4642   **17.** 2.3351   **19.** 0.7468   **21.** 0.5622; 0.5620   **23.** 18.8396; 18.8371   **25.** 0.6956

27. 16,787.5 m³   **29.** 2500   **31.** 131,787.5 m³; 132,625 m³   **33.** 230   **35.** 0.1667   **37.** 0.0026

39. 1.9100   **41.** 6.4287   **43.** 1.8446   **45.** 18.1333

Miscellaneous Exercises (page 523)

1. $(x^5/5) - (2x^3/3) + C$ **3.** $\frac{1}{4}\tan^{-1}[(x + 2)/4] + C$ **5.** $\frac{1}{2}\tan 2\theta + C$ **7.** $2(\sqrt{x} - \tan^{-1}\sqrt{x}) + C$

9. $\frac{1}{2}\ln(9 + t^2) + C$ **11.** $-(\theta + \cot\theta) + C$ **13.** $-e^{\cos x} + C$ **15.** $\frac{1}{4}\sin 2x - (x/2)\cos 2x + C$

17. $(z^5/5) + (8z^3/3) + 16z + C$ **19.** $\frac{2}{9}(2 - t)^{9/2} - \frac{12}{7}(2 - t)^{7/2} + \frac{24}{5}(2 - t)^{5/2} - \frac{16}{3}(2 - t)^{3/2} + C$

21. $\frac{1}{3}\sin^3 x - \frac{1}{5}\sin^5 x + C$ **23.** $\sqrt{2y + 1} + C$ **25.** $e^t + 2\ln|e^t - 2| + C$ **27.** $\frac{1}{2}\ln|\cosh 2v| + C$

29. $\sin x + \cos x + \ln|\csc x - \cot x| + C$ **31.** $\frac{1}{8}\ln|(x^2 - 2)/(x^2 + 2)| + C$ **33.** $\ln|\ln x| + C$

35. $\ln|\cos[(\pi/4) - \theta]| + C$ **37.** $[(ab)^x/\ln(ab)] + C$ **39.** $a\sin^{-1}(x/a) - \sqrt{a^2 - x^2} + C$

41. $(x^3/3) + (x^2/2) - x - \ln|x - 1| + C$ **43.** $\frac{1}{24}(3y^2 - 6y)^4 + C$ **45.** $-3[(x^2/2) + x + \ln|x - 1|] + C$

47. $\frac{1}{3}(\sin 3x - \frac{1}{3}\sin^3 3x) + C$ **49.** $-x - \ln|(x - 2)/(x + 2)| + C$

51. $\frac{1}{3}x^3\sin^{-1}x + \frac{1}{3}\sqrt{1 - x^2} - \frac{1}{9}(1 - x^2)^{3/2} + C$ **53.** $6\sqrt{t + 1} + 2\ln|(\sqrt{t + 1} - 1)/(\sqrt{t + 1} + 1)| + C$

55. $(x^2/4) + (x/4)\sin 2x + \frac{1}{8}\cos 2x + C$ **57.** $(1/a)\ln|x/(x + a)| + C$ **59.** $-\frac{1}{2}[w^2 + \ln|w^2 - 1|] + C$

61. Let $x = a\sin\theta$ **63.** Let $x = a\sec\theta$ **65.** Let $x = a\tan\theta$ **67.** $\ln[(x + \sqrt{x^2 - a^2})/a] + C$

69. $x^2\sin x$ **71.** $\frac{4}{9}(x + 5)^{3/2} - x - 6$ **73.** π **75.** (a) $\ln(\sqrt{2} + 1) + \sqrt{2} - 2$ (b) $\pi(2 - \pi/2)$

77. $T_n = \frac{1}{2}\Delta x[f(x_0) + 2f(x_1) + \cdots + 2f(x_{n-1}) + f(x_n)]$

$= \Delta x[f(x_0) + f(x_1) + \cdots + f(x_{n-1}) + f(x_n)] - \dfrac{\Delta x}{2}[f(x_0) + f(x_n)]$

$\displaystyle\lim_{n \to +\infty} T_n = \lim_{\Delta x \to 0}\sum_{i=1}^{n} f(x_i)\Delta x + \lim_{\Delta x \to 0}\left(\dfrac{-\Delta x}{2}\right)[f(x_0) + f(x_n)] = \int_a^b f(x)\,dx$

79. An identity in x results; $\frac{1}{4}(2x + e^{2x}) + C$ **81.** $v(t) = \frac{1}{30}(t^3 + 4)^{1/2} - \frac{1}{15}$

CHAPTER 10

Exercise 1 (page 536)

1. 5   **3.** $-\frac{12}{5}$ **5.** $\frac{3}{4}$ **7.** $\frac{1}{2}$ **9.** 2   **11.** 0   **13.** 0   **15.** 2   **17.** 0   **19.** 0   **21.** 1   **23.** $\frac{1}{4}$

25. $-\frac{1}{6}$ **27.** $\frac{2}{9}$ **29.** 0   **31.** 0   **33.** 0   **35.** 0   **37.** 2   **39.** 2   **41.** $(\ln 2)/(\ln 3)$ **43.** 0

45. 1   **47.** $\frac{1}{3}$ **49.** -2 **51.** 1   **53.** 1   **55.** 0   **57.** 0

59. $\lim\limits_{x \to +\infty} \dfrac{(\ln x)^{\beta}}{x^{\alpha}} = \lim\limits_{x \to +\infty} \dfrac{\beta(\ln x)^{\beta-1}(1/x)}{\alpha x^{\alpha-1}} = \lim\limits_{x \to +\infty} \dfrac{\beta(\ln x)^{\beta-1}}{\alpha x^{\alpha}} = \cdots = 0$

61. 1   **63.** $\lim\limits_{x \to c} \dfrac{x^{\alpha} - c^{\alpha}}{x^{\beta} - c^{\beta}} = \lim\limits_{x \to c} \dfrac{\alpha x^{\alpha-1}}{\beta x^{\beta-1}} = \dfrac{\alpha}{\beta} c^{\alpha-\beta}$ **65.** an

67. Let $x = 1/u$. Then $x \to -\infty$ implies $u \to 0^{-}$:

$$\lim_{x \to -\infty} \frac{f(x)}{g(x)} = \lim_{u \to 0^{-}} \frac{f(1/u)}{g(1/u)} = \lim_{u \to 0^{-}} \frac{(-1/u^2)f'(1/u)}{(-1/u^2)g'(1/u)} = \lim_{x \to -\infty} \frac{f'(x)}{g'(x)}$$

Exercise 2 (page 541)

1. 1   **3.** 1   **5.** 0   **7.** 0   **9.** 1   **11.** 0   **13.** 0   **15.** $\frac{2}{\pi}$ **17.** 0   **19.** 0   **21.** $\frac{1}{2}$ **23.** $\frac{1}{2}$

25. 1   **27.** 1   **29.** e **31.** 1   **33.** 1   **35.** 1

37. $y = (\cos x + 2 \sin x)^{\cot x}$; $\lim\limits_{x \to 0^{+}} \ln y = 2$; $\lim\limits_{x \to 0^{+}} y = e^2$ **39.** 0

41. $\lim\limits_{x \to 0^{+}} \dfrac{e^{-1/x^2}}{x} = \lim\limits_{x \to 0^{+}} \dfrac{1/x}{e^{1/x^2}} = \lim\limits_{x \to 0^{+}} \dfrac{x}{2e^{1/x^2}} = 0$ **43.** $f'(0) = \lim\limits_{h \to 0} \dfrac{e^{-1/h^2} - 0}{h} = 0$

Exercise 3 (page 549)

1. Improper; upper limit is $+\infty$ **3.** Not improper   **5.** Improper; $1/x$ not defined at 0

7. Improper; integrand not defined at 1   **9.** Converges; $\frac{1}{10}$ **11.** Diverges   **13.** Diverges

15. Converges; 2   **17.** Diverges   **19.** Converges; $2\sqrt{a}$ **21.** Converges; $\frac{1}{2}$ **23.** Diverges

25. Diverges   **27.** Diverges   **29.** Converges; 0   **31.** Diverges   **33.** Converges; $\pi/4$ **35.** Diverges

37. Converges; π/a **39.** Diverges   **41.** Converges; $\pi a^2/4$ **43.** Diverges   **45.** Diverges

47. Converges; $(1/2a) \ln 3$ **49.** Converges; $1 - (\pi/4)$ **51.** Converges; $\pi/2$ **53.** Converges; 4

55. Diverges   **57.** Diverges   **59.** Converges; 0   **61.** $\pi/2$ **63.** Does not exist

65. $\displaystyle\int_{0}^{+\infty} \sin x\, dx = \lim\limits_{b \to +\infty} (-\cos b + 1)$, which does not exist; $\lim\limits_{t \to +\infty} \displaystyle\int_{-t}^{t} \sin x\, dx = \lim\limits_{t \to +\infty} [\cos(-t) - \cos t] = 0$

67. 1   **69.** $\frac{1}{2}$ **71.** $\frac{1}{2}$ **73.** $\frac{1}{2}LI^2$ **75.** $Im/5r$

Exercise 4 (page 560)

1. $1 + 3(x - 1)^2$ **3.** $-6 + 5(x - 1) + 4(x - 1)^2 + (x - 1)^3$

5. $P_3(x) = 4 + 7(x - 1) + 11(x - 1)^2 + 3(x - 1)^3$; $R_4(x) = 0$

7. $P_4(x) = 7 - 25(x + 1) + 30(x + 1)^2 - 14(x + 1)^3 + 2(x + 1)^4$; $R_5(x) = 0$

9. $P_4(x) = 32 + 80(x - 2) + 80(x - 2)^2 + 40(x - 2)^3 + 10(x - 2)^4; R_5(x) = (x - 2)^5$

11. $P_5(x) = (x - 1) - \frac{1}{2}(x - 1)^2 + \frac{1}{3}(x - 1)^3 - \frac{1}{4}(x - 1)^4 + \frac{1}{5}(x - 1)^5; R_6(x) = -(x - 1)^6/6u^6$

13. $P_5(x) = 1 - (x - 1) + (x - 1)^2 - (x - 1)^3 + (x - 1)^4 - (x - 1)^5; R_6(x) = (x - 1)^6/u^7$

15. $P_6(x) = 1 - (x^2/2!) + (x^4/4!) - (x^6/6!); R_7(x) = (x^7/7!) \sin u$

17. $P_6(x) = 1 + (x^2/2!) + (x^4/4!) + (x^6/6!); R_7(x) = (x^7/7!) \sinh u$

19. $P_4(x) = 1 + x + x^2 + x^3 + x^4; R_5(x) = x^5/(1 - u)^6$

21. $P_3(x) = 1 - 2x + 3x^2 - 4x^3; R_4(x) = 5x^4/(1 + u)^6$

23. $0.095308 \ (n = 4)$ **25.** $0.909 \ (n = 3)$ **27.** $0.999847 \ (n = 2)$ **29.** $0.375 \ (n = 4)$ **31.** $1.6458 \ (n = 3)$

33. $P_4(x) = (x - 1) + \frac{1}{2}(x - 1)^2 + \frac{1}{6}(x - 1)^3 + \frac{1}{12}(x - 1)^4$ **35.** $P_2(x) = 1 - x^2$

37. $P_4(x) = 1 + 2[x - (\pi/4)] + 2[x - (\pi/4)]^2 + \frac{8}{3}[x - (\pi/4)]^3 + \frac{10}{3}[x - (\pi/4)]^4$

39. $P_3(x) = 1 + \frac{1}{2}x - \frac{1}{8}x^2 + \frac{1}{16}x^3$ **41.** $f(x) \approx -\lambda\pi - (1 + \lambda)(x - \pi) = 0$ **43.** 0.0100003

Miscellaneous Exercises (page 561)

1. 1   **3.** 9   **5.** Converges; $3/e$ **7.** Converges; $2(1 - \cos 1)$ **9.** Converges; $\frac{3}{2}(1 + \sqrt[3]{3})$

11. Converges; 1   **13.** Converges; -1 **15.** Use by-parts formula with $u = x^n$, $dv = e^{-x} dx$.

17. Let $u = \ln x$.   **19.** Let $u^8 = x$.   **21.** $V = \pi \int_0^1 x^{-4/3} dx$ diverges

23. $\lim\limits_{n \to +\infty} \dfrac{x^{1/n} - 1}{1/n} = \lim\limits_{n \to +\infty} \dfrac{(-1/n^2)(\ln x)x^{1/n}}{-1/n^2} = \ln x$ **25.** $\ln 2$

27. $1/\sqrt{2 + \sin x} \geq 1/\sqrt{3}$ and $\int_0^{+\infty} (1/\sqrt{3}) dx$ diverges   **29.** $1 - e^{-e^a}$

31. The limit of $f'(x)/g'(x)$ does not exist.

33. (a) $A_1 = \int_0^{\ln 2} e^{-x} dx = \frac{1}{2}; A_2 = \int_{\ln 2}^{+\infty} e^{-x} dx = \frac{1}{2}$ (b) $V_1 = \pi \int_0^{\ln 2} e^{-2x} dx = 3\pi/8; V_2 = \pi \int_{\ln 2}^{+\infty} e^{-2x} dx = \pi/8$

35. (a) Differentiate with respect to h; $f''(x)$ (b) $f'''(x)$ **37.** 2   **39.** $1/s$ **41.** $s/(s^2 + 1)$

43. $1/(s - 1)$ **45.** $\int_{-\infty}^{+\infty} f(x) dx = \int_a^b dx/(b - a) = 1$ **47.** $(b + a)/2$

49. $\sigma^2 = (b - a)^2/12; \sigma = (b - a)/2\sqrt{3}$

CHAPTER 11

Exercise 1 (page 578)

1. $1, \frac{1}{2}, \frac{1}{3}, \frac{1}{4}$ **3.** $0, \ln 2, \ln 3, \ln 4$ **5.** $\frac{1}{3}, -\frac{1}{5}, \frac{1}{7}, -\frac{1}{9}$ **7.** $1, 0, 1, 0$ **9.** 0   **11.** 2   **13.** 1

15. 0   **17.** Divergent   **19.** 1   **21.** 1   **23.** $-\frac{1}{2}$ **25.** 1   **27.** 2   **29.** Divergent   **31.** 0

33. 1   **35.** 0   **37.** $\pi/2$ **39.** 1   **41.** Convergent; $\lim_{n \to +\infty} s_n = 0$ **43.** Convergent; $\lim_{n \to +\infty} s_n = 0$

45. Divergent; $\lim_{n \to +\infty} s_n = +\infty$ **47.** Convergent; $\lim_{n \to +\infty} s_n = 0$ **49.** Convergent; $\lim_{n \to +\infty} s_n = 0$

51. Divergent; oscillates from -1 to 1   **53.** Convergent; $\lim_{n \to +\infty} s_n = 0$

55. Divergent; $\lim_{n \to +\infty} s_n$ does not exist.   **57.** Convergent; $\lim_{n \to +\infty} s_n = 0$

59. Convergent; $\lim_{n \to +\infty} s_n = 2\sqrt{2}$ **61.** Divergent; $\lim_{n \to +\infty} s_n = +\infty$ **63.** $34 \le n \le 99$

65. $10, 11$ **67.** $0 < r^n < 1/np$; apply the squeezing theorem   **69.** $r^n > np$; $\lim_{n \to +\infty} np = +\infty$

71. $|s_n^2 - L^2| = |s_n - L||s_n + L| \le |s_n - L|(|s_n| + |L|)$; use the fact that a convergent sequence is bounded

73. $||s_n| - |L|| \le |s_n - L|$; the converse is false; take $\{s_n\} = \{(-1)^n\}$

75. (a) $1, 1, 2, 3, 5, 8, 13, 21$ (b) Use $(1 + \sqrt{5})^2 = 2[2 + (1 + \sqrt{5})]$; $(1 - \sqrt{5})^2 = 2[2 + (1 - \sqrt{5})]$

Exercise 2 (page 592)

1. $S_4 = 1 + \frac{3}{4} + \frac{9}{16} + \frac{27}{64} = \frac{175}{64}$ **3.** $S_4 = 1 + 2 + 3 + 4 = 10$ **5.** Converges; $\frac{3}{2}$ **7.** Diverges

9. Converges; $\frac{1}{42}$ **11.** Diverges   **13.** Converges; $\frac{7}{2}$ **15.** Converges; $\frac{1}{99}$ **17.** Diverges

19. Diverges   **21.** Converges; $\frac{1}{3}$ **23.** Diverges   **25.** Diverges   **27.** Diverges   **29.** Diverges

31. Diverges   **33.** Converges; $\frac{1}{2}$ **35.** $\frac{5}{9}$ **37.** $\frac{3857}{900}$ **39.** $\left|\frac{1}{x}\right| < 1$; $\sum_{k=1}^{+\infty} \frac{1}{x^{k-1}} = \frac{1}{1 - (1/x)} = \frac{x}{x - 1}$

41. 90 ft   **43.** $\frac{2}{3}$ **45.** 31

47. S_n is the nth partial sum of $\sum_{k=1}^{+\infty} a_k$; T_n is the nth partial sum of $\sum_{k=1}^{+\infty} b_k$; $S_n + T_n$ is the nth partial sum of $\sum_{k=1}^{+\infty} (a_k + b_k)$

49. The nth partial sum of $\sum_{k=1}^{+\infty} a_k$ is $S_n + K$ if $n \ge N$, where S_n is the nth partial sum of $\sum_{k=N+1}^{+\infty} a_k$

Exercise 3 (page 606)

1. Convergent   **3.** Divergent   **5.** Divergent   **7.** Convergent   **9.** Divergent   **11.** Convergent

13. Convergent   **15.** Convergent   **17.** Convergent   **19.** Convergent   **21.** Divergent

23. Convergent   **25.** Divergent   **27.** Convergent   **29.** Convergent   **31.** Divergent

33. Convergent   **35.** Convergent   **37.** Convergent   **39.** Convergent   **41.** Convergent

43. Convergent   **45.** Divergent   **47.** Convergent   **49.** Divergent   **51.** Convergent   **53.** Divergent

55. $\int_2^{+\infty} \frac{dx}{x(\ln x)^p} = \int_{\ln 2}^{+\infty} \frac{du}{u^p}$ is convergent if and only if $p > 1$ **57.** $0 < x \le 1$ **59.** $\lim_{n \to +\infty} \frac{1/(n + 1)}{1/n} = 1$

61. $\sum_{k=1}^{+\infty} \frac{k!}{k^k}$ converges by Example 9; thus, $\lim_{n \to +\infty} \frac{n!}{n^n} = 0$

63. $\lim_{n \to +\infty} \frac{a_n}{d_n} = p > 0 \Rightarrow$ for some $\varepsilon > 0$; $\frac{a_n}{d_n} \ge \varepsilon$ for all $n >$ some $N \Rightarrow a_n \ge \varepsilon d_n$; use comparison test II;

$\lim_{n \to +\infty} \frac{a_n}{d_n} = +\infty \Rightarrow \frac{a_n}{d_n} > 1$ for all $n >$ some $N \Rightarrow a_n > d_n$; use comparison test II

65. Convergent   **67.** $\lambda = \lim_{n \to +\infty} \sqrt[n]{\frac{1}{(2n)^n}} = 0$

Exercise 4 (page 615)

1. Converges   **3.** Converges   **5.** Diverges   **7.** Converges   **9.** 0.368   **11.** 0.947   **13.** 99

15. 4   **17.** 9999   **19.** Conditionally convergent   **21.** Absolutely convergent

23. Absolutely convergent   **25.** Absolutely convergent   **27.** Absolutely convergent

29. Conditionally convergent   **31.** Absolutely convergent   **33.** Divergent   **35.** Divergent

37. Conditionally convergent   **39.** $\sum\limits_{k=1}^{+\infty} \dfrac{1}{k^{1/3}}$ diverges   **41.** This is a form of the harmonic series

43. Begin with enough positive terms so that the sum exceeds 2; then add in just enough of the negative terms for the sum to be less than 2; continuing this process gives a series that converges to 2.

45. $\left| e^{-kx} \cos kx \right| \le (e^{-x})^k = (1/e^x)^k$, a convergent geometric series

47. A rearrangement of the conditionally convergent series A does not have the sum A.

Exercise 5 (page 618)

1. Divergent   **3.** Convergent   **5.** Convergent   **7.** Divergent   **9.** Divergent   **11.** Divergent

13. Divergent   **15.** Conditionally convergent   **17.** Convergent   **19.** Divergent   **21.** Convergent

23. Divergent   **25.** Divergent   **27.** Absolutely convergent   **29.** Absolutely convergent

31. Absolutely convergent   **33.** Convergent   **35.** Convergent   **37.** Absolutely convergent

Exercise 6 (page 624)

1. $-1 \le x < 1$ **3.** $-3 < x < 3$ **5.** $-1 < x < 1$ **7.** $-\frac{3}{2} < x < \frac{3}{2}$ **9.** $-\infty < x < +\infty$

11. $-1 < x < 1$ **13.** $-4 < x < 4$ **15.** $-1 \le x < 1$ **17.** $2 < x < 4$ **19.** $-\infty < x < +\infty$

21. $1/e$ **23.** $x > 1$ or $x \le -1$ **25.** $-\infty < x < +\infty$ **27.** $-\pi/6 < x < \pi/6$

29. Convergent at $x = 2$; no

31. (a) True   (b) Cannot say   (c) Cannot say   (d) Cannot say   (e) True   (f) True

33. $\lim\limits_{n \to +\infty} \left| \dfrac{a_{n+1}}{a_n} \right| = \rho;\ |x| < R;\ \lim\limits_{n \to +\infty} \left| \dfrac{a_{n+1}x^{n+1}}{a_n x^n} \right| = |x|\rho < 1;\ \rho = \dfrac{1}{R}$

35. If either $\sum\limits_{k=1}^{+\infty} |a_k x_0^k|$ or $\sum\limits_{k=1}^{+\infty} |a_k(-x_0)^k|$ converges, so does the other.

Exercise 7 (page 632)

1. $\sum\limits_{k=0}^{+\infty} \dfrac{(-1)^k x^{2k+1}}{(2k+1)!}$ **3.** $\sum\limits_{k=0}^{+\infty} \dfrac{e(x-1)^k}{k!}$ **5.** $\sum\limits_{k=0}^{+\infty} \dfrac{\sqrt{2}}{2}(-1)^k \left[\dfrac{(x-\pi/4)^{2k+1}}{(2k+1)!} + \dfrac{(x-\pi/4)^{2k}}{(2k)!} \right]$

7. $\sum\limits_{k=0}^{+\infty} (-1)^k(x-1)^k$ **9.** $\sum\limits_{k=0}^{+\infty} (-1)^k(k+1)x^k$ **11.** $\sum\limits_{k=0}^{+\infty} (3x)^k$ **13.** $\sum\limits_{k=1}^{+\infty} \dfrac{(-1)^{k+1}x^k}{k}$

15. $-6 + 5x + 2x^2 + 3x^3$ **17.** $4 + 18(x-1) + 11(x-1)^2 + 3(x-1)^3$ **19.** $\sum\limits_{k=0}^{+\infty} x^{2k};\ -1 < x < 1$

21. 0.693   **23.** 3.1415

Exercise 8 (page 637)

1. $\sum\limits_{k=0}^{+\infty} \dfrac{(2x)^k}{k!}$ **3.** $\sum\limits_{k=1}^{+\infty} \dfrac{(-1)^{k+1}x^{2k}}{k}$ **5.** $\sum\limits_{k=0}^{+\infty} \dfrac{x^{2k+1}}{(2k+1)!}$ **7.** $x + x^2 + \frac{1}{3}x^3 - \frac{1}{30}x^5 + \cdots$

9. Integrate f; $x + \frac{1}{6}x^3 + \frac{3}{40}x^5 + \frac{5}{112}x^7$ 11. 0.3095   13. 0.995   15. 0.946   17. 0   19. 2

21. $\frac{1}{4}(\pi - 2)$(integrate $x \tan^{-1}x$)   23. $f(0) = 1; f'(0) = 0; f''(0) = -1; f'''(0) = 0$

Exercise 9 (page 640)

1. $\sum_{k=0}^{+\infty} \binom{1/2}{k} x^{2k}; -1 \leq x \leq 1$ 3. $\sum_{k=0}^{+\infty} \binom{1/5}{k} x^k; -1 \leq x \leq 1$ 5. $\sum_{k=0}^{+\infty} \binom{-3/4}{k} x^k; -1 < x \leq 1$

7. 0.2   9. 0.487

11. Use the binomial theorem to expand $(1 + 1/n)^n$. Each term is no larger than the corresponding terms of $\sum_{k=0}^{+\infty} (1/k!)$, which converges to e.

13. $\lim\limits_{n \to +\infty} \dfrac{a_{n+1}}{a_n} = \lim\limits_{n \to +\infty} \dfrac{|m - n|}{n + 1} |x| = |x|$

Miscellaneous Exercises (page 641)

1. Divergent   3. No conclusion

5. (a) $1/(1 + x) = 1 - x + x^2 - x^3 + \cdots$ (b) $\frac{1}{2} = 1 - 1 + 1 - 1 + \cdots$ (c) Series in part (a) diverges if $x = 1$

7. (a) $\ln n \leq 1 + \frac{1}{2} + \cdots + 1/(n - 1) = C_n - 1/n + \ln n \Rightarrow 1/n \leq C_n$

(b) $C_n - C_{n+1} = \ln(n + 1) - \ln n - \dfrac{1}{n + 1} > \dfrac{1}{n + 1} - \dfrac{1}{n + 1} = 0$; use (11.9); then $\lim\limits_{n \to +\infty} C_n = \gamma$

(c) 12; 31; 12,367; 2.724 × 10⁸

9. (a) 1.197532   (b) Use $f(x) = 1/x^3$; $\dfrac{1}{2} - \dfrac{1}{n^2}\left(\dfrac{1}{2} - \dfrac{1}{n}\right) < S_n < \dfrac{3}{2} - \dfrac{1}{2n^2}$ (c) $\dfrac{1}{2} < \zeta(3) < \dfrac{3}{2}$

11. $\dfrac{1}{k(k + 2)} = \dfrac{1}{2}\left(\dfrac{1}{k} - \dfrac{1}{k + 2}\right)$ 13. $\dfrac{1}{k(k + 1)(k + 2)(k + 3)} = \dfrac{1}{6}\left(\dfrac{1}{k} - \dfrac{1}{k + 3}\right) - \dfrac{1}{2}\left(\dfrac{1}{k + 1} - \dfrac{1}{k + 2}\right)$

15. Differentiate the Maclaurin series for $(e^x - 1)/x$ and put $x = 1$.

17. It is absolutely convergent by the ratio test.   19. Compare with the convergent series $\sum_{k=1}^{+\infty} \dfrac{k!}{k^k}$

21. Converges; -1 23. Diverges   25. Diverges   27. Converges   29. Converges

31. (a) $S_n \leq \int_0^n \dfrac{dx}{1 + x^2} = \tan^{-1}n$ (b) S_n is bounded and monotonic; $S_n \leq \pi/2$

(c) $S_n \geq \int_1^n \dfrac{dx}{1 + x^2} \geq \pi/4$; from part (b), $\pi/4 \leq \lim\limits_{n \to +\infty} S_n \leq \pi/2$

33. Compare to the p-series, $\sum_{k=1}^{+\infty} \dfrac{1}{k^{s-r}}$ 35. $x + (1 - x) \ln (1 - x)$ 37. 0.04267

39. (a) $\dfrac{3}{2}\left(x - \dfrac{x^3}{24} + \dfrac{x^3}{1920}\right); \dfrac{3}{2} \sum_{k=0}^{+\infty} \dfrac{(-1)^k x^{2k+1}}{4^k(2k + 1)!}$ (b) $-\infty < x < +\infty$ (c) Two terms

41. (c) 200 terms; 2 × 10¹⁰ terms

CHAPTER 12

Exercise 2 (page 653)

1. $y^2 = 4x$ **3.** $y^2 = -8x$ **5.** $y^2 = 8x$ **7.** $x^2 = 12y$ **9.** $y^2 = -8x$ **11.** $y^2 = -12x$

13. $y^2 = 32x$ **15.** $y^2 = -12x$

17. Vertex $(0, 0)$;
   focus $(2, 0)$;
   directrix $x = -2$

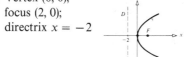

19. Vertex $(0, 0)$;
   focus $(0, -3)$;
   directrix $y = 3$

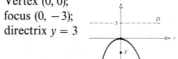

21. Vertex $(0, 0)$;
   focus $(-4, 0)$;
   directrix $x = 4$

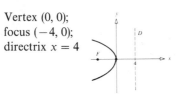

23. Vertex $(0, 0)$;
   focus $(0, 2)$;
   directrix $y = -2$

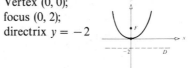

25. $y = -x + 2$ **27.** $\frac{16}{3}$ **29.** $64\pi/5$ **31.** $\frac{64}{3}$ m   **33.** $\frac{25}{16}$ ft from the vertex

35. Area of rectangle $= 2x_0 \dfrac{(x_0^2)}{4a} = \dfrac{x_0^3}{2a}$

 Area of segment $= 2 \displaystyle\int_0^{x_0} \left(\dfrac{x_0^2}{4a} - \dfrac{x^2}{4a} \right) dx = \dfrac{x_0^3}{3a}$

37. 60 tons

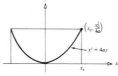

Exercise 3 (page 661)

1. $\dfrac{x^2}{36} + \dfrac{y^2}{20} = 1$ **3.** $\dfrac{x^2}{5} + \dfrac{y^2}{9} = 1$ **5.** $\dfrac{x^2}{25} + \dfrac{y^2}{16} = 1$ **7.** $\dfrac{x^2}{17} + y^2 = 1$

9. Major axis $= y$-axis;
   $a^2 = 9$; $b^2 = 4$; $c^2 = 5$;
   foci at $(0, \sqrt{5})$, $(0, -\sqrt{5})$;
   vertices at $(0, 3)$, $(0, -3)$

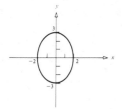

11. Major axis $= x$-axis;
   $a^2 = 16$; $b^2 = 4$; $c^2 = 12$;
   foci at $(2\sqrt{3}, 0)$, $(-2\sqrt{3}, 0)$;
   vertices at $(4, 0)$, $(-4, 0)$

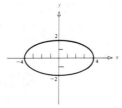

13. Major axis $= y$-axis;
   $a^2 = 4$; $b^2 = 1$; $c^2 = 3$;
   foci at $(0, +\sqrt{3})$, $(0, -\sqrt{3})$;
   vertices at $(0, 2)$, $(0, -2)$

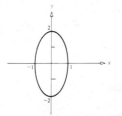

15. Circle with $R = 4$

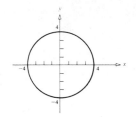

17. $y = -\frac{1}{2}x + 2$

19. (a) Circular   (b) Elliptical   (c) Flat (elongated) ellipse

21. $\dfrac{x^2}{100} + \dfrac{y^2}{64} = 1$

23. $\dfrac{x^2}{400} + \dfrac{y^2}{225} = 1$;

Interval	± 20	± 10	0
Height	0	$15\sqrt{3}/2$	15

25. $A = 4 \displaystyle\int_0^a \dfrac{b}{a} \sqrt{a^2 - x^2}\, dx = \pi ab$

27. $\dfrac{2x}{a^2} + \dfrac{2yy'}{b^2} = 0$; $y' = \dfrac{-b^2}{a^2} \dfrac{x_0}{y_0}$; $y - y_0 = \dfrac{-b^2}{a^2} \dfrac{x_0}{y_0}(x - x_0)$; $\dfrac{xx_0}{a^2} + \dfrac{yy_0}{b^2} = 1$

Exercise 4 (page 669)

1. $\dfrac{x^2}{16} - \dfrac{y^2}{20} = 1$ **3.** $\dfrac{y^2}{4} - \dfrac{x^2}{5} = 1$ **5.** $\dfrac{x^2}{9} - \dfrac{y^2}{16} = 1$ **7.** $\dfrac{5x^2}{16} - \dfrac{5y^2}{64} = 1$

9. Hyperbola;
  x-axis is transverse axis;
  center is $(0, 0)$;
  vertices at $(\pm 2, 0)$;
  foci at $(\pm\sqrt{13}, 0)$;
  asymptotes are $y = \pm\frac{3}{2}x$

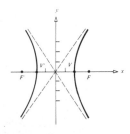

11. Hyperbola;
  x-axis is transverse axis;
  center is $(0, 0)$;
  vertices at $(\pm 4, 0)$;
  foci at $(\pm 2\sqrt{5}, 0)$;
  asymptotes are $y = \pm\frac{1}{2}x$

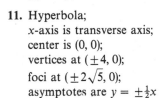

13. Hyperbola;
  y-axis is transverse axis;
  center is $(0, 0)$;
  vertices at $(0, \pm 2)$;
  foci at $(0, \pm\sqrt{5})$;
  asymptotes are $y = \pm 2x$

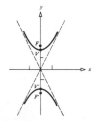

15. Hyperbola;
  x-axis is transverse axis;
  center is $(0, 0)$;
  vertices at $(\pm 4, 0)$;
  foci at $(\pm 4\sqrt{2}, 0)$;
  asymptotes are $y = \pm x$

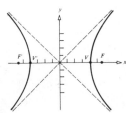

17. If e is close to 1, c is close to a, and hyperbola is very narrow. If e is very large, focus is further from vertex and hyperbola is very broad or wide.

19. $e = \sqrt{2}$

21. $a^2 = \cos^2\alpha$, $b^2 = \sin^2\alpha$; so $c^2 = a^2 + b^2 = \cos^2\alpha + \sin^2\alpha = 1$ or $c = \pm 1$, so foci are $(\pm 1, 0)$

23. (a) $y = \pm\frac{4}{5}x$ (b) $y = \pm\frac{1}{3}x$

25. (a) Tangent: $y = -x + 3$; normal: $y = x - 5$ (b) Tangent: $y = -\frac{3}{2}x + \frac{5}{2}$; normal: $y = \frac{2}{3}x - 4$

27. $\dfrac{x^2}{21{,}609} - \dfrac{y^2}{135{,}316} = 1$

29. $\displaystyle\lim_{x \to -\infty}(y_1 - y) = \lim_{x \to -\infty}\left[\frac{b}{a}\sqrt{x^2 - a^2} - \left(-\frac{b}{a}x\right)\right] = \frac{b}{a}\lim_{x \to -\infty}\left[\sqrt{x^2 - a^2} + x\right]$

$\displaystyle = \frac{b}{a}\lim_{x \to -\infty}(\sqrt{x^2 - a^2} + x)\frac{\sqrt{x^2 - a^2} - x}{\sqrt{x^2 - a^2} - x} = \frac{b}{a}\lim_{x \to -\infty}\frac{-a^2}{\sqrt{x^2 - a^2} - x} = -ab\lim_{x \to -\infty}\frac{1}{-2x} = 0;$

$\displaystyle\lim_{x \to +\infty}(y_2 - y) = \lim_{x \to +\infty}\left[-\frac{b}{a}\sqrt{x^2 - a^2} - \left(-\frac{b}{a}x\right)\right] = -\frac{b}{a}\lim_{x \to +\infty}\left[\sqrt{x^2 - a^2} - x\right] = -\frac{b}{a}\lim_{x \to +\infty}\frac{-a^2}{\sqrt{x^2 - a^2} + x}$

$\displaystyle = ab\lim_{x \to +\infty}\frac{1}{2x} = 0$

Exercise 5 (page 677)

1. Ellipse;
center at $(1, 0)$;

$(x - 1)^2 + \dfrac{y^2}{\frac{1}{4}} = 1$

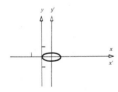

3. Hyperbola;
center at $(0, 1)$;

$\dfrac{x^2}{\frac{1}{4}} - \dfrac{(y - 1)^2}{\frac{1}{4}} = 1$

5. Parabola;
vertex at $(2, -3)$;

$(x - 2)^2 = \frac{1}{4}(y + 3)$

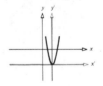

7. Ellipse;
center at $(-3, 1)$;

$\dfrac{(x + 3)^2}{16} + \dfrac{(y - 1)^2}{9} = 1$

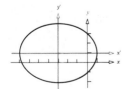

9. Hyperbola;
center at $(-3, 2)$;

$\dfrac{25(y - 2)^2}{9} - (x + 3)^2 = 1$

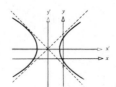

11. Hyperbola;
center at $(-1, 2)$;

$\dfrac{(x + 1)^2}{3} - \dfrac{(y - 2)^2}{4} = 1$

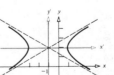

13. Hyperbola;
center at (0, 0);
$$\frac{x''^2}{8} - \frac{y''^2}{8} = 1;$$
rotation $= \pi/4$

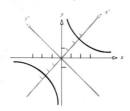

15. Ellipse;
center at (0, 0);
$$\frac{21x''^2}{20} + \frac{y''^2}{20} = 1;$$
rotation $= \pi/3$

17. Hyperbola;
center at (0, 0);
$$\frac{3y''^2}{5} - \frac{x''^2}{5} = 1;$$
rotation $= \pi/3$

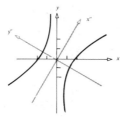

19. Parabola   **21.** Ellipse

23. $A'' + C'' = A\cos^2\theta + B\sin\theta\cos\theta + C\sin^2\theta + A\sin^2\theta - B\sin\theta\cos\theta + C\cos^2\theta$
$= A(\cos^2\theta + \sin^2\theta) + C(\sin^2\theta + \cos^2\theta) = A + C$

25. $\dfrac{(x'')^2}{4^2} + \dfrac{(y'')^2}{2^2} = 1$; an ellipse

Miscellaneous Exercises (page 678)

1. Ellipse; $\dfrac{x^2}{25} + \dfrac{y^2}{16} = 1$ **3.** Hyperbola; $\dfrac{x^2}{9} - \dfrac{y^2}{16} = 1$

5. $2yy' = 4a$; $y - y_0 = \dfrac{4a}{2y_0}(x - x_0)$; $y_0 y = 2a(x + x_0)$ **7.** $y = 1$; $2x - 3y - 5 = 0$

9. The parabola; $A_p = \frac{4}{3}bh$; $A_C = 4bh/\pi$

11. If $AC > 0$, then x^2 and y^2 have the same sign, so we have the sum of x^2 and y^2, and the equation is an ellipse.
If $A = C$, then the equation becomes $x^2 + y^2 + (D/A)x + (E/A)y + F/A = 0$, the equation of a circle.
If $AC = 0$, then one of A or C is 0; the equation is quadratic in one variable and linear in the other, hence, it is a parabola.
If $AC < 0$, then A and C are opposite in sign; thus, we have the difference of x^2 and y^2, so the graph is a hyperbola.

13. $x^2 - 2xy + y^2 - 2ax - 2ay + a^2 = 0$; $B^2 - 4AC = 0$; parabola   **15.** (a) 8π (b) 64

17. Let $d(F, P)/d(D, P) = \text{Constant} = k$; since $d(F, P) = \sqrt{(x - ea)^2 + y^2}$ and
$d(D, P) = \sqrt{[x - (a/e)]^2} = (1/e)\sqrt{(xe - a)^2}$, then

$$\frac{\sqrt{(x - ea)^2 + y^2}}{(1/e)\sqrt{(xe - a)^2}} = k \Rightarrow \frac{k^2}{e^2} = \frac{(x - ea)^2 + y^2}{(xe - a)^2}$$

But $y^2 = b^2 \left(\left(1 - \dfrac{x^2}{a^2} \right) \right.$ and $b^2 = a^2 - c^2 = a^2(1 - e^2)$, so $y^2 = (1 - e^2)(a^2 - x^2)$ and

$$\frac{k^2}{e^2} = \frac{x^2 - 2eax + e^2 a^2 + a^2 - a^2 e^2 - x^2 + x^2 e^2}{(xe - a)^2} = \frac{(xe - a)^2}{(xe - a)^2} = 1$$

$$k^2 = e^2$$

$$k = e$$

$$\frac{d(F, P)}{d(D, P)} = e$$

19. Problems 17 and 18 have shown that the definition of $e = c/a$ led to the standard equation of the ellipse $(c < a)$ and the hyperbola $(c > a)$. For the parabola, $d(F, P) = d(D, P)$ such that $d(F, P)/d(D, P) = e = 1$

21. $P_1 = (x_1, y_1)$; $P_2 = (x_2, y_2)$; $y_1^2 = 4ax_1$; $y_2^2 = 4ax_2$;

$$d(P_1, P_2) = d(P_1, F) + d(F, P_2) = \sqrt{(x_1 - a)^2 + y_1^2} + \sqrt{(x_2 - a)^2 + y_2^2} = 2a + x_1 + x_2$$

23. Length $= 2y = 2b \sqrt{1 - \dfrac{c^2}{a^2}} = \dfrac{2b}{a}\sqrt{a^2 - c^2} = \dfrac{2b^2}{a}$

25. At $x = c$, $y' = -\dfrac{b}{a} \dfrac{x}{\sqrt{a^2 - x^2}} = \dfrac{-bc}{ab} = -\dfrac{c}{a} = -e$ (upper branch). At $x = c$, $y' = e$ (lower branch).

27. Asymptote $y = \dfrac{b}{a} x$; $y = -\dfrac{a}{b}(x - c)$;

$$\frac{b}{a} x = -\frac{a}{b} x + \frac{ac}{b}$$

$$\frac{b^2 + a^2}{ab} x = \frac{ac}{b}$$

$$x = \frac{(ab)(ac)}{(c^2)(b)} = \frac{a^2}{c} = \frac{a}{e}$$

CHAPTER 13

Exercise 1 (page 686)

1.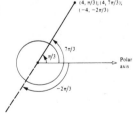

3.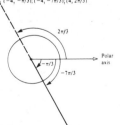

5.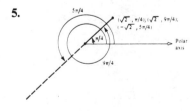

7.

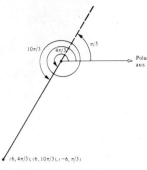

9. $(3\sqrt{3}, 3)$ **11.** $(-3\sqrt{3}, 3)$ **13.** $(0, 5)$ **15.** $(2, -2)$

17.

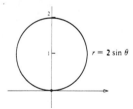

19.

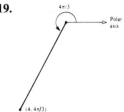

21.

23. $r^2\left(\dfrac{\cos^2\theta}{4} + \dfrac{\sin^2\theta}{9}\right) = 1$ **25.** $r = 4\cos\theta$ **27.** $r^2\cos^2\theta + 4r\sin\theta - 1 = 0$ **29.** $r^2\sin 2\theta = 2$

31. $x^2 + y^2 - x = 0$ **33.** $(x^2 + y^2)^{3/2} - y = 0$ **35.** $y^2 = 8(x + 2)$ **37.** $y = x\tan(x^2 + y^2)$

39. $x^2 + y^2 = 4$ **41.** Apply the law of cosines to the triangle OP_1P_2.

Exercise 2 (page 695)

1. Intercepts: $(0, 0)$, $(2, \pi/2)$;
symmetry: $\pi/2$ axis;
tangent at pole: $\theta = 0$

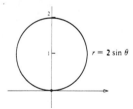

3. Intercept: $(0, 0)$;
symmetry: origin;
tangent at pole: $\theta = 11\pi/6$

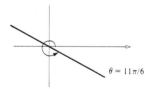

5. Intercepts: $(4, 0)$, $(4, \pi/2)$, $(4, \pi)$, $(4, 3\pi/2)$;
symmetry: polar axis, $\pi/2$ axis, pole;
tangent at pole: none

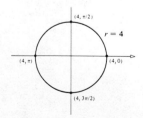

7. Intercepts: $(0, 0)$, $(1, \pi/2)$, $(2, \pi)$, $(1, 3\pi/2)$;
symmetry: polar axis;
tangent at pole: $\theta = 0$

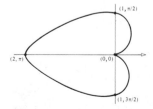

9. Intercepts: $(8, 0)$, $(4, \pi/2)$, $(0, \pi)$, $(4, 3\pi/2)$;
symmetry: polar axis;
tangent at pole: $\theta = \pi$

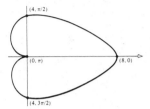

11. Intercepts: $(4, 0)$, $(1, \pi/2)$, $(4, \pi)$, $(7, 3\pi/2)$;
symmetry: $\pi/2$ axis;
tangent at pole: none

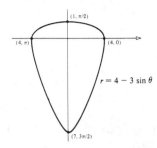

13. Intercept: $(0, 0)$;
symmetry: polar axis, $\pi/2$ axis, pole;
tangents at pole: $\theta = 0$, $\theta = \pi/2$, $\theta = \pi$, $\theta = 3\pi/2$

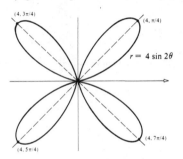

15. Intercepts: $(0, \pi/2)$, $(3, 0)$;
symmetry: polar axis;
 tangents at pole: $\theta = \pi/6$,
$\theta = \pi/2$, $\theta = 5\pi/6$,
$\theta = 7\pi/6$, $\theta = 3\pi/2$,
$\theta = 11\pi/6$

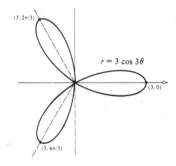

17. Intercepts: see graph;
symmetry: none;
tangent at pole: $\theta = 0$

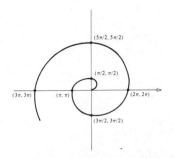

19. Intercepts: $(4, 0)$, $(0, \pi/4)$, $(-4, 0)$;
symmetry: polar axis, $\pi/2$ axis, pole;
tangents at pole: $\theta = \pi/4$, $\theta = 3\pi/4$, $\theta = 5\pi/4$,
$\theta = 7\pi/4$

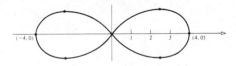

21. Intercepts: $(-1, \pi/2), (0, \pi/6)$;
symmetry: $\pi/2$ axis;
tangents at pole: $\theta = \pi/6, \theta = 5\pi/6$

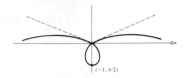

23. Intercepts: $(0, 0), (0, \pi), (2, \pi/2), (-2, \pi/2)$;
symmetry: polar axis, $\pi/2$ axis, pole;
tangent at pole: $\theta = 0$

25. Intercepts: $(1, \pi), (2, \pi/2), (2, 3\pi/2)$;
symmetry: polar axis;
tangent at pole: none

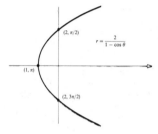

27. $\dfrac{|FP|}{|DP|} = e; |FP| = r, |DP| = d + r\cos\theta, r = e(d + r\cos\theta); r = \dfrac{ed}{1 - e\cos\theta}$

Exercise 3 (page 698)

1. $(\pi/2, \pi/8)$

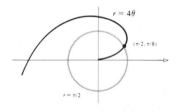

3. $(1, \pi/2), (1, 3\pi/2)$

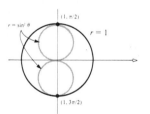

5. $(1, \pi/6), (1, 5\pi/6)$

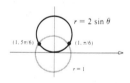

7. $(0, 0), (4, \pi)$

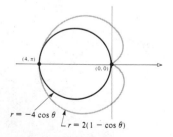

9. $(2, \pi/2), (2, 3\pi/2), (0, 0)$

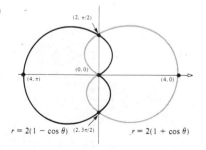

11. $(0, 0), (\sqrt{2}/2, \pi/8), (\sqrt{2}/2, 3\pi/8),$
$(\sqrt{2}/2, 5\pi/8), (\sqrt{2}/2, 7\pi/8),$
$(\sqrt{2}/2, 9\pi/8), (\sqrt{2}/2, 11\pi/8),$
$(\sqrt{2}/2, 13\pi/8), (\sqrt{2}/2, 15\pi/8)$

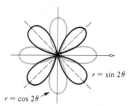

13. $(2 - \sqrt{2}, \pi/4), (2 - \sqrt{2}, -\pi/4),$
$(2 + \sqrt{2}, 3\pi/4), (2 + \sqrt{2}, -3\pi/4)$

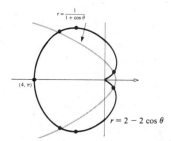

15. 4.335×10^7 miles; 2.8541×10^7 miles

Exercise 4 (page 702)

1. $A = \dfrac{3\pi}{4} + \dfrac{9\sqrt{3}}{16}$

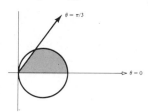

3. $A = \dfrac{4a^2\pi^3}{3}$

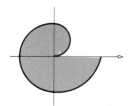

5. $A = \dfrac{3\pi}{2}$

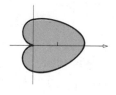

7. $A = \dfrac{3\pi}{2}$

9. $A = 2\pi$

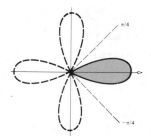

11. $A = 2$

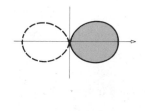

13. $A = \dfrac{\pi}{3} + \dfrac{\sqrt{3}}{2}$

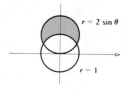

15. $A = 1 - \dfrac{\pi}{4}$

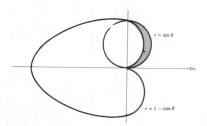

17. $A = \pi - \dfrac{3\sqrt{3}}{2}$

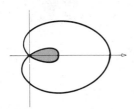

19. $A = \dfrac{32\pi}{3} + 4\sqrt{3}$

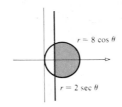

21. $A = \dfrac{9\sqrt{3}}{2} - \pi$

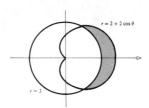

23. $A = \dfrac{7\pi}{12} - \sqrt{3}$

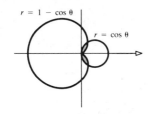

25. $A = 2\sqrt{2}$

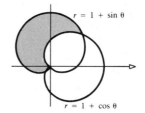

Exercise 5 (page 706)

1. $1 + \sqrt{2}$ **3.** -1 **5.** $\sqrt{3}/2$ **7.** $-\pi/2$ **9.** $-5/\sqrt{3}$

11. Horizontal: $(2, \pi/2)$, $(2, 3\pi/2)$; vertical: $(2, 0)$, $(2, \pi)$

13. Horizontal: $(\sqrt{2}, \pi/4)$, $(\sqrt{2}, -\pi/4)$; vertical: $(0, \pi/2)$, $(2, 0)$

15. Horizontal: $(0, \pi)$, $(\frac{9}{2}, -\pi/3)$, $(\frac{9}{2}, \pi/3)$; vertical: $(6, 0)$, $(\frac{3}{2}, 2\pi/3)$, $(\frac{3}{2}, 4\pi/3)$

17. Vertical: $(\frac{3}{2}, \pi/6)$, $(\frac{3}{2}, 5\pi/6)$, $(0, 3\pi/2)$; horizontal: $(2, \pi/2)$, $(\frac{1}{2}, 7\pi/6)$, $(\frac{1}{2}, 11\pi/6)$

19. The acute angle of intersection at $(\frac{3}{2}, \pi/3)$ and at $(\frac{3}{2}, 5\pi/3)$ is $\pi/6$; the graphs meet at a right angle at the pole.

21. $\tan \psi_1 = -\cot \theta$; $\tan \psi_2 = \tan \theta$; $(\tan \psi_1)(\tan \psi_2) = -1$ **23.** $\tan \psi = 1/\alpha$

Exercise 6 (page 715)

1. $x = 3$

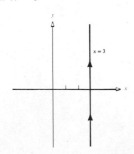

3. $x = 2$; $y \geq 4$

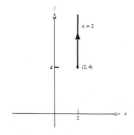

5. $y^2 = x - 5$; $x \geq 5$, $y \geq 0$

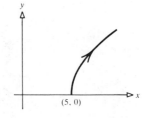

7. $y = (x - 1)^3$;
$x \geq 2, y \geq 1$

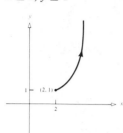

9. $y = \ln x$

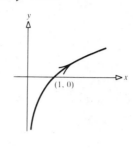

11. $x^2 - y^2 = 1; x \geq 1$

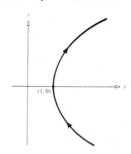

13. $\dfrac{(x + 2)^2}{9} + \dfrac{y^2}{4} = 1$

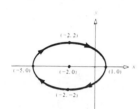

15. (*1*) $x = t$ (*2*) $x = t^{1/3}$
        $y = 4t^3$ $y = 4t$

17. (*1*) $x = t - 3$ (*2*) $x = \frac{1}{3}t^{1/2} - 3$
        $y = 9t^2$ $y = t, \quad t \geq 0$

19. $x = 3 \cos(2\pi t/3)$
     $y = 2 \sin(2\pi t/3)$

21. $\frac{4}{3}$ **23.** $e^2/2$ **25.** $-\frac{4}{3}$ **27.** (a) $(4, \pm 16\sqrt{3}/9)$ (b) $(0, 0)$ **29.** (a) None (b) None

31. (a) $(0, 0)$ (b) None   **33.** $y = 2x - 8$ **35.** $y = -\dfrac{\sqrt{2}}{4}x + \dfrac{\sqrt{2}}{2}$

37. $\dfrac{dy}{dx} = \dfrac{\sin t + \cos t}{\cos t - \sin t}; \dfrac{d^2y}{dx^2} = \dfrac{2}{e^t(\cos t - \sin t)^3}$ **39.** $\dfrac{dy}{dx} = \dfrac{t^2}{t^2 - 1}; \dfrac{d^2y}{dx^2} = \dfrac{-2t^3}{(t^2 - 1)^3}$

41. $\dfrac{dy}{dx} = \tan\theta; \dfrac{d^2y}{dx^2} = \dfrac{1}{a\theta\cos^3\theta}$

43. (a) $y = (\tan\theta)x - \dfrac{16}{v_0^2 \cos^2\theta}x^2$ (b) $v(1) = (v_0^2 - 64v_0 \sin\theta + 1024)^{1/2}; v(2) = (v_0^2 - 128v_0 \sin\theta + 4096)^{1/2}$

   (c) $y = 0 \Rightarrow t = \dfrac{v_0 \sin\theta}{16}; x = \dfrac{v_0^2 \sin 2\theta}{32}$

45. $x = 3b \cos t + b \cos 3t = a \cos^3 t; y = 3b \sin t - b \sin 3t = a \sin^3 t; (x/a)^{2/3} + (y/a)^{2/3} = 1$

47. $dy/dx = (-b^2/a^2)(x/y); d^2y/dx^2 = -b^4/a^2y^3$

Exercise 7 (page 723)

1. $\frac{8}{27}(10\sqrt{10} - 1)$ **3.** $\sqrt{5} + \frac{1}{2}\ln(2 + \sqrt{5})$ **5.** 4π **7.** 4π **9.** $\sqrt{5}(e - 1)$ **11.** 2

13. $\pi\sqrt{1 + 4\pi^2} + \frac{1}{2}\ln(2\pi + \sqrt{1 + 4\pi^2})$ **15.** $3b/2$ **17.** $5 + \frac{9}{4}\ln 3$ **19.** $\dfrac{1}{2}(3\sqrt{11} - \sqrt{3}) + \ln\dfrac{3 + \sqrt{11}}{1 + \sqrt{3}}$

21. $\sqrt{5}$ **23.** $\sqrt{0.0411} \approx 0.2028$ **25.** $a/10$ **27.** $P_n = 2n \sin(\pi/n); \lim_{n \to +\infty} P_n = 2\pi$

Exercise 8 (page 729)

1. $2/5\sqrt{5}$ **3.** $\sqrt{2}$ **5.** $\frac{2}{27}$ **7.** $30/13\sqrt{13}$ **9.** $1/7a\sqrt{7}$ **11.** 1 **13.** $1/2\sqrt{2}$

15. $3t_1^4/[2(t_1^6 + 1)^{3/2}]$ **17.** $5\sqrt{10}/3$ **19.** 1 **21.** $3\sqrt{2}$ **23.** 6

25. $\rho = [1 + (2ax + b)^2]^{3/2}/2a$ is a minimum when $x = -b/2a$ **27.** $(\sqrt{2}/2, -\ln\sqrt{2})$ **29.** $\pm(1/\sqrt[4]{5}, 1/3\sqrt[4]{125})$

31. $23/7\sqrt{7}$ **33.** $3/2\sqrt{2}a$ **35.** $\kappa = \dfrac{(1/a)\cosh(x/a)}{[1 + \sinh^2(x/a)]^{3/2}} = \dfrac{1}{a\cosh^2(x/a)}$ **37.** Zero   **39.** $\frac{1}{4}$

Exercise 9 (page 733)

1. $24\pi(2\sqrt{2} - 1)$ **3.** $6\pi a^2/5$ **5.** $(\pi/27)(10\sqrt{10} - 1)$ **7.** $(\pi a^2/4)(e^2 - e^{-2}) + \pi a^2$ **9.** $4\pi a^2$

11. $\pi[e\sqrt{1 + e^2} + \ln(e + \sqrt{1 + e^2}) - \sqrt{2} - \ln(1 + \sqrt{2})]$ **13.** $6\pi a^2/5$ **15.** $(24\pi/5)(\sqrt{2} + 1)$

17. $\pi^2/2$ **19.** $(2\sqrt{2}\pi/5)(1 + e^{2\pi})$ **21.** $S = 2\pi \int_1^{+\infty} \dfrac{\sqrt{1 + x^4}}{x^3}\, dx$ diverges; $V = \pi \int_1^{+\infty} \dfrac{1}{x^2}\, dx = \pi$ converges

23. $(\pi/2)[\sinh 2b - \sinh 2a + 2(b - a)]$

Miscellaneous Exercises (page 734)

1. $(e^2 - 1)/4e^2$ **3.** $(\pi - 1)/2\pi$ **5.** 4π **7.** $y = 2x/(x + 1), x > 0$ **9.** $y^2 = 4(x^2 - 4), x \ge 2$

11. $(x + 2)^2 + [(y - 4)^2/4] = 1$ **13.** $1/3t; -1/18t^3$ **15.** $-1/(8\sin\theta); -1/(64\sin^3\theta)$ **17.** $\frac{1}{2}\tan\theta; -\frac{1}{4}\tan^3\theta$

19. $4\pi R\sqrt{R^2 - a^2}$ **21.** (a)   (b)   (c)   (d)

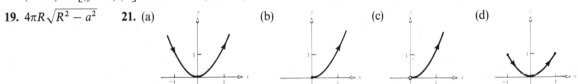

23. $\dfrac{f'g'' - f''g'}{(f')^3}$ **25.** $\tan\phi = y'$ **27.** $(\pi/2, 0)$ **29.** $(\frac{7}{4}, \frac{7}{4})$

31. Horizontal: $(\frac{10}{3}, \pm 16/3\sqrt{3})$;   **33.** $\pi + 3\sqrt{3}$ **35.** $x = m - 1; y = m^2 - 2m + 1$
vertical: $(2, 0); y = 2(x - 6), y = -2(x - 6)$

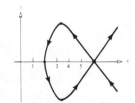

37. $(\frac{68}{3}, -8); (\frac{4}{3}, -24)$ **39.** $(a/2, b/2); \frac{1}{2}\sqrt{a^2 + b^2}$ **41.** $(\pi/6)(2\sqrt{2} - 1)$

43. $h = \dfrac{a^2 - b^2}{a}\cos^3\theta; k = \dfrac{b^2 - a^2}{b}\sin^3\theta; (ah)^{2/3} + (bk)^{2/3} = (a^2 - b^2)^{2/3}$

CHAPTER 14

Exercise 1 (page 743)

1. All are scalars except (c) and (f).   **3.** A   **5.** $-F + E - D$ **7.** $-G - H + D$ **9.** 0   **11.** 12
13. $\frac{1}{2}(v + w)$

Exercise 2 (page 748)

1. $4i - 5j$ **3.** $-i + j$ **5.** $xi + yj$ **7.** 5   **9.** $\sqrt{2}$ **11.** $\sqrt{2}\,|a|$ **13.** $3i - 8j$ **15.** $\frac{7}{6}i$
17. $\sqrt{26}$ **19.** $\frac{5}{13}i - \frac{12}{13}j$; $\frac{-5}{13}i + \frac{12}{13}j$ **21.** $(2/\sqrt{5})i + (1/\sqrt{5})j$; $(-2/\sqrt{5})i - (1/\sqrt{5})j$
23. $\frac{1}{2}i + (\sqrt{3}/2)j$; $-\frac{1}{2}i - (\sqrt{3}/2)j$ **25.** $v = 2\sqrt{3}i \pm 2j$ **27.** $v = 2\sqrt{5}i + \sqrt{5}j$; $v = -2\sqrt{5}i - \sqrt{5}j$
29. $v = -i + j$ **31.** 459.54 km/hr   **33.**

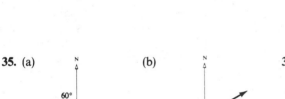

35. (a)   (b)   **37.** 217.54 km/hr   **39.** $-2 \pm \sqrt{21}$

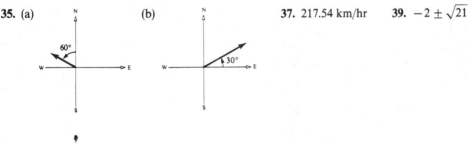

Exercise 3 (page 754)

1.   **3.**   **5.**

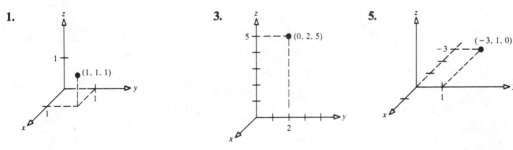

7. $(0, 0, 3)$; $(0, 1, 0)$; $(2, 0, 0)$; $(0, 1, 3)$; $(2, 1, 0)$; $(2, 0, 3)$ **9.** $(1, 2, 5)$; $(1, 4, 3)$; $(3, 2, 3)$; $(1, 4, 5)$; $(3, 4, 3)$; $(3, 2, 5)$
11. $(-1, 0, 5)$; $(-1, 2, 2)$; $(4, 0, 2)$; $(-1, 2, 5)$; $(4, 0, 5)$; $(4, 2, 2)$

13. Plane parallel to xz-plane, 3 units to the right of it   **15.** The yz-plane

17. Plane parallel to xy-plane, 5 units above it   **19.** $\sqrt{17}$ **21.** $\sqrt{57}$ **23.** $\sqrt{26}$

25. $(x-3)^2 + (y-1)^2 + (z-1)^2 = 1$ **27.** $(x+1)^2 + (y-1)^2 + (z-2)^2 = 9$ **29.** $2; (-1, 1, 0)$

31. $3; (2, -2, -1)$ **33.** $3/\sqrt{2}; (2, 0, -1)$ **35.** $x^2 + (y-3)^2 + (z-6)^2 = 17$

37. $(x+3)^2 + (y-2)^2 + (z-1)^2 = 62$ **39.** Use Pythagorean theorem and distance formula

41. Use Pythagorean theorem and distance formula   **43.** Each side has length $3\sqrt{2}$.

Exercise 4 (page 760)

1. $-3\mathbf{i} - 2\mathbf{j} + \mathbf{k}$ **3.** $3\mathbf{i} - \mathbf{k}$ **5.** $x\mathbf{i} + y\mathbf{j} + z\mathbf{k}$ **7.** $-3\mathbf{i} + \mathbf{j} - 5\mathbf{k}$ **9.** $\sqrt{13}$ **11.** $17\mathbf{i} - 3\mathbf{j} + 13\mathbf{k}$

13. $\sqrt{453}$ **15.** $\sqrt{109}$ **17.** $\pm[(-8/\sqrt{29})\mathbf{i} + (12/\sqrt{29})\mathbf{j} - (16/\sqrt{29})\mathbf{k}]$ **19.** $\sqrt{21}; 4/\sqrt{21}, 2/\sqrt{21}, -1/\sqrt{21}$

21. $\sqrt{3}; 1/\sqrt{3}, 1/\sqrt{3}, 1/\sqrt{3}$ **23.** $\sqrt{3}\,a; 1/\sqrt{3}, 1/\sqrt{3}, 1/\sqrt{3}$ **25.** $\frac{3}{2}\mathbf{i} + (3/\sqrt{2})\mathbf{j} + \frac{3}{2}\mathbf{k}$ **27.** $\sqrt{2}\mathbf{i} - \mathbf{j} + \mathbf{k}$

29. $\mathbf{c} - \mathbf{b}; \mathbf{d} - \mathbf{b}; \mathbf{d} - \mathbf{c}$ **31.** $\alpha = \pm 1$

33. (a) $\cos^2(\pi/6) + \cos^2(\pi/4) > 1$ (b) $\cos^2\beta + \cos^2\gamma = \sin^2\alpha \Rightarrow \sin\alpha \geq \cos\beta \Rightarrow \pi/4 \leq \alpha, \beta \leq \pi/2$

Exercise 5 (page 769)

1. $6; \sqrt{\frac{6}{7}}$ **3.** $-1; -\frac{1}{2}$ **5.** $-8; -8/\sqrt{66}$ **7.** $2\sqrt{3}$ **9.** $-1/\sqrt{2}$ **11.** $-8/\sqrt{6}$ **13.** 1 **15.** -2

17. $0, 4$ **19.** Remember, $\|\mathbf{i}\| = 1$ **21.** $2\sqrt{10 + 3\sqrt{3}}; 2\sqrt{10 - 3\sqrt{3}}$ **23.** 9 **25.** $\frac{8}{3}$ joules

27. $-1, 2$ **29.** Approx. $70.5°$ **31.** $-275\mathbf{i} + 75\sqrt{3}\mathbf{j}$ **33.** (a) $(\mathbf{u} + \mathbf{v}) \cdot (\mathbf{u} - \mathbf{v}) = \|\mathbf{u}\|^2 - \|\mathbf{v}\|^2 = 0$

35. $(\mathbf{v} - \alpha\mathbf{w}) \cdot \mathbf{w} = \mathbf{v} \cdot \mathbf{w} - \alpha\|\mathbf{w}\|^2 = 0$ **37.** 0 **39.** Use (14.13) and (14.18)   **41.** $\sin(\theta/2)$

43. $\mathbf{u} = a_1\mathbf{i} + b_1\mathbf{j} + c_1\mathbf{k}, \mathbf{v} = a_2\mathbf{i} + b_2\mathbf{j} + c_2\mathbf{k}, \mathbf{w} = a_3\mathbf{i} + b_3\mathbf{j} + c_3\mathbf{k};$
$\mathbf{u} \cdot (\mathbf{v} + \mathbf{w}) = a_1(a_2 + a_3) + b_1(b_2 + b_3) + c_1(c_2 + c_3) = a_1a_2 + a_1a_3 + b_1b_2 + b_1b_3 + c_1c_2 + c_1c_3;$
$\mathbf{u} \cdot \mathbf{v} = a_1a_2 + b_1b_2 + b_1b_3; \mathbf{u} \cdot \mathbf{w} = a_1a_3 + b_1b_3 + c_1c_3$

45. $\mathbf{0} \cdot \mathbf{v} = 0 \cdot a + 0 \cdot b + 0 \cdot c = 0$ **47.** $\mathbf{v} = \mathbf{0}$

Exercise 6 (page 777)

1. $-3(\mathbf{j} + \mathbf{k})$ **3.** $-2\mathbf{k}$ **5.** $-\mathbf{i} + \mathbf{j} + 5\mathbf{k}$ **7.** $-6\mathbf{i} + 21\mathbf{j} - 58\mathbf{k}$ **9.** $-8\mathbf{i} + 12\mathbf{j} + 5\mathbf{k}$ **11.** $\sqrt{1013}$

13. 58 **15.** $46\sqrt{2}$ **17.** $5\sqrt{3}$ **19.** $\sqrt{998}$ **21.** Use (14.28(c))

23. $\alpha(\mathbf{v} \times \mathbf{w}) = \alpha[(b_1c_2 - b_2c_1)\mathbf{i} - (a_1c_2 - a_2c_1)\mathbf{j} + (a_1b_2 - a_2b_1)\mathbf{k}]$
$= [(\alpha b_1)c_2 - b_2(\alpha c_1)]\mathbf{i} - [(\alpha a_1)c_2 - a_2(\alpha c_1)]\mathbf{j} + [(\alpha a_1)b_2 - a_2(\alpha b_1)]\mathbf{k} = (\alpha\mathbf{v}) \times \mathbf{w}$

25. $\mathbf{i} \times (\mathbf{i} \times \mathbf{j}) = -\mathbf{j}; (\mathbf{i} \times \mathbf{i}) \times \mathbf{j} = 0$ **27.** Use (14.26) and the definition of a third-order determinant

29. $\mathbf{u} \times (\mathbf{v} \times \mathbf{w}) = (a_1\mathbf{i} + b_1\mathbf{j} + c_1\mathbf{k}) \times [(b_2c_3 - b_3c_2)\mathbf{i} - (a_2c_3 - a_3c_2)\mathbf{j} + (a_2b_3 - a_3b_2)\mathbf{k}]$
$= [b_1(a_2b_3 - a_3b_2) + c_1(a_2c_3 - a_3c_2)]\mathbf{i} - [a_1(a_2b_3 - a_3b_2) - c_1(b_2c_3 - b_3c_2)]\mathbf{j}$
$+ [-a_1(a_2c_3 - a_3c_2) - b_1(b_2c_3 - b_3c_2)]\mathbf{k};$
$(\mathbf{u} \cdot \mathbf{w})\mathbf{v} = (a_1a_3 + b_1b_3 + c_1c_3)(a_2\mathbf{i} + b_2\mathbf{j} + c_2\mathbf{k}); (\mathbf{u} \cdot \mathbf{v})\mathbf{w} = (a_1a_2 + b_1b_2 + c_1c_2)(a_3\mathbf{i} + b_3\mathbf{j} + c_3\mathbf{k})$

31. $\pm(\frac{3}{7}\mathbf{i} - \frac{2}{7}\mathbf{j} + \frac{6}{7}\mathbf{k})$

33. $(\mathbf{a} \times \mathbf{b}) \cdot (\mathbf{c} \times \mathbf{d}) = \mathbf{a} \cdot [\mathbf{b} \times (\mathbf{c} \times \mathbf{d})] = \mathbf{a} \cdot [(\mathbf{b} \cdot \mathbf{d})\mathbf{c} - (\mathbf{b} \cdot \mathbf{c})\mathbf{d}] = (\mathbf{a} \cdot \mathbf{c})(\mathbf{b} \cdot \mathbf{d}) - (\mathbf{a} \cdot \mathbf{d})(\mathbf{b} \cdot \mathbf{c})$

35. $(\mathbf{u} - \mathbf{v}) \cdot (\mathbf{u} + \mathbf{v}) = 0$ if and only if $\mathbf{u} \cdot \mathbf{u} - \mathbf{v} \cdot \mathbf{v} = 0$ or $\|\mathbf{u}\| = \|\mathbf{v}\|$

37. $\dfrac{1}{\alpha^2 + \mathbf{a} \cdot \mathbf{a}} \left[\alpha\mathbf{b} - (\mathbf{b} \times \mathbf{a}) + \dfrac{\mathbf{b} \cdot \mathbf{a}}{\alpha} \, \mathbf{a} \right]$

Exercise 7 (page 785)

1. $\mathbf{r} = (1 + 2t)\mathbf{i} + (2 - t)\mathbf{j} + (3 + t)\mathbf{k}$; $x = 1 + 2t$, $y = 2 - t$, $z = 3 + t$; $(x - 1)/2 = (y - 2)/(-1) = (z - 3)/1$

3. $\mathbf{r} = (4 + t)\mathbf{i} + (-1 + t)\mathbf{j} + 6\mathbf{k}$; $x = 4 + t$, $y = -1 + t$, $z = 6$; $x - 4 = y + 1$, $z = 6$

5. $\mathbf{r} = (1 + 3t)\mathbf{i} + (-1 + 3t)\mathbf{j} + (3 - 2t)\mathbf{k}$; $x = 1 + 3t$, $y = -1 + 3t$, $z = 3 - 2t$;
$(x - 1)/3 = (y + 1)/3 = (z - 3)/(-2)$ **7.** $\mathbf{r} = 2t\mathbf{i} + 3t\mathbf{j} + \mathbf{k}$; $x = 2t$, $y = 3t$, $z = 1$; $x/2 = y/3$, $z = 1$

9. $(1, 6, -4)$; $\sqrt{\frac{6}{11}}$ **11.** $(-2, -3, -5)$; $-7/(3\sqrt{11})$ **13.** $x/8 = y/(-10) = z/(-7)$

15. $x = 1$, $y - 2 = (z + 1)/2$ **17.** Parallel   **19.** Intersect at $(\frac{17}{4}, \frac{11}{4}, \frac{3}{2})$ **21.** Intersect at $(\frac{25}{3}, \frac{13}{3}, \frac{19}{3})$

23. Parallel   **25.** Skew   **27.** Parallel   **29.** $x = 1 + 6t$, $y = -2 + 2t$, $z = -3 - t$ **31.** Parallel

33. $\mathbf{i} + \mathbf{j} + 2\mathbf{k}$; $(-1, -3, -4)$, $(1, -1, 0)$ **35.** $10\sqrt{78}$ m/sec   **37.** $8\sqrt{\frac{3}{14}}\, \omega$ m/sec; $\omega = 14/\sqrt{3}$ rad/sec

Exercise 8 (page 791)

1. $z = 4$ **3.** $y = -2$ **5.** $2x - y + z = 5$ **7.** $2x + y + 3z = 16$ **9.** $x + 2y - z = 12$

11. $x - y + 3z = -5$ **13.** $2x + 5y - 2z = 21$ **15.** $2x + 3y + 2z = -1$ **17.** $(2, 3, 0)$ and $(0, 0, 0)$

19. $x = 1 + 2t$, $y = 2 - t$, $z = -1 + t$ **21.** $(3, 1, 2)$ **23.** $(4, -2, 3)$ **25.** $x + y + 3z = 0$

27. $3x + 7y - 6z = 11$ **29.** $2x + 6y + 29z = 2$ **31.** Intersect; $\sqrt{2}/3$ **33.** Intersect; $17/\sqrt{406}$

35. Intersect; $5\sqrt{3}/\sqrt{106}$ **37.** Identical   **39.** $2/\sqrt{6}$ **41.** $12/\sqrt{11}$

Exercise 9 (page 803)

1. Elliptic paraboloid

3. Ellipsoid

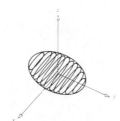

5. Elliptic cone

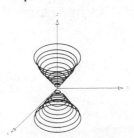

7. Parabolic cylinder

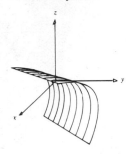

9. Hyperboloid of two sheets

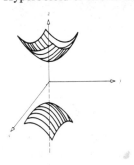

11. Hyperbolic cylinder

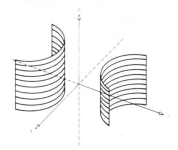

13. Elliptic paraboloid

15. Hyperbolic cylinder

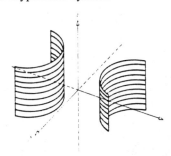

17. Elliptic cone

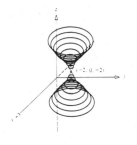

$(-2, 0, -2)$

19. Elliptic paraboloid

21. Hyperboloid of two sheets

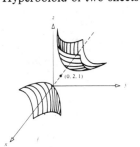

$(0, 2, 1)$

23. z is missing; yes

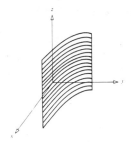

25. J   **27. H**   **29. G**   **31.** (a) **L** (b) **O** (c) **K** (d) **M** (e) **N**

Historical Exercises (page 807)

1. (a) $2 + 4i$ (b)

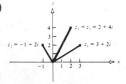

$z_1 + z_2 = 2 + 4i$
$z_2 = -1 + 2i$
$z_1 = 3 + 2i$

3. (a) $z \diamond w = (ac + bd) + (bc - ad)i$; complex

5. (a) $(0, 1, 0, 0) \times (0, 0, 1, 0) \neq (0, 0, 1, 0) \times (0, 1, 0, 0)$ (b) Yes; $(1, 0, 0, 0)$

Miscellaneous Exercises (page 808)

1. (a) $\|\mathbf{u}_r\| = 1$, $\|\mathbf{u}_\theta\| = 1$ (b) $\mathbf{u}_r \cdot \mathbf{u}_\theta = 0$

3. (a) $\sin^2\theta = 1 - \cos^2\theta$, $\cos\theta = \dfrac{\mathbf{v} \cdot \mathbf{w}}{\|\mathbf{v}\|\,\|\mathbf{w}\|}$ (b) Set $\sin\theta = 0$; it is an equality when $\mathbf{v}$ and $\mathbf{w}$ are parallel.

5. $x = 1 + t$, $y = t$, $z = -1$; the line lies in the plane $z = -1$ **7.** $y - 3z = 3$ **9.** $x + y - 2z = 1$

13. (a) Because $\mathbf{v} \cdot \mathbf{u} = 0$ and $\mathbf{w} \cdot \mathbf{u} = 0$

15. Lay the vectors end to end to form a closed figure.

17. $\|\mathbf{w}_1\| = \left\|\dfrac{\mathbf{u}_1}{\|\mathbf{u}_1\|}\right\| = 1$; $\mathbf{w}_1 \cdot \mathbf{w}_2 = \mathbf{w}_1 \cdot \dfrac{[\mathbf{u}_2 - (\mathbf{u}_2 \cdot \mathbf{w}_1)\mathbf{w}_1]}{\|\mathbf{v}_2\|} = \dfrac{\mathbf{w}_1 \cdot \mathbf{u}_2 - \mathbf{u}_2 \cdot \mathbf{w}_1}{\|\mathbf{v}_2\|} = 0$

19. $\mathbf{u} = a_1\mathbf{i} + b_1\mathbf{j}$, $\mathbf{v} = a_2\mathbf{i} + b_2\mathbf{j}$, $a_1b_2 - a_2b_1 \neq 0$; if $\mathbf{w} = w_1\mathbf{i} + w_2\mathbf{j}$, then $s = \dfrac{b_2w_1 - a_2w_2}{a_1b_2 - a_2b_1}$, $\quad t = \dfrac{a_1w_2 - b_1w_1}{a_1b_2 - a_2b_1}$

21. $(x - 4)^2 + (y - 2)^2 + (z + 8)^2 = 26$; $(x - 2)^2 + y^2 + z^2 = 26$ **23.** $x^2 + y^2 + z^2 - 2x + 4y + 2z = 3$

25. $\left(1 + \dfrac{\sqrt{2}}{2}\right)x + y + \left(-1 + \dfrac{\sqrt{2}}{2}\right)z = 0$; $\left(1 - \dfrac{\sqrt{2}}{2}\right)x + y + \left(-1 - \dfrac{\sqrt{2}}{2}\right)z = 0$ **27.** 32   **29.** 0

31. $\dfrac{x - \frac{10}{3}}{-2} = \dfrac{y + \frac{2}{3}}{7} = \dfrac{z}{3}$ **33.** Similar to the proof of (14.41)

35. $\cos\theta_{\mathbf{u}\mathbf{w}} = \dfrac{\|\mathbf{v}\|\,\|\mathbf{u}\|^2 + \|\mathbf{u}\|(\mathbf{u} \cdot \mathbf{v})}{\|\mathbf{u}\|\,\|\mathbf{w}\|} = \dfrac{\|\mathbf{v}\|\,\|\mathbf{u}\|}{\|\mathbf{w}\|} + \dfrac{\mathbf{u} \cdot \mathbf{v}}{\|\mathbf{w}\|}$; $\cos\theta_{\mathbf{v}\mathbf{w}} = \dfrac{\|\mathbf{v}\|(\mathbf{u} \cdot \mathbf{v}) + \|\mathbf{u}\|\,\|\mathbf{v}\|^2}{\|\mathbf{v}\|\,\|\mathbf{w}\|} = \dfrac{\mathbf{u} \cdot \mathbf{v}}{\|\mathbf{w}\|} + \dfrac{\|\mathbf{u}\|\,\|\mathbf{v}\|}{\|\mathbf{w}\|}$

37. $x - 2 = y/(-1) = (z + 3)/(-1)$

39. $(\mathbf{r} - \mathbf{b}) \cdot (\mathbf{r} + \mathbf{b}) = 0 \Rightarrow \|\mathbf{r}\| = \|\mathbf{b}\|$, a sphere of radius $\|\mathbf{b}\|$, center at $(0, 0, 0)$

41. $\mathbf{u} = \frac{1}{7}(5\mathbf{i} + 4\mathbf{j} - \mathbf{k})$; $\mathbf{v} = -\frac{1}{7}(3\mathbf{i} + \mathbf{j} - 2\mathbf{k})$ **43.** $2x + z = 3$

45. $\mathbf{v} - \mathbf{u}$ and $\mathbf{w} - \mathbf{u}$ are in the plane; $(\mathbf{v} - \mathbf{u}) \times (\mathbf{w} - \mathbf{u})$ is perpendicular to the plane

CHAPTER 15

Exercise 1 (page 823)

1.

3.

5.

7.

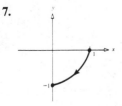

9. $\mathbf{r}'(t) = 2t\mathbf{i} + 3t^2\mathbf{j} - \mathbf{k}$; $\mathbf{r}''(t) = 2\mathbf{i} + 6t\mathbf{j}$

11. $\mathbf{r}'(t) = e^t(\cos t - \sin t)\mathbf{i} + e^t(\sin t + \cos t)\mathbf{j} + \mathbf{k}; \mathbf{r}''(t) = -2e^t \sin t\mathbf{i} + 2e^t \cos t\mathbf{j}$

13. $\mathbf{r}'(t) = (1 - 3t^2)\mathbf{i} + (1 + 3t^2)\mathbf{j} - \mathbf{k}; \mathbf{r}''(t) = -6t\mathbf{i} + 6t\mathbf{j}$

15. $\mathbf{r}(0) = \mathbf{0}$
$\mathbf{r}'(t) = \mathbf{i} + 2t\mathbf{j}$
$\mathbf{r}'(0) = \mathbf{i}$

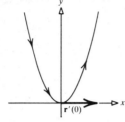

17. $\mathbf{r}(0) = \mathbf{j}$
$\mathbf{r}'(t) = \mathbf{i} + e^t\mathbf{j}$
$\mathbf{r}'(0) = \mathbf{i} + \mathbf{j}$

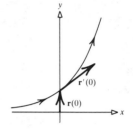

19. $\mathbf{r}(0) = -3\mathbf{j}$
$\mathbf{r}'(t) = 3 \cos t\mathbf{i} + 3 \sin t\mathbf{j}$
$\mathbf{r}'(0) = 3\mathbf{i}$

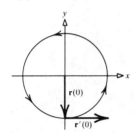

21. $\mathbf{r}(0) = 2\mathbf{i}$
$\mathbf{r}'(t) = -2 \sin t\mathbf{i} - 3 \cos t\mathbf{j}$
$\mathbf{r}'(0) = -3\mathbf{j}$

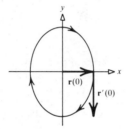

23. $-3\mathbf{i} + 2\mathbf{j} - \mathbf{k}; \mathbf{R}(u) = (1 - 3u)\mathbf{i} + 2u\mathbf{j} + (-5 - u)\mathbf{k}$ **25.** $-2\mathbf{j}; \mathbf{R}(u) = \mathbf{i} - 2u\mathbf{j} - 5\mathbf{k}$

27. $2\mathbf{k}; \mathbf{R}(u) = 2\mathbf{i} + \mathbf{j} + 2u\mathbf{k}$ **29.** $\mathbf{i} + \mathbf{j} + \mathbf{k}; \mathbf{R}(u) = (1 + u)\mathbf{i} + u\mathbf{j} + (1 + u)\mathbf{k}$

31. $12t^2; (2t + 10)\mathbf{i} - (2 + 10t)\mathbf{j} + 4(2t - t^3)\mathbf{k}$

33. $\cos 2t \sin t - \sin 2t \cos t; (2 \cos 2t - \cos t)\mathbf{i} - (\sin t - 2 \sin 2t)\mathbf{j} + (-\sin 2t \sin t - \cos 2t \cos t)\mathbf{k}$

35. $e^t - 4e^{-4t} - 2t; -2(e^{-2t} - 2te^{-2t})\mathbf{i} + (e^{2t} + e^{-t} + 2te^{2t} - te^{-t})\mathbf{j} + 3e^{-3t}\mathbf{k}$ **37.** $f''(t) = -f(t)$

39. $(0, -1, 0); (\frac{1}{4}, -\frac{3}{4}, \frac{1}{2}); (\frac{1}{4}, -\frac{3}{4}, -\frac{1}{2})$

41. (a) $\omega(-\sin \omega t\mathbf{i} + \cos \omega t\mathbf{j}); |\omega|$ (b) $2\mathbf{j} + 6t\mathbf{k}; 2\sqrt{1 + 9t^2}$ **43.** $\cos^{-1}[-\pi/(\sqrt{4 + \pi^2}\sqrt{2})]$

45. Follow the pattern of the proof of (15.4(b))   **47.** $\mathbf{u} \times \mathbf{v} = -(\mathbf{v} \times \mathbf{u})$ **49.** $\|\mathbf{r}(t)\|^2 = \mathbf{r}(t) \cdot \mathbf{r}(t)$; use (15.4(c))

Exercise 2 (page 830)

1. $\mathbf{v}(t) = \mathbf{i} + 2t\mathbf{j}$
$\mathbf{v}(0) = \mathbf{i}$
$\mathbf{a}(t) = 2\mathbf{j}$
$\mathbf{a}(0) = 2\mathbf{j}$

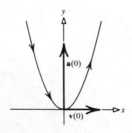

3. $\mathbf{v}(t) = \mathbf{i} + e^t\mathbf{j}$
$\mathbf{v}(0) = \mathbf{i} + \mathbf{j}$
$\mathbf{a}(t) = e^t\mathbf{j}$
$\mathbf{a}(0) = \mathbf{j}$

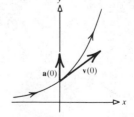

5. $v(t) = 3 \cos t\mathbf{i} + 3 \sin t\mathbf{j}$
   $v(0) = 3\mathbf{i}$
   $a(t) = -3 \sin t\mathbf{i} + 3 \cos t\mathbf{j}$
   $a(0) = 3\mathbf{j}$

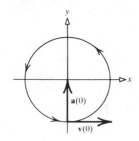

7. $v(t) = -2 \sin t\mathbf{i} - 3 \cos t\mathbf{j}$
   $v(0) = -3\mathbf{j}$
   $a(t) = -2 \cos t\mathbf{i} + 3 \sin t\mathbf{j}$
   $a(0) = -2\mathbf{i}$

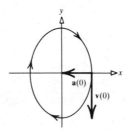

9. $2\mathbf{i} + \mathbf{j};\ 0;\ \sqrt{5}$ **11.** $-2 \sin t\mathbf{i} + 3 \cos t\mathbf{j};\ -2 \cos t\mathbf{i} - 3 \sin t\mathbf{j};\ \sqrt{4 \sin^2 t + 9 \cos^2 t}$ **13.** $\mathbf{i} + \mathbf{j} + \mathbf{k};\ 0;\ \sqrt{3}$

15. $-3 \sin t\mathbf{i} + 3 \cos t\mathbf{j};\ -3 \cos t\mathbf{i} - 3 \sin t\mathbf{j};\ 3$ **17.** $(2t - 1)\mathbf{i} + \mathbf{j} + (2t + 1)\mathbf{k};\ 2\mathbf{i} + 2\mathbf{k};\ \sqrt{8t^2 + 3}$

19. $\dfrac{1}{t}\mathbf{i} + \dfrac{1}{2\sqrt{t}}\mathbf{j} + \dfrac{3t^{1/2}}{2}\mathbf{k};\ -\dfrac{1}{t^2}\mathbf{i} - \dfrac{1}{4t^{3/2}}\mathbf{j} + \dfrac{3}{4t^{1/2}}\mathbf{k};\ \dfrac{\sqrt{9t^3 + t + 4}}{2t}$ **21.** See Problem 49, Exercise 1

23. (a) 9   (b) Less; half as much   **25.** 18,001 mph   **27.** 504.5 miles   **29.** 350 newtons

31. 41,152 newtons   **33.** $\dfrac{d}{dt}\dfrac{1}{2}m\|v(t)\|^2 = \dfrac{d}{dt}\dfrac{1}{2}m\mathbf{v}\cdot\mathbf{v} = m\mathbf{v}\cdot\mathbf{v}' = \mathbf{v}\cdot m\mathbf{a} = \mathbf{F}\cdot\mathbf{v}$

35. $\dfrac{d}{dt}\left[\mathbf{r}(t) \times m\mathbf{v}(t)\right] = \mathbf{r}' \times m\mathbf{v} + \mathbf{r} \times m\mathbf{a} = \mathbf{r} \times \mathbf{F} = \tau$

Exercise 3 (page 840)

1. $a(0) = 2\mathbf{j}$
   $a_T(0) = 0$
   $a_N(0) = 2$

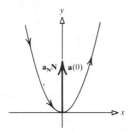

3. $a(0) = \mathbf{j}$
   $a_T(0) = 1/\sqrt{2}$
   $a_N(0) = 1/\sqrt{2}$

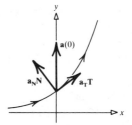

5. $a(0) = 3\mathbf{j}$
   $a_T(0) = 0$
   $a_N(0) = 3$

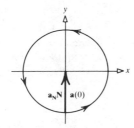

7. $a(0) = -2\mathbf{i}$
   $a_T(0) = 0$
   $a_N(0) = 2$

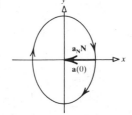

9. $0; 0$ **11.** $\dfrac{-5 \sin t \cos t}{\sqrt{4 \sin^2 t + 9 \cos^2 t}}; \dfrac{6}{\sqrt{4 \sin^2 t + 9 \cos^2 t}}$ **13.** $0; 0$ **15.** $0; 4$

17. $\dfrac{2t(9t^2 - 1)}{\sqrt{9t^4 - 2t^2 + 2}}; \dfrac{2\sqrt{9t^4 + 15t^2 + 2}}{\sqrt{9t^4 - 2t^2 + 2}}$ **19.** $\dfrac{4t + 18t^3}{\sqrt{1 + 4t^2 + 9t^4}}; \dfrac{2\sqrt{1 + 9t^2 + 9t^4}}{\sqrt{1 + 4t^2 + 9t^4}}$

21. Derivative of a constant is 0

Exercise 4 (page 845)

1. $-\cos t\mathbf{i} - \sin t\mathbf{j} + \frac{1}{2}t^2\mathbf{k} + \mathbf{c}$ **3.** $\sin t\mathbf{i} - \cos t\mathbf{j} - t\mathbf{k} + \mathbf{c}$ **5.** $-32t\mathbf{k}; 32t; -16t^2\mathbf{k}$

7. $(\sin t + 1)\mathbf{i} + (-\cos t + 1)\mathbf{j}; \sqrt{3 + 2(\sin t - \cos t)}; (-\cos t + t + 1)\mathbf{i} + (-\sin t + t + 1)\mathbf{j}$

9. $(e^t - e)\mathbf{i} - (t \ln t - t)\mathbf{j} + t^2\mathbf{k}$ **11.** 23,895 m; 53.1 sec; 3449 m

13. $\mathbf{r}(t) = \frac{1200}{13}t\mathbf{i} + \frac{500}{13}t\mathbf{j} - \frac{1}{2}(9.8)t^2\mathbf{j}; 724.55$ m   **15.** $\sin 2\theta$ is maximum when $\theta = \pi/4$

17. 25,540 m   **19.** $(-t^2/2\sqrt{2} + 3t + 1)\mathbf{i} + (-t^2/2\sqrt{2} + 4t + 2)\mathbf{j}$ **21.** 54.69 ft/sec   **23.** 318 ft; 4.5 sec

25. $\tan \alpha \approx 1.26; \alpha \approx 51.6°$ **27.** Write $\mathbf{c}$ and $\mathbf{r}$ in terms of their components and integrate.

29. Follow the same procedure as in Problem 27.

Exercise 5 (page 851)

1. $\sqrt{6}$ **3.** $\sqrt{5}\pi$ **5.** $e - e^{-1}$ **7.** 0 **9.** $\frac{4}{5}$ **11.** $\sqrt{2}/(e^{2t} + e^{-2t} + 2)$ **13.** $1/[3a(1 + t^2)^2]$

15. $6/[t^2(4 + 9t^2)^3]^{1/2}; t = 0$; no tangent line when $t = 0$ **17.** $[a(a^2 + b^2 + b^2\theta^2)^{1/2}]/[2(a^2 + b^2\theta^2)^{3/2}]$

19. $a_N = 1$ **25.** $a = 1; \kappa = a/(a^2 + 1)$

Exercise 6 (page 856)

1. 4.65×10^8 miles   **3.** 248.25 earth years

Miscellaneous Exercises (page 856)

1. $y = (1 - x)^{2/3}$ **3.** $y^2 = x/(1 - x), 0 \le x < 1$ **5.** $\cos \theta = -1/\sqrt{2}$

7. (a) $x^2 = -4(y - 2), -4 \le x \le 4$ (b)

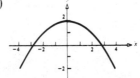

(c) $\mathbf{v} = \mathbf{0}$ when $t = k\pi$
$\mathbf{a} = 4\mathbf{i} + 8\mathbf{j}$ if k is odd
$\mathbf{a} = -4\mathbf{i} + 8\mathbf{j}$ if k is even

9. $\mathbf{a} = \mathbf{b} \times \mathbf{r}'; \mathbf{a} \cdot \mathbf{b} = 0; \dfrac{d}{dt}\|\mathbf{r}'(t)\|^2 = \dfrac{d}{dt}\mathbf{r}' \cdot \mathbf{r}' = 2\mathbf{r}' \cdot \mathbf{a} = 0$

11. $\mathbf{a} = \frac{2}{3}\mathbf{j}$ when $x = 0$; $\mathbf{a} = -\frac{2}{3}\mathbf{j}$ when $x = 2$ **13.** $a_\mathbf{T} = 0$; $a_\mathbf{N} = 4/(1 + x^2)^{3/2}$ **15.** $2 + \sqrt{2}\ln(1 + \sqrt{2})$

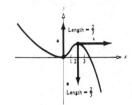

17. $(te^t - e^t)\mathbf{i} + (t^2 \sin t + 2t \cos t - 2 \sin t)\mathbf{j} + t(\ln t - 1)\mathbf{k} + \mathbf{c}$ **19.** $\mathbf{r}(s) = \dfrac{2s}{3}\mathbf{i} + \left(\dfrac{2s}{3} - 1\right)\mathbf{j} + \dfrac{s}{3}\mathbf{k}$

21. $\kappa = \frac{1}{2}$; $\mathbf{T}(\theta) = \dfrac{\cos\theta\mathbf{i} - \sin\theta\mathbf{j} + \mathbf{k}}{\sqrt{2}}$; $\mathbf{N}(\theta) = -\sin\theta\mathbf{i} - \cos\theta\mathbf{j}$; $\mathbf{B}(\theta) = \dfrac{\cos\theta\mathbf{i} - \sin\theta\mathbf{j} - \mathbf{k}}{\sqrt{2}}$

23. $\mathbf{v} = -\sin t\mathbf{u}_r + 2(2 + \cos t)\mathbf{u}_\theta$; $\mathbf{a} = (-5 \cos t - 8)\mathbf{u}_r - 4 \sin t\mathbf{u}_\theta$ **25.** (a) The $\mathbf{u}_\theta$ component is 0

27. 0.50310

CHAPTER 16

Exercise 1 (page 863)

1. (a) 3   (b) 2   (c) 10   (d) $3(x + \Delta x) + 2y + (x + \Delta x)y$ (e) $3x + 2(y + \Delta y) + x(y + \Delta y)$

3. (a) 0   (b) 0   (c) $at + a^2$ (d) $\sqrt{(x + \Delta x)y} + x + \Delta x$ (e) $\sqrt{x(y + \Delta y)} + x$

5. (a) 4   (b) -1 (c) 0   (d) $3 \sin t \cos t$ **7.** $e^{x^4 + (2y - 1)^2}$ **9.** $x \geq 0, y > 0$

11. $x \geq 0, y \geq 0$ or $x \leq 0, y \leq 0$ **13.** All x and y **15.** All x, y with $x^2 + y^2 > 4$

17. All x, y, z with $z \neq 0$ **19.** All x, y, z with $y \neq (2k + 1)\pi/2$ **21.** $S = 2xz + 2yz + xy$

23. $C = 4xy + 6yz + 4xz$

Exercise 2 (page 870)

1.

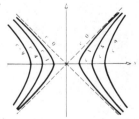

3.

5.

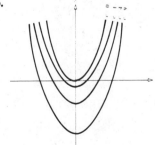

7.

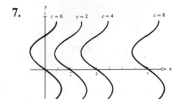

9. Sphere centered at the origin   **11.** Parallel planes
13. Circular cylinders centered on the z-axis

15. Circles centered at the origin; the surface is like a bowl with its low point (bottom) above the origin

17. Circular cylinders centered on the x-axis

Exercise 3 (page 878)

1. $f_x = 2xy$; $f_y = x^2 + 12y$ **3.** $f_x = 3x^2/y^3$; $f_y = -3x^3/y^4$

5. $f_x = -e^y \sin x + e^x \sin y$; $f_y = e^y \cos x + e^x \cos y$ **7.** $f_x = 2e^{2x+3y}$; $f_y = 3e^{2x+3y}$

9. $f_x = x/(x^2 + y^2)$; $f_y = y/(x^2 + y^2)$ **11.** $f_x = 4y \sin(2xy) \cos(2xy)$; $f_y = 4x \sin(2xy) \cos(2xy)$

13. $f_x = yx^{y-1}$; $f_y = (\ln x)x^y$ **15.** $f_{xx} = 12$; $f_{yy} = 18$; $f_{xy} = f_{yx} = -8$

17. $f_{xx} = 3x(2y^2 - x^3)/(x^3 + y^2)^2$; $f_{yy} = 2(x^3 - y^2)/(x^3 + y^2)^2$; $f_{xy} = f_{yx} = -6x^2y/(x^3 + y^2)^2$

19. $f_{xx} = 4e^{2x+3y}$; $f_{yy} = 9e^{2x+3y}$; $f_{xy} = f_{yx} = 6e^{2x+3y}$

21. $f_{xx} = -2y^3 \sin(x^2y^3) - 4x^2y^6 \cos(x^2y^3)$; $f_{yy} = -6x^2y \sin(x^2y^3) - 9x^4y^4 \cos(x^2y^3)$;
$f_{xy} = f_{yx} = -6xy^2 \sin(x^2y^3) - 6x^3y^5 \cos(x^2y^3)$

23. $f_x = y + z$; $f_y = x + z$; $f_z = y + x$

25. $f_x = y \sin z - yz \cos x$; $f_y = x \sin z - z \sin x$; $f_z = xy \cos z - y \sin x$

27. $f_x = -yz/(x^2 + y^2)$; $f_y = xz/(x^2 + y^2)$; $f_z = \tan^{-1}(y/x)$

29. $f_x = f_y = z(x + y)^{z-1}$; $f_z = [\ln(x + y)](x + y)^z$ **31.** $f_x = yzx^{yz-1}$; $f_y = z(\ln x)x^{yz}$; $f_z = y(\ln x)x^{yz}$

33. $f_x(1, 2) = 16$; $f_y(1, 2) = 12$ **35.** $f_x(0, 0) = f_y(0, 0) = 1$ **37.** $z - 5 = 2(x - 1)$, $y = 2$

39. $z - \sqrt{3}/2 = (-1/\sqrt{3})(y - 1/2)$, $x = 0$ **41.** $z - \sqrt{15} = (-1/\sqrt{15})(y - 1)$, $x = 4$

43. (a) $\frac{3}{25}$ (b) $\frac{4}{25}$ **45.** (b) $T_x(1, 0) = 200/(\ln 2)$; $T_x(0, 1) = 0$ (c) $T_y(2, 0) = 0$; $T_y(0, 2) = 100/(\ln 2)$

47. $\partial x/\partial r = \cos \theta$; $\partial x/\partial \theta = -r \sin \theta$; $\partial y/\partial r = \sin \theta$; $\partial y/\partial \theta = r \cos \theta$

49. $\partial u/\partial x = e^x \cos y = \partial v/\partial y$; $\partial u/\partial y = -e^x \sin y = -\partial v/\partial x$ **51.** $\partial u/\partial x = 2x$; $\partial u/\partial y = 8y$

53. $\partial w/\partial x = 2x$; $\partial w/\partial y = 2y - 3z$; $\partial w/\partial z = -3y$

55. $\partial^2 z/\partial x^2 = -\cos(x + y) - \cos(x - y)$; $\partial^2 z/\partial y^2 = -\cos(x + y) - \cos(x - y)$

57. $\partial^2 z/\partial x^2 = (y^2 - x^2)/(x^2 + y^2)^2$; $\partial^2 z/\partial y^2 = (x^2 - y^2)/(x^2 + y^2)^2$

59. $\partial z/\partial x = [2xe^{x^2+y^2}/(x^2 + y^2)^2](x^2 + y^2 - 1)$; $\partial z/\partial y = [2ye^{x^2+y^2}/(x^2 + y^2)^2](x^2 + y^2 - 1)$

61. $\partial V/\partial T = nr/P$; $\partial T/\partial P = V/nr$; $\partial P/\partial V = -nrT/V^2$ **63.** $\partial v/\partial p = k/2\sqrt{pd}$; $\partial v/\partial d = (-k/2)\sqrt{p/d^3}$

Exercise 4 (page 887)

1. -1 **3.** 0   **5.** $\frac{1}{4}$ **7.** 0   **9.** 0   **11.** 1   **13.** (a) 0 (b) 0 (c) $\frac{3}{14}$ (d) 0

15. Along x-axis limit is 2; along y-axis, limit is 1   **17.** Along x-axis limit is 0; along y-axis limit is -1

19. All x, y for which $x^2 = y^2$ **21.** All x, y for which $x + y + 1 \leq 0$ **23.** None

25. $f_{xy} = f_{yx} = \dfrac{6x^2 y - 2y^3}{(x^2 + y^2)^3}$ **27.** $f_{xy} = f_{yx} = \dfrac{x}{y}(\ln y)e^{x \ln y} + \dfrac{1}{y}e^{x \ln y}$

29. (a) $f_x(1, 6) = 6$; $f_y(1, 6) = 432$ (b) $\bar{f}_x(a, b) = 219\sqrt{2}$; $\bar{f}_y(a, b) = 213\sqrt{2}$

Exercise 5 (page 899)

1. $2x\,dx + 2y\,dy$ **3.** $(\sin y + y \cos x)\,dx + (x \cos y + \sin x)\,dy$

5. $(e^x \cos y - e^{-x} \sin y)\,dx + (e^{-x} \cos y - e^x \sin y)\,dy$ **7.** $\dfrac{2x}{x^2 + y^2}\,dx + \dfrac{2y}{x^2 + y^2}\,dy$

9. $(2xy + z^2)\,dx + (x^2 + 2yz)\,dy + (y^2 + 2xz)\,dz$

11. $(e^{yz} + yze^{xz} + yze^{xy})\,dx + (xze^{yz} + e^{xz} + xze^{xy})\,dy + (xye^{yz} + xye^{xz} + e^{xy})\,dz$

13. $2e^t\left(\dfrac{dx}{x} + \dfrac{dy}{y} + \dfrac{dz}{z}\right) + e^t(\ln xy + \ln xz + \ln yz)\,dt$

15. (1) $\Delta z = 2xy\Delta y + x(\Delta y)^2 + y^2\Delta x + 2y\Delta x\Delta y + \Delta x(\Delta y)^2 - 2y\Delta x - 2x\Delta y - 2\Delta x\Delta y$
    (2) Take $\eta_1 = 2y\Delta y + (\Delta y)^2 - 2\Delta y$; $\eta_2 = x\Delta y$

17. (1) $\Delta z = \dfrac{2xy\Delta y + x(\Delta y)^2 - y^2\Delta x}{x(x + \Delta x)}$ (2) $\eta_1 = \dfrac{y^2\Delta x - xy\Delta y}{x^2(x + \Delta x)}$; $\eta_2 = \dfrac{x^2\Delta y - xy\Delta x}{x^2(x + \Delta x)}$ (choices not unique)

19. 1.4   **21.** -0.3243 **23.** 4.003125   **25.** 0.1414 cm^3 **27.** 0.225 cm^2

29. $dx = \cos \theta\,dr - r \sin \theta\,d\theta$; $dy = \sin \theta\,dr + r \cos \theta\,d\theta$ **31.** 1.12%   **33.** 3%

35. (a) $dA = \frac{1}{2}c \sin \alpha\,db + \frac{1}{2}b \sin \alpha\,dc + \frac{1}{2}bc \cos \alpha\,d\alpha$ (b) 6.356%

37. $f_x(0, 0) = f_y(0, 0) = 0$; f is not continuous at $(0, 0)$ (use $y = x$ and $y = 2x$), so f is not differentiable at $(0, 0)$

39. $f_x(0, 0) = f_y(0, 0) = 0$; f is not continuous at $(0, 0)$ (use $y = x$ and $y = 2x$), so f is not differentiable at $(0, 0)$

Exercise 6 (page 907)

1. $\partial z/\partial x = 2xe^{2y} + 2y^2e^{2x}$; $\partial z/\partial y = 2x^2e^{2y} + 2ye^{2x}$

3. $\partial z/\partial x = 2xye^{x^2y}\sin[\ln(xy)] + (e^{x^2y}/x)\cos[\ln(xy)]$; $\partial z/\partial y = x^2e^{x^2y}\sin[\ln(xy)] + (e^{x^2y}/y)\cos[\ln(xy)]$

5. $\partial z/\partial x = (-y/x^2)e^{x^2+y^2} + 2ye^{x^2+y^2}$; $\partial z/\partial y = (e^{x^2+y^2}/x)(1 + 2y^2)$

7. $\partial z/\partial x = 2(5x - 3y)^2(2x + 3y)^3 + 10(2x + y)(5x - 3y)(2x + 3y)^3 + 6(2x + y)(5x - 3y)^2(2x + 3y)^2$;
$\partial z/\partial y = (5x - 3y)^2(2x + 3y)^3 - 6(2x + y)(5x - 3y)(2x + 3y)^3 + 9(2x + y)(5x - 3y)^2(2x + 3y)^2$

9. $\partial z/\partial x = 2(x^4 - y^8)/x(x^4 + y^8)$; $\partial z/\partial y = 4(y^8 - x^4)/y(y^8 + x^4)$

11. $\partial z/\partial x = \sin 2(x - y) - \sin 2(x + y)$; $\partial z/\partial y = -\sin 2(x - y) - \sin 2(x + y)$ **13.** $2te^{2t}(1 + t) + 2te^{-2t}(1 - t)$

15. $(e^{\sqrt{t}}/2\sqrt{t})(\sin \pi t + 2\pi\sqrt{t}\cos \pi t)$ **17.** $1 - 2t + (1/t)$ **19.** $e^{3t}(\sin t)(2\cos^2 t - \sin^2 t + 3\sin t\cos t)$

21. $\cos^2 t - \sin^2 t$ **23.** $y(-2x + y - 1)/x(x - 2y + 1)$ **25.** $-(\sin y + y\cos x)/(x\cos y + \sin x)$

27. $-y^{2/3}/x^{2/3}$ **29.** $\partial z/\partial x = -(z + 2xy^3)/(x + 6yz - 5)$; $\partial z/\partial y = -(3z^2 + 3x^2y^2)/(x + 6yz - 5)$

31. $\partial z/\partial x = -yz/(\cos z - y\sin z + xy)$; $\partial z/\partial y = -(\cos z + xz)/(\cos z - y\sin z + xy)$

33. $\partial z/\partial r = (\partial f/\partial x)\cos\theta + (\partial f/\partial y)\sin\theta$; $\partial z/\partial\theta = (\partial f/\partial x)(-r\sin\theta) + (\partial f/\partial y)(r\cos\theta)$

35. $\partial z/\partial x = (\partial f/\partial u)(1) + (\partial f/\partial v)(-1)$; $\partial z/\partial y = (\partial f/\partial u)(-1) + (\partial f/\partial v)(1)$

37. $\partial u/\partial x = x/u$; $\partial u/\partial y = y/u$; $\partial u/\partial z = z/u$; $\partial w/\partial x = (dw/du)(\partial u/\partial x)$ etc.

41. $\partial^2 z/\partial x^2 = (\partial^2 f/\partial u^2)(\partial g/\partial x)^2 + 2(\partial^2 f/\partial u\,\partial v)(\partial h/\partial x)(\partial g/\partial x) + (\partial^2 f/\partial v^2)(\partial h/\partial x)^2 + (\partial f/\partial u)(\partial^2 g/\partial x^2) + (\partial f/\partial v)(\partial^2 h/\partial x^2)$

43. $w = F(u, v, z)$, $u = x$, $v = y$; $\partial w/\partial y = (\partial F/\partial x)(0) + (\partial F/\partial y)(1) + (\partial F/\partial z)(\partial z/\partial x) = 0$

45. $\partial z/\partial x = -F_x/F_z$; $\partial x/\partial y = -F_y/F_x$; $\partial y/\partial z = -F_z/F_y$

47. $\partial w/\partial x = 2(3y + 4z)(2x)^{3y+4z-1}$; $\partial w/\partial y = 3\ln(2x)(2x)^{3y+4z}$; $\partial w/\partial z = 4\ln(2x)(2x)^{3y+4z}$

Miscellaneous Exercises (page 908)

1. $w_{xx} = (yz)^2e^{xyz}$; $w_{yy} = (xz)^2e^{xyz}$; $w_{zz} = (xy)^2e^{xyz}$; $w_{xy} = w_{yx} = (xyz^2 + z)e^{xyz}$; $w_{xz} = w_{zx} = (xy^2z + y)e^{xyz}$;
$w_{yz} = w_{zy} = (x^2yz + x)e^{xyz}$

3. $F_{xx} = e^x\sin y$; $F_{yy} = -e^x\sin y + e^y\sin z$; $F_{zz} = -e^y\sin z$; $F_{xy} = F_{yx} = e^x\cos y$; $F_{xz} = F_{zx} = 0$;
$F_{yz} = F_{zy} = e^y\cos z$

5. $f_x(2, 1) = 2/\sqrt{3}$; $f_y(2, -1) = 1/\sqrt{3}$ **7.** $z - 1 = 4(y + 2)$, $x = 1$; $z - 1 = 8(x - 1)$, $y = -2$

9. $\partial^2 u/\partial x^2 = -2xyz/(x^2 + y^2)^2$; $\partial^2 u/\partial y^2 = 2xyz/(x^2 + y^2)^2$; $\partial^2 u/\partial z^2 = 0$

11. Points in the upper half of a sphere of radius 1 with center at the origin that are inside the circular cone $x^2 + y^2 = z^2$

13. The difference quotients are $|h|/h$, so the limit as $h \to 0$ does not exist. The graph near the origin looks like the "point" of an ice cream cone. This is analogous to the absolute value function at $(0, 0)$.

15. Recall that $r = \sqrt{x^2 + y^2}$ and $\theta = \tan^{-1}(y/x)$ **17.** $14xy/(x - y^2)^3$ **19.** No; $f_{xy} \neq f_{yx}$

21. No; along line $y = x$, the limit is 0   **23.**

25.

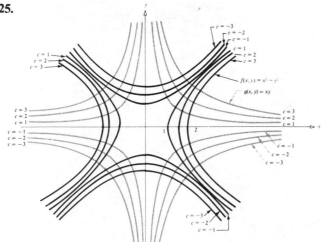

27. $f_x(0, 0) = f_y(0, 0) = 0$

29. $f_x = \dfrac{-y/x^2}{\sqrt{1-(y/x)^2}} = \dfrac{-y}{x\sqrt{x^2-y^2}}; \; f_y = \dfrac{1/x}{\sqrt{1-(y/x)^2}} = \dfrac{1}{\sqrt{x^2-y^2}}$

31. $f_{xx} = -2y^2/(x+y)^3; \; f_{xy} = 2xy/(x+y)^3; \; f_{yy} = -2x^2/(x+y)^3$ **33.** $a = -b^2 - c^2$

35. Along $x = y^2$; the limit is $\frac{3}{2}$; no; limits are different   **37.** $f_{tt} = -c^2\cos(x+ct); \; f_{xx} = -\cos(x+ct)$

39. $f_x = \dfrac{1}{2}f\left(\dfrac{1}{x+2y} - \dfrac{3}{3x-y}\right); \; f_y = \dfrac{1}{2}f\left(\dfrac{2}{x+2y} + \dfrac{1}{3x-y}\right)$

41. (a) $k = \pi/20\sqrt{\lambda}$ (b)

   (c) $\frac{10}{3}, \frac{50}{3}$ **43.** $f_x = -\ln(\cos\sqrt{x}); \; f_y = \ln(\cos\sqrt{y})$

45. $\partial u/\partial r = mr^{m-1}\cos m\theta; \; \partial^2 u/\partial r^2 = m(m-1)r^{m-2}\cos m\theta; \; \partial^2 u/\partial \theta^2 = -m^2 r^m \cos m\theta$

47. $u_{xx} = a^2 e^{ax}\cos ay; \; u_{yy} = -a^2 e^{ax}\cos ay$ **49.** $\sqrt{1-y^2}\,dx - \dfrac{xy}{\sqrt{1-y^2}}\,dy$

51. $yze^{xy}\,dx + xze^{xy}\,dy + e^{xy}\,dz$ **53.** $e^{2t}\cos(te^{2t})(2t+1) - e^{2t}\cos(te^t)(t+1) - e^t\sin(te^t)$

55. $\partial u/\partial r = 2rs - s + 2s^2; \; \partial u/\partial s = 4rs + r^2 - r - 2s$ **57.** 3.92, using $x = 0, y = 2$

59. $df = x^{y^z}[(y^z/x)\,dx + zy^{z-1}\ln x\,dy + y^z\ln y\ln x\,dz]; \; dg = (x^y)^z[(yz/x)\,dx + z\ln x\,dy + y\ln x\,dz]$

CHAPTER 17

Exercise 1 (page 923)

1. $1 - 2\sqrt{3}$ **3.** $-3 - 4\sqrt{3}$ **5.** $\frac{1}{2}(1 + \sqrt{3})$ **7.** $-\frac{7}{10}$ **9.** $-\pi\sqrt{3}/12$ **11.** $3/\sqrt{5}$

13. $(y^2 + 2x)\mathbf{i} + (2xy)\mathbf{j}$ **15.** $\left(\dfrac{-y}{x^2+y^2}\right)\mathbf{i} + \left(\dfrac{x}{x^2+y^2}\right)\mathbf{j}$ **17.** $6\mathbf{i}$ **19.** $\mathbf{i} + \mathbf{j}$ **21.** $\dfrac{1}{\sqrt{5}}\mathbf{i} - \dfrac{2}{\sqrt{5}}\mathbf{j}; \; 2\sqrt{5}$

23. $\dfrac{1}{\sqrt{2}}i + \dfrac{1}{\sqrt{2}}j;\ \sqrt{2}$　　**25.** $\dfrac{-6i + 6j + \pi k}{\sqrt{72 + \pi^2}};\ \dfrac{\sqrt{72 + \pi^2}}{4}$　　**27.** $6i + 8j$　　**29.** $6i - 4\sqrt{5}j$　　**31.** $\tfrac{2}{3}i + 9j$

33. $-\tfrac{1}{5};\ i + j;\ -i - j;\ i - j$ or $-i + j$

35. (a) $-(x^2 + y^2)^{-1/2}$　(b) 0　(c) $\dfrac{x}{x^2 + y^2}i + \dfrac{y}{x^2 + y^2}j$　(d) $\dfrac{-x}{x^2 + y^2}i + \dfrac{-y}{x^2 + y^2}j$

37. (a) $\sqrt{2}/2$　(b) $3i - 4j;\ 5$　(c) $\pm(2i + j)$　　**39.** $-j$ or $\tfrac{4}{5}i - \tfrac{3}{5}j$　　**41.** $(2xi - 2yj) \cdot (yi + xj) = 0$

43. $\dfrac{2}{\sqrt{5}}i + \dfrac{1}{\sqrt{5}}j$

Exercise 2 (page 927)

1. Tangent plane: $x - 2y + 3z - 14 = 0$; normal line: $\dfrac{x - 1}{2} = \dfrac{y + 2}{-4} = \dfrac{z - 3}{6}$

3. Tangent plane: $4x + 3y + z - 10 = 0$; normal line: $\dfrac{x - 2}{8} = \dfrac{y - 1}{6} = \dfrac{z + 1}{2}$

5. Tangent plane: $x - 3y + 2z = 0$; normal line: $\dfrac{x - 1}{2} = \dfrac{y + 1}{-6} = \dfrac{z + 2}{4}$

7. Tangent plane: $z - 5 = -4(x + 2) + 2(y - 1)$; normal line: $\dfrac{x + 2}{-4} = \dfrac{y - 1}{2} = \dfrac{z - 5}{-1}$

9. Tangent plane: $z - 2 = 4(x - 1)$; normal line: $\dfrac{x - 1}{4} = \dfrac{z - 2}{-1},\ y = 0$

11. Tangent plane: $y + z - (\pi/2) = 0$; normal line: $x = 0,\ z = y - (\pi/2)$

13. Tangent plane: $2x + y - z - 18 = 0$; normal line: $\dfrac{x - 1}{2} = y - 8 = \dfrac{z + 8}{-1}$

15. $(3, -1, 11)$　　**17.** $(x, -x, 0)$, x any real number　　**19.** $(\pm\tfrac{1}{2}, -1, \tfrac{7}{8}),\ (0, -1, 1)$

21. Each has the tangent plane $y + z - 1 = 0$

23. If (x_0, y_0, z_0) is on the sphere, the normal line obeys $x = x_0 + 2x_0 t,\ y = y_0 + 2y_0 t,\ z = z_0 + 2z_0 t$, which for $t = -\tfrac{1}{2}$ passes through $(0, 0, 0)$, the center of the sphere.

25. Tangent plane: $x + 4y - 5z - 4 = 0$; normal line: $\dfrac{x - 1}{2} = \dfrac{y - 2}{8} = \dfrac{z - 1}{-10}$　　**27.** $z = -\pi y + \dfrac{3\pi}{2},\ x = 1$

Exercise 3 (page 935)

1. $(0, 0),\ (1, 0),\ (-1, 0)$　　**3.** $(0, 0),\ (1, 1),\ (-1, -1)$

5. $(0, 0)$　　**7.** Local minimum at $(1, -2)$　　**9.** Local minimum at $(2, -1)$

11. Local minimum at $(2, 2)$　　**13.** None　　**15.** Local maximum at $(-1, -1)$; saddle point $(0, 0, 0)$

17. None　　**19.** None　　**21.** Local maximum at $(2k\pi, 2l\pi)$; local minimum at $((2k + 1)\pi, (2l + 1)\pi)$

23. Local minimum at $(k\pi, (-1)^k 3)$　　**25.** None

27. (a) Complete the square to get $(3, -2, 1)$ (b) Local maximum at $(3, -2)$, for which $z = 1$

29. S = Surface area; Base = $\sqrt{S/3} \times \sqrt{S/3}$, Height = $\frac{1}{2}\sqrt{S/3}$

31. A cube, each side of length $\sqrt{S/6}$ **33.** (a) $h = 14$ in., $w = 14$ in., $l = 28$ in.   (b) $R = 28/\pi$ in., $l = 28$ in.

35. $h = L/2\sqrt{3}$, $x = L/3$; no, semicircular area is larger   **37.** $D = (\frac{1}{3}, \frac{1}{2})$

39. Maximum profit when $x = 15,250$ tons and $y = 4100$ tons

Exercise 4 (page 944)

1. Maximum at $(\sqrt{2}, 1)$, $(-\sqrt{2}, -1)$, $3\sqrt{2}$; minimum at $(-\sqrt{2}, 1)$, $(\sqrt{2}, -1)$, $-3\sqrt{2}$

3. Maximum at $(0, 1)$, 4; minimum at $(0, -1)$, -4

5. Maximum at $(\sqrt{2}, \frac{3}{2}\sqrt{2})$, $(-\sqrt{2}, -\frac{3}{2}\sqrt{2})$, 3; minimum at $(-\sqrt{2}, \frac{3}{2}\sqrt{2})$, $(\sqrt{2}, -\frac{3}{2}\sqrt{2})$, -3

7. Minimum at $(-6, \frac{9}{2}, \frac{225}{24})$, $-\frac{225}{12}$ **9.** $(\frac{3}{5}, -\frac{9}{5})$ **11.** $(1, \sqrt{2}/2)$ and $(-1, -\sqrt{2}/2)$

13. Base = $2\sqrt[3]{3} \times 2\sqrt[3]{3}$ m; Height = $\sqrt[3]{3}$ m   **15.** (a) Maximum 50 m; minimum 20 m   (b) 3400 m² $(x = 30)$

17. Maximum $2(3 + 2\sqrt{2})$; minimum $2(3 - 2\sqrt{2})$

Miscellaneous Exercises (page 945)

1. $\nabla u(2, 1) = 16\mathbf{i} + 18\mathbf{j}$; $2\sqrt{145}$ **3.** $2\mathbf{i} - 2\mathbf{j} + \mathbf{k}$; $3/e^2$

5. $F_x = 2x_0/a^2$; $F_y = 2y_0/b^2$; $F_z = -2z_0/c^2$; the cone has no tangent plane at the origin $[\nabla F(0, 0, 0) = \mathbf{0}]$

7. $\nabla F \cdot (\mathbf{r} - \mathbf{r}_0) = 0$; $\mathbf{r}$ defines the tangent line   **9.** $m = -a/b$ **11.** $3\sqrt{2}$ **13.** $(\frac{1}{2}, \frac{1}{2}, \frac{3}{2})$

15. Use $\nabla V = \lambda \nabla g$, where $V = xyz$ and $g = 2xy + 2xz + 2zx - S = 0$; then $x = y = z = \sqrt{S/6}$

17. $m = \frac{7}{10}$, $b = \frac{6}{5}$ **19.** $y + z = \pi$; $x = 0$, $y = z$

21. $\dfrac{df}{d\theta} = \dfrac{\partial f}{\partial x}\dfrac{dx}{d\theta} + \dfrac{\partial f}{\partial y}\dfrac{dy}{d\theta} = cx(-k \sin \theta) + cy(k \cos \theta) = 0$; thus, $f(x, y)$ is constant on the circle

23. $T_x = -x/\sqrt{4 - x^2}$; the temperature is 2(maximum) when $x = 0$ and decreases very rapidly to 0 as x gets closer to 2

25. $\cos \theta = \sqrt{2}/3$; $\theta \approx 62°$ **27.** $\pm 5/\sqrt{3}$ **29.** 9

CHAPTER 18

Exercise 1 (page 953)

1. 36   **3.** 4   **5.** $\frac{47}{180}$ **7.** -2

Exercise 2 (page 960)

1. $\frac{1}{3}$ **3.** $\frac{9}{2}$ **5.** $\frac{1}{3}$ **7.** 6   **9.** $\frac{2}{27}$ **11.** $a^3/6$ **13.** $\pi/2$ **15.** $\frac{5}{6}$ **17.** $\frac{1}{2}(\ln 3)^2 - \frac{1}{2}(\ln 2)^2$

19. $\frac{36}{5}$ **21.** $\frac{423}{28}$ **23.** $\int_0^1 \left[\int_y^1 f(x, y)\, dx \right] dy$ **25.** $\int_0^a \left[\int_0^{\sqrt{a^2 - x^2}} f(x, y)\, dy \right] dx$

27. $\int_{-1}^0 \left[\int_{1-\sqrt{x+1}}^{1+\sqrt{x+1}} f(x, y)\, dy \right] dx + \int_0^8 \left[\int_{x/2}^{1+\sqrt{x+1}} f(x, y)\, dy \right] dx$

29. $\int_0^1 \left[\int_{y/3}^y xy\,dx \right] dy + \int_1^3 \left[\int_{y/3}^1 xy\,dx \right] dy = 1$ **31.** $\int_0^1 \left[\int_{1-y}^1 (x+y)\,dx \right] dy + \int_1^2 \left[\int_0^{2-y} (x+y)\,dx \right] dy = \frac{3}{2}$

33. $\frac{1}{4}(e-1)$ **35.** $(\pi/4)\ln(1+\sqrt{2})$ **37.** (a) 0 (b) $\frac{4}{3}$

Exercise 3 (page 966)

1. $\frac{1}{12}$ **3.** 72 **5.** $\frac{9}{2}$ **7.** 2 **9.** $\frac{3}{2} - 2\ln 2$ **11.** $\frac{2}{3}$ **13.** $\frac{8}{3}$ **15.** 4π **17.** 8 **19.** $\frac{56}{3}$

21. $\pi/2$ **23.** Inside the cylinder, $z^2 = 1 - x^2$, but above the triangular region, $0 \le x \le 1, 0 \le y \le x$

25. A solid of height 3, with base between $y = \sin x$, $0 \le x \le \pi$ **27.** 4 **29.** 2 **31.** $\frac{1}{2}(e-2)$

Exercise 4 (page 973)

1. $\pi/4$ **3.** 12π **5.** $(\pi/4)\sin 1$ **7.** $(\pi/4)(e^{-1} - e^{-4})$ **9.** $\pi/8$ **11.** 21π **13.** $64\pi/3$ **15.** $-\frac{1}{2}$

17. $1 - \ln 2$ **19.** $(2a^3/9)(3\pi - 4)$ **21.** $16\pi/3$ **23.** $\pi/12$ **25.** $15\pi/4$ **27.** $(5\pi/6) + (7\sqrt{3}/8)$

29. $\int_0^{1/2} \left[\int_{(1/2)-\sqrt{(1/4)-x^2}}^{\sqrt{(1/2)-x^2}} dy \right] dx$ **31.** $\int_0^{\pi/4} \left[\int_0^{\sin\theta} f(r,\theta)\,dr \right] d\theta + \int_{\pi/4}^{\pi/2} \left[\int_0^{\cos\theta} f(r,\theta)\,dr \right] d\theta$

Exercise 5 (page 981)

1. $64; (\frac{3}{2}, \frac{8}{3})$ **3.** $\frac{31}{20}; (\frac{150}{217}, \frac{44}{93})$ **5.** $k/15; (\frac{15}{28}, \frac{5}{8})$ **7.** $5k; (\frac{6}{5}, \frac{7}{10})$ **9.** $5\pi k/3; (0, \frac{21}{20})$

11. $(2ak/3)(3\sqrt{3} - \pi); (0, 3a\sqrt{3}/2(3\sqrt{3} - \pi))$ **13.** 2ρ **15.** $\rho ab^3/3$ **17.** $8\rho/15$ **19.** $\rho ab^3\pi/4$

21. $2k\pi \ln(b/a)$ **23.** $ak(\pi - 2); (a\pi/2(\pi - 2), 0)$

25. If rectangle is enclosed by $x = 0$, $x = a$, $y = 0$, $y = b$, then $\bar{x} = \dfrac{\rho ba^2/2}{\rho ba} = \dfrac{a}{2}$; $\bar{y} = \dfrac{\rho b^2 a/2}{\rho ba} = \dfrac{b}{2}$

27. $(3.58, 0)$ **29.** If triangle has vertices at $(0, 0)$, $(b, 0)$, $(0, h)$, then $m = \rho bh$; $I_x = \rho h^3 b/6 = \frac{1}{6}mh^2$

31. $(\bar{x}, \bar{y}) = \left(\dfrac{-5166}{2115}, \dfrac{-4221}{4230} \right); \dfrac{423}{28}$

Exercise 6 (page 988)

1. $(260 - 108\sqrt{3})/15$ **3.** $(\pi/6)(17\sqrt{17} - 1)$ **5.** $\pi a^2/3$ **7.** $(\pi/6)[(a^2 + 1)^{3/2} - 1]$ **9.** 16

11. $4\pi a^2$ **13.** $\frac{1}{12}(17\sqrt{17} - 1)$ **15.** $(\pi/6)(37\sqrt{37} - 5\sqrt{5})$

Exercise 7 (page 995)

1. $-\frac{2}{3}$ **3.** 72 **5.** 1

7. *(1)* $\int_0^6 \left\{ \int_0^{3-(x/2)} \left[\int_0^{2-(x/3)-(2/3)y} f(x, y, z)\,dz \right] dy \right\} dx$; *(2)* $\int_0^3 \left\{ \int_0^{6-2y} \left[\int_0^{2-(x/3)-(2/3)y} f(x, y, z)\,dz \right] dx \right\} dy$;

(3) $\int_0^6 \left\{ \int_0^{2-(1/3)x} \left[\int_0^{3-(x/2)-(3/2)z} f(x, y, z)\,dy \right] dz \right\} dx$; *(4)* $\int_0^2 \left\{ \int_0^{6-3z} \left[\int_0^{3-(x/2)-(3/2)z} f(x, y, z)\,dy \right] dx \right\} dz$;

(5) $\int_0^3 \left\{ \int_0^{2-(2/3)y} \left[\int_0^{6-3z-2y} f(x, y, z)\,dx \right] dz \right\} dy$; *(6)* $\int_0^2 \left\{ \int_0^{3-(3/2)z} \left[\int_0^{6-3z-2y} f(x, y, z)\,dx \right] dy \right\} dz$

9. $\frac{1}{8}$ **11.** $\frac{648}{5}$ **13.** $\frac{1184}{15}$ **15.** $\frac{1}{60}$ **17.** $81\pi/32$ **19.** $k\pi a^4 h/2$ **21.** $k/720$

Exercise 8 (page 1002)

1. $(-2, \pi/6, -5)$ or $(2, 7\pi/6, -5)$ 3. $(\sqrt{2}, \pi/4, \sqrt{2})$ 5. $(\sqrt{3}, 1, -5)$ 7. $(2\sqrt{3}, 2, 2)$ 9. $9\sqrt{3}$

11. $4\pi\rho a^3/3$ 13. $(0, 0, \frac{4}{3})$ 15. $(0, 0, 2.16)$ 17. $\int_{-4}^{4}\left\{\int_{-\sqrt{16-x^2}}^{+\sqrt{16-x^2}}\left[\int_{-\sqrt{x^2+y^2}}^{\sqrt{16-x^2-y^2}} z^2 x^4\, dz\right]dy\right\}dx$

Exercise 9 (page 1009)

1. $(4, 5\pi/4, \pi/6)$ 3. $(2, \pi/4, \pi/4)$ 5. $(3.74, 1.11, 0.64)$ 7. $(6, \pi/2, \pi/3)$

9. $(-1, -\sqrt{3}, -2\sqrt{3}/3)$; $(2, 4\pi/3, -2\sqrt{3}/3)$; $(4\sqrt{3}/3, 4\pi/3, 2\pi/3)$ 11. $\pi^2/64$ 13. $\sqrt{2}\pi/3 - (\pi/3)\ln(\sqrt{2}+1)$

15. $4\pi\delta a^3/3$ 17. $(0, 0, 2a/5)$ 19. $8\pi/3$ 21. $k\delta\pi a$

23. $\int_{-a}^{a}\left\{\int_{-\sqrt{a^2-x^2}}^{\sqrt{a^2-x^2}}\left[\int_{-\sqrt{a^2-x^2-y^2}}^{\sqrt{a^2-x^2-y^2}} f(x, y, z)\, dz\right]dy\right\}dx$

$$= \int_{0}^{2\pi}\left\{\int_{0}^{a}\left[\int_{-\sqrt{a^2-r^2}}^{\sqrt{a^2-r^2}} f(r\cos\theta, r\sin\theta, z) r\, dz\right]dr\right\}d\theta$$

$$= \int_{0}^{2\pi}\left\{\int_{0}^{\pi}\left[\int_{0}^{a} f(\rho\sin\phi\cos\theta, \rho\sin\phi\sin\theta, \rho\cos\phi)\rho^2\sin\phi\, d\rho\right]d\phi\right\}d\theta$$

25. $\int_{-\sqrt{8}}^{\sqrt{8}}\left\{\int_{0}^{\sqrt{8-x^2}}\left[\int_{-\sqrt{16-x^2-y^2}}^{-\sqrt{x^2+y^2}} x(x^2+y^2+z^2)\, dz\right]dy\right\}dx$

27. $\int_{0}^{\pi}\left\{\int_{0}^{\pi/2}\left[\int_{0}^{3}\rho^3\sin\theta\sin^2\phi\, d\rho\right]d\phi\right\}d\theta$ 29. 8π

Miscellaneous Exercises (page 1010)

1. $(e-1)/6e$ 3. $\sqrt{2}-1$ 5. $\pi a^3/6$ 7. (a) 0   (b) $32, -32, 0$ (c) 0

9. (a) Set $f(x, y) = f(x, y_0)$ (b) $(1/V)\iiint_S f(x, y, z)\, dV$ 11. $(1-\cos 8)/6$ 13. $\frac{1}{2}$

15. The volume of a sphere, $\frac{4}{3}\pi a^3$ 17. (a) $2k\pi\ln(b/a) \to +\infty$ as $a \to 0$ (b) $k\pi(b^2-a^2) \to k\pi b^2$ as $a \to 0$

19. $(a/2, 2b/5)$

23. $\int_{0}^{1}\int_{0}^{\sqrt{x}}\int_{0}^{1-x} dz\, dy\, dx = \int_{0}^{1}\int_{0}^{1-x}\int_{0}^{\sqrt{x}} dy\, dz\, dx = \int_{0}^{1}\int_{0}^{1-z}\int_{0}^{\sqrt{x}} dy\, dx\, dz = \int_{0}^{1}\int_{0}^{\sqrt{1-z}}\int_{y^2}^{1} dx\, dy\, dz$

$$= \int_{0}^{1}\int_{0}^{1-y^2}\int_{y^2}^{1} dx\, dz\, dy$$

25. $\int_{a}^{b}\left[\int_{a}^{y} f(x)\, dx\right]dy = \int_{a}^{b}\left[\int_{x}^{b} f(x)\, dy\right]dx = \int_{a}^{b} f(x)(b-x)\, dx$

27. (a) $I_a = \int_{0}^{a} e^{-x^2}\, dx = \int_{0}^{a} e^{-y^2}\, dy$; $I_a^2 = \int_{0}^{a} e^{-x^2}\, dx\int_{0}^{a} e^{-y^2}\, dy = \int_{0}^{a}\int_{0}^{a} e^{-(x^2+y^2)}\, dx\, dy = \iint_R e^{-r^2} r\, dr\, d\theta$

(b) $|I_a^2 - J_a| = \iint_{R_1} e^{-r^2} r\, dr\, d\theta \le \max_{R_1}(e^{-r^2})(\text{Area } R_1) = e^{-a^2}(a^2 - \pi a^2/4) = [(4-\pi)/4]a^2 e^{-a^2}$

(c) $J_a = \int_{0}^{\pi/2}\left[\int_{0}^{a} e^{-r^2} r\, dr\right]d\theta = \int_{0}^{\pi/2}(-\tfrac{1}{2}e^{-r^2}|_0^a)\, d\theta = (\pi/2)(\tfrac{1}{2} - \tfrac{1}{2}e^{-a^2})$

27. (d) $\lim\limits_{a\to+\infty} J_a = (\pi/4) - (\pi/4) \lim\limits_{a\to+\infty} e^{-a^2} = (\pi/4) - 0 = \pi/4;\ \lim\limits_{a\to+\infty} a^2 e^{-a^2} = \lim\limits_{a\to+\infty} (a^2/e^{a^2}) =$

$\lim\limits_{a\to+\infty} (2a/2ae^{a^2}) = 0;$ thus, $\lim\limits_{a\to+\infty} |I_a^2 - J_a| = 0$ and $I_{+\infty}^2 = \pi/4;$

hence, $I_{+\infty} = \displaystyle\int_0^{+\infty} e^{-x^2}\,dx = \sqrt{\pi}/2$, and finally, $\displaystyle\int_{-\infty}^{+\infty} e^{-x^2}\,dx = 2\int_0^{+\infty} e^{-x^2}\,dx = \sqrt{\pi}$

29. $\ln(1+\sqrt{2}) - \frac{1}{2}\ln[(1+\sqrt{5})/2] + (\sqrt{5}/2) - \sqrt{2}$

CHAPTER 19

Exercise 1 (page 1022)

1. $5\sqrt{2}/6$ **3.** $\sqrt{2}/4$ **5.** (a) $\frac{1}{3} - (\pi/16)$ (b) $\frac{1}{3} - (\pi/16)$ **7.** (a) 3   (b) 3   (c) 0   (d) 0

9. (a) 5   (b) $\frac{123}{20}$ **11.** $\frac{1}{2}(1 - \sin 1 - \cos 1)$ **13.** C_1 is the same as C_2

15. $19 + e^3 - e^2 + \cos 4 - \cos 9$ **17.** 0   **19.** $-\frac{3}{2}$ **21.** $\frac{1}{12}(3e^4 + 6e^{-2} - 12e + 8e^3 - 5)$

23. 0 if n is even; $2\pi ab\left[\left(\frac{1}{2}\right)\left(\frac{3}{4}\right)\cdots\cdot\left(\frac{n}{n+1}\right)\right](b^{n-1} - a^{n-1})$ if n is odd   **25.** $\dfrac{a^2}{3}\left[(4\pi^2 + 1)^{3/2} - 1\right]$

27. 3   **29.** $\frac{3}{2}$

Exercise 2 (page 1035)

1. Yes; $f(x, y) = (x^3/3) + (y^3/3)$ **3.** Yes; $f(x, y) = \frac{1}{2}x^2 e^y$

5. (a) $\partial P/\partial y = \partial Q/\partial x = 0$ (b) $f(x, y) = (x^2/2) + (y^2/2)$ (c) $\frac{19}{2}$

7. (a) $\partial P/\partial y = \partial Q/\partial x = 3$ (b) $f(x, y) = (x^3/3) + 3xy$ (c) $-\frac{181}{3}$

9. (a) $\partial P/\partial y = \partial Q/\partial x = 60xy^2 - 12y^3$ (b) $f(x, y) = x^4 + 10x^2 y^3 - 3xy^4 + y^5$ (c) 9

11. $x^3 y^2$ **13.** $(x^2/2) + 3xy$ **15.** $x^2 y - xy^2$ **17.** $(x^3/3) - (x^2/2) + xy^2 - e^y(y - 1)$

19. $y \sin x - 2x \sin y$ **21.** $x^2 + y \sin x$ **23.** $f(x, y) = \tan^{-1}(y/x)$ and $(0, 0)$ is not in R

Exercise 3 (page 1041)

1. 1   **3.** $3\pi - \frac{1}{2}$ **5.** $-\frac{19}{15}$ **7.** $\frac{2}{3}$ **9.** $(3\pi ab/4)(a^2 - b^2)$ **11.** One gives 0; the other gives -1

13. $\dfrac{-5(2^{15})}{3}\,k = -54{,}613.3k$ **15.** $1 - \dfrac{1}{\sqrt{5}}$

Exercise 4 (page 1052)

1. -12 **3.** -12 **5.** 0   **7.** $\pi/2$ **9.** $3\pi/4$ **11.** 0   **13.** $\frac{6}{35}$ **15.** $\frac{32}{9}$ **17.** $\frac{9}{2}$ **19.** 3π

21. Integrand $= \frac{1}{2}\nabla e^{x^2 + y^2}$ or $\partial P/\partial y = \partial Q/\partial x$ **23.** $\dfrac{\partial}{\partial y}(e^x \sin y) = \dfrac{\partial}{\partial x}(e^x \cos y)$ **25.** -2π **27.** -2π

29. $\displaystyle\iint_R \frac{\partial Q}{\partial x}\,dx\,dy = \int_c^d \left[\int_{g_1(y)}^{g_2(y)} \frac{\partial Q}{\partial x}\,dx\right]dy = \int_c^d Q(g_2(y), y)\,dy - \int_c^d Q(g_1(y), y)\,dy = \oint_C Q(x, y)\,dy$

Exercise 5 (page 1056)

1. $8\sqrt{3}/3$ **3.** $\sqrt{6}/4$ **5.** $\pi/2$ **7.** $\pi/4$

Exercise 6 (page 1063)

1. ρ **3.** 0 **5.** 3ρ **7.** 2π **9.** $2\pi/3$ **11.** $\frac{64}{3}$

Exercise 7 (page 1071)

1. $2x + 2y + 2z$ **3.** 4 **5.** $\frac{128}{3}$ **7.** 3 **9.** 4π **11.** 80π **13.** $\frac{2}{3}$

15. $\iint_\Sigma \mathbf{F} \cdot \mathbf{n}\, dS = 2 \iiint_T (x + y + z)\, dV$; use cylindrical coordinates   **17.** $\iint_\Sigma \mathbf{F} \cdot \mathbf{n}\, dS = 3 \iiint_T dV = 3V$

21. div $\mathbf{F}$ = div $\mathbf{G}$ = 14; use the divergence theorem

23. $\iiint_T \nabla \cdot \mathbf{F}\, dV = 4000\pi = \iint_\Sigma \mathbf{F} \cdot \mathbf{n}\, dS$

Exercise 8 (page 1080)

1. 0 **3.** $-x\mathbf{i} + xy\mathbf{j} + (1 - x)z\mathbf{k}$ **5.** $(-y^2 \cos z)\mathbf{i} + (6xyz - e^{2z})\mathbf{j} - 3xz^2\mathbf{k}$ **7.** 0 **9.** $-ze^{xz}\mathbf{j}$

11. $-(\mathbf{i} + \mathbf{j} + \mathbf{k})$ **13.** div(curl $\mathbf{F}$) $= R_{yx} - Q_{zx} - R_{xy} - P_{zy} + Q_{xz} - P_{yz} = 0$ **15.** (a), (b), (e)

17. Both equal -2π **19.** Both equal 2π **21.** 0 **23.** 4π **25.** $\nabla \times \mathbf{F} = \mathbf{0}$

27. $\nabla \times \mathbf{F} = \mathbf{0}$; $(a^2/2) + ab - bc + (c^2/2)$ **29.** π ; Hint: $\mathbf{n} = (\mathbf{i} + \mathbf{j} + \mathbf{k})\sqrt{3}$

Miscellaneous Exercises (page 1081)

1. $\frac{5}{2}$ **3.** $\sin 1$ **5.** -2 **7.** $e \sin 1$ **9.** $\mathbf{F} \cdot \mathbf{r}' = 0$ **11.** $k/\sqrt{x^2 + y^2 + z^2}$ **13.** -2 **15.** 12π

17. $\bar{x} = \dfrac{\iint_R x\, dA}{\iint_R dA} = \dfrac{1}{A} \iint_R \left(\dfrac{\partial Q}{\partial x} - \dfrac{\partial P}{\partial y} \right) dA$ (where $Q = x^2/2$, $P = 0$)

$$= \frac{1}{2A} \oint_C x^2\, dy$$

19. $\bar{x} = \bar{y} = 256a/315\pi$

21. Suppose $\mathbf{F} = P\mathbf{i} + Q\mathbf{j} + R\mathbf{k}$ is such a vector function; $R_y - Q_z = x$, $R_x - P_z = -y$, $Q_x - P_y = z \Rightarrow$ $R_{xy} = Q_{xz} + 1$, $R_{yx} = P_{yz} - 1$, $Q_{xz} = P_{yz} + 1$; this requires that $R_{xy} = Q_{xz} + 1$ and $R_{xy} = Q_{xz} - 2$

23. $2k\pi a^3$ **31.** $\frac{3}{2}$

APPENDIX I

Exercise I.1 (page A-7)

1. (a) $\pi/4 \approx 0.785$ (b) $-\pi/2 \approx -1.57$ (c) $3\pi/4 \approx 2.355$ (d) $\pi \approx 3.14$ (e) $-7\pi/6 \approx -3.66$
   (f) $5\pi/3 \approx 5.23$ (g) $-4\pi \approx -12.56$ (h) $5\pi/2 \approx 7.85$ (i) $\pi/18 \approx 0.17$ (j) $-4\pi/9 \approx -1.40$
   (k) $\pi/90 \approx 0.03$ (l) $2\pi/3 \approx 2.09$

5.

Angle	$\sin x$	$\cos x$	$\tan x$	$\cot x$	$\sec x$	$\csc x$
(a) $x = 0$	0	1	0	Undefined	1	Undefined
(b) $x = \pi/2$	1	0	Undefined	0	Undefined	1
(c) $x = \pi$	0	-1	0	Undefined	-1	Undefined
(d) $x = 3\pi/2$	-1	0	Undefined	0	Undefined	-1

7. (a) $\pi/6, 5\pi/6$ (b) $\pi/4, 7\pi/4$ (c) $0, \pi$ (d) $4\pi/3, 5\pi/3$ (e) $2\pi/3, 4\pi/3$ (f) $3\pi/4, 7\pi/4$
(g) $5\pi/6, 11\pi/6$ (h) $\pi/6, 11\pi/6$ (i) $\pi/4, 3\pi/4$ (j) $\pi/4, 5\pi/4$ (k) 0 (l) $3\pi/2$

9. (a) $\cos\theta = -\frac{4}{5}, \tan\theta = -\frac{3}{4}, \cot\theta = -\frac{4}{3}, \sec\theta = -\frac{5}{4}, \csc\theta = \frac{5}{3}$
(b) $\sin\theta = -\frac{12}{13}, \tan\theta = \frac{12}{5}, \cot\theta = \frac{5}{12}, \sec\theta = -\frac{13}{5}, \csc\theta = -\frac{13}{12}$
(c) $\sin\theta = \frac{12}{13}, \cos\theta = \frac{5}{13}, \cot\theta = \frac{5}{12}, \sec\theta = \frac{13}{5}, \csc\theta = \frac{13}{12}$
(d) $\cos\theta = 2\sqrt{2}/3, \tan\theta = -1/2\sqrt{2}, \cot\theta = -2\sqrt{2}, \sec\theta = 3/2\sqrt{2}, \csc\theta = -3$
(e) $\sin\theta = -\sqrt{21}/5, \tan\theta = -\sqrt{21}/2, \cot\theta = -2/\sqrt{21}, \sec\theta = \frac{5}{2}, \csc\theta = -5/\sqrt{21}$
(f) $\sin\theta = 1/\sqrt{101}, \cos\theta = -10/\sqrt{101}, \cot\theta = -10, \sec\theta = -\sqrt{101}/10, \csc\theta = \sqrt{101}$

11. (a) $2\pi/3$ (b) $4\pi/3$ (c) $2\pi/3$ (d) $4\pi/3$ **13.** $-\sin\theta$ **15.** $-\sin\theta$ **17.** $-\cos\theta$ **19.** $\cos\theta$

21. $\sec\alpha - \cos\alpha = \dfrac{1}{\cos\alpha} - \cos\alpha = \dfrac{1 - \cos^2\alpha}{\cos\alpha} = \dfrac{\sin^2\alpha}{\cos\alpha} = \dfrac{\sin\alpha}{\cos\alpha}\sin\alpha = \tan\alpha\sin\alpha$

23. $\cos^4(2x) - \sin^4(2x) = [\cos^2(2x) - \sin^2(2x)][\cos^2(2x) + \sin^2(2x)] = \cos^2(2x) - \sin^2(2x) = \cos(4x)$

25. $\tan x + \tan y = \dfrac{\sin x}{\cos x} + \dfrac{\sin y}{\cos y} = \dfrac{\sin x\cos y + \cos x\sin y}{\cos x\cos y} = \dfrac{\sin(x+y)}{\cos x\cos y}$

27. $\cot\theta - \tan\theta = \dfrac{\cos\theta}{\sin\theta} - \dfrac{\sin\theta}{\cos\theta} = \dfrac{\cos^2\theta - \sin^2\theta}{\sin\theta\cos\theta} = \dfrac{\cos(2\theta)}{\frac{1}{2}\sin(2\theta)} = 2\cot(2\theta)$

29. $\cos(7\alpha) + \cos(5\alpha) = 2\cos[\frac{1}{2}(7\alpha + 5\alpha)]\cos[\frac{1}{2}(7\alpha - 5\alpha)] = 2\cos(6\alpha)\cos\alpha$

31. $\sin^2 x - \sin^2 y = (\sin x + \sin y)(\sin x - \sin y) = [2\sin\frac{1}{2}(x+y)\cos\frac{1}{2}(x-y)][2\cos\frac{1}{2}(x+y)\sin\frac{1}{2}(x-y)]$
$= [2\sin\frac{1}{2}(x+y)\cos\frac{1}{2}(x+y)][2\sin\frac{1}{2}(x-y)\cos\frac{1}{2}(x-y)] = \sin(x+y)\sin(x-y)$

33. Period $= 2\pi$ **35.** Period $= \pi$ **37.** Period $= 4\pi$ **39.** Period $= \pi/2$ **41.** Period $= 4\pi$

43. Period $= 2\pi$ **45.** Period $= 8\pi$ **47.** Period $= 2$ **49.** $\dfrac{\pi}{4} + n\pi$ **51.** $\pm\dfrac{\pi}{18} + \dfrac{2n\pi}{3}$

53. $\pm\dfrac{\pi}{6} + 2n\pi; \pm\dfrac{5\pi}{6} + 2n\pi$ **55.** $2n\pi; \pm\dfrac{\pi}{3} + 2n\pi$ **57.** $\dfrac{(3n-1)\pi}{6}$

59. (a) $(f + g)(x + p) = f(x + p) + g(x + p) = f(x) + g(x) = (f + g)(x)$
(b) $(f \cdot g)(x + p) = f(x + p) \cdot g(x + p) = f(x) \cdot g(x) = (f \cdot g)(x)$
(c) $\dfrac{f}{g}(x + p) = \dfrac{f(x + p)}{g(x + p)} = \dfrac{f(x)}{g(x)} = f(g)(x)$

61. If h is the height and b is the base opposite θ, then $b/2 = a\sin(\theta/2)$ and $h = a\cos(\theta/2)$;
$\therefore A = \frac{1}{2}bh = [a\sin(\theta/2)][a\cos(\theta/2)] = \frac{1}{2}a^2\sin\theta$

63. $s = L \sin \theta;\ \cos \theta = \dfrac{L - h}{L} = 1 - \dfrac{h}{L};\ \dfrac{h}{L} = 1 - \cos \theta$

$h = L(1 - \cos \theta) = s\,\dfrac{1 - \cos \theta}{\sin \theta} = s \tan \dfrac{\theta}{2}$

$h = s\,\dfrac{1 - \sin \alpha}{\cos \alpha} = s \cot \dfrac{\alpha}{2}$

65. $L^2 = R^2 + x^2 - 2Rx \cos \theta;\ x^2 - 2Rx \cos \theta + R^2 - L^2 = 0$

$x = \dfrac{2R \cos \theta + \sqrt{4R^2 \cos^2\theta - 4R^2 + 4L^2}}{2} = R \cos \theta + \sqrt{R^2 \cos^2\theta - R^2 + L^2}$

67.

x	1.5	1	0.1	0.01	0.001	0.0001
$\sin x$	0.997	0.841	0.0998	0.00999	0.001	0.0001
$(\sin x)/x$	0.665	0.841	0.998	0.9999	0.999	1

Exercise I.2 (page A-12)

1. $x^5 + 5x^4y + 10x^3y^2 + 10x^2y^3 + 5xy^4 + y^5$ **3.** $x^3 + 9x^2y + 27xy^2 + 27y^3$

5. $16x^4 - 32x^3y + 24x^2y^2 - 8xy^3 + y^4$ **7.** $10x^2$ **9.** 405

Index

Your chance to rate
Calculus and Analytic Geometry by Mizrahi and Sullivan

In order to make this text even more responsive to your needs, it would help us to know what you, the student, thought of *Calculus and Analytic Geometry*. We would appreciate it if you would answer the following questions. Then, cut out the page, fold, seal, and mail it; no postage is required. Thank you for your help.

What Chapters in the text did you skip, or *not* cover in class? (circle)

1   2   3   4   5   6   7   8   9   10   11   12

13   14   15   16   17   18   19

Prior to taking Calculus, which courses in mathematics had you previously taken?

College Algebra _____   Analytic Geometry _____

Trigonometry _____   Algebra and Trigonometry _____

Precalculus _____   High School Algebra _____ (1 or 2 terms? _____)

Prior to taking Calculus, how long ago did you take your last Algebra course?

Within last 2 years _____   3–5 years ago _____   Over 5 years ago _____

What is your major course of study?

Engineering _____   Chemistry _____   Physics _____

Biology _____   Mathematics _____   Computer Science _____

Business _____   Social Science _____   Other _____

Would you rather see more applications to your major in college, or more detailed examples and answers?

Applications _____   More detail _____

Did the answers have any typos or misprints? If so, where?

What did you most like about the book?

FOLD HERE

What did you like least about the book?

Name _____ College _____

Home Address (optional) _____

FOLD HERE

First Class
PERMIT NO. 34
Belmont, CA

BUSINESS REPLY MAIL
No postage necessary if mailed in United States

Postage will be paid by
WADSWORTH PUBLISHING COMPANY, INC.
10 Davis Drive
Belmont, California 94002
           ATTN: Mathematics Editor